Developmental Biology

Developmental Biology

THIRD EDITION

Leon W. Browder
University of Calgary, Canada

Carol A. Erickson
University of California, Davis

William R. Jeffery
University of Texas at Austin
University of California, Davis

Saunders College Publishing

Philadelphia   Toronto
Ft. Worth   London
Chicago   Sydney
San Francisco   Tokyo
Montreal

Text Typeface: Times Roman
Compositor: Monotype Composition
Acquisitions Editor: Julie Levin Alexander
Developmental Editor: Gabrielle Goodman
Associate Developmental Editor: Dena Digilio-Betz
Managing Editor: Carol Field
Project Editor: Nancy Lubars
Copy Editor: Diane Lamsback
Manager of Art and Design: Carol Bleistine
Art & Design Coordinator: Doris Bruey
Text Designer: Tracy Baldwin
Cover Designer: Lawrence R. Didona
Text Artwork: Rolin Graphics
Layout Artist: York Production Services
Director of EDP: Tim Frelick
Production Manager: Charlene Squibb
Marketing Manager: Marjorie Waldron

About the Cover: The cover image acknowledges the fundamental importance of DNA in directing embryonic development. A fertilized egg of the popular experimental organism *Xenopus laevis* (the South African clawed frog) is shown here superimposed over the DNA. The egg contains numerous components that are synthesized during oogenesis under the control of the oocyte genome and stored in the egg cytoplasm; their utilization in initiation of embryonic development is triggered by fertilization. The egg cytoplasm has been stained with a red fluorescent dye and visualized by confocal scanning laser microscopy, a technique that enables us to gather sectional views through solid specimens. After fertilization, a swirl of alternating red fluorescent and green non-fluorescent layers (most prominent in upper right) forms on the side of the egg from which all of the dorsal embryonic structures will develop. The reorganization of cytoplasm reflects a redistribution of the products of oocyte gene expression. (DNA model by David Wagner/Phototake. The confocal image of the *Xenopus* embryo was generously provided by Dr. Michael Danilchik, Wesleyan University.)

Printed in the United States of America

DEVELOPMENTAL BIOLOGY

ISBN 0-03-013514-1

Library of Congress Catalog Card Number: 90-053487

234 039 98765432

Preface

We are privileged to be living in a period of intense and highly productive research activity in the field of developmental biology, with exciting discoveries occurring almost daily. Because change itself is the essence of the phenomena that we study, developmental biologists by temperament revel in new discoveries. We have presented developmental biology here as ongoing and evolving, rather than as a mature discipline with established dogma. This book not only details our current understanding of development but aims to give students the tools to assist them in understanding the process of discovery itself and—most important (because discovery is the life blood of any science)—to help them pose questions for further investigation.

The scope of this edition of *Developmental Biology* has been expanded substantially to provide students with a comprehensive appreciation of the developmental process in animals, presenting more detailed descriptions of development along with a conceptual framework for understanding how development occurs. As active researchers ourselves, we believe that the key to full comprehension of any experimental science is an understanding of the experimental process itself. We discuss in depth many of the crucial experiments upon which our current level of knowledge is based. This approach gives the student an appreciation for the role of experimentation in science and a better understanding of the experimental results. Familiarity with these experiments should also facilitate the understanding of new research results introduced by instructors, help students to read and understand research papers in the literature, and, we hope, stimulate them to pose questions and design experiments about the many unanswered aspects of development.

Instructional Features

Essays and Research Appendix To help students comprehend the experiments that we have analyzed, this book contains 14 essays and an extensive appendix that describe the most common methodology used in contemporary developmental biology research. The appendix has been designed as a ready reference source that summarizes the powerful techniques of molecular biology, including a discussion of nucleotide complementarity, which is the key to understanding the organization of the genome and is the basis for nucleic acid technology. The essays present supplementary material on topics such as Gene Organization and Regulation of Transcription, Growth Cone Guidance in Insects, Eukaryotic Transcription Regulatory Proteins, and DNA Methylation.

Visuals and Use of Color We believe that visual material is an important aid to the learning process and have, therefore, included numerous illustrations in the text. The entire graphic art program has been redesigned with the judicious use of color to emphasize important points and to assist in identifying key components of the drawings. In addition to the graphic art, we have included a large number of high quality photographs in both black and white and color. The extensive 24-page color insert serves two purposes: The graphic art utilizes color as a pedagogical tool to enable the reader to follow the germ layers during complex morphogenetic processes such as gastrulation. Color is also used to present research results that utilize color for analytical purposes, such as the use of fluorescent dyes to trace cells and tissues and the use of genes that encode enzymes that produce a colored product to detect the activity of that gene. The inclusion of these color figures should illustrate the power of these techniques as analytical tools. We are grateful to the many investigators and publishers who have granted permission to reproduce these figures. Complete citations for the sources of previously published figures can be found in the reference list at the end of each chapter.

Glossary This new edition also contains a comprehensive glossary. It has been designed to provide a ready reference source for the technical terms and names of organisms that are used in this book.

Biographies of Researchers We owe a debt of gratitude to those individuals who have defined the field of developmental biology, posed the questions that led to early experimentation, and designed experimental approaches to answer those questions. We salute some of these investigators with a photograph and a short biographical sketch at the beginning of each chapter. Because we believe that the literature is the foundation upon which any discipline is based, each chapter ends with an extensive reference list, which provides the student with a guide to the key sources for the material discussed in the text.

Organization of Developmental Biology, Third Edition

Two additional authors have increased the breadth of material that is covered in this edition and have added their unique perspectives, especially concerning embryogenesis, morphogenesis, patterning, and gene control in the early embryo, to enable us to present a comprehensive analysis of developmental biology. The order of presentation of the material has been revised in this edition to enable students to acquire their understanding of the developmental process in a logical sequence. In Chapter 1, we describe the origins of developmental biology and build upon this base as the story unfolds. We begin our analysis of development with gametogenesis. In addition to describing the formation of the gametes and the regulation of gametogenesis, we focus on the developmental legacy of the gametes.

The earliest events of development follow: fertilization, cleavage, embryonic polarization, and gastrulation. Fertilization is not only the physical union of gametes, but it sets into motion the entire developmental process. Consequently, we discuss both of these aspects of fertilization, beginning with the events and specializations that facilitate gamete fusion, followed by the cellular and biochemical responses of the egg to sperm entry. Fertilization endows the egg with the potential for rapid cell division (cleavage), which produces the cells that will later be molded into the embryo. The patterns of cell division during cleavage are tightly regulated and may have profound effects on later developmental processes; hence, we describe the different cleavage patterns in detail. The embryos of most species develop along axes of bilateral symmetry from eggs that are radially symmetrical. The events that establish those axes of symmetry are important developmental events and are discussed using two model systems: *Xenopus laevis* and *Drosophila melanogaster*. The formation of the embryonic body plan along the axes of symmetry is dependent upon a reorganization of the cells produced during cleavage by the process of gastrulation, which produces three embryonic germ layers, which form the tissue and organ rudiments.

A major distinction between this edition and its two predecessors is the two chapters (Chapters 7 and 8) detailing the formation of the embryonic rudiments, using both descriptive and experimental approaches. The elaboration by the embryo of diverse tissue and organ rudiments after gastrulation is a remarkable phenomenon that provides much of the fascination of the developmental process and enough unsolved questions to engage the interest of investigators for many years to come. We hope that these descriptions and discussions of morphogenesis will engender an understanding of the importance of these processes. Morphogenesis is a complex process that relies upon the concerted activities of the individual cells of the embryo, which utilize a variety of mechanisms to migrate over great distances, to adhere to one another, and to change shape. Our understanding of the nature of these mechanisms is advancing rapidly and is discussed in Chapter 9.

The role of the genome in development is one of the great challenges of contemporary molecular biology. Part V of this book provides the philosophical basis for understanding gene expression during development. We first describe the evidence that (with few exceptions) the genome remains intact during development of specialized cell types, without losing the information necessary for development of alternative cell types. The integrity of the genome implies that diversity is achieved not by loss of alternative genetic information but by differential use of the genetic material during development. There are two primary means for regulating differential gene expression to achieve cellular diversity during development: (1) the influence of distinct cytoplasms on nuclei in different regions of the early embryo and (2) extrinsic factors, such as inductive signals from neighboring cells. Our understanding of embryonic induction is increasing substantially with the discoveries that growth factors are important embryonic inducers. We describe the most recent findings in this exciting search for inducers, which began several decades ago and dominated embryological research for many years.

One of the most intriguing aspects of the onset of development is the orchestration of genetic control. Control is initially mediated by RNA that is synthesized during oogenesis and stored for delayed utilization during development. The zygote nucleus then assumes belated control over development. We describe this transition in Chapter 13. A hallmark of gene regulation in development is that it is subject to both chronological and spatial control, with different genes being expressed at distinct times during development and in certain regions of the embryo. Our understanding of the spatial regulation of gene expression has progressed rapidly in recent years owing to the judicious use of recombinant DNA technology and genetics. This fascinating story is presented in Chapter 14.

The ultimate goal of embryonic development is the production of diverse functional organs and tissues. The discussion of organogenesis in this edition has been expanded substantially. We have chosen three systems for detailed analysis: limb development and regeneration, gonad development and sex differentiation, and eye development. Among them, they provide a sampling of the complexities involved in coordinating the differentiation of multiple cell types to produce functional entities during development. Cell differentiation in differentiating organs and tissues is dependent upon the selective expression of cell-specific genes. Our understanding of the control over selective gene expression during cell differentiation has progressed rapidly in recent years with the application of the techniques of contemporary molecular biology. The exciting discoveries of factors that interact with gene regulatory elements have added new dimensions to the study of cell-specific gene expression: We are in a position not only to study the interactions of regulatory factors with their target genes but to study the regulation of the regulators themselves, which is a critical phase in cell differentiation. These topics are covered in Chapter 18.

Developmental biology is taught in a variety of formats. We have attempted to write a versatile book that should be appropriate for courses that emphasize cellular and molecular aspects of development as well as those that focus on descriptive embryology and morphogenesis. However, the book will be particularly useful for those courses that cover both the molecular and descriptive aspects of development. The approach that we have taken should be compatible with courses at various levels of instruction, from undergraduate to graduate.

Our objective while preparing this textbook has been to capture the spirit and flavor of this dynamic period in developmental biology while giving our readers an appreciation for the intellectual base embodied in the classical literature. We will have succeeded in this project if we have engendered in our readers a fascination and enthusiasm for studying the processes of change that typify development and have stimulated them to assist in unravelling the secrets of developmental biology.

Leon W. Browder
Carol A. Erickson
William R. Jeffery

March 1991

Acknowledgments

This edition of *Developmental Biology* has benefitted greatly from the assistance of many individuals, to whom the authors are indebted. The long gestation period was nurtured by Ed Murphy, the Senior Biology Editor at Saunders, and delivery was under the guidance of his successor, Julie Levin Alexander. Both of them made major contributions and commitments to the project. We are very appreciative of their support, patience, and assistance. We were given immeasurable help by Gabe Goodman. Her expertise as Developmental Editor was important in assembling the manuscript from her authors and in developing the art program, which was a major undertaking. She does her job well, and we thank her. Production of any book is a major challenge, but some are more challenging than others. The Production staff won't forget this one for a long time. The Project Editor sees the project to its completion, and we were fortunate to have the assistance of a real professional, Nancy Lubars. She coaxed, cajoled, and prodded us to complete the project while always keeping the quality of the book as her first priority. The result reflects her commitment to excellence. Thanks, Nancy!

Numerous individuals have helped us at various stages in this project. Nobody worked harder on this book than Sandy Browder, who has eased the authors' burdens while doing quality control at every stage in the process. Her knowledge of the process of producing a book, which has been gained through considerable experience over the past fifteen years, is considerable and very valuable. There is simply no way that this book could have been completed without her very able assistance. Various colleagues have given very generously of their time. We wish to make special mention of Drs. Izak Paul and Michael Pollock of Mount Royal College in Calgary. Their suggestions for improvements to the text, glossary, and art have been invaluable, and their scrutiny of the book, which was scrupulous, helped us purge it of numerous gremlins, which we have mercifully avoided as a result. Their generosity of time and their expertise have been invaluable, and we are grateful. William Jeffery was assisted in designing figures by Janet Young and in formatting text by Marsha Berkman, both of the Department of Zoology, University of Texas at Austin. In spite of all the help we have received in reviewing the manuscript, errors undoubtedly remain, for which we accept full responsibility.

Our families have given us remarkable support during this protracted ordeal—perhaps more than we deserve considering the burden that we imposed on them. In addition to Sandy Browder's contribution, Leon Browder had welcome assistance in proofreading the manuscript from his daughters, Teri and Diana (in spite of their memories of past editions), and his father- and mother-in-law, Mr. and Mrs. Lowell O'Connor. Carol Erickson is grateful to her husband, David Phillips, who edited each of her chapters in the initial stages of preparation and who assumed all household chores, including cooking, so that she could tackle this project. William Jeffery is grateful to his family, Billie, David, and Tony, who generously provided him with free time on numerous evenings and weekends to complete the book.

The following reviewers have given valuable advice on topics relating to their areas of expertise:

Dr. Lois Abbott, University of California, Davis
Dr. Robert Angerer, University of Rochester
Dr. Peter Armstrong, University of California, Davis
Dr. Karen Artzt, University of Texas at Austin
Dr. Susan Bryant, University of California, Irvine
Dr. Margaret Burns, University of California, Davis
Dr. Barbara J. Clarke, The American University
Dr. Ronald Conlon, University of Texas at Austin
Dr. Michael Danilchik, Wesleyan University
Dr. John deBanzie, Northeastern State University
Dr. R. P. Elinson, University of Toronto
Dr. Scott Fraser, University of California, Irvine
Dr. Allan Gibson, University of Calgary
Dr. Robert Grainger, University of Virginia
Dr. Hamid Habibi, University of Calgary
Dr. Albert Harris, University of North Carolina, Chapel Hill
Dr. Stephen Hauschka, University of Washington
Dr. Christine Holt, University of California, San Diego
Dr. Antone Jacobson, University of Texas at Austin
Dr. Laurinda Jaffe, University of Connecticut Health Sciences Center
Dr. Randal Johnston, University of Calgary
Dr. Ray E. Keller, University of California, Berkeley
Dr. Richard Kelley, University of Texas at Austin
Dr. Chris Kintner, Salk Institute
Dr. David Knecht, University of Connecticut
Dr. Paul Krieg, University of Texas at Austin
Dr. Paul Kugrens, Colorado State University
Dr. Paul Langer, Gwynedd-Mercy College
Dr. Jeanne Loring, University of California, Davis
Dr. George Malacinski, Indiana University
Dr. Randall Moon, University of Washington
Dr. Robert W. Nickells, California Institute of Technology
Dr. D. M. Noden, Cornell University
Dr. Richard Nuccitelli, University of California, Davis
Dr. Thomas Sargent, National Institute of Child Health and Human Development
Dr. John W. Saunders, Woods Hole, Massachusetts
Dr. Ross Shoger, Carleton College
Dr. Michael Solursh, University of Iowa
Dr. David Sonneborn, University of Wisconsin, Madison
Dr. Billie Swalla, University of California, Davis
Dr. J. P. Trinkaus, Yale University
Dr. Judith Venuti, M. D. Anderson Hospital
Dr. Matthew Winkler, University of Texas at Austin
Dr. Saul Zackson, University of Calgary

We thank these individuals for their important contribution to this book.

Contents Overview

Contents

PART *IV*

*Organizing the Multicellular
Embryo 241*

PART V

Genetic Regulation of Development   393

P A R T *VI*

*The Organized Generation of
Cell Diversity* *577*

*14 Establishment of Spatial Patterns
of Gene Expression During
Development* *578*

*15 Organogenesis: Limb
Development* *625*

16 Organogenesis: Gonad Development and Sex Differentiation 661

17 Organogenesis: Development of the Eye 684

Developmental Biology

PART I

Introduction

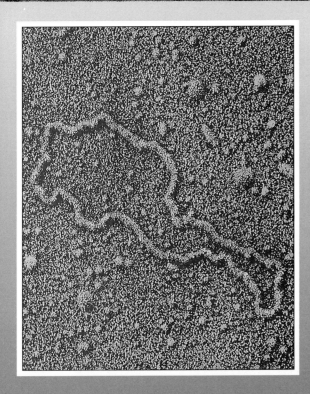

Electron micrograph of a plasmid from *E. coli*. The ability to insert DNA into plasmids is the basis for DNA cloning, which has revolutionized developmental biology. (Courtesy of Stanley N. Cohen/Science Source/PhotoResearchers, Inc.)

The Origins of Developmental Biology

Theodor Boveri (1862–1915). Boveri's most important scientific contribution was a paper published in 1902, which demonstrated that individual chromosomes have qualitatively unique effects on development. Boveri's conclusions, which were based upon cytological and embryological observations, were subsequently confirmed by genetic analysis. Boveri's work provided the foundation on which contemporary ideas on the roles of genes in development are based. (From Goldschmidt, R. B. 1956. *Portraits from Memory. Recollections of a Zoologist.* University of Washington Press, Seattle.)

The formation of complex organisms has aroused human curiosity for centuries. The question of how a new individual can arise from a seemingly formless egg has come under intense scrutiny by philosophers and scientists alike, at least since Aristotle first contemplated the formation of a chick from a hen's egg. However, no period of investigation has been as fruitful and exciting as the past decade, during which there has been an explosion of interest in studying development, resulting in an avalanche of information. The impetus for this renaissance has been the application of the power of genetics combined with the techniques of modern cellular and molecular biology to the study of development. The golden age of developmental biology is here, and the momentum continues to build. If the past decade has been exciting, the information produced during this decade should be staggering.

Although modern technology has provided the tools to answer many previously unsolved questions about development, technology is useless without an intellectual foundation that generates the questions and provides the background for interpreting the answers and for formulating new questions. The history of developmental biology is replete with the contributions of many remarkable men and women whose work has built a rich literature that challenges today's technology-blessed investigators to provide explanations for longstanding problems in terms of molecular and cellular biology.

As we shall see, a hallmark of developmental biology has been its multidisciplinary approach. The multidisciplinary approach to the study of development first emerged, before the turn of the century, as an integration of embryology with cytology and later with the new science of genetics. Literally, **embryology** means "the study of embryos," but the word cannot be rigidly defined, because it implies a point of view and is frequently used to connote the study (descriptive or experimental) of changes in the form or shape (**morphogenesis**) of animals during their embryonic phase. The cytologists of the late 1800s saw the development of the embryo as a manifestation of the changes that occur in the individual cells that compose the embryo. They believed that the cell is the key to all ultimate biological problems and that the fundamental principles underlying development would emerge by studying the properties and behavior of cells.

Every multicellular organism begins life as a single cell, the fertilized egg, which is essentially similar to all other cells but differs by its potential to divide and produce all the cells of the body. The cells diverge from one another structurally and functionally and become organized into an adult organism. Accordingly, the basic processes of development must first be understood at the cellular level before an overall understanding of development is possible.

Because the fertilized egg (or **zygote**) is derived from fusion of the male and female gametes, the cytologists assumed that the factors that control development could be identified by tracing the elements of the gametes and following their behavior in the cells of the developing organism.

The foremost proponent of the cytological approach was E. B. Wilson (see p. 162), and the most eloquent statements of this philosophy were contained in the three editions (1896, 1900, and 1925) of his classic treatise, *The Cell in Development and Inheritance* (retitled *The Cell in Development and Heredity* in 1925). He wrote in the first edition:

> Every discussion of inheritance and development must take as its point of departure the fact that the germ is a single cell similar in its essential nature to any one of the tissue-cells of which the body is composed. That a cell can carry with it the . . . heritage of the species, that it can in the course of a few days or weeks give rise to a mollusk or a man, is the greatest marvel of biological science. In attempting to analyze the problems that it involves, we must from the onset hold fast to the fact . . . that the wonderful formative energy of the germ is not impressed upon it from without, but is inherent in the egg as a heritage from the parental life of which it was originally a part. The development of the embryo is nothing new. It involves no breach of continuity, and is but a continuation of the vital processes going on in the parental body. What gives development its marvelous character is the rapidity with which it proceeds and the diversity of the results attained in a span so brief.

Thus, Wilson understood that the characteristics of the organism emerge during development by utilization of inherited information, and the only way to understand development fully is to comprehend the nature of that information and the ways in which it is utilized. These concepts have had a rocky history, both before and after Wilson's time.

1–1 Germ Cells: Bridging the Generation Gap

The idea that the form of an embryo gradually emerges during development originated with Aristotle. Nearly 2000 years later, however, in the seventeenth and eighteenth centuries, many embryologists rejected this concept and proposed that the egg contains a miniature, fully formed embryo. Development to them was analogous to the unfolding of a flower bud. Bonnet (1745) formalized this concept in the theory of *emboîtement,* or encasement. He stated that because the egg contains the complete embryo, it must also contain similar preformed eggs for all future generations, like an infinite series of Russian dolls encased one inside the other. Others of his contemporaries believed that a preformed embryo exists inside the sperm, and many microscopists of the time claimed to have seen a tiny creature (homunculus) curled up in the sperm head (Fig. 1.1).

These theories of **preformation** were not universally held, however. Caspar Wolff (1759) championed the theory of an alternative mechanism of development—**epigenesis**. Epigenesis means that the adult gradually develops from a rather formless egg as originally proposed by Aristotle. Wolff made careful observations on the development of the chick and showed that the early embryo is entirely different from the adult and that development is progressive, with new parts being formed continually. Wolff's conclusions were rejected by most of his contemporaries, and the concept of epigenesis was not accepted by a majority of biologists until the early part of the nineteenth century.

Even after epigenesis was determined to be the mechanism for the formation of the external structure of the organism during development, the nature and the significance of the germ cells remained unclear for nearly a century. The fact that the egg itself is a cell was recognized by Schwann in 1839. Likewise, the cellular nature of the sperm was determined in 1865 by Schweigger-Seidel and St. George. Another decade elapsed before Oscar Hertwig (1876) established that fertilization results from the union of the egg and sperm. Thus, each sex contributes a single cell from its own body.

These cells, which carry the complete set of instructions for production of another generation similar to the preceding one, fuse to form a single cell, the zygote, which undergoes division, or cleavage, to produce the cells that form the embryo. Within the fertilized egg, two nuclei, one derived from the egg and the other derived from the sperm, combine to form a single zygote nucleus, which gives rise by division to all nuclei of the body. In contrast to the equal nuclear contribution to the zygote made by both gametes, the cytoplasmic contributions are decidedly uneven; the egg contributes virtually all of the cytoplasm. These observations led Hertwig to the conclusion that the germ cell nuclei, and not the cytoplasm, are the vehicles of inheritance.

Hertwig's discoveries marked the beginning of a new era of investigation into the role of the nucleus in fertilization and development. Much of the attention of cytologists during this era was directed to the primary nuclear constituents, the **chromosomes**. The first detailed description of the behavior of the chromosomes in fertilized eggs was by van Beneden (1883), who used the nematode *Ascaris megalocephala*. This species was a splendid choice for chromosome study because it has only four chromosomes, which are large and stain intensely. Van Beneden observed that both the egg nucleus and the sperm nucleus provide two of the four chromosomes that align on the metaphase plate at the first cleavage division. Each chromosome splits lengthwise, and the daughter chromosomes are transported to the opposite poles of the spindle, where they are incorporated into the nuclei of the two-celled stage. Therefore, each of these nuclei receives an equal number of maternal and paternal chromosomes.

Soon after publication of van Beneden's work on *Ascaris*, reports based upon studies using several species of animals and plants established the rule that each gamete nucleus contributes one half the number of chromosomes that is characteristic of somatic cells. The behavior of chromosomes at fertilization led Hertwig and three other German scientists— Strasburger, Kolliker, and Weismann—to conclude independently that chromosomes are the means for transmission of inherited information. As Wilson stated in the 1896 edition of *The Cell,* these observations on chromosomes during fertilization could not logically lead to any other conclusion:

> These remarkable facts demonstrate the two germ-nuclei to be in a morphological sense precisely equivalent, and they not only lend very strong support to Hertwig's identification of the nucleus as the bearer of hereditary qualities, but indicate further that these qualities must be carried by the chromosomes; for their precise equivalence in number, shape, and size is the physical correlative of the fact that the two sexes play, on the whole, equal parts in hereditary transmission. And we are finally led to the view that chromatin is the physical basis of inheritance. . . .

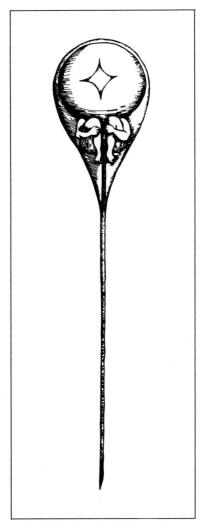

Figure 1.1

Homunculus in human sperm. (After Hartsoeker. Redrawn from J. A. Moore, 1972. *Heredity and Development,* 2nd ed. Copyright © 1972 by Oxford University Press, Inc., New York. Reprinted by permission.)

1–2 The Roux-Weismann Theory

The genetic material was beginning to take on a definite physical form in the theories of biologists. However, the absence of hard information about the nature of genetic material and its function in development led to speculation that was based on logic that Wilson characterized as bordering on the metaphysical. Wilson reserved most of his contempt for a theory of development that originated with Wilhelm Roux in 1883. Roux believed that the hereditary material represents different characteristics of the organism. He assumed that the fertilized egg receives all of these substances, which, as cell division ensues, become linearly aligned on the chromosomes. The substances are then distributed unequally to daughter cells, Roux proposed. This "qualitative division" fixes the fate of the cells and their descendants because a portion of determinants is lost to a cell at each division.

August Weismann later elaborated on Roux's ideas. Weismann developed a highly complex scheme for the hereditary material, or **germ plasm**. The primary hereditary units were biophores, which aggregated to form determinants; the determinants formed ids, and the ids formed the larger idants, which were equivalent to chromosomes. He believed that there are two kinds of division: qualitative and quantitative. Weismann proposed that the id gradually disintegrates during development, splitting into smaller and smaller groups of determinants that are isolated into different daughter cells during cell division. Finally, only one kind of determinant remains in a particular cell or group of cells. The determinant then breaks up into its constituent biophores and imparts specific characteristics on the cell. To account for the formation of germ cells that must contain the entire set of heritable information, Weismann proposed that the germ cell line is set aside early in development by quantitative, rather than qualitative, division. By equal distribution of nuclear constituents, the germ plasm remains intact.

This separate developmental pathway for the germ cells was an important concept. Weismann astutely concluded that although the body is derived from the germ plasm, the germ plasm itself is passed on without modification (the concept of mutation was unknown at that time) from one generation to another. Hence, he concluded that inheritance of acquired characteristics, as proposed by Lamarck, was impossible. The usefulness of this concept far outlived that of the Roux-Weismann theory of qualitative division, which was soon disproved.

For his part, Roux (1888) appeared to have confirmed his theories by an experiment he conducted on frog eggs. He used a hot needle to destroy one of the two cells that result from the first division after fertilization. In some cases (Fig. 1.2A), the uninjured half continued developing to form a half-embryo that lacked structures corresponding to the side that was damaged. Roux concluded that this result supported his theories because the undamaged half appeared to lack the information that was necessary to produce the other half of the embryo.

Hans Driesch (1892) approached the problem differently with sea urchin embryos, dissociating them by mechanical shaking at the two-cell stage. These half-embryos developed into normally formed dwarf larvae. Driesch subsequently modified his technique, separating the cells in calcium-free seawater, and found that isolated cells at the four-cell stage also develop normally. Thus, Driesch concluded that each cell retains all the developmental potential of the zygote.

The conflict between these two opposing views of development has been settled in favor of Driesch's interpretation by numerous cell-separation experiments on several animal species. These have included experiments on the frog embryo, the same kind of embryo that Roux used. When the

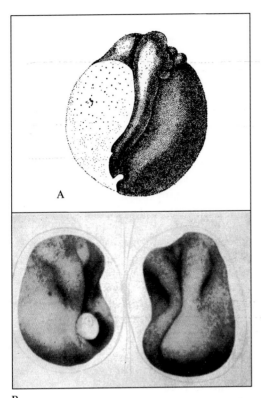

Figure 1.2

Developmental potential of half embryos of the frog. A, Half embryo produced by Roux after destruction of one blastomere with a hot needle. B, Two whole embryos produced by Schmidt after ligation at the two-cell stage. (A, From Morgan, 1927. B, From Schmidt, 1933. A and B reproduced from Balinsky, 1975.)

cells of the two-cell frog embryo are separated with a ligature, both halves proceed to develop normally (Fig. 1.2B).

The experiment conducted by Roux illustrates the importance of proper experimental design. Roux had introduced an artifact into his experiment by allowing the damaged half of the embryo to remain attached to the uninjured half, thus interfering with its development. This fact was demonstrated by Hertwig, who repeated Roux's experiment; however, instead of leaving the damaged cell attached to the normal half, Hertwig removed it. Development of the remainder of the embryo resulted in a complete, half-sized embryo. The interference caused by the death of one cell in Roux's original experiments was only temporary, for Roux himself admitted that the missing half of the damaged embryo was restored during later development, indicating that the frog embryo has the ability to regulate its development by formation of missing parts. Roux should have realized that this observation contradicted his hypothesis. Instead he attempted to rationalize the result. Perhaps this was his greatest failure. Wilson dismissed Roux's rationalization as "an artificial explanation."

Roux's work did have one major lasting effect on developmental biology, however: He had pioneered the *experimental* approach to development. For the first time, an embryologist had manipulated embryos and observed the effects of these manipulations on them. For this reason, many embryologists consider Roux to be the "Father of Experimental Embryology." Actually, Roux's contemporary, Laurent Chabry, had conducted similar experiments on tunicate embryos (see p. 432). However, Chabry's work was little known and, hence, had very little impact on embryology. Once initiated, the experimental approach attracted the attention of other embryologists, who deleted or grafted portions of embryos, centrifuged, and otherwise interfered with normal development in the belief that only by perturbing normal development could they understand how it occurs.

1–3 The Mendelian Era

It was now clear that in cell division the inherited information is equally distributed to all cells of the embryo. But the central question still remained: How does the inherited information participate in development? Frustration began to replace the optimism generated by the discoveries of the significance of the nucleus and chromosomes in inheritance. The year 1900 was the turning point. During that year, the significance of the paper that Gregor Mendel had presented to an unreceptive audience in 1865 was finally appreciated. Mendel's major new insight was that characteristics of organisms are determined by factors that retain their identities through generations of breeding. Each individual inherits a single factor for a trait from each parent. If the parents have "antagonistic" characters, the offspring (hybrid) displays the unaltered trait of one parent, whereas the trait of the other parent is missing. If the hybrid generation is bred with itself, however, the previously missing secondary character could reappear unchanged in the next generation.

Mendelism did not immediately take the scientific world by storm because existing ideas about heredity were often held tenaciously. William Bateson, a Cambridge biologist, was one of the most active proselytizers of the new faith, arguing convincingly and passionately in favor of Mendelism, which some believe was for selfish rather than for altruistic reasons. Bateson coined the term "genetics" and worked hard to establish this new biological discipline. Gradually, nebulous terms such as "biophores" and "idants" were replaced with genes and linkage groups. Mendelism made possible a new precision in thought, and soon the modern concepts of genetics were formulated.

One other experimental observation was also important in setting the stage for this new era. This was reported in a paper written by Theodor Boveri (see p. 2), published in 1902. Although published two years after the rediscovery of Mendelian inheritance, Boveri was apparently unaware of Mendelism. Boveri's work established more precisely the role of chromosomes in development.

His experiments utilized sea urchin embryos with variable numbers of chromosomes (a condition that today is called **aneuploidy**). This abnormal situation can be produced when sea urchin eggs are fertilized by two sperm. The frequency of the **dispermy** is increased by adding an excess number of sperm when fertilizing sea urchin eggs *in vitro*. The two sperm nuclei and the egg nucleus form an abnormal division figure, usually with four centrioles, so that the first mitotic division produces four cells that rarely receive the normal diploid number of chromosomes (Fig. 1.3). Because the distribution of chromosomes to these four cells is random, cells with variable numbers of chromosomes are produced from a single zygote.

Boveri separated the four cells from one another by placing the dispermic egg in calcium-free seawater and observed the development of each. Subsequent cleavage divisions were normal, thus fixing the abnormal chromosome number in the embryos developing from the separated cells. These embryos never developed normally. But *most importantly,* each typically developed in a different way, arresting at a different stage. Boveri concluded that the different developmental patterns were due to the different chromosomal combinations produced in the abnormal first division. Thus, normal development is dependent upon the *normal combination of chromosomes.* Clearly, each chromosome must have qualitatively unique effects on development.

That conclusion seems self-evident to us today—each chromosome contains a linear sequence of genes, and the gene sequences on the

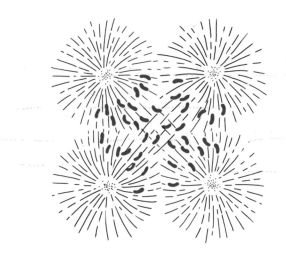

Figure 1.3

Anaphase of the first cleavage of dispermic sea urchin eggs. Each sperm contributed two centrioles, each of which attracts chromosomes at random. (Redrawn from Wilson, 1925, with permission of Macmillan Publishing Company from *The Cell in Development and Heredity*, 3rd ed. Copyright © 1925 by Macmillan Publishing Company, renewed 1953 by Anna M. K. Wilson.)

chromosomes are unique. But only a minority of biologists of the early 1900s (Wilson among them) accepted the conclusion that the hereditary factors are on the chromosomes. The evidence for the chromosome theory was based on both cytological and genetic data. Although Wilson enthusiastically promoted Boveri's conclusions, they were dependent upon indirect evidence; more convincing proof was needed.

Walter Sutton, a graduate student working with Wilson, was the first to make a clear correlation between the behavior of Mendel's genetic factors and the behavior of chromosomes. Sutton pointed out that the parallel behavior of hereditary factors and chromosomes must mean that the factors are localized on the chromosomes. Thus, in a haploid gamete, genes and chromosomes are singly represented. After fertilization, the cells of the diploid zygote contain a set of chromosomes that are derived from each parent; the chromosomes are present in homologous pairs, and the members of each chromosome pair possess the alleles, which specify the maternal and paternal traits, respectively. During the formation of gametes in the mature diploid organism, the homologous chromosomes pair, and as a result of meiosis, every gamete receives one chromosome of each homologous pair (Mendel's law of segregation). Furthermore, because the behavior of each homologous pair of chromosomes during meiosis is independent of that of every other pair, the chromosomes are distributed randomly to the gametes, resulting in several combinations of maternal and paternal chromosomes (Mendel's law of independent assortment). Sutton also noted that if two genes were located on the same chromosome, they would be inherited together and would not behave according to Mendelian principles. He had, therefore, foreshadowed the concept of linkage. The details of meiosis are outlined in Chapter 2.

The final proof of the chromosome theory of inheritance resulted from experiments conducted in the laboratory of Thomas Hunt Morgan (see p. 578), utilizing the fruit fly, *Drosophila melanogaster*. Mostly as a result of work by Morgan and his associates, A. H. Sturtevant, C. B. Bridges, and H. J. Muller, the basic concepts of transmission genetics were rapidly discovered. These concepts (which are still generally valid, although there are exceptions) have been summarized by Moore (1972). They are as follows:

1. *Inheritance is the transmission of genes from parents to offspring.* As we shall learn later, some of the characteristics of the offspring may be determined exclusively by the maternal genome during formation of the oocyte in the ovary. Genes that function in this way are called **maternal-effect genes**. The effects of the paternal alleles of maternal effect genes are exerted during formation of oocytes when their female offspring reach maturity and are manifest in the phenotype of the next generation.

2. *Genes are located on chromosomes.* Some inheritance, however, is based upon nonchromosomal structures, such as in mitochondria and chloroplasts. Evidence has recently been presented that basal bodies in the unicellular alga *Chlamydomonas reinhardtii* may also contain DNA (Hall et al., 1989).

3. *Each gene occupies a specific site (locus) on a chromosome.* Barbara McClintock (1951) has shown that genes in maize can undergo rearrangements in the genome. Her work (for which she was recently awarded the Nobel Prize) demonstrated that the genome contains mobile **transposable elements**, which have recently been exploited experimentally by investigators working with *Drosophila* to introduce exogenous genes into the genome (Rubin and Spradling, 1982).

4. *Each chromosome has many genes that are arranged in linear order.* Some chromosomes, such as the Y sex chromosome in *Drosophila*, have only a few genes.

5. *Somatic cells of diploid organisms contain two of each kind of autosomal chromosome (homologues), and thus each autosomal gene locus will be represented twice.* In some species, different kinds of individuals may have differences in chromosome number. For example, male bees (drones) are haploid, whereas female bees (queens and workers) are diploid. In other organisms, some cells may have multiple numbers of chromosomes. In *Drosophila*, for example, the nurse cells (see p. 56) are polyploid (i.e., with multiple chromosome sets).

6. *In species with nonhomologous (i.e., heteromorphic) sex chromosomes, genes may only be represented once.* In mammals and *Drosophila*, for example, males are the heterogametic sex, having one X chromosome and one Y chromosome. Females, which have two X chromosomes, are the homogametic sex.

7. *During each cell-division cycle, each gene is replicated.*

8. *Genes can exist in several alternative states (alleles). The change from one state to another is a mutation.*

9. *Genes can be transferred from one homologous chromosome to the other by crossing-over during meiosis.* Crossing-over does not occur in all instances (e.g., in the *Drosophila* male).

10. *Every gamete receives one chromosome of each homologous pair; the distribution of chromosomes to the gametes is random.* In species with the XO pattern of sex determination, half of the gametes will receive no sex chromosome.

11. *The distribution of chromosomes of one homologous pair to the gametes has no effect on the distribution of chromosomes of other pairs.* In some cases, however, chromosomes are distributed to gametes in specific groups.

12. *At fertilization, gametes unite randomly; the zygote receives one chromosome of each homologous pair from its father and one from*

its mother. As with rule 9, sex chromosomes can complicate this rule.

13. *When the cells of an organism contain two different alleles of the same gene (heterozygous condition), one allele (the dominant) usually has a greater phenotypic effect than the other (the recessive). In some cases, however, heterozygotes have intermediate phenotypes.*

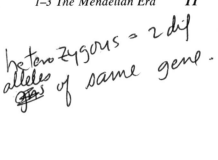

heterozygous = 2 dif alleles ~~at~~ of same gene.

The choice of *Drosophila* by Morgan was indeed a fortunate one. These remarkable creatures are tailor-made for genetics research: They are easily bred and reproduce rapidly (a single pair can produce several hundred progeny in a couple of weeks), and large numbers of them can be maintained in small vials or bottles containing simple, inexpensive "fly food." Although *Drosophila* are small, their external characteristics are simple to score with no more than a hand lens, which Morgan used in his early work.

Geneticists and cytologists were soon to discover other qualities that made *Drosophila* a nearly ideal organism for genetic research. For example, the haploid chromosome number is four, meaning that there are only four linkage groups. Furthermore, certain larval cells have giant chromosomes called **polytene chromosomes**. These structures were described in the larval salivary glands by T. S. Painter in 1934. Actually, Painter had "rediscovered" polytene chromosomes; they were known to the cytologists in 1881, and a drawing of polytene chromosomes can be found in the 1896 edition of *The Cell*. Although one would expect salivary gland nuclei to contain eight chromosomes, Painter found only four. This is due to pairing of homologous chromosomes. The most important aspect of polytene chromosomes is the presence of cross-bands on them, as shown in Painter's first published drawing of salivary gland chromosomes in Figure 1.4. The cross-banding extends across both paired chromosomes; the pairing of the homologues is so exact that each band appears as a single structure. An explanation of

— Why?

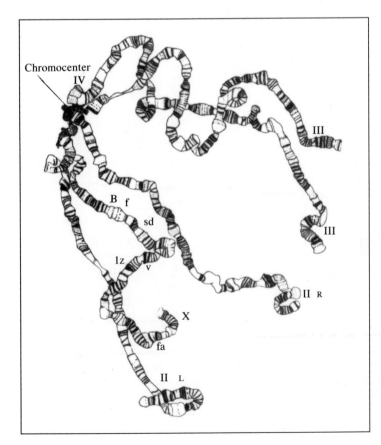

Figure 1.4

Painter's first drawing of the salivary gland chromosomes of *Drosophila melanogaster*. The chromosomes radiate out from the chromocenter. The X chromosome is attached to the chromocenter by one end so it appears as a single long structure. Both the II and III chromosomes are attached by their middle portions. Consequently, both of these chromosomes have two arms extending from the chromocenter. The tiny IV chromosome is attached by its end to the chromocenter. The approximate locations of several X chromosome genes (*B*, *f*, *sd*, etc.) are shown. (From Painter, 1934.)

these chromosomes in molecular terms can be found in Chapter 10. For the early *Drosophila* geneticists, however, the significance of polytene chromosomes was that chromosome markers were now available.

The bands are in various shapes and sizes, and interband distances are variable. In different cells of the same larva, however, and even in different larvae, the band pattern itself is constant. The pattern apparently represents the essential organization of the chromosome, and as was soon discovered, it marks the organization of genes within the chromosomes. Once again, cytology and genetics joined forces, and a new branch of genetics called **cytogenetics** emerged. For the first time, the concepts of transmission genetics, such as gene deletions and duplications, translocations, and inversions, could be correlated with chromosome markers. These correlations have resulted in very elaborate chromosome maps showing genes in a linear order and in a definite sequence (Fig. 1.5).

1–4 Embryology: Losing the Faith

One of the most important concepts to emerge from Wilson's writing concerns the nature of the relationship among genes, cells, and the developing

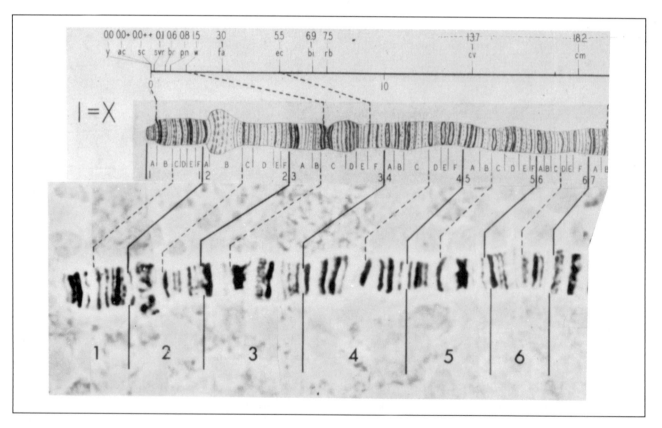

Figure 1.5

Photographic map of a portion of the *Drosophila* X chromosome, correlated with Bridges' (1935) drawings. The cytological positions of several of the common sex-linked genes are indicated. (Reproduced with permission from Lefevre, 1976. Copyright © Academic Press Inc. [London] Ltd. Bridges' chromosome map reprinted by permission of the American Genetic Association.)

embryo. Gene function generates cellular structural and functional characteristics in a process called **cell differentiation.** The characteristics of the whole organism are, in turn, the net result of the characteristics of its individual cells. Thus, genes function at the cellular level to cause development of the organism, or, as stated by Wilson, development is "the appearance of hereditary traits in regular order of space and time."

What is the specific role of the genetic material in this process? How is temporal and spatial coordination of cell differentiation regulated by the nucleus? Wilson pointed out that the nucleus itself does not cause development. Although the nucleus alone suffices to transmit inherited information, it must work in concert with the cytoplasm. Wilson proposed that the specific role of the nucleus is the regulation of "constructive metabolism" (i.e., synthetic metabolism) of the cell, which is fundamentally a problem of biochemistry and was well beyond the level of understanding of scientists of that time. Wilson and his contemporaries had expected that genetics and cytology would together lead the way to a solution of the "how" and "why" questions of development. However, Wilson admitted that they had underestimated the magnitude of the problem of development. Just as the questions of heredity required the Mendelian principles and the chromosome theory, explanation of the mechanisms of development would require the understanding of new, and as yet undiscovered, principles. Studies on the role of genes in development could not progress significantly until the nature of the gene was discovered and advances were made in understanding how genes function to influence cell biochemistry. An era had ended.

The embryologists' disillusionment with genetics was complete. Many embryologists considered genetics to be peripheral to the fundamental mechanism of development. They believed that the basic structure of an embryo is produced by *embryological* mechanisms and that the only function of genes is to add the nonessential finishing touches, such as eye and hair color, number of bristles on a leg segment, and color of flower petals. Emphasis during the 1920s, 1930s, and 1940s was on describing embryonic development and on experimentation into the interactions and rearrangements of the cells and tissues that form embryos. Although the emphasis had changed, the intensity of research had not abated. This period was truly the heyday of embryology, and it was dominated by one issue: **embryonic induction**. As we shall discuss in detail in Chapter 12, Hans Spemann (see p. 477) and his student, Hilde Mangold (see p. 196), demonstrated in 1924 that the transplantation of the dorsal lip of the blastopore of a *Triturus* (urodele amphibian) gastrula-stage embryo into the ventral region of a host embryo stimulates, or **induces**, host tissue to form a secondary embryonic axis, including neural tissue. Spemann reasoned that the behavior of an implanted dorsal lip reflects its normal function in inducing the primary axis and designated it as the embryonic **organizer**.

The remarkable properties of the organizer captured the imagination of a generation of experimental embryologists and became their overriding preoccupation. A vast amount of experimental data was collected, but the essential nature of the organizer and the process of embryonic induction remained a mystery. As we shall discuss in Chapters 12 and 14, we are now in a new era of understanding of induction and are on the threshold of major new breakthroughs in understanding how cell-cell interactions are involved in organizing the embryonic body plan.

Current progress in understanding embryonic induction has come from a realization that a class of protein molecules called **growth factors**, which were discovered by Rita Levi-Montalcini and Viktor Hamburger (see p. 242) in the 1950s, is involved in some inductive events. The first growth factor to be discovered was **nerve growth factor (NGF)**, a diffusible substance that was shown to promote the outgrowth of nerve fibers from spinal and sympathetic ganglia of chick embryos (Levi-Montalcini et al., 1954). Since the discovery of NGF, a large number of growth factors have been discovered that play important roles in a variety of developmental processes, including embryonic induction.

1–5 Genetics: Keeping the Faith

Because very few embryologists concerned themselves with gene function in development, it became incumbent upon geneticists to keep alive the concept that development occurs under genetic control. That this was truly a "rear-guard action" can be deduced from this quote by Richard Gold-schmidt (1958): "In spite of such isolationism or possessiveness [by embryologists], the geneticists will continue to worry about the problem of genetic action and take the risk of climbing over the fence erected by some jealous embryologists, who, while claiming the kingdom for themselves, do not set out to till its soil."

It is not surprising that geneticists became strong advocates of the idea that genes play an important role in basic developmental processes, because the new mutations they were discovering provided strong evidence for this fundamental principle. In *Drosophila*, for example, there are mutants that affect virtually every aspect of development from the most general down to the smallest detail. T. H. Morgan, who was an embryologist before he began studying *Drosophila* genetics, retained an interest in embryonic development and, in his book, *Experimental Embryology*, published in 1927, made the following remarkable statement concerning the roles of genes in development:

> One of the most important questions for embryology relating to the activity of the genes cannot be answered at present. Whether all the genes are active all the time, or whether some of them are more active at certain stages of development than are others, are questions of profound interest.

Thus, he was speculating about the possibility of chronological regulation of differential gene expression during development. Significantly, Morgan's laboratory was in the same department at Columbia University in which Wilson worked. In fact, Wilson as department chairman was instrumental in recruiting Morgan. It is likely that they reinforced one-another's interest in the roles of genes in development.

Some of the most striking mutations that affect *Drosophila* development are the **pattern formation genes**, which are involved in establishing the basic body plan of the fly. The body of an adult *Drosophila* consists of a head, thorax, and abdomen. The thorax is made up of three segments: the **prothorax**, the **mesothorax**, and the **metathorax**. Each of these segments bears a pair of legs; in addition, the mesothorax bears the wings, and the metathorax has a pair of rudimentary appendages called **halteres**. The halteres are balancing organs that maintain the fly in an upright position while flying. The combination of two wings and two halteres is unique to the dipterans; most insects have four wings—one pair on the mesothorax and a second pair on the metathorax. Mutations of genes in the **bithorax** gene complex can convert *Drosophila* from a typical dipteran to a four-winged creature by producing a pair of wings on the metathorax in place of the halteres (Fig. 1.6).

The detection of a mutant gene affecting a developmental process implies that the normal allele of the gene is involved in control over that

Figure 1.6

Photographs of wild-type (A) and four-winged (B) *Drosophila*. The four-winged condition is produced by combining three mutations of the bithorax gene complex. The developmental abnormality is the consequence of reduction in function of the genes *abx*, *bx³*, and *pbx*. (Courtesy of E. B. Lewis.)

developmental step. The nature of the defect often indicates which developmental process is affected. The sequence of gene-controlled steps in development of a body region can be reconstructed by careful analysis of a series of mutations that affect its development. Analysis of development in this manner is called **genetic dissection**. A very large number of pattern formation genes is known for *Drosophila*. These genes have been subjected to intensive analysis, providing us with substantial insight into the establishment of the *Drosophila* body plan. This topic, which is one of the most intensely studied and most productive aspects of contemporary developmental biology research, is discussed in detail in Chapters 6 and 14.

Many geneticists who were interested in development during the 1930s and 1940s were asking how genes produce their effects. As we have already noted, biologists had long suspected that genes are involved in regulating cellular metabolism. Therefore, the discovery of the specific function of genes could be found only by utilizing a biochemical approach. New insight into the nature of cellular metabolism began emerging from biochemical studies at approximately the same time that the new science of genetics was making rapid discoveries about the mechanism of inheritance. These critical biochemical discoveries involved the role of enzymes in biochemical reactions. Serious experimental analysis of the role of genes in metabolism was initiated by George Beadle, working initially with Boris Ephrussi and subsequently with Edward Tatum. Beadle and Tatum (1941) demonstrated the validity of the **one gene–one enzyme hypothesis**, which states that the primary function of a gene is to produce a specific enzyme. Subsequently, the rule was generalized to "one gene–one protein," because all proteins, not only enzymes, are gene-dependent. However, proteins are frequently composed of unlike subunits (polypeptides), which are specified by different genes. For example, the hemoglobin molecule is composed of four subunits representing two different polypeptide chains, each specified by a different gene. Hence, the rule can be restated as "one gene–one polypeptide."

1–6 The Modern Era

Once the function of genes in the cell was established, one of the key stumbling blocks to a clear understanding of the role of genes in development had been removed. It remained only to determine the composition of genes and the way that the genetic information is utilized. During the late 1940s, 1950s, and early 1960s, several important breakthroughs emerged that laid the groundwork for a renewed onslaught of investigations into the role of genes in development and the mechanisms involved in cell differentiation. These contributions resulted mainly from work in biochemistry and two new disciplines—cell biology and molecular biology. Biochemists gathered valuable data on the relationships between genes and proteins and the role of enzymes in cellular metabolism. Cell biologists described the structure and function of cellular components, aided by a powerful new tool, the electron microscope. Most importantly, however, molecular biologists determined the nature and structure of the genetic material, the genetic code was broken, the protein synthetic machinery was unraveled, and genomic regulatory elements were discovered.

Nearly all of the basic principles of molecular biology were obtained from work on bacteria and viruses, but these principles also apply to multicellular plants and animals, with certain variations. The rapid developments in molecular biology are too numerous to discuss here in detail, but the contemporary student of development should be intimately familiar with the discoveries that form the basic foundations of molecular biology. Some of the more important ones are the following:

1. *Genetic information is encoded in deoxyribonucleic acid (DNA) as two antiparallel polynucleotide strands in a double-helical structure (the Watson-Crick double helix).*
2. *Genetic information is stored in a linear sequence of purine and pyrimidine bases.* The genetic alphabet consists of four letters (bases): Adenine, Thymine, Guanine, Cytosine.
3. *During replication of DNA, each strand serves as a template for the formation of the complementary strand.* Each base of the template strand specifies a nucleotide bearing the complementary base: A is complementary to T, G is complementary to C, and vice versa.
4. *Genetic information in the chromosomes is expressed by transcription of the sequence of bases in DNA into a complementary sequence of bases in ribonucleic acid (RNA).* During transcription, each base of the DNA template strand specifies a ribonucleotide bearing the complementary base: A is complementary to Uracil, T is complementary to A, G is complementary to C, and C is complementary to G.
5. *The genetic information for synthesis of a specific protein is a* **structural gene**. Structural genes are transcribed into messenger RNA, which is transported into the cytoplasm, where it is translated into protein.
6. *Proteins are composed of a linear sequence of amino acids.* The placement of an amino acid in protein is designated by a triplet of bases in mRNA called a **codon**.
7. *The genetic code is (nearly) universal, nonoverlapping (except in some viruses), and degenerate (i.e., redundant), and it contains codons that punctuate protein synthesis (i.e., "start" and "stop" codons).* There are, however, some differences between the codon assignments in the "universal" code and the code in mitochondria.

Another exception to the general properties of the genetic code is the presence of overlapping genes in some viruses.

8. *Protein synthesis occurs on ribosomes, which are composed of protein and ribosomal RNA.* The genetic code is read by transfer RNA molecules. These molecules recognize a specific codon and bear the corresponding amino acid, which is added to the sequence of amino acids by formation of a peptide bond. Like messenger RNA, ribosomal and transfer RNA are transcribed from genes. Thus, not all genes code for polypeptides.

9. *Recent investigations have revealed the presence of nucleotide sequences whose function is to regulate the transcription of structural genes.* These regulatory sequences may not produce transcripts themselves.

By the late 1950s and early 1960s, enough of the critical pieces were in place. Molecular biologists began asking how gene function can be controlled to produce the wide variety of cell types in adult organisms. Biochemists began analyzing the changes in cellular biochemistry that occur during development. Cell biologists began monitoring the structural and functional changes that accompany cell differentiation and studying the cellular basis for morphogenesis. Genes that modify development were recognized as valuable tools in understanding normal developmental events. Developmental biology emerged as a vital, exciting, broadly based science. Interdisciplinary barriers fell as investigators came to realize that plant and animal development have much in common and that simple organisms such as algae and slime molds are excellent model systems for studying cell differentiation. Horizons were broadened with the realization that developmental events occur during all phases in the life span of an organism, not only during embryogenesis.

In such a broadly based science, the student is expected to understand concepts that are diverse in nature but have a common goal. That goal is to interpret the processes that produce a fly, a frog, or a human from a fertilized egg. Although the analysis is multileveled, each approach analyzes the same phenomenon from its own particular perspective. The formation of an eye is viewed differently by the biochemist, the electron microscopist, and the molecular biologist, but the eye develops nevertheless, and it is up to the developmental biologist to integrate the information provided by these diverse approaches and describe how and why the eye develops. As recently discussed by Stent (1985), development is an historical phenomenon. Thus, at any time in development, the processes occurring are both the effects of earlier developmental events and the cause of later events. Like an historian, the developmental biologist must establish the linkage between events occurring on a time scale, sorting out the sequential events that lead to cell commitment and cell differentiation. It is this dynamic aspect of developmental biology that holds such great fascination and complicates its study.

Developmental biology has made remarkable strides in a remarkably short period of time. As we mentioned at the beginning of this chapter, genetics, cell biology, and molecular biology are currently having an extraordinary impact on our understanding of development. For example, genes that control many developmental events have been identified, cloned, and manipulated. Cellular factors that regulate their expression have been identified, and the genes encoding some of these factors have also been identified and cloned. By this procedure, we are reconstructing some of the hierarchies of gene expression that allow the embryo to progress from one stage of development to another and that allow for different cell types to emerge during development in the correct locations. Morphogenesis depends upon the concerted activities of many cells that must be organized into discrete functional entities. Cells must often move relatively large distances within the embryo in order to reach their final destinations. Once they have reached their final destinations, they must establish stable multicellular structures. This involves numerous cell-cell interactions that may be necessary to construct multicellular structures or that are necessary to influence the behavior of their neighbors. Cell biology has provided the techniques to analyze these aspects of morphogenesis.

The progress made thus far in the study of development only serves to whet our appetite and to encourage us to ask even more difficult questions about the mechanisms that control development. Many unanswered questions about development remain that will require keen insight and clever new approaches. We hope that this textbook conveys to our readers the excitement and vibrancy of contemporary developmental biology and stimulates some of you to tackle the many remaining questions about development.

1–7 Chapter Synopsis

Development of complex organisms is one of the most fascinating processes in nature and has captured the imagination of scientists and philosophers alike for centuries. Speculation about the mechanisms of development, which dominated the early scientific literature, ultimately gave way to the experimental approach to development. Experimental embryologists utilized techniques to manipulate embryos and observed the effects of the manipulations on their development. One of the major accomplishments of experimental embryology was the demonstration of embryonic induction.

E. B. Wilson, who championed the multidisciplinary approach to development, provided the philosophical foundations of developmental biology. He believed that the key to understanding development would be found by studying the cells that comprise the embryo. One of the major themes running through the developmental biology literature is the role of the genome in development. Our knowledge of developmental events has recently mushroomed, due to the impact of genetics, cell biology, and molecular biology on the study of development. In particular, the application of the techiques of recombinant DNA technology has had an explosive impact on developmental biology. The ability to identify and clone genes that regulate developmental events has made it possible to begin to relate development to the action of genes, to identify individual gene products, and to investigate how gene function itself is regulated during development.

References

The history of developmental biology is thoroughly documented in a very readable book by John A. Moore, *Heredity and Development*. The authors have used Dr. Moore's book extensively in developing this chapter. The reader is referred to it for more details concerning the fascinating story of this science.

Balinsky, B. I. 1975. *An Introduction to Embryology,* 4th ed. W. B. Saunders, Philadelphia.

Beadle, G. W., and E. L. Tatum. 1941. Genetic control of biochemical control of chemical reactions in *Neurospora.* Proc. Natl. Acad. Sci. U.S.A., *27*: 499–506.

Bonnet, C. 1745. *Traité d'Insectologie*. Paris.

Bridges, C. B. 1935. Salivary chromosome maps. J. Heredity, *26*: 60–64.

Driesch, H. 1892. Entwicklungsmechanisme Studien. I. Der Werth der beiden ersten Furchungszellen in der Echinodermentwicklung. Experimentelle Erzeugen von Theil- und Doppelbildung. Zeitschrift für wissenschaftliche Zoologie, *53*: 160–178; 183–184.

Goldschmidt, R. B. 1958. *Theoretical Genetics*. University of California Press, Berkeley.

Hall, J. L., Z. Ramanis, and D. J. L. Luck. 1989. Basal body/centriolar DNA: Molecular genetic studies in Chlamydomonas. Cell, *59*: 121–132.

Hertwig, O. 1876. Beiträge zur Kenntnis der Bildung, Befruchtung und Teilung des tierischen Eies. Morphol. Jahrb., *1*: 347–434.

Lefevre, G., Jr. 1976. A photographic representation and interpretation of the polytene chromosomes of *Drosophila melanogaster* salivary glands. *In* M. Ashburner and E. Novitski (eds.), *The Genetics and Biology of Drosophila,* Vol. 1a. Academic Press, London, pp. 31–66.

Levi-Montalcini, R., H. Meyer, and V. Hamburger. 1954. *In vitro* experiments on the effects of mouse sarcomas 180 and 37 on the spinal and sympathetic ganglia of the chick embryo. Cancer Res., *14*: 49–57.

Lewis, E. B. 1963. Genes and developmental pathways. Am. Zool., *3*: 33–56.

McClintock, B. 1951. Chromosome organization and genic expression. Cold Spring Harbor Symp. Quant. Biol., *16*: 13–47.

Moore, J. A. 1972. *Heredity and Development,* 2nd ed. Oxford University Press, New York.

Morgan, T. H. 1927. *Experimental Embryology*. Columbia University Press, New York.

Painter, T. S. 1934. A new method for the study of chromosome aberrations and the plotting of chromosome maps in *Drosophila melanogaster*. Genetics, *19*: 175–188.

Roux, W. 1888. Beiträge zur Entwicklungsmechanik des Embryo. Ueber die künstliche Hervorbringung halber Embryonen durch Zerstörung einer der beiden ersten Furchungskugelin, sowie über die Nachentwicklung (Postgeneration) der fehlenden Köperhälfte. Virchows Arch. Pathol. Anat. Physiol., *114*: 113–153; 289–291.

Rubin, G. M., and A. C. Spradling. 1982. Genetic transformation of *Drosophila* with transposable element vectors. Science, *218*: 348–353.

Schmidt, G. A. 1933. Schnürungs- und Durchschneidungsversuche am Amphibienkeim. Wilhelm Roux's Archiv für Entwicklungsmechanik, *129*: 1–44.

Spemann, H., and H. Mangold. 1924. Über Induktion von Embryonalanlagen durch Implantation artfremder Organisatoren. Wilhelm Roux's Archiv für Entwicklungsmechanik, *100*: 599–638.

Stent, G. 1985. Thinking in one dimension: The impact of molecular biology on development. Cell, *40*: 1–12.

van Beneden, E. 1883. Recherches sur la maturation de l'oeuf, la fécondation et la division cellulaire. Arch. de Biol., *4*: 265–640.

Wilson, E. B. 1896. *The Cell in Development and Inheritance*. Reprinted by Johnson Reprint Corp., New York (1966).

Wilson, E. B. 1900. *The Cell in Development and Inheritance,* 2nd ed. Macmillan, New York.

Wilson, E. B. 1925. *The Cell in Development and Heredity,* 3rd ed. Macmillan, New York.

Wolff, C. F. 1759. *Theoria generationis*. Halle.

PART II

Gametogenesis

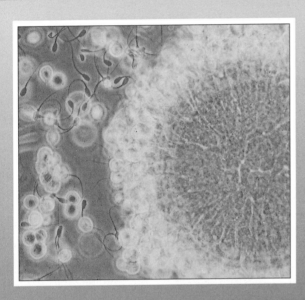

The end-products of gametogenesis in humans: an ovum surrounded by sperm. (Courtesy of SIU/Peter Arnold, Inc.)

CHAPTER 2

Spermatogenesis

D. W. Fawcett (b. 1917). Fawcett, a Professor
at Harvard University, has contributed more
than any other single investigator to our
knowledge of sperm. His descriptions of sperm
and their morphogenesis are masterful
examples of how electron microscopy can
yield valuable details on complex cells and
their formation.

The gametes (eggs and sperm) provide both the blueprint and the raw material from which the embryo is formed. Because the events of gametogenesis have such pervasive effects on embryonic development, we consider this topic in some detail. We first discuss some of the general aspects of gametogenesis and compare the process in males and females before discussing the process in detail—first in males, then (Chap. 3) in females.

SECTION ONE

Overview of Gametogenesis

Gamete formation in the two sexes is tailored to the roles of their gametes in reproduction. The male gametes are usually small and mobile. They are dispensed from the male reproductive organ—often into a hostile environment—and they must locate the female gamete, make contact, and fuse with it. The female gamete is usually less mobile than the sperm and larger, often by several orders of magnitude. The female gamete must be "competent" to be fertilized, which means that it must develop a number of specialized properties to enable it to interact with the sperm. Both classes of gametes make an equal contribution to the nucleus of the zygote, each providing a haploid genome. However, the male gamete makes a minimal contribution to the cytoplasm; the female gamete provides the zygote with virtually all of the cytoplasm, which contains the constituents from which the embryo is fashioned.

Reproduction is a prime concern for any species because it ensures the species' survival. The **germ cell line** is, therefore, a precious commodity, and its formation is an important developmental event, which is often one of the first orders of business for the embryo after fertilization. The germ cell line may derive its specificity from a specialized cytoplasmic constituent, the germ plasm, which may pre-exist in the egg before fertilization and become segregated into the germ cell line during cleavage (see Chaps. 3 and 11). Determination of the germ cell line results in two distinct categories of cells in the embryo. The nongerm cells are called the **somatic cells**. This distinction is retained throughout the life of the organism.

The initial cells in the germ line are called **primordial germ cells**. The primordial germ cells of both sexes are indistinguishable from one another. The acquisition by germ cells of sex-specific characteristics occurs at a later stage of development and is culminated by the formation of mature sex cells with distinctly different shapes and organelles. The primordial germ cells

Regular text continues on page 26.

Essay

Meiosis

Meiosis involves two consecutive cell divisions that reduce the diploid chromosome number to the haploid condition. The major events of meiosis are illustrated in Figure 1 (opposite page), which also compares meiosis to mitosis for an animal with a diploid number of four (i.e., two homologous pairs of chromosomes). Both meiosis and mitosis are preceded by DNA duplication during the S phase of interphase, and the chromosomes then consist of two chromatids, although the individual chromatids are not distinct until later in prophase.

Meiotic prophase is highly specialized to facilitate exchange of material between homologous chromosomes. This exchange occurs by **crossing-over**, which results in recombination of genes in a linkage group. Because of its complexity, prophase I is subdivided into a number of arbitrary stages. The first stage of prophase I is **leptonema**.* During leptonema, the chromosomes appear as long, thread-like structures. Homologous chromosomes then form pairs during **zygonema** in a process called **synapsis**. One member of each pair is the maternal chromosome, which was originally inherited from the mother, and the other is the paternal chromosome, which was obtained from the father. The homologues are represented by distinct colors in Figure 1; thus, each homologue can be followed through meiosis. The pairing process results in close apposition between the homologues, which is essential for crossing-over.

The next stage, **pachynema**, is characterized by a shortening of the chromosomes. During **diplonema**, the chromatids composing each homologue become individualized, and each synapsed pair of chromosomes is seen to be composed of four chromatids and is now called a **tetrad**. Diplonema is often of extremely long duration, particularly in the female. Diplotene chromosomes in some species become highly modified, with numerous loops extending laterally from the chromosome axis. These **lampbrush chromosomes** are very active synthetically and may play an important role in differentiation of the gamete. We shall discuss lampbrush chromosomes in more detail in Chapter 3. During diplonema, the homologues separate but remain joined at **chiasmata**, which are the points near the sites of physical exchange between chromatids during crossing-over, as shown in Figure 2.

The chiasmata move to the ends of the homologues during **diakinesis** in a process called terminalization. During this stage, the chromosomes shorten, the nuclear envelope breaks down, and the chromosomes begin moving toward the metaphase plate. The first division is characterized by separation of homologues from one another. A short interphase separates the two meiotic divisions. There is *no* DNA synthesis during this interphase. The second division is much more rapid and less complicated than the first. As in mitosis, the chromosomes do not pair but become aligned in tandem on the metaphase plate, and the centromere of each chromosome divides, allowing separation of sister chromatids.

The essential modification of meiosis that results in production of haploid cells is that there is only one duplication of chromosomes for the two divisions. In contrast, each mitotic division involves chromosomal duplication and division, maintaining a constant chromosome number at each division.

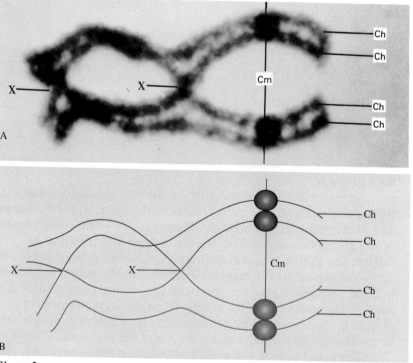

Figure 2

The meiotic tetrad. A, Photomicrograph of a tetrad from a male salamander primary spermatocyte in diplonema. Note the four chromatids (Ch). The centromeres are visible at Cm, and each chiasma is indicated by an X. B, Interpretive drawing of A. (A, Courtesy of J. Kezer, from Villee et al., 1989. B, After Villee et al., 1989.)

---

* Stages of prophase I end in ''-nema'' when used as a noun and in ''-tene'' when used as an adjective (e.g., leptonema, leptotene).

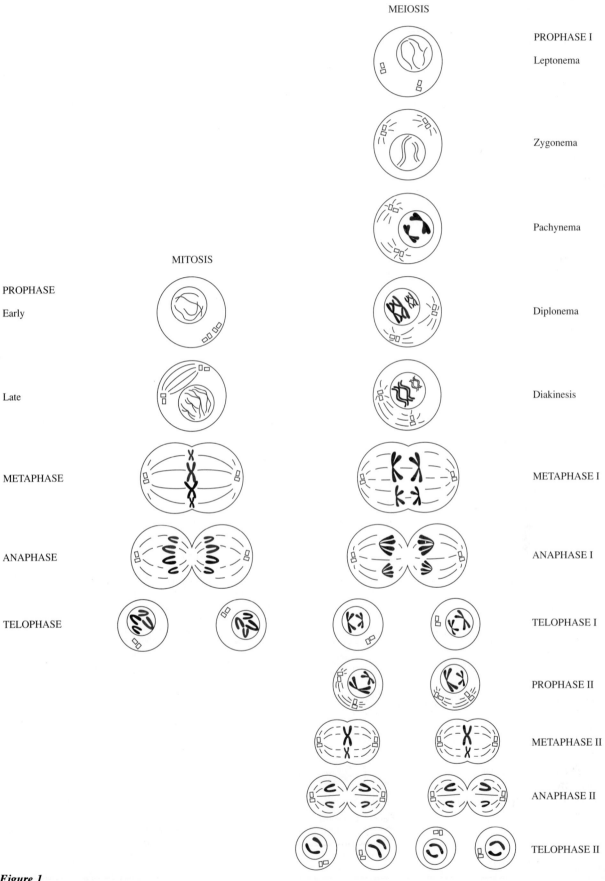

Figure 1

Meiosis and mitosis compared.

may arise at some distance from the presumptive gonads to which they migrate and become established as **stem cells**, which increase in number by mitosis. The establishment of germ cells in the gonads often involves a close association between the germ cells and the somatic cells of the gonad. These somatic cells may serve to support and protect the germ cells and to provide them with nutritive material.

In the female, the somatic cells surrounding the germ cell are called **follicle cells**. In the male, various terms have been used for them. Among the most familiar examples are the **Sertoli cells** in mammalian testes. During the proliferative phase, the germ cells are called gonia (**spermatogonia** in the testis and **oogonia** in the ovary) and act as a stem cell population for the production of cells that will differentiate into functional gametes. The gonial cell divisions may be incomplete, so that the daughter cells remain in communication with one another via intercellular bridges. Successive incomplete divisions produce very large clones of interconnected cells. This intercellular communication may serve to synchronize the development of the conjoined cells. The formation of sperm from spermatogonia is **spermatogenesis**, and the formation of ova (or eggs) from oogonia is **oogenesis**. These processes involve the reduction in chromosome number by meiosis (see essay on p. 24) and acquisition of the structural and functional characteristics of the distinct sex cells.

In the male, meiosis precedes sex cell differentiation (Fig. 2.1). A single spermatogonium enters the first meiotic division as a **primary spermatocyte**. This division produces two **secondary spermatocytes**, each of which divides to form two haploid **spermatids**. Each spermatid then differentiates (by a process called **spermiogenesis**) into a **spermatozoon** by the elaboration of structural and functional specializations that enable the sperm to fertilize the egg. Consequently, four haploid sperm result from each diploid spermatogonium. The utilization of all four haploid cells in the male is significant

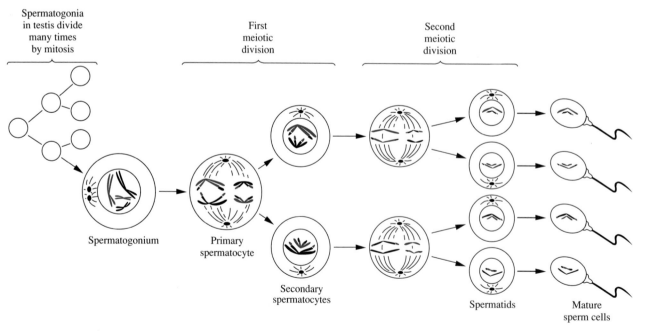

SPERMATOGENESIS

Spermatogonia in testis divide many times by mitosis

First meiotic division

Second meiotic division

Spermatogonium

Primary spermatocyte

Secondary spermatocytes

Spermatids

Mature sperm cells

Figure 2.1

Spermatogenesis. Each primary spermatocyte gives rise to four spermatids, which differentiate into mature spermatozoa. (After Villee et al., 1989.)

OOGENESIS

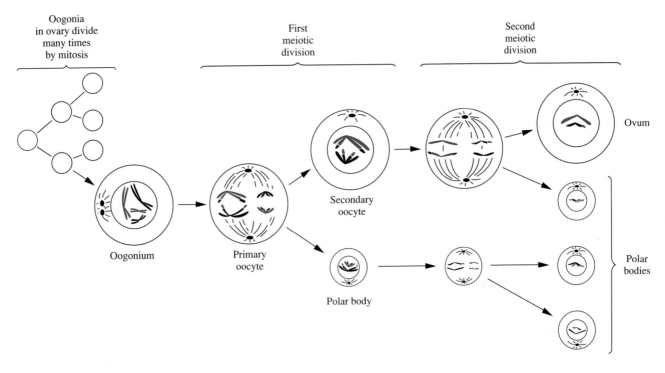

Figure 2.2

Oogenesis. A single functional ovum is produced from each primary oocyte. Polar body sizes are exaggerated in this drawing. (After Villee et al., 1989.)

because the testis must produce millions of sperm simultaneously. The loss of any of the cells during meiosis would make this task monumental. The spermatogonial mitotic divisions may be incomplete, leaving daughter cells in continuity with one another via cytoplasmic bridges. Because clusters of interconnected cells are produced from a single spermatogonium, these cells can be considered **clones**. Meiotic divisions may also be incomplete, thus enlarging the clones, which are then composed of numerous haploid spermatids. The intercellular bridges that connect members of a clone are lost in the final stages of spermiogenesis when excess cytoplasm is sloughed from the sperm (see Sect. 2–6).

In contrast to the situation in the male, germ cell differentiation in the female may occur early in meiosis (Fig. 2.2). It is also important to keep in mind the difference in the number of sex cells that result from meiosis in the male and female. Each of the meiotic divisions in the female is uneven, producing only one full-sized cell. During the first meiotic division, the **primary oocyte** divides to produce one small polar body and one **secondary oocyte**. The latter enters the second meiotic division to produce the second polar body and the haploid **ovum**, which is the only functional sex cell to result from meiotic reduction of an oogonium.

SECTION TWO

Spermatogenesis

We shall now discuss formation of sperm, which are among the most highly specialized cell types ever described. Such specialization is designed for two purposes: to get the sperm to the egg and to fuse with it. As we shall see, the testes are very efficient "sperm factories," dedicated to the production of vast numbers of these very elaborate cells.

2–1 Sperm Structure

When we think about spermatozoa, we usually envision sleek, streamlined structures with small heads and long, whip-like tails; this is indeed the form of most mammalian sperm. However, animals have evolved a variety of sperm shapes, some of which are shown in Figure 2.3. Marine and freshwater invertebrates that discharge their sperm into the water have sperm that are

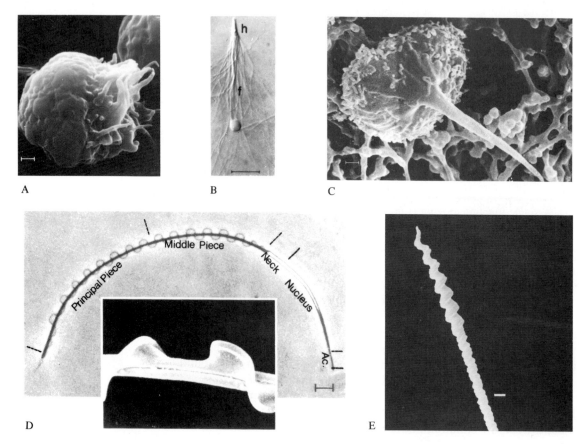

Figure 2.3

Forms of some unusual animal spermatozoa. A, Scanning electron micrograph of the amoeboid sperm of the round worm, *Ascaris*. Scale bar equals 1.0 μm. B, The multiflagellate sperm of the termite, *Mastotermes*, as seen with Nomarski optics. h: head; f: flagella. Scale bar equals 25 μm. C, Scanning electron micrograph of the tack-shaped sperm of the prawn, *Palaemonetes*. The cup-shaped basal region contains the nucleus. Scale bar equals 1.0 μm. D, Phase contrast micrograph of the sperm of the urodele amphibian, *Pleurodeles*. The sperm tail consists of a long axial fiber and a thin undulating membrane. Scale bar equals 10 μm. The inset shows details of the tail by scanning electron microscopy. E, Scanning electron micrograph of the head and midpiece of the sperm of the terrestrial snail, *Anguispira alternata*. Scale bar equals 1 μm. (A, From Abbas and Cain, 1979. B, From B. Baccetti and R. Dallai, 1978. Reprinted from *The Journal of Cell Biology*, 1978, 76: 569–576 by copyright permission of the Rockefeller University Press. C, From Koehler, 1979. D, From Picheral, 1979. E, Courtesy of J. W. Atkinson.)

considered "primitive" (see p. 34), whereas animals with internal fertilization have more elaborate sperm. Specializations found in sperm of either type are thought to be adaptations to the conditions of fertilization found in these species and may be so distinctive as to allow taxonomists to distinguish closely related species on the basis of sperm morphology (Phillips, 1983).

The primary components of most sperm are a **nucleus**, an **acrosome**, and a **flagellum**. The nucleus contains a highly condensed mass of chromatin, in which the individual chromosomes cannot be observed by light or electron microscopy. The acrosome exhibits a variety of morphologies and assists in penetration of the egg accessory layers (see Chap. 3) and in species-specific attachment of sperm to eggs. The flagellum is the locomotor organelle found in most types of sperm; however, flagella are not universal. The *Ascaris* sperm shown in Figure 2.3, for example, is ameboid rather than flagellate. Because of the variety of sperm morphologies, no single kind of sperm can be considered "typical." Because the sperm of mammals have been studied extensively, and because (as mammals ourselves) mammalian sperm are intrinsically interesting to us, a detailed description of mammalian sperm is provided and then briefly compared to primitive sperm.

Mammalian Sperm

A reconstruction of mammalian sperm structure based upon electron micrographs is shown in Figure 2.4. The cell membrane has been removed to reveal the underlying components, and representative cross-sections of the sperm are shown at various levels. The two major regions of the sperm are the **head** and the **tail**. The interior of the head consists of the nucleus. Surrounding the nucleus at the anterior end of the sperm head is the acrosome. The acrosome does not cover the nucleus entirely but forms a cap over it. The portion of the head behind the posterior margin of the acrosome is the **postacrosomal region**. The tail is subdivided into four segments: the **neck, middle piece, principal piece**, and **end piece**. The slender neck forms the articulation between the head and tail. The middle piece is characterized by a sheath of mitochondria surrounding the tail elements.

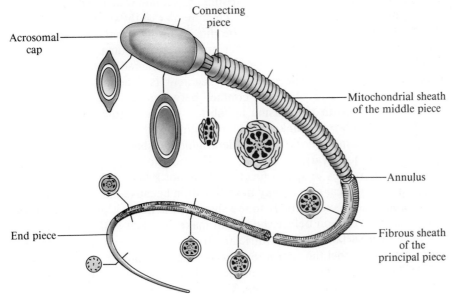

Acrosomal cap

Connecting piece

Mitochondrial sheath of the middle piece

Annulus

End piece

Fibrous sheath of the principal piece

Figure 2.4

Ultrastructure of mammalian sperm. The cell membrane has been removed. Representative cross-sections are shown at several levels. (After Bloom and Fawcett, 1975.)

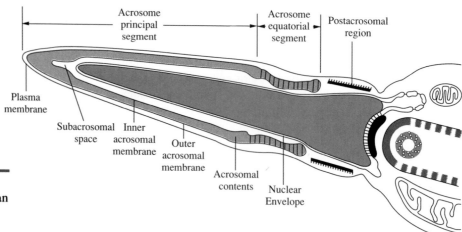

Figure 2.5

Diagrammatic representation of
sagittal section through mammalian
sperm head. Nucleus is in black.
(After Fawcett, 1975.)

The head shows little microscopic substructure because the bulk is
occupied by the nucleus, which contains highly compact chromatin with no
resolvable detail. The acrosome cap over the nucleus also has a very simple
organization. The relationship between the acrosome and nucleus is better
seen in a sagittal section of a sperm head, as shown diagrammatically in
Figure 2.5. The acrosome is sandwiched between the plasma membrane and
the nuclear envelope. The acrosome is surrounded by its own membrane.
Note that a portion of the acrosome extends anteriorly past the tip of the
nucleus. This **apical segment** of the acrosome may be quite prominent in
some species and often has a species-specific shape. In other species—
notably the human—the apical segment is small and inconspicuous.

The amorphous-appearing contents of the acrosome include several
hydrolytic enzymes. When a sperm reaches the immediate vicinity of an
egg, it undergoes the **acrosome reaction**, which causes the plasma membrane
and the outer acrosomal membrane to vesiculate and be shed, thus releasing
the enzymes of the acrosome. These enzymes apparently assist the sperm
in penetrating the accessory layers that surround the egg (see Chap. 4).

The posterior portion of the acrosome is narrow and may have a
different density from that of the anterior portions. This region of the
acrosome—the **equatorial segment**—has a unique fate during fertilization,
because it is the only part of the acrosome to remain intact. The remainder
is lost during the acrosome reaction. The integrity of the equatorial segment
during the acrosome reaction may be due to bridges that can be seen by
electron microscopy to link the inner and outer acrosomal membranes in
this region. The bridges are not present in the remainder of the acrosome
(Russell et al., 1980). The equatorial segment is functionally significant
because it is the site of initial contact between the sperm and egg at
fertilization. Beneath the plasma membrane in the postacrosomal region,
there is a dense layer of unknown composition.

The acrosome and the nucleus determine the shape of the sperm head,
which can be quite variable (Fig. 2.6). The functional significance of such
dramatic differences in sperm head shape is not known. The shape has no
apparent mechanical role to play in fertilization because the apical segment,
which contributes substantially to the shape of the head, is destroyed during
the acrosome reaction before the sperm makes contact with the egg.

The sperm tail is a very intricate structure that produces the flagellar
movement to propel the sperm toward the egg. The motor apparatus of the
sperm tail consists of two central microtubules that are surrounded by an
array of nine doublet microtubules. This structure is called the **axoneme**. At

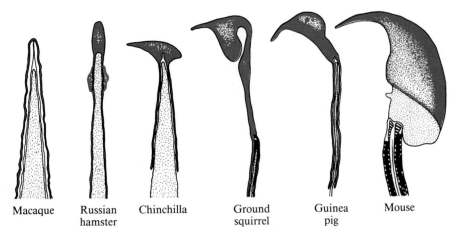

Macaque   Russian   Chinchilla   Ground   Guinea   Mouse
           hamster   squirrel pig

Figure 2.6

Drawings of sagittal sections of
sperm heads of several mammalian
species, shown to permit a
comparison of the sizes and shapes
of the apical portion of the
acrosome. (After Fawcett, 1970.)

high magnification (Fig. 2.7), the two subunits of the doublets are seen to
have different shapes. One of them is a complete tubule and, therefore,
circular in cross-section. The second is an incomplete tubule that is
C-shaped. It opens onto the wall of the cylindrical subunit. Each doublet is
associated with diffuse armlike appendages that project toward the adjacent
doublet.

As shown in Figure 2.7, each microtubule is composed of minute
protofilaments. The central pair of microtubules and the circular member

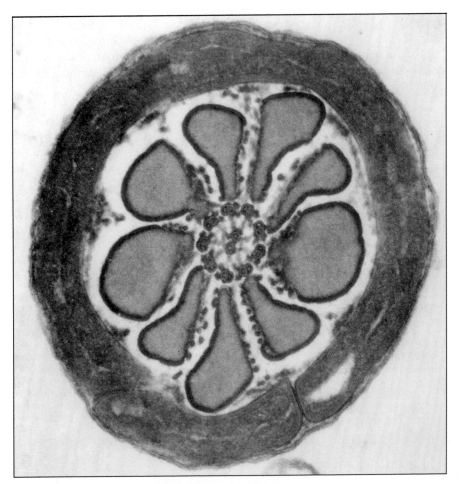

Figure 2.7

Axoneme and associated fibers in
the middle piece of a rat
spermatozoon. The protofilaments
in the walls of the central
microtubules and of the doublets
are distinctive. The dynein arms
can be seen attached to the
circular member of each doublet
and projecting toward the adjacent
doublet. Radial spokes can also be
seen. × 90,000. (From Phillips,
1983.)

of the outer doublets are each composed of 13 protofilaments, whereas the crescentric member of each outer doublet is composed of about 10 protofilaments. These protofilaments run along the entire length of a microtubule and are composed primarily of **tubulin**. The arms associated with the outer doublets are composed of another protein, **dynein 1**, which possesses ATPase activity and is responsible for converting chemical energy into mechanical movement (Ogawa et al., 1977). **Radial spokes** can also be seen to extend from the outer doublets to the central pair of microtubules.

Sperm flagellar movement probably involves localized sliding between adjacent doublets, much like the sliding filament mechanism of muscle contraction (Summers and Gibbons, 1971). Sliding is mediated by the dynein arms, which use the energy derived from the hydrolysis of ATP to attach to the adjacent doublet and cause one doublet to slide against the other. The essential role of dynein in sperm motility is suggested by the correlation between reduction in dynein arms and immotility of sperm. Afzelius (1976) described an autosomal recessive mutation in man that eliminates the dynein arms on the axonemal outer doublets (Fig. 2.8), resulting in immotile sperm that appear morphologically normal in every other respect. Men with this condition (which is called **immotile cilia syndrome**) are infertile. The syndrome is apparently inherited as an autosomal recessive mutation. The normal allele of the gene that causes immotile cilia syndrome is apparently responsible for either the synthesis of the dynein protein or the attachment of the dynein arms to the doublets. Sperm motility can also be inhibited by experimental elimination of the dynein arms of demembranated sperm. Motility can be re-established by restoring the dynein; however, dynein that has been modified to eliminate its ATPase activity is incapable of restoring motility (Gibbons et al., 1987). Clearly, the ATPase activity of dynein is essential to generate sperm motility. The bending action of the tail, which propels the sperm, is produced by an interaction of the central pair of axonemal microtubules with the radial spokes to coordinate the sliding activities of the outer doublets (Warner and Satir, 1974; Witman et al., 1978).

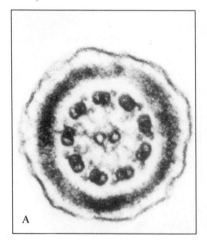
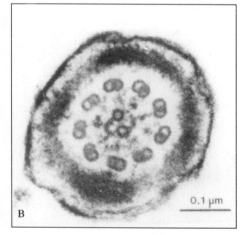

Figure 2.8

Electron micrographs of cross-sections through the human sperm tail. A, The tail of an ordinary, motile spermatozoon has nine microtubular doublets; on each of the doublets there are two dynein arms. B, The sperm tail of a sterile male (with immotile cilia syndrome) is devoid of the dynein arms. (B, From B. A. Afzelius, 1976. A human syndrome caused by immotile sperm. Science *193*: 317–319. Copyright 1976 by the American Association for the Advancement of Science.)

The axoneme runs through the entire tail, as the cross-sections in Figure 2.4 illustrate. A similar organization of microtubules is nearly universal in cilia and flagella throughout the plant and animal kingdoms. The exceptional sperm from some nonmammalian species that show deviation from the 9 + 2 tubule pattern generally have atypical swimming motion (Gibbons et al., 1983; Phillips, 1983), suggesting that this pattern is essential for generating the stereotypical motion of cilia and flagella.

The axoneme is surrounded by a row of nine **outer dense fibers**, each of which parallels an axoneme doublet. The fibers are thickest in the proximal half of the tail and progressively decrease in diameter toward the tip. The relative thickness and extent of the fibers vary considerably among mammalian species. In some they are thick and extend nearly the whole length of the tail, whereas in others they are thin and terminate in the proximal portion of the principal piece. The dense fibers are proposed to play a role in flagellar flexibility. Individuals who produce sperm with modifications of dense-fiber relationships to the axoneme are sterile due to the resultant modified beat of the sperm flagella, whose flexibility has been affected by the aberrant morphology (Serres et al., 1986).

The neck region forms the base of the tail. The major structural element in this region is the convex **connecting piece**, which articulates with a concave depression in the base of the sperm head. Very fine filaments in the space between the connecting piece and the sperm head are probably responsible for attachment of the head to the tail (see Fig. 2.5). Behind the articulation region, the connecting piece consists of nine **segmented columns** that attach to the anterior ends of the nine outer dense fibers. The sperm of some mammalian species have a centriole (the **proximal centriole**) embedded in a depression in the connecting piece. During development of the sperm tail, another centriole (the **distal centriole**) is also present, but it degenerates during development of the connecting piece.

The middle piece of the sperm is characterized by a sheath of elongated **mitochondria** that are wrapped end-to-end in a helical chain around the axoneme (Fig. 2.9). The mitochondria apparently provide the energy for sperm propulsion. The middle piece is terminated by a structure called the **annulus** (see Fig. 2.4). Immediately behind the annulus, the axoneme is encased in a **fibrous sheath**. This region of the tail is the principal piece. The sheath consists of two longitudinal columns that are connected by a

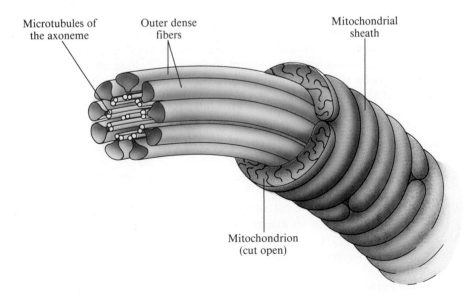

Microtubules of the axoneme

Outer dense fibers

Mitochondrial sheath

Mitochondrion (cut open)

Figure 2.9

Diagrammatic representation of a segment from the middle piece of a mammalian spermatozoon. (After Fawcett, 1975.)

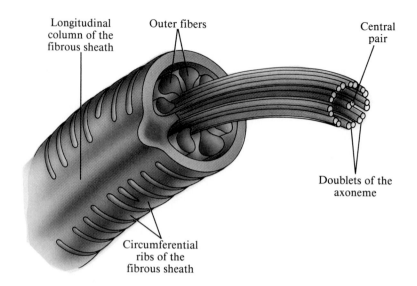

Figure 2.10

Diagrammatic representation of a segment from the principal piece of the mammalian spermatozoon illustrating one of the two longitudinal columns of the fibrous sheath and the associated ribs. Inward projections of the longitudinal columns attaching to doublets divide the tail into two unequal compartments, one containing three outer fibers and the other containing four. (After Fawcett, 1975.)

series of hemispherical ribs (Fig. 2.10). Anteriorly, the columns are attached for a short distance to two of the nine outer dense fibers. These two outer dense fibers then terminate abruptly, and along the remainder of the principal piece, the columns attach directly to the two doublets that were internal to the dense fibers. As the sperm tail tapers, the columns and ribs diminish and then end abruptly a few micrometers from the tip of the flagellum. The termination marks the junction of the principal piece and the end piece.

Primitive Sperm

As mentioned earlier, not all sperm are as morphologically elaborate as mammalian sperm. Marine and freshwater invertebrates, for example, employ simple spermatozoa. Their sperm are characterized by a head containing a rounded or conical nucleus capped by a small acrosome and a small midpiece containing a few spheroidal mitochondria surrounding the base of the tail, which consists of an axoneme with a typical 9 + 2 pattern of microtubules (Franzén, 1983). There is considerable morphological variation among these sperm, however, particularly due to variable acrosome morphology (Fig. 2.11).

The changes occurring in the acrosome during the acrosome reaction differ dramatically from those during the mammalian acrosome reaction. When sperm of marine invertebrates swim into the vicinity of the egg, the acrosome everts a long, slender process (the **acrosomal process**) that assists in egg penetration. The surface of the acrosomal process is coated with a substance (**bindin**) that binds the sperm to the egg vitelline envelope. The acrosome reaction is an important component of fertilization, and it will be discussed extensively in Chapter 4.

2–2 Germ Cell–Somatic Cell Interactions in Spermatogenesis

In most animal species, spermatogenesis occurs while the germ cells are intimately associated with specialized somatic cells. This physical relationship has been studied extensively in the testes of mammals. Mammalian testes contain numerous **seminiferous tubules**, which are shown in cross-section in Figure 2.12. Within the tubules there are a number of radially distributed Sertoli cells, with which the germ cells remain associated during

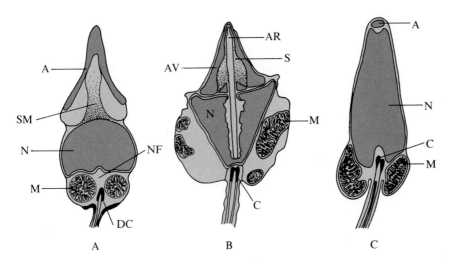

Figure 2.11

Drawings illustrating primitive sperm of some marine invertebrates. A, Longitudinal section through the head and midpiece of the sperm of the mollusk *Choromytilis meridionalis*. A: acrosome; DC: distal centriole; M: mitochondrion; N: nucleus; NF: nuclear fossa; SM: subacrosomal material. B, Longitudinal section through sperm of the polychate *Neanthes japonica*. The acrosomal vesicle (AV) is located at the periphery of the conical acrosome. The anterior half of the acrosomal rod (AR) is inserted into the subacrosomal space (S), and the posterior half is in the deep central indentation of the nucleus (N). M: mitochondrion; C: centriole. C, Longitudinal section through sperm of the sea urchin *Arbacia*. A: acrosome; N: nucleus; M: mitochondrion; C: centriole. (A, After Hodgson and Bernard, 1986. B, After Sato and Osanai, 1986. C, After Fawcett, 1970.)

the entire spermatogenic process. The Sertoli cells are columnar, with broad bases and narrow tips that extend into the lumen of the tubule. Spermatogonia are situated between the Sertoli cells and the underlying basal lamina (Fig. 2.13). Germ cells in meiosis and spermiogenesis are embedded in membrane-enclosed recesses in the Sertoli cells or are trapped in depressions between adjacent Sertoli cells. The germ cells are arranged in a very precise sequence. Spermatogonia remain at the bases of the Sertoli cells, whereas cells in progressively later stages of meiosis and spermiogenesis are situated at successively higher positions.

In any particular region of a seminiferous tubule, groups of germ cells around the circumference appear to progress toward the lumen more or less in unison (Leblond and Clermont, 1952). Thus, they move to the lumen in waves, forming circumferential zones of more advanced cells inside zones of less advanced cells. The Sertoli cells are believed to be responsible for governing the rate of spermatogenesis by controlling the translocation of germ cells from the basal lamina to the lumen. Contiguous Sertoli cells must therefore have some means for coordinating their activities in translocating germ cells. Sertoli cells and germ cells in the seminiferous tubule are joined by desmosome-like junctions. Russell (1980) and Russell and Peterson (1985) have proposed that these junctions facilitate the displacement of germ cells toward the lumen of the seminiferous tubule. Contiguous Sertoli cells are thought to undergo cooperatively conformational changes that displace their lateral margins apically. Because the germ cells are bound to the Sertoli cells by desmosome-like junctions, they are carried to the tips of the Sertoli cells by these conformational changes. At the tips of the Sertoli cells, spermatozoa occupy deep recesses or crypts and are released from there into the lumen of the seminiferous tubule.

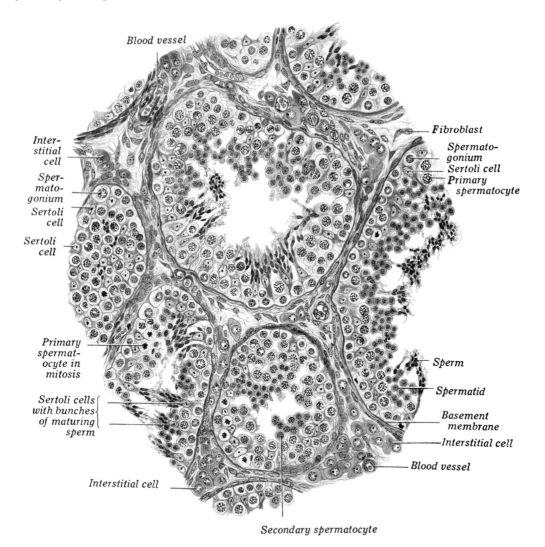

Blood vessel

Inter-
stitial
cell

Sper-
mato-
gonium

Sertoli
cell

Sertoli
cell

Primary
spermat-
ocyte in
mitosis

Sertoli cells
with bunches
of maturing
sperm

Interstitial cell

Fibroblast

Spermato-
gonium

Sertoli cell

Primary
spermatocyte

Sperm

Spermatid

Basement
membrane

Interstitial cell

Blood vessel

Secondary spermatocyte

Figure 2.12

Section of human testis. The transected tubules show various stages of
spermatogenesis. ×170. (After A. A. Maximow. From Bloom and Fawcett, 1975.)

Gap junctions are also observed between Sertoli cells and spermatocytes. Gap junctions may be involved in transfer of small molecules and
ions between Sertoli cells and the spermatocytes, which may assist in the
regulation of spermatogenesis (Russell and Peterson, 1985).

The binding of germ cells to Sertoli cells is mediated by cell surface
molecules on the germ cells that specifically recognize the Sertoli cell
surface. The biochemical basis for this specific cell adhesion is studied *in
vitro* by co-culture of immature germ cells and Sertoli cells. An important
methodology in this analysis is to prepare antibodies against germ cell
surface molecules and observe the effects of the antibodies on recognition
between germ cells and Sertoli cells. Molecules on the surfaces of spermatocytes bind the antibodies, which interfere with the binding to Sertoli
cells via the cell surface molecules (D'Agostino and Stefanini, 1987). Such
cell-cell adhesive interactions are extremely important in establishing functional interactions between cells during development. We shall discuss
adhesion in more detail in Chapter 9.

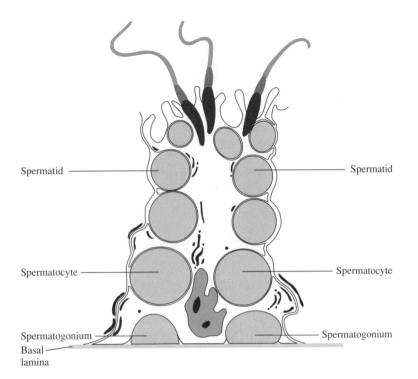

Figure 2.13

Relationship between the germ cells and a Sertoli cell. Spermatogonia occupy a basal position between the Sertoli cell and the basal lamina. Cells in progressively later stages of spermatogenesis are found at successively more apical positions. Mature spermatozoa are released into the lumen of the seminiferous tubule from the tip of the Sertoli cell. (After Dym and Fawcett, 1970.)

Interestingly, male germ cells will not thrive in culture for very long, nor will they differentiate, unless they are in contact with Sertoli cells. Such contact promotes both germ cell survival and differentiation (Palombi et al., 1979; Tres and Kierszenbaum, 1983). The *in vitro* effects of Sertoli cells on germ cells are indicative of their *in vivo* roles as regulators of germ cell metabolism and differentiation (see Sect. 2–3).

Sertoli cells form tight junctions with adjacent Sertoli cells, thus sealing the seminiferous tubule and preventing many substances of the blood plasma from entering the tubule. This constitutes the so-called **blood-testis barrier** (Dym and Fawcett, 1970; Dym, 1973). By preventing the diffusion of extrinsic substances into the intercellular spaces, the Sertoli cells are able to regulate the environment in which the germ cells differentiate. Sertoli cells secrete a fluid into the lumen of the seminiferous tubules. The fluid, which contains a number of proteins as well as small molecular weight substances, bathes the germ cells during their differentiation and carries the spermatozoa out of the testis after they are released by the Sertoli cells (Rich and de Kretser, 1983).

2–3 Hormonal Regulation of Spermatogenesis

Spermatogenesis is regulated by elegant hormonal interactions involving the pituitary gland as well as the somatic cells of the testis (Fig. 2.14). Some of these hormones are uniquely produced in males. However, as we shall discuss in Chapter 3, some of them are also involved in regulating oogenesis. The principal promoters of germ cell differentiation in male vertebrates are

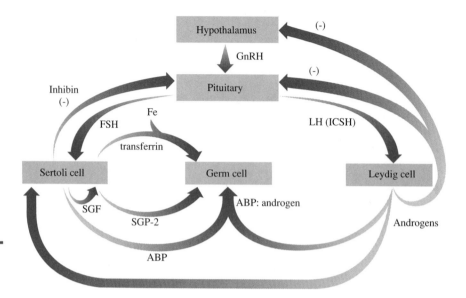

Figure 2.14

Hormonal interactions involved in regulating spermatogenesis in the adult mammal.

the **androgens**, which are synthesized by specialized somatic cells called the **interstitial cells (cells of Leydig)**. These cells are localized in the connective tissue between the seminiferous tubules. The androgens diffuse into the tubules, where they promote spermiogenesis. The production of androgens by the interstitial cells is regulated by a gonadotropic hormone released by the pituitary gland into the circulation. This hormone is **luteinizing hormone (LH)**, sometimes called **interstitial cell-stimulating hormone (ICSH)**. Another pituitary gonadotropic hormone, **follicle-stimulating hormone (FSH)**, is also involved in regulation of spermiogenesis by acting directly on the Sertoli cells. The production of these two gonadotropic hormones is regulated by **gonadotropin-releasing hormone (GnRH)**, which is produced by the hypothalamus. Androgens released into the circulation act on both the hypothalamus and the pituitary to modulate LH production. Thus, when LH production increases, androgen production is elevated. This, in turn, leads to a reduction in LH levels. This is a classical **feedback-inhibition** mechanism.

The responsiveness of a cell to hormones is dependent on the presence of hormone receptors. This is illustrated by the responsiveness of gonadal cells to FSH. Germ cells lack FSH receptors (Lyon et al., 1975); thus, FSH cannot affect the germ cells directly. However, Sertoli cells possess receptors and, as a consequence, mediate the effects of FSH in the gonad (Sanborn et al., 1977; Fritz, 1978; Means et al., 1980). These effects could be mediated either by the secretion of regulatory substances into the tubular fluid, by surface contact between Sertoli cells and germ cells (Palombi et al., 1979), or by transfer of small molecules and ions via the gap junctions. One effect of FSH on Sertoli cells in mammals is to stimulate the release of **androgen-binding protein (ABP)**, which has a high affinity for androgens. Androgen-binding protein binds androgens in the tubular fluid and presumably transports it to the germ cells to promote their differentiation. Androgens also have direct effects on Sertoli cells. Another secretory product of the Sertoli cells is **transferrin**, which is involved in local transport of iron, presumably to the germ cells (Skinner and Griswold, 1980).

Recently, mammalian Sertoli cells have been shown to produce a substance that functions as a growth factor. This substance, **seminiferous growth factor (SGF)**, stimulates somatic cell proliferation and blood vessel production in the testes during fetal and postnatal development. In the adult, Sertoli cells continue to produce SGF and respond (presumably by means of an SGF receptor in their plasma membranes) to this substance that they, themselves, secrete by secreting a number of proteins, including proteins that bind to the surface of sperm (Bellvé and Feig, 1984; Barnes, 1988).

Sertoli cells also secrete **inhibin**, which enters the circulation and functions to suppress the secretion of FSH from the pituitary (Steinberger and Steinberger, 1976). The circulating levels of FSH, in turn, regulate inhibin production, indicating that a classical feedback-inhibition mechanism operates (Ying et al., 1987).

2–4 Spermiogenesis

At the completion of meiosis, spermatids are relatively simple spherical cells with centrally located nuclei. Their differentiation into sperm requires them to undergo an extensive morphological transformation. The nucleus becomes eccentric by its presumptive acrosomal surface becoming aligned in close proximity to the plasma membrane. The acrosome and axoneme are then elaborated on opposite sides of the nucleus. The cell itself is elongated, apparently due to the action of microtubules, which orient in the long axis of the cell (Fawcett et al., 1971; Russell et al., 1983). As shown in Figure 2.13, the developing spermatozoa in mammals are oriented in Sertoli cell recesses such that their acrosomes face the base of the seminiferous tubule, and the axonemes are elaborated into the lumen of the tubule.

In addition to the elaboration of the specialized organelles that must function during sperm transport and fertilization, differentiation of spermatozoa involves extensive modifications of the nucleus and the acquisition of specialized properties by the sperm plasma membrane. We now discuss in more detail some of the morphogenetic events that occur during spermiogenesis.

Nuclear Modifications

Extensive condensation of chromatin plays an important role in helping to streamline the developing sperm, which facilitates locomotion. The tight packing of the DNA may also make it less susceptible to physical damage or mutation during storage and transport to the site of fertilization. The condensation may result from removal of chromatin-associated proteins (such as the somatic histones) or from the formation of unique DNA-protein complexes. Three categories of proteins play roles in generating the tightly condensed chromatin of the sperm head: (1) **protamines**, (2) histone-like proteins, and (3) other sperm-specific basic proteins (Bloch, 1969). The replacement of somatic histones during rooster spermatogenesis is illustrated in Figure 2.15. Figure 2.15A summarizes the events of spermatogenesis in the rooster, whereas Figure 2.15B shows the replacement of somatic histones during spermiogenesis with protamine. (See the Appendix for a discussion of the technique of gel electrophoresis, which was utilized in these experiments.) The figure shows that the protamines are initially detected as spermatids elongate and histones are gradually lost thereafter. Mature spermatozoa have no detectable histone.

Somatic histone replacement in many species (including trout and salmon) is thought to be a consequence of hyperacetylation of the histones as well as other modifications to either the histones or the DNA itself that reduce the binding affinity between DNA and histones and favor the binding between DNA and sperm-specific chromatin proteins (Christensen and

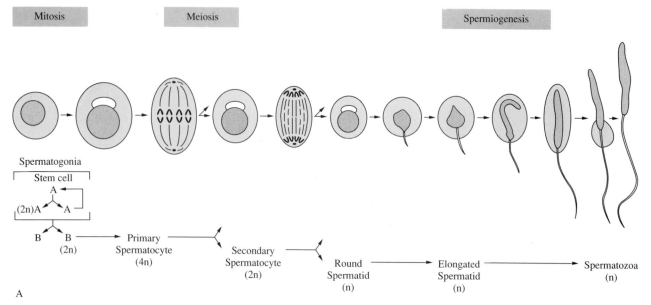

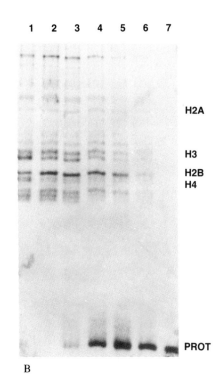

Figure 2.15

Replacement of somatic histones with protamines in the rooster. A, Diagram showing the stages of spermatogenesis in the rooster. B, Electrophoresis of nuclear proteins extracted from progressive stages of spermatogenesis (lanes 1–7). Somatic histones (H2A to H4) and protamines are labeled on the right. (A, After Oliva et al., 1988. B, From Oliva et al., 1988.)

Dixon, 1982; Oliva et al., 1987). In mammals, histone removal is thought to involve digestion of these proteins by protease enzymes that are associated with the chromatin (Marushige et al., 1976; Marushige and Marushige, 1983; Faulkner and Bhatnagar, 1987).

Sperm nuclear condensation in mammals occurs in two phases. In the first phase, histones are replaced by small, highly basic **transition proteins**. This transition is coincident with the cessation of gene transcription (Meistrich et al., 1978). The second phase involves the substitution of protamines for transition proteins. Mammalian protamines are characterized by the presence of numerous cysteine residues. The cysteines are thought to stabilize and compact the chromatin in the sperm nucleus through disulfide bond cross-linking (Bedford and Calvin, 1974).

DNA and histones in somatic cells form stereotypic complexes called **nucleosomes**. The replacement of somatic histones during spermatogenesis eliminates the nucleosomes, thus producing a tightly packed complex in which the DNA is transcriptionally inactive (Kierszenbaum and Tres, 1978). The absence of transcriptional activity in spermatids is a generalized phenomenon in the animal kingdom regardless of which proteins are complexed with DNA in the condensed chromatin. The absence of transcription after chromatin condensation means that protein synthesis, which is essential for the completion of spermiogenesis, must utilize stored messenger RNA molecules. This aspect of spermiogenesis will be discussed in more detail in Section 2–5.

Morphological changes in the nucleus occur concurrently with chromatin condensation. Fawcett et al. (1971) have suggested that particular nuclear shapes result from the pattern of DNA-protein interaction during condensation. It has also been proposed that the form of the nucleus is produced by pressure applied by microtubules outside the nucleus (McIntosh and Porter, 1967). Most investigators believe that the microtubules associated with the spermatid nucleus play a paramount role in shaping the nucleus as well as elongating the cell itself. In some species, however, nuclear shaping can occur in the complete absence of microtubules (Phillips, 1974; van Deurs, 1975). The parallel array of microtubules that surrounds the nucleus in spermatids is called the **manchette**. The manchette is a cylinder of microtubules around the nucleus in which the microtubules are oriented parallel to the long axis of the elongating sperm nucleus (Fig. 2.16). Individual microtubules of the manchette are connected by arms. These interconnecting arms presumably maintain the cylindrical nature of the manchette. A role for manchette microtubules in guiding chromatin condensation is strongly suggested by the observation that in some species individual fibers of condensing chromatin on the inside of the nuclear envelope are parallel to the microtubules attached to the outside of the nuclear envelope (Medina et al., 1986). Some investigators have reported actual physical connections between the microtubules and chromatin that pierce the nuclear envelope (Courtens and Loir, 1981).

As spermiogenesis proceeds, gaps appear in the ring of manchette microtubules as the integrity of the manchette is lost. The microtubules of the manchette are later found in isolated clusters within the excess cytoplasm that is shed from the spermatids at the completion of spermiogenesis (Rattner and Brinkley, 1972; Goodrowe and Heath, 1984; Sickels and Heath, 1986).

The function of the microtubules in nuclear shaping is elegantly illustrated by a mutant of *Drosophila* in which the nuclei fail to elongate. When spermatids of this mutant—*ms*(3)10R—are examined with the electron microscope, they are found to lack the microtubules that normally surround the nucleus during elongation (Fig. 2.17A). The chromatin condenses in the spermatid nuclei, but it does not undergo the highly ordered packing that occurs during elongation. An occasional spermatid with a partial complement of perinuclear microtubules is found in mutant testes (Fig. 2.17B). The

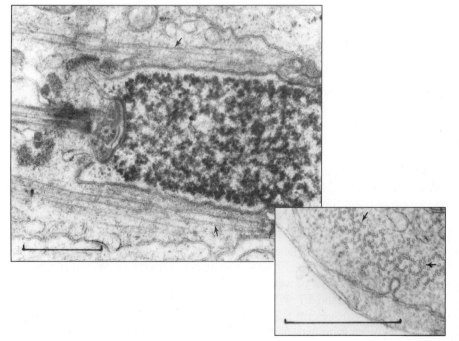

Figure 2.16

Longitudinal section of an elongating cat spermatid nucleus. Arrows identify individual microtubules of the manchette. Inset shows arms connecting neighboring microtubules (arrows), which are shown in cross-section. Scale bars equal 1.0 μm. (From Sickels and Heath, 1986. Reprinted from Anatomical Record by permission of Wiley-Liss, a division of John Wiley and Sons, Inc. Copyright © 1986 Wiley-Liss.)

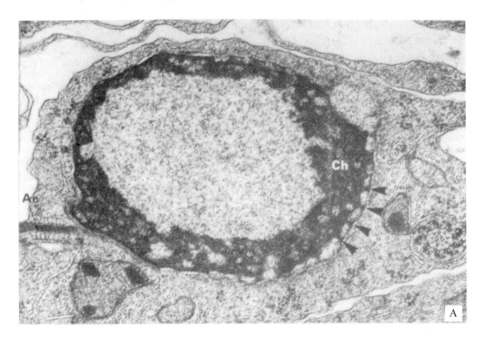
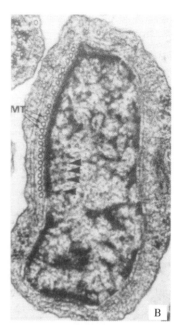

Figure 2.17

Spermatids of *ms*(3)10R *Drosophila*. A, This nucleus is devoid of surrounding microtubules, which in the wild type encircle the nucleus. Note that the chromatin (Ch) has condensed, but nuclear elongation has not occurred. Chromatin adheres to the nuclear envelope in some areas (arrows). Ac: elongate acrosome. ×52,000. B, A partial complement of perinuclear microtubules (MT) is present around the nucleus of this semi-elongated spermatid nucleus. The condensed chromatin is more ordered here (arrows) than in nuclei with no microtubules. ×48,000. (From Wilkinson et al., 1974.)

nuclei of these spermatids have partially elongated, and their chromatin is more ordered than that in nuclei that are not associated with microtubules. The correlation between the extent of nuclear elongation and the completeness of the perinuclear microtubule layer supports the hypothesis that the microtubules play a direct role in shaping the nucleus (at least in *Drosophila*).

Formation of Sperm Organelles

The near-universal presence of the acrosome in sperm heads is a remarkable example of evolutionary conservation. The presence of hydrolytic enzymes in acrosomes has led to proposals that this structure is a modified lysosome. Like lysosomes, acrosomes are derivatives of the Golgi complex. Formation of the acrosome is illustrated in Figures 2.18 and 2.19, which contain a series of electron micrographs showing successive stages of this process in the guinea pig. The first sign of acrosome formation is the appearance within the Golgi of numerous membrane-enclosed **proacrosomal granules** (see Fig. 2.18A and B). The granules coalesce into a single large **acrosomal vesicle**, which contains a dense **acrosomal granule** (see Fig. 2.18C). The acrosomal vesicle adheres to the nuclear envelope, thereby marking the future anterior tip of the sperm nucleus. The Golgi continues to form proacrosomal granules, which contribute to the enlargement of the acrosomal vesicle by fusing with it (see Fig. 2.18D and E). When the acrosomal vesicle has reached its full size, the Golgi moves into the postnuclear cytoplasm, and the acrosomal vesicle becomes modified, assuming the final shape of the mature acrosome (see Fig. 2.19).

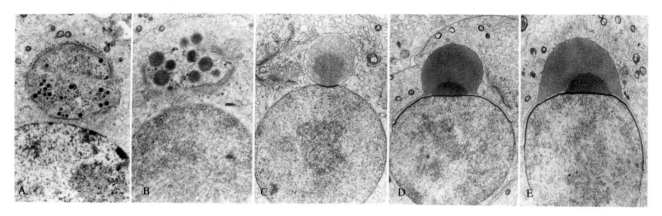

Figure 2.18

A series of electron micrographs of the juxtanuclear region of a guinea pig
spermatocyte A and spermatids B to E, showing the successive stages of
formation of the acrosomal vesicle C and its conversion to an acrosomal cap D
and E. The guinea pig is selected because the large size of its acrosome makes it
favorable material for study of the process of differentiation, which is qualitatively
similar in all mammals. (Courtesy of D. W. Fawcett.)

Formation of the middle piece is highly variable because the mitochondria in different species assume a number of configurations and undergo
a variety of structural changes. Primitive spermatozoa have very simple
middle pieces with clusters of a few enlarged mitochondria (see Fig. 2.11).
During spermiogenesis, there is a progressive decrease in the number of
mitochondria, accompanied by an increase in size of the remaining mitochondria. It is uncertain how this happens, but it is assumed that it occurs
by mitochondrial fusion (Longo and Anderson, 1974).

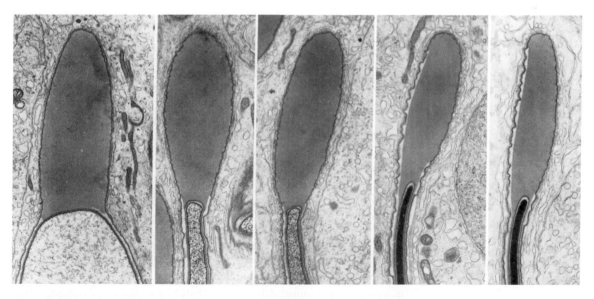

Figure 2.19

Further development of the acrosome of the guinea pig spermatozoon. When the
acrosome has attained its full size, the Golgi migrates into the postnuclear
cytoplasm, and the subsequent changes in the acrosome consist of a progressive
modification of its shape taking place concurrently with a condensation of the
chromatin and flattening of the nucleus. (From Fawcett and Phillips, 1969.)

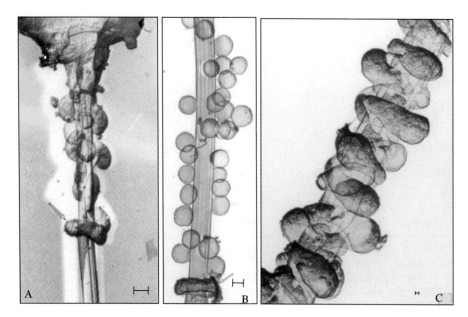

Figure 2.20

Formation of the guinea pig sperm middle piece as shown using surface replicas. In this technique, the plasma membrane is removed with a detergent. A replica of the demembranated sperm is made by coating it with platinum and carbon. The replica is then examined with the electron microscope to visualize surface details. A, Replica of spermatid showing the annulus (arrow) moving posteriorly, followed by mitochondria. Scale bar equals 1.0 μm. B, Replica of mitochondria attached to the outer dense fibers of the axoneme. Annulus (arrow) has moved farther back. Scale bar equals 0.5 μm. C, Teardrop-shaped mitochondria form helical chains around the axoneme. Scale bar equals 0.1 μm. (From Phillips, 1980.)

In mammals, the mitochondria migrate to the base of the axoneme, where they associate with the outer dense fibers. This association may be mediated by molecules of the outer mitochondrial membranes that recognize and bind to the dense fibers (Phillips, 1980). Mitochondria move down the flagellum behind the annulus. The posterior movement of the annulus and mitochondria is shown in Figure 2.20A and B. Movement of the annulus ceases when it reaches the posterior end of the middle piece. The mitochondria shown in Figure 2.20B are rounded. At a later stage (see Fig. 2.20C), they have elongated. These teardrop-shaped mitochondria wrap around the axoneme to form helical chains. The mitochondria will later associate end to end to produce the helical row of mitochondria that is depicted in Figures 2.4 and 2.9.

Differentiation of the sperm tail is basically the process of axoneme formation. Obviously, the details of tail formation may vary, because various organelles are elaborated in different species. Mammalian sperm, for example, must form outer dense fibers and a fibrous sheath. However, the axoneme is the fundamental organelle in all sperm flagella.

The axoneme is organized by one of the two sperm centrioles. The centrioles migrate to the end of the nucleus that is opposite to the acrosome and situate at right angles to one another. The centriole closest to the nucleus is called the proximal centriole, and the other member of the pair is the distal centriole. The distal centriole (from which the 9 + 2 axoneme emerges) is parallel to the long axis of the cell (Fig. 2.21). Elongation of the

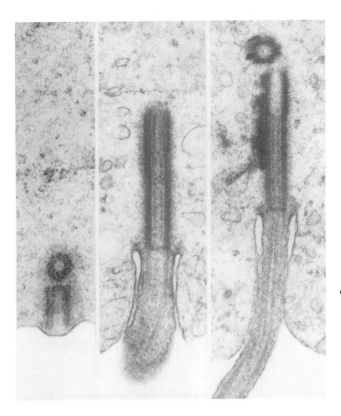

Figure 2.21

The earliest events in formation of the tail consist of migration of the centrioles to the cell surface in the postnuclear region and polymerization of microtubule protein on the template provided by the distal centriole. A typical 9 + 2 axoneme is formed, and the simple flagellum elongates by accretion of microtubule subunits to its distal end. (From Bloom and Fawcett, 1975.)

flagellum continues by growth at the distal ends of the microtubules (see Chap. 9 for a discussion of microtubular elongation).

The ultimate disposition of the two centrioles is quite variable. In mammals, the distal centriole helps organize the connecting piece of the neck before the integrity of the centriole itself is lost. The loss of this centriole indicates that it does not serve as a basal body, or kinetosome, to initiate flagellar movement in the mature spermatozoon (Fawcett, 1972). The proximal centriole is also lost in the sperm of some mammalian species (Fawcett, 1975).

These findings complicate one of the time-honored tenets of embryology. Theodor Boveri had proposed that the fertilizing sperm provides the centrioles that organize the cleavage spindle of the zygote (Wilson, 1896). According to this theory, the centrosome of the egg is either lost or incapacitated after formation of the second polar body. The theory has persisted in the literature, and it may apply for some species. In the sea urchin, for example, two centrioles enter the egg with the sperm, and the centrioles appear to participate in organizing the cleavage spindle (Longo and Anderson, 1968). As we have seen, however, the theory does not necessarily apply to mammals. Phillips (1970) also has shown that the centrioles disappear during insect spermiogenesis and cannot contribute to the mitotic spindle.

Cell Surface Changes During Spermatogenesis

One of the most important properties of a spermatozoon is its ability to interact with the egg at fertilization. The sperm surface, which is responsible for this interaction, must acquire the molecular properties that facilitate egg recognition and binding. The properties of the sperm surface undergo changes throughout spermatogenesis and continue changing during transport through the epididymis and after ejaculation. An interesting aspect of sperm surface differentiation is that some of the surface proteins that are involved in egg recognition and binding are initially inserted more or less uniformly into the plasma membrane, and, as a result of subsequent cell surface rearrangements, become localized to specific domains where they are functional (Phelps and Myles, 1987; Scully et al., 1987). This illustrates that acquisition of functional properties of cells depends not only on synthesis of cell macromolecules but also on their correct topographical distribution within the cell. Morphologically complex cells such as sperm are excellent candidates for the study of how cellular synthetic products become organized into functional cellular components.

2–5 Gene Function in Spermatogenesis

Like all differentiation processes, spermatogenesis is regulated by the genome and is dependent upon coordinated expression of selected portions of the genome. Spermatogenesis is unique, however, in that the chromatin is condensed during spermiogenesis before the sperm are completely formed. As we shall see, chromatin condensation renders the chromatin incapable of being transcribed. Consequently, transcription that is necessary for sperm differentiation must occur before the major differentiation events.

The loss of transcriptional capacity has been monitored in a number of species. These studies indicate that RNA is synthesized in spermatogonia and spermatocytes, but transcription cannot be detected during the final stages of differentiation. However, the exact stage at which the RNA synthetic capacity is lost is somewhat variable. In some species, notably *Drosophila,* transcription is terminated in the primary spermatocyte stage (Olivieri and Olivieri, 1965; Gould-Somero and Holland, 1974). In mammalian and chicken spermatogenesis, on the other hand, RNA synthesis may continue after meiosis, terminating after chromatin condensation (Kierszenbaum and Tres, 1975; Mezquita, 1985).

Although RNA synthesis stops, completion of sperm formation is dependent on the continued synthesis of protein (Monesi, 1965; Brink, 1968; Gould-Somero and Holland, 1974; O'Brien and Bellvé, 1980). Because there is no concurrent synthesis of RNA, protein synthesis must be supported by stable RNA produced during early stages of spermatogenesis and stored for translation during spermiogenesis. This is an example of cellular mechanisms acting at the *posttranscriptional level* to delay the ultimate expression of genes by storing transcripts until a later stage of development, when the stored transcripts are translated. There are a number of documented examples in the literature of genes whose expression during spermatogenesis is regulated in this fashion (Kleene, 1989). We shall use protamine genes in the mouse and rainbow trout as examples of these genes. Protamine gene expression has been particularly well characterized because large amounts of protamine are made during spermiogenesis and, hence, their messengers are abundant. Molecular probes are essential tools in detection of any mRNA. Protamine cDNA probes (see Appendix for a description of cDNA methodology) have been used to detect protamine messenger RNA during development by hybridization analysis.

In mice, protamine messengers first accumulate in early spermatid stages when the cells are round, and the messengers persist at more or less constant levels in an untranslated form for up to a week before being translated during the spermatid elongation stage (Kleene et al., 1984; Hecht et al., 1986). This strategy is essential because protamine synthesis occurs after termination of transcription, which is a consequence of the replacement of histones with transition proteins (Kierszenbaum and Tres, 1975).

A similar pattern of protamine gene expression has been found for trout, although protamine messengers begin to accumulate during meiosis— in the primary spermatocyte stage of spermatogenesis. The translationally inactive protamine messengers are stored in the germ cells until the spermatid stage. At this time, the protamine messenger is first detected on polysomes, indicating that translation has been initiated (Gedamu et al., 1977; Iatrou and Dixon, 1978; Sinclair and Dixon, 1982). The storage of long-lived messengers for utilization during terminal cell differentiation is a common developmental strategy.

2–6 Spermiation

As we noted earlier in this chapter, large numbers of developing male germ cells are joined by intercellular bridges. The formation of syncytial clones during mammalian spermatogenesis is outlined in Figure 2.22. Incomplete cytokinesis during all but the earliest spermatogonial divisions results in groups of connected spermatogonia, and incomplete meiotic divisions expand the clones even more. The maintenance of cytoplasmic continuity during male germ cell development is widespread in the animal kingdom, which suggests that this phenomenon is of fundamental importance for spermatogenesis. Although the significance of the interconnections of these cells is not known with certainty, it is likely that interconnection is responsible for the synchrony of division and differentiation of the members of the clone and the resulting production of many mature sperm at exactly the same time.

At the completion of spermiogenesis, individual sperm are released (in a process called **spermiation**) from the syncytium by the events that remove excess cytoplasm. The cytoplasm accumulates in the neck region of the spermatids in lobules called **residual bodies**. The cytoplasmic bridges are localized between residual bodies, whereas the body of each sperm is attached to its residual body by a narrow strand of cytoplasm. Constriction of these strands releases the sperm, leaving behind the syncytial chains of residual bodies. The residual cytoplasm is phagocytosed by the Sertoli cells. This process results in individual spermatozoa that emerge from the testes. In mammals, the spermatozoa are then transported to the **epididymis**, the duct through which they pass before their forcible release at ejaculation. As we shall discuss in Section 2–7, mammalian sperm undergo further changes within the epididymis that are essential to allow them to become fertile. One of these changes is the removal of a small amount of residual cytoplasm that remains attached to the neck region as a small tag called the **cytoplasmic droplet**.

2–7 Mammalian Sperm Maturation

Spermatozoa emerging from the testes are fertile in most animal species. In fact, laboratory fertilization of the eggs of some species (such as the frogs *Rana pipiens* and *Xenopus laevis*) has traditionally involved macerating the testes in a small volume of water to release the sperm. In many mammals, however, spermatozoa emerging from the testes are immature and, hence, incompetent to fertilize eggs. They must undergo a number of significant physiological, morphological, and biochemical changes that occur within the epididymis (Bedford, 1975, 1979). These changes, which constitute **sperm maturation**, are thought to be mediated by the epididymal fluids that bathe the sperm during their traverse of the epididymis. Through the process of

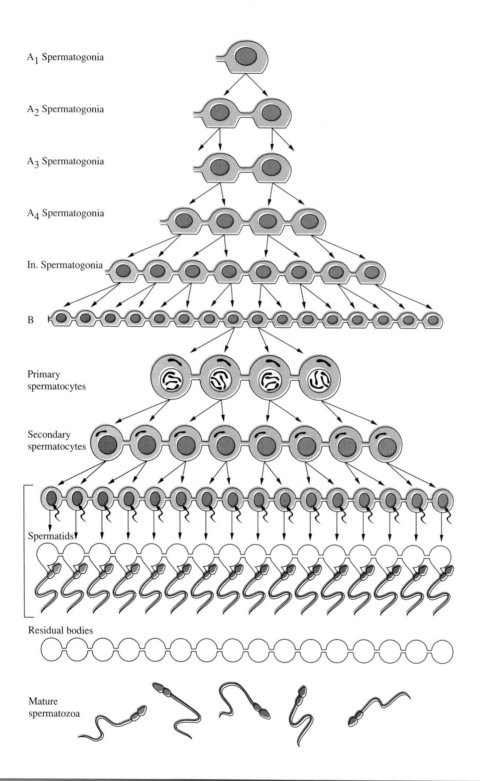

A₁ Spermatogonia

A₂ Spermatogonia

A₃ Spermatogonia

A₄ Spermatogonia

In. Spermatogonia

B

Primary spermatocytes

Secondary spermatocytes

Spermatids

Residual bodies

Mature spermatozoa

Figure 2.22

Diagram illustrating formation of syncytial clones of male germ cells and release of individual sperm. Once a spermatogonium is committed to differentiation, the daughter cells of all subsequent spermatogonial divisions and the two meiotic divisions remain connected by intercellular bridges. Six successive generations of spermatogonia are shown here. The earliest spermatogonial generations are A_1 to A_4. These are followed by intermediate (In.) spermatogonia, which divide to produce B spermatogonia. These, in turn, divide to produce the primary spermatocytes. The two meiotic divisions expand the clone. After their differentiation, individual sperm are separated from the syncytial chain of residual bodies. The actual number of interconnected cells in a clone is much larger than shown here. (After Bloom and Fawcett, 1975.)

maturation, spermatozoa acquire the potential ability to fertilize eggs. Realization of this potential does not occur, however, until the sperm have resided for some time in the female reproductive tract, in which they undergo even further changes in a process called **capacitation** before they can penetrate an egg (see Chap. 4).

The morphological changes of maturation include some remodeling of the acrosome and migration of the cytoplasmic droplet from the neck down the middle piece from which it is shed. Testicular spermatozoa are incapable of motility; progressive modifications occurring as the sperm traverse the epididymis allow them to develop the capacity to swim. The cell surface also undergoes certain changes, including the acquisition of new surface proteins and the modification, loss, and/or redistribution of some pre-existing proteins. Changes in the antigenic properties of the sperm surface provide evidence for surface modification (Bedford and Cooper, 1978). There is also evidence that sperm surface carbohydrates are modified while the sperm are in the epididymis. As we shall discuss in detail in Chapter 9, carbohydrates complexed to surface proteins may mediate the interactions between the cell and its environment. Important experimental probes to monitor cell surface carbohydrates are carbohydrate-binding proteins called **lectins**, which are specific for certain carbohydrates and allow the pattern of carbohydrate display on cell surfaces to be determined. Lectin-binding sites have been found to undergo changes in amount and distribution during maturation, suggesting that the cell surface glycoprotein configuration changes during this process (Nicolson and Yanagimachi, 1979). Because the cell surface is the region of the sperm that interfaces directly with components of the female reproductive tract after ejaculation, it is reasonable to assume that surface glycoproteins play an important role in the events that lead to fertilization. The identification of these molecules and clarification of their roles should greatly increase our knowledge of fertilization and, perhaps, suggest new targets for contraceptive procedures.

2–8 *Chapter Synopsis*

Spermatozoa are highly complex cells that are specialized for the conditions of fertilization, which vary considerably in the animal kingdom. The primary sperm components found in most sperm are a nucleus, an acrosome, and a flagellum. The acrosome, which assists in penetration of egg accessory layers and in species-specific attachment of sperm to eggs, is derived from the Golgi complex. The flagellum, which functions to propel sperm, is organized by one of the two sperm centrioles.

Sperm develop in association with specialized somatic cells. In mammals, the germ cells remain associated with Sertoli cells throughout spermatogenesis. The Sertoli cells function in translocation of the germ cells to the lumen of the seminiferous tubule and in regulating germ cell metabolism and differentiation. Spermatogenesis is also under elaborate hormonal interactions involving the pituitary gland and somatic cells of the testis.

Differentiation of sperm is called spermiogenesis. The completion of spermiogenesis is complicated in many organisms by the condensation of chromatin in the nuclei, which results in the replacement of histones with sperm-specific nuclear proteins. Chromatin condensation terminates transcription before sperm formation is complete. Because the completion of sperm differentiation depends upon protein synthesis, RNA synthesized before termination of transcription must be stored in a stable form until it is needed for protein synthesis. This is an example of posttranscriptional regulation of gene expression.

50 Chapter 2 Spermatogenesis

References

Abbas, M., and G. D. Cain. 1979. *In vitro* activation and behavior of the amoeboid sperm of *Ascaris suum* (Nematoda). Cell Tissue Res., *200*: 273–284.

Afzelius, B. A. 1976. A human syndrome caused by immotile sperm. Science, *193*: 317–319.

Baccetti, B., and R. Dallai. 1978. The spermatozoon of arthropoda. XXX. The multiflagellate spermatozoon in the termite *Mastotermes darwiniensis*. J. Cell Biol., *76*: 569–576.

Barnes, D. M. 1988. Orchestrating the sperm-egg summit. Science, *239*: 1091–1092.

Bedford, J. M. 1975. Maturation, transport, and fate of spermatozoa in the epididymis. *In* D. W. Hamilton and R. O. Greep (eds.), *Handbook of Physiology,* Vol. 5, section 7. Williams & Wilkins, Baltimore, pp. 303–317.

Bedford, J. M. 1979. Evolution of the sperm maturation and sperm storage functions of the epididymis. *In* D. W. Fawcett and J. M. Bedford (eds.), *The Spermatozoon.* Urban & Schwarzenberg, Baltimore, pp. 7–21.

Bedford, J. M., and H. I. Calvin. 1974. The occurrence and possible significance of -S-S- crosslinks in sperm heads, with particular reference to eutherian mammals. J. Exp. Zool., *188*: 137–156.

Bedford, J. M., and G. W. Cooper. 1978. Membrane fusion events in the fertilization of vertebrate eggs. *In* G. Poste and G. L. Nicolson (eds.), *Membrane Fusion. Cell Surface Reviews,* Vol. 5. Elsevier/North-Holland Biomedical Press, Amsterdam, pp. 66–125.

Bellvé, A. R., and L. A. Feig. 1984. Cell proliferation in the mammalian testis: Biology of the seminiferous growth factor (SGF). Recent Prog. Hormone Res., *40*: 531–567.

Bloch, D. P. 1969. A catalog of sperm histones. Genetics, *61* (Suppl.): 93–111.

Bloom, W., and D. W. Fawcett. 1975. *A Textbook of Histology,* 10th ed. W. B. Saunders, Philadelphia.

Brink, N. G. 1968. Protein synthesis during spermatogenesis in *Drosophila melanogaster*. Mutat. Res., *5*: 192–194.

Christensen, M. E., and G. H. Dixon. 1982. Hyperacetylation of histone H4 correlates with the terminal, transcriptionally inactive stages of spermatogenesis in rainbow trout. Dev. Biol., *93*: 404–415.

Courtens, J.-L., and M. Loir. 1981. The spermatid manchette of mammals: Formation and relation with the nuclear envelope and the chromatin. Reprod. Nutr. Dév., *21*: 467–477.

D'Agostino, A., and M. Stefanini. 1987. A rat spermatocyte surface protein is involved in adhesion of pachytene spermatocytes to Sertoli cells *in vitro*. Mol. Cell. Biol., *7*: 1250–1255.

Dym, M. 1973. The fine structure of the monkey (Macaca) Sertoli cell and its role in maintaining the blood-testis barrier. Anat. Rec., *175*: 639–656.

Dym, M., and D. W. Fawcett. 1970. The blood-testis barrier in the rat and the physiological compartmentation of the seminiferous epithelium. Biol. Reprod., *3*: 308–326.

Faulkner, R. D., and Y. M. Bhatnagar. 1987. A protease activity is associated with testicular chromatin of the mouse. Biol. Reprod., *36*: 471–480.

Fawcett, D. W. 1970. A comparative view of sperm ultrastructure. Biol. Reprod., *2* (Suppl.): 90–127.

Fawcett, D. W. 1972. Observations on cell differentiation and organelle continuity in spermatogenesis. *In* R. A. Beatty and S. Gluecksohn-Waelsch (eds.), *The Genetics of the Spermatozoan.* Published by the editors, Edinburgh, pp. 37–68.

Fawcett, D. W. 1975. The mammalian spermatozoon. Dev. Biol., *44*: 394–436.

Fawcett, D. W., W. A. Anderson, and D. M. Phillips. 1971. Morphogenetic factors influencing the shape of the sperm head. Dev. Biol., *26*: 220–251.

Fawcett, D. W., and D. M. Phillips. 1969. Observations on the release of spermatozoa and on changes in the head during passage through the epididymis. J. Reprod. Fertil., *6* (Suppl.): 405–418.

Franzén, Å. 1983. Ultrastructural studies of spermatozoa in three bivalve species with notes on evolution of elongated sperm nucleus in primitive spermatozoa. Gamete Res., *7*: 199–214.

Fritz, I. B. 1978. Sites of action of androgens and follicle stimulating hormone on cells of the seminiferous tubule. *In* G. Litwack (ed.), *Biochemical Actions of Hormones,* Vol. 5. Academic Press, New York, pp. 249–281.

Gedamu, L., P. L. Davies, and G. H. Dixon. 1977. Identification and isolation of protamine messenger ribonucleoprotein particles from rainbow trout testis. Biochemistry, *16*: 1383–1391.

Gibbons, B. H., I. R. Gibbons, and B. Baccetti. 1983. Structure and motility of the 9 + 0 flagellum of eel spermatozoa. J. Submicrosc. Cytol. *15*: 15–20.

Gibbons, I. R. et al. 1987. Photosensitized cleavage of dynein heavy chains. Cleavage at the ''V1 site'' by irradiation at 365 nm in the presence of ATP and vanadate. J. Biol. Chem., *262*: 2780–2786.

Goodrowe, K. L., and E. Heath. 1984. Disposition of the manchette in the normal equine spermatid. Anat. Rec., *209*: 177–183.

Gould-Somero, M., and L. Holland. 1974. The timing of RNA synthesis for spermiogenesis in organ cultures of *Drosophila melanogaster* testes. Wilhelm Roux's Archiv für Entwicklungsmechanik, *174*: 133–148.

Hecht, N. B. et al. 1986. Evidence for haploid expression of mouse testicular genes. Exp. Cell Res., *164*: 183–190.

Hodgson, A. N., and R. T. F. Bernard. 1986. Ultrastructure of the sperm and spermatogenesis of three species of Mytilidae (Mollusca, Bivalvia). Gamete Res., *15*: 123–135.

Iatrou, K., and G. H. Dixon. 1978. Protamine messenger RNA: Its life history during spermatogenesis in rainbow trout. Fed. Proc., *37*: 2526–2533.

Kierszenbaum, A. L., and L. L. Tres. 1975. Structural and transcriptional features of the mouse spermatid genome. J. Cell Biol., *65*: 258–270.

Kierszenbaum, A. L., and L. L. Tres. 1978. RNA transcription and chromatin structure during meiotic and postmeiotic stages of spermatogenesis. Fed. Proc., *37*: 2512–2516.

Kleene, K. C. 1989. Poly(A) shortening accompanies the activation of translation of five mRNAs during spermiogenesis in the mouse. Development, *106*: 367–373.

Kleene, K. C., R. J. Distel, and N. B. Hecht. 1984. Translational regulation and deadenylation of a protamine mRNA during spermatogenesis in the mouse. Dev. Biol., *105*: 71–79.

Koehler, L. D. 1979. A unique case of cytodifferentiation: Spermiogenesis of the prawn, *Palaemonetes paludosus*. J. Ultrastruct. Res., *69*: 109–120.

Leblond, C. P., and Y. Clermont. 1952. Definition of the stages of the cycle of the seminiferous epithelium in the rat. Ann. N. Y. Acad. Sci., *55*: 548–573.

Longo, F. J., and E. Anderson. 1968. The fine structure of pronuclear development and fusion in the sea urchin. J. Cell Biol., *39*: 339–368.

Longo, F. J., and E. Anderson. 1974. Gametogenesis. *In* J. Lash and J. R. Whittaker (eds.), *Concepts of Development*. Sinauer Associates, Sunderland, MA, pp. 3–47.

Lyon, M. F., P. H. Glenister, and M. L. Lamoreux. 1975. Normal spermatozoa from androgen-resistant germ cells of chimaeric mice and the role of androgen in spermatogenesis. Nature (Lond.), *258*: 620–622.

Marushige, K., Y. Marushige, and T. J. Wong. 1976. Complete displacement of somatic histones during transformation of spermatid chromatin: a model experiment. Biochemistry, *15*: 2047–2053.

Marushige, Y., and K. Marushige. 1983. Proteolysis of somatic type histones in transforming rat spermatid chromatin. Biochem. Biophys. Acta, *761*: 48–57.

McIntosh, J. R., and K. R. Porter. 1967. Microtubules in the spermatids of the domestic fowl. J. Cell Biol., *35*: 153–173.

Means, A. R. et al. 1980. Regulation of the testis Sertoli cell by follicle stimulating hormone. Ann. Rev. Physiol., *42*: 59–70.

Medina, A., J. Moreno, and J. L. López-Campos. 1986. Nuclear morphogenesis during spermiogenesis in the nudibranch mollusc *Hypselodoris tricolor* (Gastropoda, Opisthobranchia). Gamete Res., *13*: 159–171.

Meistrich, M. L. et al. 1978. Nuclear protein transitions during spermatogenesis. Fed. Proc., *37*: 2522–2525.

Mezquita, C. 1985. *Chromatin Composition, Structure and Function in Spermatogenesis. Revisiones Sobre Biologia Celular,* Vol. 5. Servicio Editorial, Universidad del Pais Vasco.

Monesi, V. 1965. Synthetic activities during spermatogenesis in the mouse. RNA and protein. Exp. Cell Res., *39*: 197–224.

Nicolson, G. L., and R. Yanagimachi. 1979. Cell surface changes associated with the epididymal maturation of mammalian spermatozoa. *In* D. W. Fawcett and J. M. Bedford (eds.), *The Spermatozoon.* Urban & Schwarzenberg, Baltimore, pp. 187–194.

O'Brien, D. A., and A. R. Bellvé. 1980. Protein constituents of the mouse spermatozoon. II. Temporal synthesis during spermatogenesis. Dev. Biol., *75*: 405–418.

Ogawa, K., T. Mohri, and H. Mohri. 1977. Identification of dynein as the outer arms of sea urchin sperm axonemes. Proc. Natl. Acad. Sci. U.S.A., *74*: 5006–5010.

Oliva, R. et al. 1987. Factors affecting nucleosome disassembly by protamines *in vitro*. Histone hyperacetylation and chromatin structure, time dependence, and the size of the sperm nuclear proteins. J. Biol. Chem., *262*: 17,016–17,025.

Oliva, R. et al. 1988. Haploid expression of the rooster protamine mRNA in the postmeiotic stages of spermatogenesis. Dev. Biol., *125*: 332–340.

Olivieri, G., and A. Olivieri. 1965. Autoradiographic study of nucleic acid synthesis during spermatogenesis in *Drosophila melanogaster*. Mutat. Res., *2*: 366–380.

Palombi, F. et al. 1979. Morphological characteristics of male germ cells of rats in contact with Sertoli cells *in vitro*. J. Reprod. Fertil., *57*: 325–330.

Phelps, B. M., and D. G. Myles. 1987. The guinea pig sperm plasma membrane protein, PH-20, reaches the surface via two transport pathways and becomes localized to a domain after an initial uniform distribution. Dev. Biol., *123*: 63–72.

Phillips, D. M. 1970. Insect sperm: Their structure and morphogenesis. J. Cell Biol., *44*: 243–277.

Phillips, D. M. 1974. Nuclear shaping in the absence of microtubules in scorpion spermatids. J. Cell Biol., *62*: 911–917.

Phillips, D. M. 1980. Observations on mammalian spermiogenesis using surface replicas. J. Ultrastruct. Res., *72*: 103–111.

Phillips, D. M. 1983. Analysis of sperm motility. J. Submicrosc. Cytol., *15*: 29–35.

Picheral, B. 1979. Structural, comparative, and functional aspects of spermatozoa in urodeles. *In* D. W. Fawcett and J. M. Bedford (eds.), *The Spermatozoon.* Urban & Schwarzenberg, Baltimore, pp. 267–287.

Rattner, J. B., and B. R. Brinkley. 1972. Ultrastructure of mammalian spermiogenesis. III. The organization and morphogenesis of the manchette during rodent spermiogenesis. J. Ultrastruct. Res., *41*: 209–218.

Rich, K. A., and D. M. de Kretser. 1983. Spermatogenesis and the Sertoli cell. *In* D. M. de Kretser, H. G. Burger, and B. Hudson (eds.), *The Pituitary and Testis. Clinical and Experimental Studies.* Springer-Verlag, Berlin, pp. 84–105.

Russell, L. D. 1980. Sertoli–germ cell interrelations: A review. Gamete Res., *3*: 179–202.

Russell, L. D. et al. 1983. Development of the acrosome and alignment, elongation and entrenchment of spermatids in procarbazine-treated rats. Tissue Cell, *15*: 615–626.

Russell, L. D., and R. N. Peterson. 1985. Sertoli cell junctions: Morphological and functional correlates. Int. Rev. Cytol., *94*: 177–211.

Russell, L., R. N. Peterson, and M. Freund. 1980. On the presence of bridges linking the inner and outer acrosomal membranes of boar spermatozoa. Anat. Rec., *198*: 449–459.

Sanborn, B. M. et al. 1977. Direct measurement of androgen receptors in cultured Sertoli cells. Steroids, *29*: 493–502.

Sato, M., and K. Osanai. 1986. Morphological identification of sperm receptors above egg microvilli in the polychaete, *Neanthes japonica*. Dev. Biol., *113*: 263–270.

Scully, N. F., J. H. Shaper, and B. D. Shur. 1987. Spatial and temporal expression of cell surface galactosyltransferase during mouse spermatogenesis and epididymal maturation. Dev. Biol., *124*: 111–124.

Serres, C., D. Feneux, and P. Jouannet. 1986. Abnormal distribution of the periaxonemal structures in a human sperm flagella dyskinesia. Cell Motil. Cytoskeleton, *6*: 68–76.

Sickels, K. I., and E. Heath. 1986. Disposition of the manchette and related events in the feline spermatid. Anat. Rec., *216*: 367–372.

Sinclair, G. D., and G. H. Dixon. 1982. Purification and characterization of cytoplasmic protamine messenger ribonucleoprotein particles from rainbow trout testis cells. Biochemistry, *21*: 1869–1877.

Skinner, M. K., and M. D. Griswold. 1980. Sertoli cells synthesize and secrete a transferrin-like protein. J. Biol. Chem., *255*: 9523–9525.

Steinberger, A., and E. Steinberger. 1976. Secretion of an FSH-inhibiting factor by cultured Sertoli cells. Endocrinology, *99*: 918–921.

Summers, K. E., and I. R. Gibbons. 1971. Adenosine triphosphate-induced sliding of tubules in trypsin-treated flagella of sea-urchin sperm. Proc. Natl. Acad. Sci. U.S.A., *68*: 3092–3096.

Tres, L. L., and A. L. Kierszenbaum. 1983. Viability of rat spermatogenic cells *in vitro* is facilitated by their co-culture with Sertoli cells in serum-free hormone-supplemented medium. Proc. Natl. Acad. Sci. U.S.A., *80*: 3377–3381.

van Deurs, B. 1975. Chromatin condensation and nuclear elongation in the absence of microtubules in chaetognath spermatids. J. Submicrosc. Cytol., *7*: 133–138.

Villee, C. A. et al. 1989. *Biology*. 2nd ed. Saunders College Publishing, Philadelphia.

Warner, F. D., and P. Satir. 1974. The structural basis of ciliary bend formation. Radial spoke positional changes accompanying microtubule sliding. J. Cell Biol., *63*: 35–63.

Wilkinson, R. F., H. P. Stanley, and J. T. Bowman. 1974. Genetic control of spermiogenesis in *Drosophila melanogaster*: The effects of abnormal cytoplasmic microtubule populations in mutant *ms*(3)10R and its colcemid-induced phenocopy. J. Ultrastruct. Res., *48*: 242–258.

Wilson, E. B. 1896. *The Cell in Development and Inheritance*. Reprinted by Johnson Reprint Corp., New York (1966).

Witman, G. B., J. Plummer, and G. Sander. 1978. *Chlamydomonas* flagellar mutants lacking radial spokes and central tubules. Structure, composition, and function of specific axonemal components. J. Cell Biol., *76*: 729–747.

Ying, S.-Y. et al. 1987. Secretion of follicle-stimulating hormone and production of inhibin are reciprocally related. Proc. Natl. Acad. Sci. U.S.A., *84*: 4631–4635.

Oogenesis

Jean Brachet (1909–1988), a Professor at the Université Libre de Bruxelles, Brussels, Belgium, contributed substantially to the progress of contemporary molecular biology. Using sea urchin eggs, Dr. Brachet was the first to demonstrate the presence of RNA in cells. His subsequent investigations led him to propose that RNA is involved in protein synthesis. Dr. Brachet spent much of his career studying the roles of RNA in development, especially the synthesis, storage, and utilization of RNA in oocytes. (Courtesy of Elsevier Science Publishers B.V.)

Differentiation of the female gamete, in many ways, marks the beginning of the developmental process that produces the next generation of organisms. As we have seen, the male gamete is specialized to deliver its nuclear package to the egg, but the egg contains not only a haploid nucleus but also a large dowry consisting of the materials and energy sources that are needed for construction of the embryo until it can either produce them on its own or obtain them from the environment. Furthermore, as we shall learn in this and later chapters, certain messenger RNAs and proteins that are essential for initiating development after fertilization are produced exclusively during oogenesis. In some embryos, these components may be spatially restricted to certain regions of the egg, thus producing differential effects that establish distinct embryonic regions. Thus, oogenesis has enormous significance for the course of embryonic development.

The timing of oogenesis relative to meiosis is different from that of spermatogenesis. In the male, gamete differentiation occurs *after* meiosis, but in the female, the oocyte is fully formed *before* meiosis is completed. Consequently, differentiation of the female gamete is intimately associated with meiosis. In most species the bulk of gamete differentiation occurs during prophase I of meiosis. In species producing yolky eggs, this phase is arbitrarily divided into (1) **previtellogenesis** (before yolk deposition), (2) **vitellogenesis** (yolk deposition), and (3) **postvitellogenesis** (after yolk deposition). Most oocyte growth occurs during vitellogenesis. The resumption of meiosis (**maturation**) and ovulation produce the ripe egg, or ovum. In some organisms, such as the frog *Xenopus laevis*, oocytes in these various stages of oogenesis can coexist in the ovary (Fig. 3.1).

Oogenesis can be quite prolonged in relation to the length of the life cycle of the organism. In the frog *Rana pipiens*, for example, oogenesis takes three years. After metamorphosis of tadpoles into juvenile frogs, and every year after that, oogonial divisions produce a new batch of oocytes. Thus, the ovary contains three generations of oocytes simultaneously, with one generation maturing each year and being replaced by a new one. In mammals, on the other hand, all of the oogonial divisions and transformations of oogonia into oocytes are completed either before or shortly after birth, producing a finite number of oocytes, each of which is retained in meiotic prophase I (Pinkerton et al., 1961; Franchi et al., 1962). Thus, the period of oogenesis covers virtually the entire life span from birth to ovulation. However, no oocyte growth occurs until puberty, after which a new group of oocytes resumes development during each cycle. Most of the growing oocytes fail to reach maturity during the cycle and degenerate. The major events of oogenesis in the life cycle of the mouse are diagrammatically represented in Figure 3.2.

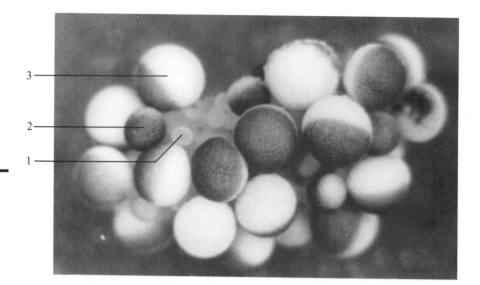

Figure 3.1

Xenopus oogenesis. Fragment of an ovary dissected from an adult female, showing oocytes at different stages of their growth. 1: previtellogenic oocyte; 2: early vitellogenic oocyte; 3: full grown, postvitellogenic oocyte.

3–1 Oocyte-Accessory Cell Interactions During Oogenesis

Like developing male germ cells, oocytes form an intimate association with nongerm cells in the ovary. These **accessory cells** may be important in steroid hormone production, in transportation of certain essential cytoplasmic components to the oocyte, and in formation of cellular and noncellular layers that surround the fully differentiated egg. The accessory cells fall into two categories: (1) follicle cells and (2) **nurse cells**. Follicle cells are somatic cells that form a layer surrounding the oocyte. This layer of cells is known as the **follicular epithelium** (Fig. 3.3A). The major distinction between follicle cells and nurse cells is that the former are derived from somatic cells, whereas the latter are derived from the germ cell line and remain associated with the oocyte via cytoplasmic bridges. Nurse cells are found in certain invertebrates, including some coelenterates, annelids, mollusks, and insects.

Figure 3.2

Diagrammatic representation of oogenesis during the life cycle of the mouse. (After Schultz and Wassarman, 1977. Reprinted by permission of Company of Biologists, Ltd.)

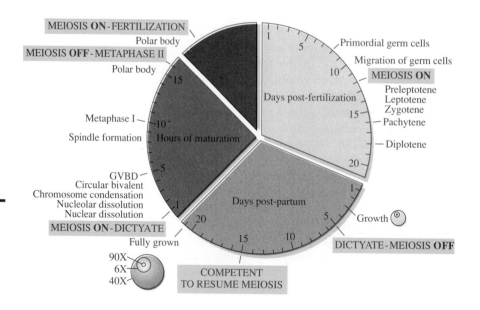

The Mammalian Follicle

During follicle growth in mammals, the follicular epithelium proliferates to become multilayered. The cells of the expanding follicle are sometimes called **granulosa cells**. The granulosa cells and the oocyte are separated by a space that widens as a result of the deposition of sulfated glycoproteins that are produced by the oocyte (Bleil and Wassarman, 1980a; Shimizu et al., 1983). This material forms a continuous coat, the **zona pellucida**, around the oocyte. Examination of the zona with the electron microscope reveals

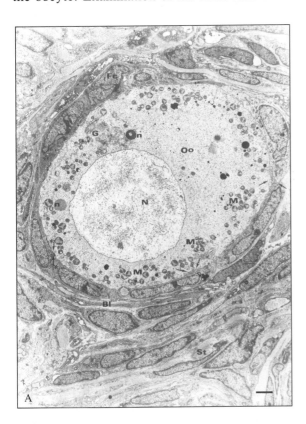
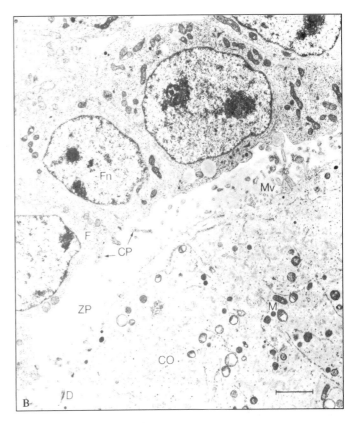

Figure 3.3

Relationship between mammalian oocyte and follicle cells. A, Electron micrograph of a primordial follicle of rabbit. The oocyte (Oo) is surrounded by a single layer of granulosa cells (Fc). The oocyte and granulosa cell membranes interdigitate extensively (arrows) and may form junctional complexes. The most prominent feature of the oocyte is the large nucleus (N). A basal lamina (Bl) surrounds the follicle and isolates it from the stroma (St) of connective tissue. G: Golgi membranes; M: mitochondria; n: nucleolus-like bodies. Scale bar equals 2 μm. B, Advanced oocyte of mouse, showing zona pellucida (ZP), microvilli (Mv) from the oocyte, and cytoplasmic processes (CP) from the follicle cells. CO: Cytoplasm of the oocyte; F: follicle cell; Fn: nucleus of follicle cell; D: desmosome at point of contact of follicle cell projection and surface of oocyte; M: mitochondria. Scale bar equals 0.5 μm. C, Scanning electron micrograph of the surface of a preovulatory rat oocyte. The zona pellucida has been removed enzymatically. Follicle cell processes are clearly evident. Note oocyte microvilli. ×1150. (A, From Van Blerkom and Motta, 1979. B, Courtesy of E. Anderson. C, From Dekel et al., 1978. Reprinted from *Gamete Research* by permission of Wiley-Liss, a division of John Wiley and Sons, Inc. Copyright © 1978, Wiley-Liss.)

that it is penetrated by numerous short microvilli from the surface of the oocyte and relatively long cytoplasmic processes from the follicle cells, which contact the oocyte surface (see Fig. 3.3B and C). Desmosomes and gap junctions are found at the sites of contact between the cytoplasmic processes and the oocyte membrane (Anderson and Albertini, 1976; Gilula et al., 1978). Nutrients and molecules that regulate oocyte development are passed into the oocyte via the gap junctions (Eppig, 1985). Before ovulation, the cytoplasmic processes and microvilli are usually withdrawn. The zona and a few granulosa cells are retained by the egg when it is ovulated (see Sect. 3–8).

As proliferation of granulosa cells nears completion, they are thought to secrete the fluid that accumulates in intercellular spaces. The fluid-filled spaces coalesce to form a large cavity—the **antrum**. Follicles with large antra are called **Graafian follicles** (Fig. 3.4). Formation of the antrum displaces the oocyte to one side of the follicle. The cluster of follicle cells around the oocyte is called the **cumulus oophorus**. As we shall discuss later, the cumulus accompanies the egg at ovulation.

The Insect Egg Chamber

Although nurse cells are found in a number of invertebrates, their formation and function are particularly well understood in insects, which will serve as our example. Oogenesis in insects with nurse cells is called **meroistic**

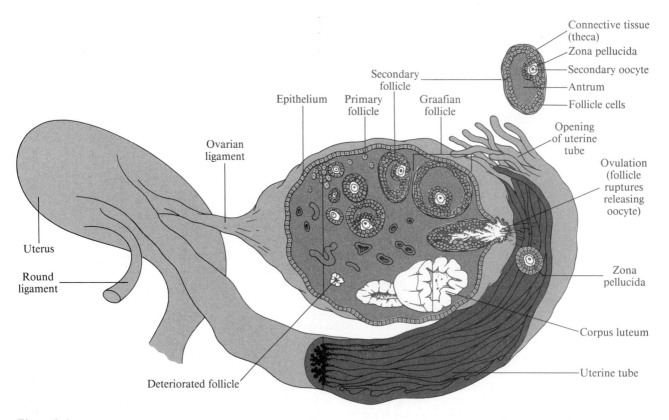

Figure 3.4

Diagram of a composite mammalian ovary. Progressive stages in the differentiation of a Graafian follicle are scattered throughout the ovary. (After Villee et al., 1989.)

oogenesis. There are, in turn, two types of meroistic ovaries: **polytrophic** and **telotrophic**. In polytrophic ovaries, the nurse cells are intimately connected to the oocyte, and the nurse cell-oocyte complex is surrounded by follicle cells. Telotrophic ovaries, on the other hand, have a cluster of nurse cells at one end of the ovary that are connected via long **trophic cords** to the oocytes, which are surrounded by follicle cells.

Polytrophic Oogenesis

Drosophila melanogaster is an example of a species with polytrophic ovaries. When *Drosophila* oogonia are undergoing their terminal divisions, cytokinesis is incomplete, resulting in continuity of the cells via cytoplasmic bridges. Although all cells in a clone appear to be equivalent, one of them becomes specialized and differentiates into the gamete; the remaining interconnected cells serve as nurse cells. The cluster of oocyte and associated nurse cells is an **egg chamber**, which is completely surrounded by follicle cells (Fig. 3.5). Nurse cells appear to provide the developing oocyte with macromolecules and even with ribosomes, which are transported from the nurse cells to the oocyte through the intercellular bridges.

Formation of the oocyte–nurse cell clone of *Drosophila* is diagrammatically represented in Figure 3.6. Mitosis of an oogonial stem cell produces two daughter cells that separate from one another. One of them continues to behave as a stem cell for the production of additional clones, whereas the other functions as a **cystoblast**, which establishes a clone by mitotic division. The daughter cells (**cystocytes**) produced by a cystoblast do not separate, thereby forming permanent **ring canals**. Mitoses involving cystocytes are also peculiar in that cell volume does not double before mitosis. Consequently, as divisions continue, the cystocytes become progressively smaller. During consecutive divisions, the spindle axes of the cystocytes shift, creating a branching chain of interconnected cells. After the fourth division has produced 16 cells, 2 cells of the clone (1 and 2) differ from all the rest in that they have four ring canals. Both of these cells (which are called **pro-oocytes**) prepare for meiosis, although the process persists to completion in only one of them. The process is aborted in the other, which develops into a nurse cell (Koch et al., 1967).

How do the pro-oocytes differ from each other such that one of them continues to develop as the germ cell, whereas the other reverts to the nurse cell pathway? Although we do not have a definitive answer to this question, there are some interesting observations that bear on the question. Koch and Spitzer (1983) reported that *Drosophila* fed colchicine (which depolymerizes microtubules; see p. 337) produce some egg chambers with 16 nurse cells and no oocyte. Thus, colchicine causes the presumptive

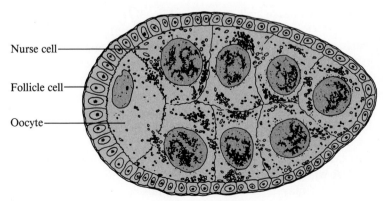

Nurse cell

Follicle cell

Oocyte

Figure 3.5

A drawing of a section through an egg chamber of *Drosophila melanogaster* as seen through the light microscope. Note the cytoplasmic continuity between cells. Organelles such as mitochondria can be seen in the intercellular bridges. The chamber is surrounded by a layer of follicle cells. The plasma membrane of the oocyte interdigitates with the membranes of the follicle cells. (After Klug et al., 1970.)

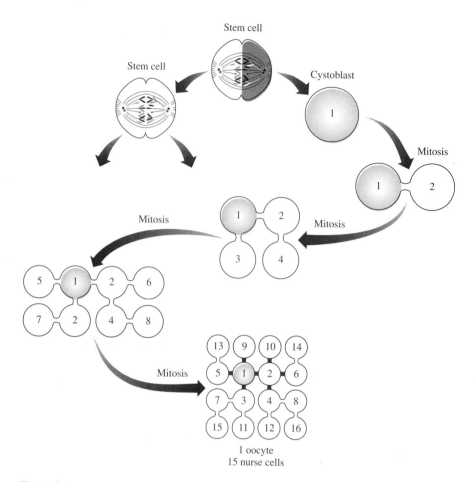

Figure 3.6

Formation of the nurse cell-oocyte cluster from a single stem cell in the
Drosophila ovary. In this drawing, the cells are represented by circles lying in a
single plane, and the ring canals have been lengthened for clarity. The area of
each circle is proportional to the volume of the cell. Note that this sequence of
mitoses is unusual in that cell volume does not double before mitosis, so that cell
volume decreases at each division. The stem cell divides into two daughters, one
of which behaves like its parent. The other differentiates into a cystoblast, which
by a series of four divisions produces 16 interconnected cystocytes. Either of the
cells connected by intercellular bridges to four others (cells 1 and 2) can become
the oocyte. In this case, cell 1 will develop as the oocyte, while cell 2 will become
one of the 15 nurse cells.

oocyte to switch to the nurse cell pathway (Fig. 3.7). Certain mutations
have the same effect (King et al., 1985). These observations indicate that
the oocyte is also capable of forming a nurse cell but is normally restrained
from doing so. Following this line of reasoning, both pro-oocytes must be
initially restrained from developing into nurse cells. In normal development,
this restraint is relieved in one pro-oocyte, whereas the restraint is maintained
in the other pro-oocyte, and it develops into the oocyte. Colchicine removes
this block, although it is uncertain whether its effects on microtubules are
responsible for this result.

The 15 nurse cells grow and become highly polyploid owing to extensive
replication of DNA without further cell division. After this **endoreplication**,
Drosophila nurse cells may have up to 1024 times as much DNA as the

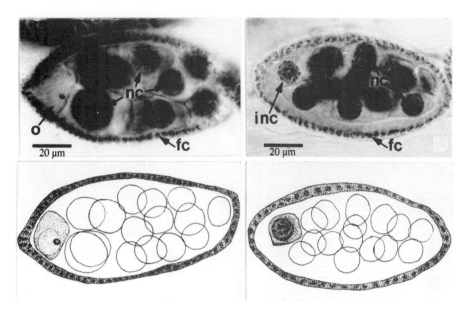

Figure 3.7

The effects of colchicine on *Drosophila* egg chambers. Feulgen-stained whole mounts are shown in the top row (Feulgen staining is specific for DNA), drawings on the bottom. Left: control chamber; right: chamber from fly fed colchicine. The tracing of the control chamber shows the positions of 15 nurse cell nuclei and the oocyte (o). The oocyte nucleus is characterized by a small, dense Feulgen-positive structure. Colchicine results in conversion of the oocyte into a nurse cell. The nuclei of the usual 15 nurse cell nuclei are shown, as is the induced nurse cell (inc), which replaced the oocyte. fc: follicle cell; nc: nurse cells. (From Koch and Spitzer, 1983.)

haploid genome (King, 1970). The polyploid nurse cell nuclei are very active in RNA synthesis. The fate of this RNA has been demonstrated by autoradiography of egg chambers exposed to labeled RNA precursors. (See the Appendix for a discussion of the technique of autoradiography.) The results of such an experiment are shown in Figure 3.8. In this experiment, egg chambers of the housefly, *Musca domestica*, were exposed to ^{3}H-cytidine for different periods of time. After short exposure to the label (see Fig. 3.8A), the nurse cell nuclei, but not the oocyte nucleus, incorporate the precursor into RNA. The fate of this RNA is seen in Figure 3.8B, which is an autoradiograph of material that was fixed five hours after injection of ^{3}H-cytidine. Here, the labeled nurse cell RNA can be seen entering the oocyte cytoplasm via ring canals. There is still no indication of oocyte nuclear RNA synthesis by this time. Because each nurse cell has the equivalent of several genomes, the oocyte can acquire vast amounts of RNA from them to promote its differentiation.

What regulates the movement of cytoplasmic constituents through the ring canals? Do they merely respond to concentration differences between the nurse cells and oocyte, or is there a means of directing particular substances in one direction or the other? There is evidence from experiments with the moth *Hyalophora cecropia* that a polarized transport of proteins and nucleic acids from the nurse cells to the oocyte is directed by differences in electrical potential between nurse cells and the oocyte. Electrical potential measurements reveal that the nurse cell cytoplasm is negative relative to the oocyte cytoplasm (Woodruff and Telfer, 1973). The voltage difference

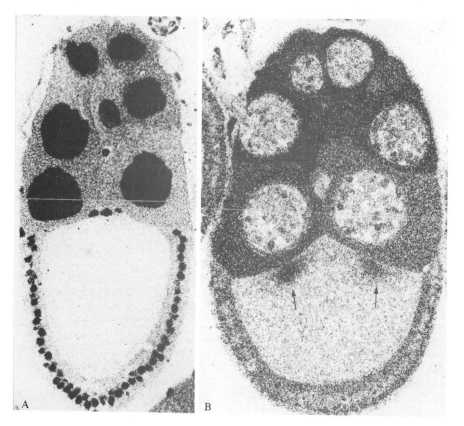

Figure 3.8

Synthesis of macromolecules in nurse cells of the housefly, *Musca domestica*, and their transport into the oocyte. A, Autoradiograph of a section of an egg chamber 1 hour after injection of ^{3}H-cytidine. Intense label in nurse cell nuclei is an indication of RNA synthesis. Some labeled RNA has been transported to nurse cell cytoplasm. B, Egg chamber 5 hours after injection of label. The labeled RNA originally synthesized in the nurse cell nuclei is now mainly localized in the nurse cell cytoplasm. Transport into the oocyte via ring canals is indicated by arrows. (Unpublished micrographs from the work of K. Bier—see Bier, 1963. Courtesy of D. Ribbert.)

is maintained by a steady flow of positive current through the ring canals from the oocyte to the nurse cells (Jaffe and Woodruff, 1979). Consistent with this electrical polarity is the observation that a negatively charged fluorescent protein will diffuse from nurse cells into the oocyte (Fig. 3.9A), but not in the opposite direction (see Fig. 3.9B). In contrast, a positively charged protein will move from the oocyte to the nurse cells (see Fig. 3.9C), but not into the oocyte from the nurse cells (see Fig. 3.9D). The behavior of charged proteins suggests that molecular movement between the oocyte and nurse cells is a kind of electrophoresis. Recent evidence for a similar voltage gradient in *Drosophila* follicles suggests that this mechanism may be a generalized one for transporting molecules from nurse cells to the oocyte (Woodruff, 1989).

Ultimately, the nurse cells inject virtually all of their cytoplasm into the oocyte. The ring canals are then severed, and the nurse cells are sloughed from the oocyte before ovulation (Cummings and King, 1969).

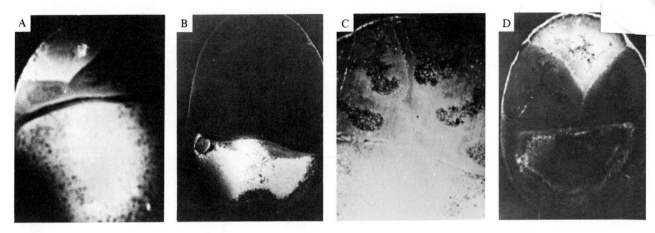

Figure 3.9

Fluorescent photomicrographs of *Hyalophora cecropia* follicles injected with fluorescent protein and incubated for 1 to 2 hours. A, Negatively charged protein injected into nurse cells. B, Negatively charged protein injected into oocyte. C, Positively charged protein injected into oocyte. D, Positively charged protein injected into nurse cells. The width of the nurse cell cap adjacent to the oocyte in each case is approximately 450 μm. (From R. I. Woodruff and W. H. Telfer, 1980. Reprinted by permission from *Nature* (Lond.), *286*: 84–86. Copyright © 1980, Macmillan Magazines Limited.)

The simultaneous support by 15 cells, each highly polyploid, enables the *Drosophila* oocyte to grow quite rapidly. Under optimal conditions, the cytoplasmic volume can increase 90,000 times in only three days (King, 1972). Such rapid growth is a considerable achievement and testifies to the remarkable advantages of this mechanism of oogenesis for facilitating rapid egg production.

Telotrophic Oogenesis

Telotrophic ovaries are found in a number of insect groups, including the Hemiptera (true bugs). One of the most extensively studied telotrophic insects is the bloodsucking bug, *Rhodnius*. The *Rhodnius* ovary consists of approximately seven lobes called **ovarioles**. At its anterior end, each ovariole contains a **tropharium**, which consists of a syncytial complex of nurse cells that project laterally and apically from a central **trophic core** that is continuous with the posteriorly projecting trophic cords, each of which connects to a growing oocyte at the posterior end of the ovariole (Fig. 3.10). As with polytrophic egg chambers, nurse cells in telotrophic ovaries are highly active in RNA and protein synthesis, whereas the oocytes are minimally active in macromolecular synthesis (Huebner, 1984). The physical separation between the nurse cells and oocytes presents an interesting problem: the polarized transport via the trophic cords of macromolecules from their sites of synthesis in the nurse cells into the oocytes.

The trophic cords of *Rhodnius* and other hemipterans are literally packed with microtubules and contain one of the most concentrated arrays of microtubules of any known tissue (Huebner, 1984). Although it has been proposed that these microtubules are involved in macromolecular transport, the cords of coleopteran insects have very few microtubules, indicating that microtubules are not essential for trophic cord transport. The trophic core within the tropharium of *Rhodnius* and other hemipterans contains an

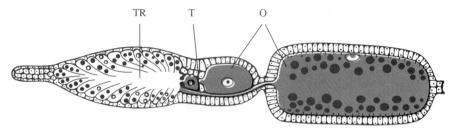

Figure 3.10

Diagram of the *Rhodnius* ovariole. A syncytial anterior tropharium consisting of lobes connected to a common trophic core (TR) is associated via long cytoplasmic bridges (trophic cords; T) to growing oocytes (O). The entire nurse cell-oocyte complex is surrounded by epithelial cells. (After Huebner and Gutzeit, 1986.)

extensive meshwork of microfilaments, which could theoretically provide the force to propel macromolecules down the trophic cords (Gutzeit and Huebner, 1986; Huebner and Gutzeit, 1986). However, there are no experimental data bearing on this possibility. As has been demonstrated for *Hyalophora* ovaries, a difference in electrical potential may be involved in transport of macromolecules from the tropharium to the oocytes of telotrophic ovaries. The oocytes are electrically positive relative to the tropharium if juvenile hormone is present. (As we discuss in Sect. 3–6, juvenile hormone is involved in regulating oocyte differentiation.) Negatively charged fluorescent proteins injected into the tropharium become dispersed within the tropharium and down the trophic cords, whereas positively charged proteins remain concentrated in the tropharium near the injection site (Telfer et al., 1981; Huebner, 1984). However, it remains to be determined whether this is a generalized means of polarized transport in telotrophic ovaries.

After cytoplasmic flow via the trophic cords is complete, the cords close, and the microtubules in them become tightly packed in crystalline-like arrays. Eventually, the cords retract and are absorbed. Thus, in contrast to polytrophic ovaries, there is no wholesale dumping of nurse cell cytoplasm into the oocytes at the end of oogenesis (Huebner, 1984).

Follicle Cells in Insects

The follicle cells that circumscribe the insect egg chamber play various roles during oogenesis. As mentioned previously, they can transfer material to the oocyte. Insect follicle cells, as is the case with mammalian follicle cells, are coupled to the oocyte via gap junctions that serve as channels for intercellular transport (Huebner, 1981). Follicle cells also play an active role in vitellogenesis; they can sequester yolk precursors from the hemolymph for transport to the oocyte, and, in some species, they may also synthesize yolk precursors (see Sect. 3–4). The follicle cells are also involved in secretion of (initially) the vitelline envelope, which is a thin glycoprotein layer that surrounds the egg plasma membrane, and (subsequently) the chorion, which forms a tough capsule surrounding the insect egg (Cummings and King, 1969; Quattropani and Anderson, 1969; Cummings et al., 1971). The deposition of these accessory layers around the entire circumference of the oocyte is facilitated by the interposition of follicle cells between the nurse cells and oocyte in the latter stages of oogenesis so that the oocyte becomes encased by these cells except for the areas occupied by the ring canals.

3–2 Organization of the Egg and Differentiation of Oocyte Constituents

In many ways, an egg is an ordinary cell with typical cell organelles. However, because of its unique roles in fertilization and in providing for the maintenance of the developing embryo, the egg acquires a number of specializations during oogenesis that are not found in somatic cells. Because oocyte structure in different species is so variable, it is not possible to discuss the structure and formation of every specialized oocyte inclusion. Instead, we discuss here a few of the better understood and most widespread oocyte constituents.

One of the most striking aspects of egg morphology is the definite spatial organization of organelles and other inclusions in the cytoplasm (often called **ooplasm**). One critical aspect of the spatial arrangement of the egg is its **polarity**. Egg constituents may be unequally distributed along the major axis of the egg. This axis is an imaginary line connecting the two poles: the animal pole and the vegetal pole. The oocyte nucleus is displaced toward the animal pole, and when meiotic maturation occurs, the polar bodies are formed at this pole. Certain cytoplasmic organelles and inclusions may also be displaced toward the animal pole. In the amphibian egg, for example, ribosomes, mitochondria, and pigment granules are prevalent at the animal pole and decrease in numbers gradually toward the vegetal pole. Yolk platelets, on the other hand, are small and loosely packed in the animal hemisphere and become progressively larger and more concentrated toward the vegetal pole (Fig. 3.11). As we shall discuss on page 74, the asymmetrical distribution of yolk platelets depends upon active translocation of platelets from the animal hemisphere to the vegetal hemisphere during late oogenesis. Translocation of cytoplasmic constituents into the vegetal hemisphere occurs during earlier stages of oogenesis as well. A cluster of mitochondria and granulofibrillar material (**mitochondrial cloud**; see p. 66) that is located near the germinal vesicle in previtellogenic *Xenopus* oocytes disperses into smaller clusters that occupy positions just below the oocyte surface in what is thought to correspond to the presumptive vegetal pole region of the

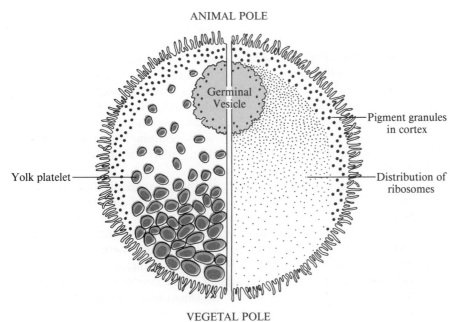

Figure 3.11

Diagrammatic representation of polarized distribution of amphibian oocyte components. Left: yolk distribution. Right: distribution of ribosomes. Note also the displacement of the germinal vesicle (oocyte nucleus) to the animal pole and the cortical location of pigment granules.

oocyte (Heasman et al., 1984). The granulofibrillar material in this complex bears a striking resemblance to the germ plasm, which is known to be in the subcortical vegetal cytoplasm of the fertilized egg. Thus, this translocation event during early oogenesis may be responsible for localizing the germ plasm to its ultimate site. As we shall discuss in Section 3–4, mRNAs for certain proteins may also have polarized distribution within the oocyte.

The unequal distribution of ooplasmic constituents plays an extremely important role in the establishment of regional specialization in the embryo (see Chap. 11). Consequently, the mechanisms that establish and maintain the spatial organization of the ooplasm set the stage for orderly embryonic development. We know very little about these mechanisms of intracellular translocation. One must also provide an explanation for the forces that drive translocation and cause its directionality. One possibility is that the distribution of particles in cells could be controlled by a flow of electrical current through the cells. In fact, an animal–vegetal current has been detected in *Xenopus* oocytes (Robinson, 1979). We have seen that current flow operates to transmit macromolecules through ring canals in the egg chambers of some insects. A similar mechanism operates in the establishment of polarity in the fertilized eggs of certain brown algae. Additional research is necessary to ascertain if this is a generalized mechanism for producing polarity.

Mitochondria and Germ Plasm

The oocyte exaggerates the production of certain components so that the egg will have a store of these components with which to begin development. This relieves the embryo of the need to produce them during the earliest stages of development. One class of components that is stockpiled during oogenesis is the mitochondria. Mitochondria possess their own self-replicating DNA. Hence, the amplification of mitochondria causes a tremendous increase in the amount of mitochondrial DNA in the cytoplasm. Whereas the ratio of nuclear DNA to mitochondrial DNA in somatic *Xenopus* cells is about 100:1, it is reversed during oogenesis and ranges from 1:1 to 1:100 in fully grown oocytes (Dawid, 1972), reflecting an amplification that results in approximately 10^7 mitochondria (Marinos, 1985).

During early oogenesis in *Xenopus*, mitochondria are arranged in clusters that form a ring around the nucleus. One of these masses is generally larger than the rest. This mass grows progressively larger, apparently by mitochondrial replication (Al-Mukhtar, 1970), and is quite distinct, even in the light microscope (Fig. 3.12). This juxtanuclear aggregate of mitochondria has been variously referred to as a "yolk nucleus," "mitochondrial cloud," or "Balbiani body." Ultrastructural analyses of the mitochondrial cloud (Fig. 3.13) reveal that, in addition to mitochondria, it contains spherical and string-like electron-dense material called **granulofibrillar material (GFM)**. The smaller perinuclear aggregates of mitochondria lack the GFM. In addition to the aggregates of mitochondria, individual mitochondria are scattered in the oocyte cytoplasm.

After its formation and enlargement, the mitochondrial cloud is dispersed into large, subcortical islands of mitochondria and GFM that are restricted to one pole of the oocyte (Fig. 3.14A and B). Interestingly, structures that are identical in appearance to the dispersed mitochondria and GFM arise in subcortical islands at the vegetal pole of later-stage oocytes and in unfertilized eggs (see Fig. 3.14D). Although it has not been possible to trace the origin of these islands, their similarity to remnants of the dispersed mitochondrial cloud led Heasman and colleagues (1984) to propose that they are derived from these remnants. These islands in

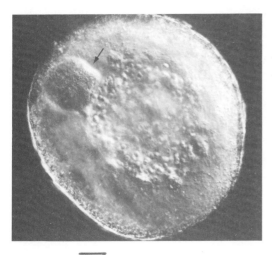

25 µm

Figure 3.12

Mitochondrial cloud of living previtellogenic oocyte of *Xenopus laevis* viewed with Nomarski interference optics. The arrow indicates the mitochondrial cloud. Large spherical structure above the mitochondrial cloud is the germinal vesicle. Note the numerous nucleoli at the periphery of the germinal vesicle. (Courtesy of J. Heasman.)

unfertilized eggs have been called **germinal granules**. In the cleavage-stage *Xenopus* embryo, the islands of mitochondria and germinal granules aggregate to form large yolk-free areas that are thought to be homologous to the germ plasm that is seen in a variety of eggs and early embryos and is considered to be the determinant of the germ cells (Czolowska, 1972; Kalt, 1973; Ikenishi and Kotani, 1975). We shall discuss the role of the germ plasm in germ cell determination in detail in Chapter 11.

Mitochondrial clouds have also been observed in oocytes of a variety of invertebrate species (Anderson and Huebner, 1968; Eckelbarger, 1979; Larkman, 1984). In some mammals, on the other hand, the mitochondria are localized to a band in the periphery of the oocytes during early stages

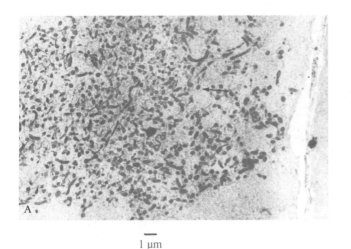

1 µm   200 nm

Figure 3.13

Ultrastructure of the mitochondrial cloud of the *Xenopus* oocyte. A, Low power electron micrograph of the cloud. The two major components of the cloud are the mitochondria and electron-dense granulofibrillar material (GFM: arrow). B, Higher power electron micrograph reveals the spherical (arrow) and string-like forms of the GFM. (A, Courtesy of J. Heasman. B, From Heasman et al., 1984.)

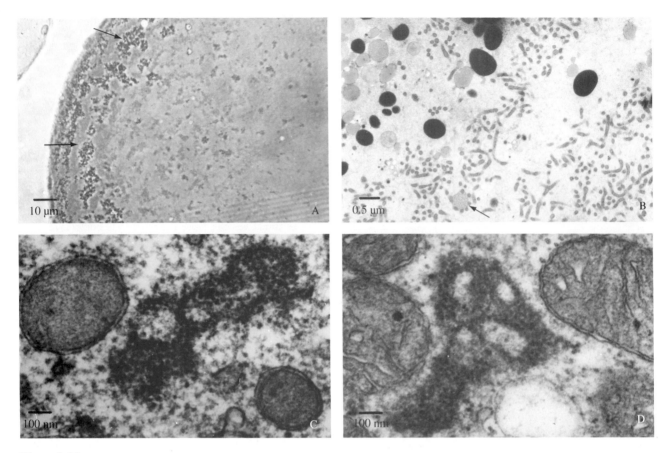

Figure 3.14

The fates of the mitochondrial cloud and GFM. A, Light micrograph showing a dispersing mitochondrial cloud. Mitochondrial islands collect at the periphery to form a distinct subcortical layer (black arrows). B, Low power electron micrograph of the region of mitochondrial cloud breakdown. The spherical form of the GFM (arrows) occurs at regular intervals. C, and D, High power electron micrographs comparing the GFM of the mitochondrial cloud of a previtellogenic oocyte (C) to a germinal granule in the subcortical region of the vegetal pole of an unfertilized egg (D). (From Heasman et al., 1984.)

of oogenesis. Thereafter, the mitochondria become dispersed throughout the ooplasm. At all times, the mitochondria are associated with cisternae of the endoplasmic reticulum (Cran et al., 1980). The mitochondria of mammals may become highly modified because of the altered orientation of cristae or the deposition of material in the mitochondria. Examples of modified mitochondria are shown in Figure 3.15.

Yolk

Amount and Distribution

The amount of yolk in animal eggs is quite variable. Early investigators devised various schemes for categorizing eggs according to yolk content and distribution. Eggs with a small amount of evenly distributed yolk have been called **oligolecithal, isolecithal,** or **homolecithal.** These eggs are found in many invertebrates, such as the sea urchin and the lower chordates (e.g., *Amphioxus* and tunicates). In these eggs, the only visible clue of egg polarity

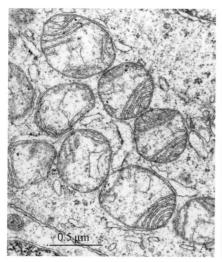

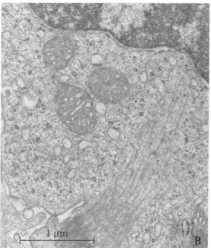

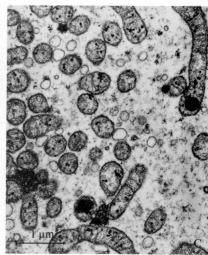

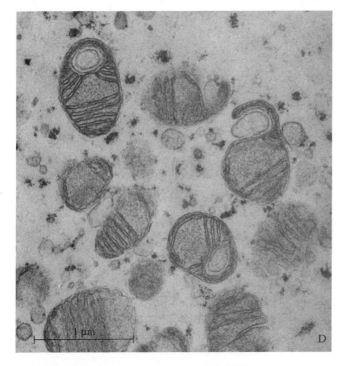

Figure 3.15

Examples of modified mitochondria found in the ooplasms of some species of mammals. A, Cristae are displaced peripherally in some mitochondria of primary rat oocytes. B, The two components of the cristae in a dividing hamster oogonium separate from each other, giving the mitochondria a vesiculated appearance. C, In oocytes of secondary follicles of the spider monkey, electron-dense granules can be seen. D, Bovine mitochondria show the development of hood-shaped extensions that surround cisternae of the endoplasmic reticulum. (A, and B, Courtesy of D. Szöllösi. C, From D. Szöllösi, 1972. Changes of some cell organelles during oogenesis in mammals. In J. D. Biggers and A. W. Schuetz (Eds.), Oogenesis. University Park Press, Baltimore, pp. 58, 61. D, From P. L. Senger and R. G. Saacke, 1970. Reprinted from *The Journal of Cell Biology*, 1970, *46*: 405–408 by copyright permission of the Rockefeller University Press.)

may be the site of polar body formation. In species with relatively large amounts of yolk, the animal and vegetal hemispheres show distinctly different organization because of the concentration of the yolk in the vegetal hemisphere. In the most extreme situations (e.g., the eggs of reptiles, bony fish, birds, and some mollusks, including cephalopods and some gastropods), there is a segregation of cytoplasm and yolk, with the cytoplasm restricted to a thin layer covering the yolk. This layer thickens at the animal pole to form a **cytoplasmic cap** that contains the nucleus (Fig. 3.16A). These eggs are called **telolecithal**. The situation is not so extreme in the amphibians, in which the yolk is concentrated in the vegetal hemisphere, whereas other cytoplasmic organelles are more numerous in the animal hemisphere. Some authors place amphibian eggs in a separate category—**mesolecithal**—whereas others classify them as moderately telolecithal. The amount and distribution of yolk have profound effects on cleavage, which will be discussed in Chapter 5.

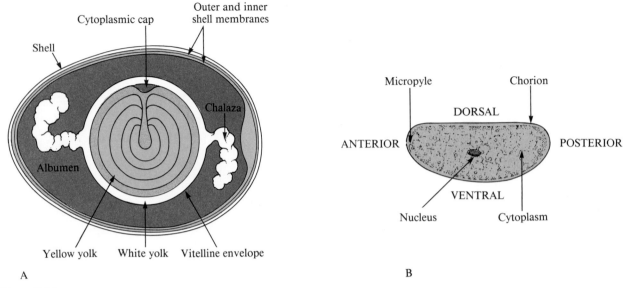

Figure 3.16

Sections of representative yolky eggs (not drawn to scale). A, Telolecithal hen's egg. B, Centrolecithal insect egg. (A, After F. R. Lillie, 1919. Redrawn from Balinsky, 1981. B, After O. A. Johannsen and F. H. Butt, 1941. *Embryology of Insects and Myriapods*, p. 10. Reprinted by permission of McGraw-Hill Book Co.)

The arthropod egg, particularly the insect egg, is unique in that the yolk assumes a central position and is surrounded by a thin coat of cytoplasm. There is also an island of cytoplasm containing the nucleus in the center of the egg. These eggs are **centrolecithal**. The animal-vegetal pole relationships are irrelevant for centrolecithal eggs. There is, however, a definite polarity in insect eggs that is reflected in the shape of the egg. The *Drosophila* egg shown in Figure 3.16B is slightly more rounded at one end. This end is destined to be the posterior portion of the embryo, whereas the opposite end will become the anterior portion. Thus, these eggs have an **anteroposterior polarity**. The dorsal and ventral aspects of the embryo are also indicated in the egg: The convex side will be the ventral side of the embryo, whereas the concave side will be the dorsal side. Dorsoventral polarity of the *Drosophila* egg will be discussed in more detail in Chapter 6.

The amount of yolk in mammalian eggs is quite variable. The eggs of primitive mammals have a great deal of yolk, but those of placental mammals have minimal yolk reserves because mammalian embryos obtain nutritive substances directly from the mother.

Vitellogenesis

Yolk is the most prominent cytoplasmic component of the egg in many species. It is a heterogeneous substance consisting of lipid, carbohydrate, and protein. The form taken by the yolk is quite variable. In some species, it is present as general reserve material with no definitive ultrastructural characteristics. In others, it may be present in definitive organelles. The amphibian **yolk platelets** are representative of yolk-containing organelles (Fig. 3.17). These are flattened ovoid structures consisting primarily of two components, **phosvitin** and **lipovitellin**, in a crystalline lattice. Phosvitin is a phosphoprotein, and lipovitellin is a lipophosphoprotein.

The raw material for vitellogenesis can come from two sources: It may be synthesized within the oocyte (**autosynthesis**) or synthesized outside the

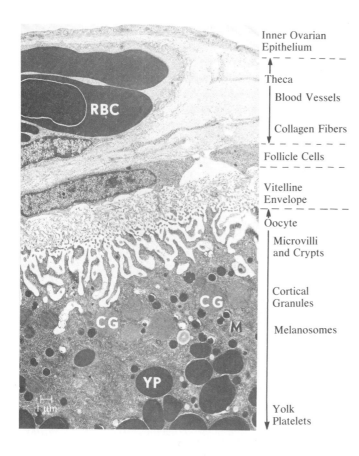

Inner Ovarian
Epithelium

Theca

Blood Vessels

Collagen Fibers

Follicle Cells

Vitelline
Envelope

Oocyte

Microvilli
and Crypts

Cortical
Granules

Melanosomes

Yolk
Platelets

Figure 3.17

Electron micrograph illustrating the organization of the cortex of the *Xenopus* oocyte. Surrounding the oocyte are (1) the ovarian epithelium; (2) a thick thecal layer that includes blood vessels, collagen fibers, and fibroblasts; (3) a single layer of follicle cells; and (4) the vitelline envelope, which is penetrated by oocyte microvilli and follicle cell processes. In addition to microvilli, the oocyte surface is contoured with deep crypts. The cortical cytoplasm contains cortical granules (CG) and melanosomes (M). Yolk platelets (YP) appear below the cortex. RBC: red blood cell. (From Dumont and Brummett, 1978. Reprinted from *Journal of Morphology* by permission of Wiley-Liss, a division of John Wiley and Sons, Inc. Copyright © 1978 Wiley-Liss.)

oocyte and then incorporated into the oocyte during vitellogenesis (**hetero-synthesis**). Organisms may use either or both of these methods to produce yolk.

Yolk production in vertebrates is predominantly heterosynthetic. Heterosynthetic yolk production has been particularly well studied in *Xenopus laevis*, which will be used to illustrate this process in vertebrates. The yolk precursor (**vitellogenin**) is synthesized in the liver and transported via the circulation to the ovary, where it is taken up into the oocytes by **receptor-mediated endocytosis**. Yolk production in the liver and its incorporation into oocytes from the blood by endocytosis is also typical of the reptiles, fishes, and birds.

The uptake of vitellogenin by amphibian oocytes has been studied extensively by Wallace, Dumont, and their associates (for review, see Wallace, 1985). Vitellogenin reaches the follicle via a capillary network located within the thecal layer of the follicle (see Fig. 3.17). After exiting the capillaries, the vitellogenin passes through channels between the follicle cells to reach the oocyte surface (Dumont, 1978), where it is incorporated by endocytosis. At the bases of the microvilli that cover the oocyte surface, numerous invaginations and pits are formed. These have a so-called bristle coat composed of proteins (the best characterized of which is **clathrin**) on their convex cytoplasmic side and a fuzzy layer of **glycocalyx** on their extracellular concave surface. The glycocalyx is generally composed of the carbohydrate side chains of glycolipids and integral membrane glycoproteins as well as extracellular proteoglycans and glycoproteins. In the developing oocyte, it contains specific vitellogenin receptors that adsorb the vitellogenin (Wallace et al., 1983). The invaginations and pits pinch off from the membrane

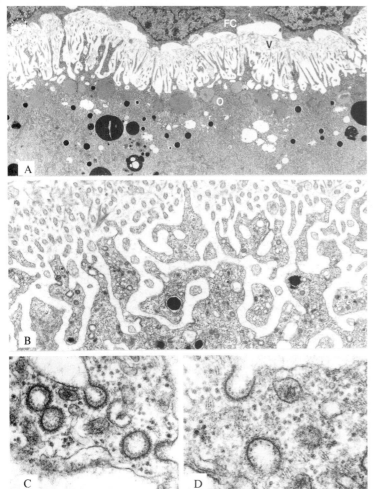

A

B

C D

Figure 3.18

Micropinocytotic pits and vesicles at the surface of vitellogenic *Xenopus laevis* oocytes. A, Overview of the surface, with wispy vitelline envelope (V) below the follicle cells (FC) and overlying the cortical cytoplasm of the oocyte (O). ×3160. B, A portion of the infolded surface showing clathrin-coated pits and vesicles. ×9570. C, and D, Pits and vesicles at the oocyte surface. Particles lining the lumenal surface have approximately the same dimension as native vitellogenin macromolecules (10 nm). C, ×55,700; D, ×73,400. (From Wallace et al., 1983.)

and form **coated vesicles** (Fig. 3.18). The vesicles carry the adsorbed vitellogenin into the oocyte. The vesicles lose the bristle coat to become endosomes, which fuse to form **multivesicular bodies (MVBs)**, which contain small membranous vesicles within a membrane-enclosed lumen (Wall and Patel, 1987). As yolk condenses and crystallizes within the MVBs, the organelles become known as **primordial yolk platelets (PYPs)**. The latter fuse to form yolk platelets. This sequence of events is summarized in Figure 3.19.

Figure 3.19

Diagrammatic representation of the sequence of events involved in formation of yolk platelets from endocytosed vitellogenin. (After Wall and Patel, 1987.)

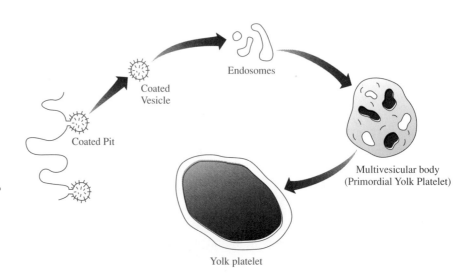

Coated Vesicle

Endosomes

Coated Pit

Multivesicular body (Primordial Yolk Platelet)

Yolk platelet

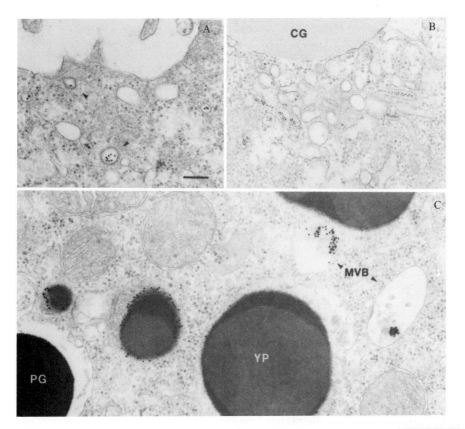

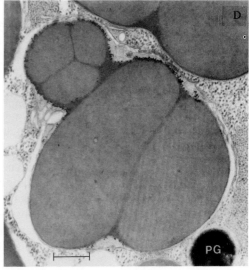

Figure 3.20

Distribution of colloidal gold-labeled vitellogenin (VG-Au) after internalization by *Xenopus* oocytes. A, After incubation of oocytes with VG-Au for 4 hour, the ligand is seen in coated pits and vesicles (arrows) and in uncoated endosomes. B, Tubular endosomes in the peripheral cytoplasm of an oocyte incubated with the conjugate for 4.5 hour. CG: cortical granule. C, After 4 hr incubation, the ligand is also seen in multivesicular bodies (MVBs), where vitellogenin condensation and crystallization occur. Electron-dense masses of condensed protein gradually grow to form small yolk platelets (YP), containing central yolk crystals surrounded by a superficial layer of dense, uncrystallized material. PG: pigment granule. D, After prolonged exposure to VG-Au, labeled yolk platelets appear to fuse to form larger yolk platelets. Scale bars equal 0.2 μm (A) and 0.4 μm (D). A, B, and C are shown at the same magnification. (From Wall and Patel, 1987.)

Experimental confirmation of this sequence has been obtained by electron microscopy of oocytes exposed to vitellogenin labeled with colloidal gold particles, which are visible as fine electron-dense granules in the electron microscope. As shown in Figure 3.20, the labeled vitellogenin is initially found in coated pits, coated vesicles, uncoated endosomes, and multivesicular bodies in the peripheral cytoplasm. After further exposure, the labeled vitellogenin is also found in small, newly formed yolk platelets and, eventually, in larger, definitive yolk platelets.

As we have discussed previously, the distribution of yolk platelets in the amphibian egg is decidedly polarized, with the largest platelets and 70%

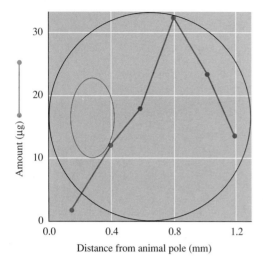

Distance from animal pole (mm)

Figure 3.21

Distribution of yolk protein in the full-grown *Xenopus* oocyte. Values on the ordinate represent the amount of yolk protein (in micrograms) in 300 µm oocyte sections. Note that the oocyte is represented with animal pole on the left. (After Danilchik and Gerhart, 1987.)

of the yolk protein localized in the vegetal hemisphere (Fig. 3.21). This animal-vegetal polarity of the amphibian egg foreshadows the polarity of the eventual embryo (see p. 198); hence, its establishment bears an important relationship to the establishment of the body plan of the embryo. The asymmetry in yolk distribution is a rather late development in oogenesis. Young oocytes are characterized by a centrally located germinal vesicle and yolk platelets that are uniformly distributed in the subcortical cytoplasm. The asymmetry in yolk platelet distribution is a consequence of internal displacement of yolk platelets from the animal hemisphere into the vegetal hemisphere. This displacement has been demonstrated by tracing the distribution of fluorescently labeled vitellogenin molecules in *Xenopus* oocytes at progressively later times after their uptake. As shown in Figure 3.22A, fluorescent vitellogenin is initially uniformly distributed around the circumference of the oocyte. In time, the labeled vitellogenin in the animal hemisphere is displaced toward the center of the vegetal hemisphere (see Fig. 3.22B and C). The center, itself, remains unlabeled because it consists of vitellogenin that was incorporated before the labeled vitellogenin was added. After asymmetry is established, newly formed yolk granules form a tree ring–like pattern, with yolk granules containing newly incorporated vitellogenin surrounding older granules (see Fig. 3.22D). Danilchik and Gerhart (1987) have estimated that yolk platelets forming in the animal hemisphere move inward as fast as 50 µm/day from the animal surface, where they originate, deflecting around the germinal vesicle, into the vegetal hemisphere. The mode of their migration is unknown. However, it is likely that the cytoskeleton is involved in yolk platelet displacement.

A mechanism for heterosynthetic yolk production analogous to that described for *Xenopus* has also been observed in insects (i.e., receptor-mediated endocytosis of exogenously produced vitellogenin). Insect vitellogenin is synthesized in the fat body and ovary (see p. 102). Recent research has characterized experimental systems for studying the endocytotic process. For example, a yolk-deficient mutant (*yolkless*) of *Drosophila* has been described in which the number of coated pits and vesicles is greatly reduced. Figure 3.23 compares the ultrastructure of the cortex of the wild-type oocyte (see Fig. 3.23A) with that of the *yolkless* mutant (see Fig. 3.23B). The deficiency of coated pits in the mutant restricts access of the blood-borne vitellogenin to the oocyte. These observations confirm the mode of yolk uptake in *Drosophila* and provide an experimental system in which to study the molecular mechanism of coated pit formation. *In vitro* systems for studying endocytosis have also been described. For example, isolated

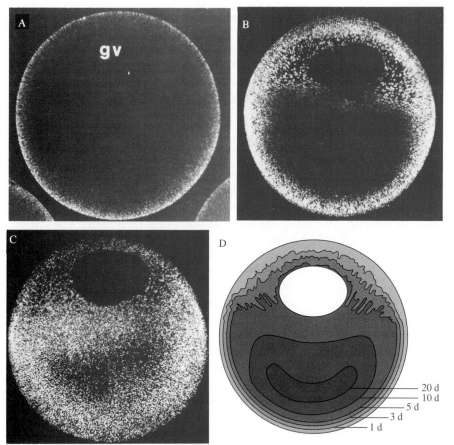

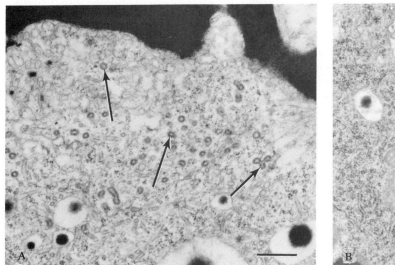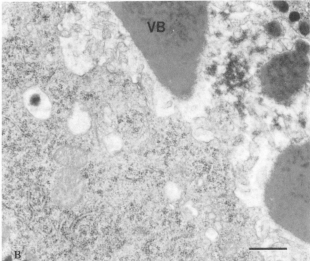

Figure 3.22

Patterns of progressive deposition of fluorescent vitellogenin. The germinal vesicle (gv in A) is oriented toward the top of the figure. A, 18 hours after injection; B, 5 days after injection; C, 20 days after injection. D, Schematic summary of the time course of the inward movement of labeled vitellogenin. The contour lines refer to the label limit at 1, 3, 5, 10, and 20 days after label injection. (From Danilchik and Gerhart, 1987.)

Figure 3.23

Ultrastructure of the cortex of (A), wild-type and (B), *yolkless Drosophila* oocytes. Arrows in A indicate coated vesicles, which are much reduced in the mutant. Scale bars equal 0.5 μm. (From P. J. DiMario and A. P. Mahowald, 1987. Reprinted from the *Journal of Cell Biology,* 1987, *105:* 199–206 by copyright permission of the Rockefeller University Press.)

ovarian follicles of the cecropia moth have been shown to incorporate radiolabeled vitellogenin into protein yolk spheres of the oocyte (Kulakovsky and Telfer, 1987). This system should be amenable to perturbation that would allow for investigation of the identity and mechanism of action of the vitellogenin receptor in binding to vitellogenin and in studying the process of internalization of vitellogenin and formation of definitive yolk spheres.

The Egg Cortex

The egg cytoplasm is organized into two definitive regions. The cytoplasmic layer just below the plasma membrane, the **cortex**, assumes physical properties that are distinct from those of the remainder of the cytoplasm. Most of the egg cytoplasm (the **endoplasm**) is in a fluid state. The cortex, however, has a higher viscosity and is a semirigid gel. These differences between the cortex and endoplasm can be demonstrated by centrifuging an egg. The components of the endoplasm are readily displaced during centrifugation, but the cortical constituents remain in place.

The cortex in the eggs of many species contains organelles that are not found in the endoplasm. The electron micrograph of the cortex of the amphibian egg (see Fig. 3.17) illustrates two common types of cortical inclusions: **cortical granules** and **pigment granules**.

Cortical granules are spherical structures that are surrounded by a membrane and contain acid mucopolysaccharides and protein. These organelles function at fertilization, when their contents are extruded into the region surrounding the egg (see Chap. 4). Cortical granules are formed within the endoplasm of the oocyte. Initially, they are randomly distributed in the egg, but near completion of oogenesis, the cortical granules migrate to the cortex. In those organisms in which their origin has been investigated, cortical granules are apparently produced by the combined efforts of the rough endoplasmic reticulum and the Golgi complex. The synthesis of cortical granule precursors apparently occurs on the ribosomes of the rough endoplasmic reticulum. The precursors are then transported within the endoplasmic reticulum to the Golgi, where they are assembled into definitive cortical granules.

In the sea urchin *Arbacia*, saccules of the Golgi containing a substance with a density similar to that of cortical granule material appear to pinch off to form membrane-enclosed vesicles. These cortical granule precursors increase in diameter, presumably by fusion of vesicles, and acquire their definitive form (Fig. 3.24). A similar pattern of cortical granule formation has been described in a wide variety of invertebrate and vertebrate species, including mammals (Selman and Anderson, 1975; Bilinski, 1981; Garwood, 1981; Gremigni and Nigro, 1983, 1984).

The nature of pigment granules in the cortex is quite variable. Those in amphibian eggs contain the dark brown or black pigment, melanin, and are called **melanosomes**. Melanosomes are found in the eggs of amphibian species that deposit their eggs in places where they are exposed to bright light. The pigmentation may protect the nucleus and cytoplasm from damage by ultraviolet radiation. Pigment granule distribution is not uniform in the amphibian egg (see Fig. 3.11). The vegetal hemisphere contains very few pigment granules and appears white, whereas the animal hemisphere (where the nucleus is located) contains many pigment granules and is darkly pigmented. Between the dark and light areas is a region of intermediate pigmentation, the **marginal zone**.

The cortex persists after fertilization and has a major role to play in development: Cortical microfilaments provide the forces that constrict cells

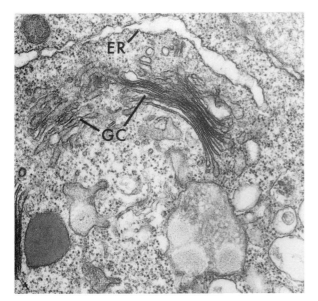

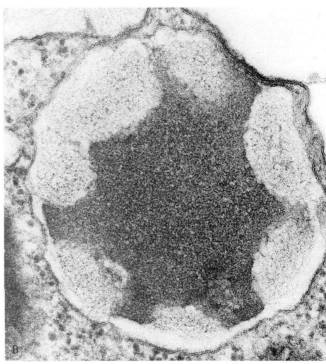

Figure 3.24

Formation of the cortical granules in the *Arbacia* oocyte. A, Stages in the initial formation of cortical granules as seen in a section through a Golgi complex (GC). Note cisternae of endoplasmic reticulum (ER). ×26,000. B, Section through a mature cortical granule. ×92,000. (From E. Anderson, 1968. Reprinted from *The Journal of Cell Biology*, 1968, *37*: 514–539 by copyright permission of the Rockefeller University Press.)

during cleavage (see Chap. 5) and alter cell shape during gastrulation (see Chap. 6).

The Oocyte Nucleus

The nucleus of the oocyte becomes highly modified during oogenesis and grows to an immense size. During early phases of oogenesis, it occupies a nearly central position in the oocyte. As the cell increases in diameter, the nucleus becomes distinctly eccentric in location and may become situated very near the plasma membrane at the animal pole. These enlarged nuclei are given a special name—**germinal vesicles**. The germinal vesicle envelope is a highly convoluted bilaminar structure. The inner and outer membranes are joined at numerous sites to form **nuclear pores** (Fig. 3.25A), which are thought to be involved in nucleocytoplasmic communication. The pores are not openings that provide nuclear and cytoplasmic continuity but are complexes of discrete elements. The hypothesis proposing a transport function is supported by micrographs that apparently show material in transit between nucleus and cytoplasm (see Fig. 3.25B). As we shall discuss in the next section, the outer membrane of the nuclear envelope may be involved in elaborating cytoplasmic membranes during oogenesis.

Cytoplasmic Membranes

The oocyte cytoplasm contains a vast array of membranous inclusions at various times during oogenesis. These include smooth and rough endoplasmic reticulum, Golgi complexes, multivesicular bodies, mitochondria, yolk bodies, and annulate lamellae. The latter are particularly prominent. Annulate lamellae consist of groups of parallel double membranes that contain pore complexes and bear a striking similarity to the nuclear envelope (Fig. 3.26A). Annulate lamellae are abundant in oocytes, but they have also been detected in spermatogonia and a variety of somatic cells as well as some tumor cells. The function of annulate lamellae is not known. They are

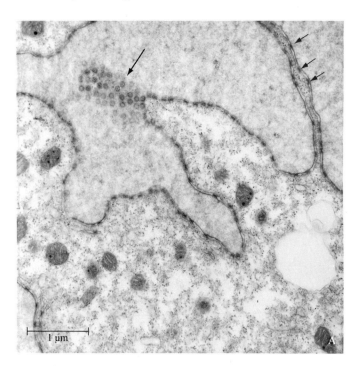
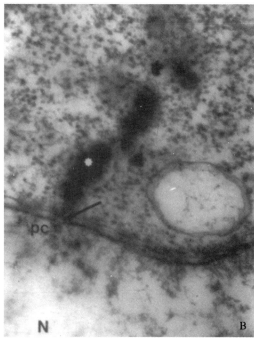

Figure 3.25

Nuclear pores in the nuclear envelopes of amphibian oocytes. A, Electron micrograph of portion of *Xenopus laevis* oocyte. Undulating nuclear envelope is penetrated by numerous regularly spaced pores (small arrows). The large arrow indicates a place where the envelope has been cut tangentially, showing cross sections of several pores. B, Electron micrograph of oocyte of the newt *Pleurodeles waltlii* showing the apparent extrusion of material (asterisk) from the nucleus (N) through a nuclear pore (arrow) into the cytoplasm. ×55,000. (A, Micrograph by Dr. G. Steinert. From Brachet, 1974. B, From Bonnanfant-Jaïs and Mentré, 1983.)

frequently accompanied by ribosomes and are often continuous with the rough endoplasmic reticulum, leading to speculation that they are involved in protein synthesis. It has also been proposed that the annulate lamellae in oocytes bear nuclear information that is dispatched to the cytoplasm, where it is stored for utilization after fertilization (Kessel, 1985).

In contrast to the extensive membranes in the full-grown oocyte, the early oocyte may be quite deficient in membranous inclusions (see Fig. 3.26B). Electron micrographs of oocytes of the urodele amphibian *Necturus maculosus* reveal that the outer membrane of the nuclear envelope appears to hypertrophy and extend into the cytoplasm (see Fig. 3.26C). Kessel and colleagues (1986) have proposed that many of the cytoplasmic membranes are derived from such elaborations.

3–3 Gene Expression During Oogenesis: Amphibians

Oogenesis is a process with dual consequences: (1) the construction of a highly complex specialized cell, and (2) the production of components that will be utilized during postfertilization development. Therefore, the utili-

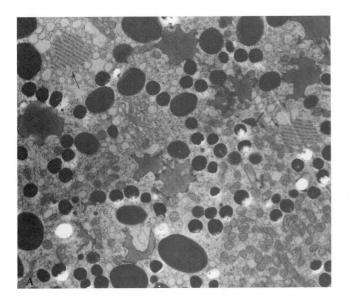

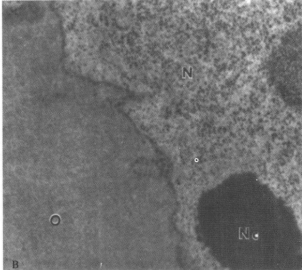

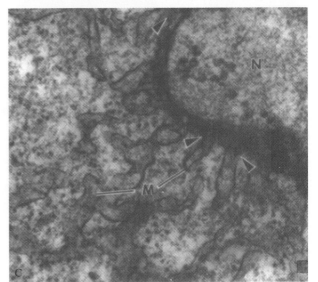

Figure 3.26

Oocyte cytoplasmic membranes. A, Portion of *Rana pipiens* oocyte. Portions of several stacks of annulate lamellae are visible (unlabeled arrows). Pores are sectioned perpendicularly. ×7,500. B, Portion of early *Necturus* oocyte. Note the scarcity of cytoplasmic membranes. N: nucleus; O: ooplasm; Nc: nucleolus. ×50,000. C, Slightly larger *Necturus* oocyte than shown in B. Note the extensive connections (arrowheads) between the nuclear envelope and cytoplasmic membranous lamellae (M). N: nucleus. ×11,700. (A, From Kessel, 1968. B, and C, From Kessel et al., 1986.)

zation of genomic information during oogenesis has important implications for the regulation of early embryonic development. The most comprehensive studies on these two aspects of oogenesis have been conducted on the oocytes of amphibians. The ready availability of experimental material, the size of oocytes, and the ability to regulate oogenesis by hormonal injection have contributed to the popularity of amphibians for research on oogenesis. Other important factors are the large size of the highly modified lampbrush chromosomes and the number of nucleoli produced during oogenesis. Because of this wealth of information, we shall concentrate on amphibians in our analysis of gene expression during oogenesis. In Section 3–4, we shall consider gene expression in other organisms.

Lampbrush Chromosomes

The process of meiosis is an integral aspect of oogenesis; consequently, the structure of oocyte chromosomes is affected by the meiotic process. The most frequently encountered situation in animal oogenesis is that in which meiosis is initiated at the beginning of oogenesis, homologous chromosomes synapse to form tetrads, and homologues begin to separate at diplonema. Then, however, meiosis is suspended, and the homologues remain attached by chiasmata. This hiatus may last for months or even years, and it is during this phase of oogenesis that the major growth and differentiation of the oocyte occur. The chromosomes at this time may become highly elongated and modified to form lampbrush chromosomes (Fig. 3.27A). Because of their extremely large size, lampbrush chromosomes of amphibians are readily amenable to experimental manipulation and have been studied in more detail than those of other groups.

To understand these chromosomes properly, we must recall that at diplonema each homologue consists of two sister chromatids arranged with their axes parallel to one another. In the lampbrush configuration, the chromosomes possess a linear sequence of globular structures (chromomeres) that are composed of compacted chromatin. The chromomeres are connected by thin parallel strands of **interchromomeric chromatin**. One or more symmetrical pairs of **loops** extend laterally from the chromomeres (see

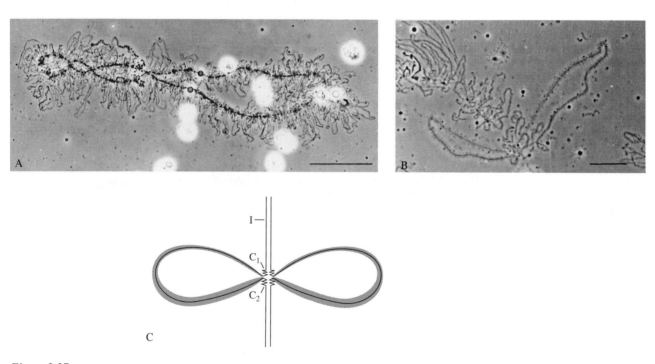

Figure 3.27

Lampbrush chromosome morphology. A, Phase contrast micrograph of lampbrush chromosome from an oocyte nucleus of the urodele *Pleurodeles waltlii*. B, Higher power micrograph of portion of a lampbrush chromosome from the newt *Triturus alpestris*, showing paired loops extending from the chromosome. C, Diagrammatic representation of the supposed general organization of the two chromatids that make up a single lampbrush chromosome. Each vertical line represents a single chromatid that consists of a single DNA duplex. Each chromatid is thought to run relatively straight through the interchromomeric region (I), become locally compacted on one side of the chromomere (C_1), pass out into a loop, where it is again in a relatively extended state, return to the compacted state on the other side of the chromomere (C_2), and then proceed along the interchromomeric fibril to the next chromomere/loop complex. Scale bars equal 50 μm (A) and 20 μm (B). (A, and B, From Scheer and Dabauvalle, 1985. C, After Macgregor, 1977.)

Fig. 3.27B), giving the chromosomes the lampbrush appearance. (Lampbrushes are brushes with looped bristles that were once used to clean oil lamps.) The presumed relationships among the various structures seen in lampbrush chromosomes are diagrammed in the model shown in Figure 3.27C.

The axis of each homologue consists of thin sister chromatid strands that span the interchromomeric region. At intervals, the chromatin becomes compacted on one side of a chromomere, extends out into loops, returns to a compacted state on the other side of the chromomere, and then extends from there through the next interchromomeric region to the next chromomere, and so on.

One of the most striking aspects of lampbrush chromosome morphology is the presence of a **matrix** on the loops. The matrix morphology of sister loops is always identical. However, the matrix on different loop pairs can be quite distinctive, with the result that lampbrush chromosomes are adorned with a variety of unusual structures (Fig. 3.28). The differences in morphology of loops originating from different chromomeres provide landmarks that

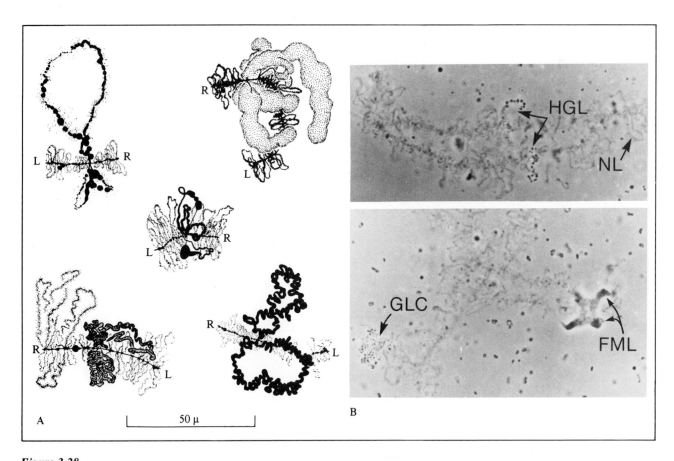

Figure 3.28

Examples of amphibian lampbrush loop matrix morphologies. A, Drawings of short segments of lampbrush chromosomes of the newt *Triturus cristatus*, illustrating some of the lampbrush loop matrix morphologies in this species. B, Photomicrographs showing portions of two lampbrush chromosomes of *Rana pipiens*. Note the homologous granular loops (HGL), occupying the same site on homologues, and the granular loop cluster (GLC) and large fusing matrix loops (FML). NL: normal loop. ×570. (A, From Callan, 1963. B, From Rogers and Browder, 1977.)

make it possible to construct "maps" of lampbrush chromosomes (Callan and Lloyd, 1975).

The matrix consists of protein associated with RNA that is being synthesized actively on the loops. The activities of loops in transcription are indicated by autoradiographs showing incorporation of radioactive precursors into nascent RNA on the loops (see Fig. 3.32). An interesting aspect of the morphology of loop matrices is that they are sometimes asymmetrical; they can be thin at one end, gradually increasing in thickness around the loop contour (Fig. 3.29). The asymmetry indicates that transcription begins on one side of the loop and proceeds until the entire loop is transcribed. These loops would appear to correspond to single transcriptional units.

Electron micrographs of asymmetrical loops prepared by a technique developed by Oscar Miller (Miller and Beatty, 1969; Miller and Hamkalo, 1972) beautifully demonstrate the directionality of transcription. Essentially, this procedure involves a gentle dispersion of nuclear contents to separate constituents without breakage of chromatin or loss of the nascent RNA. This step is followed by fixation, the spreading of the chromatin on electron microscope grids, and staining, which allows identification of the chromatin and RNA. Figure 3.30A is an electron micrograph showing transcription of a portion of a lampbrush loop. Individual transcripts can be seen because of dispersal of the loop matrix during sample preparation. RNA synthesis is proceeding in the direction of the arrow. Progressively longer RNA transcripts radiate from the loop axis as the distance from the loop origin increases. At the base of each transcript is a small granule, which is presumably the RNA polymerase molecule. Ribonucleotides are added to the base of the transcript as the polymerase moves along the loop contour,

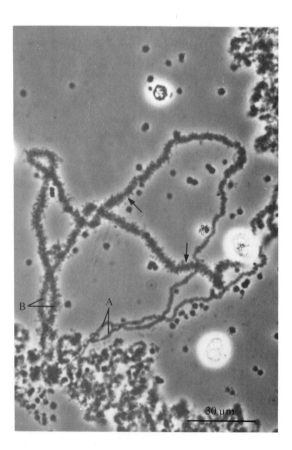

Figure 3.29

Phase contrast photomicrograph of a part of a lampbrush chromosome from the salamander *Plethodon cinereus* showing a pair of very long (240 μm) loops, both of which arise from the same chromomere. The loops have a fairly granular matrix of ribonucleoprotein, and in some places (arrows), the loop axis is faintly discernible. Each loop has a thin end or "insertion" (A) and grows gradually thicker toward the thick end (B), a phenomenon that is characteristic of most loops and is referred to as "loop asymmetry." (From Macgregor, 1977.)

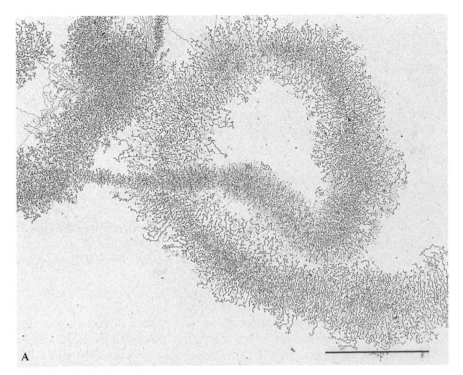

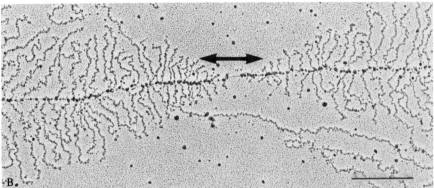

Figure 3.30

Electron microscopic visualization of transcription on loops of lampbrush chromosomes from *Pleurodeles waltlii* oocytes. A, Note the high packing density of nascent transcripts, which emanate from transcribing RNA polymerase molecules. B, This loop has two transcription units of opposite polarity. Scale bars equal 5 μm (A) and 0.5 μm (B). (From Scheer and Dabauvalle, 1985.)

progressively increasing the length of the transcript. As soon as one polymerase has moved a sufficient distance along the DNA, another polymerase attaches, and so on. Newly synthesized RNA associates with protein, resulting in deposition of ribonucleoprotein (RNP) matrix on the loops. At the termination of transcription, the completed transcript is released from the loop along with its associated protein.

Another class of loops apparently has more than one transcriptional unit. Loops with multiple transcriptional units are, in fact, quite common. On some of these loops, all the transcriptional units have the same polarity, whereas in others, transcription may be proceeding in different directions (Scheer et al., 1976a). A portion of a loop with two transcriptional units of opposite polarity is shown in Figure 3.30B.

The transcriptional units shown in Figure 3.30 illustrate that the RNA polymerases on lampbrush chromosome loops are very close to one another. The density of polymerases on DNA is the net result of initiation of transcription by attachment of polymerases to the DNA and the rate of movement of the polymerases to the termination site. Apparently, one or both of these functions is highly modified in lampbrush chromosomes as compared to chromosomes of somatic cells. The close packing of nascent transcripts, combined with the long duration of the lampbrush phase in most organisms, indicates that a tremendous amount of RNA synthesis is achieved during the lampbrush phase. However, the nature and fate of the RNA produced by the lampbrush chromosomes are only partially understood. The most useful procedure for analyzing transcription on lampbrush chromosome loops has been *in situ* hybridization (see Appendix), by which nascent transcripts on the chromosomes are hybridized to labeled probes representative of particular gene sequences. (The principles of RNA-DNA hybridization are discussed in the Appendix.) Autoradiography is then used to localize the site of radioactivity, which corresponds to the site of transcription of that sequence. For example, this technique has led to the recognition of the sites of transcription of histone gene loci on lampbrush loops (Fig. 3.31).

Additional evidence about the identities of lampbrush transcripts has been obtained from studies utilizing the inhibitor α-amanitin (Schultz et al., 1981). The three RNA polymerases are differentially sensitive to α-amanitin. Polymerase II, which synthesizes heterogeneous nuclear RNA, is inhibited by 0.5 μg/ml of α-amanitin. The results of an inhibitor experiment are illustrated in Figure 3.32. Newt oocytes were incubated *in vitro* in the presence of ^{3}H-UTP and ^{3}H-CTP, and autoradiographs were then prepared of the lampbrush chromosomes. Incorporation of label into RNA on the loops is shown in Figure 3.32A. The addition of 0.5 μg/ml of α-amanitin

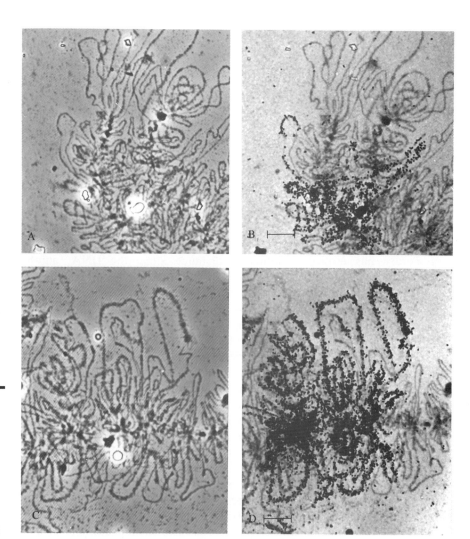

Figure 3.31

Transcription of histone genes on lampbrush loops of the newt *Notophthalmus*. A, and C, Phase contrast micrographs of histone gene loci. B, and D, The same regions after hybridization to a histone gene probe. Silver grains represent regions of hybridization to nascent RNA. Scale bars equal 7 μm. (From Diaz and Gall, 1985.)

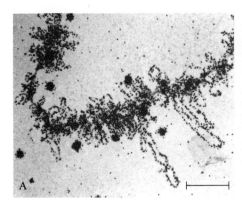

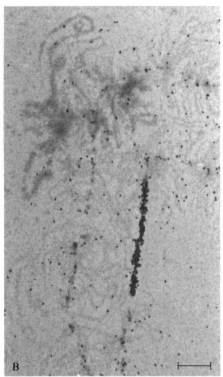

Figure 3.32

Autoradiographs of newt lampbrush chromosomes. A, Chromosomal segment showing incorporation of ³H-UTP and ³H-CTP. Note labeling of loops. Scale bar equals 20 μm. B, Short segment of chromosome after incubation in ³H-UTP and ³H-CTP in presence of α-amanitin. Label is found only on the chromosomal axis. Scale bar equals 10 μm. (From Schultz et al., 1981.)

abolishes all RNA synthesis on lampbrush loops (see Fig. 3.32B). Note that the incorporated label is confined to short stretches on the chromosomal axes. These correspond to the sites that contain the 5S RNA genes. Incubation of oocytes in 200 μg/ml of α-amanitin abolishes the label over these regions. This concentration of α-amanitin is sufficient to inhibit polymerase III as well as polymerase II. Because 5S RNA genes are normally transcribed by polymerase III, it is likely that 5S RNA is synthesized at certain sites on the axes of lampbrush chromosomes, whereas heterogeneous nuclear RNA is synthesized on lampbrush loops. It has been recently demonstrated, however, that some 5S RNA synthesis can occur on loops. Because all transcription on loops is due to polymerase II, it is likely that 5S RNA genes can be transcribed on the loops by read-through polymerase II transcription (read-through transcription will be discussed in detail on p. 87) initiated on upstream promoters (Callan et al., 1988).

Participation of RNA polymerase II in transcription on lampbrush loops is also indicated by experiments in which antibodies to the polymerase are injected into oocyte germinal vesicles (Bona et al., 1981). The results of such an experiment are illustrated in Figure 3.33. A control lampbrush chromosome from a *Pleurodeles waltlii* oocyte is shown in Figure 3.33A. Within a few minutes after microinjection of anti-RNA polymerase II, all lampbrush loops retract into the chromosome axes (see Fig. 3.33B), and all nascent transcripts are shed from the chromatin (see Fig. 3.33C). This antibody has no effect upon transcription of ribosomal RNA genes, which are transcribed by RNA polymerase I. Thus, its effects are specific to polymerase II.

The termination of the lampbrush phase is characterized by regression of the loop pairs along the length of the chromosomes and reversion of the chromosomes to their usual size. The reversibility of the lampbrush config-

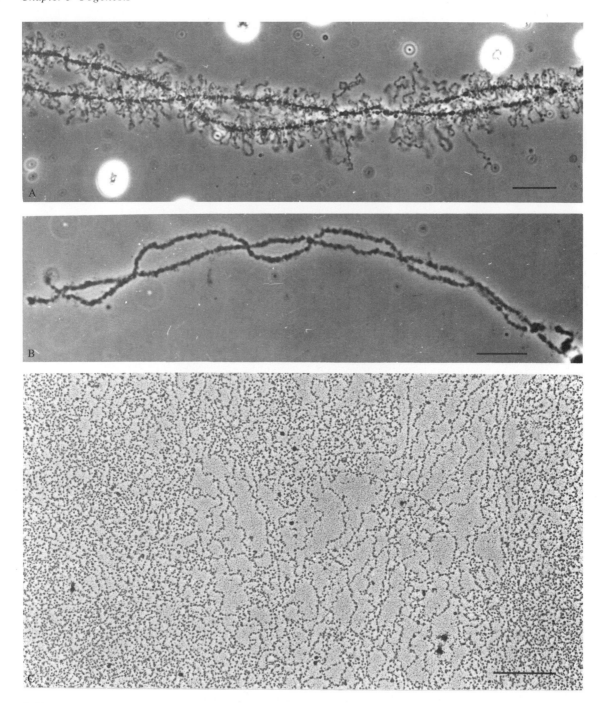

Figure 3.33

Effects of anti-RNA polymerase II on morphology of *Pleurodeles waltlii*
lampbrush chromosomes. A, Control chromosome. Scale bar equals 20 μm. B,
Loop retraction caused by injection of anti-RNA polymerase II, 5 minutes after
injection. Scale bar equals 20 μm. C, Electron micrograph of spread chromatin
from an oocyte that had been injected with anti-RNA polymerase II. Chromatin
has the configuration of transcriptionally inactive chromatin. Scale bar equals 0.5
μm. (A, Reprinted with permission from M. Bona, U. Scheer, and E. K. F.
Bautz, 1981. Journal of Molecular Biology, *151*: 81–99. Copyright by Academic
Press Inc., [London] Ltd. B, and C, Courtesy of Dr. U. Scheer.)

uration indicates that it is within the normal range of morphological variability of chromosomes.

One other noteworthy aspect of lampbrush chromosomes is that they are found during oogenesis of a wide variety of animals representing nearly the entire phylogenetic spectrum (Davidson, 1986). They are not universal, however. For example, lampbrush chromosomes are not present in oocytes of sea urchins, mammals, or meroistic insects. Thus, the lampbrush phase is not an indispensable aspect of animal oogenesis, but, it is *nearly universal*, which underscores the significance of lampbrush chromosomes and suggests that they play a major role in information transfer during oogenesis. Additional information about lampbrush chromosomes may be found in an excellent review by Macgregor (1980).

Synthesis and Accumulation of mRNA

One of the unique processes that occurs during oogenesis is the production of messenger RNA for utilization during early development. The production of mRNA during amphibian oogenesis has been analyzed by studying polyadenylated RNA and by using cloned sequences to probe specific transcripts. Polyadenylated RNA, which is usually referred to as poly (A)$^+$ RNA, is representative of the messenger RNA population, because most messengers possess tracts of 100 to 200 adenylate residues at their 3′ ends. The poly (A) tracts, which are added to transcripts by posttranscriptional modification, protect the 3′ ends of messengers from degradation, hence extending their functional lifetimes in the cell. Polyadenylation can also affect the level of translation of some mRNAs. The most extensive studies on polyadenylated RNA in oogenesis have used *Xenopus laevis* oocytes as experimental material.

The full-grown *Xenopus* oocyte contains considerable polyadenylated RNA. The majority of poly (A)$^+$ RNA is unusual in that it consists of single copy sequences that are interspersed with repetitive sequences (Anderson et al., 1982). As we shall discuss in the Appendix (see p. A-6), most single copy sequences in the genome are flanked on either side by moderately repetitive sequences. The presence of transcripts with interspersed single copy and repetitive sequences suggests that, once transcription is initiated on the promoter of a single copy sequence, it ignores normal termination signals and continues past the gene into the flanking repetitive DNA sequences. A number of repetitive DNA transcripts have, indeed, been detected on lampbrush loops (Jamrich et al., 1983; Diaz and Gall, 1985; Wu et al., 1986). If this so-called **read-through transcription** is generalized on lampbrush chromosomes, it might provide an explanation for the large size of transcripts observed on lampbrush loops (up to several kilobases). The repetitive sequences are apparently not removed from the read-through transcripts before they are transported out of the oocyte nucleus. There is no evidence that this RNA is translated, and no function for it has been demonstrated. (See Chap. 13 for further discussion of RNA containing interspersed repetitive and single copy sequences.) A portion of the poly (A)$^+$ RNA is potentially functional mRNA, however, as demonstrated by its translatability *in vitro* and after injection into *Xenopus* oocytes (Richter et al., 1984). The majority of this RNA is not active in oocyte protein synthesis but is stored in cytoplasmic RNP particles (Rosbash and Ford, 1974). Messenger RNA that is being translated (and, therefore, is on polysomes) in full-grown oocytes has been estimated to be only 20% of the total translatable mRNA (Smith et al., 1984); thus, 80% of the mRNA is unused. We shall discuss the mechanisms that may be responsible for limiting translation in the full-grown oocyte in Chapter 13. The stored mRNA is part of the information pool with which the zygote begins its development and is the mRNA that has accumulated during oogenesis; i.e., it is the net result of mRNA synthesis and degradation. Therefore, two of the prime functions of oogenesis are the synthesis and accumulation of messenger RNA.

The synthesis of polyadenylated RNA begins early in oogenesis. Net accumulation of polyadenylated mRNA plateaus at or before the beginning of vitellogenesis (Golden et al., 1980). Thereafter, however, its amount does not change significantly in spite of continued synthesis (Rosbash and Ford, 1974; Dolecki and Smith, 1979; Golden et al., 1980; Sagata et al., 1980). Experiments using recombinant DNA technology have also failed to detect any *qualitative* changes in polyadenylated RNA (Golden et al., 1980). In those experiments, *Xenopus* ovarian poly (A)$^+$ RNA was used to make cDNA. The cDNA was then made double stranded, inserted into plasmids, and cloned. These cDNA clones were used to probe for complementary RNA at different stages of oogenesis. In all cases, the RNA begins to accumulate in early oogenesis, reaches a plateau in early vitellogenesis, and then remains at the same concentration throughout the remainder of oogenesis. Thus, although synthesis continues, a constant population of polyadenylated RNA is maintained, presumably by degradation of excess RNA. Similar results have also been obtained for histone messenger RNAs, which, unlike histone mRNAs in somatic cells, are predominantly poly-adenylated (Levenson and Marcu, 1976; Ruderman and Pardue, 1977; Ruderman et al., 1979). Cloned histone H3 genes were hybridized to RNA extracted from oocytes at different stages of oogenesis, and the level of hybridization was used as a measure of the amount of histone mRNA present. As with total poly (A)$^+$ RNA, the histone mRNA accumulates in early oogenesis and does not increase in amount thereafter (van Dongen et al., 1981).

The absence of quantitative and qualitative changes in the messenger RNA population indicates that RNA degradation and synthesis are in precise balance. If RNA synthesis were occurring at a lower rate during the long period of oogenesis, mRNA levels would decline. It seems likely, therefore, that the rapid rate of transcription on lampbrush chromosomes (as evidenced by the close packing of polymerases on lampbrush loops) is an essential adaptation for maintaining a stable supply of mRNA to the end of oogenesis (Davidson, 1986). If this interpretation is correct, it provides a probable explanation for the formation of lampbrush chromosomes during oogenesis.

Localization of mRNA

As we shall discuss in detail in Chapter 11, certain cytoplasmic factors are localized to particular regions of the egg and, as a consequence, are differentially distributed to blastomeres during cleavage. These factors ultimately determine the fates of blastomeres to which they are distributed. Because messenger RNA accumulates in such large amounts during oogenesis, it is reasonable to assume that some of these mRNAs may either be the ooplasmic determinants or may encode proteins that are the determinants. Accordingly, investigators have searched for transcripts that are localized to either the animal or vegetal hemisphere of the *Xenopus* oocyte. The vast majority of mRNAs are not localized. The distribution of histone mRNAs is typical of most mRNAs, showing uniform distribution during early oogenesis, followed by a dilution of mRNA by yolk platelet accumulation during vitellogenesis. This results in a gradient of distribution from the animal to vegetal pole, reflecting the distribution of nonyolky cytoplasm in the oocyte (Jamrich et al., 1984). However, several classes of transcripts have been identified that are localized to either the animal or vegetal hemisphere and remain localized during early development (King and

Isolation of cDNA clones for localized egg RNAs

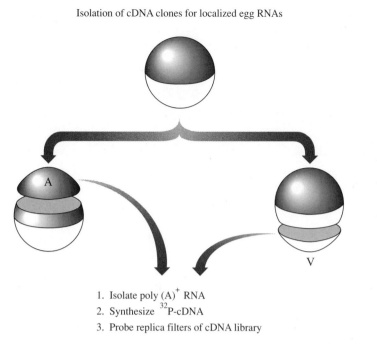

1. Isolate poly (A)$^+$ RNA
2. Synthesize ^{32}P-cDNA
3. Probe replica filters of cDNA library

Figure 3.34

Procedure used for preparing cDNA clones from RNA of either the animal (A) or vegetal (V) hemisphere of *Xenopus* eggs. (After Rebagliati et al., 1985.)

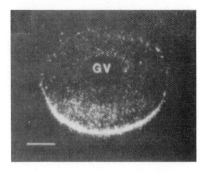

Figure 3.35

In situ hybridization, showing localization of *Vg1* RNA to the vegetal pole of a full-grown *Xenopus* oocyte. This oocyte section was observed with dark field microscopy; thus, the silver grains appear bright against a dark background. GV: germinal vesicle. Scale bar equals 200 μm. (From D. A. Melton, 1987. Reprinted by permission from *Nature* (Lond.) *328*: 80–82. Copyright © 1987, Macmillan Magazines Limited.)

Barklis, 1985; Rebagliati et al., 1985; Weeks and Melton, 1987). The strategy used to identify localized transcripts has been to screen a cDNA library that is representative of most transcripts present in eggs (see Appendix for a discussion of screening cDNA libraries) with labeled cDNA probes prepared from either animal or vegetal pole RNAs (Fig. 3.34). Autoradiographs of replica filters of the oocyte cDNA library are then examined to identify plaques that are detected exclusively by either the animal or vegetal cDNA probes. The cDNA clones identified in this way have subsequently been used in *in situ* hybridization analyses to determine the exact distribution of the localized transcripts.

One of these mRNAs has a particularly striking asymmetrical distribution. This transcript, which is called *Vg1*, was identified as a consequence of its vegetal localization in the oocyte. When oocyte sections are probed with a labeled hybridization probe for *Vg1*, the RNA is seen in a crescent at the vegetal pole of full-grown oocytes (Fig. 3.35). The ability to detect *Vg1* RNA by *in situ* hybridization allows us to examine the origin of this asymmetrical distribution. Either *Vg1* RNA is concentrated at the vegetal pole from the time of its synthesis, or it is initially uniformly distributed and becomes localized at a later stage. Figure 3.36 shows that the latter situation pertains. The mRNA is translocated into the vegetal hemisphere by a mechanism analogous to that of yolk platelet translocation, which we have discussed previously. The movement of *Vg1* RNA to the vegetal hemisphere and its anchorage in the cortex are apparently mediated by the cytoskeleton because inhibitors of microtubules prevent translocation, whereas cytochalasin B (an inhibitor of microfilaments; see p. 339) prevents anchorage (Yisraeli et al., 1990). Such a mechanism requires recognition between *Vg1* RNA and the cytoskeleton, possibly mediated by specific cellular factors that bind to both the mRNA and the cytoskeleton. Specific

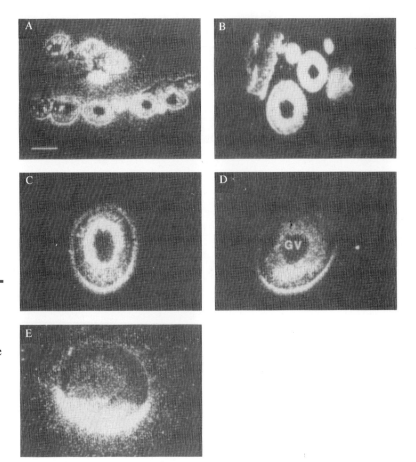

Figure 3.36

Translocation of *Vg1* RNA during oogenesis. *In situ* hybridizations were conducted on sections of oocytes at progressively later stages of oogenesis. The black hole in the center of the oocytes is the germinal vesicle (GV). Scale bar equals 200 μm. (From D. A. Melton, 1987. Reprinted by permission from *Nature* (Lond.), *328*: 80–82. Copyright © 1987, Macmillan Magazines Limited.)

sequence elements of the *Vg1* RNA that are responsible for its localization have been identified (Yisraeli and Melton, 1988); a search is presently underway for factors that might bind the RNA to the cytoskeleton.

Vg1 RNA is released from its thin cortical shell in the unfertilized egg and is inherited by most of the vegetal blastomeres during cleavage. The delocalization of *Vg1* RNA in the egg is not unusual; a similar delocalization phenomenon has also been observed for other mRNAs that are localized to the periphery of the vegetal hemisphere. Larabell and Capco (1988) have presented evidence that ion fluxes, which accompany meiotic maturation, are involved in mRNA delocalization.

Vg1 is involved in one of the most intriguing stories in contemporary developmental biology. It has been shown to encode a protein related to transforming growth factor-β (TGF-β), which is one of the growth factors that can induce development of mesoderm. The relationship of *Vg1* protein to TGF-β and its localization to the vegetal blastomeres, which are responsible for inducing cells in the marginal zone to develop into mesoderm, makes it a prime candidate to be a component of the mesoderm-inducing signals produced by the vegetal blastomeres (see Chap. 12).

Synthesis of Ribosome Components

The production and accumulation of ribosomes are major chores of oogenesis. The large store of ribosomes in the egg relieves the embryo of the need to produce its own ribosomes to support protein synthesis during early embryonic development and makes it possible for the zygote to undertake immediate protein synthesis after fertilization. Therefore, the components of ribosomes must be produced in prodigious amounts during oogenesis. These components include 18S, 28S, and 5S RNA as well as the ribosomal proteins. These components apparently are synthesized under the direction of the oocyte nucleus, assembled within the nucleoli, and transported to the oocyte cytoplasm, primarily as a storage product. The synthesis and accumulation of ribosomal components have been particularly well studied during oogenesis of *Xenopus laevis*, and the discussions that follow on individual ribosomal components will be based primarily on research using *Xenopus*. The magnitude of ribosome production during oogenesis in this species is illustrated by measurements of the rate of accumulation of ribosomes (at least 1000 times greater than in the most active somatic cells) and the total number of ribosomes produced (10^{12} ribosomes per egg) (Wormington, 1988). As we shall see, the patterns of synthesis and accumulation of the individual components of ribosomes during *Xenopus* oogenesis are quite different from one another.

5S RNA

The synthesis of 5S RNA occurs rapidly during the earliest stages of oogenesis, before the major synthetic period for 18S and 28S RNA and ribosomal proteins. The large amounts of 5S RNA made during the early stages are stored in either 7S or 42S RNP particles and later incorporated into ribosomes when the other components become available (Ford, 1971; Ford and Southern, 1973). The rapid rate of 5S RNA production and its accumulation by the previtellogenic *Xenopus* oocyte produce large stores of this class of RNA, which accounts for about 45% of the total RNA of these oocytes (Ford, 1976). By early vitellogenesis, nearly half the amount of 5S RNA needed during oogenesis has accumulated, and production continues throughout oogenesis, concurrent with the synthesis of 18S and 28S RNA, but at a lower rate.

The production of large amounts of 5S RNA is facilitated by the multiplicity of 5S RNA genes. The *Xenopus laevis* genome contains two kinds of genes that code for 5S RNA: oocyte and somatic. The oocyte 5S RNA genes are present as approximately 24,000 tandemly repeated 120 base pair coding units per haploid genome (Brown and Sugimoto, 1973). Because the chromosomes are in meiosis during oogenesis, each oocyte nucleus contains four times this many 5S RNA genes, or 96,000 of them.

Initiation of 5S RNA gene transcription is regulated by an interaction between three oocyte transcription factors and the promoter site, which, like other genes transcribed by RNA polymerase III, is located within each coding sequence, rather than in the 5′ flanking region. The formation of this transcription complex initiates transcription at the start site of the 5S RNA gene. **TFIIIA**, one of these transcription factors, has been isolated and studied extensively. TFIIIA has a prominent structural element called the **zinc finger** that is repeated nine times. Each of these elements is approximately 30 amino acids long. At both ends of the sequence are two amino acids that bind to a single Zn^{2+} ion. The metal ion draws the ends of each unit together, leaving the remaining residues to form the loop or "finger" (Fig. 3.37). The zinc fingers bind DNA in the promoter region of the 5S RNA gene. The role of the zinc fingers in DNA binding has been demonstrated experimentally by mutagenic alteration of the coding sequences of cloned TFIIIA genes, which abolishes the ability of TFIIIA to bind to DNA (Vrana et al., 1988). The presumed functional significance of the fingers is that as RNA polymerase traverses the gene (remember that the promoter of 5S RNA genes is embedded within the coding sequence), individual fingers

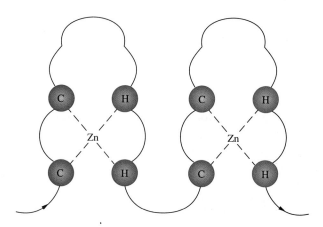

Figure 3.37

Zinc finger configuration of TFIIIA. Two consecutive fingers are shown here. The ringed residues are the amino acids that bind the zinc atom. The metal ion draws the ends of each unit together, leaving the remaining residues to form the loop or "finger." (After Miller et al., 1985.)

could transiently release the DNA sequentially to accommodate passage of the polymerase and then rebind the DNA after passage so that at any one time the majority of the contact points between TFIIIA and the DNA are maintained while the polymerase transcribes the gene (Miller et al., 1985). A number of additional zinc finger proteins have been discovered that function as transcriptional regulators of other genes. We shall discuss some of these proteins in later chapters. Thus, it would appear that the zinc finger configuration is of general importance for transcriptional regulation.

Interestingly, TFIIIA also binds to the 5S RNA after its synthesis and becomes a component of the 7S storage particle. By binding to its own transcription factor, 5S RNA can repress further transcription of the 5S RNA genes (Honda and Roeder, 1980; Pelham and Brown, 1980). This system of feedback inhibition allows us to hypothesize a mechanism for regulation of 5S RNA synthesis during oogenesis in *Xenopus*: In early stages of oogenesis, massive synthesis of transcription factor would result in accumulation of large amounts of 5S RNA. As the RNA accumulates, it binds the factor and causes transcription to level off (Pelham and Brown, 1980). Synthesis of 5S RNA could continue so long as unbound transcription factor is available to the 5S RNA genes.

As we mentioned earlier, some of the 5S RNA is also stored in 42S particles. The 42S particles are more complex than the 7S particles because they contain tRNA as well as 5S RNA. They also contain two proteins, called p50 and p43. The p50 protein associates with the tRNA. The other protein—p43—is extremely interesting because it contains nine zinc fingers that are similar to those of TFIIIA. Unlike TFIIIA, p43 binds exclusively to 5S RNA and not to the 5S gene promoter. It is presumed that the zinc finger region is responsible for the abilities of both of these proteins to bind 5S RNA (Joho et al., 1990).

After oocyte maturation, 5S RNA synthesis is no longer detectable. The amount of transcription factor has also been drastically reduced (Pelham et al., 1981). In fact, egg extracts will only support transcription of 5S RNA genes if exogenous transcription factor is added (Fig. 3.38). By contrast, oocyte extracts will transcribe 5S RNA genes without added factor. These results suggest that the reduction in the amount of transcription factor after maturation leads to a shutdown of 5S RNA synthesis (Honda and Roeder, 1980). After fertilization, the embryo reinitiates 5S RNA synthesis, but only from the somatic genes.

Ribosomal RNA

Ribosomal RNA (i.e., 18S and 28S RNA) is the major component of the oocyte RNA population. Synthesis of large amounts of ribosomal RNA in many organisms is facilitated by a mechanism that is unique to oocytes:

amplification of ribosomal RNA genes. This phenomenon occurs in many animals and has been demonstrated in some species of fish, insects, mollusks, and amphibians, although this discussion will be limited to *Xenopus*. The genetic locus that contains the ribosomal RNA genes is marked in somatic cells by the presence of the nucleolus. Somatic nuclei of *Xenopus* contain two nucleoli, one for each haploid chromosome set. Thus, the tetraploid oocyte nucleus would be expected to have four nucleoli. However, it contains literally hundreds of small nucleoli that are detached from the chromosomes and are predominantly located on the lining of the inner membrane of the nuclear envelope. The amplification process is readily monitored by incorporation of ³H-thymidine into oocyte DNA. All of the thymidine incorporation is due to replication of the nucleolar DNA because no other nuclear DNA synthesis is occurring in oocytes. As is shown in the autoradiographs in Figure 3.39, the label is incorporated into the extrachromosomal DNA that forms a cap of fibrillar material on one side of the nucleus. The major incorporation occurs during the pre-lampbrush pachytene stage.

This fibrillar material consists of numerous circular replicates of the nucleolar organizer region. During late pachynema and in diplonema, miniature nucleoli form in association with these rings of extrachromosomal DNA. The nucleoli then become more or less evenly distributed by spreading over the inner surface of the nuclear envelope. As many as 1500 nucleoli are formed during diplonema in *Xenopus*. This is about 375 times the tetraploid nucleolar number. The amount of ribosomal DNA, as measured by RNA-DNA hybridization, is increased by a comparable amount (Brown and Dawid, 1968). These multiple nucleoli provide the oocyte with numerous ribosomal coding units for the prodigious production of 18S and 28S RNA that occurs during oogenesis. It has been calculated that if amplification did not occur, it would take more than 400 years to produce the amount of ribosomal RNA found in the mature oocyte (Perkowska et al., 1968).

Synthesis of ribosomal RNA occurs at very low rates during pachynema when amplification is in progress, and the same rate prevails in the interval between amplification and the beginning of vitellogenesis, when it then undergoes a dramatic increase. There is evidence that the rate of ribosomal RNA synthesis is reduced somewhat in postvitellogenic oocytes, but it still occurs at rates exceeding those found in previtellogenesis (Davidson, 1986).

As we noted previously, maximal transcriptional rates are characterized by very close packing of polymerases on the chromatin. The proximity of these polymerases to one another, which can be monitored by observing transcription with the electron microscope, depends upon the frequency of initiation of transcription on a transcription unit and the rate of polymerase movement. Scheer and colleagues (1976b) have observed nucleolar transcription of *Triturus* oocytes, which have a ribosomal transcriptional pattern similar to that of *Xenopus*. They have found that the packing of polymerases on the ribosomal RNA genes reflects the rate of transcription. As shown in Figure 3.40, maximum packing is found in nucleoli from vitellogenic oocytes, whereas previtellogenic and postvitellogenic oocytes have fewer polymerases per unit length of nucleolar chromatin. These results clearly indicate that amplification itself is not sufficient to cause maximal rates of ribosomal RNA synthesis because amplified nucleoli may have low rates of transcription. It is apparent that transcriptional-level controls are operating to modulate the rate of rRNA production by the nucleoli.

Ribosomal Proteins

Because of the requirements for assembly of large numbers of ribosomes during oogenesis, ribosomal proteins are made in very large amounts in oocytes. Ribosome assembly occurs in the nucleoli, beginning during

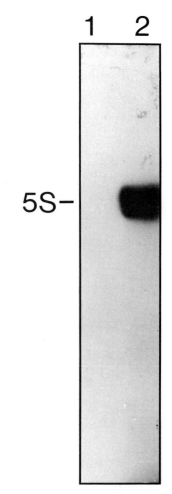

Figure 3.38

Assay for 5S RNA gene transcription factor in *Xenopus* eggs. The figure shows an autoradiograph of a polyacrylamide gel analysis of labeled RNA produced *in vitro*. The position of 5S RNA is noted on the figure. In the absence of added transcription factor (Lane 1), an egg extract will not promote synthesis of 5S RNA. This class of RNA is synthesized if transcription factor is added to the egg extract (Lane 2). (From B. M. Honda and R. G. Roeder, 1980. Association of 5S gene transcription factor with 5S RNA and altered levels of the factor during cell differentiation. Cell, *22*: 121. Copyright © Cell Press.)

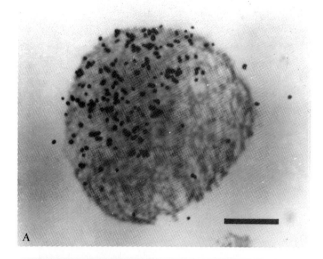

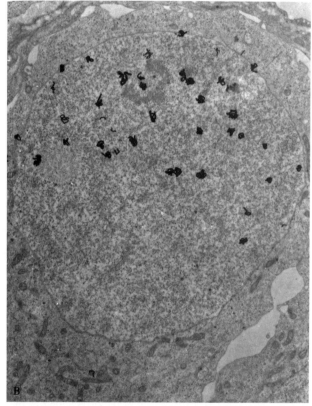

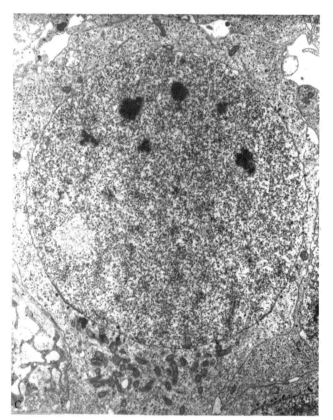

Figure 3.39

Nucleolar amplification. A, Light microscope autoradiograph, showing incorporation of ^{3}H-thymidine into nucleus of pachytene oocyte. Label is over nuclear cap. Scale bar equals 10 μm. B, EM autoradiograph, showing incorporation of ^{3}H-thymidine into fibrillar cap of pachytene nucleus. ×4,600. C, Electron micrograph of late pachytene oocyte showing formation of several small nucleoli in nuclear cap. ×6,350. (A, From L. W. Coggins and J. G. Gall, 1972. Reprinted from *The Journal of Cell Biology*, 1972, *52*: 569–576 by copyright permission of the Rockefeller University Press. B, and C, From Coggins, 1973. Reprinted by permission of Company of Biologists, Ltd.)

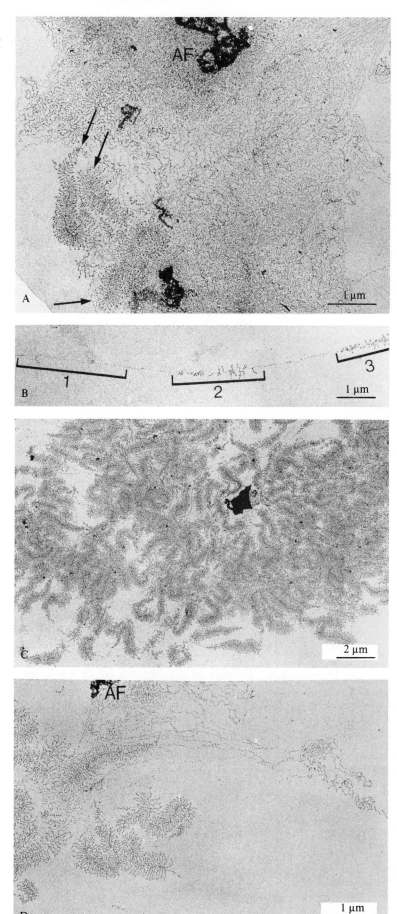

Figure 3.40

Nucleolar transcription in *Triturus* oocytes. A, and B, Previtellogenic pattern. A large proportion of the chromatin is transcriptionally silent and tends to form fibrillar aggregates (AF). Regions of low transcriptional activity and a few maximally active matrix units (arrows) are also seen in A. B shows three adjacent rRNA coding units from spread chromatin. Regions 2 and 3 show sparse packing of polymerases, while region 1 is inactive. C, Vitellogenic pattern. A typical nucleolar transcriptional pattern is seen. Virtually all of the chromatin has maximally active matrix units alternating with untranscribed spacers. D, Postvitellogenic pattern. A large part of the chromatin is transcriptionally silent and has a tendency to form fibrillar aggregates. Some regions of low transcriptional activity and maximally active matrix units are seen. (From U. Scheer, M. F. Trendelenburg, and W. W. Franke, 1976b. Reprinted from *The Journal of Cell Biology*, 1976, *69*: 465–489 by copyright permission of the Rockefeller University Press.)

vitellogenesis when the newly available 18S and 28S RNA form complexes with the 5S RNA and approximately 60 different ribosomal proteins. Much of the 5S RNA for the ribosomes is synthesized in earlier stages and stored in RNP particles for use during vitellogenesis. But what about the ribosomal proteins? Does their synthesis correlate with that of the 5S RNA or with that of 18S and 28S RNA? The available evidence suggests that ribosomal protein synthesis is occurring at higher rates during vitellogenesis than during previtellogenesis (Hallberg and Smith, 1975; Dixon and Ford, 1982; Wormington, 1988; Wormington, 1989), which indicates that their synthesis is correlated with the period when 18S and 28S RNA are synthesized at high rates and when concurrent ribosome subunit assembly is taking place.

Close examination of the chronology of ribosomal protein synthesis indicates that an interesting regulatory phenomenon is occurring. During vitellogenesis, ribosomal protein accounts for as much as 20% of total protein synthesis. As oogenesis continues, ribosomal protein synthesis continues at the same rate, but synthesis of other proteins increases substantially. Thus, ribosomal proteins account for a reduced percentage of total protein synthesis. A comparison of the behaviors of ribosomal protein mRNAs and the total oocyte mRNA population illustrates the basis for this differential synthesis (Wormington, 1988).

The vast majority of poly (A)$^+$ RNA in vitellogenic oocytes is nonpolysomal, with only 1% to 2% being associated with polysomes. Ribosomal protein mRNA, on the other hand, is at least 50% polysomal at this time. As oogenesis continues, additional poly (A)$^+$ RNA is recruited onto polysomes, whereas the amount of polysome-associated ribosomal protein mRNA remains constant. Hence, recruitment of additional mRNAs dilutes the translation of ribosomal protein mRNAs and diminishes the proportion of protein synthesis that is due to ribosomal protein synthesis. These observations indicate quite clearly that there is differential regulation of utilization of individual mRNAs by the translational apparatus in the oocyte. The basis for differential recruitment of mRNAs is unknown, but it is likely that mRNAs have signatures due to either distinct nucleotide sequences or secondary structure (i.e., the pattern of folding, which is due to base pairing of self-complementary regions in individual mRNA molecules), and these signatures are recognized by cellular factors that are developmentally regulated.

Transfer RNA Synthesis

The production of transfer RNA appears to parallel that of 5S RNA; i.e., it is synthesized in large amounts during previtellogenesis. In another similarity to 5S RNA, it has been found that there are oocyte and somatic tRNA genes. Transcripts of the oocyte-type tRNA genes are detected in oocytes and embryos, but absent in postembryonic somatic cells, whereas transcripts of the somatic-type genes are absent in oocytes, but are found in embryonic and adult cells (Stutz et al., 1989).

The tRNA that is synthesized during oogenesis accumulates in the 42S storage particles in which 5S RNA is also stored (Ford, 1976). Each 42S particle contains 12 molecules of tRNA and 4 molecules of 5S RNA (Picard et al., 1980). The 42S particles contain approximately one half the cell's 5S RNA and approximately 90% of the tRNA (Picard et al., 1980). Both of these classes of RNA are extremely long lived in the oocyte, having a lifetime of several months (Mairy and Denis, 1972).

Like 5S RNA, the accumulated tRNA accounts for about 45% of the total RNA of previtellogenic oocytes, and together they account for the vast majority of the RNA in these oocytes. The stored tRNA is released into the cytoplasm at vitellogenesis. The 42S storage particles are no longer detectable during this stage (Denis and Mairy, 1972).

3–4 Gene Expression During Oogenesis: Nonamphibian Species

In this section, we shall expand the discussion of gene expression during oogenesis by examining this process in nonamphibian species. We shall not attempt a general survey of oogenesis in the animal kingdom. Instead, we concentrate on three examples: *Drosophila*, sea urchins, and mammals. *Drosophila* has been chosen to illustrate the role of the nurse cells in macromolecular synthesis and the deposition in the oocyte cytoplasm of macromolecules that play discrete roles in controlling development of the embryo. The sea urchins are significant because of the rapidly expanding literature on the utilization of oogenic transcripts during embryonic development. The mouse is the best-studied mammal for virtually all aspects of development. Considerable research has documented the synthesis and accumulation of RNA during mouse oogenesis and its utilization in oocyte protein synthesis. Poly $(A)^+$ RNA accounts for an unusually large percentage of the RNA that accumulates during mouse oogenesis because the mouse oocyte accumulates relatively little ribosomal RNA (see Davidson, 1986, for details). However, we shall not focus on the accumulation of transcripts for utilization during development of the mouse embryo. Instead, we shall examine gene expression that is oocyte-specific (i.e., genes that are expressed exclusively during oogenesis for the production of oocyte-specific proteins).

Drosophila

As we discussed previously, oocyte chromosomes of *Drosophila* and other meroistic insects do not go through a lampbrush phase. In fact, the oocyte nucleus is of diminished functional significance in these insects, and the nurse cell nuclei assume a major role in production of oocyte cytoplasmic RNA. In contrast to those organisms with lampbrush-type oogenesis, the whole process of oogenesis in meroistic insects is extremely rapid. In the cricket (which has lampbrush chromosomes), for example, oogenesis lasts 100 days, whereas in *Drosophila* (which has nurse cells), oogenesis takes only about 8 days. Because the time allotted for RNA synthesis in meroistic insects is so short, the requirement for extensive RNA synthesis during oogenesis appears to be met by a dramatic increase in ploidy (see Sect. 3–1). Because 15 nurse cells are associated with each oocyte, the equivalent of several thousand genomes apparently satisfies the needs of the oocyte cytoplasm for RNA in a very short period of time.

As we shall discuss in detail in Chapters 6, 13, and 14, extensive genetic analyses with *Drosophila* have revealed a class of genes that are transcribed during oogenesis and exert their phenotypic effects during development. Some of these **maternal effect genes** are involved in organizing the body plan of the embryo, such as specification of the anteroposterior and dorsoventral axes. We now examine how expression of genes during oogenesis can leave a molecular imprint in the egg that directs elaboration of a polarized embryo.

If material from the anterior end of the egg is removed experimentally, the resultant embryo develops with no head or thorax (Nüsslein-Volhard et al., 1987). Thus, the components that specify anterior development must be localized at this end of the egg. The same effect is produced in eggs from females that lack a functional *bicoid* (*bcd*) gene. This indicates that *bicoid* is involved in specification of the anterior pole of the egg. Transplantation of anterior cytoplasm from wild-type embryos can normalize the phenotype of *bcd⁻* embryos (Frohnhöfer and Nüsslein-Volhard, 1986). These observations suggest that products of *bicoid* gene expression are localized during oogenesis in the anterior end of the oocyte, and their removal, either experimentally or by mutation, eliminates specification of the anterior-most components of the embryo. In fact, experimental analysis indicates that this is exactly what happens.

As shown in Figure 3.5, the nurse cells in *Drosophila* are localized at the presumptive anterior end of the oocyte. *In situ* hybridization analysis shows that when transcripts of the *bicoid* gene enter the oocyte from the nurse cells through the ring canals, these transcripts become localized in the anterior end (Fig. 3.41A to C). Thus, as with *Vg1* mRNA in the *Xenopus* oocyte, *bcd* RNA in the *Drosophila* oocyte has an address in the cell that is presumably specified by either its nucleotide sequence or its secondary structure. Another gene, *exuperantia* (*exu*), has also been shown to have maternal effects on head and thorax development. As shown in Figure 3.41D through F, *bicoid* transcripts are essentially equally distributed in the *exu* mutant oocyte. Hence, the wild-type *exu* allele must be involved in the polarized distribution of *bicoid* transcripts, possibly in elaboration of cytoskeletal elements that trap *bicoid* mRNA in the anterior end (Frohnhöfer and Nüsslein-Volhard, 1987).

As we shall discuss in Chapter 11, the localized *bicoid* mRNA is translated after fertilization, producing a protein that localizes to nuclei in the anterior half of the embryo and may influence expression of genes that are transcribed exclusively in that portion of the embryo.

An analogous mechanism is involved in specification of the posterior pole of the egg. Females that are mutant for *nanos* and *oskar* produce eggs that have a head and thorax but lack most of the abdominal segments. Likewise, removal of posterior cytoplasm produces a similar defect, and transplantation of wild-type posterior cytoplasm can induce abdominal development in *osk⁻* embryos (Lehmann and Nüsslein-Volhard, 1986). Therefore, there is localization of posterior determinants in the end of the oocyte distal to the nurse cells. Clearly, the transcripts produced during oogenesis are not randomly dumped into the cytoplasm after their synthesis, but they may be precisely localized so that the proteins they encode may be in a strategic location to perform localized functions that are necessary for proper spatial organization of the embryo.

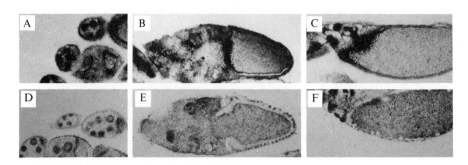

Figure 3.41

In situ hybridization analysis of *bicoid* mRNA distribution in wild-type (A-C) and *exuperantia* mutant (D-F) *Drosophila* oocytes of progressively later stages of oogenesis. The transcripts are localized to the anterior end of wild-type oocytes but are evenly distributed in the mutant oocytes. (From Berleth et al., 1988.)

Sea Urchins

A great deal of information about the messenger RNA population of sea urchin eggs and embryos has been obtained by studies with RNA-DNA hybridization technology. These hybridization studies yield valuable qualitative information about **sequence complexity** of RNA. Sequence complexity is a measure of the number of different mRNA species in a cell (see p. 552). An estimate of the complexity of a given mRNA population can be obtained by determining the extent of hybridization between total cellular mRNA and single copy DNA, the DNA class from which most mRNAs are transcribed (Davidson, 1976, 1986). They also allow comparisons to be made between RNA populations. Thus, changes in complexity can also be examined.

The sequence complexity of unfertilized sea urchin eggs has been estimated to be equivalent to approximately 11,000 different messenger RNAs (Davidson, 1986). To determine whether sequence complexity changes during oogenesis, the single copy DNA from egg RNA-DNA hybrids has been used as a probe for the presence of homologous transcripts in the cytoplasmic RNA from previtellogenic oocytes (Hough-Evans et al., 1979). Reaction of the previtellogenic oocyte RNA with this DNA measures the fraction of the egg single copy gene set that is represented in the cytoplasm of immature oocytes. Less than half (44%) of the egg single copy gene set is represented in previtellogenic oocyte cytoplasmic RNA. Thus, 56% of the mRNA sequences are added to the cytoplasm during vitellogenesis. These results contrast with those obtained with *Xenopus* oocytes, in which mRNA complexity does not increase during vitellogenesis. Thus, it would appear that the patterns of structural gene expression during oogenesis in the sea urchin and *Xenopus* differ dramatically.

One striking similarity with *Xenopus* oogenesis is that sea urchin oocytes also accumulate interspersed poly (A)$^+$ RNA (Costantini et al., 1980). (Recall that these transcripts result from transcription of adjacent single copy and moderately repetitive regions of DNA.) The synthesis of large amounts of this category of RNA in these two evolutionarily distant groups suggests that this may be a common feature of oogenesis, although its function is unclear.

The Mouse

Like the species we have previously discussed, gene expression during mouse oogenesis produces transcripts that accumulate for utilization during early development. Oogenesis in all these species also involves production of transcripts that are utilized for the synthesis of oocyte-specific proteins. Because there is an excellent example of this latter phenomenon in mice (expression of the genes encoding zona pellucida proteins), we focus our discussion of gene expression during mouse oogenesis on this aspect of oogenesis.

The zona pellucida of mammals is composed of three glycoproteins, called **ZP (zona pellucida protein)**-1, **ZP**-2, and **ZP**-3 (Bleil and Wassarman, 1980b). As we shall discuss in Chapter 4, ZP-2 and ZP-3 play important functional roles during fertilization. The zona proteins are synthesized by the oocyte, secreted from the cell, and assembled into an extracellular coat (Wassarman, 1990). The expression of the gene encoding ZP-3 has been especially well characterized and will serve as our example of expression of the zona pellucida protein genes.

The magnitude of the process of synthesizing ZP-3 during oogenesis is illustrated by the massive amount of this protein in the zona pellucida of an unfertilized mouse egg: more than 10^9 molecules (Wassarman, 1990). In order to accomplish this feat, oocytes must accumulate an impressive amount of ZP-3 mRNA. The pattern of accumulation of these transcripts during oogenesis is illustrated in Figure 3.42A. It can be seen that the accumulation of ZP-3 transcripts occurs during the period of oocyte growth (a two- to three-week period), reaching a maximum of about 300,000 copies per oocyte. For comparison, the most abundant transcripts in liver cells are present in only about 50,000 copies per cell (Wassarman, 1990). As the ZP-3 transcripts accumulate, they are being translated into ZP-3 protein. During ovulation, the ZP-3 transcript population declines dramatically, to become undetectable after fertilization. This is clearly a case of the molecular machinery providing the tools to make ZP-3 protein when it is needed and eliminating them as rapidly as possible thereafter.

Not only is ZP-3 gene expression highly regulated during oogenesis, but its expression is restricted to the oocyte. As shown in Figure 3.42B and C, no ZP-3 transcripts can be detected in a variety of other cell types, and they can never be detected in any male tissue; thus, expression of this gene is both cell-specific and sex-specific. The techniques used to detect these transcripts are those most commonly used to detect specific transcripts. The technique used in Figure 3.42B is **Northern blotting**, by which RNA is

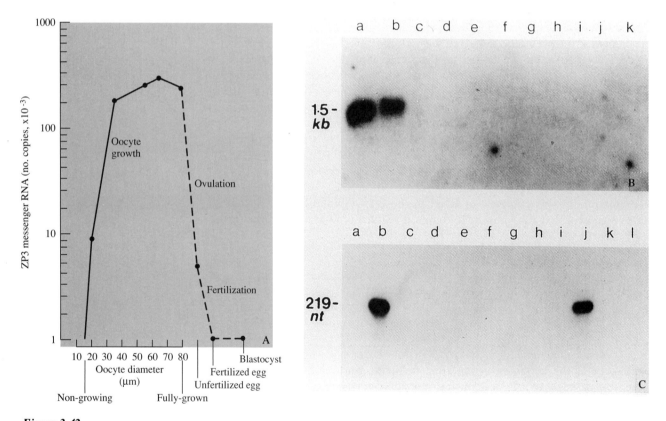

Figure 3.42

Stage- and tissue-specific expression of ZP-3 mRNA. A, Accumulation of ZP-3 messenger RNA in mouse oocytes, eggs, and preimplantation embryos. The number of copies of ZP-3 mRNA ($\times 10^{-3}$) is plotted as a function of the stage of development. B, Northern blot analysis of ZP-3 transcripts. The lanes contain RNA from: a, oocytes; b, whole ovary; c, brain; d, 13-day embryos; e, heart; f, intestine; g, kidney; h, liver; i, muscle; j, testis; and k, uterus. C, Nuclease protection analysis of ZP-3 transcripts. The lanes contain RNA from: a, *E. coli*; b, oocytes; c, brain; d, 13-day embryos; e, heart; f, intestine; g, kidney; h, liver; i, muscle; j, ovary; k, testis; and l, uterus. (A, After Wassarman, 1990. B, and C, From Wassarman, 1990. Reprinted by permission of Company of Biologists, Ltd.)

resolved by gel electrophoresis and transferred by blotting to a filter, on which it is hybridized to a labeled single-strand nucleic acid probe (see Appendix). The technique used in Figure 3.42C is **nuclease protection assay**. This assay consists of hybridizing a radioactively labeled nucleic acid probe (in this case a ZP-3 probe) to RNA derived from the tissue being tested. After the hybridization is complete, the mixture is treated with a nuclease (S1 nuclease) that specifically destroys single-stranded nucleic acids, and the remaining double-stranded pieces are separated on a gel. If ZP-3 mRNA is present in the RNA, the labeled probe hybridizes to it and becomes resistant to nuclease destruction. The piece of labeled nucleic acid can then be resolved on a gel and visualized by autoradiography.

Oocyte-specific transcripts such as the ZP-3 mRNA are obviously utilized differently than the oogenic transcripts, which are stored for translation after fertilization. The differential utilization of oocyte-specific and oogenic transcripts is an intriguing problem that awaits resolution.

3–5 Utilization of Genomic Information in Oogenesis: Conclusions

Like any differentiation process, oogenesis involves differential utilization of genomic information. The completed oocyte is composed of numerous specialized components that enable it to function in fertilization, as well as components that will be utilized during postfertilization development. Consequently, gene expression during oogenesis has both immediate and long-term implications for reproduction. A portion of the transcribed information directs the synthesis of oocyte proteins, whereas the remainder is retained after fertilization and has profound effects upon early postfertilization development (see Chap. 13). Because of this dichotomy, the full-grown oocyte can be seen either as the *result* of a complex differentiation process or an *interim stage* in a continuing process.

The oocyte is unusual in another fundamental way: Its proteins may be either synthesized endogenously or incorporated within the oocyte from exogenous sources. In some cases (as in meroistic insects), even the oocyte organelles may be a product of the coordinated expression of genes in a number of cells; the overall control of these processes is exerted by the hormones that control the reproductive cycle.

3–6 Hormonal Control of Oogenesis

Oogenesis is regulated by modulations in the concentrations of circulating hormones. As we discussed in Chapter 2, gonadal function in the vertebrates is regulated by gonadotropic hormones released from the pituitary gland. The secretory functions of the pituitary are, in turn, regulated by brain neurohormones, such as gonadotropin-releasing hormone (GnRH), released by the hypothalamus. The gonadotropins (which are peptide hormones) act on the ovary to cause oocyte growth and, ultimately, ovulation. These hormones also stimulate follicle cells in the ovary to synthesize steroid hormones.

The seasonal oogenesis cycle in amphibians illustrates the hormonal interactions that operate in **oviparous** species, in which a paramount function during oogenesis is the synthesis and deposition of yolk. The initiation of seasonal ovarian growth is triggered by environmental cues that cause the brain to stimulate the pituitary to secrete gonadotropins. The gonadotropins are carried to the ovaries in the circulation and cause the follicle cells to synthesize estrogen (Fig. 3.43). This hormone is then released into the circulation and stimulates the liver to synthesize vitellogenin. Next, the vitellogenin is released into the blood stream and is transported to the ovaries, where gonadotropins promote its uptake by oocytes. During this step, it is incorporated into the yolk platelets. When oogenesis is complete, gonadotropins then promote meiotic maturation and ovulation. These latter processes are mediated by yet another steroid hormone, **progesterone**, which is synthesized by follicle cells under gonadotropic influence.

A remarkably similar scheme for regulation of vitellogenesis is found in *Drosophila* (Fig. 3.44). In this organism, the gonadotropic hormone is the **juvenile hormone**, which is secreted by the **corpus allatum**. The juvenile hormone promotes oocyte differentiation and stimulates the ovary to produce **ecdysone**, which stimulates another organ, the **fat body**, to produce vitellogenin. The vitellogenin is released into the hemolymph and is sequestered into the oocytes under the influence of juvenile hormone. Juvenile hormone also stimulates the ovary itself to produce vitellogenin. Hence, there is a dual origin for yolk in *Drosophila*. It is important to realize that insects are a heterogeneous group of organisms, and not all of them follow this scheme. A detailed discussion of the various endocrine strategies for control of oogenesis in insects can be found in Highnam and Hill (1977).

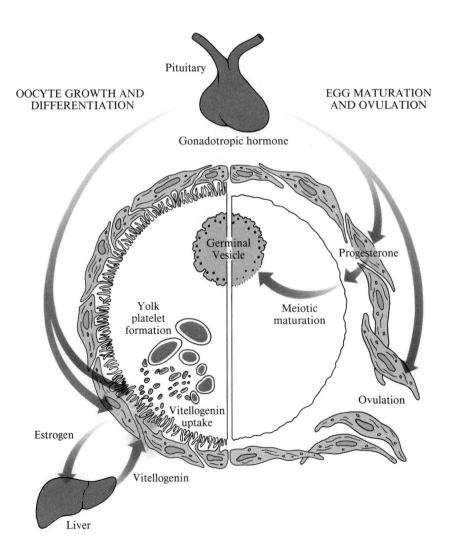

Figure 3.43

Hormonal regulation of amphibian oocyte growth and differentiation (left) and egg maturation and ovulation (right).

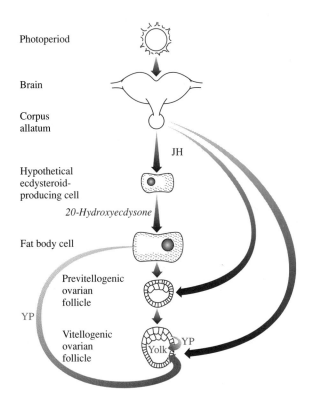

Photoperiod

Brain

Corpus
allatum

JH

Hypothetical
ecdysteroid-
producing cell

20-Hydroxyecdysone

Fat body cell

Previtellogenic
ovarian
follicle

YP

Vitellogenic
ovarian
follicle

Yolk   YP

Figure 3.44

Regulation of vitellogenesis in *Drosophila*. JH: juvenile
hormone. (After Postlethwait and Giorgi, 1985.)

Regulation of oogenesis is perhaps most complicated in the mammals,
in which there is cyclic production and ovulation of ripe eggs. The mammalian
egg develops in conjunction with its surrounding follicle cells, forming a
functional unit: the **follicle** (see Sect. 3–1). Variable numbers of follicles
develop during a cycle, each culminating in the expulsion of a mature egg
at ovulation. A single follicle usually matures in human ovaries during each
cycle, whereas in most other mammals several follicles mature simultaneously. Mammals have two gonadotropic hormones that act synergistically
to regulate oogenesis. These are the follicle-stimulating hormone (FSH) and
the luteinizing hormone (LH), which cause the growth of the follicle, prepare
the egg for ovulation, and stimulate steroid production by the follicle cells.

The menstrual cycle, which is found only in primates, is illustrated in
Figure 3.45A. The menstrual phase is considered to be the beginning of the
cycle. A notable increase in FSH production by the pituitary occurs at
menstruation. This hormone promotes growth and development of the
oocyte and, together with LH, causes the follicle cells to release increasing
amounts of estrogen, which assists in promoting follicle growth. The
consequent build-up of estrogen in the blood causes the hypothalamus to
release pulses of GnRH, which causes a large surge of gonadotropins at
mid-cycle. The surge of LH triggers ovulation and the resumption of meiosis.
After ovulation, LH promotes the conversion of the follicle—minus the
oocyte—into the **corpus luteum**, which itself is an endocrine structure that
produces progesterone and small amounts of estrogen. The progesterone
reduces the frequency of pulsatile LH secretion, thus preventing additional
LH surges during the cycle (Karsch, 1987). The corpus luteum also produces
inhibin under the control of FSH. The elevation of circulating inhibin during
this time correlates with a decline in FSH levels. It has been proposed that
the inhibin causes a decline in FSH production by the pituitary through a
classical feedback inhibition (McLachlan et al., 1987). If the egg is not

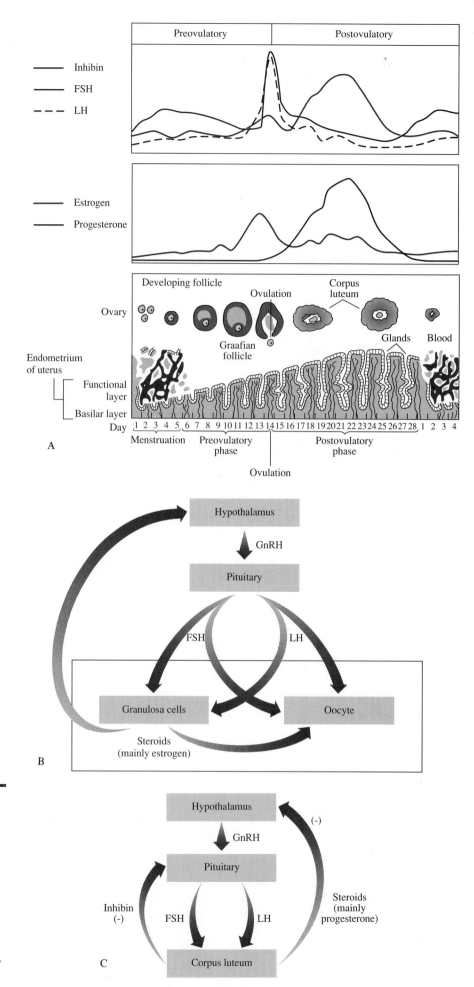

Figure 3.45

The menstrual cycle. A, Diagram of the menstrual cycle showing levels of inhibin, LH, FSH, estrogen, and progesterone, the condition of the follicle, and the condition of the uterus. B, and C, Hormonal interactions involved in regulating the menstrual cycle. B summarizes the interactions that lead to ovulation, whereas C illustrates the luteal phase of the cycle. (A, After Villee et al., 1989, and McLachlan et al., 1987.)

fertilized, the corpus luteum degenerates within eight to ten days after ovulation (possibly due to the reduced frequency of LH pulses; Soules et al., 1984), and the lining of the uterus is sloughed, causing the menstrual bleeding. Presumably, the decline in steroids and inhibin that results from regression of the corpus luteum relieves the inhibition on the pituitary and/or the brain, and FSH secretion resumes, causing the cycle to repeat (McLachlan et al., 1987), although the relative inhibitory effects of the inhibin and the steroids are unknown. The inverse relationship between FSH and inhibin during the luteal phase of the menstrual cycle is reminiscent of a similar relationship in the male in which inhibin production by the Sertoli cells regulates FSH secretion (see p. 39). The hormonal interactions that regulate the menstrual cycle are summarized in Figure 3.45B and C.

If the egg is fertilized (Fig. 3.46), production of LH is continued, and the corpus luteum expands and produces even higher levels of steroids. The corpus luteum continues to produce high levels of steroids until about the fourth month, when its steroid production begins to wane. During the latter half of pregnancy, the **placenta** becomes the principal source of steroid hormone production.

Oral contraceptives and drugs to restore fertility in infertile women are direct results of research on endocrine interactions. Synthetic steroids are employed to take advantage of feedback inhibition that curtails gonadotropin release by the pituitary, in particular the inhibition by progesterone of the LH surge that normally occurs at mid-cycle. In the absence of this surge of LH, ovulation is prevented. Gonadotropin-releasing hormone and other drugs that regulate gonadotropin secretion are commonly used to induce ovulation in infertile women.

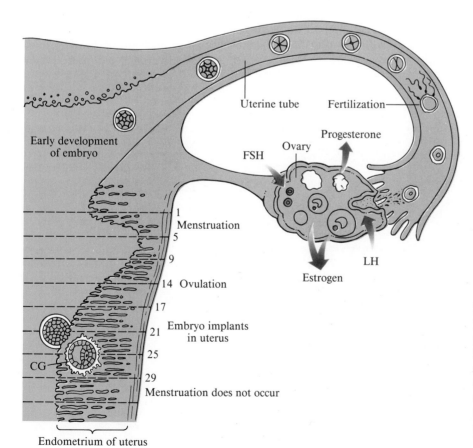

Figure 3.46

Effects of pregnancy on the menstrual cycle. The corpus luteum is maintained, and the wall of the uterus remains thickened to support implantation of the embryo, rather than being sloughed by menstruation. Numbers indicate days of the menstrual cycle. (After Villee et al., 1989.)

3–7 Conversion of the Oocyte to the Egg: Maturation and Ovulation

Differentiation of the female gamete occurs while the cell nucleus remains in a prolonged meiotic prophase (G_2 phase of the cell cycle; see p. 163 for a discussion of the phases of the cell cycle). The completion of meiosis is essential for the production of the haploid female pronucleus that associates with the male pronucleus at fertilization. The timing of meiotic completion is quite variable among animals (Table 3–1). The most familiar pattern (because it occurs in humans and most other vertebrates) is the one in which resumption of meiosis is initiated at ovulation and proceeds as far as metaphase of meiosis II. The egg remains at this stage until fertilization, which triggers the completion of meiosis II.

The phase between the initial resumption of meiosis and metaphase II is the period of oocyte maturation. The most conspicuous events of maturation are germinal vesicle breakdown (GVBD, which marks the end of prophase I), condensation of the chromosomes, formation of a spindle (at metaphase or M phase of the cell cycle), and production of the first polar body (which marks the completion of meiosis I). Thereafter, the chromosomes realign on the second M phase spindle, where they remain until fertilization. However, maturation is much more than merely the reinitiation of meiosis; it includes numerous *molecular and physiological changes* that are necessary for fertilization to occur.

Regulation of Maturation and Ovulation

As we have mentioned, both oocyte maturation and ovulation in mammals (including primates) are triggered directly by the LH surges, apparently without the direct intervention of steroids. This mechanism of control over oocyte maturation is not universal among vertebrates, however. The most extensive investigations of the events of vertebrate oocyte maturation have

Table 3–1 Stage of Egg Maturation at which Sperm Penetration Occurs in Different Animals

Young primary oocyte	Fully grown primary oocyte	First metaphase	Second metaphase	Female pronucleus
The annulate worm *Dinophilus*	The round worm *Ascaris*	The nemertine worm *Cerebratulus*	The lancelet *Amphioxus*	Coelenterates, e.g., anemones
The polychaete worm *Histriobdella*	The mesozoan *Dicyema*	The polychaete worm *Chaetopterus*	The amphibians *Siredon, Rana, Xenopus*	Echinoids, e.g., sea urchins
The flatworm *Otomesostoma*	The sponge *Grantia*	The mollusk *Dentalium*	Most mammals	
The onychophoran *Peripatopsis*	The polychaete worm *Myzostoma*	The cone worm *Pectinaria*		
The annulate worm *Saccocirrus*	The clam worm *Nereis*	Many insects		
	The clam *Spisula*			
	The echiuran worm *Urechis*			
	Dog and fox			

Modified from Austin, C.R. © 1965. *Fertilization*, p. 87. Reprinted by permission of Prentice-Hall, Englewood Cliffs, NJ.

been conducted with the amphibians *Rana pipiens* and *Xenopus laevis*, primarily because maturation and ovulation in these species can be readily induced in the laboratory and because the large size of their eggs facilitates experimental investigation. It has long been the practice to induce ovulation in amphibians by injecting gonadotropins into gravid females. Because gonadotropins also cause oocyte maturation, it has been recognized for some time that both of these events are under hormonal control. Steroid hormones are also effective inducers of maturation, indicating some sort of interaction between the gonadotropins and steroids.

Detailed investigations of the precise hormonal regulation of maturation became possible when it was discovered that maturation can be induced *in vitro* by adding either gonadotropin or the steroid hormone progesterone to culture medium containing ovarian fragments. However, gonadotropin is ineffective in inducing maturation if the oocytes are separated from their follicles, whereas subsequent addition of either ovarian tissue fragments or progesterone to the culture medium will promote maturation. These observations demonstrated that the effects of gonadotropin on the oocyte are indirect and are mediated by the follicle cells, which produce progesterone in response to gonadotropin stimulation. It is the progesterone that stimulates the oocytes to undergo maturation (Masui, 1967; Schuetz, 1967; Smith et al., 1968).

The effects of progesterone on the oocyte in bringing about maturation have been carefully documented. One surprising result is the finding that progesterone apparently acts on the oocyte surface to trigger maturation. This conclusion was drawn from attempts to induce maturation by microinjecting progesterone directly into oocytes, a treatment that is reported by most investigators to be incapable of promoting maturation, although these oocytes can subsequently be induced to mature by bathing them in the hormone (Masui and Markert, 1971; Smith and Ecker, 1971). The surface action of progesterone is also supported by the observation that oocytes will mature when exposed to a progesterone analogue that is covalently linked to a large polymer that does not enter the cells (Godeau et al., 1978). A putative progesterone receptor has been demonstrated in plasma membrane preparations of *Xenopus* oocytes (Blondeau and Baulieu, 1984).

The mode of action of progesterone in promoting oocyte maturation is quite atypical for a steroid hormone. Usually, steroid hormones enter their target cells and bind to a receptor. The resultant hormone-receptor complex then mediates the effects of the hormone by interacting with the DNA (see p. 740). The binding of progesterone to a cell surface receptor is reminiscent of the interaction between protein hormones and plasma membrane receptors. Because the complex is formed at the surface, the effects of the hormone are mediated by a second, intracellular messenger; in this case, the signal appears to be a decline in the levels of cyclic AMP (Maller, 1985).

Other than acting on the surface, there is another significant difference between the action of progesterone in inducing oocyte maturation and the classical mechanism of steroid hormone action on somatic cells. In the latter situation, the effects of steroid hormones are mediated at the level of transcription. In oocytes, however, inhibitors of RNA synthesis fail to prevent the progesterone effect. On the other hand, maturation is prevented by inhibition of protein synthesis, an indication that the induction of maturation is mediated by the cytoplasm rather than by the nucleus, apparently by components synthesized during oogenesis, stored in the cytoplasm, and activated in response to the hormone. These components must include messenger RNA, ribosomes, and other elements of the protein synthetic machinery (Wasserman and Smith, 1978a).

The Role of Maturation Promotion Factor

Recent research on maturation has focused on the central role of a cytoplasmic "factor" that mediates the hormonal effects on the oocyte. This **maturation promotion factor (MPF)** was initially demonstrated by injection of cytoplasm from progesterone-treated eggs into untreated recipients, a procedure that causes the recipient oocytes to undergo maturation (Fig. 3.47). Repeated serial transfers of cytoplasm from maturing recipient oocytes will continue to induce maturation in sequential recipients even though only the original donor oocyte was exposed to progesterone. This result indicates that injection of MPF triggers the production of still more MPF in an autocatalytic reaction.

The formation of MPF is entirely independent of the nucleus because it can be produced in enucleated oocytes (Masui and Markert, 1971; Reynhout and Smith, 1974). This can be demonstrated by treating enucleated oocytes with progesterone and injecting their cytoplasm into nucleated non–progesterone-treated oocytes; the injected cells mature, indicating the formation of MPF in the enucleated oocytes (Table 3–2).

Although protein synthesis is necessary for the initial production of MPF activity, it is not required for MPF action in promoting maturation. These conclusions stem from the observations that maturation of progesterone-treated oocytes can be inhibited with cycloheximide until the time of first appearance of MPF activity, but after this event, maturation occurs even in the presence of the inhibitor. Furthermore, cycloheximide fails to prevent the autocatalytic amplification of MPF (Wasserman and Masui,

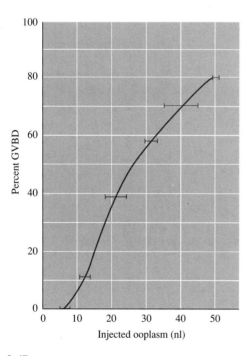

Figure 3.47

The graph reveals that the frequency of oocyte maturation is proportional to the volume of injected cytoplasm. The cytoplasm was taken from oocytes 18 to 20 hours after progesterone treatment. Abscissa: volume of injected cytoplasm. Ordinate: frequency of germinal vesicle breakdown (GVBD). (After Masui and Markert, 1971.)

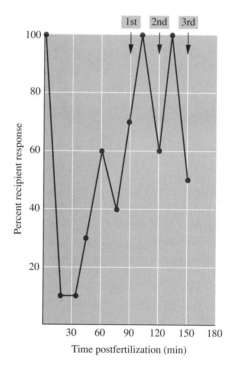

Figure 3.48

Oscillation of MPF activity in *Xenopus* embryos. The percent of recipient oocytes undergoing GVBD in response to injected embryo cytoplasm (ordinate) is plotted against the time after fertilization when the embryo cytoplasm was assayed for MPF activity (abscissa). Arrows indicate the time of each cleavage. (After Wasserman and Smith, 1978b.)

Table 3–2 Injection of Cytoplasm from Enucleated Oocytes Treated with Progesterone*

Volume (nl) injected	Enucleated donor		Nucleated donor	
	Number	GVBD (%)†	Number	GVBD (%)
5–6	18	0	12	0
11–13	34	8	54	11
19–24	49	20	54	32
35–36	40	58	45	76
50	18	67	21	71

From Masui, Y. and C. L. Markert. 1971. Cytoplasmic control of nuclear behavior during meiotic maturation of frog oocytes. J. Exp. Zool., *177*: 135.

* Injection was carried out 18 to 24 hours after progesterone treatment of donors.

† GVBD represents the percentage of injected oocytes showing germinal vesicle breakdown.

1975). These results indicate that oocytes contain a store of inactive MPF. Protein synthesis is essential to activate it; small amounts of the active MPF are then sufficient to amplify MPF activity without further protein synthesis.

Since its discovery in amphibian oocytes, MPF activity (as assayed by its ability to induce GVBD when injected into oocytes) has been found in cells of cleaving *Xenopus* embryos, in which it oscillates with the same periodicity as the cell cycle, appearing in late G_2, peaking in mitosis, and abruptly declining to undetectable levels by interphase (Fig. 3.48). Protein synthesis is essential for the reappearance of MPF activity at each division, which suggests that either MPF or an activator of MPF must be synthesized during each cycle (Wasserman and Smith, 1978b). This is an important observation; its significance will become clear later in this discussion. Maturation promotion factor activity has also been found to follow the same periodicity during the cell cycle in a variety of cells from yeast to mammals (for review, see Cyert and Kirschner, 1988). Thus, MPF may be a universal regulator of the cell cycle and should more properly be called **M-phase promoting factor**.

Although correlations between the presence of MPF and promotion of maturation are important, they do not reveal the precise role of MPF in the cell cycle. Experimental analysis of the biological activity of MPF indicates that its specific effect is to promote the $G_2 \rightarrow$ M-phase transition, and we are now beginning to understand *how* it exerts this effect. An important advance in elucidating the role of MPF was the development of a technique to assay MPF activity *in vitro* (Lohka and Maller, 1985). This assay allowed investigators to test the biological activity of putative MPF preparations and facilitated its purification.

Recently, remarkable progress has been made in clarifying the molecular nature of MPF. In its active state, MPF is a phosphoprotein that, in *Xenopus*, consists of two subunits: one of relative molecular mass of 45,000 (45K) and the other of 34,000 (34K) (Lohka et al., 1988). The small subunit (which is called **p34**) is the same size as the protein encoded by the *cdc2+* gene (cdc genes are genes affecting the cell division cycle) of fission yeast (*Schizosaccharomyces pombe*), a gene whose expression is necessary for the $G_2 \rightarrow$ M transition in these cells. The yeast protein is called p34[CDC2]. Antibodies prepared against p34[CDC2] cross-react with *Xenopus* p34, indicating that p34 is the *Xenopus* homologue of the yeast protein (Gautier et al., 1988). A p34[CDC2] homologue has also been demonstrated in cells of a variety of organisms, including mammals (Draetta et al., 1987; Lee et al., 1988), starfish (Arion et al., 1988; Labbé et al., 1988), and clams (Draetta et al., 1989), indicating that this mechanism for regulation of cell division has been conserved throughout eukaryotic evolution.

One of the best-characterized aspects of the transition to M phase in *Xenopus* maturation is protein phosphorylation (Lohka et al., 1987). p34 in yeast, *Xenopus*, mammalian cells, and starfish embryos is a protein kinase (i.e., an enzyme that phosphorylates proteins). The kinase activity of p34 remains high during M phase, resulting in the phosphorylation of a number of diverse proteins. One of these substrates for p34 kinase is histone H1 (Arion et al., 1988; Labbé et al., 1988; Lohka et al., 1988). Phosphorylation of histone H1 has been implicated in the condensation of chromatin that occurs during M phase (Matsumoto et al., 1980). The substrates for p34 kinase also include the **lamins**, which form a fibrillar network (the nuclear lamina) on the inner surface of the nuclear envelope. The lamina is thought to be important for the integrity of the nuclear envelope. The phosphorylation of the lamins by p34 kinase has led to the hypothesis that this event is responsible for dissolution of the envelope at nuclear envelope breakdown at the end of prophase (Peter et al., 1990).

The other subunit of MPF is a class of protein called **cyclins**. Cyclins derive their name from the fact that they were originally identified as proteins that oscillate in abundance during the cleavage cell cycle in marine invertebrates. Cyclins accumulate during interphase and are destroyed by proteolysis after mitosis during each cell cycle (Evans et al., 1983; Standart et al., 1987). The association between p34 and cyclins has been demonstrated by the technique of immunoprecipitation: Antibodies to cyclin precipitate not only cyclin but also p34, whereas antibodies to p34 precipitate both p34 and cyclin (Draetta and Beach, 1988; Draetta et al., 1989; Meijer et al., 1989). Thus, p34 and cyclin form a complex that is precipitated by either antibody. Furthermore, the elimination of cyclin activity in *Xenopus* egg extracts eliminates MPF activity and entry of added sperm nuclei into M phase (Minshull et al., 1989; Murray and Kirschner, 1989). Cyclin-depleted extracts can be made competent to promote entry into M phase by the addition of cyclin mRNA, which is translated and restores the cyclin (Murray and Kirschner, 1989). The apparent role of cyclins is to activate p34 kinase activity when they complex with p34. Like p34, cyclins appear to be universal cell cycle regulators in eukaryotes; they are homologous to the product of the *cdc13* gene in fission yeast (Goebl and Byers, 1988; Solomon et al., 1988).

The involvement of cyclin in meiosis was made evident by an experiment in which surf clam cyclin mRNA was injected into *Xenopus* oocytes; meiotic maturation was induced (Fig. 3.49). This result indicates that the mRNA was translated and that the resultant cyclin triggered maturation (Swenson et al., 1986). Although mitotic cells apparently have an absolute requirement for cyclin synthesis for the acquisition of MPF activity, the *Xenopus* oocyte has a substantial pre-existing store of cyclin complexed to p34 (Roy et al., 1990). In the oocyte, then, the requirement for protein synthesis for MPF activation may involve the synthesis of a protein that is necessary to activate the inactive cyclin-p34 complex. This protein may be a protein called c-MOS, which is the product of c-*mos* proto-oncogene (see the essay on p. 380 for a discussion of proto-oncogenes). c-MOS is a kinase enzyme whose synthesis is both necessary and sufficient for the induction of maturation (Sagata et al., 1989a; Freeman et al., 1989). This is shown by (1) injecting c-*mos* mRNA into *Xenopus* oocytes, which induces their maturation in the absence of progesterone (Fig. 3.50); and (2) inhibiting c-MOS synthesis, which prevents maturation. Maller and his colleagues (Roy et al., 1990) have recently shown that c-MOS can phosphorylate cyclin *in*

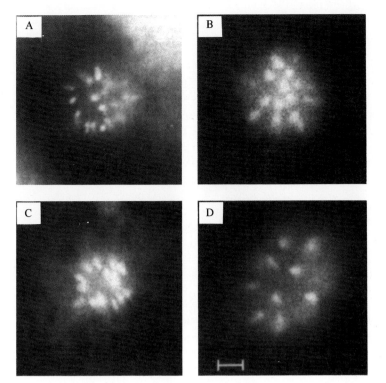

Figure 3.49

Xenopus oocyte maturation promoted by either progesterone or surf clam cyclin mRNA. A, and B, Meiosis II metaphase chromosomes from oocytes incubated with progesterone. Chromosomes are stained with Hoechst dye 33258 and visualized with fluorescent microscopy. The DNA in chromosomes is seen as bright against a dark background. C, and D, Meiosis II metaphase chromosomes from oocytes injected with cyclin mRNA. Scale bar equals 10 μm. (From Swenson et al., 1986. Copyright © Cell Press.)

vitro. Furthermore, elimination of c-MOS synthesis reduces phosphorylation of cyclin, a step that is necessary for its ability to activate MPF. As we have discussed previously, the requirement for activation of pre-existing cyclin can be countered by new cyclin synthesis from injected cyclin mRNA. This suggests that the newly synthesized cyclin combines with p34 to form fully active MPF, whereas the cyclin that pre-exists in complexes with p34 requires phosphorylation to become active.

The c-*mos* gene is normally expressed only during meiosis (in both male and female germ cells). In the oocyte, c-MOS protein is only synthesized during maturation. The evidence that we have just discussed links the newly synthesized c-MOS protein in the oocyte to the activation of cyclin. This suggests that a cascade of events is responsible for the activation of MPF: c-MOS activates cyclin, which, in turn, activates p34 kinase.

Maturation promotion factor retains its activity as long as the cyclin is present; degradation of cyclin is essential at the end of each M phase to inactivate MPF and allow the cell cycle to continue. In the mitotic cell cycle, cyclin is destroyed at the metaphase → anaphase transition, rendering MPF inactive and allowing the cell to continue the cell cycle (Murray et al., 1989). However, the mature oocyte (now called the unfertilized egg) remains in metaphase and does not continue the cell cycle. This metaphase arrest is a consequence of yet another factor, called **cytostatic factor (CSF)**. The existence of CSF was first demonstrated by Masui and Markert (1971), who showed that injection of the cytoplasm of an unfertilized egg into one blastomere of a 2-cell–stage frog embryo will cause metaphase arrest of that blastomere. This mimics the metaphase II arrest of the unfertilized egg. Does CSF prolong the life of cyclin, thus preventing release from metaphase arrest? We shall now examine evidence that this is, indeed, the case. *Xenopus* egg extracts made in the presence of the calcium chelator EGTA retain the CSF-mediated arrest; sperm nuclei added to such extracts enter (and remain in) M phase. Cytostatic factor activity can be eliminated in such extracts by the addition of calcium (which is the probable stimulus for egg activation at fertilization; see p. 153); nuclei reform, chromosomes

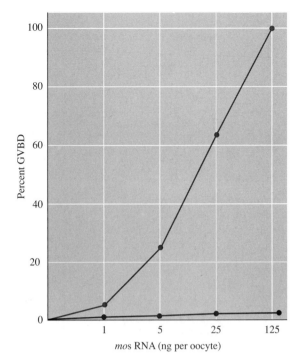

Figure 3.50

Effects of c-*mos* mRNA on oocyte maturation. Full-grown oocytes were microinjected with either functional (colored line) or non-functional (black line) c-*mos* mRNA. After 12 hours, each oocyte was examined for germinal vesicle breakdown (GVBD) to determine whether oocyte maturation had occurred. (After N. Sagata et al., 1989a. The product of the *mos* proto-oncogene as a candidate "initiator" for oocyte maturation. Science, *245:* 643–646. Copyright 1989 by the American Association for the Advancement of Science.)

decondense, and DNA synthesis occurs (M phase $\rightarrow$ G_1/S-phase transition; Lohka and Masui, 1984; Lohka and Maller, 1985). Murray et al. (1989) have shown that cyclin is stable in the presence of CSF (i.e., in the absence of Ca^{2+}), whereas it rapidly disappears when CSF activity is eliminated (after addition of Ca^{2+}).

There is now evidence that CSF is, in fact, c-MOS protein (Sagata et al., 1989b). This possibility is based upon the ability of investigators to mimic the effects of CSF with c-MOS and their ability to deplete CSF activity by depleting c-MOS, as illustrated in Figure 3.51. Injection of c-*mos* mRNA into one blastomere of a two-cell *Xenopus* embryo results in synthesis of c-MOS protein and the consequent metaphase arrest of the injected blastomere (see Fig. 3.51A). On the other hand, depletion of c-MOS protein from extracts of unfertilized egg cytoplasm with an antibody specific for c-MOS eliminates the cleavage-arrest activity of such preparations (see Fig. 3.51B).

How can we reconcile the two effects of c-MOS (i.e., activation of maturation and metaphase arrest)? One possibility is that the activation of the cyclin component of MPF also stabilizes it so as to prevent its proteolytic digestion, thus arresting the cell cycle in metaphase II.

We have discussed how a number of molecules are involved in regulating oocyte maturation and mitosis. The interactions among these molecules and the consequences of their actions are summarized in Figure 3.52.

Molecular Events in Amphibian Oocyte Maturation

The meiotic events of maturation are among a number of significant oocyte responses to hormonal stimulation, which include changes in the patterns of macromolecular synthesis. Although nuclear RNA synthesis continues in the interval between hormonal stimulation and GVBD, this RNA is apparently of no immediate consequence to the maturation process. This conclusion is based on the findings that maturation occurs in oocytes treated with inhibitors of RNA synthesis and in oocytes that are enucleated before

CT-*mos*   CN-*mos*

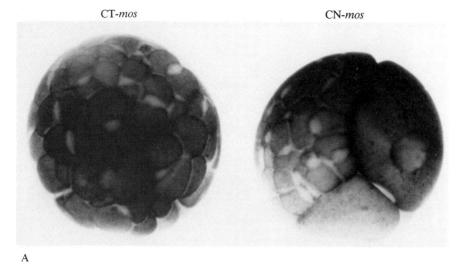

A

Control   Immunodepleted

B

Figure 3.51

Correlation between c-MOS and cytostatic factor (CSF). A, One blastomere of two-cell stage *Xenopus* embryos was injected with either truncated (non-functional; CT-*mos*) mRNA or functional (CN-*mos*) mRNA. The functional c-*mos* transcripts were translated into c-MOS protein, which arrested cleavage. Examination of the nuclei indicated that they had arrested at metaphase. B, Injection of cytoplasmic extracts of unfertilized eggs into one of two blastomeres at the two-cell stage arrests cleavage (left). Depletion of c-MOS protein with antibody eliminates the cleavage-arrest activity. (From N. Sagata et al., 1989b. Reprinted by permission from *Nature* (Lond.), *342*: 512–518. Copyright © 1989, Macmillan Magazines Limited.)

exposure to progesterone. The latter produce MPF and undergo an activation response similar to that of nucleated eggs (see Chap. 4), but they do not undergo normal cleavage. Conversely, inhibition of protein synthesis prevents maturation, an indication that the induction of maturation is mediated by mRNAs that are synthesized during oogenesis, stored in the cytoplasm, and translated in response to hormonal activation. In fact, the level of protein synthesis in *Xenopus* oocytes increases at maturation, but, in addition, certain mRNAs that accumulate during oogenesis are translated for the first time at maturation. On the contrary, other mRNAs, which are translated in oocytes, are not translated during maturation. Consequently, the mechanisms that control differential translation of these transcripts (McGrew et al., 1989) are extremely important in regulation of the progression and termination of oogenesis.

The Consequences of Germinal Vesicle Breakdown

The significance of GVBD and the resultant mixing of nucleoplasm and cytoplasm are apparent from experiments in which diploid blastula nuclei

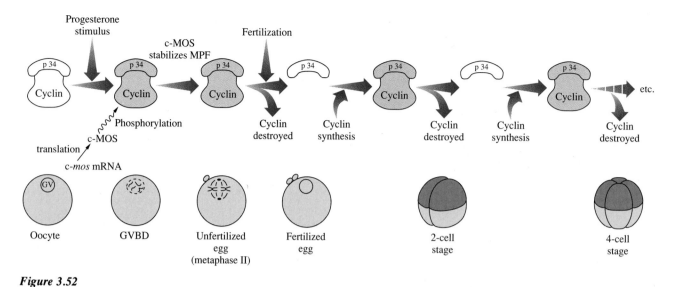

Figure 3.52

Summary of the effects of the molecules regulating oocyte maturation and mitosis in *Xenopus*.

are transplanted into *Rana pipiens* eggs after progesterone treatment. When blastula nuclei are transplanted into eggs enucleated *after GVBD*, a high percentage of the recipients will cleave and continue development. However, if blastula nuclei are transplanted into eggs enucleated *before progesterone treatment*, none of the eggs will undergo genuine cleavage after activation. Injection of germinal vesicle material into enucleated eggs before nuclear transfer can restore the ability of recipient eggs to cleave. Therefore, the cleavage restriction results solely from the absence of germinal vesicle components in the cytoplasm. Furthermore, because the germinal vesicle material used to restore the cleavage capability comes from unstimulated oocytes, the "cleavage factor" apparently pre-exists in the nucleus and is not synthesized as a result of hormonal stimulation (Smith and Ecker, 1969).

The events of maturation provide the egg with the physiological capability to become fertilized and initiate development. The acquisition of physiological maturity involves the utilization of information stored in the oocyte, by either translation or shifting components from nucleus to cytoplasm. As we see in Chapter 13, the tapping of this informational store during maturation samples only a small amount of a very large pool of information because the mature egg contains the complete program and machinery for cleavage, which is poised for utilization. Only the stimulus of fertilization is required to activate the program and utilize the machinery.

Ovulation

The mechanism of oocyte release from the mature follicle is poorly understood. One possibility is that cells surrounding the follicle contract, causing the follicle to burst and release the oocyte. For example, hamster follicles are surrounded by smooth muscle cells, whose contraction has been correlated with ovulation (Martin and Talbot, 1981; Talbot and Chacon, 1982). Another possibility is that the follicle ruptures because of degenerative changes that weaken the wall. The weakened wall ruptures, and the egg passively flows out of the follicle. The follicle could be weakened by intrinsic proteolytic enzymatic activity. There is evidence that ovulatory follicles of mammals have increased activity of the proteolytic enzyme **tissue-type**

plasminogen activator (tPA) (Reich et al., 1986a; Liu and Hsueh, 1987; Hsueh et al., 1988). Because proteolytic enzyme inhibitors block gonadotropin induction of ovulation (Reich et al., 1986a), it is possible that the activity of tPA is essential to liberate the oocyte from the follicle. The product of tPA activity is plasmin. The plasmin, in turn, may activate collagenase activity, which digests the meshwork of collagen fibers in the follicle and, thus, weakens the follicle (Reich et al., 1986b).

3–8 Egg Envelopes

Ovulated eggs are normally covered with some sort of noncellular egg envelopes. Egg envelopes are divided into the following three categories, based on their source: (1) **primary envelopes**, which are produced by the oocyte itself during oogenesis; (2) **secondary envelopes**, which are produced by the follicle cells that surround the oocyte; and (3) **tertiary envelopes**, which are added as the egg passes through the reproductive tract after ovulation. Unfortunately, the specific names given to egg envelopes cause a great deal of confusion because the same names are often given to envelopes of entirely different origin.

Examples of primary envelopes include the chorion of the fish egg, the jelly coat of the echinoderm egg, the vitelline envelope of the amphibian egg, and the zona pellucida of the mammalian egg. As we discussed in Section 3–1, the zona pellucida that surrounds the oocyte during oogenesis is penetrated by numerous microvilli from the surface of the oocyte and cytoplasmic processes from the follicle cells that make contact with the oocyte surface. During maturation, the cytoplasmic processes withdraw to facilitate subsequent ovulation. However, some of the surrounding follicle cells do not detach from the zona during maturation but remain associated with the egg to form a structure called the cumulus oophorus. This consists of an outer region containing a loose aggregate of cells and a single layer of elongated cells with fine processes that radiate toward the egg, which is called the **corona radiata**, lining the outer surface of the zona pellucida (Fig. 3.53). The cumulus cells are eventually sloughed from the zona, but in many species, they are still present at fertilization.

Although primary envelopes are produced during oogenesis, they may be subject to modification after ovulation. An example is the vitelline envelope of the *Xenopus* egg, which undergoes molecular changes during the transit of the eggs down the oviduct that are essential for the eggs to become fertilizable (Bakos et al., 1990).

Examples of secondary envelopes are the vitelline envelope and the chorion that surround insect eggs. Tertiary envelopes include such structures as the amphibian jelly coat and the albumin and shells of reptilian and avian eggs.

Egg envelopes serve various functions, such as protection and nutrition. They often provide formidable barriers through which sperm must penetrate. As we learned in Chapter 2, some types of sperm have evolved specializations that allow them to penetrate the envelopes and make contact with the egg at fertilization. Besides being barriers to sperm entry, the envelopes may also serve positive functions in fertilization (see Chap. 4).

3–9 Chapter Synopsis

The mature egg is the product of one of the most intricate developmental processes known. The constituents of the egg have both immediate and

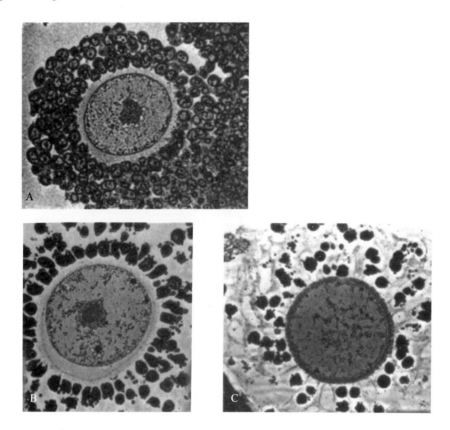

Figure 3.53

Rat cumulus-oocyte complex at various stages. A, Preovulatory complex in a
follicle. B, Preovulatory complex at a later stage. The cumulus is reduced to a
single layered corona radiata. C, Postovulatory ovum. Some corona cells remain
attached to the zona, but most have dissociated. ×420. (From N. B. Gilula, M. L.
Epstein, and W. H. Beers, 1978. Reprinted from *The Journal of Cell Biology*, 1978,
78: 58–75 by copyright permission of the Rockefeller University Press.)

long-term implications for reproduction. As a differentiated cell, the egg
contains specialized proteins and organelles that determine its functional
capabilities in fertilization. Unlike other differentiated cells, however,
fulfillment of its unique role is not a terminal function but the initial step
that leads to formation of a complete new individual.

The roles of the egg's dowry in development are far reaching. Stored
yolk fulfills the nutritional requirements of the embryo until it develops its
own mechanism for acquiring nutrients. The egg cytoplasm contains vast
amounts of messenger RNA as well as ribosomes, tRNA, and all the other
elements that are necessary for protein synthesis. These constituents not
only enable the embryo to begin immediate protein synthesis once development is activated but even determine which proteins will be made, until
the embryonic nuclei assume control over development. Furthermore, the
organization of egg constituents has significant implications for the *organization of the embryo*. The egg is a highly ordered cell, and the location of
individual components may determine the sites of specific embryonic regions.

Differentiation of the oocyte occurs while it is in prolonged meiotic
prophase. The maintenance of prophase and the termination of meiosis are
intriguing problems, and their investigation has led to the identification of

a number of factors that are involved in regulation of both the mitotic and meiotic cycles in a broad spectrum of eukaryotes. Another factor, c-MOS, has been identified whose expression is restricted to meiotic cells and whose synthesis triggers maturation.

References

Al-Mukhtar, K. A. K. 1970. Oogenesis in amphibia with special reference to the formation, replication and derivatives of mitochondria. Ph.D. thesis. Southhampton University, England.

Anderson, D. M. et al. 1982. Sequence organization of the poly (A) RNA synthesized in lampbrush chromosome stage *Xenopus laevis* oocytes. J. Mol. Biol., *155*: 281–309.

Anderson, E. 1968. Oocyte differentiation in the sea urchin, *Arbacia punctulata*, with particular reference to the origin of cortical granules and their participation in the cortical reaction. J. Cell Biol., *37*: 514–539.

Anderson, E., and D. F. Albertini. 1976. Gap junctions between oocyte and companion follicle cells in the mammalian ovary. J. Cell Biol., *71*: 680–686.

Anderson, E., and E. Huebner. 1968. Development of the oocyte and its accessory cells of the polychaete *Diopatra cuprea* (Bosc). J. Morphol., *126*: 163–197.

Arion, D. et al. 1988. *cdc2* is a component of the M phase-specific histone H1 kinase: Evidence for identity with MPF. Cell, *55*: 371–378.

Austin, C. R. 1965. *Fertilization*. Prentice-Hall, Englewood Cliffs, NJ.

Balinsky, B. I. 1975. *An Introduction to Embryology*, 4th ed. W. B. Saunders, Philadelphia.

Balinsky, B. I. 1981. *An Introduction to Embryology*, 5th ed. Saunders College Publishing, Philadelphia.

Barkos, M.-A., A. Kurosky, and J. L. Hedrick. 1990. Enzymatic and envelope converting activities of pars recta oviductal fluid from *Xenopus laevis*. Dev. Biol., *138*: 169–176.

Berleth, T. et al. 1988. The role of localization of *bicoid* RNA in organizing the anterior pattern of the *Drosophila* embryo. EMBO J., *7*: 1749–1756.

Bier, K. 1963. Autoradiographische Untersuchungen über die Leistugen des Follikelepethels und der Nährzellen bei der Dotterbildung und Eiwiessynthes im Fliegenover. Wilhelm Roux's Archiv für Entwicklungsmechanik, *154*: 552–575.

Biggers, J. D., and A. W. Schuetz (eds.). 1972. *Oogenesis*. University Park Press, Baltimore.

Bilinski, S. 1981. Ultrastructural studies on oogenesis in *Symphyla*. 2. Cortical granules and the chorion. Cell Tissue Res., *215*: 431–436.

Bleil, J. D., and P. M. Wassarman. 1980a. Structure and function of the zona pellucida: Identification and characterization of the proteins of the mouse oocyte's zona pellucida. Dev. Biol., *76*: 185–202.

Bliel, J. D., and P. M. Wassarman. 1980b. Mammalian sperm-egg interaction: Identification of a glycoprotein in mouse egg zona pellucida possessing receptor activity for sperm. Cell, *20*: 873–882.

Blondeau, J.-P., and E.-E. Baulieu. 1984. Progesterone receptor characterized by photoaffinity labelling in the plasma membrane of *Xenopus laevis* oocytes. Biochem. J., *219*: 785–792.

Bona, M., U. Scheer, and E. K. F. Bautz. 1981. Antibodies to RNA polymerase II (B) inhibit transcription in lampbrush chromosomes after microinjection into living amphibian oocytes. J. Mol. Biol., *151*: 81–99.

Bonnanfant-Jaïs, M.-L. and P. Mentré. 1983. Study of oogenesis in the newt *Pleurodeles waltlii* M. I. Ultrastructural study of the different stages of oocyte development. J. Submicrosc. Cytol., *15*: 453–478.

Brachet, J. 1974. *Introduction to Molecular Embryology*. Springer-Verlag, New York.

Brown, D. D., and I. B. Dawid. 1968. Specific gene amplification in oocytes. Science, *160*: 272–280.

Brown, D. D., and K. Sugimoto. 1973. 5S DNAs of *Xenopus laevis* and *Xenopus mulleri*: Evolution of a gene family. J. Mol. Biol., *48*: 397–415.

Callan, H. G. 1963. The nature of lampbrush chromosomes. Int. Rev. Cytol., *15*: 1–34.

Callan, H. G., J. G. Gall, and C. Murphy. 1988. The distribution of oocyte 5S, somatic 5S and 18S + 28S rDNA sequences in the lampbrush chromosomes of *Xenopus laevis*. Chromosoma, *97*: 43–54.

Callan, H. G., and L. Lloyd. 1975. Working maps of the lampbrush chromosomes of amphibia. *In* R. C. King (ed.), *Handbook of Genetics*, Vol. 4. Plenum Press, New York, pp. 57–77.

Coggins, L. W. 1973. An ultrastructural and radioautographic study of early oogenesis in the toad *Xenopus laevis*. J. Cell Sci., *12*: 71–93.

Coggins, L. W., and J. G. Gall. 1972. The timing of meiosis and DNA synthesis during early oogenesis in the toad, *Xenopus laevis*. J. Cell. Biol., *52*: 569–576.

Costantini, F. D., R. J. Britten, and E. H. Davidson. 1980. Message sequences and short repetitive sequences are interspersed in sea urchin poly (A)$^+$ RNAs. Nature (Lond.), *287*: 111–117.

Cran, D. G., R. M. Moor, and M. F. Hay. 1980. Fine structure of the sheep oocyte during antral follicle development. J. Reprod. Fertil., *59*: 125–132.

Cummings, M. R., N. M. Brown, and R. C. King. 1971. The cytology of the vitellogenic stages of oogenesis in *Drosophila melanogaster*. III. Formation of the vitelline membrane. Zeitschrift f. Zellforsch., *118*: 482–492.

Cummings, M. R., and R. C. King. 1969. The cytology of the vitellogenic stages of oogenesis in *Drosophila melanogaster*. I. General staging characteristics. J. Morphol., *128*: 427–442.

Cyert, M. S., and M. W. Kirschner. 1988. Regulation of MPF activity in vitro. Cell, *53*: 185–195.

Czolowska, R. 1972. The fine structure of the ''germinal cytoplasm'' in the egg of *Xenopus laevis*. Wilhelm Roux's Archiv für Entwicklungsmechanik, *169*: 335–344.

Danilchik, M. V., and J. C. Gerhart. 1987. Differentiation of the animal-vegetal axis in *Xenopus laevis* oocytes. I. Polarized intracellular translocation of platelets establishes the yolk gradient. Dev. Biol., *122*: 101–112.

Davidson, E. H. 1976. *Gene Activity in Early Development*, 2nd ed. Academic Press, New York.

Davidson, E. H. 1986. *Gene Activity in Early Development*, 3rd ed. Academic Press, Orlando, Florida.

Dawid, I. B. 1972. Cytoplasmic DNA. *In* J. D. Biggers and A. W. Schuetz (eds.), *Oogenesis*. University Park Press, Baltimore, pp. 215–226.

Dekel, N. et al. 1978. Cellular associations in the rat oocyte-cumulus cell complex: Morphology and ovulatory changes. Gamete Res., *1*: 47–57.

Denis, H., and M. Mairy. 1972. Recherches biochimiques sur l'oogenèse. 1. Distribution intracellulaire du RNA dans les petits oocytes de *Xenopus laevis*. Eur. J. Biochem., *25*: 524–534.

Diaz, M. O., and J. G. Gall. 1985. Giant readthrough transcription units at the histone loci on lampbrush chromosomes of the newt *Notophthalmus*. Chromosoma, *92*: 243–253.

DiMario, P. J., and A. P. Mahowald. 1987. *Female sterile (1) yolkless*: A recessive female sterile mutation in *Drosophila melanogaster* with depressed numbers of coated pits and coated vesicles within the developing oocytes. J. Cell Biol., *105*: 199–206.

Dixon, L. K., and P. J. Ford. 1982. Regulation of protein synthesis and accumulation during oogenesis in *Xenopus laevis*. Dev. Biol., *93*: 478–497.

Dolecki, G. J., and L. D. Smith. 1979. Poly(A)$^+$ RNA metabolism during oogenesis in *Xenopus laevis*. Dev. Biol., *69*: 217–236.

Draetta, G. et al. 1987. Identification of p34 and p13, human homologs of the cell cycle regulators of fission yeast encoded by *cdc2*$^+$ and *suc1*$^+$. Cell, *50*: 319–325.

Draetta, G. et al. 1989. cdc2 protein kinase is complexed with cyclin A and B: Evidence for inactivation of MPF by proteolysis. Cell, *56*: 829–838.

Draetta, G., and D. Beach. 1988. Activation of cdc2 protein kinase during mitosis in human cells: Cell-cycle dependent phosphorylation and subunit rearrangement. Cell, *54*: 17–26.

Dumont, J. N. 1978. Oogenesis in *Xenopus laevis* (Daudin). VI. The route of injected tracer transport in the follicle and developing oocyte. J. Exp. Zool., *204*: 193–218.

Dumont, J. N., and A. R. Brummett. 1978. Oogenesis in *Xenopus laevis* (Daudin). V. Relationships between developing oocytes and their investing follicular tissues. J. Morphol., *155*: 73–97.

Eckelbarger, K. J. 1979. Ultrastructural evidence for both autosynthetic and heterosynthetic yolk formation in the oocytes of an annelid (*Phragmatopoma lapidosa*: Polychaeta). Tissue Cell, *11*: 423–443.

Eppig, J. J. 1985. Oocyte–somatic cell interactions during oocyte growth and maturation in the mammal. *In* L. Browder (ed.), *Developmental Biology: A Comprehensive Synthesis*, Vol. 1. *Oogenesis*. Plenum Press, New York, pp. 313–347.

Evans, T. et al. 1983. Cyclin: A protein specified by maternal mRNA in sea urchin eggs that is destroyed at each cleavage division. Cell, *33*: 389–396.

Ford, P. J. 1971. Non-coordinated accumulation and synthesis of 5S ribonucleic acid by ovaries of *Xenopus laevis*. Nature (Lond.), *233*: 561–564.

Ford, P. J. 1976. Control of gene expression during differentiation and development. *In* C. F. Graham and P. F. Wareing (eds.), *The Developmental Biology of Plants and Animals*. W. B. Saunders, Philadelphia, pp. 302–345.

Ford, P. J., and E. M. Southern. 1973. Different sequences for 5S RNA in kidney cells and ovaries of *Xenopus laevis*. Nature (New Biol.), *241*: 7–12.

Franchi, L. L., A. M. Mandl, and S. Zuckerman. 1962. The development of the ovary and the process of oogenesis. *In* S. Zuckerman, A. M. Mandl, and P. Eckstein (eds.), *The Ovary*, Vol. 1. Academic Press, New York, pp. 1–88.

Freeman, R. S. et al. 1989. *Xenopus* homolog of the *mos* proto-oncogene transforms mammalian fibroblasts and induces maturation of *Xenopus* oocytes. Proc. Natl. Acad. Sci. U.S.A., *86*: 5805–5809.

Frohnhöfer, H. G., and C. Nüsslein-Volhard. 1986. Organization of anterior pattern in the *Drosophila* embryo by the maternal gene *bicoid*. Nature (Lond.), *324*: 120–125.

Frohnhöfer, H. G., and C. Nüsslein-Volhard. 1987. Maternal genes required for the anterior localization of *bicoid* activity in the embryo of *Drosophila*. Genes Dev., *1*: 880–890.

Garwood, P. R. 1981. Observations on the cytology of the developing female germ cell in the polychaete *Harmothoe imbricata*. Int. J. Invert. Reprod., *3*: 333–346.

Gautier, J. et al. 1988. Purified maturation-promoting factor contains the product of a Xenopus homolog of the fission yeast cell cycle control gene *cdc2*⁺. Cell, *54*: 433–439.

Gilula, N. B., M. L. Epstein, and W. H. Beers. 1978. Cell-to-cell communication and ovulation. A study of the cumulus-oocyte complex. J. Cell Biol., *78*: 58–75.

Godeau, J. F. et al. 1978. Induction of maturation in *Xenopus laevis* oocytes by a steroid linked to a polymer. Proc. Natl. Acad. Sci. U.S.A., *75*: 2353–2357.

Goebl, M., and B. Byers. 1988. Cyclin in fission yeast. Cell, *54*: 739–740.

Golden, L., U. Schafer, and M. Rosbash. 1980. Accumulation of individual pA⁺ RNAs during oogenesis of Xenopus laevis. Cell, *22*: 835–844.

Gremigni, V., and M. Nigro. 1983. An ultrastructural study of oogenesis in a marine triclad. Tissue Cell, *15*: 405–416.

Gremigni, V., and M. Nigro. 1984. Ultrastructural study of oogenesis in *Monocelis lineata* (Turbellaria: Proseriata). Int. J. Invertebr. Reprod. Dev., *7*: 105–118.

Gutzeit, H. O., and E. Huebner. 1986. Comparison of microfilament patterns in nurse cells of different insects with polytrophic and telotrophic ovarioles. J. Embryol. Exp. Morphol., *93*: 291–301.

Hallberg, R. L., and D. C. Smith. 1975. Ribosomal protein synthesis in *Xenopus laevis* oocytes. Dev. Biol., *42*: 40–52.

Heasman, J., J. Quarmby, and C. C. Wylie. 1984. The mitochondrial cloud of *Xenopus* oocytes: The source of germinal granule material. Dev. Biol., *105:* 458–469.

Highnam, K. C., and L. Hill. 1977. *The Comparative Endocrinology of the Invertebrates*, 2nd ed. Edward Arnold (Publishers) Ltd., London.

Honda, B. M., and R. G. Roeder. 1980. Association of a 5S gene transcription factor with 5S RNA and altered levels of the factor during cell differentiation. Cell, *22:* 119–126.

Hough-Evans, B. R. et al. 1979. RNA complexity in developing sea urchin oocytes. Dev. Biol., *69:* 258–269.

Hsueh, A. J. W. et al. 1988. Gonadotropin releasing hormone induces ovulation in hypophysectomized rats: Studies on ovarian tissue-type plasminogen activator activity, messenger ribonucleic acid content, and cellular localization. Endocrinology, *122:* 1486–1495.

Huebner, E. 1981. Oocyte-follicle cell interaction during normal oogenesis and atresia in an insect. J. Ultrastruct. Res., *74:* 95–104.

Huebner, E. 1984. The ultrastructure and development of the telotrophic ovary. *In* R. C. King and H. Akai (eds.): *Insect Ultrastructure*, Vol. 2. Plenum Press, New York, pp. 3–48.

Huebner, E., and H. Gutzeit. 1986. Nurse cell-oocyte association: A new F-actin mesh associated with the microtubule-rich core of an insect ovariole. Tissue Cell, *18:* 753–764.

Ikenishi, K., and M. Kotani. 1975. Ultrastructure of the "germinal plasm" in *Xenopus* embryos after cleavage. Dev. Growth Differ., *17:* 101–110.

Jaffe, L. F., and R. I. Woodruff. 1979. Large electrical currents traverse developing *Cecropia* follicles. Proc. Natl. Acad. Sci. U.S.A., *76:* 1328–1332.

Jamrich, M. et al. 1983. Transcription of repetitive sequences on *Xenopus* lampbrush chromosomes. Proc. Natl. Acad. Sci. U.S.A., *80:* 3364–3367.

Jamrich, M. et al. 1984. Histone RNA in amphibian oocytes visualized by *in situ* hybridization to methacrylate-embedded tissue sections. EMBO J., *3:* 1939–1943.

Johannsen, O. A., and F. H. Butt. 1941. *Embryology of Insects and Myriapods*. McGraw-Hill Book Co., New York.

Joho, K. E. et al. 1990. A finger protein structurally similar to TFIIIA that binds exclusively to 5S RNA in Xenopus. Cell, *61:* 293–300.

Kalt, M. R. 1973. Ultrastructural observations on the germ line of *Xenopus laevis*. Z. Zellforsch. Mikrosk. Anat., *138:* 41–62.

Karsch, F. J. 1987. Central actions of ovarian steroids in the feedback regulation of pulsatile secretion of luteinizing hormone. Ann. Rev. Physiol., *49:* 365–382.

Kessel, R. G. 1968. Annulate lamellae. J. Ultrastruct. Res. (Suppl.), *10:* 1–82.

Kessel, R. G. 1985. Annulate lamellae (porous cytomembranes): With particular emphasis on their possible role in differentiation of the female gamete. *In* L. Browder (ed.), *Developmental Biology: A Comprehensive Synthesis*, Vol. 1. *Oogenesis*. Plenum Press, New York, pp. 179–233.

Kessel, R. G. et al. 1986. Is the nuclear envelope a 'generator' of membrane? Developmental sequences in cytomembrane elaboration. Cell Tissue Res., *245:* 61–68.

King, M. L., and E. Barklis. 1985. Regional distribution of maternal messenger RNA in the amphibian oocyte. Dev. Biol., *112:* 203–212.

King, R. C. 1970. *Ovarian Development in* Drosophila melanogaster. Academic Press, New York.

King, R. C. 1972. *Drosophila* oogenesis and its genetic control. *In* J. D. Biggers and A. W. Schuetz (eds.), *Oogenesis*. University Park Press, Baltimore, pp. 253–275.

King, R. C. et al. 1985. Cytophotometric evidence for the transformation of oocytes into nurse cells in *Drosophila melanogaster*. Histochemistry, *82:* 131–134.

Klug, W. S., R. C. King, and J. M. Wattiaux. 1970. Oogenesis in the *suppressor²* *of hairy-wing* mutant of *Drosophila melanogaster*. II. Nucleolar morphology and *in vitro* studies of RNA protein synthesis. J. Exp. Zool., *174:* 125–140.

Koch, E. A., P. A. Smith, and R. C. King. 1967. The division and differentiation of *Drosophila* cystocytes. J. Morphol., *121*: 55–70.

Koch, E. A., and R. H. Spitzer. 1983. Multiple effects of colchicine on oogenesis in *Drosophila*: Induced sterility and switch of potential oocyte to nurse-cell developmental pathway. Cell Tissue Res., *228*: 21–32.

Kulakovsky, P. C., and W. H. Telfer. 1987. Selective endocytosis, *in vitro*, by ovarian follicles from *Hyalophora cecropia*. Insect Biochem., *17*: 845–858.

Labbé, J. C. et al. 1988. Activation at M-phase of a protein kinase encoded by a starfish homologue of cell cycle control gene cdc2$^+$. Nature, *335*: 251–254.

Larabell, C. A., and D. G. Capco. 1988. Role of calcium in the localization of maternal poly (A)$^+$ RNA and tubulin mRNA in *Xenopus* oocytes. Wilhelm Roux's Arch. Dev. Biol., *197*: 175–183.

Larkman, A. U. 1984. The fine structure of mitochondria and the mitochondrial cloud during oogenesis of the sea anemone *Actinia fragacea*. Tissue Cell, *16*: 393–404.

Lee, M. G. et al. 1988. Regulated expression and phosphorylation of a possible mammalian cell-cycle control protein. Nature (Lond.), *333*: 676–679.

Lehmann, R., and C. Nüsslein-Volhard. 1986. Abdominal segmentation, pole cell formation, and embryonic polarity require the localized activity of *oskar*, a maternal gene in Drosophila. Cell, *47*: 141–152.

Levenson, R. G., and K. B. Marcu. 1976. On the existence of polyadenylated histone mRNA in Xenopus laevis oocytes. Cell, *9*: 311–322.

Lillie, F. R. 1919. *The Development of the Chick*, 2nd ed. Henry Holt, New York.

Liu, Y.-X., and A. J. W. Hsueh. 1987. Plasminogen activator activity in cumulus oocyte complexes of gonadotropin-treated rats during periovulatory periods. Biol. Reprod., *36*: 1055–1062.

Lohka, M. J., M. K. Hayes, and J. L. Maller. 1988. Purification of maturation promoting factor, an intracellular regulation of early mitotic events. Proc. Natl. Acad. Sci. U.S.A., *85*: 3009–3013.

Lohka, M. J., J. L. Kyes, and J. L. Maller. 1987. Metaphase protein phosphorylation in *Xenopus laevis* eggs. Mol. Cell. Biol., *7*: 760–768.

Lohka, M. J., and J. L. Maller. 1985. Induction of nuclear envelope breakdown, chromosome condensation, and spindle formation in cell-free extracts. J. Cell Biol., *101*: 518–523.

Lohka, M. J., and Y. Masui. 1984. Effects of Ca^{2+} ions on the formation of metaphase chromosomes and sperm pronuclei in cell-free preparations from unactivated *Rana pipiens* eggs. Dev. Biol., *103*: 434–442.

Macgregor, H. C. 1977. Lampbrush chromosomes. *In* H. J. Li and R. A. Eckhardt (eds.), *Chromatin and Chromosome Structure*. Academic Press, New York, pp. 339–357.

Macgregor, H. C. 1980. Recent developments in the study of lampbrush chromosomes. Heredity, *44*: 3–35.

Mairy, M., and H. Denis. 1972. Recherches biochimiques sur l'oogenèse. 2. Assemblage des ribosomes pendant le grand accroissement des oocytes de *Xenopus laevis*. Eur. J. Biochem., *25*: 535–543.

Maller, J. L. 1985. Oocyte maturation in amphibians. *In* L. Browder (ed.), *Developmental Biology: A Comprehensive Synthesis*, Vol. 1. *Oogenesis*. Plenum Press, New York., pp. 289–311.

Marinos, E. 1985. The number of mitochondria in *Xenopus laevis* ovulated oocytes. Cell Differ., *16*: 139–143.

Martin, G. G., and P. Talbot. 1981. The role of follicular smooth muscle cells in hamster ovulation. J. Exp. Zool., *216*: 469–482.

Masui, Y. 1967. Relative roles of the pituitary, follicle cells, and progesterone in the induction of oocyte maturation in *Rana pipiens*. J Exp. Zool., *166*: 365–376.

Masui, Y., and C. L. Markert. 1971. Cytoplasmic control of nuclear behavior during meiotic maturation of frog oocytes. J. Exp. Zool., *177*: 129–146.

Matsumoto, Y. I. et al. 1980. Evidence for the involvement of H1 histone phosphorylation in chromosome condensation. Nature, *284*: 181–184.

McGrew, L. L. et al. 1989. Poly(A) elongation during *Xenopus* oocyte maturation is required for translational recruitment and is mediated by a short sequence element. Genes Dev., *3*: 803–815.

McLachlan, R. I. et al. 1987. Circulating immunoreactive inhibin levels during the normal human menstrual cycle. J. Clin. Endocrin. Metab., *65*: 954–961.

Meijer, L. et al. 1989. Cyclin is a component of the sea urchin egg M-phase specific histone H₁ kinase. EMBO J., *8*: 2275–2282.

Melton, D. A. 1987. Translocation of a localized maternal mRNA to the vegetal pole of *Xenopus* oocytes. Nature (Lond.), *328*: 80–82.

Miller, J., A. D. McLachlan, and A. Klug. 1985. Repetitive zinc-binding domains in the protein transcription factor IIIA from *Xenopus* oocytes. EMBO J., *4*: 1609–1614.

Miller, O. L., Jr., and B. R. Beatty. 1969. Portrait of a gene. J. Cell. Physiol., *74* (Suppl. 1): 225–232.

Miller, O. L., and B. A. Hamkalo. 1972. Visualization of RNA synthesis on chromosomes. Int. Rev. Cytol., *33*: 1–25.

Minshull, J., J. J. Blow, and T. Hunt. 1989. Translation of cyclin mRNA is necessary for extracts of activated Xenopus eggs to enter mitosis. Cell, *56*: 947–956.

Murray, A. W., and M. W. Kirschner. 1989. Cyclin synthesis drives the early embryonic cell cycle. Nature (Lond.), *339*: 275–280.

Murray, A. W., M. J. Solomon, and M. W. Kirschner. 1989. The role of cyclin synthesis and degradation in the control of maturation promoting factor activity. Nature (Lond.), *339*: 280–286.

Nüsslein-Volhard, C., H. G. Frohnhöfer, and R. Lehmann. 1987. Determination of anteriorposterior polarity in *Drosophila*. Science, *238*: 1675–1681.

Pelham, H. R. B., and D. D. Brown. 1980. A specific transcription factor that can bind either the 5S RNA gene or 5S RNA. Proc. Natl. Acad. Sci. U.S.A., *77*: 4170–4174.

Pelham, H. R. B., M. W. Wormington, and D. D. Brown. 1981. Related 5S RNA transcription factors in *Xenopus* oocytes and somatic cells. Proc. Natl. Acad. Sci. U.S.A., *78*: 1760–1764.

Perkowska, E., H. C. Macgregor, and M. L. Birnstiel. 1968. Gene amplification in the oocyte nucleus of mutant and wild-type *Xenopus laevis*. Nature (Lond.), *217*: 649–650.

Peter, M. et al. 1990. In vitro disassembly of the nuclear lamina and M phase-specific phosphorylation of lamins by cdc2 kinase. Cell, *61*: 591–602.

Picard, B. et al. 1980. Biochemical research on oogenesis. Composition of the 42S storage particles of *Xenopus laevis* oocytes. Eur. J. Biochem., *109*: 359–368.

Pinkerton, J. H. M. et al. 1961. Development of the human ovary—a study using histochemical technics. Obstet. Gynecol., *18*: 152–181.

Postlethwait, J. H., and F. Giorgi. 1985. Vitellogenesis in insects. *In* L. Browder (ed.), *Developmental Biology: A Comprehensive Synthesis*, Vol. 1. *Oogenesis*. Plenum Press, New York, pp. 85–126.

Quattropani, S. L., and E. Anderson. 1969. The origin and structure of the secondary coat of the egg of *Drosophila melanogaster*. Zeitschrift f. Zellforsch., *95*: 495–510.

Rebagliati, M. R. et al. 1985. Identification and cloning of localized maternal RNAs from Xenopus eggs. Cell, *42*: 769–777.

Reich, R., R. Miskin, and A. Tsafriri. 1986a. Follicular plasminogen activator: Involvement in ovulation. Endocrinology, *116*: 516–521.

Reich, R., A. Tsafriri, and G. L. Mechanic. 1986b. The involvement of collagenolysis in ovulation in the rat. Endocrinology, *116*: 522–527.

Reynhout, J. K., and L. D. Smith. 1974. Studies on the appearance and nature of a maturation-inducing factor in the cytoplasm of amphibian oocytes exposed to progesterone. Dev. Biol., *38*: 394–400.

Richter, J. D. et al. 1984. Interspersed poly(A) RNAs of amphibian oocytes are not translatable. J. Mol. Biol., *173*: 227–241.

Robinson, K. R. 1979. Electrical currents through full-grown and maturing *Xenopus* oocytes. Proc. Natl. Acad. Sci. U.S.A., *76*: 837–841.

Rogers, R. E., and L. W. Browder. 1977. Morphological observations on cultured lampbrush-stage *Rana pipiens* oocytes. Dev. Biol., *55*: 135–147.

Rosbash, M., and P. J. Ford. 1974. Polyadenylic acid-containing RNA in *Xenopus laevis* oocytes. J. Mol. Biol., *85*: 87–101.

Roy, L. M. et al. 1990. The cyclin B2 component of MPF is a substrate for the c-*mos*ˣᵉ proto-oncogene product. Cell, *61*: 825–831.

Ruderman, J. V., and M. L. Pardue. 1977. Cell-free translation analysis of messenger RNA in echinoderm and amphibian early development. Dev. Biol., *60*: 48–68.

Ruderman, J. V., H. R. Woodland, and E. A. Sturgess. 1979. Modulations of histone messenger RNA during the early development of *Xenopus laevis*. Dev. Biol., *71*: 71–82.

Sagata, N., K. Shiokawa, and K. Yamana. 1980. A study on the steady-state population of poly(A)$^+$ RNA during early development of *Xenopus laevis*. Dev. Biol., *77*: 431–448.

Sagata, N. et al. 1989a. The product of the *mos* proto-oncogene as a candidate "initiator" for oocyte maturation. Science, *245*: 643–646.

Sagata, N. et al. 1989b. The c-*mos* proto-oncogene product is a cytostatic factor responsible for meiotic arrest in vertebrate eggs. Nature (Lond.), *342*: 512–518.

Scheer, U., and M.-C. Dabauvalle. 1985. Functional organization of the amphibian oocyte nucleus. *In* L. Browder (ed.), *Developmental Biology: A Comprehensive Synthesis*, Vol. 1. Oogenesis. Plenum Press, New York, pp. 385–430.

Scheer, U. et al. 1976a. Classification of loops of lampbrush chromosomes according to the arrangement of transcriptional complexes. J. Cell Sci., *22*: 503–519.

Scheer, U., M. F. Trendelenburg, and W. W. Franke. 1976b. Regulation of transcription of genes of ribosomal RNA during amphibian oogenesis. A biochemical and morphological study. J. Cell Biol., *69*: 465–489.

Schuetz, A. W. 1967. Action of hormones on germinal vesicle breakdown in frog (*Rana pipiens*) oocytes. J. Exp. Zool., *166*: 347–354.

Schultz, L. D., B. K. Kay, and J. G. Gall. 1981. *In vitro* RNA synthesis in oocyte nuclei of the newt *Notophthalmus*. Chromosoma, *82*: 171–187.

Schultz, R. M., and P. M. Wassarman. 1977. Biochemical studies of mammalian oogenesis: Protein synthesis during oocyte growth and meiotic maturation. J. Cell Sci., *24*: 167–194.

Selman, K., and E. Anderson. 1975. The formation and cytochemical characterization of cortical granules in ovarian oocytes of the golden hamster (*Mesocricetus auratus*). J. Morphol., *147*: 251–274.

Senger, P. L., and R. G. Saacke. 1970. Unusual mitochondria of the bovine oocyte. J. Cell Biol., *46*: 405–408.

Shimizu, S., M. Tsuji, and J. Dean. 1983. *In vitro* biosynthesis of three sulfated glycoproteins of murine zonae pellucidae by oocytes grown in follicle culture. J. Biol. Chem., *258*: 5858–5863.

Smith, L. D., and R. E. Ecker. 1969. Role of the oocyte nucleus in physiological maturation in *Rana pipiens*. Dev. Biol., *19*: 281–309.

Smith, L. D., and R. E. Ecker. 1971. The interaction of steroids with *Rana pipiens* oocytes in the induction of maturation. Dev. Biol., *25*: 232–247.

Smith, L. D., R. E. Ecker, and S. Subtelny. 1968. *In vitro* induction of physiological maturation in *Rana pipiens* oocytes removed from their ovarian follicles. Dev. Biol., *17*: 627–643.

Smith, L. D., J. D. Richter, and M. A. Taylor. 1984. Regulation of translation during oogenesis. *In* E. H. Davidson and R. A. Firtel (eds.), *Molecular Biology of Development*. Alan R. Liss, New York, pp. 129–143.

Solomon, M. et al. 1988. Cyclin in fission yeast. Cell, *54*: 738–739.

Soules, M. R. et al. 1984. Progesterone modulation of pulsatile luteinizing hormone secretion in normal women. J. Clin. Endocrinol. Metab., *58*: 378–383.

Standart, N. et al. 1987. Cyclin synthesis, modification and destruction during meiotic maturation of the starfish oocyte. Dev. Biol., *124*: 248–258.

Stutz, F., E. Gouilloud, and S. G. Clarkson. 1989. Oocyte and somatic tyrosine tRNA genes in *Xenopus laevis*. Genes Dev., *3*: 1190–1198.

Swenson, K. I., K. M. Farrell, and J. V. Ruderman. 1986. The clam embryo protein cyclin A induces entry into M phase and the resumption of meiosis in Xenopus oocytes. Cell, *47*: 861–870.

Szöllösi, D. 1972. Changes of some cell organelles during oogenesis in mammals. *In* J. D. Biggers and A. W. Schuetz (eds.), *Oogenesis*. University Park Press, Baltimore, pp. 47–64.

Talbot, P., and R. S. Chacon. 1982. *In vitro* ovulation of hamster oocytes depends on contraction of follicular smooth muscle cells. J. Exp. Zool., *224*: 409–415.

Telfer, W. H., R. I. Woodruff, and E. Huebner. 1981. Electrical polarity and cellular differentiation in meroistic ovaries. Am. Zool., *21*: 675–686.

Van Blerkom, J., and P. Motta. 1979. *The Cellular Basis of Mammalian Reproduction.* Urban & Schwarzenberg, Baltimore-Munich.

van Dongen, W. et al. 1981. Quantitation of the accumulation of histone messenger RNA during oogenesis in *Xenopus laevis.* Dev. Biol., *86*: 303–314.

Villee, C. A. et al. 1989. *Biology,* 2nd ed. Saunders College Publishing, Philadelphia.

Vrana, K. E. et al. 1988. Mapping functional regions of transcription factor TFIIIA. Mol. Cell. Biol., *8*: 1684–1696.

Wall, D. A., and S. Patel. 1987. Multivesicular bodies play a key role in vitellogenin endocytosis by *Xenopus* oocytes. Dev. Biol., *119*: 275–289.

Wallace, R. A. 1985. Vitellogenesis and oocyte growth in nonmammalian vertebrates. *In* L. Browder (ed.), *Developmental Biology: A Comprehensive Synthesis,* Vol. 1. *Oogenesis.* Plenum Press, New York, pp. 127–177.

Wallace, R. A. et al. 1983. The oocyte as an endocytic cell. Ciba Found. Symp., *98*: 228–248.

Wassarman, P. M. 1990. Profile of a mammalian sperm receptor. Development, *108*: 1–17.

Wasserman, W. J., and Y. Masui. 1975. Effects of cycloheximide on cytoplasmic factor initiating meiotic maturation in *Xenopus* oocytes. Exp. Cell Res., *91*: 381–388.

Wasserman, W. J., and L. D. Smith. 1978a. Oocyte maturation: Nonmammalian vertebrates. *In* R. E. Jones (ed.), *The Vertebrate Ovary.* Plenum Publishing Corp., New York.

Wasserman, W. J., and L. D. Smith. 1978b. The cyclic behavior of a cytoplasmic factor controlling nuclear membrane breakdown. J. Cell Biol., *78*: R15-R22.

Weeks, D. L., and D. A. Melton. 1987. A maternal mRNA localized to the animal pole of *Xenopus* eggs encodes a subunit of mitochondrial ATPase. Proc. Natl. Acad. Sci. U.S.A., *84*: 2798–2802.

Woodruff, R. I. 1989. Charge-dependent molecular movement through intercellular bridges in *Drosophila* follicles. Biol. Bull., *176*: 71–78.

Woodruff, R. I., and W. H. Telfer. 1973. Polarized intercellular bridges in ovarian follicles of the cecropia moth. J. Cell Biol., *58*: 172–188.

Woodruff, R. I., and W. H. Telfer. 1980. Electrophoresis of proteins in intercellular bridges. Nature (Lond.), *286*: 84–86.

Wormington, W. M. 1988. Expression of ribosomal protein genes during *Xenopus* development. *In* L. Browder (ed.), *Developmental Biology: A Comprehensive Synthesis,* Vol. 5. *The Molecular Biology of Cell Determination and Cell Differentiation.* Plenum Press, New York, pp. 227–240.

Wormington, W. M. 1989. Developmental expression and 5S rRNA-binding activity of *Xenopus laevis* ribosomal protein L5. Mol. Cell. Biol., *9*: 5281–5288.

Wu, Z., C. Murphy, and J. G. Gall. 1986. A transcribed satellite DNA from the bullfrog *Rana catesbeiana.* Chromosoma (Berl.), *93*: 291–297.

Yamamoto, K., and I. Oota. 1967. An electron microscope study of the formation of the yolk globule in the oocyte of zebrafish, *Brachydanio rerio.* Bull. Fac. Fish. Hokkaido University, *17*: 165–174.

Yisraeli, J. K., and D. A. Melton. 1988. The maternal mRNA Vg1 is correctly localized following injection into *Xenopus* oocytes. Nature (Lond.), *336*: 592–595.

Yisraeli, J. K., S. Sokol, and D. A. Melton. 1990. A two-step model for the localization of maternal mRNA in *Xenopus* oocytes: Involvement of microtubules and microfilaments in the translocation and anchoring of Vg1 mRNA. Development, *108*: 289–298.

PART III

From Sperm and Egg to Embryo

Xenopus tadpoles with duplicated dorsal anterior structures produced by double centrifugation of eggs. (From Black, S. D., and J. C. Gerhart. 1986. *Dev. Biol. 116*: 228–240.)

CHAPTER *4*

Fertilization: The Activation of Development

Ernest E. Just (1883–1941). A professor at Howard University, E. E. Just was one of the first scientists to describe changes in the egg cortex that accompany fertilization in marine invertebrates. He also studied artificial parthenogenesis and was responsible for defining the relationship between the sperm entry point and the first cleavage plane in annelid embryos. (Photograph courtesy of the Marine Biological Laboratory Library, Woods Hole, MA.)

During gametogenesis, germ cells differentiate into sperm and eggs, which are exquisitely adapted for their subsequent roles in producing the next generation. Fertilization is the process in which sperm and egg unite to form a zygote, the predecessor of every cell in the embryo and adult. Fertilization has two important consequences. The first consequence is that the chromosome number is returned to the diploid value. This occurs when the maternal and paternal genomes combine. The second consequence of fertilization is the **activation of development**. Activation triggers a sequence of metabolic and morphological changes that result in the division of the zygote into a multicellular organism.

Much of our current understanding of fertilization and activation stems from a century or more of research on marine invertebrates, especially the sea urchin. Sea urchins are important in research on fertilization and activation because they produce large numbers of gametes and fertilization is external. Although fertilization occurs internally in mammals, they have recently joined the ranks of sea urchins in research on fertilization. Early investigations of mammalian fertilization were hampered because of the inability of sperm ejaculated directly from males to fertilize eggs *in vitro*. Eventually, it was found that in order to be competent in fertilization, mammalian sperm must undergo capacitation, a maturation step that occurs in the female reproductive tract. When sperm are collected from the female oviduct or uterus after copulation, or ejaculated sperm are bathed with fluid from these organs, they undergo capacitation and are thus competent to fertilize eggs. This discovery led to the development of *in vitro* fertilization methods for mammals (Austin, 1961; see Fig. 4.8A). Most of our information on mammalian fertilization is derived from studies conducted *in vitro* on sperm and eggs of the mouse, hamster, rabbit, and pig.

Because of their pre-eminence in the research literature, we primarily discuss fertilization in sea urchins and mammals. Although the general events of fertilization are similar in many different organisms, fertilization is a diverse process that is adapted to the different life-styles and habitats of individual species. Hence, this approach does not explore the diversity of fertilization that occurs in the animal kingdom.

4–1 Sperm and Egg Structure

Before beginning our discussion of fertilization, we return to the end of gametogenesis to review briefly the features of the sperm and egg that are important in fertilization, focusing on sea urchins and mammals.

Because sperm often travel long distances to meet an egg, they are usually streamlined into minute cells with a locomotory tail that is surmounted by a head containing the nucleus and acrosome. The sea urchin acrosome is a bulbous structure located at the anterior tip of the sperm head, whereas the mammalian acrosome forms a thin cap over the sperm nucleus that encompasses the entire anterior portion of the head. Acrosomes are membrane-enclosed organelles containing hydrolytic enzymes and other proteins. Before fertilization, the acrosome undergoes exocytosis, a process known as the acrosome reaction, releasing its contents to the outside of the sperm. The contents of the acrosome assist the sperm in penetrating the extracellular coats of the egg. Sea urchin sperm also show a specialized invagination of the nuclear membrane, the **anterior nuclear fossa**, which lies between the nucleus and acrosome. The anterior nuclear fossa contains globular actin (G-actin), which polymerizes into filamentous actin (F-actin) during the formation of the acrosomal process, a filamentous thread containing microfilaments that is extended from the anterior tip of the sperm head to the egg surface during fertilization. A plasma membrane encloses the sperm and contains a number of proteins that are important for fertilization.

The egg does not require a streamlined shape or locomotory organelles for moving long distances. Instead, it contains a stockpile of nuclear and cytoplasmic components that are used to build an embryo. It also possesses the physiological and morphological capacity to be fertilized by the sperm. The extracellular coat, the plasma membrane, and the peripheral cytoplasm, or cortex, of the egg each play important roles during fertilization. Sea urchin eggs contain an extracellular coat, or glycocalyx, consisting of two acellular layers. An outer layer, the jelly coat, contains several small peptides and large acidic polysaccharides. An inner layer, the vitelline envelope, is composed of a network of glycoprotein fibers. Mammalian eggs have an outer coat consisting of both cellular and acellular layers. In the mouse egg, for example, the corona radiata, which is a single-cell layer of follicle cells (see p. 115), is a formidable obstacle to sperm penetration. Beneath the corona radiata lies the zona pellucida, a thick meshwork of

Figure 4.1

A, Fertilization of a sturgeon egg by a sperm that has entered the micropyle and fused with a small extension of egg cytoplasm. Although a few sperm enter the micropyle, the narrow space only permits the first to fuse with the cytoplasmic extension. Contact and attachment of the sperm to the cytoplasmic extension appears to be required for the sperm to complete its journey through the micropyle. B, Attraction of sperm to the micropyle region of the isolated chorion of a herring egg. (A, Redrawn from Austin, 1965, after Ginsberg, 1959. Reprinted by permission of Prentice-Hall Inc., Englewood Cliffs, N.J. B, After Yanagimachi, 1957.)

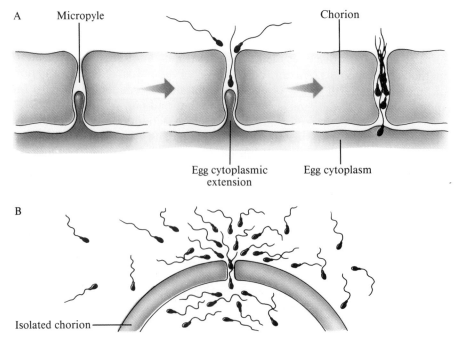

glycoprotein molecules—the mammalian counterpart to both the jelly coat and the vitelline envelope of sea urchin eggs.

Additional extracellular layers (chorions and eggshells) are deposited around the eggs of some organisms, either by the follicle cells or other somatic cells of the reproductive tract. In these organisms, sperm usually cannot penetrate these layers and must gain access to the egg through a **micropyle** (Fig. 4.1), which is a narrow channel in the chorion or eggshell through which sperm must swim to fertilize the egg.

The egg plasma membrane lies below the glycocalyx. In many eggs, the plasma membrane is folded into stout microvilli. Beneath the plasma membrane is the cortex, which has a more organized structure and is usually more viscous than the internal cytoplasm. It is also enriched in various granules, cytoskeletal networks, and endoplasmic reticulum. The largest and most abundant cortical organelle in sea urchin and mammalian eggs is the cortical granule. Cortical granules are membrane-enclosed vesicles with a unique internal structure (see Fig. 3.24) that undergo exocytosis at fertilization. They contain proteolytic enzymes, mucopolysaccharides, and structural proteins. Sea urchin eggs contain about 15,000 cortical granules, whereas mouse eggs contain about 4000 of these organelles. As we shall see, tremendous changes in the structure of the egg plasma membrane accompany cortical granule exocytosis during fertilization.

4–2 Setting the Stage for Fertilization: Events that Occur Before Sperm-Egg Fusion

Sperm must undergo a series of preparatory events to fertilize an egg successfully. First, using limited energy resources, they must travel relatively great distances to the site of fertilization. The activation of sperm motility must also be timed precisely so that it does not occur prematurely in the testes or semen. Second, sperm locomotion must be directed toward the site of fertilization. Third, once sperm reach the site of fertilization, they must be able to pass through the extracellular layers to fertilize the egg.

Activation of Sperm Motility

Sperm remain immotile for long periods of time in the testes or semen but exhibit immediate motility when they are released in the vicinity of eggs. As shown in Figure 4.2, the activation of sea urchin sperm is controlled by pH (Christen et al., 1983; Bibring et al., 1984; Clapper et al., 1985). A high

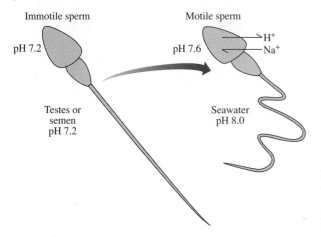

Figure 4.2

The role of intracellular pH in activation of sea urchin sperm motility. The pH values shown adjacent to sperm heads are intracellular values.

CO_2 tension in semen keeps the intracellular pH of testicular sperm at about 7.2. At this pH, cell respiration and motility do not occur. When sperm enter seawater, which has a relatively low CO_2 tension and a high pH (about 8.0), a coupled exchange of internal H^+ for external Na^+ is initiated, which quickly raises the intracellular pH to about 7.6. At this pH, respiration is activated, and sperm begin to move by beating their flagella. The importance of this change in pH in controlling sperm motility is shown by an experiment in which locomotion is examined after sperm are diluted into seawater lacking Na^+. Under these conditions, there is no Na^+/H^+ exchange, the intracellular pH remains at a level that prevents respiration, and sperm are immotile. However, if either Na^+ or NH_4OH (which also raises the intracellular pH; see Sect. 4–5) is added to immotile sperm in Na^+-free seawater, motility begins immediately. The egg jelly coat also has a role in controlling sperm motility due to the presence of **speract** (sperm-activating peptide), a 10 amino acid peptide that stimulates an increase in internal pH (Hanesbrough and Garbers, 1981). The pH change regulates sperm locomotion by activating dynein, an ATPase that is bound to the surface of microtubules in the flagellar axoneme (see p. 32). Dynein activity is very sensitive to pH (Fig. 4.3). At pH 7.2, dynein is inactive, no ATP is produced, and sperm are immotile. When the internal pH rises to 7.6, dynein is activated, and sperm begin to swim.

Compared to sea urchins, little is known about the initial activation of sperm motility in mammals. As discussed in Chapter 2, sperm maturation in the epididymis facilitates the acquisition of motility. Presumably, motility is triggered by the ejaculation of semen into the female reproductive tract. Sperm gradually become more motile after they enter the female reproductive tract (Yanagimachi, 1970). Mammals may be able to afford a gradual activation of sperm motility, because contractions of the female reproductive tract help conduct sperm to the site of fertilization in the oviduct. When sperm eventually reach the vicinity of the egg, there is an additional increase in motility. This is caused by capacitation, which will be discussed further when we examine sperm penetration through the egg surface coats.

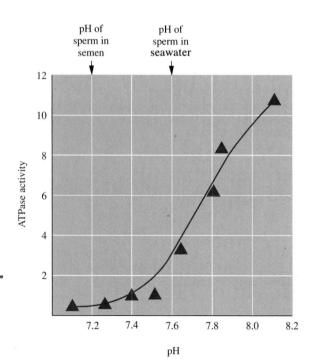

Figure 4.3

The dependence of sperm ATPase (primarily dynein ATPase) activity on the pH of the medium. After permeabilization by detergent extraction, sperm were diluted into media of different pH and assayed for ATPase activity. ATPase activity increases sharply between pH 7.2 and 7.6. (After Christen et al., 1983.)

Attraction of Sperm to the Egg

Activation of motility does not ensure that sperm will find an egg, especially when they are released into the vast expanse of an ocean, a tidepool, a pond, or even the female reproductive tract. Organisms that exhibit external fertilization have evolved specific mechanisms to attract sperm to eggs. **Chemotaxis** is defined as the locomotion of one cell toward the source of a chemical substance secreted by another cell. Chemotactic attraction of sperm to eggs is common in plants. In the seaweed *Fucus serratus*, a large number of sperm and eggs are shed into the ocean. Sperm are attracted to a specific hydrocarbon molecule, fucoserraten, which is secreted by the eggs (Müller and Jaenicke, 1973; Müller and Gassmann, 1978). Fucoserraten can be collected for experimental purposes because it is released into seawater by eggs. Placing fucoserraten in a small capillary tube and touching the tube to a suspension of sperm elicits sperm chemotaxis. After a short period of time, a cloud of sperm is attracted to the tip of the capillary tube containing fucoserraten.

Sperm chemotaxis can also occur in animals (Miller, 1977). The eggs of insects and fishes are enclosed within a chorion, a sperm-resistant extracellular coat that is elaborated during oogenesis. Sperm can enter the chorion only through the micropyle. In fish eggs, substances near the entrance of the micropyle serve as sperm attractants. These substances cause sperm to locomote rapidly in the vicinity of the micropyle. Eventually, the hyperactive sperm swim through the opening in the chorion and make contact with a thin extension of cytoplasm that protrudes into the micropyle from the surface of the egg (see Fig. 4.1A). To examine the role of the micropyle in sperm attraction, sperm were added to chorions that were dissected free of eggs. Despite the absence of an egg, sperm accumulated in the region of the micropyle and entered the opening into the empty chorion (see Fig. 4.1B). If the micropyle region is cut out of the chorion, sperm are still attracted to this region, but not to the portion of the chorion that lacks the micropyle. Thus, the substances that attract sperm must be localized in or near the micropyle.

Sperm chemotaxis also occurs in jellyfish eggs (Carre and Sardet, 1981; Cosson et al., 1986). The substance that attracts sperm is a protein localized in the **cupule**, a specialized extracellular structure surrounding the animal pole of the jellyfish egg (Fig. 4.4). This protein causes nearby sperm to alter their swimming behavior so that they fertilize the egg in the region of the cupule.

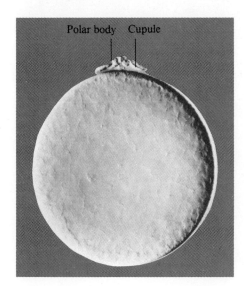

Polar body   Cupule

Figure 4.4

Unfertilized egg of the jellyfish *Muggiaea kochi* showing the cupule, which contains the sperm attractant protein, and the polar bodies at the animal pole. (From Carre and Sardet, 1981.)

For years, it was debated whether sea urchins exhibit sperm chemotaxis, but it now appears that a sperm attractant is present in the egg jelly coat. **Resact** (sperm respiratory activating peptide), a peptide consisting of only 14 amino acids, has been isolated from the jelly coat of sea urchin eggs. When a small amount of resact is injected into a drop of randomly distributed sperm, the sperm change their swimming behavior and accumulate within seconds at the position where resact was injected (Fig. 4.5). This experiment shows that resact is a sperm attractant.

Sperm Penetration of the Egg Surface Coats

Before a sperm can fuse with an egg, it must pass through the egg surface coats. The surface coats protect the egg and ensure species specificity in sperm-egg interactions. Sea urchin and mammalian eggs are not surrounded by a chorion. However, they do possess surface coats that must be penetrated before sperm can reach the plasma membrane.

Sperm Penetration in Mammals

Mammalian sperm require a period of maturation in the female reproductive tract (i.e., capacitation; Austin, 1952) before they can undergo the acrosome reaction and are capable of fertilizing the egg. Capacitation periods vary from less than 1 hour in the mouse to as much as 5 to 6 hours in the human. Depending on the species, capacitation occurs either in the uterus, the oviduct, or both. Capacitation involves alterations in the sperm plasma membrane, including removal of sperm surface components and rearrangement of intramembranous particles, but the mechanisms involved remain poorly understood. Capacitated sperm show greater respiratory activity and motility than noncapacitated sperm, which presumably assists their penetration through the egg surface coats.

Once sperm are capacitated, they are able to undergo the acrosome reaction. The acrosome reaction of mammals is initiated as sperm proceed through the cumulus oophorus or corona radiata and approach the zona pellucida. During the acrosome reaction, the outer portion of the acrosomal

Figure 4.5

The effect of resact on sea urchin sperm motility. Bright areas in the photographs represent regions of active sperm movement. Photographs were taken (A) before resact introduction (the outline of an injection pipet loaded with resact is shown), and (B) 20, (C) 40, (D) 50, (E) 70, and (F) 90 seconds after introduction of resact. Motile sperm accumulate at the position where resact was introduced. Scale bar equals 200 μm. (From Ward et al., 1986. Reprinted from *The Journal of Cell Biology*, 1986, *101*: 2324–2329 by copyright permission of the Rockefeller University Press.)

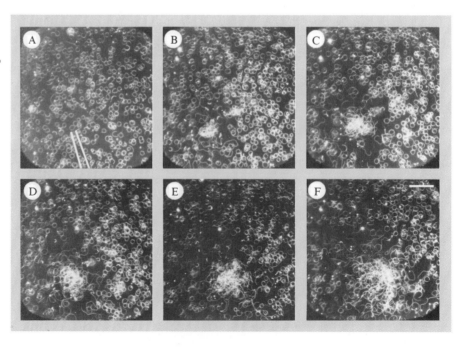

A

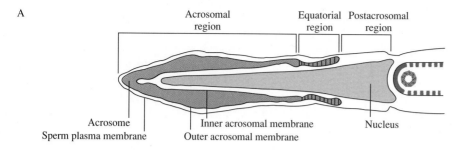

Acrosomal region   Equatorial region   Postacrosomal region

Acrosome
Sperm plasma membrane   Inner acrosomal membrane   Nucleus
Outer acrosomal membrane

B

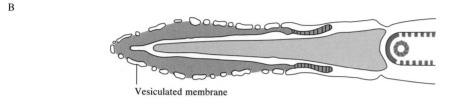

Vesiculated membrane

C

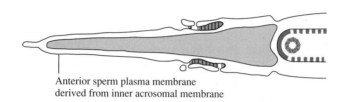

Anterior sperm plasma membrane
derived from inner acrosomal membrane

Figure 4.6

Drawing of the acrosome reaction in hamster sperm. A, A sperm head before the acrosome reaction showing an intact outer acrosomal membrane. B, A sperm head during the acrosome reaction showing vesiculation of the outer acrosomal membrane that results in release of the acrosomal contents. C, A sperm head after the acrosomal reaction showing the anterior sperm plasma membrane derived from the inner acrosomal membrane. (After Yanagimachi and Noda, 1970.)

membrane fuses with the plasma membrane, vesiculates, and disappears. In contrast, the region of the acrosomal membrane adjacent to the nucleus (inner acrosomal membrane) remains intact, reverses its orientation, and becomes part of the plasma membrane (Fig. 4.6). During exocytosis, soluble enzymes stored in the acrosome, including hyaluronidase (an enzyme that hydrolyzes hyaluronic acid, a cell surface polysaccharide) and corona-penetrating enzyme, escape into the region surrounding the sperm. These enzymes loosen the connections between cumulus oophorus or corona radiata cells and thereby help the sperm to reach the zona pellucida. Mammalian sperm penetrate the zona pellucida tangentially with respect to the egg surface (Figs. 4.7 and 4.8A), leaving behind a clear zone that is about the diameter of a single sperm head. The mechanism used by mammalian sperm to penetrate the zona pellucida is controversial. Some investigators think that sperm penetrate this layer solely by the propulsive force of their swimming (Green and Purves, 1984), whereas others believe that sperm digest their way through the zona using **acrosin**, a protease that is exposed on the sperm head when the inner acrosomal membrane becomes inverted (McRorie and Williams, 1974).

The zona pellucida plays an important role in restricting interspecific fertilization in mammals (Gwatkin, 1977). For example, when capacitated sperm of one species are placed near eggs of another species, fertilization will not occur unless the zona has first been removed from the eggs. This experiment suggests that sperm-egg binding may be mediated by species-specific sperm receptors in the zona pellucida. The zona pellucida of mouse eggs contains three glycoproteins, called ZP (zona pellucida protein)-1, ZP-2, and ZP-3 (Bleil and Wassarman, 1980). Zona pellucida protein-3 is a

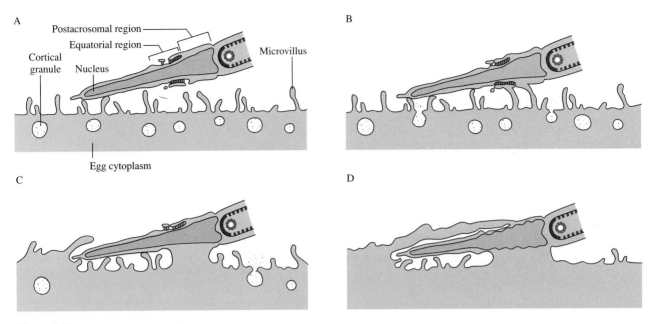

Figure 4.7

Drawings representing stages of tangential entry of sperm into the hamster egg.
(After Yanagimachi and Noda, 1970.)

sperm receptor (Bleil and Wassarman, 1986; Wassarman, 1990). This can be demonstrated by an experiment in which isolated ZP-3 is attached to glass beads, and the beads are incubated with sperm. The result is that sperm bind tightly to the ZP-3–coated beads (see Fig. 4.8B). If a ZP-3 antibody is added to the ZP-3–coated beads before the addition of sperm, however, sperm binding is prevented, showing that binding to the beads is dependent on the presence of ZP-3. The ability of ZP-3 to bind sperm is a function of its associated sugar chains. When these sugars are removed from ZP-3 by enzymatic digestion, sperm binding to the zona pellucida is blocked. ZP-2 is responsible for the maintenance of sperm binding to the egg; the function of ZP-1 is currently unknown.

An influx of Ca^{2+} into the sperm is required to initiate the acrosome reaction in mammals. This was shown by treating eggs with the ionophore

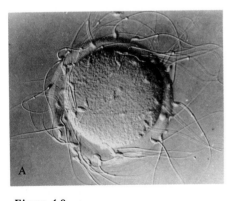
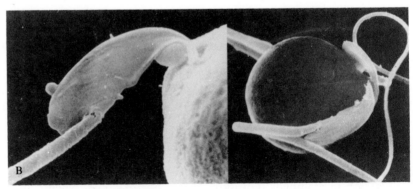

Figure 4.8

A, *In vitro* fertilization of a mouse egg showing many capacitated sperm binding to the zona pellucida. Only one sperm will penetrate the egg (inner sphere). B, Mouse sperm binding to a glass bead coated with the sperm receptor ZP-3. Right: entire bead with a single bound sperm. Left: close-up of sperm head bound to the surface of a glass bead. The glass bead is approximately 10 μm in diameter. (From Wassarman, 1990. Reprinted by permission of Company of Biologists, Ltd.)

A23187 (a molecule that inserts into the egg plasma membrane and transports Ca^{2+} into the cytoplasm) or purified ZP-3 (Wassarman, 1987). Thus, the sperm receptor ZP-3 may also function to cause the acrosomal reaction. Whereas the function of ZP-3 as a sperm receptor is mediated by its sugar chains, the ability of ZP-3 to initiate the acrosome reaction is due to its protein portion (Florman and Wassarman, 1985).

Sperm Penetration in Sea Urchins

The sperm of sea urchins and most other marine invertebrates do not require capacitation to promote the acrosome reaction. In sea urchins, two events are initiated when sperm contact the jelly coat (Fig. 4.9). First, the acrosome reaction causes the fusion of the acrosomal and sperm plasma membranes. Second, the acrosomal process, a thin protrusion, is extended from the sperm head to the vitelline envelope.

Stimulation of the acrosome reaction by the jelly coat is demonstrated by an experiment in which the acrosome reaction occurs in the absence of eggs after sperm are treated with dissolved jelly components (Dan et al., 1964). When sperm contact the jelly coat, they go through a sequence of ionic changes. First, there is a coupled exchange of external Ca^{2+} for internal K^+. A large sulfated sugar molecule in the egg jelly appears to be responsible for triggering the Ca^{2+}/K^+ exchange and subsequent acrosome-plasma membrane fusion (SeGall and Lennarz, 1979). The binding of the sulfated sugar to a glycoprotein receptor in the sperm plasma membrane also stimulates a 400-fold increase in cyclic AMP in the cytoplasm (Garbers et al., 1983; Schackmann et al., 1984). As will be discussed later, cyclic AMP stimulation is the first step in a sequence of reactions that remodel the sperm chromatin during fertilization.

The accumulation of Ca^{2+} in sea urchin sperm controls the fusion of the acrosomal and sperm membranes. This has been shown by the following experiments. First, acrosome reaction induction by the jelly coat requires Ca^{2+} in the external medium (Dan, 1954). Second, when isolated sperm are exposed to ionophore A23187, the acrosomal membrane fuses with the sperm membrane (Decker et al., 1976). Third, when sperm are treated with

Regular text continues on p. 138.

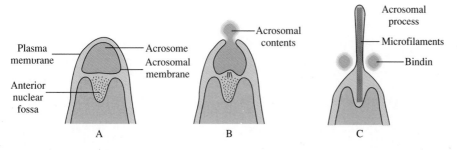

Figure 4.9

The acrosome reaction in sea urchin sperm. A, A sperm head before the beginning of the acrosome reaction showing an intact acrosome completely surrounded by the sperm plasma membrane. The depression at the tip of the nucleus is the anterior nuclear fossa. B, A sperm head at the beginning of the acrosome reaction showing the partial fusion of the plasma membrane with the acrosome membrane and release of the acrosomal contents, which include bindin. The acrosomal process is beginning to elongate by polymerization of microfilaments in the region of the anterior nuclear fossa and reversion of the remaining basal portion of the acrosomal membrane. C, A sperm head at a later stage in the acrosome reaction, in which the acrosome has disappeared and the acrosomal process has elongated further, exposing bindin at its base. (After Longo, 1987.)

Essay

Advances in the Regulation of Human Fertility: Immunocontraception and in vitro Fertilization

The steroid contraceptive pill, which inhibits ovulation, and the diaphragm and condom, which are physical barriers to sperm-egg contact, have been the mainstays of fertility control for several decades. However, recent research has focused on the possibility of perfecting immunological means of contraception. Considerable attention has also been directed toward finding means to allow women with fertility disorders to conceive and bear children. Spectacular results in this regard have been reported using in vitro fertilization techniques. Here we discuss progress in these two exciting areas of human fertility research.

Successful fertilization of the human egg is dependent on binding of sperm to the zona pellucida, a process mediated by complementary receptors on the sperm plasma membrane and the zona surface. As shown in the figures, exposure of unfertilized mouse eggs to antibodies raised against zona pellucida components in vitro will inhibit the binding of sperm to the zona (Aitken and Richardson, 1981). These results have encouraged the development of immunological approaches to contraception using the zona as a target. Indeed, some naturally occurring fertility disorders in humans may be due to the presence of zona antibodies (Shivers and Dunbar, 1977; Mori et al., 1978). It has been shown that antibodies raised against zona components of rabbits, marmosets, and pigs will prevent human sperm from binding to, and penetrating, the human zona pellucida (Henderson et al., 1987a, b). Thus, immunization of women with these zona antigens might elicit antibodies that would be directed at their zonae and prevent fertilization. Widespread use of this procedure would be facilitated if zona immunity were reversible.

Preliminary results from immunization studies with experimental animals have been encouraging. Two approaches have been used: passive and active immunization. Passive immunization involves injection of serum containing antibodies. For example, passive immunization of mice with rabbit antisera against zona pellucida antigens induces temporary infertility (Tsunoda and Chang, 1978). Active immunization involves injection of zona antigens into the target female, who would then produce antibodies. This procedure is most effective when the zona antigen is from another species. For example, mice immunized with solubilized hamster zonae become temporarily infertile; after sufficient reduction of the antibody titer, the formerly infertile mice become capable of delivering normal young (Gwatkin et al., 1977). Similar results for both passive and active immunization have been obtained in a variety of mammalian species (Aitken et al., 1981). Although the assumption inherent in these experiments has been that the antibodies inhibit fertilization, recent evidence suggests that the antibodies may be affecting ovarian function rather than (or in addition to) fertilization (Wood et al., 1981). Further investigation is necessary to establish the mode of reproductive dysfunction in immunized females.

Another possibility for immunocontraception is to induce immunity to the sperm. In vitro studies have shown that Fab fragments of rabbit anti-hamster sperm antibodies will inhibit hamster sperm binding to, and passage through, the zona, thus interfering with fertilization (Tzartos, 1979). This and similar results with anti-sperm antibodies (Lopo and Vacquier, 1980; Hamilton and Vernon, 1987) suggest the possibility of passive immunization with such antisera or of active immunization against sperm as effective strategies for contraception. Recently, 100% effective contraception has been obtained in male and female guinea pigs immunized with PH-20, a sperm surface protein that is essential for sperm adherence to the zona pellucida (Primacoff et al., 1988). The contraceptive effect was long-lasting and reversible. These results provide convincing evidence for the feasibility of developing an effective contraceptive vaccine based on sperm-specific antigens.

Successful application of the techniques of in vitro fertilization and implantation of the embryo in the mother's uterus has made the "test-tube human baby" a reality (Edwards, 1981). These procedures were perfected by Edwards and Steptoe in the United Kingdom and have since been introduced into other countries where fertility clinics have been established for the primary purpose of helping women with occluded oviducts to conceive and bear children. Preovulatory oocytes are aspirated from their follicles by the technique of laparoscopy. For the success of this procedure, it is important to predict with a high degree of accuracy the presence of preovulatory oocytes in the ovary. In the natural menstrual cycle, the approach of ovulation can be monitored by measuring estrogen levels, which assesses follicle growth, and the onset of the luteinizing hormone (LH) surge, which indicates that ovulation is imminent. Alternatively, gonadotropin-releasing hormone or the drug clomiphene can be used to regulate follicular growth and ovulation. This approach has the advantage that several follicles will develop during the cycle. Oocytes can be collected simultaneously from each of these follicles.

After collection, the oocytes are cultured to allow oocyte maturation to be completed. Mature eggs are then fertilized in droplets of medium under oil, which prevents evaporation, or in small culture tubes. A small amount of semen is introduced into the medium containing the egg, and the zygote is allowed to develop to the hatched blastocyst stage in a culture tube before it is implanted through the cervix into the uterus with a catheter. Current research is underway to improve the success of implantation, which is the most unpredictable step in this whole procedure.

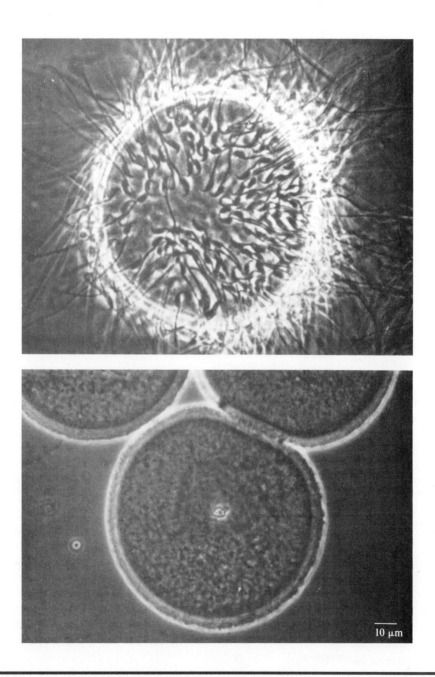

Figure 1

Sperm binding to the zona pellucida of an unfertilized mouse egg (top) and inhibition of binding after incubation of unfertilized eggs in the presence of anti-zona antibody (bottom). (From Aitken and Richardson, 1981.)

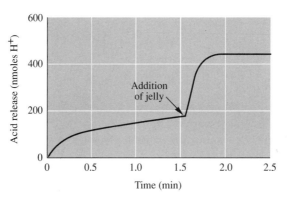

Figure 4.10

Acid release from sperm induced by the addition of egg jelly. Sperm were added to seawater and allowed to approach a steady state level of acid release. At this time, jelly was added, and the resulting shift in pH was followed. (After Schackmann et al., 1978.)

D600 or verapamil, drugs that inhibit Ca^{2+} transport into the cell, they cannot undergo acrosome-plasma membrane fusion (Schackmann et al., 1978). Fourth, radioactively labeled Ca^{2+} has been shown to enter sea urchin sperm when they contact the jelly coat (Schackmann et al., 1978).

The extension of the acrosomal process and its attachment to the vitelline envelope is also controlled by ionic changes mediated by sperm contact with the jelly coat. When sperm contact the jelly coat, there is another ionic flux: a coupled exchange of internal H^+ for external Na^+. The release of H^+ can be demonstrated by treating sperm with egg jelly and monitoring the reduction in external pH (Fig. 4.10). This escape of H^+ from the cell causes a second increase in intracellular pH (the first was associated with the release of sperm into seawater; see Fig. 4.2). This rise in intracellular pH triggers the extension of the acrosomal process by inducing G-actin stored in the anterior nuclear fossa to polymerize into F-actin (Tilney et al., 1978). Water rushes into the sperm during the acrosome reaction, suggesting that acrosomal process extension is also controlled by an increase in hydrostatic pressure.

As the acrosomal process is extended from the anterior tip of the sperm head toward the surface of the egg, it is coated by proteins originally associated with the inner side of the inverted acrosomal membrane (see Figs. 4.9C and 4.11). One of these proteins is bindin, which promotes the species-specific attachment of the acrosomal process to the vitelline envelope (Vacquier and Moy, 1977). Fertilization in sea urchins is usually species-specific. If the vitelline envelope is first removed from the egg, however, the formation of interspecific hybrids can result. The association of bindin with a sperm receptor in the egg vitelline envelope appears to account for species specificity of fertilization (Glabe and Vacquier, 1978). If sperm receptors are purified from different species of sea urchins, they each bind to sperm of their own species but not to sperm of other species (Rossignol et al., 1981). Bindin-like molecules have not been found in mammalian sperm.

After the acrosomal process attaches to the vitelline envelope, the remainder of the sperm also binds to and penetrates this layer (Fig. 4.12). Sperm penetration of the vitelline envelope is accomplished by the release of a specific chymotrypsin-like protease from the acrosome. A role for this protease in digesting the vitelline membrane was shown by experiments in

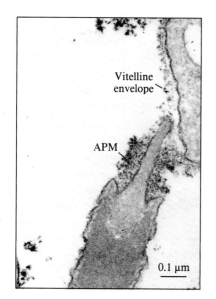

Figure 4.11

Binding of sand dollar sperm to the egg vitelline envelope. Material released from the acrosome (APM) coats the acrosomal process and the vitelline envelope. (From Summers and Hylander, 1974.)

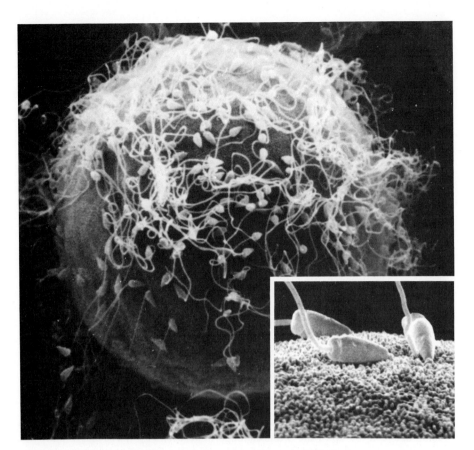

Figure 4.12

Scanning electron micrograph showing sperm bound to the vitelline envelope of the sea urchin egg. × 1900. Inset shows a close-up of the vitelline envelope with several sperm bound by their acrosomal processes. × 10,200. The vitelline envelope follows the contours of the egg surface, which is covered with microvilli. (From Epel, 1977.)

which an inhibitor of chymotrypsin-like enzymes prevented sperm penetration of the vitelline membrane (Green and Summers, 1982).

Summary of Sperm Penetration Mechanisms

Sperm interact with activator and receptor molecules before and during their penetration of the egg glycocalyx. In sea urchins, the activators are small peptides and sulfated sugars located in the jelly coat, and the receptors are glycoproteins in the vitelline envelope. In mammals, the sperm activator and receptor is a single molecule: the glycoprotein ZP-3. These activator and receptor molecules trigger ionic changes that are responsible for the acrosome reaction and species-specific sperm penetration of the extracellular coats.

4–3 Fusion of Sperm and Egg

Once the sperm has penetrated the extracellular layers of the egg, its plasma membrane fuses with the egg plasma membrane. In sea urchins, fusion begins at the tip of the acrosomal process (Fig. 4.13), whereas in mammals, which have no acrosomal process, fusion begins in the equatorial region of the sperm head immediately behind the region of the membrane that

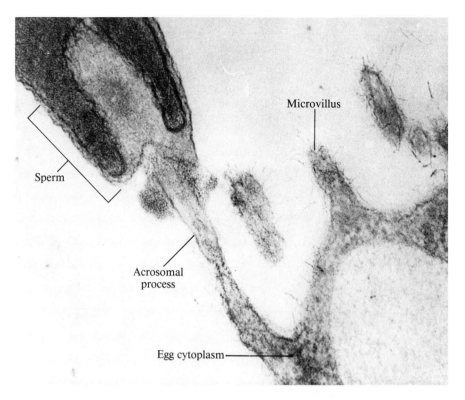

Figure 4.13

Fusion between the tip of the sea urchin sperm acrosomal process and a microvillus on the surface of the egg. The fusion results in the formation of a cytoplasmic bridge, through which the sperm enters the egg. × 54,600. (From Epel, 1977.)

originates from the acrosome (see Fig. 4.7). In either case, fusion results in the plasma membranes of the egg and sperm becoming continuous and forming a cytoplasmic bridge, which gradually widens, allowing the sperm nucleus to enter the egg cytoplasm.

Evidence obtained from experiments in which purified bindin induced the fusion of phospholipid vesicles *in vitro* suggests that bindin may have two roles in fertilization (Glabe, 1985). One role is to facilitate sperm-egg adhesion, and the other is to mediate cell fusion. Usually the entire sperm, including the nucleus, centrioles, mitochondria, and even the flagellar axoneme, enters the egg cytoplasm. The original sperm plasma membrane is also retained, becoming part of the egg plasma membrane (Gabel et al., 1979; Shapiro et al., 1981).

After the sperm fuses with the egg, the **fertilization cone**, an extension of egg cytoplasm, forms around the entering sperm head (Fig. 4.14A). The fertilization cone eventually engulfs the sperm. Microfilaments in the fertilization cone are responsible for drawing the sperm into the egg. This is shown by experiments with cytochalasin B, an inhibitor of microfilament formation. When sperm and eggs are treated with cytochalasin B at fertilization, fertilization cone formation and sperm incorporation are inhibited (Longo, 1980; Schatten and Schatten, 1980).

Figure 4.15 summarizes the events of sperm penetration and sperm-egg fusion. First, the sperm approaches the egg (see Fig. 4.15A). Then, the acrosome reaction is induced by contact with the egg jelly coat, and an acrosomal process is formed that projects to the surface of the egg (see Fig. 4.15B). Next, the tip of the acrosomal process fuses with the egg plasma

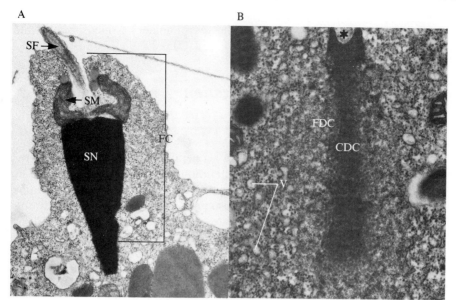

Figure 4.14

Transmission electron micrographs showing early stages of sperm incorporation into a sea urchin egg. A, Sperm nucleus (SN) incorporated into a fertilization cone (FC) shortly after sperm-egg fusion. Other sperm organelles, including a mitochondrion (SM) and the flagellum (SF), are in the process of being engulfed by the egg. × 12,600. B, Dissociation of the nuclear envelope and dispersal of sperm chromatin after the sperm has entered the egg cytoplasm. Nuclear envelope dissociation is complete except at the tip (asterisk) and the base (not seen in this section) of the nucleus. The center of the chromatin mass (CDC) remains condensed at this stage, while the peripheral chromatin is finely dispersed (FDC). Smooth vesicles (V) are evident at the periphery of the finely dispersed chromatin. The future male pronuclear envelope will be constructed by an aggregation of these vesicles. × 11,900. (A, From Longo, 1973. B, From Longo and Anderson, 1968. Reprinted from *The Journal of Cell Biology*, 1968, *39*: 339–368 by copyright permission of the Rockefeller University Press.)

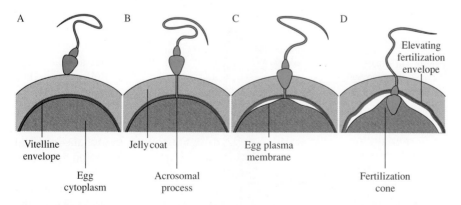

Figure 4.15

A summary of the events of sperm penetration and sperm-egg fusion in the starfish egg. (After Longo, 1987.)

membrane (see Fig. 4.15C). Finally, the sperm nucleus is incorporated into the egg via the fertilization cone (see Fig. 4.15D).

4–4 *The Response to Fertilization: Events that Occur After Sperm-Egg Fusion*

At fertilization, the zygote begins a sequence of events known as the **activation program**. The timing of the activation program has been extensively studied in sea urchin eggs and is depicted in Figure 4.16. The activation program is divided into the **early responses**, which occur within a few minutes after fertilization, and the **late responses**, which begin after the completion of the early responses (Epel, 1977). First, we examine the early responses to fertilization, which function to prevent polyspermy. Then, we discuss the late responses to fertilization, which lead directly to the formation of an embryo. We shall refer back to Figure 4.16 as we discuss the details of the egg activation program.

Prevention of Polyspermy

Although many sperm can penetrate the extracellular coats, usually only one sperm is successful in fertilizing the egg. The fusion of a single sperm

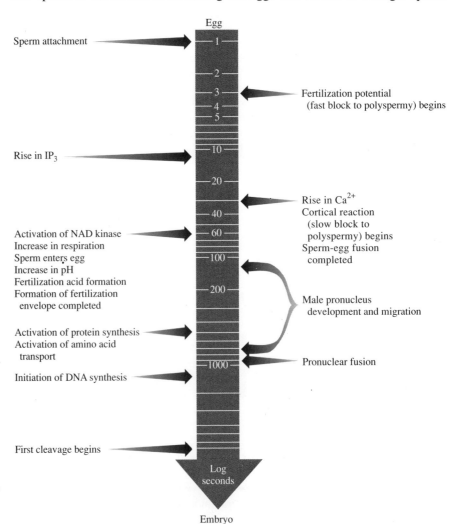

Figure 4.16

Major events of the activation program in sea urchin zygotes.

with the egg is a critical event in development that results in an equal genetic contribution from each parent and the restoration of the chromosome number to the diploid value. The abnormal condition in which more than one sperm enter an egg is called **polyspermy**. The entry of multiple sperm into the egg can result in polyploidy, abnormalities in the mechanics of chromosome separation during cell division, and ultimate death of the embryo. It is not surprising, therefore, that organisms have evolved elaborate mechanisms for preventing the formation of polyploid zygotes.

Three different strategies for preventing polyploidy have been recognized (Jaffe and Gould, 1985). The first strategy is to restrict the number of sperm that can approach the egg. We have seen how this is accomplished in fish eggs by permitting sperm to enter the chorion only through a narrow orifice, the micropyle (see Fig. 4.1A). If the chorion is removed, many sperm enter the egg, and the polyspermic embryos die (Sakai, 1961). The second strategy is to restrict the number of sperm that are able to penetrate the extracellular coats and fuse with the plasma membrane of the egg. Sea urchins and mammals have adopted this strategy. In mammals, the number of sperm that reach the site of fertilization is minimized by the requirement for migration through the female reproductive tract, but there is also a block to polyspermy involving the zona pellucida (Wolf, 1978). The third strategy restricts the number of sperm nuclei that can combine with the egg nucleus when numerous sperm enter the egg. This strategy occurs in animals with **physiological polyspermy**, in which several sperm normally enter the egg. For example, up to 20 sperm enter the salamander egg (Fankhauser, 1932). Although each sperm nucleus develops into a male pronucleus and forms a sperm aster (see p. 150), only one of the male pronuclei combines with the female pronucleus to form a zygote nucleus. The other sperm pronuclei degenerate in the egg cytoplasm before first cleavage.

We shall now start a detailed discussion of the second strategy for preventing polyploidy: the inhibition of multiple sperm entry at the level of the egg surface. Sea urchin and many other marine invertebrate eggs exhibit two successive blocks to polyspermy: a temporary or **fast block to polyspermy**, which occurs within 3 seconds after sperm contact, and a permanent or **slow block to polyspermy**, which develops by about 1 minute after fertilization (Rothschild and Swann, 1952).

The Fast Block to Polyspermy

The fast block to polyspermy is mediated by an electrical depolarization of the egg plasma membrane known as the **fertilization potential** (Steinhardt and Mazia, 1973; Jaffe, 1976; Longo et al., 1986). The fertilization potential occurs within a few seconds after the first sperm contacts the egg surface and temporarily changes the voltage across the egg plasma membrane from about -70 mV to about $+10$ mV (Fig. 4.17). The initial depolarization results from an influx of Ca^{2+}, and the positive voltage is sustained by a slow transient increase in Na^+ permeability of the egg plasma membrane. The following experiments show that the fertilization potential prevents supernumerary sperm from entering the egg (Jaffe, 1976). First, when the membrane potential is increased to a level comparable to the fertilization potential by applying a positive current to an unfertilized egg, subsequent fertilization is prevented. Second, when the membrane potential is held at a negative level by lowering the extracellular Na^+ concentration of a recently fertilized egg, multiple sperm can enter the egg.

How does the fertilization potential block supernumerary sperm from entering the egg? Two models have been proposed to explain this phenomenon. First, the egg plasma membrane may contain a sperm receptor that can only interact with sperm when the membrane potential is negative.

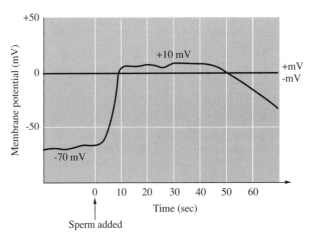

Figure 4.17

The membrane potential of sea urchin eggs before and after fertilization. The membrane potential of the unfertilized egg is about −70 mV. After the addition of sperm (arrow), the membrane potential rapidly shifts to about +10 mV and then gradually becomes negative again. (After L. A. Jaffe, 1976. Reprinted by permission from Nature, *261*: 68–71. Copyright © 1976, Macmillan Magazines Ltd.)

Second, the fertilizing sperm might insert a voltage-sensitive protein into the egg plasma membrane that promotes sperm-egg fusion. These models have been tested by experiments with interspecific hybrids (Iwao and Jaffe, 1989). These experiments depend upon the fact that various species of amphibians show different strategies for preventing polyploidy. Eggs of the amphibian *Hynobius nebulosus* have a voltage-dependent fast block to polyspermy, whereas eggs of the amphibian *Cynops pyrrhogaster* show physiological polyspermy and lack a fertilization potential. When *Cynops* eggs (normally showing voltage-independent fertilization) are fertilized with *Hynobius* sperm (normally showing voltage-dependent fertilization), fertilization is voltage-dependent. Thus, the voltage-dependent component is contributed by the sperm rather than the egg. These results support the possibility that sperm insert a positively charged fusion protein into the egg plasma membrane, which promotes sperm-egg fusion and the fertilization potential. Once the egg membrane becomes positive, the insertion of the fusion protein by additional sperm would be impossible.

In contrast to sea urchins, there is no electrical block to polyspermy in mammals (Jaffe et al., 1983). Instead, the block to polyspermy in mammals involves structural changes in the zona pellucida.

The Slow Block to Polyspermy

The fast block to polyspermy is only a temporary measure to prevent multiple sperm entry. Eventually, the fertilization potential decays to a negative level (see Fig. 4.17), and other means are used to prevent the numerous sperm that have accumulated near the egg surface from entering the egg. This is accomplished by the slow block to polyspermy.

The slow block to polyspermy is mediated by the **cortical reaction**. The term "cortical reaction" was first applied to a change in optical properties (the **fertilization wave** described by E. E. Just) that sweeps around the egg surface after fertilization (Rothschild and Swann, 1949). We now understand that this change is a wave of exocytosis that occurs as the cortical granules fuse with the egg plasma membrane and release their contents into the **perivitelline space**, the area between the plasma membrane and the vitelline envelope. The cortical reaction begins precisely at the point of sperm-egg fusion and is propagated around the surface of the egg in about 1 minute (Hinkley et al., 1986). The events of cortical granule exocytosis are summarized in Figure 4.18 (also see Figs. 4.19 and 4.20).

Whereas the fast block is an electrical event, the slow block to polyspermy is a mechanical process. As the cortical granules undergo exocytosis, they fill the perivitelline space with hydrated proteins and mucopolysaccharides, causing the vitelline envelope to elevate above the egg surface (see Fig. 4.18C). The elevated vitelline envelope is called the **fertilization envelope**. The cortical reaction blocks polyspermy in three different ways. First, fertilization envelope elevation increases the distance between the egg plasma membrane and supernumerary sperm that have attached to the vitelline envelope (see Fig. 4.18B to D). Second, the cortical granules release hydrogen peroxide and peroxidase, which cross-link glycoproteins in the fertilization envelope (Foerder and Shapiro, 1977), making it hard and resistant to digestion by sperm proteases (see Fig. 4.18D). Third, the cortical granules release proteases that destroy glycoprotein sperm

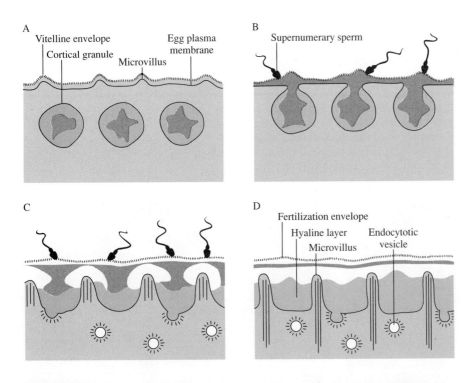

Figure 4.18

A diagram of the cortical reaction and fertilization envelope elevation in the sea urchin egg. A, Unfertilized egg with intact cortical granules, thin vitelline envelope, and plasma membrane with short microvilli. B, Cortical granule exocytosis and the beginning of fertilization envelope elevation in a recently fertilized egg. C, Completion of cortical granule exocytosis showing the elevation of the fertilization envelope with adhering supernumerary sperm. A portion of the cortical granule contents joins the hardened fertilization envelope, while another portion remains in the perivitelline space to form the hyaline layer. The microvilli begin to elongate. D, A fertilized zygote showing elevated fertilization envelope, fully elongated microvilli, and hyaline layer. The plasma membrane is beginning to be recovered by endocytosis. (After Longo, 1987.)

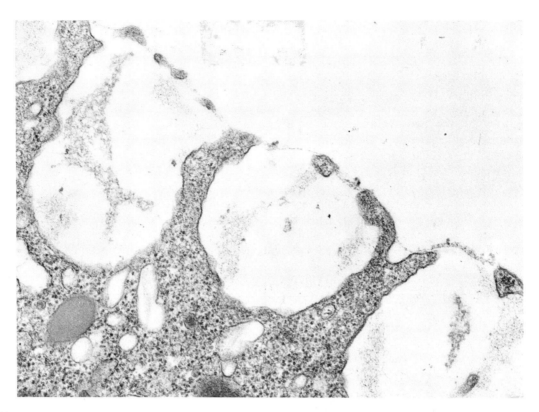

Figure 4.19

A section through the surface of a fertilized egg showing partially released cortical granules. × 30,000. (From Anderson, 1968. Reprinted from *The Journal of Cell Biology*, 1968, *37*: 514–539 by copyright permission of the Rockefeller University Press.)

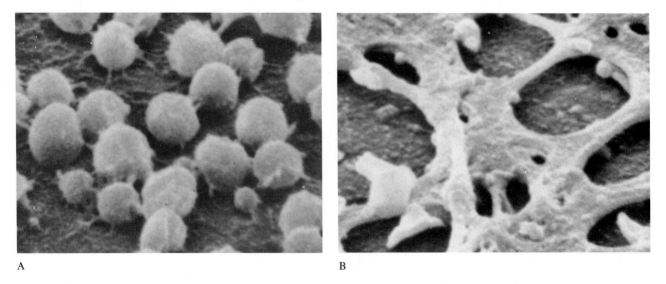

A   B

Figure 4.20

Scanning electron micrographs of ''cortical lawns'' containing intact and discharged cortical granules bound to the inner side of the plasma membrane of a sea urchin egg. A, Intact cortical granules in the unfertilized egg. × 12,000. B, Fused cortical granule material after cortical granule discharge. Discharge of cortical granules was induced by the addition of Ca^{2+} to the cortical lawns. × 18,100. (From Vacquier, 1975.)

receptors present on the vitelline envelope (Carroll and Epel, 1975), causing supernumerary sperm to detach and preventing the binding of additional sperm.

The egg surface area increases tremendously during the cortical reaction because the cortical granule membrane fuses with the egg plasma membrane (see Fig. 4.18B and C). The excess membrane is accommodated by folding the egg plasma membrane into thousands of microvilli (Fig. 4.21). Later, the excess membrane is removed from the egg surface by endocytosis (Fisher and Rebhun, 1983). Some of the cortical granule contents remain in the perivitelline layer and polymerize into the **hyaline layer** (see Fig. 4.18D). The hyaline layer maintains blastomere adherence and participates in morphogenetic changes in the developing embryo (see p. 371).

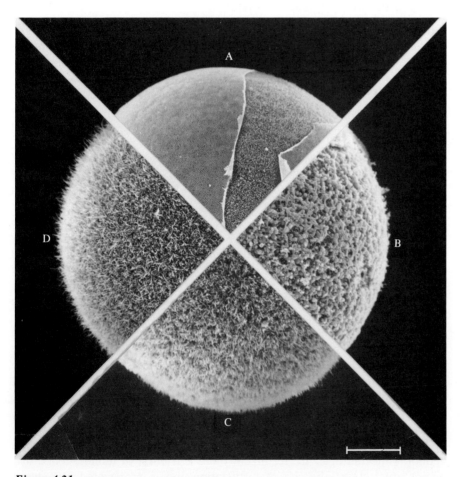

Figure 4.21

A composite of scanning electron micrographs of sea urchin egg surfaces at selected stages before and after fertilization. A, Before fertilization, with the plasma membrane revealed where the vitelline envelope is torn away. At this stage, the plasma membrane is folded into short microvilli. B, One minute after fertilization, when the surface is obscured by globular secretions from the discharged cortical granules. C, Five minutes after fertilization, following removal of materials that make up the hyaline layer. At this stage, the plasma membrane contains long microvilli. D, Thirteen minutes after fertilization, when long microvilli give the plasma membrane a "fuzzy" appearance. Scale bar equals 10 μm. (From Schroeder, 1979.)

The slow block to polyspermy in sea urchins has a counterpart in mammals and other organisms whose eggs contain cortical granules. The cortical granules of mouse eggs contain hydrolytic enzymes, which are released into the perivitelline space during the cortical reaction. Subsequently, the zona pellucida becomes hardened and refractive to sperm penetration, and its sperm receptors are inactivated (Wasserman, 1987). The alterations in the zona pellucida are called the **zona reaction**. The zona reaction may involve a chemical modification in ZP-3. After sperm-egg fusion, ZP-3 no longer behaves as a sperm receptor or induces the acrosome reaction (Wassarman, 1987). However, ZP-3 is not inactivated by a protease. Instead, it is probably modified by hydrolysis of sugar groups by glycosidases released from the cortical granules. The mechanism of zona pellucida hardening is unknown but does not appear to involve glycoprotein cross-linking.

Control of the Cortical Reaction

Unfertilized sea urchin eggs contain a large stockpile of Ca^{2+} sequestered within their cortical endoplasmic reticulum. During sperm-egg fusion, the bound Ca^{2+} is released from these organelles, resulting in a large increase in the level of free cytoplasmic Ca^{2+}. This major rise in cytoplasmic Ca^{2+} should not be confused with the relatively minor Ca^{2+} flux that initiates the fertilization potential. The minor Ca^{2+} flux involves ions that originate from outside the egg, whereas the major rise involves changes in the localization of internal Ca^{2+}.

Many different experiments illustrate the important role of Ca^{2+} in stimulating the cortical reaction: (1) sea urchin eggs treated with ionophore A23187 (Chambers et al., 1974; Steinhardt and Epel, 1974) or injected with Ca^{2+} (Hamaguchi and Hiramoto, 1981) undergo the cortical reaction in the absence of fertilization; (2) when EGTA, a Ca^{2+} chelator, is injected into sea urchin eggs, the eggs are unable to undergo the cortical reaction after fertilization (Zucker and Steinhardt, 1978); (3) when Ca^{2+} is added to isolated egg cortices that are attached to a glass slide with their cortical granule sides facing up (these preparations are called cortical lawns; see Fig. 4.20), the cortical granules are induced to fuse with the egg plasma membrane *in vitro* (Vacquier, 1975); and (4) it is possible to measure intracellular Ca^{2+} levels directly by injecting eggs with aequorin, a luminescent protein from jellyfish whose light production is very sensitive to the presence of free Ca^{2+}. When free Ca^{2+} levels are high, aequorin luminesces, but when free Ca^{2+} levels are low, there is no light produced by this protein. The first aequorin-injection experiments were done with sea urchins and the medaka, a fish with large, clear eggs that are easy to microinject. In medaka eggs, sperm-egg fusion and the cortical reaction occur at a predictable position on the egg surface, in the region immediately below the micropyle. When aequorin-injected eggs are fertilized, there is a burst of luminescence that begins precisely at the site of fertilization and is propagated as a wave around the surface of the egg (Fig. 4.22). The rate of light propagation is similar to the rate of cortical granule exocytosis, suggesting that these two processes are linked. Thus, a chain reaction of intracellular Ca^{2+} release and uptake, the **calcium wave**, accompanies the cortical reaction.

In summary, fertilization involves a sequence of membrane fusions, each of which is mediated by major increases in Ca^{2+}. The first fusion takes place between the acrosomal and sperm plasma membranes during the acrosome reaction. This fusion is mediated by an influx of Ca^{2+} into the sperm. Sperm-egg fusion triggers the major increase in Ca^{2+}, which is derived from sources located inside the egg. This rise in Ca^{2+} mediates yet

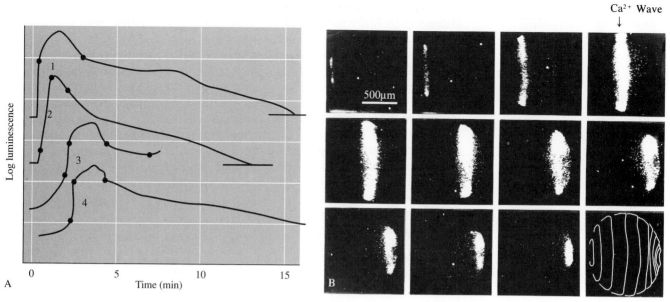

Figure 4.22

Release of calcium in fertilized medaka eggs monitored by luminescence after aequorin microinjection. A, Semilogarithmic plots of activation responses, showing a rapid rise in luminescence followed by a slower drop to the resting level. Eggs 1 and 2 were fertilized at time 0; eggs 3 and 4 were activated by A23187 treatment at time 0. B, Photographs of luminescence caused by a calcium wave propagating through the cortex of a fertilized egg. Successive photographs were taken 10 seconds apart. The last frame is a tracing showing the leading edges of 11 successive wave fronts. The egg is oriented with the micropyle (site of sperm entry) to the left. (A, After Ridgway et al., 1977. B, From Gilkey et al, 1978. Reprinted from *The Journal of Cell Biology*, 1978, 76: 448–466 by copyright permission of the Rockefeller University Press.)

a third fusion, the fusion of the cortical granule membrane with the egg plasma membrane. Later in this chapter, we shall see that the rise of free Ca^{2+} in the egg also activates the events leading to embryonic development.

Fusion of Sperm and Egg Nuclei

Sperm and egg fusion is only the beginning of fertilization. The process of fertilization is completed in the egg cytoplasm, when the sperm and egg nuclei combine to form the diploid zygotic nucleus, or **synkaryon**. The synkaryon is not formed immediately after fertilization; it is an example of a late response to fertilization that is completed between 10 and 20 minutes after sperm attachment (see Fig. 4.16). A maturation period in the egg cytoplasm, which involves a rapid reversal of the steps that were taken to remodel the sperm nucleus during spermiogenesis, is required before the sperm nucleus is competent to form a synkaryon.

As mentioned earlier, sea urchin eggs have already completed both meiotic divisions and are arrested as ootids at the time of fertilization. After the final meiotic division, the haploid egg nucleus is reorganized and is now known as the **female pronucleus**. Once the sperm nucleus enters the egg and begins its maturation period, it is known as the **male pronucleus**. The events required to form a male pronucleus from the original sperm nucleus are complicated and vary among different species. They usually involve three steps: the breakdown of the sperm nuclear envelope, the decondensation of sperm chromatin, and the formation of the male pronuclear envelope (Longo and Anderson, 1968). We now begin a detailed discussion of pronuclear fusion in the sea urchin eggs.

In sea urchins, the sperm nucleus is initially incorporated into the fertilization cone (see Fig. 4.14). Shortly after the sperm nucleus enters the egg cytoplasm, it swells and the nuclear envelope disintegrates into vesicles that at first outline the decondensing sperm chromatin but eventually are lost among the other membranous components of the egg cytoplasm. Portions of the posterior sperm nuclear membrane do not undergo this vesiculation and dispersal; they are preserved intact and are later incorporated into the new envelope that forms around the male pronucleus. The significance of this retained portion of the sperm nuclear membrane is unknown.

Chromatin decondensation begins at the periphery of the sperm nucleus and proceeds toward the center (see Fig. 4.14B). The decondensation of sperm chromatin involves the loss of sperm histone H1 and its replacement by a new histone H1 species derived from the egg cytoplasm (Green and Poccia, 1985). As described in Section 4–2, a large increase in the levels of cyclic AMP occurs when sperm attach to the jelly coat. Cyclic AMP production activates cyclic AMP-dependent phosphokinase, an enzyme that initiates the phosphorylation of sperm histone H1. This phosphorylation, which proceeds as sperm penetrate the egg glycocalyx (Porter and Vacquier, 1986), destabilizes the binding of sperm histone H1 to sperm chromatin. The new histone H1 species may promote a spatial arrangement of chromatin that is involved in forming the male pronucleus. As sperm chromatin decondensation is being completed, membranous vesicles aggregate along the periphery of the sperm chromatin. These vesicles fuse with one another and the persisting portions of the original sperm nuclear envelope to form the nuclear envelope of the male pronucleus.

At the completion of meiosis, the female pronucleus is located in the central region of the egg, whereas the male pronucleus is formed in the cortex, near the site of sperm entry. Before pronuclear fusion, the pronuclei must migrate a considerable distance through the egg cytoplasm (Fig. 4.23). In sea urchins, this movement is mediated by the **sperm aster**, a complex of long microtubules that radiate from the paired sperm centrioles. These centrioles enter the egg with the sperm nucleus and serve as part of a **microtubule organizing center (MTOC)** for the sperm aster. The astral microtubules extend into the egg cortex and into the vicinity of the male pronuclear membrane. The movements of the male and female pronucleus after fertilization are summarized in Figure 4.24. During male pronucleus

Figure 4.23

Interference contrast micrographs showing migration of the male (top) and female (bottom) pronuclei in a sea urchin egg. Note the aster radiating from the small male pronucleus. Arrows indicate time when the photographs were taken. Scale bar equals 20 μm. (From Hamaguchi and Hiramoto, 1980.)

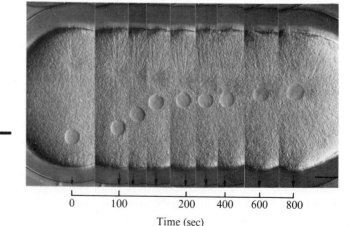

0   100   200   400   600   800

Time (sec)

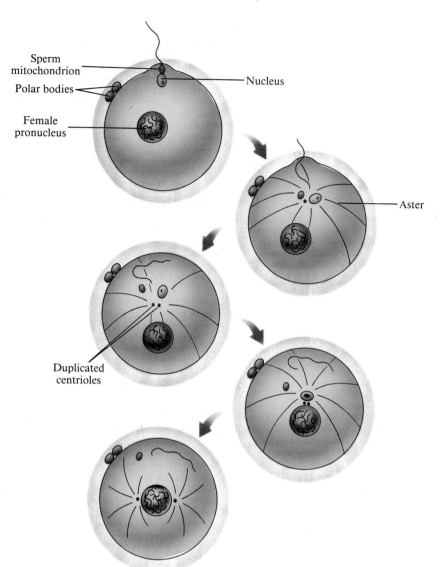

Sperm mitochondrion
Polar bodies
Female pronucleus
Nucleus
Aster
Duplicated centrioles

Figure 4.24

Diagram of pronuclear movements during sea urchin fertilization.

migration, the microtubules of the sperm aster lengthen and appear to push the male pronucleus toward the center of the egg (Wilson and Mathews, 1895; Chambers, 1939; Hamaguchi and Hiramoto, 1980). Eventually, the astral microtubules contact the female pronucleus and appear to pull the female pronucleus abruptly toward the male pronucleus. Continuing activity of the sperm aster displaces the two adjacent pronuclei to the center of the egg, where they fuse to form the synkaryon. The role of the astral microtubules in controlling pronuclear development has been demonstrated by experiments in which colchicine and other agents that depolymerize microtubules were found to inhibit pronuclear migration and fusion when added to fertilized sea urchin eggs (Zimmermann and Zimmermann, 1967; Schatten and Schatten, 1981).

Some animal eggs do not form a synkaryon by pronuclear fusion. In mammals, for example, the synkaryon forms by **approximation**, a process in which the pronuclei migrate toward one another, become closely apposed, but do not fuse (Longo and Anderson, 1969; Zamboni et al., 1972; Kunkle and Longo, 1975). They remain adjacent until the first mitotic division, when the nuclear envelopes break down and the paternal and maternal chromosomes mix at a common metaphase plate. Mammalian and sea urchin

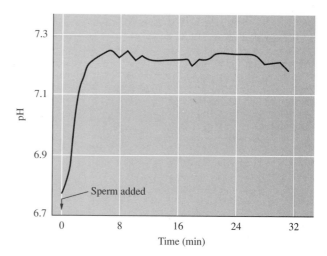

Figure 4.25

Continuous recording of intracellular pH during fertilization. (After Shen and Steinhardt, 1978. Reprinted by permission from *Nature, 272*: 253–254. Copyright © 1978, Macmillan Magazines Ltd.)

egg pronuclear movements also differ in their use of sperm or egg MTOCs to promote aster formation. In the mouse zygote, the aster is formed from many MTOCs originally derived from the egg cytoplasm (Schatten et al., 1985, 1986).

The changes that occur during male pronucleus development in the sea urchin egg are controlled by ionic signals. As shown in Figure 4.16, a major influx of Na^+ begins at about 1 minute after fertilization (Johnson et al., 1976). Unlike the previous influx of Na^+, which was responsible for the fertilization potential, the second Na^+ influx is coupled to an efflux of H^+ and is, therefore, electrically neutral. The efflux of H^+ (i.e., the release of **fertilization acid**) increases the pH of the egg cytoplasm from about 6.8 to 7.3 (Fig. 4.25). The role of the pH increase in pronuclear migration has been tested by fertilizing eggs in Na^+-free seawater. This prevents the coupled Na^+/H^+ exchange. Under these conditions, the male and female pronuclei do not develop, suggesting that pronuclear migration normally is triggered by the rise in intracellular pH (Chambers, 1976). In the next section, we shall see that increased pH also has significant effects on the initiation of sea urchin development.

4–5 The Initiation of Development

As we mentioned earlier, sperm-egg fusion initiates a series of events known as the activation program. Some of these events, including the late responses to fertilization (see Fig. 4.16), are directly related to the formation of an embryo. The late responses include metabolic changes, such as the activation of K^+ and amino acid transport, an increase in the rate of protein synthesis, and the initiation of DNA replication, as well as major regulatory events, such as the production of **inositol trisphosphate (IP₃)** and **diacylglycerol**, the release of cytoplasmic free Ca^{2+}, and the pH rise. In this section, we discuss the key regulatory events that occur during the activation program in sea urchins. Our discussion of other late responses to fertilization, including the activation of DNA, RNA, and protein synthesis and cell division, is delayed until subsequent chapters.

Research on the activation of sea urchin development has focused on the following questions: What is the relationship among the individual events in the activation program? Does the program represent a linear series of events or several parallel pathways? If there are parallel pathways, at what point in the program do they initially diverge? For example, because the fertilization potential is one of the earliest events in fertilization, it can be asked whether it stimulates any subsequent events. The fertilization potential can be mimicked by applying positive current to unfertilized eggs (Jaffe, 1976). The artificially induced fertilization potential, however, does not elicit the rise in Ca^{2+}, the cortical reaction, or any subsequent events in the program. This result indicates that the fertilization potential is not sufficient for the Ca^{2+} rise and cortical reaction. Therefore, the fertilization potential is part of a parallel pathway that terminates after fertilization and is not directly involved in the activation program. Unlike the fertilization potential, the Ca^{2+} rise triggers subsequent events in the activation program. Stimulation of Ca^{2+} release by ionophore A23187 promotes the cortical reaction and partial activation of protein and DNA synthesis in unfertilized eggs (Steinhardt and Epel, 1974). Using such experiments, the relationship among various events in the activation program is being investigated. The preliminary results of these studies are summarized in Figure 4.26.

Figure 4.26 shows that the cortical reaction and NAD kinase activation are dependent solely on the Ca^{2+} rise, but other events in the activation program are induced by the combined effects of Ca^{2+} and pH. The role of pH in the activation of protein and DNA synthesis is shown by an experiment in which unfertilized sea urchin eggs were treated with ammonia (Steinhardt

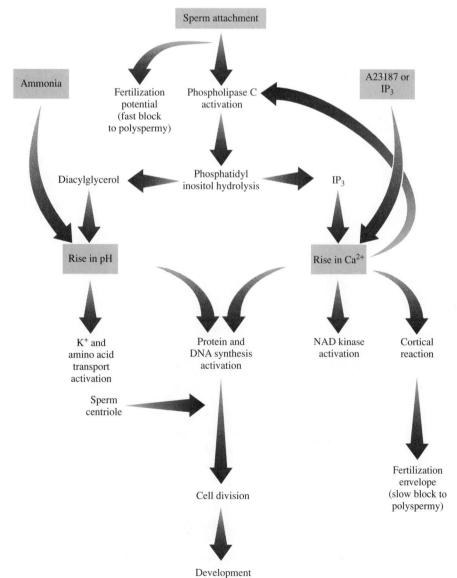

Figure 4.26

The relationships among events in the sea urchin activation program.

and Mazia, 1973). When ammonia enters the cytoplasm, it forms a complex with H^+ ($NH_3 + H^+ \rightarrow NH_4^+$) and increases the intracellular pH. This pH change leads to a stimulation of protein and DNA synthesis (Epel et al., 1974). Two additional experiments also illustrate the important role of pH in the activation program (Johnson et al., 1976; Shen and Steinhardt, 1979). First, because the pH change is dependent on a coupled Na^+/H^+ exchange, it can be prevented by removing Na^+ from the extracellular medium. Second, the pH change can also be abolished by bathing eggs in amiloride, an inhibitor of the Na^+/H^+ flux. Depletion of Na^+ or amiloride treatment before the normal pH rise blocks development. If these treatments are done after the normal pH rise, however, the egg is able to develop and undergo several divisions. These experiments show that the rise in pH is another major regulatory step in the sea urchin activation program.

How does sperm-egg fusion induce both the Ca^{2+} and pH changes? Current studies suggest that both changes result from a cascade of events involving the breakdown of **phosphatidyl inositol**, a plasma membrane lipid. This cascade was discovered in somatic cells, in which it produces intracellular signals in response to binding of a polypeptide hormone to the cell surface. The presumed pathway leading from sperm binding to the Ca^{2+} and pH rises is shown in Figure 4.27. The initial events of the cascade occur in the egg plasma membrane. Sperm binding activates a protein located on the inside of the egg plasma membrane, GTP-binding protein (G protein), which in turn stimulates the activity of phospholipase C, an enzyme located on the inner side of the plasma membrane. It is not known whether the sperm binds to a receptor in the egg plasma membrane, which then activates G protein (as indicated in Fig. 4.27), or whether the sperm inserts a G-protein activator into the membrane at the time of sperm-egg fusion. Phospholipase C causes the hydrolysis of phosphatidyl inositol to diacylglycerol and IP_3. Diacylglycerol remains in the plasma membrane, where it activates the Na^+/H^+ flux responsible for the pH rise. This was shown by treating eggs with phorbol ester, which mimics diacylglycerol and leads to an increase in pH without the accompanying rise in Ca^{2+} (Swann and Whitaker, 1985). In contrast, IP_3 is released from the membrane and diffuses into the cytoplasm, where it binds to the endoplasmic reticulum and induces the release of Ca^{2+} into the cytoplasm.

The role of IP_3 in the Ca^{2+} rise is supported by the following experiments. First, a transient rise in IP_3 was shown to be one of the earliest events in the activation program (Turner et al., 1984). Second, unfertilized

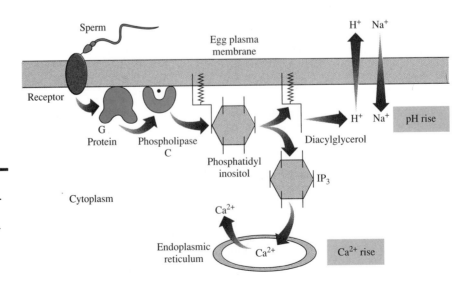

Figure 4.27

The phosphatidyl inositol breakdown pathway and its role in activating the Ca^{2+} and pH rises in fertilized sea urchin eggs. The existence of a sperm receptor that activates G protein, as shown in this diagram, is hypothetical.

eggs can be activated by injecting IP_3 (Whitaker and Irvine, 1984). Furthermore, it has been suggested that IP_3 is part of a feedback loop in the egg activation program (Swann and Whitaker, 1986). According to this idea, Ca^{2+} released by IP_3 may diffuse from the point of sperm-egg fusion to adjacent parts of the plasma membrane, stimulate phospholipase C activity, more IP_3 production, and further Ca^{2+} release (see Figs. 4.26 and 4.27). This feedback loop is thought to be responsible for propagating the calcium wave.

In summary, sperm-egg fusion appears to control both the Ca^{2+} and pH rises via the phosphatidyl inositol breakdown pathway. One breakdown product, IP_3, regulates the Ca^{2+} rise, whereas the other, diacylglycerol, controls the pH increase. The breakdown of phosphatidyl inositol represents a major branch point where parallel pathways begin to diverge in the activation program (see Fig. 4.26).

Are the events that control the sea urchin activation program the same in other organisms? When IP_3 and aequorin are injected into medaka eggs, a calcium wave is initiated (Nuccitelli, 1987) that is similar in appearance to that previously seen after fertilization (see Fig. 4.22). The activation program of most (if not all) eggs appears to include a rise in intracellular Ca^{2+} (Steinhardt et al., 1974). However, there appear to be variations in the source of Ca^{2+}. In sea urchins and vertebrates, Ca^{2+} is derived from cytoplasmic organelles, whereas in mollusks and annelids, it originates from outside the egg (Jaffe, 1983). Although Ca^{2+} appears to be the universal egg activator, the events that occur after the Ca^{2+} rise are variable in different organisms. This variation may be due to differences in timing of meiotic events. For instance, the sea urchin egg is arrested at a different stage in meiosis than most other eggs and is metabolically inert at the time of fertilization. In contrast, the amphibian egg is arrested in metaphase of the second meiotic division, and is metabolically very active. Fertilization of amphibian eggs causes a modest increase in protein synthesis (but nothing close to the 20- to 40-fold increase in sea urchin eggs; see Chap. 13) and results in alterations of certain enzyme activities (Woodland, 1974). However, these appear to be metabolic *adjustments* rather than *activations*. Clearly, considerable research needs to be done with diverse organisms before a more general picture of the activation program can be drawn.

4–6 *Parthenogenesis*

Parthenogenesis is the process in which eggs are activated and develop in the absence of a paternal genetic contribution. Parthenogenesis can be natural or artificial. **Artificial parthenogenesis** involves activation of the egg by chemical or physical means. Because the universal trigger for activation may be a Ca^{2+} flux (Jaffe, 1983), treatments that cause such a response can usually induce artificial parthenogenesis. For example, we have discussed how sea urchin and other eggs can be activated by ionophore A23187. However, this only results in a partial egg activation. DNA and protein synthesis are not stimulated to normal levels, and cell division does not occur in A23187-activated eggs, because they lack a sperm centriole and aster. The so-called **double method** for inducing artificial parthenogenesis in sea urchins involves an initial treatment with butyric acid (which probably activates the Ca^{2+} flux), followed by addition of hypertonic seawater (to induce egg asters and cell division), and can lead to complete development of parthenogenic larvae (Brandiff et al., 1975). Likewise, although pricking a frog egg with a clean needle (thereby eliciting only a Ca^{2+} flux) leads to partial activation, using a bloody needle yields embryos that develop to the tadpole stage, because the needle contains MTOCs from disrupted blood cells.

Some animals—including some flatworms, rotifers, insects, amphibians, and reptiles—reproduce by **natural parthenogenesis**. Natural parthenogenesis is the normal development of an animal without a paternal genetic contribution. These animals must overcome two important problems that are normally solved by sperm entry. First, alternate triggers must be developed to initiate their activation program. The methods used for this purpose are varied but usually involve physical perturbations of the egg. In parthenogenetic wasps, for example, activation is caused by a mechanical stimulus received by the egg as it passes out of the female reproductive tract. The development of parthenogenetic beetles is initiated by bacterial symbionts that enter the egg while it is in the reproductive tract (Peleg and Norris, 1973). Some parthenogenetic salamanders exhibit an unusual method of activation. They copulate with a male of another salamander species, and the nonspecific sperm enter and activate the egg of the parthenogenetic species (Uzzell, 1964). The sperm nucleus subsequently degenerates in the egg cytoplasm and does not contribute genetic material to the embryo. Second, parthenogenetic organisms must develop a means of attaining the diploid chromosome number in the absence of pronuclear fusion. This problem is usually solved by imposing an extra round of DNA synthesis and chromosome duplication either before meiosis or during the early cleavage of the zygote. Natural parthenogenesis has not been demonstrated in mammals. Mammals may not have evolved parthenogenesis because, unlike most other animals, genetic contributions from both the sperm and egg are required for development (see Chap. 10).

4–7 Chapter Synopsis

The sperm and egg unite at fertilization to restore the diploid chromosome number and to activate embryonic development. Gametes have evolved specific morphological features to promote efficient fertilization. The sperm is a streamlined locomotory cell specialized to deliver the paternal genome to the egg, whereas the egg is a large cell containing stockpiles of materials needed to support embryonic development and is surrounded by specialized envelopes that control the number of sperm that participate in fertilization. Most of our knowledge about fertilization comes from studies with sea urchins and mammals. Before fertilization, sperm become motile, swim toward the egg, and pass through the egg envelopes. In sea urchins, release of sperm into seawater triggers a pH rise, which activates motility. In mammals, sperm motility is activated by ejaculation into the female reproductive tract, and capacitation is required to transit the egg envelopes. The acrosome reaction is triggered by an influx of external Ca^{2+} when sperm contact the egg envelopes. Acrosome exocytosis releases hydrolytic enzymes, which digest the egg envelopes and promote sperm penetration. Mammalian sperm penetrate the zona pellucida and fuse with the egg tangentially. Sea urchin sperm penetrate the jelly coat and vitelline envelope and fuse at right angles with the egg. An acrosomal process extends from the anterior tip of the sea urchin sperm to contact and fuse with the egg plasma membrane. A fertilization cone forms at the site of sperm-egg fusion and eventually engulfs the sperm.

When the sperm nucleus enters the egg, it is remodeled to allow fusion with the egg pronucleus. Sperm-egg fusion initiates the egg activation program. The early events of this program prevent polyspermy. Two blocks to polyspermy exist in sea urchins: a fast block due to depolarization of the egg plasma membrane and a slow block due to exocytosis of cortical granules and elevation of the fertilization envelope. Cortical granule exocytosis is initiated by a wave of free cytoplasmic Ca^{2+}, which is initiated at the point of sperm entry. Other events in egg activation include a pH rise and metabolic changes leading to cell division. These events are initiated by changes in the egg plasma membrane, which involve phosphatidyl inositol breakdown to IP_3 and diacylglycerol, which activate the Ca^{2+} and pH increases, respectively. The Ca^{2+} and pH increases function synergistically to promote subsequent steps in the activation program. In some cases, egg activation can be triggered in the absence of sperm by stimuli that initiate the Ca^{2+} wave. This condition leads to parthenogenesis: the development of an animal without a paternal genetic contribution.

References

Aitken, R. J., and D. W. Richardson. 1981. Measurement of the sperm binding capacity of the mouse zona pellucida and its use in the estimation of zona antibody titres. J. Reprod. Fertil., *63*: 295–307.

Aitken, R. J. et al. 1981. The influence of anti-zona and anti-sperm antibodies on sperm-egg interactions. J. Reprod. Fertil., *62*: 597–606.

Anderson, E. 1968. Oocyte differentiation in the sea urchin, *Arbacia punctulata*, with particular reference to the origin of cortical granules and their participation in the cortical reaction. J. Cell Biol., *37*: 514–539.

Austin, C. R. 1952. The "capacitation" of the mammalian spermatozoa. Nature (Lond.), *170*: 326.

Austin, C. R. 1961. Fertilization of mammalian eggs *in vitro*. Int. Rev. Cytol., *12*: 337–359.

Austin, C. R. 1965. *Fertilization*. Prentice–Hall, Englewood Cliffs, N.J.

Bibring, T., J. Baxandall, and C. C. Harter. 1984. Sodium–dependent pH regulation in active sea urchin sperm. Dev. Biol., *101*: 425.

Bliel, J. D., and P. M. Wassarman. 1980. Mammalian sperm–egg interaction: Identification of a glycoprotein in mouse egg zona pellucida possessing receptor activity for sperm. Cell, *20*: 873–882.

Bliel, J. D., and P. M. Wassarman. 1986. Autoradiographic visualization of the mouse egg's sperm receptor bound to sperm. J. Cell Biol., *102*: 1363–1371.

Brandiff, B., R. T. Hinegardner, and R. Steinhardt. 1975. Development and life cycle of a parthenogenetically activated sea urchin embryo. J. Exp. Zool., *192*: 13–24.

Carre, D., and C. Sardet. 1981. Sperm chemotaxis in siphonophores. Biol. Cell, *40*: 119–128.

Carroll, E. J., and D. Epel. 1975. Isolation and biological activity of proteases released by sea urchin eggs following fertilization. Dev. Biol., *44*: 22–32.

Chambers, E. L. 1939. The movement of the egg nucleus in relation to the sperm aster in the echinoderm egg. J. Exp. Biol., *16*: 409–424.

Chambers, E. L. 1976. Na is essential for the activation of the inseminated sea urchin egg. J. Exp. Zool., *197*: 149–154.

Chambers, E. L., B. C. Pressman, and B. Rose. 1974. The activation of sea urchin eggs by the divalent ionophores A23187 and X-537A. Biochem. Biophys. Res. Commun., *60*: 126–132.

Christen, R., R. W. Schackmann, and B. M. Shapiro. 1983. Metabolism of sea urchin sperm. Interrelationships between intracellular pH, ATPase activity, and mitochondrial respiration. J. Biol. Chem., *258*: 5392–5399.

Clapper, D. L. et al. 1985. Involvement of zinc in the regulation of pH, motility, and acrosome reactions in sea urchin sperm. J. Cell Biol., *100*: 1817–1824.

Cosson, J., D. Carre, and M. P. Cosson. 1986. Sperm chemotaxis in siphonophores: Identification and biochemical properties of the attractant. Cell Motil. Cytoskeleton, *6*: 225–228.

Dan, J. C. 1954. Studies on the acrosome. III. Effect of calcium deficiency. Biol. Bull., *107*: 335–349.

Dan, J. C., Y. Ohori, and H. Kushida. 1964. Studies on the acrosome. VII. Formation of the acrosomal process in sea urchin spermatozoa. J. Ultrastruct. Res., *11*: 508–524.

Decker, G. L., D. B. Joseph, and W. J. Lennarz. 1976. A study of factors involved in the induction of the acrosomal reaction in sperm of the sea urchin, *Arbacia punctulata*. Dev. Biol., *53*: 115–125.

Edwards, R. G. 1981. Test tube babies, 1981. Nature (Lond.), *293*: 253–256.

Epel, D. 1977. The program of fertilization. Sci. Am., *237(5)*: 128–138.

Epel, D. et al. 1974. An analysis of the partial metabolic derepression of sea urchin eggs by ammonia; the existence of independent pathways. Dev. Biol., *40*: 245–255.

Fankhauser, G. 1932. Cytological studies on egg fragments in the salamander *Triton*. II. The history of the supernumerary sperm nuclei in normal fertilization and cleavage of fragments containing the egg nucleus. J. Exp. Zool., *62*: 185–235.

Fisher, G. W., and L. I. Rebhun. 1983. Sea urchin egg cortical granule exocytosis is followed by a burst of membrane retrieval via uptake into coated vesicles. Dev. Biol., *99*: 456–472.

Florman, H. M., and P. M. Wassarman. 1985. O–linked oligosaccharides of mouse egg ZP-3 account for its sperm receptor activity. Cell, *41*: 313–324.

Foerder, C. A., and B. M. Shapiro. 1977. Release of ovoperoxidase from sea urchin eggs hardens the fertilization membrane with tyrosine crosslinks. Proc. Natl. Acad. Sci. U.S.A., *74*: 4214–4218.

Gabel, C. A., E. M. Eddy, and B. M. Shapiro. 1979. After fertilization, sperm surface components remain as a patch in sea urchin and mouse embryos. Cell, *18*: 201–215.

Garbers, D. L. et al. 1983. Elevation of sperm adenosine 3'–5'–monophosphate concentrations by a fucose sulfate–rich complex associated with eggs: I. Structural characterization. Biol. Reprod., *29*: 1211–1220.

Gilkey, J. C. et al. 1978. A free calcium wave traverses the activating egg of the medaka, *Oryzias latipes*. J. Cell Biol., *76*: 448–466.

Ginsberg, A. S. 1959. Fertilization in the sturgeon. I. The fusion of gametes. Cytologia, *1*: 510.

Glabe, C. G. 1985. Interaction of the sperm adhesive protein, bindin, with phospholipid vesicles. I. Specific association of bindin with gel–phase phospholipid vesicles. J. Cell Biol., *100*: 794–799.

Glabe, C. G., and V. D. Vacquier. 1978. Egg surface glycoprotein receptor for sea urchin sperm binding. Proc. Natl. Acad. Sci. U.S.A., *75*: 881–885.

Green, D. P. L., and R. D. Purves. 1984. Mechanical hypothesis of sperm penetration. Biophys. J., *45*: 659–662.

Green, J. D., and R. G. Summers. 1982. Effects of protease inhibitors on sperm-related events in sea urchin fertilization. Dev. Biol., *93*: 139–144.

Green, G. R., and D. L. Poccia. 1985. Phosphorylation of sea urchin sperm H1 and H2B histones precedes chromatin decondensation and H1 exchange during pronuclear formation. Dev. Biol., *108*: 235–245.

Gwatkin, R. B. L. 1977. *Fertilization Mechanisms in Man and Mammals*. Plenum Press, New York.

Gwatkin, R. B. L., D. T. Williams, and D. J. Carlo. 1977. Immunization of mice with heat–solubilized hamster zonae: Production of anti–zona antibody and inhibition of fertility. Fertil. Steril., *28*: 871–877.

Hamaguchi, M. S., and Y. Hiramoto. 1980. Fertilization processes in the heart urchin, *Clypeaster japonicus* observed with a differential interference microscope. Dev. Growth Differ., *22*: 517–530.

Hamaguchi, Y., and Y. Hiramoto. 1981. Activation of sea urchin eggs by microinjection of calcium buffers. Exp. Cell Res., *134*: 171–179.

Hamilton, M. S., and R. B. Vernon. 1987. Inhibition of *in vitro* fertilization by mouse anti–mouse sperm sera and preliminary antigen identification. Gamete Res., *16*: 311–317.

Hanesbrough, J. R., and D. L. Garbers. 1981. Speract. Purification and characterization of a peptide associated with eggs that activates spermatozoa. J. Biol. Chem., *256*: 1447–1452.

Henderson, C. J., M. J. Hulme, and R. J. Aitken. 1987a. Analysis of the biological properties of antibodies raised against intact and deglycosylated porcine zonae pellucidae. Gamete Res., *16*: 323–341.

Henderson, C. J., P. Braude, and R. J. Aitken. 1987b. Polyclonal antibodies to a 32-KDA deglycosylated polypeptide from porcine zonae pellucidae will prevent human gamete interaction *in vitro*. Gamete Res., *18*: 251–265.

Hinkley, R. E., B. D. Wright, and J. W. Lynn. 1986. Rapid visual detection of sperm-egg fusion using the DNA–specific fluorochrome Hoechst 33342. Dev. Biol., *118*: 148–154.

Iwao, Y., and L. A. Jaffe. 1989. Evidence that the voltage–dependent component in the fertilization process is contributed by the sperm. Dev. Biol., *134*: 446–451.

Jaffe, L. A. 1976. Fast block to polyspermy in sea urchin eggs is electrically mediated. Nature (Lond.), *261*: 68–71.

Jaffe, L. A., and M. Gould. 1985. Polyspermy–preventing mechanisms. *In* C. B. Metz and A. Monroy (eds.), *Biology of Fertilization*, Vol. 3. Academic Press, New York, pp. 223–250.

Jaffe, L. A., A. P. Sharp, and D. P. Wolf. 1983. Absence of an electrical polyspermy block in the mouse. Dev. Biol., *96*: 317–323.

Jaffe, L. F. 1983. Sources of calcium in egg activation. A review and hypothesis. Dev. Biol., *99*: 265–276.

Johnson, J. D., D. Epel, and M. Paul. 1976. Intracellular pH and activation of sea urchin eggs after fertilization. Nature (Lond.), *262*: 661–664.

Kunkle, M., and F. J. Longo. 1975. Cytological events leading to the cleavage of golden hamster zygotes. J. Morphol., *146*: 197–214.

Longo, F. J. 1973. Fertilization: A comprehensive ultrastructural review. Biol. Reprod., *9*: 149–215.

Longo, F. J. 1980. Organization of microfilaments in sea urchin (*Arbacia punctulata*) eggs at fertilization: Effects of cytochalasin B. Dev. Biol., *74*: 422–433.

Longo, F. J. 1987. *Fertilization*. Chapman and Hall, New York.

Longo, F. J., and E. Anderson. 1968. The fine structure of pronuclear development and fusion in the sea urchin, *Arbacia punctulata*. J. Cell Biol., *39*: 339–368.

Longo, F. J., and E. Anderson. 1969. Cytological events leading to the formation of the two–cell stage in the rabbit: Association of the maternally and paternally derived genomes. J. Ultrastruct. Res., *29*: 86–118.

Longo, F. J. et al. 1986. Correlative ultrastructural and electrophysiological studies of sperm–egg interactions of the sea urchin, *Lytechinus variegatus*. Dev. Biol., *118*: 155–166.

Lopo, A. C., and V. D. Vacquier. 1980. Antibody to a sperm surface glycoprotein inhibits the egg jelly–induced acrosome reaction of sea urchin sperm. Dev. Biol., *79*: 325–333.

McRorie, R. A., and W. L. Williams. 1974. Biochemistry of mammalian fertilization. Ann. Rev. Biochem., *43*: 777–803.

Miller, R. L. 1977. Distribution of sperm chemotaxis in the animal kingdom. *In* K. G. Adiyodi and R. G. Adiyodi (eds.), *Advances in Invertebrate Reproduction*, Vol. 1. Peralam–Kenoth, Kerala, India, pp. 99–119.

Mori, T. et al. 1978. Possible presence of autoantibodies to zona pellucida in infertile women. Experientia, *34*: 797–799.

Müller, D. G., and G. Gassmann. 1978. Identification of the sex attractant in the marine brown alga *Fucus vesiculosus*. Naturwissenschaften, *65*: 389.

Müller, D. G., and L. Jaenicke. 1973. Fucoserraten, the female sex attractant of *Fucus serratus* L. (Phaeophyta). FEBS Lett., *30*: 137–139.

Nuccitelli, R. 1987. The wave of activation current in the egg of the medaka fish. Dev. Biol., *122*: 522–534.

Peleg, B., and N. M. Norris. 1973. Oocyte activation in *Xyleborus ferrugineus* by bacterial symbionts. Insect Physiol., *19*: 137–145.

Porter, D. C., and V. D. Vacquier. 1986. Phosphorylation of sperm histone H1 is induced by the egg jelly layer in the sea urchin *Strongylocentrotus purpuratus*. Dev. Biol., *116*: 203–212.

Primacoff, P. et al. 1988. Fully effective contraception in male and female guinea pigs immunized with the sperm protein PH–20. Nature (Lond.), *335*: 543–546.

Ridgway, E. B., J. C. Gilkey, and L. F. Jaffe. 1977. Free calcium increases explosively in activating medaka eggs. Proc. Natl. Acad. Sci. U.S.A., *74*: 623–627.

Rossignol, D. P., A. J. Roschelle, and W. J. Lennarz. 1981. Sperm–egg binding: Identification of a species–specific sperm receptor from eggs of *Strongylocentrotus purpuratus*. J. Supramol. Struct. Cell. Biochem., *15*: 347–358.

Rothschild, Lord, and M. M. Swann. 1949. The fertilisation reaction in the sea urchin egg. A propagated response to sperm attachment. J. Exp. Biol., *26*: 164–176.

Rothschild, Lord, and M. M. Swann. 1952. The fertilization reaction in the sea urchin. The block to polyspermy. J. Exp. Biol., *29*: 469–483.

Sakai, Y. 1961. Method for removal of chorion and fertilization of the naked egg in *Oryzias latipes*. Embryologia, *5*: 357–368.

Schackmann, R. W., R. Christen, and B. M. Shapiro. 1984. Measurement of plasma membrane and mitochondrial potentials in sea urchin sperm. Changes upon activation and induction of the acrosome reaction. J. Biol. Chem., *259*: 13914–13922.

Schackmann, R. W., E. M. Eddy, and B. M. Shapiro. 1978. The acrosome reaction of *Strongylocentrotus purpuratus* sperm. Ion requirements and movements. Dev. Biol., *65*: 483–495.

Schatten, H., and G. Schatten. 1980. Surface activity at the egg plasma membrane during sperm incorporation and its cytochalasin B sensitivity. Scanning electron microscopy and time–lapse video microscopy during fertilization of the sea urchin *Lytechinus variegatus*. Dev. Biol., *78*: 435–449.

Schatten, G., and H. Schatten. 1981. Effects of motility inhibitors during sea urchin fertilization. Microfilament inhibitors prevent sperm incorporation and restructuring of fertilized egg cortex, whereas microtubule inhibitors prevent pronuclear migrations. Exp. Cell Res., *135*: 311–330.

Schatten, G., C. Simerly, and H. Schatten. 1985. Microtubule configuration during fertilization, mitosis, and early development of the mouse and the requirement for egg microtubule–mediated motility during mammalian fertilization. Proc. Natl. Acad. Sci. U.S.A., *82*: 4152–4156.

Schatten, H. et al. 1986. Behavior of centrosomes during fertilization and cell division in mouse oocytes and in sea urchin eggs. Proc. Natl. Acad. Sci. U.S.A., *83*: 105–109.

Schroeder, T. E. 1979. Surface area change at fertilization: Resorption of the mosaic membrane. Dev. Biol., *70*: 306–326.

SeGall, G. K., and W. J. Lennarz. 1979. Chemical characterization of the component of the jelly coat from sea urchin eggs responsible for induction of the acrosome reaction. Dev. Biol., *71*: 33–48.

Shapiro, B. M., R. W. Schackmann, and C. A. Gabel. 1981. Molecular approaches to the study of fertilization. Ann. Rev. Biochem., *50*: 815–843.

Shen, S. S., and R. A. Steinhardt. 1978. Direct measurement of intracellular pH during metabolic derepression of the sea urchin egg. Nature (Lond.), *272*: 253–254.

Shen, S. S., and R. A. Steinhardt. 1979. Intracellular pH and the sodium requirement at fertilization. Nature (Lond.), *282*: 87–89.

Shivers, C. A., and B. S. Dunbar. 1977. Autoantibodies to zona pellucida: A possible cause for infertility in women. Science, *197*: 1082–1084.

Steinhardt, R. A., and D. Epel. 1974. Activation of sea urchin eggs by a calcium ionophore. Proc. Natl. Acad. Sci. U.S.A., *71*: 1915–1919.

Steinhardt, R. A., and D. Mazia. 1973. Development of K^+ conductance and membrane potentials in unfertilized sea urchin eggs after exposure to NH_4OH. Nature (Lond.), *241*: 400–401.

Steinhardt, R. A. et al. 1974. Is calcium ionophore a universal activator for unfertilized eggs? Nature (Lond.), *252*: 41–43.

Summers, R. G., and B. L. Hylander. 1974. An ultrastructural analysis of early fertilization in the sand dollar, *Echinarachnius parma*. Cell Tissue Res., *150*: 343–368.

Swann, K., and M. Whitaker. 1985. Stimulation of the Na/H exchanger of sea urchin eggs by phorbol ester. Nature (Lond.), *314*: 274–277.

Swann, K., and M. Whitaker. 1986. The part played by inositol triphosphate and calcium in the propagation of the fertilization wave in sea urchin eggs. J. Cell Biol., *103*: 2333–2342.

Tilney, L. G. et al. 1978. The polymerization of actin. IV. The role of Ca^{++} and H^+ in the assembly of actin and in membrane fusion in the acrosomal reaction of echinoderm sperm. J. Cell Biol., *77*: 536–550.

Tsunoda, Y., and M. C. Chang. 1978. Effect of antisera against eggs and zonae pellucidae on fertilization and development of mouse eggs *in vivo* and in culture. J. Reprod. Fertil., *54*: 233–237.

Turner, P. R., M. P. Sheetz, and L. A. Jaffe. 1984. Fertilization increases the polyphosphoinositide content of sea urchin eggs. Nature (Lond.), *310*: 414–415.

Tzartos, S. J. 1979. Inhibition of *in–vitro* fertilization of intact and denuded hamster eggs by univalent anti–sperm antibodies. J. Reprod. Fertil., *55*: 447–455.

Uzzell, T. M. 1964. Relations of the diploid and triploid species of the *Ambystoma jeffersonianum* complex. Copeia, *1964*: 257–300.

Vacquier, V. D. 1975. The isolation of intact cortical granules from sea urchin eggs: Calcium ions trigger granule discharge. Dev. Biol., *43*: 62–74.

Vacquier, V. D., and G. W. Moy. 1977. Isolation of bindin: the protein responsible for adhesion of sperm to sea urchin eggs. Proc. Natl. Acad. Sci. U.S.A., *74*: 2456–2460.

Ward, G. et al. 1986. Chemotaxis of *Arbacia punctulata* spermatozoa to resact, a peptide from the egg jelly layer. J. Cell Biol., *101*: 2324–2329.

Wassarman, P. M. 1987. The biology and chemistry of fertilization. Science, *235*: 553–560.

Wassarman, P. M. 1990. Profile of a mammalian sperm receptor. Development, *108*: 1–17.

Whitaker, M. J., and R. F. Irvine. 1984. Inositol 1,4,5–triphosphate microinjection activates sea urchin eggs. Nature (Lond.), *312*: 636–639.

Wilson, E. B., and A. P. Mathews. 1895. Maturation, fertilization, and polarity in the echinoderm egg. New light on the "quadrille of the centers." J. Morphol., *10*: 319–342.

Wolf, D. P. 1978. The block to sperm penetration in zona–free mouse eggs. Dev. Biol., *64*: 1–10.

Wood, D. M., C. Liu, and B. S. Dunbar. 1981. Effect of alloimmunization and heteroimmunization with zonae pellucidae on fertility in rabbits. Biol. Reprod., *25*: 439–450.

Woodland, H. R. 1974. Changes in the polysome content of developing *Xenopus laevis* embryos. Dev. Biol., *40*: 90–101.

Yanagimachi, R. 1957. Some properties of the sperm activating factor in the micropyle area of the herring egg. Anat. Zool. Japan, *30*: 114–119.

Yanagimachi, R. 1970. The movement of golden hamster spermatozoa before and after capacitation. J. Reprod. Fertil., *23*: 193–196.

Yanagimachi, R., and Y. D. Noda. 1970. Electron microscope studies of sperm incorporation into the golden hamster egg. Am. J. Anat., *128*: 429–462.

Zamboni, L., J. Chakraborty, and D. M. Smith. 1972. First cleavage division of the mouse zygote. Biol. Reprod., *7*: 170–193.

Zimmermann, A. M., and S. Zimmermann. 1967. Action of colcemid in sea urchin eggs. J. Cell Biol., *34*: 483–488.

Zucker, R. S., and R. A. Steinhardt. 1978. Prevention of the cortical reaction in fertilized sea urchin eggs by injection of calcium–chelating ligands. Biochim. Biophys. Acta, *541*: 459–466.

Cleavage: Becoming Multicellular

Edmund B. Wilson (1856–1939). A Professor at Williams College, the Massachusetts Institute of Technology, Bryn Mawr College, and Columbia University, E. B. Wilson was a pioneer in applying cytological methods to the study of cleavage and embryonic development. His monograph "The Cell in Development and Heredity" (Wilson, 1925) is still the classic work on the cytology of cleaving embryos. (Photograph courtesy of the Marine Biological Laboratory Library, Woods Hole, MA.)

After fertilization, the zygote begins a series of rapid and highly synchronous cell divisions that cleave it into multiple cells, or **blastomeres**. These rapid cleavages have two functions. First, cleavage reduces the volume of the blastomeres until they reach a size that is typical of somatic cells. Second, during cleavage the blastomeres begin to acquire differences that will eventually cause them to develop into different cell types. During cleavage, the foundation is being laid for regional specialization of the embryo, which will occur during later development. The developmental consequences of cleavage will be examined in Chapter 11. In this chapter, we discuss the mechanism and patterns of cleavage during the formation of the early embryo.

5–1 Cleaving Embryos Have Modified Cell Cycles

Mature, unfertilized eggs are arrested in DNA synthesis and cell division. Fertilization triggers the zygote to re-enter the **cell cycle**. The cell cycle is the period between the formation of a cell by the division of its parent cell and the time when the cell itself divides to form two new cells. During the cell cycle, DNA synthesis alternates with cell division so that the DNA content of the species is maintained. DNA synthesis and cell division are usually separated by long intervals, or "gaps." A somatic cell cycle consists of four phases (Fig. 5.1): the gap between the completion of cell division and the initiation of DNA synthesis, the G_1 phase; the period of DNA synthesis, or the **S phase**; the gap between the completion of DNA synthesis and the next cell division, the G_2 phase; and the period of cell division, the **M phase**. The somatic cell cycle also includes a growth period in which newly divided cells gradually increase in size until they become as large as the parent cell before dividing again. However, the growth phase is bypassed in the cell cycles of cleaving embryos. Without growth, the volume of the blastomeres is progressively reduced at each division until it approximates that of a typical somatic cell. The decrease in blastomere size is caused by a reduction in the amount of cytoplasm; therefore, the ratio of cytoplasmic to nuclear volume decreases during cleavage. For example, the cytoplasmic volume of an unfertilized sea urchin egg is about 550 times that of its nucleus, but by the end of the cleavage period, this ratio is only 6 to 1 (Brachet, 1950).

Once the normal cytoplasmic/nuclear volume ratio is established, both the rate and synchrony of cell division decrease to levels that are more

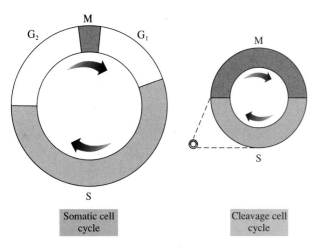

Figure 5.1

A comparison of the cell cycles of a somatic cell and a blastomere from a cleaving embryo. The drawing on the left represents the somatic cell cycle of a HeLa (human) cell showing M, G_1, S, and G_2 phases. The drawing on the right represents an enlarged version of the cleavage cell cycle of a sand dollar embryo showing only M and S phases. The very small circle from which the cleavage cell cycle is enlarged represents the sand dollar embryo cell cycle on the same scale as the HeLa cell cycle.

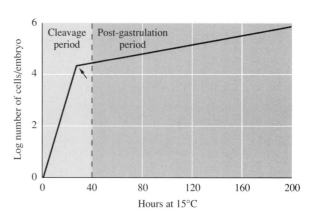

Figure 5.2

The rate of increase in cell number during early development of the frog *Rana pipiens*. Cell number increases rapidly and constantly during the cleavage period. At the end of the cleavage period, there is an abrupt mid-blastula transition (arrow) to a much slower rate of cell division. The dashed line indicates gastrulation. (After Sze, 1953.)

typical of somatic cells. In amphibians, there is an abrupt change in the duration and synchrony of the cell cycle (Fig. 5.2), which is referred to as the **mid-blastula transition (MBT)**. The MBT has been studied extensively in *Xenopus* and is an important milestone in development, which will be discussed in more detail in Chapter 13. In other animals, the transition from a cleavage to a somatic cell cycle is more gradual.

Another characteristic of cleaving embryos is that they have very short cell cycles relative to those of somatic cells. How do cleaving embryos shorten the time between cell divisions? This is accomplished by eliminating the G_1 and G_2 phases and reducing the lengths of the S and M phases. The absence of G_1 and G_2 can be demonstrated by experiments in which radioactive thymidine, a precursor of DNA, is used to determine the length of the S phase (Ito et al., 1981). The results show that S phase begins during telophase of one mitotic division and extends to prophase of the next mitotic division. Thus, S and M phases alternate without intervening gaps during the cleavage period (see Fig. 5.1).

Although S and M phases are both reduced in length during cleavage, the change in the duration of DNA synthesis is particularly dramatic. For example, *Drosophila* somatic cells have a 10-hour S phase (Dolfini et al., 1970), whereas DNA synthesis lasts only 3 to 4 minutes in cleaving embryos. The shortening of S phases during cleavage is due to simultaneous activation of multiple sites of DNA synthesis initiation in each chromosome (Kriegstein and Hogness, 1974). During the somatic S phase, there are fewer initiation sites, and they are activated less synchronously.

The short cell cycles of cleaving embryos are regulated by the cytoplasm rather than the nucleus. This can be shown by interspecific fertilization between sea urchin species that exhibit different rates of cleavage (Harvey, 1956). In these interspecific hybrids, the rate of cell division is always the same as in the maternal species, even if the egg nucleus has been removed before fertilization. In another experiment, adult brain cell nuclei, which are arrested in the G_1 phase of the cell cycle and do not synthesize DNA, were injected into amphibian eggs (Graham et al., 1966). Once within the egg cytoplasm, the brain cell nuclei are stimulated to undergo DNA synthesis and division on the same schedule as the zygotic nucleus.

As we learned in Chapter 3, cell cycles are under elaborate molecular control. Although individual cell cycles may have their own unique characteristics, they are variations of a common theme and utilize the same regulatory factors, including cdc2 and cyclin (see Chap. 3). However, research during the last decade has revealed that proteins encoded by the proto-oncogenes are also intimately involved. As we shall discuss in Chapter 9, proto-oncogenes were discovered as normal cellular homologues of viral oncogenes that can deregulate the cell cycle, resulting in cancer. It is going to be exciting to learn how the fine tuning of the cell cycle is regulated by these various factors at the molecular level. This work has important medical consequences as well, because overexpression or mutation of the proto-oncogenes can also cause normal cells to develop into cancer cells.

5–2 What Is the Mechanism of Cleavage?

Cell division consists of **karyokinesis**, the division of the nucleus, and **cytokinesis**, the division of the cytoplasm. Normally, these events occur in concert, so that daughter nuclei are distributed to separate cells. However, they can be uncoupled. For example, if the chromosomes and mitotic apparatus (see below) are removed from a fertilized sea urchin egg at late metaphase, cytokinesis will still take place, and two anucleate blastomeres will be formed. Alternatively, cytokinesis can be blocked by cytochalasin B, which disrupts the function of microfilaments, but karyokinesis will continue, forming multinucleate cells. Later in this chapter, we shall see that karyokinesis without intervening cytokinesis occurs normally during early development of insect embryos.

Karyokinesis is controlled by the **mitotic apparatus**. The mitotic apparatus is a cage-like structure consisting of **spindle fibers** with attached chromosomes, paired centrioles at either pole, and the **asters**, which emanate from the centrioles toward the periphery of the cell (Fig. 5.3A). During mitosis, the chromosomes attach to spindle microtubules, become aligned at the metaphase plate, and are withdrawn into the daughter cells. The spindle fibers and asters are composed of microtubules. The presence of

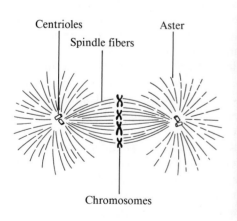

A   B

Figure 5.3

The mitotic apparatus of cleaving eggs. A, Schematic representation. B, A sea urchin egg stained with fluorescent tubulin antibody at anaphase to show the mitotic apparatus. (B, From Harris et al., 1980. Reprinted from *The Journal of Cell Biology,* 1980, *84*: 668–679 by copyright permission of the Rockefeller University Press.)

microtubules in the mitotic apparatus of sea urchin embryos can be demonstrated by staining with tubulin antibodies (see the essay on p. 344 for a discussion of immunohistochemistry), or by treating cleaving embryos with colcemid, which depolymerizes microtubules (Zimmerman and Zimmerman, 1967). Tubulin antibodies stain both the spindle fibers and the asters (see Fig. 5.3B). Colcemid treatment causes the mitotic apparatus to disappear, leaving the chromosomes stranded at the metaphase plate.

The nuclear envelope disintegrates during prophase of every mitotic division and then reforms in a two-step process at the end of karyokinesis. The first step involves the assembly of individual nuclear envelopes around each chromosome (these are called **chromosomal vesicles**). The second step involves the fusion of the chromosomal vesicles into a single interphase nucleus (Ito et al., 1981). Unlike ribosomes and mitochondria, which accumulate to high levels during oogenesis and function during early development (see Chap. 3), nuclear envelope is not stored in the egg. Instead, the egg stockpiles nuclear envelope precursors, which are rapidly assembled into envelopes during cleavage. The existence of a pool of nuclear envelope precursors can be demonstrated by injecting DNA into cleaving amphibian eggs (Forbes et al., 1983). Shortly after the DNA enters the egg cytoplasm, it is complexed with cytoplasmic proteins to form chromatin, and a complete nuclear envelope is assembled around the chromatin. Enough nuclear envelope precursors are stored in the egg cytoplasm to sustain the embryo throughout the cleavage period.

Cytokinesis does not begin until the chromosomes have initiated their movements toward the cell poles. This ensures that each daughter cell will receive a nucleus. The **cleavage furrow** is a constriction of the egg surface that splits the egg or blastomere into two parts during telophase (Fig. 5.4; also see Fig. 5.17B). The cleavage furrow is formed by the activity of a thickened region of the egg cortex, the **contractile ring** (Szollosi, 1970), which is present only during cytokinesis (Fig. 5.5). The contractile ring pinches the zygote into two parts by a mechanism similar to muscle contraction (Schroeder, 1973). The presence of actin and myosin in the contractile ring has been demonstrated by several different experiments. First, eggs treated with cytochalasin B do not form a contractile ring or cleavage furrow (Schroeder, 1973). Second, actin and myosin (Fig. 5.6) antibodies stain the contractile ring. Third, when myosin antibodies are ·injected into one of the blastomeres of a two-cell starfish embryo, the injected blastomere does not cleave, although its nucleus continues to multiply (Fig. 5.7). The results of these experiments indicate that the cleavage furrow contains an actin-myosin contractile system.

The rapid cell divisions that occur during the cleavage period result in a tremendous increase in the surface area of the zygote. Different mechanisms are used to provide the large amounts of new plasma membrane that are required for this expansion. Sea urchin zygotes use excess membrane produced during exocytosis of their cortical granules. As described in Chapter 4, excess cortical granule membrane is incorporated into numerous long microvilli that form at the surface of the egg after fertilization. During cleavage, these microvilli are progressively shortened and eventually disappear as their membrane is taken up in the surface of newly formed blastomeres (Schroeder, 1981). Amphibian embryos expand their cell surface by the insertion of cytoplasmic vesicles into the plasma membrane in the region of the cleavage furrow (de Laat and Bluemink, 1974). The new cell surface is easily distinguished in the amphibian embryo because it differs in pigmentation from the old cell surface. As shown in Figure 5.8, the cortex below the new plasma membrane is unpigmented, whereas the old membrane retains dark cortical pigment granules inherited from the egg. Thus, by following the pattern of pigmentation in cleaving amphibian embryos, it has

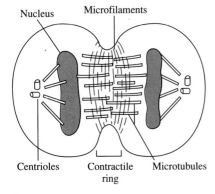

Figure 5.4

A diagram of the relationship between the cleavage furrow, contractile ring, and mitotic apparatus of a cleaving egg at telophase. The contractile ring of microfilaments constricts the egg. The microtubules of the spindle extend from the centrioles and the region of the telophase nuclei through the cleavage furrow. (After Schroeder, 1973.)

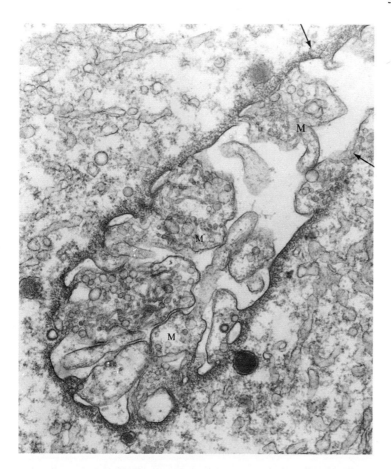

Figure 5.5

An electron micrograph of the leading edge of a cleavage furrow in the squid embryo. The furrow is caused by a band of microfilaments (the contractile ring) just below the plasma membrane. The arrows indicate the extent of the contractile ring. Short microvilli (M) extend into the cleavage furrow from the surface of the cleaving blastomeres. × 29,000. (Courtesy of J. M. Arnold.)

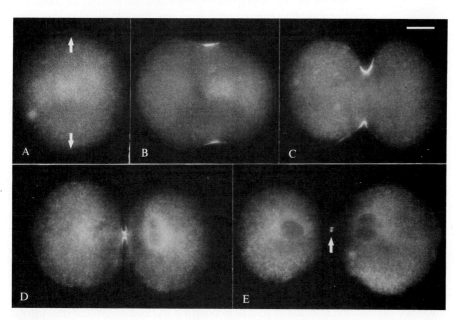

Figure 5.6

Localization of myosin in the cleavage furrow of cleaving sea urchin blastomeres. Bright regions within the cleavage furrow represent myosin detected with a fluorescent myosin antibody. A-E, Various stages in cleavage of an animal hemisphere blastomere isolated from an eight-cell embryo resulting in the formation of two daughter cells at the 16-cell stage. Arrows in A represent the future plane of the cleavage furrow before the contractile ring becomes visible by myosin staining. Arrow in E represents the remnant of the cleavage furrow after the completion of cytokinesis. Scale bar equals 10 μm. (From Schroeder, 1987.)

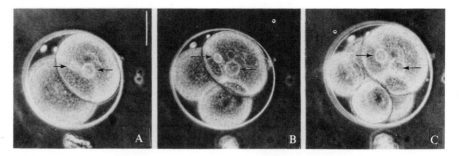

Figure 5.7

Cleavage of a starfish embryo in which one of the blastomeres was injected with myosin antibody at the two-cell stage. The uninjected blastomere cleaves twice, while the injected blastomere does not cleave. The injected blastomere is marked by two oil droplets (arrows) which were injected with the antibody. Scale bar equals 50 μm. (From Mabuchi and Okuno, 1977. Reprinted from *The Journal of Cell Biology*, 1977, *74*: 251–263 by copyright permission of the Rockefeller University Press.)

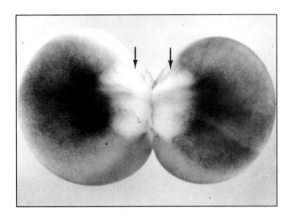

Figure 5.8

Formation of new unpigmented membrane (arrows) in the animal pole region of a cleaving amphibian egg. The old membrane is still darkly pigmented. (From de Laat and Bluemink, 1974. Reprinted from *The Journal of Cell Biology*, 1974, *60*: 529–540 by copyright permission of the Rockefeller University Press.)

been determined that the apposing walls of the blastomeres are composed of new plasma membrane.

5–3 How Is Cytokinesis Regulated?

As described in the previous section, cytokinesis is mediated by the contractile ring. The contractile ring usually cuts through the midpoint of the spindle on a plane perpendicular to the long axis of the mitotic apparatus. The role of the mitotic apparatus in determining the site of cleavage furrow formation can be demonstrated by centrifuging or deforming the egg (Harvey, 1935; Rappaport and Ebstein, 1965; Rappaport, 1976). These manipulations displace the mitotic apparatus to a new position in the cell. When the mitotic apparatus is displaced *before* metaphase, the contractile ring forms perpendicular to the new position of the displaced mitotic apparatus. By placing the mitotic apparatus in various regions of the egg, the cleavage furrow will form at virtually any site on the egg surface. In contrast, when the mitotic apparatus is displaced *after* metaphase, a cleavage furrow still forms in the expected position. Thus, the mitotic apparatus appears to dictate the site of contractile ring formation by a process that is completed during metaphase.

The Asters Regulate Cleavage Furrow Formation

Is it the spindle itself or another part of the mitotic apparatus that regulates furrowing? The following observations and experiments suggest that the asters, rather than the spindle, control the site of contractile ring formation during cleavage. First, during cellular blastoderm formation in *Drosophila* (see Fig. 5.33), cleavage furrows form across the region occupied by the spindle as well as between adjacent asters not connected by spindles. Second, parthenogenically activated sea urchin eggs lack spindles but sometimes develop cleavage furrows between their asters (Harvey, 1956). Third, when the spindle is sucked out of a sea urchin egg with a micropipette, a cleavage furrow still forms between the two remaining asters (Hiramoto, 1971). If one of the asters is removed, however, no cleavage furrow forms. Finally, by the imaginative experiment shown in Figure 5.9, Rappaport (1961) showed that cleavage furrows can form between asters that are not connected by a spindle. When a small glass ball is pressed into the center of a sand dollar egg, the shape of the egg is changed from a sphere to a torus, and the mitotic apparatus is displaced to one side of this structure. During first cleavage, a furrow appears only on the side of the torus containing the mitotic apparatus. When cytokinesis is completed, a horse-shoe-shaped egg is formed, containing two nuclei. As expected, during the second cleavage the horseshoe-shaped egg forms three cleavage furrows simultaneously. Two of these furrows form in the arms of the horseshoe

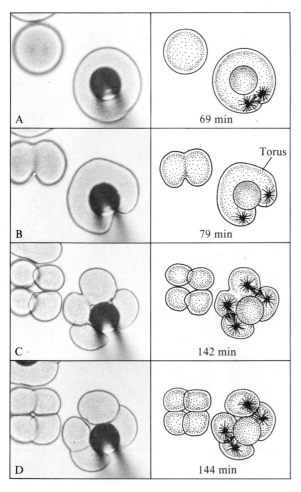

Figure 5.9

Cleavage of a torus-shaped sand dollar egg. The photographs taken during the experiment are shown on the left of each frame and are redrawn on the right of each frame. A, The position of the mitotic apparatus in a one-cell, torus-shaped egg 69 minutes after fertilization. B, The first cleavage of the torus-shaped egg occurs on one side to form a single horseshoe-shaped blastomere at 79 minutes after fertilization. The first cleavage plane is through the spindle. C, The position of each mitotic apparatus in the horseshoe-shaped embryo immediately before the second cleavage at 142 minutes after fertilization. D, The second cleavage of the horseshoe-shaped egg occurs in three positions simultaneously at 144 minutes after fertilization: between the spindles of each mitotic apparatus and between the asters on the opposite side of the egg. Four cells are eventually formed, each containing a single nucleus. Note the division synchrony of the deformed egg with the adjacent control and that the glass ball is present throughout the experiment to maintain the deformation of the egg. (After Rappaport, 1961.)

through the midpoints of each mitotic apparatus. A third cleavage furrow forms at the bend of the horseshoe between the two asters, *even though they are not connected by a spindle.*

The Asters Induce the Cleavage Furrow by Signaling the Cortex

How do the asters cause furrowing? The asters could function by physically interacting with the egg cortex. One hypothesis is that the contractile ring forms at a position where astral microtubules radiating from opposite poles of the spindle intersect in the equatorial cortex. However, this hypothesis is not confirmed by either experiment or observation. For example, if a needle is positioned in the equator and vigorously moved back and forth during cell division to disrupt any possible physical interactions between astral microtubules and the cortex, a cleavage furrow still forms perpendicular to the midpoint of the spindle (Fig. 5.10). Also, when the cortex of cleaving sea urchin eggs is examined by electron microscopy, it can be seen that astral microtubules fail to penetrate the cleavage furrow region (Asnes and Schroeder, 1979).

Another hypothesis is that the asters elicit contractile ring formation by sending a signal to the cortex. The following experiments indicate that asters do indeed determine the site of cytokinesis by signaling the cortex. The mitotic apparatus can be removed either by sucking it out of the egg with a micropipette (Hiramoto, 1956) or dissolving it by injecting sucrose or seawater into the egg (Hiramoto, 1965). As mentioned earlier, the cleavage furrow forms during telophase at a position that is predetermined at the time of metaphase. When the mitotic apparatus is removed during prometaphase, cleavage furrow formation does not occur. In contrast, when the mitotic apparatus is removed during metaphase or anaphase, a normal cleavage furrow is formed. This result shows that physical connection between the mitotic apparatus and the cortex is not required during the time the contractile ring is forming. Thus, a furrowing signal must pass from the asters to the cortex.

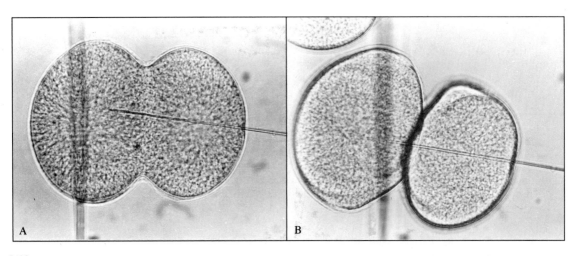

Figure 5.10

Cleavage in a sand dollar egg with a moving needle inserted through the cleavage plane. The needle was swept back and forth during the period between A and B. (From Rappaport, 1966.)

The signaling event seems to be completed in a remarkably short period of time, as demonstrated by Rappaport and Ebstein (1965) in the two part experiment illustrated in Figure 5.11. In the first part of the experiment, the mitotic apparatus is displaced to the cortex on one side of a sand dollar egg during prophase or early metaphase (see Fig. 5.11A and B). Thus, the opposite or distal side of the egg cortex is now too far from the mitotic apparatus to be stimulated to form a cleavage furrow. In the second part of the experiment, the distal egg surface is pushed toward the

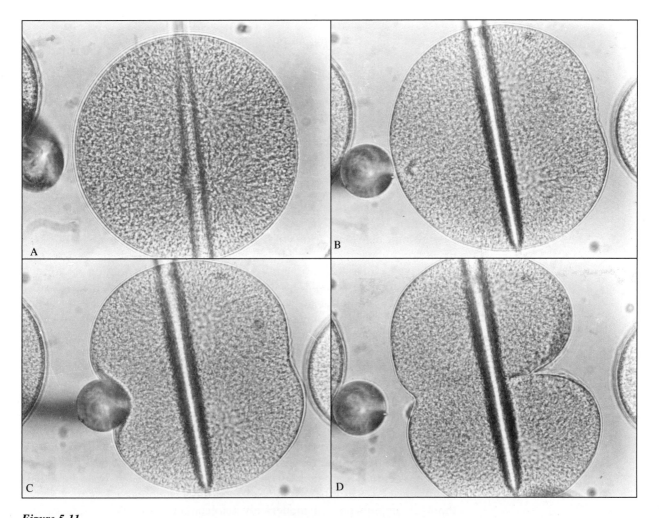

Figure 5.11

A two-part experiment in which the time required for the mitotic apparatus to determine furrowing during cytokinesis of the sand dollar egg was measured. A, and B, In the first part of the experiment, the egg is flattened by distortion under a needle. This displaces the mitotic apparatus to one side of the needle. The cleavage furrow will form unilaterally on the side of the egg near the mitotic apparatus; i.e., cutting inward from the right. C, and D, Meanwhile, in the second part of the experiment, a glass ball is used to push the distal cortical region toward the mitotic apparatus. The distal surface is held in this position for various time periods and then released to determine whether a cleavage furrow will be induced. If the distal surface is pushed toward the mitotic apparatus and held in this position for 1 minute or more, a cleavage furrow will be induced. (From Rappaport and Ebstein, 1965.)

mitotic apparatus with a small glass ball and held there for various periods to allow it to respond to the signal from the asters (see Fig. 5.11C and D). The minimal exposure necessary to initiate furrow formation was 1 minute. Unfortunately, neither the identity of the cleavage signal nor the mechanism of its relay from the asters to the egg cortex is understood at this time.

5–4 Patterns of Cleavage

Three generalizations were formulated by O. Hertwig in 1884 to describe cleavage patterns in animal eggs and are sometimes called Hertwig's rules:

1. The mitotic apparatus is positioned in the center of an uncleaved zygote or blastomere.
2. The long axis of the mitotic apparatus is oriented parallel to the long axis of the uncleaved zygote or blastomere.
3. The plane of cell division is perpendicular to the long axis of the mitotic apparatus and the uncleaved zygote or blastomere.

As we shall see, these are not rules but only generalizations that should not be pushed too far, but they will help us understand the variety of cleavage patterns that occur in animal embryos.

Types of Cleavage

The pattern of cleavage exhibited by a particular animal appears to be affected by two factors in addition to Hertwig's rules. The first factor is the amount and distribution of yolk. Large amounts of yolk tend to displace the mitotic apparatus to an off-center position and inhibit the progress of the cleavage furrow. The quantity of yolk and its distribution in the egg vary markedly between different animals. Some eggs, such as those of most mammals, have little or no yolk. These are known as **alecithal eggs**. Other eggs, such as those of many echinoderms, annelids, mollusks, and tunicates, have modest quantities of evenly distributed yolk. These are known as **isolecithal eggs**. As follows from Hertwig's rules, the mitotic apparatus is localized near the center of alecithal and isolecithal eggs. In these eggs, the first few cleavages divide the zygote into equal-sized blastomeres and the cleavage furrow extends through the entire egg. This is called **holoblastic (or complete) cleavage**. Other eggs, such as those of arthropods, cephalopod mollusks, amphibians, reptiles, and birds, have large or even enormous quantities of yolk. Depending on how the yolk is distributed, these are known as **centrolecithal** or **telolecithal eggs**. Yolk is packed into the center of centrolecithal eggs, such as those of insects, whereas it is accumulated mainly in the vegetal hemisphere of moderately telolecithal eggs, such as those of amphibians. In moderately telolecithal eggs, yolk displaces the mitotic apparatus into the animal hemisphere, and more cleavage furrows are initiated in this region. This displacement results in unequal holoblastic cleavage and retardation of cleavage furrow progress in the vegetal hemisphere. Yolk fills both hemispheres in extremely telolecithal eggs, such as those of fish, reptiles, and birds—displacing the mitotic apparatus to a small disc of cytoplasm at the animal pole. This usually prevents the formation of complete cleavage furrows. Therefore, this type of cleavage is called **meroblastic (or incomplete) cleavage**.

The second factor that affects the cleavage pattern is the orientation of the mitotic apparatus within the non-yolky cytoplasmic region of the egg. When the mitotic apparatus is perpendicular or parallel with respect to the animal-vegetal axis, cleavage is regular. Most animals exhibit regular cleavage. When the mitotic apparatus is tilted with respect to the animal-

Table 5–1 Summary of Cleavage Patterns

Cleavage pattern	Abundance and distribution of yolk	Animal groups
Radial holoblastic	Isolecithal	Echinoderms
Bilateral holoblastic	Isolecithal	Tunicates
	Moderately telolecithal	Amphibians
Bilateral meroblastic	Extremely telolecithal	Cephalopod mollusks, fish, reptiles, birds, monotreme mammals
Rotational holoblastic	Alecithal	Placental mammals
	Isolecithal	Nematodes
Spiral	Isolecithal	Flatworms, some nemertines, annelids, noncephalopod mollusks
Superficial	Centrolecithal	Most arthropods

vegetal axis, cleavage is oblique or spiral. Spiral cleavage occurs in flatworms, annelids, and non-cephalopod mollusks.

Based on further geometric considerations, cleavage is divided into five categories: **radial**, **bilateral**, **rotational**, **superficial**, and **spiral**. Within these types, cleavage can also be either holoblastic or meroblastic and equal or unequal. For example, a particular cleavage pattern may be described as either radial, holoblastic, and equal or bilateral, meroblastic, and unequal. The cleavage patterns of various animals are summarized in Table 5–1.

Radial Cleavage

Radial cleavage is distinguished by two features. First, the mitotic apparatus of each blastomere is oriented in a plane parallel or perpendicular to the plane of the egg's animal-vegetal axis, a characteristic shared with bilateral and rotational cleavage. Second, *any plane* passing through the animal-vegetal axis of a radially cleaving embryo divides it into symmetrical halves (Fig. 5.12).

Sea Cucumbers

The sea cucumber *Synapta* shows radial cleavage. In addition to its radial cleavage, the sea cucumber also exhibits equal and holoblastic cleavage throughout early development (Fig. 5.13). Before first cleavage, the mitotic apparatus becomes oriented perpendicular to the animal-vegetal axis. Thus, the first cleavage furrow cuts through the animal-vegetal axis, creating two equal-sized blastomeres. During the second cleavage, the mitotic apparatus in each blastomere is again oriented perpendicular to the animal-vegetal axis, but at right angles to the plane of first cleavage. Therefore, the second cleavage furrow also runs through the animal-vegetal axis, splitting the egg into four equal-sized cells. The first two cleavages of the sea cucumber are called meridional because the cleavage furrows pass through the animal and vegetal poles. Third cleavage, however, is equatorial; cleavage furrows pass through the equator of each blastomere. Third cleavage results in an eight-

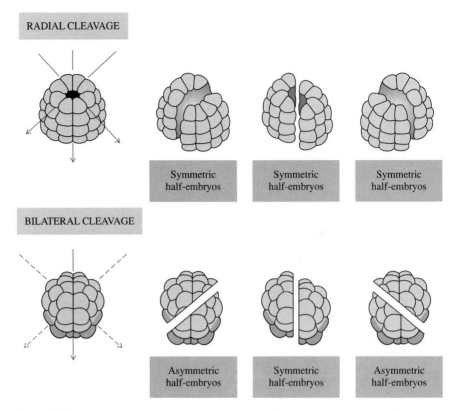

RADIAL CLEAVAGE

Symmetric half-embryos

Symmetric half-embryos

Symmetric half-embryos

BILATERAL CLEAVAGE

Asymmetric half-embryos

Symmetric half-embryos

Asymmetric half-embryos

Figure 5.12

A comparison of radial and bilateral cleavage. The diagrams show 32-cell sea cucumber (radial cleavage) and tunicate (bilateral cleavage) embryos viewed from the animal pole. In radial cleavage, every plane (solid lines) through the animal-vegetal axis bisects the embryo into symmetric halves. In bilateral cleavage, there is only one plane through the animal-vegetal axis (solid line) that bisects the embryo into symmetric halves. All other planes (dashed lines) divide the embryo into asymmetric halves.

cell embryo consisting of equal-sized blastomeres arranged in an animal and a vegetal quartet. The fourth cleavage is meridional again, dividing the quartets into animal and vegetal tiers, each consisting of eight equal-sized cells.

After several cleavages, embryos with holoblastic cleavage usually appear as a solid cluster of blastomeres. At this stage of development, the embryo is called a **morula** (Latin word for mulberry) because of its resemblance to a mulberry. The fifth cleavage is parallel to the equator, dividing each tier of eight cells into upper and lower layers of 16 cells. During subsequent divisions of the zygote into 64, 128, and 256 blastomeres, meridional cleavages alternate with equatorial cleavages.

After the morula stage, a fluid-filled cavity, the **blastocoel**, forms and gradually increases in size during the subsequent cell divisions. The cells surrounding this cavity form an epithelial layer, the **blastoderm**, and the embryo is called a **blastula**. A relatively simple blastula is found in embryos that exhibit radial holoblastic cleavage. The distinguishing feature of this blastula is that every cell in the blastoderm has an outer surface that contacts the hyaline layer (see p. 147) and an inner surface that contacts the blastocoel.

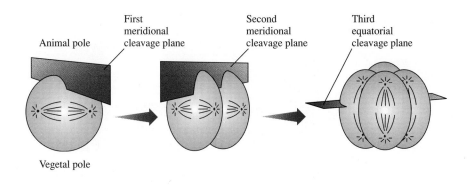

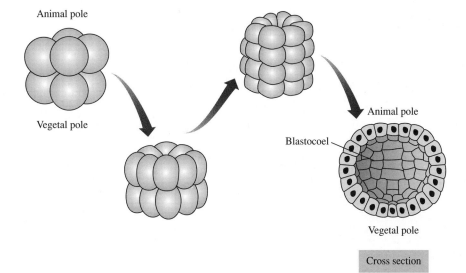

Figure 5.13

Radial holoblastic cleavage in the sea cucumber *Synapta* showing the planes of the first two meridional cleavages and the third equatorial cleavage, the formation of the morula, and a section of the blastula showing the highly regular blastocoel. (Redrawn with permission of Macmillan Publishing Company from *Developmental Biology* by J. W. Saunders, Jr. Copyright © 1982 by John W. Saunders, Jr.)

Sea Urchins

Sea urchins also show radial holoblastic cleavage, but important modifications occur during the fourth cleavage (Fig. 5.14). The first three cleavages in sea urchins are identical to those of the sea cucumber and result in an eight-cell embryo with animal and vegetal quartets of equal-sized blastomeres. During fourth cleavage, the animal quartet divides meridionally, forming eight equal-sized blastomeres: the **mesomeres**. In contrast, cleavage in the vegetal quartet is highly biased, producing an upper tier of four large cells, the **macromeres**, and a lower tier of four tiny cells, the **micromeres**.

The unequal division forming the macromeres and micromeres occurs because the mitotic apparatus shifts to a new position near the vegetal pole and becomes oriented parallel to the animal-vegetal axis. The displacement and rotation of the mitotic apparatus is controlled by the absolute time lapsed from fertilization to fourth cleavage (Hörstadius, 1939). A biological clock controlling micromere formation was discovered by placing cleaving zygotes in hypotonic seawater between the second and third cleavage. Further cleavage is inhibited while the embryos are in hypotonic seawater, but the clock continues to function. Thus, when embryos are returned to normal seawater and cleavage resumes, micromeres are formed at the third instead of the fourth cleavage. The mechanism that controls the timing of micromere formation is unknown.

At the 16-cell stage, sea urchin embryos consist of three tiers of different-sized blastomeres, the mesomeres, macromeres, and micromeres.

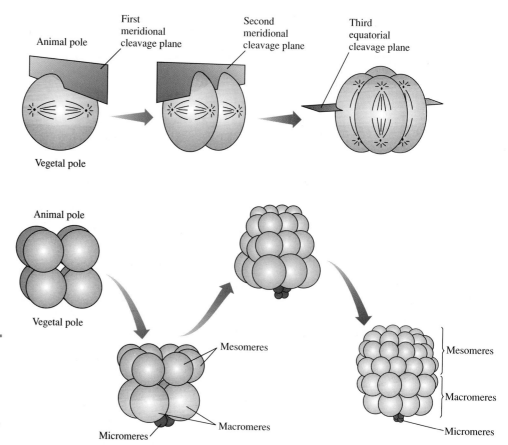

Figure 5.14

Radial holoblastic cleavage in the sea urchin *Lytechinus* showing the planes of the first three cleavages and the formation of mesomeres, macromeres, and micromeres during the unequal fourth cleavage. (After Gilbert, 1988.)

These relative size differences are maintained during subsequent cleavages as the blastula is formed. The blastoderm cells form cilia on their outer surfaces, and the embryo begins to rotate within the fertilization envelope. Embryos escape into the surrounding environment by digesting the fertilization envelope with a specialized enzyme. This process is called hatching, and the swimming embryos are known as **hatched blastulae**. The hatched blastula develops an apical tuft of long, stiff cilia at its animal pole and retains its spherical shape until the beginning of gastrulation (see p. 216).

The simple cleavage pattern found in echinoderms is not typical of most other animals. However, most animals maintain a relatively simple cleavage pattern through the third division. This pattern consists of two meridional divisions through the animal-vegetal axis followed by a third equatorial division.

Bilateral Cleavage

Most animals exhibit bilateral cleavage. Bilateral cleavage differs from radial cleavage in one important aspect: there is *only one plane* that divides a bilaterally cleaving embryo into symmetrical halves (see Fig. 5.12). Bilateral cleavage can be holoblastic or meroblastic.

Tunicates

The most striking form of bilateral holoblastic cleavage occurs in tunicates (Fig. 5.15). Eggs of the tunicate *Styela* contain different colored cytoplasmic regions (see p. 448). Even before first cleavage, the colored cytoplasmic regions are localized in a fashion that reflects the future bilateral symmetry

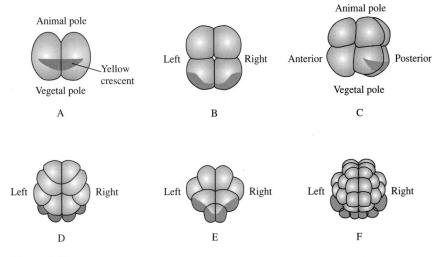

Figure 5.15

Bilateral holoblastic cleavage in the tunicate *Styela*. A, A two-cell embryo viewed from the posterior showing the position of the yellow crescent and the first meridional cleavage, which divides the egg into two blastomeres containing yellow crescent material. B, A four-cell embryo viewed from the vegetal pole showing the second meridional cleavage, which divides the embryo into two anterior blastomeres and two posterior blastomeres. Only the posterior blastomeres contain yellow crescent material. C, An eight-cell embryo viewed from the left side showing the third meridional cleavage, which divides the embryo into animal and vegetal blastomere tiers of four cells each. Only the two posterior vegetal blastomeres contain yellow crescent material. D, A 16-cell embryo viewed from the animal pole showing the irregular cleavage planes. E, A 16-cell embryo viewed from the vegetal pole showing irregular cleavage planes that are different from those in the animal hemisphere. The yellow crescent is distributed to four vegetal posterior cells. F, A 32-cell embryo viewed from the animal pole. Yellow crescent material is present in six vegetal posterior blastomeres.

of the embryo. For example, a **yellow crescent** is present below the equator in the future posterior region of the embryo. The plane of the first cleavage furrow, which is meridional, divides the yellow crescent and other colored regions into symmetrical halves (see Fig. 5.15A). The second meridional cleavage divides the zygote into four equal-sized blastomeres, and the four-cell embryo is divided into eight cells by an equatorial third cleavage. The subsequent cleavages of tunicates deviate substantially from those of echinoderms, with different patterns expressed by blastomeres in the animal and vegetal hemispheres. At the sixth cleavage, a blastula is formed consisting of 64 cells.

Amphibians

The cleavage pattern of most amphibian eggs is also holoblastic and usually bilateral. Because they are also moderately telolecithal, the mitotic apparatus is displaced into the animal hemisphere of the uncleaved zygote, and cleavage furrows are retarded by yolk in the vegetal hemisphere (Hara, 1977). Figures 5.16 and 5.17 show cleavage in amphibian embryos. The first cleavage is meridional, but because of the displacement of the mitotic apparatus, it is initiated in the animal hemisphere region and retarded in the vegetal hemisphere. The first cleavage plane usually bisects the **gray**

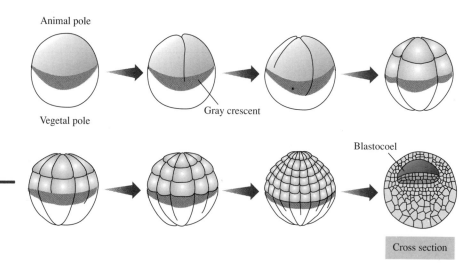

Figure 5.16

Holoblastic cleavage in an amphibian. The last frame is a section through the bastula showing the acentric blastocoel. (After Carlson, 1981.)

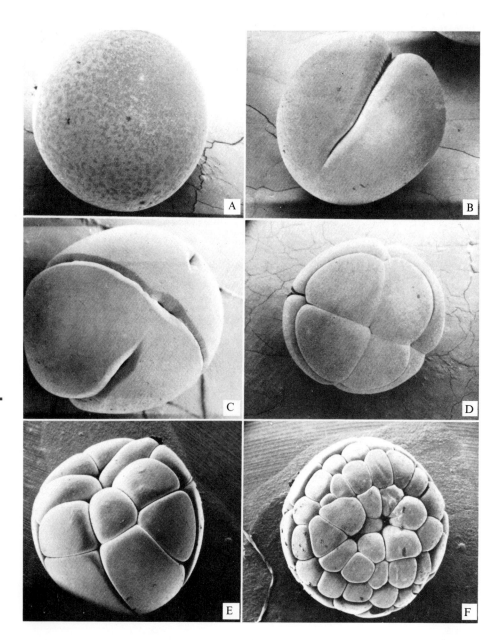

Figure 5.17

Scanning electron micrographs of cleavage in a frog egg. A, An uncleaved egg. B, First meridional cleavage viewed from a lateral side. C, Second meridional cleavage viewed from the vegetal hemisphere. D, An eight-cell embryo viewed from the animal pole. E, A 16-cell embryo viewed from a lateral side. F, An early blastula viewed from the animal pole. (From Kessel and Shih, 1974.)

crescent, a part of the egg surface that is exposed during a shift of the egg cortex with respect to the inner cytoplasm after fertilization (see p. 199). Before the first cleavage is completed, a second meridional cleavage begins in the animal hemisphere, and the third cleavage is initiated as the first and second cleavage furrows are still slowly cutting through the vegetal hemisphere. As in echinoderms and tunicates, the third cleavage is equatorial, but because the mitotic apparatus is centered in the animal hemisphere of each blastomere, the furrows are formed above the equator of the embryo. Thus, the third cleavage divides the embryo into quartets of small animal micromeres and large vegetal macromeres. Eventually, a blastula is formed with a blastocoel restricted to the animal hemisphere. The roof of the blastocoel is arched and covered with a thin layer of micromeres, whereas the floor is flat and paved with a thick layer of macromeres.

Although amphibian eggs are usually classified as having bilateral cleavage, the topographical relationship between the first cleavage plane and the embryo's plane of bilateral symmetry is actually quite variable (Danilchik and Black, 1988). In addition, because the plane of the first cleavage can be experimentally altered relative to the plane of bilateral symmetry (Black and Vincent, 1988), the frequent congruence of the two is neither obligate nor causal but simply a statistical outcome of the furrow-defining mechanism described in Section 5–3.

Cephalopods, Fish, Reptiles, and Birds

The telolecithal eggs of cephalopod mollusks (Fig. 5.18), fish, reptiles, and birds also exhibit bilateral cleavage, but the plane of bilateral symmetry is

Figure 5.18

Scanning electron micrograph of the blastodisc of a cleaving squid embryo. The arrows indicate the plane of bilateral symmetry. × 200 (Courtesy of J. M. Arnold.)

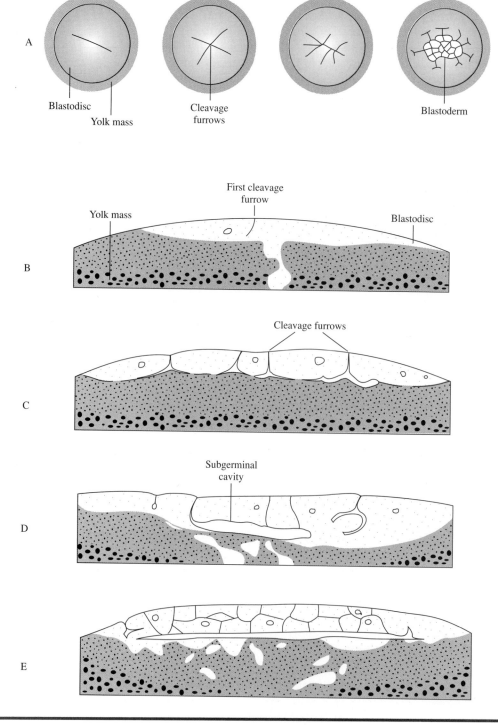

Figure 5.19

Cleavage of the chicken embryo. A, Cleavage observed in the blastodisc from
above the animal pole. B-E, Cleavage observed in sections through the blastodisc.
(After Balinsky, 1981.)

sometimes difficult to identify because the cleavage patterns are meroblastic.
The presence of large quantities of yolk in some meroblastic eggs restricts
the mitotic apparatus and cleavage furrows to a small yolk-free zone at the
animal pole, the **blastodisc**. Consequently, the egg is not completely subdivided into blastomeres (Fig. 5.19A). This form of bilateral meroblastic
cleavage is called **discoidal cleavage**. During discoidal cleavage, the embryo
proper is formed only from the blastodisc; the remainder of the zygote
becomes the yolk sac, which is digested later in development.

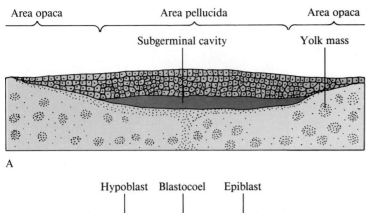

Area opaca   Area pellucida   Area opaca

Subgerminal cavity   Yolk mass

A

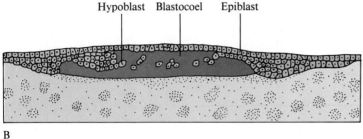

Hypoblast   Blastocoel   Epiblast

B

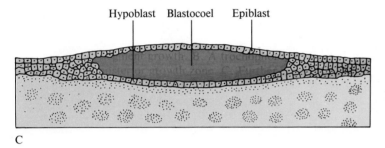

Hypoblast   Blastocoel   Epiblast

C

Figure 5.20

Formation of the blastocoel from hypoblast cells in the chicken embryo. (After Hopper and Hart, 1980.)

Figure 5.19B through E shows cross sections of early cleavages in the chicken embryo. These meridional cleavages occur within the blastodisc. The first cleavage furrow is initiated at the center of the blastodisc and extends in both directions to the edge of the underlying yolk mass. The second cleavage furrow is perpendicular to the first and also does not extend beyond the blastodisc. The first four blastomeres maintain cytoplasmic continuity with the yolk mass, both at their bases and along their lateral margins. During subsequent cleavages, the cells formed in the center of the blastodisc also retain connections with the yolk, forming a single layered blastoderm. Subsequent equatorial cleavages, however, divide the blastoderm into a layer of cells separated from the yolk mass by the **subgerminal cavity**. The subgerminal cavity forms from coalesced smaller cavities that first appear between the yolk layer and cells located at the center of the blastodisc. After the formation of the subgerminal cavity, two areas of the blastoderm can be recognized (Fig. 5.20). The cells that lie directly above the subgerminal cavity comprise the **area pellucida**, which is more translucent than the surrounding **area opaca**. The cells of the area opaca are opaque because they contact the underlying yolk mass directly. Eventually, some blastoderm cells migrate away from the area pellucida and enter the subgerminal cavity, where they continue to divide. These cells form the underlying **hypoblast**, which is separated from the **epiblast** cells above it by a blastocoel. The blastocoel is present in the extreme animal pole region of meroblastic eggs.

Fish eggs are also telolecithal and divide by discoidal cleavage (Fig. 5.21). The blastodisc is elevated as a conspicuous mound at the animal pole, where cleavage occurs by a process similar to that described for the chicken.

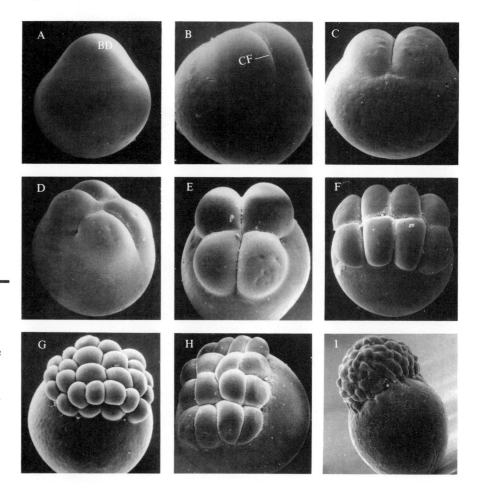

Figure 5.21

Scanning electron micrographs showing discoidal cleavage in the zebrafish embryo. A, The blastodisc (BD) appears as an uplifted portion of the animal pole region of the uncleaved egg. B, The first cleavage furrow (CF) begins to form. C, Two-cell stage. D, and E, Four-cell stage. F, Eight-cell stage. G, 16-cell stage. H, 32-cell stage. I, Blastula. A-H × 95; I × 50. (From Beams and Kessel, 1976.)

Rotational Cleavage

In radial and bilateral cleavage, the first two cleavages are meridional, whereas the third cleavage is equatorial. However, mammals (Lewis and Hartman, 1933; Gulyas, 1975) and nematodes show rotational cleavage. In nematodes, one mitotic apparatus rotates to a position perpendicular to the other and is parallel to the animal-vegetal axis during the second cleavage. In mammals, one of the first two blastomeres rotates 90 degrees with respect to the other before the second cleavage, which occurs in two perpendicular planes. Figure 5.22 compares rotational cleavage in mammals with radial

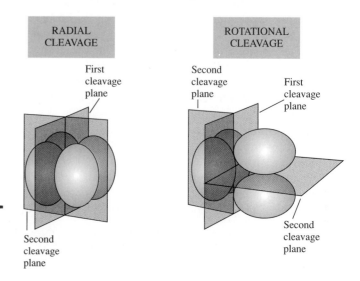

Figure 5.22

A comparison of the first two cleavage planes in (A) a radially-cleaving sea urchin embryo and (B) a rotationally-cleaving rabbit embryo. (After Gulyas, 1975.)

cleavage. In rotational cleavage, the usual situation in which two meridional cleavages are followed by an equatorial cleavage is altered to one in which the first meridional cleavage is followed by another meridional cleavage in one daughter cell and an equatorial cleavage in the other.

In addition to its rotational pattern, mammalian cleavage is holoblastic and equal (Figs. 5.23 and 5.24). Fertilization of mammalian eggs takes place in the uppermost region of the oviduct, and cleavage occurs as the zygote moves downward to the site of implantation in the wall of the uterus. Mammalian cleavages are very slow. The interphase between successive cleavages of mouse embryos lasts 12 to 24 hours. Thus, during the time it takes a mammalian egg to divide once or twice, more rapidly dividing eggs, such as those of amphibians, echinoderms, and insects, are already at advanced developmental stages. Like cleaving embryos of other animals (see Sect. 5–1), however, the cell cycle of mouse embryos lacks G_1 and G_2 phases (Gamow and Prescott, 1970).

The first cleavage divides the mammalian egg into two equal-sized blastomeres. Although the second cleavage is equatorial in one blastomere and meridional in the other, blastomeres of equal size are formed. Later stages of mammalian embryos sometimes contain odd numbers of cells because cleavage becomes slightly asynchronous during subsequent divisions (see Fig. 5.24C). During the interphase between third and fourth cleavage, mammalian blastomeres undergo a process called **compaction**, in which they become tightly packed into a compact morula (Fig. 5.25). During compaction, the cell surface of each blastomere forms intimate associations with its neighbors. These interactions are stabilized by uvomorulin, an adhesive glycoprotein localized on the cell surface (Kemler et al., 1977). When mouse embryos at the compaction stage are treated with antibodies specific for uvomorulin, they lose their adhesive properties and become decompacted (Hyafil et al., 1981) (Fig. 5.25). As compacted embryos proceed through subsequent cleavages, blastomeres become arranged on the inside and outside of the morula. The developmental significance of this arrangement will be discussed in Chapter 11.

Polar body

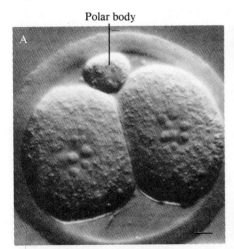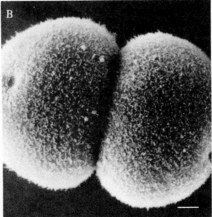

Figure 5.23

First cleavage of the mouse embryo. A, Photograph of a two-cell mouse embryo using Nomarski differential interference optics. The nuclei contain several round nucleoli. Note the relatively large polar body at the animal pole. Scale bar equals 20 µm. B, Scanning electron micrograph after removal of the zona pellucida. Scale bar equals 10 µm. (A, Photograph courtesy of P. Calarco. B, Micrograph by Dr. P. Calarco. From Epstein, 1975.)

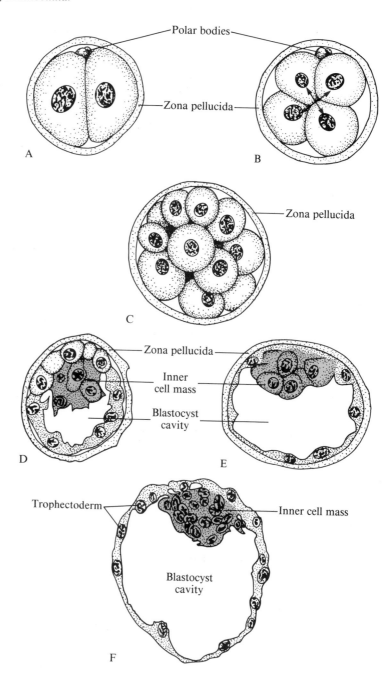

Figure 5.24

Cleavage of the pig embryo. Drawings of sectioned embryos. A, Two-cell stage. B, Four-cell stage. Arrows join sister blastomeres resulting from the second (rotational) cleavage. C, Morula of about 16 cells. D-F, Various stages in blastocyst development. (After B. M. Carlson, *Patten's Foundations of Embryology*, © 1981, McGraw-Hill Book Co.)

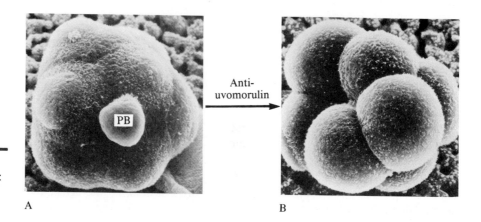

Figure 5.25

Compacted (A) and uncompacted (B) eight-cell mouse embryos. PB: polar body. (From Ducibella, 1977.)

By the 32-cell stage, the blastomeres secrete fluid into spaces within the embryo to create a blastocyst cavity (see Fig. 5.24D to F). This embryonic stage, called the **blastocyst**, is analogous to a blastula. The cells on the outside of the blastocyst form a flattened epithelial layer, the **trophectoderm**. In contrast, the cells on the inside of the blastocyst remain compact and form the **inner cell mass**, a group of cells localized on one side of the blastocyst cavity. Eventually, the inner cell mass is completely separated from the blastocyst cavity by cell processes extending from the trophectoderm cells into the region between the inner cell mass and the cavity. At the blastocyst stage, the embryo is still encased in the zona pellucida. The embryo eventually hatches by escaping through a crack in the zona and becomes implanted in the wall of the uterus, where development continues in close association with the mother.

Spiral Cleavage

Spiral cleavage is the characteristic form of cleavage in annelids and mollusks and also occurs in some flatworms and nemertine worms. The spiral cleavage pattern of these animals is one reason that they are often grouped together as spiralians. The major difference between spiral cleavage and all other forms of cleavage is that the mitotic apparatus is oriented obliquely, rather than parallel to the long axis of the egg or blastomere.

Annelids and Mollusks

Figures 5.26 and 5.27 illustrate spiral holoblastic cleavage in mollusks. The first cleavage is meridional, equal, and slightly inclined, producing two blastomeres, called the AB and CD cells. The second cleavage is also meridional, producing four equal-sized cells. Because the mitotic apparatuses of the AB and CD cells are tilted opposite to one another at the second division, one daughter cell always lies slightly higher than the other with respect to the animal pole (Fig. 5.27). In equally cleaving spiralians, it is conventional to designate the two higher cells as A and C and the two lower cells as B and D. In some spiralians, the first two cleavages are unequal (Fig. 5.28). At the two-cell stage, the larger cell is designated the CD cell. At the four-cell stage, there is one large cell, the D cell, and three smaller cells. As will be discussed in Chapter 11, the large size of the D cell permits its fate to be traced in the embryo.

The third cleavage is equatorial and highly unequal (see Figs. 5.26 to 5.28). The A blastomere, for example, gives rise to a large 1A macromere on its vegetal side and a smaller 1a micromere on its animal side. The B, C, and D blastomeres divide similarly, producing 1B, 1C, and 1D macromeres and 1b, 1c, and 1d micromeres. When the eight-cell embryo is viewed from the animal pole, the upper ends of each mitotic apparatus are tilted clockwise (see arrows in Fig. 5.26). After the next division is complete, the upper quartet of micromeres is displaced obliquely to the right side of the lower tier of sister macromeres. This right-handed or clockwise displacement of the micromeres is known as **dextral spiral cleavage**. Spiralians usually exhibit dextral spiral cleavage, although in some species **sinistral spiral cleavage** is observed. During sinistral spiral cleavage, each mitotic apparatus is tilted counterclockwise, and the micromeres are formed to the left of the macromeres at the third cleavage. As described below, the patterns of dextral and sinistral spiral cleavage are reflected in the symmetry of the adult body organization.

At fourth cleavage, each macromere divides equatorially, giving rise to another macromere on its vegetal side and a micromere on its animal side. Thus, the 1A, 1B, 1C, and 1D cells produce 2A, 2B, 2C, and 2D

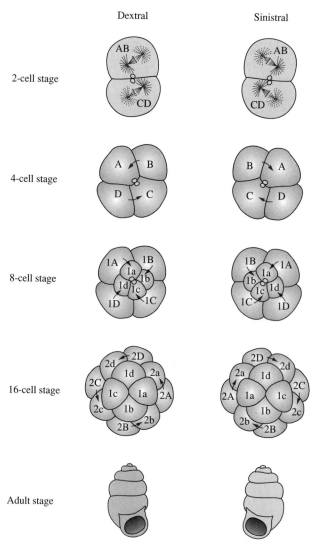

Dextral   Sinistral

2-cell stage

4-cell stage

8-cell stage

16-cell stage

Adult stage

Figure 5.26

A schematic representation of dextral (A) and sinistral (B) spiral cleavage in a snail. Note that snails with dextral coiled shells arise from embryos with a dextral (clockwise) third cleavage and that snails with sinistral coiled shells arise from embryos with a sinistral (counter-clockwise) third cleavage. The arrows represent the orientation of the mitotic apparatus during cleavage. (After Verdonk and van den Biggelaar, 1983.)

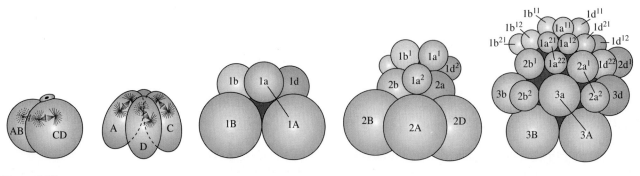

Figure 5.27

Schematic side view of spiral cleavage in a spiralian egg. (After Gilbert, 1988.)

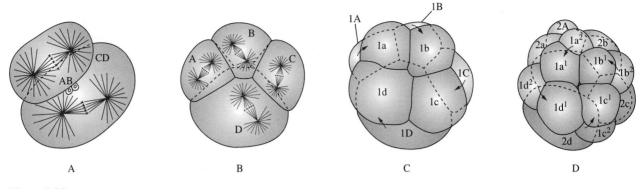

Figure 5.28

Early cleavage in zygotes of *Arenicola*, an unequally-cleaving annelid. A, 2-cell stage showing the large CD cell. B, four-cell stage showing the large D cell. C, 8-cell stage showing the large 1D cell. D, 16-cell stage showing the large 2d cell. (After Reverberi, 1971.)

macromeres and 2a, 2b, 2c, and 2d micromeres. In contrast, each micromere of the eight-cell embryo divides equally, giving rise to two micromeres (see Fig. 5.27). For example, the 1a micromere produces 1a1 and 1a2 micromeres, and other micromeres behave similarly. During fourth cleavage, the mitotic apparatus of each blastomere tilts counterclockwise, opposite to the direction that the mitotic apparatus tilted during the third cleavage (see Figs. 5.26 and 5.27). At each subsequent cleavage, the mitotic apparatuses alternate between tilting clockwise and counterclockwise. The micromeres continue to divide equally, but the macromeres divide unequally, producing additional tiers of micromeres that become stacked below the pre-existing micromeres. This rigid pattern of spiral cleavage produces a highly ordered blastula, with each cell occupying a specific position.

Genetic Control of Spiral Cleavage

Genetic experiments conducted with the freshwater snail *Lymnaea* indicate that the direction of spiral cleavage is under genetic control (Sturtevant, 1923; Boycott et al., 1930). Embryos that cleave dextrally develop into snails with dextral coiled shells, whereas embryos that cleave sinistrally develop into snails with sinistral coiled shells (see Fig. 5.26). The cleavage and shell coiling patterns are determined by a single locus, with the wild-type dextral allele (+) dominant to the less prevalent, recessive sinistral allele (s). However, the direction of coiling (the phenotype) is determined not by the genotype of the zygote but by the *genotype of the mother*. Thus, a homozygous (s/s) female will produce only sinistral offspring, even if mated with a homozygous (+/+) male (Fig. 5.29). Conversely, a heterozygous (+/s) female will produce only dextral offspring, regardless of the father's genotype. Note also that the shells of the progeny are not necessarily like the mother's shell but are determined by her genotype. For example, a sinistral +/s mother has only dextral progeny, and a dextral s/s mother has only sinistral progeny. This trait is a classic example of **maternal inheritance**, which will be discussed further in Chapter 13. The genes encoding maternal effects, such as shell coiling in snails, are usually expressed during oogenesis, and their products function during embryogenesis.

The following experiment suggests that the direction of spiral cleavage in *Lymnaea* is controlled by a substance present in the egg cytoplasm (Freeman and Lundelius, 1982). When cytoplasm taken from the eggs of dextral mothers is injected into eggs from sinistral mothers, the sinistral

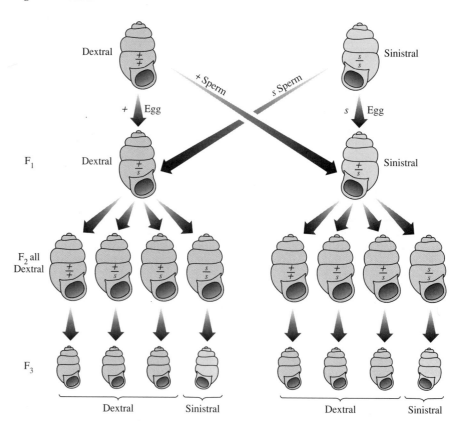

Figure 5.29

Pattern of inheritance of shell coiling in the freshwater snail, *Limnaea peregra*. (Redrawn with permission from E. Sinott, L. C. Dunn, and T. Dobzhansky, *Principles of Genetics*, 5th ed. © 1958, McGraw-Hill Book Co.)

eggs cleave dextrally. The reciprocal experiment—injection of cytoplasm from sinistral eggs into dextral eggs—has no effect on cleavage. Therefore, a dextral gene product affecting the orientation of the mitotic apparatus during cleavage must be present in dextral cleaving eggs and absent in sinistral cleaving eggs. The identity of the dextral gene product is unknown.

The Polar Lobe

An important modification in spiral cleavage occurs in some annelid and mollusk eggs that produce a **polar lobe** (Dohmen, 1983). The polar lobe is a spherical cytoplasmic protrusion that forms at the vegetal pole of the egg during cleavage (Fig. 5.30). Like cytokinesis, polar lobe formation is based on a contractile mechanism mediated by microfilaments (Conrad, 1973). Polar lobes vary in size, in the width of their attachments to the remainder of the egg, and in the timing of their appearance during early development. In some spiralians, polar lobes appear during the maturation divisions and early cleavages, whereas in others they appear only during cleavage. In the marine snail *Ilyanassa*, a large polar lobe forms during the first cleavage. After completion of cytokinesis, the polar lobe retracts into the CD blastomere. Consequently, the CD blastomere is almost twice the volume of the AB blastomere. During the next cleavage, another polar lobe forms from the CD cell and, after cytokinesis, retracts into the D blastomere. By obtaining the contents of the polar lobe, the D blastomere and its descendants become the largest cells in the embryo. As we shall see in Chapter 11, the polar lobe has special properties that cause the descendants of the CD blastomere to express particular developmental fates.

Superficial Cleavage

Superficial cleavage occurs in centrolecithal eggs. In these eggs, the yolk mass is located in the central region of the cell. Therefore, cleavage is

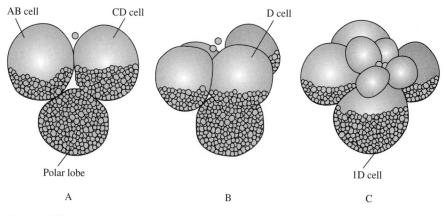

Figure 5.30

Spiral cleavage and polar lobe formation during the first and second cleavages of the snail *Ilyanassa*. A, First cleavage showing the polar lobe. B, Four-cell embryo showing the polar lobe retracting into the D cell. C, Eight-cell embryo showing polar lobe material in the 1D cell. (After Tyler, 1930.)

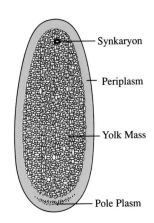

Figure 5.31

Schematic representation of an insect egg showing the synkaryon, periplasm, yolk mass, and pole plasm.

meroblastic and typically occurs only in a thin layer of superficial cytoplasm. Superficial cleavage is characteristic of most arthropods. We shall discuss the insect *Drosophila* as an example.

In insect eggs, the synkaryon is located in an **energid**, a small island of cytoplasm within the central yolk mass. The remaining cytoplasm is located in spaces surrounding the yolk granules and in the **periplasm**, the thin peripheral layer of the egg (Fig. 5.31). The periplasm is of uniform thickness and composition, except at the posterior pole, where it swells to form a specialized cytoplasmic region, the **pole plasm**. One of the unique aspects of superficial cleavage is that the synkaryon undergoes successive cycles of division in the absence of cytokinesis, forming multiple energids within the yolk mass. Thus, the egg becomes a multinucleate cell or **syncytium**.

Figure 5.32 shows the early cleavage stages of *Drosophila*. Eight nuclear divisions take place rapidly and synchronously in the central region of the egg to form a total of 256 energids. After the eighth division, most of the energids migrate into the periplasm, which is then called a **syncytial blastoderm**. Nuclear division without cytokinesis continues in the periplasm during the syncytial blastoderm stage. After each nuclear division, a partial cleavage furrow, which is oriented perpendicular to the surface of the egg, is formed between adjacent nuclei (Fig. 5.33). This causes each nucleus in the syncytial blastoderm to be elevated in a mound of surrounding cytoplasm, giving the surface of the embryo a blebbed appearance (Fig. 5.34). These partial cleavage furrows are transient, disappearing shortly after their formation. New cleavage furrows form during the next mitosis, creating mounds of cytoplasm that are more numerous and smaller than before.

The energids that enter the pole plasm behave differently from those that enter other regions of the periplasm (Turner and Mahowald, 1976). Complete cleavage furrows are formed in the pole plasm, incorporating the energids into separate cells, the pole cells (see Figs. 5.32 and 5.34G and H). As described in Chapter 11, the pole plasm contains factors that cause the pole cells to develop into germ cells.

The next phase of superficial cleavage involves the formation of the **cellular blastoderm** (see Figs. 5.32 and 5.33). Cytokinesis of the blastoderm is a dramatic event occurring simultaneously over the entire egg surface. In

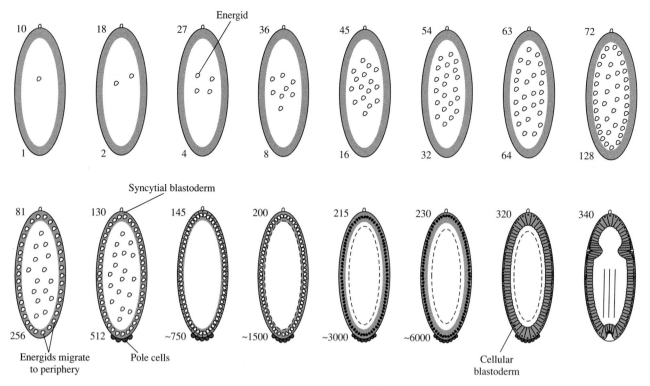

Figure 5.32

Schematic representation of nuclear multiplication during cleavage and blastoderm formation in the *Drosophila melanogaster* egg. The number above each egg corresponds to the time in minutes after the egg was deposited, whereas the number below each egg represents the number of energids at each stage. (After Zalokar and Erk, 1976.)

Drosophila melanogaster, it occurs after the fourteenth nuclear division (Zalokar and Erk, 1976). At the cellular blastoderm stage, the spherical nuclei of the syncytial blastoderm begin to enlarge and elongate in a plane perpendicular to the plasma membrane. Nuclear elongation is mediated by bundles of microtubules that surround each nucleus. As the nuclei elongate, cleavage furrows extending inward from the surface separate the nuclei into separate cells.

After cellularization, the embryo consists of three parts: the peripheral blastoderm of columnar cells, a central syncytial yolk mass containing scattered energids that did not migrate into the periplasm, and a cluster of pole cells at the posterior pole. The blastoderm cells are not uniformly distributed. More cells are concentrated on the ventral side of the yolk mass, where they form the **germ band**. As we shall discuss in Chapter 6, the cells of the germ band form most of the embryo proper, whereas many of the other blastoderm cells become extraembryonic envelopes that protect the developing embryo.

5–5 Classifying Animals According to Cleavage Patterns

Some groups of animals are strikingly uniform with regard to their cleavage patterns. Flatworms, annelids, and mollusks, for instance, cleave by spiral cleavage, and this is one of the reasons they are classified together as

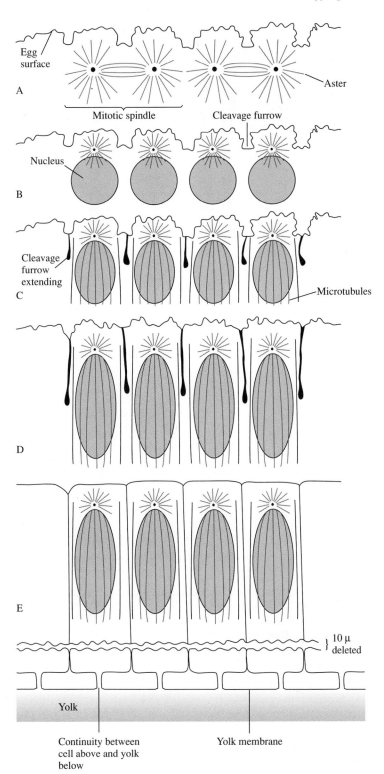

A — Egg surface, Mitotic spindle, Aster, Cleavage furrow

B — Nucleus

C — Cleavage furrow extending, Microtubules

D

E — 10 μ deleted

Yolk

Continuity between cell above and yolk below

Yolk membrane

Figure 5.33

Schematic representation of blastoderm formation during early development of *Drosophila melanogaster*. (After Fullilove and Jacobson, 1971.)

spiralians. However, classifying animals according to cleavage patterns is usually unwarranted. Similar cleavage patterns can exist in very different groups of animals. For example, the nematodes and mammals show rotational cleavage. Conversely, different patterns of cleavage can exist within the same phylogenetic groups. Although micromeres are formed by unequal division during fourth cleavage in the species of sea urchin we have examined

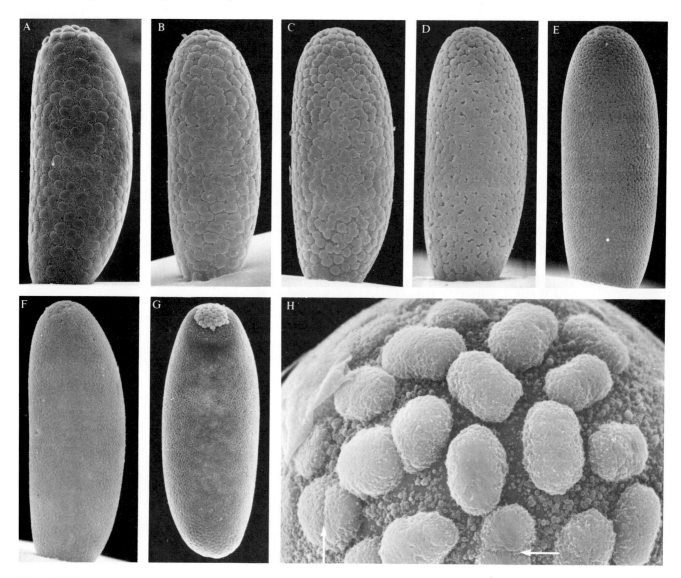

Figure 5.34

Scanning electron micrographs of *Drosophila melanogaster* embryos during the syncytial blastoderm stage. A-E, Surface views of the embryo showing the increase in numbers of nuclei within cytoplasmic mounds. F, Embryo at cellular blastoderm stage. G, Embryo at the start of gastrulation. Pole cells are evident at the posterior pole (top). H, High magnification view of pole cells. Note the division planes (arrows) in some pole cells. A-G approximately × 220; H × 1010. (From Turner and Mahowald, 1976.)

in this chapter, there are other sea urchin species that do not form micromeres at the 16-cell stage and cleave similarly to the sea cucumber (Williams and Anderson, 1975).

Cleavage patterns are highly dependent on the quantity of yolk, and yolk abundance can evolve independently of the general phylogenetic position of a species. The amphibian species we have used as an example of holoblastic cleavage in this chapter have moderately telolecithal eggs. Other amphibian species, however, have much larger quantities of yolk in their eggs, and their embryos cleave meroblastically (Del Pino, 1989). All placental mammals show holoblastic cleavage resembling that of isolecithal invertebrate eggs. In contrast, the extremely telolecithal eggs of nonplacental mammals, such as the duck-billed platypus, have meroblastic cleavage patterns resembling those of yolky reptile and bird eggs. Different cleavage patterns probably arose to distribute efficiently yolk and other cytoplasmic components to their proper places in the embryo. In Chapter 11, we shall see that these cytoplasmic components include substances that determine cell fate.

5–6 Chapter Synopsis

Rapid synchronous cleavages after fertilization convert the egg into a multicellular blastula. The cell cycles of cleaving embryos differ from those of adult somatic cells in being much shorter and lacking a growth phase. At the end of the cleavage period, the cytoplasmic to nuclear volume ratio returns to levels typical of somatic cells. The short cell cycles of cleaving embryos are regulated by cytoplasmic factors that eliminate the G_1 and G_2 phases and reduce the lengths of the S and M phases. Each cleavage division consists of karyokinesis and cytokinesis, which are normally coordinated but can be uncoupled in certain embryos. The segregation of chromosomes to daughter cells during karyokinesis is directed by the mitotic apparatus, whereas cleavage furrow formation during cytokinesis is controlled by the contractile ring. Cleavage furrow formation in the cortex is regulated by signaling from the asters located at each pole of the mitotic spindle.

Different patterns of cleavage occur in different animals depending on (1) the quantity and distribution of yolk and (2) the orientation of the mitotic apparatus with respect to the egg's animal-vegetal axis. Holoblastic cleavage occurs in eggs that lack yolk or contain equally distributed yolk, whereas meroblastic cleavage occurs in eggs that contain large quantities of yolk. Equal cleavage occurs when the mitotic apparatus is positioned in the center of a blastomere, whereas unequal cleavage occurs when it is displaced from the center. Based on geometric considerations, cleavage patterns are further distinguished as being radial, bilateral, rotational, spiral, or superficial. Breeding experiments with snails show that cleavage patterns are under genetic control. Different cleavage patterns probably evolved to distribute yolk and other cytoplasmic materials to their proper locations in the embryo.

References

Asnes, C. F., and T. E. Schroeder. 1979. Cell cleavage: Ultrastructural evidence against equatorial stimulation by aster microtubules. Exp. Cell Res., *122*: 327–338.

Balinsky, B. I. 1981. *An Introduction to Embryology*, 5th ed. Saunders College Publishing, Philadelphia.

Beams, H. W., and R. G. Kessel. 1976. Cytokinesis: A comparative study of cytoplasmic division in animal cells. Am. Sci., *64*: 279–290.

Black, S., and J.-P. Vincent. 1988. The first cleavage plane and the embryonic axis are determined by separate mechanisms in *Xenopus laevis*. II. Experimental dissociation by lateral compression of the egg. Dev. Biol., *128*: 65–71.

Boycott, A. E. et al. 1930. The inheritance of sinistrality in *Limnaea peregra* (Mollusca: Pulmonata). Philos. Trans. R. Soc. Lond. (Biol.), *219*: 51–131.

Brachet, J. 1950. *Chemical Embryology*, L. G. Barth (transl.). Facsimile edition published by Hafner Press (Macmillan), New York (1968).

Carlson, B. M. 1981. *Patten's Foundations of Embryology*. McGraw-Hill Book Co., New York.

Conrad, G. 1973. Control of polar lobe formation in fertilized eggs of *Ilyanassa obsoleta* Stimpson. Am. Zool., *13*: 961–980.

Danilchik, M. V., and S. D. Black. 1988. The first cleavage plane and the embryonic axis are determined by separate mechanisms in *Xenopus laevis*. I. Independence in unperturbed embryos. Dev. Biol., *128*: 58–64.

de Laat, S. W., and J. G. Bluemink. 1974. New membrane formation during cytokinesis in normal and cytochalasin B-treated eggs of *Xenopus laevis*. Electrophysical observations. J. Cell Biol., *60*: 529–540.

Del Pino, E. M. 1989. Modifications of oogenesis and development in marsupial frogs. Development, *107*: 169–187.

Dohmen, M. R. 1983. The polar lobe in eggs of molluscs and annelids: Structure, composition, and function. *In* W. R. Jeffery and R. A. Raff (eds.), *Time, Space and Pattern in Embryonic Development*. Alan R. Liss, New York, pp. 197–220.

Dolfini, S., A. M. Courgeon, and L. Tiepolo. 1970. The cell cycle of an established line of *Drosophila melanogaster* cells *in vitro*. Experientia (Basel), *26*: 1020.

Ducibella, T. 1977. Surface changes in the developing trophoblast cell. *In* M. H. Johnson (ed.), *Development in Mammals 1*. Elsevier/North-Holland Biomedical Press, Amsterdam, pp. 5–30.

Epstein, C. J. 1975. Gene expression and macromolecular synthesis during preimplantation embryonic development. Biol. Reprod., *12*: 82–105.

Forbes, D. J., M. W. Kirschner, and J. W. Newport. 1983. Spontaneous formation of nucleus-like structures around bacteriophage DNA microinjected into Xenopus eggs. Cell, *34*: 13–23.

Freeman, G., and J. W. Lundelius. 1982. The developmental genetics of dextrality and sinistrality in the gastropod *Lymnaea peregra*. Wilhelm Roux's Archives Dev. Biol., *191*: 69–83.

Fullilove, S. L., and A. G. Jacobson. 1971. Nuclear elongation and cytokinesis in *Drosophila montana*. Dev. Biol., *26*: 560–577.

Gamow, E. I., and D. M. Prescott. 1970. The cell life cycle during early embryogenesis of the mouse. Exp. Cell Res., *59*: 117–123.

Gilbert, S. F. 1988. *Developmental Biology*, 2nd ed. Sinauer Associates, Sunderland, MA.

Graham, C. F., K. Arms, and J. B. Gurdon. 1966. The induction of DNA synthesis by frog egg cytoplasm. Dev. Biol., *14*: 349–381.

Gulyas, B. J. 1975. A reexamination of cleavage patterns in eutherian mammalian eggs: Rotation of blastomere pairs during second cleavage in the rabbit. J. Exp. Zool., *193*: 235–248.

Hara, K. 1977. The cleavage pattern of the axolotl egg studied by cinematography and cell counting. Wilhelm Roux's Archives Dev. Biol., *181*: 73–87.

Harris, P., M. Osborn, and K. Weber. 1980. Distribution of tubulin-containing structures in the egg of the sea urchin *Strongylocentrotus purpuratus*. J. Cell Biol., *84*: 668–679.

Harvey, E. B. 1935. The mitotic figure and the cleavage plane in the egg of *Parechinus microtuberbulatus*, as influenced by centrifugal force. Biol. Bull., *69*: 287–297.

Harvey, E. B. 1956. *The American Arbacia and Other Sea Urchins*. Princeton University Press, Princeton, NJ.

Hiramoto, Y. l956. Cell division without mitotic apparatus in sea urchin eggs. Exp. Cell Res., *11*: 630–636.

Hiramoto, Y. 1965. Further studies on cell division without mitotic apparatus in sea urchin eggs. J. Cell Biol., *25*: 161–167.

Hiramoto, Y. 1971. Analysis of cleavage stimulus by means of micromanipulation in sea urchin eggs. Exp. Cell Res., *68*: 291–298.

Hopper, A. F., and N. H. Hart. 1980. *Foundations of Animal Development*. Oxford University Press, Oxford.

Hörstadius, S. 1939. The mechanics of sea urchin development, studied by operative methods. Biol. Rev., *14*: 132–179.

Hyafil, F., C. Babinet, and F. Jacob. 1981. Cell-cell interactions in early embryogenesis: A molecular approach to the role of calcium. Cell, *26*: 447–454.

Ito, S., K. Dan, and D. Goodenough. 1981. Ultrastructure and ^{3}H-thymidine incorporation into chromosome vesicles in sea urchin embryos. Chromosoma, *83*: 441–453.

Kemler, R. et al. 1977. Surface antigen in early differentiation. Proc. Natl. Acad. Sci. U.S.A., *74*: 449–452.

Kessel, R. G., and C. Y. Shih. 1974. *Scanning Electron Microscopy in Biology. A Student's Atlas on Biological Organization.* Springer-Verlag, New York.

Kriegstein, H. J., and D. S. Hogness. 1974. Mechanism of DNA replication in *Drosophila* chromosomes. Structure of replication forks and evidence for bidirectionality. Proc. Natl. Acad. Sci. U.S.A., *71*: 135–139.

Lewis, W. H., and C. G. Hartman. 1933. Early cleavage stages of the egg of the monkey (*Macacus rhesus*). Contrib. Embryol. Carnegie Inst., *24*: 187–201.

Mabuchi, I., and M. Okuno. 1977. The effect of myosin antibody on the division of starfish blastomeres. J. Cell Biol., *74*: 251–263.

Rappaport, R. 1961. Experiments concerning the cleavage stimulus in sand dollar eggs. J. Exp. Zool., *148*: 81–89.

Rappaport, R. 1966. Experiments concerning the cleavage furrow in invertebrate eggs. J. Exp. Zool., *161*: 1–8.

Rappaport, R. 1976. Furrowing in altered cell surfaces. J. Exp. Zool., *195*: 271–277.

Rappaport, R., and R. P. Ebstein. 1965. Duration of stimulus and latent period proceeding furrow formation in sand dollar eggs. J. Exp. Zool., *158*: 373–382.

Reverberi, G. 1971. Annelids. *In* G. Reverberi (ed.), *Experimental Embryology of Marine and Freshwater Invertebrates.* Elsevier/ North-Holland Biomedical Press, Amsterdam, pp. 126–163.

Saunders, J. W., Jr. 1982. *Developmental Biology.* Macmillan, New York.

Schroeder, T. E. 1973. Cell constriction: Contractile role of microfilaments in division and development. Am. Zool., *13*: 949–960.

Schroeder, T. E. 1981. Interrelations between the cell surface and cytoskeleton in cleaving sea urchin eggs. *In* G. Poste and G. L. Nicolson (eds.), *Cytoskeletal Elements and Plasma Membrane Organization.* Elsevier/North-Holland Biomedical Press, Amsterdam, pp. 169–216.

Schroeder, T. E. 1987. Fourth cleavage of sea urchin blastomeres: microtubule patterns and myosin localization in equal and unequal cell divisions. Dev. Biol., *124*: 9–22.

Sinott, E., L. C. Dunn, and T. Dobzhansky. 1958. *Principles of Genetics*, 5th ed. McGraw-Hill Book Co., New York.

Sturtevant, M. H. 1923. Inheritance of the direction of coiling in *Limnaea*. Science, *58*: 269–270.

Sze, L. C. 1953. Changes in the amount of deoxyribonucleic acid in the development of *Rana pipiens*. J. Exp. Zool., *122*: 577–601.

Szollosi, D. 1970. Cortical cytoplasmic filaments of cleaving eggs: A structural element corresponding to the contractile ring. J. Cell Biol., *44*: 192–209.

Turner, F. R., and A. P. Mahowald. 1976. Scanning electron microscopy of *Drosophila* embryogenesis. I. The structure of the egg envelopes and the formation of the cellular blastoderm. Dev. Biol., *50*: 95–108.

Tyler, A. 1930. Experimental production of double embryos in annelids and molluscs. J. Exp. Zool., *57*: 347–407.

Verdonk, N. H., and J. A. M. van den Biggelaar. 1983. Early development and the formation of germ layers. *In* K. Wilbur (ed.), *The Mollusca*, Vol. 3. *Development.* Academic Press, New York, pp. 91–122.

Williams, D. H. C., and D. T. Anderson. 1975. The reproductive system, embryonic development, larval development, and metamorphosis of the sea urchin *Heliocidaris erythrogramma*. (Val.) (Echinoidea: Echinometridae). Aust. J. Zool., *23*: 371–403.

Wilson, E. B. 1925. *The Cell in Development and Heredity*, 3rd ed. Macmillan, New York.

Zalokar, M., and I. Erk. 1976. Division and migration of nuclei during early embryogenesis of *Drosophila melanogaster*. J. Micro. Biol. Cell, *25*: 97–106.

Zimmerman, A. M., and S. Zimmerman. 1967. Action of colcemid in sea urchin eggs. J. Cell Biol., *34*: 483–488.

CHAPTER *6*

Initiating the Body Plan: Embryonic Polarization and Gastrulation

Hilde Mangold (1898–1924) was a graduate student at the University of Freiburg in Germany. Mangold (shown here with her infant son) and Hans Spemann (see p. 477) discovered the primary organizer of amphibian embryos by transplanting the dorsal lip of the blastopore from a donor to a host embryo and showing that the host developed a secondary axis. This photograph was taken shortly before her accidental death. The research initiated by Mangold resulted in a Nobel Prize (to H. Spemann). (From Hamburger, V. 1988. *The Heritage of Experimental Embryology. Hans Spemann and the Organizer.* Oxford, New York, p. 174. Reprinted by permission of Kluwer Academic Publishers.)

After the blastula is formed, the next phase of development is devoted to rearranging the embryonic cells into the larval or adult body plan. The body plan is established in discrete steps, the first of which is the development of bilateral symmetry. The eggs of most animal species show radial symmetry and are polarized along their animal-vegetal axis, which is determined during oogenesis. In contrast, the *embryos* of most animal species *develop* bilateral symmetry and become polarized along three axes: the **anteroposterior axis**, the **dorsoventral axis**, and the **left-right axis**. The body axes are determined by **embryonic polarization**, the process in which bilateral symmetry is superimposed on the egg's radial symmetry (Fig. 6.1).

The second major step in developing the body plan is gastrulation. During gastrulation, cells of the blastoderm, which are produced during cleavage, are translocated to new positions in the embryo, producing three primary germ layers: the ectoderm, mesoderm, and endoderm. K. E. von Baer (1828) was the first to recognize that the same organs come from the same germ layer in all vertebrates. Specifically, the ectoderm gives rise to the epidermis and nervous tissues, the endoderm gives rise to the organs

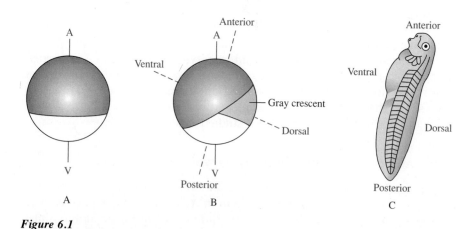

Figure 6.1

A diagram showing the relationship between the axis of the unfertilized egg (A) and that of the fertilized egg (B) and of the tailbud-stage frog embryo (C). A, The unfertilized egg is radially symmetric about the animal-vegetal axis. B, Fertilization results in the formation of the dorsoventral axis according to the point of sperm entry. C, Beginning during gastrulation, the anteroposterior axis forms perpendicular to the dorsoventral axis, and the left-right axis (not shown) is also established.

Figure 6.2

A diagram of early development in *Xenopus laevis* showing the relationship between the sperm entry site and the gray crescent, the dorsal lip of the blastopore, and the dorsoventral axis. Gastrulation is initiated at the dorsal lip of the blastopore, and dorsal structures (i.e., neural folds) are formed on the gray crescent side of the embryo opposite the sperm entry point. D: dorsal; V: ventral. Timing of development in minutes after fertilization is shown below each diagram. (After Scharf et al., 1984. Reprinted from *Molecular Biology of Development*, E. H. Davidson and R. A. Firtel, eds., by permission of Wiley-Liss, a division of John Wiley and Sons, Inc. Copyright © 1984, Wiley-Liss.)

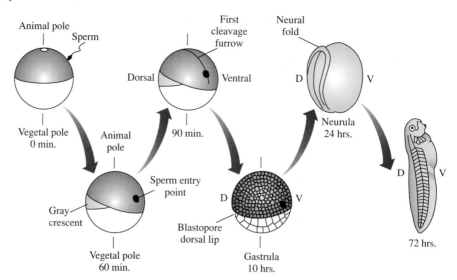

of the gut and all the accessory glands, and the mesoderm gives rise to the muscles, bones, and connective tissues of most organs. Presumptive endoderm and mesoderm cells enter the interior of the embryo, while presumptive ectoderm cells spread over the surface of the embryo. When the moving cells reach their final destinations, they begin to develop into tissues and organs. In vertebrates, the establishment of the body plan during gastrulation is coordinated by a specific region of the embryo called the **primary organizer**.

The initial steps in the formation of the body plan, including embryonic polarization, gastrulation, and the role of the primary organizer, are described in this chapter. Events that follow gastrulation, including neurulation, pattern formation, organogenesis, and cell differentiation, will be examined in subsequent chapters.

6–1 Embryonic Polarization

The timing of embryonic polarization varies considerably in the animal kingdom; it can occur during oogenesis, as a consequence of fertilization, during the cleavage period, or even later in development. In some cases, external cues such as sperm entry and gravity, are known to be important in polarizing the embryo. In most instances, however, the mechanisms of polarization are unknown. In our discussion of embryonic polarization that follows, we shall focus on amphibians and insects, which exhibit different timing of polarization and have been studied extensively.

Polarization of the Amphibian Embryo

Polarization of the amphibian embryo is initiated at fertilization. Figure 6.2 shows the development of the body plan in *Xenopus laevis* embryos. The polarity of the egg is clearly defined by differences in pigmentation along the animal-vegetal axis. The animal hemisphere contains cortical pigment granules, of which there are very few in the vegetal hemisphere. Unlike sea urchins and mammals, in which sperm-egg fusion can occur anywhere on the egg surface, fertilization is restricted to the animal hemisphere of amphibian eggs (Elinson, 1975). After sperm entry, a sperm aster is formed in the animal hemisphere. Pigment granules congregate around the sperm entry point, forming a dark spot that marks this region throughout development. About halfway between fertilization and first cleavage, the egg

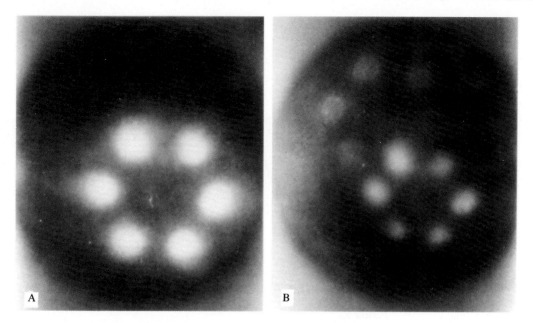

Figure 6.3

Cortical rotation in a fertilized *Xenopus* egg. A pattern of spots was imprinted on the vegetal egg surface with a grid containing a fluorescent lectin, which binds to and marks the egg cortex, and Nile blue sulfate, a fluorescent dye that stains the sub-cortical cytoplasm. The extent of rotation of the cortex over the sub-cortical cytoplasm was determined by following the movement of the fluorescent spots in the cortex relative to the Nile blue spots. A, The staining pattern before rotation. The fluorescent lectin and Nile blue spots are superimposed. B, The staining pattern after rotation. The patterns of spots have separated. The lectin pattern has moved toward the future dorsal side of the egg (upper left), whereas the Nile blue pattern has remained stationary. (From Vincent et al., 1986.)

cortex begins to rotate with respect to the internal cytoplasm. **Cortical rotation** has been demonstrated by using a grid to apply a fluorescent lectin and the vital dye Nile blue sulfate to the egg surface in a pattern of spots (Vincent et al., 1986). (**Lectins** are proteins that have multiple carbohydrate binding sites and can therefore bind to the carbohydrate moieties of cell surface glycoproteins.) The lectin binds to and marks the egg cortex, whereas the dye seeps through the cortex and stains the underlying core of internal cytoplasm. Before cortical rotation, the lectin and Nile blue spots are superimposed (Fig. 6.3A). During rotation, however, the two patterns separate: The Nile blue spots remain stationary, whereas the lectin spots move about 30 degrees toward the future dorsal side of the embryo (see Fig. 6.3B). In the animal hemisphere, the same cortical rotation shifts the pigmented egg cortex toward the site of sperm entry, revealing the **gray crescent** on the side opposite sperm entry. The gray crescent is caused by the exposure of a lightly pigmented area of cytoplasm in the **marginal zone** as the darkly pigmented cortex rotates toward the sperm entry point (Fig. 6.4). Fate mapping (see p. 226) has shown that the dorsoventral axis is determined according to the locations of the sperm entry point and the gray crescent. The site of sperm entry becomes the ventral side of the embryo, whereas the gray crescent becomes the dorsal side of the embryo (see Figs. 6.1 and 6.2). Later in development, gastrulation is initiated immediately below the gray crescent on the future dorsal side of the embryo.

How is the dorsoventral axis determined by sperm entry? Several different experiments indicate that it is not the sperm itself but the rotation

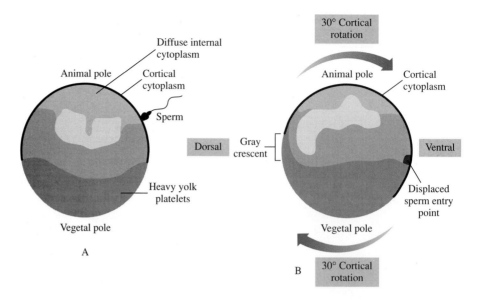

Figure 6.4

A diagram showing gray crescent formation during cortical rotation in a frog egg. A, Section through the animal-vegetal axis of an unfertilized egg. The darkly-pigmented cortical cytoplasm is positioned symmetrically with respect to the animal-vegetal axis. B, Section through the animal-vegetal axis of a fertilized egg after cortical rotation. The cortical cytoplasm has rotated, displacing the sperm entry point by 30 degrees. During rotation (arrows), the displacement of the darkly pigmented cortical cytoplasm reveals a more lightly-pigmented underlying cytoplasm on the side opposite the sperm entry point, resulting in the appearance of the gray crescent. The sizes of inclusions (represented by shades of gray) are largest in the vegetal hemisphere. (After Klag and Ubbels, 1975.)

of the egg cortex caused by sperm entry that determines polarity: (1) Polarization can be prevented by agents that inhibit cortical rotation. Eggs treated with high hydrostatic pressure, low temperature, and ultraviolet irradiation of the vegetal hemisphere lack rotation and develop into **ventralized embryos**, which lack dorsal structures and are radially symmetrical (Malacinski et al., 1975; Manes and Elinson, 1980; Scharf and Gerhart, 1980, 1983). (2) Amphibian eggs, which contain dense yolk granules, are affected by gravity and normally orient with their vegetal poles down. If eggs are tilted off-center and maintained in this position for a sufficient period of time, a rearrangement of the dense cytoplasmic inclusions occurs that mimics cortical rotation. The dorsoventral axis will then develop according to the direction of tilting, rather than to the site of sperm entry (Ancel and Vintemberger, 1948). Ultraviolet-irradiated eggs can be rescued by tilting (Fig. 6.5) or by centrifugation—procedures that promote rearrangement of cytoplasm. (3) When normal eggs are tilted or centrifuged in two different directions before first cleavage, two dorsoventral axes develop. The double-centrifuged eggs initiate gastrulation at two sites and develop into Siamese twins with duplicated dorsal structures (Fig. 6.6). The results of these experiments suggest that any part of the egg can become the dorsal side of the embryo, depending on the direction of cortical rotation.

Microtubules appear to be involved in cortical rotation. Agents that inhibit cortical rotation and promote the formation of ventralized embryos (i.e., hydrostatic pressure, low temperature, and colchicine) are known to depolymerize microtubules (Manes et al., 1978; Vincent et al., 1987). The

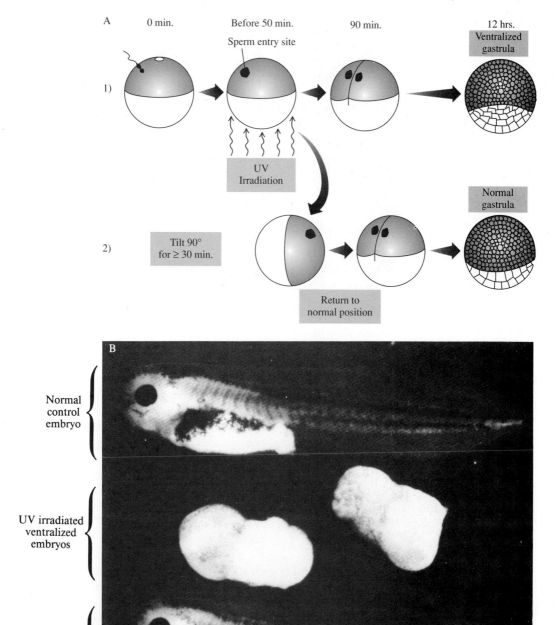

Figure 6.5

Inhibition of dorsoventral polarization by ultraviolet irradiation of the vegetal hemisphere of a fertilized *Xenopus* egg and its restoration by tilting the zygote. A, Schematic diagram of (1) the UV irradiation experiment and (2) the egg tilting procedure. B, Photographs of irradiated embryos with and without tilting. Upper embryo: An unirradiated control that has developed to the tadpole stage. Middle embryos: Two examples of irradiated ventralized embryos. No dorsoventral or left-right axes are evident in these embryos. Lower embryo: An irradiated embryo in which dorsoventral and left-right axes are formed normally after tilting as shown in A. (A, After Gerhart et al., 1983. B, From Scharf and Gerhart, 1980.)

Figure 6.6

Xenopus tadpoles with duplicated dorsal anterior structures produced by double centrifugation of eggs. (From Black and Gerhart, 1986.)

role of microtubules in polarization has also been demonstrated by experiments in which heavy water (D_2O) caused the excessive polymerization of microtubules in amphibian eggs. Fertilized eggs treated with D_2O during cortical rotation form **dorsalized embryos** with duplicated dorsal structures (Scharf et al., 1989). These dorsalized embryos are the anatomical opposites of the ventralized embryos produced by destroying microtubules. Effects of D_2O suggest that excessive polymerization of microtubules may influence the development of dorsal structures. A parallel array of microtubules has been observed in the vegetal cortex of amphibian eggs by electron microscopy and by staining with tubulin antibodies during the period of cortical rotation (Fig. 6.7). The location and orientation of these microtubules suggest that they are serving as tracks for the rotation of the cortex relative to the cytoplasm.

Dorsoventral polarization may be caused by the activation of factors (ooplasmic determinants; see Chap. 11) in the dorsal region of the embryo as a result of cortical rotation. During cleavage, these factors may be segregated to vegetal blastomeres on the dorsal side of the embryo. The descendants of the dorsal vegetal cells could initiate gastrulation themselves, or they could induce other cells to gastrulate and form dorsal structures. As we shall see in Chapter 12, vegetal blastomeres are capable of inducing marginal zone cells to become mesoderm. Transplantation experiments (shown in Fig. 6.8) were carried out to determine the role of the dorsal vegetal cells in gastrulation (Gimlich and Gerhart, 1984). In these experiments, an embryo was irradiated with ultraviolet light and allowed to develop. At the 64-cell stage, two adjacent cells were removed from the most vegetal tier of blastomeres on the dorsal side. These vegetal blastomeres were then replaced with either two cells transplanted from the same region or from the ventral side of a normal donor embryo. The transplanted dorsal vegetal cells rescued development of the ultraviolet-irradiated embryo, and a normal complement of dorsal structures formed. In contrast, ventral vegetal cells were not able to rescue development of ultraviolet-irradiated embryos. To test whether the progeny of the donor cells developed into dorsal structures or induced host cells to do so, the experiment was repeated using donor cells injected with a fluorescent tracer. The results show that fluorescent donor cells were confined to the gut and did not become dorsal

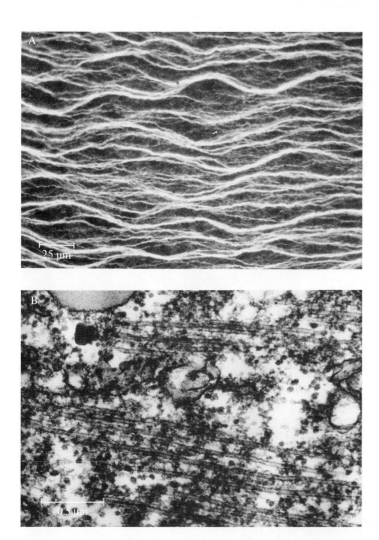

Figure 6.7

Arrays of microtubules in the vegetal cortex of *Rana pipiens* eggs. A, Parallel bundles of microtubules (wavy lines) detected by staining with a fluorescent tubulin antibody. B, Parallel arrays of microtubules observed by electron microscopy in a section cut parallel to the vegetal cortex. (From Elinson and Rowning, 1988.)

structures themselves (Fig. 6.9). Instead, dorsal structures were derived from marginal zone cells of the recipient. Thus, the transplanted vegetal cells *induced* presumptive mesoderm cells to develop into dorsal structures. In Section 6–3, we shall see that these presumptive mesoderm cells induced by underlying dorsal vegetal cells become the primary organizer of the embryo and are responsible for coordinating the development of the body plan during gastrulation.

In summary, dorsoventral polarization in amphibians occurs in several steps: (1) the sperm enters the animal hemisphere, and the egg cortex rotates toward the sperm entry point; (2) the rotation of the egg cortex activates unidentified factors on the presumptive dorsal side of the embryo, and these substances are inherited by vegetal hemisphere cells; and (3) vegetal cells induce presumptive mesoderm cells in the dorsal marginal zone to gastrulate and develop into dorsal structures. Thus, an external cue, the sperm, is

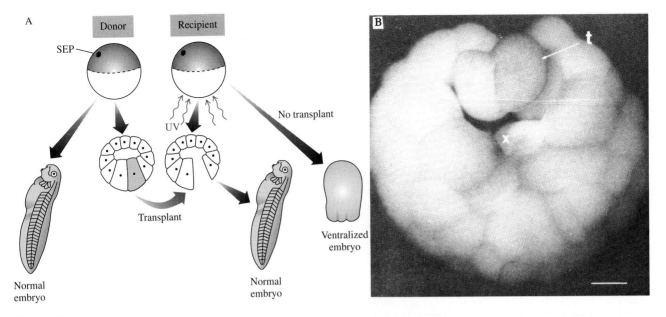

Figure 6.8

Transplantation of vegetal blastomeres in *Xenopus* embryos. A, Diagram illustrating transplantation of blastomeres from a donor to a gap caused by removal of vegetal blastomeres of an irradiated recipient embryo. SEP: sperm entry point; UV: ultraviolet irradiation. B, Photograph showing two donor blastomeres in the process of being transplanted into the vegetal hemisphere of an irradiated host embryo. X: vegetal pole; t: transplanted blastomeres. Scale bar equals 0.2 mm. (After Gimlich and Gerhart, 1984.)

Figure 6.9

Fate of transplanted vegetal cells in a UV-irradiated recipient embryo. Before being transplanted, donor blastomeres were labeled by injecting normal embryos with fluorescein-labeled dextran, a fluorescent marker that is not transferred between cells. At the tadpole stage, the host embryo was fixed and sectioned for microscopic observation. A, A fluorescence micrograph of a cross-section through the dorsoventral axis of a tadpole that developed from the recipient embryo. Fluorescent labeled cells are in the gut endoderm. B, A bright-field phase micrograph of the same section described in A showing parts of the tadpole. g: gut. No labeled cells appear in dorsal structures. Unlabeled dorsal structures include: the neural tube (nt), somites (s), and notochord (n). Scale bar equals 0.1 mm. (From Gimlich and Gerhart, 1984.)

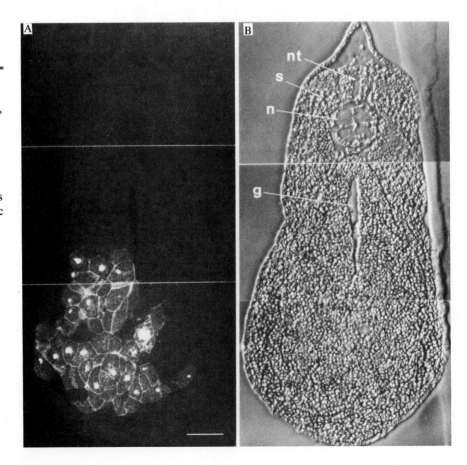

responsible for initiating a sequence of events that leads to the development of bilateral symmetry.

Polarization of the *Drosophila* Embryo

Considerable progress has recently been made in understanding embryonic polarization of *Drosophila* in terms of gene expression. In contrast to amphibian embryos, embryonic polarization begins during oogenesis in *Drosophila* and other insects. The *Drosophila* egg already shows a dorso-ventral axis at the time of ovulation; the dorsal surface is flat and the ventral surface is curved (see Color plate 1). After fertilization and ovulation, the egg begins to undergo rapid nuclear divisions without intervening cytokinesis. Eventually, the nuclei migrate into the periplasm, and after a few more divisions, the cellular blastoderm is formed (see Chap. 5). The cellular blastoderm consists of a layer of columnar cells extending around the circumference of the embryo. Cells on the dorsal surface of the blastoderm become either dorsal epidermis or amnioserosa (an extraembryonic tissue that protects the developing embryo), whereas cells on the ventral surface form the germ band, which gives rise to ventral epidermis, mesoderm, and the nervous system. Gastrulation begins by the invagination of presumptive mesoderm cells located along the ventral midline (see p. 213). After gastrulation, mesoderm and the nervous system develop ventrally, and eventually a larva is formed. The wild-type larva has belts of **denticles**, which are projections of the cuticle organized in segmentally repeated rows, on its ventral side and fine hairs on its dorsal side (Fig. 6.10A and B).

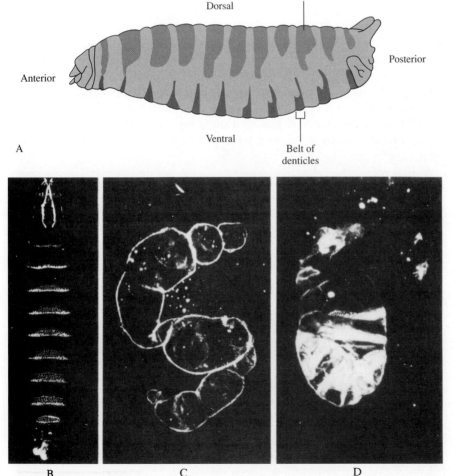

Figure 6.10

Morphology of normal and mutant *Drosophila* larvae. A, Drawing of a wild-type larva showing the pattern of hairs on the dorsal side and belts of denticles on the ventral side. B-D, Photographs of the cuticle of wild-type and mutant larvae. B, Wild-type larva showing the normal pattern of denticle bands on the ventral surface of the cuticle. C, A dorsalized *Toll*⁻ larva lacking the ventral denticle bands. D, A ventralized *Toll*ᴰ larva with denticle bands extending around the entire circumference of the cuticle. (A, After Anderson and Nüsslein-Volhard, 1984b. B-D, From Anderson, 1987.)

At least 20 different genes have been discovered in which mutations can cause abnormal or failed development of the dorsoventral axis (Table 6–1). These genes were identified by saturating the genome with mutations and isolating and characterizing all mutant alleles that affected the dorsoventral axis (Nüsslein-Volhard, 1979). Genetic crosses were then done to classify each mutant allele as belonging to a specific gene locus. Mutations that affect dorsoventral polarization can be divided into two classes: maternal-effect and zygotic. Zygotic mutations occur in genes that are expressed in the embryo (see Chap. 13). In contrast, maternal-effect mutations occur in genes that are expressed exclusively during oogenesis. There are two classes of maternal-effect genes responsible for dorsoventral polarization: (1) those that initiate the process of polarization in the oocyte and (2) those that continue polarization in the early embryo. Finally, zygotic genes begin to be expressed at the cellular blastoderm stage and complete the process of dorsoventral polarization in the *Drosophila* embryo. We shall now discuss the roles of each of the gene classes in more detail.

Dorsoventral polarization in *Drosophila* is initiated by genes that function exclusively during oogenesis. These are called the **egg-shape genes** (see Table 6–1), because mutant alleles at these loci cause changes in the shape of the egg and its shell. For example, mutations in the *gurken* and *torpedo* genes result in ventralized eggs, in which the dorsal and ventral surfaces are both curved. In contrast, mutations in the *cappuccino* and *spire* genes cause dorsalized eggs, in which the dorsal and ventral surfaces are both flattened. The shape of the insect egg is dependent on the activities of the follicle cells, which produce the egg shell (chorion). To determine the relative contributions of somatic (i.e., follicle) and germ line (i.e., oocyte and nurse cell complex) cells to the development of egg polarity, mosaic female flies were constructed containing mutant germ line cells and wild-type somatic cells and *vice versa* (Schüpbach, 1987). Mutant *spire*, *cappuccino*, and *gurken* germ lines each produce abnormally shaped eggs, even though the egg shell is produced by wild-type follicle cells. In contrast, mutant *torpedo* follicle cells result in a defective egg shape, even though the germ line is wild-type. These results suggest that follicle cells and the germ line cooperate to initiate dorsoventral polarization during oogenesis.

Although dorsoventral polarization is initiated during oogenesis, it is not irreversibly determined until later in development, because mutations

Table 6–1 Genes involved in dorsoventral axis formation in **Drosophila melanogaster**

	Maternal		Zygotic
Egg shape and embryonic polarity	Embryonic polarity (Dorsal group)		
fs (1) K10	*dorsal*		*twist*
gurken	*windbeutel*		*snail*
torpedo	*nudel*		*decapentaplegic*
cappuccino	*gastrulation defective*		*zerknüllt*
spire	*Toll*		*twisted gastrulation*
	easter		*tolloid*
	pelle		*shrew*
	spätzle		
	snake		
	tube		
	pipe		

Adapted from Anderson, K. V. 1987. Dorsal–ventral embryonic pattern genes of *Drosophila*. Trends Genet., *3*: 92.

in the second class of maternal-effect genes can disrupt dorsoventral axis formation in the embryo. This class of maternal-effect genes is called the **dorsal group** because mutations at each locus exhibit a similar dorsalized-phenotype (see Table 6–1). For example, embryos produced by $Toll^-$, $easter^-$, $pipe^-$, $dorsal^-$, or $snake^-$ mothers do not gastrulate properly and form **dorsalized larvae**, lacking ventral denticle belts and covered by dorsal hairs (see Fig. 6.10C). Recessive mutations in the other dorsal group genes have similar phenotypes. Most of the dorsal gene mutants can be rescued by injecting cytoplasm from wild-type embryos into mutant embryos, but only if microinjection is carried out before the cellular blastoderm stage (Anderson and Nüsslein-Volhard, 1984a). The rescued mutants generally develop according to the dorsoventral polarity already existing in the egg.

Several different experiments indicate that the dorsal gene *Toll* plays a central role in establishing the ventral side of the embryo. First, when $Toll^-$ mutants are rescued by injection of wild-type cytoplasm, their polarity is not necessarily the same as that of the wild-type egg (Anderson et al., 1985b). Instead, the site of gastrulation initiation (and, therefore, the ventral side of the embryo) is defined by the injection site, rather than the curved side of the egg (Fig. 6.11). Thus, embryos with reversed polarity showing curved dorsal sides and flat ventral sides can be produced by this experiment. Second, some *Toll* mutations ($Toll^D$ mutations) produce **ventralized larvae** with bands of denticles present around the entire circumference of the larva (see Fig. 6.10D). Mutations causing ventralized larvae have not been identified for most of the other dorsal group genes. Third, *Toll* is one of the last of the dorsal group genes to function in the pathway that leads to gastrulation on the ventral side of the embryo (Fig. 6.12). The order of genes that act in a sequence to produce a single event can be determined by examining the phenotypes of double mutants (i.e., $Toll^D/easter^-$, $Toll^D/pipe^-$, $Toll^D/snake^-$, etc.). For example, if the function of *Toll* is required before that of *easter, pipe, or snake*, the double mutants would show dorsalized phenotypes similar to $easter^-$, $pipe^-$, and $snake^-$ mutants; however, if *Toll* functions after *easter*, *pipe*, and *snake*, the double mutants would show a ventralized phenotype similar to $Toll^D$ mutants. In fact, the phenotypes of the double mutants are ventralized rather than dorsalized, indicating that the function of *Toll* must be required after *easter*, *pipe*, and *snake* (Anderson et al., 1985a). Presumably, the role of the *Toll* gene is to define the site of gastrulation, which dictates the ventral side of the embryo.

How is dorsoventral polarization coupled to gastrulation and the formation of the larval body plan? This may be accomplished by a factor that activates transcription of zygotic genes in the ventral cells of the embryo. Originally, it was thought that the *Toll* gene itself might encode this substance. However, by cloning the *Toll* gene, it was discovered that the *Toll* gene product is an integral membrane protein (Hashimoto et al., 1988). Therefore, the *Toll* protein cannot directly regulate the expression of genes within the nucleus. Instead, it must regulate gene expression indirectly, through one of the other dorsal group genes. The best candidate for this gene is *dorsal*, the only dorsal group gene known to function after *Toll* in the pathway leading to dorsoventral polarization (see Fig. 6.12). When $dorsal^-/Toll^D$ double mutants are constructed, their phenotype is dorsalized rather than ventralized, indicating that *dorsal* must function after *Toll*. The *dorsal* gene has been cloned, and the distribution of its protein product has been examined by staining embryos with a specific antibody (Steward et al., 1988). The *dorsal* protein is distributed in a ventral-to-dorsal gradient, with its highest concentration in the ventral midline cells that initiate gastrulation (Fig. 6.13A and B). Furthermore, the *dorsal* protein is concentrated in the nuclei of these cells (see Fig. 6.13C and D), suggesting that it

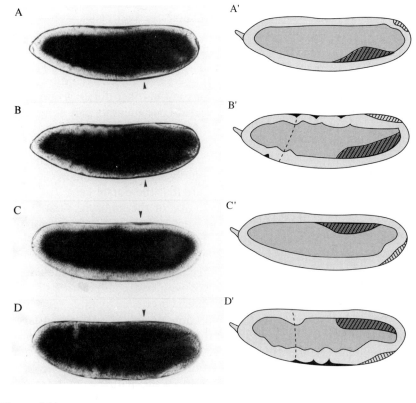

Figure 6.11

The effects of injection of wild-type cytoplasm on the position of gastrulation in *Toll⁻* embryos. A-D, Photographs of living embryos. A′D′, Drawings corresponding to the photographs in A-D, respectively. The colored cross-hatched areas on the drawings (A′-D′) represent the position of the gastrulation (i.e., ventral) furrow (A′ and C′) or the germ band (B′ and D′). The uncolored cross-hatched areas represent the position of the posterior midgut in vagination (see p. 215). The microinjection site is indicated by the arrowheads in A-D. In each embryo, the dorsal side is facing up and the ventral side is facing down. A-A′, After injection into the ventral side of an egg, gastrulation occurred precisely at the site where the wild-type cytoplasm was injected (arrowhead). In this case, this site was also the normal location of gastrulation in the wild-type embryo. B-B′, Twenty minutes later, invagination is complete, and the germ band has extended toward the dorsal side of the egg. C-C′, After injection into the dorsal side of an egg, gastrulation also begins precisely at the site of injection but on the dorsal side of the egg. D-D′, Twenty minutes later, the germ band has extended toward the ventral side of the egg, opposite its orientation in wild-type embryos. Thus, there has been a reversal of the dorsoventral axis by the injection of wild-type cytoplasm. (From Anderson et al., 1985b. Copyright © Cell Press.

Maternal-effect genes
controlling egg shape

⬇

Dorsal genes other
than *Toll* and *dorsal*

⬇

Toll

⬇

dorsal

⬇

Zygotic gene controlling
dorsoventral axis

⬇

Dorsoventral axis

Figure 6.12

A pathway for the flow of information between various genes involved in dorsoventral axis formation in *Drosophila* embryos. The continuous arrows indicate interactions that have been established by examining the phenotypes of double mutants. The broken arrow indicates possible interactions that have yet to be established experimentally.

---

activates the transcription of zygotic genes, whose products then complete the final steps of dorsoventral axis formation.

In summary, dorsoventral polarization in *Drosophila* is controlled by a group of maternal-effect and zygotic genes, whose products interact in a regulatory pathway leading to dorsoventral axis formation. As we shall see in Chapter 14, another group of maternal-effect and zygotic genes controls anteroposterior axis formation in the *Drosophila* embryo. As interactions between these genes become increasingly clear, we will be able to understand more about the determination of the larval body plan in *Drosophila*.

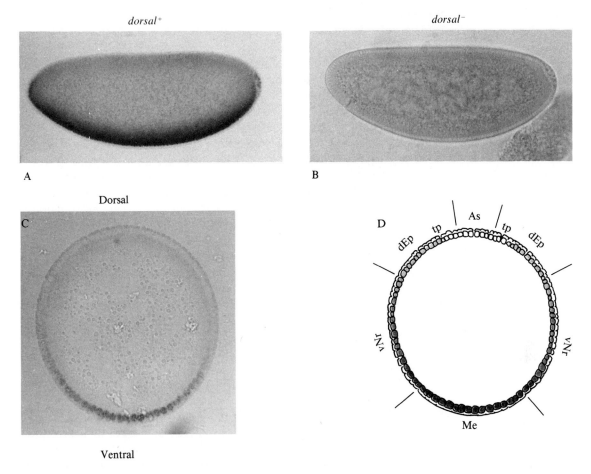

Figure 6.13

Distribution of the *dorsal* protein in the early *Drosophila* embryo as determined by staining with a specific antibody. A, and B, Flattened embryos showing nuclei in cells at the ventral midline. Note the row of stained nuclei at the ventral midline of the *dorsal+* embryo (A) and its absence in the *dorsal−* embryo (B). C, Cross section through the dorsoventral axis of an antibody-stained embryo showing the ventral-to-dorsal gradient of *dorsal* protein in the nuclei of the blastoderm cells. D: dorsal; V: ventral. D, A schematic representation of C showing blastoderm cell assignments on the fate map. As: amnioserosa; dEp: dorsal epidermis; vNr: ventral neurogenic regions; tp: tracheal pits; Me: mesoderm. (From Steward et al., 1988. Copyright © Cell Press.)

6–2 Gastrulation and Establishment of the Embryonic Body Plan

After the determination of bilateral symmetry, gastrulation makes the next major contribution to laying down the body plan. During cleavage, blastomeres usually do not change their shape or position within the embryo. After the blastula is formed, however, the cells of the blastoderm begin to show motility, which can result in their displacement into the interior of the embryo, either as individual cells or as sheets of cells. The initiation of inward displacement of blastodermal cells is defined as the beginning of gastrulation.

Gastrulation occurs by different mechanisms, even in related groups of animals. This is because the mode of cell migration is highly dependent on the quantity and distribution of yolk, which varies considerably between the eggs of different species (see Chaps. 3 and 5). In animals with isolecithal eggs, gastrulation is relatively simple and is initiated near the vegetal pole. In organisms with moderately telolecithal eggs, the vegetal cells are large and quite yolky, and their motility is minimal during gastrulation. In these embryos, gastrulation is initiated near the equator of the embryo, and alternative mechanisms have evolved to internalize cells. Finally, in organisms with telolecithal eggs, the yolk remains uncleaved, and gastrulation is initiated within the blastodisc at the animal pole of the embryo. The mode of gastrulation also depends to some extent on the number of cells in the embryo at the beginning of gastrulation. For example, annelid embryos begin to gastrulate at the 30-cell stage, and gastrulation is relatively simple. However, amphibian embryos consist of about 20,000 cells at the beginning of gastrulation and undergo a complex series of cell rearrangements.

Basic Cell Movements of Gastrulation

Gastrulation involves changes in the behavior of the blastoderm, which in some species is single layered, but in other species is a multilayered epithelium. The inner side (facing the blastocoel) of an epithelium is known as its basal surface, and the outer side (facing the environment) is its apical margin. Epithelial cells adhere to each other by junctions at their lateral surfaces. During gastrulation, some blastodermal cells undergo changes in shape that cause the entire epithelium to spread, converge, or fold. Other cells change their adhesive properties and become associated with a new set of neighboring cells or separate from the epithelium entirely and migrate individually through the embryo. The basic types of cell movements that occur during gastrulation are shown in Figure 6.14.

Three different movements cause epithelia to expand and spread on the surface or within an embryo. **Epiboly** is the process whereby epithelial cells flatten perpendicular to their apicobasal axes, accompanied by the lateral expansion of the sheet (see Fig. 6.14A). During epiboly, the superficial layers of the blastula spread and completely surround the inner portions of the embryo. **Intercalation** is also an expansion process but involving two or more layers of cells. During intercalation, cells from different layers lose contact with their neighbors and intercalate into a single, thinner layer, which consequently spreads laterally (see Fig. 6.14B). The internal spreading of sheets of cells that occurs in some embryos is a consequence of intercalation. **Convergent extension** is a process in which a sheet of cells gets narrower (converges) and longer (extends) along a single axis. During these movements, the epithelial cells lose contact with their original neighbors and form contacts with different cells (see Fig. 6.14C). Thus, convergent extension occurs by cell intercalation. The major part of the body plan of vertebrate embryos is formed by convergent extension.

Three types of movements are responsible for the displacement of cells into the interior of the embryo during gastrulation. **Invagination** occurs when epithelial cells change shape at their apical and basal surfaces but do not lose lateral contacts with other cells in the epithelium, causing it to buckle and fold into the interior of an embryo (see Fig. 6.14D). Invagination resembles the pushing in of one side of a soft, hollow rubber ball. **Ingression** occurs when individual epithelial cells change shape, lose contact with other epithelial cells, and migrate into the blastocoel (see Fig. 6.14E). During invagination and ingression, epithelial cells temporarily become bottle-shaped owing to the contraction of their apical margins and the expansion of their basal margins. Only during ingression, however, do the bottle-shaped cells actually leave the epithelium and migrate individually into the blastocoel. The third internalization mechanism is **involution**. Involution occurs when an expanding epithelium turns over on itself and continues to spread in the opposite direction along its basal margin (see Fig. 6.14F).

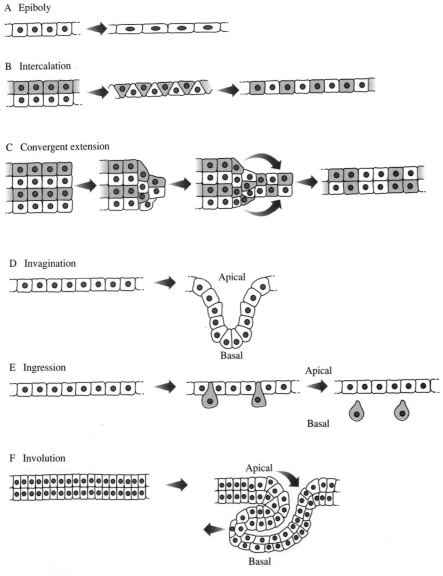

A Epiboly

B Intercalation

C Convergent extension

D Invagination

Apical

Basal

E Ingression

Apical

Basal

F Involution

Apical

Basal

Figure 6.14

Schematic diagram summarizing the basic cell movements involved in gastrulation. A, Epithelial cells flatten and spread by epiboly. B, A multiple-layered sheet of cells becomes a single sheet and spreads by intercalation. C, A sheet of cells advances, converges at a site, and extends along a single axis by convergent extension. D, An epithelium buckles and folds by invagination. E, Single cells of an epithelium terminate their association with other cells and begin to migrate during ingression. F, Cells of an expanding epithelium turn in and expand in the opposite direction on the basal surface of the same epithelium during involution.

Although the types of cell movement previously described function during gastrulation, they are not restricted to gastrulating embryos. Some of these movements are also important in the development of tissues and organs of more advanced embryos and will be discussed in more detail in Chapter 9.

Gastrulation and Formation of the Body Plan in Invertebrates

Gastrulation usually begins after relatively few cell divisions in invertebrate embryos. Consequently, the mechanisms of gastrulation in invertebrates are relatively simple. Invertebrates typically gastrulate by a combination of epiboly and ingression and/or invagination. Except for sea urchin embryos, however, the mechanisms of invertebrate gastrulation are not well characterized.

Annelids

Annelid eggs vary considerably in size and in the amount and distribution of yolk. Thus, they provide examples of alterations in cell movements that have evolved to accommodate the internalization of large, yolk-laden cells during gastrulation. Color plate 2 shows gastrulation in three different species of polychaete annelids. Most polychaetes gastrulate by epiboly and invagination, although invagination is sometimes modified or absent in species with extremely yolky eggs. The polychaete *Eupomatus* (see Color plate 2A) has small isolecithal eggs. In this species, gastrulation is initiated when presumptive endoderm cells in the vegetal hemisphere begin to elongate and invaginate into the blastocoel. The cavity formed by the invaginating vegetal hemisphere cells is called the **archenteron** and will later become the larval gut. As invagination proceeds, the archenteron becomes larger and gradually obliterates the blastocoel. The presumptive mesoderm cells are also displaced into the interior of the embryo during invagination. While invagination is occurring in the vegetal hemisphere, epiboly begins in the animal hemisphere. Eventually, the presumptive endoderm and mesoderm are completely covered by the spreading sheet of ectodermal cells. In *Scoloplos* (see Color plate 2B), which is a polychaete with larger isolecithal eggs than *Eupomatus*, the presumptive endoderm does not fold inward during gastrulation. Instead, the large endoderm cells expand at their basal margins and contract at their apical margins as they move into the blastocoel. The presumptive mesoderm cells of *Scoloplos* embryos are internalized by a unique process. They undergo multiple asymmetrical cleavages, budding off daughter cells into the interior of the embryo. The daughter cells adhere tightly to each other and are known as the **mesodermal bands**. Eventually, the endoderm and mesoderm cells that are still present on the surface are covered by presumptive ectoderm, which has spread from the animal hemisphere by epiboly. Finally, the polychaete *Neanthes* (see Color plate 2C) has moderately telolecithal eggs that cleave in the vegetal hemisphere into large yolky cells without forming a blastocoel. In this species, invagination has been discarded as a mechanism of internalization, and gastrulation occurs by epiboly. Cleavage in the animal hemisphere produces presumptive ectoderm cells that gradually spread and completely surround the large presumptive mesoderm and endoderm cells in the vegetal hemisphere. In these embryos, the archenteron is formed by a secondary process. Thus, gastrulation in polychaetes varies according to the size of the egg and the quantity of yolk. Later in this chapter, we shall see that a similar trend is found in vertebrate embryos.

After gastrulation, the larval body plan is established. In annelids, this involves the formation of a **trochophore** larva, which contains mesodermal

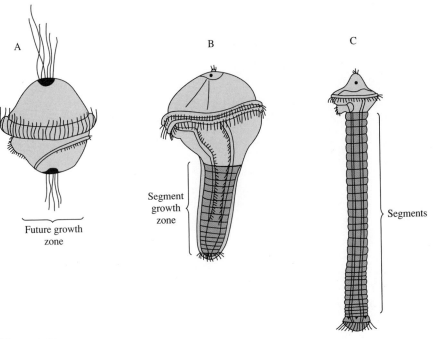

A   B   C

Future growth
zone

Segment
growth
zone

Segments

Figure 6.15

Later development and segmentation in an annelid. A, A trochophore larva before
the beginning of segment growth. B, A trochophore larva beginning to form trunk
segments in the posterior growth zone. C, Further segmentation in the trunk
region during the formation of an adult annelid. (After Kumé and Dan, 1988.)

bands and a ventral nervous system (Fig. 6.15A). Beginning at the trochophore stage, the embryo begins to feed and grow. Growth occurs in the
trochophore larva by the addition of body segments within a growth zone
at the posterior end (see Fig. 6.15B and C). Typically, only the progeny of
the 4d and 2d blastomeres of the 64-cell embryo (see Chap. 5) give rise to
the segmented region of the body (Anderson, 1973). These stem cells, or
teloblasts, compose the growth zone from which the segmented body
develops. Other embryonic cells form structures associated with the mouth,
brain, and specialized organs of the trochophore larva.

Drosophila

Insect embryos gastrulate by invagination. As previously discussed (see p.
205), the embryo proper is derived from the ventral blastoderm, whereas
cells of the dorsal blastoderm produce primarily the extraembryonic amnioserosa. In *Drosophila*, there are two phases of gastrulation (Turner and
Mahowald, 1977). During the first phase, presumptive mesoderm cells
invaginate at the ventral midline (Fig. 6.16A). During this process, the
blastoderm buckles inward to form the **ventral furrow** (Fig. 6.17). After
invagination, the folded epithelium separates from the blastoderm remaining
on the surface of the embryo and forms a tube of mesodermal cells beneath
the ventral midline (see Fig. 6.17C). The mesoderm and the overlying
blastoderm constitute the germ band. Subsequently, the ventral furrow
closes (see Fig. 6.16B), and the internal tube of mesoderm flattens and
begins to spread laterally (see Fig. 6.17D to F). The mechanisms of these
internal cell movements are poorly understood. While the ventral furrow is
closing, several shallow transverse furrows (the cephalic, anterior dorsal,
and posterior dorsal furrows) are formed by invagination (Fig. 6.16B to D).

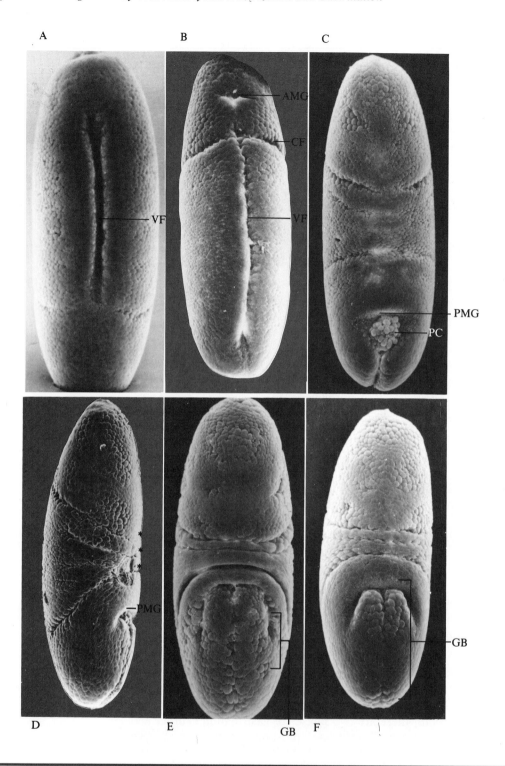

Figure 6.16

Scanning electron micrographs of gastrulation in *Drosophila*. All embryos are oriented with the anterior end pointing toward the top of the page. A, Ventral view of an early gastrula showing the ventral furrow (VF). B, Ventral view of a later gastrula showing the closing ventral furrow (VF), the cephalic furrow (CF), and the anterior midgut invagination (AMG). C, Dorsal view of an embryo during germ band extension. The pole cells (PC) have been displaced from the posterior pole and are entering the posterior midgut invagination (PMG). D, Lateral view of the same stage as C showing the transverse furrows (asterisks). E, Later stage during germ band (GB) extension. The pole cells are now entirely within the posterior midgut on the dorsal side of the embryo, and the germ band has extended anteriorly. F, Later stage during germ band extension. The germ band has extended even further anteriorly on the dorsal surface of the embryo. A-F approximately × 230. (Micrographs by Dr. F. R. Turner, from S. L. Fullilove and A. G. Jacobson. 1978. Embryonic development: Descriptive. *In* M. Ashburner and T. R. F. Wright [Eds.], *The Genetics and Biology of Drosophila*, Volume 2c. Academic Press, New York. pp. 159, 165, 166, 167, 169, 170, 173.)

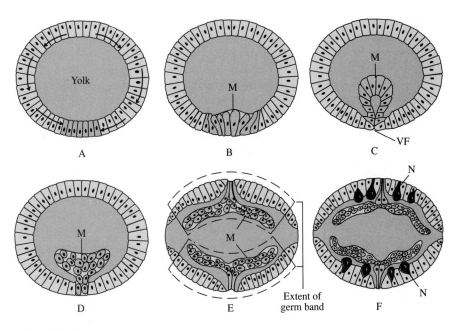

Figure 6.17

Diagrams of *Drosophila* gastrulae in cross-section showing ventral furrow invagination and mesoderm formation. A, Cellular blastoderm stage before ventral furrow formation. B, Cell movements of presumptive mesoderm cells (M) at the ventral midline initiate gastrulation. C, Invagination of presumptive mesoderm cells leads to ventral furrow (VF) formation. D, Mesoderm rudiment flattens and expands in ventral interior of gastrula. E, Extension of mesodermal rudiment to the dorsal side of the embryo during germ band extension. F, Presumptive neural cells (N) ingress from the surface of the extended germ band. (After Fullilove and Jacobson, 1978.)

These transverse furrows provide convenient markers for fate-mapping experiments (see Chap. 11) but have no apparent embryonic function and eventually disappear.

The second phase of gastrulation involves the formation of a series of pit-like invaginations, which are localized at specific positions along the ventral midline. The **anterior midgut invagination** forms anterior to the ventral furrow, the **stomodeal invagination** forms slightly anterior to the anterior midgut invagination, and the **posterior midgut invagination** forms at the posterior end of the ventral furrow near the pole cells, which gradually enter this cavity (see Fig. 6.16C to F). Initially, the posterior midgut invagination is located at the posterior end of the ventral furrow. However, as the pit deepens and the pole cells become internalized, it is translocated around the posterior pole as a result of **germ band extension**. During this process, the germ band expands posteriorly, wraps around the posterior pole, and then continues to spread in an anterior direction along the dorsal surface of the embryo (see Fig. 6.16E and F and Fig. 6.17C and D). To make room for the expanding germ band, cells in the dorsoposterior region of the embryo move laterally.

The zygotic—rather than the maternal—genome appears to control the movements of gastrulation in *Drosophila*. This has been shown by genetic experiments with two recessive lethal mutations: *folded gastrulation (fog)* and *twisted gastrulation (tsg)* (Zusman and Wieschaus, 1985). Mutant *fog* embryos do not form a posterior midgut invagination. As a result, when the germ band extends, it folds into the interior rather than moving onto the dorsal side of the embryo. Mutant *tsg* embryos have defective dorsal cells that are incapable of lateral motility and impede germ band extension. Both mutations can be rescued by fertilization with sperm from wild-type embryos, indicating that the normal alleles are expressed in the zygote.

After gastrulation is complete, the ectodermal cells on either side of the ventral furrow enter the interior of the embryo (see Fig. 6.17F) and form the ventral nerve cord, the germ band becomes segmented (see Chap. 14), and the remaining cells on the surface become either amnioserosa or epidermis. Further development results in **germ band shortening**, essentially a contraction of the extended germ band. As a result of germ band shortening, the segmented germ band returns to its original position on the ventral surface of the embryo, and the amnioserosa is stretched over the dorsal region to cover the developing midgut. The posterior midgut invagination is translocated back to the posterior tip of the embryo by these movements, where it will form the anus. Finally, the mesoderm and ectoderm expand around the midgut and fuse at the dorsal midline. The embryo is now enclosed in ectoderm, which is, in turn, surrounded by the extracellular membranes. This process is called **dorsal closure**.

Sea Urchins

Sea urchins are the only invertebrates in which cell movements during gastrulation have been analyzed by modern experimental methods. Consequently, there is a better understanding of gastrulation in this organism than in any other invertebrate, although many questions remain to be answered. The sea urchin blastula is a single-layered epithelium surrounding a blastocoel (see Color plate 3). The epithelial cells adhere to each other by junctions (Spiegel and Howard, 1983). On the apical surface of the epithelial cells are processes that attach them to the hyaline layer; their basal surface is covered by the **basal lamina**, an extracellular matrix lining the blastocoel. The blastocoel is filled with a fibrous matrix that is continuous with the basal lamina. The sea urchin basal lamina contains sulfated proteoglycans (Sugiyama, 1972; Karp and Solursh, 1974) and proteins that are biochemically and immunologically related to the vertebrate extracellular matrix components: fibronectin, collagen, and laminin (Solursh and Katow, 1982; Spiegel et al., 1983). (The extracellular matrix of vertebrates is discussed in detail in Chap. 9.) The sea urchin extracellular matrix proteins are stored in vesicles within the unfertilized egg that are distinct from the cortical granules (Wessel et al., 1984). After fertilization, exocytosis of these vesicles secretes the extracellular matrix components of the basal lamina onto the basal surface of the blastomeres.

Sea urchin gastrulation occurs in two phases (see Color plate 3). Before the first phase, cells near the vegetal pole of the swimming blastula flatten to form the **vegetal plate**. During the first phase of gastrulation, some cells ingress from the vegetal plate and enter the blastocoel. These cells, the **primary mesenchyme cells**, migrate along the basal lamina and eventually form the larval skeleton. Although gastrulation is already in progress, this stage of development is called the **mesenchyme blastula**. Other cells that remained in the vegetal plate also change their shape, and the epithelium they form invaginates into the blastocoel to form the archenteron. During the first phase of gastrulation, the archenteron is extended only about a third of the distance between the vegetal and animal poles of the embryo; then there is a pause before the beginning of the second phase of gastrulation. During the second phase, the **secondary mesenchyme cells** ingress from the blunt end of the archenteron, form long filopodia (see p. 349), and enter the blastocoel. Simultaneously, the archenteron elongates into a narrow tube that eventually reaches the wall of the blastocoel near the animal pole.

During internalization of the presumptive mesoderm and endoderm, cells of the presumptive ectoderm in the animal hemisphere spread vegetally by epiboly. These two phases of gastrulation will now be discussed in more detail.

Gastrulation is initiated by the ingression of the primary mesenchyme cells into the blastocoel. These cells, which are descendants of the micromeres, localize as a ring surrounding the vegetal pole of the swimming blastula (Fig. 6.18A and B). At the beginning of gastrulation, this ring of cells ingresses, and a depression appears on the outer surface of the vegetal plate (see Fig. 6.18C and D). The process of ingression has been studied in detail by electron microscopy (Figs. 6.18, 6.19, and 6.20). Initially, the basal lamina disappears from the inner surface of the primary mesenchyme cells. Presumably, absence of the basal lamina allows the basal region of the presumptive primary mesenchyme cells to bleb and partially extend into the blastocoel (see Fig. 6.19B). This blebbing activity is followed by coordinated expansion of the basal region and contraction of the apical region of the ingressing cells, causing them to become bottle-shaped and extend even further into the blastocoel (see Figs. 6.19C and D and 6.20). Finally, the ingressing cells withdraw entirely from the vegetal plate and enter the blastocoel (see Figs. 6.18C and D and 6.19E). The exit of primary mesenchyme cells from the vegetal plate is thought to be caused by a combination of four events: (1) disappearance of the basal lamina on the basal side of the presumptive mesenchyme cells, (2) retraction of the processes connecting the presumptive primary mesenchyme cells to the hyaline layer, (3) disappearance of junctions between the primary mesenchyme cells and adjacent cells in the vegetal plate, and (4) expansion of the adjacent vegetal plate cells into the spaces left by the primary mesenchyme cells.

Once inside the blastocoel, the primary mesenchyme cells remain in the region of the vegetal plate for a short time and then begin to migrate, using the basal lamina as a substratum. Migration appears to be facilitated by attachment, retraction, and re-attachment of their filopodia to the basal lamina. Time-lapse filming has shown that the filopodia move actively along the basal lamina (Gustafson and Wolpert, 1961). Sulfated proteoglycans in the basal lamina play an important role in the migration of the primary mesenchyme cells. In embryos treated with seawater lacking sulfate (Karp and Solursh, 1974) or with β-D-xyloside (Solursh et al., 1986)—treatments that interfere with the deposition of sulfated proteoglycans in the basal lamina—the primary mesenchyme cells complete ingression but fail to migrate up the blastocoel wall.

In normal embryos, the primary mesenchyme cells continue to move along the basal lamina until they reach an area to which they show greater adhesion. There they form a ring around the base of the archenteron (see Color plate 3 and Fig. 6.21), which by this time has invaginated into the blastocoel. Within this ring, filopodia of adjacent primary mesenchyme cells fuse and form a cable-like syncytium (Fig. 6.22). Two branches of the syncytium, one on each side of the archenteron, arise from the ring and extend into the animal hemisphere region of the blastocoel. These syncytial cables are sites of larval skeleton formation. Two triradiate spicules develop from a mineralized matrix that is deposited by the syncytial cables. The rays of the spicules grow, bend, and branch in a species-specific pattern.

Experimental evidence suggests that spicule formation is controlled by factors both intrinsic and extrinsic to the primary mesenchyme cells. The importance of factors intrinsic to the primary mesenchyme cells is shown by experiments in which micromeres from one sea urchin species were transplanted into the blastocoel of another and formed spicules resembling the donor species (Hörstadius, 1973). The importance of extrinsic

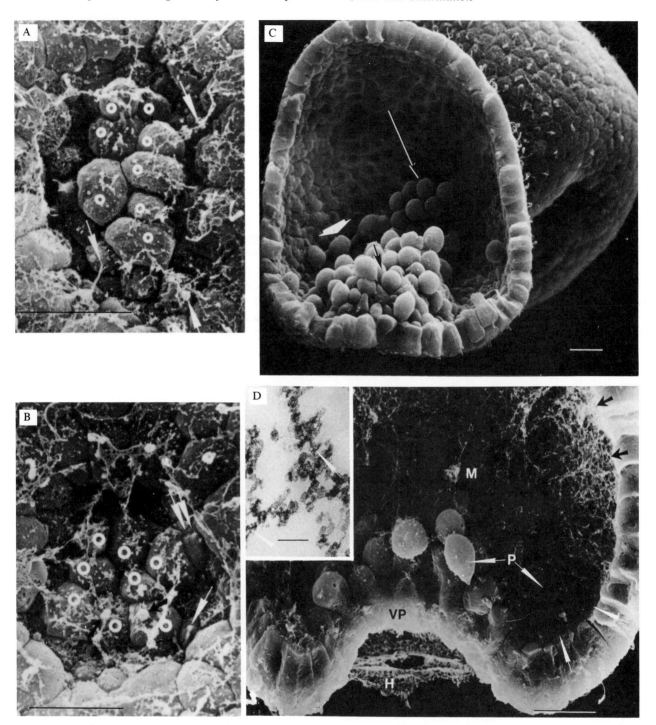

Figure 6.18

Scanning electron micrographs showing various aspects of the first phase of sea urchin gastrulation. A, The vegetal plate region of a mesenchyme blastula viewed from outside the embryo. The eight centrally-located cells (open circles) are slightly depressed but do not ingress. They are surrounded by a ring (arrows) of about 30 primary mesenchyme cells that have started to ingress. B, A slightly later stage in which ingression is continuing in the ring of primary mesenchyme cells that have almost disappeared from the surface of the vegetal plate. A and B scale bars equal 10 μm. C, Gastrulating embryos showing invagination of the vegetal plate. The left hand embryo has been fractured to observe invagination from the inside. The short black arrow indicates the inner surface of the invaginating vegetal plate cells, whereas the long black arrow shows migrating primary mesenchyme cells that are beginning to fuse and form a syncytial ring (thick white arrow). The embryo on the right shows invagination from the outside. D, An invaginating embryo showing fibrous material (M) lining the blastocoel (arrows). The blastocoel material is arranged more loosely in the vegetal (white arrows) than the animal region (black arrows). P: primary mesenchyme cells; VP: vegetal plate; H: hyaline layer. Inset: high resolution micrograph of fibrous material in the blastocoel. Scale bars for C and D equal 10 μm and 0.1 μm (inset). (A, and B, From Katow and Solursh, 1980. C, and D, From Solursh, 1986.)

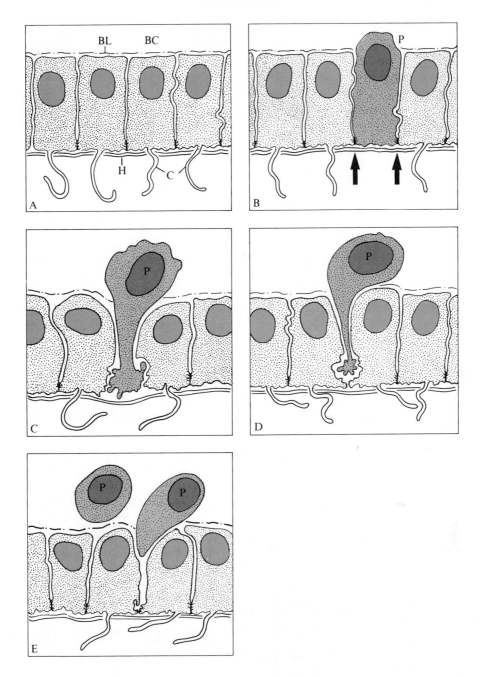

Figure 6.19

Diagram showing primary mesenchyme cell ingression reconstructed from electron micrographs. A, The initial morphology of presumptive primary mesenchyme cells is identical to that of neighboring blastomeres. They have cilia (C) and are covered by the hyaline layer (H) on their apical surfaces. The basal surface is characterized by a thin basal lamina (BL). BC: blastocoel. B, Elongation of a presumptive primary mesenchyme cell (P) into the blastocoel. The presumptive primary mesenchyme cell and the neighboring blastomeres are attached by junctions (arrows). The basal lamina has disappeared from beneath the elongated primary mesenchyme cell. C, Apical elongation and basal swelling of the primary mesenchyme cell. Junctions between the presumptive primary mesenchyme cell and neighboring blastomeres have disappeared. Microtubules are present in the peripheral cytoplasm of the neck region in the presumptive primary mesenchyme cell and in the cytoplasm of neighboring blastomeres, on the sides adjacent to the presumptive primary mesenchyme cell. D, Primary mesenchyme cells separate from the vegetal plate after the disappearance of junctions. The neighboring blastomeres extend apical cell processes toward each other, resulting in the occlusion of the intercellular spaces. E, Rounding of the primary mesenchyme cells. The neck region of the primary mesenchyme cells shortens toward the basal side, and the mesenchyme cells become round in shape and begin to migrate. (After Katow and Solursh, 1980.)

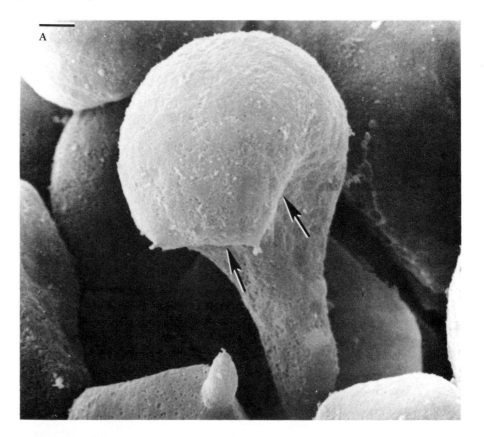

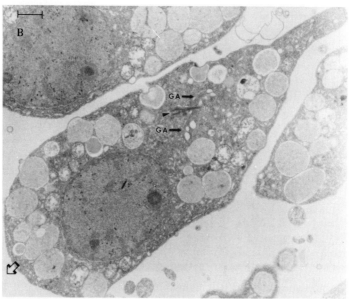

Figure 6.20

Scanning (A), and transmission (B), electron micrographs of bottle-shaped primary mesenchyme cells undergoing ingression. A, An ingressing cell viewed from inside the blastocoel. The direction of ingression is indicated by the arrows. Scale bar equals 1.0 μm. B, A section through an ingressing primary mesenchyme cell showing expanded basal and contracted apical ends. Large open arrow: direction of ingression; small black arrowhead: basal body. The nucleus is located on the basal side, and the Golgi apparatus (GA) is located on the apical side of the cell. Scale bar equals 1.0 μm. (A, From Katow and Solursh, 1980. B, From Anstrom and Raff, 1988.)

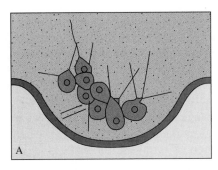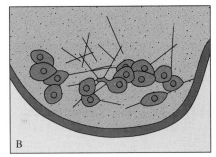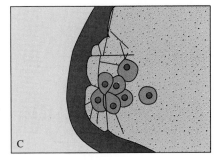

Figure 6.21

Drawings of fusion and syncytial ring formation in migrating primary mesenchyme cells of sea urchin embryos. A, Mesenchyme cells with long filopodia collecting at the level of the blastocoel where the syncytial ring will form. B, The mesenchyme ring cells become associated in a cable-like filopodial complex and begin to form attachments to the blastoderm wall. C, Syncytial ring with many filopodial attachments to the blastocoel wall. (After Gustafson and Wolpert, 1961.)

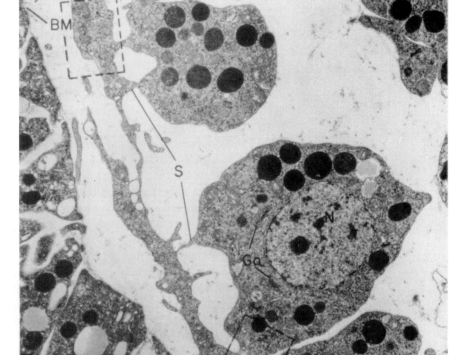

Figure 6.22

Electron micrograph of primary mesenchyme cells after fusion of filopodia to form a syncytial ring. Each cell body is connected to the cable by one or more stalks (S). Short processes extend from the cytoplasm of the cable toward the blastoderm wall. Within the cell body, the nucleus (N) is in a central position. The Golgi zone (Go) lies on one side of the nucleus. Fine extracellular fibrils are present throughout the blastocoel. BM: basal lamina. ×8300. (From Gibbins et al., 1969. Reprinted from *The Journal of Cell Biology*, 1969, *41*: 201–226 by copyright permission of the Rockefeller University Press.)

factors is shown by *in vitro* culture of micromeres and experiments with reconstituted embryos. When micromeres are removed from the embryo and cultured, they form spicules (Okazaki, 1975). However, spicules formed in culture are morphologically different from those formed within the blastocoel. When isolated primary mesenchyme cells were combined with dissociated blastula cells, the different cell types mixed and formed aggregates resembling mesenchyme blastulae (Harkey and Whiteley, 1980). The reconstituted embryos formed an outer epithelium and a blastocoel with primary mesenchyme cells organized in a syncytial ring, as in normal gastrulae. In contrast to the spicules formed in culture, the spicules that developed in the reconstituted embryos resembled normal spicules. This result suggests that cell interactions influence spicule development, perhaps with the underlying ectodermal cells assisting the primary mesenchyme cells in establishing a more normal configuration of syncytial cables before they lay down the spicule matrix.

After the primary mesenchyme cells have ingressed, neighboring cells fill the gaps left in the vegetal plate. Subsequently, these cells change from a cuboidal to a rounded shape, but they remain in contact with adjacent cells and the hyaline layer. During invagination, the vegetal plate bends inward, initially extending only about a third of the distance across the blastocoel. The archenteron that is formed opens to the outside through the blastopore (see Color plate 3). The mechanism of vegetal plate invagination is uncertain and requires further investigation. Gustafson and Wolpert (1967) proposed that invagination occurs because the rounded vegetal plate cells require a greater surface area but cannot expand laterally or apically because of their adherence to adjacent epithelial cells and the hyaline layer. Therefore, they reasoned that the only choice is for the epithelium to buckle inward toward the direction of reduced contact. In contrast, Moore and Burt (1939) showed that invagination occurs outside the embryo when the vegetal plate is removed and cultured. This result argues against invagination as a consequence of physical constraints on the vegetal plate cells and suggests that it is an intrinsic property of these cells.

Although buckling and folding of the vegetal plate initiate archenteron formation, they do not cause archenteron elongation during the second phase of gastrulation. Elongation is caused by a combination of two mechanisms: (1) intercalation of cells in the archenteron wall and (2) contractile activity of the ingressing secondary mesenchyme cells at the tip of the archenteron. During their rearrangement, cells in the wall of the archenteron lose associations with adjacent cells and become repacked into a long thin tube. This has been demonstrated by counting cells in the wall of the archenteron before and after elongation (Ettensohn, 1985). If elongation occurs by cell rearrangements, the number of cells in the archenteron wall would decrease in cross section as the archenteron elongates (Fig. 6.23A). This is precisely the result that was obtained (see Fig. 6.23B and C). Cross sections of the archenteron wall contain 20 to 25 cells before the beginning of elongation, but only about 10 cells in the same area after its completion. The second mechanism responsible for archenteron elongation is extension and contraction of filopodia (Gustafson and Kinnander, 1956; Dan and Okazaki, 1956). Filopodia are formed by the secondary mesenchyme cells at the tip of the archenteron (Fig. 6.24A and B). These filopodia extend to the inner surface of the gastrula wall and adhere. After specific adhesions are made, the filopodia contract and shorten, pulling the entire archenteron toward the animal pole.

It was believed originally that filopodial contraction was sufficient to complete archenteron formation in sea urchin embryos; however, recent experiments show that elongation is driven by a combination of filopodial contractions and cell rearrangements. In these experiments, the length of the archenteron was measured in embryos where the secondary mesenchyme cells were either prevented from participating in gastrulation or destroyed (Hardin, 1988). Treatment of sea urchin blastulae with LiCl causes **exogastrulation**, an abnormal condition in which the archenteron forms inside-out. In exogastrulae, the archenteron protrudes to the outside of the embryo,

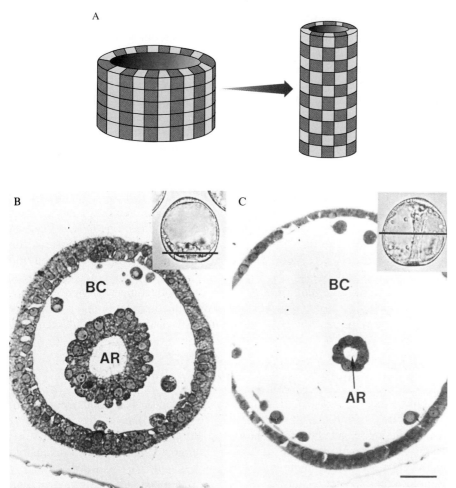

Figure 6.23

Archenteron elongation in the sea urchin gastrula. A, A diagram showing how a short, wide cylinder can be converted to a narrow tube by cell rearrangements. B, and C, Cross-sections of sea urchin gastrulae. The insets show the developmental stages examined and indicate the plane of sectioning and the level along the animal-vegetal axis at which the sections were cut. B, A cross-section passing through the middle of the gut rudiment in a gastrula before archenteron elongation. C, A cross-section passing though the middle of the gut rudiment in a gastrula after archenteron elongation. The number of cells in the wall of the archenteron has decreased from about 25–30 in A to about 10 in B. BC: blastocoel; AR: archenteron. Scale bar equals 20 μm. (From Ettensohn, 1985.)

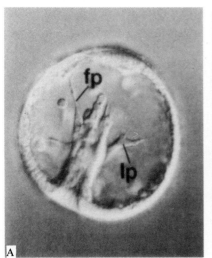
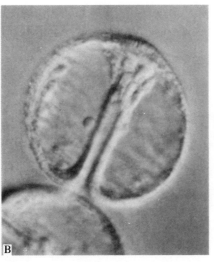
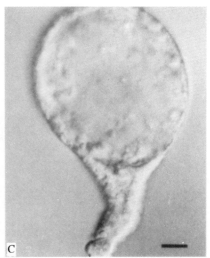

Figure 6.24

Photomicrographs of archenteron elongation during sea urchin gastrulation and exogastrulation. A, An embryo beginning the second phase of gastrulation. Note the filopodia (fp) and lamellipodia (lp) extending from the secondary mesenchyme cells. B, A late gastrula that is completing gastrulation. The archenteron has extended toward the roof of the blastocoel. C, An exogastrula with an everted archenteron. Scale bar equals 25 μm. (From Hardin, 1987.)

rather than into the blastocoel, and no secondary mesenchyme cells or filopodia are formed (see Fig. 6.24C). The protruding archenteron is only about 75% as long as a normal archenteron (see Fig. 6.25). In another experiment, the length of the archenteron was determined after the secondary mesenchyme cells were destroyed with a laser microbeam. In this case also, the archenteron elongated to only about 75% of its normal size (see Fig. 6.25). These results show that filopodial movements of the secondary mesenchyme cells and cell rearrangements in the archenteron wall are both required to complete the second phase of gastrulation.

As the tip of the archenteron nears the gastrula wall, the secondary mesenchyme cells complete ingression and migrate individually into the blastocoel. These cells differentiate into larval muscle and pigment cells (Gibson and Burke, 1985) and also form the **coelomic sac**. The coelomic sac is later constricted to form a pair of coelomic pouches, one on each side of the archenteron. The formation of the coelomic sac leaves the archenteron composed entirely of endoderm. At the point of contact between the archenteron tip and the blastocoel wall, a mouth opening is formed. At the opposite end of the embryo, the blastopore becomes the anus, and a continuous tube—the gut—is produced. With the completion of gastrulation, the basic body plan of the sea urchin **pluteus larva** is established (see Color plate 3).

Gastrulation and Formation of the Body Plan in Chordates

We have seen that gastrulation in invertebrate embryos varies according to the quantity of yolk in the egg. Although the amount and distribution of yolk influence gastrulation in chordates, the structure of the blastula also plays an important role in this process. Like most invertebrates, the blastula of lower chordates is a single-layered epithelium, and gastrulation is relatively simple. We first examine a relatively simple example of chordate gastrulation with *Amphioxus* and then proceed to examples of vertebrate gastrulation, in which the blastula is often more than one layer thick or forms a blastodisc at the animal pole of the embryo, complicating movements at gastrulation.

Amphioxus

A relatively simple form of gastrulation occurs in the cephalochordate *Amphioxus* (Conklin, 1932). Like vertebrates, *Amphioxus* contains a dorsal nerve cord, an underlying **notochord** (a stiff rod of mesodermal origin), and flanking musculature organized in segmental units called **somites**. The blastula of *Amphioxus* consists of a single-layered epithelium surrounding a blastocoel (see Color plate 4). At the beginning of gastrulation, presumptive endoderm

Figure 6.25

Comparison of final lengths of archenterons in exogastrulae (A), gastrulae whose secondary mesenchyme cells were destroyed with a laser microbeam (B), and normal gastrulae (C). The archenteron elongates to only 75% of its normal length in A and B. (After Hardin, 1988. Reprinted by permission of Company of Biologists, Ltd.)

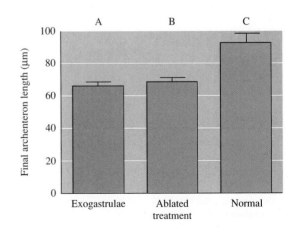

cells in the vegetal hemisphere elongate to form a flattened **endodermal plate**. *Amphioxus* gastrulates by invagination, in which the endodermal plate bends inward to form a cup-shaped embryo with a double wall. The inner wall lines the archenteron, which opens to the exterior through the blastopore. The outer wall consists of presumptive ectoderm, which will later form epidermis and neural tissues. As invagination continues, the presumptive mesoderm occupies the rim of the blastopore, with the **presumptive notochord** forming the dorsal lip and the **somitic mesoderm** (that part of the mesoderm that forms somites) forming the lateral and ventral lips. Gradually, the mesoderm is withdrawn into the interior of the embryo, but the mechanisms involved in this process have not been determined. Once inside the embryo, the mesodermal cells shift their relative positions. The somitic mesoderm converges toward the dorsal side of the embryo and straddles both sides of the presumptive notochord, which extends in the opposite direction to form the notochord (Fig. 6.26). On the outer surface, the neural ectoderm thickens over the notochord to form the **neural plate**. The notochord and the flanking mesoderm line the roof of the archenteron, whereas the endoderm forms its base and sides (see Color plate 5).

At the completion of gastrulation, the cells that will form internal structures have been displaced to the inside of the *Amphioxus* embryo. A great deal of rearrangement remains to be done, because the invaginated cells are still a continuous sheet. The notochord cells then pack into a rod flanked on either side by the somitic mesoderm. After segregation of the notochord and somitic mesoderm from the endoderm, the endoderm becomes a continuous sheet, which is the lining of the prospective digestive tract. This is illustrated by the drawings in Color plate 5A to 5D, which represent cross sections of progressively later stages of *Amphioxus* embryos. Development of the mesoderm continues as a series of invaginations from the roof of the archenteron separate to become somites, each with its own cavity, and the notochord elongates in an anteroposterior direction (see Color plate 5E). These somite cavities later expand and fuse to form the general body cavity, or **coelom**. On the surface of the embryo, the neural plate separates from the presumptive epidermis and invaginates to form the **neural tube**, from which the central nervous system is derived.

Amphibians

The structure of the amphibian embryo differs from that of the embryos we have previously described in three major ways. First, the blastula contains a greater number of cells (about 20,000) at the beginning of gastrulation. Second, large vegetal blastomeres restrict the blastocoel to the animal hemisphere (see Fig. 5.16H). Third, the blastula is a not a single-layered epithelium: Several layers of cells lie between the surface of the embryo and the wall of the blastocoel. These features necessitate a more complicated pattern of gastrulation than exists in most invertebrates and lower chordates. Because the yolky vegetal hemisphere cells are large and immobile, invagination at the vegetal pole has been discarded as a means of gastrulation.

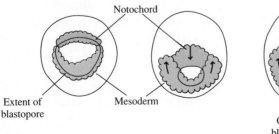

Notochord

Extent of blastopore

Mesoderm

Closing blastopore

Figure 6.26

Diagrams showing the extension of somitic mesoderm and notochord cells in opposite directions during closure of the blastopore in *Amphioxus*. (After Conklin, 1932.)

Instead, gastrulation is initiated in the marginal zone on the dorsal side of the embryo, and cells enter the interior of the embryo primarily by involution.

The multiple layers of cells in the amphibian embryo complicate the process of constructing fate maps. The classical method of making an amphibian fate map is to stain the surface of an embryo with a vital dye and follow the movement of the stained cells during gastrulation (see Chap. 11). Color plate 6A shows the fate map of the urodele amphibian *Ambystoma mexicanum*, which was constructed using the vital dye marking method (Vogt, 1929). This fate map shows three regions that correlate approximately with areas of different pigmentation in the early gastrula. The presumptive ectoderm (epidermis and neural plate) is located in the darkly pigmented animal hemisphere, the presumptive mesoderm is located in the marginal zone, and the presumptive endoderm is located in the pigment-free vegetal hemisphere. Surprisingly, fate maps of *Xenopus laevis* embryos reveal a different location for the presumptive mesoderm (Keller, 1975, 1976, 1978). In this anuran amphibian, the presumptive mesoderm maps to the marginal zone, just as it does in urodeles, but it lies *under* a superficial epithelium of endoderm (see Color plate 6B to 6D). Therefore, only the presumptive ectoderm and endoderm are on the surface in *Xenopus* blastulae. Recent research suggests that the internal position of the presumptive mesoderm may be a unique feature of *Xenopus laevis* embryos: Other anuran amphibians appear to have a fate map similar to urodele amphibians.

Using the fate maps shown in Color plate 6B to 6D, we shall now present a simplified version of gastrulation in the anuran amphibian *Xenopus laevis*. The complicated cell movements that occur during *Xenopus* gastrulation are discussed in more detail in the essay on page 228. Remembering two general rules will help us understand gastrulation in this species. First, all surface cells of the blastula will remain as surface cells throughout gastrulation. This is true even for the superficial endoderm, which forms the surface of the newly created internal cavity (the archenteron). Second, the ring of mesoderm beneath the surface turns itself inside-out, and the dorsal side elongates. These movements are shown in Color plate 7. Gastrulation begins on the surface of the embryo when endoderm cells in the dorsal marginal zone undergo invagination to form the blastopore (see Color plate 8). As endodermal cells continue to migrate into the interior of the embryo, an archenteron is formed, which is lined with endoderm. The superficial endoderm is replaced on the outside of the embryo by ectoderm, which stretches from the animal hemisphere to cover the external surface of the embryo (see Color plate 7A to 7D). The ectoderm expands by epiboly. While these movements are occurring on the surface of the embryo, the mesoderm is undergoing involution within the interior of the embryo. The sheet of mesodermal cells moves initially in a vegetal direction, rolls under itself, and then migrates in an animal direction (see Color plate 7E and 7F). Invagination of endoderm and involution of mesoderm begin on the dorsal side of the embryo, forming the **dorsal lip of the blastopore**. As lateral and ventral cells begin to invaginate and involute, the two ends of the blastopore groove gradually extend around the embryo in an increasingly greater arc (Fig. 6.27). The ends of the arc eventually meet ventrally, completing the blastopore circle. Lateral extension of the blastopore produces the **lateral lips of the blastopore**, and its completion produces the **ventral lip of the blastopore**. The large endodermal cells encircled by the blastopore form the **yolk plug**.

As gastrulation proceeds, the embryo begins to elongate along its anteroposterior axis. This elongation is mediated by convergent extension of mesoderm on the dorsal side of the embryo (see Color plate 7G to 7J). Dorsal mesoderm converges toward the point of involution while moving

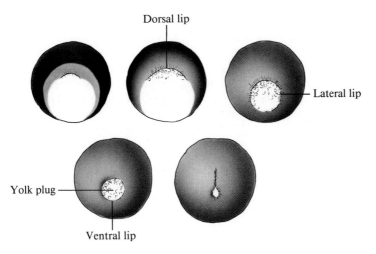

Figure 6.27

Changes in the shape of the blastopore and its closure during gastrulation in a frog. The dorsal lip forms first followed by the lateral and ventral lips, respectively. (After Balinsky, 1981.)

in a vegetal direction and then extends in an animal (anterior) direction after involuting at the dorsal lip of the blastopore. Based on its subsequent fate, the mesoderm is subdivided into three parts: (1) the presumptive notochord, which will form the notochord; (2) the somitic mesoderm, which will form somites; and (3) the **lateral plate mesoderm**, which will form ventral mesodermal derivatives. The presumptive notochord involutes over the dorsal lip, the somite mesoderm involutes over the lateral lips, and the lateral plate mesoderm moves over the ventral lip. Each of these regions involutes in succession because the dorsal lip forms first and the ventral lip forms last. Once inside the embryo, the presumptive notochord undergoes further extension to form the rod-like notochord, and the somitic mesoderm extends further and then divides to form somites. At the same time, the neural tube is forming from the ectoderm on the dorsal midline. These events combine to generate the body plan of the amphibian embryo and will be discussed in detail in Chapters 7 and 8.

Birds

Ingression is the primary mechanism of gastrulation in birds. Invagination has been abandoned as a mechanism of gastrulation because the blastodisc lies on an uncleaved mass of yolk (see Chap. 5).

Unraveling the pattern of cell movements during gastrulation in birds has proved to be a difficult problem. One difficulty is the relative inaccessibility of the lower layer of cells in the intact embryo. This layer has been made accessible through development of procedures for *in vitro* culture of the blastoderm. Classical studies of bird gastrulation used vital dyes and carbon particles to mark blastoderm cells. However, use of dyes has drawbacks because they can stain more than one cell layer. Although carbon particles specifically mark the upper surfaces of cells, they may not be displaced uniformly when the cells begin to move. Thus, either procedure can complicate the interpretation of fate maps. Recently, transplantation of identified cells from a host embryo has been used to follow cell movements during gastrulation. In one approach, cells whose nuclear DNA has been labeled with ^{3}H-thymidine are transplanted from a donor embryo to the same region of an unlabeled recipient embryo. The ultimate locations of the

Regular text continues on page 231.

Essay

Cell Movements During Xenopus *Gastrulation*

Gastrulation in *Xenopus laevis* involves a number of autonomous cell movements that occur in specific parts of the embryo (Keller, 1978, 1980, 1981, 1986; Keller and Schoenwolf, 1977; Gerhart and Keller, 1986). We shall examine five different regions of the embryo that show specific motile behaviors. Figure 1, which is a three dimensional reconstruction of the superficial (A and D) and subsurface (B and C) layers of the *Xenopus* gastrula, illustrates the locations and movements of the different regions during gastrulation. The first region is the **animal cap**, which is located in the pigmented animal hemisphere. The animal cap consists of several layers of cells that spread toward the vegetal hemisphere by epiboly and intercalation during gastrulation. The animal cap cells are presumptive ectoderm. The second region is the marginal zone. There are two subzones in this region: (1) the **noninvoluting marginal zone**, which is the animal portion of the marginal zone that spreads in front of the animal cap, but does not involute during gastrulation, and (2) the **involuting marginal zone**, which is the vegetal portion of the marginal zone that turns inside the embryo during involution. The cells of the noninvoluting and involuting marginal zones show convergent extension: convergence toward the dorsal midline and extension in an animal-vegetal (anteroposterior) direction. The noninvoluting marginal zone is presumptive ectoderm. The superficial layer of the involuting marginal zone is presumptive endoderm, whereas the subsurface layer is presumptive mesoderm. The third region, the **deep zone**, is a ring of presumptive mesodermal cells located below the surface on the vegetal edge of the involuting marginal zone. During gastrulation, the deep zone cells migrate anteriorly along the inner surface of the blastocoel roof. The migration of deep zone cells appears to initiate involution, as they adhere tightly to the involuting marginal zone and draw it inside. The fourth region is the **bottle cell zone**, which consists of a thin ring of cells on the surface of embryo between the involuting marginal zone and the fifth region, the **vegetal base** (see below). These cells are called bottle cells because they contract at their apical margins and expand at their basal margins. As they invaginate into the interior of the embryo, they are pulled anteriorly as a result of adhesion to the involuting marginal zone. The involuting marginal zone cells appear to adhere tightly to the bottle cells, dragging them along as they extend anteriorly. Contraction and invagination of bottle cells first occurs at the dorsal midline and then moves bidirectionally, eventually reaching the ventral midline (see Fig. 6.27). The bottle cells are presumptive endoderm. The fifth region, the vegetal base, contains large, yolky, vegetal blastomeres. The vegetal base cells are passively moved inward and to the ventral side of the embryo as gastrulation proceeds. The vegetal base is presumptive endoderm and forms the floor of the archenteron.

Some of the cell movements involved in gastrulation actually begin a few hours earlier: Cell motility is initiated after the midblastula transition (see p. 568). At this time, the animal cap begins to expand by epiboly and intercalation. During expansion, the superficial cells flatten and divide, with both daughter cells occupying the expanded epithelium. Meanwhile, cell layers beneath the surface intercalate with each other to form a thin layer of greater surface area, as shown in parts A through F of Figure 2. Gastrulation is initiated at the dorsal midline on the border between the involuting marginal zone and the vegetal base. Several events occur in rapid succession during the initiation of gastrulation. First, cells of the deep zone begin to migrate anteriorly using the inner roof of the blastocoel as a substratum (G and H). Second, the bottle cells on the surface of the embryo contract and invaginate to form the blastopore, which begins as a small dent in the surface of the marginal zone (see Color plate 8B). Third, involution begins when the marginal zone rolls under itself to form the dorsal lip of the blastopore and begins to extend in an anterior direction. The initiation of involution is probably caused by the migration of the deep cells, which pull the marginal zone over the dorsal lip as they migrate anteriorly. After the initiation events are completed, the remainder of gastrulation is driven by convergent extension, which occurs in the marginal zone. Convergence and extension of marginal zone cells force more cells to spread toward the blastopore, involute over the dorsal lip, and extend along the anteroposterior axis of the embryo (see Color plate 8C). Because the bottle cells adhere tightly to cells of the involuting marginal zone, they are pulled anteriorly, and the narrow space behind them enlarges to become the archenteron. The bottle cells form the leading edge of the archenteron, the involuting superficial marginal zone cells become the roof of the archenteron, and the superficial cells of the vegetal base become the floor of the archenteron. As convergent extension continues, the archenteron expands anteriorly (obliterating the blastocoel in the process), and the blastopore moves posteriorly.

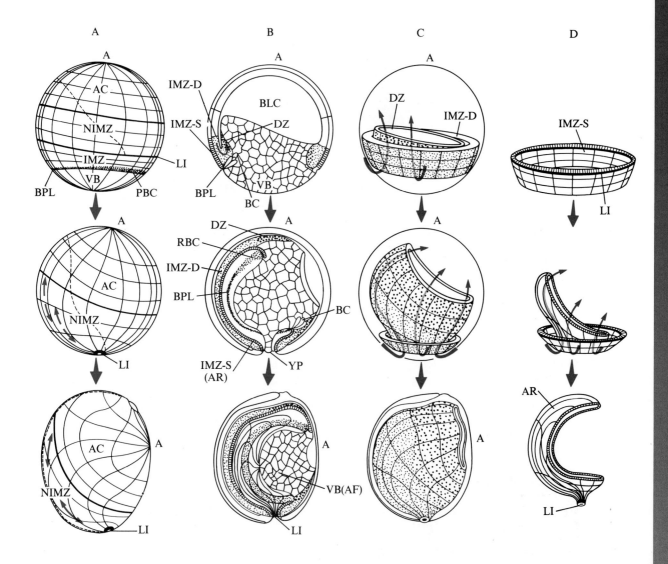

Figure 1

Schematic three-dimensional reconstructions of the behaviors of the five different regions of *Xenopus* embryos during gastrulation. All embryos are drawn with the animal pole up, the vegetal pole down, the dorsal side to the left, and the ventral side to the right. A: animal pole; AC: animal cap; AF: archenteron floor; AR: archenteron roof; BC: bottle cells; BLC: blastocoel; BPL: blastopore pigment line; DZ: deep cell zone; IMZ: involuting marginal zone; IMZ-D: IMZ deep layer; IMZ-S: IMZ superficial layer; LI: limit of involution; NIMZ: non-involuting marginal zone; PBC: prospective bottle cells; RBC: respread bottle cells; VB: vegetal base. Top row: early gastrula stage. Middle row: late gastrula stage. Bottom row: completion of gastrulation. A, Surface views of whole embryos. The surface arrows indicate sites of convergent extension. B, Midsagittal views of sectioned embryos. C, Inside view of involution (arrows) of sub-surface cells the involuting marginal zone and deep cell zone of whole embryos. D, Inside view of involution (arrows) of the superficial cells in the involuting marginal zone of whole embryos. (After Gerhart and Keller, 1986. Reproduced with permission, from the Annual Review of Cell Biology, Vol. 2. © 1986 by Annual Reviews, Inc.)

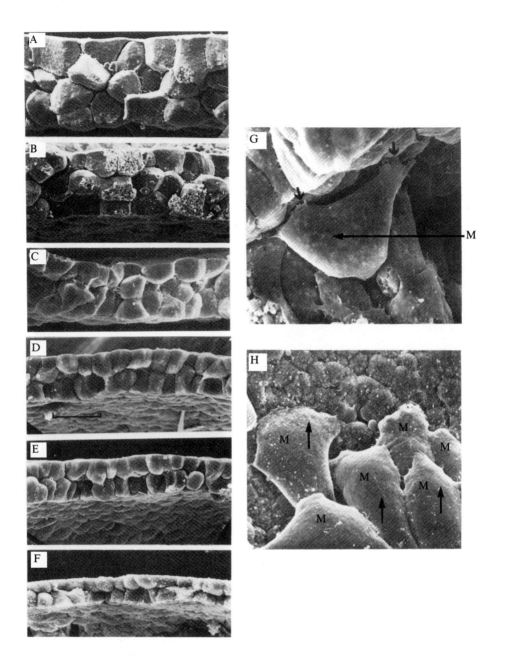

Figure 2

Cell movements during epiboly of the animal cap region (blastocoel roof) and mesodermal deep cells of *Xenopus* embryos. A-F, Scanning electron micrographs of animal cap cells at the late mid-blastula (A), late blastula (B), and early (C) through late (D-F) gastrula stages. Scale bar equals 50 μm. A-F at the same magnification. G, and H, Scanning electron micrographs of migrating mesodermal cells after their involution during amphibian gastrulation. G, A *Xenopus* gastrula showing an involuted mesodermal cell (M) attached to the inner surfaces of the deep cells of the gastrular wall by protrusions (arrows) on its margin. × 545. H, The migrating mesodermal cells (M) of an axolotl gastrula spread on the inner surface of the blastocoel wall. Arrows: direction of migration. (A-F, From Keller, 1986. G, From Keller and Schoenwolf, 1977. H, From Lundmark et al., 1984.)

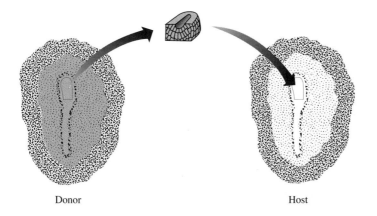

Figure 6.28

Diagram illustrating transplantation of a region from a ³H-thymidine-labeled donor to a corresponding site in an unlabeled host for fate mapping the chick blastoderm. (After Nicolet, 1970.)

labeled cells are determined by autoradiography (Fig. 6.28). Alternatively, regions of quail and chick embryos are exchanged. In this technique (see p. 265 for a discussion of the technique), cells from the two species can be distinguished from one another, and the fates of transplanted cells readily determined by light microscopy (Le Douarin, 1973).

These transplantation experiments have established that the epiblast (see p. 181) is the source of the ectoderm, mesoderm, and endoderm in the chick embryo. The presumptive mesoderm and endoderm cells of the epiblast (see p. 428 for chick fate map) appear to reach their internal locations by ingression, which results in the formation of a long cleft in the blastodisc, the **primitive streak**. The primitive streak first appears as a triangular thickening at the posterior end of the area pellucida (Fig. 6.29) and results from a convergence of epiblast cells toward this region causing the migrating cells to pile up. The streak then elongates as the anterior end of the area pellucida extends forward and the posterior end extends backward. This extension narrows and lengthens the primitive streak, and the shape of the area pellucida is modified from a rounded to a pear-shaped mass (see Fig. 6.29D). Thus, the primitive streak forms by convergent extension.

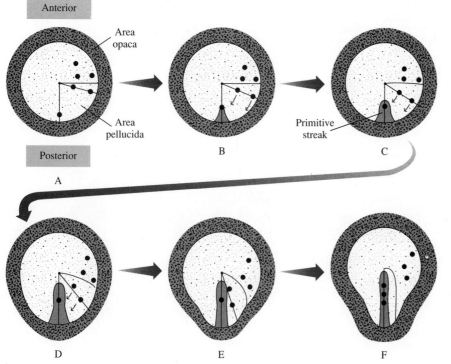

Figure 6.29

Gastrulation and primitive streak formation in the chick. A, The surface of the blastodisc before the beginning of gastrulation. B-F, Progressively later stages in gastrulation. Note that during gastrulation, marks (dark spots) applied to regions of the area pellucida converge (arrows) toward the primitive streak. (After Spratt, 1946.)

Ingression of cells through the primitive streak is shown in Figure 6.30. A depression, the **primitive groove**, is located at the midline of the streak. Cells at the bottom of the groove are elongated and similar in shape to ingressing primary mesenchyme cells of sea urchins. Their basal surfaces are highly motile and have reduced adhesion to adjoining cells, whereas their apical surfaces adhere tightly to one another. Consequently, they are thought to pull neighboring cells into the groove as a spreading sheet. Once in the groove, the apical ends of the cells separate from their neighbors, and the cells migrate into the blastocoel. Inward displacement from the surface epithelium is facilitated by the absence of a basal lamina in the streak region. Once on the inside of the embryo, the cells round up and enter a stream of migrating mesenchymal cells (Solursh and Revel, 1978). The direction of movement is mainly from the epiblast down to the hypoblast, but the cells also spread laterally and forward from the anterior end of the primitive streak. Some of the ingressed cells form a sheet of mesoderm, whereas others enter the hypoblast and displace the original hypoblast cells to the periphery. The cells that colonize the hypoblast form most of the embryonic endoderm, whereas the original hypoblast cells form extraembryonic endoderm (Vakaet, 1962; Fontaine and Le Douarin, 1977). One of the most intriguing questions to emerge from this analysis of chick gastrulation is how the internalized cells become sorted into presumptive mesoderm

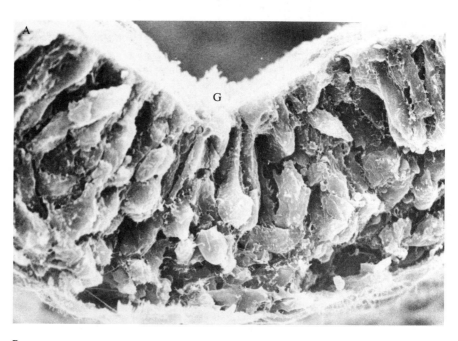

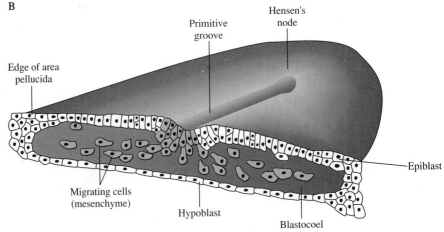

Figure 6.30

Ingression of cells at the primitive groove during chick gastrulation. A, Scanning electron micrograph of a transversely-fractured area pellucida. The columnar cells of the epiblast, visible laterally, form bottle cells in the primitive groove (G). ×2100. B, Diagrammatic representation of A. The anterior half of the area pellucida is shown in cross section. (A, From Solursh and Revel, 1978. B, After Balinsky, 1981.)

and endoderm. Unfortunately, we do not understand how this sorting is accomplished.

During gastrulation, a thickening called **Hensen's node** develops at the anterior end of the primitive streak. Presumptive notochord cells accumulate within this region. After the primitive streak reaches its maximal length, Hensen's node recedes toward the posterior end of the area pellucida. During this process, the presumptive notochord cells, which are located in front of the retreating node, extend posteriorly. As the node retreats, the body is taking shape anterior to it. Thus, in the chick embryo, the anterior end of the body begins to form before gastrulation is completed posteriorly. The anteroposterior progression of chick embryo development is illustrated in Figure 6.31. This figure shows the formation of typical vertebrate dorsal structures, including neural tube, notochord, and somites. Anterior sections

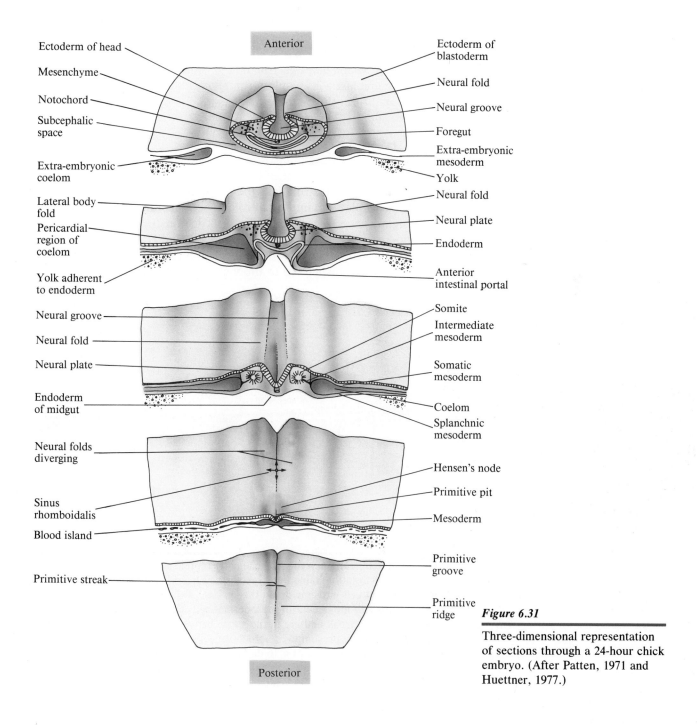

Figure 6.31

Three-dimensional representation of sections through a 24-hour chick embryo. (After Patten, 1971 and Huettner, 1977.)

show these structures in advanced stages, whereas the posterior sections show progressively earlier stages in body formation.

Because the amphibian gastrula is spherical, formation of a cylindrical body is a relatively straightforward process (see Color plate 7I and 7J). However, a much more complicated scenario unfolds in the chick embryo, because the body cylinder is constructed from a flattened sheet. The elaboration of the body plan from this sheet will be discussed in detail in Chapters 7 and 8.

Mammals

Before gastrulation, mammalian embryos bear a remarkable resemblance to bird embryos. As described in Chapter 5, a bilayered disc forms from the inner cell mass. The upper layer is the epiblast, and the lower layer, the hypoblast. The disc undergoes a change in shape, much like that of the chick embryo, and forms the **embryonic shield**, which possesses a primitive streak at the posterior end of the epiblast (Fig. 6.32A). A three-layered embryo is formed by ingression of epiblast cells through the primitive streak (see Fig. 6.32B). Mammalian gastrulation is similar to that of the chick. Thus, migrating epiblast cells would form the mesoderm and may also enter the hypoblast and form the endoderm.

6–3 The Primary Organizer

In addition to its role in gastrulation, the dorsal lip of the blastopore of the amphibian gastrula plays an important role in organization of the amphibian body. The significance of the dorsal lip was demonstrated by the classic transplantation experiments done by H. Spemann and H. Mangold in 1924. They showed that the ability to organize the embryo is retained by the dorsal lip when it is transplanted to the ventral or lateral region of another

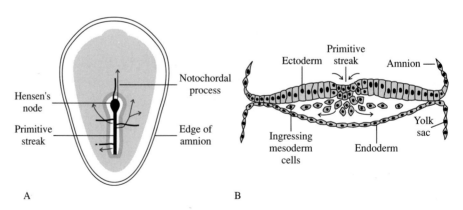

Figure 6.32

Gastrulation in the human embryo. A, Schematic representation of cell movements in a gastrulating human embryo viewed from the dorsal surface. The black lines represent cell movements on the surface of the embryo, and the colored lines represent movement of mesodermal cells that have already passed through the primitive streak and are in the space between the ectoderm and endoderm. Note that the presumptive notochord cells move anteriorly. B, Drawing of a transverse section of a 16-day embryo showing lateral spreading of mesodermal cells from the primitive streak. (After J. Langman, 1969. *Medical Embryology*, 2nd ed. © 1969, The Williams & Wilkins Co., Baltimore.)

embryo. As a result, a secondary embryo is formed, complete with gut, neural tube, notochord, and somites. Because the neural tube is formed from the ventrolateral ectoderm (which would normally form epidermis), their results showed that the dorsal lip induced the host ectoderm to develop into neural ectoderm. The role of the dorsal lip in establishing the antero-posterior axis led Spemann to designate it as the primary organizer. The intrinsic properties of the dorsal lip cause it to involute and form the chordamesoderm. After involution, the chordamesoderm extends along the inner surface of the dorsal ectoderm. Because neural induction occurs during this period, Spemann proposed that the chordamesoderm, which we now know is derived from the subsurface layers of the dorsal marginal zone in *Xenopus*, induces the overlying dorsal ectoderm to develop as neural ectoderm instead of epidermis. We shall discuss neural induction in more detail in Chapter 12.

6–4 Chapter Synopsis

This chapter describes the formation of the embryonic body plan during polarization and gastrulation. Although the radially symmetrical egg has only one axis, bilaterally symmetrical embryos have three axes. In different animals, embryonic polarization can occur either in the oocyte, the fertilized egg, or the cleaving embryo. Our current knowledge of embryonic polarization comes primarily from studies on amphibian and *Drosophila* embryos. In amphibians, polarization occurs between fertilization and first cleavage. An external cue, the sperm, promotes cortical rotation, which leads to gray crescent formation. The point of sperm entry later becomes the ventral side of the embryo, whereas the gray crescent becomes the dorsal side. Cortical rotation probably activates unidentified factors in the egg's vegetal hemisphere, which are later inherited by vegetal blastomeres, that cause dorsal marginal zone cells to begin gastrulation. In *Drosophila*, embryonic polarization is initiated in the oocyte and completed in the fertilized egg. At least 20 different genes have been identified that play a role in dorsoventral polarization, beginning with the egg shape genes, which function in the oocyte and follicle, and ending with the embryonic polarity genes, which function in the embryo. The products of these genes interact in a regulatory pathway leading to dorsoventral polarization.

During gastrulation, the embryonic body plan is established by a specific set of movements in which cells are displaced from the surface to the interior of the embryo, where they subsequently create the embryonic body plan. Gastrulation occurs by different mechanisms in different animals and depends to a large extent on the quantity and distribution of yolk. In relatively non-yolky embryos, such as those of many invertebrates, gastrulation is simple and is initiated near the vegetal pole. In yolky eggs, such as those of most vertebrates, gastrulation is more complex and is initiated either in the equatorial zone or in the blastodisc. In vertebrates, the cell movements that establish the body plan are coordinated by the primary organizer. In amphibians, the primary organizer is the dorsal lip of the blastopore.

References

Ancel, P., and P. Vintemberger. 1948. Recherches sur le déterminisme des la symetrie bilaterale dans l'oeuf des amphibiens. Bull. Biol. Fr. Belg., *31* (Suppl.): 1–182.

Anderson, D. T. 1973. *Embryology and Phylogeny in the Annelids and Arthropods*. Pergamon Press, Oxford.

Anderson, K. V. 1987. Dorsal-ventral embryonic pattern genes of *Drosophila*. Trends Genet., *3*: 91–97.

Anderson, K. V., and C. Nüsslein-Volhard. 1984a. Information for the dorsal-ventral pattern of the *Drosophila* embryo is stored as maternal mRNA. Nature (Lond.), *311*: 223–227.

Anderson, K. V., and C. Nüsslein-Volhard. 1984b. Genetic analysis of dorsal-ventral embryonic pattern in *Drosophila*. *In* G. M. Malacinski and S. V. Bryant (eds.), *Pattern Formation: A Primer in Developmental Biology*. Macmillan, New York, pp. 269–289.

Anderson, K. V., G. Jürgens, and C. Nüsslein-Volhard. 1985a. Establishment of dorsal-ventral polarity in the Drosophila embryo: Genetic studies on the role of the Toll gene product. Cell, *42*: 779–789.

Anderson, K. V., L. Bokla, and C. Nüsslein-Volhard. 1985b. Establishment of dorsal-ventral polarity in the Drosophila embryo: The induction of polarity by the Toll gene product. Cell, *42*: 791–798.

Anstrom, J. A., and R. A. Raff. 1988. Sea urchin primary mesenchyme cells: Relation of cell polarity to the epithelial-mesenchymal transformation. Dev. Biol., *130*: 57–66.

Balinsky, B. I. 1975. *An Introduction to Embryology*, 4th ed. W. B. Saunders, Philadelphia.

Balinsky, B. I. 1981. *An Introduction to Embryology,* 5th ed. Saunders College Publishing, Philadelphia.

Berrill, N. J. 1987. Early chordate evolution. Part 2. Amphioxus and ascidians: To settle or not to settle. Int. J. Invert. Reprod. Dev., *11*: 15–28.

Black, S. D., and J. C. Gerhart. 1986. High frequency twinning of *Xenopus laevis* embryos from eggs centrifuged before first cleavage. Dev. Biol., *116*: 228–240.

Conklin, E. G. 1932. The embryology of *Amphioxus*. J. Morphol., *54*: 69–118.

Dan, K., and K. Okazaki. 1956. Cyto-embryological studies of sea urchins. III. Role of the secondary mesenchyme cells in the formation of the primitive gut in sea urchin larvae. Biol. Bull., *110*: 29–42.

Elinson, R. P. 1975. Site of sperm entry and a cortical contraction associated with egg activation in the frog *Rana pipiens*. Dev. Biol., *47*: 257–268.

Elinson, R. P., and B. Rowning. 1988. A transient array of parallel microtubules in frog eggs: Potential tracks for a cytoplasmic rotation that specifies the dorso-ventral axis. Dev. Biol., *128:* 185–197.

Ettensohn, C. A. 1985. Gastrulation in the sea urchin embryo is accompanied by the rearrangement of invaginating epithelial cells. Dev. Biol., *112*: 383–390.

Fontaine, J., and N. M. Le Douarin. 1977. Analysis of endoderm formation in the avian blastoderm by the use of quail-chick chimaeras. The problem of the neuroectodermal origin of the cells of the APUD series. J. Embryol. Exp. Morphol., *41*: 209–222.

Fullilove, S. L., and A. G. Jacobson. 1978. Embryonic development: Descriptive. *In* M. Ashburner and T. R. F. Wright (eds.), *The Genetics and Biology of Drosophila*, Vol. 2c. Academic Press, New York, pp. 106–227.

Gerhart, J. C., and R. E. Keller. 1986. Region-specific cell activities in amphibian gastrulation. Ann. Rev. Cell Biol., *2*: 201–229.

Gerhart, J. C., S. Black, and S. Scharf. 1983. Cellular and pancellular organization of the amphibian embryo. Mod. Cell Biol., *2*: 483–507.

Gibbins, J. R., L. G. Tilney, and K. R. Porter. 1969. Microtubules in the formation and development of the primary mesenchyme in *Arbacia punctulata*. I. Distribution of microtubules. J. Cell Biol., *41*: 201–226.

Gibson, A. W., and R. D. Burke. 1985. The origin of pigment cells in embryos of the sea urchin *Strongylocentrotus purpuratus*. Dev. Biol., *107*: 414–419.

Gimlich, R. L., and J. C. Gerhart. 1984. Early cellular interactions promote embryonic axis formation in *Xenopus laevis*. Dev. Biol., *104*: 117–130.

Gustafson, T., and H. Kinnander. 1956. Micro-aquaria for time-lapse cinemicrographic studies of morphogenesis in swimming larvae and observations on gastrulation. Exp. Cell Res., *22*: 437–449.

Gustafson, T., and L. Wolpert. 1961. Studies on the cellular basis of morphogenesis in the sea urchin embryo. Directed movements of primary mesenchyme cells in normal and vegetalized larvae. Exp. Cell Res., *24*: 64–79.

Gustafson, T., and L. Wolpert. 1967. Cellular movement and contact in sea urchin morphogenesis. Biol. Rev., *42*: 442–498.

Hamburger, V. 1988. *The Heritage of Experimental Embryology. Hans Spemann and the Organizer.* Oxford University Press, New York.

Hardin, J. 1987. Archenteron elongation in the sea urchin embryo is a microtubule independent process. Dev. Biol., *121*: 253–262.

Hardin, J. 1988. The role of secondary mesenchyme cells during sea urchin gastrulation studied by laser ablation. Development, *103*: 317–324.

Harkey, M. A., and A. H. Whiteley. 1980. Isolation, culture, and differentiation of echinoid primary mesenchyme cells. Wilhelm Roux's Archives Dev. Biol., *189*: 111–122.

Hashimoto, C., K. L. Hudson, and K. V. Anderson. 1988. The Toll gene of Drosophila, required for dorsal-ventral polarity, appears to encode a transmembrane protein. Cell, *52*: 269–279.

Hörstadius, S. 1973. *Experimental Embryology of Echinoderms.* Clarendon Press, Oxford.

Huettner, A. F. 1949. *Fundamentals of Comparative Embryology of the Vertebrates.* Macmillan, New York.

Karp, G. C., and M. Solursh. 1974. Acid mucopolysaccharide metabolism, the cell surface, and primary mesenchyme cell activity in the sea urchin embryo. Dev. Biol., *41*: 110–123.

Katow, H., and M. Solursh. 1980. Ultrastructure of primary mesenchyme cell ingression in the sea urchin *Lytechinus pictus*. J. Exp. Zool., *213*: 231–246.

Keller, R. E. 1975. Vital dye mapping of the gastrula and neurula of *Xenopus laevis*. I. Prospective areas and morphogenetic movements of the superficial layer. Dev. Biol., *42*: 222–241.

Keller, R. E. 1976. Vital dye mapping of the gastrula and neurula of *Xenopus laevis*. II. Prospective areas and morphogenetic movements of the deep layer. Dev. Biol., *51*: 118–137.

Keller, R. E. 1978. Time-lapse cinemicrographic analysis of superficial cell behavior during and prior to gastrulation in *Xenopus laevis*. J. Morphol., *157*: 223–248.

Keller, R. E. 1980. The cellular basis of epiboly: An SEM study of deep-cell rearrangement during gastrulation in *Xenopus laevis*. J. Embryol. Exp. Morphol., *60*: 201–234.

Keller, R. E. 1981. An experimental analysis of the role of bottle cells and the deep marginal zone in gastrulation of *Xenopus laevis*. J. Exp. Zool., *216*: 81–101.

Keller, R. E. 1986. The cellular basis of amphibian gastrulation. *In* L. Browder (ed.), *Developmental Biology: A Comprehensive Synthesis*, Vol. 2. Plenum Press, New York, pp. 241–327.

Keller, R. E., and G. C. Schoenwolf. 1977. An SEM study of cellular morphology, contact, and arrangement, as related to gastrulation in *Xenopus laevis*. Wilhelm Roux's Archives Dev. Biol., *182*: 165–186.

Keller, R. E. et al. 1985. The function and mechanism of convergent extension during gastrulation in *Xenopus laevis*. J. Embryol. Exp. Morphol., *89* (Suppl.): 185–209.

Klag, J. J., and G. A. Ubbels. 1975. Regional morphological and cytochemical differentiation of the fertilized egg of *Discoglossus pictus* (Anura). Differentiation, *3*: 15–20.

Korschelt, E. 1936. *Vergleichende Entwicklungsgeschichte der Tiere*, Band I. Gustav Fischer Verlag, Jena.

Kumé, M., and K. Dan (eds). 1988. *Invertebrate Embryology.* Garland, New York.

Langman, J. 1969. *Medical Embryology*, 2nd ed. Williams & Wilkins, Baltimore.

Le Douarin, N. 1973. A biological cell labelling technique and its use in experimental embryology. Dev. Biol., *30*: 217–222.

Løvtrup, S. 1975. Fate maps and gastrulation in amphibia—A critique of current views. Can. J. Zool., *53*: 473–479.

Lundmark, C. et al. 1984. Amphibian gastrulation as seen by scanning electron microscopy. Scanning Electron Microsc., *3*: 1289–1300.

Malacinski, G. M., H. Benford, and H. M. Chung. 1975. Association of an ultraviolet irradiation sensitive cytoplasmic localization with the future dorsal side of the amphibian egg. J. Exp. Zool., *191*: 97–110.

Manes, M., and R. Elinson. 1980. Ultraviolet light inhibits grey crescent formation in the frog egg. Wilhelm Roux's Archives Dev. Biol., *189*: 73–76.

Manes, M., R. P. Elinson, and F. D. Barbieri. 1978. Formation of the amphibian grey crescent: Effects of colchicine and cytochalasin B. Wilhelm Roux's Archives Dev. Biol., *185*: 99–104.

Moore, A. R., and A. S. Burt. 1939. On the locus and nature of forces causing gastrulation in the embryos of *Dendraster excentricus*. J. Exp. Zool., *82*: 159–171.

Nicolet, G. 1970. Determination et contrôle de la différenciation des somites. Méd. Hygiene, *28*: 1433–1437.

Nüsslein-Volhard, C. 1979. Maternal effect mutations that alter the spatial coordinates of the embryo of *Drosophila melanogaster*. Symp. Soc. Dev. Biol., *37*: 195–211.

Okazaki, K. 1975. Spicule formation by isolated micromeres of the sea urchin embryo. Am. Zool., *15*: 567–581.

Patten, B. M. 1971. *Early Embryology of the Chick*, 5th ed. McGraw-Hill Book Co., New York.

Scharf, S. R., and J. C. Gerhart. 1980. Determination of the dorso-ventral axis in eggs of *Xenopus laevis*: Complete rescue of UV-impaired eggs by oblique orientation before first cleavage. Dev. Biol., *79*: 181–198.

Scharf, S. R., and J. C. Gerhart. 1983. Axis determination in eggs of *Xenopus laevis*: A critical period before first cleavage identified by the common effects of cold, pressure, and ultraviolet irradiations. Dev. Biol., *99*: 75–87.

Scharf, S. R., J.-P. Vincent, and J. C. Gerhart. 1984. Axis determination in the *Xenopus* egg. *In* E. H. Davidson and R. A. Firtel (eds.), *Molecular Biology of Development*. Alan R. Liss, New York, pp. 51–73.

Scharf, S. R. et al. 1989. Hyperdorsoanterior embryos from *Xenopus* eggs treated with D$_2$O. Dev. Biol., *134*: 175–188.

Schüpbach, T. 1987. Germ and soma cooperate during oogenesis to establish the dorsoventral pattern of egg shell and embryo in Drosophila melanogaster. Cell, *49*: 699–707.

Solursh, M. 1986. Migration of sea urchin primary mesenchyme cells. *In* L. Browder (ed.), *Developmental Biology: A Comprehensive Synthesis*, Vol. 2. Plenum Press, New York, pp. 391–431.

Solursh, M., and H. Katow. 1982. Initial characterization of sulfated macromolecules in blastocoels of mesenchyme blastulae of *Strongylocentrotus purpuratus* and *Lytechinus pictus*. Dev. Biol., *94*: 326–336.

Solursh, M., and J. P. Revel. 1978. A scanning electron microscope study of cell shape and cell appendages in the primitive streak region of the rat and chick embryos. Differentiation, *11*: 185–190.

Solursh, M., S. L. Mitchell, and H. Katow. 1986. Inhibition of cell migration in sea urchin embryos by ß-D-xyloside. Dev. Biol., *118*: 325–332.

Spemann, H. 1936. *Experimentelle Beiträge zu einer Theorie der Entwicklung*. Julius Springer, Berlin.

Spemann, H., and H. Mangold. 1924. Induction of embryonic primordia by implantation of organizers from a different species. English translation of original German. *In* B. H. Willier and J. M. Oppenheimer (eds.), *Foundations of Experimental Embryology*, 2nd ed. Hafner Press, New York, pp. 144–184 (1974).

Spiegel, E., and L. Howard. 1983. Development of cell junctions in sea urchin embryos. J. Cell Sci., *62*: 27–48.

Spiegel, E., M. Burger, and M. Spiegel. 1983. Fibronectin and laminin in the extracellular matrix and basement membrane of sea urchin embryos. Exp. Cell. Res., *144*: 47–55.

Spratt, N. T., Jr. 1946. Formation of the primitive streak in the explanted chick blastoderm marked with carbon particles. J. Exp. Zool., *103*: 259–304.

Steward, R. et al. 1988. The dorsal protein is distributed in a gradient in early Drosophila embryos. Cell, *55*: 487–495.

Sugiyama, K. 1972. Occurrence of mucopolysaccharides in the early development of the sea urchin embryo and its role in gastrulation. Dev. Growth Differ., *14*: 63–73.

Turner, F. R., and A. P. Mahowald. 1977. Scanning electron microscopy of *Drosophila melanogaster* embryogenesis. II. Gastrulation and segmentation. Dev. Biol., *57*: 403–416.

Vakaet, L. 1962. Some new data concerning the formation of the definitive endoblast in the chick embryo. J. Embryol. Exp. Morphol., *10*: 38–57.

Vincent, J.-P., G. F. Oster, and J.C. Gerhart. 1986. Kinematics of grey crescent formation in *Xenopus* eggs: The displacement of subcortical cytoplasm relative to the egg surface. Dev. Biol., *113*: 484–500.

Vincent, J.-P., S. R. Scharf, and J. C. Gerhart. 1987. Subcortical rotation in *Xenopus* eggs: A preliminary study of its mechanochemical basis. Cell Motil. Cytoskeleton, *8*: 143–154.

Vogt, W. 1929. Gestaltungsanalyse am Amphibienkeim mit örtlicher Vitalfärbung. II. Gastrulation und Mesodermbildung bei Urodelen und Anuren. Wilhelm Roux's Archiv. für Entwicklungsmechanik, *120*: 385–706.

von Baer, K. E. 1828. *Ueber Entwicklungsgeschichte der Tiere, Beobachtung, und Reflexion*. Königsberg.

Wessel, G. M., R. B. Marchase, and D. R. McClay. 1984. Ontogeny of the basal lamina in the sea urchin embryo. Dev. Biol., *103*: 235–245.

Zusman, S. B., and E. F. Wieschaus. 1985. Requirements for zygotic gene activity during gastrulation in *Drosophila melanogaster*. Dev. Biol., *111*: 359–371.

PART IV

Organizing the Multicellular Embryo

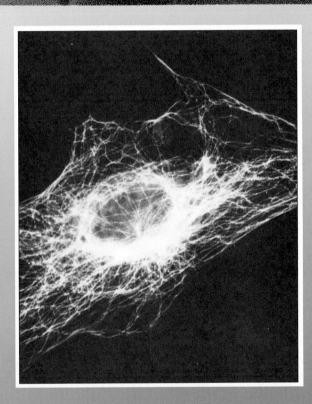

Microtubules are seen to emanate from the microtubule organizing center and swirl throughout this tissue culture cell. The microtubules are visualized with immunofluorescence microscopy using antibodies against tubulin. (From Osborn, M., and K. Weber. 1976. Proc. Natl. Sci. U.S.A., 73:867–871.)

CHAPTER 7

Laying Down the Vertebrate Body Plan: The Generation of Ectodermal Organ Rudiments

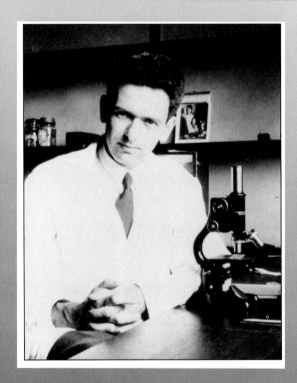

Viktor Hamburger (b.1900) was a student of Spemann and is one of the great experimental embryologists. He has made fundamental contributions to our understanding of the development of the nervous system and, with Rita Levi-Montalcini, discovered nerve growth factor. He is also well known for his careful description of the normal development of the chick. This staged series of morphological landmarks was published with Hamilton in 1951 and provide the standard by which all chick development is described today. (Courtesy of Marine Biological Laboratory, Woods Hole, MA.)

As we have just discussed, the process of gastrulation in the vertebrate embryo produces three germ layers, from which all the individual tissues and organs will arise. The three germ layers have approximately the same spatial relationships to one another that their descendants will have in the adult organism, so that the essence of the body plan is really laid down during the process of gastrulation. This is readily seen in the amphibian (Fig. 7.1A), in which gastrulation results in a sphere consisting of three layers: an enveloping layer of ectoderm that will eventually become the skin, an inner archenteron lined with endoderm that will become the gut, and a sheet of mesoderm in between that will form the intervening tissues. This body plan has been referred to as a "tube within a tube."

A different scheme occurs in the chick embryo, in which the body cylinder must be constructed from a flattened sheet (see Fig. 7.1B). Gastrulation does not produce a completely enclosed archenteron cavity in the chick. Instead, the forerunner of the gut is a space lying below the endoderm and above the yolk. An enclosed gut must be formed after gastrulation by a lateral infolding of the endoderm to enclose and surround this space (see Fig. 7.1C to E). As the lateral infolding occurs, the endoderm is accompanied by an overlying layer of mesoderm that closely adheres to it. Ventrally, the gut initially remains in open communication with the yolk. The ventral connection between the yolk and the gut is gradually constricted as the **body folds** come together. Eventually, the chick body plan resembles the amphibian "tube within a tube."

During the elaboration of the vertebrate body plan, the continuity of the germ layers is lost as their cells reorganize to form clusters that are the rudiments from which definitive tissues and organs will be formed. In this chapter and in Chapter 8, we explore the various tissues and organs that arise from the three germ layers, beginning with the ectodermal derivatives in this chapter and continuing with mesodermal and endodermal derivatives in Chapter 8.

K. E. von Baer (1828), and later E. Haeckel (1868), recognized that the early stages of tissue and organ development are virtually identical in most vertebrates; i.e., postgastrula- and neurula-stage embryos of various vertebrates are structurally similar. Furthermore, the more general developmental features appear first, with specializations appearing later (Fig. 7.2). Thus, we consider in these chapters the earliest developmental stages in the laying down of the basic structure of embryos, and because most vertebrates are similar during this time period, we focus on the embryos of amphibians, birds, and the human.

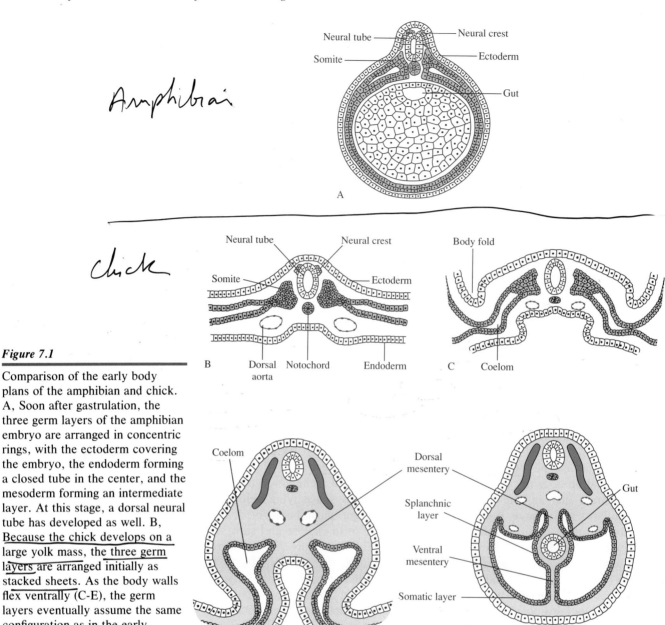

Amphibian

chick

Figure 7.1

Comparison of the early body plans of the amphibian and chick. A, Soon after gastrulation, the three germ layers of the amphibian embryo are arranged in concentric rings, with the ectoderm covering the embryo, the endoderm forming a closed tube in the center, and the mesoderm forming an intermediate layer. At this stage, a dorsal neural tube has developed as well. B, Because the chick develops on a large yolk mass, the three germ layers are arranged initially as stacked sheets. As the body walls flex ventrally (C-E), the germ layers eventually assume the same configuration as in the early amphibian.

7–1 Neurulation: Establishment of the Body Axis and Separation of Ectoderm into Subpopulations

After gastrulation, ectoderm envelops the embryo, but soon this germ layer will become partitioned into three separate cell populations that will have different fates. These are the epidermal ectoderm, which will become primarily **epidermis**, the **neural ectoderm**, which comprises the central nervous system (CNS), and the **neural crest**, which will form a portion of the peripheral nervous system as well as a variety of non-neural cell types. This partitioning of ectoderm is accomplished by the formation and inward displacement of the neural tube in a process known as **neurulation**.

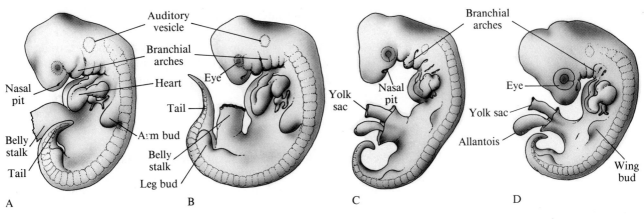

Figure 7.2

K. E. von Baer recognized that the earliest stages of tissue and organ development are virtually identical in most vertebrates. This is amply demonstrated in these drawings of (A) man, (B) pig, (C) reptile, and (D) bird at corresponding developmental stages.

The process of neurulation so dominates the shape and appearance of the embryo that the embryo is known as a neurula at this time. The initial indication of neurulation is the flattening and thickening of the dorsal ectoderm to form the neural plate (Fig. 7.3). The edges of the neural plate then rise above the surface to form the neural folds and flank a central depression called the **neural groove** that extends along the entire mid-dorsal line of the embryo. The neural folds eventually meet above the deepening neural groove, where they fuse to form the neural tube. This tube is the rudiment of the central nervous system. At its anterior end, the neural tube expands and elaborates the brain, whereas in the trunk, the neural tube becomes the spinal cord.

Fusion of the neural folds results in the separation of the neural ectoderm from the presumptive epidermis, which now completely envelops the embryo. Also, during the process of neurulation, a second population of cells separates from the dorsal neural tube and comes to lie on top of it (see Fig. 7.3). Because of their location on top of the neural tube, these cells are called the neural crest cells. They do not remain in this location

during neurulation, embryo known as neurula.

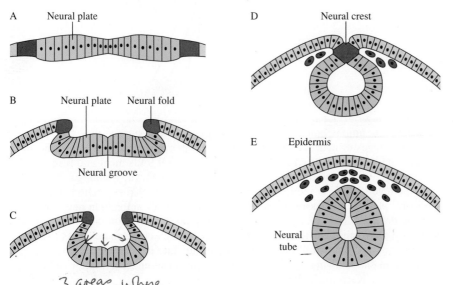

3 areas where cells bend.

Figure 7.3

Diagrammatic view of neurulation. Initially, the dorsal ectoderm forms a thickened plate, which is contiguous with the epidermal ectoderm. The edges of the plate then elevate to form the neural folds. The folds eventually come together and fuse at their tips. Upon fusion, the epidermal ectoderm separates from the neural epithelium, which has now formed a hollow tube. After fusion, a population of non-cohesive cells known as the neural crest emigrates from the neural epithelium.

long; instead, they migrate laterally and ventrally to give rise to a variety of cell types scattered throughout the body (see p. 265).

Because the developing nervous system dominates the external morphology of the early vertebrate embryo, we briefly review neurulation in two vertebrates with two contrasting patterns of neurulation: amphibians and birds.

Amphibians

Initially, the neural plate of an amphibian, such as a salamander, constitutes an oval disc encompassing the entire dorsal hemisphere of the gastrula (Fig. 7.4A). The plate cells can be distinguished from the adjacent ectoderm cells because the former have changed shape and are, at this stage, more columnar. Eventually, the disc as a whole changes shape so that it becomes narrower and longer and is distorted into a keyhole configuration (see Fig. 7.4B).

Neurulation

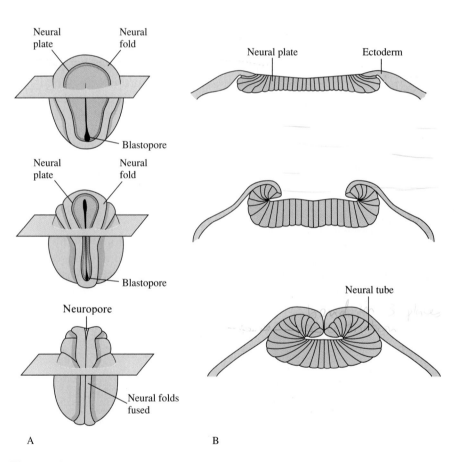

A   B

Figure 7.4

Neurulation in amphibians. A, These drawings represent surface views of the sequential stages of neurulation. Note that the anterior neural plate is significantly broader and will eventually form the brain. The neural folds meet and fuse at approximately the same time along the length of the embryo. Rectangular planes show the levels of cross sections that are drawn in B. B, Cross sections through the neural plate of successively more advanced stages of amphibian neurulation, showing the cell shape changes that accompany the process. Initially, the cells of the neural plate are taller than the adjacent epidermal ectoderm cells. As the plate begins to roll up, the neural plate cells where the plate is bending become pyramidal in shape owing to constriction of their apical ends. (A, After Balinsky, 1981. B, After Karfunkel, 1974.)

Finally, the neural plate elongates rapidly and folds up almost simultaneously along its length to form the neural tube (see Fig. 7.4C). This rolling up of the neural plate to form the neural tube is accompanied by another change in cell shape. Specifically, neuroepithelial cells at the most lateral edges of the neural plate develop constrictions at their apical edges so that they are no longer columnar, but rather are pyramidal. As the apical surface area of the neural epithelium becomes smaller relative to the basal surface, the neural tube rolls up (see Fig. 7.4B).

Chick

Although the basic process of neurulation in the chick is similar to that in amphibians, it differs in several important ways. In amphibians, the neural tube forms almost simultaneously all along its length. In the chick (as well as in all reptiles and mammals), however, neurulation and the accompanying process of body plan organization (such as the formation of somites from the mesoderm) begin at the anterior end, while the posterior end is still undergoing gastrulation. The postgastrulation processes progressively spread to the posterior end as the primitive streak recedes. The anteroposterior progression of neurulation in the chick embryo is illustrated in Figure 7.5. Note that at the anterior end, fusion of the neural folds is imminent, whereas posteriorly the neural folds are just beginning to elevate in front of the

Anterior

— Neural fold

200 μm

Posterior

Figure 7.5

Scanning electron micrograph (SEM) of a chick embryo during the process of neurulation. At the anterior end, in the head region, the neural folds have nearly fused, whereas at the posterior end the neural plate has only just formed. (From Watterson and Schoenwolf, 1984. Courtesy of G. Schoenwolf.)

receding primitive streak. An interesting consequence of this spatial-temporal gradient is that if you look at serial sections through the trunk of a chick embryo at this stage sequentially from posterior to anterior, the effect is as if you were moving forward in time.

As in amphibians, at least two important changes in cell shape accompany neurulation in the chick (Fig. 7.6). First, as the neural plate develops, the neural ectoderm cells change from a cuboidal to a columnar shape, resulting in a narrowing and thickening of the neural plate. Second, as the neural plate rolls up to form the neural tube, the apical ends of some neural epithelial cells narrow and constrict, whereas their basal surfaces broaden to form pyramidal cells. This shape change occurs in two regions of the neural plate in the chick embryo (Schoenwolf, 1982; Schoenwolf and Franks, 1984): in the portion of the neural plate above the notochord and along the dorsomedial side of the two neural folds, and it is in these three

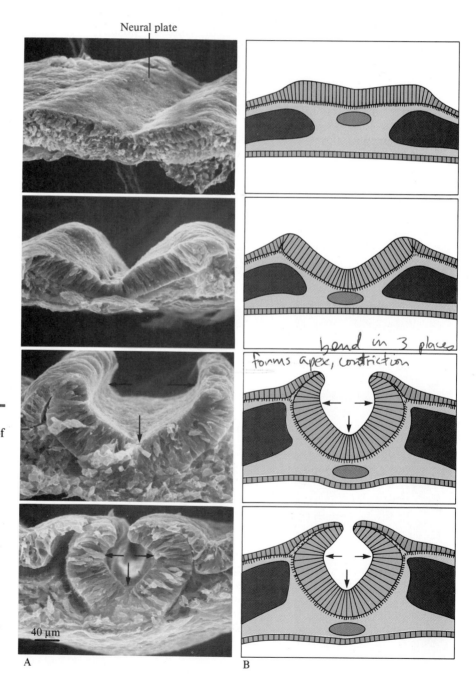

Figure 7.6

Cross-sections through a chick embryo during successive stages of neurulation. A, SEMs of the cut surfaces of neural plates during their successive stages of bending during neurulation. The plate bends in three locations: at the center of the plate just above the notochord and on each side of the neural plate, just below the tips of the neural folds (arrows). B, Line drawings of the same stages as in "A," elucidating the cell shape changes that accompany neurulation in the chick. (A, From Schoenwolf, 1982. B, After Martins-Green, 1988.)

regions that the neural plate flexes. Thus, whereas the amphibian neural tube appears to bend only at the dorsal edges, the chick appears to have "hinge points" above the notochord and along the sides of the tube.

7–2 *Mechanisms of Neurulation*

Although neurulation has been studied for many years, the mechanisms responsible for the development of the neural plate and the elevation of the neural folds remain controversial. It has generally been assumed that changes in cell shape drive the process of neurulation (Fig. 7.7), and there is certainly good evidence to support this notion.

Neural Plate Formation

As the neural plate forms, the cells become dramatically taller. During this elongation, cell volume stays constant, so that their apical surface area must concomitantly decrease, and the neural plate becomes narrower (Jacobson and Gordon, 1976; Schoenwolf, 1985). Thus, this elongation, and its associated apical shrinkage, is a critical component for the shaping of the neural plate from an oval to a long, narrow plate.

The lengthening of individual neural epithelial cells and the maintenance of the elongated condition are generally believed to depend upon microtubules. This assumption is based upon the observation that neural epithelial cells contain bundles of microtubules oriented in the direction of elongation

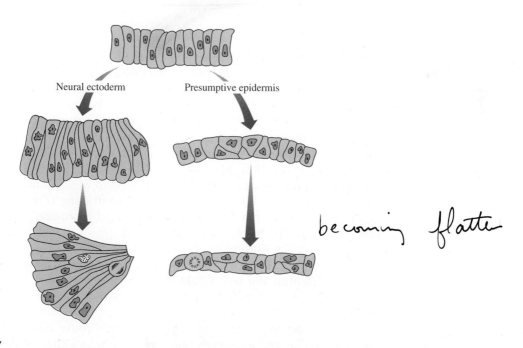

Neural ectoderm   Presumptive epidermis

becoming flatter

Figure 7.7

Drawings summarizing the cell shape changes that are associated with neurulation. The cells in the ectoderm that will form the future neural tube first become taller to create the thickened neural plate and then constrict at their apical surfaces at the locations where the neural plate bends. The cells in the future epidermis, on the other hand, become flatter as the surface area of the embryo increases. (After Burnside, 1971.)

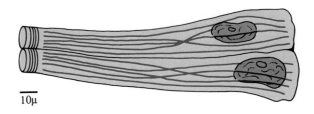

10μ

Figure 7.8

Schematic illustration of the orientations of microtubules and microfilaments in elongating neural plate cells. Numerous microtubules are aligned parallel to the cell's long axis, and circumferential bundles of microfilaments encircle the cell's apex in a drawstring fashion. (After Burnside, 1973.)

(Fig. 7.8) (e.g., amphibian neural plate, Baker and Schroeder, 1967; Burnside, 1971; Karfunkel, 1971; chick neural plate, Karfunkel, 1972) and that disruption of these microtubules with drugs, such as colchicine, converts the columnar cells to rounded cells (amphibian neural plate, Karfunkel, 1971; Burnside, 1973; chick neural plate, Karfunkel, 1972). A serious fault with such studies is that drugs often have additional confounding effects. Colchicine, for example, has also been shown to affect cell volume and may consequently change cell shape (Beebe et al., 1979). Another argument against the supposition that microtubules are responsible for elongation is that serial sectioning along the length of an elongated neural epithelial cell shows no change in the length of microtubules as the cells elongate, nor do the microtubules appear to slide against each other to cause elongation (Burnside, 1973). Thus, microtubules may not *cause* elongation; perhaps they are more important in stabilizing the elongated state.

Bending of the Neural Plate

Numerous factors have been hypothesized to cause bending of the neural plate (see Schoenwolf, 1982; Gordon, 1985). We shall discuss only two alternatives. The first hypothesis suggests that apical surfaces of neural epithelial cells constrict at the sites of bending because of the contraction of actin microfilaments. This notion is based on several observations: (1) Cells in the region of bending become narrower at their apical surfaces, thereby reducing the surface area and presumably causing a buckling (Karfunkel, 1971; Burnside, 1973; Schoenwolf and Franks, 1984); (2) circumferential bands of microfilaments develop at the apical surfaces of the epithelial cells whose apical ends are constricting (Fig. 7.9), and the bands of microfilaments thicken as the apices get narrower, suggesting that contraction of the microfilaments constricts the cell apices (Baker and Schroeder, 1967; Karfunkel, 1972; Burnside, 1973); and (3) when the microfilaments are disrupted with the drug cytochalasin, the microfilaments are depolymerized and neural fold elevation is prevented or even reversed if neurulation was complete when the drug was applied (Karfunkel, 1971, 1972; Burnside, 1973).

The evidence is strong that microfilaments cause constriction of the apices of neural epithelial cells and thereby produce the rolling up of the neural tube. There are several cautionary notes, however: (1) Like colchicine, cytochalasin also has side effects, so such drug studies must be interpreted cautiously; and (2) the cells may become wedge-shaped because the plate bends due to external forces; as a result, the apices would be passively deformed. In fact, Schoenwolf (1988) has produced evidence that forces extrinsic to the neural plate may cause it to bend and fold. If he surgically

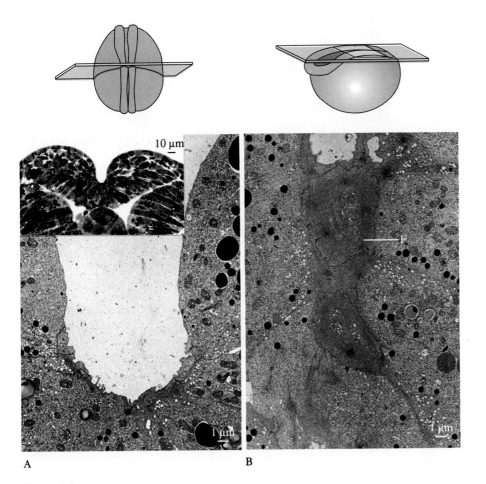

Figure 7.9

Cross-section (A) and frontal section (B) through a mid-neurula stage amphibian embryo visualized in the transmission electron microscope. Rectangular planes through the whole embryos indicate the planes of section reproduced below them. A, Cross-section showing the apices of approximately 10 flask-shaped cells at the base of the neural groove and one or two unconstricted cells lining the walls on each side. The apices of the flask-shaped cells contain arrays of microfilaments, which cannot be seen clearly at this magnification. Inset: Light micrograph of a comparable section. B, This frontal section grazes through the apices of the cells in the neural groove. A circular array of microfilaments (F) can be seen in two cells. (From Baker and Schroeder, 1967. Courtesy of T. Schroeder.)

cuts very young chick embryos along the border between the prospective neural plate and epidermal ectoderm so that the two tissues are separated, the neural plate assumes its proper shape (i.e., it thickens and narrows), but no elevation and closure of the folds occur. If the forces responsible for the elevation of the folds were due to cell shape changes intrinsic to the neural plate cells, one would have expected neurulation to occur even in the cut embryos. This study does not define the nature of these extrinsic forces, but it does provide strong evidence that the cellular basis of neurulation is complex and that the mechanism remains unclear.

An alternative mechanism for elevation of the neural folds has been suggested by numerous studies by A. G. Jacobson and his colleagues. Their suggestion is that stretching along the length of the neural plate would cause the neural tube to roll up. This phenomenon, called **Poisson buckling**, is

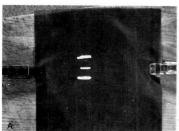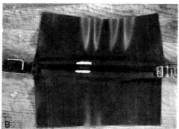

Figure 7.10

Model of neural tube formation by stretching an elastic sheet. A rubber sheet (A) is stretched along the midline, causing a fold to be raised on each side of the line. B, Additional stretching causes the folds to roll toward one another until they meet and form a tube. C, The white lines provide landmarks showing that lateral material is moved to the midline and into the neural tube. (From Jacobson, 1978.)

graphically demonstrated in Figure 7.10. One could visualize the neural plate as an elastic sheet that when stretched would buckle the plate lateral to the midline and roll it into a tube. Jacobson and his colleagues have demonstrated that the neural plate in amphibians does indeed elongate very rapidly just at the time of neural fold elevation and then stops dramatically after neural fold closure (Jacobson and Gordon, 1976; A. G. Jacobson, 1978). This elongation is due, in part, to cell rearrangements in the base of the neural tube, called the **notoplate** (Fig. 7.11) (Burnside and Jacobson, 1968; Jacobson and Gordon, 1976). The correlation between elongation and rolling up of the neural tube is even more dramatic in the chick, in which neurulation occurs in an anteroposterior wave. Only those regions in which the tube is closing show rapid elongation, whereas other areas do not (Jacobson, 1980, 1984). Furthermore, if elongation is prevented by X-irradiation of the chick neural epithelium, neurulation will not occur (Jacobson, 1984). More experimentation is needed to understand the cause of neural plate elongation and to determine whether elongation actually causes neurulation or is merely a coincident event.

7–3 Development of the Central Nervous System

Before neurulation is complete, the anterior portion of the neural tube can already be distinguished from the posterior neural tube by the development anteriorly of distinct swellings or vesicles. This anterior expansion is the precursor to the brain, whereas the posterior tube, which does not develop these expansions, becomes the spinal cord.

We briefly examine the regional differentiation of the brain and spinal cord using a variety of vertebrates, including the human, as examples. However, most of the experimental work has been done with the chick, and we shall emphasize these studies.

Cell Derivatives of the Neural Ectoderm

It is impossible to discuss the development of the nervous system without first describing the cells that compose it. We shall therefore digress for a moment to identify the cell types that differentiate from the neural epithelium.

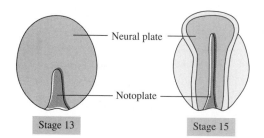

Stage 13

Stage 15

Neural plate

Notoplate

Figure 7.11

Drawings of stage 13 and stage 15 *Taricha torosa* embryos, showing how the notoplate elongates during the process of neurulation. (After A. G. Jacobson, 1981. Morphogenesis of the neural plate and tube. *In* T. G. Connelly et al. (eds.), *Morphogenesis and Pattern Formation.* Raven Press, New York.)

Two basic cell types make up the nervous system: the **neurons** and the **glial cells.** Neurons carry out the essential function of the nervous system—that of communication via electrical impulses. Neurons consist of a cell body, containing a nucleus, and processes emanating from it (Fig. 7.12). Typically, the cell body is drawn out at one end into several short processes known as the **dendrites.** The dendrites generally receive nervous impulses from other cells. At the other end of the neuron is usually one longer process known as the **axon.** The axon transmits nervous impulses either toward other neurons or to effector organs such as muscles. This stereotypic plan has been modified in hundreds of ways, and many shapes and sizes of neurons are found throughout the central and peripheral nervous systems (Fig. 7.13).

A neuron communicates with other neurons or with effector organs at specialized junctions known as **synapses.** Typically at these junctions, electrical impulses pass from one cell to another across a gap by way of **chemical neurotransmitters.** The nature of these transmitters varies among different cell types.

Neuroglia cells are the second cell type derived from the neural epithelium; some glial cells are also derived from the neural crest (Weston, 1970; Le Douarin, 1982). The name "glial" literally means "cement" in Greek and implies that these cells have no more than a supportive function in the nervous system. However, the role of the glial cells is now known to be a more active one. They can act as supportive cells of the nervous system or as phagocytes of the central nervous system. They can also

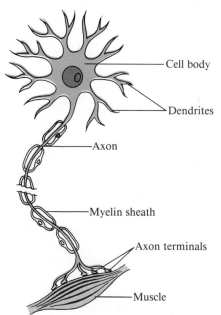

Cell body

Dendrites

Axon

Myelin sheath

Axon terminals

Muscle

Figure 7.12

A diagrammatic representation of a stereotyped motor neuron, showing the cell body or perikaryon, the axon, and dendrites. (After Villee et al., 1989.)

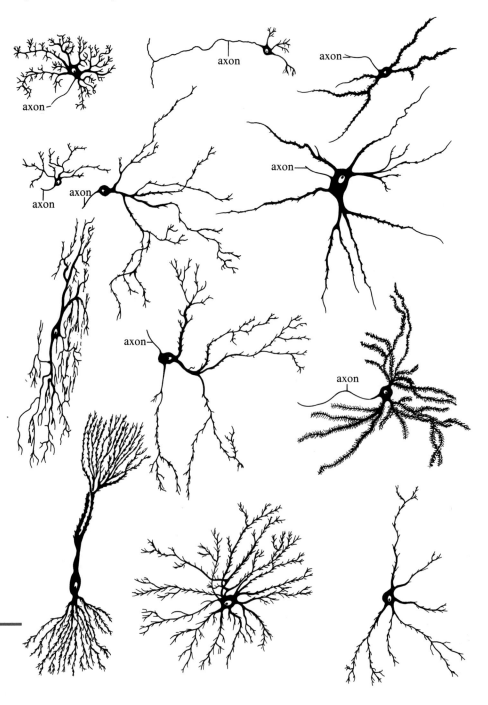

Figure 7.13

Neurons take on a variety of shapes depending upon their location and function. (After Bloom and Fawcett, 1975.)

wrap themselves around axons, completely ensheathing them in multiple layers of a specialized plasma membrane known as **myelin** (Fig. 7.14). These membranes are not of the typical cell membrane composition; rather, they are high in lipids (>75%) and contain specialized proteins, including myelin basic protein and proteolipid protein. This sheathing makes transmission of electrical signals along the axons more efficient. The glial cells that produce myelin membranes in the central nervous system are the **oligodendrocytes**, and the cells in the peripheral nervous system that accomplish the same function are the **Schwann cells**.

We now return to the earliest appearance of the nervous system and observe how these two cell types arise and become organized.

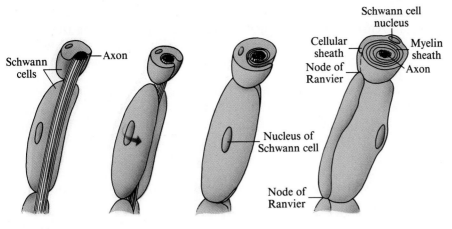

Figure 7.14

Formation of the myelin sheath in the peripheral nervous system. Schwann cells wrap around peripheral nerve axons numerous times to form the insulating sheath. The Schwann cells are periodically arranged along the axon, leaving intercellular spaces known as the nodes of Ranvier. (After Villee et al., 1989.)

Gross Morphology of the Early Central Nervous System

The early brain is composed of three distinct swellings (Fig. 7.15A) known, from the anterior to the posterior, as the **prosencephalon** (forebrain), the **mesencephalon** (midbrain), and the **rhombencephalon** (hindbrain). The prosencephalon is distinguished by lateral outgrowths known as the **optic vesicles**, which are the precursors of the eyes (see Fig. 7.15B and C). As the neural tube expands to form the brain vesicles, it also bends or flexes at the borders of the brain vesicles (see Fig. 7.15C). This flexure is due to the differential growth of the dorsal side of the brain over the ventral side.

   With time, this initial simple segmentation of the brain becomes further subdivided (see Fig. 7.15B and C). The prosencephalon becomes subdivided

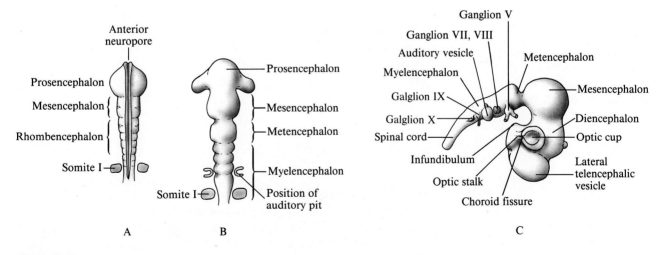

Figure 7.15

Diagrams showing the topography and segments of the chick brain at 27 hours (dorsal view, A), 36 hours (dorsal view, B), and 4 days (lateral view, C) of development. By day 4, cranial flexures result in the ventral displacement of the brain. (After Patten, 1951.)

into the anterior **telencephalon** and the posterior **diencephalon**, the mesencephalon remains undivided, and the rhombencephalon gives rise to the anterior **metencephalon** and the posterior **myelencephalon**. This developmental sequence and the ultimate derivatives of the brain vesicles are outlined in Table 7–1.

The obvious demarcations of the brain are rapidly lost as selective growth of various regions radically changes the relationships of the areas of the brain. Especially dramatic is the growth of the telencephalon to give rise to the cerebral hemispheres, which extend back over the diencephalon. The tremendous growth and intricate rearrangements of the mammalian brain are diagrammed in Figure 7.16.

The spinal cord remains undivided but continues to elongate during this time.

Histological Differentiation of the Central Nervous System

The described changes in the external features of the brain and spinal cord are due to the proliferation of cells of the neural epithelium and the migration of these cells to form new cell layers.

Initially, the cells of the neural tube are arranged as an epithelium, in which the cells extend from the inner (or luminal) surface to the outer (or basal) surface of the neural tube. The cells within this epithelium are rapidly dividing. Consequently, it is also known as the **germinal epithelium**. For reasons that are obscure, the nuclei of the cells that are dividing are found

Table 7–1 Divisions and Functions of the Higher Vertebrate Brain

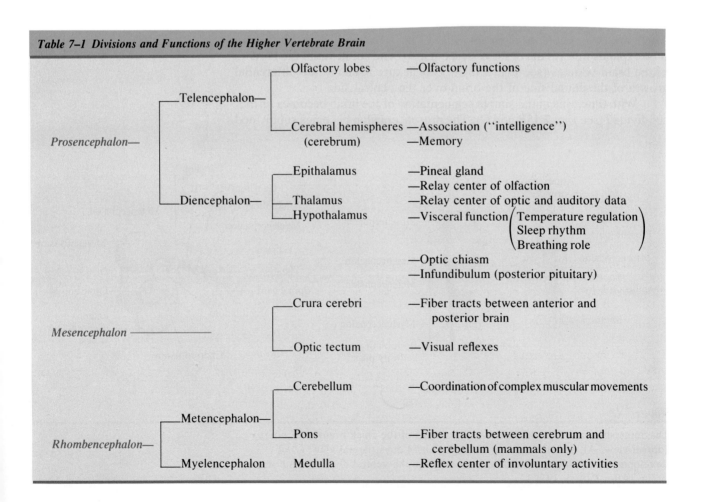

Prosencephalon—	Telencephalon—	Olfactory lobes	—Olfactory functions
		Cerebral hemispheres (cerebrum)	—Association ("intelligence") —Memory
	Diencephalon—	Epithalamus	—Pineal gland —Relay center of olfaction
		Thalamus	—Relay center of optic and auditory data
		Hypothalamus	—Visceral function (Temperature regulation / Sleep rhythm / Breathing role) —Optic chiasm —Infundibulum (posterior pituitary)
Mesencephalon		Crura cerebri	—Fiber tracts between anterior and posterior brain
		Optic tectum	—Visual reflexes
Rhombencephalon—	Metencephalon—	Cerebellum	—Coordination of complex muscular movements
		Pons	—Fiber tracts between cerebrum and cerebellum (mammals only)
	Myelencephalon	Medulla	—Reflex center of involuntary activities

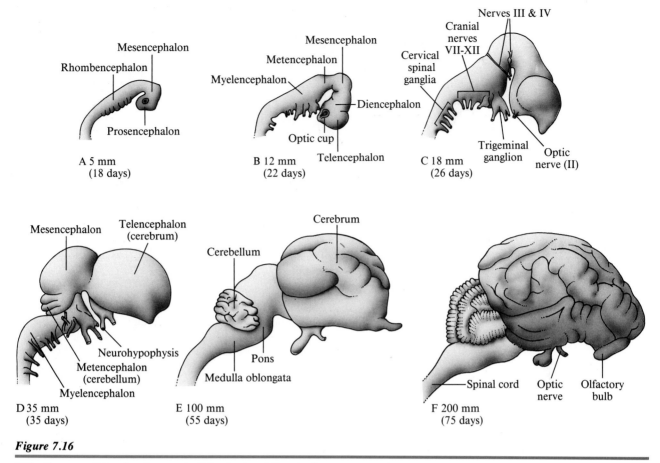

Figure 7.16

This diagram of the development of the brain of the pig demonstrates the differences in the relative growth of various regions of the brain. (After D. M. Noden and A. de Lahunta. 1985. *The Embryology of Domestic Animals: Developmental Mechanisms and Malformations.* © 1985, the Williams & Wilkins Co., Baltimore.)

at the luminal surface. Sauer (1935) was the first to show that as cells pass through the cell cycle and enter mitosis, the nuclei migrate toward the luminal surface. After they have completed mitosis, the nuclei migrate back toward the basal surface of the epithelium. This migration of nuclei through the cell cycle is depicted in Figure 7.17.

The cells of the germinal epithelium eventually cease dividing, and, as they do so, they detach from the luminal surface and migrate peripherally in the neural tube to form other layers of the central nervous system. The initial detachment and migration of these cells are shown in Figure 7.18. Both neurons and glial cells are generated from the germinal epithelium in this manner.

As more cells are recruited from the germinal layer, they form a second layer external to the germinal epithelium, which is known as the **mantle layer** (Fig. 7.19). The former germinal epithelium is now known as the **ependymal layer.** Cells in the mantle layer that differentiate into neurons extend axons further peripherally, whereas their cell bodies remain in the mantle zone. This new outer zone that is rich in axons is known as the **marginal layer.** Some glial cells from the mantle layer eventually cover the axons in the marginal layer and produce myelin membranes, which give the axons a glistening white appearance. The marginal layer is therefore often

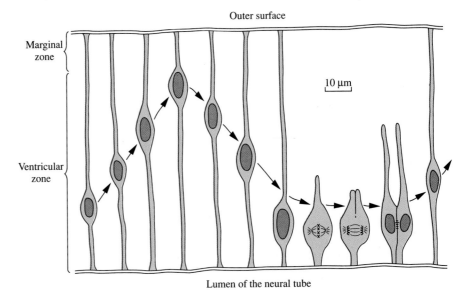

Figure 7.17

A diagrammatic section through the neural tube of a chick embryo demonstrating the position of a single cell nucleus as it progresses through the cell cycle. The nuclei of all dividing cells are found at the luminal surface of the neural tube. (After Sauer, 1935.)

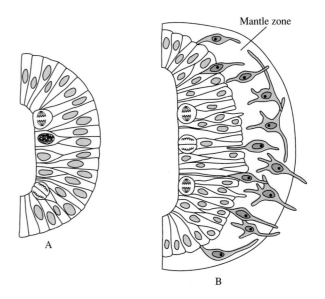

Figure 7.18

Two stages in the development of the central nervous system. A, Initially, the neural tube is a pseudostratified epithelium, in which cells are actively dividing. B, Once a cell ceases all mitotic activity, it detaches from the luminal surface and migrates into the mantle zone. These cells then produce axons that become myelinated and give rise to a third layer, the marginal layer. (After Balinsky, 1981.)

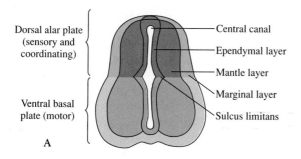

Dorsal alar plate (sensory and coordinating)

Ventral basal plate (motor)

Central canal

Ependymal layer

Mantle layer

Marginal layer

Sulcus limitans

A

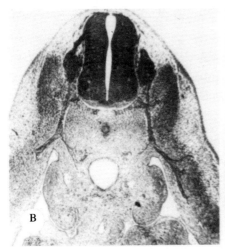

B

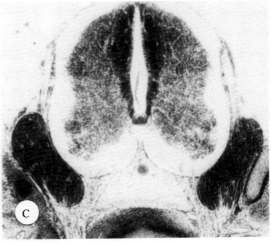

C

Figure 7.19

The layered organization of the neural tube. A, Drawing of a cross-section through a 5-day chick neural tube illustrating the three layers of the neural tube: the inner ependymal layer, the middle mantle layer, and the outer marginal layer. B, and C, Cross-sections through the spinal cords of calves of 32 and 42 days of gestation, respectively. By 32 days, the mantle layer is well developed, but the marginal layer is quite small. By 42 days, the marginal zone—especially in the basal plate—is established. (A, After J. Langman. 1985. *Medical Embryology,* 5th ed. © 1985, the Williams & Wilkins Co., Baltimore. B, and C, After D. M. Noden and A. de Lahunta. 1985. *The Embryology of Domestic Animals: Developmental Mechanisms and Malformations.* © 1985, the Williams & Wilkins Co., Baltimore.)

known as the **white matter**, whereas the mantle layer, which is devoid of myelin and is darker in appearance, is known as the **gray matter**.

Thus, early in development, three layers arise from the original germinal layer. These layers are retained throughout development in the spinal cord and the posterior brain, but very complicated rearrangements of these cell layers occur in the anterior portions of the brain. In the cerebellum and cerebrum (in which the greatest deviation in the pattern of the neural tube occurs), many of the gray matter cell bodies become aggregated to form functional nuclei, and other neurons migrate peripherally past the white matter to form additional external layers of gray matter.

The student interested in the intricate arrangement of cells within the various regions of the brain can seek this information in a number of excellent treatises devoted to the development of the nervous system (see Crelin, 1974; M. Jacobson, 1978; Purves and Lichtman, 1985).

7–4 Development of the Peripheral Nervous System

The central nervous system consists of the brain and spinal cord, whereas the **peripheral nervous system** comprises all the nervous tissue exterior to the central nervous system. The peripheral nervous system mediates communication of all tissues of the body with the central nervous system, and the central nervous system integrates activity of the whole organism.

The peripheral nervous system is composed of nerves and glial cells that are derived from the central nervous system, the neural crest, and epidermal placodes. In this section, we examine the structure of the peripheral nervous system and how it develops from these three sources of cells.

Structure and Function of the Peripheral Nervous System

Before discussing the development of the peripheral nervous system, we first describe the basic structure and function of this component of the nervous system. This discussion is necessarily brief and generalized, and the interested student is urged to examine physiology and anatomy textbooks for a fuller appreciation of the complexity of the organization and function of the peripheral nervous system.

The nervous system, both central nervous system and peripheral nervous system, can be divided into two components: the **autonomic nervous system** and the **somatic nervous system** (Fig. 7.20).

The autonomic nervous system contains those motor fibers that innervate the smooth muscles of the visceral organs and various glands. Thus, it is also known as the **visceral nervous system.** This system controls the functions of the heart, the abdominal organs, sweat glands, the muscles of the eye, and hair follicles. Generally, the functions controlled by the autonomic nervous system are mediated below the conscious level and therefore are thought of as being under involuntary control.

The autonomic nervous system can be further divided into the **sympathetic** and the **parasympathetic** systems. These two systems reflect anatomical, functional, and biochemical differences that will be described in more detail later. In general, the two systems act antagonistically in the same organ. For example, the sympathetic system innervating the heart will increase the heart beat, whereas the parasympathetic system will decrease the heart rate.

The **somatic nervous system**, on the other hand, includes the motor neurons that innervate and control the skeletal muscles, and one can generalize to say that this system is under voluntary control.

Both the somatic and autonomic nervous systems contain two types of nerve fibers: **sensory (afferent) neurons** and **motor (efferent) neurons** (see Fig. 7.20). Sensory fibers carry sensations such as pain, touch, and proprioception from the periphery. The sensory fibers enter the spinal cord at its dorsal side and synapse on cells in the gray matter. The cell bodies of these sensory neurons lie just outside the spinal cord and are aggregated into structures known as ganglia (from the Greek, meaning ''swelling''). Because the sensory neurites enter the spinal cord via the dorsal root nerve, their ganglia are known as the **dorsal root,** or **sensory, ganglia.** The sensory ganglia are arranged in a periodic fashion along the length of the spinal cord (Fig. 7.21).

Once a message has been received from the periphery via the sensory neurons, the message is integrated by complicated pathways within the central nervous system. Finally, messages are conducted toward the pe-

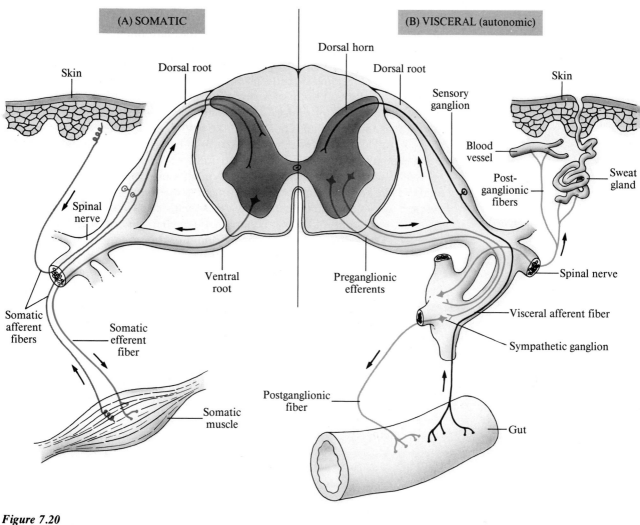

(A) SOMATIC

(B) VISCERAL (autonomic)

Figure 7.20

A schematic drawing of the various elements of the somatic (A) and autonomic (B) peripheral nervous systems.

riphery to reach a muscle or gland. These messages are transmitted via the motor, or efferent, neurons.

Motor fibers in the somatic nervous system have their cell bodies in the ventral gray matter of the spinal cord and exit the spinal cord to form the **ventral root** of the **spinal nerves** (see Figs. 7.20 and 7.21). These neurons synapse directly on the appropriate striated muscle. Thus, somatic motor innervation is accomplished by long axons that pass without intervening synapses from the neural tube directly to the neuromuscular junction.

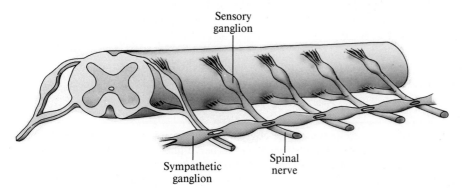

Figure 7.21

A surface view of a portion of the trunk neural tube showing the relationships of the sensory and sympathetic ganglia and the spinal nerves. Note the segmental arrangement of the sensory and sympathetic ganglia. (After Yip, 1986.)

The motor fibers of the autonomic nervous system are more complicated in that, unlike somatic peripheral innervation, autonomic peripheral innervation requires two neurons arranged in series. The first of these fibers has its cell body in the ventral neural tube and also exits via the ventral root. Rather than synapsing directly on the muscle or gland, however, this fiber synapses on the cell body of a second neuron, which in turn synapses on the smooth muscle or gland. The cell bodies of the second fiber are also aggregated into ganglia. Thus, the first fiber that originates in the neural tube is called the **preganglionic fiber**, and the second is designated the **postganglionic fiber** (see Figs. 7.20 and 7.22).

The preganglionic axons of the sympathetic system emerge from the spinal cord and, together with the sensory fibers and motor fibers of the somatic nervous system, make up the **spinal nerves** (see Fig. 7.20). The preganglionic fibers of the parasympathetic system generally arise from the brain stem and, together with sensory and motor fibers there, make up the **cranial nerves**.

The ganglia of the autonomic nervous system are found in different places depending upon their function. Those fibers that are part of the sympathetic autonomic system have their cell bodies clustered in ganglia next to the spinal cord and are known as the **sympathetic ganglia** (see Figs. 7.21 and 7.22). These sympathetic ganglia form a chain along the spinal cord in register with the periodic sensory ganglia. The ganglia that make up the parasympathetic autonomic nervous system are found near or in the organ they innervate. An example of such a parasympathetic ganglion is the **ciliary ganglion** (Fig. 7.23), which is found in the posterior portion of the eye near the iris and ciliary muscle.

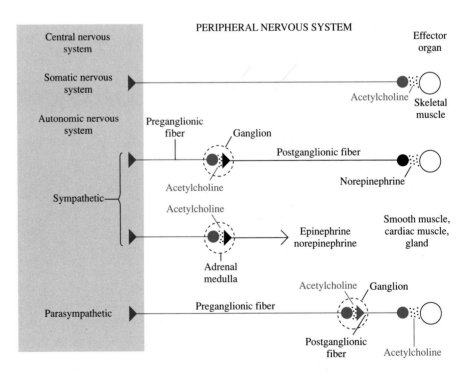

Figure 7.22

Diagrammatic representation of the efferent (motor) division of the peripheral nervous system, outlining the arrangement of nerve fibers, the composition and positions of ganglia, and the dominant neurotransmitter released at each site. (After Vander et al., 1985.)

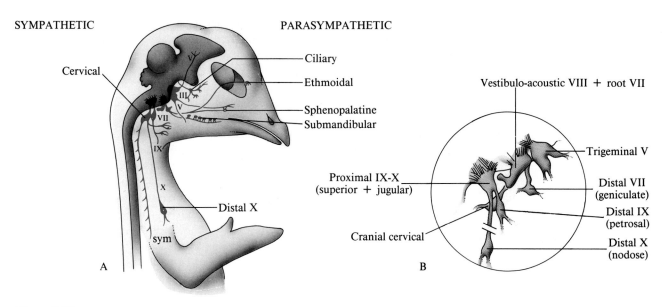

Figure 7.23

A, Schematic illustration of the autonomic and sensory ganglia of the head and some of the cranial nerves in a 12-day chick embryo. B, An enlarged and more detailed view of those sensory ganglia found closest to the head. (After D'Amico–Martel and Noden, 1983. Reprinted from American Journal of Anatomy by permission of Wiley-Liss, a division of John Wiley and Sons, Inc. Copyright © 1983, Wiley-Liss.)

For reasons of simplicity, we shall discuss further the development of the peripheral nervous system emanating from the spinal cord. It is important to remember, however, that there is also a portion of the peripheral nervous system associated with the head. These are the cranial nerves and the associated sensory and parasympathetic cranial ganglia (see Table 7–2; Fig. 7.23). Organization of the peripheral nervous system associated with the

Table 7–2 Cranial Nerves and Associated Ganglia

	Cranial Nerve	Sensory Ganglia		Parasympathetic Ganglia
		Proximal	Distal	
I	Olfactory			
II	Optic			
III	Oculomotor			Ciliary
IV	Trochlear			
V	Trigeminal	Trigeminal		
VI	Abducens			
VII	Facial	Root VII	Geniculate	Submandibular Sphenopalatine Ethmoidal
VIII	Acoustic	Vestibulo-acoustic		
IX	Glossopharyn-geal	Superior	Petrosum	Otic
X	Vagus	Jugular	Nodosum	
XI	Accessory			
XII	Hypoglossal			

head is rather more complicated than that associated with the spinal cord, and we shall not deal with its development here.

Neural Crest Cells

All of the motor (afferent) fibers of the somatic nervous system (i.e., those that innervate skeletal muscle) and the preganglionic fibers of the autonomic nervous system are derived from the neural tube and have their cell bodies there. Much of the remainder of the peripheral nervous system—the sensory nerves and ganglia as well as the postganglionic fibers and the ganglia of the autonomic nervous system—is derived from the neural crest. In order to understand how the peripheral nervous system develops, with its extraordinary order and pattern, it will be necessary to discuss first the origin and migration of the neural crest cells.

Shortly before the neural tube fuses in the head, or after the neural tube fuses in the trunk, a population of cells (the neural crest cells) separates from the neural epithelium as a group of individual cells (Fig. 7.24).

Soon after their appearance, the neural crest cells begin to migrate and move extensively throughout the embryo. This migration occurs first in the head region and then progresses along the trunk in an anteroposterior wave (see Fig. 7.24). The neural crest cells ultimately come to reside in distant parts of the embryo and give rise to a multitude of cell types (Weston, 1970; Le Douarin, 1982). In the chick embryo, in which neural crest cell migration and differentiation have been studied most extensively, the head neural crest cells are known to give rise to portions of the cranial nerves and ganglia, pigment cells, and parts of the connective tissue and skeleton of the head. In the trunk, the neural crest cells give rise to the sensory ganglia, the sympathetic ganglia, the medulla of the adrenal gland, pigment cells of the skin, and the enteric ganglia of the gut (Table 7–3).

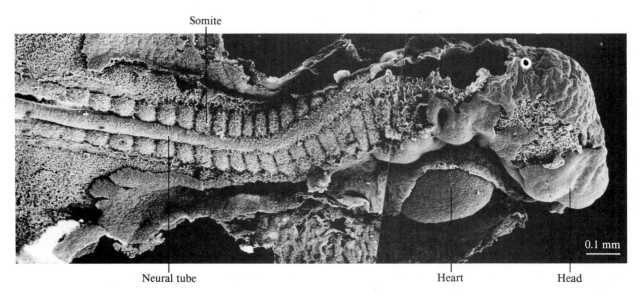

Somite

Neural tube   Heart   Head

0.1 mm

Figure 7.24

SEM of a 2.5-day-old chick embryo with its ectoderm removed to reveal the underlying neural crest cells escaping from the neural tube. Such preparations show that neural crest cells begin migrating first in the head and then initiate their migration progressively posteriorly. The blocks of tissue adjacent to the neural tube are the somites, which provide convenient landmarks when studying the timing of neural crest migration. (From Tosney, 1978. Courtesy of K. Tosney.)

Table 7–3 Neural Crest Derivatives

Pigment Cells	Sensory Nervous System	Autonomic Nervous System	Skeletal and Connective Tissue	Endocrine
Trunk Crest (includes vagal crest)				
Melanophores Xanthophores Iridophores	Spinal ganglia	Sympathetic ganglia Superior cervical ganglia Prevertebral ganglia Paravertebral (sympathetic) ganglia Adrenal medulla Parasympathetic ganglia Remak's ganglion Pelvic plexus Enteric ganglia of gut	Walls of aortic arch Connective tissue of parathyroid, thyroid, thymus	Adrenal medulla Calcitonin- producing cells Type I cells of carotid body Parafollicle cells of thyroid
		Supportive (glial cells)		
Cranial Crest				
Melanophores Xanthophores Iridophores	Trigeminal ganglia (V) Facial root (VII) Superior (IX) Jugular (X)	Parasympathetic ganglia Ciliary Submandibular Sphenopalatine Ethmoidal Otic Intrinsic ganglia of viscera	Connective tissue in: Arterial walls Odontoblasts (teeth) Dermis of face Eye Skeletal tissue (carti- lage and bone) Floor of skull Face Jaw	
		Supportive cells		

Recent studies have shown that neural crest cells that arise from different axial levels of the head and trunk (i.e., at different locations along the neural tube) give rise to different derivatives. This was first demonstrated in detail by Nicole Le Douarin and her colleagues (Le Douarin and Teillet, 1974). One of her many important contributions to our understanding of the neural crest was to discover a way to identify neural crest cells as they migrate. After the neural crest cells have left the neural tube, they cannot be distinguished morphologically from the rest of the embryonic tissue surrounding them. However, Le Douarin discovered that Feulgen stain, which stains DNA, will distinguish chick cells from the closely related quail cells because of the presence of condensed nucleolar-associated hetero-chromatin found in quail, but not in chick, nuclei (Fig. 7.25). She, therefore, removed a portion of a chick neural tube and replaced it with a piece of quail neural tube before the time when the neural crest cells would migrate. The host chicken embryo was then allowed to develop further, until it was fixed, sectioned, and then stained with Feulgen. With the Feulgen stain, all the grafted quail cells, including the neural crest cells that migrated from the grafted neural tube, could be distinguished from the surrounding chick tissues.

In a series of experiments, Le Douarin and Teillet (1974) replaced all the segments of the chick neural tube or the head folds with quail neural tubes from the same axial level and of the same age. The result was a fate

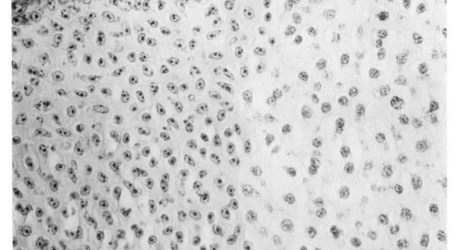

Figure 7.25

A section of cartilage from a chick/quail chimera. Cells derived from grafted quail mesenchyme are on the left; these are clearly recognized by the presence of a distinct intranuclear marker. Chick cells, on the right, do not have the marker. The quail-specific marker is due to nucleolar-associated DNA that remains condensed throughout the cell cycle. (Courtesy of D. Noden.)

map of the origins of the different neural crest derivatives and their ultimate destinations. Their discoveries are summarized in Figure 7.26. In this diagram, the axial level (i.e., the position along the neural tube) is marked by the somite number. They found that the neural crest cells that give rise to the sympathetic ganglia originate from the neural tube at somite levels 7 through 28. Sensory ganglia, on the other hand, come from all levels of the trunk neural tube. Neural crest cells that will form the adrenal medulla migrate from somite levels 18 through 24, and those that will develop into the enteric ganglia of the gut come from somite levels 1 through 7, which

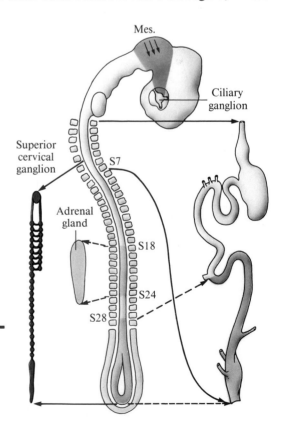

Figure 7.26

A fate map showing the origins of the autonomic ganglia and adrenal medulla in the avian embryo. The avian gut is enlarged on the right and the sympathetic chain and adrenal medulla on the left. Arrows indicate from which axial levels the neural crest cells arise and produce these derivatives. Axial levels are defined by the somite (S) number. Mes.: mesencephalon. (After Le Douarin, 1982.)

is really the most posterior portion of the brain. Finally, neural crest cells from all axial levels will give rise to pigment cells of the skin. Similar studies have also established fate maps for neural crest cells arising from the brain (Noden, 1975).

Because the neural crest produces so many different phenotypes, it is of interest to know what controls the differentiation of neural crest cells. Are they predetermined to be a particular phenotype before they leave the neural tube and simply migrate to the appropriate spot? Or are the neural crest cells pluripotent (i.e., capable of forming many or all derivatives of the neural crest), and subsequently differentiate according to the migratory pathways taken and where they eventually come to rest? In order to answer this question, Le Douarin and Teillet (1974) replaced pieces of neural tubes containing neural crest cells from one axial level of a chick embryo with neural tube segments from a different axial level of a quail embryo (Fig. 7.27). For example, they took a piece of neural tube from somite levels 1 through 7 and grafted it into somite levels 18 through 24. Normally, the neural crest cells from somite levels 1 through 7 migrate to the gut and form enteric ganglia that produce the neurotransmitter acetylcholine (see Fig. 7.22). When such an experiment was done, they found instead that the grafted neural crest cells form sympathetic ganglia that produce the neurotransmitter norepinephrine. In addition, these grafted neural crest cells give rise to the adrenal medulla, which produces the transmitter epinephrine (adrenalin). Neural crest cells from the reciprocal graft (i.e., when a quail neural tube from somite levels 18 through 24 is placed into a chick at somite

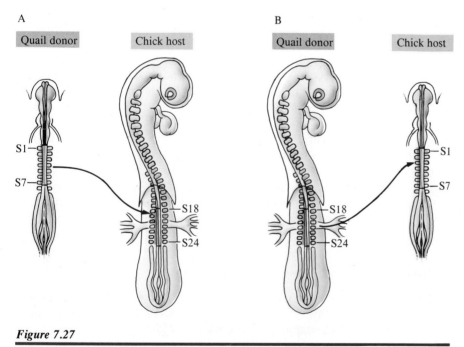

Figure 7.27

Diagrammatic representation of the heterotopic grafting experiments between the "vagal" (somite levels 1–7) and "adrenomedullary" (somite levels 18–24) levels of the neural tube. In A, a piece of neural tube from the vagal level of a quail embryo was removed just before the time when the neural crest cells would migrate and was grafted into somite level 18–24 of a chick embryo. In B, the reverse graft was done. Such experiments demonstrate that the phenotypes derived from the neural crest cells are dependent upon the environments through which they move. (After Le Douarin, 1982.)

levels 1 through 7) give rise to enteric ganglia of the gut. Clearly, the fates of the population of neural crest cells are not predetermined by their axial level of origin.

An even more dramatic demonstration of the role of the environment in controlling neural crest cell differentiation is the following: Prosencephalic neural crest cells, which do not normally produce neurons, can be grafted into the trunk, and these will give rise to sensory, sympathetic, and parasympathetic neurons (Noden, 1978; Ayer-Le Lièvre and Le Douarin, 1982).

Although it is generally the case, this pluripotency is not strictly true for all neural crest cells. For example, only neural crest cells from the head can give rise to skeletal elements (Le Douarin and Teillet, 1974), and some of the bones in the head are prepatterned before their immigration (Noden, 1983). Thus, it seems clear that the fates of some neural crest cells are fixed at the time they arise from the neural tube.

It is not clear from such heterotopic grafting experiments whether each individual neural crest cell can have multiple fates and differentiate according to cues in its environment. These experiments can only demonstrate the developmental potential of the entire *population* of neural crest cells from a particular axial level. Therefore, it is possible that subpopulations of the neural crest cells at any particular axial level have fixed fates and that certain of these will be targeted for survival by the environment in particular regions, whereas the rest of the neural crest cells either die or change their fate.

Recent experiments by Bronner-Fraser and Fraser (1988) suggest that at least some individual neural crest cells are, in fact, pluripotent when they leave the neural tube. These investigators injected a fluorescent vital dye, rhodamine dextran, into single neural epithelial cells in the dorsal neural tube, the region from which neural crest cells will migrate. This technique allowed them to mark all the descendants of that injected cell as it divided. After one to two days, the embryos were fixed and sectioned, and the distribution of labeled cells (which are all descendants of the single labeled cell) was examined. These results clearly show that a variety of phenotypes differentiate from a single cell, including sensory neurons, sympathetic neurons, pigment cells, glial cells, and adrenal medulla cells (Fig. 7.28). Thus, at least some neural crest cells are pluripotent when they leave the neural tube. These results still cannot rule out the possibility that some premigratory neural crest cells are predetermined. Whatever the case, the environment through which the cells move and their ultimate destinations play important roles in the organization and differentiation of the peripheral nervous system. The pathways taken by the neural crest cells are considered next.

Pathways of Neural Crest Cell Migration

The segmental organization of the peripheral nervous system is established by the patterns and pathways of neural crest cell migration. This has recently become apparent after mapping the pathways of neural crest migration in some detail.

Pathways of neural crest cell migration have been identified by developing markers for the neural crest cells to distinguish them from the surrounding embryonic tissue. The chick/quail chimera technique developed by Le Douarin as well as radioisotope labeling (Weston, 1963) are two such methods used to identify neural crest cells.

Recently, an antibody has been developed called NC–1, which primarily recognizes migrating neural crest cells (Vincent and Thiery, 1984). When

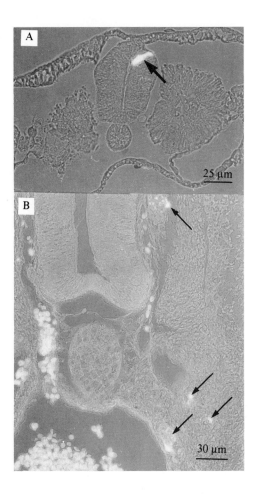

25 μm

30 μm

Figure 7.28

A, An individual dorsal neural epithelial cell was injected with rhodamine dextran and immediately fixed and sectioned to confirm that only a single cell was targeted. B, A single neural tube cell was filled and the embryo fixed 2 days after injection. In this section, fluorescent cells derived from the original filled cell are found in the sensory ganglion and around the dorsal aorta where adrenergic cells localize. (From Bronner-Fraser and Fraser, 1988. Reprinted by permission from *Nature* Vol. 335 pp. 161–164. Copyright © 1988 Macmillan Magazines Ltd.)

this antibody is fluorescently tagged and is applied to sections of embryos of various ages, it will identify the neural crest cells at progressive positions in their migration (see the essay on p. 344 for a discussion of the technique of immunofluorescence microscopy).

Such studies using markers have identified the various migratory pathways taken by the neural crest cells. For the sake of simplicity, we discuss only the pathways in the trunk of chick embryos and relate these pathways to the pattern of the peripheral nervous system. There is some evidence that the pathways and patterns in the chick are similar to those in other vertebrates.

In the chick, some neural crest cells migrate laterally between the ectoderm and the somite, and these neural crest cells will ultimately enter the ectoderm and become pigment cells of the skin (Teillet and Le Douarin, 1970; Teillet, 1971).

The rest of the neural crest cells migrate ventrally and move into the space between the somite and the neural tube. These are the neural crest cells that will give rise to the neurons and glial cells of the peripheral nervous system. The organization of the elements of the peripheral nervous system is based upon the pattern of migration taken by the neural crest cells. As the neural crest cells migrate from the neural tube, they soon reach a neighboring somite and migrate into it. They move into the somite along the interface between the dermamyotome and sclerotome (see p. 296 for a description of the somite) and spread along the undersurface of the dermamyotome until they reach the dorsal aorta (Fig. 7.29). Neural crest cells do not disperse throughout the whole somite, however: They are found

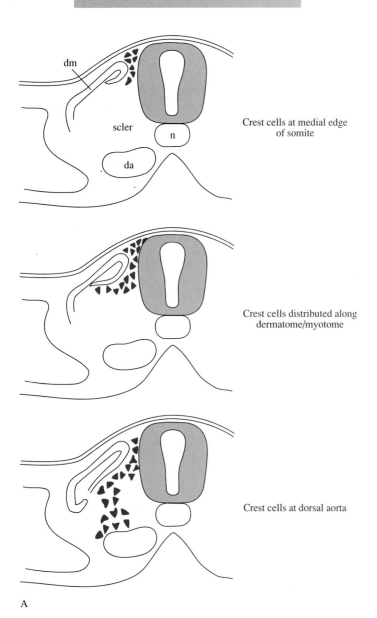

STAGE OF NEURAL CREST MIGRATION

Crest cells at medial edge
of somite

Crest cells distributed along
dermatome/myotome

Crest cells at dorsal aorta

A

Figure 7.29

A, A schematic diagram of the progressive migration of chick neural crest cells under the dermamyotome (dm) to the dorsal aorta (da). n: notochord; Scler: sclerotome. B, A representative section through a chick embryo stained with an antibody that recognizes migratory neural crest cells. NT: neural tube; DM: dermamyotome; PCV: posterior cardinal vein; PD: pronephric duct; Sc: sclerotome, marked by open arrowhead; DA: dorsal aorta. Scale bar equals 100 μm. (From Loring and Erickson, 1987.)

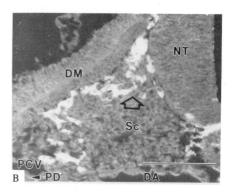

only in the anterior portion of the somite (Rickmann et al., 1985; Bronner-Fraser, 1986; Loring and Erickson, 1987; Teillet et al., 1987). Some neural crest cells stop migrating as soon as they enter the somite, and these cells give rise to the sensory ganglion. Other neural crest cells continue to migrate along the undersurface of the dermamyotome until they reach the dorsal aorta. These neural crest cells then aggregate to form the sympathetic ganglia of the autonomic nervous system (see Figs. 7.29 and 7.30). At the level of the developing kidney, some of the neural crest cells that migrate through the somite also cluster near the kidney, and these develop into the medulla of the adrenal gland.

Those neural crest cells that migrate from the neural tube opposite the posterior border of the somite will migrate either anteriorly or posteriorly along the neural tube and enter the anterior segment of their own (or the

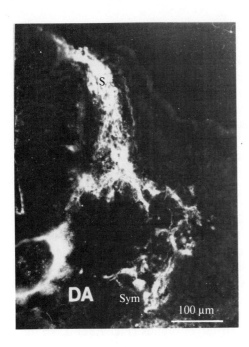

Figure 7.30

Section through a 4-day chick embryo stained with an antibody that recognizes neural crest cells. The pattern of the peripheral nervous system can already be distinguished at this early stage. S: sensory ganglion; Sym: sympathetic ganglion; DA: dorsal aorta. (From Loring and Erickson, 1987.)

adjacent) somite (Teillet et al., 1987). Thus, the somite immediately imposes a periodic, segmented organization on the neural crest cells. Furthermore, the sensory ganglia and sympathetic ganglia develop in register with each other. *Therefore, the basic organization of the peripheral nervous system is established by the migratory pathways taken by the neural crest cells* (Fig. 7.31).

The importance of the somite in the patterning of neural crest cell migration is dramatically demonstrated in the experiments of Lehmann (1927) and Detwiler (1937), in which they removed somites of the axolotl, and one large, unsegmented sensory ganglion developed, whereas if extra somites were added, additional ganglia developed.

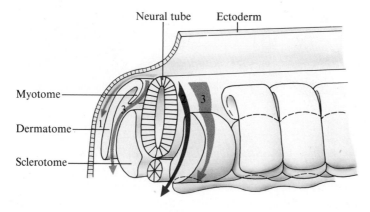

Figure 7.31

A diagrammatic view of an avian embryo with its ectoderm peeled back to reveal the migratory pathways of neural crest cells. These are: (1) between the ectoderm and somite to give rise to the pigment cells of the skin; (2) in the intersomitic space, possibly to give rise to portions of the sympathetic ganglia; and (3) through the anterior half of the somite to form the sensory and sympathetic ganglia. (After Le Douarin et al., 1984.)

The pathways taken during neural crest cell migration are determined primarily by elements in their environment, and many factors undoubtedly control the patterns of neural crest cell migration. These include extracellular matrix molecules that serve as adhesive molecules for migration, spaces that allow the neural crest cells to move, and barriers created by embryonic structures and extracellular matrix molecules. Several of these may, in fact, act together to form a "fail-safe mechanism" for guiding these very important cells to their precise destination. We discuss the mechanisms involved in directing neural crest cell migration in Chapter 9.

Connections of Neurons with Their Targets

The peripheral nervous system consists of neurons whose cell bodies are fixed in place either in the neural tube or in ganglia, with their axons extending to their targets. The distances between the cell bodies and targets can be substantial, and the possibilities for incorrect connections would seem enormous. When and how do the axons extend to the appropriate target organs, and what guides the axons to their targets?

The axons of the motor fibers of the ventral roots exit the neural tube and are the first axons to extend to their target organs (Fig. 7.32). Like the neural crest cells, these fibers grow through only the anterior portion of the somites (Keynes and Stern, 1984). Soon thereafter, the neural crest cells that have aggregated to form the sensory ganglia send processes in two directions: (1) toward the dorsal neural tube, where they will synapse on cell bodies of the mantle layer; and (2) toward the periphery, where they will synapse on sensory receptors of the skin, muscles, and visceral organs.

Finally, the afferent fibers of the autonomic nervous system begin to send their axons toward the appropriate targets. Preganglionic axons of the sympathetic and parasympathetic motor neurons exit the ventral root and synapse at the appropriate ganglia. Meanwhile, the postsynaptic neurons

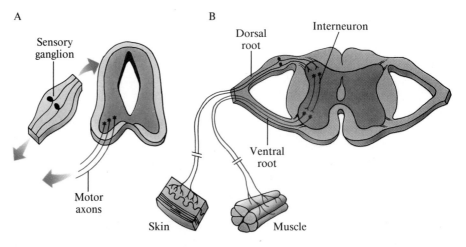

Figure 7.32

Development of axons in the peripheral nervous system. The first axons to appear are the ventral root motor fibers, which extend from cell bodies in the ventral portion of the neural tube (A) to striated muscles (B). About the same time, neural crest cells that have aggregated to form the sensory ganglia extend axons in two directions: toward the neural tube, where they synapse with cells in the dorsal portion of the neural tube, and toward the periphery, where they innervate sensory structures.

derived from the neural crest extend axons from their ganglion to their targets in the viscera and skin.

The outgrowth of neurons is accomplished by the migratory tip of the axon known as a **growth cone** (Fig. 7.33). Growth cones were first observed by the renowned neurohistologist S. Ramón y Cajal (1890). The true function of the growth cone was not proved until Ross G. Harrison (see p. 625) (1910) grew frog neural tubes in tissue culture and was able to observe directly the extension of the neuron because of the migratory activity of the growth cone. In fact, until this remarkable experiment was reported, many believed that axons were formed from chains of cells that fused to make one long process or that they were somehow secreted along the pathway by the cells lining the paths. Harrison, incidentally, is credited with first developing the technique of growing living tissues in culture. **Tissue culture** has since become an extremely powerful technique for cell and developmental biologists alike.

The growth cone is similar to the lamellipodium of the fibroblast, which is discussed in detail in Chapter 9. The growth cone is the locomotory appendage of the axon, and its motility is dependent upon microfilaments (Yamada et al., 1971; Letourneau, 1985). Unlike the fibroblast, however, whose trailing end periodically detaches and snaps forward in order for the cell to advance, the neuron cell body generally remains fixed in the same spot as the axon extends (Wessells et al., 1973). Therefore, the neuron must also grow as it extends outward. Consequently, a second important process occurring in the growth cone is that new plasma membrane is inserted in this region to accomplish the growth necessary for continuous extension, hence its name.

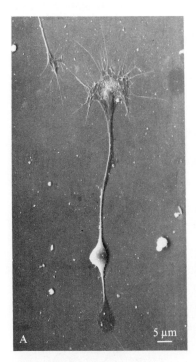
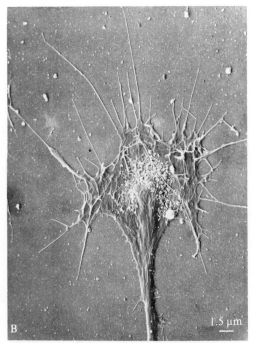

Figure 7.33

A, SEM of a neuron in tissue culture, with its cell body and growth cone at the end of the extending axon. B, An enlargement of the growth cone. The long, spiky processes are filopodia, which may have a substratum-sensing function. (From N. K. Wessells, 1977. *Tissue Interactions and Development.* Benjamin/ Cummings, Menlo Park, California.)

An important question that has been addressed experimentally by neurobiologists is the following: What controls the directional outgrowth of an axon so that it hooks up with the appropriate target, either in the central nervous system or in the peripheral nervous system? It should be pointed out that generalizations about neuronal pathfinding are difficult to make because the results of many experiments seem contradictory. Perhaps two generalizations can be made, however: (1) Early in neurogenesis, as the axons are extending, the neurons already appear to possess the specificities for the target with which they will connect; and (2) as the axons extend to their target, they make very few errors in pathway navigation or connectivity that they will have to adjust later.

In many cases, normal neuronal pathways can be obstructed, and yet the nerves still find their ways to the appropriate target. For example, a barrier of the mineral mica can be placed between the spinal cord and limb of a developing salamander embryo, and the motor axons will grow around the barrier and still innervate the limb appropriately. Similarly, Lance-Jones and Landmesser (1980) reversed a small segment of chick spinal cord so that the emerging motor fibers were displaced by a few somite segments. The motor fibers still found their way into the limb, albeit by different pathways (Fig. 7.34), and innervated the appropriate targets. Such studies suggest that these motor fibers are predetermined in the spinal cord to hook up with particular muscles. If, however, the neurons are displaced too far from their appropriate target, they can innervate an inappropriate muscle (Hollyday, 1981).

The apparent lack of many major mistakes by neurons in the peripheral nervous system has been most graphically demonstrated in insects. In the grasshopper, for example, limb sensory neurons arise at the distal tip of the limb bud and send axons proximally so that they will eventually connect with the central nervous system. Because these are the first neurons to develop, they are called **pioneer neurons**. Pioneer neurons can be stained

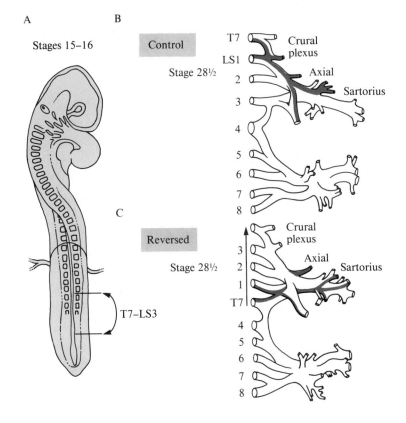

Figure 7.34

When neurons are slightly displaced, they are able to innervate the appropriate muscles— even though they may take different pathways than usual. A, 2½-day-old chick embryo, in which the neural tube between segments T7 and LS3 was reversed. B, The normal pattern of nerves that form the crural and sciatic plexus near the hind limb. The normal projection of the neurons from segments T7 and LS1 within the crural plexus is highlighted in color. C, The pattern of neurons in the crural plexus after segments T7 to LS3 are reversed. Note that the pattern of the plexus itself is normal, but the neurons from segments T7 and LS1 take different pathways within the plexus. Even though the nerves follow different pathways than usual, they still are able to innervate their appropriate muscle. (After Lance-Jones and Landmesser, 1980. Reprinted by permission of Cambridge University Press.)

with specific antibodies at their earliest stages of outgrowth, and the pathways they take can be traced (Bentley and Keshishian, 1982; Keshishian and Bentley, 1983). The extraordinary observation is that these axons grow along stereotyped pathways that do not vary from embryo to embryo; furthermore, the growth cone guiding the axon makes choices to grow in certain directions with virtually no error (Fig. 7.35). Such precision in pathfinding has been observed directly in the zebra fish as well (Eisen et al., 1986).

Higher vertebrates' axons also display precise pathfinding, although the techniques for studying pathfinding in these organisms are more indirect. Landmesser (1978) looked for the distribution of motor neuron cell bodies in the spinal cord in relationship to the muscles they innervated in the limb. She was able to trace the neurons from the spinal cord to the limb by labeling the neurons with horseradish peroxidase (HRP) (see the essay on p. 280). For example, she injected HRP into a muscle mass of the chick

Regular text continues on page 279.

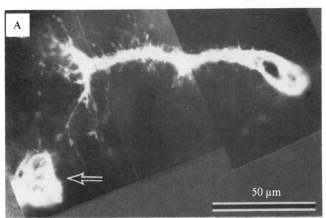

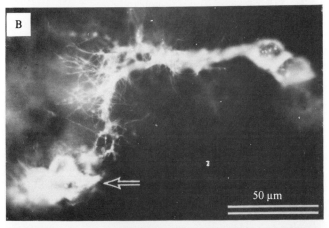

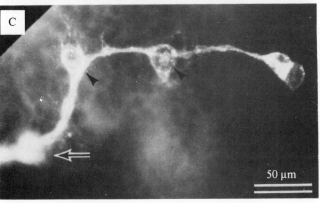

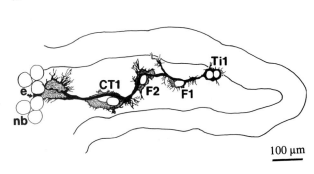

Figure 7.35

The first neuron to arise at the tip of the grasshopper limb and migrate toward the CNS (toward the left in this figure) is followed in this series of micrographs after staining the neuron with a fluorescently-labeled antibody. A, The Ti1 neuron extends medially toward the CNS owing to the activity of its growth cone. B, The neuron now takes a sharp turn posteriorly. C, The axon takes yet another sharp turn, this time toward the CNS, which is outside the field of this micrograph. D, This migratory pattern is consistent between embryos and is summarized in this figure. The points at which the Ti1 neuron changes direction are coincident with distinctive cells that are named F1, F2 (arrowheads in micrographs), and CT1 (arrows in micrographs). These cells may act as "guideposts" to direct the migration of the axon. (From Bentley and Caudy, 1983. Reprinted by permission from *Nature* Vol. 304 pp. 62–65. Copyright © 1983 Macmillan Magazines Ltd. Courtesy of D. Bentley.)

Essay

Growth Cone Guidance in Insects

Growth cone guidance is often difficult to study in vertebrates because the embryos are opaque, which precludes direct observation, and even the earliest stages of outgrowth are already quite complex. Insects, on the other hand, have provided an attractive system for studying neuronal guidance because they have relatively simple nervous systems and neurons can often be visualized directly in the living embryo. Furthermore, at least in *Drosophila,* classical and molecular genetic approaches are used for studying axon guidance. In addition to the attractive practical features of the insect model, the gradually emerging picture from studies of the nervous system of insects suggests that the molecular mechanisms that control the development of insect nervous systems are similar to those implicated in the development of the vertebrate nervous system. Thus, what we learn from insects is very likely to have broad implications for nerve guidance in general.

The development of the segmental ganglia along the ventral surface of the grasshopper embryo will be cited here as one example of how growth cone guidance can be studied in insects. In grasshoppers, 17 segmental ganglia develop from the ventral ectoderm. Each ganglion develops from 30 neuroblasts (Fig. 1) on each side, one median neuroblast, and 7 midline precursors. Each neuroblast (we shall use the so-called 7-4 neuroblast for our discussion) divides repeatedly to form **ganglion mother cells**, which, in turn, divide to form two daughter neurons. These neurons then project axons laterally across the segment until the growth cones reach bundles of nerves running longitudinally. The axons then turn either anteriorly or posteriorly and migrate along the existing nerve bundles. For example, projecting and growing neurons Q1 and Q2 traverse the segment until they reach neuron dMP2, which they then follow posteriorly. These axons come together to form bundles, or **fascicles**, in a process known as **fasciculation**. In contrast to neurons Q1 and Q2, neurons C and G, traveling together, pass by dMP2 and instead follow other longitudinal axons: C follows axons X1 and X2 posteriorly, and G follows axons P1 and P2 anteriorly. Such precise patterning has been demonstrated for many hundreds of neurons within the grasshopper. As a result of this fasciculation, the nervous system resembles a ladder, with transverse tracts, or **commissures**, and longitudinal tracts extending from the ganglia (see also Figs. 2, and 3A).

The precision with which the embryo is "wired" suggests that a particular axon recognizes the surface of another axon that has already grown out. To test this idea, several investigators have selectively removed specific neurons and observed the behavior of other axons that would normally fasciculate with them. Because the grasshopper embryo is relatively transparent, individual neurons and their cell bodies and growth cones can be directly observed with Nomarski optics, enabling investigators to ablate individual neurons with laser surgery so that particular axons within the fascicles no longer exist. When a particular

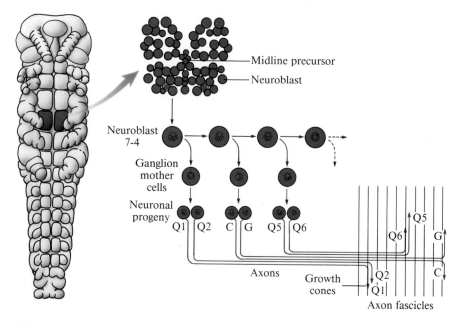

Figure 1

A grasshopper embryo has 17 segmental ganglia, formed from 60 neuroblasts, seven midline precursors, and one median neuroblast. Each neuroblast divides many times, each time generating a ganglion mother cell, which in turn divides to form two progeny neurons. In this figure, neuroblast 7-4 and its first six progeny are indicated. Each neuron sends out an axon until it reaches a particular longitudinal axon, with which it will fasciculate. The pathway taken by each neuron is different and specific for that neuron. (After C. S. Goodman and M. J. Bastiani. 1984. How embryonic nerve cells recognize one another. Scientific American, *251* (6): 58–66. Copyright © by Scientific American, Inc. All rights reserved.)

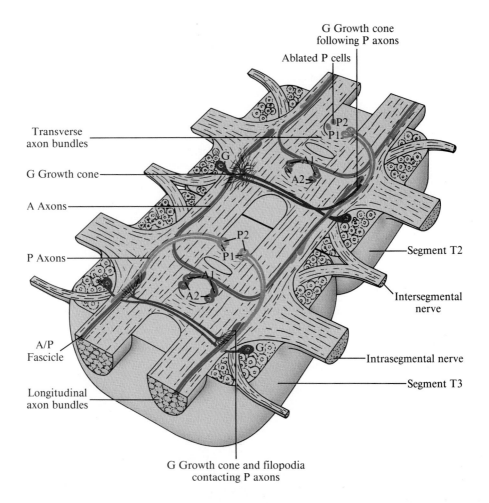

G Growth cone following P axons

Ablated P cells

Transverse axon bundles

G Growth cone

A Axons

P Axons

A/P Fascicle

Longitudinal axon bundles

Segment T2

Intersegmental nerve

Intrasegmental nerve

Segment T3

G Growth cone and filopodia contacting P axons

Figure 2

Two segments (T2 and T3) of a grasshopper embryo, showing only the A, P, and G neurons and their axons. The more anterior segment is slightly more advanced developmentally. Normally, the G neuron sends its axon to the A/P bundle, with which it will fasciculate. If the P1/P2 sibling pairs are ablated (so that their axons will not grow out into the A/P fascicle), the growth cone from the G neuron will no longer fasciculate with the remaining A neuron (shown in upper left). If, on the other hand, the A cells are ablated, the G growth cone will fasciculate as usual with the remaining P axons. (After C. S. Goodman and M. J. Bastiani. 1984. How embryonic nerve cells recognize one another. Scientific American, *251* (6): 58–66. Copyright © by Scientific American, Inc. All rights reserved.)

axon is removed, the behavior of neurons that usually fasciculate with it can then be observed. Figure 2 represents two segments of a grasshopper embryo with only certain pairs of daughter neurons and their axons indicated. Normally, the axons of the G neuron fasciculate with the A/P fascicle. In fact, the growth cone of the G neuron will crawl across, and apparently ignore, at least 25 other fascicles containing over 100 axons before it gets to the A/P fascicle. Upon arrival, it crawls onto the surface of the fascicle and extends anteriorly. Electron microscopy suggests that the G axon associates with the P axons within the fascicle, rather than with the A axons. When a G growth cone arrives at an A/P fascicle from which P axons had been specifically removed by laser ablation, the G growth cone branches abnormally and wanders aimlessly. Such specificity has been demonstrated with many neurons using such laser ablation experiments.

These observations led Corey Goodman and his associates to formulate the **labeled pathway hypothesis**, which states that different axons possess different chemical cues on their surfaces and these direct the fasciculating behavior of each and every axon. To test this idea, they raised antibodies against grasshopper neurons and showed that different antibodies do indeed recognize different axon bundles in different regions. In Figure 3, for example, two different antibodies label different portions of the axon tracts (Bastiani et al., 1987). In Figure 3B, an antibody recognizes bundles that traverse the segment, the so-called commissures. In Figure 3C, another antibody recognizes only the longitudinal bundles. Not only do these antibodies stain different fascicles, but their staining patterns change with time, suggesting that they recognize developmentally regulated cell surface molecules.

Incubation of grasshopper embryos in these antibodies disrupted the nervous system patterning, suggesting that cell surface molecules play a role in axonal recognition (Harrelson et al., 1988). The antibodies were then used to isolate two cell surface glycoproteins, which were named **fasciclin** I and **fasciclin** II. With so many axons that must follow specific pathways, many more adhesion molecules will likely be discovered in the future.

Grasshopper fasciclin I and II, whose genes have now been cloned and sequenced, bear a surprising similarity to vertebrate adhesion molecules N-CAM and L1, which

we shall discuss in detail in Chapter 9. Because of such similarities, the insect model, with its relative simplicity and accessibility, will undoubtedly continue to shed light on the more complex vertebrate nervous system and controls for growth cone pathfinding.

More recently, it has been shown that the nervous system of *Drosophila* embryos is similar, if not identical, to that of the grasshopper. Because *Drosophila* has many mutations in the nervous system, we may now use genetic tools to make rapid progress in our understanding of the factors that control connections within the nervous system.

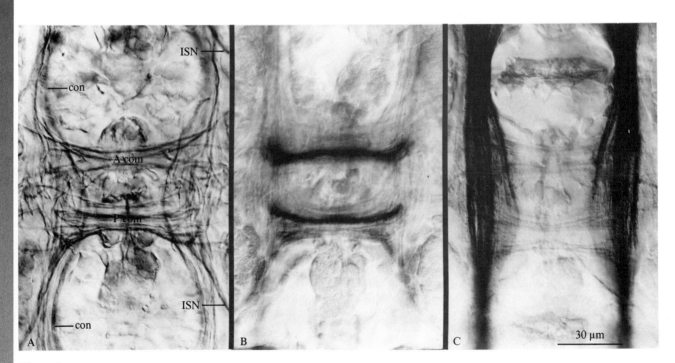

Figure 3

Different antibodies recognize unique subsets of axons in the grasshopper nervous system. A, A whole-mount preparation of the grasshopper neuroepithelium viewed with Nomarski optics and stained with an antibody to HRP, which stains all the neurons. The axons form a ladder pattern composed of transverse, or commissural, axons (A com, P com) and longitudinal axons. con: longitudinal connectives; ISN: intersegmental nerve. B, If an embryo of similar age is stained with antibody 3B11 (against fasciclin I), a subset of the commissural axons is recognized by the antibody. C, In contrast, antibody 8C6 (against fasciclin II) stains many of the longitudinal bundles—but not the commissural bundles. (From Bastiani et al., 1987; courtesy of C. Goodman. Reprinted by permission of Cell Press.)

limb, and all the neurons in the injection site took up the HRP at their tips and transported the HRP back to the neural tube. Because the HRP technique produces a brown stain, she was able to trace the neurons from the location of their cell bodies in the spinal cord to the particular muscle mass that they innervated. Her results, which are summarized in Figure 7.36, show that motor neurons whose cell bodies are localized in particular regions of the spinal cord innervate the same muscle masses on day 11 as they do on day 6; i.e., a neuron from a particular region of the neural tube sends its axons to a particular muscle mass by day 6 and is still innervating that muscle at day 11. Thus, if mistakes are made, they are below the resolution of this technique. Similarly, sensory nerve fibers innervate the appropriate skin receptors at the time of their initial outgrowth (Scott, 1982).

The environmental factors that actually determine the migratory pathways taken by axons are not yet known in detail, but the study of motor and sensory innervation of the limbs provides an informative model system.

Regular text continues on page 282.

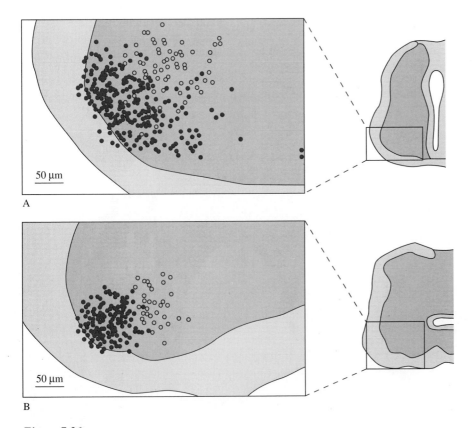

Figure 7.36

The projection patterns of motor neurons that arise in the ventral neural tube to the chick limb were determined in 6-day-old (A) and 11-day-old (B) embryos. The fiber pathways were traced by injecting HRP into either the dorsal (filled circles) or ventral (open circles) muscle mass on day 6. The neurons innervating the muscles then transported the HRP back toward the neural tube, allowing the investigator to trace the projection of any given neuron from its cell body in the neural tube to the muscle. The distribution of motor neurons in the neural tube that innervate either the dorsal or ventral muscles appears to be the same at early and later stages, suggesting that as soon as the neurons enter the leg, they connect up with the appropriate muscle mass. (After Landmesser, 1978. Reprinted by permission of Cambridge University Press.)

Essay

Marking Neurons with Horseradish Peroxidase

During development, axons grow out from neuron cell bodies to targets that can often be considerable distances away. To understand what controls these patterns of directed growth, the entire axon must be traced along its pathway. The technique of marking neurons with the enzyme horseradish peroxidase (HRP) has made these important studies possible. Since its introduction in the early 1970s, the HRP marking technique has also been applied to other developmental systems as a method of determining cell lineage.

Horseradish peroxidase is a heme protein that has been isolated from the roots of the horseradish. By itself, the protein is not visible; however, when it is reacted with a suitable substrate, such as diaminobenzidine (DAB) or tetramethylbenzidine (TMB), the resulting reaction product is readily visible as a dark precipitate.

A number of properties of HRP enhance its value as a marker. First, only a few HRP molecules are required to yield a large amount of reaction product. The resulting amplification makes HRP a particularly sensitive marker. Second, once taken up by neurons or other cells, the HRP generally stays within the cells and does not leak out, so it can be considered a specific cell marker.

Horseradish peroxidase can be used to mark neurons in a number of ways. Although intact neuronal terminals will also take up HRP, especially if they are electrically stimulated, generally better labeling is obtained if the axons are slightly injured (Tosney and Landmesser, 1986). The most common method is to inject the HRP on or near an injured axon. The injured axon takes up the HRP, presumably where a break has occurred in the plasma membrane, allowing easy entrance of the marker. Horseradish peroxidase is then transported in a retrograde fashion (i.e., toward the cell body). Retrograde transport thus fills the axon with HRP, which—when "developed" with DAB or TMB—marks the entire neuron, including the axon and its cell body. These labeled neurons can be visualized in sectioned material, both in the light and electron microscopes, or in whole-mount preparations. Such a retrograde marking technique al-

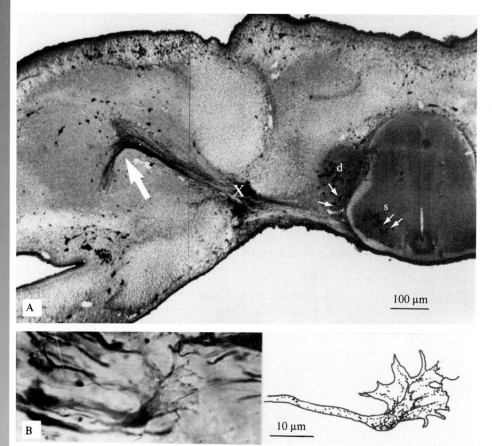

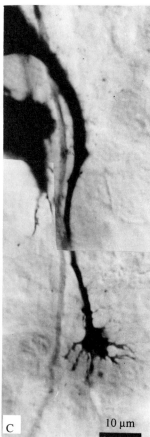

lows one to trace the projections of neurons from their site of innervation back to the cell bodies.

A second method takes advantage of the neuron's ability to transport materials in an anterograde fashion as well (i.e., from the cell body outward toward the growth cone or axon terminus). When HRP is injected near a slightly injured cell body, it is taken up and transported along the axon toward the growth cone or axonal terminus. Such applications have been particularly useful in visualizing the growth cones of growing axons and the terminal arborization of neurons.

Finally, HRP can be introduced into individual neurons, if they are large enough, by injecting the solution directly into the cell body. This gives us a remarkable picture of the entire neuron, including its very intricate growth cone. Although the neurobiologist has made special use of HRP, the intracellular injection technique has also been applied to other cell types as a method of labeling individual cells and their progeny as they divide. For example, if an individual blastomere of an embryo is labeled with HRP, the reaction product can still be detected in that cell's progeny after many rounds of divi-

sion, thus indicating what cells arise from the particular injected blastomere (see Fig. 11.3).

Other labeling techniques have been introduced since the development of the HRP technique, but HRP still remains one of the most useful tools in the arsenals of neural and developmental biologists.

Figure 1 (opposite)

A, This chick embryo has been injected with HRP in the plexus of nerves that enter the limb (injection site marked with X). HRP has been carried in an anterograde direction to the site of neuron termination in the muscle mass (large white arrow) and in a retrograde direction to the cell bodies (small white arrows) in the dorsal root ganglion (d) and spinal cord (s). B, High magnification of a growth cone of a motor neuron that had been labeled in an anterograde direction. Note the thin lamellipodium that is visible with this technique. The drawing on the right depicts the injected neuron. C, A peripheral nerve exiting the grasshopper nerve cord has been injected directly with HRP. The entire nerve, including the cell body, axon, and growth cone is labeled. (A, and B, From Tosney and Landmesser, 1985. Reprinted with permission of Oxford University Press. C, From Harrelson et al., 1988. Reprinted with permission of Sinauer Associates. Courtesy of C. Goodman.)

Motor neurons that innervate the hind limb are derived from nine segments (i.e., nine somite levels; see Fig. 7.34). Those neurons from the first four anterior segments (thoracic level 7 and lumbosacral levels 1 through 3) migrate toward the limb and aggregate to form a plexus of nerves known as the **crural plexus**, whereas those motor axons that emerge from more posterior parts of the neural tube (lumbosacral levels four through eight) give rise to the **sciatic plexus**.

The patterns of nerve projections that make up these plexes are consistent, suggesting that cues in the extracellular environment establish highways along which axons extend. It is generally believed that the factors that determine these pathways are nonspecific cues that could be followed by many types of neurons. For example, nerves from other axial levels can be grafted to the site of the hind limb, and they will slavishly trace out the same highway system as the appropriate nerves would have done. In the experiments of Lance-Jones and Landmesser (1980) outlined in Figure 7.34, the motor fibers still traced out the exact pattern of the crural plexus, even though the individual fibers took portions of the highway they ordinarily would not have taken. One can even graft spinal cord segments that normally give rise to the wing neurons to the site of the hind limb, and they will form a normal crural and sciatic plexus (Hollyday, 1981). This suggests, then, that some of the cues that guide the neurons can be relatively nonspecific.

Besides the cues that are relatively nonspecific and that determine the highways, there also must be other cues that sort out axons and tell them to exit the major highways to innervate specific muscles or sensory organs. For example, once a neuron is in the plexus, what directs it into specific portions of the limb? Again, in the experiments of Lance-Jones and Landmesser (1980), when a neural tube segment was reversed, the appropriate axons still got to their appropriate target, even though they took different parts of the highway (see Fig. 7.34). Furthermore, Ferguson (1983) demonstrated that after rotation of the chick hind limb around the dorso-ventral axis 180 degrees, the motor neurons still innervated their appropriate muscles. Surprisingly, many of these cues seem to reside somewhere in or near the plexus region because this is where the sorting seems to occur. Furthermore, an intact limb bud does not seem necessary for proper sorting out, because Tosney and Landmesser (1985) removed the limb bud and yet the axons still sorted out appropriately in the plexus.

Numerous cues could direct axon extension. The same guidance mechanisms probably hold for neurons as for fibroblasts. These cues are discussed in detail in Chapter 9. Cues that could provide relatively nonspecific tracks include adhesive "lanes" and extracellular spaces. More specific cues to direct a particular axon to a specific muscle group could include chemotaxis or specific adhesion molecules that regulate cell-cell interactions. It is likely that these cues may overlap or even be redundant, so that if neurons are slightly displaced and miss their adhesive pathways, they still may find their way to the target, perhaps by chemotactic means. To date, however, these possibilities remain unproven.

7–5 Ectodermal Placodes

There is another important source of neurons in addition to the neural tube and the neural crest cells. This other source is the thickenings of the ectoderm, known as placodes. Ectodermal placodes are found only in the head and are regions of the ectoderm that have become a columnar epithelium (Fig. 7.37). These placodes include the **nasal placode** at the anterior-most end of the embryo, which will give rise to the **nasal epithelium** (the sensory

PLACODES

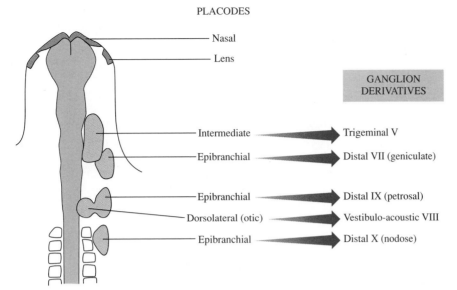

Figure 7.37

Diagram of a 2-day chick embryo showing the locations of the head placodes and their derivatives in the adult. (After D'Amico-Martel and Noden, 1983.)

lining of the nose), the **lens placode**, which will give rise to the lens, and the **otic placode**, from which the inner ear will develop. In addition, a series of smaller placodes, the **epibranchial placodes**, the **intermediate placode**, as well as the otic placode, give rise to some of the neurons in the cranial sensory ganglia (D'Amico-Martel and Noden, 1983; Le Douarin et al., 1986).

The morphogenetic behavior of the cells within these placodes varies. Some of these placodes eventually invaginate and pinch off entirely from the ectoderm to form an epithelial ball. For example, as the eye develops, outpockets of the brain known as the optic vesicles approach the overlying ectoderm, and where they contact it, the ectoderm thickens to become the lens placode. The lens placode then inpockets to form a cup. Eventually, the rim of the cup closes up like a purse string, and the placode separates entirely from the surface epithelium to form an internal vesicle. By a similar mechanism, the otic placode gives rise to the **otic vesicle**, which will undergo complicated changes to form the inner ear (Fig. 7.38).

The nasal placode, on the other hand, invaginates but never completely detaches from the ectoderm. These pouches of thickened ectoderm, called the **olfactory pits**, eventually form the olfactory epithelium of the nose, which contains all the specialized sensory nerves for smell. The neurons that make up the cranial nerve I (olfactory nerve) are also derived from the nasal epithelium and synapse in the olfactory bulb of the telencephalon.

A different morphogenetic behavior is seen in the cells that are derived from the epibranchial, intermediate, and otic placodes, which contribute to the sensory ganglia of the head (Fig. 7.39). Soon after these placodes form, single cells stream from the thickened epithelium (D'Amico-Martel and Noden, 1983; Nichols, 1986). As the placode cells detach as single cells or

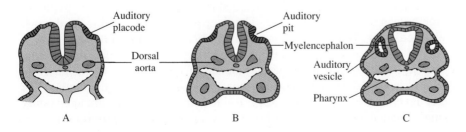

Figure 7.38

Three stages in the development of the otic placode, which will give rise to the inner ear. (After Hopper and Hart, 1985.)

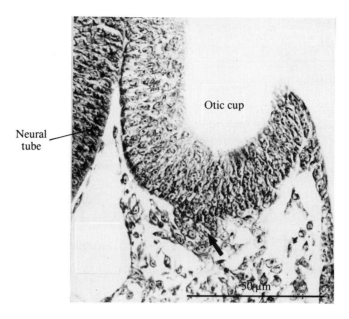

Figure 7.39

Section through a chick otic placode showing the thickened epithelium, which will give rise to the inner ear, and the separation of single cells at the base of the placode (arrow), which will contribute neurons to the vestibulo-acoustic ganglion (VIII). (From D'Amico-Martel and Noden, 1983. Reprinted from *American Journal of Anatomy* by permission of Wiley-Liss, a division of John Wiley and Sons, Inc. Copyright © 1983, Wiley-Liss. Courtesy of D. Noden.)

in clusters, they join with neural crest cells to form the cranial sensory ganglia. The placode-derived cells always differentiate into neurons, whereas the neural crest cells in these same ganglia can develop into either neurons or glial cells. The relative contributions of the neural crest and ectodermal placodes to the various cranial ganglia are outlined in Figure 7.37.

Thus, many of the sensory organs and ganglia are composed of cells from a variety of sources, including the ectodermal placodes. It is difficult to imagine how these complicated morphogenetic events are orchestrated and even more amazing to consider how such a scheme arose during evolution.

7–6 Epidermis and Cutaneous Appendages

The rest of the ectoderm remains on the surface of the embryo and differentiates into the epidermis of the skin. We shall use the development of the human skin as an example. The deeper layer of the skin, the **dermis**, is derived from the mesoderm and will be discussed in Chapter 8.

The adult epidermis consists of a multilayer of skin cells, or **keratinocytes**, so named because of the prevalence of intermediate filaments composed of keratin (see p. 339 for a discussion of intermediate filaments). The outer layers of the epidermis are cornified dead cells that are constantly being sloughed off, whereas the inner layers of cells are proliferating and progressively differentiating to produce new surface cells.

The innermost layer of the epidermis, or the **basal layer**, is continuously proliferating to give rise to cells that are subsequently pushed outward and begin to differentiate (Fig. 7.40A). The next layer, moving peripherally, is

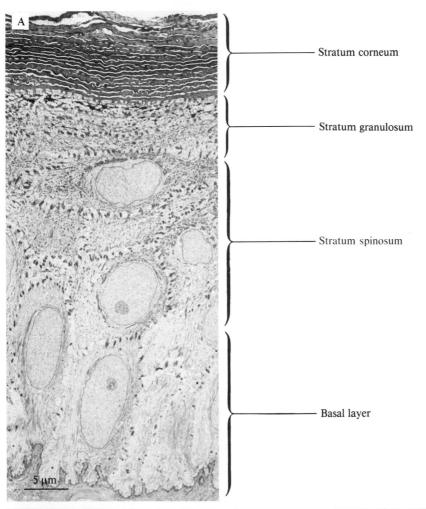

Stratum corneum

Stratum granulosum

Stratum spinosum

Basal layer

5 μm

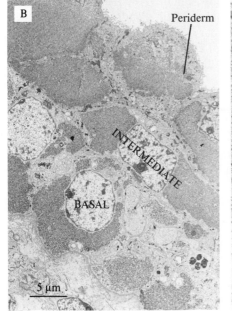

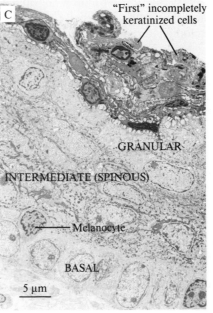

Figure 7.40

Development of the human skin. A, Electron micrograph through human epidermis, showing the four major layers of the adult skin. B, Section through the epidermis of a 62-day human embryo showing the basal layer, the transient periderm, and the first appearance of the intermediate layers. C, Section through the epidermis of a 21-week-old human fetus at a stage when all the definitive cell layers, including a keratinized layer, have formed. (A, From Odland, 1983. B, and C, From Holbrook, 1983. A, B, and C, Reprinted by permission from L. A. Goldsmith (ed.), *Biochemistry and Physiology of the Skin,* Copyright © 1983 by Oxford University Press, Inc., New York. Courtesy of K. Holbrook.)

known as the **stratum spinosum**. This consists of polyhedral cells that are held together by developing desmosomes; the layer is so named because of the spiny appearance of these junctions and their associated keratin filaments. Next to the stratum spinosum is the **stratum granulosum**, whose cells have acquired granular material known as **keratohyalin granules**. These granules contain the highly phosphorylated protein **profilaggrin**, which will be used eventually to aggregate the keratin filaments (Dale et al., 1978). Finally, the outer five or six layers form the **stratum corneum**. The keratinocytes of this layer are flattened squamous cells that have lost their nuclei and are tightly packed with bundles of keratin intermediate filaments. These cells are constantly being sloughed from the surface of the skin.

In the two-week human embryo, the epidermis is a single cell layer. Shortly thereafter, however, it begins to proliferate rapidly and forms a second, outer cell layer known as the **periderm** (see Fig. 7.40B). The periderm is characteristic of primates only and is a temporary protective covering for the skin. The periderm may also facilitate glucose transport to the fetus *in utero*. By six months postfertilization, when the hairs begin to develop and the skin begins to keratinize, the periderm is pushed away from the epidermis. This cast-off periderm mixes with sebaceous secretions from the skin and desquamatized epidermis to make the **vernix**, which is spread on the fetal skin. The vernix protects the developing epidermis from abrasion as well as damage from immersion in amniotic fluid.

By the third month of development, the basal cells have produced another cell layer through proliferation, which is called the **intermediate layer** because of its position between the basal layer and the periderm (see Fig. 7.40B). Additional intermediate layers are added between the fourth and fifth months. These intermediate layers will become the stratum spinosum and granulosum. By the fifth month, some of the intermediate cells have begun to differentiate (or keratinize) to produce the first cornified cells of the stratum corneum (see Fig. 7.40C).

The epidermis also contains pigment cells, called **melanocytes**, which are derivatives of the neural crest. These cells migrate from the dermis, through the basal lamina of the epidermis, and come to lie in the basal layer. The pigment cells in humans produce the brown pigment melanin, which is packed in membrane-enclosed vesicles called **melanosomes**. The melanosomes are then transferred through cytoplasmic processes from the melanocytes into the keratinocytes of the basal layer.

Numerous structures, or appendages, can develop from the epidermis, including hair, fingernails, feathers, horns, and glands of various sorts, including sweat, apocrine, and mammary glands. These structures will generally form from thickenings of basal epidermis that inpocket.

7–7 Chapter Synopsis

After gastrulation, the embryo is composed of three germ layers, from which all the tissues and organs of the body will arise. One of these germ layers, the ectoderm, is initially an epithelium that covers the embryo. As a result of neurulation, the ectoderm is partitioned into three separate cell populations having different fates: the epidermal ectoderm (which forms the epidermis), the neural ectoderm (which comprises the central nervous system), and the neural crest (which forms most of the peripheral nervous system and gives rise to a variety of other cell types as well). The developmental fates of each of these cell populations diverge to produce a diverse array of structures.

The epidermal ectoderm gives rise primarily to the keratinocytes of the epidermis, but these cells also differentiate to produce a variety of cutaneous structures, including hair, scales, nails and claws, and glands of many types. In addition, the epidermal ectoderm can thicken to form placodes that give rise to neuronal elements, such as the nasal epithelium.

The neural tube, which forms during neurulation, is initially a columnar epithelium. These cells ultimately will undergo complex morphogenetic movements to produce the many layers of the brain and spinal cord. Two types of cells are derived from this original epithelium: the neurons, which are the excitatory cells of the central nervous system, and a variety of glial cell types, which are the supportive cells of the central nervous system.

The neural crest cells, which are derived from the neural epithelium, migrate extensively as individual cells to many positions in the embryo, where they give rise to the neurons and glial cells of the peripheral nervous system. In addition, the neural crest cells produce a wide variety of non-nervous tissue cell types, including pigment cells, connective tissue in the head, and teeth. These cells provide an excellent model system to study the factors that control directional migration of cells during embryogenesis and the differentiation of diverse cell types from a single precursor.

References

Ayer-Le Lièvre, C. S., and N. M. Le Douarin. 1982. The early development of cranial sensory ganglia and the potentialities of their component cells studied in quail-chick chimeras. Dev. Biol., *94*: 291–310.

Baker, P. C., and T. E. Schroeder. 1967. Cytoplasmic filaments and morphogenetic movement in the amphibian neural tube. Dev. Biol., *15*: 432–450.

Balinsky, B. I. 1981. *An Introduction to Embryology*, 5th ed. Saunders College Publishing, Philadelphia.

Bastiani, M. J. et al. 1987. Expression of fasciclin I and II glycoproteins on subsets of axon pathways during neuronal development in the grasshopper. Cell, *48*: 745–755.

Beebe, D. C. et al. 1979. Lens epithelial cell elongation in the absence of microtubules: Evidence for a new effect of colchicine. Science, *206*: 836–838.

Bentley, D., and M. Caudy. 1983. Pioneer axons lose directed growth after selective killing of guide post cells. Nature (Lond.), *304*: 62–65.

Bentley, D., and H. Keshishian. 1982. Pathfinding by peripheral pioneer neurons in grasshoppers. Science, *218*: 1082–1088.

Bloom, W., and D. W. Fawcett. 1986. *A Textbook of Histology*, 11th ed. W. B. Saunders, Philadelphia.

Bronner-Fraser, M. 1986. Analysis of the early stages of trunk neural crest cell migration in avian embryos using monoclonal antibody HNK–1. Dev. Biol., *115*: 44–53.

Bronner-Fraser, M., and S. E. Fraser. 1988. Cell lineage analysis reveals multipotency of some avian neural crest cells. Nature, *335*: 161–164.

Burnside, B. 1971. Microtubules and microfilaments in newt neurulation. Dev. Biol., *26*: 416–441.

Burnside, B. 1973. Microtubules and microfilaments in amphibian neurulation. Am. Zool., *13*: 989–1006.

Burnside, M. B., and A. G. Jacobson. 1968. Analysis of morphogenetic movements in the neural plate of the newt *Taricha torosa*. Dev. Biol., *18*: 537–552.

Carlson, B. M. 1988. *Patten's Foundations of Embryology*, 5th ed. McGraw-Hill Book Co., New York.

Crelin, E. S. 1974. Development of the nervous system. Clin. Sym., *26*: 1–32.

Dale, B. A., K. A. Holbrook, and P. M. Steinert. 1978. Assembly of stratum corneum basic protein and keratin filaments in macrofibrils. Nature (Lond.), *276*: 223–227.

D'Amico-Martel, A., and D. M. Noden. 1983. Contributions of placodal and neural crest cells to avian cranial peripheral ganglia. Am. J. Anat., *166*: 445–468.

Detwiler, S. R. 1937. Observations upon the migration of neural crest cells, and upon the development of the spinal ganglia and vertebral arches in *Ambystoma*. Am. J. Anat., *61*: 63–94.

Eisen, J. S., P. Z. Myers, and M. Westerfield. 1986. Pathway selection by growth cones of identified motoneurons in live zebra fish embryos. Nature (Lond.), *320*: 269–271.

Ferguson, B. A. 1983. Development of motor innervation of the chick following dorsal-ventral limb bud rotations. J. Neurosci., *3*: 1760–1772.

Goodman, C. S., and M. J. Bastiani. 1984. How embryonic nerve cells recognize one another. Sci. Am., *251*(6): 58–66.

Gordon, R. 1985. A review of the theories of vertebrate neurulation and their relationship to the mechanics of neural tube birth defects. J. Embryol. Exp. Morphol., *89* (Suppl.): 229–255.

Haeckel, E. 1868. *Natürliche Schöpfungsgeschichte*. Berlin.

Harrelson, A. L. et al. 1988. From cell ablation experiments to surface glycoproteins: Selective fasciculation and the search for axonal recognition molecules. *In* S. S. Easter, Jr., K. F. Barald, and B. M. Carlson (eds.), *From Message to Mind*. Sinauer Associates, Sunderland, MA, pp. 96–109.

Harrison, R. G. 1910. The outgrowth of the nerve fiber as a mode of protoplasmic movement. J. Exp. Zool., *9*: 787–846.

Holbrook, K. A. 1983. Structure and function of the developing human skin. *In* L. A. Goldsmith (ed.), *Biochemistry and Physiology of the Skin*. Oxford University Press, New York, pp. 64–101.

Hollyday, M. 1981. Rules of motor innervation in chick embryos with supernumerary limbs. J. Comp. Neurol., *202*: 439–465.

Hopper, A. F., and N. H. Hart. 1985. *Foundations of Animal Development*. Oxford University Press, New York.

Huettner, A. F. 1949. *Fundamentals of Comparative Embryology of the Vertebrates*, rev. ed. Macmillan, New York.

Jacobson, A. G. 1978. Some forces that shape the nervous system. Zoon, *6*: 13–21.

Jacobson, A. G. 1980. Computer modeling of morphogenesis. Am. Zool., *20*: 669–677.

Jacobson, A. G. 1981. Morphogenesis of the neural plate and tube. *In* T. G. Connelly et al. (eds.), *Morphogenesis and Pattern Formation*. Raven Press, New York, pp. 233–263.

Jacobson, A. G. 1984. Further evidence that formation of the neural tube requires elongation of the nervous system. J. Exp. Zool., *230*: 23–28.

Jacobson, A. G., and R. Gordon. 1976. Changes in the shape of the developing vertebrate nervous system analyzed experimentally, mathematically and by computer simulation. J. Exp. Zool., *197*: 191–246.

Jacobson, M. 1978. *Developmental Neurobiology*. Plenum Press, New York.

Karfunkel, P. 1971. The role of microtubules and microfilaments in neurulation in *Xenopus*. Dev. Biol., *25*: 30–56.

Karfunkel, P. 1972. The activity of microtubules and microfilaments in neurulation in the chick. J. Exp. Zool., *181*: 289–302.

Karfunkel, P. 1974. The mechanisms of neural tube formation. Int. Rev. Cytol., *38*: 245–271.

Kelly, D. E., R. L. Wood, and A. C. Enders. 1984. *Bailey's Textbook of Microscopic Anatomy/Bailey's Textbook of Histology*. Williams & Wilkins, Baltimore.

Keshishian, H., and D. Bentley. 1983. Embryogenesis of peripheral nerve pathways in grasshopper legs. I. The initial nerve pathway to the CNS. Dev. Biol., *96*: 89–102.

Keynes, R. J., and C. D. Stern. 1984. Segmentation in the vertebrate nervous system. Nature (Lond.), *310*: 786–789.

Lance-Jones, C., and L. Landmesser. 1980. Motoneurone projection patterns in chick hind limb following partial reversals of the spinal cord. J. Physiol. (Lond.), *302*: 581–602.

Landmesser, L. 1978. The development of motor projection patterns in the chick hind limb. J. Physiol. (Lond.), *284*: 391–414.

Langman, J. 1985. *Medical Embryology*, 5th ed. Williams & Wilkins, Baltimore.

Le Douarin, N. 1982. *The Neural Crest*. Cambridge University Press, Cambridge.

Le Douarin, N. M. et al. 1984. Nuclear, cytoplasmic and membrane markers to follow neural crest cell migration: A comparative study. *In* R. L. Trelstad (ed.), *The Role of the Extracellular Matrix in Development*. Alan R. Liss, New York, pp. 373–398.

Le Douarin, N. M., J. Fontaine-Pérus, and G. Couly. 1986. Cephalic ectodermal placodes and neurogenesis. TINS, *9*: 175–180.

Le Douarin, N. M., and M.-A. Teillet. 1974. Experimental analysis of the migration and differentiation of neuroblasts of the autonomic nervous system and of neuroectodermal mesenchymal derivatives, using a biological cell marking technique. Dev. Biol., *41*: 162–184.

Lehmann, F. 1927. Further studies on the morphogenetic role of the somites in the development of the nervous system of amphibians. J. Exp. Zool., *49*: 93–131.

Letourneau, P. C. 1985. Axonal growth and guidance. *In* G. M. Edelman, W. E. Gall, and W. M. Cowan (eds.), *Molecular Bases of Neural Development*. John Wiley & Sons, New York, pp. 269–294.

Loring, J. F., and C. A. Erickson. 1987. Neural crest cell migratory pathways in the trunk of the chick embryo. Dev. Biol., *121*: 220–236.

Martins-Green, M. 1988. Origin of the dorsal surface of the neural tube by progressive delamination of epidermal ectoderm and neuroepithelium: Implications for neurulation and neural tube defects. Development, *103*: 687–706.

Nichols, D. H. 1986. Mesenchyme formation from the trigeminal placodes of the mouse embryo. Am. J. Anat., *176*: 19–31.

Noden, D. M. 1975. An analysis of the migratory behavior of avian cephalic neural crest cells. Dev. Biol., *42*: 106–130.

Noden, D. M. 1978. The control of avian cephalic neural crest cytodifferentiation. II. Neural tissues. Dev. Biol., *67*: 313–329.

Noden, D. M. 1983. The role of the neural crest in patterning of avian cranial skeletal, connective, and muscle tissues. Dev. Biol., *96*: 144–165.

Noden, D. M., and A. de Lahunta. 1985. *The Embryology of Domestic Animals: Developmental Mechanisms and Malformations*. Williams & Wilkins, Baltimore.

Odland, C. F. 1983. Structure of the skin. *In* L. A. Goldsmith (ed.), *Biochemistry and Physiology of the Skin*. Oxford University Press, New York, pp. 3–63.

Patten, B. M. 1951. *Early Embryology of the Chick*. McGraw-Hill Book Co., New York.

Patten, W. 1922. *Evolution*. Dartmouth College Press, Hanover, NH.

Purves, D., and J. W. Lichtman. 1985. *Principles of Neural Development*. Sinauer Associates, Sunderland, MA.

Ramón y Cajal, S. 1890. Sur l'origine et les ramifications des fibres nerveuses de la moelle embryonaire. Anatomischer Anzeiger, *5*: 609–613.

Rickmann, M., J. W. Fawcett, and R. J. Keynes. 1985. The migration of neural crest cells and the growth cones of motor axons through the rostral half of the chick somite. J. Embryol. Exp. Morphol., *90*: 437–455.

Sauer, F. C. 1935. Mitosis in the neural tube. J. Comp. Neurol., *62*: 377–405.

Saunders, J. W., Jr. 1982. *Developmental Biology. Patterns/Problems/Principles*. Macmillan Press, New York.

Schoenwolf, G. C. 1982. On the morphogenesis of the early rudiments of the developing central nervous system. Scanning Electron Microsc. *I*: 289–308.

Schoenwolf, G. C. 1985. Shaping and bending of the avian neuroepithelium: Morphometric analysis. Dev. Biol., *109*: 127–139.

Schoenwolf, G. C. 1988. Microsurgical analyses of avian neurulation: Separation of medial and lateral tissues. J. Comp. Neurology, *276*: 498–507.

Schoenwolf, G. C., and M. V. Franks. 1984. Quantitative analyses of changes in cell shapes during bending of the avian neural plate. Dev. Biol., *105*: 257–272.

Scott, S. A. 1982. The development of the segmental pattern of skin sensory innervation in the embryonic chick hind limb. J. Physiol. (Lond.), *303*: 203–220.

Teillet, M.-A. 1971. Recherches sur le mode de migration et la différenciation des mélanoblastes cutanés chez l'embryon d'oiseau: Étude expérimentale par la méthode des greffes hétérospécifiques entre embryons de Caille et de Poulet. Ann. Embryol. Morphogenet., *4*: 95–109.

Teillet, M.-A., C. Kalcheim, and N. M. Le Douarin. 1987. Formation of the dorsal root ganglia in the avian embryo: Segmental origin and migratory behavior of neural crest progenitor cells. Dev. Biol., *120*: 329–347.

Teillet, M.-A., and N. Le Douarin. 1970. La migration des cellules pigmentaires étudiée par la methode des greffes hétérospécifiques de tube nerveux chez l'embryon d'Oiseau. C. R. Acad. Sci., *270*: 3095–3098.

Tosney, K. W. 1978. The early migration of neural crest cells in the trunk region of the avian embryo: An electron microscopic study. Dev. Biol., *62*: 317–333.

Tosney, K. W., and L. T. Landmesser. 1985. Growth cone morphology and trajectory in the lumbosacral region of the chick embryo. J. Neurosci., *5*: 2345–2358.

Tosney, K. W., and L. T. Landmesser. 1986. Neurites and growth cones in the chick embryo. J. Histochem. Cytochem., *34*: 953–957.

Vander, A. J., J. H. Sherman, and D. S. Luciano. 1985. *Human Physiology. The Mechanisms of Body Function*, 4th ed. McGraw-Hill Book Co., New York.

Villee, C. A. et al. 1989. *Biology*, 2nd ed. Saunders College Publishing, Philadelphia.

Vincent, M., and J. P. Thiery. 1984. A cell surface marker for neural crest and placodal cells: Further evolution in peripheral and central nervous system. Dev. Biol., *103*: 468–481.

von Baer, K. E. 1828. Ueber Entwicklungsgeschichte der Tiere, Beobachtung, und Reflexion. Königsberg.

Watterson, R. L., and G. C. Schoenwolf. 1984. *Laboratory Studies of Chick, Pig and Frog Embryos*. Burgess Publishing Co., Minneapolis.

Wessells, N. K. 1977. *Tissue Interactions and Development*. Benjamin/Cummings Publishing Co., Menlo Park, CA.

Wessells, N. K., B. S. Spooner, and M. A. Ludueña. 1973. Surface movements, microfilaments and cell locomotion. In *Locomotion of Tissue Cells*, Ciba Foundation Symposium *14*: new series, Elsevier, Amsterdam, pp. 53–82.

Weston, J. A. 1963. A radioautographic analysis of the migration and localization of trunk neural crest cells in the chick. Dev. Biol., *6*: 279–310.

Weston, J. A. 1970. The migration and differentiation of neural crest cells. Adv. Morphogenet., *8*: 41–114.

Yamada, K. M., B. S. Spooner, and N. K. Wessells. 1971. Ultrastructure and function of growth cones and axons of cultured nerve cells. J. Cell Biol., *49*: 614–635.

Yip, J. W. 1986. Specific innervation of neurons in the paravertebral sympathetic ganglia of chick. J. Neurosci., *6*: 3459–3464.

CHAPTER 8

Laying Down the Vertebrate Body Plan: Mesodermal and Endodermal Organ Rudiments

Stephen P. Meier (1947–1986) studied many aspects of morphogenesis and the role of the extracellular matrix in developmental processes. He is perhaps best known for his discovery that the paraxial mesoderm is prepatterned into "somitomeres," which are the precursors of somites in the trunk, but which also represent the fundamental unit of segmentation of the head mesoderm.

We now consider the development of the two inner germ layers. At the completion of gastrulation, the mesoderm is a loosely associated layer of cells between the ectoderm and endoderm. In the amphibian, this mantle of mesoderm nearly envelops the tubular gut, whereas in the amniotes, such as humans and birds, the mesoderm is a sheet on the dorsal surface of the endoderm. Eventually, as a result of formation of the head, tail, and body folds, the mesoderm attains the same spatial relationship to the endoderm in amniotes as it does in the early amphibian, while the endoderm forms a tube that is open mid-ventrally to the yolk sac through much of its development (Fig. 8.1).

Initially, the endodermal tube ends blindly at both ends. The anterior end of the endoderm is in close contact with an inpocketing of the ectoderm, known as the **stomodeum**; these two tissues fuse to form the **oral plate** (see Fig. 8.1). At the posterior extremity, the endoderm contacts the ectodermal

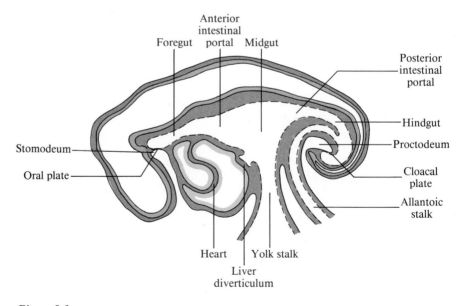

Figure 8.1

Sagittal section through an amniote embryo illustrating the opening of the gut to the yolk sac. At its anterior and posterior ends, the gut ends blindly but will eventually open into the stomodeum and proctodeum. (After Torrey and Feduccia, 1979.)

inpocket, which is called the **proctodeum**, to form the **cloacal plate**. Eventually, these sites of ectoderm-endoderm contact perforate, opening the endodermal tube to the exterior, at the mouth and anus, respectively. Thus, the openings of the gut at either end are lined with ectoderm.

After discussing the development of the mesoderm and endoderm, we turn our attention to the extraembryonic membranes of amniotes. The four extraembryonic membranes that form during embryonic development play important protective and functional roles during embryonic life.

8–1 The Mesoderm

Before neurulation, a mid-dorsal strip of mesoderm separates from the rest of the mesodermal sheet to form the notochord. In some amphibians, such as *Xenopus*, the notochord arises by simply splitting from the rest of the mesoderm. In the chick, the notochord is laid down from cells from Hensen's node as the node moves posteriorly during primitive streak regression. Thus, Hensen's node splits the existing mesoderm in half and leaves the notochord cells in its wake (see Figs. 6.31 and 8.2A and B).

The remaining non-notochordal mesoderm comprises three subdivisions in the trunk of the embryo (see Fig. 8.2C). The mesoderm immediately adjacent to the notochord forms a thick band known as the **paraxial mesoderm.** It will eventually segment into blocks of tissues known as the **somites** (see Fig. 8.2D). Connecting the paraxial mesoderm to the rest of the mesodermal sheet is a thin stalk known as the **intermediate mesoderm** (see Fig. 8.2C and D). It will eventually give rise to components of the urogenital system. Finally, the rest of the trunk mesoderm forms a broad sheet known as the **lateral plate mesoderm** (see Fig. 8.2C). Initially, the lateral plate mesoderm is a loose epithelium, but eventually it splits to form two layers: the **splanchnic mesoderm**, which is closely associated with the endoderm, and the **somatic mesoderm**, which becomes closely applied to the ectoderm

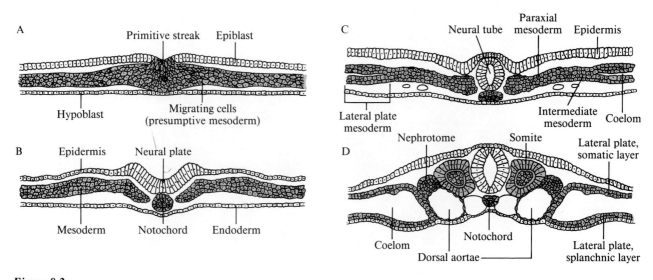

Figure 8.2

Transverse sections through the trunk level of chick embryos illustrating four stages in the development of the mesoderm. Note the progressive separation of the early unsegmented mesoderm into somites, intermediate mesoderm, and lateral plate mesoderm. (After Balinsky, 1981.)

(see Fig. 8.2C and D). The cavity created by the separation of the lateral plate mesoderm into these two layers is the future **coelomic cavity**, in which all the viscera will be suspended.

In the head, the mesoderm is similarly organized into paraxial mesoderm and lateral plate mesoderm, but there is no intermediate mesoderm (Fig. 8.3). We now consider the early derivatives of each of these portions of the mesoderm, beginning with the paraxial mesoderm.

Paraxial Mesoderm—Trunk

Initially, the mesoderm adjacent to the neural tube and notochord is unsegmented. Soon after neurulation is complete, however, the paraxial mesoderm begins to condense into somites (Fig. 8.4). Somites first segregate from the paraxial mesoderm in the anterior end of the embryo adjacent to the myelencephalon. The stretch of paraxial mesoderm that reaches from the most recently formed somite to the end of the primitive streak and is initially not visually segmented is known as the **segmental plate** (see Figs. 8.4 and 8.5). All the future somites in amniotes will develop from the segmental plate in an anteroposterior wave. This segmentation is accomplished by the formation of segmental clefts at the cranial and caudal sides of the prospective somites.

The mechanism of this remarkable transformation of the unsegmented paraxial mesoderm into a series of somites and the specification of where and how many somites will eventually develop is unknown. Recent morphological studies have now shown, however, that the pattern of prospective somites is laid down during gastrulation. This segmented pattern can be seen as circular swirls of the mesodermal cells about a center when viewed by stereoimaging of scanning electron micrographs (Fig. 8.6A); these have

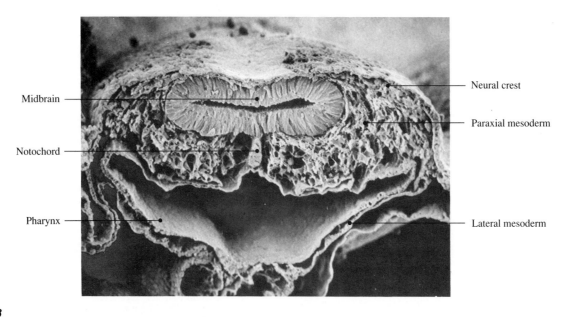

Midbrain

Notochord

Pharynx

Neural crest

Paraxial mesoderm

Lateral mesoderm

Figure 8.3

A scanning electron micrograph (SEM) of a 10-somite quail embryo cut transversely at the level of the midbrain showing the distribution of the paraxial mesoderm, lateral plate mesoderm, and neural crest cells. (Courtesy of D. Noden and K. Reiss.)

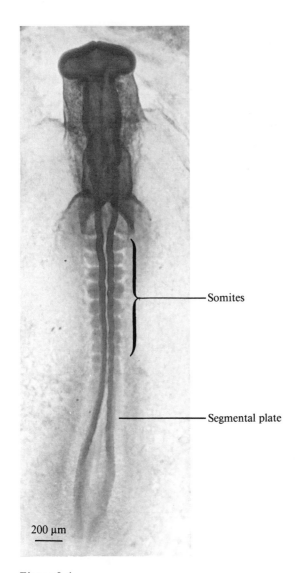

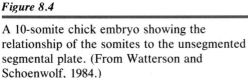

Figure 8.4

A 10-somite chick embryo showing the relationship of the somites to the unsegmented segmental plate. (From Watterson and Schoenwolf, 1984.)

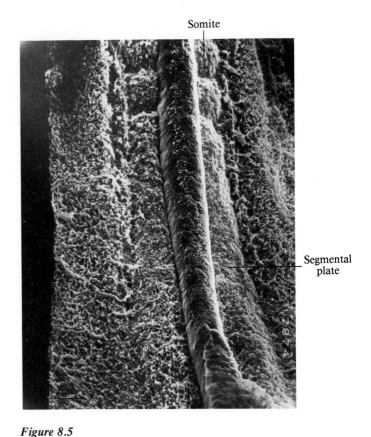

Figure 8.5

A scanning electron micrograph of the posterior portion of the chick trunk with the ectoderm removed to reveal the underlying somites and contiguous segmental plate. (From Jacobson and Meier, 1986.)

been named **somitomeres** by their discoverer (Meier, 1979). The somitomeres are known to turn into somites because when the segmental plate is excised from a chick or mouse and placed in tissue culture, each somitomere can be observed directly to compact and then form cranial and caudal segmental clefts to become a somite (Packard and Meier, 1983).

The paraxial mesoderm is believed to be organized into somitomeres by Hensen's node and the regressing primitive streak, because grafting additional nodes onto a blastoderm results in the induction of more somites (Hornbruch et al., 1979). Conversely, the removal of the node and primitive streak results in cessation of somite formation (Bellairs, 1963). Furthermore, somitomeres always appear in the paraxial mesoderm just behind the regressing node (see Fig. 8.6B). The nature of this induction of the paraxial mesoderm by the regressing node and disappearing primitive streak is still unknown.

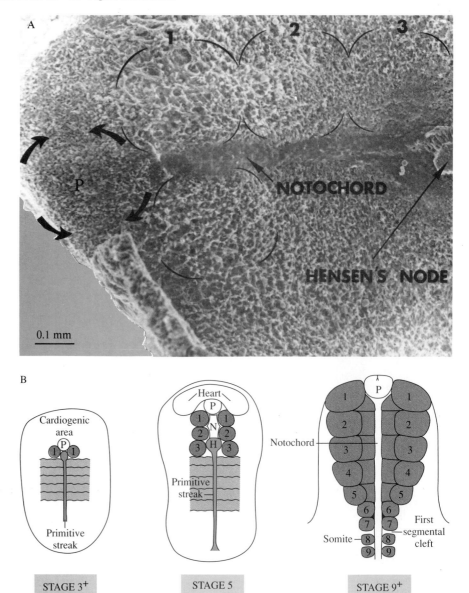

Figure 8.6

Development of somitomeres in the chick embryo. A, SEM of a stage 5 chick embryo with the ectoderm removed to reveal the underlying pattern of mesoderm in the somitomeres. The somitomeres are outlined in brackets. Anterior to the left. B, Diagrams showing the positions and progressive development of somitomeres in the head of a chick embryo. Note that the somitomeres develop in the wake of the receding Hensen's node (H) and primitive streak. Anterior on top. P: prechordal plate; N: notochord. (A, From Meier, 1981; B, After Meier, 1982.)

Initially, the somite in amniotes (birds, reptiles, and mammals) is an epithelial ball with a few loose cells filling the cavity. However, shortly after the somite forms, it begins to break up into three components (Fig. 8.7). The ventromedial wall of the somite loses its epithelial organization and becomes a loose association of single cells known as a mesenchyme. This mesenchymal portion of the somite is known as the **sclerotome**; these cells are the precursors of the vertebrae and ribs. The remaining portion of the somite is an epithelial sheet, but this too undergoes a change. The craniomedial corner of the sheet begins to curl under and extend along its

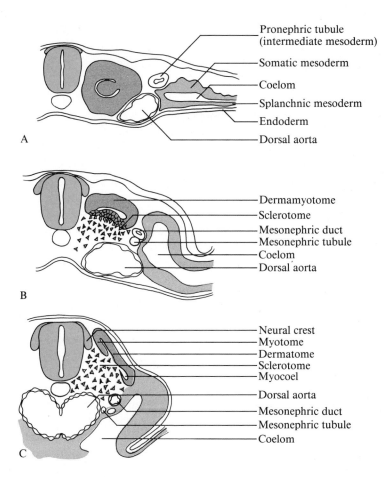

Figure 8.7

Stages in somite formation. A, At first, the somite is an epithelial ball, whose lumen is filled with a few loose mesenchyme cells. B, Within a few hours, the ventral-medial portion of the epithelium begins to disperse to form a mesenchyme, known as the sclerotome. C, The remaining dorsal portion of somitic epithelium is the future dermatome. Its cranio-medial lip turns under and spreads on the undersurface of the dermatome to form the myotome. (After Carlson, 1988.)

underside (Christ et al., 1978; Kaehn et al., 1988). Eventually, a two-layered epithelial sandwich develops. The upper epithelium is the **dermatome**, which will become the dermis of the dorsal skin. The ventral layer is the **myotome**, which will develop into the musculature of the back, abdominal walls, and the limbs.

Sclerotome Derivatives

The cells of the sclerotome surround the neural tube and give rise to the vertebrae (Fig. 8.8). Those cells that surround the dorsal portion of the neural tube form the neural arch of the vertebrum, whereas those mesodermal cells underneath the neural tube give rise to the centrum, to which the intervertebral muscles of the back attach. In the thoracic region, the sclerotome cells form the ribs as well.

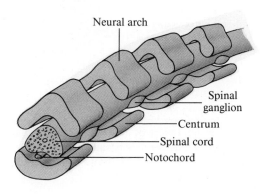

Figure 8.8

Development of the vertebral column. Sclerotome cells surround the neural tube and notochord to form the neural arch and centrum of the vertebrae. (After Carlson, 1988.)

Once the sclerotome cells are in place, they begin to differentiate into cartilage, the tough but pliant connective tissue common in the embryo. Eventually, the cartilage model is replaced by bone.

Myotome Derivatives

Soon after the myotomes form, they begin to expand ventrally (Figs. 8.9 and 8.10). The dorsal portion of the myotome is known as the **epimere**, and the larger ventral expansion is known as the **hypomere** (see Fig. 8.9). These become the **epaxial** and **hypaxial** columns of muscles, respectively. The derivatives of the trunk myotomes are summarized in Figure 8.11. Although many of the deeper muscles retain their metameric arrangement in the adult, especially those that become intervertebral and intercostal (rib) muscles, the more superficial muscles frequently undergo substantial rearrangements and lose the metamerism seen in the embryo or in the more primitive vertebrates.

Paraxial Mesoderm—Head

The paraxial mesoderm adjacent to the brain also becomes organized into somitomeres, but unlike those in the trunk, the paraxial mesoderm of the head eventually disperses without ever dividing into somites. The derivatives of these head somitomeres have been studied in great detail by Noden (1983), who replaced individual chick head somitomeres with identical ones from quail. The derivatives of the grafted quail somitomeres were then identified with Feulgen stain (see p. 265). A summary of the results of his experiments identifying the muscle structures derived from the somitomeres is shown in Figure 8.12. The skeletal structures of the human head are illustrated in Figure 8.13.

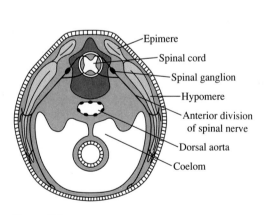

Figure 8.9

The myotome is initially in a dorsal position in the embryo but eventually spreads ventrally and splits into two portions, the dorsal epimere and the ventral hypomere. The dermatome has dispersed by this time, and its derivatives occupy the space between the myotome and presumptive epidermis. (After Hopper and Hart, 1985.)

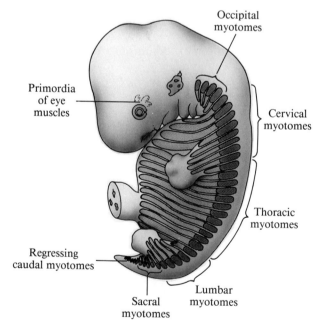

Figure 8.10

Lateral view of a human embryo. The original position of the myotomes is shown as being darker than the regions into which they will ultimately extend. (After Carlson, 1981.)

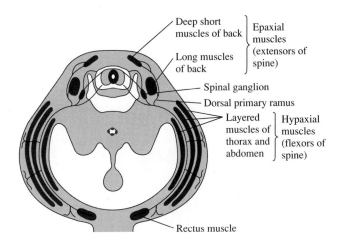

Deep short
muscles of back ⎫ Epaxial
⎬ muscles
Long muscles ⎭ (extensors of
of back   spine)

Spinal ganglion

Dorsal primary ramus

Layered ⎫ Hypaxial
muscles of ⎬ muscles
thorax and ⎭ (flexors of
abdomen   spine)

Rectus muscle

Figure 8.11

Diagram of the primitive muscle masses derived from the epimere and hypomere in a section through the trunk. (After Hopper and Hart, 1985.)

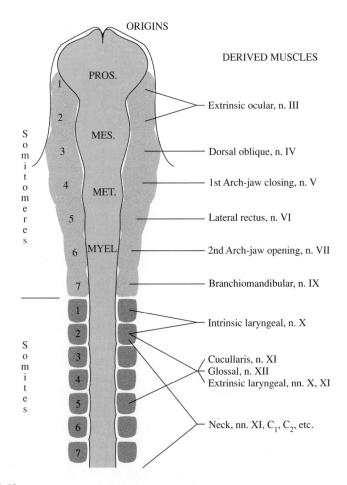

ORIGINS

DERIVED MUSCLES

Extrinsic ocular, n. III

Dorsal oblique, n. IV

1st Arch-jaw closing, n. V

Lateral rectus, n. VI

2nd Arch-jaw opening, n. VII

Branchiomandibular, n. IX

Intrinsic laryngeal, n. X

Cucullaris, n. XI
Glossal, n. XII
Extrinsic laryngeal, nn. X, XI

Neck, nn. XI, C₁, C₂, etc.

Figure 8.12

Summary of the origins of the muscles in the head from the paraxial mesoderm and the nerves that innervate these muscles. A few muscles in the head may also be derived from the lateral plate mesoderm. (After D. M. Noden and A. de Lahunta. *The Embryology of Domestic Animals: Developmental Mechanisms and Malformations.* © 1985, the Williams & Wilkins Co., Baltimore.)

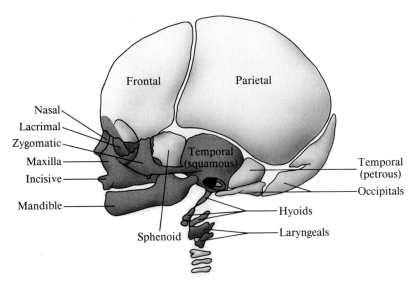

Figure 8.13

Skeletal elements in the head are derived from the paraxial mesoderm (white) and neural crest cells (in color). Only some of the cartilages around the larynx are derived from the lateral plate mesoderm (shaded). (After Noden, 1986. Reprinted from Journal of Craniofacial Genetics and Developmental Biology by permission of Wiley-Liss, a division of John Wiley and Sons, Inc. Copyright © 1986, Wiley-Liss.)

Intermediate Mesoderm

The intermediate mesoderm is present, in the trunk only, as a stalk of cells connecting the somites with the lateral plate mesoderm (see Fig. 8.2). The cells of this region give rise to the organs of the urinary system and part of the reproductive system. We shall defer our discussion of sex determination and development of the gonads until Chapter 16.

The kidneys consist of two elements: (1) the kidney tubules, or nephrons, which are the functional filtering units of the kidney; and (2) the paired kidney ducts, which collect the filtrate from the tubules and transport it to the cloaca. Both the tubules and ducts form from the strip of intermediate mesoderm that runs the length of the trunk.

Three different kidneys develop in successive stages in a spatial and temporal progression from the anterior to posterior end of the embryo, although it should be pointed out that the borders between these kidneys are often a blurred continuum. The first kidney to develop does so from the most anterior intermediate mesoderm and is consequently known as the **pronephric** (from the Greek *pro*, meaning ''before'') **kidney**. This kidney is functional only in the adults of the evolutionarily lowest vertebrates, such as hagfishes and some teleosts, and in the larvae of amphibians. In most vertebrates, it has a brief and transitory appearance and then begins to regress as a second kidney form develops posterior to it. This new kidney is known as the **mesonephric** (from the Greek *mesos*, meaning ''middle'') **kidney**, and it is the functional kidney of the adult amphibians and fishes and in the embryos of higher vertebrates. In birds, reptiles, and mammals, it too begins to regress and is replaced by the **metanephric** (from the Greek *meta*, meaning ''behind'' or ''after'') **kidney** or permanent kidney. Kidney development thus represents a continuum of progressive differentiation from the anterior to posterior end of the embryo that may also reflect the

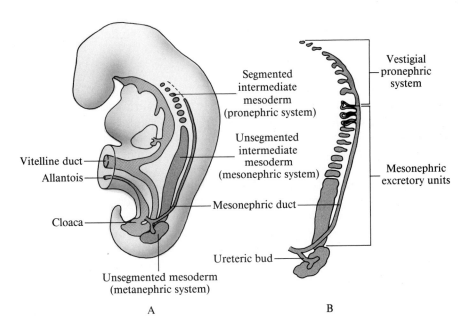

Figure 8.14

A, Diagram of a mammalian
embryo depicting the position of
the kidney-forming mesoderm and
the sequential development of the
pronephric, mesonephric, and
metanephric kidneys. B, Expanded
view of the kidney-forming tissue
showing the position of the three
kidneys and their relationship to
the mesonephric duct. (After
Langman, 1985.)

evolutionary history of the vertebrates. This continuum is schematically
represented in Figure 8.14.

Pronephric Kidney

The typical **pronephric tubule** opens at one end into the coelomic cavity
and at the other end into a duct known as the **pronephric duct** (Fig. 8.15).
The pronephric duct initially is a solid cord of cells, but it eventually hollows
out, and the posterior end then begins to migrate back to the cloaca, where
it fuses, thereby providing an opening to the exterior of the organism. The
migration and guidance of the pronephric duct are considered in Chapter 9.
In those organisms in which this kidney is functional, waste products are
filtered from the blood in a vascular ridge near the pronephric tubule known
as the **glomus**. This filtrate is swept from the coelom into the tubule and
then into the pronephric duct and back to the cloaca.

While the most caudal pronephric tubules are forming, the more
anterior tubules are already beginning to degenerate in most species. The

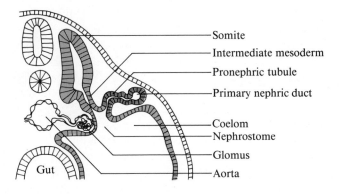

Figure 8.15

Cross-section through an early chick embryo showing the morphology and
position of a primitive pronephric kidney tubule and its connection to both the
coelom and pronephric duct (primary nephric duct). (After Carlson, 1988.)

pronephric duct itself persists, however, and becomes the functional excretory duct of the mesonephric kidney.

Mesonephric Kidney

Like the pronephric tubules, the **mesonephric tubules** form in a craniocaudal progression. They differ from the pronephric tubules chiefly in their relationship to the blood vessels, although the transition between the pronephric and mesonephric tubules is a gradual one, and the demarcation between the two kidneys is not always distinct.

Generally, the mesonephric tubules elongate and develop a concave, cup-shaped end (the **glomerular capsule**), which surrounds a knot of capillaries known as the glomerulus (Fig. 8.16). This glomerulus, which is a tufted segmented outgrowth of the dorsal aorta, differs from the glomus of the pronephric tubule, which is a long vascular ridge. The distal end of the mesonephric tubule connects with the former pronephric duct. Because this duct is now associated with the mesonephric kidney, it is known as the **mesonephric** (or **Wolffian**) **duct**. Filtrate from the glomerulus is passed into the mesonephric tubule, where some ion resorption occurs. The unconcentrated urine then passes into the mesonephric duct.

Like the pronephric kidney before it, the mesonephric kidney also begins to degenerate, and its functions are assumed by the final kidney, the metanephric kidney in birds, reptiles, and mammals. All of the tissue is not lost, however, and the Wolffian duct and tubules will form part of the male reproductive organs and ducts (see Chap. 16).

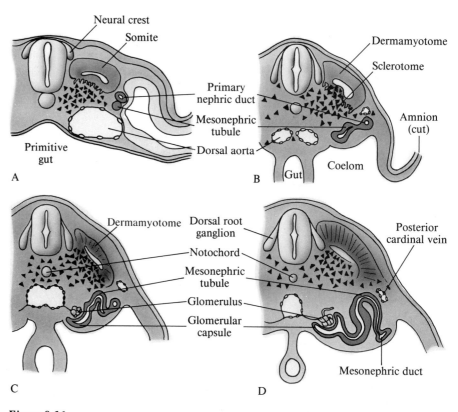

Figure 8.16

Diagrams depicting the progressive development of a mesonephric tubule in successively older chick embryos. The mesonephric tubule opens into the mesonephric duct at its distal end and forms a glomerular capsule at its proximal end. (After Carlson, 1988.)

Metanephric Kidney

The development of the metanephric or permanent adult kidney of the reptiles, birds, and mammals departs from that of the previous two. This kidney develops from two sources: The collecting ducts arise from an outgrowth of the mesonephric duct, and the tubules differentiate from a capsule of nephrogenic mesenchyme posterior to the mesonephric tubules.

The first appearance of the metanephric kidney is an outgrowth of the mesonephric duct near the point where it fuses with the cloaca; this outgrowth is known as the **metanephric diverticulum** (Fig. 8.17A). As it elongates, it forms the future **ureter** and ends in an expanded blind pocket that will ultimately give rise to the **renal pelvis**. As the diverticulum pushes its way into the posterior nephrogenic mesenchyme, this latter mesoderm condenses around the pelvis to form the **metanephric blastema**, from which the tubular excretory units of the kidney arise. The result of the reciprocal interaction between these two tissues is the condensation of the metanephric blastema to form the renal tubules, which is diagrammed in Figure 8.17B, and the dichotomous branching of the metanephric duct to form the system of collecting ducts.

Lateral Plate Mesoderm

Originally, the mesoderm of the lateral plate is a solid mass of cells (see Fig. 8.2). It eventually splits into the somatic and splanchnic mesoderm, and the intervening space becomes the coelomic cavity (see Figs. 8.2 and 8.18). The specific derivatives of the lateral plate mesoderm are dependent upon the axial level from which they arise. Anteriorly, the lateral plate mesoderm forms the heart and the pericardial cavity; further posteriorly, the coelomic space forms the pleural cavity that surrounds the lungs; and finally, most caudally, the coelomic space develops into the peritoneal cavity, in which the stomach, intestines, and the various digestive glands are found.

We shall now examine in more detail the development of the structures within the peritoneal and pericardial cavities.

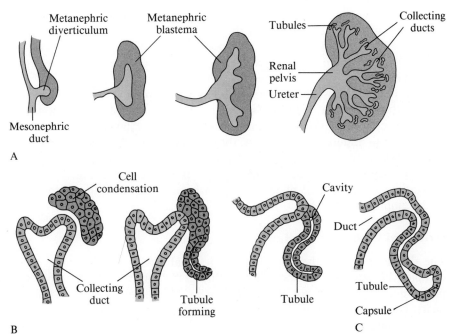

Figure 8.17

A, Diagrammatic representation of stages of metanephric kidney development. An outpocket of the mesonephric duct near the point where it fuses with the cloaca begins to push its way into the intermediate mesoderm. As it does so, the mesoderm condenses around the ureteric bud and is known as the metanephric mesenchyme or capsule. The ureteric bud branches at its distal end to form the collecting ducts of the kidney. B, The metanephric mesenchyme in association with the collecting ducts begins to condense and form the kidney tubules. C, Eventually, the tubules and collecting ducts fuse so that the urine formed in the tubules can make its way into the ducts. (After Wessells, 1977.)

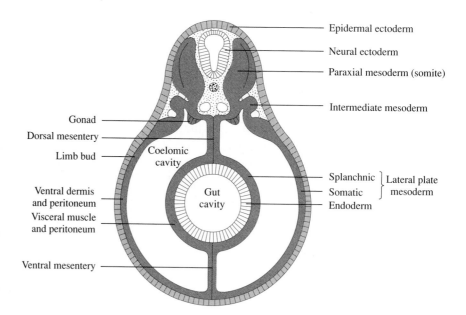

ADULT STRUCTURES FROM LATERAL PLATE MESODERM

GERM LAYER

Epidermal ectoderm
Neural ectoderm
Paraxial mesoderm (somite)
Intermediate mesoderm

Gonad
Dorsal mesentery
Limb bud
Coelomic cavity

Ventral dermis and peritoneum
Visceral muscle and peritoneum

Gut cavity

Splanchnic ⎱ Lateral plate
Somatic ⎰ mesoderm
Endoderm

Ventral mesentery

Figure 8.18

A schematic representation of the distribution of the somatic and splanchnic mesoderm in amphibians and a summary of their derivatives.

Peritoneal Cavity

As we have seen, in the amphibian, the gut is already tubular at the termination of gastrulation, and the mesoderm is sandwiched between the ectoderm and endoderm. The mesodermal mantle begins to split first in the dorsal portion of the lateral plate and gradually spreads ventrally, thus forming the coelomic cavity (see Fig. 7.1). This coelomic cavity expands so that the splanchnic mesoderm becomes tightly apposed to the endoderm and will form the connective tissue and muscle of the gut organs (see Fig. 8.18). It will also give rise to the gonadal ridges (precursors of the ovaries and testes) in the dorsal portion of the coelomic cavity. The somatic mesoderm will form the dermis of the belly skin, some of the ventral body wall musculature, and the lining (the peritoneum) of the peritoneal cavity. It will also contribute all of the connective tissue and appendicular skeleton to the limb buds.

In the amniotes, the gut is still a flat endodermal sheet overlying the yolk at the end of gastrulation (see Fig. 8.2). The mesoderm lies above it and soon splits to form the somatic and splanchnic layers. As the body folds begin to undercut the embryo, the right and left sides of the endoderm and the associated splanchnic mesoderm are brought together and eventually fuse to form a tubular gut surrounded by mesoderm (see Fig. 7.1). The derivatives of the lateral plate mesoderm in amniotes are similar to those of the amphibian.

Pericardial Cavity and the Heart

The development of the heart and pericardial cavity is somewhat more complex because the mesoderm not only gives rise to the linings of the pericardial cavity itself but also contributes to the major components of the heart. By contrast, within the peritoneal cavity, the functional linings of the major organs are of endodermal origin.

The development of the heart and the associated pericardial cavity is relatively simple in the amphibian. At the end of gastrulation, the mesodermal mantles continue to spread anteriorly and ventrally. The future heart will develop at the tips of these mantles.

The first evidence of heart development is the proliferation of cells from the tips of the mesodermal mantles (Fig. 8.19A). These cells form a solid cord (see Fig. 8.19B) that eventually hollows out to form an endothelial tube. This tube is the **endocardium**, or inner lining of the heart. The endocardium is continuous anteriorly and posteriorly with similar endothelial tubes that will become the major blood vessels to and from the heart.

After the formation of the endocardium, the mesodermal mantles meet and fuse in the midline ventral to the endocardium (see Fig. 8.19C). The splanchnic layer of the mesoderm then spreads dorsally on either side of the endocardium, enveloping the tube, and eventually fuses above the heart as well (see Fig. 8.19D and E). The portion of the splanchnic mesoderm that surrounds the heart will become the muscle layer, or **epimyocardium**; the regions above and below the heart where the mesoderm from each side meets and fuses become the mesenteries that suspend the heart in the pericardial cavity. These pericardial mesenteries will eventually perforate and disappear.

The coelomic cavity enveloping the heart then expands and is known as the **pericardial cavity**. The somatic layer of lateral plate mesoderm that lines the pericardial cavity is known as the **pericardium**.

Development of the heart in the amniotes is complicated by the organization of the germ layers as flattened sheets on the yolk. Consequently, the heart in these animals develops initially as paired primordia that then come together and fuse as the body folds swing toward the midline and fuse.

The heart begins to develop in the head-fold stage, when cells of the splanchnic mesoderm detach from the epithelium, migrate along the under-

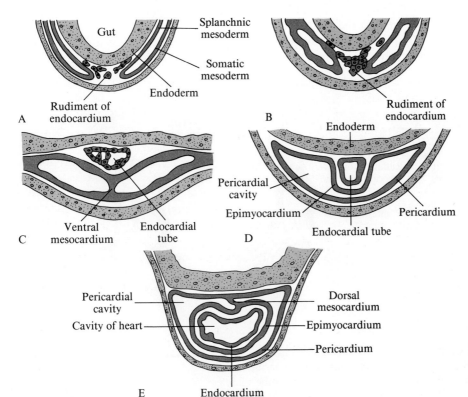

Figure 8.19

Stages in the development of the amphibian heart seen in cross-sections. (After Balinsky, 1981.)

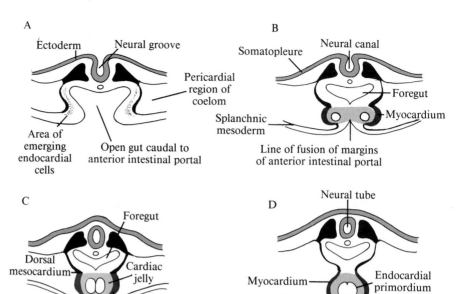

Figure 8.20

Stages in the development of the chick heart from the splanchnic mesoderm. Since the chick develops on top of a large yolk mass, the heart primordia are initially paired and must then come together as the body folds move toward the dorsal midline and fuse. (After Hopper and Hart, 1985.)

lying endoderm, and coalesce near the midline to form thin-walled tubes (Fig. 8.20; DeHaan, 1963). These tubes will eventually fuse and become the endocardium of the heart. The rest of the splanchnic mesoderm meets and fuses in the midline as well, enveloping the endocardium with the future myocardium. The chick heart develops in an anteroposterior wave, so that the paired heart tubes fuse first anteriorly and then progressively fuse caudally (Fig. 8.21). By 30 hours of incubation, the chick has a beating tubular heart, as yet uncompartmentalized. Blood flows into the heart at its posterior end, which is the future atrium, and is pumped into the ventricle and out through the developing aortic arches.

Traditionally, it has been believed that the outer enveloping layer of the heart, the epicardium, develops from the myocardium, but recent work

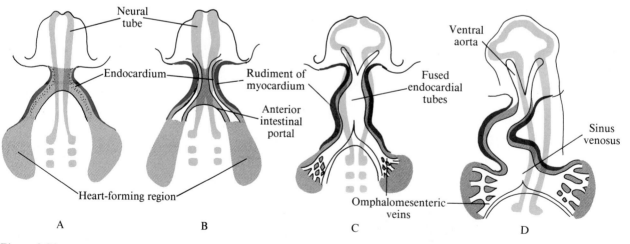

Figure 8.21

Ventral views of a chick embryo showing the origin and anterior-to-posterior fusion of the paired cardiac primordia. Note the bending of the cardiac tube that begins in D. (After DeHaan, 1965.)

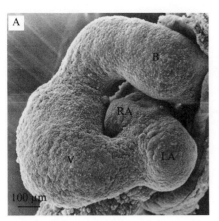
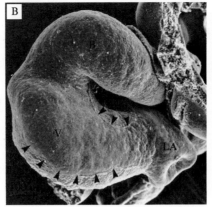
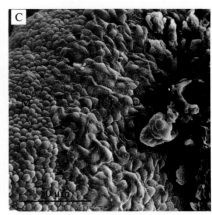

Figure 8.22

Scanning electron micrographs showing the development of the epicardium in the chick embryo. A, An early stage in the development of the heart, which has just begun to bend but is not yet invested with an epicardium. LA: left atrium; RA: right atrium; V: ventricle; B: bulbus cordis. B, At a later stage, the epicardium (arrowheads) begins to spread over the myocardium. Scale bar equals 100 μm. C, Higher magnification showing the spreading epicardium (E). M: myocardium. (From Ho and Shimada, 1978.)

(Fig. 8.22) shows that the epicardium migrates as an epithelial sheet from the mesoderm surrounding the posterior-most region of the heart (the sinus venosus; see Fig. 8.21). As in amphibians, the heart tube of amniotes is surrounded by a coelom that is lined with a pericardium.

The adult heart, which is essentially formed in the chick by five days of incubation, is a complicated four-chamber affair, reflecting the separation of the pulmonary and systemic blood streams in the higher vertebrates. The simple tube is converted into the adult form by at least two important morphogenetic events: (1) It undergoes looping and bending to bring the future atria dorsal to the ventricles; and (2) septa form to divide the tube into chambers.

The formation of the heart loops is diagrammed in Figure 8.23. The forces behind this remarkable event are not well understood but probably involve at least cell shape changes (Manasek et al., 1972) in the myocardial epithelium (see Chap. 9). We do know that most of this event is due to endogenous changes in the heart itself, because the heart tube can be

Figure 8.23

Ventral views of the human heart during progressive stages of cardiac looping. (After Hopper and Hart, 1985.)

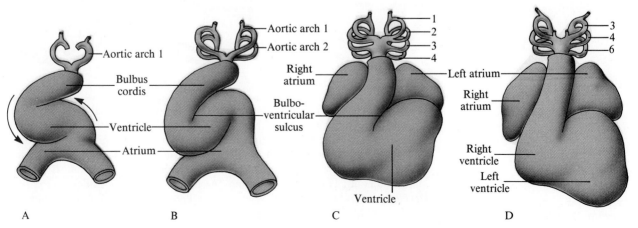

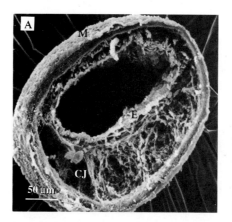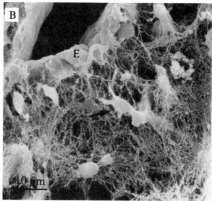

Figure 8.24

A, Scanning electron micrograph of a cross-sectional view of a chick heart illustrating the fibrous cardiac jelly (CJ), which fills the space between the endocardium (E) and myocardium (M). B, Higher magnification showing the first endocardial cushion cells (arrow) detaching from the endocardium and migrating into the cardiac jelly. (From Fitzharris and Markwald, 1982.)

explanted into culture and will undergo looping independent of the rest of the embryo.

Some of the septa and valves that divide the heart into chambers form as a result of the remarkable proliferation of a population of cells that detach from the endocardium and are known as **endocardial cushion cells**. After detaching, they migrate toward the myocardial wall in a large extracellular space filled with a gelatinous matrix called **cardiac jelly** (Fig. 8.24). This matrix is rich in hyaluronic acid (see p. 355 for a discussion of this proteoglycan), which expands the space (Markwald et al., 1978). It also contains other glycoproteins that induce the cushion cells to separate from the endocardium and migrate (Krug et al., 1985). As a result of their

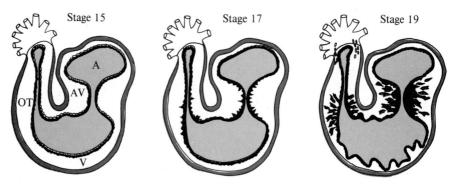

Figure 8.25

Diagrammatic sagittal sections through a chick heart at successive stages, demonstrating the migration of the cardiac cushion cells from the endothelium into the cardiac jelly. The accumulation of these cells produces a bulge in the heart wall that occludes the lumen of the heart and gives rise to the valves and septa. OT: outflow tract; V: ventricle; AV: atrioventricular node; A: atrium. (From Markwald et al., 1984. Reprinted from *The Role of the Extracellular Matrix in Development* by permission of Wiley-Liss, a division of John Wiley and Sons, Inc. Copyright © 1984, Wiley-Liss.)

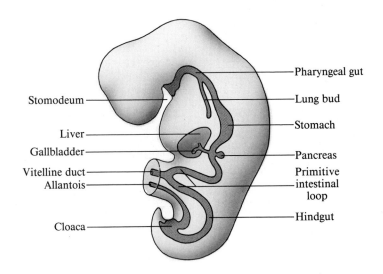

Stomodeum

Liver

Gallbladder

Vitelline duct
Allantois

Cloaca

Pharyngeal gut

Lung bud

Stomach

Pancreas

Primitive
intestinal
loop

Hindgut

Figure 8.26

Schematic representation of the
regions of the gut in a sagittal view
of a human embryo. (After
Langman, 1985.)

migration and proliferation, the cushion cells occlude the lumen of the heart
and eventually become the valves and septa (Fig. 8.25).

8–2 The Endoderm

The simple tube of endoderm becomes expanded at its anterior end to form
the **pharynx** (Fig. 8.26). The rest of the tube can be subdivided into the
future **esophagus, stomach, intestines**, and, finally, the **cloaca**, into which the
ducts of the excretory and reproductive systems will also empty. In the
simple tube, outpockets soon develop that will give rise to all the digestive
organs as well as the respiratory organs. We now examine in more detail
the development of the pharynx, the digestive organs, and the respiratory
system.

Pharynx

The anterior portion of the endodermal tube is expanded to form a flattened
cavity, the pharynx. This cavity has evaginations or outpockets called
pharyngeal pouches that contact the overlying ectoderm in regions where
the overlying ectoderm inpockets to form the **pharyngeal grooves**. In lower
vertebrates, the regions of contact perforate to form the gill slits (or
pharyngeal clefts), which provide a passage for water from the pharynx to
the outside (Fig. 8.27). In higher vertebrates, this perforation does not occur
or occurs only transiently; however, formation of the pharyngeal pouches
and grooves in early development clearly reflects the embryo's phylogenetic
history.

Between the regions of ectodermal and endodermal contact are pockets
of mesenchyme that are derived from the neural crest (Noden, 1975). This
accumulation of mesenchyme forms columns of material known as the
pharyngeal (or visceral) arches (see Figs. 8.27 and 8.28). Blood vessels
develop within this mesenchyme, so that the arches in lower vertebrates
function as respiratory organs. The number of visceral arches varies among
the vertebrates: Fishes have six, amphibians generally have five, and the
amniotes have four.

In the higher vertebrates, in which the respiratory function is assumed
by the lungs, the endodermal pouches are exploited for other roles (Fig.
8.29). The **first** pharyngeal pouches become the paired eustachian tubes and
tympanic cavities of the ears. The epithelium of the **second** pharyngeal

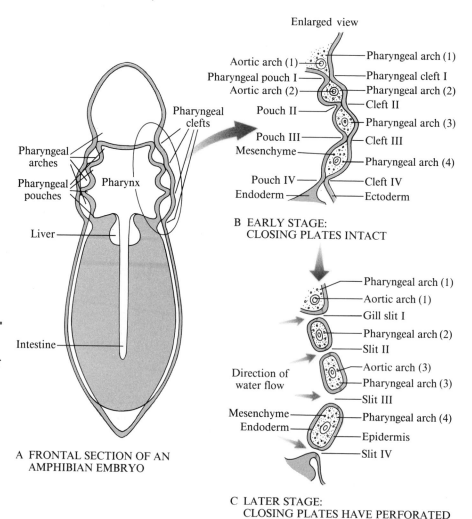

A FRONTAL SECTION OF AN AMPHIBIAN EMBRYO

B EARLY STAGE: CLOSING PLATES INTACT

C LATER STAGE: CLOSING PLATES HAVE PERFORATED

Figure 8.27

Development of the pharynx. A, Frontal section of an amphibian embryo showing the relationship of the pharyngeal pouches (endoderm), the pharyngeal clefts (ectoderm), and the pharyngeal arches (neural crest). The sites of the future gill slits (pharyngeal clefts) are indicated. B, Enlarged view of the pharynx. C, Later stage in the development of the gill slits. (After P. B. Armstrong, by permission.)

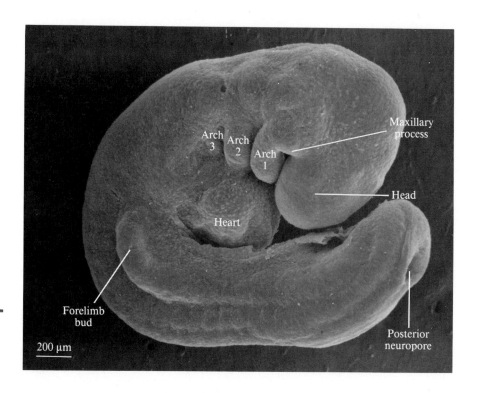

Figure 8.28

Scanning electron micrograph of a 9½-day mouse embryo showing an external view of the pharyngeal arches. (Courtesy of K. Sulik.)

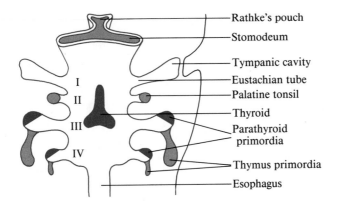

Figure 8.29

Diagrammatic summary of the derivatives of the pharyngeal endoderm in higher vertebrates. I-IV: pharyngeal pouches. (After Torrey and Feduccia, 1979.)

pouch produces the tonsils. The thymus, which is the original source of some lymphocytes, is derived from the fusion of portions of the **third** pharyngeal pouch with a limited contribution from the fourth pouch. Finally, the parathyroid glands are derived from portions of the third and **fourth** pouches.

One other evagination of the pharyngeal endoderm gives rise to a significant structure in the adult: A diverticulum forms in the floor of the pharynx between the first and second pharyngeal pouches that eventually detaches from the endoderm and migrates to the neck, where it will become the **thyroid gland**.

Digestive Tube and Associated Glands

The digestive tube, which consists of the esophagus, stomach, intestines, and cloaca, is lined by a **mucosa**, which is derived from the endoderm and is the tissue of the gut tube that functions in digestion. The smooth muscles and connective tissues of the digestive organs, as well as the visceral peritoneum that covers them, are derived from the splanchnic mesoderm. Associated with the gut and derived from it as outpockets are the digestive glands: the liver, the pancreas, and the gallbladder.

In the human, the earliest evidence of the liver is an endodermal outgrowth from the gut in the region that will become the duodenum. This diverticulum is known as the **hepatic diverticulum** (Fig. 8.30). The liver bud grows into the surrounding mesoderm, which is instrumental in the further development of the liver. The diverticulum develops into highly branched cords of cells. The distal cords become the functional cells of the liver, and the cords closest to the gut become the hepatic ducts (Fig. 8.31). A second outgrowth develops from the diverticulum at the point where the hepatic ducts arise; this is the primordium of the **gallbladder** (see Figs. 8.30 and 8.31). Closer to the gut, a third outgrowth develops from the hepatic diverticulum, which will be the ventral primordium of the **pancreas**. The dorsal primordium of the pancreas arises directly from the duodenum. As the two buds grow and as the intestine begins to twist, the two primordia of the pancreas eventually meet and fuse into one structure.

The glands that develop from the endoderm undergo extensive branching that is under the control of the mesoderm, into which they grow. The control of branching behavior during gland morphogenesis will be discussed in detail in Chapter 12.

Respiratory Organs

The endoderm also gives rise to the organs of respiration, which include the lungs and the trachea. The first indication of lung formation is the appearance of a midventral furrow in the floor of the pharyngeal endoderm

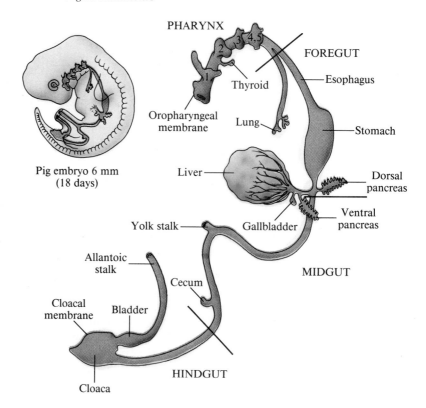

Figure 8.30

Sagittal section through an 18-day-old pig embryo showing the positions of the organs derived from the gut. (After D. M. Noden and A. de Lahunta. *The Embryology of Domestic Animals: Developmental Mechanisms and Malformations.* © 1985, the Williams & Wilkins Co., Baltimore.)

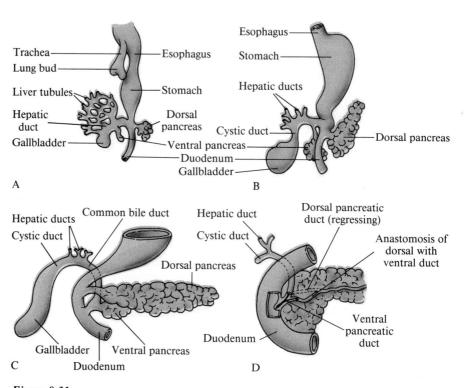

Figure 8.31

Progressive development of the hepatic and pancreatic primordia in the pig. Note how the twisting of the gut brings the dorsal and ventral pancreatic primordia together so that they can fuse. (After B. M. Carlson, 1988. *Patten's Foundations of Embryology*, 5th ed. Reprinted by permission of McGraw-Hill Book Co.)

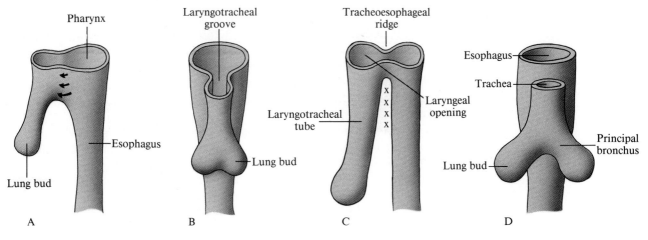

Figure 8.32

The progressive stages in the formation of the trachea and lungs in the human embryo. A, A groove (arrows) forms in the ventral surface of the pharynx, known as the laryngotracheal groove (end of third week, lateral view). B, End of third week, ventral view. C, The groove gradually constricts (x x x x), separating the future esophagus from the trachea. Meanwhile, the most posterior portion of the trachea lengthens posteriorly and expands to form the two lung buds (fourth week, lateral view). D, Fourth week, ventral view. (After D. M. Noden and A. de Lahunta. *The Embryology of Domestic Animals: Developmental Mechanisms and Malformations.* © 1985, the Williams & Wilkins Co., Baltimore.)

at its most posterior border (see Figs. 8.30 and 8.32). This furrow is known as the **laryngotracheal groove.** As this groove deepens, it eventually constricts altogether from the gut, except for a small opening at its cranial end, which will become the future glottis.

The tracheal outgrowth lengthens caudally, parallel to the esophagus, and eventually bifurcates to form the two lobes of the lungs. These lobes will grow into the associated mesodermal mesenchyme, where they branch repeatedly under its influences. The tracheal mesoderm, on the other hand, represses branching (Wessells, 1970). The branching of the endodermally derived lung buds gives rise to the bronchial trees, the tips of which end in blind air sacs. The mesodermal derivatives of the lungs include the muscle, connective tissues, and pleural coverings.

8–3 *The Extraembryonic Membranes*

The development of embryos in an aquatic environment is relatively simple because the surrounding water protects and keeps the embryo moist and allows for rapid exchange of gases and metabolic wastes. In order for vertebrates to move onto land, however, a number of important evolutionary adaptations were necessary. For the reptiles and birds (who lay eggs), **extraembryonic membranes** were acquired that accomplish the vital functions of respiration and removal of embryonic waste. Mammals have lost their egg shells in favor of internal development, but the extraembryonic membranes are retained with basically the same functions.

Chick Extraembryonic Membranes

Because the functions and development of the extraembryonic membranes are similar in most vertebrates, we shall examine the chick as the traditional example. The progressive development of the four extraembryonic membranes is depicted in Figure 8.33. The chick embryo and the fully formed extraembryonic membranes are depicted in Figure 8.33D. The **amnion** surrounds the embryo as a fluid-filled sac and is responsible for keeping the embryo moist and protecting it from shock. Because of the importance of this membrane, the birds, reptiles, and mammals are collectively known as **amniotes**.

The **chorion** is the outermost extraembryonic membrane, which abuts the egg shell. It functions as the major vehicle of gas exchange between the embryo and the environment. The chorion becomes a portion of the placenta in mammals, in which it acquires the functions of supplying nutrition and waste removal in addition to respiration.

The **allantois** is an extension of the gut; in birds and reptiles, it serves as a repository for excretory wastes that, of necessity, must accumulate within the bounds of the shell. The allantois, together with the chorion, also has a respiratory role. It expands enormously until it touches and fuses with the chorion. This fused membrane is known as the **chorioallantoic membrane**. The allantois becomes highly vascularized and establishes a vascular connection with the embryo.

The **yolk sac** is involved in the nutrition of the embryo. It intimately surrounds and digests the yolk, which is then transported to the embryo through the **omphalomesenteric veins**, which develop in the walls of the yolk sac.

The four extraembryonic membranes develop from the lateral extension of the embryonic germ layers. Each extraembryonic membrane is composed of two germ layers: either (1) a combination of ectoderm and somatic mesoderm known as the **somatopleure**, which forms the amnion and chorion; or (2) a combination of endoderm and splanchnic mesoderm known as the **splanchnopleure**, which forms the allantois and yolk sac (see Fig. 8.33C and D). Initially, the boundary between the embryo and the extraembryonic

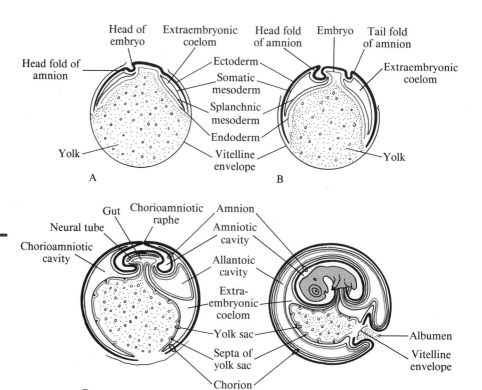

Figure 8.33

Sagittal views of the progressive development of the four extraembryonic membranes from the contiguous embryonic germ layers in a chick embryo. See Color plate 9 for color version. (After B. M. Carlson, 1988. *Patten's Foundations of Embryology*, 5th ed. Reprinted by permission of McGraw-Hill Book Co.)

membranes cannot be discerned. Eventually, the head, tail, and body folds will undercut the embryo, separating it from the yolk, at which time the embryo can be distinguished from the extraembryonic regions.

As the head and tail folds begin to undercut the embryo, and as the embryo's head and tail grow and begin to sink into the yolk, the somatopleure is thrown into folds above the head and tail of the chick, forming the amniotic folds. The amniotic folds of the head and tail approach each other and fuse. This fusion results in two extraembryonic membranes: the outer chorion and inner amnion (see Fig. 8.33).

As the body folds gradually undercut the embryo, they convert the flat splanchnopleure into a tube and lift the embryo off the yolk. As this folding progresses, the splanchnopleure also spreads over the yolk and virtually envelops it. Initially, the tubular gut is formed only at the head and tail regions of the embryo, with the remainder open to the underlying yolk at the midgut. As folding progresses, the gut becomes longer and the region open to the yolk narrows until the embryo and yolk sac are only connected by a narrow **yolk stalk** (Fig. 8.34). The boundary of the splanchnopleure of the gut and of the yolk sac is thus established at the yolk stalk.

The allantois develops as a diverticulum of the gut (see Fig. 8.34). Thus, it is also a derivative of the splanchnopleure. As the allantois extends, it fills with fluid and balloons out over the embryo, filling the space under

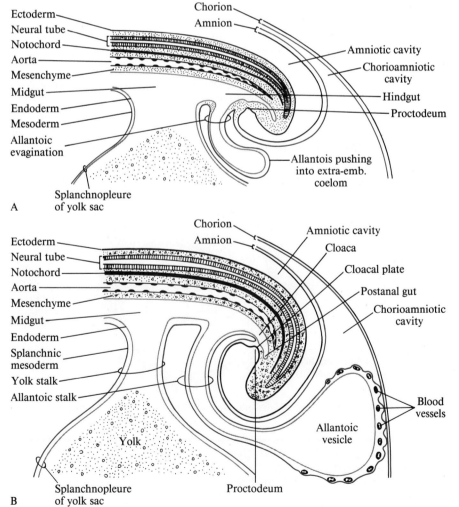

Figure 8.34

Sagittal views of the caudal end of a chick embryo showing in detail the relationship of the germ layers comprising the extraembryonic membranes with the germ layers of the embryo proper. Note in particular the extensive outgrowth of the allantois. See also Color plate 10. (After B. M. Carlson, 1988. *Patten's Foundations of Embryology*, 5th ed. Reprinted by permission of McGraw-Hill Book Co.)

the chorion. As described previously, the allantois eventually fuses with the chorion to form the chorioallantoic membrane.

Human Placenta

In mammalian embryos, the nutritional and excretory functions are assumed by an extraordinary organ, the placenta. The placenta has a dual origin: (1) a maternal component derived from the endometrium of the uterus; and (2) an embryonic component composed of the extraembryonic membranes, primarily the chorion. In this section, we explore the origin of the human extraembryonic membranes and later show how these associate with the uterus to form the placenta (see Boyd and Hamilton, 1970).

Human Extraembryonic Membranes

The four extraembryonic membranes in mammals are named after those in the chick because they have analogous origins and functions (Fig. 8.35). As we shall see, however, they arise in a very different fashion. Unlike the chick, in which the embryo and the extraembryonic membranes develop at the same time, the extraembryonic membranes in the mammal form well ahead of the embryo, reflecting the fact that the mammalian embryo is dependent on the mother for all its needs and must develop an exchange system with the maternal blood supply before embryonic development can proceed very far.

The human zygote gives rise to a blastocyst by day 4, which is composed of two cell layers: an outer epithelium called the trophectoderm, which will ultimately give rise to portions of the placenta, and an inner cell mass, which will form the embryo proper. By day 8, the embryo begins to invade the uterine wall; as it does so, the trophectoderm differentiates into two cell layers, an inner cellular layer known as the **cytotrophoblast** and an outer syncytial layer, the **syntrophoblast**, which develops from the cytotrophoblast wherever the embryo is in contact with the uterus (Fig. 8.36). Thus, as the embryo continues to invade and bury itself in the lining of the uterus, the syntrophoblast is progressively generated around the embryo.

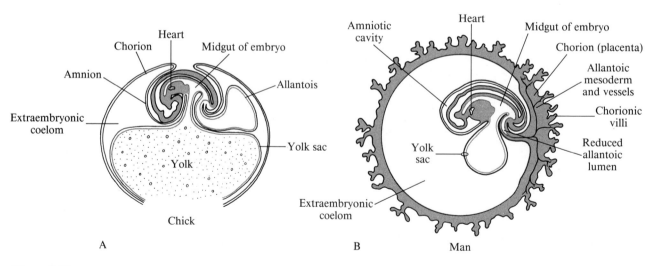

Figure 8.35

Arrangement of the four extraembryonic membranes in early chick (A) and human (B) embryos. See also Color plate 11. (After B. M. Carlson, 1988. *Patten's Foundations of Embryology*, 5th ed. Reprinted by permission of McGraw-Hill Book Co.)

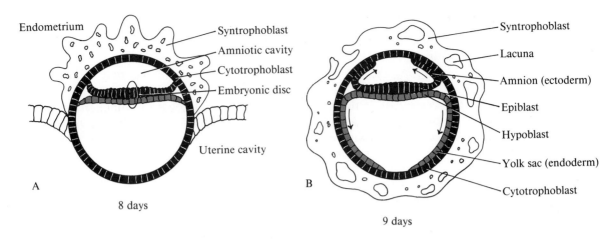

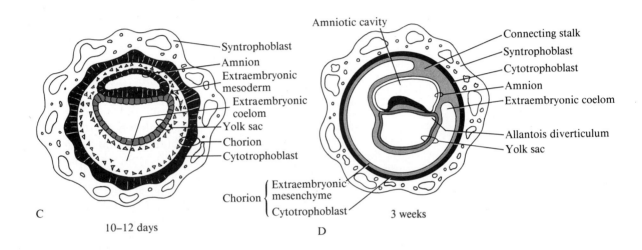

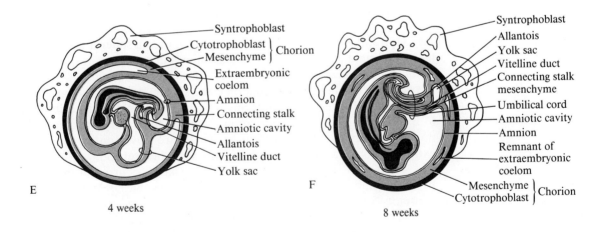

Figure 8.36

Diagrams of the development of the extraembryonic membranes in 8-day (A), 9-day (B), 12-day (C), 3-week (D), 4-week (E), and 8-week (F) human embryos. Note in particular the germ layer components of each extraembryonic membrane and the changes in the relationships of the membranes to give rise to the umbilical cord and placenta. See also Color plate 12. (After Tuchmann-Duplessis et al., 1972.)

During invasion, the inner cell mass forms the epiblast and hypoblast. Gastrulation does not begin at this point, however, but is delayed until day 16.

As the blastocyst invades the uterine wall, the first of the extraembryonic layers begins to take form. The edges of the epiblast spread on the inner wall of the cytotrophoblast above the embryo until it forms a roof over the epiblast (see Fig. 8.36B; Luckett, 1975). This roof is the future ectoderm of the **amnion**, and it encloses the **amniotic cavity**. This layer is so named by analogy to the chick amnion, because eventually it will be composed of both ectoderm and mesoderm that are continuous with the embryonic germ layers and will have a similar function. Its development is radically different from the chick, however, and the extraembryonic germ layers are not yet contiguous with those of the embryo.

Similarly, the hypoblast spreads below the embryo and forms the future endoderm of the **yolk sac**. The human embryo has no yolk food reserves, however, and this is a remarkable example of a developmental process that recapitulates evolutionary history. The yolk sac may have an early absorptive function but is vestigial during later development.

By day 12, many changes have occurred in the extraembryonic layers, whereas there is virtually no change in the embryo itself. This is because an exchange system must develop between the embryo and mother before the embryo itself can grow much larger. The most obvious change is that the entire embryo is embedded in the uterine wall and is now completely surrounded by a syntrophoblast layer. As the syntrophoblast continues to erode its way through the uterine wall, it branches and forms sinuous processes that then anastomose to enclose cavities known as **lacunae** (see Fig. 8.36B and C). These fill up with maternal blood and lymph that have leaked from eroded endometrial blood vessels and glands. The contents of the lacunae provide the initial nourishment and oxygen to the embryo.

The other major change by day 12 is the appearance of the extraembryonic mesoderm. Until recently, it was thought that this layer was derived from the inner trophoblast wall. Recent studies by Luckett (1978) of fixed human tissue and carefully staged Rhesus monkey embryos suggest that the extraembryonic mesoderm is derived from the epiblast via the primitive streak (see Fig. 8.36C). As the extraembryonic mesoderm begins to stream out of the caudal end of the embryo, it spreads along the inner surface of the cytotrophoblast. At a slightly later stage, the mesoderm also surrounds the ectoderm and endoderm of the amnion and yolk sac, respectively, to form the mesodermal component of these two extraembryonic layers. Between the two layers of mesoderm is the so-called **extraembryonic coelom**.

By day 16, the embryo begins to gastrulate, as described in Chapter 6. By the completion of gastrulation (day 21), the embryo is composed of three germ layers (see Fig. 8.36D), which are now contiguous with the extraembryonic layers that formed many days beforehand.

By day 21, several additional changes have occurred in the extraembryonic membranes. For example, the extraembryonic coelom has expanded enormously so that the future embryo is no longer attached to the cytotrophoblast except at the future posterior end of the embryo, where a mass of mesoderm forms the **connecting stalk**. Ultimately, the connecting stalk will contribute to a portion of the connective tissue of the umbilical cord. In addition, a bulge from the yolk sac endoderm pushes into the connecting stalk mesoderm and is called the **allantois**, by analogy to the similar structure in the chick. It will give rise to the vasculature of the umbilical cord and connect the fetal blood vessels with those of the placenta.

By day 21, then, four extraembryonic membranes are clearly established: the amnion, composed of extraembryonic ectoderm and mesoderm;

the chorion, composed of cytotrophoblast (ectoderm by analogy with the chick, although it is never contiguous with the embryonic ectoderm) and mesoderm; and the yolk sac and allantois, composed of extraembryonic endoderm and mesoderm.

By day 24, the neural tube undergoes flexure, as it does in chick development, and the amniotic cavity expands (see Fig. 8.36E). As a result, the amnion begins to surround the embryo, obliterating the extraembryonic coelom and eventually fusing with the chorion (see Fig. 8.36F). As the amnion envelops the embryo, it compresses the yolk sac and allantois together between the folds of the amnion, thereby forming the umbilical cord (see Fig. 8.36F).

In the preceding discussion, we concentrated on the development of the extraembryonic membranes. We now show how these membranes, in association with the uterus, form the placenta.

Placenta Development

The placenta is composed of two sources of tissue: the chorion and a portion of the lining of the uterus known as the decidua. We shall examine the development of each.

As the embryo burrows into the uterus, owing to the invasive behavior of the trophoblast, it erodes the endometrium and generates pools of maternal blood that fill the lacunae (Fig. 8.37A). Shortly thereafter, the cytotrophoblast extends fingers of tissue through the syntrophoblast. These are called the **primary villi** (see Fig. 8.37B), which are the forerunners of the placental villi. The cytotrophoblast eventually penetrates the syntrophoblast and spreads over the endometrial wall, thereby anchoring the villus to the uterus. Eventually, the villi are invaded by chorionic mesoderm and

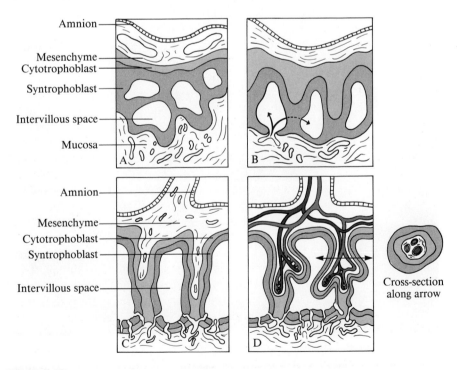

Figure 8.37

Diagrams of the development of the chorionic villi in 13-day (A), 15-day (B), 18-day (C), and 21-day (D) old placentas. (After Tuchmann-Duplessis et al., 1972.)

are now known as **secondary villi** (see Fig. 8.37C). Finally, the villi form many branches that increase the amount of surface area that is exposed to the maternal blood in the greatly expanded lacunae (now known as the **intervillous spaces**). These are the **tertiary villi** (see Fig. 8.37D), which cover the chorion and give it a bushy appearance (Fig. 8.38). The tertiary villi have well-developed blood vessels that fuse with the embryonic vascular system. Thus, nutrients from the maternal blood in the intervillous spaces diffuse across the villi into the fetal blood, which is circulated by the fetal heart. Exchange between the mother and fetus is now well established.

Meanwhile, the endometrium (known as the **decidua** during pregnancy), in which the embryo is embedded, also changes as development proceeds (Fig. 8.39). The portion of the uterine decidua that is unperturbed by the invading embryo is known as the **parietal decidua**. The decidua that lies beneath the embryo is known as the **basal decidua**, and that portion of the

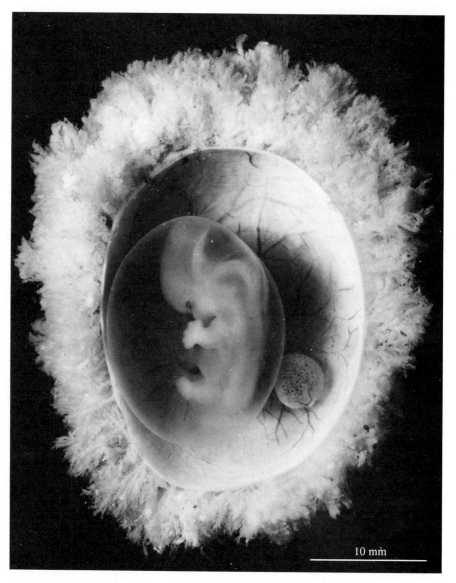

10 mm

Figure 8.38

The chorion has been broken open to expose a 7-week old fetus in its amnion. The yolk sac is the vesicle to the right. (Embryo No. 8537A from the Carnegie Collection. Photograph courtesy of R. O'Rahilly.)

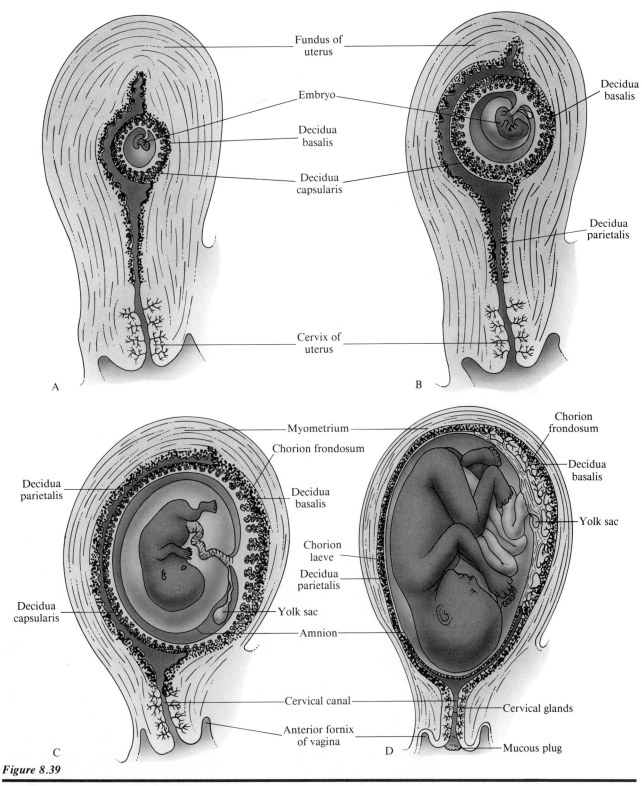

Figure 8.39

Diagrams showing the relationship of the fetus and its extraembryonic membranes to the uterus at three weeks (A), five weeks (B), eight weeks (C), and five months (D). Note in particular the gradual loss of chorionic villi beneath the capsular decidua (decidua capsularis), the progressive expansion of the capsular decidua, and its eventual fusion with the parietal decidua (decidua parietalis). The basal decidua (decidua basalis) comprises the maternal portion of the placenta, and the chorion frondosum makes up the fetal component of the placenta. Decidua represented by stipples in color. (After Carlson, 1988.)

decidua that covers the embryo and lines the uterine cavity is the **capsular decidua**. As the embryo grows, it pushes the capsular decidua into the lumen and obliterates the uterine cavity (see Fig. 8.39C). The capsular decidua eventually fuses with the parietal decidua by the fourth month (see Fig. 8.39D). As the capsular decidua is pushed out and thins, it loses much of its blood supply. As a result, the chorionic villi in this region eventually are lost; this portion of the chorion is known as the **chorion laeve** (see Fig. 8.39D). A discoid portion of the chorion around the end of umbilical cord, in association with the highly vascularized basal decidua, retains its lawn of villi; this region is known as the **chorion frondosum** (from the Latin *frons*, meaning "foliage"). The chorion frondosum forms the fetal component of the placenta, and the basal decidua makes up the maternal component.

Full-Term Placenta

In the mature placenta (Fig. 8.40), the chorionic villi, composed of syntrophoblast, cytotrophoblast, and mesenchyme filled with blood vessels, are suspended in large pools of blood that fill the intervillous spaces. Nutrients and oxygen diffuse from the maternal blood across the villi and into the fetal blood, which is being continually pumped by the fetal heart. Likewise, fetal metabolic wastes and carbon dioxide pass out of the fetal blood and into the intervillous space.

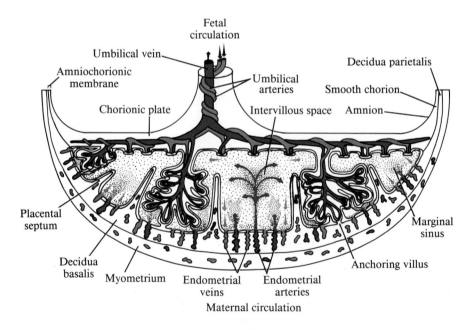

Figure 8.40

A schematic drawing through a mature placenta depicting the blood circulation in the fetal placenta (chorionic villi) and the maternal placenta (basal decidua). As the highly oxygenated blood spurts into the intervillous spaces, it flows around the villi, where exchange of O_2 and nutrients takes place with the fetal blood in the chorionic villi. CO_2 and waste metabolites are removed from the fetal blood into the intervillous space, and this blood settles into the endometrial veins and passes to the maternal circulation. Note that in the full-term placenta the chorionic villi dangle in large intervillous spaces that have been excavated from the decidua. See also Color plate 13. (After Moore, 1988.)

The blood in the intervillous space is constantly being recirculated because of differences in hydrostatic pressure. The blood from the placental spiral arteries is propelled into the intervillous space at very high pressure (70 mm Hg). As it squirts out of the arteries, it bathes the villi in oxygen and nutrient-rich blood. The pressure of the intervillous space and the veins that drain these cavities is only 10 mm Hg, so that after fetal exchange, the oxygen-depleted blood settles into the veins and is carried away into the maternal circulation.

As development proceeds and the efficiency of exchange accelerates to accommodate increased fetal demand, there is a progressive loss of the cell layers covering the villi. Near term, the villi consist of little more than fetal blood vessels with a cover of syntrophoblast dangling in pools of maternal blood. But never, even at this extreme, do the maternal and fetal blood supplies merge.

8–4 Chapter Synopsis

In this chapter, we considered the tissue and organ derivatives of the mesoderm and endoderm. The mesoderm in the trunk can be divided into three segments: (1) the dorsal-most paraxial mesoderm; (2) the intermediate mesoderm; and (3) the lateral plate mesoderm. The paraxial mesoderm forms blocks of tissue, the somites, that ultimately give rise to the vertebrae and ribs, the dermis of the skin, and the axial musculature. The intermediate mesoderm forms the kidney tubules and the ureter and also contributes to portions of the reproductive system. The lateral plate mesoderm splits at an early stage to form the somatic mesoderm in contact with the epidermis and the splanchnic mesoderm associated with the endoderm separated by a cavity, the coelom. The exact derivatives of the lateral plate mesoderm depend upon the axial level examined.

In the head, the mesoderm comprises two segments: (1) the paraxial mesoderm and (2) the lateral plate mesoderm. These give rise to the muscle, bone, and other connective tissues of the head and neck.

The endoderm produces most of the functional portion of all the organs of the gut, including the esophagus, stomach, intestines, and cloaca. In addition, the endoderm expands and branches to form gut-associated glands: the liver, pancreas, and gallbladder. Finally, the endoderm branches to produce the organs of respiration, the paired lungs and the trachea. All of the endoderm-derived organs develop in association with the mesoderm, which contributes the connective tissue component.

Besides giving rise to embryonic tissues and organs, the germ layers also form the extraembryonic membranes, which accomplish a variety of functions including nutrition, respiration, and protection of the embryo. In the mammal, these membranes are adapted to construct a remarkable structure, the placenta.

We have now completed a review of the major organs and tissues that develop from the three germ layers of the vertebrate embryo and are in a position to explore the mechanisms that are responsible for their development. In Chapter 9, we shall examine the mechanisms that are responsible for shaping the tissues and organs. In Part Six, we shall examine the differentiation of the varied cell types that make up the embryo.

References

Balinsky, B. I. 1981. *An Introduction to Embryology*, 5th ed. Saunders College Publishing, Philadelphia.

Bellairs, R. 1963. The development of somites in the chick embryo. J. Embryol. Exp. Morphol., *11*: 697–714.

Boyd, J. D., and W. J. Hamilton. 1970. *The Human Placenta*. W. Heffer and Sons, Ltd., Cambridge.

Carlson, B. M. 1981. *Patten's Foundations of Embryology*. McGraw-Hill Book Co., New York.

Carlson, B. M. 1988. *Patten's Foundations of Embryology*, 5th ed. McGraw-Hill Book Co., New York.

Christ, B., H. J. Jacob, and M. Jacob. 1978. On the formation of the myotomes in avian embryos. An experimental and scanning electron microscope study. Experientia, *34*: 514–516.

DeHaan, R. L. 1963. Migration patterns of the precardiac mesoderm in the early chick embryo. Exp. Cell Res., *29*: 544–560.

DeHaan, R. L. 1965. Morphogenesis of the human heart. *In* R. L. DeHaan and H. Ursprung (eds.), *Organogenesis*. Holt, Rinehart and Winston, New York, pp. 377–419.

Fitzharris, T. P., and R. R. Markwald, 1982. Cellular migration through the cardiac jelly matrix: A stereoanalysis by high-voltage electron microscopy. Dev. Biol., *92*: 315–329.

Ho, E., and Y. Shimada. 1978. Formation of the epicardium studied with the scanning electron microscope. Dev. Biol., *66*: 579–585.

Hopper, A. F., and N. H. Hart. 1985. *Foundations of Animal Development*. Oxford University Press, New York.

Hornbruch, A., D. Summerbell, and L. Wolpert. 1979. Somite formation in the early chick embryo following grafts of Hensen's node. J. Embryol. Exp. Morphol., *51*: 51–62.

Jacobson, A. G., and S. Meier. 1986. Somitomeres: The primordial body segments. *In* R. Bellairs, P. A. Ede, and J. W. Lash (eds.), *Somites in Developing Embryos*. Plenum Press, New York, pp. 1–16.

Kaehn, K. et al. 1988. The onset of myotome formation in the chick. Anat. Embryol., *177*: 191–201.

Krug, E. L., R. B. Runyan, and R. R. Markwald. 1985. Protein extracts from early embryonic hearts initiate cardiac endothelial cytodifferentiation. Dev. Biol., *112*: 414–426.

Langman, J. 1985. *Medical Embryology*, 5th ed. Williams & Wilkins, Baltimore.

Luckett, W. P. 1975. The development of primordial and definitive amniotic cavities in early Rhesus monkey and human embryos. Am. J. Anat., *144*: 149–168.

Luckett, W. P. 1978. Origin and differentiation of the yolk sac and extraembryonic mesoderm in presomite human and Rhesus monkey embryos. Am. J. Anat., *152*: 59–98.

Manasek, F. J., M. B. Burnside, and R. E. Waterman. 1972. Myocardial cell shape changes as a mechanism of embryonic heart looping. Dev. Biol., *29*: 349–371.

Markwald, R. R. et al. 1978. Structural analyses on the matrical organization of glycosaminoglycans in developing endocardial cushions. Dev. Biol., *62*: 292–316.

Markwald, R. R. et al. 1984. Use of collagen gel cultures to study heart development: Proteoglycan and glycoprotein interactions during formation of endocardial cushion tissue. *In* R. L. Trelstad (ed.), *The Role of the Extracellular Matrix in Development*. Alan R. Liss, New York, pp. 323–350.

Meier, S. 1979. Development of the chick embryo mesoblast. Formation of the embryonic axis and establishment of the metameric pattern. Dev. Biol., *73*: 25–45.

Meier, S. 1981. Development of the chick embryo mesoblast: Morphogenesis of the prechordal plate and cranial segments. Dev. Biol., *83*: 49–61.

Meier, S. 1982. Development of segmentation in the cranial region of vertebrate somites. SEM/1982/III: 1269–1282.

Moore, K. L. 1988. *The Developing Human*, 4th ed. W. B. Saunders, Philadelphia.

Noden, D. M. 1975. An analysis of the migratory behavior of avian cephalic neural crest cells. Dev. Biol., *42*: 106–130.

Noden, D. M. 1983. The embryonic origins of avian cephalic and cervical muscles and associated connective tissue. Am. J. Anat., *168*: 257–276.

Noden, D. M. 1986. Origins and patterning of craniofacial mesenchyme tissues. J. Craniofacial Gen. Dev. Biol., *2* (Suppl.): 15–31.

Noden, D. M., and A. de Lahunta. 1985. *The Embryology of Domestic Animals: Developmental Mechanisms and Malformations*. Williams & Wilkins, Baltimore.

Packard, D. S., Jr., and S. Meier. 1983. An experimental study of the somitomeric organization of the avian segmental plate. Dev. Biol., *97*: 191–202.

Patten, B. M. 1951. *Early Embryology of the Chick*. McGraw-Hill Book Co., New York.

Torrey, T. W., and A. Feduccia. 1979. *Morphogenesis of the Vertebrates*. John Wiley & Sons, New York.

Tuchmann-Duplessis, H., G. David, and P. Haegel. 1972. *Illustrated Human Embryology*, Vol. 1. *Embryogenesis*. Masson et Cie, Paris.

Watterson, R. L., and G. C. Schoenwolf. 1984. *Laboratory Studies of Chick, Pig and Frog Embryos*. Burgess Publishing Co., Minneapolis.

Wessells, N. K. 1970. Mammalian lung development: Interactions in formation and morphogenesis of tracheal buds. J. Exp. Zool., *175*: 455–466.

Wessells, N. K. 1977. *Tissue Interactions and Development*. Benjamin/Cummings Publishing Co., Menlo Park, CA.

CHAPTER *9*

The Cellular Basis of Morphogenesis

Michael Abercrombie (1912–1979), seated on the left, was a British developmental biologist whose early work was on the behavior of the primitive streak and Schwann cell migration on peripheral nerves. However, his most indelible mark was made through his work on cell behavior in tissue culture, rather than in experimental embryology. He greatly expanded our understanding of cell locomotion, not only by observing cells in culture directly but by introducing quantitative methods and experimental procedures. The phenomenon of contact inhibition, which he and his colleagues characterized in detail, remains today an important concept in the understanding of morphogenetic cell behavior in the embryo. (Courtesy of R. Bellairs.)

At the completion of the blastula stage, the embryo is little more than a mass of cells, in many species forming a sphere and in others a sheet, bearing no resemblance to the future organism. With the onset of gastrulation, cells actively rearrange themselves and take up positions approximating those that they will occupy in the adult: Cells that remain on the outside of the embryo (the ectoderm) will become the skin; other cells organize into a tube in the center of the embryo (the endoderm) to become the gut and respiratory system; and those cells that lie in between (the mesoderm) will give rise to muscle, skeleton, connective tissue, and the urogenital system. Embryonic form becomes increasingly more complex as the various organ rudiments later emerge. As a result of various cell movements, the diverse organs and tissues of the embryo are formed, and the future external and internal form of the adult is achieved.

The process by which embryonic form and structure are achieved is called **morphogenesis** (from the Greek, meaning "generation/organization of form"). It involves a rearrangement of cell positions relative to each other, as well as the shaping and alignment of individual cells. The cell is the basic unit of morphogenesis, and we shall see in this chapter how individual cells or groups of cells can change the shape of the embryo and the composition of its various tissues and organs.

There are several ways in which changes in the form of embryonic structures are generated. Cells can literally migrate from one region to another in the embryo. We have already observed several examples of this sort of movement, including outgrowth of neurons and migrations of the neural crest in vertebrate embryos and primary mesenchyme cells during sea urchin gastrulation. In some cases, this locomotion occurs among cells that are tightly bound together to form epithelia. These sheets can migrate or spread; in addition, individual cells within the sheet can change shape, producing a deformation in the epithelium. An example of morphogenesis due to cell shape changes is the rolling up of the neural tube during neurulation owing to the conversion of columnar cells to pyramidal cells, which we discussed in detail in Chapter 7. These changes are collectively referred to as **morphogenetic movements**.

In addition to morphogenetic movements, embryonic cells can produce a change in the form of the developing embryo by two other means. Cells in one region of the embryo can begin to proliferate faster in relationship to surrounding cells and produce an outgrowth. The limb is an example of a structure that develops as a result of **selective cell proliferation**. Alternatively, cells in specific regions will predictably die. Such **morphogenetic cell death** can also sculpt the embryonic form. The individual digits of the hand

and foot emerge from a paddle-shaped appendage as a result of the death of cells between the presumptive digits.

In this chapter, we examine the cellular basis for morphogenetic cell movement, selective cell growth, and morphogenetic cell death.

9–1 Morphogenetic Movements

Modes of Epithelial Morphogenesis

At the onset of gastrulation, which is one of the first major sets of embryonic morphogenetic events, the embryo consists of a series of epithelial layers. Thus, the earliest changes in embryonic form and organization involve epithelial cells. Later, during the establishment of the major organ rudiments, epithelia undergo a variety of folding and spreading motions that continue to sculpt the embryo. Transformations in epithelia therefore represent major forces in shaping the embryo.

Cells within an epithelium are tightly bound to one another with cell junctions and have a distinct polarity. The **apical** side of an epithelial cell generally faces a lumen or the outside of the animal, whereas the **basal** side rests on a basal lamina and is associated with a mesenchymal cell population. Epithelial cells are held together by a series of junctions that form a **junctional complex** near their apical surface (Fig. 9.1). These junctions include (1) the most apical **zonula occludens** or **tight junctions**, so named because these junctions prevent any molecules from diffusing into the paracellular spaces; (2) **zonula adherens** or **adherens junctions**, which encircle the apical borders of the epithelial cells; and (3) the **maculae adherens** or **desmosomes**, which are punctate weld spots along the lateral sides of epithelial cells. Thus, the forces of morphogenesis can be transduced across the entire epithelium, even if they occur in only a few of the epithelial cells, because of the tight associations among all the component cells. **Gap junctions** also occur on the lateral surfaces of epithelial cells, but they have a more important role in cell-cell communication than in adhesion.

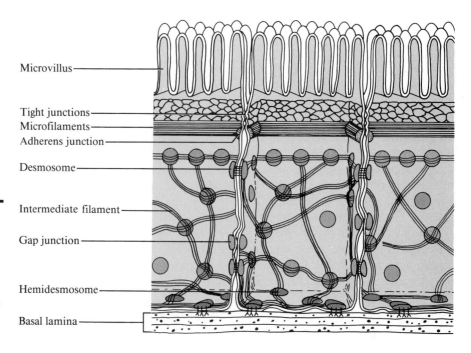

Microvillus——

Tight junctions——
Microfilaments——
Adherens junction——

Desmosome——

Intermediate filament——

Gap junction——

Hemidesmosome——

Basal lamina——

Figure 9.1

Diagram of the principal types of cell junctions between epithelial cells, represented here in intestinal epithelial cells. Note the association of microfilaments with the adherens junctions and the association of intermediate filaments with the desmosomes. (After Darnell et al., 1990.)

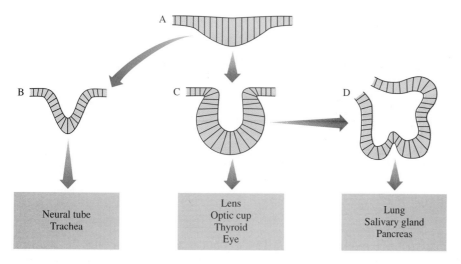

Figure 9.2

A diagrammatic representation of the principal changes in cell shape that accompany epithelial morphogenesis. A, The elongation of epithelial cells, called palisading, generally precedes any other morphogenetic event in epithelia. This thickened epithelium can then either fold along a line to form a groove (B) or inpocket to generate a pouch (C). In some instances, the pouch can also branch (D). (After Hilfer and Searls, 1986.)

Epithelial sheets undergo several major forms of deformation in the embryo: They can (1) fold; (2) branch; or (3) the entire sheet can spread. Most deformations of epithelial sheets are preceded by a thickening of the epithelium, known as **palisading** (Fig. 9.2A), due to the elongation of the cells within it from a cuboidal to columnar shape. Such thickened epithelia can be quite extensive, such as the broad neural plate, or they can be more punctate, such as the head placodes (see Chap. 7). These thickenings then often bend or fold.

Folding of an epithelium can be directed either inward or outward. When an epithelium bends away from a lumen or the surface of the embryo, the process is known as **evagination**, whereas inpocketing into the embryo or into a cavity is **invagination**.

Folding along a line will give rise to a groove (see Fig. 9.2B). The neural tube and the laryngotracheal tracheal groove form in this manner. In some instances, the fold can separate entirely from the surrounding epithelium to form a tube, an event that occurs during neurulation. In other situations, the folds retain their connection with the original epithelium, as in the case of one end of the laryngotracheal groove, which results in the opening of the trachea from the pharynx.

Folding of epithelia can also produce round depressions or pouches (see Fig. 9.2C). The lens, otic, and nasal placodes, for example, will invaginate to form pouches. In some instances, the pouches retain their connection with the surface, such as the nasal placodes, which form the lining of the nares. In other circumstances, the pouches pinch off to form hollow vesicles, such as the lens or the otic vesicles.

In addition, folds and pouches can be modified to form branched structures (see Fig. 9.2D). All the glands, for example, result from a cleft or fold appearing at the tip of an epithelial outpocket. Such a cleft subdivides the bulbous outpocket into two new end buds, a process that then repeats itself. The control of branching will be considered in detail in Chapter 12.

Folding or bending of a cell sheet is accompanied by cell shape changes within the epithelium. The columnar cells narrow, generally at the apical

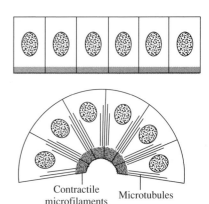

Figure 9.3

Diagram illustrating that a narrowing of the apices of cells in an epithelial sheet will produce a curvature in the sheet. (After Balinsky, 1981.)

end, to form pyramidal cells. This produces a change in surface area on one side of the epithelial sheet compared with the other, and the epithelium consequently bends (Fig. 9.3). Because all of the cells within the epithelium are tightly associated, any changes in cell shape, even in a few cells, can be translated over the whole epithelium. Furthermore, cell shape changes in a thickening epithelium will be more effective than in a flattened one because the taller the component cells are, the more pronounced the effect of a change in the diameter of an apical surface will be on the curvature of a cell sheet. The basis for these cell shape changes will be considered in Section 9–2.

Besides bending, epithelia also spread. During gastrulation in *Xenopus*, for example, the future ectoderm spreads to cover the embryo as endoderm disappears from the surface, a process known as epiboly (see p. 210). Or, during neurulation, the epidermal ectoderm must spread to cover the area vacated by the removal of the neural epithelium. This process of spreading is also often accompanied by a cell shape change, notably a thinning and flattening of individual cells, which results in an increase in surface area of the epithelium (see Fig. 7.7). In some instances, such as in *Xenopus laevis* gastrulation, cells from deeper layers are sequestered and intercalated into the more superficial layers, which also results in an increase in surface area (Keller, 1986). The forces that drive the process of sheet spreading have not yet been identified.

Modes of Individual Cell Movement

Individual cells also move from one point to another during many developmental events. They accomplish this translocation by a peculiar crawling motion that involves extension of the cell body, adhesion to the substratum, and changes in the cytoskeleton. Because cell movement is so basic to morphogenesis, we shall examine it in some detail.

Ideally, cell movement would be observed directly within the embryo. However, because most embryos are opaque, movement in the underlying tissues is usually obscured, and in only a few instances have motile cells actually been observed as they migrate within an embryo. To understand the mechanisms of cell motility, investigators have studied cells moving in tissue culture, where conditions permit observations at the highest possible magnification and resolution and where motility can be perturbed in specific ways to investigate the molecular mechanisms of cell movement.

The most intensely studied cells are those cells called **fibroblasts** (Fig. 9.4), which give rise to connective tissues. Fibroblast motility has been studied since the turn of the century, but we owe much to the detailed work of Michael Abercrombie (see p. 326) and his associates (Abercrombie et

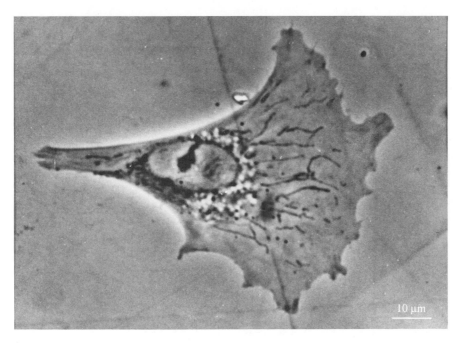

Figure 9.4

Phase contrast micrograph of a migrating fibroblast. This cell is moving to the right due to the extension of its lamellipodium. (Courtesy of N. K. Wessells.)

al., 1970), who used time-lapse cinemicroscopy. They observed that cells propel themselves forward by generating regions that are quite thin and fan-shaped and are called **lamellae**. Each lamella bears a very thin, flat extension that is not yet adherent to the substratum and is called a **lamellipodium**. As the lamellipodium protrudes forward, parts of it may lift up off the substratum and fold back on itself, giving a ruffled appearance (Fig. 9.5). Therefore, these leading edges of moving cells are often referred to as **ruffling lamellipodia**. As lamellipodia continue to be produced at the leading edge, the cell gets longer and longer until it is so stretched that the back or trailing

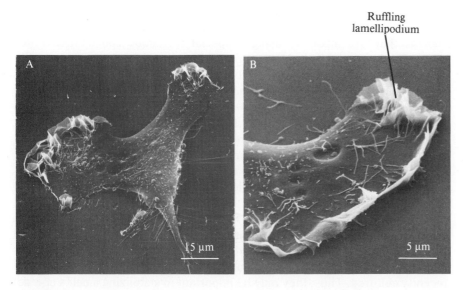

Ruffling
lamellipodium

Figure 9.5

Scanning electron micrographs of glial cells in tissue culture producing ruffling lamellipodia. A, A glial cell with two lamellipodia that are ruffling extensively. B, High magnification of a lamellipodium showing it in the process of uplifting from the substratum and folding backwards. (From Collins et al., 1977.)

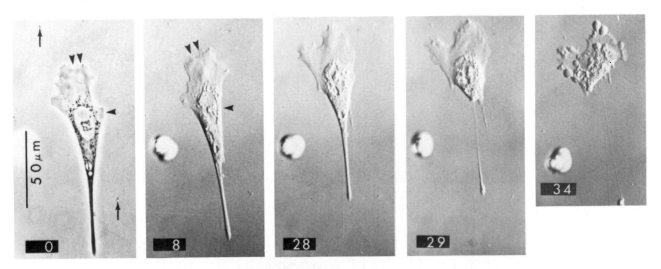

Figure 9.6

A chick heart fibroblast making net forward progress owing to the extension of lamellipodia (arrowheads) and retraction of its trailing edge. The cell is moving toward the top of the page. Bits of debris adherent to the substratum serve as fixed reference points (arrows). Numbers refer to minutes after the initiation of filming. At time 0 the cell is extending in the direction of the arrows. Its trailing edge gradually begins to pull away from the substratum, and the cell snaps forward. Note the bit of cell left behind: Cells often detach by tearing away from their adhesion to the substratum. (From Chen, 1981a. Reprinted by permission of Company of Biologists, Ltd.)

end detaches from the substratum and the tail snaps forward. Thus, movement of the cell consists of two phases: an extension phase, in which the cell extends itself to the limits of its plasma membrane, and a detachment phase, in which the posterior end of the cell is pulled forward (Fig. 9.6).

In order for the cell to move, it must adhere to its substratum and develop "traction." Contacts between the cell and the substratum can be seen as they are made by using a special technique called **interference reflection microscopy**. This microscopic technique reveals how far the cell's plasma membrane is from the underlying substratum. In general, the closer the cell comes to the substratum, the darker the image of that contact will be. Two major classes of adhesions have been revealed by this technique (Izzard and Lochner, 1976): (1) the **close contact**, in which the cell's plasma membrane comes within 30 to 50 nm of the substratum and appears gray in the interference reflection image; and (2) the **focal contact**, in which the membrane is 10 to 20 nm from the substratum and appears black (Fig. 9.7). Close contacts are generated at the leading end of the cell as a broad band; these contacts appear to be crucial in anchoring the lamellipodium to the substratum as it protrudes forward. The focal contact is more punctate than the close contact and is found in more posterior (i.e., older) portions of the lamellipodium and in the trailing end. Focal contacts are not always found in cultured cells, especially those that move rapidly or those that have been recently cultured. Thus, they may be important in stabilizing a cell's contact with the substratum, rather than being critical in the motile process itself.

Because a cell adheres to the substratum, it can then exert a tractional force on that substratum. This has been elegantly demonstrated by Harris and his co-workers (1980, 1981) by growing cells on a deformable rubber sheet. As the cell generates traction, it compresses the underlying rubber

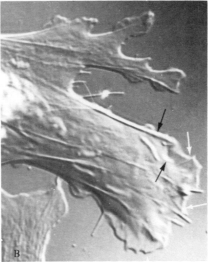

Figure 9.7

A living chick heart fibroblast observed with interference reflection optics (A) to visualize the distribution of contacts with the substratum. The same cell is also shown in differential interference optics (B) to show the lamellipodium. Two types of contacts can be seen: The dark black streaks represent the focal contacts, and the broader gray areas represent the close contacts. Regions that are white, such as under the tip of the lamellipodium, are much further away from the substratum. (From Izzard et al., 1985. Reprinted by permission of S. Karger AG, Basel, Switzerland.)

to form wrinkles, the amount of wrinkling being proportional to the traction developed (Fig. 9.8). The amount of traction exerted varies widely among cell types and is strongest in cells that have both close and focal contacts and weakest in those with only close contacts.

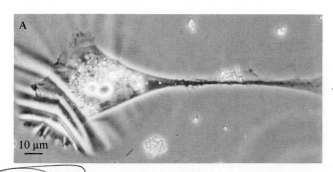
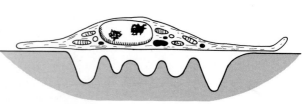

Figure 9.8

A, A fibroblast moving on an elastic silicone rubber sheet produces compression wrinkles in the substratum beneath the lamellipodium, revealing the tractional force it is exerting. B, A diagrammatic side view of a cell wrinkling its rubber substratum by compression as it crawls along. (A, From Harris et al., 1981. Reprinted by permission from *Nature* Vol. 290 pp. 249–251. Copyright © 1981 Macmillan Magazines Ltd. B, After A. K. Harris et al., 1980. Silicone rubber substrata: A new wrinkle in the study of cell locomotion. Science *208*: 177–179. Copyright 1980 by the American Association for the Advancement of Science.)

9–2 The Cytoskeletal Components

Changes in cell shape and the production of cell movement are mediated by a complex array of fibers in the cytoplasm that are referred to collectively as the **cytoskeleton**. Although the term "cytoskeleton" accurately defines the role of the fibers in maintaining cell shape, it also has the unfortunate connotations of rigidity and permanence. On the contrary, the cytoskeleton constitutes a dynamic system that is capable of *considerable* change.

The cytoskeleton is composed of three types of fibers: **microtubules, microfilaments,** and **intermediate filaments** (Fig. 9.9). There is evidence that each of these is involved in mediating cell motility and changes in cell shape. To understand the roles played by the cytoskeleton, we shall first discuss each of the cytoskeletal elements present in embryonic cells and then observe how their distribution in the cell produces motility.

Microtubules

Microtubules are present in all animal cells as vital parts of the basic cellular structure and machinery. They are essential functional elements in the mitotic spindle, are involved in the translocation of cellular organelles from one place to another in the cytoplasm, and are responsible for the movement of cilia and flagella. In addition to these general functions, microtubules also play a specific role in the generation and maintenance of cell asymmetry of both palisading cells and migratory cells.

Microtubules are hollow cylindrical rods (Fig. 9.10), 25 nm in diameter, formed of 13 rows of solid protofilaments that run parallel to the microtubule long axis (Fig. 9.11). The protofilaments are composed of two similar kinds of protein subunits, called α- and β-tubulins. *In vitro* studies of microtubules have revealed how the tubulin subunits are assembled into protofilaments and microtubules. One molecule of α-tubulin combines with one molecule of β-tubulin to form a **tubulin dimer**. The dimers, in turn, polymerize end to end to form the protofilaments. Concomitant side-by-side assembly of the 13 protofilaments produces the hollow microtubule.

Addition and loss of tubulin dimers can occur only at the ends of microtubules. The two ends of a microtubule are structurally different, and,

Figure 9.9

Electron micrograph of a portion of a motile tissue culture cell showing microtubules (MT), bundles of microfilaments (MF), and intermediate filaments (F). ×40,900. (From Goldman, 1971. Reprinted from *The Journal of Cell Biology*, 1971, *51*: 752–762 by copyright permission of the Rockefeller University Press.)

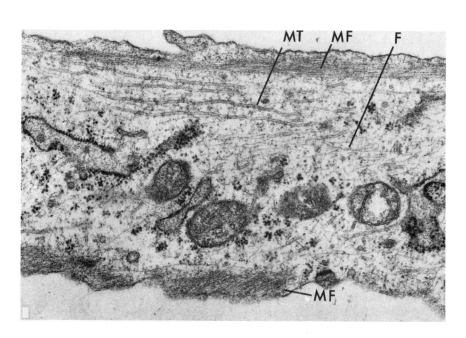

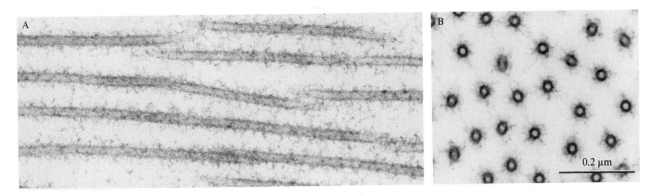

Figure 9.10

Longitudinal (A) and transverse (B) sections of microtubules assembled *in vitro*. The fuzzy material attached to the microtubules is composed of MAPs. (Courtesy of Helen Kim and Lester I. Binder.)

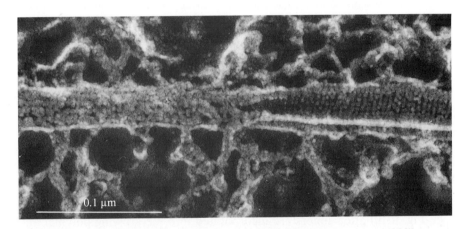

Figure 9.11

A purified microtubule fractured and deep-etched after quick-freezing. The microtubule is broken open on the right to reveal the protofilaments consisting of longitudinally arranged tubulin dimers. (From Heuser and Kirschner, 1980. Reprinted from *The Journal of Cell Biology*, 1980, *86*: 212–234 by copyright permission of the Rockefeller University Press. Courtesy of John Heuser.)

as a consequence, the microtubule is said to be polar. One end of a microtubule is kinetically fast (+ end), and the other end is kinetically slow (− end); i.e., both assembly and disassembly occur rapidly at the + end of a microtubule and slowly at the − end.

Several factors in cells exert control over microtubule assembly. First, the − end of the microtubule is embedded in a structure known as the **microtubule organizing center (MTOC)**. The MTOC serves as the point of initiation of growth for the microtubules and may specify the number, distribution, and length of the microtubules that emerge (Brinkley et al., 1981; Fig. 9.12). Once microtubules are initiated, their continued growth occurs by tubulin dimer addition to their distal (+ end) tips (Borisy, 1978; Bergen and Borisy, 1980). A critical concentration of tubulin subunits is a prerequisite for microtubule growth.

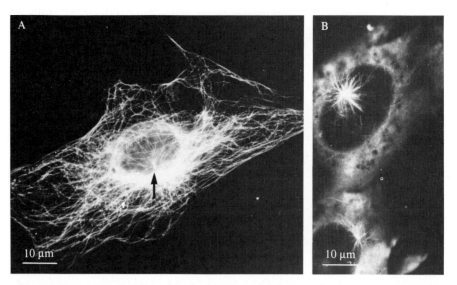

Figure 9.12

Microtubules visualized in a tissue culture cell using antibodies against tubulin and immunofluorescence microscopy (see essay on page 344). A, Microtubules emanate from microtubule organizing centers (MTOC; arrow) and swirl throughout the cell. B, The MTOC can be seen particularly well if the microtubules are first depolymerized and then allowed to regrow. These microtubules begin to grow radially from the MTOC 30 minutes after recovery from colchicine, which has depolymerized the microtubules. (From Osborn and Weber, 1976.)

Other factors also may be involved in the control of microtubule growth, although we are less certain of these. One possibility is that **microtubule associated proteins (MAPs),** which have been shown to promote microtubule assembly *in vitro*, are involved in the growth control of microtubules. These proteins associate with the surface of the microtubule (see Fig. 9.10) after dimers are incorporated into the growing polymer and stabilize the microtubule against sudden disassembly. However, it remains to be seen whether MAPs play a similar role in regulating microtubule growth in the intact cell.

Another possible control for microtubule growth is the capping of microtubules with bound guanosine triphosphates (GTPs). Each tubulin dimer binds two molecules of GTP, and, *in vitro*, if one of the GTPs is hydrolyzed to form GDP at the end of a microtubule, the microtubule becomes unstable and will rapidly depolymerize. If the GTP is not hydrolyzed, it forms a so-called **GTP cap.** This capped microtubule will be stable and will serve as a primer for the addition of more tubulin dimers (Mitchison and Kirschner, 1984). Microtubules have been observed to undergo rapid shortening and lengthening both *in vitro* (Horio and Hotani, 1986) and in living cells (Sammak and Borisy, 1988). This phenomenon is known as **dynamic instability,** and it underscores the dynamic nature of the cytoskeleton.

The results of *in vitro* studies have important implications for the behavior of microtubules *in vivo* and the regulation of their assembly. Clearly, elongation of microtubules can occur only by the addition of dimers to the ends of microtubules, and then only if the equilibrium favors assembly over disassembly. Conversely, microtubules will shorten if disassembly is favored. Although polymerization of dimers will occur spontaneously *in vitro* by the addition of the correct components at favorable temperature, microtubule assembly *in vivo* is a highly regulated process, with both the length and orientation of microtubules under close control.

A number of drugs affect the assembly and disassembly of microtubules and are excellent probes to determine whether microtubules are involved in a particular morphogenetic process. **Colchicine** has been the most commonly used drug. Originally used to control the medical condition known as gout, we now know that colchicine binds to tubulin monomers and prevents their assembly into microtubules (Margolis and Wilson, 1977). Because microtubules are continually being disassembled and reassembled, eventually all microtubules will disappear as the monomers become complexed with the drug. Recently, a new drug called **nocodazole** has been added to the cell biologist's arsenal. This drug also binds to tubulin subunits and prevents polymerization (Hoebeke et al., 1976) but appears to have fewer nonspecific side effects than colchicine. Either of these drugs, when applied to a developing system, serves as an excellent probe for determining whether microtubules are involved in the developmental process.

Microfilaments

Microfilaments, like microtubules, are nearly universal constituents of animal cells and are involved in a variety of basic cellular functions. Resolved with the electron microscope as 6-nm diameter fibers, microfilaments can be found singly throughout a cell, organized as a meshwork directly beneath the plasma membrane, or gathered into thick bundles that run the length of the cell (Buckley and Porter, 1967; Fig. 9.13).

Like microtubules, microfilaments are polymers—in this case formed from subunits of the contractile protein actin. The assembled polymers are

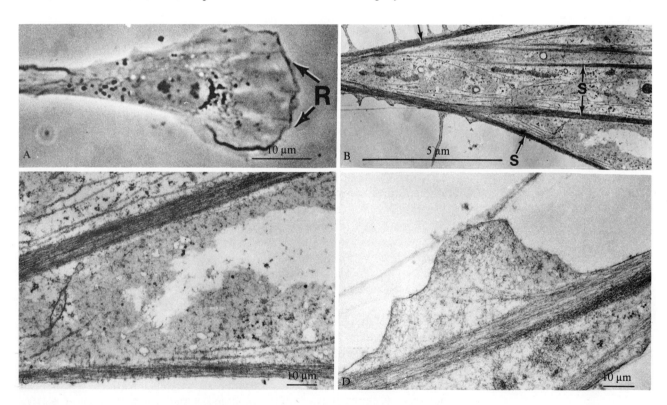

Figure 9.13

Distribution of microfilaments in a migratory cell. A, A living cell locomoting to the right owing to the protrusion of a ruffling lamellipodium (R). B, The rear end of a cell similar to that in A sectioned for electron microscopy. Sheaths of microfilaments (S) extend in the direction of locomotion. C, Higher magnification of B showing the bundles of microfilaments. D, High magnification of a ruffling lamellipodium showing that the microfilaments are organized in a meshwork. A sheath of microfilaments is seen below the meshwork. (Courtesy of N. K. Wessells.)

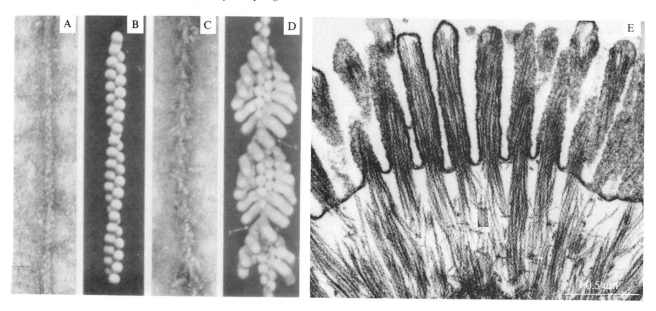

Figure 9.14

Electron micrographs of a negatively stained actin filament (A) and a negatively stained actin filament decorated with HMM (C). B, and D, Representative models of the electron microscope images. E, Electron micrograph of the actin filaments in an intestinal cell brush border decorated with HMM. Actin filaments are attached to the plasma membrane with the same polarity. Here, all the arrowheads point away from the plasma membrane. (From Pollard, 1981. Reprinted from *The Journal of Cell Biology*, 1981, *91*: 156s–165s by copyright permission of the Rockefeller University Press. E, Courtesy of D. Begg.)

frequently called **F-actin** (for filamentous actin), whereas the unpolymerized subunits are called **G-actin** (for globular actin). The equilibrium between actin subunits and polymeric filaments imparts dynamic properties to the actin network and allows it to adapt rapidly to the changing requirements of cells during morphogenesis. As with tubulin, *in vitro* polymerization of actin subunits occurs in a head-to-tail fashion onto both ends of microfilaments (Wegner, 1976). One end of the filaments is favored for polymerization, and this end is also called the + end. The polarity of an actin filament can be revealed by treating the microfilaments with heavy meromyosin (HMM) (Woodrum et al., 1975), which is a proteolytic fragment of myosin that contains the actin-binding portion of the molecule. When HMM is added to actin microfilaments, it binds in an arrowhead configuration (Fig. 9.14A to D). Not only can HMM be used as a specific probe to identify a filament as actin, but the arrowhead also defines the polarity of the molecules: The barbed end of the HMM arrowhead marks the + end of the microfilaments, whereas the pointed end marks the − end. In most cells, the + end of the microfilaments, where polymerization is favored, is attached to the plasma membrane (see Fig. 9.14E), and the − end, where assembly is not favored, is free in the cytoplasm (Tilney et al., 1981).

An array of proteins is associated with actin (for reviews, see Weeds, 1982; Stossell et al., 1985) and contributes to the function of actin in nonmuscle cells during development. One of the most important of these is **myosin**. The generation of contractile forces by microfilaments is thought to be based upon a mechanism of actin-myosin sliding filaments similar to that in smooth muscle, although the details of their interaction are unknown. Low levels of myosin are associated with microfilaments in nonmuscle cells. The actin-myosin interaction is regulated by Ca^{2+} interaction with the calcium-binding protein calmodulin (for review, see Adelstein, 1982).

In addition to myosin, a large number of other actin-binding proteins are found in nonmuscle cells, in which they have a variety of functions (Schliwa, 1981; Weeds, 1982). Some molecules, such as **filamin** and **α-actinin,** form flexible cross-links between actin microfilaments, producing a lattice or meshwork. Others, such as **fimbrin,** bundle microfilaments together into parallel arrays. Finally, molecules such as **villin** and **gelsolin** are Ca^{2+}-dependent actin-fragmenting proteins, which can break microfilaments into smaller molecules and accomplish changes in actin organization in the cell.

These are strongly suspected of controlling motility, but how this is done is not yet clear.

As with microtubules, the microfilament network can be altered experimentally by exposing cells to a chemical probe. The drug **cytochalasin B** blocks actin polymerization (Lin et al., 1980); thus, it can be used to determine whether microfilaments are involved in such developmental processes as shape changes or in cell motility. Unfortunately, the effects of cytochalasin on the cell are not so specific. Cytochalasin causes several side effects, including alterations of membrane transport functions (Tanenbaum, 1978), so that with this drug the investigator must consider the possibility of confounding nonspecific side effects.

Intermediate Filaments

The intermediate filaments are the third category of cytoskeletal components (Fig. 9.15). Intermediate filaments have a mean diameter of 10 nm and are therefore "intermediate" in size between microfilaments and microtubules. Unlike microfilaments and microtubules, which are assembled uniformly from their respective subunits of actin and tubulin, the intermediate filaments are much more heterogeneous in composition.

There are five classes of intermediate filaments, all of which are structurally and biochemically related, but each of which is made up of a distinct set of subunit proteins (Steinert et al., 1985). These five classes are found in different cell types (Lazarides, 1982; Osborn et al., 1982). The classes of intermediate filaments and the cells in which they are found are: (1) **keratin filaments**, which are restricted to epithelial cells arising from both ectoderm and endoderm; (2) **vimentin filaments**, found primarily in mesodermally derived connective tissues such as bone and cartilage; (3) **neurofilaments**, found in nerve cells; (4) **glial filaments**, which are found in glial cells that form the supportive tissues of the nervous system; and (5) **desmin**

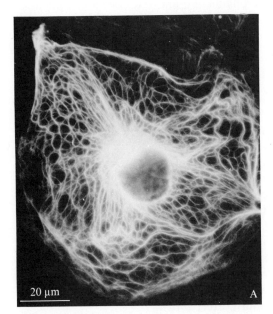

20 μm

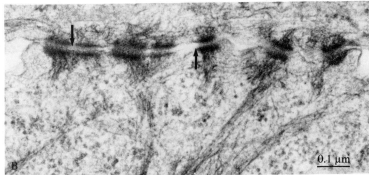

0.1 μm

Figure 9.15

Intermediate filaments in tissue culture cells. A, Intermediate filaments are visualized in this cell by immunofluorescence microscopy using antibodies against mouse epidermal keratin. The filaments are dispersed in the cell cytoplasm and form a cage around the nucleus. B, An electron micrograph of a cultured mouse epidermal cell showing the wavy bundles of intermediate filaments. Many of these terminate at cell junctions that connect neighboring cells to each other (arrows). (A, Courtesy of Ms. Karen Vikstrom and Dr. Robert Goldman. B, Courtesy of Drs. Jonathan Jones and Robert Goldman.)

filaments, which appear in skeletal, cardiac, and smooth muscle. Generally, one intermediate filament type characterizes a given adult differentiated tissue, but sometimes several can co-exist in a single cell type. It is particularly interesting to note that the kind of intermediate filaments present in a cell can change as the cell differentiates (e.g., Paranko and Virtanen, 1986).

Unlike the microfilaments and microtubules, the intermediate filaments appear to be very stable *in vivo*, and there is apparently no balanced equilibrium between polymerized and unpolymerized subunits. In fact, when intermediate filaments are isolated, they can be depolymerized *in vitro* using only the strongest denaturing agents, such as 8M urea. Even during mitosis, when microfilaments and microtubules break down, the intermediate filaments remain.

Such stability has made difficult the search for a suitable chemical probe to study their function. However, intermediate filaments can be broken down by either injecting antibodies to intermediate filaments into a cell or by treating cells with acrylamide (Eckert, 1985). Such treatments do not affect cell motility or any other observed cell function, at least in culture. The mechanical conditions inside embryos are likely to be quite different, however, so these results should not be taken as evidence for lack of function. Thus, the function of the several types of intermediate filaments remains obscure for the moment, although it has been proposed that the intermediate filaments interact with cellular organelles to help organize and maintain their three-dimensional arrangement in the cytoplasm (Klymkowsky et al., 1989).

9–3 Organization of Cytoskeletal Elements in Epithelia and Fibroblasts

To bring about appropriate changes in cell shape and cell motility, cytoskeletal elements—microtubules, microfilaments, and intermediate filaments—must be arranged and organized in some manner. This organization varies from cell type to cell type.

Role of Cytoskeleton in Epithelial Folding

The changes in cell shape that accompany epithelial folding are thought to drive this process. The cytoskeleton is undoubtedly involved in cell shape changes. We consider the organization of the cytoskeleton and its role in epithelial morphogenesis in this section.

Microfilaments, which are composed of actin, are organized as a meshwork beneath the plasma membrane. In addition, they insert as bundles into the zonula adherens junctions and encircle the apical tips of epithelial cells like a purse string (see Fig. 9.1). Keratin intermediate filaments are intimately associated with the maculae adherens (see Fig. 9.1) and are known as **tonofilaments**. The tonofilaments often stretch across an epithelial cell from one junction to another and form a superstructure within the cell. Epithelial cells also contain microtubules that are short and randomly arranged in cuboidal cells but are often organized in bundles running the length of columnar cells (Byers and Porter, 1964).

Role of Microtubules in Palisading Cells

There are several observations indicating that microtubules have a mechanical role in the elongation or palisading of epithelial cells: (1) Microtubules are often randomly arranged in cuboidal epithelia but can become organized along the length of a cell as it elongates (Byers and Porter, 1964; Burnside, 1971). Observations of this kind suggest that their appearance at the appropriate time and their distribution play some causal role in the elongation process. (2) Drugs or other treatments, such as cold shock and increased hydrostatic pressure, that prevent the polymerization of microtubules also prevent or reverse cell elongation (e.g., Karfunkel, 1971, 1972; see also p. 337).

A few notes of caution, however: (1) Drugs such as colchicine have side effects that could disrupt cell functions other than polymerization of microtubules. For example, colchicine can also change cell volume, and, at least in the case of lens epithelial cells, an increase in cell volume is responsible for cell elongation (Beebe et al., 1979). (2) Burnside (1971) measured the length and examined the distribution of microtubules in the developing amphibian neural tube and found no evidence that the microtubules were getting any longer or that they were sliding against each other to generate a force. Thus, the role of polymerization of microtubules in cell elongation remains an interesting idea, but one that is not yet proved. It could be, for example, that other forces are responsible for elongation, but that the microtubules act to stabilize the elongated state, as they apparently do in fibroblast locomotion.

Role of Microfilaments in Folding

As epithelia fold, the apical surfaces of cells become narrower. Early studies demonstrated that there is a circumferentially arranged band of microfilaments at the cell apices that attaches to the plasma membrane at the zonula adherens. These bands of microfilaments become thicker as the apices constrict in a variety of epithelia, including neural plate (see Fig. 7.9), uterine glands (Wrenn and Wessells, 1970), salivary gland (Spooner, 1973), and lens (Wrenn and Wessells, 1969), to name a few. When the microfilaments are depolymerized with cytochalasin B, the epithelia stop folding and revert to flat sheets. Because these microfilaments are composed of the contractile protein actin, this supports the idea that they are the mechanical cause of the narrowing of the apical surfaces, resulting in folding or branching.

Although the evidence is strong, we must again recall that drugs may have effects other than those on microfilaments. Ettensohn (1984), for example, showed that cytochalasin B decreases the adhesion between cells of the sea urchin gastrula, an effect that could also alter epithelial shape. Also, it is possible that other forces are responsible for the deformation of the epithelial sheet and that because the cells of the epithelium are tightly associated, they are simply passively deformed as the sheet bends.

Convergent Extension

Evagination and invagination can also result from the epithelium narrowing and lengthening through rearrangement of cells in the epithelium. As we discussed in Chapter 6, involution movements during gastrulation in the *Xenopus* embryo and elongation of the sea urchin archenteron both involve the process of convergent extension (i.e., convergence of a sheet of cells toward a central site, followed by its extension along a single axis through forceful intercalation of the cells of the epithelium [Fig. 9.16]). Active intercalation of cells is involved in numerous other folding events, including the morphogenesis of the insect imaginal disc (Fristrom, 1976) and the elongation of the notochord (Keller et al., 1989), but it is not known what directs this "cell shuffle." It is likely that many other morphogenetic events that involve epithelial sheet folding involve convergent extension as well and deserve re-examination by investigators.

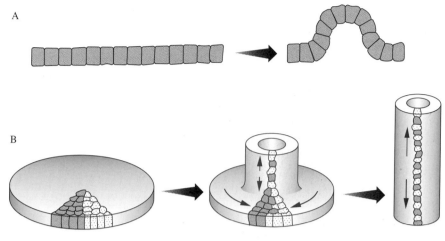

Figure 9.16

The traditional view of invagination of an epithelial sheet is illustrated in A. Here, the sheet folds owing to changes in shape of individual cells in the sheet. An alternative mechanism to account for folding of epithelial sheets is illustrated in B. Here, the forceful repacking of cells within the epithelium as they converge toward each other results in the elongation and bending of the epithelium. (After Keller, 1986.)

Distribution of the Cytoskeleton in Fibroblasts

Motility of individual cells is generated by cytoskeletal elements that are composed of contractile proteins. Indeed, forms of all of the major contractile proteins of muscle cells can be found in the motile fibroblast. Their distribution has been demonstrated using immunofluorescent techniques (see essay on p. 344) and with electron microscopy.

Actin, which is one of the basic constituents of muscle cells, has been identified in motile cells. Over half of the actin in cells is in the G-actin form, and the rest is arranged in microfilaments (Lazarides and Weber, 1974). The microfilaments are organized in two major ways: (1) as a meshwork that lies just beneath the plasma membrane in the cell cortex and fills the lamellae (Fig. 9.17); and (2) as bundles of microfilaments that extend along the length of the cell, known as **stress bundles** (Buckley and Porter, 1967; see Figs. 9.13, 9.18A).

Myosin is also found in nonmuscle cells, where it is localized primarily in a periodic distribution in the stress bundles in association with actin (see Fig. 9.18B). Curiously, myosin is not found in the lamellipodium, where the cell is extending forward (Willingham et al., 1981).

Many other actin-binding proteins are also found in the locomoting fibroblast (Stossell et al., 1985). Among these are tropomyosin (Lazarides, 1975) and α-actinin (Lazarides and Burridge, 1975), which are found in a periodic distribution along the stress bundles (see Fig. 9.18C and D).

In order to generate motility, the microfilaments must be linked in some way to the plasma membrane. It is not yet known precisely how they are so anchored, but they are attached consistently at their + end, as revealed by HMM binding (Small et al., 1978). Currently, we are learning much about the attachment of the cytoskeleton to at least one portion of the fibroblast, the focal contact. Actin stress bundles are known to end in focal contacts (Fig. 9.19). The attachment of the actin bundles to the focal contact appears to be mediated by several molecules, including α-actinin, vinculin, and talin, although the specific interactions are not presently

Regular text continues on page 346.

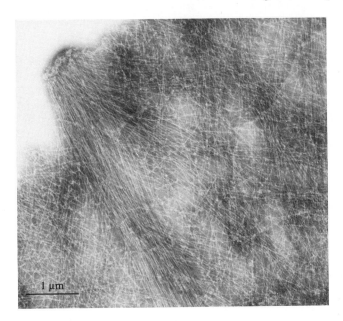

Figure 9.17

The meshwork of actin filaments can be seen clearly in the lamellipodium of this chick heart fibroblast, which had first been extracted with the detergent Triton X-100 and negatively stained and viewed in the electron microscope. (From Small, 1981. Reprinted from *The Journal of Cell Biology*, 1981, *91*: 695–705 by copyright permission of the Rockefeller University Press.)

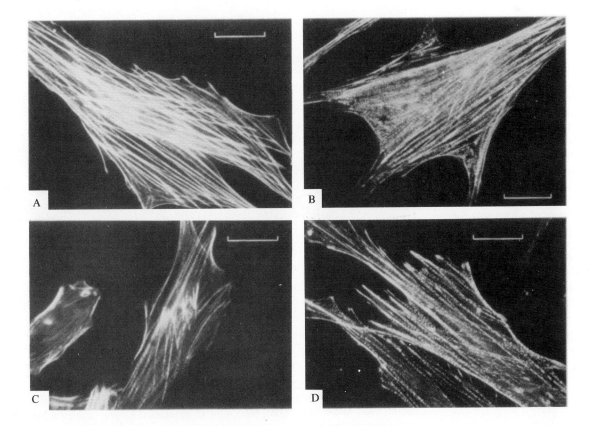

Figure 9.18

Rat mammary cells grown in tissue culture and stained with fluorescent-labeled antibodies to various specific contractile proteins: (A) actin, (B) myosin, (C) tropomyosin, and (D) α-actinin. The actin antibodies stain the stress bundles continuously, whereas the other antibodies stain only periodically along the stress bundles. Scale bars equal 25 μm. (From Abercrombie, 1980.)

Essay:

Visualization of the Cytoskeleton by Immunofluorescence Microscopy

The electron microscope is a powerful tool for studying the intricate details of cellular components. However, it is not always the instrument of choice for studying the overall pattern of distribution of a particular component so that one may correlate its distribution with its function. This is particularly true for components that may be responsible for generating *changes* in cellular configuration, because the components themselves are in a dynamic state while generating these changes. The cytoskeletal components are prime examples. Correlations of cytoskeletal configurations with changing cell morphology could assist in understanding the cytoplasmic basis for cellular dynamics. A technique that makes such an analysis possible is **immunofluorescence microscopy.**

Immunofluorescence microscopy is a technique for detecting a cell protein by visualizing with the fluorescence microscope antibodies to that protein that have been coupled to a fluorescent dye. The fluorescence microscope uses ultraviolet light, which causes bound fluorescent molecules to appear bright against a dark background. In *indirect* immunofluorescence, two kinds of antibodies are employed. One is an antibody that has been raised against a single cell protein (the so-called primary antibody), whereas the one that is actually visualized is a fluorescent-tagged antibody directed against primary antibody; the fluorescent

Figure 1

A section through a 54-hour quail embryo stained with an antibody recognizing the intermediate filament protein vimentin. Some cell layers lack staining, especially the ectoderm, endoderm (En) and pronephric duct (P). Other layers stain heavily with the antibody, such as the neural tube (NT), sclerotome (Sc), and intermediate mesoderm (IM). The change in the distribution of intermediate filaments during embryogenesis has been correlated with a role for intermediate filaments in morphogenetic movement. NC: neural crest; D: dermatome; M: myotome; Som: somatic mesoderm; C: coelom; Spl: splanchnic mesoderm. (From Erickson et al., 1987.)

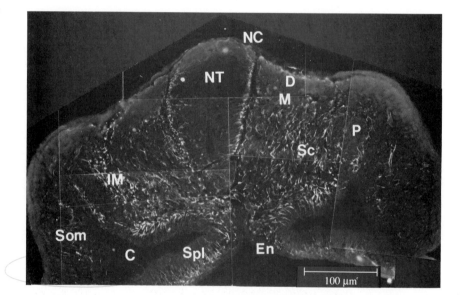

Figure 2 (opposite)

This fibroblast was injected with a fluorescent-labeled actin and then photographed at intervals of t = 0(A), 6(B), 9(C), and 15 minutes (D). The labeled actin is incorporated into the stress bundles. In this sequence, the development of an actin bundle can be observed (arrow). From such studies, it has been learned that stress bundles develop from a focal contact and then grow back toward the nucleus. In addition, once a stress bundle is assembled, it remains stationary relative to the substratum even though the cell continues to move forward. (From Wang, 1984. Reprinted from *The Journal of Cell Biology*, 1971, *99*:1478–1485 by copyright permission of the Rockefeller University Press.)

antibody is thus the secondary antibody. In one possible research strategy, the antibody to the cell protein (primary antibody) is elicited in rabbits, whereas the fluorescent antibody (secondary antibody) is an anti-rabbit antibody that is prepared in goats and coupled to the dye. A major advantage to the indirect method is that fluorescence is amplified owing to the binding of several fluorescent antibodies to a single primary antibody.

Immunofluorescence reveals the presence and locations of all major cytoskeleton constituents. The distributions of microtubules, intermediate filaments, actin microfilaments, and actin-associated proteins are visualized using this technique in Figure 1 and in Figures 9.12, 9.15, and 9.18.

In immunofluorescence techniques, cells are generally fixed first with formaldehyde or ethanol. Consequently, the dynamics of the cytoskeleton cannot be appreciated. A recently devised approach for overcoming the disadvantages of fixation is to inject into living cells cytoskeletal proteins that have been fluorescently labeled (Fig. 2). These proteins become incorporated into the cell's cytoskeleton, and the dynamics of the cytoskeletal components can then be observed as the cell moves. Because the amount of light emitted from such an injected cell is very weak, special cameras that can amplify low light are generally used. This approach has been used successfully for many of the cytoskeletal components (Taylor and Wang, 1980; Kreis and Birchmeier, 1981) and should eventually allow us to understand how changes in the molecular organization of the contractile machinery of a cell generate its spectacular movements.

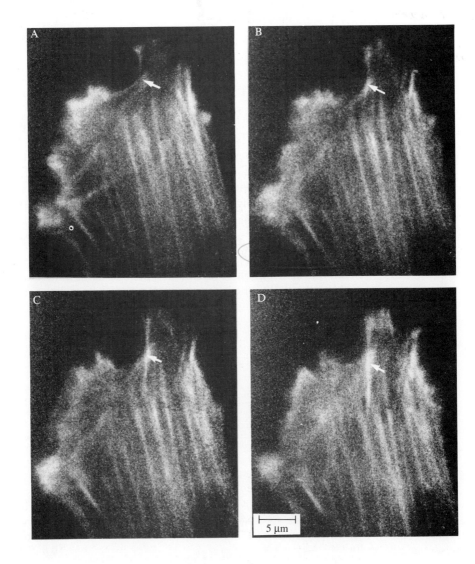

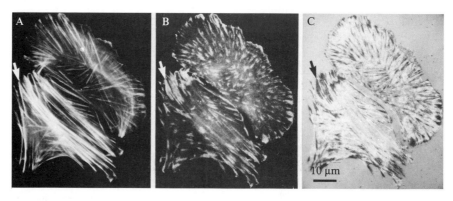

10 µm

Figure 9.19

The organization and relationship of actin bundles and focal contacts in two chick embryo fibroblasts. A, Actin filaments and bundles are revealed by treating the cells with rhodamine phalloidin. B, The same cells were treated with antibodies to the protein talin; its distribution is revealed with a fluorescein-conjugated secondary antibody. C, Focal contacts observed with interference reflection microscopy. Note that the distribution of talin coincides with the focal contacts and that the actin bundles also terminate in these contacts. (From Burridge et al., 1988. Reproduced, with permission, from the Annual Review of Cell Biology, Vol. 4 © 1988 by Annual Reviews Inc.)

[handwritten note: Seems to only localize at focal adhesion pts. to stress bundles]

understood (Burridge et al., 1988). None of these molecules are integral membrane proteins, however; consequently, they must in turn attach to another molecule in the plasma membrane itself. One such integral protein linking actin to the plasma membrane is a glycoprotein complex called **integrin** (Pytela et al., 1985; Damsky et al., 1985). Integrin is found localized in, or at the edge of, focal contacts and has a strong affinity for the vinculin-talin complex. Furthermore, the domain of this molecule that lies external to the cell binds to many extracellular matrix molecules and therefore mediates cell attachment to the substratum as well (see Sect. 9–5). Our current understanding of the organization of cytoskeletal proteins in cell contact sites is summarized in Figure 9.20.

How Do Cells Move?

Given this distribution of microfilaments and actin-associated proteins, how might cell motility be generated? Because all of the muscle contractile

Figure 9.20

Schematic drawings of the contacts a cell makes with its substratum and the organization of the cytoskeletal elements in those contacts. See text for description. Note that we are not entirely sure which elements are present in cell contacts; these are marked with ? in the figure. FN: fibronectin.

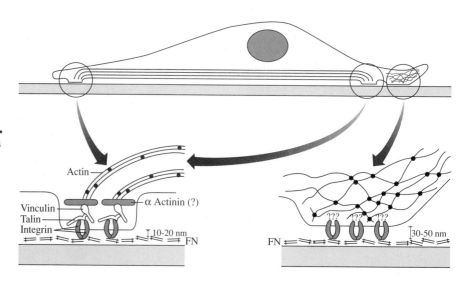

proteins are present in motile cells, it seems reasonable that some sort of contractile machinery linked to the plasma membrane generates cell motility. When tissue culture cells are treated with cytochalasin B so that microfilaments are eventually broken down, cell motility ceases (Spooner et al., 1971; Wessells et al., 1973). Furthermore, if antibodies against actin are injected into a motile cell, all movement ceases. Actin, then, is critical to movement.

Myosin, the other major contractile protein in muscle cells that generates force by sliding against actin, may not be so important in fibroblast movement. First, no myosin is present in the lamellipodium, where cell extension is occurring. Second, recent studies have shown that if the myosin gene is mutated or prevented from being expressed, the amoebae of the slime mold *Dictyostelium* can continue to locomote, although they cannot divide (Knecht and Loomis, 1987; De Lozanne and Spudich, 1987). This evidence suggests that at least the extension phase is not dependent upon myosin.

How, then, do the various cytoskeletal elements produce extension of the leading edge and retraction of the trailing end? One hypothesis for controlling extension at the leading edge of a cell is that actin polymerization generates a protrusive force. This would be similar to the force thought to be generated by actin polymerization coupled with osmotic pressure changes (Tilney and Inoué, 1985) in the extension of the acrosomal process (Tilney and Kallenbach 1979; see p. 138). Alternatively, a local increase in osmotic pressure in the lamellipodium may be the force for extension, whereas polymerization of microfilaments provides the structure (Oster, 1984; Trinkaus, 1985). We have not yet been able to distinguish between these two alternatives. New plasma membrane, which is needed to form the extending lamellipodium, could be recruited from a more posterior region of the cell, or it could be newly synthesized and inserted into the leading edge (Singer and Kupfer, 1986). This has not yet been established, however. In any case, as the lamellipodium is extended, new close contacts (and sometimes focal contacts) are generated that anchor the lamellipodium in its new position, thus allowing for net forward progress.

Our models for explaining formation of the lamellipodium are continually being modified as new data are acquired. A recent finding that a modified myosin is present in the leading edge of *Dictyostelium* may radically change our ideas concerning cell motility. A form of myosin is present in at least some motile cells such as *Acanthamoeba* (Pollard and Korn, 1973) and *Dictyostelium* (Cote et al., 1985) that has a head group almost identical with the well-characterized myosin that was discussed previously but with a greatly shortened tail that is capable of binding both to actin in an ATP-insensitive manner (Lynch et al., 1986) and to plasma membrane (Adams and Pollard, 1989). These have been named mini-myosins or myosin I. The role of myosin I in motility is far from clear, but the recent observation that myosin I is found in the leading edge of *Dictyostelium* and the conventional form of nonmuscle myosin is found in the trailing end of the same cell suggests that this smaller myosin may be critical in the protrusive process (Fig. 9.21). To date, the distribution of myosin I in other eukaryotic cells is not known, but its potential importance has spurred efforts to find it in other cell types.

A cell can extend and stretch only so far before eventually the trailing end snaps forward. This retraction of the trailing edge involves both an ATP-dependent contraction of the cytoskeleton and also a passive recoil due to the elasticity of the cytoskeleton and its associated plasma membrane (Chen, 1981b). This contraction event is likely to be due to the contraction of the stress bundles that contain both actin and myosin and have been shown to shorten visibly during cell movement (Isenberg et al., 1976). After

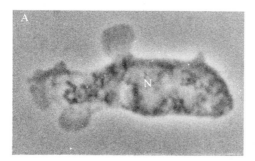

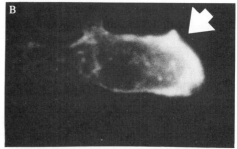

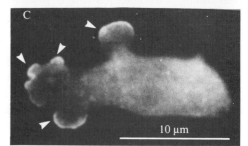

Figure 9.21

Localization of myosin I and myosin II in *Dictyostelium*. A, A *Dictyostelium* amoeba viewed in phase contrast migrating from right to left. N: nucleus. Myosin I (C) and myosin II (B) have been localized in the same cell using specific antibodies. Note that myosin I is found in the leading edge (arrowheads in C), whereas myosin II is more concentrated in the tail (large arrow in B). (From Fukui et al., 1989. Reprinted by permission from *Nature*, Vol. 341, pp. 328–331. Copyright © 1989 Macmillan Magazines Ltd.)

the tail end snaps forward, new focal contacts are established at the trailing end, and the cycle begins again.

Although progress has been made, much remains unknown about nonmuscle cell motility. We are still ignorant regarding aspects of the molecular organization of the microfilaments and their associated proteins and how these elements are linked to the plasma membrane. Furthermore, we still do not know precisely what forces produce the extension of the lamellipodium, although actin is clearly implicated somehow, nor do we know what triggers such an extension. Finally, the generation of adhesions, which is critical to motility, and the organization of the proteins in those adhesion sites remain mysteries.

Cell forces and contractions applied at random do not produce cell translocation; for this to occur, motility must be directional. A dominant lamellipodium must form in order for the cell to make net forward movement. When several lamellipodia are generated around the cell perimeter, the cell will simply tug in all directions at once and go nowhere until one of the lamellae becomes larger and more dominant.

Microtubules are apparently responsible for organizing the direction of protrusive activity. Microtubules are generally arrayed along the length of a cell and, as we have seen, emanate from an MTOC. If the microtubules are depolymerized with drugs such as colchicine, the cell will begin to produce lamellipodia around its entire circumference (Vasiliev et al., 1970). Furthermore, if cells are forced to change directions (such as with a chemotactic agent), the MTOC shifts so that it is on the side of the nucleus that faces the new lamellipodium (Kupfer et al., 1982). The Golgi apparatus also shifts to the position of the MTOC. Singer and Kupfer (1986) have hypothesized that the MTOC aligns the Golgi apparatus and that new plasma membrane needed for the extending lamellipodium is transported directionally from the Golgi apparatus along a conduit of microtubules.

Intermediate filaments are generally distributed in the same pattern as microtubules. However, depolymerization of intermediate filaments with either antibodies against them or with acrylamide has no effect on motility. It is generally thought that intermediate filaments have little role in cell movement but perhaps are involved in the stabilization of cytoplasmic components or the biomechanical properties of tissues.

9–4 Cell Motility in the Embryo

We have discussed in some detail the mode and mechanisms of cell motility in tissue culture, but we should not lose sight of the fact that as developmental biologists we are interested in cell movement during embryogenesis. Is motility in the embryo comparable to that studied *in vitro*? Fortunately, cell behavior, where it can be observed *in situ*, is similar, if not identical, to that which we have studied in such detail in culture.

A number of embryos are crystal clear and afford the opportunity to study cell movement and behavior in their natural habitat. Perhaps the most carefully studied of all *in vivo* cell movements are the deep cells of fish embryos. Like the sea urchin, fish embryos are exquisitely clear. This system has been exploited to great advantage in particular by J. P. Trinkaus in the fish *Fundulus*. The fish blastula consists of a mass of individual cells called **deep cells**, sandwiched between an overlying **enveloping layer** and the underlying **yolk syncytial layer** (Fig. 9.22). During ensuing gastrulation, the deep cells begin to migrate on the yolk syncytial layer and converge to form the embryonic shield, which will become the embryo proper. Because the embryos are clear, the movements of the deep cells have been studied in great detail using time-lapse cinemicrography.

Deep cells display several modes of motility. In some instances, finger-like protrusions, rather like sausages, extend from the rounded cell bodies (Fig. 9.23). These have been called **lobopodia** (Trinkaus, 1973). These lobopodia can shorten, thereby pulling the rounded cell body forward (Fig. 9.23), or the cytoplasm from the cell body can pour into the lobopodium. Alternatively, filopodia and lamellipodia can form on the ends of short lobopodia (Fig. 9.24). The latter have even occasionally been caught in the process of ruffling (Fig. 9.25; Trinkaus, 1973). Furthermore, filopodia and lamellipodia can transform into one another (Trinkaus and Erickson, 1983). This interconvertibility suggests that they are simply morphological variants of the same motile machinery.

Such studies suggest that cell motility *in vivo* is similar to that in tissue culture because morphologically the cells appear alike and their modes of movement resemble each other.

In higher vertebrates, a particularly good structure in which to observe cell motility is the developing cornea of the chick embryo. At early stages, the corneal matrix consists of many layers of collagen fibrils. Into this collagenous matrix, corneal fibroblasts of neural crest origin migrate and ultimately take up residence (Fig. 9.26). The cornea can be removed from the embryo and placed in organ culture, where the migration of these cells can be observed using high-resolution differential interference contrast optics (Bard and Hay, 1975). This cell movement is of particular interest because it occurs relatively late in development and represents the movement of genuine tissue cells, rather than blastula or gastrula cells.

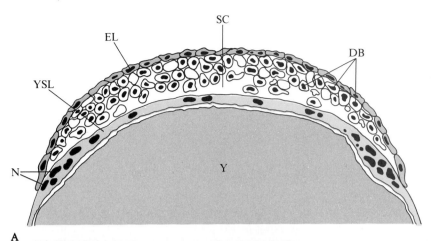

A

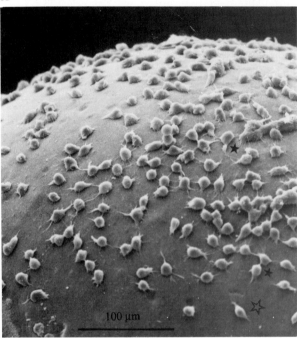

B

Figure 9.22

A, Diagram of an early *Fundulus* gastrula revealing the overlying blastoderm (EL), the underlying yolk syncytial layer (YSL) with scattered nuclei (N), and the deep blastomeres (DB) in the segmentation cavity (SC). Y: yolk. ×140. B, A scanning electron micrograph of *Fundulus* deep cells that have been revealed after the removal of the blastoderm. (A, After Lentz and Trinkaus, 1967. B, From Trinkaus and Erickson, 1983.)

Figure 9.23

Frames from a time-lapse film showing deep cells moving in an intact *Fundulus* gastrula. A flattened lamellipodium (arrowhead) extends from the more rounded lobopodium, and the cell makes forward progress owing to the shortening of the lobopodium. Time elapsed is in minutes and seconds. Use the nucleus (N) of cell in the underlying layer as a reference point. Cell labeled FC is a flattened stationary cell. (From Trinkaus and Erickson, 1983.)

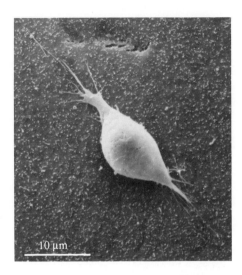

Figure 9.24

The locomotory process of a deep cell can be seen in more detail in this higher resolution scanning electron micrograph. Note that the flattened lamellipodium has long filopodia extending from it. These filopodia are often not visible in the living embryo. (From Trinkaus and Erickson, 1983.)

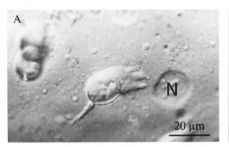
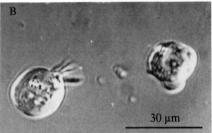

Figure 9.25

A, A deep cell photographed in the living embryo in the process of ruffling. B, A deep cell that has been isolated and is locomoting in culture. Note the similarity of its ruffling lamellipodium to the one shown in A. N: nucleus of cell in underlying layer. (From Trinkaus and Erickson, 1983.)

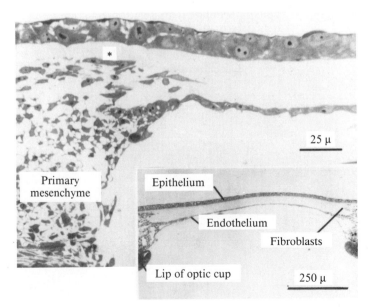

Figure 9.26

Light micrograph of a section through a 5½-day chick embryo eye revealing the beginning of the invasion of corneal fibroblasts (cells marked by *) into the corneal stroma. The inset shows an eye whose lens has been removed in order to film the migration of the corneal fibroblasts in a culture chamber. (From Bard and Hay, 1975. Reprinted from *The Journal of Cell Biology*, 1975, *67*: 400–418 by copyright permission of the Rockefeller University Press.)

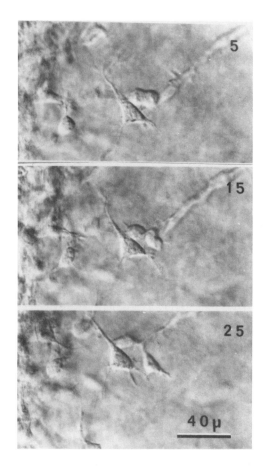

Figure 9.27

Frames from a time-lapse film (time elapsed recorded in minutes) of corneal fibroblasts migrating in the corneal stroma. The cells move primarily using long, thin filopodia. (From Bard and Hay, 1975. Reprinted from *The Journal of Cell Biology*, 1975, *67*: 400–418 by copyright permission of the Rockefeller University Press.)

The corneal fibroblasts locomote in a manner similar to fibroblasts on a flat petri dish, except that the advancing end is morphologically less flattened and narrower, more like a filopodium than a fibroblast lamellipodium (Fig. 9.27). Ruffling activity is not observed. The function appears to be the same, however, which is to advance the cell. These embryonic fibroblasts have even been observed to undergo contact inhibition in the cornea (contact inhibition will be discussed on p. 365), as fibroblasts do in culture (Bard and Hay, 1975).

If these same fibroblasts are isolated and grown on a petri dish, they assume the morphology of a typical fibroblast with a flattened lamellipodium at its advancing edge (Fig. 9.28). If, instead, these cells are cultured in

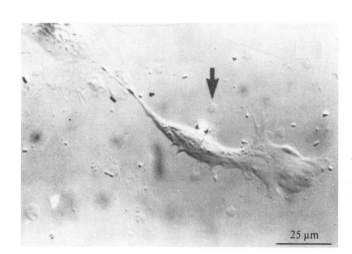

Figure 9.28

A corneal fibroblast that has been isolated and cultured on plastic. Note the large lamellipodium on the right, rather than the filopodia that form *in situ*. (From Bard and Hay, 1975. Reprinted from *The Journal of Cell Biology*, 1975, *67*: 400–418 by copyright permission of the Rockefeller University Press.)

three-dimensional collagen gels, which more closely approximate the extracellular matrix through which they move in the cornea, they then develop filopodia, typifying cells *in vivo*.

The preceding studies raise two important points: (1) The morphology of cells depends on the substratum upon which they move; and (2) filopodia and lamellipodia are readily interconvertible and therefore probably use the same cytoplasmic machinery for motility. Thus, experiments in culture are directly applicable to the situation in the embryo.

The major difference between the modes of cell movement *in vivo* and *in vitro* is that ruffling activity is generally not seen *in vivo*. The reason for this is probably trivial. It is generally assumed that in the embryo there is not space above the cell for the lamellipodium to lift up, as exists in culture. *Fundulus* deep cells, which do form ruffling lamellipodia, locomote in the blastocoel cavity, which clearly does provide room for ruffles to develop.

9–5 Cell-Matrix Adhesions

In order for cells to move or to change shape, they must be able to adhere to each other or to substrata in their environment. In this section, we consider adhesion of cells to molecules in the extracellular environment.

Although much has been gained from studying the movements of cells on plastic culture dishes, cells in an embryo do not move on a compositionally uniform, two-dimensional plastic. Rather, moving cells adhere to complex environmental surfaces composed of molecules known as **extracellular matrix (ECM) molecules** (Hay, 1981a, b). These molecules can be divided into three categories: (1) collagen, (2) proteoglycans, and (3) other glycoproteins. Most extracellular matrix molecules are secreted into the spaces between cells and assemble into a meshwork (Fig. 9.29). In addition to forming the interstitial matrix, some matrix molecules form a dense sheet on the basal

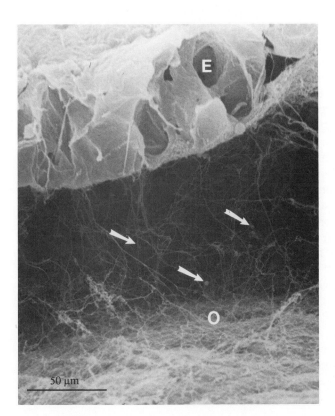

50 µm

Figure 9.29

SEM of the extracellular matrix found in a chick embryo between the ectoderm (E), and the optic vesicle (O). The matrix is composed of fibrils of various sizes that are probably collagen I, III, and fibronectin. The matrix forms a dense mat known as a basal lamina on the undersurface of the ectoderm and on the optic epithelium. The arrows point to spheres called interstitial bodies, which are composed of fibronectin and proteoglycans. (From Tosney, 1982.)

Epithelium

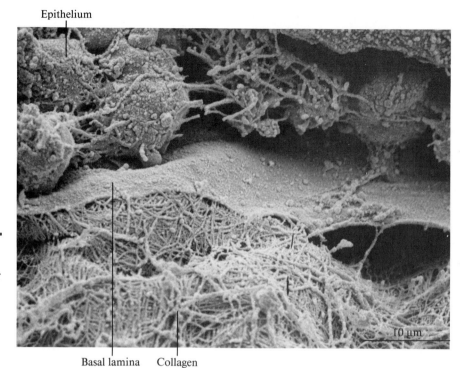

Basal lamina   Collagen

Figure 9.30

SEM showing the position of the basal lamina. The basal lamina is a dense condensation of extracellular matrix molecules deposited beneath an epithelium, on which it rests. The basal lamina separates the epithelium from the rest of the interstitial matrix. (Courtesy of R. Trelstad.)

surface of epithelial cells called a **basal lamina** (Fig. 9.30). The ability to self-assemble into extracellular matrices is due to the fact that most of the ECM molecules can self-aggregate; furthermore, they can bind to each other at specific binding domains.

Collagen

Collagen, which constitutes over half the total body protein, is the most abundant matrix molecule. Collagen molecules are distinguished from other glycoproteins by their unusual amino acid composition (33% glycine, 10% proline, 10% hydroxyproline), their resistance to proteases, and their special microscopic structure of striated fibrils (Hay, 1981a, b).

A collagen molecule is composed of three polypeptide chains that wrap around each other to form a triple helix. These chains associate end to end and side to side with other chains to form fibrils that are generally striated when viewed in the electron microscope (Fig. 9.31) and are relatively resistant to lateral bending. Many types of collagen have been isolated (12 at last count; Burgeson, 1988), and different tissues have different combinations of each type (see Table 9–1). For example, tendons are made up of bundles of collagen fibrils composed of collagen types I and III, and those in the cornea are composed of types I and V. A few of the collagen types do not form fibrils, but instead assemble in a fine network. Type IV collagen is one such example of a network, well suited to its role in assembly of the sheet-like basal lamina, where it is concentrated (Timpl et al., 1981).

Proteoglycans

A second major constituent of the ECM is a class of highly charged sugars known as **glycosaminoglycans** (Lindahl and Höök, 1978). The glycosaminoglycans are negatively charged sugar chains that are made up of dimeric subunits, consisting of two alternating classes of monosaccharides,

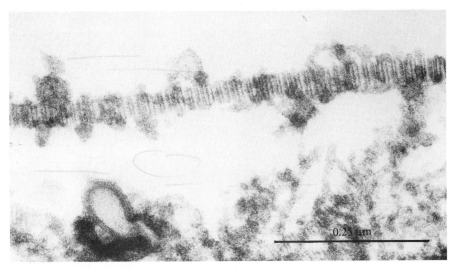

0.25 μm

Figure 9.31

Collagen fibrils appear striated when stained and observed in the electron microscope. These fibrils are type I collagen. (From Hay et al., 1978.)

the amino sugars, and uronic acids. The different glycosaminoglycans are distinguished by their precise monomeric composition, the type of linkage between the monomers, and the amount or location of added sulfate groups that can modify these sugars. The common glycosaminoglycans (GAGs) found in developing tissues are shown in Figure 9.32.

With the exception of hyaluronic acid (HA), which can exist by itself in the interstitial matrix, the rest of the GAGs are covalently bound to a core protein to form **proteoglycans** (Hascall and Hascall, 1981). At one end of the core protein is a linker region that binds the proteoglycan to HA. Many such proteoglycans can complex with HA to form large complexes or aggregates of relative molecular mass up to 10^8 M_r. Binding to HA is stabilized by a linker protein. These interactions are summarized in Figure 9.32.

Because of their high density of negative charge, the proteoglycans do not form compact tertiary structures like proteins, but rather are extended, space-filling molecules. Furthermore, they bind water tightly in microdomains. Therefore, these molecules occupy much space and are resistant to compression. These properties, as we shall see, are critical for their role in development.

Table 9–1 Major Collagen Types and Locations

Type	Chain Composition	Structure	Location in Tissues
I	$[\alpha1(I)]_2\alpha2(I)$	Fibril	Skin, bone, tendon, ligaments, cornea, organs (90% of total collagen)
II	$[\alpha1(II)]_3$	Fibril	Cartilage
III	$[\alpha1(III)]_3$	Fibril	Skin, blood vessels, internal organs
IV	$[\alpha1(IV)]_2\alpha2(IV)$	Network	Basal laminae
V	A family composed of various combinations of $\alpha1(V)$, $\alpha2(V)$, $\alpha3(V)$, $\alpha4(V)$	Unknown	In all connective tissues (10% or less of total collagen)

From Mayne, R. 1984. The different types of collagen and collagenous peptides. *In* R. L. Trelstad (ed.), *The Role of the Extracellular Matrix in Development*. Alan R. Liss, New York, pp. 33–42.

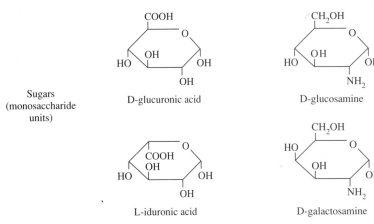

Sugars
(monosaccharide
units)

D-glucuronic acid   D-glucosamine

L-iduronic acid   D-galactosamine

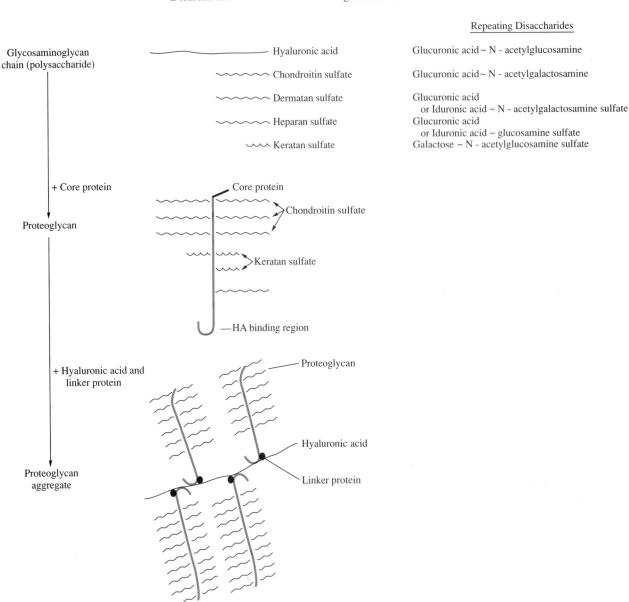

Repeating Disaccharides

Glycosaminoglycan
chain (polysaccharide)

Hyaluronic acid   Glucuronic acid ~ N - acetylglucosamine

Chondroitin sulfate   Glucuronic acid~ N - acetylgalactosamine

Dermatan sulfate   Glucuronic acid
    or Iduronic acid ~ N - acetylgalactosamine sulfate

Heparan sulfate   Glucuronic acid
    or Iduronic acid ~ glucosamine sulfate

Keratan sulfate   Galactose ~ N - acetylglucosamine sulfate

+ Core protein

Core protein

Proteoglycan

Chondroitin sulfate

Keratan sulfate

HA binding region

+ Hyaluronic acid and
linker protein

Proteoglycan

Hyaluronic acid

Linker protein

Proteoglycan
aggregate

Figure 9.32

Summary of the organization of glycosaminoglycans into proteoglycan aggregates.
(After Hascall and Hascall, 1981.)

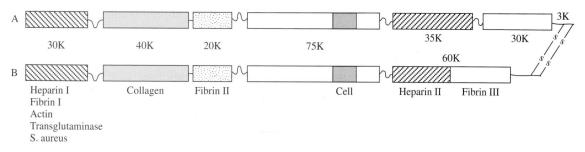

FUNCTIONAL DOMAINS OF FIBRONECTIN

Figure 9.33

Schematic diagram of a fibronectin molecule, showing the distribution
of binding sites to other extracellular matrix molecules and to cells.
The multiple binding sites allow these molecules to cross-link various
extracellular matrix molecules and to mediate cell attachment to the
extracellular matrix. (After Yamada et al., 1984.)

Adhesive Glycoproteins

A variety of glycoproteins occur in the extracellular matrix, and the list
seems to grow on a monthly basis. These glycoproteins have several roles,
including regulating the assembly of the three-dimensional ECM and mediating the attachment of cells to the ECM (Hewitt and Martin, 1984;
Ruoslahti et al., 1985).

Most is known about the glycoprotein **fibronectin** and how it mediates
ECM assembly and cell attachment (Yamada et al., 1984; Hynes, 1985;
Ruoslahti, 1988). Fibronectin is a glycoprotein dimer of 400,000 M_r. Each
polypeptide chain consists of a linear array of binding domains, which
include several cell-binding domains and binding sites for collagen, heparin,
and HA (Fig. 9.33). The two polypeptide chains are linked together at their
carboxyl termini by disulfide bonds. The dimeric nature of this molecule
means that fibronectin can cross-link various elements of the ECM into
stable arrays, can mediate attachment of a cell to the ECM, or even cross-
link several cells together.

Another ECM molecule that mediates cell adhesion to the ECM and
cross-links the ECM is **laminin** (Kleinman, 1986), which is found primarily,
but not exclusively, in the basal lamina. Like fibronectin, laminin also
consists of a linear array of binding domains (Fig. 9.34). Laminin specifically
mediates epithelial cell adhesion to the ECM of the basal lamina and is also
thought to stimulate motility of a number of embryonic cells, including the
neural crest cells and neurons (Gundersen, 1987).

Cell Adhesion Receptors

The binding of cells to the extracellular matrix is mediated by receptors
that are integral membrane proteins. So far, receptors that mediate adhesion

**FUNCTIONAL DOMAINS
OF LAMININ**

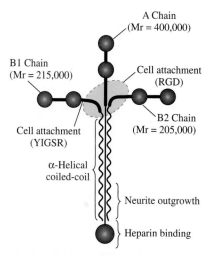

Figure 9.34

Schematic diagram of a laminin
molecule, showing the distribution
of the binding sites to other
extracellular matrix molecules and
to cells. (After Martin and Timple,
1987.)

to collagen, laminin, fibronectin, and another ECM molecule, vitronectin, have been isolated. Others will undoubtedly also be discovered.

The fibronectin receptor is currently the best characterized extracellular matrix receptor (Damsky et al., 1985; Ruoslahti, 1988). It has been isolated using two different approaches. In one, the plasma membranes of cells are solubilized and poured over a column that has the fibronectin molecules bound to an inert substrate (a so-called affinity column). Any protein in the solubilized solution that can bind to fibronectin will then stick to the column, whereas the rest of the mixture will wash out. What sticks to the column is a possible fibronectin receptor. In the other approach, antibodies are raised against cell surface molecules, and these are then tested to find those antibodies that can inhibit cell attachment to a fibronectin substratum. These antibodies are then used as affinity reagents to bind cell receptors as described previously. These two techniques isolate essentially the same proteins.

The fibronectin receptor isolated from a variety of cell types consists of two distinct subunits, termed α and β. The polypeptides that make up each subunit are approximately 140,000 M_r, although there is some heterogeneity depending upon the source of the receptors. Based upon their amino acid sequence and the ability of the isolated receptor to integrate into the artificial membrane of a liposome, these proteins are almost certainly transmembrane proteins. For this reason, these receptors have been termed integrins. The integrin molecules are oriented with the amino terminus facing the exterior and the carboxyl terminus embedded in the cytoplasm, where they interact with elements of the cytoskeleton, as we discussed in Section 9–3 (Fig. 9.35).

The integrins bind to the fibronectin molecule at an attachment site domain that is composed minimally of only three amino acids, Arg-Gly-Asp (also termed the **RGD sequence**, based upon the standard single-letter designations for these amino acids), as demonstrated by the fact that cells can attach to RGD-derivatized substrata. Furthermore, cell adhesion to fibronectin (and subsequent migration on it) can be broken by adding peptides containing the RGD tripeptide sequence to the medium. These presumably occupy the integrin receptor and prevent the cell from binding to fibronectin (Pierschbacher and Ruoslahti, 1984).

Recently, it has been discovered that cell receptors for other extracellular matrix proteins, such as collagen, laminin (Gehlsen et al., 1988), and vitronectin, are very similar to the receptor for fibronectin, being composed of α- and β-subunits that are very similar (and in some cases identical to) the subunits of the fibronectin receptor. Therefore, these

Figure 9.35

A schematic drawing of the integrin molecule, demonstrating the two subunits, their orientation in the plasma membrane, and their presumed binding site to the RGD sequence of an extracellular matrix molecule. (After Ruoslahti, 1988.)

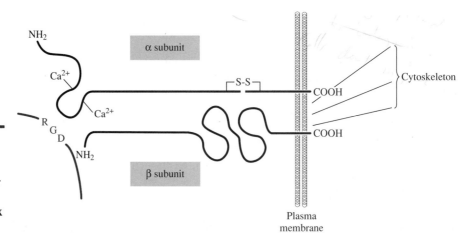

receptors have been classified as members of an **integrin superfamily** (Hynes, 1987). Not only are these receptors alike in their structure, many of them also recognize an RGD peptide sequence in their respective adhesion molecule. For example, the collagen receptor binds to RGD in collagen, the vitronectin receptor binds to RGD in vitronectin, and so on. The specificity of the receptor for its particular matrix molecule depends upon the three-dimensional configuration of the RGD sequence in a particular matrix protein (Ruoslahti and Pierschbacher, 1987).

9–6 *Control of Directional Migration*

During development, cells migrate—often over large distances—to reach their final destinations. Many factors control such cell migrations and rearrangements. These include factors in the extracellular environment, such as the extracellular matrix, that can direct locomoting cells. Even after the initial, long-range migrations are complete and the cells have stopped in a particular location, they will often undergo rearrangement with neighboring cells during the histogenesis of a particular organ. The final positions of cells in an organ may be determined by their surface properties, a topic that will be discussed in Section 9–7. In this section, however, we focus on how the extracellular environment can direct morphogenetic movements.

Determination of Migratory Pathways

Cells often take very precise pathways to their final destinations. There are many ways in which these pathways can be determined by extracellular factors.

Contact Guidance

A number of investigators (Harrison 1912; Weiss, 1934) demonstrated that cells can detect discontinuities in their substratum and will align and migrate along such features as fibers or scratches on the bottom of a petri dish (Fig. 9.36). Weiss (1958) termed this behavior **contact guidance**. Although the mechanism of this phenomenon remains obscure, its relevance to directed motility in the embryo is clear. Perhaps orientations in the substratum along the path of migratory cells could create tracks on which the cells could move.

In at least one case, migratory cells do appear to be guided by oriented fibrils. Nakatsuji et al. (1982) observed in amphibian embryos that the

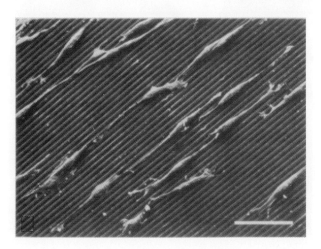

Figure 9.36

Contact guidance. These mammalian cells are growing on a grooved surface. The cells preferentially align on the grooves. Scale bar equals 50 μm. (From Dow et al., 1987. Reprinted by permission of Company of Biologists, Ltd.)

mesodermal cells move in a directed fashion from the blastopore to the future anterior end of the embryo along the roof of the blastocoel cavity. Extracellular matrix fibrils are oriented along the blastopore–anterior axis as well. If the roof of the blastocoel is removed and placed in culture, the tissue will deposit on the tissue culture substratum a pattern of fibrils identical to those seen in the gastrula. Mesoderm cells seeded onto these fibrils will adhere and migrate along them (Nakatsuji and Johnson, 1983). If the orientation of these fibrils is artificially manipulated by exerting tension on the roof of the blastocoel, the directional migration of the mesodermal cells is correspondingly altered (Nakatsuji and Johnson, 1984). Thus, it appears that pathways may be determined by fibrils on the substratum and presumably by other structural irregularities as well.

It is of considerable interest to understand how fibrils become aligned. One hypothesis is that fibroblasts, which exert a great deal of traction on their substratum (see Fig. 9.8), may use this force to align the matrix to which they adhere. Stopak and Harris (1982) tested this notion by embedding chick heart fibroblasts in a three-dimensional collagen matrix in which the collagen fibrils were randomly arranged. Very rapidly, the cells attach to, and rearrange, the collagen fibrils into bundles owing to the tractional force they exert on them (Fig. 9.37). Such tractional force may very well be used by fibroblasts in the embryo to organize the extracellular matrix into patterns that, in turn, guide the movement of other cells.

Adhesive Differences

Numerous studies have demonstrated that cells adhere differentially to various ECM molecules. For example, neural crest cells adhere strongly to fibronectin and laminin, but only weakly to chondroitin sulfate proteoglycan (Erickson and Turley, 1983). Perhaps pathways are created by very adhesive molecules being laid down in tracks.

Letourneau (1975) addressed the question of whether a cell could detect adhesive boundaries and whether this could pattern their migration. In a clever experiment with neurons in tissue culture, he tested this possibility by coating plastic petri dishes with various substances to which the neurons could adhere either tightly or weakly. In Figure 9.38A, which is taken from his work, we see a neuron migrating along a lane of a very adhesive molecule, polyornithine, and avoiding the surrounding squares of the less adhesive heavy metal palladium. Therefore, some cells can detect boundaries

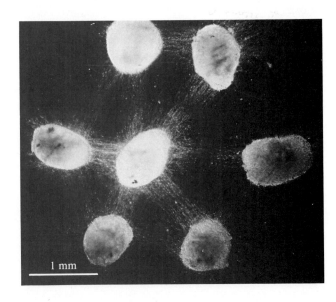

Figure 9.37

Aggregates of chick heart cells have been cultured in a collagen gel and viewed with dark-field illumination. The cells exert enough tractional force to align the collagen fibrils between the individual explants. (From Stopak and Harris, 1982.)

1 mm

where there is a difference in adhesion and move only along the more adhesive substratum. Such boundaries appear at numerous places in the embryo (see essay on p. 362).

Extracellular Spaces

The interior of the embryo does not consist entirely of densely packed, abutting tissues. Rather, tissue masses are separated by extracellular spaces, and some tissues consist of cells loosely associated with extracellular matrix. There are many instances in development in which the pathways taken by cells coincide with the appearance of spaces. It seems logical that cells might utilize such spaces as pathways for migration, providing there is something there to which they can adhere, because such spaces represent corridors that might offer less resistance to cell movement.

A particularly good example of the role of spaces in directing cell movement is the development of the optic nerve. In the mouse, the optic nerve develops when neurons from the retina migrate through the optic stalk. The migratory path of the neurons through the optic stalk is a precise one, and Silver and Robb (1979) observed that before the neurons entered the optic stalk, channels develop as a result of the selective death of some of the stalk cells. The neurons then migrate through these channels. Silver and Robb discovered that in a mouse mutant named *or^J*, these channels never develop and that the optic nerve fibers never enter the optic nerve stalk. These studies suggest that in some instances spaces are necessary to allow cells to pass and the timely and precise appearance of these corridors is likely to control certain pathways of migration. The role of spaces has been implicated in several other developmental events, including migration of neural crest cells, the formation of the sternum primordium, and movement of neurons in the brain.

Control of Directionality of Migration

In all of the preceding examples, extracellular matrix molecules or spaces create tracks along which cells can move. However, these tracks do not impose directionality on the cells. If we think of the tracks as a street, the tracks certainly determine where the roadway is, but not which direction to travel on the roadway. Clearly, other factors must be superimposed on the roadbed to determine directionality.

Adhesive Gradients

Carter (1967) showed that cells will migrate along an adhesive gradient in tissue culture, from the less adhesive end to the more adhesive end, owing—at least in part—to the more frequent detachment from the less adhesive end of the substratum (Harris, 1973). Therefore, adhesive gradients could provide directional cues to embryonic cells. However, only sparse evidence exists for the directed movement of cells along an adhesive gradient (called **haptotaxis**) in the embryo.

Indirect evidence for an adhesive gradient comes from studies of the migration of the amphibian pronephric duct (Poole and Steinberg, 1982). The amphibian pronephric duct arises as an ovoid structure from the anterior mesoderm. The posterior end of the rudiment then begins to migrate along the ventral border of the somites in a posterior direction toward the cloaca, with which it fuses, whereas the rest of the rudiment remains anchored in place. The cues for this posterior migration appear to reside in the underlying mesoderm. If a second pronephric duct rudiment is grafted to the flank mesoderm, it will migrate caudally to fuse with the host duct. Ventral or

Regular text continues on page 364.

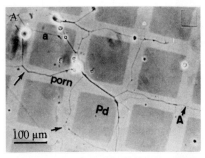

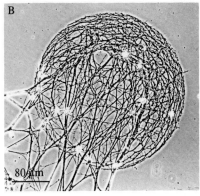

Figure 9.38

Preferential migration of neurons on more adhesive substrata. A, Chick embryo sensory neurons are shown here migrating on lanes covered with the highly adhesive molecule polyornithine (porn), but avoiding the less adhesive squares of palladium (Pd). a: axon on palladium; A: axon on polyornithine. B, A dot of the extracellular matrix molecule laminin was placed on a collagen substratum. Neurons growing on the collagen adhere preferentially to the laminin and will migrate endlessly within the dot but do not venture across the border back onto the collagen. These experiments suggest that laminin may be an important determinant of cell migratory pathways in the embryo. (A, From Letourneau, 1975. B, From Gundersen, 1987.)

Essay:

The Role of the Extracellular Matrix in Neural Crest Cell Migration

Neural crest cell migration is among the most dramatic of morphogenetic movements in the developing embryo. These cells migrate relatively long distances from their origin in the dorsal neural tube and follow precise pathways. Current evidence suggests that the extracellular matrix (ECM) is responsible for the pattern of neural crest cell migration.

As we saw in Chapter 7, the neural crest cells in the trunk of the embryo take several pathways. Some crest cells migrate under the ectoderm, and these become pigment cells in the skin. Most of the crest cells migrate ventrally between the neural tube and somite and eventually enter the somite, but only in its anterior half. It is believed that a combination of extracellular matrix molecules is responsible for the pattern of migration.

Fibronectin is found in the crest pathways (Fig. 1A; Newgreen and Thiery, 1980; Thiery et al., 1982). Several studies have shown that

Staining y fibronectin

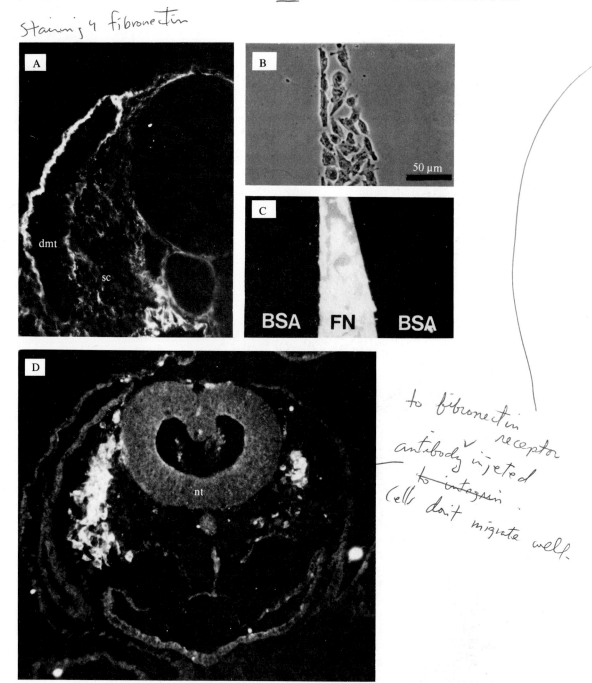

to fibronectin receptor antibody injected to integrin. Cells don't migrate well.

neural crest cells adhere strongly to fibronectin (Fig. 1B and C; Newgreen et al., 1982; Erickson and Turley, 1983; Rovasio et al., 1983), and when either antibodies to the fibronectin receptor (Fig. 1D; Bronner-Fraser, 1985) or a decapeptide (10 amino-acid) fragment of fibronectin, which inhibits cell attachment to fibronectin (Boucaut et al., 1984), is injected into the embryo, crest cell migration is in-

terfered with to some degree, at least in the head region. Another ECM molecule, laminin, is found in the crest pathways and in the basal laminae that border the neural crest cell pathways (Fig. 2A; Duband and Thiery, 1987). Neural crest cells also adhere strongly to this molecule (Fig. 2B and C; Newgreen, 1984), and antibodies to a receptor for a laminin-heparan sulfate complex

(Bronner-Fraser and Lallier, 1988) also perturb neural crest cell migration.

Another category of ECM molecules that may influence neural crest cell migration is the glycosaminoglycans (GAGs). Hyaluronic acid (HA) is one such GAG that is found in the neural crest pathways (Derby, 1978; Pintar, 1978). Neural crest cells cannot adhere to this molecule, but when HA is added

continues

Figure 1 (opposite)

A, The distribution of fibronectin revealed by fluorescein-labeled antibodies in a cross-section of a 3-day chick embryo. Note that fibronectin is found in the sclerotome (sc) and on the basal surface of the dermamyotome (dmt), both regions where crest cells will migrate. B, and C, Stripes of fibronectin (FN) alternating with bovine serum albumin (BSA) were made on a glass surface. When neural crest cells were cultured on such a surface, they attached and spread more avidly on the fibronectin. D, Antibodies that recognize the cell-surface fibronectin receptor were injected into one side of a chick embryo's head at the time when cranial neural crest cells had just begun to migrate. In this micrograph, the neural crest cells are revealed with a labeled antibody. Note that on the right side, where the antibody was injected, neural crest migration has been perturbed, whereas on the left side, the distribution is normal. nt: neural tube. (A, From Thiery et al., 1982. B, and C, From Newgreen, 1984. D, From Bronner-Fraser, 1985. Reprinted from *The Journal of Cell Biology*, 1985, *101*: 610–617 by copyright permission of the Rockefeller University Press.)

importance y integrins on der embo in vivo. Indir Immunofluorescence

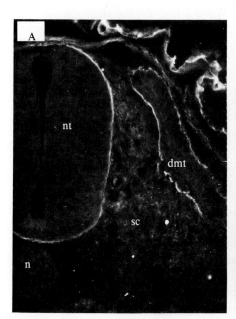

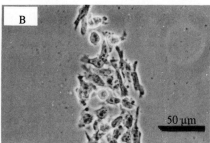

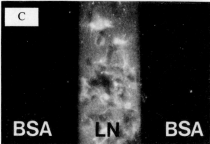

Figure 2

A, Distribution of laminin is revealed by labeled antibodies in a cross-section of a 3-day chick embryo. Laminin is found in regions similar to fibronectin (see Fig. 1). nt: neural tube; n: notochord; sc: sclerotome; dmt: dermamyotome. B, and C, When lanes of laminin (LN) are laid down on a glass substratum, neural crest cells prefer to migrate on the laminin. (A, From Duband and Thiery, 1987. Reprinted with permission of Company of Biologists, Ltd. B, and C, From Newgreen, 1984.)

to three-dimensional collagen matrices, it opens up spaces between the collagen molecules and speeds up the movement of neural crest cells embedded in the matrices, compared to their movement in collagen alone (Tucker and Erickson, 1984). This suggests that HA may act to open up spaces and create paths of lesser resistance (Toole and Trelstad, 1971). Another GAG, chondroitin sulfate, is not very adhesive for neural crest cells (Erickson and Turley, 1983) and is found in regions of the embryo where neural crest cells do not migrate (Newgreen et al., 1986). Thus, such weakly adhesive molecules may function to prevent neural crest cell migration into particular regions.

Neural crest cell pathways are also bordered by basal laminae that are extracellular matrix-containing structures. Neural crest cells adhere to and migrate on these basal laminae but do not go through them (Erickson, 1987). Therefore, basal laminae act as impenetrable barriers that constrain the crest cells to migrate along particular paths.

Thus, there are many factors in the extracellular environment that control neural crest cell migration. This redundancy is probably important in providing a fail-safe mechanism to guide these very important cells to their destinations.

anterior migration is never observed, regardless of the orientation of the grafted rudiment (Fig. 9.39). There is some evidence suggesting that alkaline phosphatase is arrayed in a gradient along the axolotl flank and may be the adhesive substratum utilized by the pronephric duct (Zackson and Steinberg, 1988).

A second example of adhesive gradients directing migration of embryonic cells is migration of the precursor cells of the chick heart. As we saw in Chapter 8, the precursor cells of the heart in the chick embryo are derived from the lateral plate mesoderm, from which they migrate anteriorly as single cells on the endoderm to form the paired endocardial tubes. Fibronectin is associated with the endoderm in a gradient increasing in the anterior direction that is correlated with the direction of heart cell migration. Linask and Lash (1988a) have shown that if a chick embryo is incubated in antibodies against fibronectin, or in RGD peptides that interfere with cell adhesion to fibronectin, presumptive heart cell migration stops. Thus, these cells are likely to use at least fibronectin to mediate their attachment and migration. Furthermore, if the endoderm on which the heart cells migrate is removed and rotated 180 degrees so that the fibronectin gradient is now reversed, heart cell migration halts prematurely when the cells arrive at the boundary that represents the high end of the gradient (Linask and Lash, 1988b). Migration is normal when this same area is surgically removed and reinserted in its original orientation. These experiments strongly suggest that the gradient of fibronectin not only sustains but also directs heart mesenchyme migration.

Chemotaxis

A variety of cell types can orient their locomotion along a concentration gradient of certain diffusible molecules by extending their lamellipodia toward the source. This movement toward the source of a diffusible chemical is known as chemotaxis. Chemotaxis is distinct, although practically difficult to distinguish, from movement in relation to a gradient of substances attached to a substratum (haptotaxis). Numerous examples of chemotaxis by cells in culture have been identified (e.g., Zigmond, 1982). There is some evidence suggesting that this phenomenon plays a role in directing cell movements during embryogenesis as well.

Perhaps the clearest demonstration of chemotaxis in development involves the migration of neurons. Studying mouse embryos, Lumsden and Davies (1983) examined the cues that direct the migration of neurons of the trigeminal ganglion to the whiskers of the maxillary process. They were able to show that in tissue culture, explants of the trigeminal ganglion will grow toward mesoderm taken from the maxillary process but will not grow

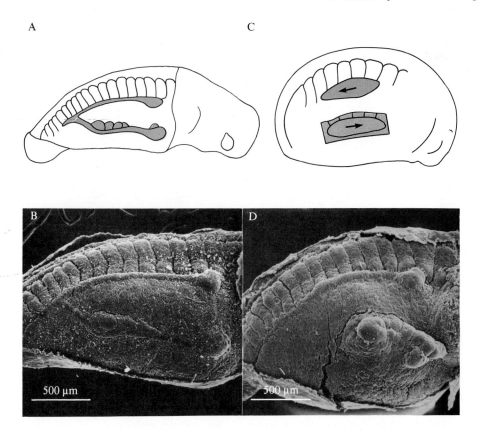

Figure 9.39

Adhesive gradients may direct pronephric duct migration in amphibians. A, and B, A camera lucida drawing and a scanning electron micrograph, respectively, of an amphibian embryo showing the migration of the pronephric duct from the anterior region of the embryo to the cloaca. A second duct grafted to the belly will also migrate posteriorly and fuse with the host duct. C, and D, A camera lucida drawing and a scanning electron micrograph, respectively, of an amphibian embryo in which a second duct has been grafted to the belly with its anteroposterior axis reversed. The portion of the duct that should extend has rounded up and does not migrate onto the underlying mesoderm. This suggests that the duct cannot adhere to the underlying mesoderm. (From Poole and Steinberg, 1982.)

toward mesoderm taken from an inappropriate target (mesoderm from the limb bud) (Fig. 9.40). This directed growth occurred in culture only when the tissue was taken from the mouse at the time when this directed migration was occurring *in vivo*. Nerve growth factor (NGF), a protein often implicated in neurite extension and neuronal survival, was not responsible for this migration because antibodies against NGF, which should inactivate NGF, had no effect on the directed migration. These results suggest that some diffusible agent in solution is released from the target tissue and is attracting neurons in culture. This chemotactic agent, which has not yet been isolated or characterized, may therefore have a role during normal development in directing trigeminal neurons to the developing whiskers.

Contact Inhibition

When one cell contacts another in culture, the ruffling lamellipodium of the contacting cell ceases protrusive activity, forms a new process elsewhere, and migrates away from the cell it contacted (Fig. 9.41). This phenomenon is known as **contact inhibition** (Abercrombie, 1970). One of the outcomes of this process is directional migration: Cells will move away from areas of high cell density toward regions where there are fewer cells.

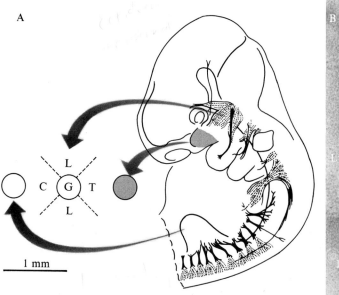

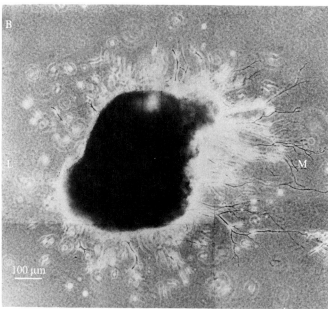

Figure 9.40

Chemotaxis in trigeminal ganglion outgrowth. A, Camera lucida drawing of a 10-day mouse embryo showing the position of the trigeminal ganglion and the maxillary process (colorized), to which the trigeminal neurons grow. Trigeminal ganglia were removed from the embryo and co-cultured with both mesoderm from the maxillary process and mesoderm from a nontarget tissue, in this case the limb bud. The four quadrants surrounding the ganglion (G) are labeled T (the target quadrant), C (the control quadrant), and L (two lateral quadrants). B, A phase contrast micrograph of an experimental culture as outlined above. The trigeminal ganglion neurons have grown out toward the maxillary process mesoderm (M) but have not extended toward the limb bud mesoderm (L). These results suggest that there is an agent produced by the maxillary tissue that directs the growth of early trigeminal neurons. (From Lumsden and Davies, 1983. Reprinted by permission from *Nature*, Vol. 306, pp. 786–788. Copyright © 1983 Macmillan Magazines Limited.)

One possible example of directional migration being produced by contact inhibition is in the case of the dispersal of neural crest cells away from the neural tube. Neural crest cells arise at the dorsal surface of the neural tube and then disperse from their point of origin. In a variety of experiments, haptotaxis and chemotaxis have been ruled out as the basis of this directional migration (Erickson, 1985). However, neural crest cells demonstrate contact inhibition in tissue culture, and indirect evidence suggests that contact inhibition may drive these cells from their origin into the periphery, where they are found at lower concentrations. Despite suggestive evidence from tissue culture studies, the role of contact inhibition in directional migration can only be proved by observing it in the living embryo.

9–7 Intercellular Adhesion

Besides adhering to extracellular matrix molecules in the environment, cells also adhere to each other. In fact, one of the most important factors governing the behavior of cells as they migrate and as they rearrange themselves within tissues to form organs is the nature of a cell's interaction with its neighbors.

The importance of cell associations in directing morphogenesis was first demonstrated by experiments in which tissues were dissociated and allowed to reassemble into aggregates. H. V. Wilson (1907) was the first to perform this type of experiment, using sponges as the experimental organisms. Wilson dissociated sponges to form a suspension of cell fragments, individual cells, and small clusters of cells. The dissociated cells sink to the bottom of the culture dish, where they initially adhere. Then the cells actively migrate until they adhere to one another to form clusters that increase in size by the addition of single cells and by fusion with adjacent clusters. The aggregates eventually reform into "new" individual sponges, complete with spicules, flagellated chambers, and branched canals. This amazing phenomenon was termed **reconstitution**. It was not clear from these results whether individual cells redifferentiated to form the appropriate cell type in the correct position, or whether different cell types sorted out to assume the correct position.

The ability to reconstitute the original form turned out not to be the province of only lower invertebrates; similar experiments showed that when cells of developing vertebrate embryos are mixed, they too will segregate according to cell type and re-establish their former associations. These studies were initiated by Johannes Holtfreter using amphibian embryos. In particular, Townes and Holtfreter (1955) were the first to present convincing evidence that embryonic cells that originate from different germ layers **sort out**. Their experiments took advantage of the fact that cells from amphibian gastrulae can be dissociated in solutions of alkaline pH. They could then mix single cell suspensions from ectoderm, mesoderm, and endoderm and follow the movements of these cells in aggregates owing to the fact that cells from each germ layer have a distinctive size and pigmentation. Their experiments produced several important results (Fig. 9.42). First, the various cell types initially adhere to each other to form aggregates in which the cells are randomly arranged. Soon, however, the cells begin to shift their positions within the aggregates, so that each cell type sorts out from the other. Thus, in a mixed aggregate of mesoderm and presumptive epidermis, for example, the epidermis is eventually found as a layer on the outside, with the mesoderm forming a cohesive mesenchyme in the center (see Fig. 9.42A).

The second important observation, which is most extraordinary, is that the cells take up the approximate positions they would have normally occupied had they been allowed to stay in the embryo and go through their normal gastrulation movements. For example, when cells of all three germ layers are combined, the endoderm forms a compact ball in the center, the ectoderm forms an epidermis on the surface of the aggregate, and the mesoderm forms a loose arrangement of cells sandwiched between these two layers (see Fig. 9.42B). Even cells from the same germ layer, but of different developmental fates, will segregate from each other and take up positions appropriate to their normal relationship during development. For example, if presumptive neural plate ectoderm and epidermis are mixed in an aggregate, the epidermis will form a surface epithelium, whereas the neural ectoderm will sink into the interior to form structures resembling brain vesicles (see Fig. 9.42C).

These experiments demonstrate not only that cells will sort out but that cells of the three germ layers assume positions in the aggregate that reflect their *in vivo* relationship. Any explanation of intercellular adhesion that is relevant to embryonic development must account for these two aspects. We shall now discuss two hypotheses of the mechanism of intercellular adhesion.

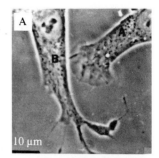
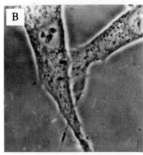
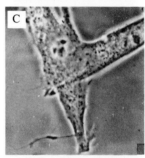
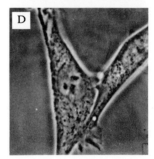

Figure 9.41

Contact inhibition. The fibroblast on the right approaches (A) and contacts (B) a second cell, to which it adheres. Ruffling lamellipodium activity ceases, and the contacting cell retracts and moves off in another direction (C and D). (From Erickson, 1978.)

[Handwritten margin notes: "Ectodermal cells / mesodermal cells" (top left); "like-cells neighbouring ea other" (top right)]

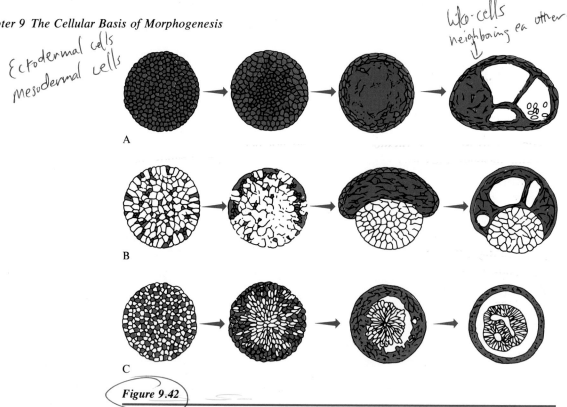

A

B

C

Figure 9.42

The pattern of sorting out of disaggregated and reaggregated embryonic cells. A, Combination of ectodermal and mesodermal cells. B, Combination of epidermal, mesodermal, and endodermal cells. C, Combination of epidermal and neural plate cells. (After Townes and Holtfreter, 1955.)

Specific Cellular Adhesiveness

Moscona (1957) extended the sorting out experiments of Townes and Holtfreter to cells of birds and mammals. He took advantage of the fact that the nuclei of cells derived from chick and mice stain differentially with hematoxylin. Therefore, when he mixed two different cell types derived from these animals (e.g., mouse mesonephric tissue and chick chondrocytes), he could identify the derivation of the cells in the mixed aggregate, whether from chick or mouse. In such an experiment, the mouse cells all form tubules, whereas the chick cells differentiate into cartilage, and they sort out from each other. Moscona's experiments were historically important for several reasons: (1) He developed a method of dissociating tissue from higher vertebrates using trypsin (Holtfreter used a pH shift), thus enabling him to extend sorting out behavior to higher vertebrates; (2) he showed, using a permanent nuclear marker, that without a doubt different cell types sort out in aggregates, rather than differentiate according to their location in the aggregate; and (3) he showed that cells from different tissues in higher vertebrates also possess some means of distinguishing self from non-self. Now sorting behavior had been observed across many species and was regarded as an important phenomenon that, if understood, might possibly explain organogenesis in the embryo.

Later, Moscona (1960) proposed that **specific cellular adhesiveness** is an important factor that determines cellular associations. According to this hypothesis, cell adhesion is a property of specific cell surface macromolecules that allow cells to recognize like cells and adhere. Thus, cells belonging to different tissues will segregate from one another in a mixed aggregate because they recognize and preferentially associate with their own kind.

Specificity of cell recognition would be determined by the molecular characteristics of the cell surface receptors.

Differential Adhesion Hypothesis

In contrast to Moscona, Steinberg (1963) did not believe that the sorting behavior in aggregates was necessarily governed by specific ligands, but rather by relative differences in adhesiveness. In a series of experiments, different tissue types were dissociated with trypsin and recombined in aggregates. From such aggregates, the cells will sort out according to tissue type, usually with one cell type surrounding another. His hypothesis was that the relative positions of cells in mixed aggregates result from random motility of the cells and quantitative differences in the adhesiveness between them. Thus, he predicted that cells with the stronger mutual adhesion will aggregate in the center, whereas cells with weaker attraction will remain at the surface. This he named the **differential cellular adhesiveness hypothesis** (Steinberg, 1970).

An example of such a sorting-out event is shown in Figure 9.43. In a mixture of limb-bud cartilage and heart cells, the heart cells eventually

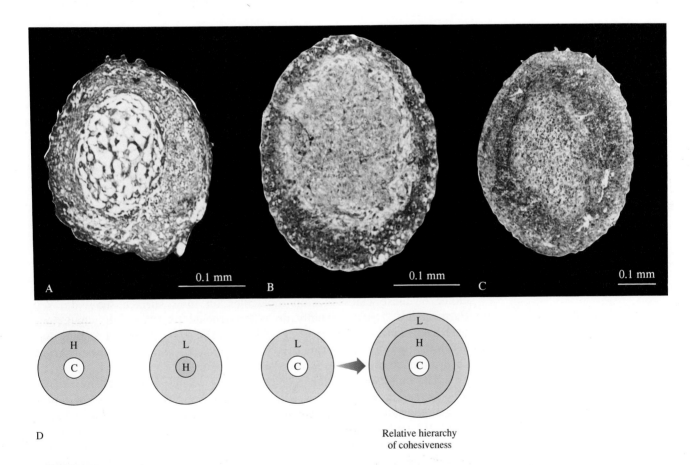

Figure 9.43

Hierarchy of sorting out in reaggregated chick embryo cells. Heart sorts out externally to limb bud chondrocytes (A) but internally to liver cells (B). When liver and chondrocytes are mixed (C), the liver cells sort externally to the chondrocytes. From such various combinations, a hierarchy of cohesiveness can be constructed (D). (A-C, From M. S. Steinberg. 1963. Reconstruction of tissues by dissociated cells. *Science, 141*: 401–408. Copyright 1963 by the American Association for the Advancement of Science.)

surround the cartilage (see Fig. 9.43A). When heart and liver cells are combined, the liver cells surround the heart cells (see Fig. 9.43B). Steinberg proposed that heart cells are less adhesive than cartilage, and liver is less adhesive than heart, accounting for the rearrangements of the cells in these two combinations. He concluded that, of these three tissues, liver is the least adhesive and cartilage is the most adhesive. Consequently, he predicted that if he were to combine liver and cartilage cells, the liver cells would surround the strongly adhesive cartilage cells. As shown in Figure 9.43C, this is exactly what he observed. Thus, there is an **adhesive hierarchy** in which cells at the top of the scale will segregate internal to those below it, whereas cells at the bottom of the scale segregate external to all others.

Phillips and Steinberg (1969) attempted to demonstrate different adhesive strengths by centrifuging reaggregates of cells derived from various tissues against a flat substratum. They believed that the degree of flattening of the round aggregate could be inversely correlated with the strength of intercellular adhesiveness by assuming that cell-cell adhesions had to be broken in order to flatten against the substratum. Thus, an aggregate with strong intercellular adhesiveness would resist flattening, whereas one with relatively weak adhesions would readily deform. The results using this assay are consistent with adhesive hierarchies derived from the original sorting-out studies.

Steinberg's model attempts to explain sorting-out behavior based on cells rearranging themselves into the most thermodynamically stable patterns. As such, the hypothesis does not deal with the particular chemical or physical mechanisms of cell-cell adhesions. It does, however, explain the consequences if there were differences in the strengths of adhesion between various kinds of cells.

Molecular Basis of Sorting Out

The preceding studies suggest that formation of aggregates and sorting out are mediated by cell adhesion molecules, but until recently the molecular details of this behavior were not known. Takeichi and colleagues (Nose et al., 1988) showed directly in a series of elegant experiments employing molecular cloning techniques that specific adhesion molecules can regulate cell aggregation and cell sorting. They transfected genes for either E- or P-cadherin, two different cell adhesion molecules (see Table 9–2), into a cell line known as L cells. The L cells, which previously had not expressed either of these cell adhesion molecules on their surface, now did so. Furthermore, L cells usually do not aggregate together, but when they express either E- or P-cadherin on their surface, they formed aggregates. When the two cell lines expressing E- or P-cadherin were mixed, they sorted out from each other (Fig. 9.44).

The biological relevance of this sorting was further tested by mixing transfected L cells with dissociated lung tissue. When recombined, the dissociated lung tissue sorts out into mesenchymal and epithelial components. If L cells labeled with a fluorescent dye (so that they can be recognized later) are mixed with the lung cells, they sort out with the mesenchyme. If, however, L cells expressing E-cadherin are mixed with dissociated lung, the L cells now sort out with the epithelium. These results show directly that (1) cells can sort out from each other when there is a difference in only a single type of cell adhesion molecule (L cells with either E- or P-cadherin); and (2) if a single cell adhesion molecule is altered, it can change the sorting behavior (in this case, from association with the lung mesenchyme to the lung epithelium).

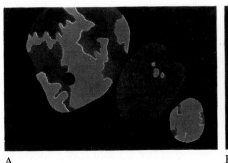
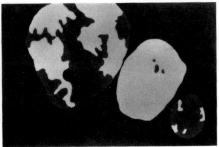

A   B

Figure 9.44

L cells transfected with the genes for P- or E-cadherin were mixed in a 1:1 ratio and cultured for 24 hours, during which time they formed aggregates. The aggregates were sectioned, incubated with antibodies to E- and P-cadherin, and the primary antibodies localized with secondary antibodies tagged with either rhodamine or fluorescein. Cells expressing P-cadherin are represented in (A) by colored regions in the aggregates, and those expressing E-cadherin are shown as light gray regions in (B). The two cell types are clearly segregated and show a complementary staining pattern. ×65. As shown in Color plate 14, the regions staining with rhodamine appear red, whereas those regions staining with fluorescein are green. (After Nose et al., 1988. Copyright © Cell Press.)

Changes in Cell Adhesion Accompany Changes in Morphogenetic Behavior

The early work of Holtfreter suggested that cell affinities could direct morphogenetic behavior. We now have direct evidence for this hypothesis, perhaps most graphically seen in the early gastrulation movements of the sea urchin embryo.

Just before gastrulation, the sea urchin blastula consists of a single epithelium enclosing the blastocoel. The epithelium tightly adheres to an external layer called the hyaline layer, and the epithelium rests on an intact basal lamina that lines the blastocoel (see Fig. 6.19).

The first gastrulation movement is the ingression of the primary mesenchyme cells (PMC). These cells separate from the vegetal plate by detaching from the hyaline layer, breaking through the basal lamina into the blastocoel, and then migrating along the basal lamina inside the blastocoel (Fig. 9.45). Research by McClay and his colleagues has shown that ingression of the primary mesenchyme cells is due to changes in at least three

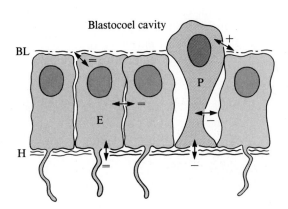

Figure 9.45

A summary of the changes in cell affinity that occur between the primary mesenchyme cells (P) and the external hyaline layer (H), the vegetal plate epithelium (E), and the internal basal lamina (BL) during sea urchin gastrulation. =: no change in affinity; +: an increase in affinity; −: a decrease in cell adhesivity. (After Fink and McClay, 1985.)

independent cell adhesions. First, the PMCs lose their affinity to the hyaline layer (McClay and Fink, 1982). This change in adhesiveness is specific for the PMCs because the future ectoderm and endoderm do not lose their attachment to hyalin. Second, the PMCs also detach from the other vegetal plate epithelial cells, and it can be demonstrated that there is a reduction in adhesiveness of the PMCs to the surrounding cells and to each other (Fink and McClay, 1985). Finally, at the same time that the PMCs lose affinity to the hyaline layer and other cells, they also show an increased affinity to the basal lamina (Fink and McClay, 1985). During this same time period, no change takes place in the adhesion of the ectoderm or endoderm to the basal lamina. Thus, three simultaneous changes in cell adhesion, either to other cells or to components of the extracellular matrix, can account for the early separation of PMCs from the vegetal plate.

Once the PMCs have separated from the vegetal plate, they migrate on the basal lamina that lines the blastocoel until they reach a point at the equator of the gastrula where they organize into a ring pattern. At this time, the PMCs demonstrate a renewed adhesiveness toward each other, which had been lost at the time of ingression. Thus, changes in cell affinities are carefully timed and integrated in order to generate the appropriate patterns of PMC migration. The molecular basis for these changes in affinity is not yet known in sea urchins.

What Are the Adhesive Molecules?

We have shown that embryonic cells have different affinities for each other and that changes in those affinities, for each other as well as for the extracellular matrix, can direct morphogenetic processes and histogenesis. What is the molecular nature of these affinities? Recently, much effort has been made to identify the molecules that mediate adhesive interactions between cells; several different approaches have been taken to identify such molecules with the assumption that "affinities" represent at least in part, adhesiveness.

The first adhesion molecules to be identified were those substances that could promote cell aggregation when cells were dissociated under a variety of conditions and then recombined. These aggregation factors were isolated from the dissociation medium or were detected in conditioned medium (i.e., medium in which the cells had been incubated and into which they shed adhesive molecules).

Several problems are inherent in trying to identify cell adhesion molecules through their presence in a dissociation medium. Most notably, there is no guarantee that particular cell adhesion molecules will be released from the membranes into the medium. Secondly, there is some doubt that the factors involved in cell aggregation *in vitro* are the same factors that are responsible for cell adhesion in the embryo. Recently, two quite different methods have been developed that have resulted in the identification of many cell adhesion molecules in a relatively short period of time.

One approach for identifying cell adhesion molecules is based upon the observations, first made by Takeichi (1977), that adhesion in aggregation systems can be either Ca^{2+}-dependent (CD) or Ca^{2+}-independent (CID) (i.e., some adhesion molecules require the presence of Ca^{2+} to be functional, whereas others do not). Generally, any one cell type has both Ca^{2+}-dependent and independent adhesion systems. What Takeichi noted was that one or the other of these adhesion systems could be removed from the cell surface depending upon the digestion conditions used. Specifically, if cells are digested with trypsin in the presence of Ca^{2+}, the CID system is removed, but the CD system remains. This is thought to be because the Ca^{2+} protects the CD system proteins from digestion. On the other hand, if cells are digested with low concentrations of trypsin in the absence of Ca^{2+}, the CD system is digested, whereas the CID system remains. This is because the CD system is much more sensitive to trypsin than the CID system, and once the protective Ca^{2+} is removed, the CD proteins are easily digested. By selectively removing these different adhesion systems and identifying the proteins remaining in the membranes, it has been possible to identify membrane glycoproteins that are potentially important in adhesion. Takeichi, in particular, has been instrumental in isolating many of the Ca^{2+}-dependent adhesion molecules, which are called **cadherins** (see Table 9–2). To make a strong case for their being adhesion molecules, however, the molecules have been purified and antibodies have been made that specifically bind to these molecules and perturb adhesion.

Another successful approach has used immunological tools to identify previously unknown adhesion molecules. The rationale is that if antibodies can be made to cell surface molecules, the antibodies should be able to block adhesion if they bind adhesion molecules. Once the adhesion-blocking ability of an antibody has been demonstrated, that antibody can then be used to purify the adhesion molecule itself. This approach was first used by Gerisch and his colleagues (for review, see Gerisch, 1980) to identify molecules that mediate the adhesion of slime molds. The approach has now been successfully applied to the isolation of a number of cell-cell adhesion molecules in vertebrate embryos by a number of laboratories, notably by Edelman and his colleagues. Probably the best characterized adhesion molecule isolated in this fashion is called **N-CAM** (**neural-cell adhesion molecule**); because it is so well characterized, we shall discuss it in some detail.

N-CAM

Brackenbury et al. (1977) identified adhesion molecules in the neural retina by first injecting chicken retina cells into a rabbit and allowing the rabbit to produce antibodies against retina cell surface components. The rabbit serum was then tested until serum was found that contained antibodies that bound to retina cell surfaces. These antibodies were then added to a suspension of retina cells to observe whether they would inhibit the adhesion of the retina cells to each other. The antibodies used were not the usual intact, divalent molecules, but rather were cleaved so that each molecule had only one antigen binding site. These cleaved univalent antibodies are known as **Fab′ fragments**. When Fab′ fragments were added to a suspension of retina cells, adhesion was blocked. Uncleaved antibodies, being divalent, would have agglutinated the cells rather than acting to block adhesion.

Once an antibody is developed that can block cell adhesion, it can be used for two additional purposes. First, it can be used to isolate the adhesion molecule itself. Thiery and other members of Edelman's laboratory (Thiery et al., 1977) solubilized the components of the retina cell plasma membranes and used the antibody to purify the adhesion molecule by affinity chromatography. In this method, an antibody is bound to a solid substrate in a column, and the membrane components are poured through the column. The antigen that is recognized by the immobilized antibody binds to the antibody, whereas the rest of the components flow through. Finally, the antigen is eluted from the column, providing a pure sample of the protein. The first molecule from Edelman's group purified using these techniques was N-CAM, which is a glycoprotein of 160,000 M_r and is unusual in that its sugar component is made up of a highly negatively charged sugar called **sialic acid.** N-CAM has three domains: the amino-terminal end of the molecule that contains the cell-binding region, a middle domain that carries the polysialic acid-rich sugar, and the carboxyl-terminal end that is embedded in the plasma membrane (Edelman, 1986; 1988). Two cells containing this protein bind together because the amino-terminal end of an N-CAM molecule on one cell binds to the complementary amino-terminal end of an N-CAM molecule on the other cell (Fig. 9.46). This binding of two like molecules to each other is known as **homophilic binding**.

The second important purpose of having specific antibodies against adhesion molecules is that they can be used to perturb normal developmental processes in the embryo and, thus, to determine directly if the particular putative adhesion molecule controls a developmental event. Such a study was done on the chick retina (Buskirk et al., 1980). Normally by day 6, the neuronal fibers and cell bodies have rearranged themselves in the chick retina to form two layers, known as the inner and outer plexiform layers. This same cell rearrangement can occur if the retina is explanted into organ culture. However, if antibodies to N-CAM are added to the culture medium, this normal rearrangement does not occur, and instead the arrangement of neuronal cell bodies and fibers is random (Fig. 9.47). Clearly, N-CAM has some role in the normal histogenesis of the retina. However, the mechanism by which N-CAM actually produces and directs this rearrangement is not known.

One problem with trying to perturb development with antibodies or other specific chemicals is that these substances may affect more than just the specific adhesion molecule being investigated. However, nature provides its own experimental animals in the form of developmental mutants, which can yield powerful evidence for the precise roles of individual adhesion molecules. One example is a particular mutation in mouse called *staggerer*, which has provided evidence that N-CAM is critical in the normal development of the cerebellum (Edelman and Chuong, 1982).

The *staggerer* mice have defects in the connection of some of the neurons in the cerebellum, and these result in abnormalities in their motion and balance. It was also discovered that there is a defect in the processing of N-CAM in these mice. In the normal mouse brain, much of the sialic acid (see Fig. 9.46) is removed from the sugar chains of N-CAM at around day 14. As a result of this removal of a highly negatively charged polysac-

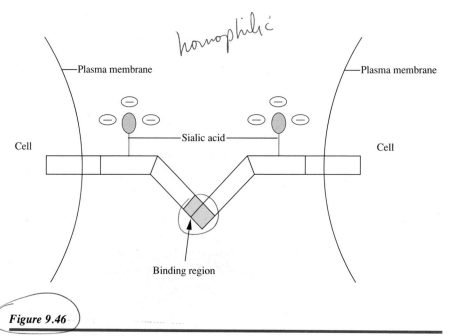

Figure 9.46

Model for homophilic binding interaction for N-CAM. The N-CAM molecule has three domains (see text). (After Edelman, 1983.)

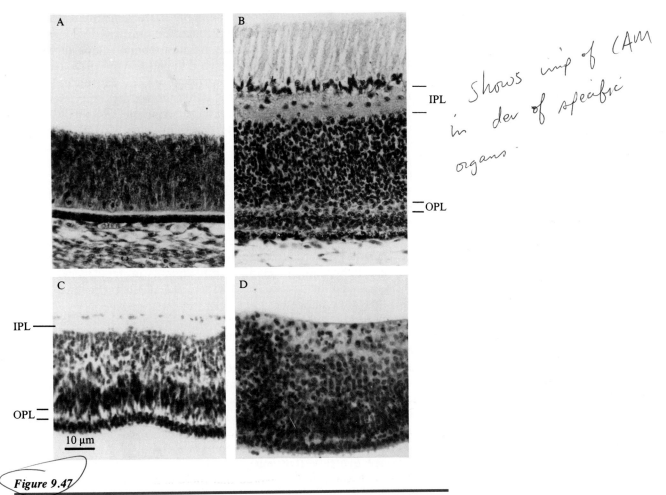

shows imp of CAM in dev of specific organs.

Figure 9.47

Histogenesis of the chick retina can be disrupted by antibodies against N-CAM. A, and B, Sections through retinas of 6- and 9-day-old chick embryos. Between these days, the retinal cells sort out to form the inner (IPL) and outer plexiform layers (OPL). C, The retina can also develop in culture and forms the OPL and IPL, although the retina is thinner than *in vivo*. D, When the retina is cultured with Fab' fragments against N-CAM, normal histogenesis is inhibited. (From Buskirk et al., 1980. Reprinted by permission from *Nature*, Vol. 285, pp. 488–489. Copyright © 1980 Macmillan Magazines Limited.)

ie cells can't homophilic bind & aggregate.

charide, the N-CAM molecules adhere more tightly to each other; consequently, so do the nerve cells (Hoffman and Edelman, 1983). This gradual conversion of N-CAM from an embryonic to an adult form is known as the E (Embryonic) → A (Adult) conversion. In the mouse brain, it is reasonable to expect that changes in cell adhesiveness may direct morphogenetic events, such as migration or sorting of neurons in the cerebellum. In *staggerer*, the normal conversion from the E to the A form of N-CAM has not occurred, even by day 21. The exact mechanism whereby this failure of the E → A conversion brings about the morphological defects still remains to be established, but the correlation is strong evidence linking N-CAM modulation to a developmental phenomenon.

Role of CAMs in Development

Using the two isolation procedures described earlier, many cell adhesion molecules (CAMs) have been identified that appear to be developmentally regulated (i.e., when the distribution of these molecules is determined in an embryo using antibodies, they are found in precisely defined areas, and the changes in the distribution of these molecules are correlated with morphogenetic movements and differentiation). A list of the most important CAMs and their distribution in the early embryo is found in Table 9–2. Note that these molecules can be either Ca^{2+}-independent or Ca^{2+}-dependent and that they mediate cell-cell attachment by either homophilic or heterophilic mechanisms. Several of these molecules have been found even in the earliest stages of development, and almost all tissues throughout development have at least one of these early CAMs (Fig. 9.48). Other CAMs appear at somewhat later stages.

It is important to recognize that the CAMs have precise distributions that change during embryogenesis and that are correlated with morphogenetic movements or cell differentiation. For example, neural crest cells lose A-CAM as they migrate out of the neural tube but regain it as they aggregate to form ganglia, and E-cadherin appears on mouse blastomeres just before the time when they undergo compaction at the 16-cell stage. Such fluctuations in CAM distribution are too numerous to discuss in detail here, but the student is referred to several reviews on the subject (Edelman, 1986; Takeichi, 1988).

It is presently believed that these molecules play important roles in developmental events. Clearly, perturbation experiments using antibodies against CAMs have shown that these molecules can be critical for normal development to proceed. What is not yet known is how these adhesion molecules mediate developmental processes. Are they functioning merely to hold like cells together, while other cues direct development? Or do they provide the directions to carry out normal developmental programs? Until more is understood of the molecular events that these molecules mediate, we can offer little more than speculation at this time.

Another question that must be answered is how many adhesive molecules are necessary to control all the developmental processes. Some

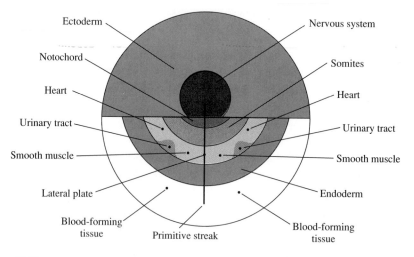

Figure 9.48

A fate map (see Chap. 11) of a chick embryo superimposed upon the blastoderm shows the origins of all the embryonic tissues. The distribution of CAMs from 5–14 day embryos based on immunofluorescence localization has been superimposed on the same fate map. N-CAM is represented by color, L-CAM by gray, and tissues where neither is present by white. Mesodermal cells that form the kidneys express both CAMs. (After Edelman, 1984.)

Table 9–2 Cell-Cell Adhesion Molecules

Molecule (+ synonyms)	Binding Mechanism	Ion Dependence	Tissue Distribution	
N-CAM (D₂, BSP-2)	Homophilic	Ca²⁺-independent	*Early:*	Epiblast Neural plate Placodes Mesoderm
			Late:	Nervous system Striated muscle Adrenal cortex Gonad cortex Heart Kidney epithelia Somatopleural and splanchnopleural elements
Ng-CAM (L1, NILE)	Heterophilic	Ca²⁺-independent	*Late:*	Neural epithelium
L-CAM (E-cadherin, Uvomorulin, Cell CAM 120/80, Arc-1)	Homophilic	Ca²⁺-dependent	*Early (Preimplantation):*	Blastomeres Inner cell mass Trophectoderm
			Early (Postimplantation):	Ectoderm Endoderm
			Late:	Most epithelia
A-CAM (N-cadherin, C-Cal-CAM)	Homophilic	Ca²⁺-dependent	*Early:*	Mesoderm Notochord
			Late:	Neural tissues Lens Cardiac and skeletal muscles Nephric primordia Mesothelium Primordial germ cells
P-cadherin	Homophilic	Ca²⁺-dependent	*Early:*	Extraembryonic ectoderm Endoderm Lateral plate mesoderm Notochord
			Late:	Placenta Epidermis Pigmented retina Mesothelium
gp140k		Ca²⁺-dependent	*Xenopus* epithelial cell lines	

From Edelman, G. M. 1986. Cell adhesion molecules in the regulation of animal form and tissue pattern. Ann. Rev. Cell Biol., *2:* 81–116; and Takeichi, M. 1988. The cadherins: Cell-cell adhesion molecules controlling animal morphogenesis. Development, *102:* 639–655.

investigators (e.g., Sperry, 1963) have proposed that very specific molecules exist that govern each and every adhesive event. For example, adhesive molecules involved in directing liver histogenesis would be different from those that direct neural retina development. This view would, of necessity, predict that there are nearly countless numbers of adhesive molecules. On the other hand, some biologists believe that only a few adhesive molecules are necessary to direct all the morphogenetic events in development (Edelman, 1986). Changes in affinity could possibly come about by modulating the number of adhesive molecules on the cell, the distribution of the molecules on the cell, or a modulation in the molecule itself that would affect its adhesive strength (such as the E → A conversion of N-CAM), but the evidence is by no means conclusive. Presumably, timing and location are important, too; thus, liver and retinal cells could have the same adhesive molecule because they would never encounter each other.

The truth probably lies somewhere in between. For example, modulation of only a few CAMs during early development might be involved in the earliest morphogenetic movements, whereas at later times more refined movements, such as those that must take place during tissue histogenesis or connectivity between axons and their target organs, might be directed by a greater array of molecules. This question will be answered only when the developmental roles of each of these adhesive molecules have been identified.

9–8 Cell Death and Cell Proliferation

In addition to cells moving from one place to another to shape the embryo, it is clear that selective death of cells plays a crucial role in sculpting various structures of the emerging embryo (Saunders, 1966). Among these embryonic structures are the brain, limbs, and the palate. Why this should be such a prevalent mechanism isn't clear, because it seems that self-destruction of embryonic cells is a waste of resources and information.

Cell death is seen quite prominently in the developing limb. Areas of cell death can be visualized in the chick limb by staining the embryos *in ovo* with the vital stain Nile Blue sulfate (Saunders et al., 1962). This stain becomes concentrated in dead and dying cells as well as in the macrophages involved in a "mop up" campaign. Four areas of cell death in the mesoderm are noted (Fig. 9.49): at the anterior and posterior margins of the limb in regions known as the **anterior (ANZ)** and **posterior necrotic zones (PNZ)**, in a central region known as the "opaque patch" because it is opaque to transmitted light, and in the **interdigital necrotic zones (INZ)** separating the chondrifying digits.

The role of the necrotic zones between the developing digits seems clearly to shape and separate the digits. This idea is supported by studies of various mutants such as *talpid* in the chick embryo, in which the necrotic zones do not appear between the digits, with the consequence that a web of mesenchyme continues to connect the digits (Hinchliffe and Thorogood, 1974).

The roles of the ANZ and PNZ have not been so easy to establish experimentally. Again, mutants that do not develop an ANZ or PNZ provide us with some clues. In the *talpid* mutant, the ANZ and PNZ are absent, and the resulting limbs have up to eight digits. The ANZ is absent in the mole as well, unlike in the related rat and mouse, and these animals have an extra digit. The emerging evidence suggests that the ANZ and PNZ are responsible for regulating the amount of limb mesenchyme and consequently determining the correct size of the limb and the number of digits.

The control of selective cell death is not understood, although this process seems to be programmed into the cells destined for destruction. This assertion is based upon two facts: First, cells such as those in the PNZ can be explanted into tissue culture and will subsequently undergo death at the appropriate time, and, second, single gene mutations can change the developmental potentials of necrotic zones, as we have seen in *talpid*.

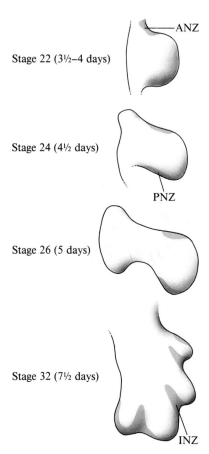

Stage 22 (3½–4 days) — ANZ

Stage 24 (4½ days) — PNZ

Stage 26 (5 days)

Stage 32 (7½ days) — INZ

Figure 9.49

Necrotic zones in chick limbs of increasing age, identified with Nile blue sulfate. The numbers refer to the developmental stage using the Hamburger and Hamilton staging system. (After Saunders, 1969.)

Cell death is also critical in the development of the nervous system (Silver, 1978). Cell death is necessary for separating the developing neural tube from the contiguous epidermis. We have already seen that cell death is essential in the optic stalk for the passage of retinal fibers on their way to the optic tectum (Silver and Robb, 1979).

Cell Death in the Nematode

Although cell death has been studied in vertebrates, the underlying genes controlling this morphogenetic event are not understood. One model system in which the genetics of **programmed cell death** is being approached with success is in the nematode *Caenorhabditis elegans*. Over 130 instances of cell death have been observed in *C. elegans,* and some of the genes responsible have been cloned. One of the interesting systems in which cell death is responsible for shaping an organ is in the gonad. In the *C. elegans* hermaphrodite, the uterus is a two-armed structure attached at the base of the arms to the vulva. In a second genus of nematode, *Panagrellus redivivus* (Fig. 9.50A), the uterus has only one arm. This difference in structure of

Regular text continues on page 383.

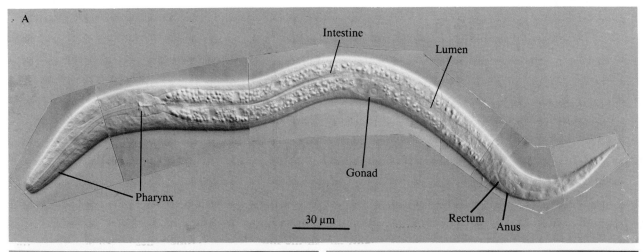

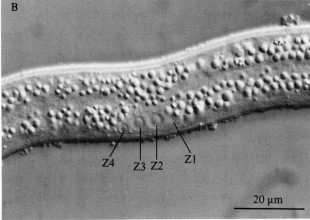

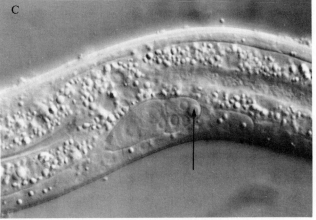

Figure 9.50

A, Photograph of a single living *Panagrellus* worm. Note that with special optics all the internal cells can be observed and identified. B, Higher magnification to show the four cells that will give rise to the gonad and the associated germ cells. C, An arrow points to a dying Z4.pp cell in a *Panagrellus* female. This programmed cell death will result in a one-armed uterus. (A, From Sternberg and Horvitz, 1982. B, and C, From Sternberg and Horvitz, 1981.)

Essay:

Roles of Oncogenes in Controlling Cell Proliferation

Recently, much has been learned about the roles of growth factors in controlling cell proliferation, especially in tissue culture model systems. Furthermore, evidence is rapidly mounting that growth factors play key roles in regulating various aspects of embryonic development, including cellular growth control. Although the effects of adding exogenous growth factors to embryonic cells have been studied in detail, little is known about their precise roles in normal development or how their production may be controlled. An unexpected source of new information on the control of cell proliferation comes from studies of tumor viruses, which release cells from growth control originally established during embryogenesis, transforming them into cancer cells. An understanding of how normal cells are released from growth control during transformation may well elucidate how this control is exerted on embryonic cells in the first instance.

Malignant transformation by tumor viruses includes a wide variety of changes, including alterations in morphology and adhesion to the substratum, an increase in invasive behavior, and a loss of growth control. It was first demonstrated in 1979 (Copeland et al.) that a single gene in the Rous sarcoma virus called the *src* gene can cause transformation. Since then, many such individual genes have been isolated from a wide variety of tumor viruses. These genes have been called **viral oncogenes (v-oncs)**. In the last ten years, the very exciting finding has been made that these same or similar genes are present in all normal cells and that either the overexpression of, or mutation in,

these genes is responsible for the development of cancer cells from normal cells (Bishop, 1987). The cellular equivalents of the viral oncogenes are known as **cellular oncogenes (c-oncs)** or **proto-oncogenes**. Most of the oncogenes that have been recognized to date appear to have some role in the control of growth (see Table 9–A). Furthermore, it is now becoming clear that many (if not all) of the proto-oncogenes are important regulators of a variety of developmental processes (some of which are discussed elsewhere in this book).

For a clear understanding of growth control, we must examine the functional relationship between growth factors and proto-oncogenes (the so-called **growth control cascade**). Growth factors are secreted into the extracellular medium by one cell type and cause changes in other cell types after binding to receptors on their plasma membranes. When growth factors bind to their receptors, a whole sequence of events mediated by a variety of proteins occurs to transmit the cell division signal through the cytoplasm to the nucleus. Many of the proteins in these cascades are similar to known cellular oncogene products. One such cascade is summarized in Figure 1. (Different growth factors have variations of this sequence.) For example, when platelet-derived growth factor (PDGF) is added to a cell, it first binds to a PDGF receptor at the cell surface. This binding can stimulate tyrosine kinase activity in some growth factor receptors that will result in the phosphorylation of many proteins in the cell. Growth factor binding in turn also causes the activation of an associated G-protein. In the continuing cascade, G-protein activation in turn activates the enzyme phospholipase C, which cleaves the membrane lipid phosphatidylinositol-4, 5-bisphosphate (PIP_2) into two components: inositol trisphosphate (IP_3) and diacylglycerol (DAG). These newly generated components have numerous effects in the cell, many of which stimu-

late cell mitosis. For example, IP_3 releases Ca^{2+} from the endoplasmic reticulum. Diacylglycerol, on the other hand, activates **protein kinase C**, which has numerous effects, including phosphorylating proteins and activating the Na^+/H^+ pump, which increases cytoplasmic pH. The increases in cytoplasmic calcium and pH are known to stimulate cell division. (You will recall that a similar sequence of events is involved in the activation of the egg at fertilization; see p. 152.) Phosphorylation of cytoplasmic proteins likewise sends a message to the nucleus to divide, although it is not clear how.

Viral oncogenes have been identified that substitute for, and mimic, various elements in this growth response pathway. They fall into five categories: those that mimic (1) growth factors, (2) growth factor receptors, (3) G-proteins, (4) protein kinases, and (5) nuclear proteins. Their cellular counterparts are shown in Table 9–A and Figure 2. For example, the oncogene isolated from the Simian sarcoma virus, v-*sis*, is similar to PDGF and binds to actual PDGF receptors. Thus, when this gene is turned on in a tumor, it may grow rapidly as a result of the increase in the amount of the growth factor produced. Another example is v-*erb*-B, the oncogene found in the avian erythroblastosis virus. The v-*erb*-B gene product is identical to the cytoplasmic and membrane domains of the epidermal growth factor (EGF) receptor. It only lacks the EGF binding portion of the normal EGF receptor. When EGF binds to its normal receptor, it activates the receptor's tyrosine kinase activity, which is known to be a signal for mitosis. The *erb*-B receptor, without the binding site, appears to be in a constantly stimulated state of high kinase activity and thus will continuously stimulate a cell to divide, even in the absence of EGF itself. The oncogene v-*ras* is similar to other G-proteins; its role in cell division is indicated by an experiment in which an antibody to *ras* is

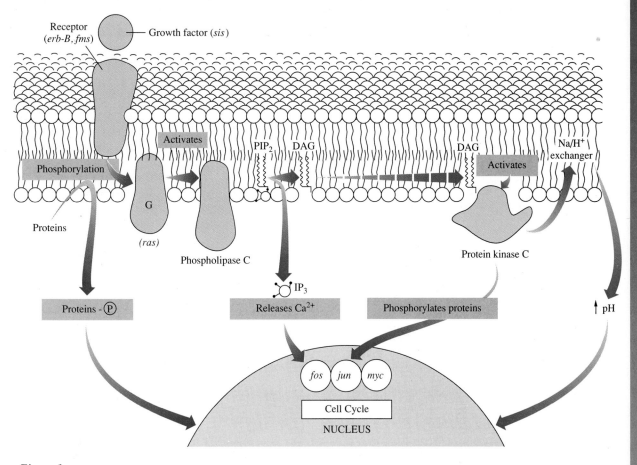

Figure 1

Diagrammatic summary of signaling events that mediate growth factor-induced cell division. In this cascade, a series of phosphorylation events by activated kinases, an increase in cytoplasmic Ca^{2+}, and an increase in pH all signal mitosis. Oncogene products can substitute for normal cellular proteins and stimulate or short-circuit this cycle. In addition, the nuclear oncogenes *fos*, *jun*, and *myc* can bypass the cytoplasmic events entirely. (After Berridge, 1985.)

injected into amphibian embryos, thus blocking cell division (Baltus et al., 1988). The nuclear oncogenes, which are at the end of the cascade, are particularly interesting because they can modulate gene expression itself: *fos* and *jun* are transcriptional regulators, whereas *myc* is apparently involved in the processing of nuclear RNA molecules (Prendergast and Cole, 1989).

We are currently learning much about what activates or turns on these genes in tumor cells and the roles that the individual oncogenes play in the malignant process. In addition, the discovery of new oncogenes has increased our understanding of steps in the pathway of normal growth control. An exciting consequence of this research is newly emerging information on the roles of proto-oncogenes in development that is allowing us to look at growth control during development from a new perspective.

continues

Table 9–A Proto-oncogenes: Localization and Possible Function

c-oncs	Identity	Localization	Activity	Putative Function
Tyrosine kinases				
c-*erb*-B	EGF receptor	Plasma membrane	EGF binding TGFα-binding	Signal transduction
c-*neu*	homology to EGF-receptor	Plasma membrane	Tyr kinase	?
c-*fms*	CSF-1 receptor	Plasma membrane	CSF-1 binding Tyr kinase	Signal transduction
c-*abl*		Plasma membrane	Phosphorylates vinculin	B-cell differentiation
c-*src*		Cytoplasmic, membrane associated	Tyr kinase (phosphorylates vinculin, vimentin, filamin, p36)	Neuron and muscle development?
c-*fes* } c-*fps* }		Cytoplasmic, membrane associated	Tyr kinase	Macrophage development?
c-*mos*	CSF	Cytoplasmic	Ser/thr kinase	Meiotic inhibitor
GTPases				
c-Ki-*ras* } c-Ha-*ras* } N-*ras* }		Cytoplasmic, membrane associated	GTP binding	Adenylate cyclase regulation?
Nuclear proteins				
c-*myc*		Nucleus	Nuclear RNA processing	Proliferation?
N-*myc*				?
L-*myc*				?
c-*fos* } c-*jun* }	transcription factor AP-1	Nucleus	DNA-binding	Transcriptional regulation
c-*myb*		Nucleus	DNA-binding	Differentiation of hematopoietic cell
Others				
c-*sis*	PDGF B chain	Secreted protein	PDGF-R binding	Mitogenesis
c-*erb*-A	T3 receptor	Cytoplasmic and nuclear	Thyroxine binding	Metabolic regulator

Abbreviations: EGF, epidermal growth factor; Tyr, tyrosine; TGF-α, transforming growth factor-α; CSF-1, colony stimulating factor-1; CSF, cytostatic factor; Ser/Thr, serine/threonine; PDGF, platelet-derived growth factor; T3, thyroid hormone.

From Rempel, S. 1990. Steroid hormone regulation of c-*myc* gene expression in proliferating chick oviduct. Ph.D. thesis, University of Calgary, Calgary, Alberta, Canada.

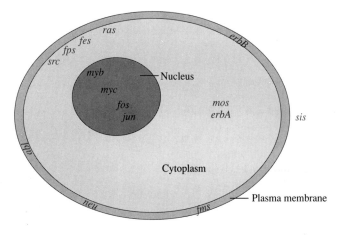

Figure 2

Subcellular localization of proto-oncogene proteins. (After Rempel, 1990.)

the uterus between the two worms is due to the programmed cell death in one cell that makes up the gonad.

In both worms, the gonad and the germ cells arise from four cells, termed Z1, Z2, Z3, and Z4 (see Fig. 9.50B). The germ cells are all derived from division of the Z2 and Z3 cells. The somatic cells of the gonad are derived from the Z1 and Z4 cells. One of the descendants of the Z1 and Z4 cells, named Z1.aa and Z4.pp respectively, is called the distal tip cell. There are two distal tip cells in the early gonad, one at the tip of each arm of the uterus in *C. elegans*. If one of these is destroyed by a laser, that arm fails to grow out, and the uterus is similar in shape to the one-armed *P. redivivus* uterus (Kimble, 1981).

Sternberg and Horvitz (1981) discovered that *P. redivivus* similarly has two distal tip cells, which are also designated Z1.aa and Z4.pp. The Z1.aa remains at the anterior tip of the primordium and leads the outgrowth of that arm of the uterus. The Z4.pp cell, on the other hand, undergoes cell death, producing the truncation of the second arm of the uterus (see Fig. 9.50C). This is a dramatic demonstration of the effect of cell death on the shape of an organ.

The genetic basis of this programmed cell death is being intensively investigated in these organisms. A number of mutant genes have been isolated that control various phases of cell death. For example, the genes *ced-3* and *ced-4* control the first known step in the developmental pathway for programmed cell death (Ellis and Horvitz, 1986) and may be required for the initiation of the cell death program. *Caenorhabditis elegans* mutants for either of these genes produce an embryo in which virtually all the cells that normally die will instead survive and may differentiate. Other genes are involved in other steps of the cell death pathway. For example, the wild-type genes *ced-1*[+] and *ced-2*[+] are required for the phagocytosis of the dying cells, and gene product nuc-1 is a nuclease that degrades the DNA of the dead cells (Hedgecock et al., 1983). The identification of additional genes in the cell death cascade (Fig. 9.51) and the isolation of the gene products will ultimately elucidate the mechanisms by which cell death is programmed into a cell.

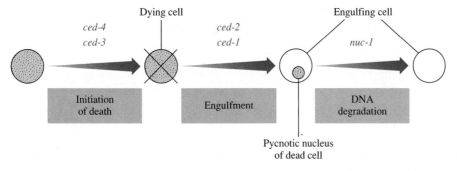

Figure 9.51

A proposed scheme for control of programmed cell death in *C. elegans*. The wild-type genes *ced-3* and *ced-4* are probably those that initiate the process of cell death. In their absence, cells that would normally die survive, instead. In the normal embryo, once cells die, they are phagocytosed by neighboring cells. This function is controlled by at least two wild-type genes: *ced-2*[+] and *ced-1*[+]. Finally, the *nuc-1* gene product is a nuclease that degrades the DNA of the dying cells. Other genes may be necessary to complete this pathway. (After Ellis and Horvitz, 1986.)

Cell Proliferation

The reciprocal process of selective cell proliferation, or **hyperplasia**, is also an important morphogenetic mechanism. This process is also aptly demonstrated in the limb. The limb itself develops initially as a site of rapid proliferation in the somatic mesoderm that results in the protrusion of the limb bud. Furthermore, in some vertebrates, such as amphibians, there is no interdigit necrotic zone, and the digits form by differential growth. In anurans, for example, the digits form by the proliferative outgrowth of "digital rays" (Hinchliffe, 1972). The different developmental strategies for the formation of digits pose a fascinating evolutionary problem.

9–9 Conclusions

We have, of necessity, dealt separately with the various mechanisms of morphogenesis in this chapter. It is important to remember, however, that morphogenesis is actually a complicated orchestration of all of the mechanisms that have been discussed. Consider primary mesenchyme cell migration in sea urchin embryos, for example. Migration of these cells is directed by changes in the affinity of cells to each other as well as to their migratory substrata. Changes in cell affinities are carefully timed and integrated in order to generate the appropriate patterns of primary mesenchyme cell migration. Migration itself is mediated by elements of the cytoskeleton. In Chapters 15, 16, and 17, we shall discuss how these morphogenetic processes are utilized and integrated to generate some of the individual organ systems that comprise the embryo.

9–10 Chapter Synopsis

Embryonic form is achieved by the complex movement and rearrangement of cells known as morphogenetic movements. These movements can involve translocation of individual cells from one area to another, or epithelial sheets can deform, bend, or spread. These morphogenetic movements are dependent upon changes in single cells and, for the most part, these changes are accomplished by the cytoskeleton. In order for the cytoskeleton to cause morphogenesis, the cells must adhere either to a substratum in the case of single cell locomotion, or to each other, as in the case of epithelial sheet bending. The biochemistry of these adhesive junctions is now beginning to be understood.

A variety of factors control morphogenetic movements. Cells can be directed in their migration by molecules in their environment, the so-called extracellular matrix. Alternatively, the final positions of cells within a tissue or organ can be determined by cell surface properties, especially the cell surface adhesion molecules known as CAMs.

Two other mechanisms can sculpt the embryo. Cells in a particular region will predictably die to change embryonic form. The digits develop as a result of death of cells between the presumptive digits. Alternatively, a group of cells may divide more rapidly than their neighbors, creating an outgrowth. The limb is an example of a structure that develops as a result of selective cell proliferation.

References

Abercrombie, M. 1970. Contact inhibition in tissue culture. In Vitro, *6*: 128–142.

Abercrombie, M. 1980. The Croonian Lecture, 1978. The crawling movement of metazoan cells. Proc. R. Soc. Lond. B., *207*: 129–147.

Abercrombie, M., G. A. Dunn, and J. P. Heath. 1977. The shape and movement of fibroblasts in culture. *In* J. W. Lash and M. M. Burger (eds.), *Cell and Tissue Interactions*. Raven Press, New York, pp. 57–70.

Abercrombie, M., J. E. M. Heaysman, and S. M. Pegrum. 1970. The locomotion of fibroblasts in culture. I. Movements of the leading edge. Exp. Cell Res., *59*: 393–398.

Adams, R. J., and T. D. Pollard. 1989. Binding of myosin I to membrane lipids. Nature (Lond.), *340*: 565–568.

Adamson, E. D. 1987. Oncogenes in development. Development, *99*: 449–471.

Adelstein, R. S. 1982. Calmodulin and the regulation of the actin-myosin interaction in smooth muscle and nonmuscle cells. Cell, *30*: 349–350.

Balinsky, B. I. 1981. *An Introduction to Embryology*, 5th ed. Saunders College Publishing, Philadelphia.

Baltus, E. et al. 1988. Injection of an antibody against a p21 c-Ha-*ras* protein inhibits cleavage in axolotl eggs. Proc. Natl. Acad. Sci. U.S.A., *85*: 502–506.

Bard, J. B. L., and E. D. Hay. 1975. The behavior of fibroblasts from the developing avian cornea. J. Cell Biol., *67*: 400–418.

Beebe, D. C. et al. 1979. Lens epithelial cell elongation in the absence of microtubules: Evidence for a new effect of colchicine. Science, *206*: 836–838.

Bergen, L. G., and G. G. Borisy. 1980. Head to tail polymerization of microtubules *in vitro*: Electron microscope analysis of seeded assembly. J. Cell Biol., *84*: 141–150.

Berridge, M. J. 1985. The molecular basis of communication within the cell. Sci. Am., *253* (10): 142–152.

Bishop, J. M. 1987. The molecular genetics of cancer. Science, *235*: 305–311.

Borisy, G. G. 1978. Polarity of microtubules of the mitotic spindle. J. Mol. Biol., *124*: 565–570.

Boucaut, J. C. et al. 1984. Biologically active synthetic peptides as probes of embryonic development: A competitive peptide inhibitor of fibronectin function inhibits gastrulation in amphibian embryos and neural crest cell migration in avian embryos. J. Cell Biol., *99*: 1822–1830.

Brackenbury, R. et al. 1977. Adhesion among neural cells of the chick embryo. I. An immunological assay for molecules involved in cell-cell binding. J. Biol. Chem., *252*: 6835–6840.

Brinkley, B. R. et al. 1981. Tubulin assembly sites and the organization of cytoplasmic microtubules in cultured mammalian cells. J. Cell Biol., *90*: 554–562.

Bronner-Fraser, M. 1985. Alterations in neural crest migration by a monoclonal antibody that affects cell adhesion. J. Cell Biol., *101*: 610–617.

Bronner-Fraser, M., and T. Lallier. 1988. A monoclonal antibody against a laminin-heparan sulfate proteoglycan receptor perturbs cranial neural crest cell migration *in vivo*. J. Cell Biol., *106*: 1321–1329.

Buckley, I. K., and K. R. Porter. 1967. Cytoplasmic fibrils in living cultured cells. A light and electron microscope study. Protoplasma, *64*: 349–380.

Burgeson, R. E. 1988. New collagens, new concepts. Ann. Rev. Cell Biol., *4*: 551–577.

Burnside, B. 1971. Microtubules and microfilaments in newt neurulation. Dev. Biol., *26*: 416–441.

Burridge, K. et al. 1988. Focal adhesions: Transmembrane junctions between the extracellular matrix and cytoskeleton. Ann. Rev. Cell Biol., *4*: 487–525.

Buskirk, D. R. et al. 1980. Antibodies to a neural cell adhesion molecule disrupt histogenesis in cultured chick retinae. Nature (Lond.), *285*: 488–489.

Byers, B., and K. R. Porter. 1964. Oriented microtubules in elongating cells of the developing lens rudiment after induction. Proc. Natl. Acad. Sci. U.S.A., *52*: 1091–1099.

Carter, S. 1967. Haptotaxis and the mechanisms of cell motility. Nature (Lond.), *213*: 256–260.

Chen, W.-T. 1981a. Surface changes during retraction-induced spreading of fibroblasts. J. Cell Sci., *49*: 1–13.

Chen, W.-T. 1981b. Mechanisms of the retraction of the trailing edge during fibroblast movement. J. Cell Biol., *90*: 187–200.

Collins, V. P. et al. 1977. Cell proliferation and plasma membrane motility. Stationary and proliferating human glia and glioma cells at various densities. Scanning Electron Microsc./1977, *2*: 1–11.

Copeland, N. G., A. D. Zelenetz, and G. M. Cooper. 1979. Transformation of NIH/3T3 cells by DNA of Raus sarcoma virus. Cell, *17*: 993–1002.

Cote, G. P. et al. 1985. Purification from *Dictyostelium discoideum* of a low-molecular weight myosin that resembles myosin I from *Acanthamoeba castellanii*. J. Biol. Chem., *260*: 4543–4546.

Damsky, C. H. et al. 1985. Distribution of cell substratum attachment (CSAT) antigen on myogenic and fibroblastic cells in culture. J. Cell Biol., *100*: 1528–1539.

Darnell, J., H. Lodish, and D. Baltimore. 1990. *Molecular Cell Biology*, 2nd ed. Scientific American Books, New York.

De Lozanne, A., and J. A. Spudich. 1987. Disruption of the *Dictyostelium* myosin heavy chain gene by homologous recombination. Science, *236*: 1086–1091.

Derby, M. A. 1978. Analysis of glycosaminoglycans within the extracellular environments encountered by migrating neural crest cells. Dev. Biol., *66*: 321–336.

Dow, J. A. T. et al. 1987. Novel methods for the guidance and monitoring of single cells and simple networks in culture. J. Cell Sci. Suppl., *8*: 55–79.

Duband, J.-L., and J. P. Thiery. 1987. Distribution of laminin and collagens during avian neural crest development. Development, *101*: 461–478.

Eckert, B. S. 1985. Alteration of intermediate filament distribution in PtK$_1$ cells by acrylamide. Eur. J. Cell Biol., *37*: 169–174.

Edelman, G. M. 1983. Cell adhesion molecules. Science, *219*: 450–457.

Edelman, G. M. 1984. Cell-adhesion molecules: A molecular basis for animal form. Sci. Am., *250* (4): 118–129.

Edelman, G. M. 1986. Cell adhesion molecules in the regulation of animal form and tissue pattern. Ann. Rev. Cell Biol., *2*: 81–116.

Edelman, G. M. 1988. Morphoregulatory molecules. Biochemistry, *27*: 3533–3543.

Edelman, G. M., and C.-M. Chuong. 1982. Embryonic to adult conversion of neural cell adhesion molecules in normal and *staggerer* mice. Proc. Natl. Acad. Sci. U.S.A., *79*: 7036–7040.

Ellis, H. M., and H. R. Horvitz. 1986. Genetic control of programmed cell death in the nematode C. elegans. Cell, *44*: 817–829.

Erickson, C. A. 1978. Analysis of the formation of parallel arrays by BHK cells *in vitro*. Exp. Cell Res., *115*: 303–315.

Erickson, C. A. 1985. Control of neural crest cell dispersion in the trunk of the avian embryo. Dev. Biol., *111*: 138–157.

Erickson, C. A. 1987. Behavior of neural crest cells on embryonic basal laminae. Dev. Biol., *120*: 38–49.

Erickson, C. A., and E. A. Turley. 1983. Substrata formed by combinations of extracellular matrix components alter neural crest cell motility *in vitro*. J. Cell Sci., *61*: 299–323.

Erickson, C. A, R. P. Tucker, and B. F. Edwards. 1987. Changes in the distribution of intermediate-filament types in Japanese quail embryos during morphogenesis. Differentiation, *34*: 88–97.

Ettensohn, C. A. 1984. Primary invagination of the vegetal plate during sea urchin gastrulation. Am. Zool., *24*: 571–588.

Fink, R. D., and D. R. McClay. 1985. Three cell recognition changes accompany the ingression of sea urchin primary mesenchyme cells. Dev. Biol., *107*: 66–74.

Fristrom, D. 1976. The mechanism of evagination of imaginal discs of *Drosophila melanogaster*. III. Evidence for cell rearrangements. Dev. Biol., *54*: 163–171.

Fukui, Y. et al. 1989. Myosin I is located at the leading edges of locomoting *Dictyostelium* amoebae. Nature (Lond.), *341*: 328–331.

Gehlsen, K. R. et al. 1988. The human laminin receptor is a member of the integrin family of cell adhesion receptors. Science, *241*: 1228–1229.

Gerisch, G. 1980. Univalent antibody fragments as tools for the analysis of cell interactions in *Dictyostelium*. *In* M. Friedlander (ed.), *Immunological Approaches to Embryonic Development and Differentiation. Part II*. Curr. Top. Dev. Biol., *14:* 243–270.

Goldman, R. D. 1971. The role of three cytoplasmic fibers in BHK–21 cell motility. I. Microtubules and the effects of colchicine. J. Cell Biol., *51*: 752–762.

Gundersen, R. W. 1987. Response of sensory neurites and growth cones to patterned substrata of laminin and fibronectin *in vitro*. Dev. Biol., *121*: 423–431.

Harris, A. K. 1973. Behaviour of cultured cells on substrata of variable adhesiveness. Exp. Cell Res., *77*: 285–297.

Harris, A. K., D. Stopak, and P. Wild. 1981. Fibroblast traction as a mechanism for collagen morphogenesis. Nature (Lond.), *290*: 249–251.

Harris, A. K., P. Wild, and D. Stopak. 1980. Silicone rubber substrata: A new wrinkle in the study of cell locomotion. Science, *208*: 177–179.

Harrison, R. G. 1912. The cultivation of tissues in extraneous media as a method of morphogenetic study. Anat. Rec., *6*: 181–193.

Hascall, V. C., and G. K. Hascall. 1981. Proteoglycans. *In* E. D. Hay (ed.), *Cell Biology of Extracellular Matrix*. Plenum Press, New York, pp. 39–63.

Hay, E. D. 1981a. Introductory remarks. *In* E. D. Hay (ed.), *Cell Biology of Extracellular Matrix*. Plenum Press, New York, pp. 1–4.

Hay, E. D. 1981b. Extracellular matrix. J. Cell Biol., *91*: 205s–223s.

Hay, E. D., D. L. Hasty, and K. L. Kiehnau. 1978. Fine structure of collagens and their relation to glucosaminoglycans (GAG). *In* K. Kuhn and R. Marx (eds.), *Collagen-Platelet Interaction*. F. K. Schattauer Verlag, Stuttgart, pp. 129–151.

Hedgecock, E., J. E. Sulston, and N. Thomas. 1983. Mutations affecting programmed cell deaths in the nematode *Caenorhabditis elegans*. Science, *220*: 1277–1279.

Heuser, J. E., and M. W. Kirschner. 1980. Filament organization revealed in platinum replicas of freeze-dried cytoskeletons. J. Cell Biol., *86*: 212–234.

Hewitt, A. T., and G. R. Martin. 1984. Attachment proteins and their role in extracellular matrices. *In* R. J. Ivatt (ed.), *The Biology of Glycoproteins*. Plenum Press, New York, pp. 65–93.

Hilfer, S. K., and R. L. Searls. 1986. Cytoskeletal dynamics in animal morphogenesis. *In* L. W. Browder (ed.), *Developmental Biology: A Comprehensive Synthesis*, Vol. 2, *The Cellular Basis of Morphogenesis*. Plenum Press, New York, pp. 3–29.

Hinchliffe, J. R. 1972. Cell death in relation to the genesis of form and pattern in the developing chick and amphibian limb. *In* P. Sengel (ed.), *Colloque International sur le Développement du Membre*. Grenoble, p. 26.

Hinchliffe, J. R., and P. V. Thorogood. 1974. Genetic inhibition of mesenchymal cell death and the development of form and skeletal pattern in the limbs of *talpid³* mutant chick embryos. J. Embryol. Exp. Morphol., *31*: 747–760.

Hoebeke, J., G. Van Nijen, and M. De Brabander. 1976. Interaction of oncodazole (R 17934), a new antitumoral drug, with rat brain tubulin. Biochem. Biophys. Res. Commun., *69*: 319–324.

Hoffman, S., and G. M. Edelman. 1983. Kinetics of homophilic binding by E and A forms of the neural cell adhesion molecule. Proc. Natl. Acad. Sci. U.S.A., *80*: 5762–5766.

Horio, T., and H. Hotani. 1986. Visualization of the dynamic instability of individual microtubules by dark-field microscopy. Nature, *321*: 605–607.

Hynes, R. O. 1985. Molecular biology of fibronectin. Ann. Rev. Cell Biol., *1*: 67–90.

Hynes, R. O. 1987. Integrins: A family of cell surface receptors. Cell, *48*: 549–554.

Isenberg, G. et al. 1976. Cytoplasmic actomyosin fibrils in tissue culture cells. Direct proof of contractility by visualization of ATP-induced contraction in fibrils isolated by laser microbeam dissection. Cell Tissue Res., *166*: 427–443.

Izzard, C. S., S. L. Izzard, and J. A. De Pasquale. 1985. Molecular basis of cell-substrate adhesions. Exp. Biol. Med., *10*: 1–22.

Izzard, C. S., and L. R. Lochner. 1976. Cell-to-substrate contacts in living fibroblasts: An interference-reflexion study with an evaluation of the technique. J. Cell Sci., *21*: 129–160.

Karfunkel, P. 1971. The role of microtubules and microfilaments in neurulation in *Xenopus*. Dev. Biol., *25*: 30–56.

Karfunkel, P. 1972. The activities of microtubules and microfilaments in neurulation in the chick. J. Exp. Zool., *181*: 289–302.

Keller, R. E. 1986. The cellular basis of amphibian gastrulation. *In* L. W. Browder (ed.), *Developmental Biology: A Comprehensive Synthesis*, Vol. 2, *The Cellular Basis of Morphogenesis*. Plenum Press, New York, pp. 241–328.

Keller, R. E. et al. 1989. Cell intercalation during notochord development in *Xenopus laevis*. J. Exp. Zool., *251*: 134–154.

Kimble, J. 1981. Alterations in cell lineage following laser ablation of cells in the somatic gonad of *Caenorhabditis elegans*. Dev. Biol., *86*: 286–300.

Kleinman, H. K. 1986. Biological activities of laminin. J. Cell. Biochem., *27*: 317–325.

Klymkowsky, M. W., J. B. Bachant, and A. Domingo. 1989. Functions of intermediate filaments. Cell Motil. Cytoskel., *14*: 309–331.

Knecht, D. A., and W. F. Loomis. 1987. Antisense RNA inactivation of myosin heavy chain gene expression in *Dictyostelium discoideum*. Science, *236*: 1081–1086.

Kreis, T. E., and W. Birchmeier. 1981. Microinjection of fluorescently labeled proteins into living cells with emphasis on cytoskeletal proteins. Int. Rev. Cytol., *75*: 209–227.

Kupfer, A., D. Louvard, and S. J. Singer. 1982. Polarization of the Golgi apparatus and microtubule-organizing center in cultured fibroblasts at the edge of an experimental wound. Proc. Natl. Acad. Sci. U.S.A., *79*: 2603–2607.

Lazarides, E. 1975. Tropomyosin antibody: The specific localization of tropomyosin in non-muscle cells. J. Cell Biol., *65*: 549–561.

Lazarides, E. 1982. Intermediate filaments: A chemically heterogeneous, developmentally regulated class of proteins. Ann. Rev. Biochem., *51*: 219–250.

Lazarides, E., and K. Burridge. 1975. α-Actinin: Immunofluorescent localization of a muscle structural protein in non-muscle cells. Cell, *6*: 289–298.

Lazarides, E., and K. Weber. 1974. Actin antibody: The specific visualization of actin filaments in non-muscle cells. Proc. Natl. Acad. Sci. U.S.A., *71*: 2268–2272.

Lentz, T. L., and J. P. Trinkaus. 1967. A fine structural study of cytodifferentiation during cleavage, blastula, and gastrula stages of *Fundulus heteroclitus*. J. Cell Biol., *32*: 121–138.

Letourneau, P. C. 1975. Cell-to-substratum adhesion and guidance of axonal elongation. Dev. Biol., *44*: 92–101.

Lin, D. C. et al. 1980. Cytochalasins inhibit nuclei-induced actin polymerization by blocking filament elongation. J. Cell Biol., *84*: 455–460.

Linask, K. K., and J. W. Lash. 1988a. A role for fibronectin in the migration of avian precardiac cells. I. Dose-dependent effects of fibronectin antibody. Dev. Biol., *129*: 315–323.

Linask, K. K., and J. W. Lash. 1988b. A role for fibronectin in the migration of avian precardiac cells. II. Rotation of the heart-forming region during different stages and its effects. Dev. Biol., *129*: 324–329.

Lindahl, U., and M. Höök. 1978. Glycosaminoglycans and their binding to biological macromolecules. Ann. Rev. Biochem., *47*: 385–417.

Lumsden, A. G. S., and A. M. Davies. 1983. Earliest sensory nerve fibers are guided to peripheral targets by attractants other than nerve growth factor. Nature (Lond.), *306*: 786–788.

Lynch, T. J. et al. 1986. ATPase activities and actin binding properties of subfragments of *Acanthamoeba* myosin IA. J. Biol. Chem., *261*: 17,156–17,162.

McClay, D. R., and R. D. Fink. 1982. The role of hyalin in early sea urchin development. Dev. Biol., *92*: 285–293.

Margolis, R. L., and L. Wilson. 1977. Addition of colchicine-tubulin complex to microtubule ends: The mechanism of substoichiometric colchicine poisoning. Proc. Natl. Acad. Sci. U.S.A., *74*: 3466–3470.

Martin, G. R., and R. Timple. 1987. Laminin and other basement membrane components. Ann. Rev. Cell Biol., *3*: 57–86.

Mayne, R. 1984. The different types of collagen and collagenous peptides. *In* R. L. Trelstad (ed.), *The Role of the Extracellular Matrix in Development.* Alan R. Liss, New York, pp. 33–42.

Mitchison, T., and M. Kirschner. 1984. Dynamic instability of microtubule growth. Nature (Lond.), *312*: 237–242.

Moscona, A. A. 1957. Development *in vitro* of chimaeric aggregates of dissociated embryonic chick and mouse cells. Proc. Natl. Acad. Sci. U.S.A., *43*: 184–194.

Moscona, A. A. 1960. Patterns and mechanisms of tissue reconstruction from dissociated cells. *In* D. Rudnick (ed.), *Developing Cell Systems and Their Control.* Ronald Press, New York, pp. 45–70.

Nakatsuji, N., A. C. Gould, and K. Johnson. 1982. Movement and guidance of migrating mesodermal cells in *Ambystoma maculatum* gastrulae. J. Cell Sci., *56*: 207–222.

Nakatsuji, N., and K. Johnson. 1983. Conditioning of a culture substratum by the ectodermal layer promotes attachment and oriented locomotion by amphibian gastrula mesodermal cells. J. Cell Sci., *59*: 43–60.

Nakatsuji, N., and K. Johnson. 1984. Experimental manipulation of a contact guidance system in amphibian gastrulation by mechanical tension. Nature (Lond.), *307*: 453–455.

Newgreen, D. 1984. Spreading of explants of embryonic chick mesenchymes and epithelia on fibronectin and laminin. Cell Tissue Res., *236*: 265–277.

Newgreen, D., and J.-P. Thiery. 1980. Fibronectin in early avian embryos: Synthesis and distribution along migration pathways of neural crest cells. Cell Tissue Res., *211*: 269–291.

Newgreen, D. F., M. Scheel, and V. Kastner. 1986. Morphogenesis of sclerotome and neural crest in avian embryos. *In vivo* and *in vitro* studies on the role of notochordal extracellular material. Cell Tissue Res., *244*: 299–313.

Newgreen, D. F. et al. 1982. Ultrastructural and tissue culture studies on the role of fibronectin, collagen and glycosaminoglycans in the migration of neural crest cells in the fowl embryo. Cell Tissue Res., *221*: 521–549.

Nose, A., A. Nagafuchi, and M. Takeichi. 1988. Expressed recombinant cadherins mediate cell sorting in model systems. Cell, *54*: 993–1001.

Osborn, M., and K. Weber. 1976. Cytoplasmic microtubules in tissue culture cells appear to grow from an organizing structure towards the plasma membrane. Proc. Natl. Acad. Sci. U.S.A., *73*: 867–871.

Osborn, M. et al. 1982. Intermediate filaments. Cold Spring Harbor Symp. Quant. Biol., *46*: 413–429.

Oster, G. F. 1984. On the crawling of cells. J. Embryol. Exp. Morphol., *83* (Suppl.): 329–364.

Paranko, J., and I. Virtanen. 1986. Epithelial and mesenchymal cell differentiation in the fetal rat genital ducts: Changes in expression of cytokeratin and vimentin type of intermediate filaments and desmosomal plaque proteins. Dev. Biol., *117*: 135–145.

Phillips, H. M., and M. S. Steinberg. 1969. Equilibrium measurements of embryonic chick cell adhesiveness. I. Shape equilibrium in centrifugal fields. Proc. Natl. Acad. Sci. U.S.A., *64*: 121–127.

Pierschbacher, M. D., and E. Ruoslahti. 1984. Cell attachment activity of fibronectin can be duplicated by small synthetic fragments of the molecule. Nature (Lond.), *309*: 30–33.

Pintar, J. E. 1978. Distribution and synthesis of glycosaminoglycans during quail neural crest morphogenesis. Dev. Biol., *67*: 444–464.

Pollard, T. D. 1981. Cytoplasmic contractile proteins. J. Cell Biol., *91*: 156s–165s.

Pollard, T. D., and E. D. Korn. 1973. *Acanthamoeba* myosin I. Isolation from *Acanthamoeba castellanii* of an enzyme similar to muscle myosin. J. Biol. Chem., *248:* 4682–4690.

Poole, T. J., and M. S. Steinberg. 1982. Evidence for the guidance of pronephric direct migration by a craniocaudally traveling adhesion gradient. Dev. Biol., *92:* 144–158.

Prendergast, G. C., and M. D. Cole. 1989. Posttranscriptional regulation of cellular gene expression by the c-*myc* oncogene. Molec. Cell. Biol., *9:* 124–134.

Pytela, R., M. Pierschbacher, and E. Ruoslahti. 1985. Identification and isolation of a 140 kd cell surface glycoprotein with properties of a fibronectin receptor. Cell, *40:* 191–198.

Rempel, S. 1990. Steroid hormone regulation of c-*myc* gene expression in proliferating chick oviduct. Ph.D. thesis, University of Calgary, Calgary, Alberta, Canada.

Rovasio, R. A. et al. 1983. Neural crest cell migration: Requirements for exogenous fibronectin and high cell density. J. Cell Biol., *96:* 462–473.

Ruoslahti, E. 1988. Fibronectin and its receptors. Ann. Rev. Biochem., *57:* 375–413.

Ruoslahti, E., E. G. Hayman, and M. D. Pierschbacher. 1985. Extracellular matrices and cell adhesion. Arteriosclerosis, *5:* 581–594.

Ruoslahti, E., and M. D. Pierschbacher. 1987. New perspectives in cell adhesion: RGD and integrins. Science, *238:* 491–497.

Sammak, P. J., and G. G. Borisy. 1988. Detection of single fluorescent microtubules and methods for determining their dynamics in living cells. Cell Motil. Cytoskel., *10:* 237–245.

Saunders, J. W., Jr. 1966. Death in embryonic systems. Science, *154:* 604–612.

Saunders, J. W., Jr. 1969. The interplay of morphogenetic factors. *In* C. A. Swinyard (ed.), *Limb Development and Deformity: Problems of Evaluation and Rehabilitation.* Charles C. Thomas, Springfield, IL, pp. 84–100.

Saunders, J. W., Jr. 1982. *Developmental Biology.* Macmillan Publishing Co., New York.

Saunders, J. W., Jr., M. T. Gasseling, and L. C. Saunders. 1962. Cellular death in morphogenesis of the avian wing. Dev. Biol., *5:* 147–178.

Schliwa, M. 1981. Proteins associated with cytoplasmic actin. Cell, *25:* 587–590.

Silver, J. 1978. Cell death during development of the nervous system. *In* M. Jacobson (ed.), *Handbook of Sensory Physiology IX: Development of Sensory Systems.* Springer Verlag, Berlin, pp. 419–436.

Silver, J., and R. M. Robb. 1979. Studies on the development of the eye cup and optic nerve in normal mice and in mutants with congenital optic nerve aplasia. Dev. Biol., *68:* 175–190.

Singer, S. J., and A. Kupfer. 1986. The directed migration of eukaryotic cells. Ann. Rev. Cell Biol., *2:* 337–365.

Small, J. V. 1981. Organization of actin in the leading edge of cultured cells. J. Cell Biol., *91:* 695–705.

Small, J. V., G. Isenberg, and J. E. Celis. 1978. Polarity of actin at the leading edge of cultured cells. Nature (Lond.), *272:* 638–639.

Solursh, M. 1986. Migration of sea urchin primary mesenchyme cells. *In* L. W. Browder (ed.), *Developmental Biology: A Comprehensive Synthesis,* Vol. 2, *The Cellular Basis of Morphogenesis.* Plenum Press, New York, pp. 391–431.

Sperry, R. W. 1963. Chemoaffinity in the orderly growth of nerve fiber patterns and connections. Proc. Natl. Acad. Sci. U.S.A., *50:* 703–710.

Spooner, B. S. 1973. Microfilaments, cell shape changes, and morphogenesis of salivary epithelium. Am. Zool., *13:* 1007–1022.

Spooner, B. S., K. M. Yamada, and N. K. Wessells. 1971. Microfilaments and cell locomotion. J. Cell Biol., *49:* 595–613.

Steinberg, M. S. 1963. Reconstruction of tissues by dissociated cells. Science, *141:* 401–408.

Steinberg, M. S. 1970. Does differential adhesion govern self-assembly processes in histogenesis? Equilibrium configurations and the emergence of hierarchy among populations of embryonic cells. J. Exp. Zool., *173:* 395–434.

Steinert, P. M. et al. 1984. Intermediate filaments. J. Cell Biol., *99*: 22s–27s.

Steinert, P. M., A. C. Steven, and D. R. Roop. 1985. The molecular biology of intermediate filaments. Cell, *42*: 411–419.

Sternberg, P. W., and H. R. Horvitz. 1981. Gonadal cell lineages of the nematode *Panagrellus redivivus* and implications for evolution by the modification of cell lineage. Dev. Biol., *88*: 147–166.

Sternberg, P. W., and H. R. Horvitz. 1982. Postembryonic nongonadal cell lineages of the nematode *Panagrellus redivivus*: Description and comparison with those of *Caenorhabditis elegans*. Dev. Biol., *93*: 181–205.

Stopak, D., and A. K. Harris. 1982. Connective tissue morphogenesis by fibroblast traction. I. Tissue culture observations. Dev. Biol., *90*: 383–398.

Stossell, T. et al. 1985. Nonmuscle actin-binding proteins. Ann. Rev. Cell Biol., *1*: 353–402.

Takeichi, M. 1977. Functional correlation between cell adhesive properties and some cell surface proteins. J. Cell Biol., *75*: 464–474.

Takeichi, M. 1988. The cadherins: Cell-cell adhesion molecules controlling animal morphogenesis. Development, *102*: 639–655.

Tanenbaum, S. W. (ed.). 1978. *Cytochalasins-Biochemical and Cell Biological Aspects*. Elsevier/North-Holland, Amsterdam.

Taylor, D. L., and Y.-L. Wang. 1980. Fluorescently labeled molecules as probe of the structure and function of living cells. Nature (Lond.), *284*: 405–410.

Thiery, J.-P., J. L. Duband, and A. Delouvée. 1982. Pathways and mechanisms of avian trunk neural crest cell migration and localization. Dev. Biol., *93*: 324–343.

Thiery, J.-P. et al. 1977. Adhesion among neural cells of the chick embryo. II. Purification and characterization of a cell adhesion molecule from neural retina. J. Biol. Chem., *252*: 6841–6845.

Tilney, L. G., E. M. Bonder, and D. J. DeRosier. 1981. Actin filaments elongate from their membrane-associated ends. J. Cell Biol., *90*: 485–494.

Tilney, L. G., and S. Inoué. 1985. Acrosomal reaction of the *Thyone* sperm. III. The relationship between actin assembly and water influx during the extension of the acrosomal process. J. Cell Biol., *100*: 1273–1283.

Tilney, L. G., and N. Kallenbach. 1979. Polymerization of actin. VI. The polarity of actin filaments in the acrosomal process and how it might be determined. J. Cell Biol., *81*: 608–623.

Timpl, R. et al. 1981. A network model for the organization of type IV collagen molecules in basement membranes. Biochem. J., *120*: 203–211.

Toole, B. P., and R. L. Trelstad. 1971. Hyaluronate production and removal during corneal development in the chick. Dev. Biol., *26*: 28–35.

Tosney, K. W. 1982. The segregation and early migration of cranial neural crest cells in the avian embryo. Dev. Biol., *89*: 13–24.

Townes, P. L., and J. Holtfreter. 1955. Directed movements and selective adhesion of embryonic amphibian cells. J. Exp. Zool., *128*: 53–120.

Trinkaus, J. P. 1973. Surface activity and locomotion of *Fundulus* deep cells during blastula and gastrula stages. Dev. Biol., *30*: 68–103.

Trinkaus, J. P. 1985. Protrusive activity of the cell surface and the initiation of cell movement during morphogenesis. Exp. Biol. Med., *10*: 130–173.

Trinkaus, J. P., and C. A. Erickson. 1983. Protrusive activity, mode and rate of locomotion, and pattern of adhesion of *Fundulus* deep cells during gastrulation. J. Exp. Zool., *228*: 41–70.

Tucker, R. P., and C. A. Erickson. 1984. Morphology and behavior of quail neural crest cells in artificial three-dimensional extracellular matrices. Dev. Biol., *104*: 390–405.

Vasiliev, J. M. et al. 1970. Effect of colcemid on the locomotory behaviour of fibroblasts. J. Embryol. Exp. Morphol., *24*: 625–640.

Wang, Y.-L. 1984. Reorganization of actin filament bundles in living fibroblasts. J. Cell Biol., *99*: 1478–1485.

Weeds, A. 1982. Actin-binding proteins—regulators of cell architecture and motility. Nature (Lond.), *296*: 811–816.

Wegner, A. 1976. Head-to-tail polymerization of actin. J. Mol. Biol., *108*: 139–150.

Weiss, P. 1934. *In vitro* experiments on the factors determining the course of the outgrowing nerve fiber. J. Exp. Zool., *68*: 393–448.

Weiss, P. 1958. Cell contact. Int. Rev. Cytol., *7*: 391–423.

Wessells, N. K., B. S. Spooner, and M. A. Ludueña. 1973. Surface movements, microfilaments and cell locomotion. *In* R. Porter and D. W. Fitzsimons (eds.), *Locomotion of Tissue Cells*. Ciba Found. Symp., *14*: 53–77.

Willingham, M. C. et al. 1981. Ultrastructural immunocytochemical localization of myosin in cultured fibroblastic cells. J. Histochem. Cytochem., *29*: 1289–1301.

Wilson, H. V. 1907. On some phenomena of coalescence and regeneration in sponges. J. Exp. Zool., *5*: 245–258.

Woodrum, D. T., S. A. Rich, and T. D. Pollard. 1975. Evidence for biased bidirectional polymerization of actin filaments using heavy meromyosin prepared by an improved method. J. Cell Biol., *67*: 231–237.

Wrenn, J. T., and N. K. Wessells. 1969. An ultrastructural study of lens invagination in the mouse. J. Exp. Zool., *171*: 359–368.

Wrenn, J. T., and N. K. Wessells. 1970. Cytochalasin B: Effects upon microfilaments involved in morphogenesis of estrogen-induced glands of oviduct. Proc. Natl. Acad. Sci. U.S.A., *66*: 904–908.

Yamada, K. M. et al. 1984. Fibronectin and cell surface interactions. *In* R. L. Trelstad (ed.), *The Role of Extracellular Matrix in Development*. Alan R. Liss, New York, pp. 89–121.

Zackson, S. L., and M. S. Steinberg. 1988. A molecular marker for cell guidance information in the axolotl embryo. Dev. Biol., *127*: 435–442.

Zigmond, S. H. 1982. Polymorphonuclear leucocyte response to chemotactic gradients. *In* R. Bellairs, A. Curtis, and G. Dunn (eds.), *Cell Behavior*. Cambridge University Press, Cambridge, pp. 183–202.

Color Plates

*Complete Literature Citations Can Be Found in
Corresponding Chapters*

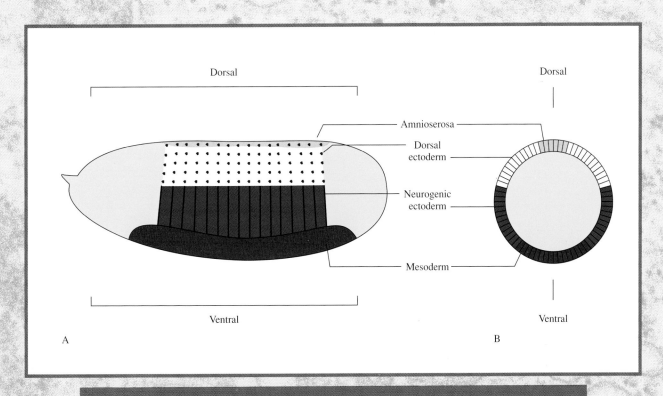

Dorsal

Dorsal

Amnioserosa

Dorsal
ectoderm

Neurogenic
ectoderm

Mesoderm

Ventral

Ventral

A

B

Color plate 1 Fate map of a *Drosophila* embryo at the cellular blastoderm stage. A, Side view. B, Cross-section about halfway through the egg. (After Anderson, 1987.)

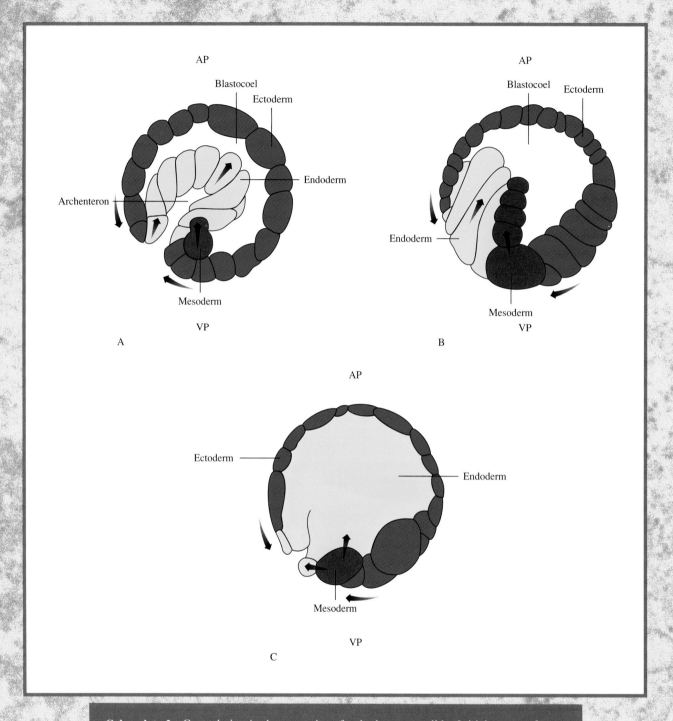

Color plate 2 Gastrulation in three species of polychaete annelids. Initial stages of gastrulation are shown in drawings of sectioned embryos. A, *Eupomatus*. B, *Scoloplos*. C, *Neanthes*. AP: animal pole; VP: vegetal pole. (After Anderson, 1973.)

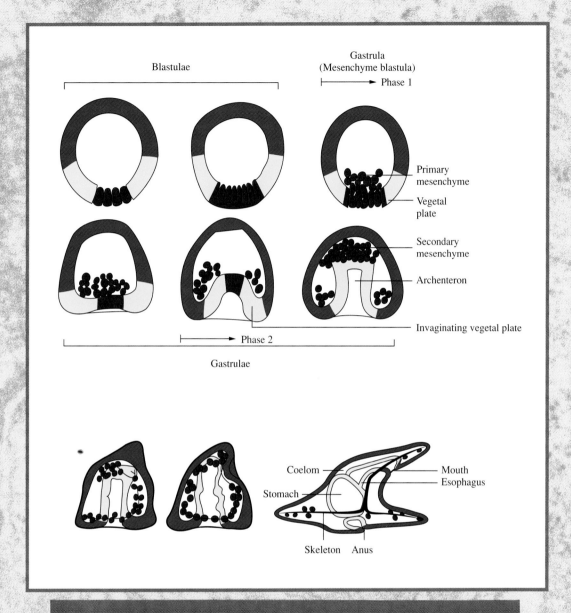

Blastulae

Gastrula
(Mesenchyme blastula)
Phase 1

Primary mesenchyme

Vegetal plate

Secondary mesenchyme

Archenteron

Invaginating vegetal plate

Phase 2

Gastrulae

Coelom

Mouth

Esophagus

Stomach

Skeleton   Anus

Color plate 3   Sea urchin gastrulation. Drawings of the pre-gastrula, gastrula, and post-gastrula stages up to pluteus larva. Blue: presumptive ectoderm; red: presumptive mesoderm; yellow: presumptive endoderm. (After Spemann, 1936.)

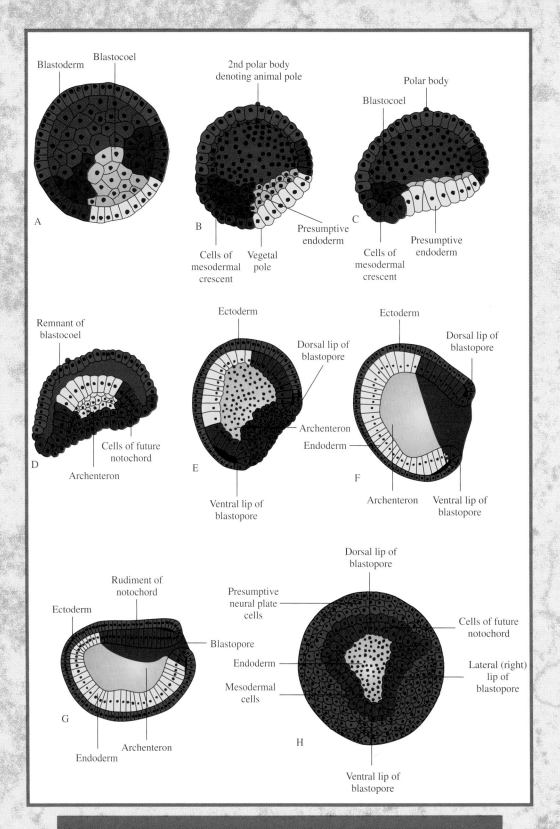

Color plate 4 Gastrulation in the cephalochordate *Amphioxus*. The embryos in A to G are cut in the median plane. A, Blastula. B, and C, Formation of the endodermal plate. D, As invagination begins, the embryo attains the structure of a double-walled cup with a broad opening to the exterior. E, and F, Constriction of the blastopore. G, Completed gastrula. H, Whole mid-gastrula, viewed from the blastopore side. Blue: presumptive ectoderm; red: presumptive mesoderm; yellow: presumptive endoderm. (After Conklin, 1932, and Balinsky, 1981.)

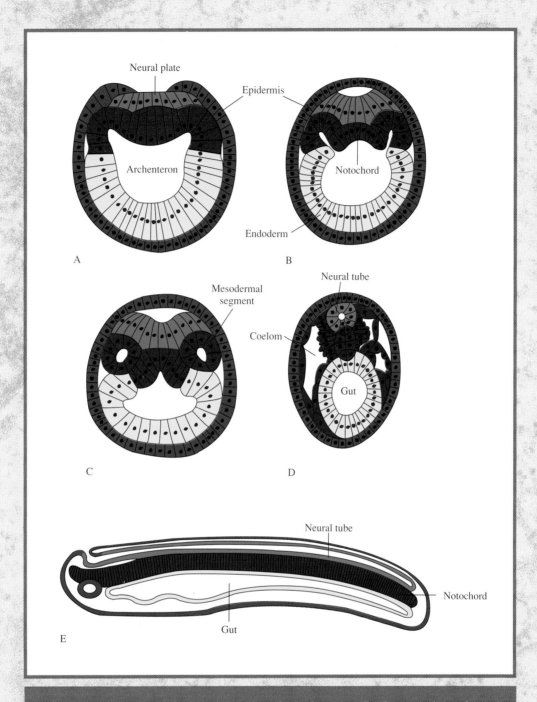

Color plate 5 Elaboration of the body plan of *Amphioxus*. A-D, Cross-sections of progressively later stages. Blue: epidermis; green: neural plate and tube; red: mesoderm; yellow: endoderm. E, Sagittal section of a later stage showing anteroposterior elongation of the embryo. (After Korschelt, 1936, Balinsky, 1981, and Berrill, 1987.)

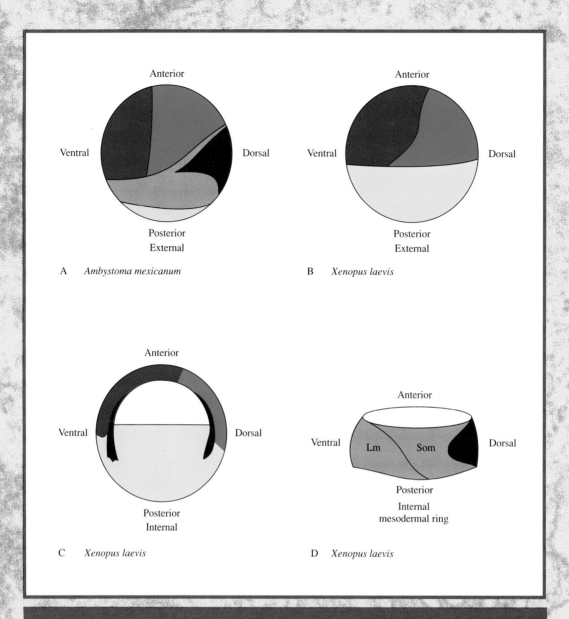

Color plate 6 Fate maps of amphibian embryos. A, Fate map of an *Ambystoma mexicanum* (urodele) early gastrula. All presumptive germ layer cells are on the surface. B. Fate map of the superficial cell layer of a *Xenopus laevis* (anuran) early gastrula. C, Cross-section of B showing the fate map of the deep layer of a *Xenopus laevis* early gastrula. D, Fate map of the mesodermal ring of a *Xenopus laevis* embryo. Blue: presumptive epidermal ectoderm; green: presumptive neuroectoderm; red: presumptive mesoderm (in C), notochord (in A and D); orange: somitic and lateral plate mesoderm (in A and D); yellow: presumptive endoderm. (A, After Løvtrup, 1975. B. After Keller, 1975. D, After Keller, 1976.)

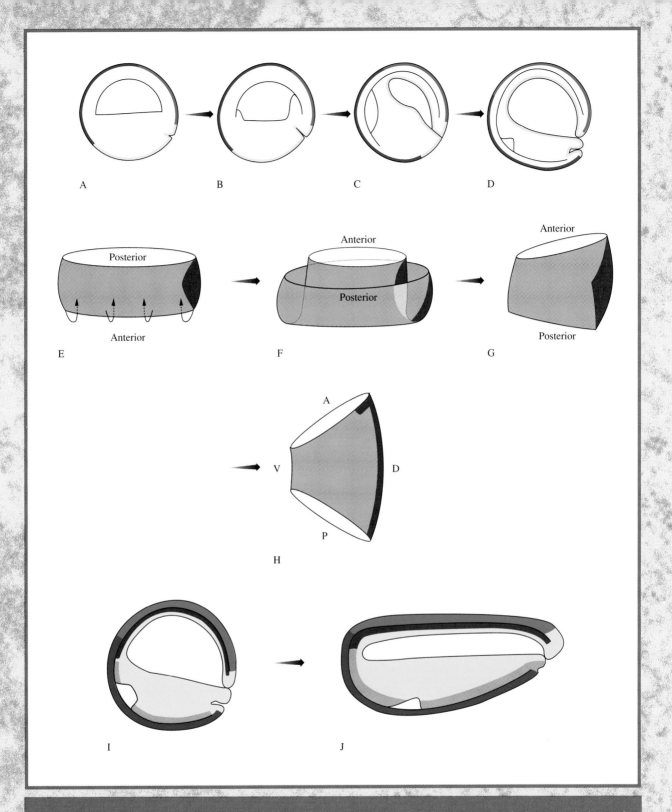

Color plate 7 Diagrams of presumptive ectoderm, mesoderm, and endoderm movements during gastrulation and anteroposterior axis elongation in *Xenopus*. A-D, Sagittal sections of embryos during gastrulation showing the movements of superficial cells during gastrulation (i.e., presumptive endoderm, epidermis, and neural ectoderm). E-H, Involution and dorsal elongation of the mesodermal ring. Red: presumptive notochord. I, and J, Sagittal sections of the late gastrula and neurula stages showing behavior of germ layers during the establishment of the anteroposterior axis. Blue: epidermal ectoderm; green: neuroectoderm; red: presumptive notochord; orange: presumptive somitic and lateral plate mesoderm; yellow: presumptive endoderm.

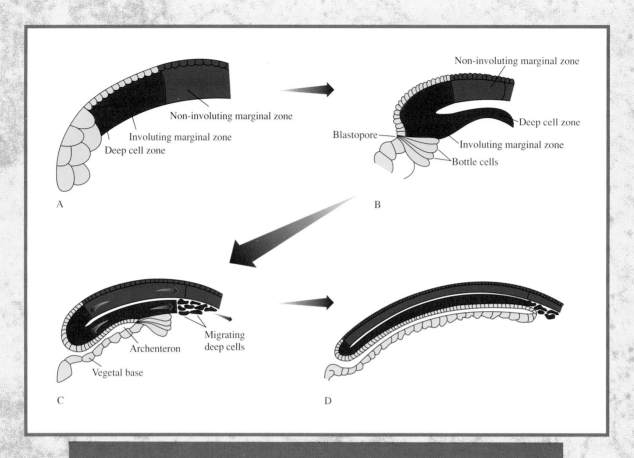

Color plate 8 Relationship between cell migration, involution, invagination, and convergent extension in the marginal zone during gastrulation in *Xenopus* embryos. Diagrams of mid-sagittal sections through the marginal region on the dorsal side of a late blastula (A), an early gastrula (B), a mid-gastrula (C), and a late gastrula (D). The arrows represent convergent extension in the involuting and noninvoluting marginal zones. (After Keller et al., 1985.)

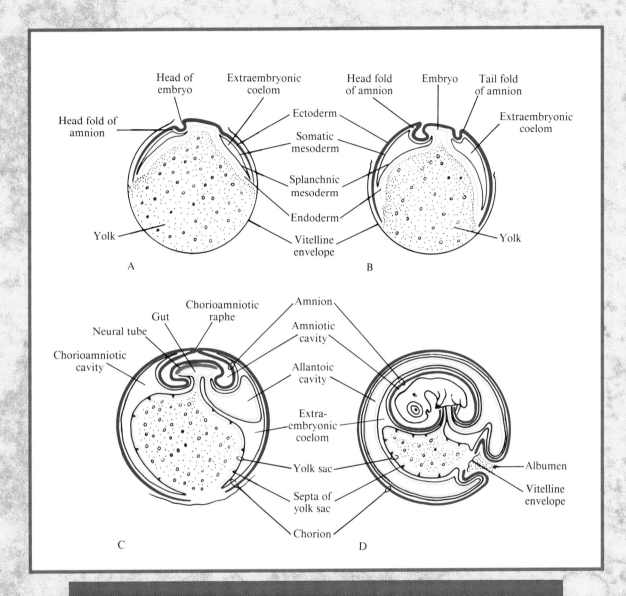

Color plate 9   Color version of Figure 8.33. Sagittal views of the progressive development of the four extraembryonic membranes from the contiguous embryonic germ layers in a chick embryo. Blue: ectoderm; red: mesoderm; yellow: endoderm; green: neural tube. (After B. M. Carlson, 1988, *Patten's Foundations of Embryology*, 5th ed. Reprinted by permission of McGraw-Hill Book Co.)

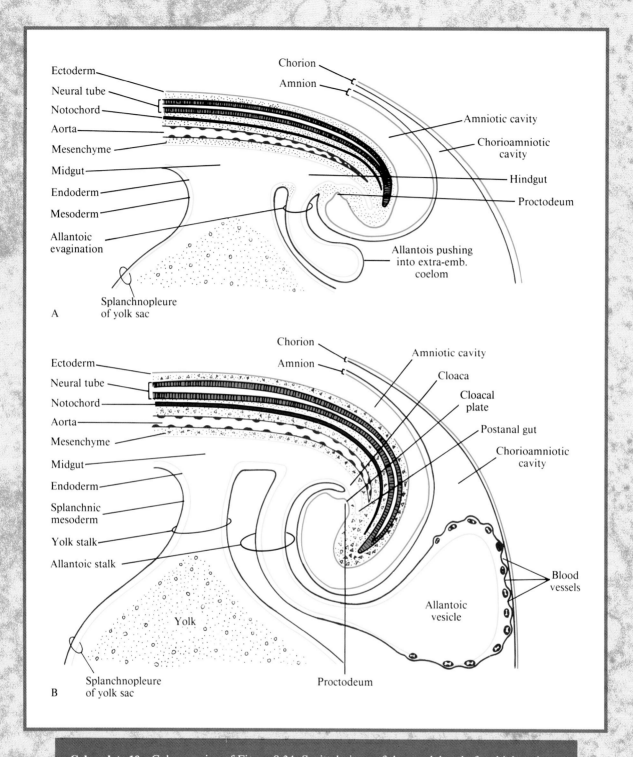

Color plate 10 Color version of Figure 8.34. Sagittal views of the caudal end of a chick embryo showing in detail the relationship of the germ layers comprising the extraembryonic membranes with the germ layers of the embryo proper. Note in particular the extensive outgrowth of the allantois. Blue: ectoderm; red: mesoderm; yellow: endoderm; green: neural tube. (After B. M. Carlson, 1988, *Patten's Foundations of Embryology*, 5th ed. Reprinted by permission of McGraw-Hill Book Co.)

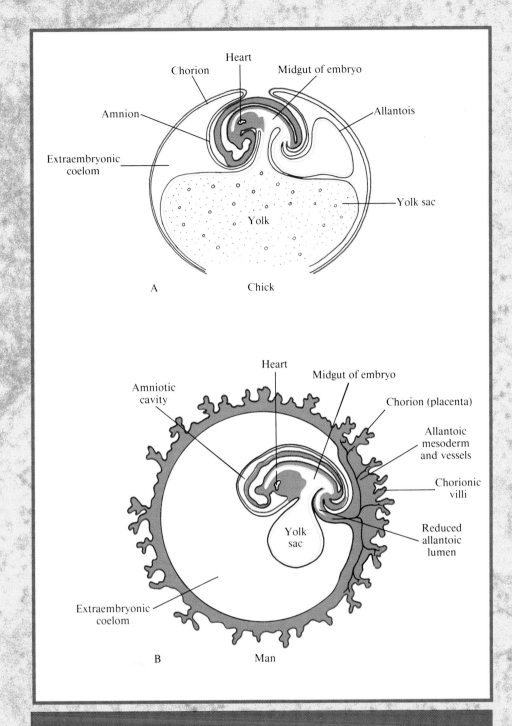

Color plate 11 Color version of Figure 8.35. Arrangement of the four extraembryonic membranes in early chick (A) and human (B) embryos. Blue: ectoderm; red: mesoderm; yellow: endoderm; green: neural tube. (After B. M. Carlson, 1988, *Patten's Foundations of Embryology*, 5th ed. Reprinted by permission of McGraw-Hill Book Co.)

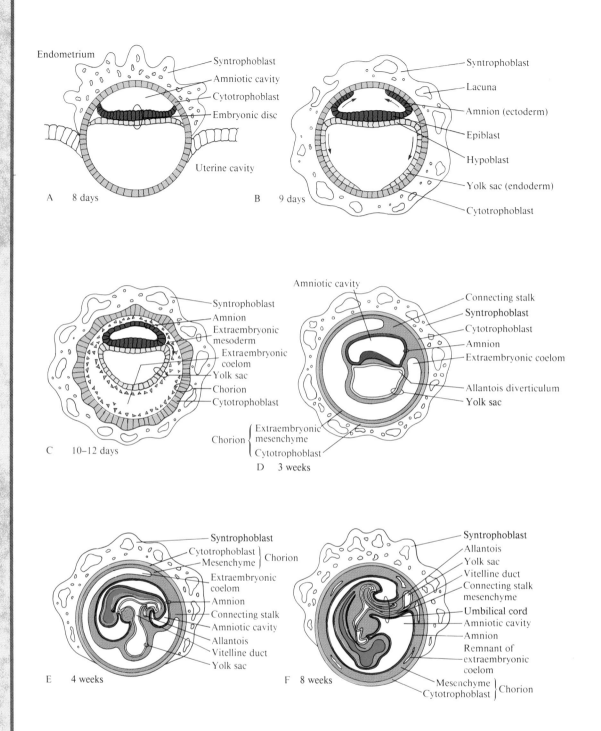

Color plate 12 Color version of Figure 8.36. Diagrams of the development of the extraembryonic membranes in 8-day (A), 9-day (B), 12-day (C), 3-week (D), 4-week (E) and 8-week (F) human embryos. Note in particular the germ layer components of each extraembryonic membrane and the changes in the relationships of the membranes to give rise to the umbilical cord and placenta. Blue: ectoderm; red: mesoderm; yellow: endoderm; green: neural tube. (After Tuchmann-Duplessis et al., 1972.)

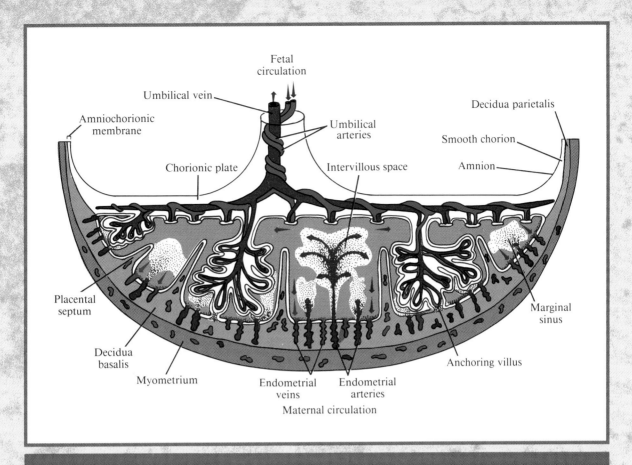

Fetal
circulation

Umbilical vein

Amniochorionic
membrane

Decidua parietalis

Umbilical
arteries

Smooth chorion

Chorionic plate

Amnion

Intervillous space

Placental
septum

Marginal
sinus

Decidua
basalis

Anchoring villus

Myometrium

Endometrial
veins

Endometrial
arteries

Maternal circulation

Color plate 13 Color version of Figure 8.40. A schematic drawing through a mature placenta depicting the blood circulation in the fetal placenta (chorionic villi) and the maternal placenta (basal decidua). As the highly oxygenated blood spurts into the intervillous spaces, if flows around the villi, where exchange of O_2 and nutrients takes place with the fetal blood in the chorionic villi. CO_2 and waste metabolites are removed from the fetal blood into the intervillous space, and this blood settles into the endometrial veins and passes to the maternal circulation. Note that in the full-term placenta the chorionic villi dangle in large intervillous spaces that have been excavated from the decidua. Oxygenated blood is shown in red, whereas the blood after exchange of gases and metabolites is shown in blue. (After Moore, 1988.)

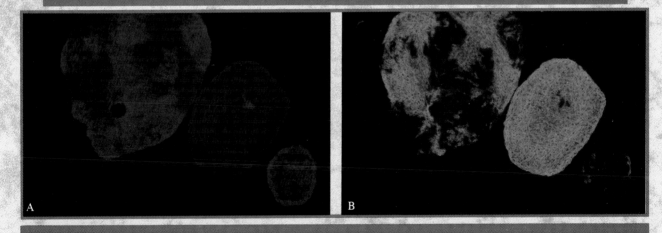

Color plate 14 Color version of Figure 9.44. L cells transfected with the genes for P- or E-cadherin were mixed in a 1:1 ratio and cultured for 24 hours, during which time they formed aggregates. The aggregates were sectioned, incubated with antibodies to E- and P-cadherin, and the primary antibodies localized with secondary antibodies tagged with either rhodamine or fluorescein. Cells expressing P-cadherin are visualized in A, and those expressing E-cadherin are seen in B. The two cell types are clearly segregated and show a complementary staining pattern. The regions staining with rhodamine appear red, whereas those regions staining with fluorescein are green. (From Nose et al., 1988. Copyright © Cell Press.)

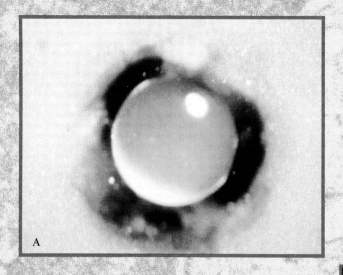

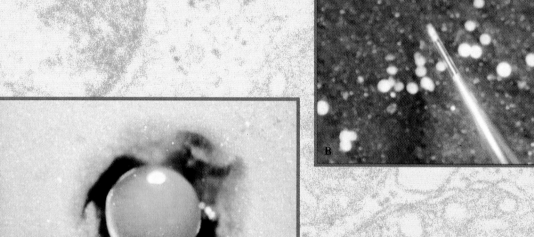

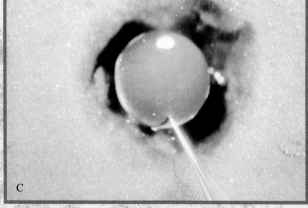

Color plate 15 Nuclear transplantation in *Rana pipiens*. A, Enucleated egg. B, Dissociated blastula cells, from which donor nuclei are to be obtained. C, Nuclear transplant recipient egg. D, Nuclear transplant frog. (Photographs courtesy of R. G. McKinnell.)

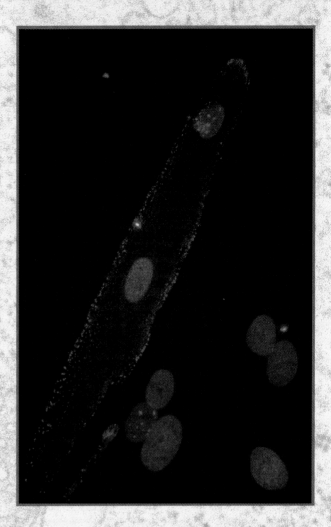

Color plate 16   Color version of Figure 10.23. Activation of the human muscle gene for 5.1H11 in heterokaryons. Live cells were incubated with a monoclonal antibody that recognizes a muscle-specific protein that is found on the surfaces of human (but not mouse) myotubes, followed by treatment with a secondary antibody that can be visualized by fluorescence microscopy. Heterokaryons are shown in fluorescence microscopy at two different wavelengths. Nuclei fluoresce blue, whereas antigen-antibody complexes fluoresce red. The binucleate heterokaryon containing one punctate mouse muscle nucleus and one uniformly stained human hepatocyte nucleus expresses the antigen, which is uniformly distributed on the cell surface. The trinucleate heterokaryon (lower center) has not activated the gene for 5.1H11. (From H. M. Blau et al., 1985. Plasticity of the differentiated state. Science *230:* 758–766. Copyright 1985 by the American Association for the Advancement of Science.)

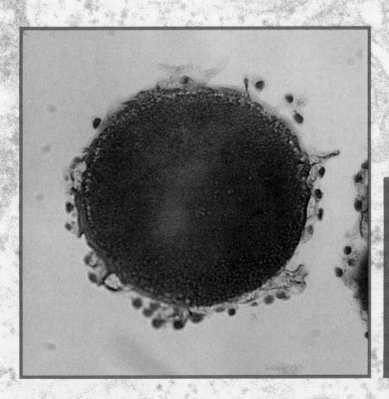

Color plate 17   Photomicrograph of a *Styela* egg that has been stained to reveal the myoplasm (dark red), ectoplasm (pink), and yolk (blue). In this orientation, only a portion of the myoplasm (which would appear yellow without staining) can be seen. (Courtesy of W. R. Jeffery from the front cover of Development *102* (2), February, 1988. Reprinted by permission of Company of Biologists, Ltd.)

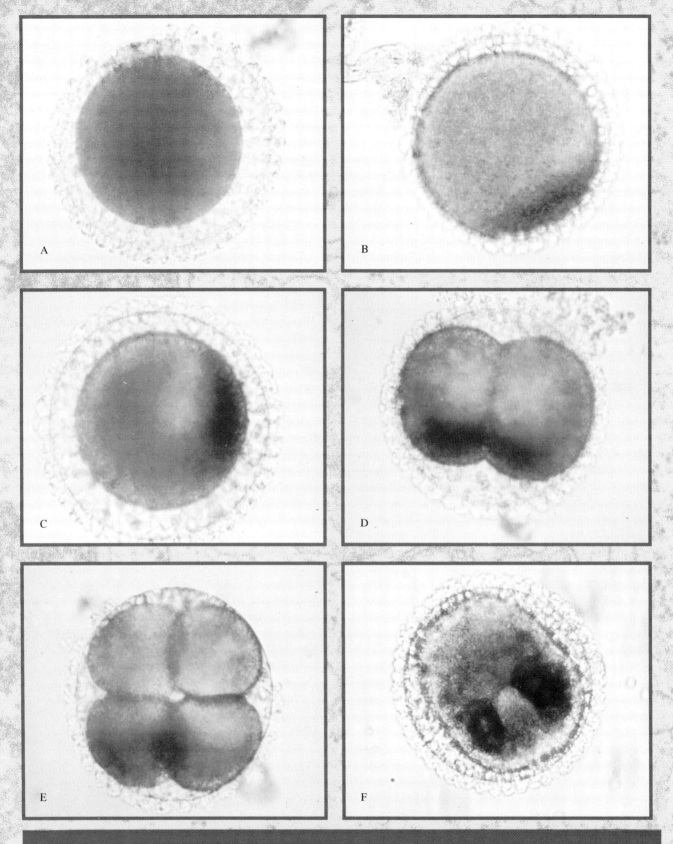

Color plate 18   Photomicrographs of living eggs and embryos of the tunicate *Boltenia villosa* showing segregation and partitioning during cleavage of the orange myoplasm. A, An unfertilized egg with myoplasm throughout the cortex. B, A fertilized egg, in which myoplasm has moved to the vegetal pole region during the first phase of ooplasmic segregation. C, A fertilized egg, in which the myoplasm has moved from the vegetal pole to the posterior region forming a crescent. The clear ectoplasm can be seen adjacent to the orange myoplasm. D, A two-cell embryo with myoplasm in both blastomeres. E, A four-cell embryo with myoplasm in two posterior blastomeres. F, A late gastrula showing two rows of myoplasm-containing presumptive muscle cells surrounding the lateral lips of the blastopore (center). (Courtesy of D. Hursh.)

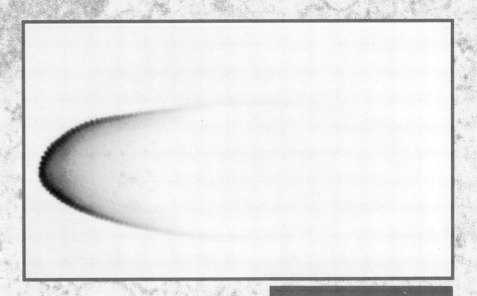

Color plate 19 Color version of Figure 11.37A. Whole mount of a syncytial blastoderm stage *Drosophila* embryo stained with antibody to bicoid protein. A gradient of staining is seen with its highest concentration at the anterior pole (left). (From Driever and Nüsslein-Volhard, 1988. Copyright © Cell Press.)

Color plate 20 Spemann's Nobel Prize Certificate. The prize was awarded October 24, 1935. (Courtesy of K. Sander.)

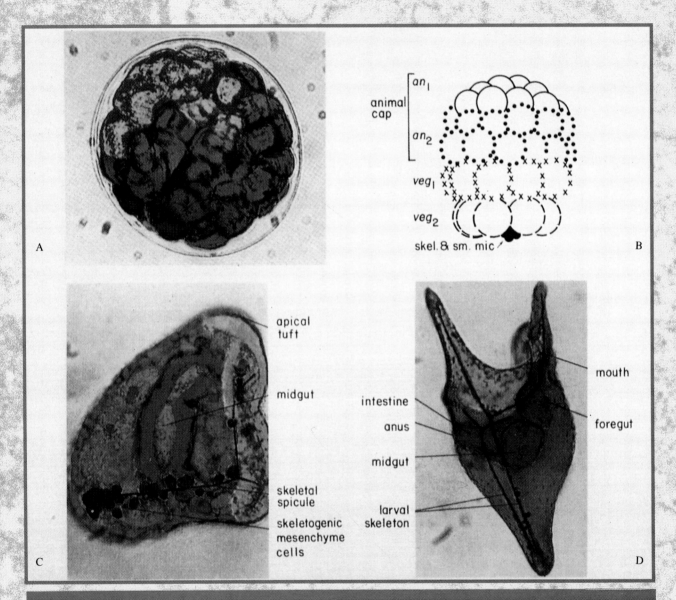

Color plate 21 Development of the *Paracentrotus purpuratus* embryo. A, Photograph of early blastula-stage embryo, which has been colorized to show the positions of territories. In this photograph and in those of the gastrula (C) and the pluteus larva (D), green denotes aboral ectoderm, yellow indicates oral ectoderm, blue indicates vegetal plate, red indicates skeletogenic mesenchyme, and pink indicates the small, non-skeletogenic micromeres. B, A Hörstadius tier diagram of an early sea urchin blastula. (From Davidson, 1989. Reprinted by permission of Company of Biologists, Ltd.)

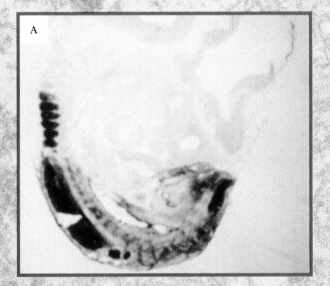

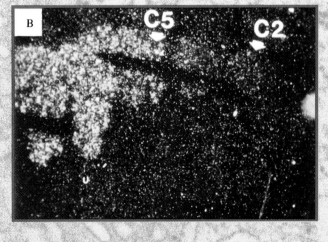

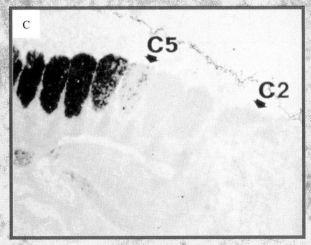

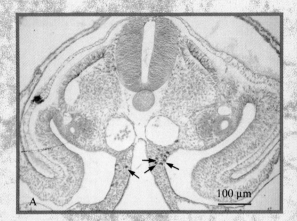

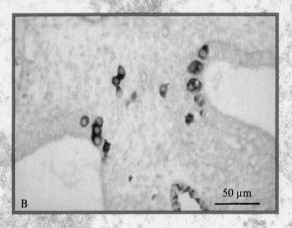

Color plate 24 Color version of Figure 16.13. Migration of primordial germ cells from the dorsal aorta to the genital ridge. A, Section through a 3-day chick embryo stained with PAS (deep pink), showing germ cells (arrows) in the mesentery, one cell leaving the dorsal aorta, and others embedded in the genital ridge. B, High magnification of germ cells identified by PAS staining in the mesentery and genital ridge epithelium. (Courtesy of L. Urven.)

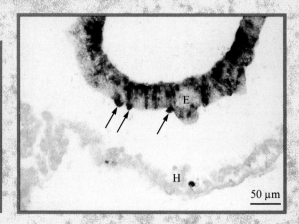

Color plate 25 Color version of Figure 16.14. A section through the epiblast (E) and hypoblast (H) of an early 12-hour-old embryo stained with EMA-1 antibody, which recognizes germ cells (arrows) at a stage earlier than is possible with PAS. (From Urven et al., 1988. Reprinted by permission of Company of Biologists, Ltd.)

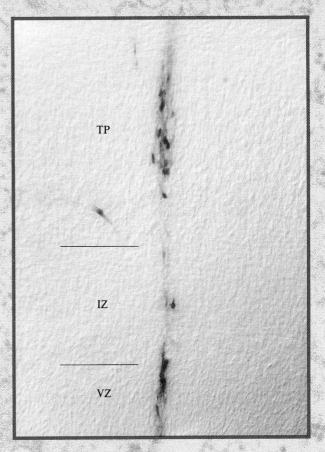

Color plate 26 Color version of essay Figure 1, p. 703. A chicken embryo was injected in the optic tectum with virus carrying the gene for β-galactosidase at stage 16 (51–56 hours of development) and fixed, sectioned, and stained at stage 34. A clone of cells derived from a single infected cell is revealed after staining with X-gal. The clone is arrayed radially across the tectum, suggesting that progeny are displaced vertically through the various layers, with minimal mixing laterally. (From Gray et al., 1988.)

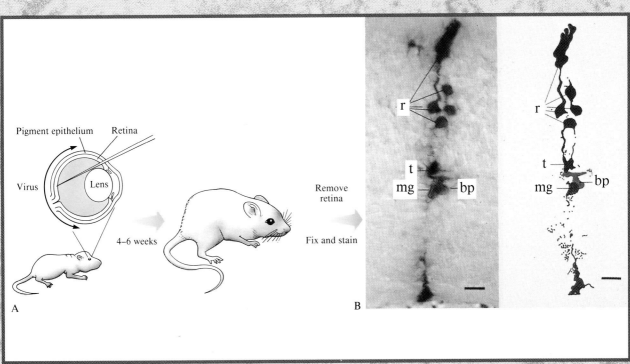

Color plate 27 Color version of Figure 17.18. Retinal cell origins in the rat. A, A retroviral vector that expresses the bacterial ß-galactosidase gene was injected into the space between the neural and pigmented retina in neonatal rats. The animals were killed at 4–6 weeks of age and the injected retinas fixed and processed for ß-galactosidase activity. B, A section through a retina that had been infected with a vector carrying a bacterial β-galactosidase gene. The section has been processed for enzyme activity. A single infected cell has given rise to this radially-arrayed clone, which includes five rods (r), a bipolar cell (bp), and a Müller glial cell (mg). t: rod terminals. Scale bars equal 10 μm. (From Turner and Cepko, 1987. Reprinted by permission from *Nature* Vol. 328 pp. 131–136. Copyright © 1987 Macmillan Magazines Ltd.)

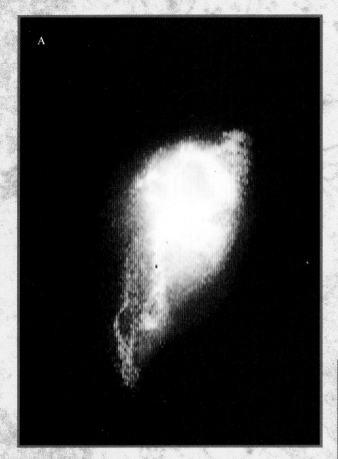

A

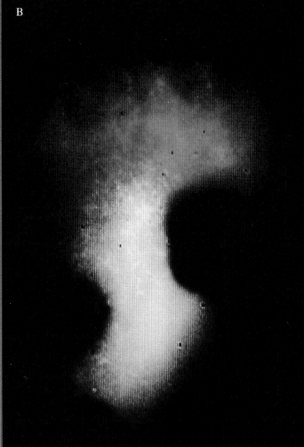

B

Color plate 28 The distribution of retinal fibers on the tectum can be observed in a living animal by using fluorescent vital dyes. In this case, a few nasal retinal cells of a *Xenopus* embryo were labeled with fluorescein dextran (green), and some temporal retinal cells were labeled with rhodamine dextran (red). A, When the fibers first reach the tectum, their terminal arbors overlap considerably. B, With time, fibers from the two poles of the retina sort out and occupy different regions of the tectum. (Courtesy of S. Fraser.)

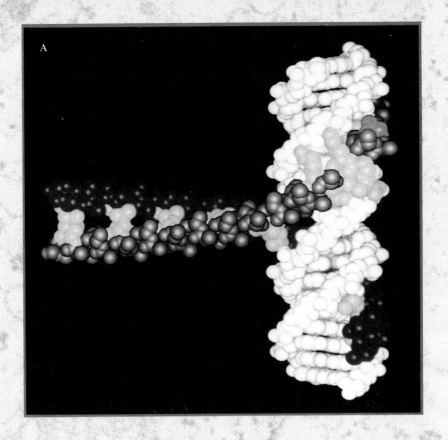

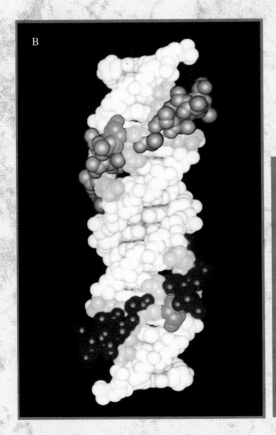

Color plate 29 Hypothetical model of a dimeric leucine zipper protein interacting with its target DNA. The DNA is shown in light gray. One subunit of the leucine zipper protein is shown in red, and the other is in blue. The leucines that interact to hold the dimer together and the basic residues in the DNA-binding region are shown in yellow. Two views of the protein-DNA complex are shown here. A, is a side view, in which the zipper portion of the protein projects to the left and the DNA-binding regions wrap around the DNA molecule. B, is a back view in which the DNA-binding regions of the two monomers are seen to lie in the major groove of the DNA. The zipper region would project back from the plane of the page. The dimer appears to have a "scissors grip" on the DNA double helix. (From C. R. Vinson et al., 1989. Scissor-grip model for DNA recognition by a family of leucine zipper proteins. Science, *246*: 911–916. Copyright 1989 by the American Association for the Advancement of Science.)

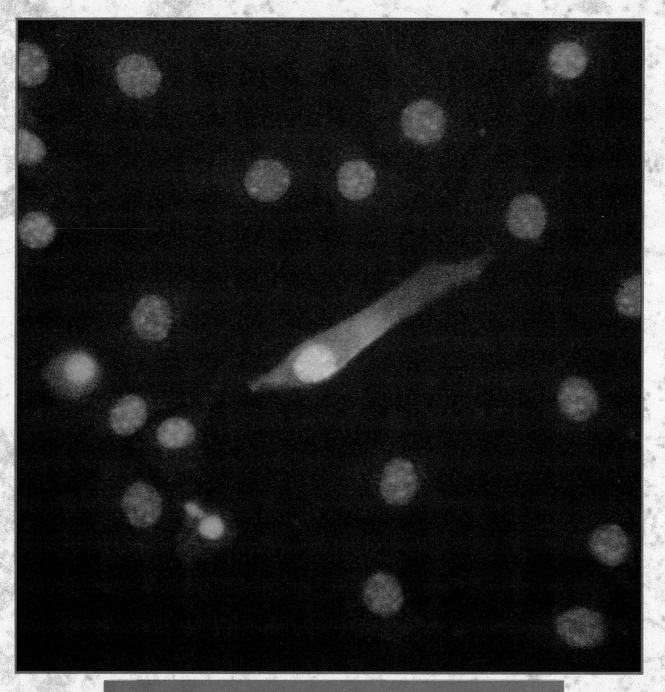

Color plate 30 Color version of Figure 18.5. Conversion of a fibroblast to muscle by transfection with the *MyoD1* gene. A single transfected cell is shown against a background of untransfected cells. The nuclei of the untransfected cells are stained blue with the DNA-specific stain DAPI. Expression of MyoD1 and myosin heavy chain protein in the transfected cell are demonstrated by immunofluorescent staining. The MyoD1 is localized to the nucleus of the transfected cell, whereas myosin heavy chain protein is present in the cytoplasm. Note the elongate appearance of the cell. (Courtesy of S. J. Tapscott from the front cover of Science, 4 August, 1989. Copyight 1989 by the American Association for the Advancement of Science.)

PART V

Genetic Regulation of Development

Ommatidia of the *Drosophila* eye as seen by
scanning electron microscopy. (From Tomlinson and
Ready, 1987.)

CHAPTER 10

Genome Constancy and Its Implications for Development

Robert Briggs (1911–1983), who—with Thomas J. King—pioneered the technique of nuclear transplantation. Briggs and King discovered that as development proceeds, there is a progressive decrease in the percentage of transplanted nuclei that are capable of promoting normal development of enucleated eggs. The early nuclear transplantation experiments were conducted at the Institute for Cancer Research in Philadelphia, Pennsylvania. Later, Briggs moved to Indiana University, where he and Dr. Rufus Humphrey pioneered the use of the Mexican axolotl (*Ambystoma mexicanum*) for the study of developmental genetics. He was particularly interested in genes that exert maternal effects on development; i.e., genes that are transcribed during oogenesis but exert their phenotypic effects after fertilization. (Photograph by Jerry Mitchell, Courtesy of Indiana University News Bureau.)

During the development of a multicellular organism, numerous functionally distinct cell types are produced from a single cell—the fertilized egg, or zygote. The magnitude of this feat and the challenge of understanding how a multitude of diverse cell types is generated in time and space during development become apparent if we examine the starting material that the zygote utilizes to form a complex organism: a single diploid nucleus, which is embedded in the egg cytoplasm and contains the genes that designate the properties of each and every cell type in the body.

How does the embryo use a single genome to produce the various cell types that constitute the completed organism? The cell separation experiments by Driesch (see p. 6) clearly demonstrate that there is no loss of genetic information by early cleavage-stage nuclei. However, the Driesch experiments did not completely quell the beliefs of many developmentalists that modifications in the genetic material itself could occur during development. The task of the nucleus would be much simpler if unnecessary genetic information could be dispensed with or modified in some way so that it becomes nonfunctional. In other words, the nucleus itself would become highly adapted to perform a singular role in development (e.g., the production of a blood cell or a bone cell or a skin cell). Such adaptation would commit a nucleus permanently to a restricted developmental role. In the jargon of embryology, the nucleus would lose **potency**. The zygote nucleus is said to be **totipotent**; it has complete potential to support the development of the entire range of cell types that make up the organism. The potency of nuclei of differentiated cells can be evaluated either by functional analyses of nuclear potential or by analyzing the genetic material itself to determine whether it remains intact after cell differentiation. We now examine the evidence on nuclear potency from these two perspectives.

10–1 Functional Analyses of Genome Equivalence

One way to determine whether the genomes of differentiated cells retain genetic potential that they are not using is to challenge their nuclei experimentally to reinitiate development and promote differentiation of alternate cell types. For plants, potency can be demonstrated directly: Isolated plant cells—even from mature plants—may be induced to form an entire plant (Vasil and Hildebrandt, 1965). However, it has proved much more difficult to resolve this problem for animals because the cell separation experiments that have been used to demonstrate the totipotency of early cleavage-stage nuclei are impractical in later stages of animal development. As cleavage proceeds, not only does the volume of cytoplasm become progressively reduced, but localized cytoplasmic constituents may become segregated into different blastomeres. As we shall discuss later in this chapter, the function of the nucleus in development is highly dependent upon the cytoplasm. Hence, nuclear potential can only be demonstrated if the nuclei are in competent cytoplasm. Consequently, techniques had to be developed to combine later stage nuclei with early embryonic cytoplasm containing all essential cytoplasmic constituents.

Hans Spemann (1938) suggested a radical method for testing nuclear potency by isolating the nuclei of progressively older embryos and implanting them into enucleated eggs. Spemann was suggesting the possibility of **nuclear transplantation** from a somatic cell to an enucleated egg but could see no way to perform the experiment. However, 14 years later, Briggs and King (1952) reported successful transplantation of nuclei of the leopard frog,

Figure 10.1

A, Eggs are activated with a clean glass needle. B, Activated eggs rotate and are enucleated. C, Exovate containing the nucleus is trapped by the vitelline envelope. D, Donor cells are dissociated. An individual cell is drawn into a micropipette of slightly smaller diameter than the cell, bursting it and releasing the nucleus. E, Donor nucleus is inserted into enucleated egg. F, Leakage is minimized by severing the connection that forms between the egg surface and the vitelline envelope. (After King, 1966.)

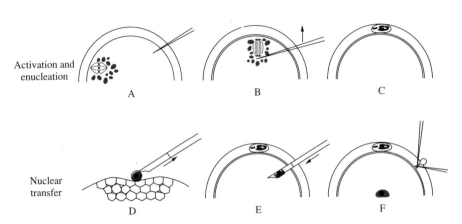

Rana pipiens. This species and other amphibia have proved to be particularly amenable to nuclear transplantation, resulting in an extensive literature dealing with the question of nuclear potency in development. Nuclear transplantation has also been tried with varying degrees of success using a variety of other organisms, including mammals (for review, see DiBerardino, 1980). These ''cloning'' experiments have prompted considerable confusion among both biologists and the lay public alike, particularly concerning the possibility of cloning human beings. We first discuss what the amphibian nuclear transplantation experiments have revealed about nuclear potency and then discuss recent mammalian nuclear transplantation experiments.

Nuclear Transplantation: Amphibians

The transplantation procedure developed by Briggs and King for *Rana pipiens* (Figs. 10.1, 10.2, and Color plate 15) involves two main steps: (1) preparation of the recipient egg; and (2) isolation of a donor cell and transfer of its nucleus. The recipient eggs are obtained from a female by induced ovulation. Because normal fertilization has two components (activation of development and formation of a diploid nucleus), successful nuclear transplantation must also mimic these events. Activation of the egg is normally

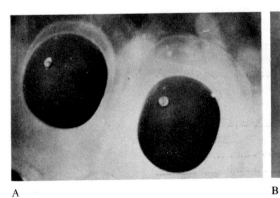

A   B

Figure 10.2

Nuclear transplantation in *Rana pipiens*. A, Enucleated eggs approximately 30 minutes after activation and enucleation. Exovate containing nucleus is visible above the eggs. B, Nuclear transplant recipient at the four-cell stage. (From King, 1966.)

caused by the interaction of the sperm with the egg surface. Activation for nuclear transplantation is achieved by pricking the surface of the egg with a clean glass needle. As a result, the second meiotic spindle moves to the animal pole in preparation for the formation of the second polar body, thus facilitating its removal with a second glass needle. (Recall that the egg is arrested at the second meiotic metaphase.) Donor cells are obtained by dissociating the cells of an embryo in a cell dissociation solution. Normal cell adhesion requires the presence of calcium ions. The dissociation solution is depleted of Ca^{2+}, causing cells to separate from one another. An individual cell is drawn into a micropipette that has a slightly smaller diameter than the cell itself, causing the cell membrane to break. The nucleus, together with its surrounding cytoplasm, is then injected into the activated enucleated egg with the same pipette.

When nuclei from blastulae were transplanted to enucleated eggs, approximately 55% promoted normal cleavage and blastula formation, and 80% of the resulting blastulae continued to develop to the tadpole stage. Of these, 75% reached metamorphosis, at which stage the experiment was terminated (Briggs and King, 1960). McKinnell (1962) obtained normal postmetamorphic frogs from blastula nuclear transplants, providing additional evidence that there is no reduction of developmental potential in blastula nuclei. In a series of investigations, Briggs and King used donor nuclei from endodermal cells of progressively later stages of development in order to examine the possible loss of developmental potential in nuclei (King and Briggs, 1956; Briggs and King, 1957, 1960). They indeed found that the percentage of transplants showing normal development declined as the developmental stage of the donor nuclei increased. Early gastrula nuclei were nearly as effective in promoting normal development as the cleavage nuclei, but nuclei from late gastrula stages showed a definite restriction in developmental potential, as indicated by a dramatic reduction in the number of nuclear transplant recipients that developed normally. Transplants with nuclei from postgastrula stages exhibited a progressive loss of the ability to develop past the gastrula stage. In those transplants receiving late gastrula and postgastrula endodermal cell nuclei that did develop further, the most pronounced deficiencies were found in the size and extent of differentiation of ectodermal and mesodermal derivatives, whereas endodermal derivatives were hardly affected. This condition is termed the "endoderm syndrome." Tailbud-stage endoderm nuclei were incapable of supporting normal development. These data suggest a progressive restriction in the ability of endodermal nuclei to promote normal development and further indicate that the nuclei have become irreversibly specialized as "endodermal" nuclei.

The early Briggs and King experiments with postgastrula nuclei were conducted with nuclei from only one type of cell of one species, endoderm of *Rana pipiens*. Perhaps some peculiarity of these endoderm nuclei causes this restriction, which may not be typical of other nuclei. This may be partially true because occasionally nuclei from other postgastrula cells have been found to promote normal development. DiBerardino and King (1967), for example, obtained a low percentage of normal larvae from neural plate cell nuclei derived from *Rana pipiens* neurulae. It is necessary to point out, however, that neural nuclear transplants are similar to endodermal nuclear transplants in that the percentage of normal development decreases dramatically when the nuclei are derived from progressively older donor embryos. Furthermore, the developmental defects found in some abnormal neural nuclear transplants reflect the origin of the nuclei because the extent of ectodermal development exceeds the development of endodermal and mesodermal derivatives. This "ectoderm syndrome" is the exact opposite of the pattern seen in transplants with nuclei of endodermal origin.

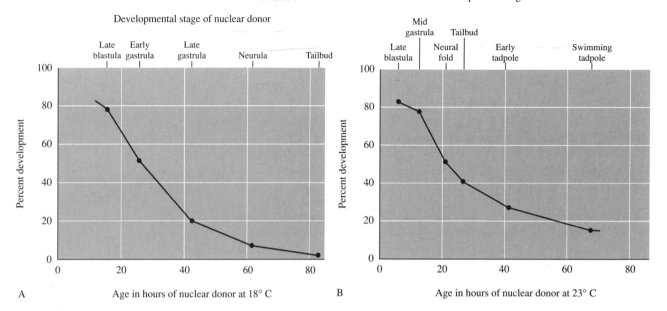

Figure 10.3

A comparison of the capacities of *Rana pipiens* (A) and *Xenopus laevis* (B) nuclei to promote normal development as the ages of donor nuclei increase. *Rana pipiens* develops much more slowly than does *Xenopus laevis*, but the loss of potency occurs at approximately the same rate in both species. (After McKinnell, 1972.)

The results of nuclear transplantation experiments with embryonic nuclei of a number of amphibian species have confirmed that as cells become more highly differentiated, their nuclei progressively become more restricted in their ability to promote development of recipient eggs (Fig. 10.3).

Transplantation of Postembryonic Nuclei

In spite of the general tendency for progressive reduction in nuclear potency as embryos develop, experiments have shown that nuclei of some cells from postembryonic stages can promote considerable development of enucleated eggs. One of the most instructive series of experiments of this sort involved transplantation of nuclei of the germ cell line. Primordial germ cell nuclei of young tadpoles were demonstrated to be highly competent in promoting development of enucleated eggs: 40% of complete blastulae developed into normal tadpoles (Smith, 1965). A high degree of potency makes sense if one considers that primordial germ cells are the precursors of the gametes. Clearly, they are *genetically totipotent* (i.e., they retain an intact genome that has the potential to participate in formation of an embryo). This does not mean, however, that germ cell nuclei are *developmentally totipotent* at all stages during their development. This question was addressed by transplanting germ cell nuclei from later stages of germ cell development. When spermatogonial nuclei of juvenile and adult frogs were transplanted, most of the complete blastulae arrested before finishing gastrulation (DiBerardino and Hoffner, 1971). Of 13 complete blastulae, 3 developed past the gastrula stage, and 1 formed an abnormal larva that commenced feeding but did not survive.

The germ cell nuclear transplantation experiments are important for interpreting the results of somatic cell nuclear transplantation. They demonstrate that genetic totipotence is not sufficient for developmental totipotence. The nucleus apparently can undergo changes that affect the utilization of the genetic material while leaving the genome intact. Although the germ cell line obviously retains the complete genome, germ cell nuclei apparently undergo a developmental restriction after the primordial germ cell stage, limiting the utilization of the genome for development after nuclear transplantation.

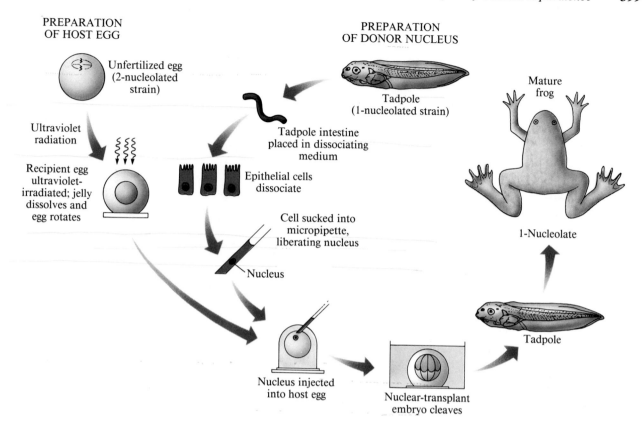

PREPARATION
OF HOST EGG

Unfertilized egg
(2-nucleolated
strain)

Ultraviolet
radiation

Recipient egg
ultraviolet-
irradiated; jelly
dissolves and
egg rotates

PREPARATION
OF DONOR NUCLEUS

Tadpole
(1-nucleolated strain)

Tadpole intestine
placed in dissociating
medium

Epithelial cells
dissociate

Cell sucked into
micropipette,
liberating nucleus

Nucleus

Nucleus injected
into host egg

Nuclear-transplant
embryo cleaves

Mature
frog

1-Nucleolate

Tadpole

Figure 10.4

Nuclear transplantation in *Xenopus laevis*. Donor tadpole is from a mutant strain with one nucleolus per nucleus. Host egg nucleus contains two nucleoli. Presence of nuclei with one nucleolus in nuclear transplant individual proves that development is due to donor nucleus. (After Gurdon, 1966.)

The developmental potency of nuclei from postembryonic cells of *Xenopus laevis* has also been examined by extensive nuclear transplantation studies. In the initial experiments of this type, Gurdon (1962) transplanted nuclei from *Xenopus* tadpole intestinal epithelium cells to enucleated *Xenopus* eggs. The techniques used for transplantation of *Xenopus* nuclei (Fig. 10.4) are somewhat different from those used for *Rana*. Instead of mechanical enucleation, the nucleus is destroyed by ultraviolet irradiation. The irradiation treatment also activates the egg. The donor nucleus is derived from a cell obtained by chemical dissociation of tadpole intestinal epithelium. As with *Rana*, the cell containing the donor nucleus bursts as it is sucked into the transplant pipette. The liberated nucleus is then injected into the enucleated host egg. A genetic marker found in certain mutant *Xenopus* allows the investigator to determine whether the nuclei of the nuclear transplant individuals are indeed derived from the donor nucleus (the reliability of this marker has been challenged by Du Pasquier and Wabl, 1977). Donor nuclei are obtained from a strain of *Xenopus* that has only one, rather than the normal two, nucleoli per nucleus. The host eggs are from a wild-type strain with two nucleoli. Microscopic examination of cells of the nuclear transplant individuals is conducted routinely to confirm the origin of the nucleus. (For a detailed discussion of this mutation, see Chap. 13.) A very low percentage (approximately 1.5%) of intestinal cell nuclear transplants actually developed into normal feeding tadpoles. Some of these nuclear transplant tadpoles were successfully reared in the laboratory into adult frogs.

Gurdon (1962) observed that a large percentage of *Xenopus* nuclear transplant recipients underwent partial cleavage. This could result from damage to eggs during transplantation or from incomplete chromosomal replication in some transplant nuclei or their daughter nuclei during cleavage. (As we shall discuss on p. 401, incomplete chromosomal replication is a

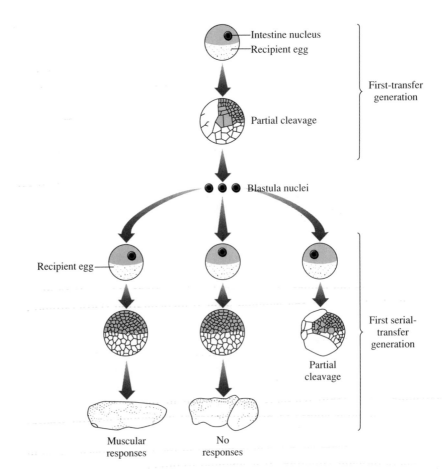

Figure 10.5

Serial nuclear transplantation. Abnormal first-transfer embryos were dissociated, and nuclei from individual cells were transplanted to enucleated eggs. Many serial-transfer embryos developed more normally than the first-transfer donor; some developed functional muscle, as indicated by muscular responses. (After Gurdon, 1968.)

common problem in transplanted nuclei.) The potential of such nuclei to promote differentiation would never be realized because gastrulation would be abnormal. Therefore, the percentage of normal development would not reflect the developmental capabilities of the transplanted nuclei accurately. For this reason, **serial nuclear transplantation** (Fig. 10.5) has been conducted to maximize the opportunities for transplanted nuclei to promote normal development (Gurdon, 1962). This technique was developed earlier by King and Briggs (1956) with *Rana pipiens*.

Abnormal first-transfer *Xenopus* embryos (blastula stage or earlier) were selected at random and dissociated. Several nuclei from each embryo were then injected singly into enucleated eggs. All serial-transfer embryos produced from a single first-transfer embryo are a clone of genetically identical individuals descended from a single intestinal nucleus. In every case, some of the serial-transfer embryos developed more normally than their first-transfer embryo nuclear donor, indicating that totipotent nuclei can be present in embryos that exhibit limited development. Some of the serial-transfer embryos developed into feeding tadpoles. When the results of the first-transfer generation were combined with those of the first serial-transfer generation, the percentage of intestinal cell nuclei that were capable of promoting development to the tadpole stage was raised to 7%. The most normal serial-transfer blastula was then used to provide nuclei for a second serial-transfer generation. This generation did not show any further improvement in development over that of the first serial-transfer generation. Thus, one serial-transfer generation is sufficient to demonstrate the maximal developmental capacity of the original intestinal cell nucleus. If one of the serial-transfer embryos is used subsequently as a source of nuclei for transplantation, and this process is repeated in each generation, the nuclear clone could be perpetuated indefinitely.

Nuclei from another larval cell type of *Xenopus* also have been shown to be totipotent. Kobel et al. (1973) obtained a mature adult frog after serial transplantation of nuclei from epidermal cells of hatching tadpoles. Interestingly, only the nuclei from nonciliated epidermal cells promoted development; nuclei from ciliated epidermal cells, which appear to be more highly differentiated, never promoted development beyond the blastula stage. This suggests that the nuclei of the nonciliated cells are much less differentiated than those of their ciliated cohorts.

Two generalizations emerge from the experiments we have discussed thus far: (1) As cells differentiate, their nuclei become progressively restricted in their ability to promote the development of enucleated eggs; and (2) cells from *some* postembryonic tissues contain nuclei that can promote at least partial development of enucleated eggs. What is the basis for the developmental restriction, and what is the explanation for the exceptions?

Mitotic Incompatibility

One reason for the failure of some nuclear transplant recipients to develop normally is damage to the genome after transplantation (DiBerardino and Hoffner, 1970; DiBerardino, 1979). Consider the difficult task that faces a nucleus of a postgastrula stage when transplanted to an egg. As we discussed in Chapter 5, the cleavage cell cycle is very rapid and highly modified, with no G_1 or G_2 stages. At the mid-blastula transition, the mitotic rate is reduced considerably. The continued reduction in the normal mitotic rate during development roughly parallels the decrease in the ability of nuclei from progressively later stages to promote normal development after transplantation. Furthermore, many differentiated cells never undergo mitosis. Thus, when a nucleus from such a cell is placed in the egg cytoplasm, it is in an extremely difficult situation. It is adapted to directing the specialization of one particular cell type that may no longer divide, but it is suddenly called upon to replicate its DNA and divide at the very rapid rate demanded during cleavage. Chromosomal replication may be incomplete under such circumstances, resulting in chromosomal abnormalities.

The hypothesis that mitotic adaptability of donor nuclei affects their capacity to promote development is supported by experiments with adult *Xenopus* erythroblast and erythrocyte nuclei. Erythroblasts proliferate extensively by mitosis and are the precursors of erythrocytes, which are no longer proliferative. (Unlike mammalian erythroid cells, amphibian erythrocytes are nucleated.) Thus, the effects of mitotic capacity of nuclei on their developmental potential can be tested by comparing the results of transplantation of nuclei of these two cell types. In fact, erythrocyte nuclei are incapable of promoting development beyond the early gastrula stage, whereas some erythroblast nuclei promote the development of eggs to abnormal early tadpoles (Brun, 1978). Although these nuclei differ in ways other than their mitotic activity, the correlation between mitotic activity and developmental potential is striking.

Is the failure of mitotic adaptability an explanation for the general failure of nuclear transplantation from postgastrula-stage embryonic cells? Numerous investigators have reported that abnormal nuclear transplants often have an aberrant karyotype (i.e., the chromosome number may be modified, or the chromosomes themselves may appear abnormal when studied with the microscope). Furthermore, the severity of chromosomal abnormalities appears to be directly correlated with the extent of the developmental restriction shown by nuclear transplants (DiBerardino, 1979). Thus, those with the most severe chromosomal abnormalities arrest the earliest, whereas those with less severe aberrations can develop to larval stages. Careful studies of nuclei transplanted from late gastrula endoderm cells demonstrate that chromosomes of these nuclei frequently are unable to replicate properly when transferred to the cytoplasm of eggs (DiBerardino and Hoffner, 1970). The most obvious cause of this restriction appears to be the failure of some of the chromatin to decondense sufficiently to allow replication of the entire genome.

The endoderm and ectoderm syndromes previously described appear to be exceptions to the correlation between abnormal development of nuclear transplant recipients and aberrant chromosome constitution. Briggs et al. (1961), Subtelny (1965), and DiBerardino and King (1967) observed that nuclear transplant recipients displaying either the endoderm or ectoderm syndrome apparently had normal karyotypes, whereas only those with generalized developmental abnormalities could be correlated with chromosome abnormalities. This observation implies that there can be developmental restriction without chromosomal damage. There is, however, the possibility that subtle damage to the genome has occurred that has not been detected by examining karyotypes.

Mitotic Adaptation

One way to overcome the problem of mitotic incompatibility is to allow nuclei to adapt to a more rapid division rate before transplantation. This procedure attempts to maximize the opportunities for nuclei to make the transition to the rapid mitotic rate of cleavage and thus minimize or eliminate posttransplantation genetic damage. In these experiments, cells obtained from adult frogs are grown in tissue culture, where they undergo mitosis before nuclear transfer.

This approach has been used with cells of the adult *Xenopus* kidney, heart, lung, and skin. Their nuclei have been shown to promote development of abnormal larvae in serial nuclear transplantation studies, provided the cells were first grown in tissue culture before nuclear transplantation (Laskey and Gurdon, 1970). An extensive series of experiments has been conducted with cultured adult *Xenopus* skin cells. In order to rule out the possibility that nuclei from undifferentiated cells were being transferred, the differentiated state of these cells was confirmed by the detection of keratin, the protein that is specific for differentiated epidermal cells. Abnormal swimming tadpoles have been obtained in serial transfers of nuclei of these cells. Histological examination of the tadpoles has revealed that even though these larvae are not completely normal, the transplanted nuclei have supported the development of a number of diverse tissues and organs, including the heart, striated muscle, brain, nerve cord, notochord, pronephros, intestine, and eye (Gurdon et al., 1975). Similar studies have shown that adult *Xenopus* lymphocyte nuclei can support the development of abnormal swimming tadpoles (Du Pasquier and Wabl, 1977).

These various experiments illustrate that mitotic preadaptation of donor nuclei can increase their developmental potential, but the recipients are still not *completely* normal. Thus, mitotic incompatibility is not the only explanation for developmental restrictions on nuclei.

Exceptions to Developmental Restrictions on Nuclear Potency

As we have discussed, some tissues from advanced embryonic and young larval stages contain cells with nuclei that can promote normal development in nuclear transplantation experiments without the necessity for mitotic adaptation. Except for the primordial germ cell nuclear transplants, the frequency of successful nuclear transplantation from these stages is very low. (Recall that only 7% of intestinal cell nuclei could promote development to the tadpole stage.) In none of these experiments have the donor cells been unequivocally shown to be differentiated. Therefore, it is possible that the exceptional successful nuclear transplantations have resulted from nuclei of uncommitted stem cells, rather than differentiated cells (DiBerardino, 1989). In the case of *Xenopus* intestinal epithelium, the nuclei might have been derived from primordial germ cells, which migrate through the gut on their way to the gonads (see p. 446).

Specific Nuclear Differentiation

In spite of nearly four decades of intense investigation, nuclear transplantation has yet to provide an unequivocal answer to the question of nuclear totipotency. The weight of evidence suggests that there is, indeed, *specific nuclear differentiation* during development. Indeed, no nucleus of a proven differentiated somatic cell nor of any adult cell *of any kind* has been shown to be totipotent. Genomes apparently become *imprinted* during development to express a limited portion of the original developmental repertoire (DiBerardino, 1989). Transplantation of nuclei to eggs may be an ineffectual means for releasing the nuclei from their limitations and allowing them to express their full genetic potential.

Nuclear Reprogramming

If we accept the conclusion that developmental restrictions are imposed upon most nuclei during development, are there ways to reprogram nuclei (including nuclei from differentiated cells) to express more—if not all—of their genetic potential? Recently, experiments have been initiated to attempt to reprogram nuclei by exposing them to oocyte, rather than egg, cytoplasm. These experiments have utilized erythrocytes of *Rana pipiens* as the sources of nuclei. Among the experimental advantages of erythrocytes are their ease of collection and the certainty that they are, indeed, fully differentiated cells, rather than stem cells. Erythrocyte nuclei are injected into oocytes at first meiotic metaphase (Fig. 10.6). After the completion of meiotic maturation, the eggs are activated by pricking with a glass needle. Some of the resultant blastulae have been used to provide donor nuclei, which were injected into enucleated eggs. Some of these second transfer-generation

Figure 10.6

Transplantation of *Rana pipiens* erythrocyte nuclei into oocytes. Erythrocytes were broken by osmotic shock and microinjected into oocytes at first meiotic metaphase. After maturation, the matured oocyte (i.e., egg) was activated with a glass needle, and the nucleus was removed by microsurgery. Prehatching tadpoles were derived from the original transplant generation. Retransfer of nuclei from dissociated nuclear transplant blastulae resulted in swimming tadpoles. (After DiBerardino et al., 1984.)

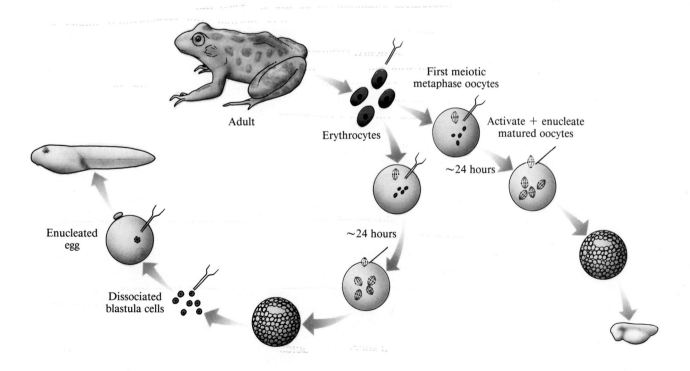

First meiotic metaphase oocytes

Adult

Erythrocytes

Activate + enucleate matured oocytes

~24 hours

~24 hours

Enucleated egg

Dissociated blastula cells

Figure 10.7

Triploid *Rana pipiens* nuclear transplant tadpole from a triploid blastula nucleus derived from an erythrocyte of a triploid juvenile frog. Arrow denotes one of a pair of hind limb buds. (From DiBerardino et al., 1986.)

embryos developed into tadpoles (Fig. 10.7). Parallel experiments in which erythrocyte nuclei were transplanted in the first instance into unfertilized eggs produced second transfer generation embryos that never developed further than the gastrula stage. Clearly, the oocyte cytoplasm is able to elicit nuclear potential from nuclei of these differentiated cells that remains repressed without this interaction. Thus, the imprinting of the genome with the singular properties of an erythrocyte phenotype can be erased to allow the nuclei to specify the multiple cell phenotypes of tadpoles (DiBerardino, 1989). Erythrocyte nuclei that were originally passaged through oocyte cytoplasm have been serially transplanted through eight transplant generations with no loss of developmental potential, indicating that the erasure of the previous pattern of imprinting is stable (Hoffner Orr et al., 1986).

Recent attempts to reprogram nuclei have produced promising results. However, nuclear *totipotency* must remain an uncertainty until some means is found to produce a high frequency of perfectly normal nuclear transplants from nuclei of adult differentiated cells. But, *at the very least*, amphibian nuclear transplantation experiments demonstrate that the nuclei of some differentiated cells can promote gastrulation, organization of the basic body plan, and elaboration of a variety of diverse organs and tissues, and are therefore **pluripotent** (i.e., can be partially reprogrammed during development).

Nuclear Transplantation: Mammals

Nuclear transplantation experiments with mammals also demonstrate that nuclei from early cleavage-stage embryos are totipotent. However, there are considerable differences among different mammalian species in the success of transplantation of nuclei from later stages of development. As we shall see, developmental restrictions on nuclei are observed quite early in mouse development, whereas nuclei from somewhat later stages of sheep and cattle embryos are totipotent.

Mouse Nuclear Transplantation

The nuclear transfer techniques used for mammals are quite different from those used for amphibians because mammalian embryos are quite sensitive to penetration by micropipette. Procedures to transfer nuclei to mouse embryos without using a micropipette to penetrate embryos have been developed by McGrath and Solter (1983). Their technique, which is outlined in Figure 10.8, combines microsurgery with cell fusion. As shown in the drawings, the pronuclei of one-cell–stage embryos are aspirated into a micropipette within a membrane-enclosed karyoplast. In the presence of the cytoskeletal inhibitors cytochalasin B and colcemid, the karyoplast will separate from the embryo. The pronuclei of karyoplasts removed in this way may be introduced into enucleated recipient embryos by using Sendai viruses that have been killed with ultraviolet light. The inactivated viruses, although noninfective, bind to cell membranes and cause them to adhere to one another. At the points of contact, the membranes fuse, allowing the contents of the karyoplast to enter the cytoplasm of the previously enucleated embryo. Over 90% of nuclear transplant recipients develop normally when this technique is used (McGrath and Solter, 1986).

An extension of the McGrath and Solter technique involves the fusion of karyoplasts containing diploid nuclei from progressively later stages of development. As shown in Figure 10.9, a dramatic reduction in the success of nuclear transplantation occurs during the two-cell stage. Transfer of early two-cell nuclei was as effective as transfer of pronuclei. However, transfer of late two-cell nuclei produced only a few morulae (22%), whereas four-cell and eight-cell nuclei could only rarely direct more than one cleavage division.

In spite of the failure to observe development after the transfer of older nuclei into one-cell embryos, nuclei from as late as the eight-cell stage have been transplanted into enucleated late two-cell–stage recipients, resulting in normal blastocysts (Robl et al., 1986; Howlett et al., 1987) or even live young (Tsunoda et al., 1987). This paradoxical result points to the difficulties in interpreting the results of nuclear transplantation experiments: Nuclei may have potential that is not revealed in the experiment. An essential nucleocytoplasmic interaction apparently occurs at the mid-two-cell stage of mouse development (an important transitional stage; see page 565). Once they have undergone this interaction, nuclei apparently lose the ability to repeat this interaction, leaving the cytoplasm deficient. Hence, if older nuclei are used, the cytoplasm must be from an embryo in which such an interaction has already occurred. The results of mouse nuclear transplantation experiments thus provide evidence for both a restriction of nuclear potential in all nuclei at the two-cell stage as well as the retention of significant developmental capacity—at least to the eight-cell stage.

Parental Imprinting of Genes in Mouse Gametes

In most organisms, both parents contribute equally to the zygotic genome. Thus, the paternal and maternal genes should be interchangeable. In support of this hypothesis, eggs of many species can produce viable progeny by parthenogenesis (i.e., no sperm nucleus; see Chap. 4), and fertilized anucleate sea urchin merogones (no egg nucleus) can develop into adults. However, sperm and egg do not appear to make identical contributions to the zygotic genome in mammals. Natural parthenogenesis has not been documented in mammals, and most artificially activated mammalian eggs die after implantation (Kaufman et al., 1977).

The requirement for both female and male pronuclei for development can be shown directly by nuclear transplantation experiments done in the mouse. In these experiments, embryos are constructed containing only two female (**gynogenetic embryos**) or two male (**androgenetic embryos**) pronuclei (McGrath and Solter, 1984; Barton et al., 1984). This is accomplished by microsurgically removing either the male or female pronucleus from one zygote and replacing it with a nucleus of the opposite type from another zygote (see Fig. 10.8). After the operation, the two pronuclei fuse into a single synkaryon and the zygotes cleave, but neither androgenetic nor gynogenetic embryos develop beyond implantation. As shown in Figure

A

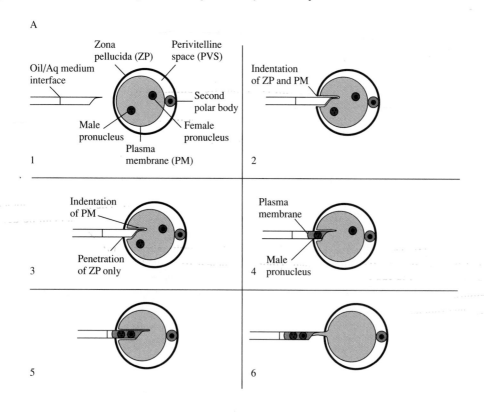

B

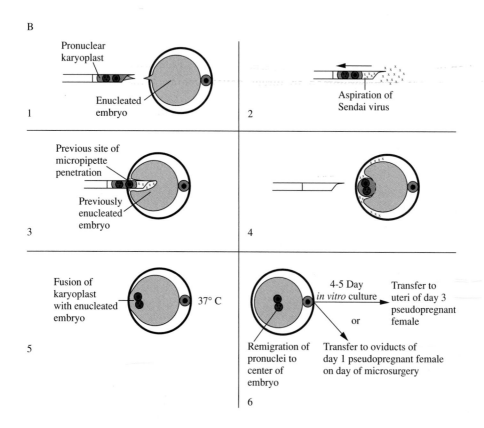

Figure 10.8 (opposite)

Technique of nuclear transplantation in mouse embryos. A, Diagrammatic representation of the removal of the pronuclei from the one-cell stage mouse embryo. The embryo, held in place by a suction pipette (not shown), is oriented with the second polar body opposite to the approach of the enucleation pipette (1). This orientation facilitates the relocation of the original site of pipette penetration when a second nucleus is introduced after enucleation. The zona pellucida (ZP) and embryo plasma membrane are indented by the advancing pipette (2) until the ZP is penetrated (3). Considerable deformation of the embryo may be necessary. The pipette tip is subsequently moved to overlie a pronucleus, and the latter—preceded by a portion of the embryo plasma membrane—is aspirated into the pipette (4). The second pronucleus is similarly aspirated (5), and the pipette is withdrawn. Upon continued withdrawal (6) and in the presence of cytoskeletal inhibitors as described in the text, the pronuclear karyoplast (nucleated vesicle) will separate from the embryo cytoplasm. B, Diagrammatic representation of the introduction of pronuclei into a previously enucleated one-cell stage mouse embryo. The pipette containing the pronuclear karyoplast is placed in a drop containing inactivated Sendai virus, and a small volume of the virus suspension is aspirated into the pipette (2). The pipette is then inserted into a previously enucleated one-cell stage embryo through the previous enucleation site (3), and the contents of the pipette are injected into the perivitelline space (4). After transfer to a 37°C incubator, the pronuclear karyoplast fuses with the enucleated embryo (5). The number of enucleated embryos that incorporated the donor nuclei can be determined using a dissecting microscope. The newly-introduced pronuclei subsequently migrate to the center of the embryo (6). Nuclear transplant embryos may be cultured *in vitro* or transferred directly to the oviducts of pseudopregnant females. (After McGrath and Solter, 1986.)

10.10, gynogenetic embryos are normal but small, and their extraembryonic tissue is very poorly developed. Androgenetic embryos, on the other hand, are quite retarded, whereas their extraembryonic tissue is well developed. This result is not caused by injury to the pronuclei during the operation because normal fetuses can be obtained from control zygotes containing a female pronucleus and an implanted male pronucleus. Thus, the two parental genomes apparently function in a complementary fashion during development: Regions of paternal chromosomes are apparently preprogrammed to direct proliferation of the extraembryonic tissues, whereas certain maternal chromosomal regions are preprogrammed to direct development of the embryo proper.

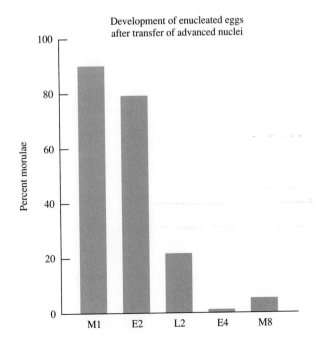

Development of enucleated eggs after transfer of advanced nuclei

Figure 10.9

Development of enucleated mouse eggs after transfer of nuclei from progressively older stages of development. The graph shows the percentage of nuclear transplant recipients that reached the morula stage after enucleation and transfer of pronuclei (M1), early (E2) or late (L2) two-cell nuclei, early four-cell (E4) nuclei, or eight-cell (M8) nuclei. (After Howlett et al., 1989.)

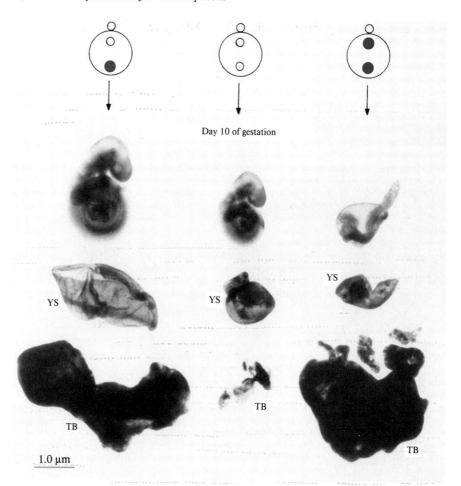

Day 10 of gestation

1.0 μm

Figure 10.10

The appearance on day 10 of gestation of mouse conceptuses derived from fertilized, gynogenetic, or androgenetic eggs. Embryo, yolk sac (YS), and trophoblast (TB) have been dissected in each case. Maternal pronuclei are shown in the drawing as open circles, and paternal pronuclei are colorized. (From Howlett et al., 1989.)

Differential programming of the maternal and paternal genomes apparently occurs during gametogenesis in a process known as **parental imprinting** (for review, see Howlett et al., 1989). Although the imprinting mechanism is unknown, it has been proposed that modifications to the DNA during gametogenesis mark the chromosomes as to their origin, thus directing the expression of distinct sets of genes. In this regard, there is evidence that the pattern of <u>DNA methylation</u> may be different in female and male pronuclear genomes (Sapienza et al., 1987; Swain et al., 1987). As we shall discuss in the essay on p. 732, methylation is a modification to DNA that can be transmitted to daughter cells at cell division and has been <u>implicated</u> in regulation of <u>gene expression</u>. Accordingly, it has been proposed that differential methylation of portions of the genome in sperm and eggs influences their utilization during development (Monk et al., 1987; Monk, 1988).

Sheep and Cow Nuclear Transplantation

In contrast to the results of mouse nuclear transplantation, there is no evidence for a restriction of nuclear potential at the two-cell stage in sheep and cattle. In fact, nuclei from stages as late as the 16-cell stage in sheep and the 64-cell stage in cattle are fully totipotent when transferred to enucleated eggs. Nuclear transplantation is now being used commercially for the cloning of embryos from prized cattle. Typically, cattle nuclei are transferred at the 16- to 32-cell stage. Investigators have found no evidence for developmental restriction on any nuclei at this stage (Willadsen, 1989). These results point to the danger in drawing general conclusions for all mammals from experiments with mice. Mice are valuable experimental animals with many advantages, but results obtained from murine studies should not necessarily be extrapolated to all mammals.

Unlike the kinds of studies that have been reported for amphibians, there has been no systematic investigation of potency of nuclei from later stages of sheep and cattle, including nuclei from differentiated cells. Consequently, we still do not know whether there are developmental restrictions on advanced nuclei in these species.

10–2 Integrity of the Genome During Development

It is clear from the amphibian nuclear transplantation experiments that loss of genetic information does not normally accompany cell differentiation. Another convincing example of the retention of the complete genome is the constancy of the polytene chromosome map of *Drosophila* larvae (see pp. 11 and 411). The band pattern (which reflects genome organization) is identical in diverse differentiated cells, indicating that genetic composition remains constant during development of those cell types.

genome equivalence

There is also a vast amount of molecular data that offers support for the concept of genomic constancy. Genetic information is encoded in DNA in the sequence of its nucleotides, which does not vary from cell to cell in an adult organism (i.e., there is no loss or gain of genetic information during cell differentiation). This is even true for genes that encode tissue-specific proteins, such as the globin gene: These genes are retained, and their nucleotide sequence is identical in all cell types—those in which they are expressed (red blood cells in the case of the globin gene) and those in which they remain silent. Likewise, with very few documented exceptions, specialization for protein synthesis does not involve amplification of genes encoding those proteins.

10–3 Differential Gene Expression

Let us now consider some of the ramifications of the constancy of the genome during development. The primary function of the genome is to produce the transcripts that direct the synthesis of proteins. Theoretically, every cell has the capacity to produce the same proteins. However, the capacity to synthesize certain proteins becomes restricted to specialized cells during development. For example, red blood cells are specialized to synthesize globin, a protein that is not found in other cell types in the body.

Restrictions on protein synthesis must mean that certain mechanisms are operating to select genes to be expressed in different cells. Therefore, the variability in protein composition in differentiated cells must result from **differential gene expression**. Thus, the pattern of RNA synthesis varies among differentiated cells. For example, regulated production of globin transcripts (from which globin molecules are produced) occurs in chick embryo erythropoietic tissue, but production does not occur in nonerythropoietic tissue (Groudine et al., 1974; Lois et al., 1990). As we shall discuss in detail in the following chapters, gene expression is subject to extensive spatial and temporal regulation throughout development.

The mechanisms that control development must exert their effects by controlling differential gene expression. When the developmental pathway of a cell is specified (**cell determination**), control mechanisms must select

Essay

Exceptions to the Gene Constancy Rule

Although the genome is *normally* retained intact during development, there are some important exceptions to this rule. These include DNA deletion, DNA amplification, and DNA rearrangement. These exceptions do not nullify the rule of genome constancy but emphasize how rules can occasionally be bypassed when it is advantageous to the organism.

DNA Deletion

The entire somatic cell line of some animals may lose a portion of its genome during early cleavage. This chromatin diminution is known to occur during development of some nematodes and crustaceans. The germ cell line in these embryos retains the complete genome. Some determinative event in very early development discriminates between the presumptive somatic cells and germ cells, enabling the latter cells to retain the complete genome while causing selective loss of a portion of the genome in somatic cells. An example of such a mechanism occurs in the nematode *Ascaris megalocephala* and will be discussed in Chapter 11.

Chromatin diminution is a singular and dramatic developmental event that lacks the subtlety of dif-ferential gene expression. As described in Chapter 11, it appears that repetitive DNA sequences, rather than structural genes, are lost during chromatin diminution. Furthermore, there is *no evidence* that further selective loss of genetic material occurs in cell determination in species exhibiting this phenomenon. They presumably must resort to more conventional means of information selection for subsequent cell determination.

DNA Amplification

Multiple copies of certain genes and surrounding regions of the DNA may be selectively produced in some cells to provide additional templates for the exaggerated production of transcripts from them. This is one mechanism that is employed by cells that require a large amount of a given gene product in a very short time. An example of gene amplification is the differential replication of 18S and 28S ribosomal RNA genes in amphibian oocytes, which was discussed in Chapter 3.

Another example of specific structural gene amplification has been discovered in *Drosophila melanogaster*. Genes for the chorion proteins are amplified in ovarian follicle cells (Spradling and Mahowald, 1980). As with ribosomal RNA gene amplification, the multiple copies of the chorion protein genes satisfy the need to produce a great deal of product in a very short time. The polytene chromo-somes of *Drosophila* and other dipterans are examples of total genome amplification. Polytene chromosomes of the midge *Rhynchosciara* undergo selective amplification of a few genes in addition to the generalized amplification associated with polytenization (Pavan and Da Cunha, 1969). This mechanism presumably allows for the enhanced production of the proteins encoded by the selectively amplified genes.

DNA Rearrangement

In 1965, Dryer and Bennett proposed that the means for the incredible diversity of immunoglobulins produced by the vertebrate immune system is generated during development by rearrangement of separate genetic elements into functional immunoglobulin genes. Contemporary molecular biology research has confirmed that this is, indeed, one of the mechanisms for generation of immunoglobulin diversity (Hozumi and Tonegawa, 1976; Seidman and Leder, 1978). Clearly, immunoglobulin-producing cells undergo irreversible changes in their genomes during their development—a definite exception to the gene constancy rule. We shall discuss rearrangement of immunoglobulin genes in more detail in Chapter 18. A similar phenomenon has been demonstrated for genes encoding antigen-specific receptors on mammalian T lymphocytes (Chien et al., 1984; Malissen et al., 1984; Siu et al., 1984).

which genes in that cell are to be expressed and which genes are to remain dormant. Cell differentiation (i.e., acquisition of specialized structure and function) is a consequence of the utilization of those restricted portions of the genome.

There is no single mechanism that restricts gene expression during development; regulation of gene expression can be exerted at each step in cellular information processing. Indeed, we have discussed examples of posttranscriptional regulation of gene expression in Chapters 2 and 3, and the significance of this mode of regulation will become even more apparent in Chapter 13. However, the primary step in information processing is transcription. Obviously, if a gene were not transcribed, there would be no need for posttranscriptional regulation. One of the most dramatic demonstrations of differential transcription is the morphological change occurring in certain loci of dipteran polytene chromosomes when they become functionally active.

Polytene chromosomes are found in certain enlarged cells of the larvae of some dipterans, including *Drosophila* and *Chironomus*. Dipterans are somewhat unusual because homologous chromosomes pair and remain perfectly aligned in somatic cells during interphase. In larval salivary glands, midgut, and malpighian tubules of *Drosophila* and *Chironomus*, repeated DNA replication of both homologues follows somatic pairing. The numerous replicates (chromatids) do not separate but remain attached side by side, forming one extremely large structure—a polytene chromosome. Polytene chromosomes can attain a cross-sectional width as much as 10,000 times that of a normal interphase chromosome. The increased size and the parallel alignment of the chromatids reveal structural details that cannot be seen in normal metaphase chromosomes or interphase chromatin.

Figure 10.11 shows the polytene chromosomes of the *Drosophila* larval salivary glands. The four chromosomes radiate out from a structure called

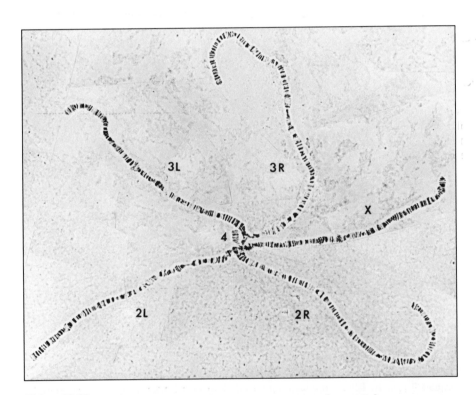

Figure 10.11

Composite photomicrograph of *Drosophila* salivary gland polytene chromosomes. The four chromosomes radiate from the chromocenter, which is formed by an aggregation of centromeres. The centromere of the X chromosome is at one end of the chromosome, so it appears as a single element. The centromeres of the second and third chromosomes are centrally located, so these chromosomes each have two arms—designated L (left) and R (right). The small fourth chromosome is located adjacent to the centromere. (From G. Lefevre, Jr., 1976, with permission, from *The Genetics and Biology of Drosophila*, Vol. 1a. M. Ashburner and E. Novitski, Eds. Copyright: Academic Press Inc. [London] Ltd.)

the **chromocenter**, which results from an aggregation of the centromeres of all the chromosomes. The chromosomes have a striated appearance of bands alternating with interband regions. The bands are visible because the concentration of DNA is greater than in the interbands owing to local folding of each chromatid. Each of these folded domains is called a **chromomere**.

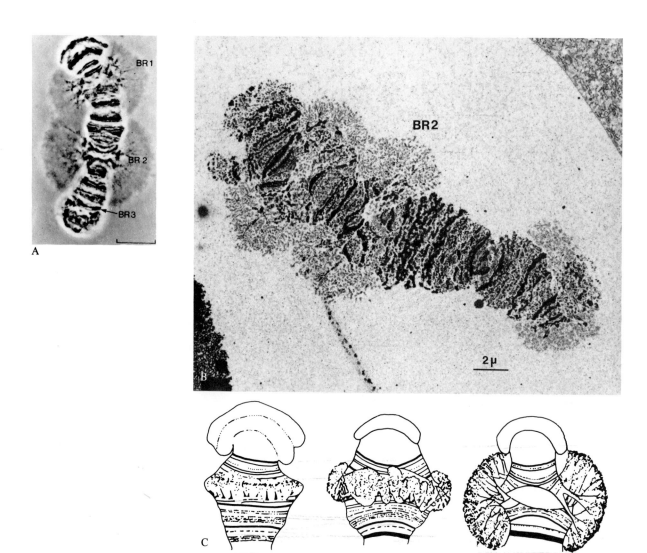

Figure 10.12

Polytene chromosome puffing. A, Phase contrast micrograph of chromosome 4 from the salivary gland of *Chironomus tentans*. The three Balbiani rings (BR) are indicated on the micrograph. Scale bar equals 7 μm. B, Electron micrograph of chromosome 4 from a salivary gland cell of *Chironomus tentans*. The individual longitudinal strands, the chromatids, of the polytene chromosome are not resolved in this micrograph, but the transverse bands, each formed from homologous chromomeres, are easily observed. Moreover, the three Balbiani rings (BR 2 being the intermediate one) are readily recognized. C, Diagrammatic representation of one of the three large puffs (Balbiani rings) of chromosome 4 characteristic of all salivary gland nuclei of *Chironomus tentans* and *Chironomus pallidivittatus*. Three different stages of puffing are shown. Magnification approximately ×890. (A, From Sass, 1981. B, From Daneholt, 1975. Copyright © Cell Press. C, From Beermann, 1963.)

The parallel alignment of chromomeres on the individual strands produces the dense bands.

Beermann (1952) observed that some bands may exhibit a swollen or **puffed** appearance under certain conditions. A puff results from an unfolding of the chromomeres that constitute a band (Fig. 10.12). Puffing may be slight, with the only apparent change being a small enlargement and diffusion of the band, or it may be extreme, with the uncoiling of the chromomeres to form large loops. The larger puffs found in *Chironomus* are called **Balbiani rings**. Beermann proposed that the puffs represent sites of intense transcriptional activity. This hypothesis is supported by several lines of evidence. For example, examination of *Chironomus* salivary gland chromosomes, using the Miller technique for observing transcription by electron microscopy (see Chap. 3), reveals regions that are assumed to be Balbiani rings that show chromatin loops bearing highly active transcription units (Fig. 10.13). These loops are presumably formed by the unfolding of individual chromomeres. Also, puffs actively incorporate radioactive RNA precursors, as demonstrated by autoradiography (Fig. 10.14). Furthermore, one can cor-

Puff = transcr.

Figure 10.13

Electron micrograph of active transcription units from chromosome 4 of *Chironomus tentans*. Scale bar equals 1 μm. (From Lamb and Daneholt, 1979. Copyright © Cell Press.)

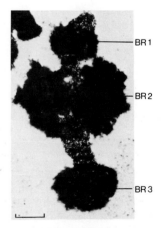

Figure 10.14

Autoradiograph of chromosome 4 of *Chironomus tentans* illustrating ^{3}H-uridine incorporation by Balbiani rings (BR). Scale bar equals 7 μm. (From Sass, 1981.)

Is RNA p'ase inhibitor.
If use, then regression
of Balbiani Rings

relate the amount of incorporation of label with the extent of puffing; a puff site that has undergone experimentally induced puff regression demonstrates a commensurate reduction in ³H-uridine incorporation. Regression of puffing can be induced with a variety of agents. This is illustrated by incubation of *Chironomus* salivary glands with the RNA polymerase II and III inhibitor, α-amanitin, which causes regression of Balbiani rings. As shown in Figure 10.15, incorporation of ³H-uridine has been eliminated in the regions of puff regression (as for all nonnucleolar chromosomal regions). These results suggest that these regions are transcriptionally active when they are decondensed (i.e., puffed) and suppressed when they are condensed.

Confirmation of this interpretation is provided by experiments in which puffing is induced, and this morphological change is correlated with the synthesis of specific transcripts. An experimental procedure that facilitates this kind of analysis is treatment of *Drosophila* with high temperatures (Ritossa, 1962). Briefly shifting the organisms from normal temperature (approximately 25°C) to 37°C elicits the formation of a discrete set of puffs (Fig. 10.16) and the synthesis of a specialized set of mRNAs and proteins. These proteins are called **heat shock proteins**. At the same time, the expression of other gene loci is inhibited; all pre-existing puffs regress, most non-heat shock RNAs are no longer produced, and the translation of pre-existing messengers is halted. The same effects on RNA and protein synthesis are also observed in other *Drosophila* tissues and cells as well as in tissue culture cells. The induction of heat shock protein synthesis by a sudden increase in temperature (which is called the **heat shock response**) is a universal homeostatic mechanism in organisms from bacteria to man.

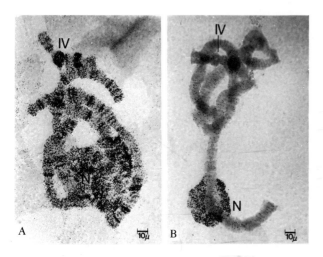

Figure 10.15

Effects of α-amanitin on ³H-uridine incorporation by *Chironomus pallidivittatus* salivary gland chromosomes. A, Control. B, Chromosomes from α-amanitin treated salivary gland. Note that the nucleolus (N) remains enlarged and incorporates label, while the Balbiani rings on chromosome 4 (IV) have regressed and have ceased incorporation. The nucleoli are sites of ribosomal RNA synthesis, which depends upon RNA polymerase I. This polymerase is insensitive to α-amanitin. (From Beermann, 1971.)

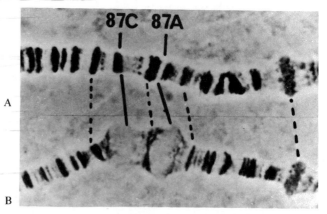

Figure 10.16

Induction of puffs on a *Drosophila melanogaster* salivary gland chromosome by heat shock (40 minutes at 37°C). A, Control. B, Heat-shocked. Note puffs formed by bands at 87C and 87A. (From Ashburner and Bonner, 1979. Copyright © Cell Press.)

The role of heat shock puffs in the synthesis of heat shock RNA in salivary glands is demonstrated by *in situ* hybridization of a cloned heat shock gene to nascent RNA on the puffs. As shown in Figure 10.17, two of the heat shock–induced puffs contain RNA that hybridizes with this specific cloned probe. No hybridization is detected without heat shock. Consequently, heat shock induces transcription at these loci.

Comparisons of puffing patterns of polytene chromosomes during normal development illustrate that transcription is subject to both spatial and temporal regulation. Examination of puffs in different cell types reveals that each type has a unique puffing pattern. For example, Balbiani ring 2 is formed in salivary gland chromosomes of *Chironomus tentans* but not in other tissues having polytene chromosomes, such as the malpighian tubules (Daneholt et al., 1978). Temporal regulation is illustrated by examination of puffs of *Drosophila* in the same tissue at different stages of development (Fig. 10.18); the pattern of transcription changes as development proceeds.

Developmental regulation of puffing is due to the molting hormone ecdysone, which is produced by the prothoracic gland. The role of ecdysone in regulation of puffing has been demonstrated by experiments in which the hormone was injected into larvae, resulting in premature appearance of stage-specific puffs (Clever and Karlson, 1960; Clever, 1961). More recently, isolated *Drosophila melanogaster* salivary glands have been exposed to ecdysone *in vitro*. The chromosomes of hormone-treated glands apparently progress through a normal puffing pattern. Figure 10.19 shows a chromosomal region from salivary glands treated with ecdysone *in vitro*. The region

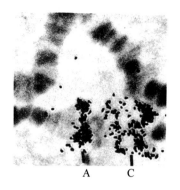

Figure 10.17

Hybridization *in situ* of a cloned heat shock gene to RNA on a salivary gland chromosome from a *Drosophila melanogaster* larva kept at 36°C for 15 minutes. This autoradiograph shows hybridization at puffs formed by bands at 87A and 87C (labeled A and C, respectively). × 1175. (From Livak et al., 1978.)

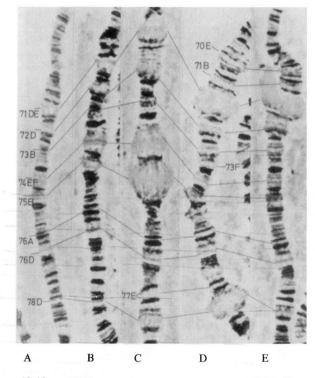

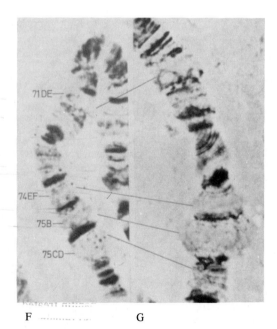

Figure 10.18

Puffing sequence of a portion of salivary gland chromosome 3 of *Drosophila melanogaster*. A, and B, 110 hour larva. C, 115 hour larva. D, 0 hour pre-pupa. E, 4 hour pre-pupa. F, 8 hour pre-pupa. G, 12 hour pre-pupa. (From Ashburner, 1972b.)

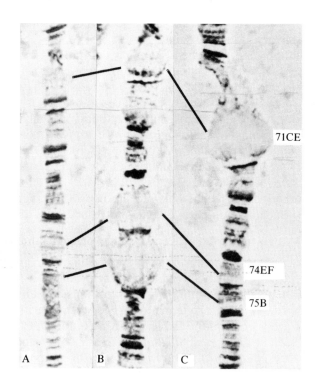

71CE

74EF

75B

A   B   C

Figure 10.19

Ecdysone-induced puffing cycle in cultured *Drosophila* salivary glands. This chromosomal region corresponds to that shown in Figure 10.18. A, Unincubated control; B, 5 hours; C, 10 hours. (From Ashburner, 1990. Copyright © Cell Press.)

corresponds to one whose normal *in vivo* puffing pattern was shown in Figure 10.18. A comparison of these two figures shows that hormone treatment *in vitro* results in a pattern that is virtually identical to that found *in vivo*. Note in particular the early response of bands 74EF and 75B, whose puffs later regress as band 71CE reaches its peak response. Interestingly, the early puffs have been cloned and shown to encode DNA binding proteins, which are apparently involved in regulating transcription of other genes (see review by Ashburner, 1990).

Among the gene loci known to undergo transient puffing cycles during development in response to ecdysone are those encoding salivary secretory products called **glue proteins**, which attach the pupal case to the substrate (Korge, 1975, 1977; Akam et al., 1978; Velissariou and Ashburner, 1980, 1981). Experimental analysis has shown that puffing of these loci and accumulation of glue protein mRNA are coordinately regulated by ecdysone (Crowley and Meyerowitz, 1984). Although there is a strong correlation between puffing and high levels of transcription, recent evidence indicates that puffing is not *essential* for transcription of the glue protein genes (McNabb and Beckendorf, 1986; Crosby and Meyerowitz, 1986). Thus, although puffing appears to provide an index of very high levels of transcriptional activity of certain genes, transcription may occur without the necessity of puff formation.

Evidence for differential transcription is also provided by facultative heterochromatin, in which one chromosome of a pair becomes highly condensed during development, and, as a consequence, inactive in transcription. Once heterochromatization occurs, this chromosome and all of its mitotic derivatives remain transcriptionally inactive throughout development. As a consequence, only one member of each pair of alleles on these chromosomes is active in transcription.

A classical example of facultative heterochromatin involves one of the X chromosomes of female mammals. Although females contain two X

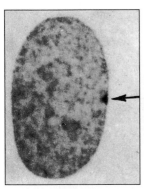

Figure 10.20

Barr bodies (arrows) in nuclei of human fibroblasts cultured from the skin of a female. ×2200. (From Villee et al., 1989. Courtesy of U. Mittwoch.)

chromosomes (and hence two copies of each X-linked gene) and males carry only one, both sexes produce similar amounts of proteins that are coded by genes located on the X chromosome. This is achieved by a mechanism called **dosage compensation**. Mary Lyon (1961) proposed that dosage compensation occurs by random heterochromatization of either the paternal or maternal X chromosome during development (the **Lyon hypothesis**). After the initial inactivation occurs in an embryonic cell, the same X chromosome remains inactive in that cell and all its mitotic derivatives during the lifetime of the organism. The heterochromatic X can be seen as a darkly staining structure called the **Barr body** lying near the nuclear envelope in interphase nuclei (Fig. 10.20). The maintenance of the heterochromatic state is believed to involve methylation of cytosine residues in the DNA by the process known as DNA methylation.

As a consequence of random heterochromatization of parental X chromosomes, females that are heterozygous for genes on the X chromosome have a mosaic phenotype, with the maternal X chromosome gene being expressed in some cells and the paternal gene being expressed in others.

10–4 Nucleocytoplasmic Interactions

We shall now return to the question that we posed at the beginning of this chapter: How does the fertilized egg generate the diverse cell types that constitute an adult organism? Because most cells of a developing organism apparently have identical genomes, the pathway of differentiation to be followed in any particular cell must be imposed upon the genome. This implicates the cytoplasm in regulation of gene expression because the nucleus is, in a sense, the captive of the cytoplasm, which completely surrounds it. E. B. Wilson referred to the cell as a "reaction system" in which the nucleus and cytoplasm each play a role in determination of cellular characteristics. Nucleus and cytoplasm are in dynamic interaction, resulting in an altered physiological state (differentiation) in which new nucleocytoplasmic interactions come into operation.

One way to examine the role of nucleocytoplasmic interactions in regulation of gene expression is to combine nuclei and cytoplasms from different cells. The technique of **somatic cell hybridization** is particularly useful for such experiments. Fusion of somatic cells may be facilitated by use of inactivated Sendai virus (see p. 405). When the fusion is between cells from different species, the hybrid cell that is formed is called a **heterokaryon**. A particularly interesting heterokaryotic combination is between a human tissue culture cell, called a HeLa cell, and a hen erythrocyte. These cells were selected for fusion because of their contrasting patterns

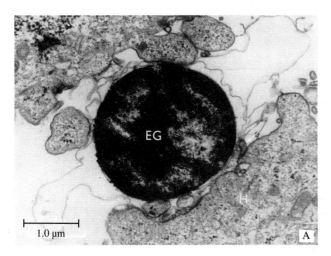

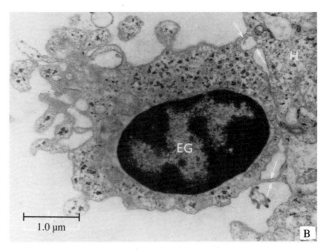

Figure 10.21

Fusion of an erythrocyte ghost (EG) and a HeLa cell (H). A, The arrow shows a virus particle wedged between the two cell membranes. B, The HeLa cytoplasm has flowed into the erythrocyte ghost. Note the highly condensed chromatin in the erythrocyte nucleus. (From Schneeberger and Harris, 1966. Reprinted by permission of Company of Biologists, Ltd.)

of DNA and RNA synthesis. HeLa cells are highly active in both DNA and RNA synthesis, whereas hen erythrocytes are inactive in DNA synthesis and synthesize very little RNA. Hen erythrocytes are particularly useful for cell fusion experiments because the Sendai virus causes the cells to rupture, releasing the cytoplasm. When these karyoplasts fuse with a HeLa cell, all of the cytoplasm of the heterokaryon is derived from the HeLa cell (Fig. 10.21). Therefore, the effects on the erythrocyte nuclei are caused solely by the HeLa cytoplasm. After fusion, the erythrocyte nuclei increase in volume, and their chromatin disperses (Fig. 10.22A to C). These enlarged nuclei become highly active in both RNA and DNA synthesis, apparently in response to signals from the HeLa cytoplasm that regulate the level of RNA and DNA synthesis (Harris, 1965, 1968). RNA synthesis in the erythrocyte nuclei is shown in Figure 10.22D, which is an autoradiograph of a heterokaryon after incorporation of ^{3}H-uridine.

Recent work with heterokaryons has involved monitoring cell-specific gene expression in naive nuclei in response to cytoplasmic signals from differentiated cells. Blau et al. (1985) have shown that nuclei of human nonmuscle cells will express muscle-specific genes when fused with mouse muscle cells. The advantage of fusing cells from different species is that the messengers produced by the two nuclei, or the proteins they encode, can be distinguished from one another. An elegant example of this is shown in the fluorescent micrograph in Figure 10.23. The nuclei have been stained with the fluorescent dye Hoechst 33258, which reveals heterokaryons. Human nuclei stain uniformly blue, whereas mouse nuclei appear punctate. This preparation was also treated with a monoclonal antibody that recognizes a muscle-specific protein (5.1H11) that is found on the surfaces of human (but not mouse) myotubes. The antibody-antigen complexes are, in turn, detected by a secondary antibody that is labeled such that it appears red in ultraviolet light. As shown in this figure, the human gene for muscle-specific protein is activated in hepatocyte (liver cell) nuclei in heterokaryons. These

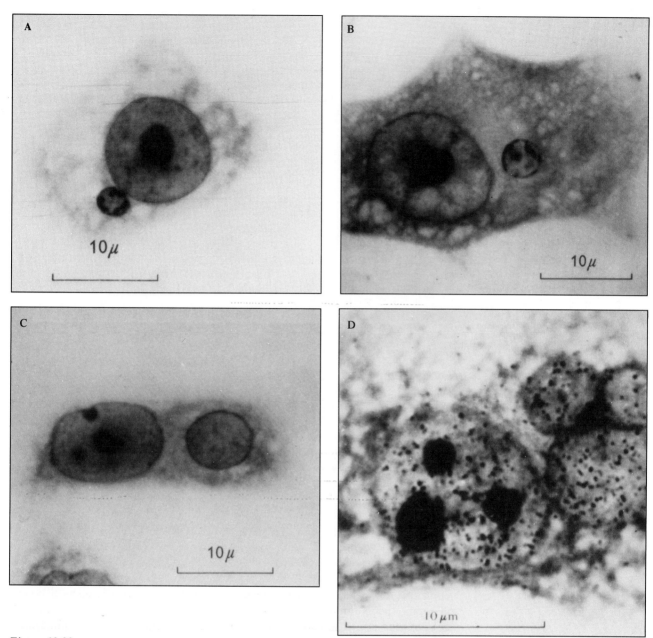

Figure 10.22

Heterokaryons with HeLa and erythrocyte nuclei. A to C, Morphological changes in erythrocyte nuclei. A, A dikaryon immediately after fusion. B, Erythrocyte nucleus has begun to enlarge. C, Further stage of enlargement. Note dispersion of erythrocyte chromatin. D, Autoradiograph of heterokaryon after incorporation of ³H-uridine. The silver grains represent synthesis of RNA. The cell contains one HeLa nucleus and three erythrocyte nuclei in various stages of enlargement. Note that the labeling of erythrocyte nuclei increases as they enlarge. (A to C, From Harris, 1967. Reprinted by permission of Company of Biologists, Ltd. D, From Harris, 1968.)

experiments clearly indicate that cytoplasmic regulatory factors can activate expression of dormant genes.

Cytoplasmic regulation of gene expression provides an answer to the question of how numerous cell types are differentiated from the same basic genetic information. However, it also points to an apparent enigma. Multicellular organisms begin life as a single cell, the zygote. If the cytoplasm

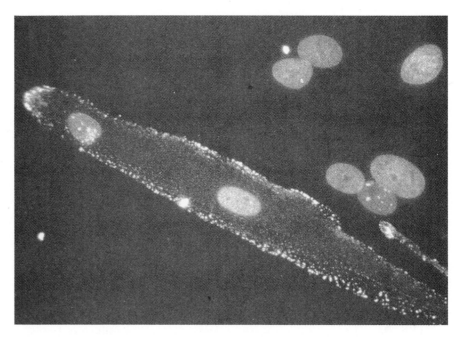

Figure 10.23

Activation of the human muscle gene for 5.1H11 in heterokaryons. Live cells were incubated with a monoclonal antibody that recognizes a muscle-specific protein that is found on the surfaces of human (but not mouse) myotubes, followed by treatment with a secondary antibody that can be visualized by fluorescence microscopy. Heterokaryons are shown in fluorescence microscopy at two different wave-lengths. The binucleate heterokaryon containing one punctate mouse muscle nucleus and one uniformly stained human hepatocyte nucleus expresses the antigen, which is uniformly distributed on the cell surface. The trinucleate heterokaryon (lower center) has not activated the gene for 5.1H11. See Color plate 16 for color version. (From H. M. Blau et al., 1985. Plasticity of the differentiated state. Science *230*: 758–766. Copyright 1985 by the American Association for the Advancement of Science.)

controls gene expression, how can the derivatives of this single cell ever generate the cytoplasmic diversity that will produce distinct patterns of gene expression? There are apparently two general solutions to this problem. The first is the **intrinsic heterogeneity** in the egg cytoplasm itself. If the egg cytoplasm consists of regions that differ from one another, cleavage would cause fixation of these differences, producing cells with distinct cytoplasmic constituents. For example, gene regulatory factors or messengers encoding such factors can be localized to a particular region of the egg cytoplasm and therefore parcelled into a subset of cells during cleavage. The second solution involves **extrinsic influences**. In this developmental strategy, the different regions of the embryo receive positional information that establishes their fates. Both modes of determination are discussed in detail in Chapter 11.

Extrinsic factors also include signals that originate from other embryonic cells. Embryonic induction is the process in which cell fate is determined by cell-cell interactions. These interactions involve the production of an inducing substance by one cell type, the extracellular diffusion of the inducer, and the stimulation of cell determination in another cell type. Many successive episodes of embryonic induction occur during embryonic development, beginning during the cleavage period and continuing during the development of the embryonic body plan (morphogenesis) and specific organ systems (organogenesis). Embryonic induction will be discussed in Chapter 12. Another type of extrinsic influence on cells during development is that mediated by hormones (e.g., the induction of salivary chromosome gene expression by ecdysone). However, those interactions that occur during later development are all dependent upon the initial events that establish cytoplasmic heterogeneity during early development. Because heterogeneity often results from inherent regional differences within the egg, its establishment is in turn dependent upon processes that occur during oogenesis.

10–5 Chapter Synopsis

In this chapter, we posed the question: How does the embryo use a single genome to produce the various cell types that constitute the completed organism? One possibility is that nuclei could dispense with genetic information that they do not use in the formation of specialized cells. Evidence provided by nuclear transplantation studies and molecular analyses generally precludes the latter possibility, with some very interesting exceptions. Nuclear transplantation studies have revealed, however, that nuclei do become progressively restricted in their ability to promote development of enucleated eggs. Mouse nuclear transplantation studies have revealed that gamete nuclei of this species are imprinted during gametogenesis so as to restrict their function during development: Paternal chromosomal regions are restricted to directing proliferation of the extraembryonic tissues, whereas maternal chromosomal regions are programmed to direct development of the embryo itself.

If nuclei do not dispense with irrelevant genetic information, how is the appropriate genetic information selected for utilization in cell differentiation? Differential gene expression can be regulated at both the transcriptional and posttranscriptional levels. Polytene chromosome puffing and facultative heterochromatin provide visual evidence for differential transcription. The regulation of differential gene expression is dependent upon interactions between the nucleus and cytoplasm. Since the zygote is a single cell, diversity of cells during development can be generated by the segregation of egg cytoplasmic constituents during cleavage, which packages nuclei in distinct cytoplasm so that nuclei in different regions of the embryo can come under different influences. Nuclei can also come under diverse influences as a result of their different positions in the embryo or from signals emitted from neighboring cells. In the following chapters, we shall see how these strategies are used to direct embryonic development.

References

Akam, M. E. et al. 1978. Drosophila: The genetics of two major larval proteins. Cell, *13*: 215–225.

Ashburner, M. 1972a. Patterns of puffing activity in the salivary gland chromosomes of Drosophila. VI. Induction by ecdysone in salivary glands of *D. melanogaster* cultured *in vitro*. Chromosoma, *38*: 255–281.

Ashburner, M. 1972b. Puffing patterns in *Drosophila melanogaster* and related species. *In* W. Beermann (ed.), *Developmental Studies on Giant Chromosomes*. Springer-Verlag, Berlin, pp. 101–151.

Ashburner, M. 1990. Puffs, genes, and hormones revisited. Cell, *61*: 1–3.

Ashburner, M., and J. J. Bonner. 1979. The induction of gene activity in Drosophila by heat shock. Cell, *17*: 241–254.

Barton, S. C., M. A. H. Surani, and M. L. Norris. 1984. Role of paternal and maternal genomes in mouse development. Nature (Lond.), *311*: 374–376.

Beermann, W. 1952. Chromomerenkonstanz und spezifische Modifikationen der Chromosomenstruktur in der Entwicklung und Organdifferenzierung von *Chironomus tentans*. Chromosoma, *5*: 139–198.

Beermann, W. 1963. Cytological aspects of information transfer in cellular differentiation. Am. Zool., *3*: 23–32.

Beermann, W. 1971. Effect of α-amanitine on puffing and intranuclear RNA synthesis in *Chironomus* salivary glands. Chromosoma, *34*: 152–167.

Blau, H. M. et al. 1985. Plasticity of the differentiated state. Science, *230*: 758–766.

Briggs, R., and T. J. King. 1952. Transplantation of living nuclei from blastula cells into enucleated frogs' eggs. Proc. Natl. Acad. Sci. U.S.A., *38*: 455–463.

Briggs, R., and T. J. King. 1957. Changes in the nuclei of differentiating endoderm cells as revealed by nuclear transplantation. J. Morphol., *100*: 269–312.

Briggs, R., and T. J. King. 1960. Nuclear transplantation studies on the early gastrula (*Rana pipiens*). I. Nuclei of presumptive endoderm. Dev. Biol., *2*: 252–270.

Briggs, R., T. J. King, and M. A. DiBerardino. 1961. Development of nuclear-transplant embryos. *In* S. Ranzi (ed.), *Symposium on the Germ Cells and Earliest Stages of Development*. Fond. A. Baselli, Milan, pp. 441–477.

Brun, R. B. 1978. Developmental capacities of *Xenopus* eggs, provided with erythrocyte or erythroblast nuclei from adults. Dev. Biol., *65*: 271–284.

Chien, Y.-h. et al. 1984. Somatic recombination in a murine T-cell receptor gene. Nature (Lond.), *309*: 322–326.

Clever, U. 1961. Genaktivitäten in den Riesenchromosomen von *Chironomus tentans* und ihre Beziehungen zur Entwicklung. I. Genaktivierungen durch Ecdyson. Chromosoma, *12*: 607–675.

Clever, U., and P. Karlson. 1960. Induktion von Puff-veranderüngen in den Speicheldrüsenchromosomen von *Chironomus tentans* durch Ecdyson. Exp. Cell. Res., *20*: 623–626.

Crosby, M. A., and E. M. Meyerowitz. 1986. *Drosophila* glue gene *Sgs–3*: Sequences required for puffing and transcriptional regulation. Dev. Biol., *118*: 593–607.

Crowley, T. E., and E. M. Meyerowitz. 1984. Steroid regulation of RNAs transcribed from the *Drosophila* 68C polytene chromosome puff. Dev. Biol., *102*: 110–121.

Daneholt, B. 1975. Transcription in polytene chromosomes. Cell, *4*: 1–9.

Daneholt, B. et al. 1978. The 75S RNA transcription unit in Balbiani ring 2 and its relation to chromosome number. Philos. Trans. R. Soc. Lond. (Biol.), *283*: 383–389.

DiBerardino, M. A. 1979. Nuclear and chromosomal behavior in amphibian nuclear transplants. *In* J. F. Danielli and M. A. DiBerardino (eds.), *Nuclear Transplantation*. Int. Rev. Cytol. Suppl. *9*: 129–160.

DiBerardino, M. A. 1980. Genetic stability and modulation of metazoan nuclei transplanted into eggs and oocytes. Differentiation, *17*: 17–30.

DiBerardino, M. A. 1989. Genomic activation in differentiated somatic cells. *In* M. A. DiBerardino and L. D. Etkin (eds.), *Developmental Biology: A Comprehensive Synthesis*, Vol. 6. *Genomic Adaptability in Somatic Cell Specialization*. Plenum Press, New York, pp. 175–198.

DiBerardino, M. A., and N. Hoffner. 1970. Origin of chromosomal abnormalities in nuclear transplants—a reevaluation of nuclear differentiation and nuclear equivalence in amphibians. Dev. Biol., *23*: 185–209.

DiBerardino, M. A., and N. Hoffner. 1971. Development and chromosomal constitution of nuclear-transplants derived from male germ cells. J. Exp. Zool., *176*: 61–72.

DiBerardino, M. A., N. J. Hoffner, and L. D. Etkin. 1984. Activation of dormant genes in specialized cells. Science, *224*: 946–952.

DiBerardino, M. A., N. Hoffner Orr, and R. G. McKinnell. 1986. Feeding tadpoles cloned from *Rana* erythrocyte nuclei. Proc. Natl. Acad. Sci. U.S.A., *83*: 8231–8234.

DiBerardino, M. A., and T. J. King. 1967. Development and cellular differentiation of neural nuclear transplants of known karyotype. Dev. Biol., *15*: 102–128.

Dryer, W. J., and J. C. Bennett. 1965. The molecular basis of antibody formation. A paradox. Proc. Natl. Acad. Sci. U.S.A., *54*: 864–869.

Du Pasquier, L., and M. R. Wabl. 1977. Transplantation of nuclei from lymphocytes of adult frogs into enucleated eggs. Special focus on technical parameters. Differentiation, *8*: 9–19.

Groudine, M. et al. 1974. Lineage-dependent transcription of globin genes. Cell, *3*: 243–247.

Gurdon, J. B. 1962. The developmental capacity of nuclei taken from intestinal epithelium cells of feeding tadpoles. J. Embryol. Exp. Morphol., *10*: 622–641.

Gurdon, J. B. 1966. The cytoplasmic control of gene activity. Endeavour, *25*: 95–99.

Gurdon, J. B. 1968. Transplanted nuclei and cell differentiation. Sci. Am., *219* (6): 24–35.

Gurdon, J. B., R. A. Laskey, and O. R. Reeves. 1975. The developmental capacity of nuclei transplanted from keratinized skin cells of adult frogs. J. Embryol. Exp. Morphol., *34*: 93–112.

Harris, H. 1965. Behaviour of differentiated nuclei in heterokaryons of animal cells from different species. Nature (Lond.), *206*: 583–588.

Harris, H. 1967. The reactivation of the red cell nucleus. J. Cell Sci., *2*: 23–32.

Harris, H. 1968. *Nucleus and Cytoplasm*. Clarendon Press (Oxford University Press), New York.

Hoffner Orr, N., M. A. DiBerardino, and R. G. McKinnell. 1986. The genome of frog erythrocytes displays centuplicate replications. Proc. Natl. Acad. Sci. U.S.A., *83*: 1369–1373.

Howlett, S. K., S. C. Barton, and A. Surani. 1987. Nuclear cytoplasmic interactions following nuclear transplantation in mouse embryos. Development, *101*: 915–923.

Howlett, S. K. et al. 1989. Genomic imprinting in the mouse. *In* M. A. DiBerardino and L. D. Etkin (eds.), *Developmental Biology: A Comprehensive Synthesis*, Vol. 6: *Genomic Adaptability in Somatic Cell Specialization*. Plenum Press, New York, pp. 59–77.

Hozumi, N., and S. Tonegawa. 1976. Evidence for somatic rearrangement of immunoglobulin genes coding for variable and constant regions. Proc. Natl. Acad. Sci. U.S.A., *73*: 3628–3632.

Kaufman, M. H. et al. 1977. Normal postimplantation development of mouse parthenogenetic embryos to the forelimb bud stage. Nature, *265*: 53–55.

King, T. J. 1966. Nuclear transplantation in amphibia. *In* D. J. Prescott (ed.), *Methods in Cell Physiology*, Vol. 2. Academic Press, New York, pp. 1–36.

King, T. J., and R. Briggs. 1956. Serial transplantation of embryonic nuclei. Cold Spring Harbor Symp. Quant. Biol., *21*: 271–290.

King, T. J., and M. A. DiBerardino. 1965. Transplantation of nuclei from the frog renal adenocarcinoma. I. Development of tumor nuclear-transplant embryos. Ann. N. Y. Acad. Sci., *126*: 115–126.

Kobel, H. R., R. B. Brun, and M. Fischberg. 1973. Nuclear transplantation with melanophores, ciliated epidermal cells, and the established cell-line A-8 in *Xenopus laevis*. J. Embryol. Exp. Morphol., *29*: 539–547.

Korge, G. 1975. Chromosome puff activity and protein synthesis in larval salivary glands of *Drosophila melanogaster*. Proc. Natl. Acad. Sci. U.S.A., *72*: 4550–4554.

Korge, G. 1977. Direct correlation between a chromosome puff and the synthesis of a larval saliva protein in *Drosophila melanogaster*. Chromosoma, *62*: 155–174.

Lamb, M. M., and B. Daneholt. 1979. Characterization of active transcription units in Balbiani rings of Chironomus tentans. Cell, *17*: 835–848.

Laskey, R. A., and J. B. Gurdon. 1970. Genetic content of adult somatic cells tested by nuclear transplantation from cultured cells. Nature (Lond.), *228*: 1332–1334.

Lefevre, G., Jr. 1976. A photographic representation and interpretation of the polytene chromosomes of *Drosophila melanogaster* salivary gland. *In* M. Ashburner and E. Novitski (eds.), *The Genetics and Biology of Drosophila*, Vol. 1a. Academic Press, London, pp. 31–66.

Livak, K. J. et al. 1978. Sequence organization and transcription at two heat shock loci in *Drosophila*. Proc. Natl. Acad. Sci. U.S.A., *75*: 5613–5617.

Lois, R. et al. 1990. Active β-globin transcription occurs in methylated, DNase I-resistant chromatin of nonerythroid chicken cells. Mol. Cell. Biol., *10*: 16–27.

Lyon, M. 1961. Gene action in the X-chromosome of the mouse (*Mus musculus* L.). Nature (Lond.), *190*: 372–373.

Malissen, M. et al. 1984. Mouse T cell antigen receptor: Structure and organization of constant and joining gene segments encoding the β polypeptide. Cell, *37*: 1101–1110.

McGrath, J., and D. Solter. 1983. Nuclear transplantation in the mouse embryo by microsurgery and cell fusion. Science, *220*: 1300–1302.

McGrath, J., and D. Solter. 1984. Completion of mouse embryogenesis requires both maternal and paternal genomes. Cell, *37*: 179–183.

McGrath, J., and D. Solter. 1986. Nuclear and cytoplasmic transfer in mammalian embryos. *In* R. B. L. Gwatkin (ed.), *Developmental Biology: A Comprehensive Synthesis*, Vol. 4. *Manipulation of Mammalian Development*. Plenum Press, New York, pp. 37–55.

McKinnell, R. G. 1962. Intraspecific nuclear transplantation in frogs. J. Hered., *53*: 199–207.

McKinnell, R. G. 1972. Nuclear transfer in *Xenopus* and *Rana* compared. *In* R. Harris, P. Allin, and D. Viza (eds.), *Cell Differentiation*. Munksgaard International Publishers, Copenhagen, pp. 61–64.

McNabb, S. L., and S. K. Beckendorf. 1986. *Cis*–acting sequences which regulate expression of the *Sgs-4* glue protein gene of *Drosophila*. EMBO J., *5*: 2331–2340.

Monk, M. 1988. Genomic imprinting. Genes and Development, *2*: 921–925.

Monk, M., M. Boubelik, and S. Lehnert. 1987. Temporal and regional changes in DNA methylation in the embryonic, extraembryonic and germ cell lineages during mouse embryo development. Development, *99*: 371–382.

Pavan, C., and A. B. Da Cunha. 1969. Chromosomal activities in *Rhynchosciara* and other *Sciaridae*. Ann. Rev. Genet., *3*: 425–450.

Ritossa, F. M. 1962. A new puffing pattern induced by heat shock and DNP in *Drosophila*. Experientia, *18*: 571–573.

Robl, J. M. et al. 1986. Nuclear transplantation in mouse embryos: Assessment of recipient cell stage. Biol. Reprod., *34*: 733–739.

Sapienza, C. et al. 1987. Degree of methylation of transgenes is dependent on gamete of origin. Nature (Lond.), *328*: 248–251.

Sass, H. 1981. Effects of DMSO on the structure and function of polytene chromosomes of *Chironomus*. Chromosoma, *83*: 619–643.

Schneeberger, E. E., and H. Harris. 1966. An ultrastructural study of interspecific cell fusion induced by inactivated Sendai virus. J. Cell Sci., *1*: 401–405.

Seidman, J. G., and P. Leder. 1978. The arrangement and rearrangement of antibody genes. Nature (Lond.), *276*: 790–795.

Siu, G. et al. 1984. The human T cell antigen receptor is encoded by variable, diversity, and joining gene segments that rearrange to generate a complete V gene. Cell, *37*: 393–401.

Smith, L. D. 1965. Transplantation of the nuclei of primordial germ cells into enucleated eggs of *Rana pipiens*. Proc. Natl. Acad. Sci. U.S.A., *54*: 101–107.

Spemann, H. 1938. *Embryonic Development and Induction*. Yale University Press, New Haven, CN. Reprinted by Hafner Press (Macmillan, Inc.), New York (1962).

Spradling, A. C., and A. P. Mahowald. 1980. Amplification of genes for chorion proteins during oogenesis in *Drosophila melanogaster*. Proc. Natl. Acad. Sci. U.S.A., *77*: 1096–1100.

Subtelny, S. 1965. On the nature of the restricted differentiation-promoting ability of transplanted *Rana pipiens* nuclei from differentiating endoderm cells. J. Exp. Zool., *159*: 59–92.

Swain, J. L., T. A. Stewart, and P. Leder. 1987. Parental legacy determines methylation and expression of an autosomal transgene: A molecular mechanism for parental imprinting. Cell, *50*: 719–727.

Tsunoda, Y. et al. 1987. Full-term development of mouse blastomere nuclei transplanted into enucleated two-cell embryos. J. Exp. Zool., *242*: 147–151.

Vasil, V., and A. C. Hildebrandt. 1965. Differentiation of tobacco plants from single isolated cells in microcultures. Science, *150*: 889–892.

Velissariou, V., and M. Ashburner. 1980. The secretory proteins of the larval salivary gland of *Drosophila melanogaster*. Cytogenetic correlation of a protein and a puff. Chromosoma, *77*: 13–27.

Velissariou, V., and M. Ashburner. 1981. Cytogenetic and genetic mapping of a salivary gland secretion protein in *Drosophila melanogaster*. Chromosoma, *84*: 173–185.

Villee, C. A. et al. 1989. *Biology*. 2nd ed. Saunders College Publishing, Philadelphia.

Willadsen, S. M. 1989. Cloning of sheep and cow embryos. Genome, *31*: 956–962.

The Progressive Determination of Cell Fate

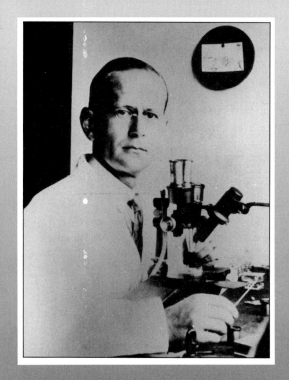

Sven Hörstadius (b. 1898). A professor at the Universities of Stockholm and Uppsala in Sweden, S. Hörstadius has done classic experiments on the determination of cell fate in sea urchin embryos and devised a model to explain these results that continue to provide a conceptual framework for further investigation. (Photograph courtesy of the Marine Biological Laboratory Library, Woods Hole, MA.)

In Chapter 10, we defined cell determination as the process by which portions of the genome are selected for expression in different embryonic cells. As we shall now discuss, cell determination is a progressive process involving a sequence of decisions that gradually restrict cell fate. Before the initial decision, blastomeres are totipotent; they have the capacity to form every cell type in the larva or adult. As more divisions take place, blastomeres become pluripotent; they are able to form fewer of the various cell types. Finally, after further divisions, blastomeres are determined; they can form only one cell type.

In some instances, cell determination is controlled intrinsically by factors that reside within the determined cells. These intrinsic factors, called **ooplasmic determinants**, are present in the egg and are partitioned unequally to different blastomeres during cleavage. In other instances, cell fate is controlled extrinsically by factors that originate from outside the determined cells. The extrinsic factors include instructions obtained as a function of the positions of blastomeres within the embryo as well as signals that are transmitted between different blastomeres. Embryonic induction is the process in which cell fate is determined by signalling between different cells. In this chapter, we examine the role of ooplasmic determinants and the position of blastomeres in determining cell fate. The role of embryonic induction in determination of cell fate will be examined in Chapter 12.

11–1 Methods of Studying Cell Determination

Before considering examples of cell determination, we discuss some of the methods used to study this process.

Fate Maps

A **fate map** shows what each part of an egg or early embryo will become at a later stage in development (Fig. 11.1). Fate maps allow one to trace the embryonic origins of specific cells and to design experiments to learn how cell fates are established. There are several ways to make a fate map; the simplest is to watch different regions of an egg or embryo develop into particular larval or adult structures. Although some fate maps have been constructed in this way (Poulson, 1950), it is painstaking work because specific regions of the egg or embryo must be followed throughout devel-

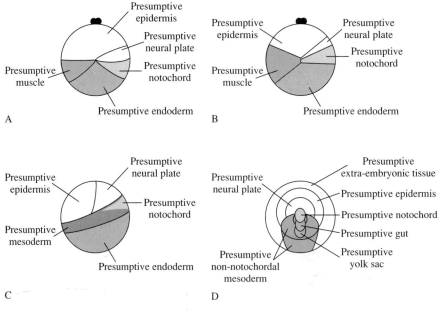

Figure 11.1

Fate maps of various chordate embryos. A, Tunicate one-cell zygote (lateral view). B, *Amphioxus* one-cell zygote (lateral view). C, Late blastula of a urodele amphibian (lateral view). D, Chick epiblast at early gastrula stage. (After Oppenheimer, 1980.)

opment. Natural markers exist in some eggs. For example, E. G. Conklin (1905) produced a fate map of a tunicate (*Styela*) egg by tracing the distribution of colored pigment granules during development (Fig. 11.2). Unfortunately, most eggs (including those of most tunicates) lack natural markers, so other ways must be devised to construct fate maps. One method is to mark the surface of an egg or early embryo with a vital dye (Vogt, 1929; Keller, 1975) or a carbon particle (Ortolani, 1955) and then determine the positions of the marks during later development. A disadvantage of this method is that only those blastomeres that receive the marked surface during cleavage can be mapped. Most dyes also have a tendency to spread and fade during the course of development and therefore are of limited use in constructing a fate map. Another method involves transplantation of genetically or radioactively labeled cells from a donor embryo to the same position in a host embryo after the corresponding host cells have been removed (Nicolet, 1971; Gardner, 1978; Janning, 1978). The fates of the labeled cells in the host can then be determined by assaying for their phenotypes or radioactivity at a later stage of development. As we discussed in Chapter 7, cytological markers, combined with interspecific grafting of cells, have been used to make fate maps of avian embryos (Le Douarin, 1971). These fate maps have been made by following the distribution of quail cells transplanted into chick embryos. The quail cells are not rejected by the chick host and can be distinguished from chick tissues by the morphology of their nucleoli. Recently, new methods for making fate maps have been introduced that use antibody or nucleic acid probes to examine the spatial distribution of cell- and tissue-specific gene products (see Chap. 14 and the essay on p. 703).

The fate maps of several chordate eggs are shown in Figure 11.1. Note that, although each chordate egg has a distinct fate map, the fate maps are

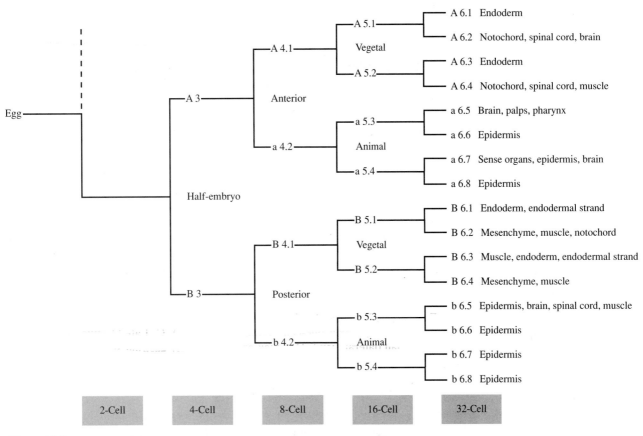

Derivatives

Figure 11.2

The cell lineage of the *Styela* embryo up to the 32-cell stage. The lineage of one half of the bilaterally symmetric embryo is shown. (Adapted from a number of sources including Conklin, 1905; Ortolani, 1955; Nishida and Satoh, 1983, 1985.)

quite similar because overall development of chordates is similar. Other groups of organisms also have characteristic fate maps. Such fate maps, however, are relatively imprecise because they usually label large numbers of cells.

The most precise fate maps are obtained by examining the fates of individual cells. The history of an individual embryonic cell is its **cell lineage**. A lineage map of the entire animal is obtained when the histories of all its cells are known. Information obtained from studying cell lineages may include: the region of cytoplasm that a blastomere inherits from the egg; the number and variety of tissue types a specific blastomere forms; the number of contacts with other cells that a blastomere makes during development; and the number and symmetry of divisions that occur in a blastomere lineage before the formation of a specific cell type.

The first cell lineage was determined for the leech *Clepsine* by C. O. Whitman (1878). Whitman's pioneering studies were soon followed by extremely detailed determinations of cell lineages in other embryos. These early cell lineage studies were done solely by microscopic observations. Thus, they rarely extended beyond the blastula stage, when embryonic cells become small and difficult to trace. Figure 11.2 shows a portion of the cell lineage of the tunicate *Styela* (Conklin, 1905). As described previously,

Conklin was able to determine the cell lineage of *Styela* because of the presence of different colored pigment granules in the egg (see Sect. 11–3). Recently, new methods of cell lineage analysis have been developed. These consist of injecting substances, such as the enzyme horseradish peroxidase (Fig. 11.3) or fluorescent dyes, which cannot pass through gap junctions and therefore are not readily exchanged by animal cells (Weisblat et al., 1978; Jacobson and Hirose, 1981). Embryos with single identifiable blastomeres injected are then allowed to develop to a later stage, and the enzyme activity is assayed or the fluorescence is visualized. These advances have allowed more accurate cell lineages to be constructed. For instance, errors in Conklin's original tunicate cell lineage have been corrected using the horseradish peroxidase injection method (Nishida and Satoh, 1983, 1985).

Development of powerful microscopical methods during the last few decades has also advanced cell lineage analysis. For example, the complete cell lineage of the nematode worm *Caenorhabditis elegans* has been determined using Nomarski optics (Sulston et al., 1983). Cell lineage analysis is aided by the transparency of the *C. elegans* embryo as well as by the relatively small and constant number of cells present in adult hermaphrodite or male worms (precisely 959 and 1031 cells, respectively). By studying cell lineages in *C. elegans*, it has been determined that the cleavage pattern leading to the adult worm is essentially the same in every embryo. In contrast, cell lineage studies in amphibians and other vertebrates have shown that there is some variation among different embryos (Dale and Slack, 1987; Kimmel and Warga, 1988).

How Is Cell Determination Assayed?

Although determined cells have entered an exclusive pathway of development, they are usually indistinguishable from each other morphologically.

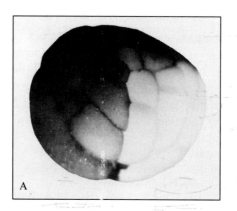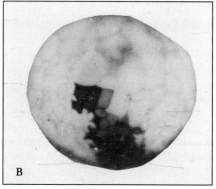

Figure 11.3

An example of the use of horseradish peroxidase microinjection to follow cell lineages in *Xenopus laevis* embryos. The horseradish peroxidase enzyme (which oxidizes a substrate and produces a colored precipitate) is restricted to the descendants of the injected cell. A, One blastomere was microinjected with horseradish peroxidase at the two-cell stage and examined at the morula stage. The dark areas of the morula represent cells containing the precipitate formed by the injected horseradish peroxidase. One half of the morula is labeled. B, One blastomere was microinjected with horseradish peroxidase at the 32-cell stage and examined at the blastula stage. The photograph shows the blastula viewed from the vegetal hemisphere with a specific group of cells labeled. (From Slack, 1983. Reprinted by permission of Cambridge University Press.)

How is it established whether a blastomere is determined or not? This can be accomplished in several different ways. First, a specific blastomere can be destroyed or removed from the embryo. If the embryo with the missing blastomere develops lacking a particular cell type, this suggests that the missing cell was determined to produce that cell type. This approach was used by L. Chabry to study cell determination in tunicate embryos. Second, a blastomere can be isolated from an embryo to see whether it forms a specific cell type in culture. If an isolated blastomere forms only the cells or tissues that are expected from the normal fate map or cell lineage, it is considered to be determined at the time of isolation. In contrast, if it forms an entire embryo or more cell types than expected from the fate map or cell lineage, it is not determined. For example, the trochophore larva of the limpet *Patella* contains 16 ciliated **trochoblast** cells (Fig. 11.4A to C). Cell lineage studies indicate that these cells arise from four blastomeres in the 16-cell embryo. The four precursor cells cleave two more times, then cease dividing and differentiate into trochoblast cells. To investigate whether the trochoblast precursor cells are already *determined* at the 16-cell stage, E. B. Wilson (1904a) removed them from 16-cell embryos and cultured them *in vitro*. The isolated cells divided twice and formed ciliated trophoblast cells, exactly as if they had remained in the embryo (see Fig. 11.4D to G). This experiment shows that the four trochoblast precursor cells are already determined by the 16-cell stage. Third, blastomeres can be transplanted from one part of an embryo to another part of the same or a different embryo. If the transplant develops as it would have without transplantation, this is evidence that it contains determined cells. If the transplant develops differently, however, it is not determined. H. Spemann conducted reciprocal transplantation experiments in amphibian embryos and demonstrated that the neural ectoderm is determined between the early and late gastrula stages (see pp. 485–486, especially Fig. 12.7).

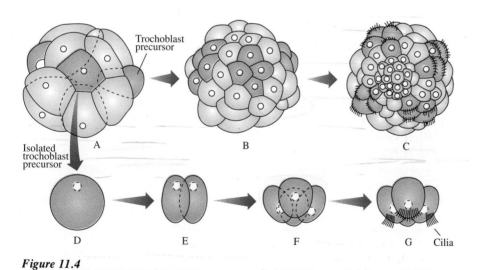

Figure 11.4

Development of the intact embryo (A–C) and of isolated trochoblast precursor cells (D–G) in the limpet *Patella*. A, A 16-cell embryo with four trochoblast precursor cells (three can be seen in this view). B, A 48-cell stage embryo showing undifferentiated descendants of the trochoblast precursor cells. C, A trochopore larva showing the 16 ciliated trochoblast cells. D–G, A trochoblast precursor cell isolated at the 16-cell stage divides twice in culture, and four cells differentiate a row of cilia typical of differentiated trochoblast cells. (After Wilson, 1904a.)

In some cases, blastomeres become restricted even before the 16-cell stage. This was demonstrated by destruction or isolation of blastomeres. The first blastomere destruction experiments were published by L. Chabry in 1887. When Chabry killed one blastomere of a two-cell tunicate embryo, the surviving blastomere continued to develop as if it were a part of the whole embryo, eventually forming a half-larva. Similarly, when a blastomere was killed at the four-cell stage, a defective embryo that lacked specific structures was formed. At the same time, similar experiments were being conducted by W. Roux (1888) on frog embryos. When Roux destroyed a single blastomere at the two- or four-cell stage, he obtained half- or three-quarter-embryos, respectively (see Chap. 1). These results led Roux to consider the early frog embryo as a mosaic of determined cell types. Shortly after Chabry's and Roux's studies were published, H. Driesch (1892) conducted similar experiments with sea urchin embryos. Instead of destroying blastomeres, however, Driesch studied the development of isolated blastomeres. By vigorously shaking embryos, he was able to isolate blastomeres without seriously injuring them. Considering the results of the previous experiments, Driesch was surprised to find that some blastomeres isolated from two- or four-cell sea urchin embryos produced complete pluteus larvae in culture. As discussed in Chapter 1, this result led Driesch to reject Roux's mosaic theory, especially after it was shown that isolated blastomeres from two-cell frog embryos were actually able to form entire tadpole larvae in culture (McClendon, 1910). McClendon's experiment was a direct repeat of Roux's, except that the dead blastomere was removed. In Roux's experiment, the dead blastomere had been left in association with the living cells and affected their ability to develop normally. In subsequent experiments, however, it was shown that isolated blastomeres of other animals, especially tunicates and other invertebrates, may indeed have restricted fates as early as the two- or four-cell stage. Thus, it is clear that the embryos of various animals differ with respect to the potential of isolated blastomeres to develop into complete embryos.

11–2 Mosaic and Regulative Embryos

Embryos of different animals are sometimes divided into two groups: **mosaic** and **regulative** embryos. In mosaic (or **determinative**) embryos, such as those of tunicates, blastomeres become restricted during the first few cleavages and often as soon as they are formed. Because cell fates are established early in mosaic embryos, they cannot compensate for blastomeres that are removed or destroyed, so the embryo lacks the structures derived from the missing blastomeres. In regulative (or **indeterminate**) embryos, such as those of sea urchins and amphibians, the restriction of blastomeres begins later. Because cell fates are established later, regulative embryos can compensate for blastomeres that are removed or destroyed early in development. As shown in the following examples, regulative embryos differ from mosaic embryos in the orientation of cleavage planes with respect to the distribution of ooplasmic substances involved in cell determination. This relationship is illustrated in Figure 11.5.

Sea urchin embryos are classified as regulative because each blastomere can form a complete larva when isolated at the two- or four-cell stage. However, restrictions in developmental potency can be seen when these experiments are extended to the eight-cell stage (Hörstadius, 1939). As described in Chapter 5, at third cleavage the sea urchin zygote divides equatorially, forming two tiers of equal-sized animal and vegetal cells. When

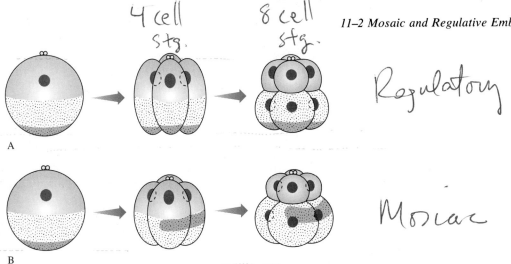

4 cell stg.   8 cell stg.

Regulatory

Mosiac

Figure 11.5

A schematic diagram showing the roles of cleavage plane orientation in the distribution of localized cytoplasmic regions in (A) a regulatory and (B) a mosaic embryo. In (A), a localized region (shown in color) is segregated at the third cleavage. When blastomeres are separated at the four-cell stage, each cell can form a complete larva because it contains every cytoplasmic region. When blastomeres are separated at the eight-cell stage, however, they do not form a complete larva because the cytoplasmic regions have been segregated to different cells. In (B), a localized region is segregated at the second cleavage. When blastomeres are separated at the four-cell stage, only the two cells containing each cytoplasmic region develop into complete larvae. When blastomeres are separated at the four- or eight-cell stages, none form complete larvae because the various cytoplasmic regions have been segregated to different cells. (After Wilson, 1925.)

an eight-cell embryo is bisected through the animal and vegetal pole with a glass needle, halves containing two animal and two vegetal blastomeres are formed. Despite this operation, each half-embryo develops into a normal larva (Fig. 11.6A). Different results are obtained, however, when eight-cell embryos are bisected through the equator, and halves containing animal and vegetal quartets are formed. In contrast to the earlier experiment, only the vegetal half-embryo forms a pluteus larva. The larva derived from the vegetal half, which shows reduced oral arms (which normally arise from animal hemisphere cells) and an enlarged gut, is called a **vegetalized pluteus** (see Fig. 11.6B). The animal half-embryo forms a blastula-like sphere of ciliated cells that is called an **animalized embryo**. Similar results are obtained when unfertilized eggs are divided into halves, and each half is fertilized (Hörstadius, 1939). When a cut is made meridionally, bisecting the unfertilized egg through the animal and vegetal pole, and both halves are subsequently fertilized, each half develops into a complete larva (see Fig. 11.6C). When a cut is made through the equator of the egg, and both halves are fertilized, the animal half forms an animalized embryo and the vegetal half forms a vegetalized larva (see Fig. 11.6D). Although only one half of the egg contains the nucleus, this does not affect the experiment, because when the anucleate half is inseminated it is able to develop as a haploid embryo. These experiments show that restrictions in cell fate leading to cell determination begin after third cleavage in the sea urchin embryo.

Amphibian embryos are regulative, based on the ability of blastomeres separated during the early cleavages to form complete embryos. Observations on newt eggs suggest that the orientation of the first cleavage plane is important in determining the developmental potential of the first two

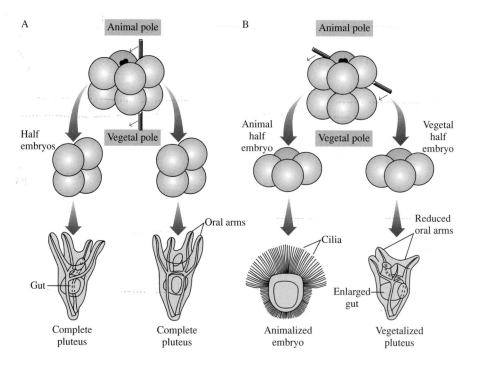

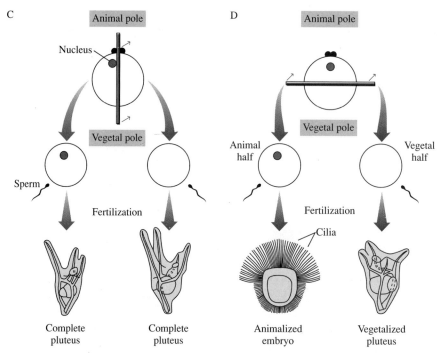

Figure 11.6

Blastomere separation and egg fragmentation experiments with sea urchin embryos and eggs. A, Meridional separation of blastomeres at the eight-cell stage gives rise to two complete plutei. B, Equatorial separation of blastomeres at the eight-cell stage gives rise to a vegetalized pluteus from the vegetal cells and an animalized embryo from the animal cells. C, Meridional division of an egg leads to two normal plutei (one diploid and one haploid) after fertilization of both halves. D, Equatorial division of an egg and subsequent fertilization of both halves results in the formation of a vegetalized pluteus from the vegetal half and an animalized embryo from the animal half. (After Hörstadius, 1939.)

blastomeres (Spemann, 1938). Amphibian eggs contain dark pigment granules in the animal hemisphere cortex, whereas the vegetal hemisphere has much less pigmentation. During cortical rotation, the egg cortex shifts toward the sperm entry site, forming the gray crescent on the opposite side of the egg. The significance of this shift of cortical cytoplasm for the development of the amphibian body plan was discussed in Chapter 6. Normally, the plane of first cleavage cuts through the gray crescent dividing the embryo into blastomeres of equal developmental potential (Fig. 11.7A). In rare cases, however, the first cleavage does not cut through the gray crescent, and this region is distributed to only one of the blastomeres. If the blastomeres are now separated and cultured, only the cell containing the gray crescent forms a complete embryo (see Fig. 11.7B), whereas the other blastomere forms a ventralized embryo that lacks dorsal structures (see p. 200). However, the regulative capacity of the amphibian embryo is demonstrated by the fact that normal larvae are produced in intact embryos in the rare cases in which the first cleavage furrow does not pass through the gray crescent.

Some developmental biologists also believe that mosaic and regulative embryos differ in the extent of their usage of intrinsic and extrinsic mechanisms for cell determination. Mosaic embryos may rely mostly on ooplasmic determinants for cell fate determination, whereas regulative embryos may rely mostly on cell interactions for this process. We now examine examples in which cell determination is regulated by either intrinsic or extrinsic factors.

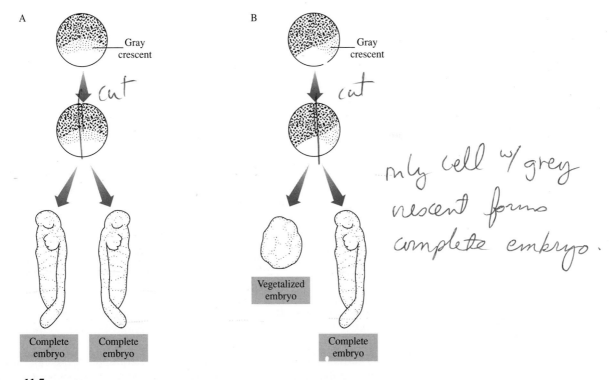

Figure 11.7

The effects of the gray crescent on the development of amphibian eggs. A, Blastomeres isolated from the two-cell stage of embryos in which the first cleavage plane cuts through the midpoint of the gray crescent. Both blastomeres give rise to complete larvae. B, Two-cell embryos in which the first cleavage plane segregated the gray crescent into one blastomere. The blastomeres are separated and cultured individually. Only one blastomere (the gray crescent-containing blastomere) gives rise to a complete larva.

11–3 *Regulation of Cell Determination by Ooplasmic Determinants*

The Roux-Weismann theory (see Chap. 1) stated that segregation of nuclear determinants during cleavage causes differences in cell fate. This required that nuclei become qualitatively different during development. The blastomere separation studies of Driesch, however, demonstrate that nuclei do not lose genetic information during early cleavage divisions because blastomeres have the capacity to develop more cell types than expected from the fate map. These experiments and the more recent nuclear transplantation experiments (see Chap. 10) indicate that each nucleus has the genetic potential to produce the entire range of cell types of the organism, although that potential may not be realized.

Embryologists working near the turn of the century found evidence that the egg cytoplasm is not uniform. Visible differences in cytoplasmic regions of the egg are known as **cytoplasmic localizations**. Although cytoplasmic localizations sometimes reflect the presence of ooplasmic determinants, this is not always the case. The cytoplasmic localizations simply allow us to identify reproducibly specific regions of the egg cytoplasm. The existence of ooplasmic determinants must be shown by experimental demonstration of functional differences in regions of the egg cytoplasm. In this section, we discuss examples in which the initial restriction of cell fate is controlled by ooplasmic determinants. There is evidence that ooplasmic determinants are important in the determination of both germ and somatic cell lineages. As we discussed in Chapter 6, ooplasmic determinants may also play a role in defining the general embryonic body plan.

Germ Cell Determination

In some organisms, germ cell and somatic cell lineages become separate during early development. The germ cell precursors form **primordial germ cells,** which eventually migrate to the gonads where they multiply and become germ cells. As we discussed in Chapter 1, A. Weismann proposed in 1893 that germ cell development is due to the presence of the **germ plasm**. He believed that germ plasm contains the hereditary material of the species and that somatic cells lose most of this material during cleavage. These ideas were the basis for W. Roux's theory of preformation, which was subsequently discredited. However, the concept of the germ plasm has persisted, albeit in a highly modified form, and has gained credibility during the early part of this century.

Investigators in the 1890s and early 1900s reported that primordial germ cells are indeed set aside from somatic cells during early development. Sometimes the primordial germ cells could be distinguished from other cells in the embryo by their distinct morphology. The morphological differences include distinct sizes, shapes, and, most importantly, cytoplasmic granules. In many different animals, there are specialized granules in the egg that are segregated only to the primordial germ cells during development (Beams and Kessel, 1974). The localization of these granules suggests that they may contain ooplasmic determinants that are responsible for germ cell determination.

The Germ Plasm of Nematode Eggs

The existence of a germ plasm was first demonstrated by Boveri's (1887) observations and experiments on the origin of germ cells in the nematode *Ascaris*. As we discussed in the essay on page 410, *Ascaris* shows chromatin diminution, the process in which specific parts of chromosomes are lost from somatic cells during the early cleavages. Chromatin diminution is an exceptional process that occurs in only a few organisms. In these animals, only the germ cells retain a complete set of chromosomes and the full amount of nuclear DNA. Although this process might at first appear to support the Roux-Weismann hypothesis of nuclear restriction, it is unlikely that chromatin diminution involves the loss of structural genes from somatic cell nuclei (Bennett and Ward, 1986). What is lost from the genome appears to be repeated DNA sequences, which do not contain structural genes. In Boveri's experiments, chromatin diminution provided a distinct cytological marker for distinguishing between primordial germ cells and somatic cells.

Cleavage and early development of *Ascaris* are shown in Figure 11.8. Nematodes exhibit an unusual cleavage pattern in which the first cleavage is equatorial, rather than meridional. First cleavage divides the egg into an animal hemisphere blastomere, known as the AB cell, and a vegetal hemisphere blastomere, known as the P1 cell (Fig. 11.8A; see also Fig. 11.10, which contains a summary of early cleavages in *C. elegans*). Both the AB and P1 cells have intact chromosomes. Before second cleavage, one mitotic apparatus rotates to a position perpendicular to the other mitotic

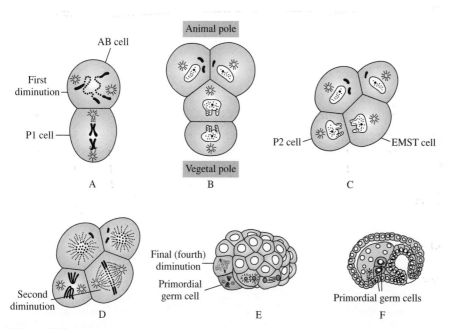

Figure 11.8

Early development of *Ascaris megalocephala*, showing chromatin diminution. A, Beginning with the second cleavage, diminution is occurring (dotted regions of chromosomes) in the upper blastomere (AB), which is a somatic cell precursor. Conventional division occurs in the lower blastomere (P1), which is a precursor to both somatic and germ cells. B, Later during the second cleavage, eliminated chromatin remains at the equator of the spindle of the upper blastomere, which cleaves into two somatic cell precursors. The lower blastomere also undergoes conventional cleavage. C, At the completion of the second division, the embryo changes shape. The vegetal-most blastomere (P2) will be protected from diminution at the next division, whereas its sister blastomere (EMST) will undergo diminution. D, Third cleavage. Diminution now occurs in EMST. Another diminution occurs before the 16-cell stage. At the conclusion of the fourth cleavage (i.e., the 16-cell stage; not shown), only two cells retain an intact genome. E, As the embryo prepares for the fifth division, the final diminution occurs in one cell. The primordial germ cell will divide to produce two primordial germ cells, neither of which will undergo diminution. F, A section of a gastrula showing the two primordial germ cells, which have sunk into the interior of the embryo, where they will enter the gonad and multiply to form the germ cells. (After Wilson, 1925.)

apparatus. This rotation results in meridional cleavage of the AB blastomere and equatorial cleavage of the P1 blastomere and temporarily forms a T-shaped four-cell embryo (see Fig. 11.8B). Later, the embryo becomes shaped like a trapezoid. During second cleavage, chromatin diminution occurs in the AB blastomere, and its descendants form somatic cells. The chromosomes of the P1 blastomere and its two daughter cells remain intact. During third cleavage, however, one descendant of the P1 blastomere (somatic cell precursor, designated the EMST cell) undergoes chromatin diminution, whereas the chromosomes remain intact in its sister cell (the P2 cell). During third cleavage, the P2 cell divides into a somatic cell (which undergoes diminution) and another germ cell precursor (which retains an intact genome), the P3 cell. This process is subsequently continued until a P4 cell is formed after four cleavages. At fifth cleavage, the P4 cell divides into two cells, which are the primordial germ cells. Both of these primordial germ cells (and their descendants) retain an intact genome. The primordial germ cells migrate into the gonad, where they continue to multiply and eventually form the germ cells.

Is germ cell determination in nematodes controlled by the nucleus or cytoplasm? Realizing that ooplasmic determinants localized in the posterior cytoplasm of the egg might account for germ cell determination, Boveri (1910) shifted the orientation of the mitotic apparatus with respect to various cytoplasmic regions by centrifuging *Ascaris* eggs before first cleavage. The results of this experiment are shown in Figure 11.9. When the mitotic apparatus was shifted to a position 90 degrees from its former location, the first cleavage was meridional rather than equatorial, and each blastomere

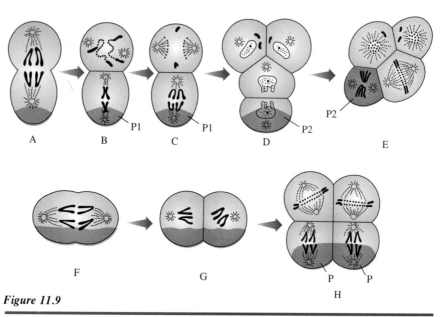

Figure 11.9

Distribution of posterior (vegetal) cytoplasm (shown in color) during cleavage of normal (A–E) and centrifuged (F–H) *Ascaris* zygotes. In the former, the distribution of posterior cytoplasm to germ-line blastomeres during the first and second cleavages protects them from diminution as the next divisions begin. At the four-cell stage, the normal embryo has one germ cell precursor (P2). After centrifugation (F), the spindle has been displaced 90°, and cleavage is vertical, causing posterior cytoplasm to be distributed to both blastomeres at the first division. Thus, neither cell undergoes diminution. After the second cleavage (H), both vegetal cells retain posterior cytoplasm and function as potential germ cell precursors. (After Waddington, 1966.)

received a part of the granular posterior cytoplasm, which usually enters only the P1 blastomere. Chromatin diminution did not occur during the second cleavage, presumably because the posterior cytoplasm was present in both blastomeres of the two-cell embryo. During the second cleavage, however, the posterior cytoplasm is partitioned to only two of the four blastomeres, and chromatin diminution occurs in the blastomeres lacking the posterior cytoplasm during the subsequent division. Because two P cells are formed, the resulting embryo develops twice the normal number of germ cells. These experiments show that the posterior cytoplasm protects nuclei from chromatin diminution.

The formation and history of the P1 cell lineage have been confirmed in *C. elegans* (Fig. 11.10). This nematode does not show chromatin diminution, but the germ cell lineage does contain specific cytoplasmic granules. As shown in Figure 11.11, these structures, called **P granules,** can be detected by using a specific antibody. (See the essay on p. 344 for a discussion of immunofluorescent microscopy.) P granules are segregated into the posterior cytoplasm of the egg, where they are partitioned first to the vegetal blastomere, then to the P cell lineage, and finally into primordial germ cells. Although the distribution of P granules correlates with germ cell development in *C. elegans* embryos, it is still uncertain whether these organelles actually contain ooplasmic determinants needed for germ cell determination.

The Pole Plasm of Insect Eggs

The pole plasm is a striking example of a germ plasm in insect embryos. As discussed in Chapter 5, the pole plasm is incorporated into the pole cells, large blastomeres that form precociously at the posterior pole of the insect embryo. The pole cells are precursors of primordial germ cells.

Evidence that pole plasm contains ooplasmic determinants for germ cell development has been obtained from several different experiments. In one of the earliest experiments, the posterior region of beetle eggs was

pole plasm = germ plasm (but in insect)

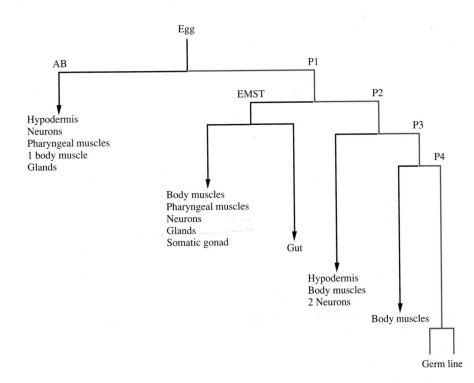

Figure 11.10

The P cell lineage of the *C. elegans* embryo. (After Strome and Wood, 1982.)

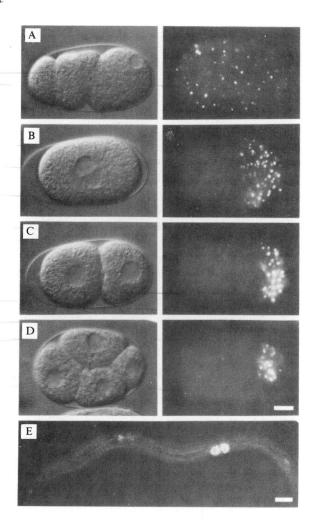

Figure 11.11

Fluorescent micrographs showing the segregation and partition of the P granules during early development of *C. elegans*. The left-hand frames (A–D) show Nomarski images of living embryos. The right-hand frames (A–D) show embryos at the same stage of development stained with an antibody specific for P granules. Anterior is left, and posterior is right. A, One-cell embryo at the pronuclear stage showing shape changes that occur before first cleavage. The P granules are dispersed throughout the cytoplasm. B, One-cell embryo after pronuclear fusion showing P granule segregation to the posterior cytoplasm. C, Two-cell embryo showing P granules partitioned to the P1 cell. D, Four-cell embryo showing P granules partitioned to the P2 cell. E, A larva showing P granules in the two primordial germ cells. Scale bars equal 10 μm. (From Strome, 1989.)

destroyed with a hot needle before pole cell formation (Hegner, 1911). Despite destruction of the pole plasm, nuclei entered the posterior region of the embryo and a cellular blastoderm was formed. No pole cells developed, however, and the adult beetles were sterile. Many subsequent experiments on germ cell determination have been conducted on *Drosophila* eggs. For instance, it was demonstrated that germ cell formation can be prevented by irradiating the posterior tip of the egg with ultraviolet light (Geigy, 1931). The irradiated eggs do not form pole cells (Figs. 11.12 and 11.13) and develop into flies lacking germ cells. Experiments combining ultraviolet irradiation and microinjection have also established a causal relationship between the posterior region and germ cell determination in *Drosophila* (Okada et al., 1974). These experiments show that pole plasm from a normal embryo microinjected into the posterior pole region of an ultraviolet-irradiated embryo allows pole cells to form in the host (see Fig. 11.13).

Because the presence of pole cells alone is insufficient proof of the capacity to form functional germ cells, an experiment was designed to determine whether pole cells that form as a result of transplanted pole plasm develop into functional germ cells (Illmensee and Mahowald, 1974, 1976). This experiment included a genetic marker for pole cells and is illustrated in Figure 11.14. Initially, pole plasm was removed from the posterior pole of a donor embryo and microinjected into the anterior pole region of a genetically marked preblastoderm recipient embryo (see Fig. 11.14A and B). Normally, the anterior region of the embryo does not form pole cells.

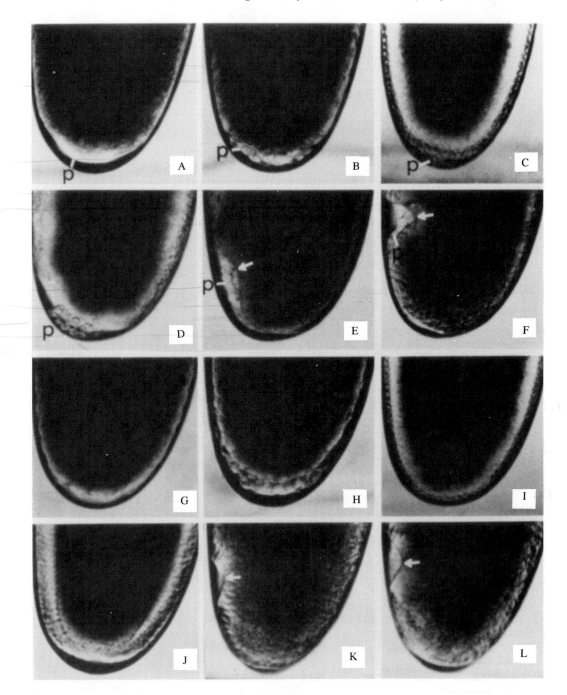

Figure 11.12

Photomicrographs showing the posterior region of normal (A–F) and UV-irradiated (G–I) embryos. A–C, Pole cell (p) formation beginning between the eighth and eleventh nuclear divisions. D, Gastrulation starts on the ventral side of the embryo. E, Posterior midgut invagination starts. F, Pole cells are entering the posterior midgut invagination. G–I, No pole cells are formed between the eighth and eleventh nuclear divisions. J, Gastrulation begins. K, and L, The posterior midgut invagination forms without pole cells. Arrows indicate site of posterior midgut formation. (From Okada et al., 1974.)

However, cells having the appearance of pole cells formed in the anterior region of the recipient that received injected pole plasm (see Fig. 11.14C). Because it is unlikely that pole cells formed ectopically in the anterior region of the embryo could migrate to the gonads (which is required for germ cell

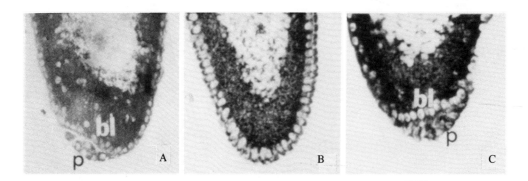

Figure 11.13

Longitudinal sections of the posterior region of *Drosophila* embryos at the cellular blastoderm stage. A, A normal, unirradiated embryo. Blastoderm (bl) and pole cells (p) are present. B, An embryo that was irradiated before the blastoderm stage. No pole cells have formed, and the blastoderm at the posterior pole has cleaved into smaller somatic cell precursors. C, An embryo that was irradiated before the blastoderm stage and subsequently injected with pole plasm from an unirradiated host. Blastoderm and pole cells appear to be similar to those present in normal embryos. (From Okada et al., 1974.)

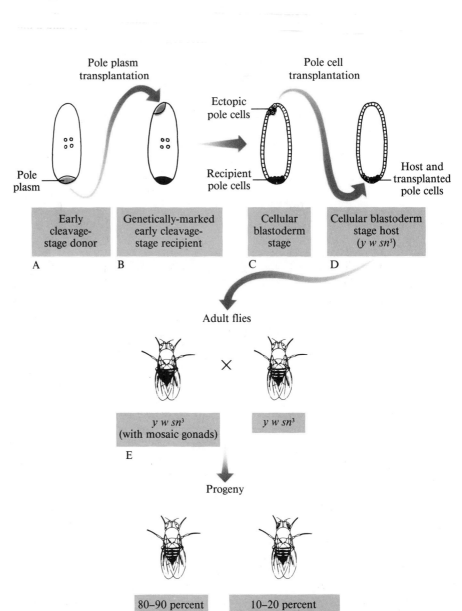

Figure 11.14

Experiment demonstrating germ cell determination by transplantation of pole plasm in *Drosophila*. A, and B, Pole plasm from the egg of donor fly is withdrawn into a micropipette and injected into the anterior pole of a genetically-marked recipient egg. C, and D, The anterior pole cells formed in this region are removed and transplanted to the posterior region of a mutant *y w sn³* embryo. E, After reaching adulthood, the *y w sn³* recipients were mated to other *y w sn³* flies. F, The progeny of the cross shown in E, indicating the presence of two genotypes; 10–20% of the flies were derived from the transplanted pole cells. This experiment demonstrates that pole plasm is sufficient to cause the formation of pole cells that can develop into gametes. (After Illmensee and Mahowald, 1974.)

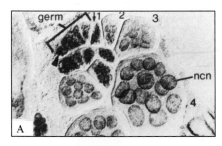

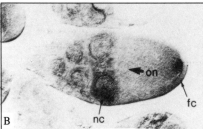

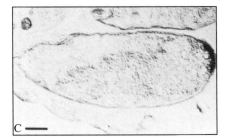

Figure 11.15

The distribution of polar granule antigens in *Drosophila* ovaries as determined by staining with an antibody specific for polar granules. A, Polar granule antigen localization in germ-line stem cells (germ) and developing egg chambers (1–4). In egg chamber 4, staining can be seen near the nuclear envelope (ncn) of nurse cells. B, Polar granule antigen localization in nurse cells (nc) and a thin cortical region at the posterior pole of the growing oocyte. The oocyte nucleus (on) is indicated by the arrow. fc: follicle cells. C, Polar granule antigen localization in the posterior region (pole plasm) of a mature oocyte. Scale bar equals 200 μm. (From Hay et al., 1988. Reprinted by permission of Company of Biologists, Ltd.)

differentiation), the anterior pole cells were transplanted into the posterior region of genetically distinct hosts to test their potency in germ cell development (see Fig. 11.14C and D). The host embryo exhibited the homozygous recessive genotype: yellow body (*y/y*), white eyes (*w/w*), and singed bristles (*sn³/sn³*). After reaching sexual maturity, the *y w sn³* flies (which should also contain gametes bearing the genotype of the transplanted pole cells if they became functional germ cells) were mated to *y w sn³* flies (see Fig. 11.14E). If the transplanted pole cells form functional germ cells, the progeny of this cross should include flies of both genotypes. As shown in Figure 11.14F, 10% to 20% of the progeny were derived from gametes resulting from the transplanted pole cells.

The pole plasm contains specifically staining granules, the **polar granules.** The polar granules can be visualized either by electron microscopy (Mahowald, 1971b) or by staining eggs and embryos with an antibody that is specific for these organelles (Hay et al., 1988). Antibody staining indicates that the polar granule antigen is originally derived from the germ cells, accumulates in the nurse cells, enters the oocyte with other nurse cell components during oogenesis, and becomes localized in a thin cortical layer at the posterior pole of the egg (Fig. 11.15). After nuclear migration and pole cell formation, the polar granules are enclosed within the pole cells and are converted into **nuage** (from French, meaning "cloud"), a fibrous substance associated with the nuclear envelope of primordial germ cells and gametes. Later, the pole cells migrate through the posterior midgut to the gonad (Fig. 11.16).

By electron microscopy, polar granules appear as dense granular organelles (Fig. 11.17). Specific staining procedures have shown that the polar granules contain RNA and protein. Large polar granules, closely associated with mitochondria, are found in mature eggs (Mahowald, 1968,

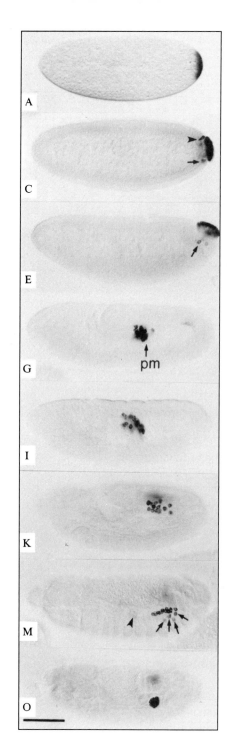
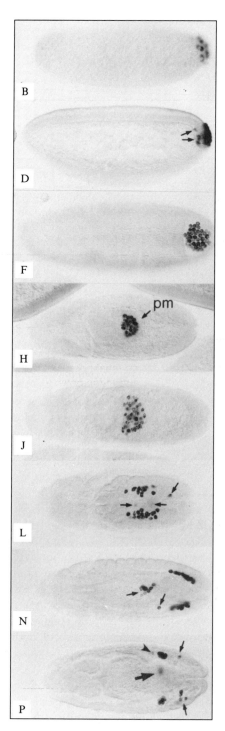

Figure 11.16

The distribution of polar granule antigens during *Drosophila* development as determined by staining with an antibody specific for polar granules. A, A syncytial embryo during the early stages of nuclear multiplication showing antibody staining within a cortical cap at the posterior pole. B, A syncytial embryo in the process of forming pole cells. Antibody staining is concentrated around the nuclei within the forming pole cells. C, A cellular blastoderm stage embryo showing stained pole cells and unstained blastoderm cells. D–F, Various views of post-blastoderm embryos showing stained pole cells within and beneath (arrows) the blastoderm layer. G, and H, Various views of post-gastrula stage embryos showing stained pole cells near the posterior midgut (pm). I, and J, Various views of stained pole cells migrating through the posterior midgut at a later stage of development. K–N, Various views of embryos undergoing germ band shortening showing migration and bilateral disposition of stained pole cells. Some pole cells (arrows) remain in the yolk. O, and P, Two views of embryos during gonad formation showing the stained pole cells encapsulated by mesoderm in the region of the developing gonad (arrowhead). Arrows indicate pole cells that remain outside the gonads. Scale bar equals 100 μm. (From Hay et al., 1988. Reprinted by permission of Company of Biologists, Ltd.)

444

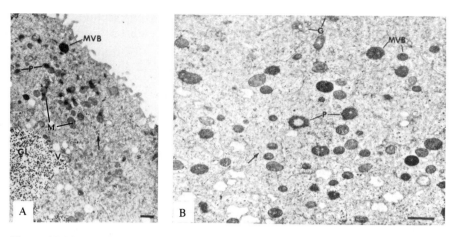

Figure 11.17

Sections through the pole plasm of *Drosophila* embryos. A, *Drosophila melanogaster* embryo shortly after oviposition, showing polar granules (P) associated with mitochondria (M) near the cell surface. Arrow indicates glycogen granules. B, *Drosophila willistoni* embryo shortly before nuclei have reached the pole plasm. Polar granules (P) are no longer associated with mitochondria. Helical polyribosomes (arrows) accumulate near the polar granules. G: golgi; GL: glycogen yolk; MVB: multivesicular body; V: vacuole. Scale bars equal 1 μm. (From Mahowald, 1968.)

1971a), and this configuration is retained for a short time after fertilization. Before pole cell formation, however, the polar granules lose their association with mitochondria and fragment into many smaller structures. The smaller structures disperse in the pole plasm, lose their RNA, and are surrounded by polyribosomes (Mahowald, 1971b). If radioactive amino acids are added to embryos at this time, it can be shown that high levels of protein synthesis occur only in the pole cells (Zalokar, 1976). The fragmentation of the polar granules and the accompanying burst of protein synthesis in pole cells suggest that these structures release mRNA molecules into the surrounding cytoplasm. This has led to the hypothesis that mRNA stored in the polar granules may be involved in germ cell determination. Later in this chapter, we shall discuss evidence that mRNA is also an ooplasmic determinant for the anterior region of the insect embryo.

The behavior of polar granules in the egg and early embryo suggests that they may be involved in germ cell determination. If polar granules function in germ cell determination, it should be possible to isolate mutants that do not form germ cells, and these mutants would also be expected to lack polar granules. Recently, mutations with these characteristics have been identified in *Drosophila* (Boswell and Mahowald, 1985; Lehmann and Nüsslein-Volhard, 1986; Schupbach and Wieschaus, 1986). They are members of a general class of maternal-effect mutations known as *grandchildless* because homozygous females show deficiencies in pole cell formation that affect germ cell development. Thus, a first generation of progeny is produced, but it is sterile, and there is no second generation. Cytological examination of the posterior region of eggs from females homozygous for mutations at certain *grandchildless* loci show that they lack a pole plasm and polar granules. Thus, sterility is correlated closely with the lack of polar granules, strongly supporting the idea that these organelles may function as ooplasmic determinants for germ cell formation. Because most *grandchildless* mutants do not completely obliterate polar granules, however, it is unlikely that polar granules are the only substances required for germ cell determination. Some of the *grandchildless* mutations also affect the formation of the abdominal region of the larva, suggesting that they correspond to gene loci that determine the entire posterior pattern of the embryo.

The Germ Plasm of Frog Eggs

A germ plasm is also present in frog eggs. Initially, it was noted that granular materials were localized in the vegetal cytoplasm of frog eggs and that these materials showed staining properties similar to the polar granules of insect eggs (Bounoure, 1934). Therefore, they were called germinal granules. The fate of the germinal granules has been traced in several different species of frogs, and although details vary, a generalized pattern has emerged (Fig. 11.18). The germinal granules are located in a rim of cytoplasm below the cortex near the vegetal pole of the egg. During cleavage, this material is distributed to large cells destined to form endoderm in this region. Later, descendants of these cells migrate dorsally around the posterior gut region and enter the mesodermal regions from which the gonads will be formed

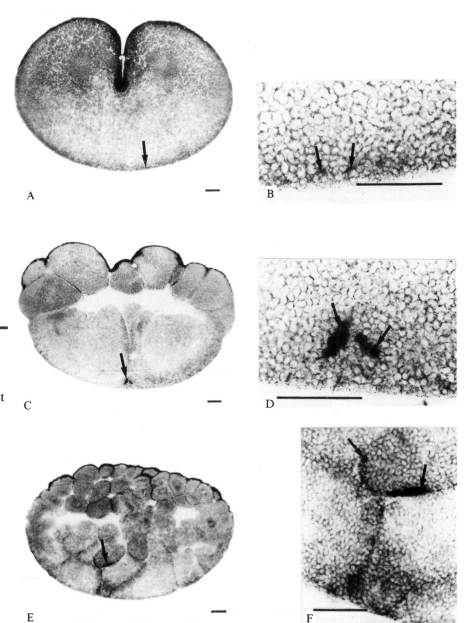

Figure 11.18

The germ plasm of the *Xenopus* embryo. A, The germ plasm (arrow) is located in a rim of cytoplasm immediately inside the vegetal cortex of the zygote at first cleavage. B, Enlargement of the germ plasm region in A. C, The germ plasm is located near the vegetal pole of two large endodermal blastomeres in this morula-stage embryo. D, Enlargement of the germ plasm region in C. E, The germ plasm moves anteriorly in endodermal cells by the blastula stage. F, Enlargement of the germ plasm region in E. Scale bars equal 100 μm. (From Ressom and Dixon, 1988. Reprinted by permission of Company of Biologists, Ltd.)

(the genital ridges). In the gonads, these cells proliferate as gonia, enter meiotic phases, and eventually differentiate into functional gametes.

The evidence that the amphibian germ plasm contains germ cell determinants is similar to that for *Drosophila*. First, organelles located in the amphibian germ plasm bear a striking morphological resemblance to insect polar granules. In eggs and early embryos, these organelles appear as a mass of fibrillar and granular bodies associated with mitochondria and surrounded by ribosomes (Fig. 11.19). By the time the primordial germ cells migrate into the genital ridges, the granular material can no longer be detected. Instead, the cells have complexes of fibrous material and mitochondria located adjacent to the nuclear envelope, similar to the nuage of *Drosophila* germ cells. Second, like the insect pole plasm, the amphibian germ plasm is sensitive to ultraviolet irradiation (Bounoure, 1934). When vegetal (but not animal) cytoplasm from nonirradiated donors is microinjected into the vegetal hemisphere of ultraviolet irradiated eggs, primordial germ cells are again seen in the genital ridges (Smith, 1966). The usual interpretation of these experiments is that the ability of the germ plasm to determine germ cells is destroyed by ultraviolet irradiation. Other studies of embryos and tadpoles that have been irradiated during cleavage, however, suggest that primordial germ cells are present in the endoderm of these embryos but that germ cell migration to the genital ridges of the tadpoles is

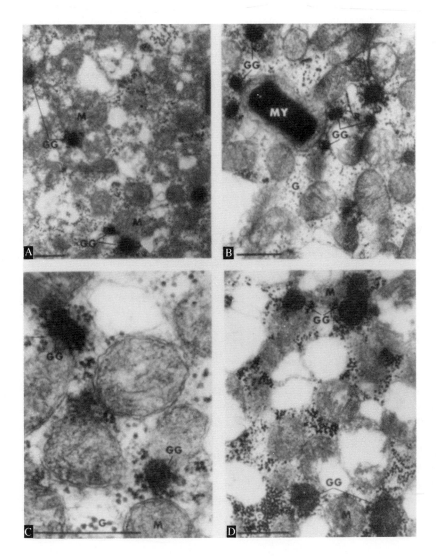

Figure 11.19

Ultrastructure of germ plasm region of *Rana pipiens* embryos. A, 1.5 hours after fertilization. B, Two-cell stage. C, Four-cell stage. D, 16-cell stage. These areas contain germinal granules (GG), mitochondria (M), mitochondria-containing yolk (MY), and glycogen (G). Scale bars equal 0.5 μm. (From Williams and Smith, 1971.)

delayed or inhibited (Züst and Dixon, 1977; Ikenishi and Kotani, 1979). Thus, the role of the germ plasm in amphibians may be to determine the ability of primordial germ cells to migrate from the endoderm to the genital ridges.

Somatic Cell Determination

As in germ cell determination, there are several classic examples in which ooplasmic determinants are implicated in somatic cell determination.

Ooplasmic Determinants in Tunicates

As mentioned earlier, colored cytoplasmic regions are present in eggs of some tunicate species. The most brilliantly colored region is the **myoplasm**, which is yellow in *Styela* eggs and orange in *Boltenia* eggs (see Color plates 17 and 18). Other major cytoplasmic regions are the **ectoplasm** and the **endoplasm**. These regions become localized after fertilization by a series of cytoplasmic movements known as **ooplasmic segregation** (Fig. 11.20). Although ooplasmic segregation occurs in eggs of many different animals, it is more evident in *Styela* and *Boltenia* eggs because of their colored cytoplasmic regions (Jeffery and Bates, 1989). The ectoplasm is derived primarily from the germinal vesicle contents when the germinal vesicle breaks down during oocyte maturation. In the unfertilized egg, the ectoplasm is located in the animal hemisphere, whereas the endoplasm is localized mainly in the vegetal hemisphere, and the myoplasm occupies the egg cortex (see Fig. 11.20A). During ooplasmic segregation, the myoplasm moves vegetally, collecting in the vegetal pole region. Next, the ectoplasm flows into the vegetal hemisphere, and, at the same time, the endoplasm is displaced entirely into the animal hemisphere (see Fig. 11.20B). Thus, at the end of the first phase of ooplasmic segregation, the endoplasm, ectoplasm, and myoplasm are stratified perpendicular to the animal–vegetal axis from the animal to the vegetal pole, respectively (see Fig. 11.20C). The three cytoplasmic regions subsequently enter a second phase of movements that establish bilateral symmetry in the egg (see p. 176). The sperm aster is the major organizer of the second movement (Sawada and Schatten, 1989). The myoplasm moves into the subequatorial region and is eventually extended into a crescent in the posterior region of the egg. In *Styela* eggs, the myoplasm at this stage is known as the **yellow crescent**. The ectoplasm streams into the animal hemisphere (see Fig. 11.20D), after temporarily forming a clear crescent immediately above the crescent of myoplasm (see Fig. 11.20C). Opposite the yellow crescent, a fourth cytoplasmic region, the **chordoplasm**, appears in the anterior vegetal region of the egg. By the completion of the second phase of ooplasmic segregation, the ectoplasm enters the animal hemisphere, and the endoplasm returns to the vegetal hemisphere.

During cleavage, the colored cytoplasmic regions in eggs are distributed to different blastomeres. The first cleavage furrow passes through the plane of bilateral symmetry, dividing the cytoplasmic regions into two equal parts. The second cleavage is also meridional and oriented perpendicular to the first, so that the two anterior blastomeres receive the chordoplasm and the two posterior blastomeres obtain the crescent of myoplasm. Because the third cleavage is equatorial, it further segregates the cytoplasmic regions (see Fig. 11.20E). The ectoplasm is distributed primarily to four animal blastomeres, the endoplasm enters four vegetal blastomeres, the chordoplasm enters two anterior vegetal cells, and the yellow crescent is segregated to two posterior vegetal cells. This segregation of cytoplasmic regions continues during the subsequent cleavages. By following the fates of colored blastomeres, E. G. Conklin (1905) determined that cells containing ectoplasm

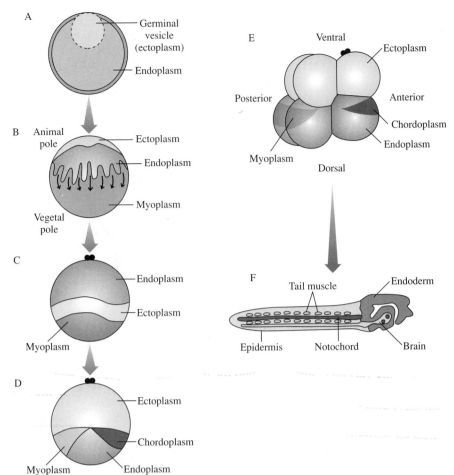

ooplasmic segregation

Figure 11.20

Schematic diagram of the distribution of colored cytoplasmic regions during ooplasmic segregation and early cleavage in *Styela*. A, Cross-section of an unfertilized egg. Myoplasm (color) is in the cortex. B, Fertilized egg during the first phase of ooplasmic segregation. C, Egg that has completed the first phase of ooplasmic segregation. D, Egg that has completed the second phase of ooplasmic segregation. E, Eight-cell zygote. F, Sagittal section of a tadpole larva.

become epidermis and neural tissues, cells containing endoplasm become gut, cells containing chordoplasm become notochord, and cells containing myoplasm become tail muscle (see Fig. 11.20F).

Cell determination in tunicate embryos has been examined by blastomere destruction and isolation experiments. The pioneering experiments of L. Chabry, who showed that the prospective fates of the blastomeres are determined during the early cleavages, have already been discussed. When pairs of blastomeres from the bilaterally symmetrical embryo are isolated at the eight-cell stage and cultured, they continue to divide and differentiate as expected from the cell lineage (Reverberi and Minganti, 1946). Descendants of ectoplasm-containing blastomeres differentiate into ectodermal tissue, endoplasm-containing blastomeres differentiate into gut tissue, chordoplasm-containing cells form notochord tissue, and myoplasm-containing cells become muscle tissue (Fig. 11.21). Enzyme markers, rather than morphological criteria, have also been used to assess the determination of isolated blastomeres (Whittaker et al., 1977). In these experiments, acetylcholinesterase, an enzyme that appears primarily in tail muscle cells (Fig. 11.22A), is used as a histological marker for muscle cell differentiation. When myoplasm-containing vegetal blastomeres are removed from eight-cell embryos, these cells continue to divide and eventually produce acetylcholinesterase. In contrast, very few partial embryos lacking these cells produce acetylcholinesterase (Deno et al., 1985) (see Fig. 11.22B). These results suggest that ooplasmic determinants involved in muscle cell determination are segregated to the myoplasm-containing blastomeres during cleavage.

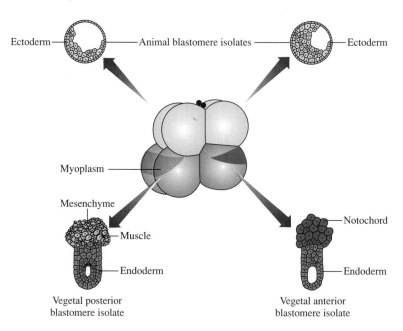

Figure 11.21

Schematic representation of the results of blastomere isolation experiments in an eight-cell tunicate embryo. Isolated animal hemisphere blastomeres (upper part of diagram) form ectoderm. Isolated anterior vegetal blastomeres (lower right) form endoderm and notochord. Isolated posterior vegetal blastomeres (lower left) form endoderm, muscle, and mesenchyme. (After Reverberi, 1971.)

Further evidence for the segregation of ooplasmic determinants has been obtained by examining acetylcholinesterase expression in **cleavage-arrested embryos**. Cleavage-arrested embryos are prepared by treating embryos with cytochalasin B. As discussed in Chapter 9, cytochalasin B inhibits cytokinesis, but nuclei continue to divide within the cytoplasm of the cleavage-arrested cells. When embryos are treated with cytochalasin B at various times during development and cultured until the time when they would normally form muscle cells, the cleavage-arrested embryos develop acetylcholinesterase in cells corresponding to the muscle cell lineage (Whittaker, 1973). For example, two distinctly located cells produce acetylcholinesterase in embryos treated with cytochalasin B at the two-, four-, or eight-cell stages, whereas four and six distinctly located cells produce this enzyme in embryos arrested at the 16- or 32-cell stages, respectively (Fig. 11.23). The numbers and locations of these cells are exactly as predicted from the tunicate egg fate map (see Fig. 11.2). These experiments suggest that ooplasmic determinants are segregated to the muscle precursor cells at each cleavage.

To test whether the muscle determinants are nuclear or cytoplasmic, the normal distribution of myoplasm was changed by compressing embryos between glass coverslips at third cleavage (Fig. 11.24). The third cleavage of tunicates is equatorial, partitioning myoplasm into two vegetal cells. When the four-cell embryo is compressed in a plane perpendicular to the animal–vegetal axis, the third cleavage becomes meridional, and myoplasm is distributed to four, rather than two, cells. The distribution of nuclei to various blastomeres is not changed in this experiment. The results of this experiment are striking. When the compressed embryos are cleavage-arrested at the eight-cell stage by treatment with cytochalasin B, acetylcholinesterase is sometimes produced in all four myoplasm-containing blastomeres (see Fig. 11.24H). Because the two extra acetylcholinesterase-producing cells obtained the same nuclei they would have received during normal cleavage, this experiment shows that muscle cell determinants are localized in the cytoplasm of tunicate embryos and that these determinants have imposed a new fate on the nuclei in the cells to which they have been diverted.

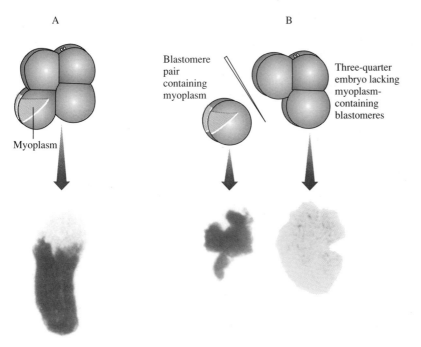

Figure 11.22

Diagram and results of blastomere isolation experiments with eight-cell tunicate embryos, in which muscle cell differentiation was assayed by staining for acetylcholinesterase. A, A normal eight-cell embryo gives rise to a larva containing acetylcholinesterase (dark area) in its tail muscle cells. B, An eight-cell embryo, in which the blastomere pair containing the myoplasm was separated from the remainder of the embryo with a glass needle. The myoplasmic blastomeres give rise to a partial embryo (left) that stains positively for acetylcholinesterase, whereas no acetylcholinesterase staining is seen in the partial embryo lacking these blastomeres (right). Note that in other experiments, some acetylcholinesterase production is seen in the partial embryo (right) lacking the myoplasmic blastomeres (see Deno et al., 1985). (After Whittaker et al., 1977.)

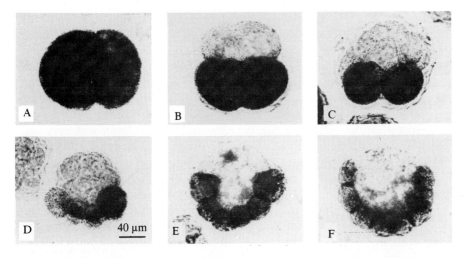

Figure 11.23

Acetylcholinesterase development in the blastomeres of cleavage-arrested tunicate embryos. Acetylcholinesterase is formed in two cells of two-cell (A), four-cell (B), and eight-cell (C) cleavage-arrested embryos, four cells of 16-cell cleavage-arrested embryos (D), six cells of 32-cell cleavage-arrested embryos (E), and eight cells of 64-cell cleavage-arrested embryos (F). (From Whittaker, 1973.)

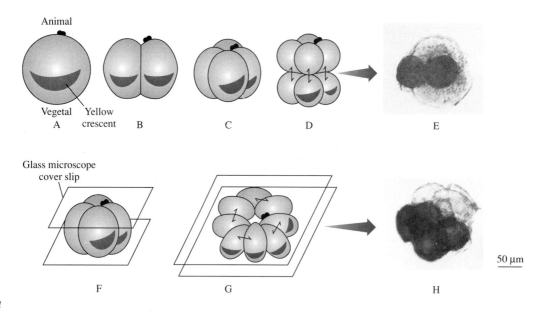

Figure 11.24

Diagram and results of experiment in which the yellow crescent of a tunicate embryo was redistributed by compression. A–D, Normal development through the eight-cell stage. Polar bodies are shown at the animal pole. Sister blastomeres formed at third cleavage are indicated by a barbed line. The yellow crescent material is partitioned to two cells of the eight-cell embryo during normal development. E, The normal embryo shows acetylcholinesterase activity in two blastomeres at the eight-cell stage. F, Four-cell stage embryo compressed between glass cover slips. G, A compressed embryo divides to form a flat plate of cells lying in the same plane. Sister blastomeres are indicated by a barbed line. Four blastomeres now contain the yellow crescent material. H, The cleavage-arrested compressed embryo shows acetylcholinesterase activity in four blastomeres. (After Whittaker, 1980. Reprinted by permission of Company of Biologists, Ltd.)

The muscle cell determinants may be attached to the egg cytoskeleton, which promotes their specific segregation into the presumptive muscle cells during cleavage. Information on the nature of the cytoskeleton has been obtained by treating *Styela* eggs and embryos with the nonionic detergent Triton X-100 (Jeffery and Meier, 1983). This detergent removes soluble and membranous components from cells but does not affect the cytoskeleton and associated components, which remain as an insoluble residue. When eggs and embryos were treated with Triton X-100, most of the cytoplasmic structures disappeared, but a highly organized cytoskeleton was revealed in the myoplasm. Scanning electron microscopy showed that this cytoskeleton consists of two parts: a network of actin filaments located immediately beneath the plasma membrane (Fig. 11.25A) and a deeper lattice of cytoskeletal filaments containing the pigment granules (see Fig. 11.25B). Contraction of the actin network may be responsible for segregation of the myoplasm after fertilization, whereas the cytoskeletal lattice may be the framework to which the muscle cell determinants are attached (Jeffery, 1984). In the following sections, we shall see that the cytoskeleton is also implicated in the localization and segregation of ooplasmic determinants in other organisms.

Ooplasmic Determinants in Ctenophores

Ctenophores are marine animals that are known for their ability to produce light. Their cydippid larvae contain light-producing **photocytes** and loco-

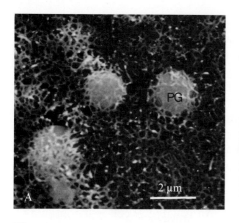
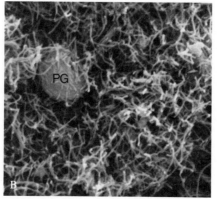

Figure 11.25

The cytoskeleton of the myoplasm in *Styela* eggs visualized by Triton X-100 extraction and scanning electron microscopy. A, The actin network lies immediately beneath the egg plasma membrane. Pigment granules (PG) of the myoplasm can be seen immediately below this layer. (From Jeffery and Meier, 1983). B, The deep lattice of cytoskeletal filaments contains embedded pigment granules. (From Jeffery, 1983.)

motory organs called **comb plates** (Fig. 11.26E). Ctenophore embryos are especially useful for experiments on the determination of somatic cell fate because differentiated comb plate cells are easy to detect and photocyte differentiation can be measured by the emission of light.

Figure 11.26A through D shows the cleavage pattern of the ctenophore embryo. The first two cleavages are meridional, dividing the egg into four equal-sized cells. The third cleavage is oblique, forming four inner blastomeres, called **M cells**, and four outer blastomeres, called **E cells**. The outer blastomeres are slightly smaller than the inner blastomeres. The fourth cleavage is highly unequal, forming micromeres and macromeres in both M

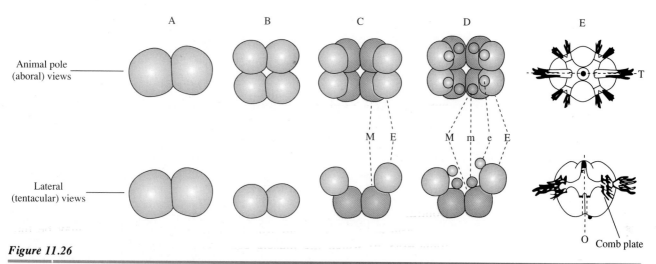

Figure 11.26

Diagrammatic representation of two-, four-, eight-, and 16-cell embryos (A–D) and the cydippid larva (E) of the ctenophore *Mnemiopsis*. The top part of the figure depicts these stages from above (animal pole or aboral plane views). The bottom part of the figure shows the same stages from the side (lateral or tentacular plane views). M and E indicate the macromeres, whereas m and e indicate the micromeres. The oral–aboral (O) and tentacular planes (T) are shown in E. (After Freeman, 1976.)

and E cells. The **m micromeres** are derived from M cells, and the **e micromeres** are derived from E cells. During subsequent cleavages, the m and e micromeres divide and are joined by new tiers of micromeres contributed by the underlying macromeres.

A fate map of the ctenophore embryo was made by applying colored chalk particles to blastomeres during the early cleavages (Reverberi and Ortolani, 1963). Particles applied to the e micromeres eventually appeared on the comb plate cells, whereas particles applied to the m micromeres labeled the photocytes. These experiments indicate that the E cell lineages form the comb plate and the M cell lineages form the photocytes. Before the third cleavage, isolated blastomeres produce both kinds of cells. When blastomeres are isolated at the eight-cell stage (i.e., after the third cleavage), however, different results are obtained. An isolated E cell will continue to cleave as if it were part of the intact embryo, and comb plate cells, but not photocytes, will differentiate. In contrast, photocytes, but not comb plate cells, will differentiate from isolated M cells. These experiments demonstrate that comb plate and photocyte determinants are segregated from each other between the second and third cleavages.

Comb plate and photocyte determinants intermingle before the four-cell stage but then become localized in specific cytoplasmic regions of each blastomere. This can be demonstrated by experiments in which anucleate portions of the outer or inner cytoplasmic regions of blastomeres are cut off at the two- or four-cell stage (Freeman, 1976). The first experiment was done at the two-cell stage, and its effect on development was examined by subsequent isolation of blastomeres. Regardless of whether the outer or inner cytoplasm was removed, the isolated blastomeres behaved normally, the E cells forming comb plates and the M cells forming photocytes. When the experiment was done at the four-cell stage, however, the isolated blastomeres showed specific defects: E cells that formed from blastomeres whose outside cytoplasm was removed failed to develop comb plates, whereas M cells forming from blastomeres whose inner cytoplasm had been removed did not differentiate photocytes.

Ooplasmic Determinants in Nemertines

In the nemertine worm *Cerebratulus*, ooplasmic determinants are localized during the maturation divisions and the early cleavages. Hörstadius (1937) demonstrated that the four animal blastomeres yield an apical tuft of cilia, whereas the four vegetal blastomeres yield gut tissue. After removing the animal or vegetal blastomeres of eight-cell embryos, embryos lacking the animal cells form guts but not apical tufts, and embryos lacking the vegetal cells form apical tufts but not guts. These experiments show that apical tuft and gut determinants are localized in specific regions (animal versus vegetal hemisphere) by the eight-cell stage. The localization process also was examined by cutting unfertilized eggs and one-, two-, or four-cell embryos into animal and vegetal hemisphere fragments and testing the ability of these fragments to form embryos with apical tufts and guts (Freeman, 1978). The results show that apical tuft determinants do not become localized in the animal hemisphere region of the zygote until between the first and second cleavages, whereas gut determinants began to be concentrated in the vegetal hemisphere much earlier, after the second maturation division (Fig. 11.27). Thus, in *Cerebratulus*, apical tuft and gut determinants become localized in the zygote at different times during early development.

What mechanisms are responsible for localization of ooplasmic determinants during the maturation divisions or cleavage? Several experiments indicate that the aster plays an important role in localizing apical tuft and gut determinants (Freeman, 1978). First, after exposure of *Cerebratulus*

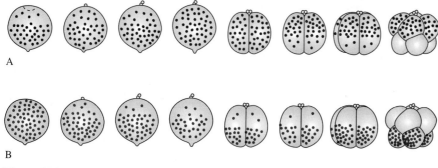

A

B

Figure 11.27

Diagrammatic representation of the localization of apical tuft (A) and gut (B) determinants during the maturation divisions and early cleavages of *Cerebratulus lacteus* eggs. (After Freeman, 1978.)

eggs to ethyl carbamate (an inhibitor of aster formation) at the second maturation division, aster formation is inhibited but development proceeds. In ethyl carbamate–treated eggs, the apical tuft and gut determinants are not localized in the animal and vegetal blastomeres at the eight-cell stage. Second, when the meiotic apparatus is cut out of eggs, asters are not formed during the maturation divisions. However, asters develop during the early cleavages. In these embryos, localization of gut determinants occurs, but apical tuft determinants are not localized. Third, when extra asters are induced by treating eggs with hypertonic seawater, apical tuft determinants are precociously segregated into the animal hemisphere region of the one-cell zygote with multiple asters. Although these experiments suggest that cytoskeletal elements are involved in localizing determinants in *Cerebratulus*, the mechanism of localization is unknown.

Ooplasmic Determinants in the Spiralian Polar Lobe

As described in Chapter 5, the eggs of some spiralians form a polar lobe during the early cleavages. The polar lobe is a transient cytoplasmic bulge that shunts vegetal cytoplasm specifically into the CD blastomere and then into the D blastomere (see Figs. 11.29 and 11.31). Blastomere separation and destruction experiments have been conducted with embryos containing polar lobes (Wilson, 1904b; Rattenbury and Berg, 1954; Clement, 1956; Cather and Verdonk, 1979). These experiments show that the restriction of developmental potential begins at first cleavage: Only the isolated CD and D blastomeres are capable of forming some of the mesodermal derivatives normally contributed by the D quadrant (Fig. 11.28). The incomplete embryos that develop from isolated AB, A, B, or C blastomeres are radially (rather than bilaterally) symmetrical and lack some of the tissues and organs normally derived from the mesoderm. This suggests that ooplasmic determinants for the missing structures are localized in the polar lobe and successively shunted to the CD and D blastomeres.

An experiment showing that ooplasmic determinants are present in the polar lobe of the scaphopod mollusk *Dentalium* was conducted by E. B. Wilson (1904a). In *Dentalium*, a large polar lobe forms during the early cleavages (Fig. 11.29A). During telophase of first cleavage, the polar lobe is connected to the future CD blastomere only by a narrow neck of cytoplasm. This is called the **trefoil stage** because of the three-lobed appearance of the embryo. The narrow neck of cytoplasm separating the polar lobe from the CD blastomere is readily severed by shaking or by treating trefoil stage embryos with Ca^{2+}-free seawater. When Wilson

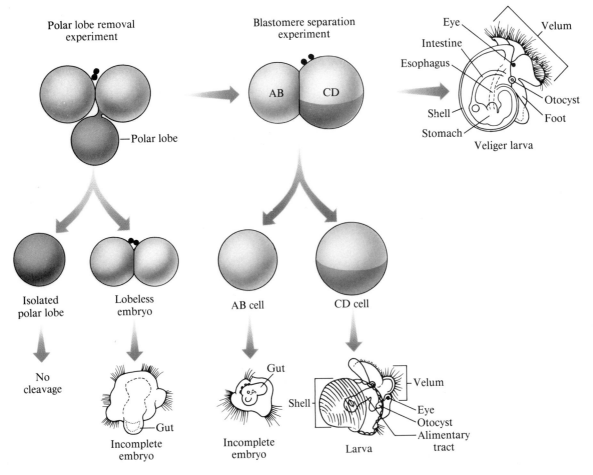

Figure 11.28

Diagrammatic representation of polar lobe removal and blastomere separation experiments in *Ilyanassa* embryos.

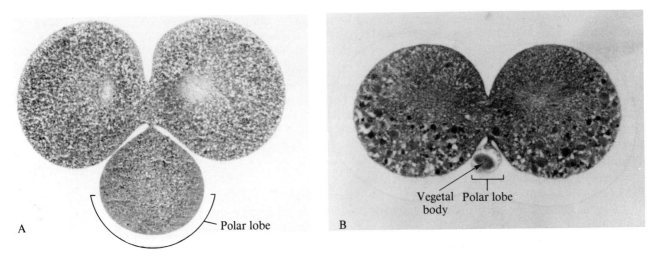

Figure 11.29

Cross-sections of (A) *Dentalium* and (B) *Bithynia* eggs at the trefoil stage. The vegetal body can be seen in the small polar lobe of *Bithynia*. (From Dohmen and Verdonk, 1979.)

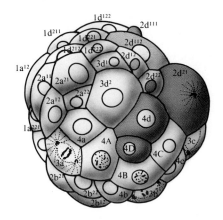

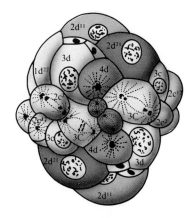

A   Normal embryo

B   Double embryo

Figure 11.30

Normal (A) and double (B) *Dentalium* embryos at about the sixth cleavage. Note duplication of some of the micromeres in the double embryo, which has been caused by treatment with cytochalasin B. (After Guerrier et al., 1978.)

removed the polar lobe at the trefoil stage, the **lobeless embryo** continued to cleave and develop but formed a radially symmetrical larva deficient in some mesodermal derivatives as well as other structures that are induced by tissues derived from the mesoderm. The defects obtained by removing the polar lobe of molluskan embryos are similar to those observed in embryos that develop from separated AB, A, B, or C blastomeres (see Fig. 11.28). The influence of the polar lobe on development is also demonstrated by experiments in which its formation is suppressed during first cleavage (Guerrier et al., 1978). *Dentalium* eggs treated with low concentrations of cytochalasin B do not form polar lobes (because cytoskeletal structures required for polar lobe constriction are inhibited), and when first cleavage occurs, the polar lobe cytoplasm is distributed to both AB and CD blastomeres. This results in the formation of a double embryo (Fig. 11.30), which develops into a trochophore larva with two shells.

Detailed studies on embryonic defects resulting from removal of the polar lobe have also been conducted on the gastropod mollusk *Ilyanassa*. The polar lobe of *Ilyanassa* behaves in a fashion similar to that of *Dentalium* during the first two cleavages (Fig. 11.31). The *Ilyanassa* embryo develops into a larva called a **veliger**, which contains a shell, digestive tract, pigment cells, foot, otocyst, heart, eyes, and the velum (a ciliated structure required for larval locomotion and feeding; see Fig. 11.28). Larvae that develop from lobeless embryos possess cilia, an endodermal mass, pigment cells, the nervous system, muscle, and stomach, but lack the velum, heart, intestine, otocyst, and eyes (see Fig. 11.28). The missing structures are either mesodermal derivatives or structures induced by mesoderm during later development. The results of these experiments suggest that the polar lobe contains ooplasmic determinants that enter the D cell lineage during cleavage. Later, these substances direct the formation of some of the mesodermal derivatives. Because the polar lobe contains no nucleus, the polar lobe determinants must be present in the cytoplasm.

As discussed in Chapter 5, spiral cleavage produces tiers of micromeres stacked above a vegetal quartet of macromeres. Once segregated into the D cell lineage, the polar lobe determinants appear to be parceled out to the lineage of micromeres during subsequent cleavages. The distribution of polar lobe determinants has been studied in an elegant series of blastomere destruction experiments (Clement, 1962). By destroying the D macromere after successive cleavages, it is possible to determine the developmental role of determinants passed to the d lineage micromeres (Fig. 11.32). The first quartet of cells is produced at the third division. At this time, the D blastomere divides into the 1d micromere and the 1D macromere. When

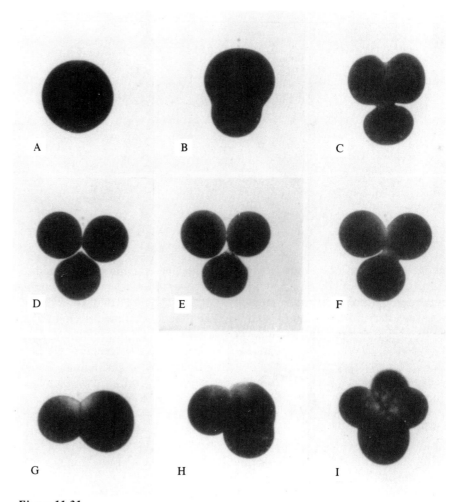

Figure 11.31

Early cleavage and polar lobe formation in the *Ilyanassa* embryo. The animal pole is toward the top, and the vegetal pole is toward the bottom. A–C, Early stage of polar lobe formation in the one-cell zygote. D–F, The trefoil stage. F, Polar lobe material enters the CD blastomere. G, A two-cell embryo showing the difference in size between the AB (left) and CD (right) blastomeres. H, A two-cell embryo showing the second polar lobe. I, A four-cell embryo showing the difference in size between the D (lower) and A, B, and C (upper) blastomeres. (From Clement, 1976.)

the 1D macromere is destroyed, the resulting larva has nearly the same set of defects as a lobeless embryo, indicating that there is no qualitative segregation of determinants between the 1D and 1d cells. After the next two cleavages, however, removal of the D macromere has less drastic effects on development. For example, destruction of the 2D macromere results in partial embryos that can develop a shell, destruction of the 3D macromere results in embryos that can develop a shell, velum, foot, and eyes; and destruction of the 4D macromere (which is of no unique developmental significance) results in the formation of a complete larva. This suggests that the last of the polar lobe determinants are passed to the 4d micromere at the sixth cleavage. The 4d micromere is the precursor cell to the mesodermal derivatives. These results show that each d micromere, except 1d, is involved in the development of specific larval structures (Table 11–1).

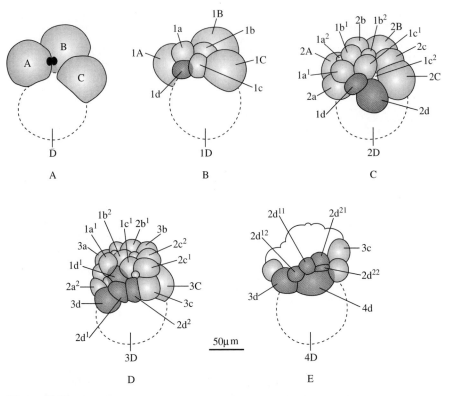

Figure 11.32

Staged removal of the D quadrant macromere of *Ilyanassa*. The cell removed is shown in broken outline. All D quadrant micromere derivatives are shown in color. A, D blastomere removed, leaving the ABC combination. B, 1D macromere removed, leaving the ABC + 1d combination. C, 2D macromere removed, leaving the ABC + 1d + 2d combination. D, 3D macromere removed, leaving the ABC + 1d + 2d + 3d combination. E, 4D macromere removed, leaving the ABC + 1d + 2d + 3d + 4d combination. (After Clement, 1962.)

The results of these experiments illustrate a clear case in which ooplasmic determinants are involved in programming specific blastomeres to yield progeny cells that adopt specific fates. However, the d micromeres themselves do not always form a particular structure. In some cases, the d micromeres induce other micromeres arising from the A, B, and C macromeres to differentiate into certain structures. For instance, eyes develop from the descendants of a and c micromeres after interacting with descendants of d micromeres. Thus, polar lobe determinants affect cell determination indirectly, by causing cells to induce other cells to follow specific developmental fates.

Centrifugation experiments suggest that the polar lobe determinants are absent from the fluid cytoplasm of the polar lobe (Clement, 1968). In these experiments, *Ilyanassa* eggs were centrifuged vegetal pole up, so the yolk-filled vegetal cytoplasm that normally enters the polar lobe is moved into the animal hemisphere in exchange for animal cytoplasm. When the polar lobe of a centrifuged embryo was removed at the trefoil stage, the same defects were found as in uncentrifuged lobeless embryos. This suggests that determinants are not displaced from the polar lobe by centrifugation and, therefore, could be attached to the cortical cytoskeleton or plasma membrane. Are there any differences in the cell surface (cortex and plasma membrane) of the polar lobe relative to other regions of the embryo? In

Table 11–1 Effects of Removal of the D Macromere of Ilyanassa at Different Stages

Partial Embryo	Defects Expected from Fate Map	Defects Found
ABC	Heart, shell, part of foot and gut	Intestine, heart, shell, foot, statocysts, eyes, velum
ABC + 1d	As above	As above
ABC + 1d + 2d	Heart, part of foot and gut	As above
ABC + 1d + 2d + 3d	Heart, part of gut	Intestine, heart
ABC + 1d + 2d + 3d + 4d	Part of gut	None

fact, the surface of the polar lobe has been shown by scanning electron microscopy (Fig. 11.33) to be morphologically distinct from the remainder of the egg (Dohmen and van der Mey, 1977).

Polar lobes have been examined by transmission electron microscopy for unique structures. So far, unique organelles have not been found in the large polar lobe of *Ilyanassa*. However, the much smaller polar lobe of the snail *Bithynia* (see Fig. 11.29B) contains a granular organelle (Fig. 11.34). This structure, called the **vegetal body**, is a cup-shaped mass of small vesicles, most of which are filled with a darkly staining substance. The contents of these vesicles have not been completely identified, although staining experiments suggest that the vegetal body contains RNA (Dohmen and Verdonk, 1974).

11–4 What Is an Ooplasmic Determinant?

T. H. Morgan first suggested that ooplasmic determinants may be substances that activate specific genes in restricted parts of the embryo (Morgan, 1934). Morgan, of course, was unaware that genes are made of DNA and produce protein via an mRNA intermediate. The expression of genes is controlled by proteins called transcription factors, which bind to regulatory sites in genes and modulate their ability to synthesize RNA. Therefore, proteins

Figure 11.33

Scanning electron micrographs of mollusk embryos showing the morphology of the cell surface of the polar lobe (A, *Buccinium*) and the CD blastomere, which has inherited polar lobe surface (B, *Crepidula*). (From Dohmen and van der Mey, 1977.)

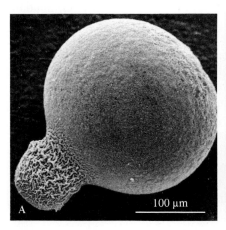
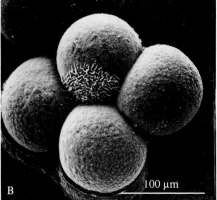

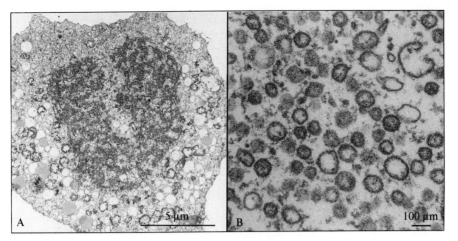

Figure 11.34

Electron micrographs of the polar lobe of *Bithynia* showing the structure of the vegetal body. A, The cup-shaped vegetal body is present in the polar lobe. B, Detail of the vegetal body showing the distribution of RNA-staining vesicles. (From Dohmen and Verdonk, 1974. Reprinted by permission of Company of Biologists, Ltd.)

that regulate differential gene activity are excellent candidates for ooplasmic determinants. It is also possible that mRNAs encoding transcription factors, rather than the factors themselves, are present in the egg cytoplasm and therefore serve as ooplasmic determinants. As we shall see in Chapter 13, many different oogenic mRNAs are stored in the egg cytoplasm and their translation is initiated after fertilization.

The possibility that ooplasmic determinants may be oogenic mRNAs is supported by detection of RNA in localized organelles, such as the polar granules of *Drosophila* and the vegetal body of *Bithynia*, and by a number of experiments. For example, we have described the rescue of pole cell formation in ultraviolet-irradiated *Drosophila* eggs by the microinjection of pole plasm from a normal donor into the posterior region of the egg. These experiments have been extended by microinjecting RNA purified from donor embryos (Tobashi et al., 1986). Microinjection of mRNA into the posterior region of irradiated eggs causes pole cells to be formed, suggesting that the pole cell determinant may be mRNA. However, functional gametes do not form, indicating that additional factors may be necessary for germ cell development.

Several different experiments support the possibility that the anterior body pattern of the chironomid insect *Smittia* is mediated by RNA localized in the anterior pole of the egg. Ultraviolet irradiation prevents the development of anterior larval structures such as the head and thorax (Kalthoff and Sander, 1968). Irradiated embryos show a mirror-image duplication of the posterior segments of the larva—except for pole cells—in the anterior portion of the embryo, and thus are called **double abdomens** (Fig. 11.35). The evidence that double abdomens are caused by the inactivation of oogenic RNA is as follows: First, the wavelength of ultraviolet light (260 nm) that is most effective in producing double abdomens is also most efficient for the inactivation of nucleic acids (Kalthoff, 1973). Second, the effect of ultraviolet light can be reversed by **photoreactivation**, in which irradiated embryos are treated with visible light (Kalthoff, 1973). Because photoreactivation reverses damage to nucleic acids, but not proteins, the ultraviolet-sensitive component is likely to be nucleic acid. Third, microinjection of RNase into the anterior pole of the egg causes formation of a double abdomen (Kandler-Singer and Kalthoff, 1976). Finally, it has been shown that the anterior structures of ultraviolet-irradiated eggs can be rescued by microinjecting RNA from unirradiated eggs (Elbetieha and Kalthoff, 1988).

Anterior determinants are also being studied in *Drosophila*, in which their identification is facilitated by genetic analysis. Mutations have been

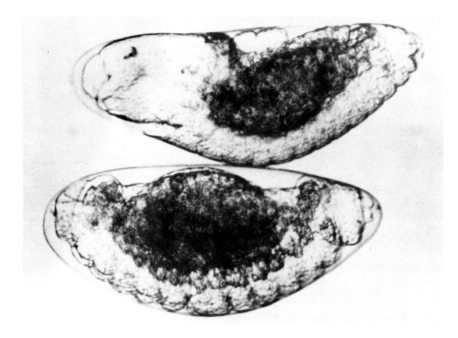

Figure 11.35

Normal (top) and double abdomen (bottom) embryos of the midge *Smittia*. Double abdomen embryos are induced by exposure to ultraviolet irradiation. (Photograph courtesy of K. Kalthoff.)

isolated in a small number of gene loci whose phenotypes are the absence of anterior structures from the embryo (Table 11–2). For example, embryos from *bicoid* (*bcd*) females are missing the head and thorax regions, which are replaced by another telson—a posterior structure. Defects in anterior embryonic pattern are also caused by the *exuperantia* (*exu*), *swallow* (*sww*), and *staufen* (*stau*) mutations (Fig. 11.36). Transplantation experiments suggest that *bcd*⁺ gene products are responsible for determining the anterior larval structures (Frohnhöfer and Nüsslein-Volhard, 1986). When cytoplasm from the anterior tip of a wild-type egg is microinjected into a *bcd* egg, a head and thorax develop in the *bcd* embryo. The anterior structures are formed at the site of the injection, *even if it is the posterior region of the egg*. Thus, substances in the wild-type cytoplasm substitute for mutant *bcd* gene products in determining the anterior embryonic structures. Cytoplasm

---

Table 11–2 Maternal-Effect Mutations Affecting Anterior Pattern in **Drosophila**

Mutant Gene	Phenotype
bicoid	Head and thorax absent, telson duplicated anteriorly
exuperantia	Anterior reduced, gnathal and thoracic regions enlarged
swallow	Same as *exuperantia*
bicaudal	Anterior defects (some alleles exhibit mirror-image duplication of abdomen)
staufen	Same as *exuperantia* and *swallow* but weaker; also affects posterior pattern

Adapted from Driever, W. and C. Nüsslein-Volhard. 1988. The bicoid protein determines position in the Drosophila embryo in a concentration-dependent manner. Cell, *54*: 95–104. Copyright © Cell Press.

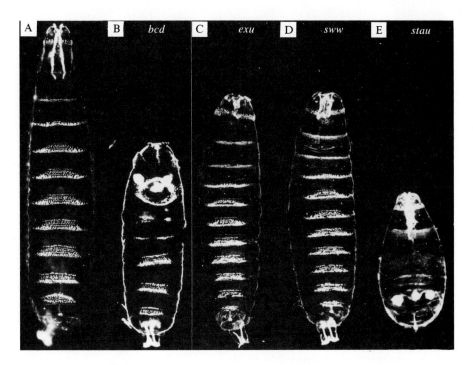

Figure 11.36

Maternal-effect mutations affecting the anterior region of the *Drosophila* embryo. Each frame shows larval cuticle with head and thorax region at the top and the posterior telson at the bottom. A, Normal larva. The head and thorax are present, and wide belts of denticles are located in the middle and posterior regions of the larva. B, A *bicoid* larva. Head and thorax are replaced by a duplicated telson, and the anterior abdomen is defective. C, An *exuperantia* larva showing anterior defects. D, A *swallow* larva showing anterior defects. E, A *staufen* larva showing severe anterior and abdominal defects. (From Driever and Nüsslein-Volhard, 1988b. Copyright © Cell Press.)

taken from the anterior region of the egg is more effective at forming head and thorax than cytoplasm taken from the middle or posterior regions of the egg. Therefore, *bcd* rescue activity is found primarily in the anterior region of the egg but is not restricted to the anterior tip.

The *bcd*⁺ gene has been cloned, and gene-specific probes have been constructed. The *bcd* gene is one of the class of genes that contains a **homeobox**, a short sequence of nucleotides that encodes a DNA-binding domain in gene regulatory proteins. Homeoboxes were originally identified in homeotic genes that control pattern formation in insect embryos and will be discussed in detail in Chapter 14. Using a gene-specific probe, the distribution of *bcd* mRNA has been followed in the egg by *in situ* hybridization (Berleth et al., 1988). As described in Chapter 3 (see Fig. 3.41), *bcd* mRNA is produced in the nurse cells and is deposited in the anterior tip of the oocyte during oogenesis, where it remains as an oogenic mRNA until it is translated after fertilization. Thus, in *Drosophila*, the ooplasmic determinant controlling development of anterior embryonic structures appears to be an mRNA molecule.

The *bcd* mRNA itself does not function directly in promoting anterior pattern formation; *bcd* mRNA is translated during early cleavage, and bcd protein becomes distributed along a concentration gradient that is highest in the anterior region of the syncytial embryo (Driever and Nüsslein-Volhard, 1988a). This gradient can be quantified by staining early embryos

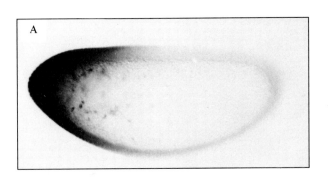

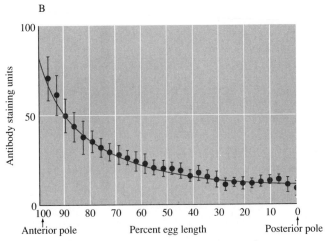

Figure 11.37

Gradient of bicoid protein revealed by staining *Drosophila* embryos with bicoid antibody. A, Whole mount of a syncytial blastoderm stage embryo showing a gradient of staining with its highest concentration at the anterior pole (left). See Color plate 19 for color version. B, Quantification of staining shows a steep gradient of bicoid protein with its highest concentration at the anterior pole (left). (From Driever and Nüsslein-Volhard, 1988a. Copyright © Cell Press.)

with an antibody specific for bcd protein. As shown in Figure 11.37, the protein gradient is very steep and extends from the anterior pole to about one third of the distance to the posterior pole; this is precisely the region in which the anterior structures of the embryo develop. The gradient of bcd protein in the embryo explains the results of the microinjection experiments discussed previously, in which rescue of *bcd* eggs was obtained by microinjection of middle-region cytoplasm from wild-type eggs, even though bcd mRNA is only present in the anteriormost region: The bicoid protein must be providing the rescue activity.

After cellular blastoderm formation, the bcd protein enters the nuclei, in which it modulates transcription of other genes involved in forming the anterior structures of the embryo (the gap genes; see Chap. 14). Genetic experiments demonstrate that the blastoderm cells can respond to small differences in the shape of the bcd protein gradient by altering the position where anterior embryonic structures are determined (Driever and Nüsslein-Volhard, 1988b). These experiments entail increasing the *bcd*[+] gene dosage by constructing flies with multiple *bcd*[+] genes. The bcd protein gradient becomes steeper and extends further toward the posterior end of the egg as the number of *bcd*[+] genes is increased from one to four (Fig. 11.38). This shift of the bcd gradient causes a corresponding change in the fate map of the embryo, such that anterior structures begin to be seen at more posterior positions. For example, the cephalic furrow (see p. 213), which is an early indicator of the posterior limit of the head, is gradually moved posteriorly when the dose of *bcd*[+] genes is increased (see Fig. 11.38A). These experiments suggest that the anterior portion of the *Drosophila* embryo is determined by an ooplasmic determinant consisting of a localized mRNA that directs the synthesis of a gene regulatory protein. It is this protein that is the causal factor in determining the anterior pattern of the *Drosophila* embryo.

If the *bcd* gene encodes the ooplasmic determinant for anterior structures, why do mutations in other gene loci (see Table 11–2) also result

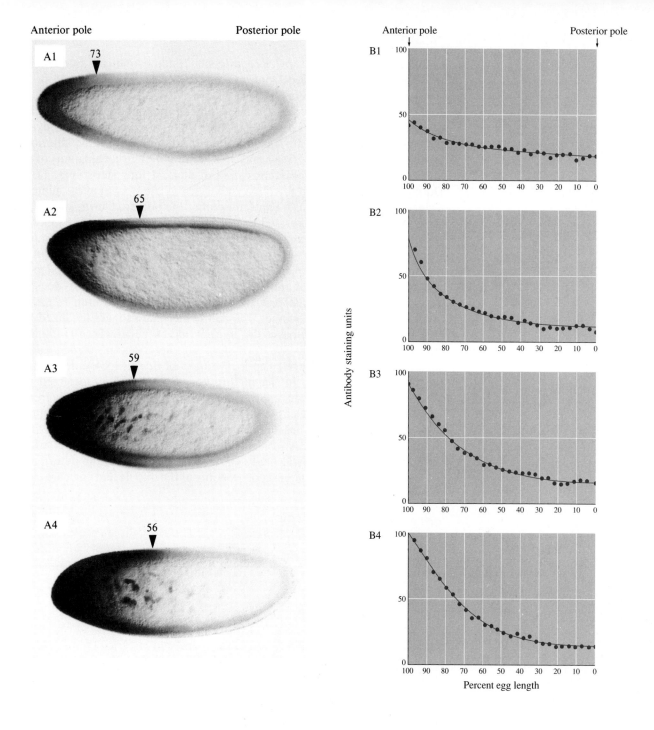

Figure 11.38

Dependence of the bicoid protein gradient and the fate map of the *Drosophila* embryo on the dosage of *bicoid*[+] genes. A, Photographs of blastoderm-stage embryos with 1–4 copies of the *bicoid*[+] gene. Note that the bicoid protein gradient, shown by staining with a bicoid-specific antibody, becomes steeper (also see B) as the number of *bicoid*[+] genes increases. Arrowheads indicate the position of the cephalic fold when these embryos gastrulate, a marker of the posterior extent of the head, and the percent egg length. The percent egg length is determined from the posterior to the anterior pole; hence, decreasing percentages reflect a movement of the marker to positions nearer the posterior pole. A1–A4 represent embryos carrying 1–4 copies of the *bicoid*[+] gene, respectively. B, Dependence of the bicoid protein gradient on the number of *bicoid*[+] gene copies. The gradient becomes steeper and extends more posteriorly as the gene dosage increases. B1–B4 represent gradients determined by antibody staining of embryos carrying 1–4 copies of the *bicoid*[+] gene, respectively. (From Driever and Nüsslein-Volhard, 1988b. Copyright © Cell Press.)

in defective anterior pattern? Recent genetic studies indicate that some of these genes function in localizing *bcd* mRNA to the anterior tip of the egg (Frohnhöfer and Nüsslein-Volhard, 1987; Stephenson et al., 1988; St. Johnston et al., 1989). When *exu*, *sww*, and *stau* mutant eggs are examined by *in situ* hybridization, *bcd* mRNA is either uniformly distributed throughout the egg (*exu* mutants) or forms a very shallow gradient from the anterior to posterior end of the egg (*sww* and *stau* mutants). The localization of *bcd* mRNA is a gradual process consisting of at least four steps (Fig. 11.39). During the first step, *bcd* mRNA synthesized in the oocyte accumulates in a ring-shaped structure at the anterior end of the oocyte. During the second step, *bcd* mRNA is synthesized and accumulates in the apical periphery in each nurse cell. *Exu* mutants appear to be defective in the first and second steps of localization: The ring of *bcd* mRNA appears to be more diffuse in the oocyte, there is no apical localization of *bcd* mRNA in the nurse cells, and *bcd* mRNA released into the oocyte from the nurse cells does not localize in the anterior pole of the egg. During the third step (after *bcd* mRNA has entered the oocyte), the messenger is no longer localized as a ring, but instead covers most of the anterior end of the oocyte. *Sww* mutants appear to be defective in the third step: The anterior ring of *bcd* mRNA slips posteriorly, and there is no localization of *bcd* mRNA entering from the nurse cells. Finally, in the mature egg, *bcd* mRNA is localized in a spherical region on the presumptive dorsal side of the anterior pole. In *stau* mutants, *bcd* mRNA is released from the anterior pole late during oogenesis.

At present, there is little information about how *bcd* mRNA localization is achieved at the molecular or cellular levels. As research on *bcd* mRNA advances, however, we will likely discover how an ooplasmic determinant is localized and what structural components of the oocyte and egg participate in this process.

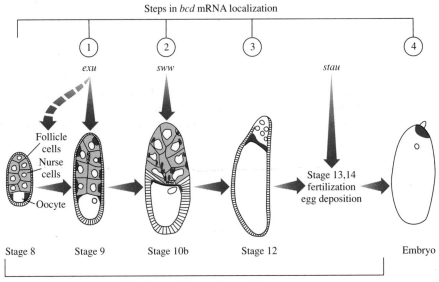

Figure 11.39

A drawing showing the four phases of *bcd* mRNA localization during oogenesis and the points at which the wild-type alleles of *exu*, *sww*, and *stau* are required. (After St. Johnston et al., 1989. Reprinted by permission of Company of Biologists, Ltd.)

11–5 Regulation of Cell Determination by Extrinsic Factors

Cell determination can also be controlled by extrinsic factors, such as environmental stimuli or cell-cell interactions. In this section, we discuss examples of the role of blastomere position in determining cell fate. Consideration of inductive interactions related to the determination of tissues and organs during later development will be discussed in Chapter 12.

Extrinsic Determination of Trophectoderm and Inner Cell Mass in Mammals

An early step in cell determination in mammals involves the formation of two different tissues: the trophectoderm and inner cell mass. As discussed in Chapter 5, trophectoderm and inner cell mass have different fates during mammalian development. The trophectoderm forms most of the extraembryonic tissues, whereas the inner cell mass forms the embryo proper. Compacted eight-cell mammalian embryos contain four outer and four inner blastomeres. Fate maps indicate that the outer cells form trophectoderm and the inner cells give rise to the inner cell mass (Tarkowski and Wróblewska, 1967).

When are the trophectoderm and inner cell mass determined during development? This question can be approached by traditional blastomere-isolation experiments. Investigators have shown that blastomeres isolated at the two-, four-, or eight-cell stage have the capacity to form both trophectoderm and inner cell mass (Herbert and Graham, 1974). These results suggest that each blastomere in the eight-cell mammalian embryo is totipotent. Blastomere isolation experiments are difficult to do at later developmental stages because isolated mammalian blastomeres have a low capacity for survival in culture. Another approach is to separate blastomeres at various stages of development, allow them to reaggregate, and follow the development of the reaggregated embryos (Stern, 1972). Despite complete disorganization, reaggregated embryos form blastocysts containing trophectoderm and inner cell mass, even if dissociated up to the early blastocyst stage. Because blastomeres tend to reaggregate without regard to their original position in the embryo, it is clear from these results that determination depends on the location of a cell during a particular stage of development.

It has been proposed that the determination of mammalian blastomeres is dependent on whether they are located on the inside or outside of the embryo (Tarkowski and Wróblewska, 1967). This **inside-outside hypothesis** proposes that cells located inside the embryo become inner cell mass and cells located on the outside of the embryo become trophectoderm. The inside-outside hypothesis can be tested by experiments with chimeric mouse embryos. **Chimeras** are experimental embryos containing cells from two or more animals, usually different genetically marked strains of mice. A favorite genetic marker is coat color because the experimental result is easily determined by examining the color of adult animals. A chimera experiment has been used to determine the developmental potential of blastomeres at the four-cell stage (Hillman et al., 1972). In this experiment, single blastomeres obtained from four-cell mouse embryos are placed on the outside of a mass of aggregated blastomeres from another embryo. The transplanted cells always develop into trophectoderm. These results suggest that outer cells become trophectoderm but do not indicate the fate of the inner cells. A more elaborate experiment, used to examine the fate of the inside cells, is shown in Figure 11.40. Four-cell embryos of an albino mouse strain were dissociated, the individual blastomeres were allowed to divide one more time, and then each of the eight blastomeres was combined with four blastomeres derived from the eight-cell stage of an embryo derived from a black strain of mice. Chimeric embryos derived from these albino and black strains of mice can be distinguished biochemically as well as by their coat

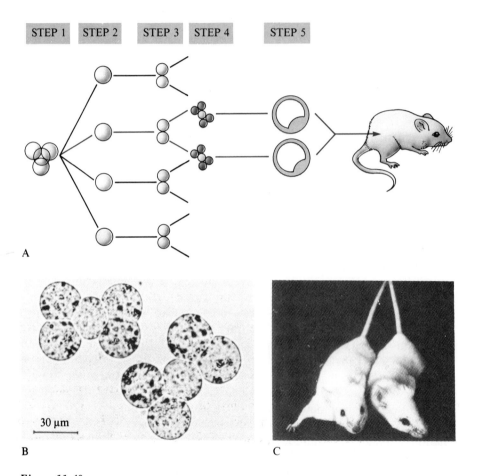

STEP 1 **STEP 2** **STEP 3** **STEP 4** **STEP 5**

A

30 μm

B   C

Figure 11.40

Experiment to test the "inside-outside" hypothesis. A, Experimental design. The donor embryo is from an albino strain (designated "A"), and the host blastomeres are from a strain (designated "B") with black pigmentation. Step 1. The donor, "A"-type embryo is obtained at the four-cell stage, and the zona pellucida is removed. Step 2. The blastomeres are dissociated. Step 3. The blastomeres divide to the eight-cell stage, giving four pairs of "octet" blastomeres. Step 4. Each pair of blastomeres is dissociated, and each individual blastomere is surrounded by four eight-cell stage type "B" blastomeres by micromanipulation. Step 5. The pairs of chimeric embryos are cultured to the blastocyst stage. Each pair of blastocysts is transferred to one of the uterine horns of a pseudopregnant recipient. B, A pair of composite blastomeres. The "A"-type donor blastomeres are each surrounded by four "B"-type blastomeres. C, Two albino mice recovered from this experiment. Each mouse was derived from one of the eight-cell stage blastomeres of an "A" type embryo, combined with four eight-cell stage "B" type blastomeres. (From Kelly, 1977.)

color because they also have different electrophoretic variants of the enzyme glucose phosphate isomerase. Therefore, samples of tissue can be taken during development to determine whether the inner cell mass is derived from the inner (albino strain) blastomere or the outer (black strain) blastomeres. The variant of glucose phosphate isomerase characteristic of the inner cell (albino) was expressed by the majority of chimeric embryos. In addition, most of the mice that developed from these embryos were albino. Because cells placed at the outside or inside of chimeric mouse embryos do not change their positions during development (Mintz, 1965), the results of these experiments provide further evidence that blastomeres are totipotent in early mammalian embryos. The results also support the inside-outside hypothesis and indicate that the trophectoderm and inner cell mass are determined extrinsically by their spatial position in the embryo. However, the precise mechanism through which the position of a mammalian blastomere affects its fate is unknown.

Cell Position and Gradients in Sea Urchin Development

Classic experiments conducted by Hörstadius (1935, 1939) show that cell position is also an important factor in sea urchin development. As discussed in Chapter 5, sea urchin zygotes cleave to form a 64-cell embryo with five tiers of blastomeres (Fig. 11.41A). The 32 animal blastomeres (the animal cap) are descendants of the mesomeres, the first and second tiers of 16 vegetal blastomeres are descendants of the macromeres, and the third (bottom) vegetal tier consists of 16 derivatives of the micromeres. Fate maps of the sea urchin embryo show that the animal cap and upper tier of vegetal blastomeres are precursors of ectoderm, the second vegetal tier becomes endoderm and secondary mesenchyme, and the micromeres form the primary mesenchyme from which the larval skeleton is derived (see Chap. 6).

To determine how the primary germ layers form in sea urchin embryos, Hörstadius isolated tiers of blastomeres from 64-cell stage sea urchin embryos with a glass needle and recombined them in various ways. These experiments are illustrated in Figure 11.41. In the first part of the experiment, animal caps were isolated and cultured alone. These animal caps became animalized embryos (see Sect. 11–2), consisting of spheres of ectodermal cells with an abundance of long apical cilia but no mesoderm or endoderm (see Fig. 11.41B). When isolated animal caps are recombined with the first tier of vegetal cells, the resulting larva is less animalized; it now contains fewer long apical cilia and a portion of the gut (see Fig. 11.41C). More striking are the results obtained when the animal caps were combined with the second tier of vegetal cells or the micromeres (see Fig. 11.41D and E). These combinations develop into complete plutei containing ectoderm, endoderm, and mesoderm. In recombinates of the animal cap with micromeres, the gut must be formed from animal cells, rather than its normal precursors—the second tier of vegetal cells. This result shows that animal cap cells retain the potential to become gut cells until the 64-cell stage in sea urchins. However, the most important conclusion that can be drawn from these experiments is that endoderm determination is mediated by cell-cell interactions.

Hörstadius used J. Runnström's double-gradient model to rationalize the results of his recombination experiments. According to this model, the ectoderm, endoderm, and mesoderm are determined by apposing gradients of animalizing and vegetalizing factors. The animalizing factors would be most concentrated at the animal pole and gradually decrease toward the vegetal pole. The vegetalizing factors would show the opposite distribution. Therefore, cells that contain a high concentration of animalizing factors and

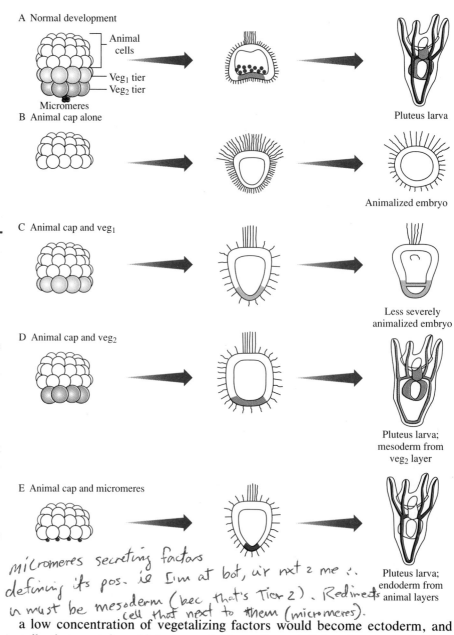

A Normal development

Animal cells

Veg₁ tier
Veg₂ tier

Micromeres

Pluteus larva

B Animal cap alone

Animalized embryo

C Animal cap and veg₁

Less severely animalized embryo

D Animal cap and veg₂

Pluteus larva; mesoderm from veg₂ layer

E Animal cap and micromeres

Pluteus larva; endoderm from animal layers

Figure 11.41

Schematic representation of blastomere layer isolation and recombination experiments carried out by Hörstadius. The nature of the experimental embryo is shown on the left of each row, the swimming blastula derived from this embryo is shown in the middle of each row, and the pluteus larva or equivalent structure formed from this embryo is shown at the right of each row. A, Cell fate in a normal embryo. B, Development of isolated animal tiers into animalized larvae. C, Development of isolated animal tiers recombined with the first vegetal tier into less severely animalized embryos. D, Development of isolated animal tiers recombined with the second vegetal tier into a complete pluteus larva. E, Development of isolated animal tiers recombined with micromeres into complete pluteus larvae. (After Hörstadius, 1939.)

[handwritten annotation: Micromeres secreting factors defining its pos. ie I'm at bot, uir nxt 2 me ∴ u must be mesoderm (bec that's Tier 2). Redirects cell that next to them (micromeres).]

[handwritten annotation: ie pos imp in det what cell forms-]

a low concentration of vegetalizing factors would become ectoderm, and cells that contain a high concentration of vegetalizing factors and a low concentration of animalizing factors would become mesoderm. Cells in the equatorial region of the embryo would form ectoderm or endoderm, depending on the relative concentration of animalizing and vegetalizing factors, respectively. The equatorial animal cells must remain undetermined until late during development and may be induced to become endoderm by cells containing higher levels of vegetalizing factors. The gradient model was tested by a variation of the experiment shown in Figure 11.41E, in which different numbers of micromeres were successively added to isolated animal caps (Hörstadius, 1939). As progressively more micromeres were added to an isolated animal tier, a more complete embryo was formed. This result is consistent with the gradient model.

Clearly, the double-gradient model predicts the existence of both intrinsic and extrinsic factors controlling cell determination in sea urchin development. The intrinsic factors would be the putative animalizing and vegetalizing substances, whereas the extrinsic factors would be inductive events promoted by these factors in the various tiers of blastomeres. Although the animalizing and vegetalizing factors have not been isolated and identified, they can be mimicked by treating embryos with a number of different organic and inorganic substances. For example, heavy metals and certain dyes are examples of substances that promote the development of animalized embryos, whereas lithium and carbon monoxide are examples of substances that form vegetalized embryos.

The double-gradient model is useful as an interpretive framework for the results of Hörstadius' experiments but is falling into disfavor as a mechanism to explain cell determination (Angerer and Davidson, 1984; Wilt, 1987; Davidson, 1989). There is no experimental evidence requiring two gradients rather than a sequence of inductive events starting from the vegetal pole. Although the determined state of the micromeres can be demonstrated by differentiation in culture (see Chap. 6), animal cap cells do not differentiate in culture, as might be expected for blastomeres that are the source of an animalizing gradient. Furthermore, recent studies suggest that animalization by heavy metals represents an arrest of development rather than the selection of a specific developmental pathway (Nemer, 1986). For these reasons, current research is focussed on how vegetalizing agents function to produce the inductive factors that are important in normal development. Chapter 12 will examine the role of embryonic induction in the determination of cell fate.

11–6 Chapter Synopsis

This chapter describes how the fates of totipotent blastomeres become progressively restricted by intrinsic and extrinsic mechanisms during embryonic development. Intrinsic mechanisms involve ooplasmic determinants, whereas extrinsic mechanisms are based on embryonic induction or information obtained from the position of blastomeres in the embryo. Since cell determination is not accompanied by visible changes in embryonic cells, it is studied using procedures such as cell isolation and transplantation, which require knowledge of the fate map or cell lineage of the embryo. Embryos are distinguished as being either mosaic or regulative depending on how rapidly blastomere fates are decided during embryogenesis. In mosaic embryos, blastomeres become restricted during the first few cleavages, whereas in regulative embryos blastomeres remain totipotent until later in development.

Germ cell and some somatic cell fates are controlled by ooplasmic determinants localized in the egg and segregated to specific blastomeres during cleavage. In nematode eggs, ooplasmic determinants segregated to the P cells cause them to become primordial germ cells. The nematode germ cell determinants may be localized in P granules. Germ cell determinants may also be localized in polar granules in the pole plasm of *Drosophila* eggs and the germinal granules of frog eggs. Muscle cell determination in tunicates is caused by ooplasmic determinants localized in the egg's yellow crescent, which is segregated to presumptive muscle cells during cleavage. The fates of photocytes and comb plate cells of ctenophore embryos and gut cells of nemertine embryos are also established by segregated ooplasmic determinants. All ooplasmic determinants do not establish cell fate function intrinsically. Ooplasmic determinants localized in the polar lobe of spiralian embryos initiate determination of cells that subsequently induce other cells to become mesoderm.

Ooplasmic determinants may be proteins that regulate transcription or localized mRNA molecules that are translated during embryogenesis into gene regulatory proteins. An example of the latter is *bicoid* mRNA in *Drosophila*, which encodes a transcription factor distributed in an anterior to posterior gradient and is required for development of the anterior region of the embryo.

Somatic cell fates are also controlled extrinsically by environmental stimuli or cell-cell interactions. Embryonic cells of mammals are determined by their position during late cleavage: Inside cells become inner cell mass, and outside cells become trophectoderm. The primary germ layers of sea urchin embryos are determined by a vegetal to animal gradient of cell-cell interactions. In the next chapter, we shall examine the determination of cell fate by cell-cell interactions.

References

Angerer, R. C., and E. H. Davidson. 1984. Molecular indices of cell lineage specification in sea urchin embryos. Science, *226*: 1153–1160.

Beams, H. W., and R. G. Kessel. 1974. The problem of germ cell determinants. Int. Rev. Cytol., *39*: 419–479.

Bennett, K. L., and S. Ward. 1986. Neither a germ line-specific nor several somatically expressed genes are lost or rearranged during embryonic chromatin diminution in the nematode *Ascaris lumbricoides* var. *suum*. Dev. Biol., *118*: 141–147.

Berleth, T. et al. 1988. The role of localization of *bicoid* RNA in organizing the anterior pattern of the *Drosophila* embryo. EMBO J., *7*: 1749–1756.

Boswell, R. E., and A. P. Mahowald. 1985. tudor, a gene required for assembly of the germ plasm in Drosophila melanogaster. Cell, *43*: 97–104.

Bounoure, L. 1934. Recherches sur la lignée germinale chez la grenouille rousse aux premier stades au développement. Ann. Sci. Natur. Zool., 10e Ser., *17*: 67–248.

Boveri, T. 1887. Über Differenzierung der Zellkerne wahrend der Furchung des Eis von *Ascaris megalocephala*. Anat. Anz., *2*: 688–693.

Boveri, T. 1910. Über die Teilung centrifugierter Eier von *Ascaris megalocephala*. Wilhelm Roux's Archiv für Entwicklungsmechanik, *30*: 101–125.

Cather, J. N., and N. H. Verdonk. 1979. Development of *Dentalium* following removal of D quadrant blastomeres at successive cleavage stages. Wilhelm Roux's Archives Dev. Biol., *187*: 355–366.

Clement, A. C. 1956. Experimental studies on germinal localization in *Ilyanassa*. II. The development of isolated blastomeres. J. Exp. Zool., *132*: 427–446.

Clement, A. C. 1962. Development of *Ilyanassa* following removal of the D macromere at successive cleavage stages. J. Exp. Zool., *149*: 193–216.

Clement, A. C. 1968. Development of the vegetal half of the *Ilyanassa* egg after removal of most of the yolk by centrifugal force compared with the development of animal halves of similar visible composition. Dev. Biol., *17*: 165–186.

Clement, A. C. 1976. Cell determination and organogenesis in molluscan development: A reappraisal based on deletion experiments in *Ilyanassa*. Am. Zool., *16*: 447–453.

Conklin, E. G. 1905. The organization and cell-lineage of the ascidian egg. J. Acad. Nat. Sci. Phila., Ser. 2., *13*: 1–119.

Dale, L., and J. M. W. Slack. 1987. Fate map for the 32-cell stage of *Xenopus laevis*. Development, *99*: 527–551.

Davidson, E. H. 1989. Lineage-specifc gene expression and the regulative capacities of the sea urchin embryo: A proposed mechanism. Development, *105*: 421–445.

Deno, T., H. Nishida, and N. Satoh. 1985. Histospecific acetylcholinesterase development in quarter ascidian embryos derived from each blastomere pair of the eight-cell stage. Biol. Bull. Mar. Biol. Lab. Woods Hole, 168:239–248.

Dohmen, M. R., and J. C. A. van der Mey. 1977. Local surface differentiation of the vegetal pole of the eggs of *Nassarius reticulatus*, *Buccinium undatum*, and *Crepidula fornicata* (Gastropoda, Prosobranchia). Dev. Biol., *61*: 104–113.

Dohmen, M. R., and N. H. Verdonk. 1974. The structure of a morphogenetic cytoplasm, present in the polar lobe of *Bithynia tentaculata* (Gastropoda, Prosobranchia). J. Embryol. Exp. Morphol., *31*: 423–433.

Dohmen, M. R., and N. H. Verdonk. 1979. The ultrastructure and role of the polar lobe in development of molluscs. *In* S. Subtelny and I. R. Konigsberg (eds.), *Determinants of Spatial Organization*. Academic Press, New York, pp. 3–27.

Driesch, H. 1892. The potency of the first two cleavage cells in echinoderm development. Experimental production of partial and double formations. *Reprinted in* B. H. Willier and J. M. Oppenheimer (eds.), *Foundations of Experimental Embryology*. Hafner, New York.

Driever, W., and C. Nüsslein-Volhard. 1988a. A gradient of bicoid protein in Drosophila embryos. Cell, *54*: 83–93.

Driever, W., and C. Nüsslein-Volhard. 1988b. The bicoid protein determines position in the the Drosophila embryo in a concentration-dependent manner. Cell, *54*: 95–104.

Elbetieha, A., and K. Kalthoff. 1988. Anterior determinants in embryos of *Chironomus samoensis*: Characterization by rescue bioassay. Development, *101*: 61–75.

Freeman, G. 1976. The role of cleavage in the localization of developmental potential in the ctenophore *Mnemiopsus leidyi*. Dev. Biol., *49*: 143–177.

Freeman, G. 1978. The role of asters in the localization of factors that specify the apical tuft and the gut of the nemertine *Cerebratulus lacteus*. J. Exp. Zool., *206*: 81–107.

Frohnhöfer, H. G., and C. Nüsslein-Volhard. 1986. Organization of anterior pattern in the *Drosophila* embryo by the maternal gene *bicoid*. Nature (Lond.), *324*: 120–125.

Frohnhöfer, H. G., and C. Nüsslein-Volhard. 1987. Maternal genes required for the anterior localization of *bicoid* activity in the embryo of *Drosophila*. Genes Dev., *1*: 880–890.

Gardner, R. L. 1978. The relationship between cell lineage and differentiation in the extraembryonic endoderm of the mouse embryo. *In* W. J. Gehring (ed.), *Genetic Mosaics and Cell Differentiation*. Springer-Verlag, Berlin, pp. 205–241.

Geigy, R. 1931. Action de l'ultra-violet sur le pole germinal dans l'oeuf de *Drosophila melanogaster* (Castration et mutabilité). Rev. Suisse Zool., *38*: 187–288.

Guerrier, P. et al. 1978. Significance of the polar lobe for the determination of dorso-ventral polarity in *Dentalium vulgare* (da Costa). Dev. Biol., *63*: 233–242.

Hay, B. et al. 1988. Identification of a component of *Drosophila* polar granules. Development, *103*: 625–640.

Hegner, R. W. 1911. Experiments with chrysomelid beetles. III. The effects of killing parts of the eggs of *Leptinotarsa decemlineata*. Biol. Bull., *20*: 237–251.

Herbert, M. C., and C. F. Graham. 1974. Cell determination and biochemical differentiation of the early mammalian embryo. *In* A. A. Moscona and A. Monroy (eds.), *Current Topics in Developmental Biology*, vol. 8. Academic Press, New York, pp. 151–178.

Hillman, N., M. I. Sherman, and C. Graham. 1972. The effect of spatial arrangement on cell determination during mouse development. J. Embryol. Exp. Morphol., *28*: 263–278.

Hörstadius, S. 1935. Über die Determination im Verlaufe der Eiasche bei Seeigeln. Pubb. Staz. Zool. Napoli, *14*: 251–479.

Hörstadius, S. 1937. Experiments on the early development of *Cerebratulus lacteus*. Biol. Bull., *73*: 317–342.

Hörstadius, S. 1939. The mechanics of sea urchin development studied by operative methods. Biol. Rev., *14*: 132–179.

Ikenishi, K., and M. Kotani. 1979. Ultraviolet effects on presumptive primordial germ cells (pPGCs) in *Xenopus laevis* after the cleavage stage. Dev. Biol., *69*: 237–246.

Illmensee, K., and A. P. Mahowald. 1974. Transplantation of posterior polar plasm in *Drosophila*. Induction of germ cells in the anterior pole of the egg. Proc. Natl. Acad. Sci. U.S.A., *71*: 1016–1020.

Illmensee, K., and A. P. Mahowald. 1976. The autonomous function of germ plasm in a somatic region of the *Drosophila* egg. Exp. Cell Res., *97*: 127–140.

Jacobson, M., and G. Hirose. 1981. Clonal organization of the central nervous system of the frog. II. Clones stemming from individual blastomeres of the 32- and 64-cell stages. J. Neurosci., *1*: 271–284.

Janning, W. 1978. Gynandromorph fate maps in *Drosophila*. *In* W. J. Gehring (ed.), *Genetic Mosaics and Cell Differentiation*. Springer-Verlag, Berlin, pp. 1–28.

Jeffery, W. R. 1983. Messenger RNA localization and cytoskeletal domains in ascidian eggs. *In* W. R. Jeffery and R. A. Raff (eds.), *Time, Space and Pattern in Embryonic Development*. A. R. Liss Press, New York, pp. 241–259.

Jeffery, W. R. 1984. Pattern formation by ooplasmic segregation in the ascidian egg. Biol. Bull., *166*: 277–298.

Jeffery, W. R., and W. R. Bates. 1989. Ooplasmic segregation in the ascidian *Styela*. *In* H. Schatten and G. Schatten (eds.), *The Molecular Biology of Fertilization*. Academic Press, New York, pp. 341–367.

Jeffery, W. R., and S. Meier. 1983. A yellow crescent cytoskeletal domain in ascidian eggs and its role in early development. Dev. Biol., *196*: 125–143.

Kalthoff, K. 1973. Action spectrum for UV induction and photoreversal of a switch in the developmental program of the egg of an insect (*Smittia*). Photochem. Photobiol., *18*: 355–364.

Kalthoff, K., and K. Sander. 1968. Der Entwicklungsgang der Missbildung "Doppelabdomen" im partiell UV-bestrahlen Ei von *Smittia parthenogenetica* (Dipt., Chironomidae). Wilhelm Roux's Archiv für Entwicklungsmechanik, *161*: 129–146.

Kandler-Singer, I., and K. Kalthoff. 1976. RNase sensitivity of an anterior morphogenetic determinant in an insect egg (*Smittia* sp., Chironomidae, Diptera). Proc. Natl. Acad. Sci. U.S.A., *73*: 3739–3743.

Keller, R. E. 1975. Vital dye mapping of the gastrula and neurula of *Xenopus laevis*. I. Prospective areas and morphogenetic movements of the superficial layer. Dev. Biol., *42*: 222–241.

Kelly, S. J. 1977. Studies on the developmental potential of 4- and 8-cell stage mouse blastomeres. J. Exp. Zool., *200*: 365–376.

Kimmel, C. B., and R. M. Warga. 1988. Cell lineage and developmental potential of cells in the zebrafish embryo. Trends Genet., *4*: 68–74.

Le Douarin, N. 1971. Characteristiques ultrastucturales du noyau interphasique chez la caille et chez le poulet et utilization de cellules de caille comme "marquers biologiques" en embryologie experimentale. Ann. Embryol. Morphogenet., *4*: 125–135.

Lehmann, R., and C. Nüsslein-Volhard. 1986. Abdominal segmentation, pole cell formation, and embryonic polarity require the localized activity of oskar, a maternal gene in Drosophila. Cell, *47*: 141–152.

McClendon, J. F. 1910. The development of isolated blastomeres of the frog's egg. Am J. Anat., *10*: 425–430.

Mahowald, A. P. 1968. Polar granules of *Drosophila*. II. Ultrastructural changes during early embryogenesis. J. Exp. Zool., *167*: 237–262.

Mahowald, A. P. 1971a. Polar granules of *Drosophila*. III. The continuity of the polar granules during the life cycle of *Drosophila*. J. Exp. Zool., *176*: 329–343.

Mahowald, A. P. 1971b. Polar granules of *Drosophila*. IV. Cytochemical studies showing loss of RNA from polar granules during early stages of embryogenesis. J. Exp. Zool., *176*: 345–352.

Mintz, B. 1965. Experimental genetic mosaicism in the mouse. *In* G. E.

Nishida, H., and N. Satoh, 1983. Cell lineage analysis in ascidian embryos by intracellular injection of a tracer enzyme. I. Up to the eight-cell stage. Dev. Biol., *99*: 382–394.

Nishida, H., and N. Satoh, 1985. Cell lineage analysis in ascidian embryos by intracellular injection of a tracer enzyme. II. The 16- and 32-cell stages. Dev. Biol., *110*: 440–454.

Okada, M., I. A. Kleinman, and H. A. Schneiderman. 1974. Restoration of fertility in sterilized *Drosophila* eggs by transplantation of polar cytoplasm. Dev. Biol., *37*: 43–54.

Oppenheimer, S. B. 1980. *Introduction to Embryonic Development*. Allyn and Bacon, Boston.

Ortolani, G. 1955. The presumptive territory of the mesoderm in the ascidian germ. Experientia, *11*: 445–446.

Poulson, D. F. 1950. Histogenesis, organogenesis and differentiation in the embryo of *Drosophila melanogaster*. *In* M. Demerec (ed.), *The Biology of Drosophila*. John Wiley & Sons, New York, pp. 168–174.

Rattenbury, J. C., and W. E. Berg. 1954. Embryonic segregation during early development of *Mytilis edulis*. J. Morphol., *95*: 393–414.

Ressom, R. E., and K. E. Dixon. 1988. Relocation and reorganization of germ plasm in *Xenopus* embryos after fertilization. Development, *103*: 507–518.

Reverberi, G. 1971. Ascidians. *In* G. Reverberi (ed.), *Experimental Embryology of Marine and Freshwater Invertebrates*. North Holland-Elsevier, Amsterdam, pp. 507–550.

Reverberi, G., and A. Minganti. 1946. Fenomeni di evocazione nello sviluppo dell' uovo di Ascidie. Resulati dell' indagine spermentale sull' uovo di *Ascidiela aspersa* e di *Ascidia malaca* allo stadio di 8 blastomeri. Pubb. Staz. Zool. Napoli, *20*: 199–252.

Reverberi, G., and G. Ortolani. 1963. On the origin of the ciliated plates and the mesoderm in ctenophores. Acta Embryol. Morphol. Exp., *6*: 175–190.

Roux, W. 1888. Contribution to the developmental mechanics of the embryo. On the artificial production of half-embryos by destruction of one of the first two blastomeres and the later development (postgeneration) of the missing half of the body. *Reprinted in* B. H. Willier and J. M. Oppenheimer (eds.), *Foundations of Experimental Embryology*, 2nd ed., 1974. Hafner, New York, pp 2–37.

Sawada, T., and G. Schatten. 1989. The effects of cytoskeletal inhibitors on ooplasmic segregation and microtubule organization during fertilization and early development in the ascidian *Molgula occidentalis*. Dev. Biol., *132*: 331–342.

Schupbach, T., and E. Wieschaus. 1986. Maternal-effect mutations altering the anterior-posterior pattern of the *Drosophila* embryo. Wilhelm Roux's Archives Dev. Biol., *195*: 302–317.

Slack, J. M. W. 1983. *From Egg to Embryo: Determinative Events in Early Development*. Cambridge University Press, London.

Smith, L. D. 1966. The role of a "germinal plasm" in the formation of primordial germ cells in *Rana pipiens*. Dev. Biol., *14*: 330–347.

Spemann, H. 1938. *Embryonic Development and Induction*. Yale University Press, New Haven, CN.

St. Johnston, D. et al. 1989. Multiple steps in the localization of *bicoid* RNA to the anterior pole of the *Drosophila* oocytes. Development, *107* (Suppl.): 13–19.

Stephenson, E. C., Y. Chao, and J. D. Fackenthal. 1988. Molecular analysis of the *swallow* gene of *Drosophila melanogaster*. Genes Dev., *2*: 1655–1665.

Stern, M. S. 1972. Experimental studies on the organization of the preimplantation mouse embryo. II. Reaggregation of disaggregated embryos. J. Embryol. Exp. Morphol., *18*: 155–180.

Strome, S. 1989. Generation of cell diversity during early embryogenesis in the nematode *Caenorhabditis elegans*. Int. Rev. Cytol., *114*: 81–123.

Strome, S., and W. B. Wood. 1982. Immunofluorescence visualization of germ-line specific cytoplasmic granules in embryos, larvae, and adults of *Caenorhabditis elegans*. Proc. Natl. Acad. Sci. U.S.A., *79*: 1558–1562.

Sulston, J. E. et al. 1983. The embryonic cell lineage of the nematode *Caenorhabditis elegans*. Dev. Biol., *78*: 577–597.

Tarkowski, A. K., and J. Wróblewska. 1967. Development of blastomeres of mouse eggs isolated at the 4- and 8-cell stage. J. Embryol. Exp. Morphol., *18*: 155–180.

Tobashi, S., S. Kobayashi, and M. Okada. 1986. Function of maternal mRNA as a cytoplasmic factor responsible for pole cell formation in *Drosophila* embryos. Dev. Biol., *118*: 352–360.

Vogt, W. 1929. Gestaltungsanalyse am Amphibienkeim mit ortlicher Vitalfarbung. II. Gastrulation und Mesodermbildung bei Urodelen und Anuran. Wilhelm Roux's Archiv für Entwicklungsmechanik, *120*: 384–706.

Waddington, C. H. 1966. *Principles of Development and Differentiation*. Macmillan, New York.

Weisblat, D. A., R. T. Sawyer, and G. Stent. 1978. Cell lineage analysis by intracellular injection of a tracer enzyme. Science, *202*: 1295–1298.

Whitman, C. O. 1878. The embryology of *Clepsine*. Q. J. Microsc. Sci., *18*: 215–315.

Whittaker, J. R. 1973. Segregation during ascidian embryogenesis of egg cytoplasmic information for tissue specific enzyme development. Proc. Natl. Acad. Sci. U.S.A., *70*: 2096–2100.

Whittaker, J. R. 1980. Acetylcholinesterase development in extra cells caused by changing the distribution of myoplasm in ascidian embryos. J. Embryol. Exp. Morphol., *55*: 343–354.

Whittaker, J. R., G. Ortolani, and N. Farinella-Ferruzza. 1977. Autonomy of acetylcholinesterase differentiation in muscle-lineage cells of ascidian embryos. Dev. Biol., *55*: 196–200.

Williams, M. A., and L. D. Smith. 1971. Ultrastructure of the "germinal plasm" during maturation and early cleavage in *Rana pipiens*. Dev. Biol., *25*: 568–580.

Wilson, E. B. 1904a. Experimental studies on germinal localization. II. Experiments on the cleavage-mosaic in *Patella* and *Dentalium*. J. Exp. Zool., *1*: 197–268.

Wilson, E. B. 1904b. Experimental studies on germinal localization. I. The germ regions in the egg of *Dentalium*. J. Exp. Zool., *1*: 1–72.

Wilson, E. B. 1925. *The Cell in Development and Heredity*, 3rd ed. Macmillan, New York, pp 1072–1073.

Wilt, F. 1987. Determination and morphogenesis in the sea urchin embryo. Development, *100*; 559–575.

Zalokar, M. 1976. Autoradiographic study of protein and RNA formation during early development of *Drosophila* eggs. Dev. Biol., *49*: 425–437.

Züst, B., and K. E. Dixon. 1977. Events in the germ cell lineage after entry of the primordial germ cells into the genital ridges in normal and UV-irradiated *Xenopus laevis*. J. Embryol. Exp. Morphol., *41*: 33–46.

CHAPTER 12

The Role of Induction in Determination of Cell Fate

Hans Spemann (1869–1941). A Professor at the University of Rostock, the Kaiser Wilhelm Institute for Biology, and the University of Freiburg in Germany, Spemann was responsible for a number of important discoveries involving cell determination and induction in amphibian embryos. In 1935, he received the Nobel Prize for his discovery of the primary organizer. (See Color plate 20 for a reproduction of the Nobel Prize certificate.) Viktor Hamburger (see p. 242), himself a superb experimental embryologist, was a student in Spemann's laboratory and recently wrote a spellbinding account of experimental embryology, including the experiments done in Spemann's laboratory (Hamburger, 1988). For students interested in the history of science, this is an extraordinary book. (Photograph courtesy of the Marine Biological Laboratory Library, Woods Hole, MA.)

As we have learned in Chapter 11, specialized regions of the egg cytoplasm may determine the fates of cells: In this scenario, the pattern of cell division establishes the distribution of ooplasmic determinants and, hence, the locations of presumptive tissues and organs. Most differentiated tissues, however, arise by a more complex route that requires interactions with neighboring cells of the embryo. The early embryo is a complex reaction system in which the possible fates of localized regions become progressively narrowed or restricted as a result of these interactions. This ability to alter the eventual fate of a population of cells as a result of interactions with neighboring cells is known as induction. In this chapter, we explore the nature of inductive interactions.

Hans Spemann, the noted German embryologist, and his colleagues were the first to study embryonic induction. Through a series of grafting experiments using amphibian embryos, which we shall discuss later, they showed that the fate of the neural ectoderm is fixed during gastrulation by an interaction with the underlying chordamesoderm (i.e., the presumptive notochord and somitic mesoderm), which is derived from the cells that pass over the dorsal lip of the blastopore. More recently, however, an earlier inductive event during amphibian development has been discovered. This is the induction of mesoderm during the blastula stage by cells of the vegetal hemisphere. This discovery has also led to a breakthrough in the understanding of the molecular nature of cellular interactions during development. Therefore, we discuss mesoderm induction before returning to a discussion of Spemann's experiments.

12–1 Mesoderm Induction in Amphibians

At the 32-cell stage, the amphibian embryo consists of only two cell types: small ectodermal cells in the animal hemisphere and larger endodermal cells in the vegetal hemisphere. A third cell type, from which the mesoderm is derived, first becomes distinct in the early blastula stage. These are the equatorial cells, so named because of their position in the blastula. Fate mapping of the amphibian embryo has shown that equatorial cells (excluding the surface layer in *Xenopus*; see Chap. 6), which contact both animal and vegetal cells in the equator (or marginal zone) of the blastula, are precursors of the mesoderm (Fig. 12.1).

How are the equatorial cells determined to become mesoderm? One possibility is that they contain intrinsic factors (ooplasmic determinants; see Chap. 11) that specify mesoderm determination. Another possibility is

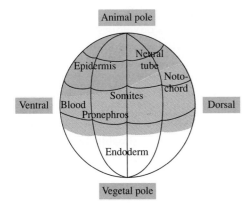

Figure 12.1

A fate map of the *Xenopus* embryo at the 32-cell stage. (After Smith, 1989.)

.that mesoderm is induced by extrinsic factors derived from neighboring cells. These possibilities were tested in the experiment shown in Figure 12.2. In this experiment, blastulae were divided into explants of animal, equatorial, and vegetal cells (Nieuwkoop, 1969). Subsequently, each type of explant was cultured either alone or in combination with one of the other types of explants, and the identities of the tissues that developed were

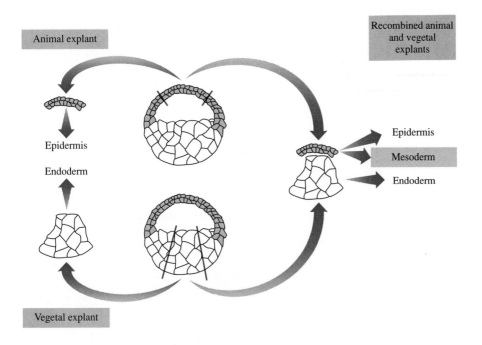

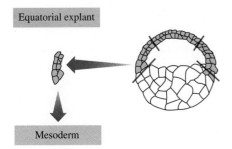

Figure 12.2

Schematic diagram showing the design of the explant recombination experiments conducted by Nieuwkoop on amphibian blastulae. (After Smith, 1987.)

assessed by their morphology. After several days of culture, animal explants developed into epidermis, marginal zone explants developed primarily into mesodermal tissues, and vegetal explants developed into endodermal cells. However, animal explants will also form mesoderm if they are combined with vegetal explants, ruling out the possibility that mesoderm formation is caused exclusively by ooplasmic determinants. These results suggest that equatorial cells develop into mesoderm because an inducer is secreted by the adjacent vegetal cells. In the normal embryo, the animal hemisphere cells become ectoderm, rather than mesoderm, presumably because they are too far from the source of the inducer and are separated from the vegetal cells by the blastocoel.

Recently, Nieuwkoop's explant experiment has been repeated using a molecular marker for amphibian mesoderm development (Gurdon et al., 1985). Because muscle cells differentiate from mesoderm, gene products made specifically in muscle (such as the mRNA encoding muscle-specific α-cardiac actin) can be used as markers for mesoderm differentiation. α-Cardiac actin is expressed at high levels in developing muscle and is only weakly expressed in adult skeletal muscle. It is also the major actin in adult hearts. Nuclease protection assays have been used to examine the expression of cardiac actin mRNA in the blastula. An assay conducted with RNA obtained from animal, equatorial, vegetal, and combined animal and vegetal explants is shown in Figure 12.3. It can be seen that muscle-specific actin mRNA is present in the equatorial cells and in combined animal and vegetal explants, but not in the animal and vegetal explants alone, thus confirming the earlier morphological studies.

Regulation of cardiac actin gene transcription during *Xenopus* development has also been studied by injection of fusion genes containing the cardiac actin gene promoter into fertilized eggs. The injected DNA becomes distributed throughout the embryo and is regulated in the same way as the

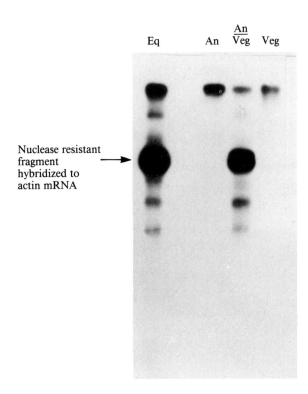

Figure 12.3

A nuclease protection assay for the presence of muscle-specific (cardiac) actin mRNA in RNA isolated from marginal zone (equatorial) explants (Eq), animal explants (An), vegetal explants (Veg), and explants in which animal and vegetal cells are recombined (An/Veg). The muscle-specific actin mRNA (arrow) is detected only in equatorial explants and explants in which animal and vegetal cells are recombined. (From Gurdon et al., 1985. Copyright © Cell Press.)

endogenous cardiac actin genes; i.e., it is first transcribed during gastrulation and then only in the somitic mesoderm (Mohun et al., 1986; Wilson et al., 1986). Mohun et al. (1989) have demonstrated that the injected fusion genes are subject to the same regulatory signals as are the endogenous cardiac actin genes: The injected fusion genes are not expressed in cultured animal or vegetal fragments from blastulae. However, they are expressed in cultured equatorial zone fragments and animal-vegetal recombinants. These results are shown in Figure 12.4. Transcription is monitored in these experiments by the nuclease protection technique. Figure 12.4A shows the gene constructs used in these experiments. A 603-nucleotide fragment containing the *Xenopus* cardiac actin gene promoter was fused to the human β-globin gene. The transcription of this gene was assayed by probing its transcripts with a plasmid containing 303 nucleotides comprising the 5′ portion of the human β-globin cDNA fused to the *Xenopus* cardiac actin gene promoter and

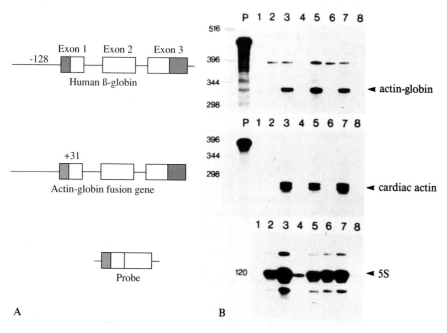

A   B

Figure 12.4

Muscle-specific expression of a cardiac actin-globin fusion gene in *Xenopus* embryo fragments. A, Diagram of the structures of the human β-globin gene, the construct derived from fusing it with the *Xenopus* cardiac actin gene promoter (see p. 582), and the cDNA probe that is used to detect the presence of fusion gene transcripts. The cDNA protects an RNA fragment that is 341 nucleotides long. Boxes represent exons (see p. 583), gray shading represents non-coding portions of the human β-globin gene, and color represents 5′ untranslated sequences from the *Xenopus* cardiac actin gene. B, Nuclease protection assay. Each sample was analyzed for fusion gene transcription as well as transcription of the endogenous cardiac actin and 5S RNA genes. Numbers on the left indicate fragment sizes in number of nucleotides. Lane P: undigested probe; lane 1: 5 μg tRNA; lanes 2-4: animal, equatorial, and vegetal fragments, respectively, from embryos injected in the animal portion of each blastomere at the two-cell stage; lane 5: recombined animal and vegetal fragments; lanes 6-8: animal, equatorial, and vegetal fragments, respectively, from embryos injected in the vegetal portion of each blastomere at the two-cell stage. These lanes give the same results as lanes 2-4, indicating that the site of injection does not affect the results. (From Mohun et al., 1989.)

transcription start site. Digestion of hybrids between this probe and fusion gene transcripts would produce a 341 nucleotide fragment. Endogenous cardiac actin transcripts and 5S RNA (the latter served as a control) were also detected with specific probes. Figure 12.4B shows the results of the nuclease protection assays. The fusion gene transcripts are detected only in equatorial fragments or in conjugates formed by recombining animal and vegetal fragments. The endogenous cardiac actin fragments show the same distribution. Clearly, cardiac actin gene expression is restricted to mesodermal cells, and the expression of this gene is a consequence of the inductive interaction.

The mesoderm is composed of several different types of tissue, including notochord, muscle, and blood. Another series of recombination experiments has shown that the induction of different kinds of mesoderm is dependent on the origin of the vegetal inducing cells (Boterenbrood and Nieuwkoop, 1973; Dale and Slack, 1987). For instance, vegetal cells from the dorsal side of the blastula tend to induce notochord and muscle, whereas vegetal cells from the lateral and ventral side of the blastula tend to induce blood cells. Curiously, ventral vegetal blastomeres induce little or no muscle from animal pole cells, yet a significant proportion of muscle is formed from marginal zone blastomeres on the ventral side of the embryo (see Fig. 12.1). This puzzling feature of amphibian mesoderm induction can be explained by the **three signal model** (Smith and Slack, 1983), which postulates that there are three distinct inductive signals involved in mesoderm formation (Fig. 12.5). The first signal, on the dorsal side of the embryo (where the gray crescent is located), induces notochord and a small amount of muscle. The second signal, on the ventral side of the embryo, induces primarily blood. The first two signals originate from the vegetal cells, but a third signal is derived from the newly induced dorsal mesoderm. This third signal acts within the presumptive mesoderm to cause ventral mesoderm adjacent to dorsal mesoderm to form muscle instead of blood. Evidence supporting the three signal model comes from experiments in which dorsal and ventral presumptive mesodermal regions of the early gastrula are juxtaposed (Dale and Slack, 1987). In isolation, dorsal mesoderm forms mainly notochord, whereas ventral mesoderm forms blood but no muscle. In combinations, however, the ventral mesoderm also forms large amounts of muscle.

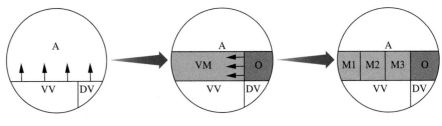

Figure 12.5

The three signal model. Two mesoderm-inducing signals are postulated to be derived from the vegetal blastomeres (vertical arrows). The dorsal-vegetal signal (DV) induces notochord and other dorsal mesoderm (i.e., the Spemann organizer, O), whereas the ventral vegetal signal (VV) induces blood and other ventral mesodermal derivatives. The ventral mesoderm (VM) then receives a signal from the organizer (probably during gastrulation) that results in the induction of muscle (M3) and possibly pronephros (M2). Only the most ventral mesoderm (M1) differentiates into blood. A: animal hemisphere. (After Smith, 1989.)

How do vegetal cells induce equatorial cells to become mesoderm? A possibility already mentioned is that inducers may be secreted by the vegetal cells, diffusing to the equatorial cells and causing the latter to become mesoderm. Another possibility is that induction involves the formation of intimate contacts between the inducing and responding cells. These possibilities, although not mutually exclusive, can be tested by experiments in which embryos are cultured either intact or disaggregated in medium lacking Ca^{2+} and Mg^{2+} (Gurdon et al., 1984; Sargent et al., 1986). In the absence of divalent cations, blastomeres lose their adhesive qualities, and when the vitelline envelope is removed, the embryo collapses into a heap of disorganized cells. The cells in this heap remain contiguous but do not adhere to each other. Despite their disorganized state, nuclease protection assays show that muscle-specific actin mRNA accumulates within the heap of cells. This result indicates that cell adhesion and maintenance of normal spatial arrangement of blastomeres are not required for mesodermal induction. Different results, however, are obtained if the heap of embryonic cells is dispersed so that none of the cells are in physical proximity. Under these conditions, no muscle actin mRNA is transcribed, indicating that mesoderm is not induced. These results indicate that *proximity*, but not *adhesive cell contact*, is necessary for mesoderm induction in *Xenopus*. Therefore, inducers may be substances secreted by vegetal cells that are unstable and only show regional inductive activity in the embryo. This would explain why vegetal cells induce the equatorial cells, but not the overlying animal cells, to become mesoderm in the intact embryo.

Considerable progress has recently been made in identifying the mesoderm-inducing substances in amphibian embryos. For many years, it was known that mesoderm induction could be mimicked by various macromolecules isolated from diverse types of late embryonic or adult cells and tissues (Saxén and Toivonen, 1958; Tiedemann and Tiedemann, 1959), including chick embryos, carp swim bladder, guinea pig bone marrow, and cultured mammalian cells. More recently, inductive activity was also discovered in XTC-MIF (*Xenopus* tissue culture cell mesoderm-inducing factor), a substance secreted into the culture medium by XTC cells, a *Xenopus laevis* cell line (Fig. 12.6). Although all of these substances are of heterogeneous origin, they have one feature in common: They each contain growth factors. As we have previously discussed, growth factors have properties that conform to the attributes expected of inducer substances; e.g., they are released by one cell type and cause changes in other cell types. The realization that heterogeneous inducers contain growth factors has led to direct tests of the inducing activity of these substances. The test for mesoderm-inducing activity involves the addition of purified growth factors to animal pole explants and assaying for mesoderm induction. Using these procedures, two classes of mesoderm-inducing growth factors have been identified (Kimelman and Kirschner, 1987; Slack et al., 1987; Smith, 1987; Rosa et al., 1988). The first class of mesoderm-inducing factors is members of the transforming growth factor-β (TGF-β) family. The TGF-β class induces a variety of mesodermal cell types, including notochord, muscle, and pronephros, when added to animal pole explants (Smith et al., 1988). The second class of mesoderm-inducing factors is related to the acidic and basic fibroblast growth factors (aFGF and bFGF); aFGF and bFGF are capable of inducing all types of mesoderm except notochord, the most dorsal type. This has led to the suggestion that the dorsal signal in the three signal model (see Fig. 12.5) is caused by a growth factor in the TGF-β family, whereas the ventral signal is caused by a growth factor related to FGF.

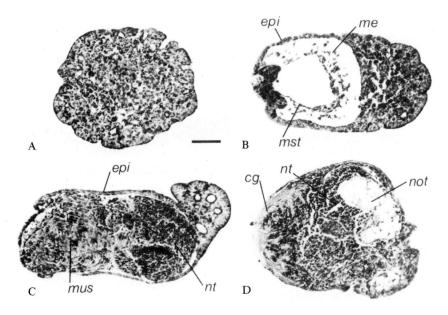

Figure 12.6

Sections of animal explants of blastulae cultured in the presence of increasing concentrations of the purified mesoderm-inducing factor secreted by a *Xenopus* cell line in culture (XTC-MIF). A, No inducing factor added. The explant develops into epidermal tissue. B, 0.4 ng/ml of inducing factor added. The explant develops an internal region of mesothelium (mst) and mesenchyme (me) within the epidermis (epi). C, 3 ng/ml of inducing factor added. The explant develops muscle (mus) and neural (nt) tissue. These explants sometimes also develop notochord (not shown). D, 25 ng/ml of inducing factor added. The explant develops notochord (not), neural tissue (nt), and cement gland (cg). Scale bar equals 100 μm. (From Smith et al., 1988. Reprinted by permission of Company of Biologists, Ltd.)

Recently, progress has been made in identifying the growth factors in the TGF-β family that have mesoderm-inducing activity in *Xenopus* embryos. These factors appear to be **activins**, which are growth factors that also have a number of regulatory roles in adult mammals. One of the activins is activin A, which can mimic the effects of the dorsal signal. Treatment of animal pole explants with purified activin A leads to differentiation of mesodermal derivatives such as mesenchyme, notochord, blood cells, and muscle (Asashima et al., 1990). The inducing activity in XTC-MIF has been purified and found to be a homologue of activin A (Smith et al., 1990; van den Eijnden-Van Raaij et al., 1990). The *Xenopus* activin A gene has been cloned, and the cloned gene was used to construct a probe to monitor the level of activin A mRNA in the embryo. These transcripts are not detected until the gastrula stage, which implies that activin A is not the endogenous mesoderm inducer. However, transcripts for another activin, activin B, are detected at the blastula stage, when mesoderm induction occurs. Thus, activin B, which has been shown to be a potent mesoderm inducer, may be the actual mesoderm inducer in *Xenopus* (Thomsen et al., 1990).

If growth factors are the actual inducers, they should be found in the *Xenopus* embryo itself. Their isolation has been a difficult task, however, probably because they are present at very low levels in the embryo. As we discussed in Chapter 3, transcripts of a gene encoding one of the TGF-β-related growth factors (*Vg1* mRNA) are present in the vegetal hemisphere of *Xenopus* oocytes (Weeks and Melton, 1987). During cleavage, *Vg1* mRNA is distributed to the vegetal hemisphere cells and is in the correct spatial position to be translated into a growth factor responsible for mesoderm induction. Current investigations indicate that both the mRNA and protein corresponding to a bFGF are present in the embryo (Kimelman et al., 1988; Slack and Isaacs, 1989). Future work in this area will be very interesting, for it may not only reveal the molecular basis of the signaling systems between vegetal and equatorial cells that result in the induction of mesoderm, but also provide a molecular model for induction in general.

12–2 Determination of the Central Nervous System

As mentioned previously, experiments by Spemann and his colleagues (particularly Hilde Mangold) led to the discovery that the dorsal mesoderm induces ectoderm to differentiate into neural tissue. Spemann (1918) exchanged bits of presumptive epidermis and neural ectoderm tissue between two species of newt whose cells differed in their degree of pigmentation (Fig. 12.7A). In this way, he could follow the grafted tissue with a pigment marker as it developed in the unpigmented host. When Spemann grafted a piece of prospective neural ectoderm from an early gastrula-stage embryo into a region where belly skin should develop, he found that the graft developed according to its surroundings; i.e., it also became belly ectoderm. Conversely, if he grafted prospective epidermis into the region fated to be neural plate, the graft differentiated into neural tissue. Thus, if the exchange

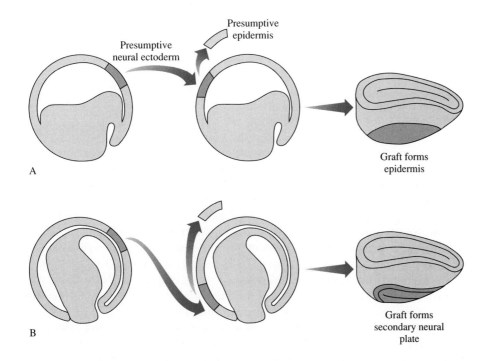

Figure 12.7

A, When prospective neural ectoderm is grafted into a region of prospective epidermal ectoderm in an early amphibian gastrula, the graft develops according to its new site and becomes epidermal ectoderm. B, When the same experiment is repeated during late gastrulation, the prospective ectoderm forms neural tissue, showing that at this stage its fate is fixed. (After Saxén and Toivonen, 1962.)

is made during the early gastrula stage, the implants developed according to their new surroundings. This means that the fates of these two regions are not fixed at the early gastrula stage.

Exchange of presumptive epidermis and neural ectoderm at the end of gastrulation yielded an entirely different result, however: Neural ectoderm transplanted into the prospective belly region differentiated into neural tissue, although it failed to roll up into a neural tube (see Fig. 12.7B), whereas presumptive epidermis continued to develop as epidermis regardless of its new position. These results indicate that the fate of the neural ectoderm becomes fixed during the process of gastrulation.

What interactions are responsible for the determination of the neural ectoderm during gastrulation? Hilde Mangold's elegant experiments showed that the neural epithelium is induced to form by the underlying chordamesoderm (Spemann and Mangold, 1924). Spemann and Mangold first tested the ability of the cells that compose the dorsal lip of the blastopore to self-differentiate. They grafted dorsal lips from the unpigmented newt embryo *Triturus cristatus* into the ventral region of a second species, *Triturus taeniatus*, which is darkly pigmented (Fig. 12.8A). Normally, dorsal lip tissue undergoes involution during gastrulation and differentiates into dorsal mesoderm. Spemann and Mangold found that the ability to involute is retained even when the dorsal lip is transplanted to an inappropriate region. Furthermore, an entire secondary axis forms—complete with gut, neural tube, notochord and somites (see Fig. 12.8B to E). The formation of a secondary gut is not surprising because the process of involution of the transplanted dorsal lip would produce a secondary archenteron. A notochord and somites might also be expected because these structures are normally derived from the dorsal lip mesoderm. However, the neural tube is not a derivative of the transplanted tissue because the pigmentation marker revealed that the neural tube differentiated from the host ectoderm that had been destined to form belly skin. Spemann proposed that the dorsal lip material had *induced* the overlying ectoderm to become neural tissue. As the chordamesoderm involutes during gastrulation, it comes into contact with the overlying ectoderm, the fate of which is altered as a result of this contact (Fig. 12.9). The rest of the ectoderm that does not contact the chordamesoderm becomes skin.

The inductive event between the chordamesoderm and ectoderm has been termed **primary induction** because it was originally believed to be the first inductive event during embryogenesis. This is a misnomer now because we know that at least one other inductive event (mesoderm induction) has occurred previously. The term still persists in the literature, however.

Other important inductive interactions occur in the early gastrula as well. Smith and Slack (1983) have demonstrated that the dorsal mesoderm (i.e., the chordamesoderm) induces the differentiation and organization of the lateral mesoderm adjacent to it. In Spemann and Mangold's experiments, for example, the second set of somites were induced to form from what should have been lateral plate mesoderm. This has been termed the "dorsalizing" influence of the chordamesoderm (see Fig. 12.5). In addition, the chordamesoderm organizes the anteroposterior axis of the underlying gut. As a result of the ability of the dorsal lip tissue to organize the anteroposterior axis, Spemann designated this tissue as the **primary organizer**, and it was for this work that he was awarded the Nobel Prize in 1935 (see Spemann, 1938, for review).

Regional Determination by the Chordamesoderm

As we discussed in Chapter 7, the neural epithelium is not uniform along the length of the embryo. Rather, it is regionally differentiated into the

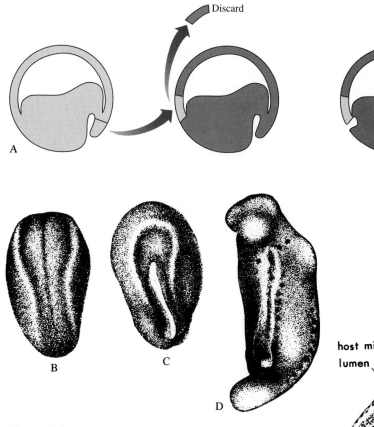

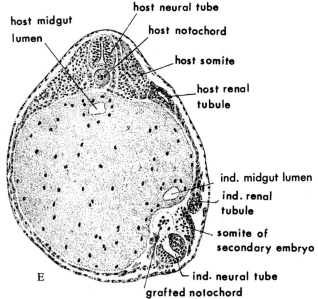

Figure 12.8

Induction of a secondary embryo by transplantation of a piece of the dorsal lip of the blastopore. A, Transplantation of the dorsal lip of an unpigmented *Triturus cristatus* gastrula to the ventral side of a pigmented *Triturus taeniatus* gastrula. B, The pigmented neural plate of the host embryo. C, Ventral view of the host, showing the induced neural plate. D, Same embryo in the tail-bud stage, showing the induced secondary axis on the lower left side of the host embryo. E, Cross-section of the host embryo at the tail-bud stage, showing induced dorsal structures (ind.). (After Spemann, 1936.)

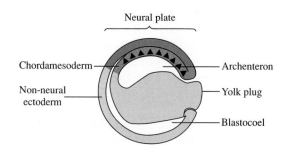

Figure 12.9

Late gastrula stage of the urodele showing the relationship between the chordamesoderm and future neural ectoderm. The inductive signal is indicated by arrowheads.

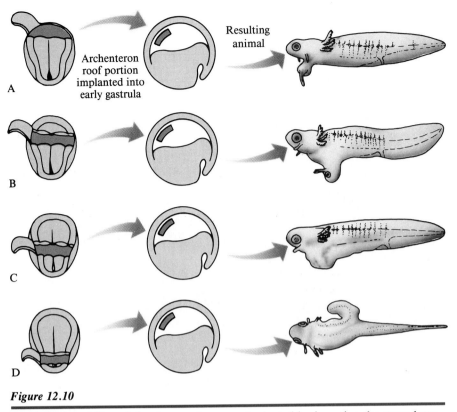

Figure 12.10

The inducing ability of chordamesoderm can be tested by inserting the mesoderm into the blastocoel cavity of an early gastrula embryo and allowing it to complete gastrulation. As the blastocoel cavity becomes diminished, the grafted tissue is pressed against the overlying ectoderm. Anterior pieces of chordamesoderm induce anterior brain structures (A and B), whereas more posterior pieces induce tail structures (C and D). (After Mangold, 1933.)

forebrain (also known as **archencephalon**), hindbrain (or **deuterocephalon**), and spinal–caudal regions. The underlying chordamesoderm is responsible for this regional differentiation. Otto Mangold (1933), who was Hilde Mangold's husband, removed four successive regions of chordamesoderm from *Triturus* embryos that had just completed gastrulation and implanted them separately into the blastocoel cavities of early *Triturus* gastrulae (Fig. 12.10). As gastrulation proceeds and the archenteron cavity expands, the implanted tissue is gradually compressed against the wall of the gastrula and therefore can interact with the overlying prospective ventral ectoderm. The most anterior piece of chordamesoderm induced the formation of mouth parts and suckers. The next region produced anterior head structures, such as nose, eyes, and otic vesicles; the third induced otic structures only; and the most posterior region gave rise to trunk and tail structures, including neural tube and mesodermal derivatives.

A similar regional differentiation results from transplants of dorsal lips that are excised at progressively later times in their involution. We know that the first chordamesoderm to involute contains the anterior chordamesoderm and the last chordamesoderm to involute contains the posterior chordamesoderm. Transplantation of the former to a host gastrula tends to induce a secondary head, whereas the latter will induce a secondary trunk and tail (Fig. 12.11).

This regionalization was also confirmed by a series of experiments done in tissue culture. Holtfreter (1936) removed dorsal lips from embryos

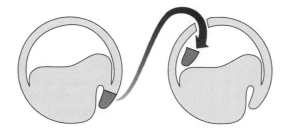

A

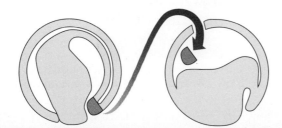

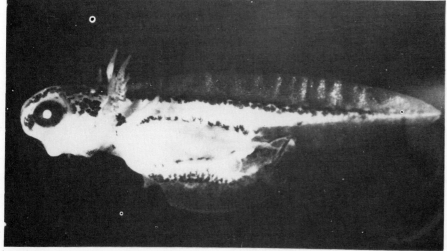

2 μm

B

Figure 12.11

Dorsal lip material taken from an early gastrula and inserted into the blastocoel cavity will induce head structures (A), whereas dorsal lip from an older gastrula stage will induce spinal-caudal structures only (B). (From Saxén and Toivonen, 1962. Reprinted by permission of Prentice Hall, Inc., Englewood Cliffs, New Jersey.)

at progressively later stages of gastrulation and sandwiched them between pieces of prospective ectoderm from early gastrula embryos (Fig. 12.12). (These explants have been subsequently dubbed "Holtfreter sandwiches"!) The early dorsal lips induced the ectoderm to form archencephalic structures, whereas the oldest dorsal lips induced spinal–caudal structures.

What is responsible for the regional neural induction properties of the chordamesoderm? As we have learned, growth factors are clearly important in mesoderm induction, and recent evidence suggests that they are also involved in conferring a graded anteroposterior character on the inductive signals emanating from the mesoderm. Ruiz i Altaba and Melton (1989, 1990) treated isolated animal caps with either XTC-MIF (the TGF-β-related growth factor produced by *Xenopus* XTC cells; see p. 483) or bFGF and implanted them into the blastocoels of gastrulating hosts. As shown in Figure 12.13, implants of animal caps treated with XTC-MIF mimic the effects of implanted anterior mesoderm, inducing head structures, whereas animal caps treated with bFGF induce tail structures. These results are almost exactly those achieved by Otto Mangold in 1933 when he grafted bits of either anterior or posterior chordamesoderm into the blastocoel cavity (see Fig. 12.10). Thus, even though we do not know how the

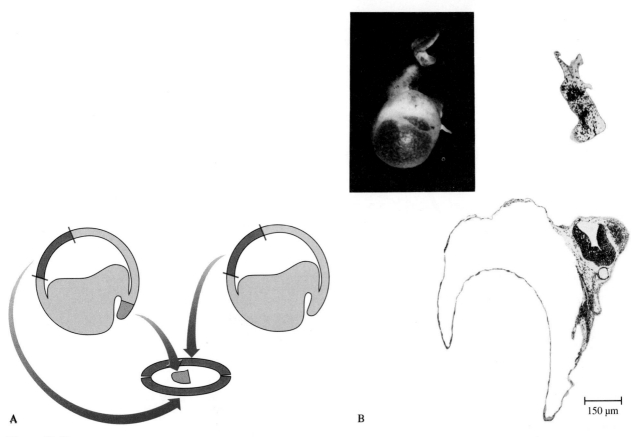

A　　　　　　　　　　　　　　　　**B**

Figure 12.12

Holtfreter showed that pieces of dorsal lip sandwiched between two pieces of presumptive ectoderm (A) are capable of inducing the ectoderm to form neural structures. B, A piece of dorsal lip from an early gastrula has induced the ectoderm in a Holtfreter sandwich to form head, trunk, and tail tissue, as seen in this section. Scale bar equals 150 μm. The inset shows the intact, unsectioned tissue. (From Saxén and Toivonen, 1962. Reprinted by permission of Prentice Hall, Inc., Englewood Cliffs, New Jersey.)

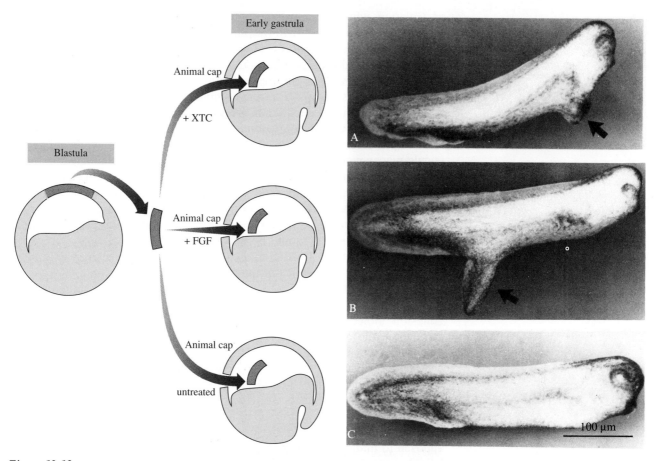

Figure 12.13

Different growth factors induce mesoderm with different anteroposterior character as shown by the ability of animal caps treated with different growth factors to induce different neural/epidermal structures. The positional values of animal caps treated with either XTC-MIF or bFGF were determined by implanting them into the blastocoels of gastrulating embryos. The implants are pushed ventrally during gastrulation and induce overlying ectoderm. XTC-MIF-treated animal caps induce head structures (A), whereas bFGF-treated animal caps induce tail structures (B) and untreated animal caps have no effect (C). Arrows indicate induced structures in A and B. (From Ruiz i Altaba and Melton, 1989. Reprinted by permission from *Nature*, Vol. 341, pp. 33–38. Copyright © 1989 Macmillan Magazines Ltd.)

chordamesoderm induces regionalization, it seems probable that growth factors establish the anteroposterior properties of the chordamesoderm itself.

It is likely, based on recent data from a number of laboratories, that the induction of neural tube by the underlying chordamesoderm during gastrulation is probably an overly simplified view of the process. It is currently hypothesized that neural induction is at least a two-step process, with one of these inductive events occurring before the onset of gastrulation. One study that supports this view will be discussed in detail. Dixon and Kintner (1989) examined the effectiveness of two kinds of interactions between mesoderm and ectoderm on neural differentiation. One interaction would occur just before gastrulation, when the mesoderm induces the contiguous ectoderm to a neural fate, whereas the other would occur after gastrulation, when the chordamesoderm underlies the ectoderm (Fig. 12.14). In the first instance, the signal would diffuse in the plane of the epithelium. In the second instance, the signal would pass from the involuting mesoderm to the ectoderm during gastrulation.

Dixon and Kintner cultured explants of ectoderm and mesoderm from *Xenopus* embryos in three different combinations (Fig. 12.15) and measured the amount of neural-specific mRNA produced in the ectoderm. In the first combination, blocks of contiguous ectoderm and mesoderm of the dorsal lip were dissected from two blastula-stage embryos and sandwiched together (single pieces would roll up into a ball and obscure morphological details). In the second combination, dorsal mesoderm from a midgastrula-stage embryo was removed and sandwiched between two sheets of ectoderm removed from a late blastula (a Holtfreter sandwich). Thus, in the first case, any interaction would be within the plane of the epithelium, whereas in the second, the interaction would be between two cell layers across an extracellular space. In the third combination, the two types of explants were combined so that both types of interactions occurred.

The accumulation of two neural-specific mRNAs was monitored by nuclease protection assays. One of these transcripts encodes the neural cell adhesion molecule N-CAM (see p. 373), which is normally expressed throughout the neural ectoderm soon after induction. The second encodes a neurofilament-like protein called NF-3, which is normally expressed after neural tube closure. Together, these two transcripts provide a measure of the amount and extent of neural tissue differentiation.

When epithelial sheets were removed from blastulae before involution, both messengers were expressed at a somewhat lower level than in normal development (50% to 70% of the levels of intact embryos), whereas ectoderm apposed to midgastrula-stage mesoderm (as it normally does during gastrulation) expresses very few of these transcripts (1% to 10% of the levels in normal embryos). When *both* kinds of interaction were allowed to occur, however, messenger levels approximated those in intact embryos. Thus, generation of the full complement of neural tissue appears to depend upon signals transmitted both within the epithelium and between mesoderm and ectoderm coming into apposition during gastrulation. This supports a model proposed by Nieuwkoop (1973), which predicts two steps in the neural

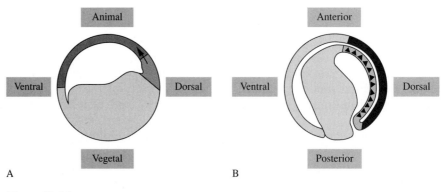

A   B

Figure 12.14

Induction of the ectoderm by the mesoderm to form neural tissue could potentially occur at two times: (A) before gastrulation, when the mesoderm contiguous with the ectoderm would induce in the plane of the epithelium (arrow); or (B) during gastrulation, as the involuting mesoderm contacts the ectoderm above (arrowheads). Both kinds of interaction appear to occur.

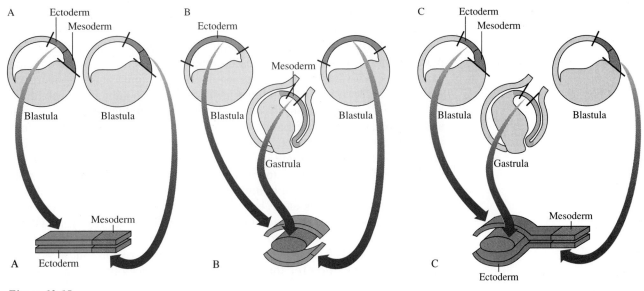

Figure 12.15

Ectoderm and mesoderm were explanted in three combinations, and expression of neural-specific transcripts (N-CAM and NF-3) was measured by nuclease protection assay. As discussed in the text, neural induction was more efficient if the two tissues interacted in the plane of the epithelium than if they were simply juxtaposed in a Holtfreter sandwich. However, when both types of contacts occur, the induction is synergistic. (After Dixon and Kintner, 1989.)

induction process. The first step occurs in the early gastrula between the mesoderm that will form the dorsal lip and the contiguous ectoderm. This step primes the ectoderm to be neural tissue. The second step occurs after gastrulation between the involuted chordamesoderm and primed neural ectoderm that overlies it. This step regionalizes the neural tube into anterior versus posterior structures, as we have previously discussed.

What Is the Nature of the Neural Inducer?

The discovery of induction by Spemann triggered a flurry of research activity aimed at analyzing the "inductive factor" (for an excellent review, see Witkowski, 1985). However, neural induction proved to be a very difficult problem, and it remains today one of the major unsolved problems of developmental biology. One of the chief reasons that neural induction has been intractable is that it is very difficult to isolate a prospective inducing molecule that is undoubtedly present in minute amounts from chordamesoderm, which is difficult to collect in quantity. Secondly, the assumption that there is only one inducer and inductive event is probably misleading. Even so, considerable research activity has yielded a number of important results. One of the first was the discovery that the dorsal lip can still function as an inducer after it has been killed (Bautzmann et al., 1932). The simplest interpretation of this experiment is that a stable inducing substance diffuses from the dead tissue. The hypothesis that inducing tissue produces a stable, diffusible inducer is supported by a number of observations. For example, nonliving cell-free extracts of dorsal lip material are induction-competent (Waddington et al., 1933). Also, preventing cell-cell contact between living dorsal lip tissue and ectoderm by inserting a permeable membrane with pores of 0.1 mm between them allows passage, presumably by diffusion, of the inductive influence (Toivonen et al., 1976).

To overcome the problem of limited amounts of tissue, many investigators tried to extract inducing substances from adult tissues, from which they hoped to obtain a sufficient quantity for analysis. The surprising and puzzling outcome was that one could obtain diverse extracts with inducing capabilities. Extracts from sources as diverse as guinea pig bone marrow, liver cells, and cultured human tumor cell lines all have inducing capabilities. Even *Hydra* body sections have weak powers of induction (Waddington and Wolsky, 1936). As disparate a group of substances as methylene blue, toluene, steroids, and slightly acid or basic salt solutions could also elicit an inductive response. Some investigators believed that all these inducing substances function merely to release the true inducing molecule that is already in the ectoderm in a bound or inactive form. Others suggested that these were not inducers at all but were, in fact, eliciting a toxic response in the cells that caused them to develop into neural tissue. Holtfreter (1947) suggested that "sublethal cytolysis" caused ectoderm cells to realize their neurogenic potential. This is hardly a satisfying explanation, but at present there is no explanation for the multitude of substances that have inducing ability.

One of the useful outcomes of the experiments just described was that the artificial inducers were generally found to have one of two possible inducing capacities: Factors such as guinea pig liver can induce anterior brain structures (these were coined "neuralizing" factors), whereas others, such as guinea pig bone marrow, can induce mesoderm and occasionally posterior neural tube structures (coined "mesodermalizing" factors). Toivonen and Saxén (1955b) demonstrated that if they implanted *both* types of inducers into the ventral region of a newt embryo at the same time, such as guinea pig bone marrow and liver, then the full range of neural structures would be induced, including forebrain, midbrain, hindbrain, and spinal cord (Fig. 12.16). The actions of these inducers mimic those of the two signals that have been proposed by Nieuwkoop and others (Dixon and Kintner, 1989) to be responsible for complete neural induction. Saxén and Toivonen's

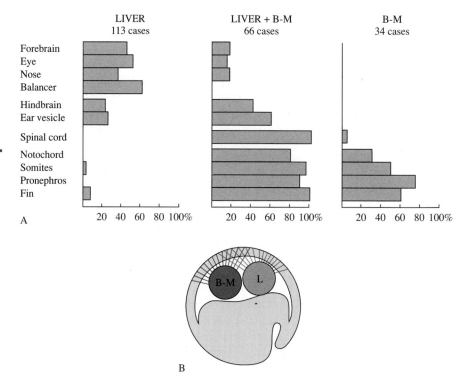

Figure 12.16

A, When implanted into an early gastrula, guinea pig liver (L) induces primarily anterior head structures, whereas guinea pig bone marrow (B-M) induces tail and mesodermal tissue. B, When these two tissues are implanted together, they induce the whole range of neural derivatives, including posterior brain structures, which would not develop from either inducer individually. (A, After Toivonen and Saxén, 1955a. B, After Toivonen and Saxén, 1955b.)

neuralizing factors mimic the signal that apparently comes from the contiguous dorsal lip before gastrulation and uniformly induces the neural plate, whereas the mesodermalizing factors mimic the graded signal that comes from the chordamesoderm and induces the regional characteristics of the central nervous system.

To date, the intercellular signals responsible for neural tube differentiation have not been isolated, and the search continues. Attention will obviously be focused on growth factors, in view of their apparent role in mesoderm induction, but recent evidence supports the notion that retinoic acid, which, as we shall see, is an important regulator of pattern in the chick and amphibian limb, may also be important in anteroposterior patterning of the neural tube.

Durston et al. (1989) treated *Xenopus* embryos with retinoic acid and found that anterior brain structures were repressed, producing microcephalic embryos with no eyes (Fig. 12.17). Similar effects could be produced in tissue culture with isolated neural tissue as well. These workers suggest that a gradient of retinoic acid may exist in the embryo that transforms anterior structures to posterior structures and could thus produce the anteroposterior axis of the central nervous system. The behavior of retinoic acid correlates very well with the mesodermal gradient predicted by Saxén and Toivonen. If a gradient of retinoic acid confers regional properties on the neural ectoderm, it will be interesting to learn whether growth factors play a role in establishing this gradient.

Tremendous progress has been made in the last several years in understanding the nature of the signals responsible for mesoderm and neural induction. With the availability of molecular markers for genes that are turned on early during induction and the recent identification of candidate inducing molecules, rapid progress is likely to be made on a problem that was first described over 60 years ago.

Primary Organizers in Higher Vertebrates

Primary organizers have also been demonstrated in other vertebrates. The primary organizer of the bird embryo is Hensen's node, as shown by

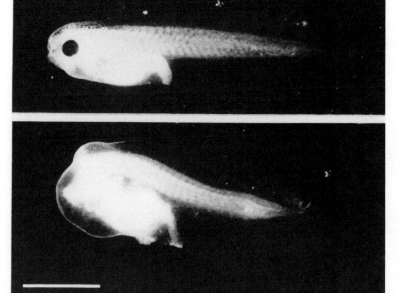

Figure 12.17

The effects of retinoic acid on anterior neural development. The top panel shows a normal *Xenopus* feeding-stage tadpole, and the bottom panel shows an embryo of the same age that had been treated with retinoic acid for 30 minutes at stage 10. The treated embryo has no eye, and most anterior head structures are missing. (From Durston et al., 1989. Reprinted by permission from *Nature*, Vol. 340, pp. 140–144. Copyright © 1989 Macmillan Magazines Ltd.)

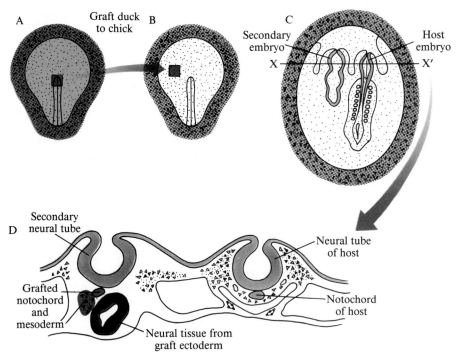

Figure 12.18

Induction of a secondary embryo as a result of transplanting Hensen's node from a duck embryo to a chick embryo. Origin (A) and site of placement (B) of graft. C, Embryo cultured for 31½ hours after transplantation of node. Dorsal view showing the development of the secondary embryo. D, Cross-section from level indicated by the line X-X' in C showing the secondary neural tube. (After Waddington and Schmidt, 1933.)

transplanting the anterior portion of the primitive streak from a duck embryo to the region below the epiblast of a chick embryo. The transplant forms a secondary embryo consisting of a notochord and mesoderm derived from duck tissue and a neural tube derived from chick ectoderm (Fig. 12.18). Thus, Hensen's node appears analogous to the dorsal lip of the amphibian embryo because tissue derived from it (i.e., the chordamesoderm) induces the ectoderm to undergo neurulation. Recent evidence indicates that the inducer is an activin (Mitrani et al., 1990).

The dorsal structures of the mammalian embryo may also be organized by the anterior region of the primitive streak. This was shown by transplantation experiments in which a chick primitive streak was implanted into a rabbit embryo (Waddington, 1934, 1936). A secondary neural plate developed in the rabbit embryo as a result of this transplantation. Likewise, in the reciprocal experiment, Hensen's node from a rabbit embryo induced a secondary neural plate when transplanted to a chick embryo.

12–3 Secondary Induction

The initial inductive events that produce mesoderm and the neural axis are critical for the patterning of the early embryo, but these are only the first of many signaling events that will occur over the life of the organism. These secondary induction events are important for at least two reasons:

(1) Induction allows the embryo to position differentiated cells in precise locations in the embryo. For example, we shall learn that the optic cup directs a lens to form from head ectoderm that is already predisposed to form a lens (see p. 498). As a result, the lens is situated in the center of the optic cup. A displaced lens would be of no visual use to an organism. (2) Sequential inductive events contribute to producing diverse cell types from relatively few distinct precursor cells. Thus, once a tissue is induced to differentiate, it becomes capable of inducing other tissues to differentiate. The eye (which we shall discuss in detail in Chap. 17) is an excellent example of sequential induction.

We now discuss how secondary induction contributes to the generation of cell type diversity during development.

Instructive Versus Permissive Interactions

What is the nature of the cellular interactions between the inducing and responding tissues during induction? Holtzer (1968) categorized the induction process into two major types based upon experimental evidence. These have been elaborated upon by Saxén (1977), Wessells (1977), and Gurdon (1987). The two types of induction have been termed **instructive** and **permissive**. In an instructive interaction, the inducing tissue apparently gives precise information or "instructions" to *commit* cells to a new pathway of development. Such an interaction presumes that the responding tissue is relatively undetermined. Alternatively, induction may be permissive. In this case, responding cells are already determined and poised to differentiate, simply requiring a signal from the inducing tissue to *allow* them to express their potential. Permissive interactions are relatively nonspecific, and they are generally identified experimentally by the fact that many different tissues can elicit the same inductive response from a particular responding tissue.

The preceding definitions stress the importance of the inducing tissue, but it is also critical that the responding tissue be competent to receive the inducing cues. As we shall see, the responding tissue is generally competent only during a narrow window of time during development.

The differences between instructive and permissive interactions can be best clarified with some examples.

Cutaneous Structures: Instructive Interactions Between the Mesenchyme and Ectoderm

The skin is composed of two cell layers: (1) the overlying epithelial ectoderm, which differentiates into the epidermis bearing such differentiated products as hair, feathers, and scales; and (2) the underlying mesoderm, which is initially organized as a loose aggregation of cells (generally referred to as a mesenchyme) that will become the dermis. The cell types that differentiate from the epidermis depend upon instructions from the underlying dermis. The inductive events that occur in the skin are typical of interactions that occur between an epithelium and a mesenchymal cell population in most organ rudiments. These have been termed **epitheliomesenchymal interactions**, and we shall examine many other examples of this kind of interaction in this chapter.

In the chick, broad feathers form on the wing, whereas thigh feathers are narrower, and the lower leg has scales and claws. Cairns and Saunders (1954) transplanted mesoderm from different regions of the leg or wing into foreign regions and showed that the overlying ectoderm developed according to the *origin of the mesoderm* (Fig. 12.19). For example, if a piece of thigh mesoderm is grafted under the ectoderm of the wing, the developing wing ectoderm will form thigh feathers above the grafted mesoderm. Likewise,

Figure 12.19

Schematic representation showing the effects of combining ectoderm and mesoderm from different regions of the limb; i.e., thigh mesoderm with wing ectoderm (A), feather-area mesoderm with nonfeather ectoderm (B), and scale-area mesoderm with feather-area ectoderm (C). In all cases, the mesoderm controls the differentiation of structures from the ectoderm. (After Wessells, 1977.)

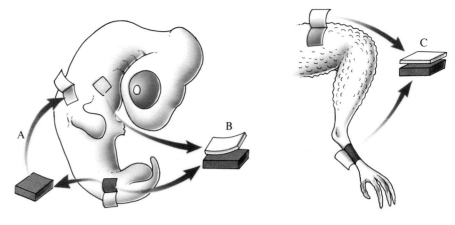

if foot mesoderm is combined with wing ectoderm, scales and claws will form (Fig. 12.20). Conversely, leg ectoderm combined with wing mesoderm will result in broad feathers typical of wing. Such studies show that the mesoderm can direct regional differentiation of the overlying ectoderm, presumably by directing the expression of specific genes in the ectoderm.

Another example of an instructive interaction between epithelium and mesenchyme is found in the developing tooth. The tooth arises from two tissues: the jaw ectoderm and the underlying mesenchyme, which is derived from the neural crest (Lumsden, 1987). The ectoderm invaginates to form a cup-shaped structure, known as the enamel organ, which produces the outer enamel layer of the tooth, whereas the underlying mesenchyme condenses to form the dental papilla, which ultimately produces the hard dentine core of the tooth (Fig. 12.21). The classic studies of Kollar and Baird (1969) demonstrate that the dental papilla (the mesenchyme) can instruct the overlying ectoderm to differentiate. Kollar and Baird isolated tooth buds from young mouse embryos (11 to 14 days) and with proteolytic enzymes separated the ectoderm from mesenchyme. When cultured alone, neither the ectoderm nor mesenchyme would differentiate. When combined together, a morphologically normal molar tooth composed of dentine and enamel developed. If the tooth mesenchyme were combined with another source of ectoderm, such as plantar (foot) or limb ectoderm, the mesenchyme would induce the ectoderm to differentiate into an enamel organ (Fig. 12.22). Conversely, any other mesenchyme, such as limb, when combined with jaw ectoderm, resulted only in keratinized cells characteristic of skin. Thus, the mesenchyme can instruct the ectoderm from numerous sources to differentiate into an enamel organ—an instructive interaction.

Furthermore, the underlying mesoderm determines the shape of the tooth. Kollar and Baird separated the mesenchyme from overlying ectoderm of incisor and molar tooth germs and interchanged the ectoderm and mesenchyme. In all cases, the eventual shape of the tooth was determined by the source of the mesenchyme, rather than the ectoderm.

Induction of the Amphibian Lens: Progressive Determination

The above experiments and discussion have suggested that a single induction is sufficient to cause a tissue to differentiate. This classic view of induction is likely to be too simple, however. Recent work examining the induction of the lens suggests that induction of this tissue is a multistep process, and in fact there appears to be a series of inductive events that progressively

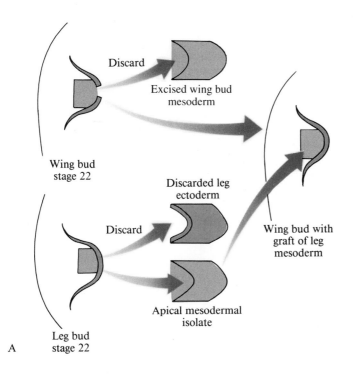

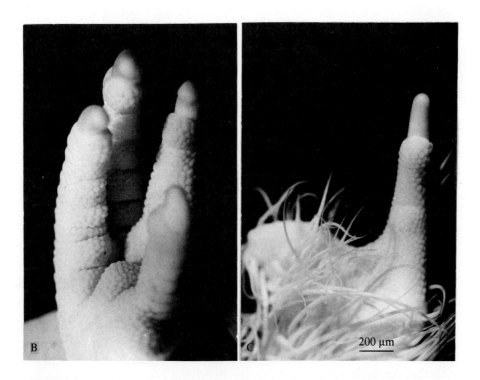

Figure 12.20

Experiment demonstrating the effects of foot mesoderm on forelimb ectoderm. A, Mesoderm is removed from the tip of a host chick forelimb bud and is replaced with equivalent mesoderm from a hind limb bud. The resulting forelimb develops part of a claw at its tip (C). A claw of a normal 12-day chick embryo is seen in (B) for comparison. (From Cairns and Saunders, 1954.)

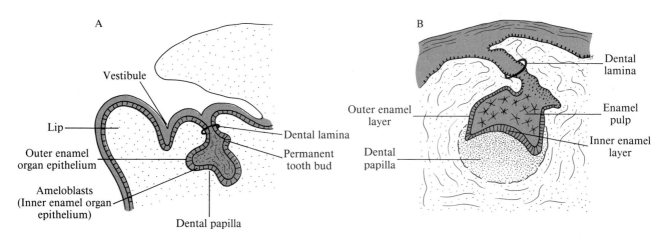

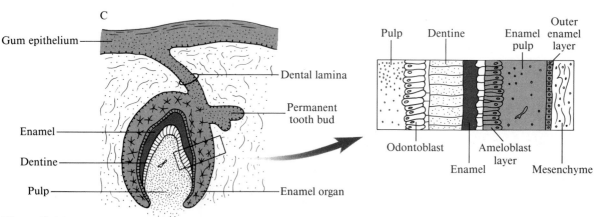

Figure 12.21

Development of human teeth. A, Drawing showing a sagittal section of the lip and
jaw of an 11-week-old human fetus, illustrating the invagination of the oral
epithelium to form the tooth bud. B, A developing tooth in a three-month-old
fetus, showing the inner enamel layer, which is derived from the oral ectoderm,
and the condensing dental papilla, which is derived from neural crest cells. C, At
seven months, the tooth has clearly differentiated enamel and dentine layers,
which are deposited from the ameloblast and odontoblast layers, respectively.
(After Hopper and Hart, 1985.)

bias the head ectoderm to form lens, rather than a single instructive event
precipitating the formation of the lens from the head ectoderm.

Early work from Spemann (1901), Lewis (1904), and others led to the
idea that the optic vesicle contacts the overlying ectoderm as it grows out
from the diencephalon and induces the formation of a lens from the overlying
ectoderm. If Spemann removed the eye rudiment of a neurula-stage frog
embryo by microcautery, the lens did not appear. Furthermore, when Lewis
grafted an eye rudiment to the flank of a frog neurula-stage embryo, a lens
developed in association with the grafted optic vesicle, suggesting that the
optic vesicle induced a lens from the flank ectoderm. Such experiments led
to the long-held belief that the optic vesicle produces an "instructive" signal

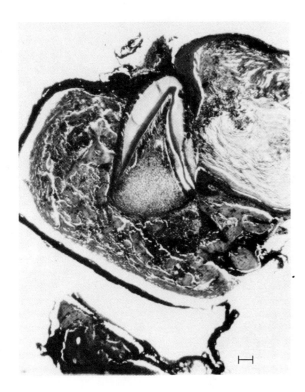

Figure 12.22

The result of combining a mouse dental papilla with foot epidermis. The epidermis forms an enamel organ. Scale bar equals 100 μm. (From Kollar and Baird, 1970. Reprinted by permission of Company of Biologists, Ltd.)

and that the optic vesicle was not only necessary, but sufficient, for induction. Recent experiments by Jacobson (1966) and by Grainger and his colleagues (1988) have now drawn our attention to the fact that a series of interactions is necessary for the lens to appear and that the optic vesicle is, in reality, a very weak inducer.

Grainger et al. (1988) repeated the experiments of Lewis, except they labeled the donor optic vesicles with horseradish peroxidase (see essay on p. 280 for technique) in order to distinguish donor from host tissue. Their results were remarkable: When a lens formed, it was from the donor tissue, not the host flank ectoderm, showing that the optic vesicle did not *induce* a new lens. Rather, the lens came from tissue contaminated with presumptive lens from the donor (Fig. 12.23). This result suggests that either the optic vesicle is not a strong inducer and cannot induce flank ectoderm or that the flank ectoderm is not competent to respond to the inductive cues. In fact, both situations appear to be the case, as evidenced by the following experiments.

In a series of grafting experiments, Henry and Grainger (1987) showed that, if ectoderm is taken from any region of a young *Xenopus* neurula and grafted into the prospective lens region of another *Xenopus* early neurula, the grafted ectoderm will form a lens. When posterior ventral ectoderm is obtained from progressively older embryos (late neurula or tail-bud stages) and grafted to the presumptive lens-forming region of a neurula-stage embryo, however, the posterior ectoderm progressively loses it ability to form lens, whereas any head ectoderm progressively increases in its ability to form a lens when grafted to an early neurula embryo. These experiments make two important points: (1) The nonhead ectoderm is gradually restricted in its potential so that it can no longer respond to the optic vesicle; and (2) because head ectoderm responds better when obtained from progressively older embryos, it is likely that a number of earlier inductive events that are necessary for lens formation occur before contact with the optic vesicle during the late neurula stage.

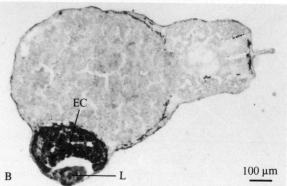

Figure 12.23

Repeat of the classical optic vesicle transplantation experiment. A, A young *Xenopus* larva that had received a transplant of an optic vesicle to its flank at the gastrula stage. The transplanted optic vesicle has a well-developed lens associated with it. B, When such a transplanted vesicle, which had been labeled with HRP before transplantation, is sectioned, the lens (L) as well as the eye cup (EC) exhibit HRP staining, showing that the lens was derived from donor tissue and not induced from the unlabeled flank ectoderm. (From Grainger et al., 1988. Reprinted by permission of Company of Biologists, Ltd.)

What might be the inductive cues that predispose head ectoderm to form lens? During gastrulation, two tissue layers come into contact with the presumptive lens ectoderm. These are the endoderm of the pharynx and the heart mesoderm, which sweeps by this region (Fig. 12.24). Jacobson (1966) showed in a series of tissue culture experiments that isolated presumptive lens ectoderm will not form lens. On the other hand, if he included either endoderm or heart mesoderm in the cultured explants, lens tissue would differentiate from ectoderm. However, if he included *both* endoderm and heart mesoderm, he achieved the highest percentage of lens induction. Thus, Jacobson believes that the prior induction by heart and endoderm is necessary for the lens to develop.

Another possibility that needs to be tested experimentally is that the presumptive eye-forming region of the brain during the neural-plate stage may induce the contiguous epidermal ectoderm to become lens, or at least predispose it in that direction (see Fig. 12.24A). Hence, during gastrulation, head ectoderm becomes predisposed to form lens, whereas trunk ectoderm loses this ability—probably because it becomes predisposed to form something else. Endoderm and heart mesoderm may provide the later inductive cues, although these inductive interactions need to be explored in detail.

What, then, is the role of the optic vesicle in lens development? It is likely to be the last inductive interaction and a weak one at that, but it is probably important in centering the lens in the center of the optic cup (see Fig. 12.24D). Thus, the vesicle serves to further restrict the lens' potential to develop from the entire head ectoderm in the early embryo to the more narrow area above the optic cup. In addition, the retina is necessary for the continued well-being of the lens because lenses will eventually regress in the absence of a retina.

The story of lens induction is an interesting one for several reasons. It is a graphic demonstration that developmental processes are often complex

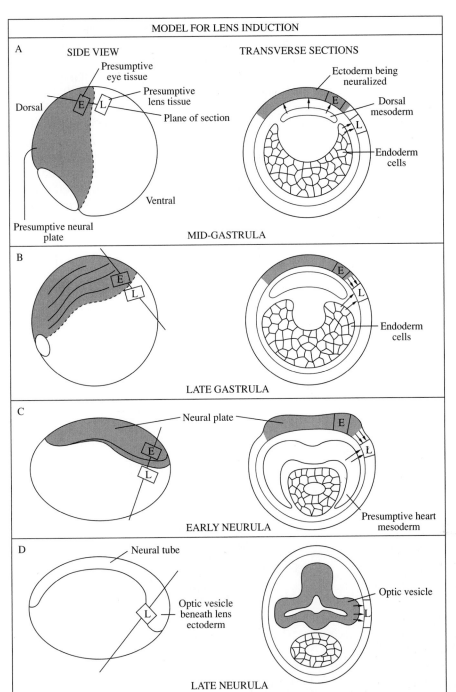

MODEL FOR LENS INDUCTION

A
SIDE VIEW
TRANSVERSE SECTIONS

Presumptive
eye tissue

Presumptive
lens tissue

Dorsal

Plane of section

Ventral

Presumptive neural
plate

MID-GASTRULA

Ectoderm being
neuralized

Dorsal
mesoderm

Endoderm
cells

B

LATE GASTRULA

Endoderm
cells

C

Neural plate

EARLY NEURULA

Presumptive heart
mesoderm

D

Neural tube

Optic vesicle
beneath lens
ectoderm

Optic vesicle

LATE NEURULA

Figure 12.24

The various steps believed to
induce the lens are outlined in this
diagram. Each inductive step
between two tissues is indicated by
arrows. Presumptive lens is
indicated by L, and presumptive
optic vesicle is indicated by E (eye
rudiment). The colored area is
presumptive neural plate. (After
Saha et al., 1989.)

and carefully orchestrated in time and position in order to achieve the final
result. Consequently, the process of lens induction may indicate that other
inductive interactions are similarly complex and involve a series of steps
(Jacobson and Sater, 1988; Saha et al., 1989). It also stands as a reminder
to the student of development that "classic experiments" such as those of
Lewis may lead to the wrong conclusions when the experiment is not
carefully controlled.

Pancreas Development: A Series of Permissive Interactions

All of the glands and many internal organs develop as outpockets of an
epithelium in conjunction with a mesenchyme. The extent of dependence

upon the mesenchyme varies considerably among these organs. For example, the salivary gland epithelium requires a specific mesenchyme in order to develop its branched morphology and acquire its glandular function, whereas the pancreatic diverticulum can continue to develop *in vitro* if it is cocultured with virtually any mesenchyme.

As we discussed in Chapter 8, the pancreas is first observable as a diverticulum of the embryonic gut in a nine-day mouse (Fig. 12.25A). As the diverticulum continues to grow, the endoderm pushes into the mesenchyme and begins to branch, forming blind pockets known as the **pancreatic acini** (see Fig. 12.25B to D). The cells of these acini differentiate into **exocrine cells** and produce the various pancreatic enzymes, including proteases, peptidases, nucleases, and amylases. In addition, cells separate from the pancreatic epithelium to form clusters of cells surrounded by mesoderm,

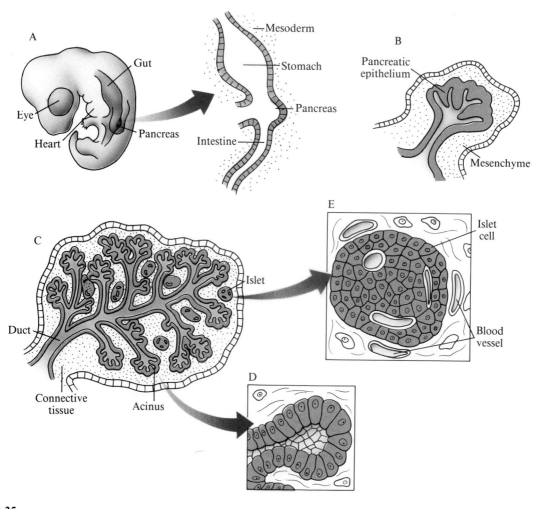

Figure 12.25

The development of the pancreas in the mouse. A, The position of the gut is indicated in the whole embryo with a higher magnification showing the pancreas as it forms an outpocket in a nine-day-old mouse. B, By day 12, the pancreatic epithelium has grown into the adjacent mesoderm and begun to branch. C, By day 15, the endoderm is highly branched with well developed acinar cells of exocrine function and islet cells of endocrine function. D, and E, Acinar and islet cells, respectively, at higher magnification. (After Wessells, 1977.)

known as the pancreatic **islets of Langerhans** (see Fig. 12.25C and E). These cells are endocrine in function and produce the hormones insulin, glucagon, and somatostatin. Thus, both types of glandular cells arise from the epithelial portion of the embryonic gut.

Pancreatic development also requires a contribution from the mesoderm. If the pancreatic mesoderm and endoderm from a nine-day embryo (15-somite stage) are separated from one another and cultured separately, they will not develop further. However, if they are recombined in culture, the endoderm will differentiate into the normal exocrine and endocrine cells of the pancreas (Fig. 12.26) (Golosow and Grobstein, 1962; Wessells and Cohen, 1967). Rutter and his colleagues (1964) have shown that if they substituted many different types of mesenchyme from other organs, such as the salivary gland, for the pancreatic mesenchyme, the pancreatic endoderm would still differentiate. Somites, which normally produce muscle and cartilage cells, will also induce pancreatic endoderm to differentiate as evidenced by amylase activity in the endoderm (Wessells, 1968). Even embryo extract (purée of young embryos) will suffice. Therefore, at the time when pancreatic endoderm can first be isolated, it is already determined to the extent that a relatively nonspecific cue will suffice to complete the differentiation process. This is a good example of a permissive interaction. It is suspected that the inductive interaction that is critical for continued development (Rutter et al., 1964; Wessells, 1968) is little more than a stimulus to divide.

At some point in the earlier developmental history of the pancreas, a more specific signal must predispose this portion of the endoderm to become pancreas. Wessells and Cohen (1967) discovered that the pancreatic endoderm will differentiate *in vitro* when combined with a variety of foreign mesoderms as long as the endoderm is taken from a 15-somite or older mouse embryo (Fig. 12.27C). If the endoderm is isolated from a seven- to eight-somite embryo, however, it will differentiate only when combined with pancreatic mesoderm (see Fig. 12.27B). Finally, even if the whole midsection of a zero- to six-somite mouse embryo is explanted into culture,

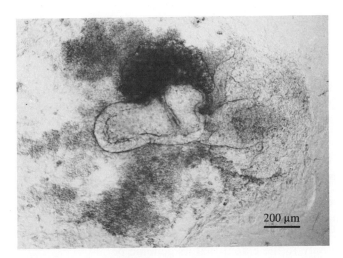

200 μm

Figure 12.26

Pancreatic endoderm will differentiate in culture when it is explanted with associated mesoderm. After six days in culture, the acinar cells differentiate to produce zymogen granules, which give the pancreas a darkened appearance. (From Wessells and Rutter, 1969.)

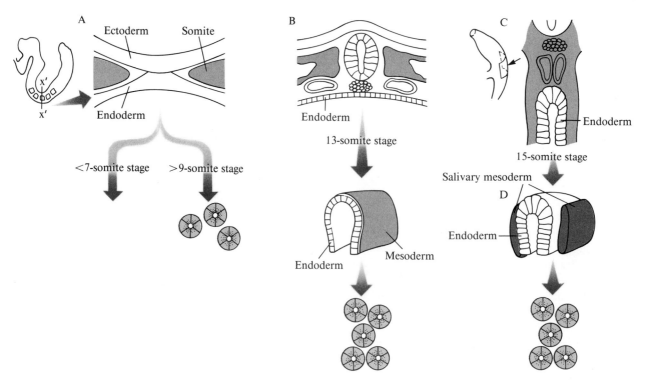

Figure 12.27

The dependence of the pancreatic endoderm on mesoderm for continued development varies depending upon the age of the endoderm. The top row shows cross sections through the pancreatic region of embryos that are 7-somites, 13-somites, and 15-somites old, respectively. A, If the whole mid-gut region of a 7-somite embryo is explanted into culture, the pancreas will not develop further. By the 9-somite stage, the pancreatic endoderm will branch and differentiate. B, By the 13-somite stage, the pancreatic endoderm plus mesenchyme can be separated from the rest of the embryo and will differentiate when placed in culture, as long as the pancreatic mesoderm is present. C, and D, By the 15-somite stage, the pancreatic endoderm (arrow in drawing on left and shown enlarged on the right and in D) can be recombined with foreign mesenchyme, such as salivary mesoderm, and will still continue to branch. Note that even by the 15-somite stage, there is still no obvious pancreatic diverticulum. (After Wessells, 1977.)

the pancreatic endoderm never differentiates (see Fig. 12.27A). Clearly, at one of these earlier stages, a more specific or "instructional" cue must limit or direct pancreatic endoderm into the appropriate developmental pathway.

Induction Across Species Boundaries: The Role of the Ectodermal Genome

Many of the inductive cues appear to have been conserved during evolution, because tissue from one species can often be substituted for another in inductive situations. Generally, however, the induced tissue can respond and will differentiate only to the extent that its genome permits.

One of the first demonstrations of cross-species induction came from the experiments of Spemann and Schotté (1932). They examined the ability of mouth mesoderm to induce head ectoderm structures in the salamander and the frog. The head structures in these two animals differ significantly. The frog possesses two mucus-secreting suckers on either side of a toothless

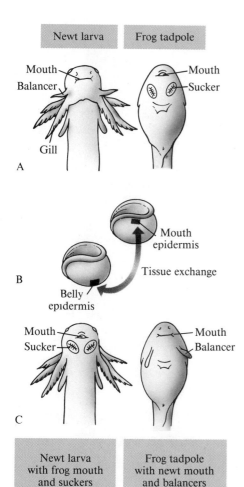

Figure 12.28

Experiment demonstrating cross-species induction in amphibia. A, The mouth parts of urodeles and anurans are different. The urodeles have tooth-bearing jaws and balancer organs, whereas the frogs have toothless mouths and suckers. B, When belly epidermis from one species is grafted to the future head region of the other, mouth parts will differentiate from the grafted belly skin, but the particular mouth parts are typical of the donor, not the host, embryo (C). (After Walbot and Holder, 1987.)

jaw, whereas the salamander has two balancers adjacent to a jaw that contains calcareous teeth (Fig. 12.28A). When they grafted skin from the flank of a frog embryo onto the head of a salamander embryo, the host embryo developed a frog-like mouth including suckers, whereas when salamander flank ectoderm was grafted onto a frog embryo, balancers and a tooth-bearing jaw developed (see Fig. 12.28B and C). Clearly, the mesoderm of both species could induce mouth parts in the skin of the other. Although the ectoderm could be induced across species boundaries, it was also clear from these studies that the ectoderm could respond only insofar as its genome would permit. Frog skin cannot produce balancers, so it makes what it can (i.e., suckers).

Species specificity has also been examined in the skin. Recall from our earlier discussion that the mesoderm induces the overlying ectoderm to form different structures (such as hair, scales, or feathers) and specifies the kind of ectoderm-derived structure. Coulombre and Coulombre (1971) demonstrated that the chick corneal epithelium, which under usual circumstances forms the clear cornea, will produce cutaneous structures if skin mesoderm is introduced into the anterior chamber of the eye. Moreover, when mouse skin mesoderm is introduced into the chick eye, feather germs will appear in the corneal epithelium (Fig. 12.29). Not only is the corneal epithelium inducible to form cutaneous structures, but it will do so under the influence of mesoderm from other species. As expected, however, the chick cornea does not produce hair when combined with mouse mesoderm;

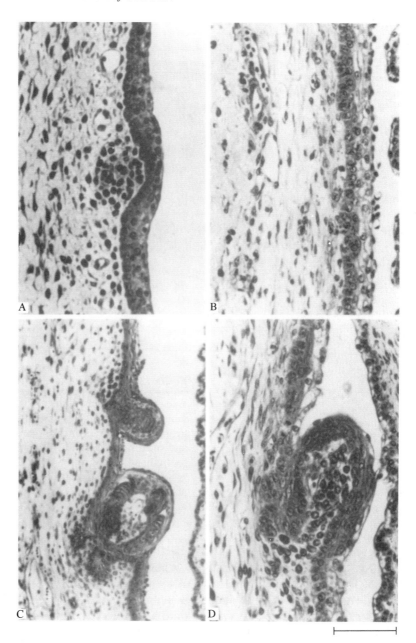

Figure 12.29

When mouse skin mesoderm is grafted into the anterior chamber of the chick eye, it induces the overlying cornea to form a feather germ. In A, the mouse cells lie just below the chick epithelium. B, C, and D show progressive stages in the formation of feather primordia. The grafted mouse cells compose the feather pulp. Scale bar equals 100 μm (A), 50 μm (B), 30 μm (C), and 50 μm (D). (From Coulombre and Coulombre, 1971.)

it can only produce what is endowed in its genetic repertoire (i.e., feathers). Similar recombinations of mesoderm and ectoderm among reptiles, chick, and mammals have also revealed the universality of the inductive response in skin (Dhouailly, 1975).

Possibly the most extraordinary demonstration of the cross-species inductive ability of mesoderm is in tooth development. Normally, chick embryos do not develop teeth. Yet, when Kollar and Fisher (1980) combined chick oral epithelium with mouse mesodermal tooth germ, the chick ectoderm differentiated into ameloblasts and formed enamel (Fig. 12.30). Clearly, the

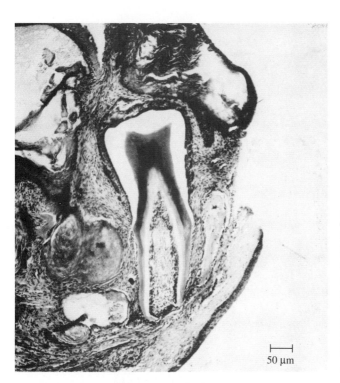

50 μm

Figure 12.30

Genes that have been silenced long ago in evolutionary history are reawakened when chick oral epithelium is recombined with mouse dermal papilla. The mouse mesenchyme induces the chick oral epithelium to form ameloblasts and enamel, although it would not do so during normal development. (From E. J. Kollar and C. Fisher, 1980, Tooth induction in chick epithelium: Expression of quiescent genes for enamel synthesis. Science, *207*: 993–995. Copyright 1980 by the American Association for the Advancement of Science.)

enamel genes have been retained in the chick genome, even though teeth have long since been lost during evolution.

Nature of the Inductive Interaction

How are inducing factors passed from one cell layer to another? Grobstein (1955) proposed a number of years ago that there are at least three means by which the induction signal could be transferred: (1) by direct cell-cell contact; (2) by the diffusion of specific molecules over a short distance from the inducing cell to the responding cell; or (3) by the extracellular matrix produced by the inducing cell, which remains closely associated with the cell surface and does not diffuse. There is, in fact, evidence that all three occur.

To test the ability of inducing factors to diffuse, Grobstein (1953b, 1956) developed a technique for culturing inducing and responding tissue while separated from each other by a porous Millipore filter barrier (Fig. 12.31). These filters have tortuous pores that extend through the filters (Fig. 12.32); the pore diameter can be chosen by the investigator. When the pore sizes are quite large, cell processes from the explanted tissue can grow into the pores and make direct contact with the tissue on the other side. When the pore sizes fall below 0.1 mm, cell processes would presumably no longer invade the filter, although diffusible molecules could still pass from one side to another. Using such a filter system, Grobstein asked if induction could occur in a variety of developmental systems with or without direct contact. During the development of the pancreas, the endoderm is induced by the surrounding mesoderm and branches to form pancreatic acini and pancreatic islets. If the mesoderm and endoderm are cultured on either side of a Millipore filter to limit contact between the two tissues, induction of the pancreatic endoderm will still occur. More recently, an enriched extract from mesoderm that appears to contain a glycoprotein has been shown to substitute for mesodermal cells in the induction process, again showing that cell contact may not be necessary for passage of the inductive signal (reviewed in Rutter et al., 1978).

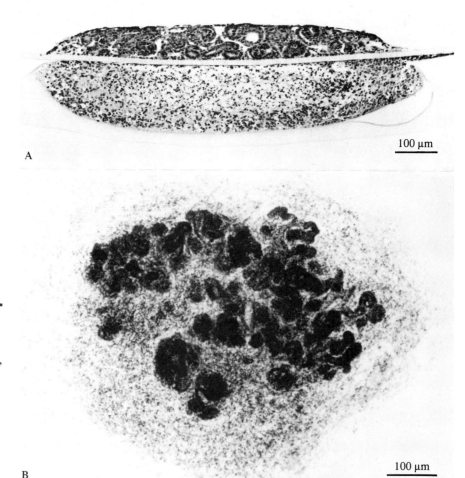

A

100 μm

B

100 μm

Figure 12.31

Kidney tubules differentiate after 65 hours when metanephrogenic mesoderm is cultured across a Millipore filter from spinal cord. A, Cross-sectional view of this trans-filter experiment, with metanephrogenic mesenchyme above the filter. B, View of the differentiating tubules seen from above. (From Saxén, 1987. Reprinted by permission of Cambridge University Press.)

Until recently, no inductive interactions that require cell-cell contact had been demonstrated. However, changes in experimental protocols have substantially changed this view. Nucleopore filters have replaced Millipore filters in the induction assays (Wartiovaara et al., 1974). Nucleopore filters have pores that are quite straight, rather than tortuous, like the Millipore filters. When one of these filters is sectioned so that the entire length of the pore can be seen in one section, the presence of cell processes can be identified unequivocally (Fig. 12.33A). In addition, the methods used to preserve cell morphology that are available now are vastly superior to any that Grobstein had at his disposal in 1956 (Lehtonen et al., 1975). It was discovered using these improved techniques that, in fact, cell processes in many cases do extend all the way through the filters and contact each other (see Fig. 12.33B).

One example of an induction process that appears to need cell contact is kidney induction. As we discussed in Chapter 8, the adult kidney develops from two sources of tissue: (1) the metanephric mesenchyme, which will eventually form kidney tubules; and (2) the metanephric diverticulum, which will form the kidney ducts (see Fig. 8.17). The presence of both tissue types is needed for either of them to differentiate. When Millipore filters were used in combination with older fixation techniques, no penetration of processes was observed. However, when Nucleopore filters were employed

Figure 12.32

Transmission electron micrograph through metanephrogenic mesenchyme (top) and spinal cord (bottom) cultured across a Millipore filter. This micrograph reveals the tortuous pores of the filter and also demonstrates that cytoplasmic processes from the cultured tissue (C) have invaded the filter. (From Lehtonen et al., 1975. Reprinted by permission of Company of Biologists, Ltd.)

for the same experiment, cell contact between processes were observed between the two tissues (Wartiovaara et al., 1974). Furthermore, as the size of the pores was steadily decreased, the point at which no induction occurred corresponded with the pore size that no longer permitted penetration of cell processes. Thus, contact appears to be necessary for induction in the developing kidney, although no differentiated cell junctions have ever been observed.

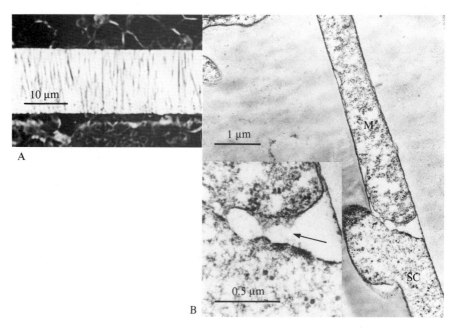

Figure 12.33

Demonstration that cytoplasmic processes can penetrate filters. A, A section through a Nucleopore filter that has kidney mesenchyme cultured above and spinal cord below. This section shows how straight the pores are; cytoplasmic processes can be clearly seen to penetrate through the entire thickness of the filter. B, An electron micrograph of a similar filter demonstrating that the mesenchyme (M) and spinal cord (SC) cell processes contact each other. The insert shows that extracellular matrix material (arrow) is also present on the surfaces of the processes. (From Wartiovaara et al., 1974. Reprinted by permission of Company of Biologists, Ltd.)

Now it appears that many inductive interactions need cell contact of some type. It is not always clear if direct plasma membrane contact is necessary, or if the inductive interaction is mediated by components of the extracellular matrix, which are closely associated with the cell surface. When inductive tissues are observed *in situ*, one sees a tight association between the two cell types, suggesting that the cells are intimately associated (Fig. 12.34).

A few inductive signals are clearly conveyed by diffusible factors. These include induction of the neural epithelium from the ectoderm (Toivonen et al., 1976), induction of *Xenopus* mesoderm by the vegetal cells (Grunz and Tacke, 1986), and, probably, pancreatic induction. In these cases, induction occurs in the absence of cell process penetration when the cells are separated by a filter, and induction can be mediated by soluble factors isolated from the inductive tissue.

What Are the Inducing Molecules?

Ever since the identification of the induction mechanism by Spemann, numerous investigators have pursued the elusive goal of isolating and characterizing the inducing molecules from a variety of cell types. In general, this has been a nearly impossible task because it is very difficult to isolate enough of the inductive tissue to actually characterize a molecule that is probably present in small quantities. As we have discussed already in this

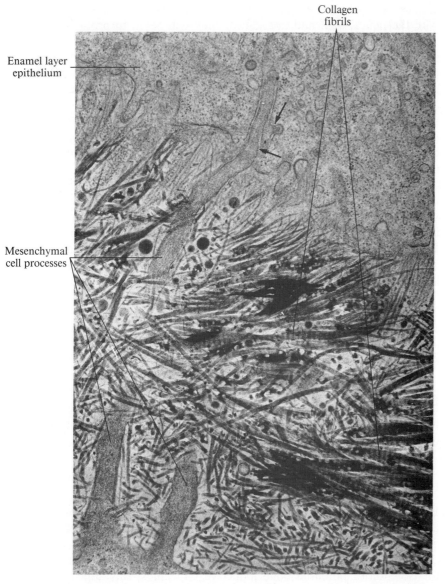

Collagen
fibrils

Enamel layer
epithelium

Mesenchymal
cell processes

Figure 12.34

In the embryo, there is often intimate contact between the epithelial and mesenchymal components during induction. In this electron micrograph of a developing tooth germ, the mesenchymal cells of the dental papilla extend through a dense region of collagen and other extracellular matrix components to contact (at arrows) the inner enamel layer epithelium. (From Slavkin and Bringas, 1976.)

chapter, considerable evidence is accumulating that growth factor–like substances are involved in mesoderm induction in amphibians and retinoic acid is a candidate for involvement in conferring regional properties on the neural ectoderm. We now discuss briefly some other developmental systems in which the inducing molecule is plentiful enough that it has been reasonably well characterized.

Extracellular Matrix as an Inducer

There are several situations in which collagen, perhaps in association with proteoglycans, has been shown to have inductive properties. These include the differentiation of somite cells into cartilage, the differentiation of myoblasts into muscle cells, and the differentiation of the cornea. It is likely in all these events that collagen is a permissive inducer, allowing these cells to acquire characteristics already programmed by earlier events.

As we discussed in Chapter 8, the sclerotome portion of the somites normally surrounds the neural tube and notochord and differentiates into the cartilaginous vertebrae. If the sclerotome cells are grown in tissue culture, they will rarely become cartilage. If, however, they are cocultured with the neural tube and notochord, they will begin to produce cartilage-specific molecules—namely collagen type II and a unique form of chondroitin sulfate proteoglycan. If notochords are pretreated with proteolytic enzymes to remove their extracellular matrix molecules before coculturing the notochords with sclerotome, the notochords lose their inductive ability (Kosher and Lash, 1975). Lash and Vasan (1978) showed that if the sclerotome cells were cocultured with collagen or collagen plus proteoglycan in the absence of neural tube, they would differentiate into cartilage. Thus, collagen and proteoglycans are clearly implicated in promoting cartilage differentiation.

Similarly, Hay and her coworkers showed that the chick cornea will not produce a normal corneal stroma when cultured by itself. If, however, the cornea is cultured on top of a lens primordium, it will differentiate, even if the lens is first freeze-killed (Hay and Dodson, 1973). When purified type I or type II collagen is substituted for the lens in culture, normal induction occurs (Meier and Hay, 1974). This induction event appears to be mediated by direct contact between the corneal epithelium and the lens extracellular matrix stroma because in transfilter experiments, induction occurs only when cell processes from the cornea completely penetrate the filter (Fig. 12.35). In fact, the amount of extracellular matrix produced by the cornea decreases as pore size decreases and is directly proportional to the number of cell processes that are able to contact the lens.

The rate of differentiation of myoblasts into myotubes is enhanced by culturing them on a collagen substratum (Hauschka and Konigsberg, 1966). It is likely that other extracellular matrix molecules will also be found to have a role in inductive processes. For example, the basal lamina glycoprotein, laminin (see Chap. 9), has been shown to cause the pigmented retina to transdifferentiate into neuronal retina (Reh et al., 1987).

Bone Morphogenetic Protein

Bone is an unusual tissue in that, even in postfetal life, it continues to grow and remodel. Presumably, some of this continuous generation of new bone cells is the result of differentiation of undifferentiated mesoderm cells into osteoprogenitor cells. Because large quantities of adult bone are available for isolation of potential bone inductive molecules, this appears to be an ideal system in which to search for a putative bone inducer.

An acidic molecule of 17,500 M_r has been partially purified from demineralized bone matrix. When a pellet of this isolated protein is implanted into the cheek pouch of a hamster (which contains only cartilage and loose connective tissue), mesenchymal cells in the cheek differentiate into bone cells. This protein, which is related to TGF-β, has been named bone morphogenetic protein (BMP; for review, see Urist et al., 1983). Although BMP is presently being used clinically to stimulate bone differentiation and healing, it remains to be identified in embryos as a factor in fetal bone differentiation.

Stalk Cell Differentiation in *Dictyostelium*

Dictyostelium, the cellular slime mold, dwells on the forest floor (and now also in the biologist's laboratory). During part of its life cycle, it is a free-living amoeba, until starvation conditions produce a remarkable transformation: The amoebae stream together to form a mound of cells. Each mound produces a nipple-like tip, which then elongates to form a "finger." The finger eventually falls over on its side to produce a migratory slug. Within 24 hours, the slug is transformed into a fruiting body, a collection of spores that sit atop a slender stalk (Fig. 12.36).

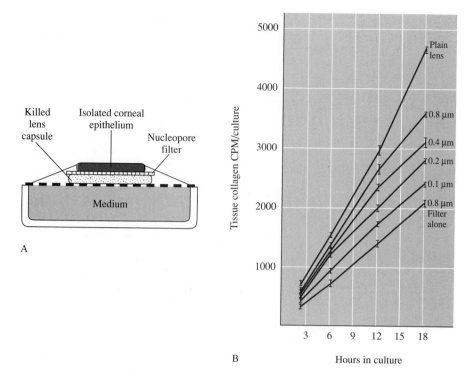

Figure 12.35

Differentiation of corneal epithelium upon contact with the lens capsule. A, When corneal epithelium is cultured opposite lens capsule on a Nucleopore filter, as shown in this drawing, epithelial processes can be found in the filter and contacting the lens. B, The degree of corneal differentiation, as measured by the amount of collagen produced by cornea, is proportional to the extent of contact between the two tissues is extensive, large amounts of collagen are produced, whereas when there is minimal contact between the tissues, collagen production is reduced. Collagen synthesis is measured by relative radioactivity in collagen (CPM, ordinate) from incorporation of radiolabeled precursor. (After Meier and Hay, 1975. Reprinted from *The Journal of Cell Biology*, 1975, 66: 275–291 by copyright permission of the Rockefeller University Press.)

It has been long known that the prespore cells differentiate in the posterior part of the migrating slug, whereas the prestalk cells differentiate from the anterior slug cells (Raper, 1940). The signals for differentiation had remained elusive until recently, when a crude factor was isolated from slugs, called **DIF (differentiation induction factor)**, that induces isolated amoebae in culture to differentiate into stalk cells. Furthermore, DIF also inhibits spore cell formation, thus switching the cells to the stalk differentiative pathway (Kay and Jermyn, 1983). A variety of analytical techniques, especially mass spectroscopy, has identified DIF as a phenyl hexone with di-chloro, di-hydroxy, and methoxy substitutions of the benzene ring (Morris et al., 1987). At least five different forms of DIF have been isolated, with DIF-1 accounting for 96% of activity.

Differentiation induction factor also redirects differentiation in the intact slug as well. If slugs are grown in DIF, a larger percentage of cells are converted from the spore to stalk phenotypes (Fig. 12.37).

At present, it is not known what the origin or distribution of DIF is in the mound stage or where differentiation must be occurring, nor is it completely understood how DIF generates the prespore and prestalk patterns (for reviews, see Williams, 1988; Kay et al., 1989). However, rapid strides

Figure 12.36

Two stages in the development of *Dictyostelium*. On the right is a mature fruiting body, which stands about 1 mm. The spores are contained in the mass at the top of the slender stalk. On the left is the initial aggregation stage, in which the amoebae have streamed together to form a mound with a nipple-shaped tip. It is at this stage of development that prespore and prestalk cells begin to differentiate. (Courtesy of R. Kay from the front cover of Development 103 (1), 1988. Reprinted by permission of Company of Biologists, Ltd.)

0.2 mm

are likely to be made in the near future in understanding how DIF controls the genome of *Dictyostelium*. First, the molecule can now be manufactured in large and pure quantities, alleviating the brute-force tactics previously needed to obtain DIF from amoebae. Second, two genes have now been isolated and characterized whose expression in stalk cells is dependent upon DIF (Williams et al., 1987; Williams et al., 1989). With stalk-specific genes at last in hand, investigators can now monitor when stalk cells differentiate and how and when pattern is generated.

Use of Mutations to Study the Molecular Events in Induction

The search for molecules involved in induction has been primarily a random one in which the inducing molecule has been serendipitously discovered. Direct approaches to identifying the molecular events in induction have been difficult because of the limitations that we have discussed previously. However, a new approach using organisms in which mutations can be generated and isolated efficiently has the potential of overcoming this handicap, leading to the identification of molecules involved in these interactions.

Mutants of the fruit fly *Drosophila* are now providing powerful tools in the study of cell-cell interactions. An example can be found in mutants that affect eye development. *Drosophila* eyes are comprised of identical repeating units called **ommatidia** (Fig. 12.38). Each ommatidium is composed of 20 cells arranged in a characteristic way and has a well-characterized morphology. The first eight ommatidal cells to differentiate are the photo-receptor cells. The first of these is receptor cell 8 (R8; Fig. 12.39). Next, the pair of receptor cells 2/5 forms in association with cell 8. These are

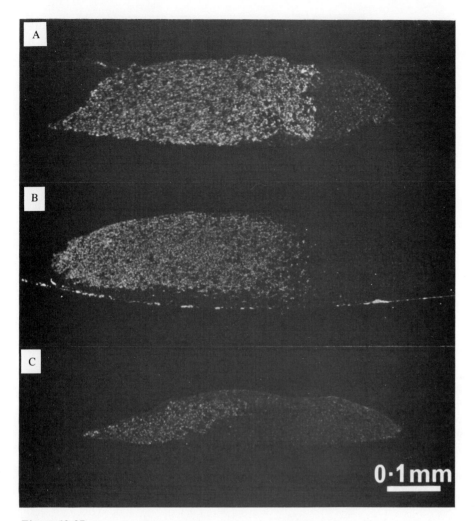

Figure 12.37

Effects of differentiation induction factor on cell differentiation in the
Dictyostelium slug. A, The normal distribution of prespore and prestalk cells is
revealed in this slug stained with an antibody that recognizes prespore cells. If the
slugs are treated with DIF-1 (B and C), the zone of prestalk cells increases and
the number of prespore cells decreases. (From Kay et al., 1989. Reprinted by
permission of Company of Biologists, Ltd.)

followed by the cell pair 3/4, then cell pair 1/6, and finally cell 7. This
pattern of development suggested to Tomlinson and Ready (1987) that the
ommatidium forms as a developmental cascade, such that once cell 8 arises,
its contact with neighboring cells in the epithelium induces cells 2 and 5 to
differentiate. These, in turn, induce cells 3 and 4, and so on. Such a model
also predicts that the inducing cells produce a signal and that the responding
cells must have signal reception and transduction mechanisms. Mutational
analysis has identified genes whose products are involved in both signal
generation and signal reception and transduction.

To date, three genes have been isolated that have a role in cell-cell
communication in the ommatidium. The mutation *rough* (*ro*) interrupts eye
development after the differentiation of cells 2 and 5, suggesting that they
are involved in the induction of subsequent cells 3 and 4. Mutations in two

Figure 12.38

The *Drosophila* eye is composed of 700–800 ommatidia, shown here by scanning electron microscopy. Each of the individual facets is made up of an invariant number of cells, eight of which are photoreceptor cells that depend upon interactions with each other to differentiate appropriately. Scale bar equals 10 μm. (From Tomlinson and Ready, 1987.)

other genes, *sevenless* (*sev*) and *bride of sevenless* (*boss*), are so named because cell 7 fails to differentiate (see Fig. 12.39A). **Mosaic analysis** has allowed investigators to determine if the mutation in each case affects production of the signal itself, or if it affects the reception mechanism (reviewed by Tomlinson, 1989).

Mosaic analysis involves inducing a patch of mutant tissue in an otherwise normal (i.e., wild-type) eye. When the two patches of tissue meet, wild-type and mutant cells will contact each other. If wild-type cells can "rescue" mutant cells (i.e., if the mutant cells differentiate normally), it suggests that the mutation is in the signaling (or inducer) cell rather than in the receiving (or induced) cell. Conversely, if the wild-type cell cannot rescue the mutant cells, the defect must be in the cell being induced. Such mosaic analysis shows that the *sev* gene is required in cell 7 to receive a signal from cell 8, whereas *boss* is required in cell 8 to send the signal. The *ro* gene product is required in cell 2 and 5 to signal cells 3 and 4 to differentiate (Fig. 12.40). What is intriguing about all these genes is that in the mosaic analysis, normal ommatidia can be found directly adjacent to defective ones. This suggests that the signaling mechanism between the cells operates at very close range and probably involves cell-cell contact.

Which gene products are responsible for communication between inducing and responding cells? The powerful techniques of molecular biology have in these cases allowed investigators to pursue the identities and functions of the proteins encoded by these genes and to begin to unravel the chain of molecular events in signaling.

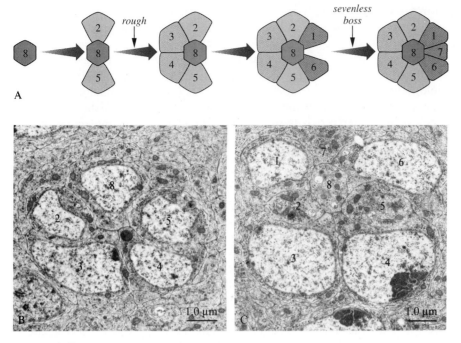

Figure 12.39

Development of ommatidia in *Drosophila*. A, The individual photoreceptor cells are the first to differentiate from the eye imaginal disc epithelium. They differentiate in a fixed sequence, as shown here, with the first cell to differentiate being R8, followed by pairs 2/5, 3/4, 1/6, and finally 7. Mutations in the genes *rough*, *sevenless*, and *bride of sevenless* interrupt development at the points indicated in the diagram. B, and C, Electron micrographs through the differentiating ommatidium, showing the progressive appearance of the receptor pairs in two stages of development about eight hours apart. Pair 1/6 and receptor 7 have appeared in C, but are not yet differentiated in B. (A, After Rubin, 1989. B, From Tomlinson, 1988. Reprinted by permission of Company of Biologists, Ltd.)

The *sev* gene has been isolated, its sequence determined, and the presumed amino acid sequence of its protein product deduced. By comparison with other known protein sequences, this protein has been identified as a transmembrane protein with a large extracellular domain and a cytoplasmic domain with tyrosine kinase activity (Hafen et al., 1987). This structure is very similar to that of receptors of external signals, such as the platelet-derived growth factor receptor, in which a ligand binds to the receptor and tyrosine kinase is modulated. The kinase phosphorylates certain proteins, which modulate their function, thus mediating the effects of the signaling molecule (see the essay on p. 380). Basler and Hafen (1988) have shown that a single amino acid substitution in the tyrosine kinase domain shuts off *sev* gene function, showing that signal transduction operates through the kinase. Thus, the structure and putative function of the *sev* gene agree with the mosaic analysis, which indicated that the *sev* gene is part of the response system in cell 7. In addition, antibody localization of the *sev* protein product shows that it accumulates in cell 7 only in regions where it contacts cell 8, further reinforcing the idea that the *sev* protein receives a signal by contact with cell 8.

Mutations in the *boss* gene result in the same phenotype as mutations in the *sev* gene, but the mutation is in cell 8 (i.e., the signaling cell), as

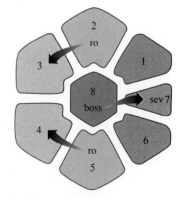

Figure 12.40

Roles of *rough*, *sevenless*, and *bride of sevenless* in photoreceptor cell differentiation. Mosaic analysis has revealed that the *ro* gene is expressed in R2/5 and is needed to induce the differentiation of R3/4. Both the *boss* and *sev* genes are necessary for the differentiation of R7, but *boss* is required in R8, whereas *sev* is required in R7 for normal development. (After Rubin, 1989.)

shown by mosaic analysis (Reinke and Zipursky, 1988). The *ro* gene apparently affects signaling by cells 2 and 5, which cause cells 3 and 4 to develop (Tomlinson, 1989). Therefore, the *boss* and *ro* genes have the potential to reveal details about the inducing signals themselves. The combined approaches of genetic and mosaic analysis with modern molecular biology tools facilitate a more direct approach to understanding induction than is possible from using the classical, indirect approaches. We look forward to further progress in resolving the sequence of molecular events in production, reception, and transduction of the signals that cause the receptor cells to differentiate.

Drosophila is not the only organism in which a combination of genetic analysis and the techniques of molecular biology has facilitated the study of induction. Recently, a number of mutations in the nematode *Caenorhabditis elegans* have allowed us to make significant strides in understanding what genes are involved in cell-cell interactions and identifying their gene products. As we discussed in Chapter 11, many studies have shown that the fate of each cell of the *C. elegans* embryo is invariant; i.e., this nematode is a classic example of mosaic development, where the fate of each cell is fixed throughout development. However, recently a number of investigators have shown that if individual cells are destroyed at different times during development, the fates of neighboring cells are altered. Therefore cell-cell interactions also occur in a manner similar to the inductive interactions documented in vertebrates.

One good example of such a regulative interaction is the development of the gametes within the gonad. As we have already discussed in Chapter 9 (see p. 383), the gonad of the hermaphrodite is derived from four precursor cells; two of them produce gonadal somatic tissue, and two give rise to germ cells. A cell at the tip of the developing gonad (the distal tip cell) is responsible for its elongation. If this cell is ablated, the gonad will cease elongation (Fig. 12.41). Ablation of the distal tip cell also affects gametogenesis.

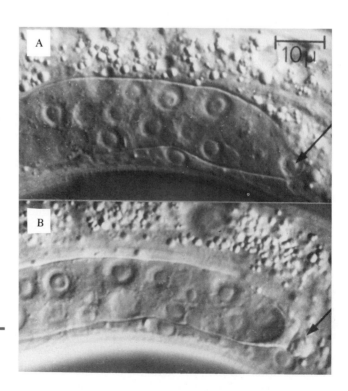

Figure 12.41

A distal tip of the gonad of a *C. elegans* hermaphrodite is indicated by an arrow in A and after laser ablation in B. (From Kimble and White, 1981.)

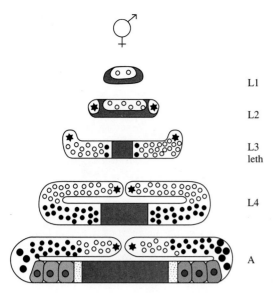

Figure 12.42

Diagram depicting the development of the gonad in a *C. elegans* hermaphrodite at the four larval stages of postembryonic development (Larval stages 1,2,3,4) and in the adult (A). The asterisks represent the distal tip cells, open circles are germ-cell nuclei in mitotic division, closed circles are germ-cell nuclei in meiotic division, and oocytes are shown in color, whereas sperm are represented by stippling. The somatic tissue of the gonad is shaded. leth: lethargus; period just before molting. (After Austin and Kimble, 1987.)

Normally, a gradient of germ cell differentiation is seen along the length of the hermaphrodite gonad (Fig. 12.42). At the distal tip, the germ cell nuclei go through several rounds of mitosis, but as they move away from the distal tip, they enter meiosis. Finally, at the proximal end of the gonad near the vulva, they differentiate into sperm and eggs. If the distal tip is removed at any time during embryonic or adult development, the germ cells shift into meiosis. Thus, some signal appears to emanate from the distal tip that either stimulates mitosis or represses meiosis. Once a germ cell has moved out of range of this signal, it goes into meiosis. At least one gene required for this interaction has been identified.

In screening many mutations that affect germ cell development, Austin and Kimble (1987) found eight recessive alleles, which they named *glp-1* (*germ line proliferation defective*), in which the germ cell precursors undergo meiosis but not mitosis, giving rise to only four to eight germ cells (Fig. 12.43). The effects of this mutation are reminiscent of the defect caused by laser ablation of the distal tip cell (see p. 383). Does this mutation affect the signaling component of the interaction or the receiving end? Mosaic analysis has revealed that the *glp-1* activity is produced by the germ line and not the distal tip cell, showing that the gene is part of the receiving end of the signaling interaction.

The *glp-1* gene has recently been cloned and sequenced, which has yielded some clues about its function (reviewed in Austin et al., 1989). It encodes a large transmembrane protein, which one might expect of a receptor that would transduce an extracellular signal. In its external domain, it contains many epidermal growth factor–like repeating units, although the function of such units in the variety of molecules in which they have been found is not known. However, with the *glp-1* gene in hand, answers should come quickly to the role of the *glp-1* molecule in the signaling between distal tip cell and germ line cells.

Role of Induction in Morphogenesis: Branching of Glands

Until now, we have considered the role of induction in controlling determination of cell fate. It is important at this point to recognize that inductive events can control aspects of morphogenesis as well. One example of an epitheliomesenchymal interaction directing morphogenesis is the branching

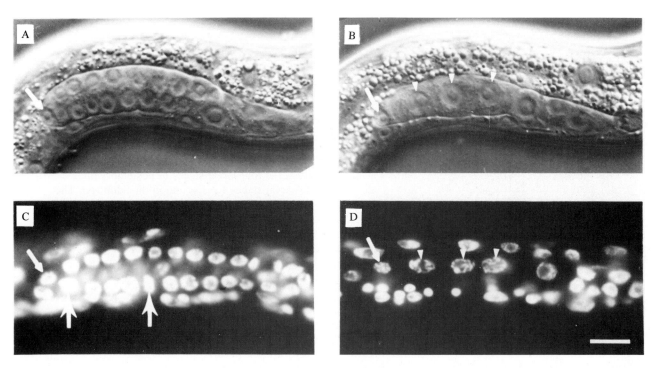

Figure 12.43

When *C. elegans* embryos are mutant for *glp-1*, the germ cells no longer respond
to the inductive cues emanating from the distal tip cells and do not go through
mitosis, resulting in no proliferation of the germ cell precursors. Rather, they
proceed directly into meiosis. The top micrographs are Nomarski images of a
normal (A) and a *glp-1* mutant (B), showing the distal tip cell (arrow) in the gonad
of a hermaphrodite. Note the normal complement of germ cells in the normal
gonad, but only three (arrowheads) in the mutant. The bottom micrographs show
a normal (C) and a mutant (D) animal stained with DAPI, which identifies
chromatin. Many mitotic figures (arrows) are seen in the normal animal (C),
whereas only germ cells in the pachytene stage of meiosis are found in the
mutants. Scale bar equals 10 μm. (From Austin and Kimble, 1987. Copyright ©
Cell Press.)

of various glands. The interaction between salivary gland epithelium and
mesenchyme has been studied extensively, particularly with respect to the
morphogenesis of the epithelial component.

The initial rudiments of the mammalian submandibular salivary glands
are a pair of club-shaped epithelial buds that grow downward on either side
of the tongue into the connective tissue layer. The buds soon become
surrounded by mesenchyme, which condenses around them to form a
connective tissue capsule. The epithelial portions of the glands then begin
to branch and form knob-like thickenings (or **lobes**) at the tips of each branch
(Fig. 12.44). The lobes divide repeatedly to form the secretory acini, whereas
the branches form the duct system. The branching of the lobes begins with
the appearance of clefts in the rounded outer surface. As the clefts gradually
deepen, the adjacent portions of the lobe expand and become bifurcated by
new clefts, producing a tree-like structure. It has been postulated that two
morphogenetic mechanisms assist in branching. Expansion of the tips of
the lobes appears to result from an elevated rate of mitosis (Wessells, 1977),
whereas the formation of clefts may be due to contraction of microfilaments

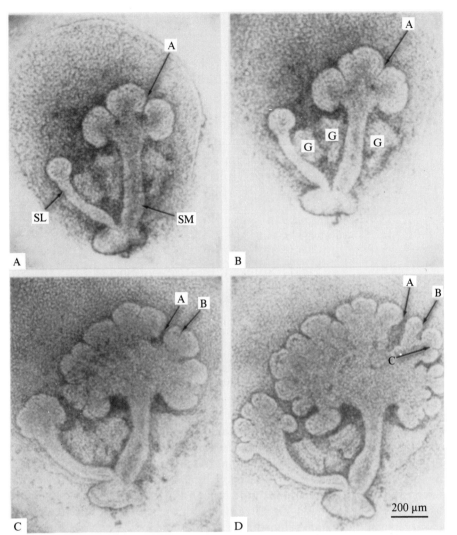

Figure 12.44

In vitro morphogenesis of a mouse salivary gland. Both the submandibular (SM) and sublingual (SL) glands are present. The branching process can be followed in the progressively older glands (A-D). Note the progressive deepening of cleft A, the appearance of cleft B, its deepening, and finally the appearance of cleft C. G: ganglia. (From N. K. Wessells, 1977. *Tissue Interactions and Development.* Reprinted by permission of Benjamin/Cummings Publishing Co.)

(Spooner and Wessells, 1972), because treatment with cytochalasin will relax the clefts.

The need for mesenchyme in promoting epithelial morphogenesis is evident when the two components are separated and cultured *in vitro*. Both epithelium and mesenchyme are viable when cultured separately, but they do not undergo normal morphogenesis: The epithelium fails to branch, and the mesenchyme forms a sheet of proliferating cells. If, however, the components are positioned adjacent to one another in a culture dish, the mesenchyme surrounds the epithelial bud, which proceeds to branch and form secretory acini (Fig. 12.45). The capacity of salivary mesenchyme to promote the specialized branched morphology of the salivary glands is indicated by an experiment in which mouse salivary mesenchyme was combined *in vitro* with mouse mammary gland epithelium. The mammary epithelium normally develops a branched morphology quite distinct from that of the salivary epithelium (Fig. 12.46A and B). However, in combination with salivary mesenchyme, it branches and forms knobs reminiscent of salivary tissue (see Fig. 12.46C). Although the mammary tissue assumes the *form* of salivary tissue, it is not known whether the cells undergo salivary *differentiation*. In other words, does salivary mesenchyme induce the mammary cells to produce salivary-specific proteins instead of milk proteins?

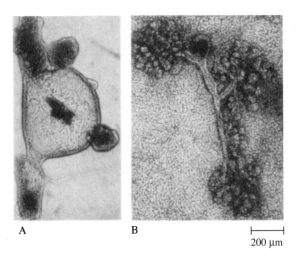

A   B

|———|
200 µm

Figure 12.45

The effects of homologous mesenchyme on morphogenesis of the mouse submandibular gland *in vitro*. A, When the epithelium is cultured alone, the gland fails to branch and instead spreads on the culture dish. B, When the epithelium is cultured with homologous mesenchyme, normal branching occurs. (From Grobstein, 1953a.)

The ability of salivary mesenchyme to promote branching morphogenesis of heterologous epithelium suggests that this mesenchyme has a critical formative role in salivary development. Is this mesenchyme unique in its ability to promote salivary morphogenesis, or can other mesenchymes substitute for it? The specificity of the mesenchymal requirement has been examined by combining the submandibular epithelial bud with mesenchyme of various origins. Most mesenchymes fail to promote the growth and differentiation of the epithelium (Grobstein, 1953a). On this basis, the salivary mesenchyme had long been considered to be a specific promoter of salivary morphogenesis. More recent evidence, however, indicates that the requirement of salivary epithelium for its homologous mesenchyme is not so specific as was previously thought. Lawson (1974) has shown that mouse submandibular epithelium will branch and form acini when cocultured with mouse lung mesenchyme. However, lung mesenchyme must be present in larger amounts than submandibular mesenchyme to achieve the same

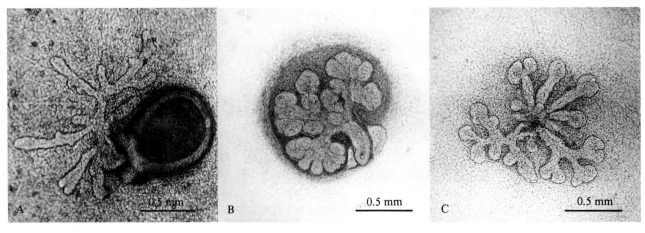

A   0.5 mm   B   0.5 mm   C   0.5 mm

Figure 12.46

The effects of salivary mesenchyme on morphogenesis of mouse mammary epithelium. Mammary gland (A) and salivary gland (B) have very different branching morphologies. When mammary gland epithelium is combined with salivary gland mesenchyme (C), morphogenesis resembles that of a typical salivary gland. (From Kratochwil, 1969.)

degree of morphogenesis. This result suggests that the ability to branch in a characteristic way is inherent to salivary epithelium, and the role of the mesenchyme is to *permit* morphogenesis.

The nature of the interaction between salivary mesenchyme and epithelium has been studied extensively in efforts to clarify how epithelial morphogenesis is influenced by the mesenchyme. One of the possibilities to be considered is that a diffusible inducing substance is transmitted to the epithelium from the mesenchyme. However, transfilter experiments (see p. 509) indicate that induction is not due to a diffusible substance (Saxén et al., 1976). Recent experiments have implicated the basal lamina in maintenance of the branched morphology and the mesenchyme in modification of the basal lamina to facilitate branching morphogenesis. In the intact embryo, the basal lamina is located at the interface between the epithelium and mesenchyme. However, it is not distributed uniformly over the epithelial surface. Instead, it is thinner at the distal tips of the lobes than in the clefts. Ultrastructurally, there are many discontinuities of the basal lamina at the tips (Coughlin, 1975), and mesenchymal cells are occasionally observed to contact the epithelial cells in these regions. To evaluate the morphogenetic roles of the basal lamina and mesenchyme, investigators have conducted experiments examining the *in vitro* morphogenesis of epithelia either in their absence or with one or both of these components present.

If mesenchyme alone is removed from the epithelium (leaving the basal lamina intact), the epithelium retains its branched morphology, although it will not branch further. Treatment of the mesenchyme-free epithelium with hyaluronidase degrades glycosaminoglycans (GAGs), thus removing the basal lamina. In the presence of mesenchyme, the basal lamina–free epithelium rounds up, and *after considerable delay*, the basal lamina is regenerated, after which branching is reinitiated (Fig. 12.47).

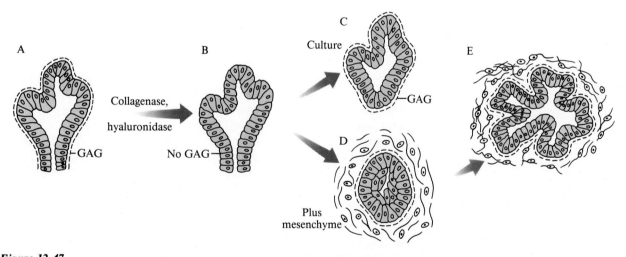

Figure 12.47

The combined effects of GAG and mesenchyme on salivary morphogenesis. A, Salivary epithelium with its GAG-containing basal lamina intact. B, Enzymatic treatment has removed the GAG and thereby disrupted the basal lamina. One of two results is obtained when this epithelium is cultured. In the absence of mesenchyme (C), GAG is replenished, but no morphogenesis takes place. In the presence of mesenchyme, the epithelium initially rounds up (D), but eventually the GAG is replenished and branching morphogenesis resumes shortly thereafter (E). (After Wessells, 1977.)

The effects of basal lamina removal can be reversed by first culturing the epithelium in the absence of mesenchyme for a short time. Under these conditions, the basal lamina is rapidly regenerated. When this epithelium is combined with mesenchyme, the branched morphology is maintained. These results indicate that (1) the integrity of the basal lamina is dependent upon its glycosaminoglycans; (2) maintenance of branched morphology is dependent upon the basal lamina; and (3) mesenchyme delays the formation of basal lamina. As a consequence of this delay, the epithelium rounds up and requires a long latent period before branching can resume (Bernfield, 1981). It is seemingly enigmatic that mesenchyme, which is required for branching morphogenesis, causes loss of branching and delays both basal lamina formation and branching morphogenesis.

What, then, is the role of mesenchyme in morphogenesis? Not surprisingly, if epithelium cleaned of both mesenchyme and basal lamina is cultured without mesenchyme, branching will not occur, even though basal lamina will eventually accumulate at the epithelial surface (see Fig. 12.47). Thus, although basal lamina seems to be necessary for morphogenesis, it is not **sufficient**; the mesenchyme provides the proper conditions.

The presumed roles of the mesenchyme in salivary morphogenesis are summarized in the model in Figure 12.48. The mesenchyme appears to function in the generation of the branching morphology of the epithelium and in stabilizing the morphology once it appears. As we discussed previously, the epithelium expands by mitosis at the tips of the lobes. The GAGs in the basal laminae covering the tips are undergoing rapid turnover (the rate of degradation at the tips actually exceeds the rate of synthesis; Bernfield and Banerjee, 1982). The mesenchyme's role in GAG turnover apparently involves producing hyaluronidase, which degrades the existing GAGs at the tips (Bernfield et al., 1984). Restoration of GAGs then occurs by the synthetic activity of the epithelial cells. However, before producing a new basal lamina, the epithelium (which has become free of this restrictive layer) invaginates by the contractile action of microfilaments within the cells. A cleft is thus produced.

The beginning cleft deepens by expansion of the flanking lobes (due to cell division). As it deepens, the mesenchyme helps to stabilize the cleft by depositing bundles of collagen. Collagen apparently stabilizes the cleft by rendering the basal lamina more stable to mesenchyme-induced degradation (David and Bernfield, 1981). As a consequence, the basal lamina plus associated collagen maintains the morphology of the epithelium in the cleft region (Bernfield, 1981). Meanwhile, at the tips of the expanding lobes, the GAGs are being degraded so that additional clefts can be formed.

Does mesenchyme also play a role in promoting cell differentiation as well as morphogenesis? The answer to this question might be found if investigators could ascertain whether the cells of mammary epithelium, in combination with salivary mesenchyme, synthesize salivary—rather than mammary—proteins.

12–4 Signaling and Pattern

The organization of differentiated cells in the organism is extremely complex and ordered. Thus, it is not sufficient for us to understand what factors make a particular cell become bone or another muscle; we also wish to understand what controls the spatial organization of these cells within a particular structure. For example, in the limb, we would like to know not only what causes bone to form, but why a particular group of bone cells becomes a humerus, whereas another group becomes radius or ulna. This

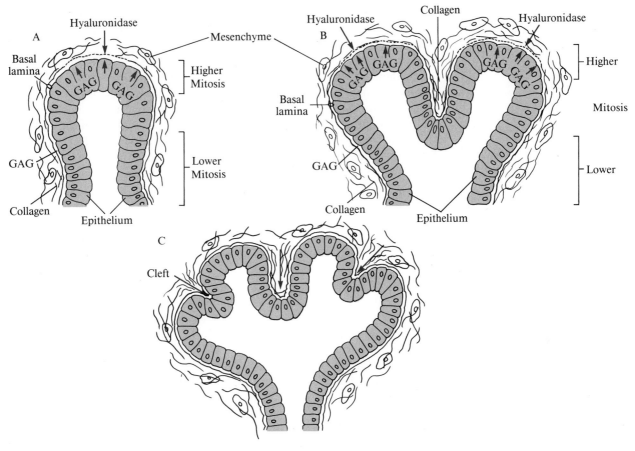

Figure 12.48

Model depicting the control of salivary gland morphogenesis. A, As a result of hyaluronidase secretion by the mesenchyme at the tip region, basal lamina is degraded locally. Clefting is initiated at this site. B, The GAG-containing basal lamina is stabilized in the cleft by the deposition of collagen, whereas basal lamina degradation resumes at the new tips, resulting in another round of branching (C). (After Wessells, 1977.)

ordered spatial arrangement of differentiated cells is known as **pattern formation**. As we shall discuss later, intercellular signaling appears to play an important role in pattern formation.

We have a much more thorough understanding of the *rules* that govern patterning than of the mechanisms that are involved. For example, as we shall see on page 528, regeneration in *Hydra* follows a set of rules that reliably predict the pattern of regenerating tissues. It is hoped that an understanding of these rules will lead to a better understanding of the mechanisms that are involved. The state of our understanding of pattern formation might be likened to that of genetics before the elucidation of the structure of DNA and the genetic code. By the middle of the twentieth century, we had a thorough understanding of the rules for predicting inheritance of specific characteristics, even though we had only a vague notion that the genetic information was carried in DNA and were not to know until 1953 how the structure of DNA with its base complementarity could account for genetic *rules* that had already been discerned.

Numerous models have been proposed to account for the many patterns that arise during development. A model is an hypothesis that proposes a mechanism of how a developmental phenomenon occurs. Central to most of these is the concept of **positional information** that was proposed by Lewis Wolpert (1969) and that was responsible for a resurgence of interest in the subject of pattern. Wolpert reasoned that cells act as though they "sense" where they are in a developing structure. Each group of cells that is to become a specific structure, such as a wing or a leg, is known as a **developmental field**. Such fields have discrete boundaries or borders, and cells that make up the field are believed to know what position they occupy within the boundaries of the field. There are at least two coordinate systems that may specify a cell's position within a field: the polar and Cartesian coordinate systems. Once a cell knows what its position is with respect to other cells within the field and in relationship to the field's boundaries, it must **interpret** its position appropriately (i.e., differentiate into the correct cell type for its position within the field). Thus, a mesoderm cell will differentiate into bone, be it a humerus or wrist element.

A major problem addressed by many of the theories of pattern is what establishes the coordinate system. By far the most prevalent idea is that regulatory molecules known as **morphogens** are distributed in some gradient within the field that can direct differentiation (Slack, 1987). Morphogens are signaling molecules that are generally supposed to be diffusible but do not necessarily have to be. Furthermore, some of these diffusing molecules may exist as a gradient in the milieu around individual cells or, if they are sufficiently small, could be passed directly from one cell to the next through gap junctions. There is evidence that all of these possibilities may exist in different developmental systems.

The best-characterized diffusible gradient system operates to organize the anteroposterior axis in the embryonic chick wing. As we shall discuss in detail in Chapter 15, retinoic acid emanates from the posterior wing bud and diffuses to the anterior side. The resultant gradient is sufficient to account for organizing the anteroposterior pattern of wing structures.

Morphogens need not exist as a simple monotonic gradient, like the retinoic acid gradient in the limb. In the early 1950s a class of models was described (Turing, 1952) in which spatial patterns arise after minor perturbations in morphogen gradients that had been essentially homogeneous. Developmental biologists who were interested in pattern generation saw the usefulness of these models, and they have been extensively adapted, particularly in the work of Gierer and Meinhardt (1972). Their models involve two morphogens that diffuse from a single source to form separate gradients. One gradient consists of an **activator**, which stimulates its own production and is responsible for differentiation in specific local regions. The other consists of an **inhibitor**, which will repress the formation of the activator and thus prevent the spread of the activator to wider areas. As the two substances diffuse from their initial source, they are removed at a rate proportional to their concentrations. In their model, Gierer and Meinhardt have imposed two important parameters on the activator and inhibitor: (1) The inhibitor diffuses faster and further than the activator (this is to ensure that the activator induces locally, and the inhibitor will prevent release of the activator beyond this local area); and (2) the inhibitor is much more labile, so that if its source is removed, the concentration of the inhibitor in the rest of the tissue will fall rapidly, whereas the self-catalyzing activator will remain much longer. This model is known as a reaction-diffusion model.

In at least one situation—the regeneration of *Hydra*—there is now considerable evidence that activator and inhibitor gradients exist. *Hydra* is a coelenterate whose simple body plan (Fig. 12.49) is composed of a two-layered epithelial tube, with a **basal disc**, or foot, and a head composed of

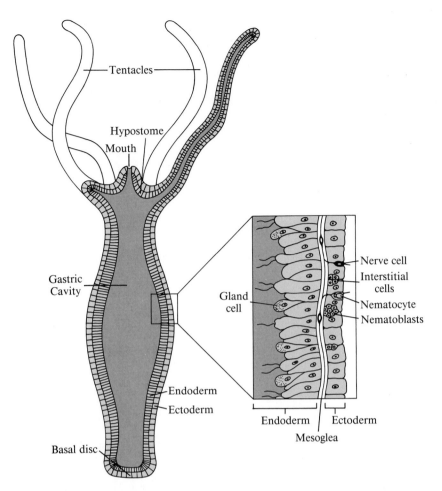

Figure 12.49

Diagram of a *Hydra* with boxed area of the epithelium expanded on the right to show the various epithelial cell types. (After Schaller et al., 1989.)

a central mouth and a ring of tentacles known as a **hypostome**. *Hydra* generally reproduces by asexual budding of offspring from the body wall of the parent. If the head of *Hydra* is removed, the former middle of the organism will form a new head, and all the cells in the *Hydra* will become respecified to form a normally proportioned, although smaller, animal. When a *Hydra* is cut in half, the cut edge of the piece retaining the head forms a new foot, whereas the cut end of the piece containing the original foot forms a new head. Also, if a slice of tissue is removed from the center of the body, a new head and foot will regenerate, reflecting the original polarity of the intact organism. Such results suggest that the cells in the *Hydra* have acquired positional information that allow them to reconstruct their original pattern.

A variety of grafting experiments suggests that both a head activator and a head inhibitor gradient arise from the head (reviewed by Bode and Bode, 1984; Schaller et al., 1989). The presence of a head inhibitor can be demonstrated by the following experiments (Fig. 12.50). If a section of the body wall of *Hydra* is taken from the gastric region and grafted to a region of a host near the middle of the body column, a second head will form less than 20% of the time. However, the frequency of head formation can be increased more than fourfold if the host head is cut off at the time of

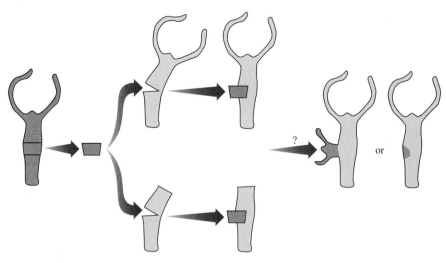

Figure 12.50

Diagrammatic representation of the transplantation experiments used to demonstrate the existence of the head inhibitor gradient. A piece of body column inserted into the body column of a host will form a second head in less than 20% of cases, whereas if the host is decapitated, the frequency of head formation increases more than four-fold. (After Bode and Bode, 1984.)

transplantation. This experiment suggests that the head produces an inhibitor of head production that is removed when the head is decapitated. This inhibitor is found in a gradient with its source at the head. For example, one can take pieces of body column from directly below the head and graft them into the body wall at increasing distances from the host head (Fig. 12.51). The percentage of heads generated increases the further the graft is placed from the head. Put another way, the closer the graft is to the host head, the greater the inhibition of head generation.

Apparently, the head inhibitor factor is very labile (i.e., it is degraded rapidly): In decapitation experiments, the inhibitor was found to be at a minimum level 6 hours after decapitation. Schaller et al. (1979, 1989) have reported the partial purification of a small molecule that appears to have head inhibition properties. Its small size is suited to its need to diffuse rapidly.

The qualities of the head inhibitor correspond to those predicted by the Gierer-Meinhardt model. How is the gradient of the inhibitor molecule propagated down the body column? Because the *Hydra* body column is an epithelium, it had been hypothesized that these molecules may diffuse from cell to cell through epithelial gap junctions. In order to do so, a molecule must be fairly small, which the inhibitor molecule, at least, is known to be. Recently, some elegant experiments by Fraser et al. (1987) suggest that the patterning gradients of *Hydra* are indeed propagated through gap junctions.

Fraser et al. (1987) first showed that epithelial cells in *Hydra* can communicate with each other by injecting into a single cell a dye, Lucifer yellow, that can diffuse only through gap junctions. The dye was found some time later in many of the adjacent epithelial cells. Next, they showed that they could inactivate the gap junctions (i.e., render them nonfunctional) with an antibody against a gap junction protein. When an animal was treated with this antibody, injected Lucifer yellow no longer diffused to adjacent cells. When the junctions were inactivated, Fraser and colleagues were also

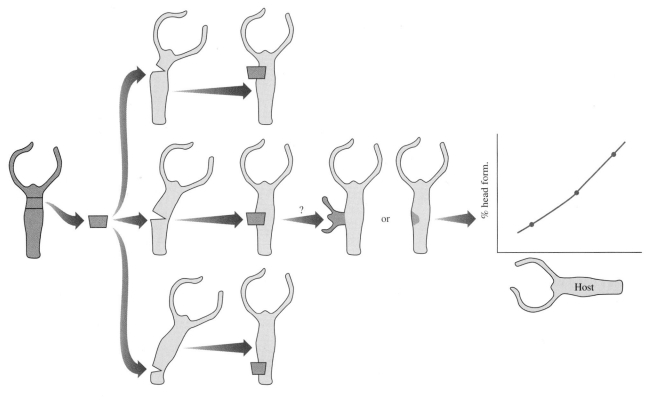

Figure 12.51

On the left, a piece of *Hydra* is explanted from the most anterior region, which contains the highest level of head activator substance. It is then grafted into another host *Hydra* at varying distances from the head. The graph on the right shows that the further the graft is placed from the head, the greater the chance of inducing a new head. Thus, there is a gradient of head inhibition emanating from the head and decreasing along the length of the *Hydra*. (After Bode and Bode, 1984.)

able to demonstrate by grafting experiments that the inhibition gradient was lost. For example, they took a piece of epithelium from near the head region and grafted it near the middle of the body cavity. As described previously, only a low level of head formation results from such grafts because the head inhibitor represses new head formation. When the same experiment is done using antibody-treated host animals, a significant increase in head formation over the control animals results (22% increase). This suggests that the head inhibitor has not been able to diffuse from the head owing to the inactivation of the gap junctions. This is the first experimental evidence that gap junctions can function in the maintenance of gradients that carry patterning information.

A gradient of a head activator substance has also been identified experimentally by grafting rings of epithelium from various levels of the *Hydra* to the middle of a host body column (Fig. 12.52). The ability of the grafts to initiate a new head decreases as their level of origin is progressively more posterior. If the grafted pieces are placed in decapitated hosts, the frequency of head formation is higher because the inhibitor gradient is gone, but the same gradation in head activation is still observed. The activator gradient differs from the inhibitor gradient in that it decays much less rapidly after decapitation, and it diffuses less down the body column.

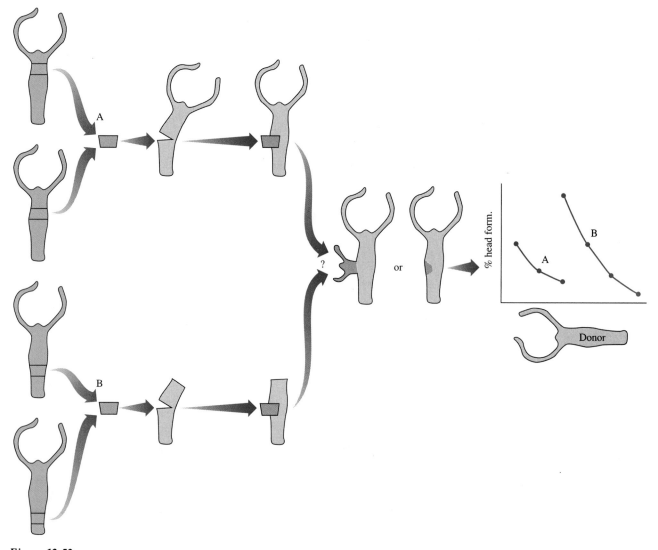

Figure 12.52

Grafting procedures to demonstrate the existence of a head activator gradient. A, Pieces of *Hydra* are taken from various anterior levels of the animal and grafted into the middle of a host. The graph illustrates that the more anterior the origin of the graft, the greater the percentage of head formation. B, The gradient is more striking if the host animal is first decapitated so that there is no head inhibitor gradient. Under these circumstances, even pieces of *Hydra* taken from posterior levels (which contain low levels of head activator) are capable of inducing a new head. (After Bode and Bode, 1984.)

Head activator has been isolated and identified biochemically (Schaller and Bodenmüller, 1981). It is a small peptide that is prevented from diffusing rapidly by binding to a large carrier molecule. If activator is not bound to the carrier, it dimerizes rapidly and becomes inactive (Bodenmüller et al., 1986). This rapid inactivation fulfills the requirement of limited diffusion.

Head activator has been shown to have at least two effects on *Hydra* cells (reviewed in Schaller et al., 1989): (1) All cells in *Hydra* arrested in G_2 phase are stimulated to divide within one to two hours after treatment; and (2) head activator directs the differentiation of epithelial cells to head-type and increases the number of head-type nerve cells that differentiate from stem cells, which are called interstitial cells. It is these nerve cells that produce head activator in the first place, thus creating an autocrine feedback loop that ensures continued head development. It is interesting to note that head activator is also found in mammalian tissue and functions as a growth factor in mammals as well.

The Role of Specific Genes in Patterning

The factors that specify pattern must exert their effects on development by regulating gene expression. In at least one developmental system, the *Drosophila* embryo, we are beginning to understand pattern at the level of the gene thanks to the extraordinary number of mutants that affect pattern and also to the techniques of molecular biology that allow us to probe gene function. The genes that are particularly valuable for understanding pattern formation are those concerned with segmentation of the early *Drosophila* embryo. We shall discuss this fascinating topic in Chapter 14.

12–5 Chapter Synopsis

This chapter describes the role of induction in cell determination. The first inductive interaction during vertebrate development involves the formation of mesoderm. In amphibian embryos, mesoderm is formed by the interaction of vegetal hemisphere blastomeres, which produce inducing substances, with equatorial blastomeres, which are induced to become mesoderm. Qualitatively distinct inductive events are generated by vegetal hemisphere blastomeres depending on their position in the embryo. Vegetal cells on the dorsal side of the embryo induce notochord and muscle, whereas vegetal cells on the lateral and ventral side of the embryo induce blood cells. A third inductive signal from the dorsal mesoderm induces ventral mesoderm to form muscle. Recent research suggests that mesoderm inducers may be growth factors belonging to the TGF-β and FGF families.

Perhaps the most intensively studied of all inductive interactions is neural induction by the chordamesoderm. Spemann and Mangold showed that if they grafted a second dorsal lip from one urodele embryo into the ventral region of another early gastrula embryo, it developed a second anteroposterior axis, including a duplicated neural tube, somites, and gut. By using labeled cells for their grafts, they showed that the grafted cells contributed only to the notochord and portions of the somites, whereas the neural tube was induced to form from the host's overlying ectoderm. The identities of the inductive signals are currently being sought.

Many other cell types originate as a result of inductive interactions during development, a process termed secondary induction. Thus, induction is a major mechanism whereby different cell types are generated from a few undifferentiated precursors. Inductive interactions have been classified as either instructive or permissive. In an instructive interaction, the inducer provides the induced tissues with instructions as to what it will become. In permissive interactions, the induced tissue is already poised to differentiate into a particular tissue type and only needs the start signal to begin the process of differentiation.

Few inducing molecules have been identified, because they are present in such minute amounts in tissue that can only be isolated in small amounts for biochemical analysis. Most inducers have been identified in lower organisms, such as hydra or *Dictyostelium*, from which large amounts of tissue are available.

In recent years, our understanding of induction and cell-cell interactions has been increased enormously by studies using organisms in which mutants that affect induction are available. Many such mutants have been identified in *Drosophila* and *C. elegans*. Because the mutant genes and their normal counterparts can be cloned and analyzed, investigators are now beginning to identify some of the molecules responsible for induction in these organisms. It is believed that many of the cell-cell interactions in lower organisms will be comparable to those that occur in vertebrates.

We have not only considered the factors that control differentiation of specific cell types, but we have also discussed the mechanisms by which these cell types acquire their specific spatial organization within the organism. This ordered spatial arrangement of differentiated cells is known as pattern formation. Some of the rules and specific molecules that govern patterning in the embryos have been discussed.

References

Asashima, M. et al. 1990. Mesoderm induction in early amphibian embryos by activin A (erythroid differentiation factor). Wilhelm Roux's Archives Dev. Biol., *198*: 330–335.

Austin, J., and J. Kimble. 1987. *glp-1* is required in the germ line for regulation of the decision between mitosis and meiosis in C. elegans. Cell, *51*: 589–599.

Austin, J., E. M. Maine, and J. Kimble. 1989. Genetics of intercellular signaling in *C. elegans*. Development, *107* (Suppl.): 53–57.

Balinsky, B. I. 1975. *An Introduction to Embryology*, 4th ed. W. B. Saunders, Philadelphia.

Basler, K., and E. Hafen. 1988. Control of photoreceptor cell fate by the sevenless protein requires a functional tyrosine kinase domain. Cell, *54*: 299–311.

Bautzmann, H. et al. 1932. Versuche zur Analyse der Induktionsmittel in der Embryonalentwicklung. Naturwissenschaften, *20*: 971–974.

Bernfield, M. R. 1981. Organization and remodeling of the extracellular matrix in morphogenesis. *In* T. G. Connelly, L. L. Brinkley, and B. M. Carlson (eds.), *Morphogenesis and Pattern Formation*. Raven Press, New York, pp. 139–162.

Bernfield, M. R., and S. D. Banerjee. 1982. The turnover of basal lamina glycosaminoglycan correlates with epithelial morphogenesis. Dev. Biol., *90*: 291–305.

Bernfield, M. R. et al. 1984. Remodeling of the basement membrane as a mechanism of morphogenetic tissue interaction. *In* R. L. Trelstad (ed.), *The Role of Extracellular Matrix in Development*. Alan R. Liss, New York, pp. 545–572.

Bode, P. M., and H. R. Bode. 1984. Patterning in hydra. *In* G. M. Malacinski and S. V. Bryant (eds.), *Pattern Formation*. Macmillan Publishing Co., New York, pp. 213–241.

Bodenmüller, H. et al. 1986. The neuropeptide head activator loses its biological activity by dimerisation. EMBO J., *5*: 1825–1829.

Boterenbrood, E. C., and P. D. Nieuwkoop. 1973. The formation of mesoderm in urodelean amphibians. V. Its regional induction by the endoderm. Wilhelm Roux's Archives Dev. Biol., *173*: 319–322.

Cairns, J. M., and J. W. Saunders, Jr. 1954. The influence of embryonic mesoderm on the regional specification of epidermal derivatives in the chick. J. Exp. Zool., *127*: 221–248.

Coughlin, M. D. 1975. Early development of parasympathetic nerves in the mouse submandibular gland. Dev. Biol., *43*: 123–139.

Coulombre, J. L., and A. J. Coulombre. 1971. Metaplastic induction of scales and feathers in the corneal anterior epithelium of the chick embryo. Dev. Biol., *25*: 464–478.

Dale, L., and J. M. W. Slack. 1987. Regional specification within the mesoderm of early embryos of *Xenopus laevis*. Development, *100*: 279–295.

David, G., and M. R. Bernfield. 1981. Type I collagen reduces the degradation of basal lamina proteoglycan by mammary epithelial cells. J. Cell Biol., *91*: 281–286.

Dhouailly, D. 1975. Formation of cutaneous appendages in dermo-epidermal recombinations between reptiles, birds and mammals. Wilhelm Roux's Archives Dev. Biol., *177*: 323–340.

Dixon, J. E., and C. R. Kintner. 1989. Cellular contacts required for neural induction in *Xenopus* embryos: Evidence for two signals. Development, *106*: 749–757.

Durston, A. J. et al. 1989. Retinoic acid causes an anteroposterior transformation in the developing central nervous system. Nature (Lond.), *340*: 140–144.

Fraser, S. E. et al. 1987. Selective disruption of gap junctional communication interferes with a patterning process in *Hydra*. Science, *237*: 49–55.

Gierer, A., and H. Meinhardt. 1972. A theory of biological pattern formation. Kybernetik, *12*: 30–39.

Gilbert, S. F. 1985. *Developmental Biology*. Sinauer Associates, Sunderland, MA.

Golosow, N., and C. Grobstein. 1962. Epitheliomesenchymal interaction in pancreatic morphogenesis. Dev. Biol., *4*: 242–255.

Grainger, R. M., J. J. Henry, and R. A. Henderson. 1988. Reinvestigation of the role of the optic vesicle in embryonic lens induction. Development, *102*: 517–526.

Green, J. B. A., and J. C. Smith. 1990. Graded changes in dose of a *Xenopus* activin A homologue elicit stepwise transitions in embryonic cell fate. Nature (Lond.), *347*: 391–394.

Grobstein, C. 1953a. Epithelio-mesenchymal specificity in the morphogenesis of mouse sub-mandibular rudiments *in vitro*. J. Exp. Zool., *124*: 383–413.

Grobstein, C. 1953b. Morphogenetic interaction between embryonic mouse tissues separated by a membrane filter. Nature (Lond.), *172*: 869–871.

Grobstein, C. 1955. Tissue interaction in the morphogenesis of mouse embryonic rudiments *in vitro*. *In* D. Rudnick (ed.), *Aspects of Synthesis and Order in Growth*. Princeton University Press, Princeton, pp. 233–256.

Grobstein, C. 1956. Transfilter induction of tubules in mouse metanephrogenic mesenchyme. Exp. Cell Res., *10*: 424–440.

Grunz, H., and L. Tacke. 1986. The inducing capacity of the presumptive endoderm of *Xenopus laevis* studied by transfilter experiments. Wilhelm Roux's Archives Dev. Biol., *195*: 467–473.

Gurdon, J. B. 1987. Embryonic induction—molecular prospects. Development, *99*: 285–306.

Gurdon, J. B. et al. 1984. Transcription of muscle-specific actin genes in early Xenopus development: Nuclear transplantation and cell dissociation. Cell, *38*: 691–700.

Gurdon, J. B. et al. 1985. Activation of muscle-specific actin genes in Xenopus development by an induction between animal and vegetal cells of a blastula. Cell, *41*: 913–922.

Hafen, E. et al. 1987. *sevenless*, a cell-specific homeotic gene of *Drosophila*, encodes a putative transmembrane receptor with a tyrosine kinase domain. Science, *236*: 55–63.

Hamburger, V. 1988. *The Heritage of Experimental Embryology: Hans Spemann and the Organizer*. Oxford University Press, Oxford.

Hauschka, S., and I. Konigsberg. 1966. The influence of collagen on the development of muscle clones. Proc. Natl. Acad. Sci. U.S.A., *55*: 119–126.

Hay, E. D., and J. W. Dodson. 1973. Secretion of collagen by corneal epithelium. I. Morphology of the collagenous products produced by isolated epithelia grown on frozen-killed lens. J. Cell Biol., *57*: 190–213.

Henry, J. J., and R. M. Grainger. 1987. Inductive interactions in the spatial and temporal restriction of lens-forming potential in embryonic ectoderm of *Xenopus laevis*. Dev. Biol., *124*: 200–214.

Holtfreter, J. 1936. Regionale Induktionen in xenoplastisch zusammengesetzten Explantaten. Wilhelm Roux's Archiv für Entwicklungsmechanik, *134*: 466–550.

Holtfreter, J. 1947. Neural induction in explants which have passed through a sublethal cytolysis. J. Exp. Zool., *106*: 197–222.

Holtzer, H. 1968. Induction of chondrogenesis: A concept in terms of mechanisms. *In* R. Fleischmajer and R. E. Billingham (eds.), *Epithelial-Mesenchymal Interactions.* 18th Hahnemann Symposium. Williams & Wilkins, Baltimore, pp. 152–164.

Hopper, A. F., and N. H. Hart. 1985. *Foundations of Animal Development.* Oxford University Press, New York.

Jacobson, A. 1966. Inductive processes in embryonic development. Science, *152*: 25–34.

Jacobson, A. G., and A. K. Sater. 1988. Features of embryonic induction. Development, *104*: 341–359.

Kay, R. R., M. Berks, and D. Traynor. 1989. Morphogen hunting in *Dictyostelium.* Development, *107* (Suppl.): 81–90.

Kay, R. R., and K. A. Jermyn. 1983. A possible morphogen controlling differentiation in *Dictyostelium.* Nature (Lond.), *303*: 242–244.

Kimble, J. E., and J. G. White. 1981. On the control of germ cell development in *Caenorhabditis elegans.* Dev. Biol., *81*: 208–219.

Kimelman, D., and M. Kirschner. 1987. Synergistic induction of mesoderm by FGF and TGFβ and the identification of an mRNA coding for FGF in the early Xenopus embryo. Cell, *51*: 869–877.

Kimelman, D. et al. 1988. The presence of FGF in the frog egg: Its role as a natural mesoderm inducer. Science, *242*: 1053–1056.

Kollar, E. J., and G. R. Baird. 1969. The influence of the dental papilla on the development of tooth shape in embryonic mouse tooth germs. J. Embryol. Exp. Morphol., *21*: 131–148.

Kollar, E. J., and G. R. Baird. 1970. Tissue interactions in embryonic mouse tooth germs. II. The inductive role of the dental papilla. J. Embryol. Exp. Morphol., *24*: 173–186.

Kollar, E. J., and C. Fisher. 1980. Tooth induction in chick epithelium: Expression of quiescent genes for enamel synthesis. Science, *207*: 993–995.

Kosher, R. A., and J. W. Lash. 1975. Notochord stimulation of *in vitro* chondrogenesis before and after enzymatic removal of perinotochordal materials. Dev. Biol., *42*: 362–378.

Kratochwil, K. 1969. Organ specificity in mesenchymal induction demonstrated in the embryonic development of the mammary gland of the mouse. Dev. Biol., *20*: 46–71.

Lash, J. W., and N. S. Vasan. 1978. Somite chondrogenesis *in vitro*: Stimulation by exogenous extracellular matrix components. Dev. Biol., *66*: 151–171.

Lawson, K. A. 1974. Mesenchyme specificity in rodent salivary gland development: The response of salivary epithelium to lung mesenchyme *in vitro.* J. Embryol. Exp. Morphol., *32*: 469–493.

Lehtonen, E. et al. 1975. Demonstration of cytoplasmic processes in Millipore filters permitting kidney induction. J. Embryol. Exp. Morphol., *33*: 187–203.

Lewis, W. H. 1904. Experimental studies on the development of the eye in amphibia. I. On the origin of the lens in *Rana palustris.* Am. J. Anat., *3*: 505–536.

Lumsden, A. G. S. 1987. The neural crest contribution to tooth development in the mammalian embryo. *In* P. F. A. Maderson (ed.), *Developmental and Evolutionary Aspects of the Neural Crest.* John Wiley & Sons, New York, pp. 261–300.

Mangold, O. 1933. Über die Induktionsfähigkeit der verschiedenen Bezirke der Neurula von Urodelen. Naturwissenschaften, *21*: 761–766.

Meier, S., and E. D. Hay. 1974. Control of corneal differentiation by extracellular materials. Collagen as a promotor and stabilizer of epithelial stroma production. Dev. Biol., *38*: 249–270.

Meier, S., and E. D. Hay. 1975. Stimulation of corneal differentiation by interaction between cell surface and extracellular matrix. I. Morphometric analysis of transfilter ''induction.'' J. Cell Biol., *66*: 275–291.

Mitrani, E. et al. 1990. Activin can induce the formation of axial structures and is expressed in the hypoblast of the chick. Cell, *63*: 495–501.

Mohun, T. J., N. Garrett, and J. B. Gurdon. 1986. Upstream sequences required for tissue-specific activation of the cardiac actin gene in *Xenopus laevis* embryos. EMBO J., *5*: 3185–3193.

Mohun, T. J. et al. 1989. The CArG promoter sequence is necessary for muscle-specific transcription of the cardiac actin gene in *Xenopus* embryos. EMBO J., *8*:1153–1161.

Morris, H. R. et al. 1987. Chemical structure of the morphogen differentiation inducing factor from *Dictyostelium discoideum*. Nature (Lond.), *328*: 811–814.

Nieuwkoop, P. D. 1969. The formation of the mesoderm in urodelan amphibians. IV. Qualitative evidence for purely "ectodermal" origin of the entire mesoderm and of the pharyngeal endoderm. Wilhelm Roux's Archiv für Entwicklungsmechanik, *169*: 185–199.

Nieuwkoop, P. D. 1973. The "organization center" of the amphibian embryo: Its origin, spatial organization and morphogenetic action. Adv. Morphol., *10*: 1–39.

Raper, K. B. 1940. Pseudoplasmodium formation and organization in *Dictyostelium discoideum*. J. Elisha Mitchell Sci. Soc., *56*: 241–282.

Reh, T. A., T. Nagy, and H. Gretton. 1987. Retinal pigmented epithelial cells induced to transdifferentiate to neurons by laminin. Nature (Lond.), *350*: 68–71.

Reinke, R., and S. L. Zipursky. 1988. Cell-cell interactions in the Drosophila retina: The bride of sevenless gene product is required in photoreceptor cell R8 for R7 cell development. Cell, *55*: 321–330.

Rosa, F. et al. 1988. Mesoderm induction in amphibians: The role of TGF-β-like factors. Science, *239*: 783–785.

Rubin, G. M. 1989. Development of the Drosophila retina: Inductive events studied at single cell resolution. Cell, *57*: 519–520.

Ruiz i Altaba, A., and D. A. Melton. 1989. Interaction between peptide growth factors and homeobox genes in the establishment of antero-posterior polarity in frog embryos. Nature (Lond.), *341*: 33–38.

Ruiz i Altaba, A., and D. A. Melton. 1990. Axial patterning and the establishment of polarity in the frog embryo. Trends Genet., *6*: 57–64.

Rutter, W. J., N. K. Wessells, and C. Grobstein. 1964. Control of specific synthesis in the developing pancreas. Natl. Cancer Inst. Mon. (13): 51–61.

Rutter, W. J. et al. 1978. An analysis of pancreatic development: Role of mesenchymal factor and other extracellular factors. *In* J. Papaconstantinou and W. J. Rutter (eds.), *Molecular Control of Proliferation and Differentiation*. Academic Press, New York, pp. 205–227.

Saha, M. S., C. L. Spann, and R. M. Grainger. 1989. Embryonic lens induction: more than meets the optic vesicle. Cell Differ. Dev., *28*: 153–172.

Sargent, T. D., M. Jamrich, and I. B. Dawid. 1986. Cell interactions and the control of gene activity during early development of *Xenopus laevis*. Dev. Biol., *114*: 238–246.

Saxén, L. 1977. Directive versus permissive induction: A working hypothesis. *In* J. Lash and M. Burger (eds.), *Cell and Tissue Interactions*. Raven Press, New York, pp 1–9.

Saxén, L. 1987. *Organogenesis of the Kidney*. Cambridge Univ. Press, Cambridge.

Saxén, L., and S. Toivonen. 1958. The dependence of the embryonic induction action of HeLa cells on their growth media. J. Embryol. Exp. Morphol., *6*: 616–633.

Saxén, L., and S. Toivonen. 1962. *Primary Embryonic Induction*. Logos Press, London.

Saxén, L. et al. 1976. Are morphogenetic tissue interactions mediated by transmissible signal substances or through cell contacts? Nature (Lond.), *259*: 662–663.

Schaller, H. C., and H. Bodenmüller. 1981. Isolation and amino acid sequence of a morphogenetic peptide from hydra. Proc. Natl. Acad. Sci. U.S.A., *78*: 7000–7004.

Schaller, H. C., S. A. H. Hoffmeister, and S. Dübel. 1989. Role of the neuropeptide head activator for growth and development in hydra and mammals. Development, *107* (Suppl.): 99–107.

Schaller, H. C., T. Schmidt, and C. J. P. Grimmelikhuijzen. 1979. Separation and specificity of action of four morphogens from hydra. Wilhelm Roux's Archives Dev. Biol., *186*: 139–149.

Slack, J. M. W. 1987. Morphogenetic gradients—past and present. TIBS, *12*: 200–204.

Slack, J. M. W., and H. Isaacs. 1989. Presence of basic fibroblast growth factor in the early *Xenopus* embryo. Development, *105*: 147–153.

Slack, J. M. W. et al. 1987. Mesoderm induction in early *Xenopus* embryos by heparin-binding growth factors. Nature (Lond.), *326*: 197–200.

Slavkin, H. C., and P. Bringas, Jr. 1976. Epithelial-mesenchymal interactions during odontogenesis. IV. Morphological evidence for direct heterotypic cell-cell contacts. Dev. Biol., *50*: 428–442.

Smith, J. C. 1987. A mesoderm-inducing factor is produced by a *Xenopus* cell line. Development, *99*: 3–14.

Smith, J. C. 1989. Mesoderm induction and mesoderm-inducing factors in early amphibian development. Development, *105*: 666–677.

Smith, J. C., and J. M. W. Slack. 1983. Dorsalization and neural induction: Properties of the organizer in *Xenopus laevis*. J. Embryol. Exp. Morphol., *78*: 299–317.

Smith, J. C., M. Yaqoob, and K. Symes. 1988. Purification, partial characterization and biological effects of the XTC mesoderm-inducing factor. Development, *103*: 591–600.

Smith, J. C. et al. 1990. Identification of a potent *Xenopus* mesoderm-inducing factor as a homologue of activin A. Nature (Lond.), *345*: 729–731.

Spemann, H. 1901. Über Korrelationen in die Entwicklung des Auges. Verh. Anat. Ges., *15*: 61–79.

Spemann, H. 1912. Zur Entwicklung des Wirbeltierauges. Zool. Jahrb. Abt. f. allg. Zool. u. Phys. D. Tiere, *32*: 1–98.

Spemann, H. 1918. Über die Determination der ersten Organanlagen des Amphibienembryo. Wilhelm Roux's Archiv für Entwicklungsmechanik, *43*: 448–555.

Spemann, H. 1936. *Experimentelle Beiträge zu einer Theorie der Entwicklung*. Julius Springer, Berlin.

Spemann, H. 1938. *Embryonic Development and Induction*. Yale University Press, New Haven, CN. Reprinted by Hafner Press (Macmillan, Inc.), New York (1962).

Spemann, H., and H. Mangold. 1924. Über Induktion von Embryonalanlagen durch Implantation artfremder Organisatoren. Wilhelm Roux's Archiv für Entwicklungsmechanik., *100*: 599–638.

Spemann, H., and O. Schotté. 1932. Über xenoplastische Transplantation als Mittel zur Analyse der embryonalen Induktion. Naturwissenschaften, *20*: 463–467.

Spooner, B. S., and N. K. Wessells. 1972. An analysis of salivary gland morphogenesis: Role of cytoplasmic microfilaments and microtubules. Dev. Biol., *27*: 38–54.

Thomsen, G. et al. 1990. Activins are expressed early in *Xenopus* embryogenesis and can induce axial mesoderm and anterior structures. Cell, *63*: 485–493.

Tiedemann, H., and H. Tiedemann. 1959. Versuche zur gewinnung eines mesodermalen induktionsstoffes aus huhnerembryonen. Hoppe-Seyler's Z. Physiol. Chem., *314*: 156–176.

Toivonen, S., and L. Saxén. 1955a. Über die Induktion des Neuralohrs bei Trituruskeimen als simultane Leistung des Leber- und Knochenmarkgewebes vom Meerschweinchen. Ann. Acad. Sci. fenn ser., A, IV, *30*: 1–29.

Toivonen, S., and L. Saxén. 1955b. The simultaneous inducing action of liver and bone marrow of the Guinea-pig in implantation and explantation experiments with embryos of *Triturus*. Exp. Cell Res., *3*: 346–357.

Toivonen, S., D. Tarin and L. Saxén. 1976. The transmission of morphogenetic signals from amphibian mesoderm to ectoderm in primary induction. Differentiation, *5*: 49–55.

Tomlinson, A. 1988. Cellular interactions in the developing *Drosophila* eye. Development, *104*: 183–193.

Tomlinson, A. 1989. Short-range positional signals in the developing *Drosophila* eye. Development, *107* (Suppl.): 59–63.

Tomlinson, A., and D. F. Ready. 1987. Neuronal differentiation in the *Drosophila* ommatidium. Dev. Biol., *120*: 366–376.

Turing, A. M. 1952. The chemical basis of morphogenesis. Philos. Trans. R. Soc. Lond. (Biol.), *641*: 37–72.

Urist, M. R., R. J. DeLange, and G. A. M. Finerman. 1983. Bone cell differentiation and growth factors. Science, *220*: 680–686.

van den Eijnden-Van Raaij, A. J. M. et al. 1990. Activin-like factor from a *Xenopus* cell line responsible for mesoderm induction. Nature (Lond.), *345*: 732–734.

Waddington, C. H. 1934. Experiments on embryonic induction. III. A note on inductions by chick primitive streak transplanted to the rabbit embryo. J. Exp. Biol., *11*: 224–226.

Waddington, C. H. 1936. Organizers in mammalian development. Nature (Lond.), *138*: 125.

Waddington, C. H., J. Needham and D. M. Needham. 1933. Physico-chemical experiments of the amphibian organizer. Nature (Lond.), *132*: 239.

Waddington, C. H., and G. A. Schmidt. 1933. Induction by heteroplastic grafts of the primitive streak in birds. Wilhelm Roux's Archiv für Entwicklungsmechanik, *128*: 522–563.

Waddington, C. H., and A. Wolsky. 1936. The occurrence of the evocator in organisms which possess no nerve cord. J. Exp. Biol., *13*: 92–94.

Walbot, V., and N. Holder. 1987. *Developmental Biology*. Random House, New York.

Wartiovaara, J. et al. 1974. Transfilter induction of kidney tubules: Correlation with cytoplasmic penetration into Nucleopore filters. J. Embryol. Exp. Morphol., *31*: 667–682.

Weeks, D. L., and D. A. Melton. 1987. A maternal mRNA localized to the vegetal hemisphere in Xenopus eggs codes for a growth factor related to TGFβ. Cell, *51*: 861–867.

Wessells, N. K. 1968. Problems in the analysis of determination, mitosis, and differentiation. *In* R. Fleischmajer and R. E. Billingham (eds.), *Epithelial-Mesenchymal Interactions*. 18th Hahnemann Symposium. Williams & Wilkins, Baltimore, pp. 132–151.

Wessells, N. K. 1977. *Tissue Interactions and Development*. Benjamin/Cummings Publishing Co., Menlo Park, CA.

Wessells, N. K., and J. H. Cohen. 1967. Early pancreas organogenesis: Morphogenesis, tissue interactions, and mass effect. Dev. Biol., *15*: 237–270.

Wessells, N. K., and W. J. Rutter. 1969. Phases in cell differentiation. Sci. Am., *220*: 36–44.

Williams, J. G. 1988. The role of diffusible molecules in regulating the cellular differentiation of *Dictyostelium discoideum*. Development, *103*: 1–16.

Williams, J. G. et al. 1987. Direct induction of Dictyostelium prestalk gene expression by DIF provides evidence that DIF is a morphogen. Cell, *49*: 185–192.

Williams, J. G., K. A. Jermyn, and K. T. Duffy. 1989. Formation and anatomy of the prestalk zone of *Dictyostelium*. Development, *107* (Suppl.): 91–97.

Wilson, C., G. S. Cross, and H. R. Woodland. 1986. Tissue-specific expression of actin genes injected into Xenopus embryos. Cell, *47*: 589–599.

Witkowski, J. 1985. The hunting of the organizer: An episode in biochemical embryology. TIBS, *10*: 379–381.

Wolpert, L. 1969. Positional information and the spatial pattern of cellular differentiation. J. Theoret. Biol., *25*: 1–47.

CHAPTER 13

Maternal and Zygotic Control During the Initiation of Development

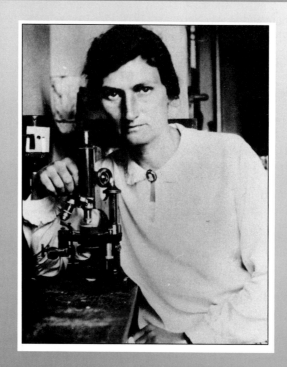

Ethel Browne Harvey (1885–1965). E. B. Harvey studied the effects of high-speed centrifugation on sea urchin eggs at the Marine Biological Laboratory in Woods Hole, Massachusetts, and at Princeton University. These studies provided evidence for the maternal control of early development. (Photograph courtesy of the Marine Biological Laboratory Library, Woods Hole, MA.)

After fertilization, the union of the sperm and egg nuclei produces a diploid zygotic genome that directs the development of the embryo into an adult. However, the zygotic genome usually does not have an immediate impact on development. Instead, the developmental program is initiated by the maternal genome through products accumulated in the egg during oogenesis. Messenger RNA transcribed by the maternal genome during oogenesis is known as **oogenic mRNA**, whereas mRNA transcribed by the zygotic genome after fertilization is called **zygotic mRNA**. In many organisms, oogenic mRNA directs protein synthesis during the early cleavages and until the blastula stage, when it is gradually replaced by zygotic mRNA. In this chapter, we discuss the maternal control of early development, the activation of translation after fertilization, the translational control of gene expression during early development, and the onset of zygotic control of development.

13–1 Maternal Control of Early Development

In many animals, early development is controlled by the products of the maternal genome that are synthesized during oogenesis, stored in the egg, and activated after meiotic maturation or fertilization. Interspecies hybridization experiments provided the first evidence for control by the maternal genome during the initiation of development (Tennent, 1914). Although fertilization is usually species-specific (see Chap. 4), hybridization can sometimes occur between different species of sea urchins, especially when the extracellular coats of the egg are removed before fertilization. In an interspecies hybridization experiment, eggs of one species are fertilized by sperm of another species, and the contribution of the zygotic genome is assessed by determining when paternal features first appear during development. Specific features of sea urchin hybrids that serve as markers for gene expression are the rate of cleavage and the time and place of appearance of the mesenchyme cells. For instance, in *Lytechinus variegatus*, cleavage is rapid, and descendants of the micromeres (primary mesenchyme cells) enter the blastocoel before invagination of the endoderm. Later, these cells give rise exclusively to the skeletal system of the pluteus larva (see Chap. 6). In the "primitive" sea urchin *Eucidaris tribuloides*, cleavage is slower, and mesenchyme cells enter the blastocoel after the beginning of archenteron formation (Table 13–1). The mesenchyme cells of *Eucidaris* embryos give rise to the larval skeleton as well as other mesodermal derivatives (Wray

Table 13–1 Development of **Eucidaris** *(♀) ×* **Lytechinus** *(♂) hybrids*

	Beginning of archenteron invagination (hr)	Primary mesenchyme formation (hr)	Site of origin of primary mesenchyme cells
Eucidaris (♀)	20	23–26 (after invagination)	Archenteron tip
Lytechinus (♂)	9	8 (precedes invagination)	Archenteron base and sides
Eucidaris (♀) × *Lytechinus* (♂)	20	24 (after invagination)	Archenteron base and sides

From Davidson, E. H. 1976. *Gene Activity in Early Development*, 2nd ed. Academic Press, New York; collated from the data of Tennent, 1914.

and McClay, 1988). When *Eucidaris* eggs are fertilized by *Lytechinus* sperm, the hybrid embryos *initially* develop like *Eucidaris*, the maternal species. Cleavage is slow, and the mesenchyme cells do not enter the blastocoel until after archenteron formation. There are two ways of interpreting these experiments: Either the maternal genome is exclusively responsible for early sea urchin development, or both maternal and paternal genomes are involved, but the latter does not function correctly in egg cytoplasm from the maternal species. Although more recent experiments indicate that paternal genes sometimes fail to be expressed in sea urchin interspecies hybrids (Tufaro and Brandhorst, 1982), it is clear that the paternal genome can be expressed in the *Eucidaris/Lytechinus* hybrids because paternal features appear *later* in development (see Sect. 13–4). These experiments suggest that the maternal genome directs early sea urchin development. The earliest point at which a paternal influence on development can be detected is after the blastula stage.

Experiments with anucleate sea urchin eggs provide further evidence for maternal control of early development. E. B. Harvey (1936, 1940) showed that anucleate egg fragments could be produced by centrifugation. Centrifugation of unfertilized eggs initially leads to stratification of organelles into zones perpendicular to the direction of centrifugal force (Fig. 13.1). During continued centrifugation, the eggs elongate and split into two fragments, or **merogones.** The nucleus always enters the centripetal merogone after the eggs split, whereas the centrifugal merogone is anucleate. Artificial activation of anucleate merogones can be achieved with hypertonic seawater (see Chap. 4). Even though the anucleate merogones lack nuclei, after activation they are able to cleave, form cilia, and hatch into embryos that resemble swimming blastulae. However, they arrest at the hatched blastula stage.

Cleavage without a functional nucleus has also been reported in amphibians. Frog embryos that lack a functional paternal genome can be produced by fertilizing eggs with sperm whose chromatin was previously destroyed by irradiation with x-rays. Subsequently, the female pronucleus can be manually removed from the activated eggs (see Chap. 10). Despite the absence of a functional genome, these eggs are able to cleave and form partial blastulae (Fig. 13.2). The development of anucleate frog eggs indicates that the zygotic genome is not required for cleavage.

Nuclear function can also be blocked by treating embryos with inhibitors of transcription, such as actinomycin D or α-amanitin. After actinomycin D treatment, sea urchin eggs cleave and sometimes develop as

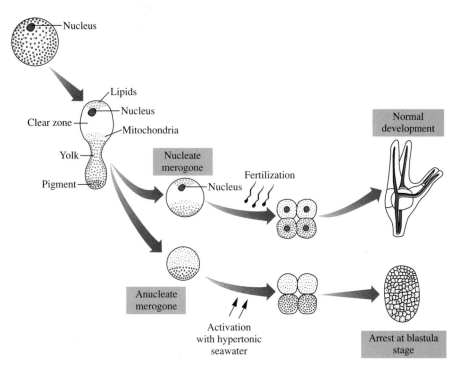

Figure 13.1

Production of nucleate and anucleate merogones of sea urchin eggs by centrifugation. A centrifuged egg is initially stratified into layers of organelles. Note that the nucleus is present in the upper (centripetal) end of the egg, and it will be distributed to the centripetal or nucleate fragment during further centrifugation. As centrifugation is continued, the stratified egg gradually separates into nucleate and anucleate merogones. The nucleate merogone develops normally. Some of the anucleate merogones cleave, form a blastula, and hatch after parthenogenic activation.

far as the blastula stage (Gross et al., 1964). Similar results have been obtained for many other organisms (Table 13–2). Transcriptionally inhibited embryos arrest at different stages of development, ranging from early cleavage (mammals) to gastrula (the gastropod mollusk *Ilyanassa*). Each organism is arrested at a stage that approximates the switch from the utilization of maternal to zygotic gene products.

The significance of the maternal genome for early development is most clearly demonstrated by studies on maternal-effect genes. Maternal-effect genes are expressed during oogenesis and produce products that are required

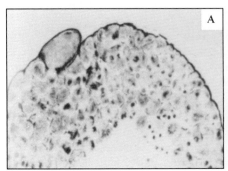
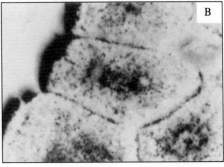

Figure 13.2

Cleavage of an activated anucleate frog egg. A, Section through the cleaved animal hemisphere of a partial blastula produced from an egg that lacked a functional nucleus. ×75. B, Higher magnification of a portion of A. ×570. (From Briggs et al., 1951.)

Table 13-2 Effects of Inhibitors of RNA Synthesis on Early Development

Organism	Observations	References
Coleopteran insects	Arrest at blastoderm stage with either actinomycin D or α-amanitin	Lockshin, 1966; Maisonhaute, 1977
Ilyanassa obsoleta (marine snail)	Arrest at gastrulation with actinomycin D	Collier, 1966; Feigenbaum and Goldberg, 1965
Sea urchin	Arrest at mesenchyme blastula stage with actinomycin D	Gross et al., 1964; Summers, 1970
Rana pipiens	Arrest at blastula stage with actinomycin D	Wallace and Elsdale, 1963
Fundulus heteroclitus (marine teleost)	Arrest at blastula stage with actinomycin D	Wilde and Crawford, 1966

for early postfertilization embryogenesis. An egg produced by a female homozygous for a recessive mutation in a maternal-effect gene shows the mutant phenotype as an embryo even if it is fertilized by wild-type sperm. This indicates that the defect is in a maternal gene product that cannot be replaced by transcription from the zygotic genome. We have already examined some developmental processes controlled by maternal-effect genes, including shell coiling in snails (see Chap. 5) and establishment of embryonic axes in *Drosophila* (see Chap. 6).

A large number of maternal-effect genes have been detected by genetic tests in *Drosophila*. In fact, maternal-effect mutants exist for each gene locus known to be involved in cellular blastoderm formation (Bakken, 1973; Fullilove and Woodruff, 1974; Rice and Garen, 1975), indicating that this process may be entirely under maternal control. Some maternal-effect genes function in adults as well as in eggs and embryos. An example is *rudimentary* (*r*), a gene required for normal wing morphology in adult *Drosophila*. The results of genetic crosses involving wild-type and *rudimentary* flies are shown in Table 13-3. When +/*r* females are mated to *r*/*r* males, embryonic development is normal and half the adult progeny are *rudimentary*. In contrast, all embryos that develop from *r*/*r* females die during early development, even if they mate with +/*r* males. This survival pattern is characteristic for the inheritance of a maternal-effect gene. Eggs produced by *r*/*r* females will develop normally if they are microinjected with pyrimidines before the cellular blastoderm stage, showing that *rudimentary* gene products are required for pyrimidine metabolism during the cleavage period (Okada et al., 1974).

Table 13-3 Effects of the Rudimentary Mutation on Embryonic Development of Drosophila melanogaster

Genotypes of parents	Genotypes of progeny	Phenotypes of progeny
+/*r* ♀ × *r*/*r* ♂	+/*r*	Wild-type
	r/*r*	Rudimentary
r/*r* ♀ × +/*r* ♂	+/*r*	Embryonic lethal
	r/*r*	Embryonic lethal

A maternal-effect gene called *ova deficient* (*o*) has been described in the axolotl, an amphibian. Embryos derived from eggs of *o/o* females arrest during gastrulation, regardless of the genotype of the sperm. However, eggs from mothers of the genotype *o/o* can be rescued by microinjecting the contents of the germinal vesicle (or the oocyte cytoplasm after germinal vesicle breakdown) from oocytes of wild-type or +/*o* females (Briggs and Cassens, 1966). Apparently, the germinal vesicle from oocytes of wild-type and +/*o* females contains a gene product required for normal development (Briggs and Justus, 1968). This substance is produced by the maternal genome during oogenesis and cannot be provided by transcription from the zygotic genome.

In summary, four different kinds of evidence demonstrate that early development is controlled by the maternal genome: (1) early maternal dominance in interspecies hybrids; (2) early cleavage in the absence of a functional nucleus; (3) development to blastula stage (except in mammals) when nuclear transcription is blocked by an inhibitor; and (4) the phenotypes of mutant maternal-effect genes. In the next section, we describe in detail one of the early events of development that is under maternal control: the activation of protein synthesis.

13–2 *Activation of Protein Synthesis*

Although transcription is not essential immediately after fertilization, translational activity is essential. This has been demonstrated by treating unfertilized eggs with drugs that interfere with protein synthesis. In sea urchins (Hultin, 1961) and amphibians (Ecker and Smith, 1971), treatment with puromycin immediately after fertilization causes arrest of development before first cleavage. Thus, the proteins required for early development must result from *de novo* translation of mRNA. We now examine this important topic.

Our knowledge of protein synthesis during early development is derived mainly from studies on sea urchin eggs. Unfertilized sea urchin eggs exhibit a very low level of translational activity, but protein synthesis increases dramatically after fertilization (Hultin, 1952). An experiment designed to determine the timing of translational activation in sea urchins is shown in Figure 13.3. In this experiment, the incorporation of radioactive amino acids into newly synthesized proteins was used to assess the rate of translation before and after fertilization. Labeled amino acids were added to unfertilized eggs in seawater, sufficient time was allowed for these amino acids to enter the eggs, and the excess amino acids were removed from the seawater before fertilization. This procedure preloads unfertilized eggs with radioactive amino acids and is necessary to be certain that enhanced amino acid incorporation into protein reflects increased protein synthesis, rather than the activation of amino acid uptake that also accompanies fertilization (see Chap. 4). The results indicate that the rate of protein synthesis begins to increase by 7 to 8 minutes after fertilization. Protein synthesis usually increases from 10- to 40-fold during the interval between fertilization and first cleavage. Subsequently, the rate of protein synthesis continues to rise; by the gastrula stage, it reaches a level about 100-fold higher than that of the unfertilized egg (Regier and Kafatos, 1977).

Sea urchins are exceptional in that they are arrested before fertilization as ootids that have completed the meiotic divisions. In contrast, unfertilized eggs of most organisms are arrested before the completion of meiosis. In these organisms, the major period of translational activation occurs during oocyte maturation, and fertilization elicits a supplementary increase in

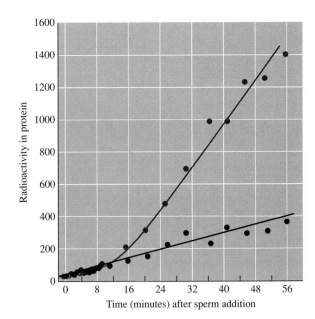

Figure 13.3

Protein synthesis in unfertilized (black line) and fertilized (colored line) eggs measured by incorporation of radioactive leucine. (After Epel, 1967.)

protein synthesis that is usually much less than in sea urchin eggs. For example, only a twofold stimulation in protein synthesis occurs after fertilization in *Xenopus* eggs (Wasserman et al., 1982).

Mechanism of Translational Activation after Fertilization

Protein synthesis involves three major steps: initiation, elongation, and termination (Fig. 13.4). Although many different components are involved in synthesizing a polypeptide, the major requirements are an energy source (ATP and GTP), ribosomes, tRNA, mRNA, and initiation, elongation, and termination factors. What is the mechanism of activation of protein synthesis after fertilization in sea urchin eggs? There are two alternatives for increased translation in fertilized eggs (Fig. 13.5). First, the *efficiency* of protein synthesis (the number of polypeptides synthesized per ribosome per minute) could be increased. The mechanism could be an increase in the rate of initiation (leading to larger polysomes), the rate of elongation (leading to smaller polysomes), or both (with no change in polysome size) (see Fig. 13.5A to C). Second, the *amount* of protein synthesis could be increased because of the activation of more translational components (e.g., mRNAs, ribosomes, and initiation factors) (see Fig. 13.5D). The mechanism of translational activation has been determined by measuring the number of ribosomes engaged in protein synthesis (Fig. 13.6; also see Fig. 13.15) and polysome size (Fig. 13.7) before and after fertilization. These measurements indicate that there is a 40- to 60-fold increase in the number of ribosomes engaged in protein synthesis during the first 6 hours after fertilization but no change in polysome size (Humphreys, 1971). Although these results indicate that the amount of protein synthesis increases after fertilization, the *magnitude* of this change is insufficient to account for the entire 100-fold increase in protein synthesis that occurs between fertilization and the gastrula stage. Therefore, the efficiency of protein synthesis must also be increased after fertilization. Changes in the efficiency of protein synthesis have been measured by determining the **mRNA transit time**, which is the interval required for an average-sized mRNA to be translated into a single polypeptide. Estimates indicate that mRNA transit time decreases two- to three-fold after fertilization (Brandis and Raff, 1978; Hille and Albers, 1979), suggesting that there is an equivalent increase in the efficiency of protein

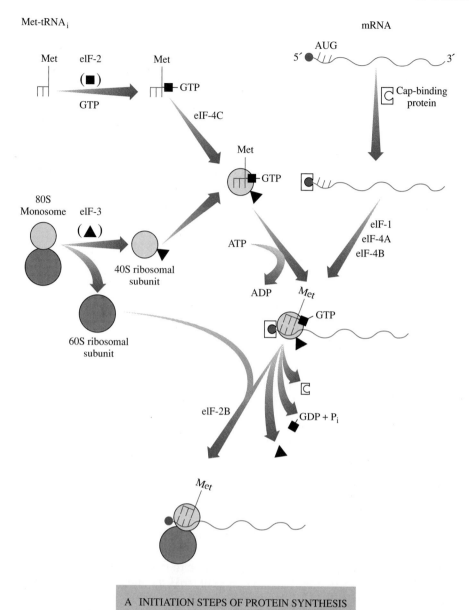

A   INITIATION STEPS OF PROTEIN SYNTHESIS

Figure 13.4A (Part B is on the following page.)

The steps of eukaryotic protein synthesis. A, Initiation. Three initiation steps occur simultaneously: A cap-binding protein is associated with the 5′ terminus of mRNA, a complex of eukaryotic initiation factor 2 (eIF-2) and GTP bind the initiator Met-tRNA (Met-tRNA$_i$), and 80S monosomes are dissociated into 40S and 60S ribosomal subunits by the action of eIF-3, which binds to the 40S ribosomal subunit. The 40S ribosomal subunit-eIF-3 complex then binds to the Met-tRNA$_i$–eIF-2–GTP complex, and hydrolysis of ATP occurs as this complex in turn binds to mRNA. Subsequently, the 60S ribosomal subunit is added to this complex, and the initiation factors and cap-binding protein are released. The involvement of the other initiation factors (including eIF-2B, eIF-4C, eIF-1, eIF-4A, and eIF-4B) is summarized in the figure.

synthesis. Together, the large increase in the number of translational components engaged in protein synthesis (40- to 60-fold) and the small increase in the efficiency of protein synthesis (two- to three-fold) account for the 100-fold activation of translation during early development.

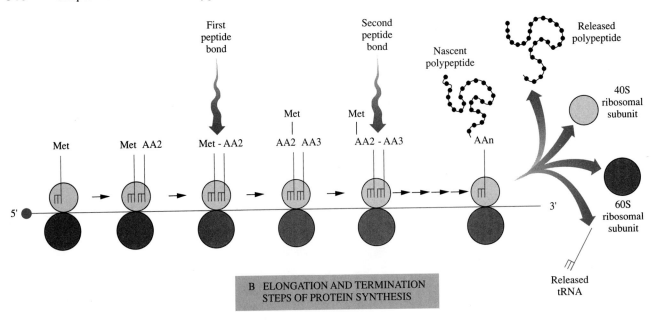

B ELONGATION AND TERMINATION STEPS OF PROTEIN SYNTHESIS

Figure 13.4B (Part A is on the preceding page.)

The steps of eukaryotic protein synthesis (continued). B, Elongation and termination. The first ribosome accepts the tRNA carrying the second amino acid (AA2), the first peptide bond is formed between methionine and AA2, the ribosome accepts a tRNA carrying the third amino acid (AA3), with concomitant release of Met-tRNA$_i$, and the second polypeptide bond is formed. This process continues until a termination site is reached in the mRNA; the nascent polypeptide is then released, and the ribosome dissociates into 40S and 60S subunits. Elongation and termination factors are involved in these processes (not shown).

As we learned in Chapter 4, the increase in protein synthesis after fertilization is preceded by rises in intracellular pH and Ca^{2+}. The following experiments show that pH and Ca^{2+} cooperate to stimulate protein synthesis after fertilization. In the first experiment, pH is increased by treating unfertilized eggs with ammonia. As shown in Figure 13.8A, increased pH causes an enhancement of protein synthesis. Transfer of the ammonia-activated eggs to seawater lacking ammonia (*arrow* in Fig. 13.8A) results in a subsequent drop both in pH and the rate of protein synthesis. In the second experiment, pH is reduced by treating fertilized eggs with sodium acetate (see Fig. 13.8B). Sodium acetate is a weak acid that enters the cytoplasm and lowers the pH. When sodium acetate is added 20 minutes after fertilization (*arrow*), the intracellular pH begins to drop and eventually is lowered to the level of unfertilized eggs. The effects of these pH shifts on protein synthesis are shown in Figure 13.8C. The addition of sodium acetate (indicated by *arrow 1*) causes protein synthesis to decrease to a level almost as low as that of unfertilized eggs. This effect is reversible because protein synthesis returns to normal levels after sodium acetate is removed (*arrow 2* in Fig. 13.8C). Although elevation of intracellular pH mimics the effect of fertilization on protein synthesis rate, the responses are not quantitatively identical. The natural increase in protein synthesis can be mimicked exactly, however, by using combinations of different activation methods (Fig. 13.9). Although protein synthesis in ammonia-activated eggs is less than that in fertilized eggs (Fig. 13.9, *curves b* and *c*), it can be increased to normal levels by simultaneous treatment with ionophore

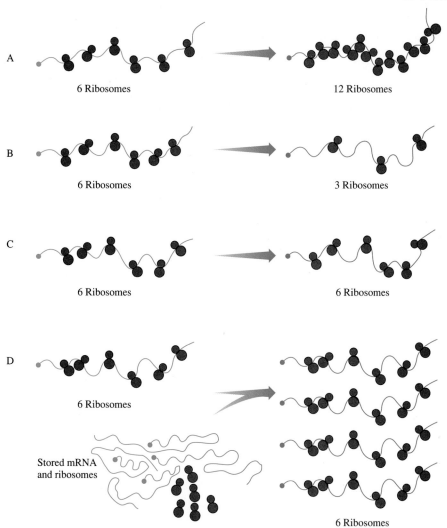

Figure 13.5

Diagram showing some of the possibilities for the mechanism of translational activation after fertilization in sea urchin eggs and expected consequences with respect to polysome size. A, A two-fold increase in the efficiency of initiation with no increase in elongation efficiency would increase polysomes from 6 to 12 ribosomes. B, A two-fold increase in the efficiency of elongation with no increase in initiation would decrease polysomes from 6 to 3 ribosomes. C, Equivalent increases in the efficiency of initiation and elongation have no effect on polysome size. In A-C, there is no increase in the amount of mRNA engaged in protein synthesis or of other translational components. D, An increase in the amount of translational components (e.g., stored mRNAs and ribosomes) engaged in protein synthesis without changes in initiation or elongation efficiency would lead to more polysomes of the same size.

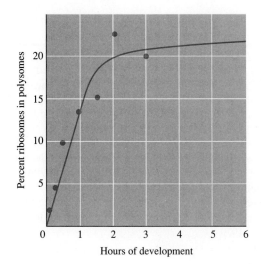

Figure 13.6

The increase in percent of ribosomes in polysomes after fertilization of the sea urchin egg. The proportion of ribosomes was measured at various times after fertilization. (After Humphreys, 1971.)

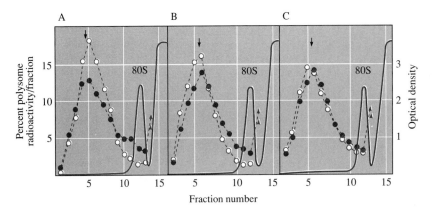

Figure 13.7

Comparison of polysome sizes in sea urchin eggs and embryos. Unfertilized eggs were labeled with ^{3}H-leucine (open circles). Embryos at 20 (A), 60 (B), and 360 (C) minutes after fertilization were labeled with ^{14}C-leucine (closed circles) and mixed with an aliquot of ^{3}H-leucine-labeled unfertilized eggs. Polysomes were extracted from each mixture and centrifuged on sucrose density gradients. The gradients were monitored for radioactivity (dashed lines), which represents the positions of polysomes, and optical density (solid lines), which represents the positions of 80S monosomes. The average size of polysomes (arrow) does not change in unfertilized and fertilized eggs. (After Humphreys, 1971.)

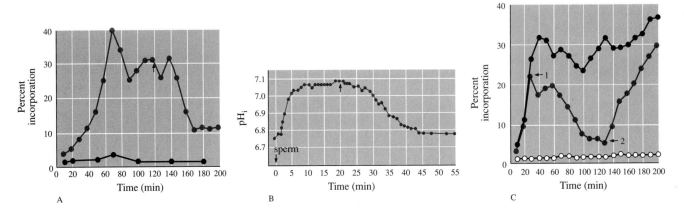

Figure 13.8

Experiments examining the relationship between pH and protein synthesis in sea urchin eggs. A, The rate of protein synthesis in ammonia-activated eggs. Unfertilized eggs were placed in either seawater (black line) or seawater containing NH_4Cl (colored line), and the rate of protein synthesis was determined at 10-minute intervals. At 120 minutes (arrow), the eggs were removed from NH_4Cl-containing seawater and resuspended in normal seawater. Note the subsequent drop in the rate of protein synthesis. B, Measurement of pH in sodium acetate-treated eggs. Eggs were fertilized and maintained in seawater. Internal pH (pH_i) was measured with a microelectrode. At 20 minutes after fertilization (arrow), eggs were exposed to sodium acetate. Internal pH declined, dropping to the unfertilized egg level within 25 minutes. C, The effect of sodium acetate on the rate of protein synthesis (colored line). At 30 minutes after fertilization (arrow no. 1), eggs were resuspended in seawater containing sodium acetate. The rate of protein synthesis dropped to the unfertilized egg level. At 130 minutes (arrow no. 2), the eggs were returned to normal seawater. Protein synthesis accelerated. For comparison, the rates of protein synthesis for unfertilized eggs (line with open circles) and fertilized eggs maintained in seawater (black line) are shown. Percent incorporation is the percent amino acid precursor taken up by eggs that is incorporated into protein. (After Grainger et al., 1979.)

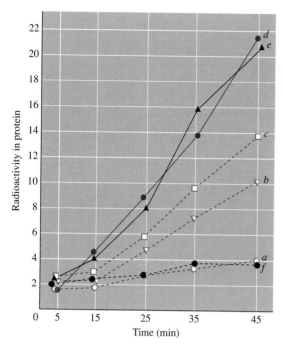

Figure 13.9

The role of increased pH and Ca^{2+} in the activation of protein synthesis in sea urchin eggs. In this experiment, the incorporation of radioactive valine into newly-synthesized protein was followed over a 45-minute period beginning 5 minutes after activation in various ionic media. The plotted data points represent the mean incorporation. The time points have been spread out slightly over the x axis for clarity. a, Eggs activated by treatment with ionophore A23187 in Na^+-free seawater. This treatment stimulates Ca^{2+} release without a pH rise. b, Eggs activated by treatment with ammonia in Ca^{2+}-free seawater. This treatment stimulates the pH rise without a Ca^{2+} release. c, Eggs activated by treatment with ammonia in Na^+-free seawater. This treatment stimulates the pH rise with some Ca^{2+} influx. d, Eggs activated by treatment with ammonia and ionophore A23187. Both the rise in pH and Ca^{2+} release occur. e, Eggs fertilized in normal seawater. f, Unfertilized eggs in normal seawater. (After Winkler et al., 1980. Reprinted by permission from *Nature* Vol. 287 pp. 558–560. Copyright © 1980 Macmillan Magazines Ltd.)

A23187 (Fig. 13.9, *curve d*), which induces a rise in Ca^{2+}. This experiment suggests that Ca^{2+} and pH function together to activate translation after fertilization.

Maternal Control of Translational Activation

We have seen that early development is independent of the zygotic genome. However, protein synthesis is essential for the onset of development. How can we resolve this apparent paradox? We must conclude that the activation of protein synthesis is under maternal control. Supporting evidence for this conclusion comes from experiments in which protein synthesis is activated in enucleated frog eggs (Smith and Ecker, 1965), anucleate sea urchin merogones (Denny and Tyler, 1964; Raff et al., 1972), or in sea urchin zygotes treated with actinomycin D (Gross et al., 1964). These results show that substantial quantities of each of the components required for protein synthesis (i.e., mRNA, tRNA, ribosomes, initiation factors, and so forth) must pre-exist in unfertilized eggs. Mammalian eggs also contain oogenic mRNA and other translation components. However, their slower rate of development and small size make them dependent on components synthesized under the direction of the zygotic genome much earlier in development relative to other animals (see Sect. 13–4).

mRNA

A direct way of demonstrating the presence of mRNA in unfertilized eggs is to isolate total RNA from them and determine whether it can be translated into protein in a **cell-free translation system**. Cell-free translation systems contain all the components required for protein synthesis except mRNA. When mRNA (or total cellular RNA) and a labeled amino acid are added to a cell-free translation system, protein synthesis is initiated, and newly synthesized proteins can be identified by gel electrophoresis. An example of such a two-dimensional gel pattern (see the Appendix for a discussion of two-dimensional gel electrophoresis) is shown in Figure 13.10. This gel contains more than 400 radioactive proteins, including actin in the region

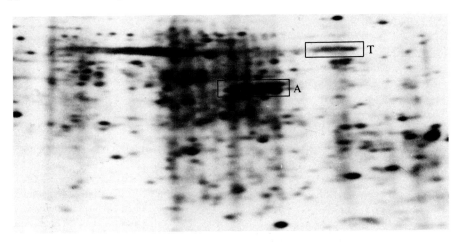

Figure 13.10

Cell-free translation products of mRNA extracted from unfertilized sea urchin eggs. This autoradiograph of a two-dimensional gel shows about 400 different polypeptides. Boxes labeled T and A show the regions of the gel containing tubulins and actin, respectively. (From Infante and Heilmann, 1981. Reprinted with permission from Biochemistry, *20*: 1–8. Copyright 1981 American Chemical Society.)

labeled A and tubulin in the region labeled T. Another way of identifying different oogenic mRNAs is by hybridization. Northern blot and *in situ* hybridization experiments (see Appendix) have shown that actin (Crain et al., 1981; Scheller et al., 1981), tubulin (Alexandraki and Ruderman, 1985), and histone (see Figs. 13.16 and 13.26) mRNAs are present in sea urchin eggs, and fibronectin mRNA is present in amphibian eggs (Fig. 13.11). Sea urchin eggs also contain mRNA species coding for all of the five classes of histones (Davidson et al., 1982). These oogenic mRNAs direct the synthesis of specialized histones that are incorporated exclusively into the chromatin of cleaving embryos (see Sect. 13–4).

The oogenic messenger population has also been characterized based upon transcript abundance. Any eukaryotic mRNA population can be divided into prevalent and rare classes based on transcript abundance. The **prevalent mRNA class** contains several hundred different mRNA species, including transcripts coding for actin, tubulin, histone, and many other structural proteins. These messengers, which are present in 1000 or more copies per cell, account for about 50% to 60% of the total mass of oogenic mRNA (Hough-Evans et al., 1977). Sequence complexity measurements (see Ap-

Figure 13.11

Northern hybridization showing the level of oogenic mRNA encoding fibronectin in eggs and early embryos of *Xenopus laevis*. The blot was hybridized with a labeled fibronectin gene probe. MBT: midblastula transition; st 9–18: stages 9–18 (see p. 564). (From Kimelman and Kirschner, 1987. Copyright © Cell Press.)

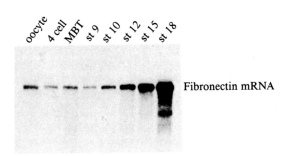

pendix) indicate that about 11,000 different mRNAs exist in sea urchin eggs (Davidson, 1986), whereas *Drosophila* eggs contain about 6000 different mRNAs (Hough-Evans et al., 1980). Although approximately one half of the mass of oogenic mRNA is accounted for by only several hundred prevalent transcripts, the remainder of this population contains the majority of different mRNAs that are present in only a few copies each per cell. This is the **rare mRNA class**, which numbers only about two copies per cell in sea urchin pluteus larvae. The relationship between mRNA abundance and complexity is illustrated in Figure 13.12. Rare mRNAs are impossible to detect by methods such as cell-free translation and two-dimensional electrophoresis of translation products, which are only effective in detecting the prevalent mRNA species (as in Fig. 13.10).

Ribosomes and tRNA

The storage of ribosomes in unfertilized sea urchin eggs can be demonstrated by extracting RNA and fractionating it by sucrose density gradient centrifugation. Most of the RNA seen in these gradients is 18S and 28S rRNA, indicating that ribosomes must be present in the unfertilized egg. The presence of ribosomes can also be shown by subjecting extracts of egg

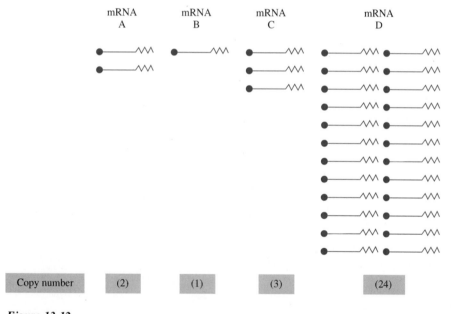

Figure 13.12

Schematic diagram illustrating the relationship between transcript abundance and complexity in a mRNA population. Each mRNA molecule is shown as a vertical line beginning with a filled circle (the 5' cap structure) and ending with a jagged line (the 3' poly (A) tail). The total number of mRNA molecules in this population is 30, which is the sum of two copies of mRNA species A, one copy of mRNA species B, three copies of mRNA species C, and 24 copies of mRNA species D. Since transcripts of mRNA species D make up 80% (24 of 30 copies) of the molecules in the population, species D is an abundant mRNA. In contrast, mRNA species A, B, and C, which make up 6.7% (2 of 30 copies), 3.3% (1 of 30 copies) and 10% (3 of 30 copies) of the total transcript population, respectively, are rare mRNAs. The complexity of this transcript population is 4 (since there is a total of only 4 different mRNAs), and the rare mRNA class makes up 75% (3 of 4 different species) of the complexity.

cytoplasm to density gradient centrifugation. In these gradients, most egg ribosomes appear as 80S monosomes (see Fig. 13.7), indicating that they are not engaged in protein synthesis.

Experiments involving an anucleolate mutant of *Xenopus laevis* (Elsdale et al., 1958) indicate that enough ribosomes are stored in eggs of this species to promote development up to the tadpole stage. Wild-type frogs (*2-nu*) exhibit two nucleolar organizers and nucleoli in each diploid cell. The anucleolate mutant (*0-nu*) lacks nucleolar organizers and nucleoli, cannot manufacture ribosomes, and is lethal in the homozygous condition. Heterozygotes (*1-nu*) contain only one nucleolar organizer and nucleolus per diploid cell. The anucleolate mutation is inherited in a typical Mendelian fashion. When two heterozygotes are mated, there is a 1:2:1 ratio of progeny that exhibit two (*2-nu*), one (*1-nu*), or no (*0-nu*) nucleoli per diploid cell, respectively. The anucleolate (*0-nu*) progeny appear perfectly normal as embryos, but development is retarded after hatching, and they die as tadpoles before the feeding stage. In the experiment shown in Figure 13.13, anucleolate (*0-nu*) and control (*1-nu* and *2-nu*) embryos were allowed to incorporate radioactive uridine into RNA until the tadpole stage, when RNA was extracted and fractionated by sucrose density gradient centrifugation. The amount of total RNA (oogenic and zygotic RNA) in the gradient was determined by measuring optical density, whereas the RNA synthesized by the zygote is indicated by radioactivity. The control embryos synthesize 18S rRNA, 28S rRNA, 5S rRNA, and tRNA. In contrast, the anucleolate mutants synthesize 5S rRNA and tRNA, but not 18S or 28S rRNA. The only 18S and 28S rRNA seen in the gradient is unlabeled; therefore, it must have been derived from the egg. Because the *0-nu* embryos develop from eggs of heterozygous mothers, they are able to form tadpoles using the ribosomes already present in the egg.

As we discussed in Chapter 3, an impressive amount of 4S RNA (mainly tRNA) and 5S rRNA accumulates during oogenesis. For example, late vitellogenic stage *Xenopus* oocytes have accumulated more than five 4S RNA molecules and seven 5S rRNA molecules for each molecule of 18S and 28S rRNA. Large amounts of 5S rRNA and tRNA are retained in the

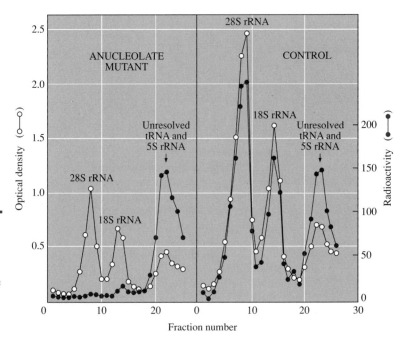

Figure 13.13

Sucrose density gradient centrifugation of RNA isolated from anucleolate mutant (*0-nu*) and control embryos. Total RNA is represented by optical density measurements (black lines), whereas RNA synthesized by the embryos is represented by the radioactivity measurements (colored lines). (After Brown and Gurdon, 1964.)

Xenopus egg. The presence of oogenic 4S RNA and 5S rRNA in eggs can be demonstrated by labeling RNA during oogenesis and measuring the amount of radioactive RNA recovered in the embryo (Fig. 13.14).

Mechanisms of Translational Suppression in the Unfertilized Egg

Although all of the necessary translational components are present in unfertilized sea urchin eggs, they exhibit a very low level of protein synthesis. As we have discussed, the increase in protein synthesis after fertilization depends primarily on the activation of stored translational components. Thus, one or more of these translational components must be inactive in the unfertilized egg. Although this question has been studied for several decades, the inactive component(s) have yet to be precisely identified.

One popular idea is that oogenic mRNA is limiting before fertilization because it is complexed with inhibitory proteins that prevent its interaction with initiation factors, ribosomes, or other translational components. After fertilization, the inhibitory proteins could be removed or degraded, leading to translational activation of the mRNA. This idea, known as the **masked mRNA hypothesis**, is supported by the fact that mRNA molecules are normally complexed with proteins in cells, forming **messenger ribonucleo- protein** or **mRNP particles** (Spirin, 1966). In principle, it seems simple to test the masked mRNA hypothesis. If this idea is correct, mRNPs isolated from fertilized eggs—but not from unfertilized eggs—should direct protein synthesis when added to a cell-free translation system. Another expectation would be that different proteins might be present in mRNP particles isolated from unfertilized eggs than from fertilized eggs. Unfortunately, mRNP particles are extremely difficult to isolate and characterize because their structure is very sensitive to the composition of the isolation medium, and there are no good assays for "correct" mRNP composition. Although it has been demonstrated that some proteins are different in mRNP particles isolated from unfertilized and fertilized eggs (Moon et al., 1980), conflicting results have been obtained on the translation of isolated mRNPs (Jenkins et al., 1978; Moon et al., 1982). Thus, although the masked mRNA hypothesis

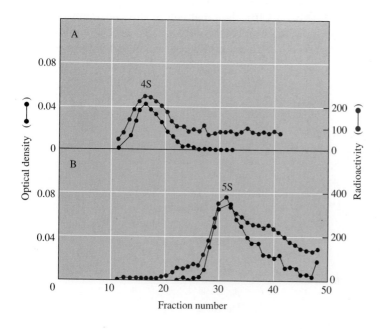

Figure 13.14

Measurements of oogenic tRNA (4S) and 5S rRNA in mature *Xenopus* eggs. Eggs were obtained from a female that had been injected with radioactive uridine. Consequently, RNA synthesized during oogenesis is radioactive. RNA was isolated from egg supernatant (A) or ribosomal fraction (B), and low molecular weight RNA from these fractions was analyzed by column chromatography. Optical density: black lines. Radioactivity: colored lines. (After Brown and Littna, 1966 with permission from Journal of Molecular Biology. Copyright: Academic Press [London] Ltd.)

is an attractive idea, other schemes for the inhibition of mRNA translation are also deserving of consideration.

One possible alternative is that mRNA structure may be different in unfertilized and fertilized eggs. After transcription, nuclear mRNA precursors normally undergo a number of processing steps, including the addition of a methylated cap structure to the 5' end, intron removal and exon splicing in the interior of the molecule, and the addition of a long poly (A) tail to the 3' end, before they enter the cytoplasm. The 5' cap, and possibly the 3' poly (A) tail, serve to protect the mRNA from degradation and may also aid in translation. Some of these nuclear events might be skipped during oogenesis, resulting in the accumulation of incompletely processed mRNA in the egg cytoplasm. After fertilization, the specific mRNA-processing steps could be completed rapidly, permitting the mRNA to be translated. In support of the existence of unprocessed oogenic mRNA, very large polyadenylated RNAs containing interspersed repetitive sequences, such as those that might be found in introns, have been reported in the cytoplasm of sea urchin (Costantini et al., 1980) and amphibian (Anderson et al., 1982) eggs. In Chapter 3 we found no evidence that these large transcripts are translatable (Richter et al., 1984; McGrew and Richter, 1989) or that they are processed into functional mRNA later in development. There are also indications that some oogenic mRNAs have unmethylated cap structures and very short poly (A) tails. After fertilization, the caps of these mRNAs are methylated (Caldwell and Emerson, 1985) and their poly (A) tails are elongated (Dolecki et al., 1977; Wilt, 1977) in the cytoplasm, but there is no experimental evidence indicating that these modifications are *necessary* for translation (Mescher and Humphreys, 1974; Showman et al., 1987).

A second alternative for suppressing mRNA translation is to prevent its interaction with other translational components. This could be accomplished by sequestering mRNA in the unfertilized egg. After fertilization, the sequestered mRNA could be released and interact with the other translational components. There is evidence for sequestration of oogenic histone mRNA in the female pronucleus of sea urchin eggs (DeLeon et al., 1983). In sea urchins, translation of mRNA encoding early histones (see Sect. 13–4) does not increase immediately after fertilization but is delayed until first cleavage (Fig. 13.15). *In situ* hybridization (see Appendix) shows that early histone mRNA is confined to the female pronucleus until its breakdown before first cleavage, when it is released into the cytoplasm and begins to be translated (Fig. 13.16). Nuclear sequestration cannot be a general method for translational suppression of oogenic mRNA, however, because most mRNA species are stored in the egg cytoplasm (Angerer and

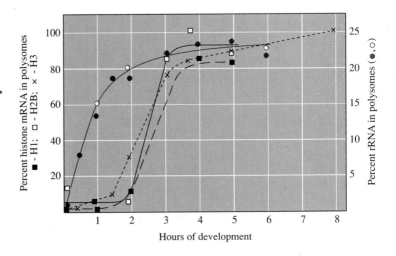

Figure 13.15

Delayed recruitment of three different early histone mRNAs into polysomes of fertilized sea urchin eggs. The graph shows the amount of histone mRNAs incorporated into polysomes as a function of time after fertilization as determined by hybridization with cloned probes for H1, H2B, and H3 histones. Also shown is the incorporation of ribosomes into polysomes, which is a measure of the general mobilization of mRNA after fertilization. (After Raff, 1983.)

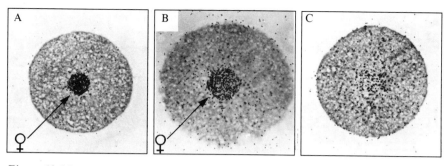

Figure 13.16

Demonstration of early histone mRNA sequestration in the female pronucleus (arrow) of sea urchin eggs and its release before first cleavage. A to C illustrate sectioned eggs that have been hybridized *in situ* with a specific probe that recognizes all five classes of early histone mRNAs. A, 70 minutes after fertilization. Histone mRNA (indicated by autoradiographic grains) is present in the nucleus. B, 80 minutes after fertilization. Most histone mRNA is still in the nucleus. C, 90 minutes after fertilization. Histone mRNA has escaped to the cytoplasm after nuclear envelope breakdown. (From DeLeon et al., 1983.)

Angerer, 1981; Showman et al., 1982) and are translationally activated immediately after fertilization. Another means of limiting interaction between mRNA and the translational machinery involves the egg cytoskeleton. Most protein synthesis in fertilized sea urchin eggs occurs on polysomes that are associated with the cytoskeleton (Moon et al., 1983). Translation of mRNA could be restricted by a mechanism that limits mRNA access to the cytoskeleton in unfertilized eggs. This is an intriguing idea because the cytoskeleton of sea urchin eggs is extensively reorganized at the time of fertilization (Begg et al., 1982), which may promote interaction between mRNA and translational components.

Messenger RNA is not necessarily the translational component that is limiting in the unfertilized egg. Indeed, other translational components may be suppressed and activated in concert with mRNA at the time of fertilization. This is shown by an experiment in which globin mRNA is microinjected into fertilized sea urchin eggs and protein synthesis is examined by gel electrophoresis (Fig. 13.17). If mRNA is the limiting component, total protein synthesis would be greatly enhanced after globin mRNA injection. Instead, total protein synthesis does not increase, although injection of sufficient globin mRNA eventually makes it the predominant protein synthesized in the microinjected eggs. Injected globin mRNA must compete with oogenic mRNA for a restricted number of active translational components. Thus, even in the presence of excess mRNA, there is little or no spare translational capacity in the unfertilized egg. This suggests that mRNA is not the limiting factor in protein synthesis. Similar results were obtained when mRNA was injected into *Xenopus* oocytes (Laskey et al., 1977).

Protein synthesis could be limited by ribosomes, initiation factors, elongation factors, tRNA, or any other of the many factors involved in translation (see Fig. 13.4). When the translational capabilities of ribosomes from unfertilized and fertilized sea urchin eggs are compared in a cell-free translation system, the ribosomes of unfertilized eggs are less active in protein synthesis than those of embryos (Danilchik and Hille, 1981). This experiment suggests that ribosomes are also suppressed in the unfertilized egg and activated after fertilization. Similarly, there is evidence that unfertilized sea urchin eggs have limited quantities of functional initiation factors. When initiation factors are added to cell-free translation systems prepared from unfertilized sea urchin eggs, protein synthesis is enhanced

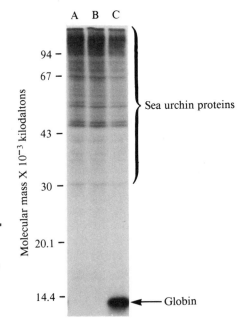

A B C

Molecular mass X 10^{-3} kilodaltons

94 —

67 —

} Sea urchin proteins

43 —

30 —

20.1 —

14.4 — ← Globin

Figure 13.17

Translation of globin mRNA after microinjection into fertilized sea urchin eggs. This experiment compares proteins synthesized in (A) uninjected eggs, (B) eggs injected with buffer, and (C) eggs injected with globin mRNA. Note in C that the level of synthesis of endogenous proteins is reduced when globin mRNA is injected and translated into protein, showing that mRNA is not the limiting factor in protein synthesis. (From Colin and Hille, 1986.)

(Winkler et al., 1985; Colin et al., 1987; Huang et al., 1988). Although the identity of the initiation factor (or factors) that is rate-limiting in the unfertilized egg is still unclear, the best candidates appear to be eIF-2, eIF-2B, and eIF-4A (see Fig. 13.4). Thus, suppression of protein synthesis may be caused by inactivation of many different translational components including mRNA, ribosomes, and one or more initiation factors.

Recent work with *Xenopus* oocytes and eggs suggests that one of the factors limiting translation before fertilization in that species is a "translational potentiator" that is necessary to unwind mRNA so that translation can proceed (Fu et al., 1991). Complementary regions in RNA molecules can bind to one another by base-pairing to form double-stranded regions that would inhibit translation unless they could be unwound. Potentiator activity apparently increases dramatically at fertilization, unwinding folded mRNA molecules and facilitating their translation. Current research is aimed at determining whether this mechanism operates on a small subset of mRNA molecules or is a mechanism for regulating translation of oogenic mRNA in general.

13–3 Specific Translational Control During Early Development

Orderly embryonic development requires that specific proteins be produced at particular times during development and in different regions of the embryo. During later development, changes in protein synthetic patterns are modulated primarily by transcriptional control: Different genes are transcribed at different times and in different cells (see Chaps. 14 and 18). We have learned, however, that transcription has little impact on early development, and there is already a large number of different mRNAs present in the unfertilized egg. Thus, differential protein synthesis during early development must be mediated by posttranscriptional mechanisms, which can occur at the levels of mRNA translation or mRNA and protein turnover.

Translational control is the process by which gene expression is modulated at the level of mRNA translation. We have discussed *general*

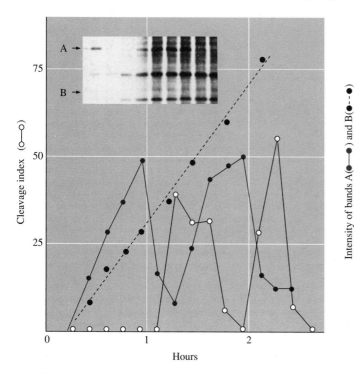

Figure 13.18

Fluctuation of cyclin levels during early cleavage of sea urchin embryos. Graphed data were obtained from the series of one-dimensional gels (inset), in which the position of cyclin (A) is indicated by an arrow. Another protein that does not fluctuate like cyclin is indicated by B in the gels and graph. Also indicated in the graph are the timing of first and second cleavages (cleavage index). It can be seen that cyclin levels are high during interphase and low during cleavage. (After Evans et al., 1983. Copyright © Cell Press.)

translational control, in which translation rates of the entire population of mRNA molecules vary. We shall now discuss *differential*, or *specific*, translational control, in which some mRNAs are translated at higher rates than others.

The sea urchin embryo shows very few changes in the pattern of protein synthesis as compared to the unfertilized egg (Brandhorst, 1976). An exception, however, is cyclin (a protein component of MPF, which is involved in the regulation of meiosis and mitosis; see Chap. 3), which is not synthesized in unfertilized eggs but appears after fertilization (Fig. 13.18). The synthesis of cyclin after fertilization is due to specific translational activation of cyclin mRNA (Evans et al., 1983).

Differential translational control appears to be widespread in organisms that, unlike the sea urchin, exhibit a major activation of protein synthesis during oocyte maturation. Examples include starfish, the clam *Spisula* (actually, in this species, fertilization triggers maturation, but the latter is thought to be the critical step with regard to translational control), amphibians, and the mouse. In these organisms, a variety of different proteins are synthesized before and after maturation (Schultz and Wassarman, 1977; Rosenthal et al., 1980, 1982; Rosenthal and Ruderman, 1987). In *Spisula*, proteins X, Y, and Z (see Fig. 13.19A) are synthesized primarily before maturation, whereas proteins A, B, and C (see Fig. 13.19B) are synthesized after maturation. However, synthesis of both sets of proteins can be directed by mRNA from either mature or immature oocytes in a cell-free translation system (Fig. 13.19C and D). This demonstrates that X, Y, and Z mRNAs and A, B, and C mRNAs are present in both mature and immature oocytes. However, X, Y, and Z mRNAs are preferentially translated in immature oocytes, and A, B, and C mRNAs are preferentially translated in mature oocytes. Thus, the qualitative switch in protein synthesis at maturation is mediated by changing the relative translational efficiencies of various oogenic mRNAs.

These organisms may also exhibit specific changes in the pattern of protein synthesis at fertilization. For example, the small subunit of the multimeric enzyme ribonucleotide reductase is synthesized exclusively after fertilization of clam eggs. This enzyme catalyzes the conversion of uridine

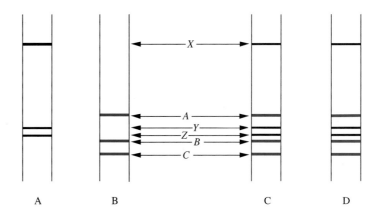

A   B   C   D

Figure 13.19

Drawing showing results of gel electrophoresis of proteins synthesized *in vivo* (A, B) and in a cell-free translation system (C, D) before and after maturation of *Spisula* oocytes. A, *In vivo* synthetic pattern in immature oocytes. Note the predominance of proteins *X*, *Y*, and *Z*. B, *In vivo* synthetic pattern in mature oocytes. Note the predominance of proteins *A*, *B*, and *C*. C, and D, Synthetic patterns obtained when RNA extracted from immature (C) or mature (D) oocytes was added to a cell-free translation system. Note the presence of high levels of proteins *X*, *Y*, *Z* and *A*, *B*, *C* in each gel. (After Rosenthal et al., 1980.)

to deoxyuridine, which is subsequently converted to thymidine, a precursor of DNA. The production of ribonucleotide reductase is controlled at the translational level (Standart et al., 1986). The ribonucleotide reductase large subunit protein, but not the small subunit protein, is stored in the egg. In contrast, the egg stores the small subunit mRNA in a translationally inactive form (Fig. 13.20). After fertilization, ribonucleotide reductase mRNA is activated, enters polysomes, and directs the synthesis of the small subunit. Only then are the small and large subunits assembled into active enzyme. The reason that clam eggs store one subunit of ribonucleotide reductase as protein and the other as mRNA is unknown. Perhaps this is a safeguard against premature enzyme assembly and unscheduled DNA synthesis in the unfertilized egg.

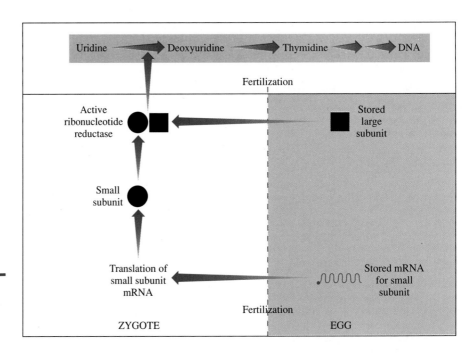

Figure 13.20

A schematic diagram for ribonucleotide reductase synthesis and assembly after fertilization in clam eggs.

Differential translational control continues to operate as cleavage proceeds. In fertilized *Xenopus* eggs, actin and other mRNAs show delayed entry into polysomes (Ballantine et al., 1979; Woodland and Ballantine, 1980); i.e., their utilization in protein synthesis is deferred until later in development. Another example of delayed recruitment involves the mRNAs encoding two glycoproteins that function in sea urchin gastrulation (Lau and Lennarz, 1983). These mRNAs are stored in the egg but remain translationally inactive until before gastrulation when they begin to direct glycoprotein synthesis. Finally, differential translational control also occurs during early development of the gastropod mollusk *Ilyanassa* (Brandhorst and Newrock, 1981). *Ilyanassa* development is accompanied by major changes in the spectrum of polypeptides synthesized between the early cleavages and gastrulation (Fig. 13.21A and B). As described in Chapter 5, *Ilyanassa* embryos form an anucleate polar lobe at first cleavage, which can be removed from the rest of the embryo. Isolated polar lobes are metabolically active and undergo a similar sequence of changes in polypeptide synthesis as occurs in intact embryos (Fig. 13.21C and D). Because polar lobes lack a nucleus, however, these differences in protein synthesis must be due to selective translation of different mRNAs rather than to differential transcription. These experiments show that changes in the pattern of protein synthesis during early development depend to a large degree on differential translational control.

Posttranscriptional control can also occur at the level of mRNA or protein degradation. An example of selective mRNA turnover occurs during maturation in *Xenopus* oocytes. As discussed earlier, ribosomes used by the early embryo are produced during oogenesis and stored in the mature egg. *Xenopus* oocytes actively translate oogenic mRNAs encoding about 60 different ribosomal proteins. In contrast, only three of these ribosomal proteins are synthesized during cleavage (Pierandrei-Amaldi et al., 1982; Baum and Wormington, 1985). The oogenic mRNAs encoding these three ribosomal proteins are among a minority of oogenic transcripts that survive turnover during oocyte maturation, when there is a major burst of mRNA degradation. Cyclin is an example of a protein whose levels are regulated by selective protein degradation. In sea urchin and clam embryos, cyclin is destroyed during mitosis and resynthesized during interphase of every cell cycle (see Fig. 13.18). The destruction of cyclin is accomplished by protein degradation, and its resynthesis is directed by stable oogenic mRNA.

13–4 Zygotic Control of Development

As embryonic development proceeds, there is a switch from control by the maternal genome to the zygotic genome. In some organisms, this change is abrupt, whereas in others it is gradual. As discussed earlier, the delayed function of the zygotic genome was first appreciated because of the results of interspecies hybridization experiments.

The sea urchins *Lytechinus variegatus* and *Eucidaris tribuloides* differ in their modes of mesenchyme formation (see Table 13–1). *Eucidaris* does not have a distinct population of mesenchyme cells that ingress before the beginning of archenteron invagination (see Chap. 6). Instead, all mesenchyme cells originate at the tip of the archenteron after invagination has been initiated. In *Lytechinus*, however, primary mesenchyme cells originate before invagination of the archenteron both from the tip and base of the presumptive archenteron. When *Eucidaris* eggs are fertilized with *Lytechinus* sperm, mesenchyme formation follows invagination, just as it does in the maternal species (see Table 13–1). However, the mesenchyme cells originate

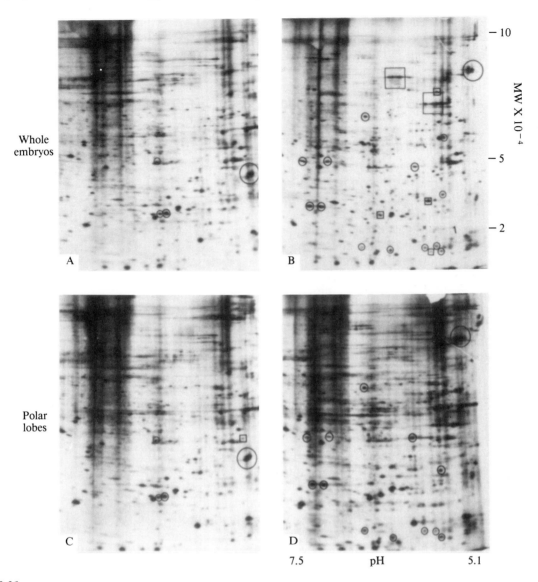

Whole
embryos

Polar
lobes

MW X 10⁻⁴

7.5   pH   5.1

Figure 13.21

Two-dimensional gel electrophoresis of proteins synthesized during short intervals
in whole embryos (A and B) and isolated polar lobes (C and D) of the marine snail
Ilyanassa. A, Embryos labeled for four hours from the one-cell to eight-cell stage.
B, Embryos labeled for four hours during the gastrula stage. C, Isolated polar
lobes labeled at the same time as in A. D, Isolated polar lobes labeled at the same
time as in B. The circled and boxed proteins change both in whole embryos and
isolated polar lobes and therefore represent proteins whose synthesis is controlled
at the translational level. (From Brandhorst and Newrock, 1981.)

from both the tip and sides of the archenteron, as occurs in the paternal
species. This result indicates that the zygotic genome has begun to influence
development by the time of gastrulation.

During later development, the primary mesenchyme cells form spicules,
crystalline structures that constitute the branched skeleton of the pluteus
larva. Spicule morphology differs among sea urchin species and can be used
as a marker in interspecies hybridization experiments (Fig. 13.22A and B).
For instance, the portion of the spicule that extends into the anal arms of

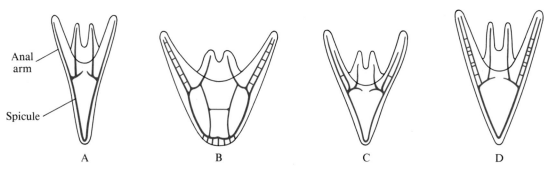

Figure 13.22

A diagram of normal and hybrid sea urchin pluteus larvae showing the morphology of the spicules (rod-like structures). A, A normal *Strongylocentrotus lividus* larva with spicules exhibiting a single unbranched rod in the anal arms. B, A normal *Sphaerechinus granularis* larva with spicules exhibiting two rods with many interconnecting bars in the anal arms. C, and D, Examples of hybrid larvae produced from a cross between *Sphaerechinus* eggs and *Strongylocentrotus* sperm showing spicules with various degrees of duplication, branching, and cross-bars in the anal arms. (After Vernon, 1900.)

Strongylocentrotus lividus plutei is a single unbranched rod. In contrast, the corresponding part of the spicule in *Sphaerechinus granularis* plutei consists of two rods, with multiple interconnecting bars. When *Sphaerechinus* eggs are fertilized with *Strongylocentrotus* sperm, hybrid plutei develop with spicules resembling both parental types (see Fig. 13.22C and D). The hybrid characteristics of these spicules indicate that they are formed by the combined activity of the paternal and maternal genomes. The involvement of the zygotic genome in spicule formation is confirmed by treatment of sea urchin embryos with actinomycin D. Although primary mesenchyme cells form in the presence of actinomycin D, they do not differentiate to form a skeleton. Thus, the zygotic genome begins to influence primary mesenchyme cell development by the time sea urchin embryos reach the mesenchyme blastula stage.

Activation of Transcription

As described in Chapter 3, oocytes are very active in RNA synthesis. Whether transcription continues after the completion of oogenesis, however, varies in different organisms. After oocyte maturation or fertilization, some organisms enter a transcriptional quiescent period, whereas others continue to synthesize RNA.

In sea urchins, transcription occurs in the unfertilized egg, the post-fertilization zygote, and during the early cleavages (Wilt, 1964; Kedes and Gross, 1969; Brandhorst, 1980; Poccia et al., 1985). The major class of RNA transcribed during early development is mRNA. This is demonstrated by the heterogeneous size of the newly synthesized transcripts and their presence in polysomes. The transcription of mRNA appears to increase gradually during the cleavage period. For example, gradual accumulation of zygotic mRNA encoding early histone H2B during the cleavage period of sea urchin embryos has been demonstrated by Northern hybridization experiments (see Fig. 13.26). In contrast to early histone mRNA, transcription of tRNA does not begin until the 64-cell stage (O'Melia and Villee, 1972), and rRNA synthesis is initiated at the blastula stage (Emerson and

Humphreys, 1970). When rRNA transcription first appears in blastulae, transcript accumulation is low relative to the rate of rRNA synthesis during later development (Griffith et al., 1981). The rRNA genes remain relatively inactive until the pluteus stage, when there is a major increase in rRNA synthesis (Humphreys, 1973).

In amphibians, there is a transcriptionally quiescent period during early development. In *Xenopus* embryos, radioactive uridine incorporation into nuclear RNA is initially detected at about the twelfth cleavage (Fig. 13.23). The period between the eleventh and twelfth cleavages is the developmental period referred to as the **mid-blastula transition** (Newport and Kirschner, 1982a). Transfer RNA and some kinds of mRNA (Krieg and Melton, 1987) first appear at this stage of development, whereas rRNA and many other zygotic mRNA species are first synthesized somewhat later (Shiokawa et al., 1981). *Drosophila* embryos also exhibit a general activation of zygotic transcription at the cellular blastoderm stage (Anderson and Lengyel, 1979). Unlike the situation in amphibians, however, there is evidence that some genes are transcribed earlier. As described in Chapter 3, gene transcription can be visualized directly by disrupting nuclei and viewing the spread chromatin with an electron microscope (McKnight and Miller, 1976). Electron microscopy of chromatin from nuclei of pre-blasto-derm stage *Drosophila* embryos clearly shows genes in the act of being transcribed (Fig. 13.24). The histone mRNAs are among the first to be transcribed and can be detected as early as the syncytial blastoderm stage (Anderson and Lengyel, 1980). The early transcription of some of the *Drosophila* segmentation genes will be discussed in Chapter 14.

Transcription is activated in mammals between fertilization and first cleavage. This can be demonstrated in two different ways. First, when mouse embryos are treated with inhibitors of transcription after fertilization, they arrest at the two-cell stage. Second, when radioactive uridine is added to mouse embryos after fertilization, incorporation into RNA can be detected as early as pronuclear fusion. The transcripts synthesized at the two-cell stage include mRNA (Levey et al., 1978) and rRNA (Clegg and Pikó, 1983), and the four-cell stage marks the first appearance of tRNA (Woodland and Graham, 1969). Although mammals activate transcription at an early de-

Figure 13.23

The initiation of RNA synthesis at the mid-blastula transition in *Xenopus* embryos. The lower plot (colored line) represents nuclear RNA synthesis as determined by autoradiography. The upper plot (black line) represents the duration of the cell cycle, which begins to increase at the mid-blastula transition. (After Bachvarova and Davidson, 1966.)

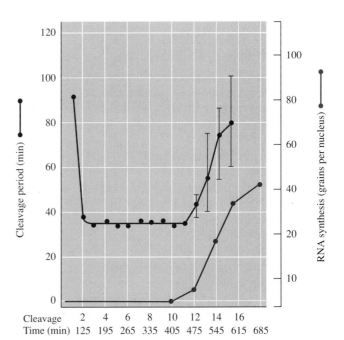

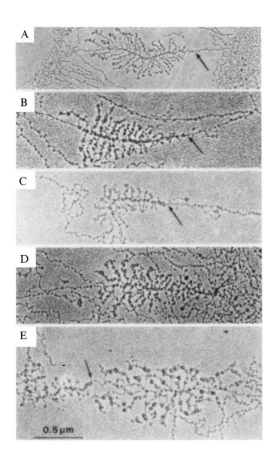

Figure 13.24

Electron micrographs of spread chromatin showing transcription complexes with nascent RNAs from pre-blastoderm *Drosophila* embryos. Various non-ribosomal transcription complexes are shown. The arrows mark the transcriptional initiation sites for each complex. In E, the arrow indicates the initiation sites of two oppositely oriented transcription complexes. (From McKnight and Miller, 1976. Copyright © Cell Press.)

velopmental stage, the actual time interval between fertilization and the beginning of RNA synthesis is much longer because mammals develop so slowly.

The Switch from Oogenic to Zygotic mRNA Translation

All organisms eventually switch from the translation of oogenic to zygotic mRNA. This can happen through an abrupt loss of oogenic mRNA or via a gradual attrition of oogenic messengers. In the mammalian embryo, for example, the amount of oogenic mRNA decreases dramatically during the two-cell stage. Northern hybridization experiments (Fig. 13.25) show that oogenic actin and histone mRNAs are abundant in one-cell mouse embryos but disappear before first cleavage. Later, some of these mRNA species are replaced by transcription from the zygotic genome. Although most of the newly synthesized zygotic mRNAs encode the same proteins as the oogenic mRNAs being replaced, new species of transcripts also appear at this time, as reflected by changes in the pattern of protein synthesis. For example, treatment of mouse embryos with the transcriptional inhibitor α-amanitin before first cleavage has no effect on protein synthesis. However, if two-cell embryos are treated with this inhibitor, a distinct group of proteins that normally appears at this stage of development is missing (Flach et al., 1982). Human embryos appear to undergo the same transition, but at the four-cell stage. No differences are apparent in the pattern of protein synthesis in human embryos exposed to radioactive amino acids between the unfertilized egg and the two-cell stage (Braude et al., 1988). However, at the four-cell stage, some of the proteins characteristic of earlier stages disappear

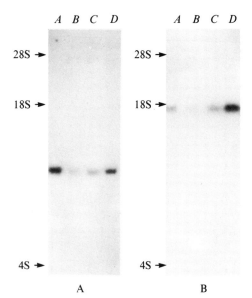

Figure 13.25

Changes in the abundance of histone H3 and actin messengers during early mouse development. A, Northern blot hybridization of histone H3 RNA from different stages of early mouse development. Lane *A*: RNA from 1000 unfertilized eggs; Lane *B*: RNA from 1000 two-cell embryos; Lane *C*: RNA from 1000 eight-cell embryos; Lane *D*: RNA from 1000 blastocysts. B, Northern blot hybridization of actin RNA from stages of early mouse development. Stages are identical to those in A. (From Giebelhaus et al., 1983.)

and a unique group of proteins is synthesized. When unfertilized human eggs are treated with α-amanitin and then inseminated, the unique proteins are missing, and development arrests at the four-cell stage. The abrupt switch from oogenic to zygotic mRNA translation during early cleavage in mammals is an extreme example of a more gradual, but fundamentally similar, phenomenon that occurs in other organisms.

The switch from translation of oogenic to zygotic mRNAs is less dramatic in sea urchin embryos. It occurs gradually, between the early cleavage and mesenchyme blastula stages, and specific mRNA species are affected to different extents. In some instances, the switch includes an intermediate stage characterized by the appearance of zygotic mRNA species transcribed exclusively during cleavage stages, as exemplified by histone gene transcription. Sea urchins have specialized histone gene families that are transcribed at different times during development. Each gene family contains a set of genes for each of the five types of histones (H1, H2A, H2B, H4, and H5). Genes in one family differ in nucleotide sequence from the homologous genes in other families. The histones assembled into chromatin during the very early cleavage stages are called the **cleavage-stage**, or **CS**, histones. The CS histones are synthesized by oogenic mRNAs and, for a short time after fertilization, by mRNAs transcribed from the zygotic genome. The zygotic CS histone genes are subsequently turned off during the early cleavages. Synthesis of the **early histones** is initiated by the two-cell stage. Like CS histones, early histones are encoded by both oogenic and zygotic mRNAs. However, the oogenic CS and early histone mRNAs are stored in two separate locations in the egg. As discussed previously, early histone mRNA is stored in the female pronucleus, whereas CS histone mRNA is stored in the egg cytoplasm, along with other oogenic mRNAs (Showman et al., 1982). A third set of histone variants, the **late histones**, is synthesized from the 16-cell stage through the pluteus stage and in adult sea urchins (Maxson et al., 1983; Halsell et al., 1987). By the mid-gastrula stage, late histones are the only histones being synthesized in the embryo. The changing levels of early and late forms of histone 2B mRNA are shown by Northern hybridization experiments using probes specific for each transcript (Fig. 13.26). Messenger RNAs transcribed from the early H2B genes begin to accumulate by the 16-cell stage and reach maximal levels by the 200-cell stage, after which they decline in amount (Mauron et al., 1982;

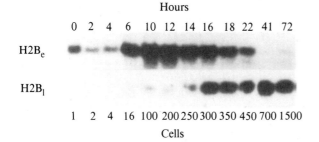

Hours

0 2 4 6 10 12 14 16 18 22 41 72

H2B$_e$

H2B$_l$

1 2 4 16 100 200 250 300 350 450 700 1500

Cells

Figure 13.26

Northern blot hybridization of early and late histone H2B mRNA during *S. purpuratus* development. The amount of early H2B (H2B$_e$) RNA is reduced as late H2B (H2B$_l$) RNA increases in amount. (From R. E. Maxson et al., 1983. Reprinted by permission from *Nature* Vol. 301 pp. 120–125. Copyright © 1983 Macmillan Magazines Ltd.)

Maxson and Wilt, 1982). After about the 450-cell stage, the late histone gene transcripts are the most prevalent histone mRNAs in the embryo. In general, it appears that gradual degradation of oogenic mRNAs leads to their eventual disappearance from the embryo by the gastrula stage.

Although many of the zygotic mRNA species that appear during development encode the same proteins as the degraded oogenic mRNAs do, new species of zygotic mRNA also appear between late cleavage and the pluteus stage. As we shall discuss in detail in Chapter 14, zygotic mRNAs transcribed from most of the *Drosophila* segmentation genes appear for the first time during early embryogenesis. In sea urchins, new zygotic mRNAs include those encoding a muscle form of actin (Davidson, 1986), matrix proteins of the spicule (Benson et al., 1987), and a family of Ca^{2+}-binding proteins known as the Spec (*Strongylocentrotus purpuratus* ectoderm specific) family, whose expression is restricted to the aboral ectoderm (see Chap. 14). The gradual accumulation of two of the Spec mRNAs during sea urchin embryogenesis is shown in Figure 13.27.

13–5 Control of Transcriptional Activation

The appearance of new mRNA, tRNA, and rRNA during early development indicates that a general activation of transcription has occurred in the embryo. As discussed previously, this activation occurs early in sea urchin embryos, whereas it occurs later and abruptly in amphibian embryos. Transcriptional activation in amphibian embryos (at the mid-blastula transition) appears to be controlled by factors located in the egg cytoplasm. This is demonstrated by microinjecting nuclei obtained from transcriptionally

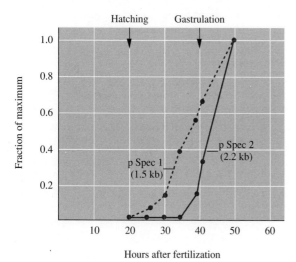

Figure 13.27

Accumulation of Spec 1 and 2 mRNAs during sea urchin embryogenesis. (From Klein et al., 1983.)

active embryonic cells into enucleated eggs and examining subsequent RNA synthesis (Gurdon and Woodland, 1969). The injected nuclei immediately stop transcription. Later, they are transcriptionally activated at the mid-blastula transition. Cytoplasmic factors controlling transcription do not appear to be species-specific. Nuclei from other organisms, even bacterial plasmids containing a cloned yeast tRNA gene (Fig. 13.28), become transcriptionally inactive after injection into the amphibian egg and are reactivated on the same schedule as the endogenous nuclei.

In addition to the stimulation of transcription, other changes occur at the mid-blastula transition in amphibian embryos (Newport and Kirschner, 1982a): The rate of cleavage slows, blastomeres in various regions of the embryo begin to divide asynchronously, and cells become motile for the first time. It appears that these events may be controlled coordinately. There are several ways in which the timing of transcriptional activation and other changes that occur at the mid-blastula transition could be triggered: First, these events may respond to a clock-like mechanism that counts the absolute time lapsed between fertilization and the mid-blastula transition. Second, the timing of these events could be controlled by the number of DNA replications or cleavages that occur between fertilization and the mid-blastula transition. Third, because the nuclear/cytoplasmic ratio gradually increases during the early cleavages (see Chap. 5), the transition could be triggered when a critical nuclear/cytoplasmic ratio is reached.

A series of experiments was conducted by Newport and Kirschner (1982a, b) to test these possibilities. To determine whether the number of cleavages controls transcriptional activation, zygotes were centrifuged or treated with cytochalasin B before first cleavage. Cytokinesis is blocked by these treatments, but nuclear division and DNA synthesis continue, the uncleaved zygote becomes multinucleate, and the nuclear/cytoplasmic ratio increases accordingly. Transcription is activated on schedule in these syncytial embryos, however, indicating that the mechanism of activation is not regulated by the number of cleavages. To examine the role of absolute time lapsed from fertilization and the number of DNA synthetic cycles in

Figure 13.28

Kinetics of transcriptional inactivation and activation after a plasmid containing a cloned yeast tRNA gene is microinjected into a *Xenopus* egg. Initially, the gene is active in transcription, then it is transcriptionally suppressed after being assembled into chromatin in the egg cytoplasm, and finally it is reactivated when the endogenous *Xenopus* genes become transcriptionally active. (After Newport and Kirschner, 1982b. Copyright © Cell Press.)

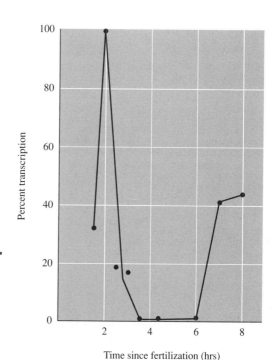

Time since fertilization (hrs)

activating RNA synthesis, Newport and Kirschner conducted the experiment shown in Figure 13.29. Using a procedure first developed by Spemann (1928), a fertilized egg was constricted into two parts by ligation with a loop of human hair. Initially, the zygote nucleus is too large to pass through the narrow constriction separating the ligated parts of the egg, and only the portion of the egg containing the nucleus cleaves. However, as nuclei continue to divide, they become smaller and some are eventually able to slip through the constriction and initiate cleavage in the uncleaved portion of the egg. Even though the same absolute time has elapsed, and the same number of nuclear DNA synthetic cycles have occurred, the portion of the egg that receives a nucleus later in development is retarded in reaching the mid-blastula transition by one to two cleavages. This result suggests that transcriptional activation is controlled neither by absolute time nor the number of DNA synthetic cycles after fertilization. However, because the nuclear/cytoplasmic ratio is decreased in the retarded portion of the egg, the results suggest that the mid-blastula transition is triggered by an increase in the nuclear/cytoplasmic ratio. Further evidence for control of the mid-blastula transition by the nuclear/cytoplasmic ratio has been obtained as follows: First, when the nuclear/cytoplasmic ratio is increased by removing cytoplasm from an uncleaved egg or by increasing the number of nuclei in the egg cytoplasm by polyspermy, the mid-blastula transition is initiated prematurely. Second, when the mid-blastula transition is examined in haploid embryos (which have a lower nuclear/cytoplasmic ratio because haploid nuclei are smaller than diploid nuclei), the mid-blastula transition is retarded by one cleavage cycle.

Based upon the results described previously, Newport and Kirschner (1982b) proposed a model to explain the activation of transcription at the mid-blastula transition. They believe that amphibian eggs contain a factor that inhibits transcription by binding to nuclear DNA. When the nuclear/cytoplasmic ratio is low, the inhibitor would be abundant enough to saturate all the DNA binding sites, and suppression of transcription would result. When the nuclear/cytoplasmic ratio increases to a critical level, the inhibitory factor is no longer able to occupy all of the DNA binding sites, and

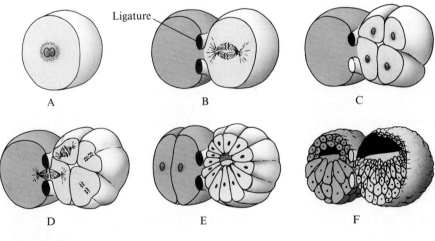

Figure 13.29

Schematic diagram of the egg constriction experiment used to demonstrate that the onset of the mid-blastula transition is related to nuclear/cytoplasmic ratio rather than absolute time after fertilization, number of cleavages, or DNA replication cycles in *Xenopus* eggs. (After Newport and Kirschner, 1982a.)

transcription would be initiated. As new nuclei are produced during development, more and more unoccupied regulatory sites would appear in DNA, and transcription would gradually increase.

According to the Newport and Kirschner model, it should be possible to activate transcription prematurely by microinjecting excess DNA into the egg. Thus, increasing amounts of bacterial plasmid DNA have been injected into cleavage-arrested *Xenopus* eggs that had been injected previously with lesser quantities of an unrelated plasmid containing a yeast tRNA gene. The yeast tRNA gene was used as a marker for transcriptional activation because its specific transcripts can be assayed using a cloned probe. As described earlier, the injected yeast tRNA genes become quiescent shortly after microinjection and are normally transcriptionally activated at the mid-blastula transition along with the amphibian genes (see Fig. 13.28). However, when small amounts of bacterial plasmid DNA were microinjected into the same egg, the yeast tRNA gene remained transcriptionally inactive. However, when the amount of injected DNA exceeded a critical level, the tRNA gene continued to be transcribed. The amount of DNA required to sustain transcription is similar to that present in an uninjected embryo at the mid-blastula transition. This experiment supports the idea that transcription is controlled by the concentration of an inhibitor that can be titrated out of the cytoplasm by the addition of exogenous DNA. Recent experiments suggest that this inhibitor may function by preventing the lengthening of the cell cycle, which normally occurs at the mid-blastula transition (Kimelman et al., 1987).

13–6 *Chapter Synopsis*

This chapter describes the relative roles of the maternal and zygotic genomes in controlling early development. Development is initiated by maternal gene products stored in the egg. This is shown by development of anucleate merogones, early dominance of maternal features in interspecies hybrids, and the existence of maternal-effect genes. Protein synthesis after fertilization in sea urchin eggs is an example of maternal regulation of early development. Although the mechanism of activation of protein synthesis is unknown, it relies on a large stockpile of oogenic mRNAs, ribosomes, and other translational factors in the unfertilized egg. After fertilization, one or more of these factors becomes available for translation, and protein synthesis is increased 100-fold in the absence of appreciable zygotic transcription. Other organisms exhibit similar, but less extensive, changes in the intensity of protein synthesis during oocyte maturation or fertilization.

As development proceeds, there is a switch from control by the maternal genome to the zygotic genome; this switch is timed differently in various animals. In sea urchin embryos, low levels of zygotic transcription continue throughout early development, and zygotic gene products accumulate gradually beginning at the blastula stage. In amphibian embryos, zygotic transcription begins abruptly at the mid-blastula transition and may be dependent on reducing the titer of DNA-binding factors stored in the egg cytoplasm. In mammalian embryos, which exhibit very slow development, the switch from maternal to zygotic control begins as early as the two-cell stage. Activation of zygotic transcription includes genes that had been active during oogenesis as well as formerly inactive genes that function at specific times in development and in particular regions of the embryo. Differential zygotic gene expression will be examined in the next chapter.

References

Alexandraki, D., and J. Ruderman. 1985. Multiple polymorphic α tubulin and β tubulin messenger RNAs are present in sea urchin eggs. Proc. Natl. Acad. Sci. U.S.A., *82*: 134–138.

Anderson, D. M. et al. 1982. Sequence organization of the poly (A) RNA synthesized in lampbrush chromosome stage *Xenopus laevis* oocytes. J. Mol. Biol., *155*: 281–309.

Anderson, K. V., and J. A. Lengyel. 1979. Rates of synthesis of the major classes of RNA in *Drosophila* embryos. Dev. Biol., *70*: 217–231.

Anderson, K. V., and J. A. Lengyel. 1980. Changing rates of histone mRNA synthesis and turnover in Drosophila embryos. Cell, *21*: 717–727.

Angerer, L. M., and R. C. Angerer. 1981. Detection of poly (A)+ RNA in sea urchin eggs and embryos by quantitative *in situ* hybridization. Nucleic Acids Res., *9*: 2819–2840.

Bachvarova, R., and E. Davidson. 1966. Nuclear activation at the onset of amphibian gastrulation. J. Exp. Zool., *163*: 285–296.

Bakken, A. H. 1973. A cytological and genetic study of oogenesis in *Drosophila melanogaster*. Dev. Biol., *33*: 100–122.

Ballantine, J. E. M., H. R. Woodland, and E. A. Sturgess. 1979. Changes in protein synthesis during the development of *Xenopus laevis*. J. Embryol. Exp. Morphol., *51*: 137–153.

Baum, E. Z., and W. M. Wormington. 1985. Coordinate expression of ribosomal protein genes during *Xenopus* development. Dev. Biol., *111*: 488–489.

Begg, D. A., L. I. Rebhun, and H. Hyatt. 1982. Structural organization of actin in the sea urchin egg cortex: Microvillar elongation in the absence of actin filament bundle formation. J. Cell Biol., *93*: 24–32.

Benson, S. et al. 1987. A lineage-specific gene encoding a major matrix protein of the sea urchin embryo spicule. I. Authentication of the cloned gene and its developmental expression. Dev. Biol., *120*: 499–506.

Brandhorst, B. P. 1976. Two-dimensional gel patterns of protein synthesis before and after fertilization of sea urchin eggs. Dev. Biol., *52*: 310–317.

Brandhorst, B. P. 1980. Simultaneous synthesis, translation, and storage of mRNA including histone mRNA in sea urchin eggs. Dev. Biol., *79*: 139–148.

Brandhorst, B. P., and K. M. Newrock. 1981. Post-transcriptional regulation of protein synthesis in *Ilyanassa* embryos and isolated polar lobes. Dev. Biol., *83*: 250–254.

Brandis, J. W., and R. A. Raff. 1978. Translation of oogenetic mRNA in sea urchin eggs and early embryos: Demonstration of a change in translational efficiency following fertilization. Dev. Biol., *67*: 97–113.

Braude, P., V. Bolton, and S. Moore. 1988. Human gene expression first occurs between the four- and eight-cell stages of preimplantation development. Nature (Lond.), *332*: 459–461.

Briggs, R., and G. Cassens. 1966. Accumulation in the oocyte nucleus of a gene product essential for development beyond gastrulation. Proc. Natl. Acad. Sci. U.S.A., *55*: 1103–1109.

Briggs, R., E. U. Green, and T. J. King. 1951. An investigation of the capacity for cleavage and differentiation in *Rana pipiens* eggs lacking "functional" chromosomes. J. Exp. Zool., *116*: 455–499.

Briggs, R., and J. T. Justus. 1968. Partial characterization of the component from normal eggs which corrects the maternal-effect in gene *o* in the Mexican axolotl (*Ambystoma mexicanum*). J. Exp. Zool., *167*: 105–116.

Brown, D. D., and J. B. Gurdon. 1964. Absence of ribosomal RNA synthesis in the anucleolate mutant of *Xenopus laevis*. Proc. Natl. Acad. Sci. U.S.A., *51*: 139–146.

Brown, D. D., and E. Littna. 1966. Synthesis and accumulation of low molecular weight RNA during embryogenesis of *Xenopus laevis*. J. Mol. Biol., *20*: 95–112.

Caldwell, D. C., and C. P. Emerson. 1985. The role of cap methylation in the translational activation of stored maternal histone mRNA in sea urchin embryos. Cell, *42*: 691–700.

Clegg, K. B., and L. Pikó. 1983. Poly (A) length, cytoplasmic adenylation and synthesis of poly (A)⁺ RNA in early mouse embryos. Dev. Biol., *95*: 331–341.

Colin, A. M., and M. B. Hille. 1986. Injected mRNA does not increase protein synthesis in unfertilized, fertilized, or ammonia-activated sea urchin eggs. Dev. Biol., *115*: 184–192.

Colin, A. M. et al. 1987. Evidence for simultaneous derepression of messenger RNA and the guanine nucleotide exchange factor in fertilized sea urchin eggs. Dev. Biol., *123*: 354–363.

Collier, J. R. 1966. The transcription of genetic information in the spiralian embryo. *In* A. A. Moscona and A. Monroy (eds.), *Current Topics in Developmental Biology*, Vol. 1. Academic Press, New York, pp. 39–59.

Costantini, F. D., R. J. Britten, and E. H. Davidson. 1980. Message sequences and short repetitive sequences are interspersed in sea urchin egg poly (A)⁺ RNAs. Nature (Lond.), *287*: 111–117.

Crain, W. R., Jr., D. S. Durica, and K. Van Doren. 1981. Actin gene expression in developing sea urchin embryos. Mol. Cell. Biol., *1*: 711–720.

Danilchik, M. V., and M. B. Hille. 1981. Sea urchin egg and embryo ribosomes: Differences in translational activity in a cell-free system. Dev. Biol., *84*: 291–298.

Davidson, E. H. 1976. *Gene Activity in Early Development*, 2nd ed. Academic Press, New York.

Davidson, E. H. 1986. *Gene Activity in Early Development*, 3rd ed. Academic Press, New York.

Davidson, E. H., B. R. Hough-Evans, and R. J. Britten. 1982. Molecular biology of the sea urchin embryo. Science, *217*: 17–26.

DeLeon, D. V. et al. 1983. Most early variant histone mRNA is contained in pronuclei of sea urchin eggs. Dev. Biol., *100*: 197–206.

Denny, P. C., and A. Tyler. 1964. Activation of protein biosynthesis in non-nucleate fragments of sea urchin eggs. Biochem. Biophys. Res. Commun., *14*: 245–249.

Dolecki, G. J., R. F. Duncan, and T. Humphreys. 1977. Complete turnover of poly (A) on maternal mRNA of sea urchin embryos. Cell, *11*: 339–344.

Ecker, R. E., and L. D. Smith. 1971. The nature and fate of *Rana pipiens* proteins synthesized during maturation and early cleavage. Dev. Biol., *24*: 559–576.

Elsdale, T. R., M. Fischberg, and S. Smith. 1958. A mutation that reduces nucleolar number in *Xenopus laevis*. Exp. Cell Res., *14*: 642–643.

Emerson, C. P., and T. Humphreys. 1970. Regulation of DNA-like RNA and the apparent activation of ribosomal RNA synthesis in early sea urchin embryos: Quantitative measurements of newly-synthesized RNA. Dev. Biol., *23*: 86–112.

Epel, D. 1967. Protein synthesis in sea urchin eggs: A "late" response to fertilization. Proc. Natl. Acad. Sci. U.S.A., *57*: 899–906.

Evans, T. et al. 1983. Cyclin: A protein specified by maternal mRNA in sea urchin eggs that is destroyed at each cleavage division. Cell, *33*: 389–396.

Feigenbaum, L., and E. Goldberg. 1965. Effect of actinomycin D on morphogenesis in *Ilyanassa*. Am. Zool., *5*: 198.

Flach, G. et al. 1982. The transition from maternal to embryonic control in the 2-cell mouse embryo. EMBO J., *1*: 681–686.

Fu, L. et al. 1991. Translational potentiation of mRNA with secondary structure during oogenesis and early development of *Xenopus*. Science, *251*: 807–810.

Fullilove, S. L., and R. C. Woodruff. 1974. Genetic, cytological, and ultrastructural characterization of a temperature-sensitive lethal in *Drosophila melanogaster*. Dev. Biol., *38*: 291–307.

Giebelhaus, D. H., J. J. Heikkila, and G. A. Schultz. 1983. Changes in the quantity of histone and actin messenger RNA during the development of preimplantation mouse embryos. Dev. Biol., *98*: 148–154.

Grainger, J. L. et al. 1979. Intracellular pH controls protein synthesis rate in the sea urchin egg and early embryo. Dev. Biol., *68*: 396–406.

Griffith, J. K., B. B. Griffith, and T. Humphreys. 1981. Regulation of ribosomal RNA synthesis in sea urchin embryos and oocytes. Dev. Biol., *87*: 220–228.

Gross, P. R., L. E. Malkin, and W. A. Moyer. 1964. Templates for the first proteins of embryonic development. Proc. Natl. Acad. Sci. U.S.A., *51*: 407–414.

Gurdon, J. B., and H. R. Woodland. 1969. The influence of the cytoplasm on the nucleus during cell differentiation, with special reference to RNA synthesis during amphibian cleavage. Proc. R. Soc. Lond., Ser. B., *173*: 99–111.

Halsell, S. R., M. Ito, and R. Maxson. 1987. Differential expression of early and late histone embryonic genes in adult tissues of the sea urchin *Strongylocentrotus purpuratus*. Dev. Biol., *119*: 268–274.

Harvey, E. B. 1936. Parthenogenetic merogony or cleavage without nuclei in *Arbacia punctulata*. Biol. Bull., *71*: 101–121.

Harvey, E. B. 1940. A comparison of the development of nucleate and non-nucleate eggs of *Arbacia punctulata*. Biol. Bull., *79*: 166–187.

Hille, M. B., and A. A. Albers. 1979. Efficiency of protein synthesis after fertilization in sea urchin eggs. Nature (Lond.), *278*: 469–471.

Hough-Evans, B. R. et al. 1977. Appearance and persistence of maternal RNA sequences in sea urchin development. Dev. Biol., *60*: 258–277.

Hough-Evans, B. R. et al. 1980. Complexity of RNA in eggs of *Drosophila melanogaster* and *Musca domestica*. Genetics, *95*: 81–94.

Huang, W-I. et al. 1988. Inactivator of eIF-4F in unfertilized sea urchin eggs. Proc. Natl. Acad. Sci. U.S.A., *84*: 6359–6363.

Hultin, T. 1952. Incorporation of N^{15}-labeled glycine and alanine into the proteins of developing sea urchin eggs. Exp. Cell Res., *3*: 494–496.

Hultin, T. 1961. The effect of puromycin on protein metabolism and cell division in fertilized sea urchin eggs. Experientia, *17*: 410–411.

Humphreys, T. 1971. Measurements of messenger RNA entering polysomes upon fertilization in sea urchin eggs. Dev. Biol., *26*: 201–208.

Humphreys, T. 1973. RNA and protein synthesis during early animal embryogenesis. *In* S. J. Coward (ed.), *Developmental Regulation: Aspects of Cell Differentiation*. Academic Press, New York, pp. 1–22.

Infante, A. A., and L. J. Heilmann. 1981. Distribution of messenger ribonucleic acid in polysomes and nonpolysomal particles of sea urchin embryos: Translational control of actin synthesis. Biochemistry, *20*: 1–8.

Jenkins, N. A. et al. 1978. A test for masked message: The template activity of messenger ribonucleoprotein particles isolated from sea urchin eggs. Dev. Biol., *63*: 279–298.

Kedes, L. H., and P. R. Gross. 1969. Synthesis and function of messenger RNA during early embryonic development. J. Mol. Biol., *42*: 559–575.

Kimelman, D., and M. Kirschner. 1987. Synergistic induction of mesoderm by FGF and TGF-ß and the identification of an mRNA coding for FGF in the early Xenopus embryo. Cell, *51*: 869–877.

Kimelman, D., M. Kirschner, and T. Scherson. 1987. The events of midblastula transition in Xenopus are regulated by changes in the cell cycle. Cell, *48*: 399–407.

Klein, W. H. et al. 1983. A family of genes expressed in the embryonic ectoderm of sea urchins. *In* W. R. Jeffery and R. A. Raff (eds.), *Time, Space and Pattern in Embryonic Development*. Alan R. Liss Press, New York, pp. 87–100.

Krieg, P. A., and D. A. Melton. 1987. An enhancer responsible for activating transcription at the midblastula transition in *Xenopus* development. Proc. Natl. Acad. Sci. U.S.A., *84*: 2331–2335.

Laskey, R. et al. 1977. Protein synthesis in oocytes of Xenopus laevis is not regulated by the supply of messenger RNA. Cell, *11*: 345–351.

Lau, J. T. Y., and W. J. Lennarz. 1983. Regulation of sea urchin glycoprotein mRNAs during embryonic development. Proc. Natl. Acad. Sci. U.S.A., *80*: 1028–1032.

Levey, I. L., G. B. Stull, and R. L. Brinster. 1978. Poly(A) and synthesis of polyadenylated RNA in the preimplantation mouse embryo. Dev. Biol., *64*: 140–148.

Lockshin, R. A. 1966. Insect embryogenesis: Macromolecular synthesis during early development. Science, *154*: 775–776.

Maisonhaute, C. 1977. Comparison des effets de deux inhibiteurs de la synthese d'ARN (Actinomycin D et α-amanitine) sur le développement embryonnaire d'un insecte déterminisme génique du début de l'embryogénèse de *Leptinotarsa decemlineata* (Coleptera). Wilhelm Roux's Archives Dev. Biol., *183*: 61–77.

Mauron, A. et al. 1982. Accumulation of individual histone mRNAs during embryogenesis of the sea urchin *Strongylocentrotus purpuratus*. Dev. Biol., *94*: 425–434.

Maxson, R. E., Jr., and F. H. Wilt. 1982. Accumulation of the early histone messenger RNAs during the development of *Strongylocentrotus purpuratus*. Dev. Biol., *94*: 435–440.

Maxson, R. E. et al. 1983. Distinct organizations and patterns of expression of early and late histone gene sets in the sea urchin. Nature (Lond.), *301*: 120–125.

McGrew, L. L., and J. D. Richter. 1989. *Xenopus* oocyte poly(A) RNAs that hybridize to a cloned interspersed repeat sequence are not translatable. Dev. Biol., *134*: 267–270.

McKnight, S. L., and O. L. Miller. 1976. Ultrastructural patterns of RNA synthesis during early embryogenesis of Drosophila melanogaster. Cell, *8*: 305–319.

Mescher, A., and T. Humphreys. 1974. Activation of maternal mRNA in the absence of poly (A) formation in fertilized sea urchin eggs. Nature (Lond.), *249*: 138–139.

Moon, R. T., M. L. Danilchik, and M. B. Hille. 1982. An assessment of the masked message hypothesis: Sea urchin egg messenger ribonucleoprotein complexes are efficient templates for *in vitro* protein synthesis. Dev. Biol., *93*: 389–403.

Moon, R. T., K. D. Moe, and M. B. Hille. 1980. Polypeptides of nonpolyribosomal messenger ribonucleoprotein complexes of sea urchin eggs. Biochemistry, *19*: 2723–2730.

Moon, R. T. et al. 1983. The cytoskeletal framework of sea urchin eggs and embryos: Developmental changes in the association of messenger RNA. Dev. Biol., *95*: 447–458.

Newport, J., and M. Kirschner. 1982a. A major developmental transition in early Xenopus embryos: I. Characterization and timing of changes at the midblastula stage. Cell, *30*: 675–686.

Newport, J., and M. Kirschner. 1982b. A major developmental transition in early Xenopus embryos: II. Control of the onset of transcription. Cell, *30*: 687–696.

Okada, M., I. A. Kleinman, and H. A. Schneiderman. 1974. Repair of a genetically-caused defect in oogenesis of *Drosophila melanogaster* by transplantation of cytoplasm from wild-type eggs and injection of pyrimidine nucleosides. Dev. Biol., *37*: 55–62.

O'Melia, A. F., and C. A. Villee. 1972. *De novo* synthesis of transfer RNA and 5S RNA[if] in cleaving sea urchin embryos. Nature (Lond.) New Biol., *239*: 51–53.

Pierandrei-Amaldi, P. et al. 1982. Expression of ribosomal protein genes in Xenopus laevis. Cell, *30*: 163–171.

Poccia, D. et al. 1985. RNA synthesis in male pronuclei of the sea urchin. Biochim. Biophys. Acta, *824*: 349–356.

Raff, R. A. 1983. Localization and temporal control of expression of maternal histone mRNA in sea urchin embryos. *In* W. R. Jeffery and R. A. Raff (eds.), *Time, Space and Pattern in Embryonic Development*. Alan R. Liss Press, New York, pp. 65–86.

Raff, R. A. et al. 1972. Oogenic origin of messenger RNA for embryonic synthesis of microtubule proteins. Nature (Lond.), *235*: 211–214.

Regier, J. C., and F. C. Kafatos. 1977. Absolute rates of protein synthesis in sea urchins with specific activity measurements of radioactive leucine and leucyl-tRNA. Dev. Biol., *57*: 270–283.

Rice, T. B., and A. Garen. 1975. Localized defects of blastoderm formation in maternal-effect mutants of *Drosophila*. Dev. Biol., *93*: 277–286.

Richter, J. D. et al. 1984. Interspersed poly (A) RNAs of amphibian oocytes are not translatable. J. Mol. Biol., *173*: 227–241.

Rosenthal, E. T., B. P. Brandhorst, and J. V. Ruderman. 1982. Translationally mediated changes in patterns of protein synthesis during maturation of starfish oocytes. Dev. Biol., *91*: 215–220.

Rosenthal, E. T., T. Hunt, and J. V. Ruderman. 1980. Selective translation of mRNA controls the pattern of protein synthesis during early development of the surf clam, Spisula solidissima. Cell, *20*: 487–494.

Rosenthal, E. T., and J. V. Ruderman. 1987. Widespread changes in the translation and adenylation of maternal messenger RNAs following fertilization of *Spisula* oocytes. Dev. Biol., *121*: 237–246.

Scheller, R. H. et al. 1981. Organization and expression of multiple actin genes in the sea urchin. Mol. Cell. Biol., *1*: 609–628.

Schultz, R. M., and P. M. Wassarman. 1977. Specific changes in the pattern of protein synthesis during meiotic maturation of mammalian oocytes *in vitro*. Proc. Natl. Acad. Sci. U.S.A., *74*: 538–541.

Shiokawa, K., Y. Misumi, and K. Yamana. 1981. Demonstration of rRNA synthesis in pre-gastrular embryos of *Xenopus laevis*. Dev. Growth Differ., *23*: 579–587.

Showman, R. M. et al. 1982. Message-specific sequestration of maternal histone mRNA in the sea urchin egg. Proc. Natl. Acad. Sci. U.S.A., *79*: 5944–5947.

Showman, R. M. et al. 1987. Translation of maternal histone mRNAs in sea urchin embryos: A test of control by 5′ cap methylation. Dev. Biol., *212*: 284–287.

Smith, L. D., and R. E. Ecker. 1965. Protein synthesis in enucleated eggs of *Rana pipiens*. Science, *150*: 777–779.

Spemann, H. 1928. Die Entwicklung seitlicher und dorso-ventraler Keimhälften bei verzögerter Kernversorgung. Zeitschrift für wissenschaftliche Zoologie, *132*: 105–134.

Spirin, A. S. 1966. On "masked" forms of messenger RNA during early embryogenesis and in other differentiating systems. Curr. Topics Dev. Biol., *1*: 1–38.

Standart, N., T. Hunt, and J. V. Ruderman. 1986. Differential accumulation of ribonucleotide reductase subunits in clam oocytes: The large subunit is stored as a polypeptide, the small subunit as untranslated mRNA. J. Cell Biol., *103*: 2129–2136.

Summers, R. G. 1970. The effect of actinomycin D on demembranated *Lytechinus variegatus* embryos. Exp. Cell Res., *59*: 170–171.

Tennent, D. H. 1914. The early influence of the spermatozoan upon the characters of echinoid larvae. Carnegie Inst. Wash. Publ., *182*: 129–138.

Tufaro, F., and B. P. Brandhorst. 1982. Restricted expression of paternal genes in sea urchin interspecies hybrids. Dev. Biol. *92*: 209–220.

Vernon, H. M. 1900. Cross-fertilization among echinoids. Wilhelm Roux's Archiv für Entwicklungsmechanik, *9*: 464–478.

Wallace, H. and T. R. Elsdale. 1963. Effects of actinomycin D on amphibian development. Acta Embryol. Morphol. Exp., *6*: 275–282.

Wassarman, W. J., J. D. Richter, and L. D. Smith. 1982. Protein synthesis during maturation promoting factor- and progesterone-induced maturation in *Xenopus* oocytes. Dev. Biol., *89*: 152–158.

Wilde, Ch. E., and R. B. Crawford. 1966. Cellular differentiation in the anamniota. III. Effects of actinomycin D and cyanide on the morphogenesis of *Fundulus*. Exp. Cell Res., *44*: 471–488.

Wilt, F. H. 1964. Ribonucleic acid synthesis during sea urchin embryogenesis. Dev. Biol., *9*: 299–313.

Wilt, F. H. 1977. The dynamics of maternal poly (A)-containing mRNA in fertilized sea urchin eggs. Cell, *11*: 673–681.

Winkler, M. M. et al. 1980. Dual ionic controls for the activation of protein synthesis at fertilization. Nature (Lond.), *287*: 558–560.

Winkler, M. M. et al. 1985. Multiple levels of regulation of protein synthesis at fertilization in sea urchin eggs. Dev. Biol., *107*: 290–300.

Woodland, H. R., and J. E. M. Ballantine. 1980. Parental gene expression in developing hybrid embryos of *Xenopus laevis* and *Xenopus borealis*. J. Embryol. Exp. Morphol., *60*: 359–372.

Woodland, H. R., and C. F. Graham. 1969. RNA synthesis during early development of the mouse. Nature (Lond.), *221*: 327–332.

Wray, G. A., and D. R. McClay. 1988. The origin of spicule-forming cells in a "primitive" sea urchin (*Eucidaris tribuloides*) which appears to lack primary mesenchyme cells. Development, *103*: 305–315.

PART VI

The Organized Generation of Cell Diversity

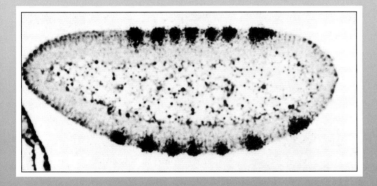

The expression of *fushi tarazu* mRNA in the *Drosophila* embryo as demonstrated by *in situ* hybridization. The transcripts are restricted to seven stripes in the blastoderm. (From Ingham, P., and P. Gergen. 1988. Development, *104* (Suppl.): 51–60.)

CHAPTER *14*

Establishment of Spatial Patterns of Gene Expression During Development

Thomas Hunt Morgan (1866–1945), an embryologist and geneticist at Bryn Mawr College, Columbia University, and the California Institute of Technology, was the first to suggest that development may be controlled by differential gene expression. T. H. Morgan received the Nobel Prize for his discovery of mutations in *Drosophila* and their use in understanding the role of genes in heredity. (Photograph courtesy of Marine Biological Laboratory Library, Woods Hole, MA.)

In the last chapter, we discussed the initiation of development, which is dependent on maternal information, and the onset of transcription, which allows the embryonic genome to take control over the development of the embryo. As development proceeds, gene expression is regulated to ensure that the correct developmental events occur at the right time and *in the right cells* to produce the orderly development of the embryo. We shall now examine the establishment of regionally distinct patterns of gene expression, which occurs during the initial diversification of cells during development. As we have discussed in earlier chapters, the establishment of body axes, germ layers, and—in all higher organisms—body segments is a prerequisite for the later production of functionally distinct, fully differentiated cell types in the correct spatial arrangement. Thus, regionalization of the embryo has important ramifications for embryonic development, and the molecules that control these events are among the most important regulatory molecules produced during development.

In this chapter, we discuss examples of regional patterns of gene expression and the regulation of gene expression during regionalization. We have chosen three systems that have proved to be particularly productive for the experimental analysis of regionalization: sea urchins, *Drosophila*, and vertebrates. As we shall see, the generation of diversity and its coordination are fascinating stories that have been yielding to experimental analysis and constitute one of the most active and exciting topics of research in contemporary developmental biology.

14–1 Region-Specific Gene Expression in the Sea Urchin Embryo

As we have discussed numerous times in this book, the sea urchin embryo has long been one of the most important subjects of embryological investigation. The literature on sea urchin development is particularly rich with insights on the events of early development, thanks to the pioneering work of individuals such as Driesch, Boveri, and Hörstadius. In recent years, the classical experimental embryology literature has been used as a foundation for investigating the roles of the genome in directing both temporal and spatial aspects of development. As we discussed in Chapter 13, much of the information we have on regulation of temporal aspects of gene expression is based on work with sea urchin embryos. These embryos have also proved to be valuable for the study of spatial patterns of gene expression during early development, thanks to refinements in the technique of *in situ* hybridization by Angerer and Angerer (1981) and the use of cloned genes as probes to monitor gene expression (Angerer and Davidson, 1984). Now, the powerful and sensitive techniques of contemporary molecular biology are being utilized to study the *regulation* of spatially restricted gene expression.

Cell lineage analyses have revealed the establishment of five distinct embryonic regions, or **territories**, in sea urchin blastulae (Davidson, 1989). Each territory is a contiguous region of the blastoderm that is characterized by a distinct pattern of gene expression and unique cell fates. The territories are the prospective **aboral ectoderm**, the prospective **oral ectoderm**, the prospective **skeletogenic mesenchyme**, the **vegetal plate**, and the **small micromeres**.

Color plate 21A shows the relationships between the cells of the blastula and their derivatives in the gastrula and pluteus-stage embryos of the sea urchin *Paracentrotus lividus*. Color coding distinguishes cells of individual territories. The presumptive ectoderm, which constitutes the two tiers of animal blastomeres and the upper tier of vegetal blastomeres, gives rise to two distinct ectoderm territories in the pluteus larva: the aboral ectoderm (a single layer of cells forming the body wall on the side opposite the mouth), which is colored green in the figure, and the oral ectoderm (ectoderm on the side where the mouth is located), which is colored yellow. The oral ectoderm also gives rise to the larval nervous system. The lower tier of vegetal blastomeres (blue) gives rise to the vegetal plate of the swimming blastula. These cells invaginate during gastrulation and form the gut and secondary mesenchyme (see Chap. 6). Micromeres constitute two distinct territories: One group (the skeletogenic mesenchyme; red) enters the blastocoel to form the skeleton of the pluteus larva, whereas the small micromeres (pink) may contribute to formation of the coelomic sacs (Pherson and Cohen, 1986).

Recent investigations have characterized patterns of gene expression within each territory and have begun to analyze the factors involved in regulating gene expression within them. There are four general patterns of gene expression in the territories (Table 14–1; Davidson, 1989):

1. Genes that are activated early and are expressed in a single territory. One example is **SM50**, which encodes a 50 kDa spicule matrix protein that is expressed early in the skeletogenic mesenchyme (Benson et al., 1987). The aboral ectoderm has a number of very nice examples of early, territory-restricted gene expression. One is the gene for the enzyme arylsulfatase, which is expressed at around the hatching blastula stage (Yang et al., 1989). One of the cytoskeletal actin genes, which is called **CyIIIa** is expressed even earlier than the arylsulfatase gene—at the 7th to 8th cleavage (Davidson, 1989). Genes belonging to the Spec 1 family (see p. 567), which encodes a group of calcium-binding proteins related to calmodulin, are also expressed concurrently with CyIIIa in aboral ectoderm in the species *Strongylocentrotus purpuratus*. The expression pattern of Spec 1 is illustrated in Figure 14.1. Silver grains (which indicate hybridization between the labeled probes and cellular transcripts) are dense over the aboral ectoderm of the larvae and the presumptive aboral side of the blastulae. Grain density over other lineages is not significantly different from background density.

   Diverse, unlinked genes that are co-expressed, such as CyIIIa and Spec 1, constitute what are known as **gene batteries**. This term was used first by Morgan (1934) and more recently by Britten and Davidson (1969, 1971). According to Britten and Davidson, coordinate regulation of genes is most logically explained by the response of diverse genes to the same diffusible transcriptional factors. The simplest way for multiple genes to respond coordinately to the same

Table 14–1 Lineage-Specific Gene Expression in the Sea Urchin Embryo

Gene	Product or function	Where expressed
Spec 1 } Spec 2 }	Members of calmodulin family	Aboral ectoderm
CyIIIa Arylsulfatase	Cytoskeletal actin Enzyme	
Hbox 1	Protein containing homeodomain	General aboral ectoderm early; restricted to a portion of aboral ecto- derm later
SM50 msp130	50 kDa spicule matrix protein Cell-surface glycoprotein	Skeletogenic mesen- chyme
CyIIa	Cytoskeletal actin	Vegetal plate derivatives and skeletogenic mesenchyme
CyI } CyIIb }	Cytoskeletal actins	Ubiquitous early; oral ectoderm and gut late
Mt	Metallothionein	Generally expressed early; aboral ectoderm in pluteus stage

Adapted from Davidson, E. H. 1989. Lineage-specific gene expression and the
regulative capacities of the sea urchin embryo: a proposed mechanism. Development, *105*: 421–445.

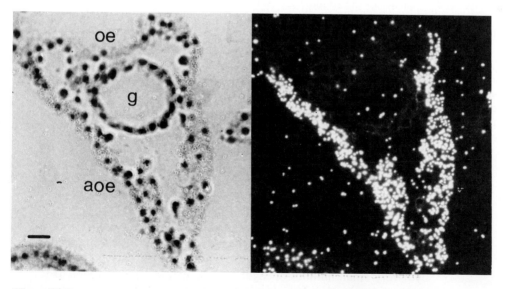

Figure 14.1

Expression of Spec 1 in the aboral ectoderm of a *Strongylocentrotus purpuratus*
pluteus larva as demonstrated by *in situ* hybridization. A phase-contrast
micrograph is shown on the left, and a dark-field micrograph is shown on the
right. The light dots in the dark-field micrograph are silver grains, the density of
which indicates the extent of hybridization between a labeled Spec 1 probe and
Spec 1 cellular transcripts. aoe: aboral ectoderm; oe: oral ectoderm; g: gut. Scale
bar equals 10 μm. (Courtesy of L. M. Angerer and R. M. Angerer.)

Essay

Gene Organization and Regulation of Transcription

Three classes of genes are recognized in eukaryotes, based upon the polymerase that transcribes that class of gene. RNA polymerase I is found preferentially in the nucleolus and is utilized in transcription of rRNA. Polymerase III transcribes the genes for transfer RNA and 5S RNA, whereas polymerase II transcribes genes that code for proteins. The latter genes are often referred to as Pol II genes. In this essay, we examine the organization of Pol II genes and the relationship between gene organization and regulation of transcription.

Polymerase molecules transcribe DNA in a linear fashion by sequential addition of ribonucleotides to a growing RNA chain. Each nucleotide in RNA has the same orientation, giving the chain a definite polarity, which is opposite that of the DNA template. The end terminated by the 5'-carbon atom and its associated $(PO_4)_3$ group is called the 5' end, whereas the end containing the 3'-carbon atom and its OH group is the 3' end. During transcription, a polymerase molecule attaches to the beginning of the region of DNA to be transcribed (a transcriptional unit). The site at which the polymerase initiates transcription is called the **promoter**. The promoter signals where transcription should begin and also influences the efficiency of initiation. Attachment of the polymerase to the DNA causes the double helix in the attachment region to unwind. The polymerase then moves along the DNA, causing local unwinding and reformation of the helix in its wake. Only one DNA strand is transcribed by the polymerase, which always moves in the $3' \rightarrow 5'$ direction of this strand. This produces an RNA molecule with opposite polarity—it begins at the 5' end and terminates at the 3' end. The growing RNA strand remains attached to the polymerase during the entire transcriptional process, and when this complex reaches the 5' end of the transcriptional unit, both the polymerase and the completed transcript are released from the DNA.

The nucleotide sequence of DNA is always given in terms of the sequence of the RNA it produces and therefore is the sequence of the strand that is complementary to the coding strand itself. This convention is followed because the RNA sequence can be obtained directly from the DNA sequence by substituting uracil (U) for thymine (T). Thus, the nucleotide sequence of DNA is represented from 5' to 3'. The 5' flanking region, which is involved in transcription initiation, precedes the gene, and the 3' flanking region follows the gene.

The promoters of Pol II genes contain a number of consensus sequences that have been shown experimentally to be important functional elements. These include the **TATA box** and the **CCAAT box**. Analysis of the roles of the eukaryotic consensus sequences in transcription of polymerase II genes has involved observing the effects on transcription of modification or deletion of these sequences. Genes with aberrant 5' flanking regions are cloned, and transcription of the cloned sequences is assessed by either *in vitro* transcription with polymerase II or the technique of **surrogate genetics**. The latter procedure involves introducing the cloned genes into the nucleus of another cell and allowing the endogenous RNA polymerase II of the host nucleus to transcribe them. Exogenous genetic material may either be injected into the nucleus of a living cell (such as a *Xenopus* oocyte) or introduced into tissue culture cells by transformation (Birnstiel and Chipchase, 1977).

Experimental analysis of the TATA box indicates that it directs the polymerase to initiate transcription about 30 nucleotides downstream from it. Thus, it specifies the RNA start site. Regions located farther upstream—among them the CCAAT box—exert an influence on the efficiency of transcription. Yet another upstream transcriptional factors is if they share the same gene regulatory sequences that interact with these factors to promote transcription of the genes (see the essay above for a discussion of transcriptional regulation). Gene batteries may be quite large but may be regulated by a very small number of regulatory molecules. Those regulatory molecules are, of course, gene products themselves. In the case of the early embryo, the regulatory molecules could either be synthesized from zygotic transcripts or maternal transcripts. Thus, the sequestration of maternally produced transcriptional factors or their transcripts within a distinct spatial domain of the zygote could result in the activation of constellations of genes that would have unique developmental consequences.

2. Genes that are initially expressed in all cells within a territory, but

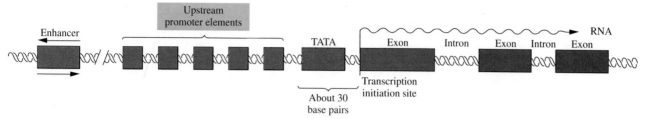

Figure 1

A typical developmentally-regulated eukaryotic gene. Transcription is regulated by the binding of transcription factors to the various promoter elements and to the enhancer. Transcription is initiated about 30 nucleotides downstream of the TATA box. The gene consists of both exons and introns. The latter are spliced out of the transcript during processing of the transcript in the nucleus before it is transported to the cytoplasm to function as a messenger RNA. (After Villee et al., 1989.)

element, the **GC box** (GGGCGG) may be present one or more times. GC boxes are not always present and may be in inverse orientation. Therefore, the eukaryotic polymerase II promoter is a multisite entity in the 5′ flanking region (Groschedl and Birnstiel, 1980; Wasylyk et al., 1980; Dierks et al., 1981; Sassone-Corsi et al., 1981; Grosveld et al., 1982).

Genes that are subject to developmental regulation possess a number of gene-specific regulatory elements in addition to the generalized promoter elements (see Fig. 1). As we discuss in this chapter, developmentally regulated genes may have a complex array of these regulatory elements, usually located upstream of the generalized promoter elements. This class of sequence elements may include those that are either gene-specific or gene battery-specific. These specialized regulatory elements are of two types. One type is functional only if it is located near the polymerase initiation site and is, therefore, a component of the promoter, and the other type (**enhancer**) can be located at a distance from the coding sequence, either upstream or downstream of the coding unit and in either orientation (i.e., either oriented 5′ to 3′ or 3′ to 5′). As their name implies, enhancers enhance transcription from the gene promoter.

The effectiveness of the promoter and enhancers in regulating transcription of a gene depends upon sequence-specific DNA-binding proteins called **transcription factors**, which bind to individual sequence elements and regulate transcription of the gene by RNA polymerase II. Some of these factors recognize the general promoter elements. These include a TATA-binding factor (TFIID), CCAT-binding factor, and SP1 (which binds to GC boxes). However, others specifically recognize the gene-specific sequence elements. Some of these activate transcription, and others inhibit it. The combination of general and specific transcription factors bound to the promoter and enhancer of a gene regulates the level of transcription of that gene. As we learn in this chapter and discuss again in Chapter 18, this mechanism is an elegant and sophisticated control mechanism that regulates both temporal and spatial patterns of gene expression during development.

whose expression is later confined to a subset of those cells. **Hbox1**, a gene containing a homeobox sequence related to the *Antennapedia* homeobox of *Drosophila* (see p. 609), is initially expressed throughout the aboral ectoderm, but later (between the blastula and pluteus stages) its expression is restricted to a small area of ectoderm that lies at the extreme aboral end of the pluteus (Dolecki et al., 1986; Angerer et al., 1989).

3. Genes that are precluded from expression in a territory. Another cytoskeletal actin gene, **CyIIa**, is never expressed in the aboral ectoderm. It is expressed relatively late in development, exclusively in the vegetal plate derivatives and in skeletogenic mesenchyme.

4. Genes that are initially expressed everywhere but later become restricted in their pattern of expression. Examples are the **CyI** (Fig.

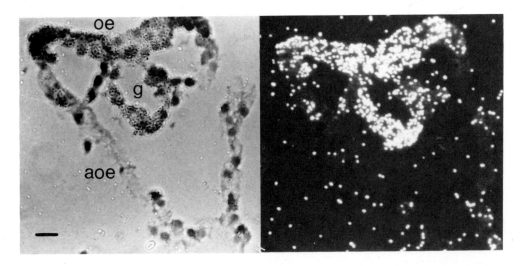

Figure 14.2

Expression of CyI cytoskeletal actin in a *S. purpuratus* pluteus larva as demonstrated by *in situ* hybridization using a CyI-specific probe. Left: phase-contrast micrograph; right: dark-field micrograph. The oral ectoderm (oe) and the gut (g) are heavily labeled, whereas there is no labeling over the aboral ectoderm (aoe). Scale bar equals 10 μm. (Courtesy of L. M. Angerer and R. M. Angerer.)

14.2) and **CyIIb** cytoskeletal actin genes, which are initially expressed throughout the embryo but continue to be expressed in oral ectoderm and gut while becoming silenced in aboral ectoderm. Another example is the genes encoding metallothioneins, which are small cysteine-rich proteins that bind heavy metals. Metallothionein mRNA is distributed throughout the embryo in early development and becomes restricted primarily to the aboral ectoderm during the pluteus stage (Angerer et al., 1986).

5. Genes that are expressed relatively late and in the derivatives of a single territory. Examples include **msp130**, which encodes a cell surface glycoprotein that is restricted to skeletogenic mesenchyme cells only after they have ingressed into the blastocoel. Another group of Spec genes, **Spec 2**, is expressed in the aboral ectoderm several hours after Spec 1 gene expression was initiated in those same cells.

The abrupt, aboral ectoderm-restricted expression of CyIIIa has provided an opportunity to study the molecular mechanisms by which regional patterns of gene expression are controlled. These experiments, which have been conducted by Eric Davidson and his colleagues, rely on the ability to introduce into sea urchin eggs gene regulatory sequences that are fused to a "reporter" gene. In these experiments the reporter gene encodes a bacterial enzyme, **chloramphenicol acetyl transferase (CAT)**, which is not found in eukaryotes. These **fusion genes** are microinjected into eggs, and the exogenous DNA is stably incorporated into the sea urchin genome and replicates along with the embryonic nuclear DNA. Transcription of the CAT genes is directed by the CyIIIa regulatory DNA, and the resultant transcripts are translated into CAT enzyme. Highly sensitive techniques are available to detect CAT enzyme activity, which reflects the level of CAT transcripts and thus provides a measure of the role of the regulatory

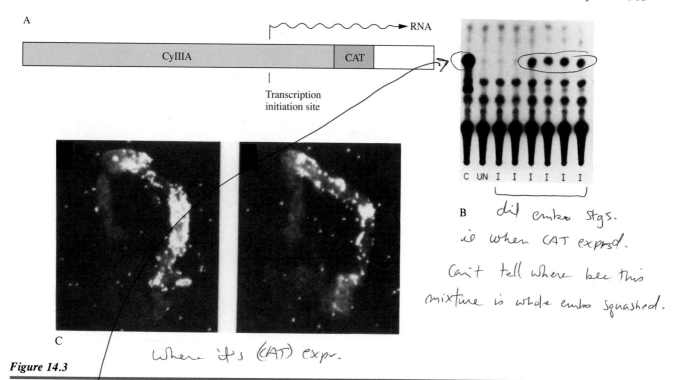

A

CyIIIA   CAT   RNA

Transcription
initiation site

C UN I I I I I

B *did embryo stgs.*
is when CAT exprsd.
Can't tell where bec this
mixture is whole embryo squashed.

C   *When it's (CAT) expr.*

Figure 14.3

Developmentally-regulated expression of a CyIIIa·CAT fusion gene. A, Diagram of the fusion gene. The transcript produced by expression of this gene is shown above the gene. B, Appearance of CAT activity during development of embryos from eggs injected with CyIIIa·CAT fusion genes. CAT acetylates chloramphenicol; the acetylated product (upper spot) migrates further during chromatography. Both the substrate and product are radioactively labeled so that the conversion can be detected by autoradiography after chromatography. C: control CAT activity; UN: uninjected; I: injected. C, *In situ* hybridization of consecutive sections of a pluteus embryo that developed from an egg injected with CyIIIa·CAT fusion genes using CAT antisense RNA as a probe. Significant labeling is seen only over the aboral ectoderm (right side of each section). (A, After Davidson, 1989. B, From Davidson et al., 1985. C, From Hough-Evans et al., 1987.)

sequences in directing gene expression during development. Alternatively, CAT transcripts can be detected by hybridization technology.

A CyIIIa·CAT fusion gene is shown in Figure 14.3A. It contains the CyIIIa 5′ flanking region, 5′ noncoding sequences, and a few codons of CyIIIa protein coding sequence fused to the CAT coding sequence. Transfer of this construct into sea urchin eggs results in the appearance of CAT enzyme in the correct temporal and spatial pattern for CyIIIa; i.e., CAT activity appears abruptly in late cleavage, and CAT mRNA appears only in aboral ectoderm, which is exactly the pattern that is observed for accumulation of CyIIIa transcripts (see Fig. 14.3B and C). Thus, the 5′ sequences have all the information that is both necessary and sufficient for temporal and spatial regulation of the CyIIIa gene. As the number of fusion genes injected into embryos is increased beyond a certain amount, a maximum amount of CAT enzyme production is reached; injecting more DNA fails to result in more CAT production (Livant et al., 1988). These results suggest that there is a limiting number of diffusible transcription factors that regulate CyIIIa gene transcription in the embryo. Consequently, excess CyIIIa 5′ flanking sequences will compete with one another for the limited number of these transcription factors. Davidson and his colleagues have strengthened this interpretation by coinjecting excess molecules of the putative regulatory region itself along with the CyIIIa·CAT fusion constructs. As predicted, the 5′ flanking region reduces CAT activity almost stoichiometrically, presumably by titrating diffusible transcriptional factors present in the embryonic

Figure 14.4

In vivo competition between the CyIIIa regulatory domain and CyIIIa·CAT fusion genes. The two elements were coinjected in the molar ratios indicated on the abscissa. Each data point represents an experiment carried out with a single batch of eggs. The ideal competitive stoichiometry is indicated by the dashed line, whereas the solid line indicates the best fit for the experimental points if the regulatory DNA competes with approximately 76% of ideal efficiency or if about 76% of the injected competing sequences reside in the same nuclei as the CyIIIa·CAT fusion constructs. The curve plotted from the data points conforms very closely to the ideal competition curve. (After Franks et al., 1990. Reprinted by permission of Company of Biologists, Ltd.)

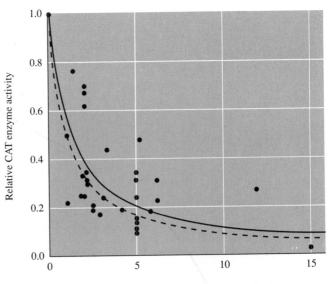

Molar ratio: competitor DNA/CyIIIa-CAT DNA

nuclei, thus preventing them from interacting with the CyIIIa·CAT fusion constructs (Fig. 14.4).

Deletion of different portions of the 5′ regulatory region in CyIIIa·CAT fusion constructs decreases the activity of the construct to different extents, suggesting that there are a number of independent regulatory domains within it (Flytzanis et al., 1987). The functions of these regulatory domains in promoting CyIIIa gene transcription have been studied by coinjecting an excess of different subfragments of the regulatory region along with the complete CyIIIa·CAT fusion construct into eggs before fertilization. As we shall discuss presently, these fragments contain one or more specific binding sites for transcription factors. A number of these domains compete as strongly as does the entire regulatory domain, thus reducing CAT activity stoichiometrically (examples of the effects of three subfragments are shown in Figure 14.5). This suggests that simultaneous interactions at several sites within the 5′ flanking region are necessary for maximal expression of CyIIIa. Similar competition experiments have demonstrated that spatial regulation of CyIIIa is under *negative* regulation (Hough-Evans et al., 1990). Co-injection of CyIIIa·CAT genes with fragments representing two distinct sites within the 5′ regulatory region causes dramatic ectopic expression of the CyIIIa·CAT gene: CAT mRNA is detected in oral ectoderm, gut, and mesenchyme cells. These results suggest that these fragments contain sequences that normally function to repress CyIIIa transcription, except in the aboral ectoderm cells. Competitive binding of transcription factors by the fragments prevents the normal regulatory interactions that repress CyIIIa gene expression in oral ectoderm, gut, and mesenchyme cells. As a consequence, inappropriate expression of CyIIIa·CAT genes can occur.

The functional significance of interactions between transcription factors and gene regulatory elements is abundantly obvious from the data obtained from the gene injection experiments that we have just discussed. We shall now examine these interactions in more detail. The technique used for these experiments is called **mobility-shift** (or **gel retardation**) **analysis**. These *in vitro* experiments involve incubating radiolabeled subfragments of the 5′ flanking region with nuclear extracts from sea urchin embryos. The DNA is then subjected to electrophoresis, followed by autoradiography. If the subfragment contains sequences that interact stably with proteins in the

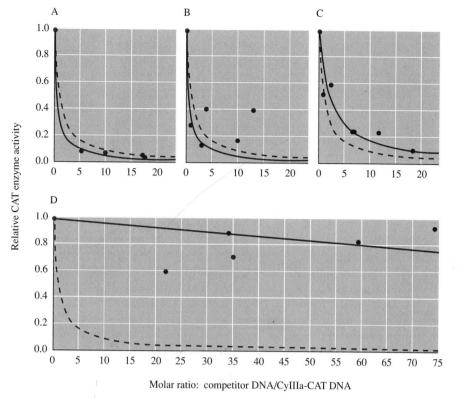

Figure 14.5

In vivo competition between subfragments of the CyIIIa regulatory domain and CyIIIa·CAT fusion genes. Three of these subfragments (data shown in A, B, and C) compete as strongly as does the entire regulatory domain (see Fig. 14.4), whereas the subfragment used in D shows no significant competition. (After Franks et al., 1990. Reprinted by permission of Company of Biologists, Ltd.)

nuclear extract, the mobility of the DNA in the gel will be retarded by the protein. The specificity of the interaction is determined by adding the same subfragment that is unlabeled. Specific interactions are those in which the unlabeled DNA will compete with the labeled probe as a site for binding by the protein, thus eliminating the DNA-protein complex on the autoradiograph. An example of such an experiment is shown in Figure 14.6. At least 15 individual sites within the CyIIIa regulatory domain have been identified where specific DNA-protein interactions take place (Calzone et al., 1988; Davidson, 1989).

The picture of gene regulation during development that emerges from these experiments is that transcription factors in the nuclei of embryonic

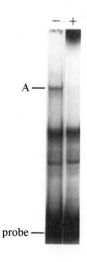

Figure 14.6

Mobility-shift assay for embryo nuclear DNA-binding factors. Nuclear extracts were reacted with a subfragment of the CyIIIa 5′ upstream region with (+) and without (−) unlabeled competitor DNA containing the same sequence. Such reactions also contain large excesses of non-specific competitor DNA, which would interact with general DNA-binding proteins (e.g., histones), which are present in the extracts. The DNA was subjected to electrophoresis; an autoradiograph of the gel is shown. The locations of the free probe and the DNA-protein complexes (A) are indicated. (From Calzone et al., 1988.)

cells must interact with multiple regulatory sites to control the temporal and spatial pattern of gene expression. Thus, it is important that investigators focus their attention on such factors to monitor changes in either their amount or activity during development (i.e., to study regulation of the regulators). Such an analysis has begun on the factors affecting CyIIIa gene expression. These studies reveal that only one of these factors is present in an active form in unfertilized eggs (these experiments cannot distinguish whether the remaining factors are absent or present in inactive forms). Significant levels of activity of the remaining factors become detectable by the mid-cleavage stage. Of these, some remain at constant levels, whereas a number of them increase dramatically in amount or activity by the time that the CyIIIa gene is being expressed in aboral ectoderm nuclei. The rise in the concentrations or activities of these factors may be responsible for activation of this gene at the early blastula stage (Calzone et al., 1988). In addition, the spatial specificity of CyIIIa expression may be caused by lineage-specific differences in either the amounts or activity of the factors that negatively regulate CyIIIa expression; i.e., these factors must either be absent or inactive in the aboral ectoderm cells.

As discussed earlier, CyIIIa is a member of a battery of genes whose expression is restricted to the aboral ectoderm. It is probable (as predicted by Britten and Davidson) that members of this battery share binding sites that confer the specificity of their co-expression. (These binding sites would be prime candidates for those that confer spatial specificity of expression.) This possibility has been tested by intergenic gel shift competitions (i.e., using unlabeled regulatory regions from other genes as competitor). Interestingly, Spec 1 (a member of the same battery as CyIIIa) has certain upstream subfragments that compete with CyIIIa binding sites, whereas SM50 and CyI lack competitor sites (Davidson, 1989). It is thus likely that common protein binding sites provide complete specificity for the temporal and spatial control of gene expression and that the onset of expression of members of a battery is regulated entirely by the abundance and functional competence of the protein factors that bind to these sites.

The experiments that we have discussed demonstrate that the establishment of region-specific patterns of gene expression during development is clearly dependent upon interactions between gene regulatory elements and diffusible transcription factors in the nuclei of the early embryo. This means that we must now look for the events and embryonic components that govern such regulatory interactions. We have learned in Chapters 3, 11, and 13 that the oocyte accumulates a vast number of transcripts and proteins that are present in the fertilized egg, some of which have been demonstrated to have clear effects on the course of regionalization of the embryo. It is going to be quite exciting to unravel the complete story of how the oogenic dowry is utilized in the process of synthesis and/or mobilization of the transcription factors at the right time and in the correct lineages. Treatments that perturb the specificity of gene expression during early development will be valuable tools in this process. As we discussed in Chapter 11, lithium causes vegetalization of the sea urchin embryo; i.e., isolated animal blastomeres will form vegetal structures in the presence of LiCl. It has recently been reported that such vegetalized "embryoids" express the SM50 gene, which is normally restricted in its expression to the skeletogenic mesenchyme cells (Livingston and Wilt, 1989). Thus, LiCl elicits inappropriate expression of this gene and may do so by mimicking the normal SM50 regulatory process. Further exploitation of this system should improve our understanding of the sequence of events in the regulation of SM50 gene expression, including the synthesis or mobilization of the transcription factors that initiate expression of that gene in its normal lineage.

14–2 Gene Expression Controlling Segmentation in Drosophila

As we shall now discuss, research on *Drosophila* has yielded a bonanza of information about regulation of the spatial patterns of gene expression during early development. In *Drosophila*, a number of mutations in developmentally important genes have been isolated, and their mutant phenotypes allow one to predict the developmental functions of their wild-type alleles. These studies have revealed a class of developmental control genes that function to define the anteroposterior axis and are expressed in a distinct temporal and spatial pattern in the embryo. As we shall learn in Section 14–3, some of the genetic elements that regulate development in *Drosophila* have been conserved during evolution and modified to generate the additional complexity in form and function that characterizes more advanced organisms, such as the vertebrates.

   Before discussing gene expression during *Drosophila* development, we shall review the phases of the *Drosophila* life cycle, which are summarized in Figure 14.7. During the first phase of the life cycle, the egg develops into an embryo, and the embryo eventually hatches into a larva. As discussed in previous chapters, *Drosophila* development is initiated when the egg is laid and then undergoes a series of rapid nuclear divisions without intervening cytokinesis to become a syncytium. When the nuclei generated in

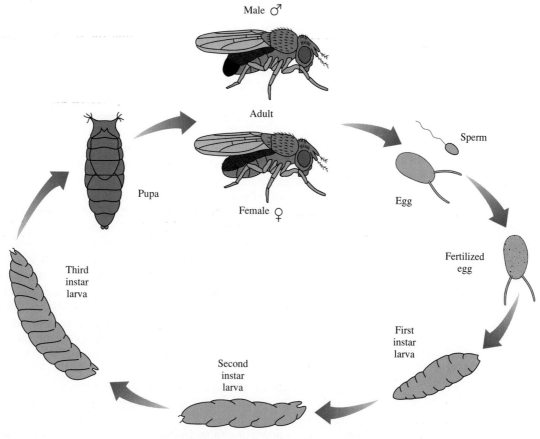

Figure 14.7

A summary of the *Drosophila* life cycle. (After Villee et al., 1989.)

the interior migrate to the egg periphery, the embryo is at the syncytial blastoderm stage. After another series of nuclear divisions, a cellular blastoderm is created by inward extension of the egg plasma membrane around each nucleus. Gastrulation occurs after cellularization, and the germ band appears on the ventral surface of the embryo. As the germ band elongates posteriorly, it becomes separated into a series of bulges or **parasegments**. These parasegments divide the embryo into 14 distinct regions along its anteroposterior axis and reflect different domains of gene expression (see p. 598). Later, as the embryo develops into a larva, the parasegments disappear and are replaced by **segments**, the repetitive body units of *Drosophila* and other arthropods. (The relationship between parasegments and segments is described in more detail on p. 594.)

The second phase of the life cycle begins when the larva hatches and begins to feed and grow. The larval growth period is divided into three feeding stages called **instars**. Each instar is separated by molting, a process in which the old larval exoskeleton is shed, the body grows, and a new exoskeleton is elaborated to cover the larger larva. After the end of the third instar, the larva forms a **pupa**. During the pupal stage, the larva is converted into an adult by **metamorphosis**, a process in which larval structures are destroyed and replaced by cells destined to form the adult tissues and organs. Metamorphosis is triggered by **ecdysone**, the adult molting hormone. During the adult phase, primordial germ cells differentiate into sperm and eggs, and the life cycle is eventually repeated.

In the following pages, we shall present evidence that early *Drosophila* development is regulated by a network of developmental control genes—some that are transcribed during oogenesis and others that are transcribed in the early embryo.

Imaginal Discs and Transdetermination

The *Drosophila* embryo contains presumptive larval cells and **imaginal cells** (i.e., adult cells; *imago* is a term used to describe an adult insect). The larval cells differentiate early, forming larval tissues and organs, whereas the imaginal cells delay their differentiation until metamorphosis. As their name implies, imaginal cells remain undifferentiated in the larva but differentiate to form specific adult structures after metamorphosis. There are two kinds of imaginal cells. Adult abdominal structures develop from imaginal cells located in the **abdominal histoblast nests**. The imaginal cells in the histoblast nests have been difficult to study because they are small and hard to identify. Imaginal cells destined to form structures in other parts of the adult are located in the **imaginal discs** (Ursprung and Nothiger, 1972). Imaginal discs are discrete packages of undifferentiated cells in the larva. There are 19 imaginal discs: 18 of these are arranged in pairs on either side of the larva, and a single fused disc is located at the midline (Fig. 14.8). Although both larval and imaginal cells are determined at the cellular blastoderm stage (Chan and Gehring, 1971), the imaginal cells remain undifferentiated until metamorphosis, when they form specific elements of the adult exoskeleton. For instance, a pair of discs in the anterior region of the larva forms the eyes and antennae, a pair of discs in the middle region develops into the wings, and the disc at the posterior midline forms the genitalia.

The first evidence for developmental control genes came from experiments with imaginal discs. To study the developmental fate of imaginal discs, the German embryologist Ernst Hadorn developed the experimental procedure shown in Figure 14.9. In this procedure, a specific imaginal disc was removed from a larva and cut into two parts. One part was transplanted

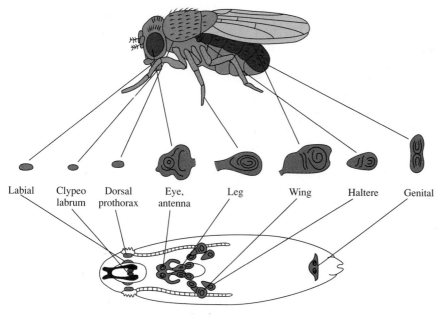

Figure 14.8

Imaginal discs and their developmental fates in *Drosophila*. Each imaginal disc has a specific morphology and is located in a typical region of the head, thorax, or abdomen of the larva. (After Villee et al., 1989.)

into the abdominal cavity of an adult fly (see Fig. 14.9A),whereas the other was transplanted into a third instar larva (see Fig. 14.9B). Because ecdysone is present only in the pupa, the imaginal disc cells transplanted into adults multiplied but did not differentiate. By serial transplantation into successive adult hosts, these cells could be propagated in an undifferentiated state. In contrast, the part of the disc transplanted into a third instar larva differentiated into a specific structure during metamorphosis. Thus, Hadorn was able to demonstrate that each imaginal disc is predetermined to form a specific structure in the adult (as shown in Fig. 14.8).

The next experiment was to determine whether the fate of an imaginal disc was established irreversibly. The stability of imaginal disc determination was tested by transplanting a disc that had been propagated by serial transplantation into a third instar larva (see Fig. 14.9C). If the same adult structure was obtained after many cycles of growth and cell division, then it could be concluded that determination was irreversible. However, if different structures were obtained after serial transplantation, it would suggest that imaginal discs could switch their developmental fates after proliferation during serial transplantation. The results of these experiments showed that the adult structure that formed usually corresponded to the original fate of the imaginal disc. Thus, the fate of an imaginal disc established during embryogenesis is normally very stable, even after many cycles of transplantation. In exceptional cases, however, a transplanted disc suddenly switched its fate (e.g., a genital disc developed into a leg, or an antennal disc formed a wing). Although these changes were rare, their frequency was high enough to distinguish them from spontaneous mutations. Therefore, it was learned that imaginal disc determination is not irreversible; an alteration in cell fate, or **transdetermination**, had taken place. Transdetermination causes imaginal disc cells to change the original developmental pathway that was selected during embryogenesis (Hadorn, 1978).

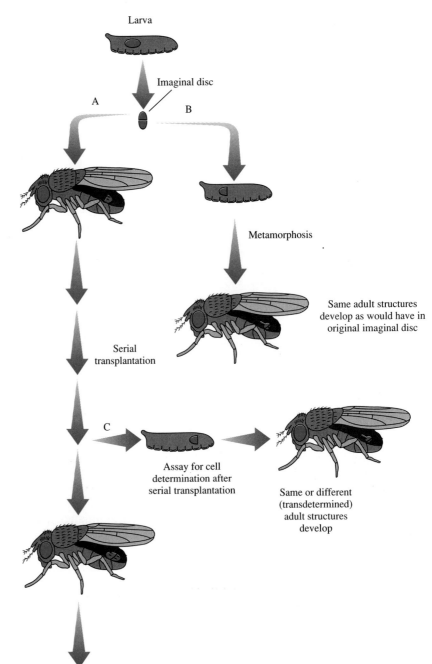

Figure 14.9

Transplantation experiments conducted to determine the stability of imaginal disc determination. An imaginal disc is excised from a larva and divided into two parts. A, One part is transplanted into the abdomen of an adult host to propagate the disc in its undifferentiated state. B, The other part is transplanted to a third instar larva to test the developmental potential of the imaginal disc cells. C, After many cycles of serial transplantation, an imaginal disc is transplanted into a third instar larva to determine whether its developmental potential has been changed.

Hadorn's experiments also showed that particular imaginal discs did not have the option to switch to any new fate. Instead, there were restrictions in the type of switches that could occur during transdetermination (Fig. 14.10). For example, the formation of legs or antennae from a genital disc occurred more frequently than the formation of genitalia from an antennal disc, and some transdeterminations, such as the formation of a leg from an antennal disc, were never observed. These results have been explained by assuming that imaginal disc determination involves a developmental pathway consisting of a defined sequence of steps. It would be relatively easy to reverse one of these steps, but reversing more than one would be very difficult. Thus, the more steps that occur during the determination of a particular imaginal disc, the less likely it is that a single spontaneous event

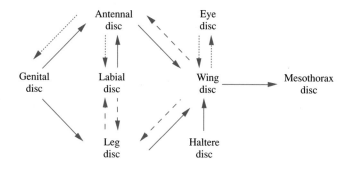

```
────▶   Frequent transdeterminations
- - - ▶   Less frequent transdeterminations
·······▶   Rare or unobserved transdeterminations
```

Figure 14.10

The frequencies of transdetermination in imaginal discs. (After Hadorn, 1978.)

during later development could completely reverse this process. Developmental biologists believe that transdetermination is caused by changes in the expression of a special class of genes controlling cell determination.

Homeotic Mutations and Segmentation

Further evidence for the existence of genes that regulate *Drosophila* development was provided by the discovery of **homeotic mutations**. Homeotic mutations disrupt the function of the **homeotic selector genes** (see p. 606) and result in bizarre substitutions of one body part for another. For example, in the dominant homeotic mutation *Antennapedia (Antp)*, legs arise from the region of the head that would normally produce antennae (Fig. 14.11). Another example of a homeotic mutation is *Ultrabithorax (Ubx)*, in which the **halteres** (balancing organs that function in flight) are replaced by a second pair of wings (see Fig. 14.12; see also Fig. 1.6). Interpreting the changes that occur in homeotic mutants requires an understanding of the segmental organization of the *Drosophila* body, which we now describe.

The body of *Drosophila* larvae and adults is divided into three parts: head, thorax, and abdomen (Fig. 14.13). The head consists of the **acron**, an anterior unsegmented region, followed by at least three fused segments (H1, H2, and H3), which form the eyes, antennae, and mouth parts. The thorax also consists of three segments: The prothorax (or segment T1) contains

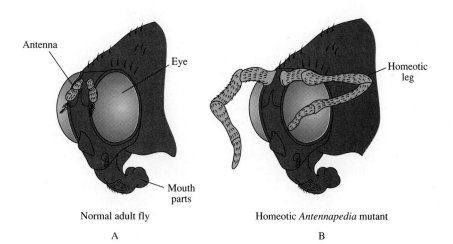

Normal adult fly

A

Homeotic *Antennapedia* mutant

B

Figure 14.11

The phenotypes of wild-type and *Antennapedia* mutants. A, The head of a wild-type fly, showing the paired antennae. B, The head of a dominant *Antennapedia* mutant, showing a pair of legs protruding from where the antennae would normally be formed. (After Villee et al., 1989.)

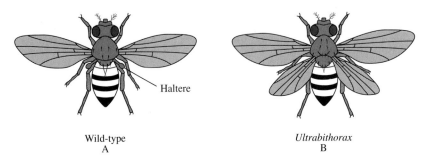

Wild-type
A

Ultrabithorax
B

Figure 14.12

The phenotypes of wild-type and *Ultrabithorax* mutants. A, A wild-type fly viewed from the dorsal side, showing the mesothorax with paired wings and the metathorax with paired halteres. Note that the metathorax is very small relative to the mesothorax. B, An *Ultrabithorax* mutant fly viewed from the dorsal side, showing the normal mesothorax and the enlarged metathorax bearing a second set of paired wings instead of halteres. (After Raff and Kaufman, 1983.)

the first pair of legs on its ventrolateral margin, the mesothorax (or segment T2) contains the second pair of legs on its ventrolateral margin and the wings on its dorsolateral margin, and the metathorax (or segment T3) contains the third pair of legs on its ventrolateral margin and halteres on its dorsolateral margin. The abdomen consists of eight superficially similar segments (A1 to A8), followed by a terminal unsegmented region, the **telson**.

The first sign of segmentation appears during germ band extension when the germ band is divided into 14 parasegments (P1 to P14), marking anteroposterior divisions in the ectoderm and the underlying mesoderm (Martinez-Arias and Lawrence, 1985). Each parasegment contains an anterior and a posterior portion that become parts of different adjacent segments during subsequent development. The relationship between the order of parasegments and segments is shown in Figure 14.14. The anterior portion of a given parasegment becomes the posterior portion of the preceding segment, which develops a half-segmental unit anterior to the parasegment, whereas the posterior portion forms the anterior portion of the succeeding segment, which develops a half-segmental unit posterior to the parasegment. Thus, every parasegment is a half-segment out of register with a segment. The significance of parasegments will be apparent later in this chapter, when we describe the expression of the segmentation genes in *Drosophila*.

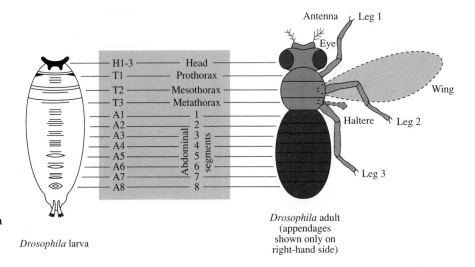

Figure 14.13

Segmentation in a *Drosophila* larva and adult. (After De Pomeroi, 1985.)

Drosophila larva

Drosophila adult (appendages shown only on right-hand side)

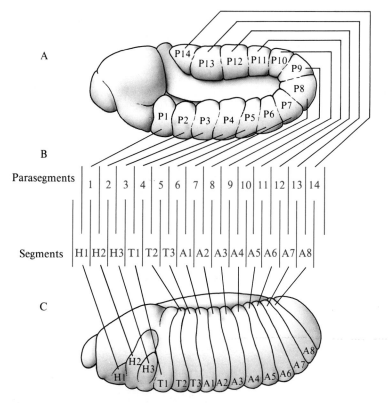

Figure 14.14

The relationship between parasegments and segments during *Drosophila* development. A, A diagram showing parasegments in the elongated germ band of an early embryo. B, A diagram showing the relationship between parasegments and segments. Each parasegment is a half segment out of register with the corresponding segments. The anterior compartment of each parasegment becomes the posterior compartment of the preceding (out-of-register) segment, while the posterior compartment of each parasegment becomes the anterior compartment of each (out-of-register) succeeding segment. C, A diagram showing the pattern of segments in a later stage embryo. (After Martinez-Arias and Lawrence, 1985.)

How are homeotic mutations related to the pattern of normal segmentation? To answer this question, we must consider the origin of segmentation during arthropod evolution. As shown in Figure 14.15A, the ancestors of the arthropods are thought to have been worm-like organisms made up of a series of identical segments. Later, each segment developed a pair of appendages (see Fig. 14.15B). Gradually, the anterior segments fused to form a head, and their appendages were converted into antennae and mouth parts, whereas the abdominal appendages degenerated (Fig. 14.15C to E). Thus, the thorax was the only part of the body where appendages (legs) were retained. During insect evolution, exoskeletal elements of the mesothorax and metathorax gained the capacity to form wings. Indeed, two pair of wings are found in some extant groups of insects (e.g., dragonflies). Finally, in dipterans (flies), wings on the metathorax were converted to halteres, leaving a single pair of wings on the mesothorax. After considering these events, it is clear that homeotic mutations result in the conversion of structures in one segment to their homologous counterparts in another segment. In *Antp* mutants, the antennae are replaced by their homologues,

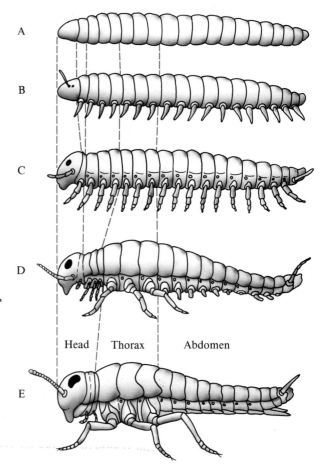

Figure 14.15

Hypothetical changes in segmentation during the evolution of arthropods. A, Putative arthropod ancestor consisting of a series of similar segments. B, A more advanced form, in which each segment bears a pair of appendages, and the head is beginning to be specialized. C-E, Subsequent stages during evolution, in which the anterior segments progressively fuse to form the head, with mouthparts evolving from appendages, and the posterior segments progressively losing appendages to form the abdomen. These modifications leave only the three thoracic segments with legs. The evolution of wings is not shown in this diagram. (After Raff and Kaufman, 1983.)

the legs (Fig. 14.16), whereas in *Ubx* mutants, the halteres are replaced by their homologues, the wings.

The properties of homeotic mutations show similarities to transdetermination. First, both abnormalities affect the developmental fate of imaginal disc cells. Second, the fate of an imaginal disc switches from structures of one segment to those of another. Finally, there is a limited range of options available for changes affected by these abnormalities. Thus, it is believed that homeotic mutations and transdetermination reflect changes in the expression of the same set of developmental control genes. These genes, which are known as the homeotic selector genes, are members of a larger class of regulatory genes that control *Drosophila* development (Nüsslein-Volhard and Wieschaus, 1980). We shall return to the function of the homeotic selector genes later in this chapter.

The Coordinate Genes Determine the Anteroposterior Axis

In Chapter 6, we learned that the dorsoventral axis of *Drosophila* embryos is controlled by a class of genes that functions in the egg and early embryo. Here, we describe another class of developmental control genes that organizes the anteroposterior axis by specifying the pattern of segmentation. Many of these genes were first identified using the mutagenesis and screening methods described in Chapter 6. Segmentation is a progressive process in the *Drosophila* embryo. First, the anterior and posterior compartments are established in the egg. Next, the embryo is subdivided into 3 different regions (corresponding to head, thorax to anterior abdomen, and posterior

Antennal segments
Arista
ANTENNA
Claws
Tarsal segments
LEG
Coxa
Trochanter
Tibia
Femur

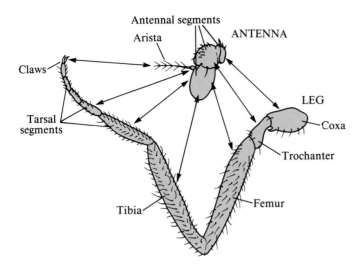

Figure 14.16

Homologies (shown by arrows) between parts of the antenna of a wild-type *Drosophila* and the leg that develops in the antennal region of a dominant *Antennapedia* mutant. (After Postlethwait and Schneiderman, 1971.)

abdomen), then into 7 regions about two parasegments wide, and finally into 14 regions corresponding to the parasegments.

The first step in segment development is initiated during oogenesis and involves the activity of the **coordinate genes** (Fig. 14.17A), which divide the egg into anterior and posterior regions (Nüsslein-Volhard et al., 1987). We previously described *bicoid*, which is a coordinate gene that determines the anterior region of the egg (Frohnhöfer and Nüsslein-Volhard, 1986). The head and thorax are absent in *bicoid* mutants, which consist almost entirely of abdominal segments. The *bicoid* gene encodes an mRNA determinant that is localized in the anterior region of the oocyte (see Fig. 14.19, Column 1). During early embryogenesis, *bicoid* mRNA is translated into a protein

help set up ant-post axis

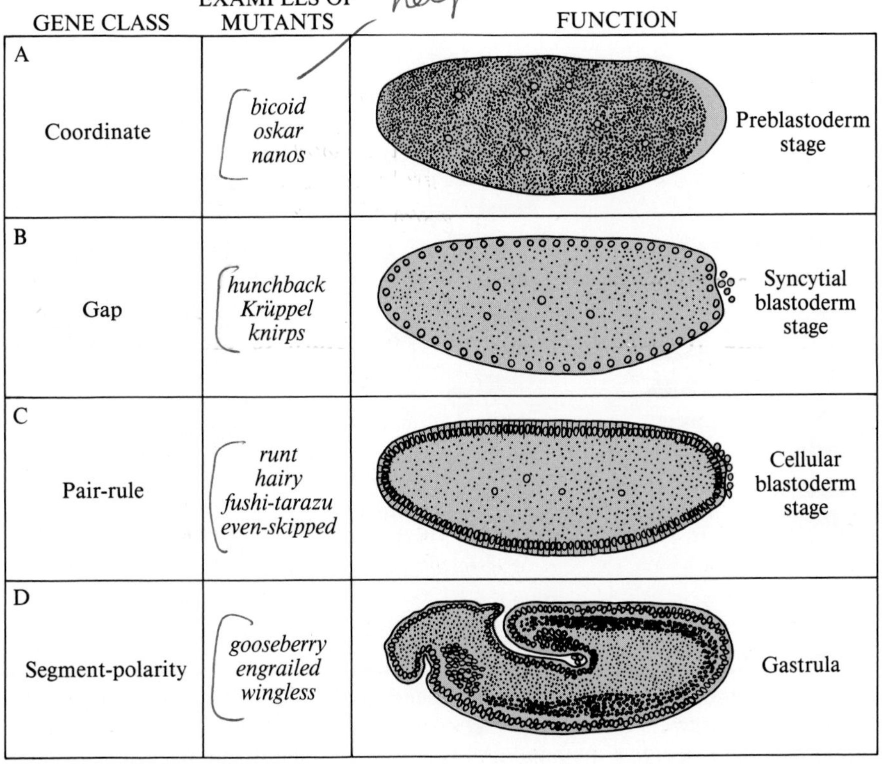

GENE CLASS
EXAMPLES OF MUTANTS
FUNCTION
A
Coordinate
bicoid
oskar
nanos
Preblastoderm stage
B
Gap
hunchback
Krüppel
knirps
Syncytial blastoderm stage
C
Pair-rule
runt
hairy
fushi-tarazu
even-skipped
Cellular blastoderm stage
D
Segment-polarity
gooseberry
engrailed
wingless
Gastrula

Figure 14.17

A summary of the different classes of genes and the time of their function in regulating segmentation along the anteroposterior axis of *Drosophila* embryos. The drawings on the right show the stages of development in which the genes are activated in the embryo. A, The coordinate genes divide the embryo into anterior and posterior regions. B, The gap genes divide the embryo into anterior, middle, and posterior regions. C, The pair-rule genes divide the embryo into seven regions spanning the width of two parasegments. D, The segment-polarity genes divide the embryo into 14 regions corresponding to each parasegment.

that is distributed in an anteroposterior gradient with its highest concentration at the anterior pole (Driever and Nüsslein-Volhard, 1988). The *bicoid* protein is a morphogenetic substance (or morphogen) that causes anterior structures to form where it is most concentrated. Later (see p. 603), we shall see that the *bicoid* protein regulates the transcription of genes that determine the head and thorax.

Other coordinate genes appear to be responsible for organizing the posterior region of the egg. Examples are the maternal effect genes *nanos* and *oskar*. The phenotypes of these mutants are essentially the opposite of *bicoid*: embryos develop head and thorax but lack most of the abdominal segments (Lehmann and Nüsslein-Volhard, 1986). The *nanos* and *oskar* genes are thought to produce products that are localized in the posterior region of the egg, but their identities and functions are unknown.

The Segmentation Genes Divide the Embryo into Segments

After the anterior and posterior regions of the embryo are established, further development along the anteroposterior axis is mediated by the **segmentation genes** (see Fig. 14.17B to D). In contrast to the coordinate genes, the segmentation genes are primarily zygotic-effect genes, which are transcribed in the early embryo. Examination of mutant phenotypes have allowed the segmentation genes to be divided into three categories: (1) **gap genes**; (2) **pair-rule genes**; and (3) **segment-polarity genes**. Each category of segmentation genes is activated sequentially in the early embryo.

The gap genes (see Fig. 14.17B) are expressed beginning at the syncytial blastoderm stage. Mutations in gap genes eliminate a large block of adjacent segments in the anterior, middle, or posterior regions of the embryo (Fig. 14.18A). For example, segments T1 through A5 (extending from the posterior compartment of P3 through the anterior compartment of P10) are missing in *Krüppel* mutants (see also Fig. 14.20). Likewise, the gap mutant *hunchback* lacks head segments, and the *knirps* mutant lacks posterior abdominal

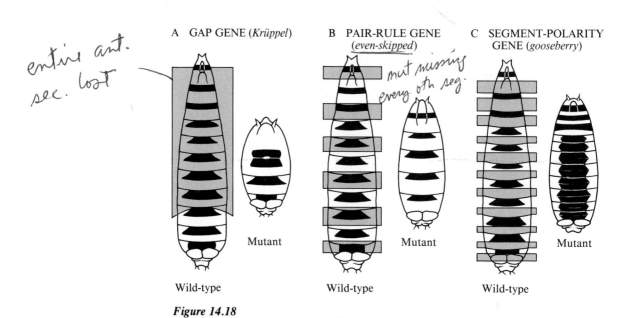

A GAP GENE (*Krüppel*) B PAIR-RULE GENE (*even-skipped*) C SEGMENT-POLARITY GENE (*gooseberry*)

Mutant Mutant Mutant

Wild-type Wild-type Wild-type

Figure 14.18

A comparison of the phenotypes of representative gap (*Krüppel*), pair-rule (*even-skipped*), and segment-polarity (*gooseberry*) mutations. Shadowed segments are missing in the respective mutants. (After Nüsslein-Volhard and Wieschaus, 1980.)

segments. From these phenotypes, it has been established that the gap genes function in subdividing the embryo into anterior, middle, and posterior regions. The *Krüppel* gene has recently been cloned, and its mRNA and protein products have been located in the embryo by *in situ* hybridization (Fig. 14.19, Column 2) and antibody staining (Jäckle et al., 1986; Gaul et al., 1987), respectively. The results show that *Krüppel* mRNA and protein are expressed in P4, P5, and P6 of wild-type embryos—parasegments that are missing in mutant embryos (Fig. 14.20). Thus, by combining genetic and molecular approaches, developmental biologists have been able to relate the expression pattern of a specific gene to its developmental function. However, there may be differences between the domains of gene function and expression. For example, *Krüppel* mutants also show abnormalities in regions beyond the normal domain of gene expression (see Fig. 14.20). A possible explanation for this is that expression of the *Krüppel* gene may result in production of an extracellular signal that affects the development of segments adjacent to those in which the gene itself is transcribed.

The pair-rule genes (see Fig. 14.17C) begin to be expressed just before the cellular blastoderm stage. Mutations in these genes show defects in all

[handwritten: later onset of pair rule gn. notice 7 diff sections. These 7 segments become 4 polarized. Hence get 14 Sections.]

[handwritten: bcd prot]
[handwritten: becomes less]

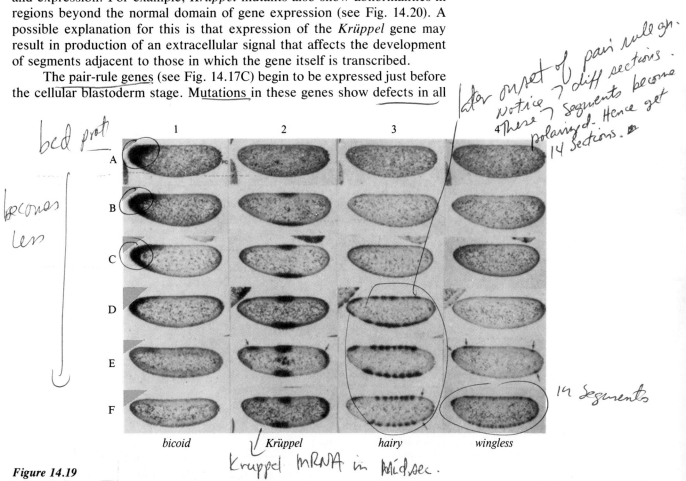

[handwritten: 14 Segments]

Figure 14.19

[handwritten: Kruppel mRNA in Mid sec.]

Sequential accumulation and localization of mRNA corresponding to examples of coordinate (*bicoid*), gap (*Krüppel*), pair-rule (*hairy*), and segment-polarity (*wingless*) genes as determined by *in situ* hybridization. Columns 1–4 correspond to adjacent sections of the same embryo hybridized with cloned probes recognizing *bicoid*, *Krüppel*, *hairy*, or *wingless* mRNA, respectively. Row A: A newly-laid embryo. *Bicoid* mRNA is localized at the anterior pole in section 1, but in sections 2–4 the other mRNAs have yet to be synthesized. Row B: An embryo about 10 minutes after laying. *Krüppel* mRNA accumulation is beginning in the middle of the embryo in section 2. Row C: An embryo about 20 minutes after laying. The level of *Krüppel* mRNA has increased in section 2, and *hairy* mRNA is appearing throughout the periphery of the embryo in section 3. Row D: An embryo just before the cellular blastoderm stage, showing the seven major stripes of *hairy* transcript accumulation in section 3. Row E: An embryo at 10–15 minutes after cellularization, showing two new regions of *Krüppel* mRNA at the anterior and posterior regions of the embryo in section 2 (arrows) and the beginning of *wingless* transcript accumulation in section 4 (arrows). Row F: A gastrulating embryo showing a new region of *hairy* transcript accumulation (arrows) and 14 stripes of *wingless* transcript accumulation in section 4. PC: pole cells. (From Ingham, 1988. Reprinted by permission from *Nature* Vol. 335 pp. 25–34. Copyright © 1988 Macmillan Magazines, Ltd.)

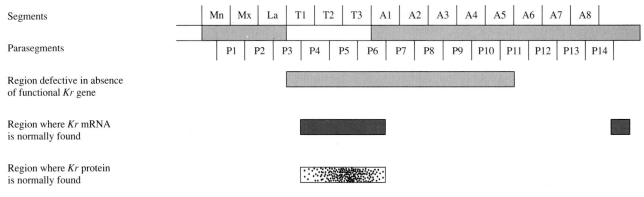

Figure 14.20

The relationship between defective segments in *Krüppel* (*Kr*) mutants and regions of *Krüppel* mRNA and protein synthesis at the syncytial blastoderm stage. Mn: mandible; Mx: maxilla; La: labium (mandible, maxilla, and labium are mouthparts); T1–3: thoracic segments; A1–8: abdominal segments; P1–14: parasegments. (After Alberts et al., 1989.)

or part of alternate parasegments, leaving the larva with half the number of normal segments. For example, *even-skipped* mutants lack the even-numbered parasegments (see Fig. 14.18B), and *fushi tarazu* (a Japanese word meaning "too few segments") mutants lack the odd-numbered parasegments. Mutant pair-rule genes are not necessarily missing entire parasegments. The pair-rule mutants *hairy* and *runt* are defective in parasegment-sized regions that are out of register with the parasegments. The pattern of pair-rule gene expression has been defined by *in situ* hybridization with cloned gene probes and antibody staining. The results indicate that the domains of gene expression are the same as the region deleted by the corresponding mutation. For example, *fushi tarazu* mRNA and protein are present in seven stripes around the circumference of the cellular blastoderm (Fig. 14.21). Each stripe is found in the odd-numbered parasegments that are missing in *fushi tarazu* mutants. The expression of the *hairy* gene is more complicated (see Fig. 14.19, Column 3); it is initially expressed throughout the periphery of the embryo, but expression is later confined to cells within the seven major regions that are missing in *hairy* mutants. Thus, there must be a mechanism that sharpens the pattern of *hairy* gene expression. Later, we shall show that this mechanism involves interactions between different segmentation genes and their products.

The segment-polarity genes (see Fig. 14.17D) begin to be transcribed between the cellular blastoderm stage and gastrulation. Mutations in these genes remove a part of each segment and replace it with a duplication of the remaining part of the segment (see Fig. 14.18C). For instance, in *gooseberry* and *engrailed* mutants, the posterior region of each segment (i.e., the anterior region of each parasegment) is replaced by a mirror image of the anterior region of the adjacent segment. Thus, the segment-polarity genes must be expressed in only a part of every segment (or parasegment). This was demonstrated by examining the distribution of *engrailed* transcripts and protein as described previously (Kornberg et al., 1985; O'Farrell et al., 1985; Howard and Ingham, 1986). The results showed that wild-type embryos contain 14 stripes of *engrailed* transcripts and proteins (Fig. 14.22)—one stripe for each parasegment. Another segment-polarity gene, *wingless*, is also expressed in a pattern of 14 stripes (see Fig. 14.19, Column 4).

In summary, the mutant phenotypes and expression patterns of the segmentation genes suggest that they function sequentially during devel-

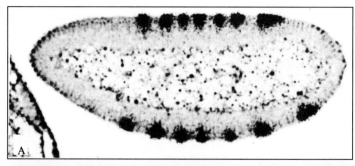

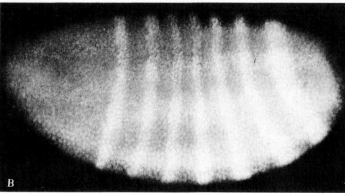

Figure 14.21

The expression of *fushi tarazu* mRNA (A) and protein (B) at the cellular blastoderm stage as determined by *in situ* hybridization and antibody staining, respectively. Note the presence of seven stripes of mRNA and protein in the blastoderm. (A, From Ingham and Gergen, 1988. Reprinted by permission of Company of Biologists, Ltd. B, From Carroll and Scott, 1986. Copyright © Cell Press.)

Indirect Immunofluorescense.

14 Segments

Figure 14.22

The expression of engrailed protein at (A) the extended germ band stage and (B) later during embryogenesis as determined by antibody staining. Expression is seen in 14 stripes, each corresponding to a single region of a parasegment. See Color plate 22 for a color version of A. (Courtesy of T. Kornberg.)

opment to subdivide the embryo into smaller and smaller regions along the anteroposterior axis. We now consider the mechanisms controlling the sequential expression of the segmentation genes.

Regulation of the Segmentation Genes

The segmentation genes are part of a gene control network in which gene products at one tier regulate the expression of genes at the next tier (Fig. 14.23A). Thus, the coordinate genes, which are in the top tier, regulate the gap genes, the gap genes control the pair-rule genes, and the pair-rule genes regulate the segment-polarity genes. In addition, segmentation genes at various tiers regulate the homeotic selector genes, which are in the bottom tier. Finally, the homeotic selector genes regulate genes that may be directly

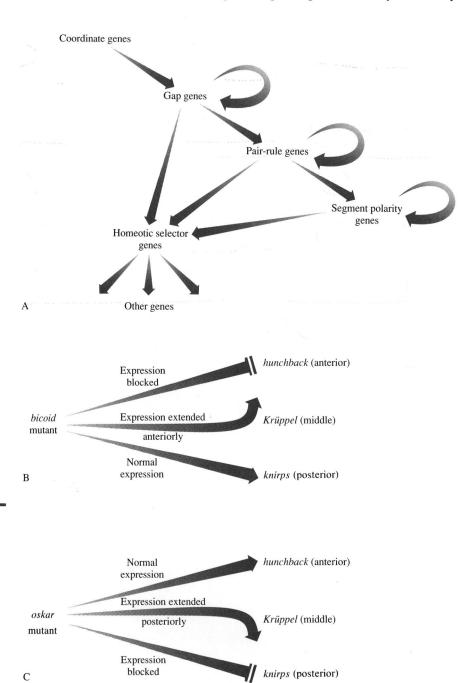

Figure 14.23

The hierarchy of gene control involving the coordinate, segmentation, and homeotic selector genes. A, Genes at one level of the hierarchy regulate genes at a lower level or at the same level. B, Effects of *bicoid* mutations on the expression of *hunchback*, *Krüppel*, and *knirps*. C, Effects of *oskar* mutations on the expression of *hunchback*, *Krüppel*, and *knirps*.

responsible for cell differentiation. Another feature of this hierarchy is that regulatory interactions occur among genes within the same tier (see Fig. 14.24).

Interactions Between Genes in Different Tiers of the Hierarchy

The hierarchy shown in Figure 14.23A has been established by a combination of molecular and genetic methods. The method used to determine whether a certain gene is regulated by the product of another gene is to examine the expression of the first gene in a mutant in the second gene. If expression of the first gene is altered in the mutant, this shows that the first gene is dependent on the activity of the second gene. As described previously, the *hunchback* gene is expressed primarily in the anterior region, *Krüppel* primarily in the middle region, and *knirps* in the posterior region of the early embryo (Jäckle et al., 1986). In recessive *bicoid* mutants, the *hunchback* gene is not expressed, *Krüppel* expression is extended anteriorly, and *knirps* expression is normal (see Fig. 14.23B; Tautz, 1988). Thus, the *bicoid* gene product must stimulate *hunchback* expression and suppress *Krüppel* activity in the anterior region of the embryo without affecting *knirps* activity in the posterior region. Similarly, in recessive *oskar* mutants, only *hunchback* and *Krüppel* are expressed, and *Krüppel* expression extends posteriorly (see Fig. 14.23C). By analogy to *bicoid*, these experiments suggest that the *oskar* product activates *knirps* transcription and suppresses *Krüppel* transcription in the posterior region of the embryo.

These positive and negative regulatory interactions between the co-ordinate and gap genes suggest a mechanism for the establishment of the three gap gene domains: *bicoid* activates *hunchback* and suppresses *Krüppel* expression in the anterior compartment (Fig. 14.24), whereas *oskar* (or a related posterior coordinate gene) activates *knirps* and suppresses *Krüppel* in the posterior compartment. Thus, *Krüppel* expression is restricted to the middle compartment, where the *bicoid* and *oskar* gene products are presumably at their lowest concentrations. Similar experiments have been conducted to demonstrate that the gap gene products control the expression of the pair-rule genes and the pair-rule gene products regulate the segment-polarity genes. If the expression of the pair-rule gene *fushi tarazu* is examined in a mutant *Krüppel* embryo, the characteristic *fushi tarazu* stripes are missing in the region where *Krüppel* is normally expressed (Carroll and Scott, 1986; Ingham et al., 1986), suggesting that *Krüppel* gene products regulate *fushi tarazu* gene activity. Likewise, mutations in any of the pair-rule genes alter the expression pattern of the segment polarity genes (Harding et al., 1985; MacDonald et al., 1986).

Interactions Between Genes in the Same Tier of the Hierarchy

Mutual inhibition of expression between the *hunchback* and *Krüppel* genes illustrates the importance of interactions between genes within the same tier of the hierarchy (see Fig. 14.24). There is normally a boundary between the domain of *hunchback* expression in the anterior part of the blastoderm and *Krüppel* expression in the middle. In *hunchback* mutants, however, the

Figure 14.24

Diagram illustrating positive and negative interactions among the *bicoid*, *hunchback*, and *Krüppel* genes.

Krüppel domain extends anteriorly, whereas in *Krüppel* mutants, the *hunchback* domain extends posteriorly. Thus, the boundary between *hunchback* and *Krüppel* expression must be sharpened by their mutual inhibition during normal development. A similar mechanism is probably used to sharpen the expression patterns of other segmentation genes (Carroll and Scott, 1986; Howard and Ingham, 1986).

Segmentation Genes Encode DNA-Binding Proteins that Modulate Transcription

Research is currently underway to determine the molecular mechanisms involved in regulating segmentation gene expression. Thus far, the results indicate that regulation is primarily at the transcriptional level. The *bicoid* protein is a transcription factor that binds to six different regulatory sequences in the 5′ flanking region of the *hunchback* gene (Driever and Nüsslein-Volhard, 1989; Driever et al., 1989; Struhl et al., 1989). These multiple binding sites regulate the intensity of *hunchback* gene expression according to the anterior to posterior gradient of *bicoid* protein concentration. Thus, in the anterior region of the embryo (where *bicoid* protein is most concentrated), all six sites in the 5′ flanking region are occupied, and the *hunchback* gene is transcribed strongly. Proceeding from anterior to posterior, fewer and fewer of the binding sites are occupied, resulting in progressively weaker *hunchback* gene expression. Eventually, a threshold of *bicoid* protein concentration is reached, below which there is no *hunchback* gene transcription. The interaction of *bicoid* protein with DNA may be mediated by the homeodomain, a polypeptide sequence that serves as a DNA-binding site (see p. 609). Similarly, the gap gene products are transcription factors, but their DNA-binding regions are different from *bicoid*. The *hunchback* and *Krüppel* proteins contain zinc fingers (Fig. 14.25; Rosenberg et al., 1986; Tautz et al., 1987), which are DNA-binding motifs originally discovered in the *Xenopus* transcription factor IIIA (Miller et al., 1985; see also p. 91). In contrast, the *knirps* gene encodes a DNA-binding protein that is similar to a steroid hormone receptor (Nauber et al., 1988).

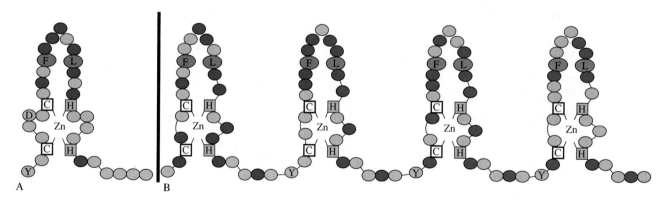

Figure 14.25

Diagram showing zinc-finger motifs in (A) the *Xenopus* transcription factor IIIA and (B) the *Drosophila Krüppel* protein. The marked circles, ellipses, and squares indicate conserved amino acid residues in the proteins. Y: tyrosine; C: cysteine; D: aspartic acid; H: histidine; F: phenylalanine; L: leucine. The darkly colored circles mark the positions of DNA-binding amino acids. The binding of zinc (Zn) to four conserved amino acids in the motif allows the other amino acids to form a DNA-binding loop or finger. The *Krüppel* protein contains four zinc fingers. (After Rosenberg et al., 1986. Reprinted by permission from *Nature* Vol. 319 pp. 336–339. Copyright © 1986 Macmillan Magazines Ltd.)

Further evidence that gap gene products are transcription factors comes from molecular studies showing that *hunchback* and *Krüppel* proteins bind to promoters in the 5' flanking region of the *hunchback* gene (Stanojevic et al., 1989; Treisman and Desplan, 1989). This is probably the mechanism by which the *Krüppel* protein suppresses *hunchback* activity and may also allow *hunchback* to influence its own expression. Finally, the *fushi tarazu* protein is known to activate transcription of the homeotic selector genes *Antennapedia* and *Ultrabithorax* (Winslow et al., 1989). This was demonstrated by the experiment outlined in Figure 14.26, in which plasmids containing either *Antennapedia*, *Ultrabithorax*, or heat-shock gene promoters were introduced into *Drosophila* tissue culture cells (by a process

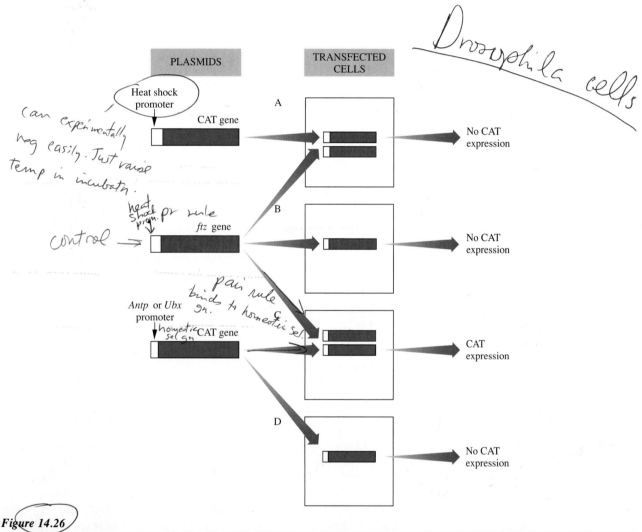

Figure 14.26

Regulation of *Antennapedia* and *Ultrabithorax* gene transcription by the *fushi tarazu* gene product. **A,** *Drosophila* tissue culture cells maintained at normal laboratory temperatures do not express the CAT gene to high levels when transfected with a plasmid consisting of a heat-shock gene promoter fused to the CAT gene and a plasmid containing a *fushi tarazu (ftz)* gene. **B,** Cells transfected with only the plasmid containing the *fushi tarazu* gene also do not express the CAT gene. This control experiment shows that there is no CAT produced by these cells unless the CAT-containing plasmid is present. **C,** Cells co-transfected with a plasmid containing an *Antennapedia (Antp)* or *Ultrabithorax (Ubx)* promoter fused to the CAT gene and a plasmid containing the *fushi tarazu* gene express the CAT gene. In this experiment, the *fushi tarazu* gene is transcribed, and a protein is formed that specifically activates transcription by binding to the homeotic selector gene (*Antp* and *Ubx*) promoters. **D,** Cells transfected with the plasmid containing an *Antennapedia (Antp)* or *Ultrabithorax (Ubx)* promoter fused to the CAT gene without the *fushi tarazu* plasmid do not express CAT to high levels, showing that the *fushi tarazu* gene product is required for expression of CAT protein.

known as **transfection**) with or without another plasmid containing the *fushi tarazu* gene. The *Antennapedia* and *Ultrabithorax* promoters were used to examine the transcriptional control of these genes, whereas the heat-shock promoter was used as a control because it should not be functional under the conditions used in these experiments. The test promoters were fused with a CAT gene (see p. 584), which is used to evaluate transcription by the promoters. When plasmids containing *Antennapedia* or *Ultrabithorax* promoters linked to the CAT gene were introduced into cells without the *fushi tarazu* plasmid, very little CAT protein was synthesized (see Fig 14.26D). However, when either plasmid was introduced into cells along with the *fushi tarazu* gene, high levels of CAT protein were produced (see Fig 14.26C). Because the *fushi tarazu* plasmid did not stimulate CAT when transfected into cells by itself (see Fig. 14.26B) or with a plasmid containing a heat-shock promoter fused with the CAT gene (see Fig. 14.26A), these experiments show that *fushi tarazu* is a specific transcriptional activator of *Antennapedia* and *Ultrabithorax*.

The Homeotic Selector Genes Specify Segment Identity

After the segments are formed, they receive a specific identity according to their position along the anteroposterior axis. For instance, the T1, T2, and T3 segments differentiate into prothorax, mesothorax, and metathorax, respectively. Segment identity is specified by the segment-polarity genes and the homeotic selector genes. We have seen that mutations in homeotic selector genes result in transformations among homologous body parts of the fly. In genetic mapping studies (Sànchez-Herrero et al., 1985), six of the homeotic selector genes were shown to be clustered in two regions of the third chromosome, the **antennapedia complex** and the **bithorax complex**.

The Antennapedia Complex

The antennapedia complex contains three homeotic selector genes—*deformed, sex combs reduced*, and *Antennapedia*—as well as the *bicoid* and *fushi tarazu* genes (Scott et al., 1983). The homeotic genes located in the antennapedia complex control head and anterior thorax development (through the anterior compartment of the mesothorax; see Fig. 14.28). Here, we describe the function of the *Antennapedia* gene, which has been determined by a combination of genetic and molecular studies. These studies suggest that the *Antennapedia* gene controls mesothorax development. When the entire *Antennapedia* gene is genetically deleted, all thorax segments tend to develop like the head and prothorax (Struhl, 1981; Wakimoto and Kaufman, 1981). Furthermore, in wild-type flies, *Antennapedia* mRNA (Levine et al., 1983; Martinez-Arias, 1986) and proteins (Wirz et al., 1986) appear primarily in segment T2 (parts of parasegments 4 and 5) during development. Finally, *Antennapedia* is inappropriately expressed in the head segments of *Antp* mutants (Schneuwly et al., 1987b), resulting in replacement of antennae by mesothoracic legs (see Fig. 14.11). The role of the *Antennapedia* gene in mesothorax development is dramatically demonstrated by an experiment in which extra *Antennapedia* genes were introduced into the *Drosophila* genome by **germ-line transformation** (Spradling and Rubin, 1982) and expressed ectopically during embryogenesis (Fig. 14.27). Germ-line transformation involves the fusion of a cloned gene with the P-element, which is a small DNA transposon that contains sequences allowing it (and any DNA linked to it) to be inserted into the *Drosophila* genome. After microinjection into eggs, some of the plasmids containing the P-element may enter the pole plasm and be incorporated into germ-line DNA. Ectopic expression of the extra *Antennapedia* genes was obtained

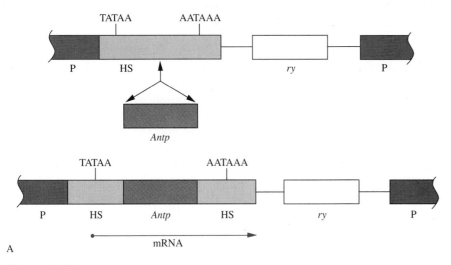

Figure 14.27A

The production of homeotic legs in the head of *Drosophila* by ectopic expression of *Antennapedia* genes introduced into the genome by germ-line transformation. A, Diagrams showing how the plasmid was constructed to introduce *Antennapedia* genes into the germ line. As shown in the upper drawing, a cloned *Antennapedia* (*Antp*) cDNA was inserted (arrow) into a plasmid containing a P-element (P), part of a heat shock gene (HS) containing a promoter (TATAA) and mRNA processing sequences (AATAAA), and the *rosy* gene (*ry*). The *rosy* gene is used as a marker to follow the incorporation of the plasmid into the germ line. Fine lines: plasmid DNA. The plasmid containing the inserted *Antennapedia* gene is shown in the lower drawing. The orientation of the *Antennapedia* coding sequence within the plasmid is such that its expression comes under the control of the heat shock promoter, and its mRNA transcripts are processed correctly owing to the heat shock processing sequences. The transcript produced by the engineered *Antennapedia* gene is represented by the horizontal arrow.

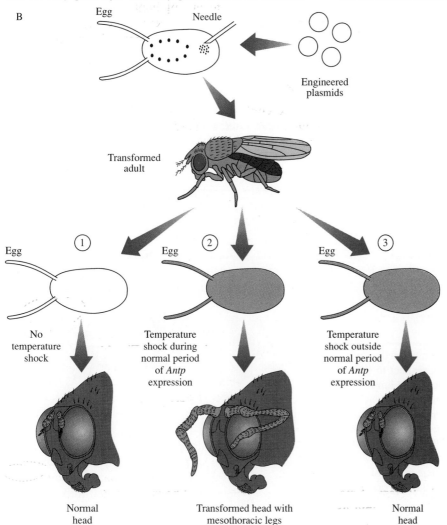

Figure 14.27B

A diagram showing the experiments to determine the effect of ectopic *Antennapedia* expression on development. After injection into eggs, some of the engineered *Antennapedia* plasmids become incorporated into germ-line DNA, and a strain of flies is formed containing extra *Antennapedia* genes. 1. When embryos from the transformed strain are grown at normal temperature, they develop into normal flies. 2. When embryos from the transformed strain are exposed to elevated temperature during the time of normal *Antennapedia* expression, they develop into flies in which a pair of legs replaces the antennae. 3. When embryos from the transformed strain are exposed to elevated temperature outside the time of normal *Antennapedia* expression, they develop into normal flies. (After Schneuwly et al., 1987a.)

607

by removing their natural promoters and replacing them with the promoters of a heat-shock gene before transformation (see Fig. 14.27A). The heat-shock promoter allows the *Antennapedia* genes to be activated by elevating the temperature. When embryos were exposed to elevated temperature during the normal period of *Antennapedia* gene expression, a pair of mesothoracic legs developed in place of the antennae (see Fig. 14.27B). In contrast, normal development was obtained either when the transformed embryos were maintained at normal temperature or when the extra genes were expressed outside the normal interval of *Antennapedia* activity. This experiment shows that the *Antennapedia* gene controls the development of the mesothorax and must be *inactivated* in the head during normal development to permit antenna formation.

The Bithorax Complex

The bithorax complex contains the homeotic selector genes *Ultrabithorax, abdominal A*, and *abdominal B*. These genes specify segments posterior to T1 (beginning in the posterior compartment of the mesothorax; Lewis, 1978, 1985). This was demonstrated by deleting the entire bithorax complex, which caused all segments posterior to T2 to develop as replicas of the mesothorax. Further analysis of mutations in the bithorax complex have revealed a relationship between their position on the chromosome and the spatial ordering of segments along the body axis (Lewis, 1978). *Ultrabithorax,* the first gene in the complex, specifies the posterior thorax segments, *abdominal A* (the second gene) specifies the anterior abdominal segments, and *abdominal B* (the third gene) is responsible for development of the posterior abdominal segments (Fig. 14.28). *In situ* hybridization experiments indicate that the genes of the bithorax complex are expressed in the regions of the embryo corresponding to the segments they specify during development. However, the control of segment identity by these genes is complex and cannot be revealed simply by examining their domains of expression. The complexity of this process can be appreciated by considering that the three homeotic genes comprise only about 10% of the DNA sequence in the bithorax complex. Genetic analysis has shown that the remaining DNA

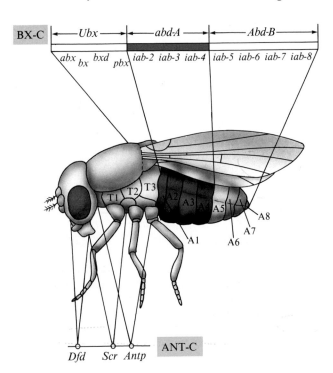

Figure 14.28

The regions in which the genes of the antennapedia and bithorax complexes are transcribed with respect to the adult body plan. The vertical lines indicate the limits of gene transcription. BX-C: bithorax complex; *Ubx: ultrabithorax; abd-A: abdominal-A; Abd-B: Abdominal-B; abx: anteriobithorax; bx: bithorax; bxd: bithoraxoid; pbx: postbithorax; iab: interabdominal*; ANT-C: antennapedia complex; *Dfd: Deformed; Scr: Sex combs reduced; Antp: Antennapedia*. The *bicoid* and *fushi tarazu* genes of the antennapedia complex are not shown in the diagram. (After Harding et al., 1985.)

sequence is also involved in determining segment identity (Bender et al., 1983). For example, the *anteriobithorax* and *bithorax* mutations, which map outside the three gene loci of the bithorax complex, cause the anterior compartment of T3 to develop like the anterior compartment of T2. Thus, in mutant flies, the metathorax contains the anterior half of a wing and the posterior half of a haltere (Crick and Lawrence, 1975). These homeotic mutations represent changes in enhancers (see essay on p. 582) and other regulatory sequences located in introns and regions flanking the structural genes of the bithorax complex. The mechanisms by which these regulatory sequences control transcription of the homeotic selector genes is a subject of intense contemporary research in developmental biology.

Homeoboxes and Homeodomains

The homeobox, a 180 nucleotide DNA sequence, was discovered during the cloning and sequencing of the homeotic selector genes (McGinnis et al., 1984; Scott and Weiner, 1984). This DNA sequence encodes an amino acid sequence called the **homeodomain** and is also found in other genes that control *Drosophila* development, such as *bicoid, even-skipped, engrailed*, and *fushi tarazu*. The homeodomain is a DNA-binding site (Desplan et al., 1985), and proteins containing this sequence are thought to be transcription factors. An important characteristic of the homeodomain is its conservation among widely divergent organisms—from yeast to humans (Levine et al., 1984): Homeoboxes have been identified in the DNA of yeast, nematodes, arthropods, sea urchins, and vertebrates. Figure 14.29 shows the remarkable

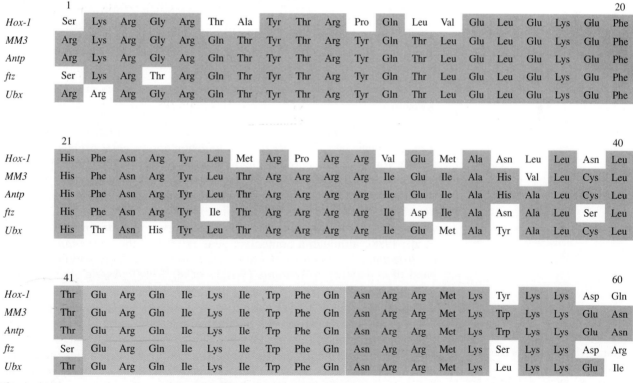

Figure 14.29

The amino acid sequences of homeodomains encoded by homeobox-containing genes in the mouse (*Hox–1*), *Xenopus* (*MM3*), and the *Drosophila* genes *Antennapedia* (*Antp*), *fushi tarazu* (*ftz*), and *Ultrabithorax* (*Ubx*). The shaded box encloses the homologous region in each homeodomain. The colorized box corresponds to the putative DNA-binding region of the proteins. (After Gehring, 1985.)

degree of conservation in amino acids among some of the *Drosophila* homeobox-containing genes, the mouse homeobox-containing gene *Hox-1*, and the *Xenopus* homeobox-containing gene *MM3*. As we shall see in Section 14–3, homeobox genes may also be involved in regulating development in vertebrates.

14–3 Regional Patterns of Gene Expression During Early Development in Vertebrates

Now that we have examined the establishment of regional patterns of gene expression in invertebrates, we turn our attention to this topic in the vertebrates. Most of the research on regional gene expression in vertebrates has focused on the vertebrate homologues of the insect homeobox genes. We have seen that the expression of homeobox genes is regionally restricted in both insects and sea urchins. The former are segmented, and expression of these genes in insects is thought to be related to the establishment of segmental boundaries. Sea urchins are, however, not segmented animals. Hence, expression of homeobox genes in sea urchins is not related to segmentation. In vertebrate embryos (which are segmented), homeobox gene expression is regionally restricted, primarily in dorsal tissues along the anteroposterior axis. There is some evidence for segmentally restricted expression of homeobox genes in vertebrates, but in most cases the domains of homeobox gene expression bear no apparent histological or morphological boundaries. Thus, expression of these genes is not necessarily related specifically to segmentation. Their expression, however, may be quite important in the hierarchy of regulatory steps that establish the vertebrate body plan.

Xenopus

A number of *Xenopus* homeobox genes have been identified by screening *Xenopus* gene libraries with *Drosophila* homeobox gene probes (see Appendix for description of the technique of library screening). When the spatial patterns of expression of these genes are examined, they are typically found to be spatially restricted in their expression. As is the case with other vertebrates, most homeobox genes are expressed within the neuroectoderm and/or mesoderm (De Robertis et al., 1988; Harvey and Melton, 1988; Oliver et al., 1988), although a homeobox gene related to the *Antennapedia* gene of *Drosophila* has been shown to be expressed exclusively within a narrow band of endoderm in *Xenopus* (Wright et al., 1988). As we shall discuss, not only has homeobox gene expression been characterized during *Xenopus* development, but experimentation with *Xenopus* embryos has provided clues as to the possible roles of homeobox genes in vertebrate development and the regulation of their expression.

We shall focus this discussion of *Xenopus* homeobox genes on *Xhox3*, which was isolated by screening a *Xenopus* genomic library with the homeobox portion of the *Drosophila* pair-rule gene *even-skipped (eve)*. As we shall see, expression of *Xhox3* appears to be involved in mediating anteroposterior polarity during *Xenopus* development.

Analyses of the chronological expression of *Xhox3* using nuclease protection assays (see p. 101) reveal that transcripts are first detected at the mid-blastula transition and peak at the early neurula stage (Fig. 14.30). The *spatial* pattern of *Xhox3* expression was initially demonstrated by

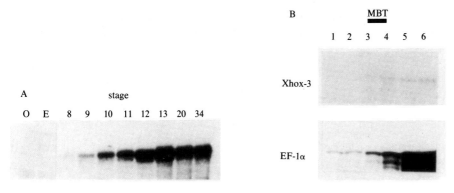

Figure 14.30

Nuclease protection profiles of *Xhox3* transcripts during early *Xenopus* development. A, *Xhox3* transcripts are absent in oocytes (O) and eggs (E) and are first detected at the mid-blastula transition (MBT). Stages shown are mid and late blastulae (8–9), early to late gastrulae (10–13), early to late neurulae (13–20), and swimming tadpole (34). B, More exact staging reveals that *Xhox3* transcripts accumulate precisely at the MBT. The nuclease protection profile for *Xhox3* is compared to that for the translational factor *EF-Iα*, whose transcripts accumulate extensively beginning at the MBT. The first sample of embryos was at the early blastula stage, and samples were collected at hourly intervals (hourly intervals are indicated above each lane). The bar indicates the MBT, as determined by the increased expression of the *EF-Iα* gene. (From Ruiz i Altaba and Melton, 1989a. Reprinted by permission of Company of Biologists, Ltd.)

dissecting embryos into anterior, middle, and posterior pieces and conducting nuclease protection assays. As shown in Figure 14.31, *Xhox3* transcripts show a graded distribution along the anteroposterior axis, with highest levels at the posterior end. Are *Xhox3* transcripts found in all germ layers, or are they germ-layer restricted? Separation of the germ layers indicates that *Xhox3* transcripts accumulate preferentially (perhaps exclusively) in the mesoderm during this time (Fig. 14.32), thus providing an index of the positional values of mesodermal cells along the anteroposterior axis. Experimental treatments that respecify the anteroposterior fates of *Xenopus* mesodermal cells (Kao and Elinson, 1988) have been shown to alter *Xhox3* expression (Ruiz i Altaba and Melton, 1989a): Treatment of embryos with ultraviolet light, which causes cells to assume more posterior fates, elevates *Xhox3* transcript levels, whereas lithium, which anteriorizes embryos, drastically reduces *Xhox3* transcript levels (Fig. 14.33).

In view of the roles of homeobox genes in conferring positional information on cells in the *Drosophila* embryo, Ruiz i Altaba and Melton (1989a, 1990) have suggested that the level of *Xhox3* expression (as reflected in *Xhox3* transcript levels and, presumably, *Xhox3* protein levels) may impart positional information to the mesodermal cells that could influence the fates of these cells and of the lineages that they establish (i.e., the extent of anterior development is inversely related to the level of *Xhox3* expression). As a test of this hypothesis, Ruiz i Altaba and Melton (1989b) have injected synthetic *Xhox3* mRNA into the anterior region of *Xenopus* embryos (where *Xhox3* messengers are usually present at their lowest levels). This results in a uniform, high concentration of *Xhox3* mRNA along the anteroposterior axis, obliterating the normal graded transcript distribution. This resultant overabundance of *Xhox3* transcripts in the anterior region causes defects in anterior (i.e., head) development (Fig. 14.34). Because the primary effect of overabundance of *Xhox3* transcripts is to cause anterior deficiencies,

Figure 14.31

Graded distribution of *Xhox3* transcripts along the anteroposterior axis during early *Xenopus* development. Drawings of stages 11–19 are shown at the top. St. 11: mid-gastrula; St. 13: late gastrula–early neurula; St. 15: mid-neurula; St. 19: late neurula; D: dorsal; V: ventral; A: anterior; M: middle; P: posterior. The same RNA samples were probed with both *Xhox3* and *EF-Iα* probes in a nuclease protection assay. The latter provides an index of cell numbers to ensure that equivalent amounts of RNA are loaded for samples from each stage of development. (From Ruiz i Altaba and Melton, 1989a. Reprinted by permission of Company of Biologists, Ltd.)

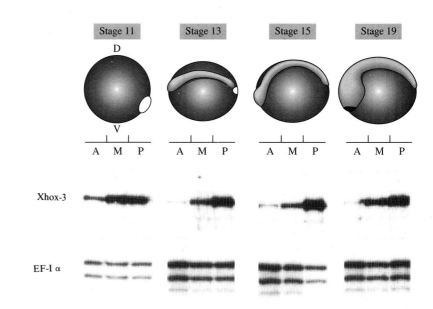

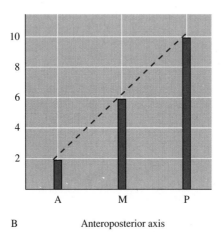

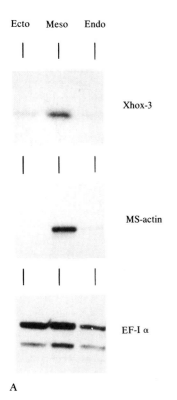

Figure 14.32

Xhox3 expression in separated germ layers of *Xenopus* embryos. A, Embryos were incubated in high salt medium, which caused the mesoderm and endoderm to flow outside rather than inside, during gastrulation. The germ layers of the resultant exogastrulae are readily dissected. RNA was extracted and assayed for the presence of *Xhox3*, muscle-specific actin (MS-actin; to confirm the identity of the mesodermal fraction), and *EF-Iα* by nuclease protection assays. B, Diagram showing the relative abundance of *Xhox3* mRNA in regions of the axial mesoderm of late gastrula–early neurula embryos. (A, From Ruiz i Altaba and Melton, 1989a. Reprinted by permission of Company of Biologists, Ltd. B, After Ruiz i Altaba and Melton, 1989c. Reprinted by permission of *Nature* Vol. 341 pp. 33–38. Copyright © 1989 Macmillan Magazines, Ltd.)

rather than to transform anterior structures to posterior structures, it would appear that the level of *Xhox3* is necessary, but not sufficient, to specify anteroposterior development. Presumably, the expression of other genes (perhaps other homeobox genes) is essential to refine positional values and thus specify exact cell fate.

Because of the possible involvement of *Xhox3* in regulating proper anteroposterior development in the mesoderm, it is important to understand how the levels of *Xhox3* are regulated. Ruiz i Altaba and Melton (1989a) have proposed that a diffusible graded signal from the posterior pole is responsible for the graded distribution of *Xhox3* transcripts; this would, in

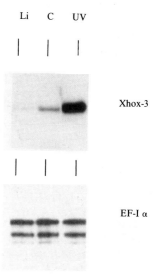

Figure 14.33

The effects of lithium (Li) and ultraviolet light (UV) treatment on *Xhox3* expression in *Xenopus* embryos. Embryos were treated with lithium or ultraviolet and allowed to develop to the neurula stage. Lithium-treated embryos with exaggerated anterior development and ultraviolet light-treated embryos with exaggerated posterior development were selected. RNA from these embryos was assayed by nuclease protection assay for the presence of *Xhox3* and *EF-Iα* transcripts. There is a clear excess of *Xhox3* transcripts in the posteriorized embryos and a deficiency in the anteriorized embryos over control (C) levels, whereas *EF-Iα* transcript levels are constant. (From Ruiz i Altaba and Melton, 1989a. Reprinted by permission of Company of Biologists, Ltd.)

turn, specify positional information in the mesoderm along the antero-posterior axis (such as allowing anterior differentiation in the mesoderm). Possible candidates for such a signal would be the growth factors that are responsible for mesoderm induction. As we discussed in Chapter 12, mesoderm is induced by members of the transforming growth factor-β (TGF-β) and fibroblast growth factor (FGF) families. Indeed, Ruiz i Altaba and Melton (1989c) have shown that incubating isolated animal caps of *Xenopus* blastulae in either basic fibroblast growth factor (bFGF) or XTC-MIF (a member of the TGF-β family; see p. 483) induces expression of *Xhox3* mRNA. To test whether these growth factors might have differential effects on *Xhox3* expression in the mesoderm, animal caps were incubated in levels of either growth factor that are sufficient to induce dorsal mesoderm, and *Xhox3* mRNA levels were determined. As shown in Figure 14.35, bFGF induces five- to ten-fold higher levels of *Xhox3* mRNA at 15 (neurula) stage than does XTC-MIF. This difference is comparable to the endogenous difference in *Xhox3* mRNA levels between anterior and posterior mesoderm in the intact neurula. Interestingly, raising the dose of XTC-MIF decreases *Xhox3* mRNA levels, whereas higher doses of bFGF increase them. These results suggest that the complementary action of endogenous members of the FGF and TGF-β families regulates anteroposterior patterning in the mesoderm during development. High doses of an FGF family member would induce high levels of *Xhox3* expression in the mesoderm (thus specifying posterior character), whereas high doses of a TGF-β–like molecule result in low levels of *Xhox3* expression (hence allowing anterior mesoderm development). As we have discussed, lithium reduces *Xhox3* levels and anteriorizes embryos. As also shown in Figure 14.35, lithium appears to exert its effect by attenuating the induction of *Xhox3* mRNA by FGF.

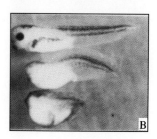

Figure 14.34

The effects of overabundance of *Xhox3* transcripts on *Xenopus* development. The embryo on the top in both A and B was injected with β-globin mRNA as a control, and the rest were injected with *Xhox3* mRNA. Embryos in A were injected with RNAs at concentrations of 100 μg/ml, whereas those in B were injected with 200 μg/ml. A range of axial defects is seen in the embryos injected with *Xhox3* RNA—the most obvious and consistent being deficiencies in anterior development, which are more severe at high RNA levels. (From Ruiz i Altaba and Melton, 1989b. Copyright © Cell Press.)

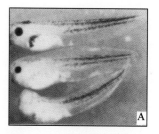

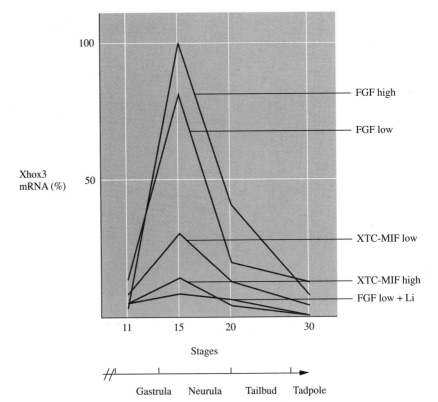

Figure 14.35

Different growth factors induce different levels of *Xhox3* mRNA. Groups of isolated animal caps of *Xenopus* blastulae were treated with bFGF, bFGF plus lithium (Li), or XTC-MIF at different concentrations, and RNA levels were determined at different stages by nuclease protection assay. (After Ruiz i Altaba and Melton, 1989c. Reprinted by permission from *Nature* Vol. 341 pp. 33–38. Copyright © 1989 Macmillan Magazines, Ltd.)

A consequence of anteroposterior polarity in the mesoderm is in the nature of the neural and epidermal elements that it induces in the overlying ectoderm. Thus, anterior mesoderm induces anterior structures, whereas posterior mesoderm induces posterior structures (see Fig. 12.10, p. 488, and Fig. 12.11, p. 489). The results of recent experiments further suggest that growth factors confer a graded anteroposterior character on the inductive signals emanating from the mesoderm (see Fig. 12.13, p. 491). Because the positional values of the mesoderm correlate with *Xhox3* expression, it is intriguing to consider the possibility that the level of *Xhox3* expression mediates the inducing properties of mesoderm. Indeed, as shown in Figure 14.34, overabundance of *Xhox3* in the anterior end of the embryo results in head deficiencies—an indication of faulty anterior neural induction.

Because homeobox genes encode DNA-binding proteins that (in *Drosophila*) regulate expression of other genes, it will be instructive to learn whether the protein encoded by *Xhox3* has a direct effect on expression of genes that are involved in induction of ectoderm. In any case, we have now entered a new era in investigating the establishment of the body plan of amphibians and can look forward not only to learning the identity of Spemann's organizer but learning how it is regulated.

At the late neurula to early tailbud stages, the spatial pattern of *Xhox3* gene expression changes, reflecting expression in the nervous system and tailbud. When somite formation is complete, *Xhox3* expression in the tailbud disappears, but neural expression continues (Ruiz i Altaba and Melton, 1989a). Thus, germ layer specificity has changed. *In situ* hybridization analyses of the expression of *Xhox3* in the developing nervous system reveal that transcripts are restricted to the anterior neural tube and neural crest (Fig. 14.36). In the adult, *Xhox3* transcripts can be detected in the brain. The expression of *Xhox3* in the developing central nervous system is not

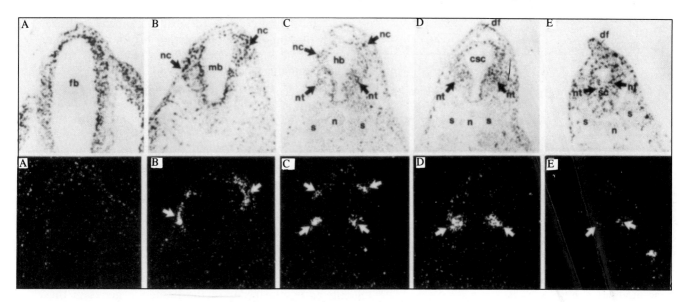

Figure 14.36

Expression of *Xhox3* mRNA in tailbud/early tadpole stages of *Xenopus* development. Sections of embryos were examined for the presence of *Xhox3* transcripts by *in situ* hybridization. The panels show the dorsal portion of cross-sections at progressively more posterior positions along the anteroposterior axis. The top row shows bright-field micrographs, whereas the bottom row shows dark-field micrographs. Arrows show areas of specific hybridization. csc: cervical spinal cord; df: dorsal fin; fb: forebrain; hb: hindbrain; mb: midbrain; n: notochord; nc: neural crest; nt: neural tube; s: somites; sc: spinal cord. (From Ruiz i Altaba, 1990. Reprinted by permission of Company of Biologists, Ltd.)

surprising because most homeobox gene expression that has been detected in vertebrates has been localized to the central nervous system. These results suggest that this later expression of the *Xhox3* gene may have a different consequence for development than the first; perhaps it is responsible for establishing and maintaining cell lineages in the brain.

Mammals

The basic body plan of the mammalian embryo is established after implantation. The anteroposterior axis is manifest with the appearance of the primitive streak at the posterior end of the embryo. The emergence of the primitive streak marks the beginning of gastrulation, which—as the process continues—generates the three embryonic germ layers (see Chap. 6).

Of the three germ layers, portions of two of them undergo segmentation (Rossant and Joyner, 1989). The dorsal mesoderm undergoes progressive segmentation in an anteroposterior progression to form somites. The intermediate mesoderm also segments in register with the somite mesoderm. Meanwhile, the mid-dorsal ectoderm forms the neural plate, which folds up to form the neural tube—the precursor of the nervous system. The anterior neural tube becomes segmented into periodic structures known as **neuromeres**. Recent evidence from studies with chick embryos suggests that neuromeres in the hindbrain (which are called **rhombomeres**) are responsible for establishing the repeating pattern of cranial nerves in this region. Furthermore, the neural crest derivatives of the rhombomeres are responsible for patterning the skeletal and muscular components of each branchial arch.

(Recall that we discussed the neural crest origin of the branchial arches in Chap. 7.) Hence, the segmentation of the hindbrain has profound effects on development of the head (Lumsden and Keynes, 1989). Presumably, similar relationships hold for the mammalian hindbrain as well.

Much of the remainder of the embryo (including the lateral plate mesoderm, surface ectoderm, and endoderm) remains unsegmented. As development proceeds, both the segmental and nonsegmented structures are subject to regional specialization as diverse elements are generated along the body axis. Do homeobox genes have any role to play in either the segmentation or regionalization of the mammalian embryo? This is a more difficult question to answer for mammals than for *Drosophila* or *Xenopus*, because mammalian embryos are not as amenable to experimental manipulation as amphibian embryos are. However, considerable evidence has been obtained that correlates the expression of homeobox genes in mice with patterning along the anteroposterior axis.

As we have discussed previously, homeobox genes in *Drosophila* are clustered in two separate and adjacent complexes: antennapedia and bithorax (see Sect. 14–2). Significantly, the order of the homeotic genes in these clusters in *Drosophila* corresponds with the order in which these genes are expressed along the anteroposterior axis of the embryo (see Fig. 14.28). A similar situation occurs in the mouse. For example, there are a large number of *Antp*-related genes organized into four clusters, which are designated *Hox-1, Hox-2, Hox-3*, and *Hox-5*. There is a high degree of homology among members of the clusters, but also in the alignment within clusters. There is evidence that these clusters originated from a single ancestral cluster by duplication and divergence (Akam, 1989). The organization and pattern of expression of members of the *Hox-2* cluster have recently been analyzed in detail and reveal a clear pattern of expression along the anteroposterior axis of the mouse embryo.

The 5′ to 3′ order of eight members of the *Hox-2* complex is shown in Figure 14.37 (the numbers assigned to these genes reflect the order in which they were discovered). The patterns of expression of members of the complex have been determined by Northern blotting and *in situ* hybridization. As with most homeobox genes, the *Hox-2* genes are expressed in both the central and peripheral nervous systems and in the mesoderm (Graham et al., 1989). A detailed analysis of expression of these members of the *Hox-2* complex using *in situ* hybridization reveals that most of these genes have no clear posterior limit of expression, but each has a discrete anterior limit of expression. This is illustrated in Figure 14.38 for three members of this complex. Reference to Figure 14.37 will reveal that the limit of expression of these genes becomes progressively more anterior as one progresses through the gene complex in the 5′ to 3′ direction. This trend is typical of this complex, as summarized in Figure 14.39, which shows the anterior limits of expression within the CNS of individual members of the *Hox-2* complex.

In contrast to the absence of posterior limits of expression of the *Hox-2* genes that we have described previously, one newly discovered

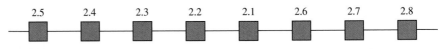

Figure 14.37

The relative positions of eight members of the *Hox-2* gene complex in the mouse. 5′ is on the left, 3′ is on the right.

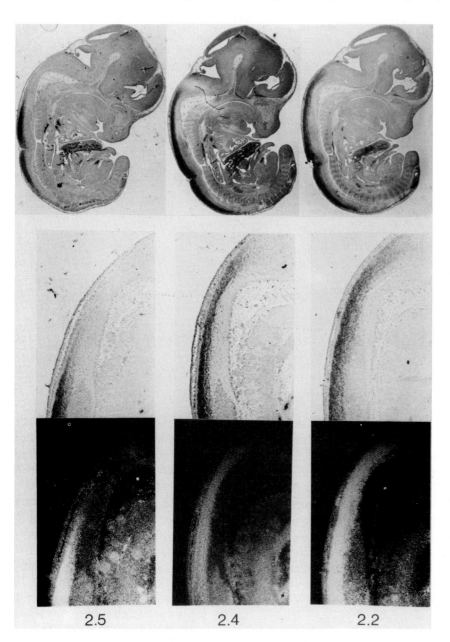

Figure 14.38

Comparison of the patterns of expression of the genes *Hox-2.5*, *Hox-2.4*, and *Hox-2.2* in the 12.5-day mouse embryo. The top row represents bright field images of *in situ* hybridizations to near-adjacent sagittal sections of entire embryos. Below are high-power micrographs of these sections, focusing on the regions of the central nervous system in which the anterior limits of expression are observed. These are represented in both bright field and dark field. The probe for each of these sections is indicated at the bottom of each column of pictures. (From Graham et al., 1989. Copyright © Cell Press.)

Hox-2 gene (*Hox-2.9*) has very precise anterior and posterior boundaries of expression in the developing hindbrain: It is expressed in a single rhombomere (Fig. 14.40). This observation suggests that *Hox-2.9* may be involved in specification of the identity of this segment and that other *Hox* genes (either alone or in combination with other *Hox* genes) may be involved in specification of other neural segments.

Because homeobox gene expression in mammals responds to positional cues, it is important to determine what the cues are and how they are used to regulate expression of the genes. The most likely scenario is that expression of these genes is regulated by interactions between regulatory elements in the promoters of these genes and regionally restricted transcriptional factors. An initial step in examining the regulatory mechanism is to characterize the regulatory sequences for these genes that specify their spatial patterns of expression as has been done for the CyIIIa gene of sea urchins. In a typical experiment, putative regulatory sequences of individual

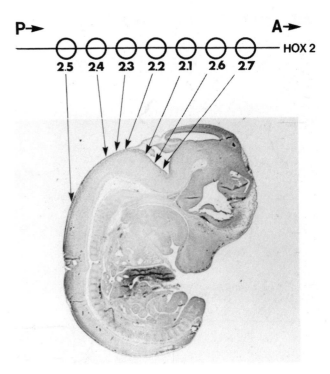

Figure 14.39

Anterior boundaries of expression of individual *Hox-2*
genes within the central nervous system and the
correlation of those boundaries with gene position within
the *Hox-2* cluster. A: anterior; P: posterior. (From Graham
et al., 1989. Copyright © Cell Press.)

homeobox genes are fused to reporter gene sequences, and the fusion gene
constructs are injected into the male pronuclei of fertilized eggs to produce
transgenic embryos (see Appendix). The exogenous genes may become
stably integrated into the zygotic genome, and the pattern of expression
during development can be examined.

The transgenic embryos are transferred to the reproductive tracts of
foster mothers and allowed to develop. The pattern of expression of the
transgene is determined by discerning the spatial pattern of the product of
the reporter gene. If the pattern of expression corresponds to that of the
endogenous homeobox gene, the test sequences must contain the information
for specifying the regional pattern of expression of the homeobox gene. The
results of a representative experiment of this sort are shown in Figure 14.41.

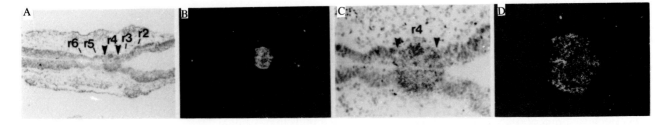

Figure 14.40

Expression of *Hox-2.9* in the 9.5-day mouse embryo hindbrain as shown by *in situ*
hybridization. A, and C, bright-field micrographs; B, and D, dark-field
micrographs. A and B are low power micrographs (50×) showing rhombomeres
2–6, whereas C and D are slightly higher power images (66×). The boundaries of
rhombomere 4 are indicated by arrows. Silver grains are restricted to rhombomere 4.
(From Wilkinson et al., 1989. Reprinted by permission from *Nature* Vol. 341 pp.
405–409. Copyright © 1989 Macmillan Magazines Ltd.)

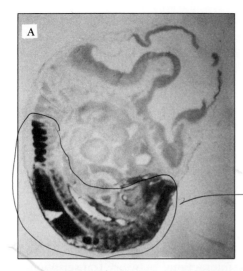

(handwritten: (actually Hox 1.1·lacZ transgene Hox 1 gn expres here.)

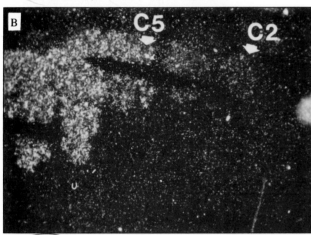

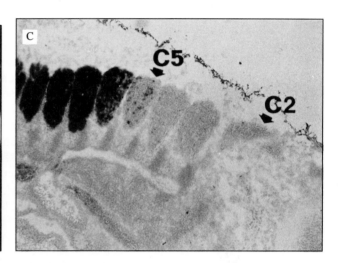

Figure 14.41

Expression of *Hox-1.1* and the *Hox-1.1·lac-Z* transgene in the 12-day mouse embryo. A, Mid-sagittal section stained for β-galactosidase activity. B, *In situ* *(handwritten: ie lacZ.)* hybridization of section using a probe for *Hox-1.1*. Section is oriented with the rostral (anterior) end of the embryo to the right. The positions of cervical ganglia 2 and 5 are indicated by C2 and C5, respectively. No *Hox-1.1* expression above background is detected anterior to C5. C, Comparable section to that shown in B stained for β-galactosidase. The staining pattern in cervical ganglia (which represents transgene expression) reflects the expression of *Hox-1.1* shown in B. See Color plate 23 for color version. (From Püschel et al., 1990. Reprinted by permission of Company of Biologists, Ltd.)

The promoter region of *Hox-1.1* was fused to the bacterial β-galactosidase (*lac-Z*) gene. The pattern of expression of the transgene is determined by staining for β-galactosidase activity and is compared with the pattern of expression of the endogenous *Hox-1.1* gene, which is determined by *in situ* hybridization. Figure 14.41A shows a sagittal section of a transgenic embryo stained for β-galactosidase activity. The limits of transgene expression can be seen readily by the dark staining in the dorsal tissues in the posterior half of the embryo. *Hox-1.1* expression in cervical ganglia is shown by *in situ* hybridization in Figure 14.41B. Expression is moderate in the fifth cervical ganglion and becomes more intense in the more posterior ganglia. A similar pattern of expression is seen for the *Hox-1.1·lac-Z* transgene in Figure 14.41C. Thus, the promoter of the transgene contains the information responsible for restricting the expression of this gene to a particular region of the embryo. Further analysis of this system should reveal which promoter elements convey the positional information and lead to identification of the regulatory factors that interact with them.

14–4 Chapter Synopsis

Positional information provides a molecular blueprint of the body plan of the embryo. Cells interpret these molecular cues to establish their patterns of gene expression. Distinct gene expression patterns are established initially in embryonic regions consisting of cells of a variety of cell fates and ultimately in homogeneous cell populations. The CyIIIa gene of sea urchins provides an excellent model system for investigating the molecular interactions involved in regulating expression of a regionally restricted gene during early development. Complex interactions between transcriptional factors and CyIIIa regulatory sequences control both the temporal and spatial patterns of expression of this gene.

Development along the anteroposterior axis of the *Drosophila* embryo is regulated by a cascade of regulatory interactions involving developmental control genes. Those genes that are expressed earliest in development encode proteins that regulate the expression of genes that are expressed later. One class of developmental control genes, called homeobox genes, encodes proteins containing a DNA-binding region called the homeodomain.

Expression of the homologues of the homeobox genes in vertebrates appears to be regulated by positional information in the early vertebrate embryo. In *Xenopus*, growth factors appear to be involved in regulating expression of one of the homeobox genes, *Xhox3*. The homeobox genes, in turn, specify more precise positional information along the anteroposterior axis, presumably by regulating the patterns of expression of specific genes.

References

Akam, M. 1989. Hox and HOM: Homologous gene clusters in insects and vertebrates. Cell, *57*: 347–349.

Alberts, B. et al. 1989. *The Molecular Biology of the Cell*, 2nd ed. Garland Publishing, New York.

Angerer, L. M. et al. 1986. Spatial patterns of metallothionein mRNA expression in the sea urchin embryo. Dev. Biol., *116*: 543–547.

Angerer, L. M. et al. 1989. Progressively restricted expression of a homeobox gene within the aboral ectoderm of developing sea urchin embryos. Genes Dev., *3*: 370–383.

Angerer, L. M., and R. C. Angerer. 1981. Detection of poly(A)$^+$ RNA in sea urchin eggs and embryos by quantitative *in situ* hybridization. Nucl. Acids Res., *9*: 2819–2840.

Angerer, R. C., and E. H. Davidson. 1984. Molecular indices of cell lineage specification in the sea urchin embryo. Science, *226*: 1153–1160.

Bender, W. et al. 1983. Molecular genetics of the bithorax complex in *Drosophila melanogaster*. Science, *221*: 23–29.

Benson, S. et al. 1987. A lineage-specific gene encoding a major matrix protein of the sea urchin embryo spicule. I. Authentication of the cloned gene and its developmental expression. Dev. Biol., *120*: 499–506.

Birnstiel, M. L., and M. Chipchase. 1977. Current work on the histone operon. Trends Biochem. Sci., *2*: 149–152.

Britten, R. J., and E. H. Davidson. 1969. Gene regulation for higher cells: A theory. Science, *165*: 349–357.

Britten, R. J., and E. H. Davidson. 1971. Repetitive and nonrepetitive DNA sequences and a speculation on the origins of evolutionary novelty. Q. Rev. Biol., *46*: 111–138.

Calzone, F. J. et al. 1988. Developmental appearance of factors that bind specifically to *cis*-regulatory sequences of a gene expressed in the sea urchin embryo. Genes Dev., *2*: 1074–1088.

Carroll, S. B., and M. P. Scott. 1986. Zygotically active genes that affect the spatial expression of the fushi tarazu segmentation gene during early Drosophila embryogenesis. Cell, *45*: 113–126.

Chan, L. N., and W. J. Gehring. 1971. Determination of blastoderm cells in *Drosophila melanogaster*. Proc. Natl. Acad. Sci. U.S.A., *68*: 2217–2221.

Crick, F. H. C., and P. A. Lawrence. 1975. Compartments and polyclones in insect development. Science, *189*: 340–347.

Davidson, E. H. 1989. Lineage-specific gene expression and the regulative capacities of the sea urchin embryo: a proposed mechanism. Development, *105*: 421–445.

Davidson, E. H. et al. 1985. Lineage-specific gene expression in the sea urchin embryo. Cold Spring Harbor Symp. Quant. Biol., *50*: 321–328.

De Pomeroi, D. 1985. *From Gene to Animal*. Cambridge University Press, New York.

De Robertis, E. M. et al. 1988. Families of vertebrate homeodomain proteins. *In* J. Gralla (ed.), *DNA-Protein Interactions in Transcription*. UCLA Symp., *95*: 107–115.

Desplan, C., J. Theis, and P. H. O'Farrell. 1985. The *Drosophila* developmental gene *engrailed* encodes a sequence-specific DNA binding activity. Nature (Lond.), *318*: 630–635.

Dierks, P. et al. 1981. DNA sequences preceding the rabbit β-globin gene are required for formation in mouse L cells of β-globin RNA with the correct 5′ terminus. Proc. Natl. Acad. Sci. U.S.A., *78*: 1411–1415.

Dolecki, G. J. et al. 1986. Stage-specific expression of a homeo box-containing gene in the non-segmented sea urchin embryo. EMBO J., *5*: 925–930.

Driever, W., and C. Nüsslein-Volhard. 1988. A gradient of bicoid protein in Drosophila embryos. Cell, *54*: 83–93.

Driever, W., and C. Nüsslein-Volhard. 1989. The *bicoid* protein is a positive regulator of *hunchback* transcription in the early *Drosophila* embryo. Nature (Lond.), *337*: 138–143.

Driever, W., G. Thoma, and C. Nüsslein-Volhard. 1989. Determination of spatial domains of zygotic gene expression in the *Drosophila* embryo by the affinity of binding sites for the *bicoid* morphogen. Nature (Lond.), *340*: 363–367.

Flytzanis, C. N., R. H. Britten, and E. H. Davidson. 1987. Ontogenic activation of a fusion gene introduced into the sea urchin egg. Proc. Natl. Acad. Sci. U.S.A., *84*: 151–155.

Franks, R. R. et al. 1990. Competitive titration in living sea urchin embryos of regulatory factors required for expression of the CyIIIa actin gene. Development, *110*: 31–40.

Frohnhöfer, H. G., and C. Nüsslein-Volhard. 1986. Organization of anterior pattern in the *Drosophila* embryo by the maternal gene *bicoid*. Nature (Lond.), *324*: 120–123.

Gaul, U. et al. 1987. Analysis of Krüppel protein distribution during early Drosophila development reveals posttranscriptional regulation. Cell, *50*: 639–647.

Gehring, W. J. 1985. Homeotic genes, the homeobox, and the genetic control of development. Cold Spring Harbor Symp. Quant. Biol., *50*: 243–251.

Graham, A., N. Papalopulu, and R. Krumlauf. 1989. The murine and Drosophila homeobox gene complexes have common features of organization and expression. Cell, *57*: 367–378.

Grosschedl, R., and M. L. Birnstiel. 1980. Identification of regulatory sequences in the prelude sequences of an H2A histone gene by the study of specific deletion mutants *in vitro*. Proc. Natl. Acad. Sci. U.S.A., *77*: 1432–1436.

Grosveld, G. C. et al. 1982. DNA sequences necessary for transcription of the rabbit β-globin gene *in vitro*. Nature (Lond.), *295*: 120–126.

Hadorn, E. 1978. Transdetermination. *In* M. Ashburner and T. R. F. Wright (eds.), *The Genetics and Biology of Drosophila*, Vol. 2c. Academic Press, New York, pp. 556–617.

Harding, K., C. Wedeen, and M. Levine. 1985. Spatially regulated expression of homeotic genes in *Drosophila*. Science, *229*: 1236–1242.

Harvey, R. P., and D. A. Melton. 1988. Microinjection of synthetic Xhox-1A homeobox mRNA disrupts somite formation in developing Xenopus embryos. Cell, *53*: 687–697.

Hough-Evans, B. R. et al. 1987. Correct cell type-specific expression of a fusion gene injected into sea urchin eggs. Dev. Biol., *121*: 576–579.

Hough-Evans, B. R. et al. 1990. Negative spatial regulation of the lineage specific CyIIIa actin gene in the sea urchin embryo. Development, *110*: 41–50.

Howard K., and P. Ingham. 1986. Regulatory interactions between the segmentation genes fushi tarazu, hairy and engrailed in the Drosophila blastoderm. Cell, *44*: 949–957.

Ingham, P., and P. Gergen. 1988. Interactions between the pair-rule genes *runt, hairy, even-skipped*, and *fushi-tarazu* and the establishment of the periodic pattern in the *Drosophila* embryo. Development, *104* (Suppl.): 51–60.

Ingham, P., D. Ish-Horowicz, and K. Howard. 1986. Correlative changes in homeotic and segmentation gene expression in *Krüppel* mutant embryos in *Drosophila*. EMBO J., *5*: 1659–1665.

Ingham, P. W. 1988. The molecular genetics of embryonic pattern formation in *Drosophila*. Nature (Lond.), *335*: 25–34.

Jäckle, H. et al. 1986. Cross regulatory interactions among the gap genes of *Drosophila*. Nature (Lond.), *324*: 668–670.

Kao, K. R., and R. P. Elinson. 1988. The entire mesodermal mantle behaves as Spemann's organizer in dorsoanterior enhanced *Xenopus laevis* embryos. Dev. Biol., *127*: 64–77.

Kornberg, T. et al. 1985. The engrailed locus of Drosophila: In situ localization of transcripts reveals compartment-specific expression. Cell, *40*: 45–53.

Lehmann, R., and C. Nüsslein-Volhard. 1986. Abdominal segmentation, pole cell formation, and embryonic polarity require the localized activity of oskar, a maternal gene in Drosophila. Cell, *47*: 141–152.

Levine, M. et al. 1983. Spatial distribution of *Antennapedia* transcripts during *Drosophila* development. EMBO J., *2*: 2037–2046.

Levine, M., G. M. Rubin, and R. Tijian. 1984. Human DNA sequences homologous to a protein-coding region conserved between homeotic genes of Drosophila. Cell, *38*: 667–673.

Lewis, E. B. 1978. A gene complex controlling segmentation in *Drosophila*. Nature (Lond.), *276*: 565–570.

Lewis, E. B. 1985. Regulation of the genes of the bithorax complex in *Drosophila*. Cold Spring Harbor Symp. Quant. Biol., *50*: 155–164.

Livant, D. et al. 1988. Titration of the activity of a fusion gene in intact sea urchin embryos. Proc. Natl. Acad. Sci. U.S.A., *85*: 7607–7611.

Livingston, B. T., and F. H. Wilt. 1989. Lithium evokes expression of vegetal-specific molecules in the animal blastomeres of sea urchin embryos. Proc. Natl. Acad. Sci. U.S.A., *86*: 3669–3673.

Lumsden, A., and R. Keynes. 1989. Segmental patterns of neuronal development in the chick hindbrain. Nature (Lond.), *337*: 424–428.

MacDonald, P. M., P. Ingham, and G. Struhl. 1986. Isolation, structure, and expression of even-skipped: A second pair-rule gene of Drosophila containing a homeobox. Cell, *47*: 721–734.

Martinez-Arias, A. 1986. The *Antennapedia* gene is required and expressed in parasegments 4 and 5 of the *Drosophila* embryo. EMBO J., *5*: 135–141.

Martinez-Arias, A., and P. A. Lawrence. 1985. Parasegments and compartments in the *Drosophila* embryo. Nature (Lond.), *313*: 639–642.

McGinnis, W. et al. 1984. A homologous protein-coding sequence in Drosophila homeotic genes and its conservation in other metazoans. Cell, *37*: 403–408.

Miller, J., A. D. McLachlan, and A. Klug. 1985. Repetitive zinc-binding domains in the protein transcription factor IIIA from *Xenopus* oocytes. EMBO J., *4*: 1609–1614.

Morgan, T. H. 1934. *Embryology and Genetics*. Columbia University Press, New York.

Nauber, U. et al. 1988. Abdominal segmentation of the *Drosophila* embryo requires a hormone-like protein encoded by the gap gene *knirps*. Nature (Lond.), *336*: 489–492.

Nüsslein-Volhard, C., and E. Wieschaus. 1980. Mutations affecting segment number and polarity in *Drosophila*. Nature (Lond.), *287*: 795–801.

Nüsslein-Volhard, C., H. G. Frohnhöfer, and R. Lehmann. 1987. Determination of anteroposterior polarity in *Drosophila*. Science, *238*: 1675–1681.

O'Farrell, P. H. et al. 1985. Embryonic pattern in *Drosophila*: The spatial distribution and sequence-specific DNA binding of *engrailed* protein. Cold Spring Harbor Symp. Quant. Biol., *50*: 235–242.

Oliver, G. et al. 1988. Differential antero-posterior expression of two proteins encoded by a homeobox gene in *Xenopus* and mouse embryos. EMBO J., *7*: 3199–3209.

Pherson, J. R., and L. H. Cohen. 1986. The fate of the small micromeres in sea urchin development. Dev. Biol., *113*: 522–526.

Postlethwait, J. H., and H. A. Schneiderman. 1971. Pattern formation and determination in the antennae of the homeotic mutant *Antennapedia* of *Drosophila melanogaster*. Dev. Biol., *25*: 606–640.

Püschel, A. W., R. Balling, and P. Gruss. 1990. Position-specific activity of the Hox1.1 promoter in transgenic mice. Development, *108*: 435–442.

Raff, R. A., and T. C. Kaufman. 1983. *Embryos, Genes, and Evolution*. Macmillan Publishing Co., New York.

Rosenberg, U. B. et al. 1986. Structural homology of the product of the *Drosophila Krüppel* gene with *Xenopus* transcription factor IIIA. Nature (Lond.), *319*: 336–339.

Rossant, J., and A. L. Joyner. 1989. Towards a molecular-genetic analysis of mammalian development. Trends Genet., *5*: 277–283.

Ruiz i Altaba, A. 1990. Neural expression of the *Xenopus* homeobox gene Xhox3: Evidence for a patterning neural signal that spreads through the ectoderm. Development, *108*: 595–604.

Ruiz i Altaba, A., and D. A. Melton. 1989a. Bimodal and graded expression of the *Xenopus* homeobox gene *Xhox3* during embryonic development. Development, *106*: 173–183.

Ruiz i Altaba, A., and D. A. Melton. 1989b. Involvement of the Xenopus homeobox gene Xhox3 in pattern formation along the anterior-posterior axis. Cell, *57*: 317–326.

Ruiz i Altaba, A., and D. A. Melton. 1989c. Interaction between peptide growth factors and homeobox genes in the establishment of anterior-posterior polarity in frog embryos. Nature (Lond.), *341*: 33–38.

Ruiz i Altaba, A., and D. A. Melton. 1990. Axial patterning and the establishment of polarity in the frog embryo. Trends Genet., *6*: 57–64.

Sànchez-Herrero, E. et al. 1985. Genetic organization of *Drosophila* bithorax complex. Nature (Lond.), *313*: 108–113.

Sassone-Corsi, P. et al. 1981. Promotion of specific *in vitro* transcription by excised "TATA" box sequences inserted in a foreign nucleotide environment. Nucleic Acids Res., *9*: 3941–3958.

Schneuwly, S., R. Klemenz, and W. J. Gehring. 1987a. Redesigning the body plan of *Drosophila* by ectopic expression of the homeotic gene *Antennapedia*. Nature (Lond.), *325*: 816–818.

Schneuwly, S., A. Kuroiwa, and W. J. Gehring. 1987b. Molecular analysis of the dominant homeotic *Antennapedia* phenotype. EMBO J., *6*: 201–206.

Scott, M. P., and A. J. Weiner. 1984. Structural relationships among genes that control development: Sequence homology between the *Antennapedia*, *Ultrabithorax*, and *fushi tarazu* loci of *Drosophila*. Proc. Natl. Acad. Sci. U.S.A., *81*: 4115–4119.

Scott, M. P. et al. 1983. The molecular organization of Antennapedia locus of Drosophila. Cell, *35*: 763–776.

Spradling, A. C., and G. M. Rubin. 1982. Transposition of cloned P elements in *Drosophila* germ line chromosomes. Science, *218*: 341–347.

Stanojevic, D., T. Hoey, and M. Levine. 1989. Sequence-specific binding activities of the gap proteins encoded by *hunchback* and *Krüppel* in *Drosophila*. Nature (Lond.), *341*: 331–335.

Struhl, G. 1981. A homeotic mutation transforming leg to antenna in *Drosophila*. Nature (Lond.), *292*: 635–638.

Struhl, G. et al. 1989. The gradient morphogen bicoid is a concentration dependent transcriptional activator. Cell, *57*: 1259–1273.

Tautz, D. 1988. Regulation of the *Drosophila* segmentation gene *hunchback* by two maternal morphogenetic centres. Nature (Lond.), *332*: 281–284.

Tautz, D. et al. 1987. Finger domain of novel structure encoded by *hunchback*, a second member of the gap gene class of *Drosophila* segmentation genes. Nature (Lond.), *327*: 383–389.

Treisman, J., and C. Desplan. 1989. The products of the *Drosophila* gap genes *hunchback* and *Krüppel* bind to the *hunchback* promoters. Nature (Lond.), *341*: 335–337.

Ursprung, H., and R. Nothiger. 1972. *The Biology of Imaginal Disks. Results and Problems in Cell Differentiation*, Vol. 5. Springer-Verlag, Heidelberg.

Villee, C. A. et al. 1989. *Biology*, 2nd ed. Saunders College Publishing, Philadelphia.

Wakimoto, B. T., and T. C. Kaufman. 1981. Analysis of larval segmentation in the lethal genotypes associated with the *Antennapedia* gene complex in *Drosophila melanogaster*. Dev. Biol., *81*: 51–64.

Wasylyk, B. et al. 1980. Specific *in vitro* transcription of conalbumin gene is drastically decreased by single-point mutation in T-A-T-A box homology sequence. Proc. Natl. Acad. Sci. U.S.A., *77*: 7024–7028.

Wilkinson, D. G. et al. 1989. Segmental expression of Hox–2 homeobox-containing genes in the developing mouse hindbrain. Nature (Lond.), *341*: 405–409.

Winslow, G. M. et al. 1989. Transcriptional activation by the Antennapedia and fushi tarazu proteins in cultured Drosophila cells. Cell, *57*: 1017–1030.

Wirz, J., L. I. Fessler, and W. J. Gehring. 1986. Localization of the *Antennapedia* protein in the *Drosophila* embryo and imaginal discs. EMBO J., *5*: 3327–3334.

Wright, C. V. E., P. Schnegelsberg, and E. M. De Robertis. 1988. XlHbox 8: A novel *Xenopus* homeo protein restricted to a narrow band of endoderm. Development, *104*: 787–794.

Yang, Q., L. M. Angerer, and R. C. Angerer. 1989. Structure and tissue-specific developmental expression of a sea urchin arylsulfatase gene. Dev. Biol., *135*: 53–65.

Organogenesis: Limb Development

Ross G. Harrison (1870–1959), shown here in his laboratory at Yale University, where he was a professor from 1907. Harrison is best known for the experiment in which he explanted a frog neural tube into clotted lymph and observed directly that axons extend owing to the activity of the growth cone. In so doing, he not only settled the controversy concerning how axons extend, but also invented the technique of tissue culture, which is extensively used in modern cell biology research. He also made fundamental observations on the establishment of axis polarity in the limb and developed the concept of the limb morphogenetic field. (Courtesy of Marine Biological Laboratory Library, Woods Hole, MA.)

The elaboration of functional organs and tissues from their rudiments is the crowning achievement of the developmental process. The formation of these functional entities involves the causally interrelated events of (1) morphogenesis, which is dependent upon individual cell movement, cell proliferation, and cell death, and (2) differentiation of the component cells, whereby they attain the ability to perform specialized functions. The chronological and spatial regulation of morphogenesis and cell differentiation in individual organs are, in turn, responsible for the orderly acquisition of adult form and function. The challenge in understanding the formation of any organ is to integrate multiple developmental processes that are occurring simultaneously at many levels of organization. To examine organogenesis, we have chosen three organ systems that display different strategies to produce functional organs. The systems selected for discussion in this chapter and the following two chapters are among those most amenable to study, either because they can be easily manipulated experimentally or because aspects of their development can be studied in culture. Although there are, of course, many other events in development, these three will serve to illustrate the complexity of development and the intricacy with which developmental mechanisms are orchestrated.

15–1 The Vertebrate Limb as a Developmental System

The vertebrate limb is an outgrowth of the embryonic body wall and is composed of a central core of loose mesenchyme derived from the somatic part of the lateral plate mesoderm and the somites, encased in an epithelial jacket derived from ectoderm. The predominant structural elements of the limb (cartilage, muscle, and connective tissue) are formed from the mesenchyme. Although all vertebrate limbs are composed of these tissues, their final forms are variable, and the special form of each limb is due primarily to a unique spatial pattern of cell differentiation. Another reason for studying the limb is that muscle, cartilage, and connective tissue also differentiate in other regions of the body. Hence, the implications of pattern formation in the limb go far beyond the formation of the limb itself.

15–2 The Chick Limb: Illustration of Principles for Limb Development

Much of our understanding of development of the vertebrate limb derives from extensive analyses of the chick limb. The attractiveness of the chick limb for study by developmental biologists is due partially to the ready accessibility of the limb bud, which juts out from the body axis, for experimental manipulation *in ovo*. In addition, the component tissues of the bud can be disassembled and readily reconstructed, and the limb can be grafted to a host embryo in a variety of orientations and positions. Finally, the organization of the mesodermal derivatives is clearly evident in the limb by the positions of the muscle, cartilage, and connective tissue. The cartilage, most of which is later replaced by bone, is particularly easy to monitor because it readily takes up the dye alcian blue in fixed preparations. Such a stained and cleared preparation is shown in Figure 15.3.

Early Outgrowth of the Limb

The first evidence of limb outgrowth is proliferation of somatic mesoderm cells along the length of the embryo. These cells gradually accumulate under the ectoderm to form a longitudinal ridge known as the **Wolffian ridge**. The ridge begins to enlarge in the pectoral and pelvic regions by day 3 (stage 17) to form the definitive limb buds (Fig. 15.1A). This enlargement is due to more rapid proliferation of cells in these regions than in the rest of the Wolffian ridge (Searls and Janners, 1971).

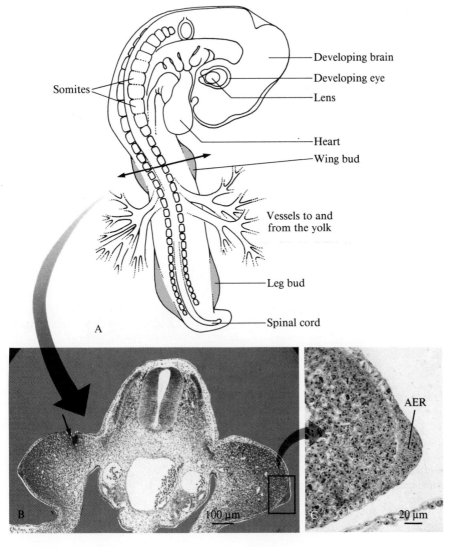

Somites

Developing brain
Developing eye
Lens
Heart
Wing bud
Vessels to and from the yolk
Leg bud
Spinal cord

A

B   100 μm

AER

C   20 μm

Figure 15.1

A, Diagram of a chick embryo at approximately 3 days of development illustrating the positions of the wing and leg buds. The arrow through the wing buds marks the plane of section shown in B. B, Section through a wing bud of a 3-day chick embryo, which reveals the proliferation of somatic mesoderm to form the bud and the thickening of the ectoderm at the limb margin, known as the AER. C, A higher magnification of the boxed area in B shows the pseudostratified columnar epithelium of the AER in more detail. (B, and C, From Saunders, 1948.)

The distal tip of the limb is covered with a ridge of thickened ectoderm that runs along the length of its anteroposterior margin (see Fig. 15.1B and C). This thickening of the ectoderm is known as the **apical ectodermal ridge (AER)**. The AER is found in most higher vertebrates, including the amniotes, but is not present in urodele amphibians. As we shall see, the AER is crucial for the continued outgrowth of the chick limb. The AER is a transient structure and disappears after the time the digits have formed on day 8.

The limb bud elongates by the continued proliferation of undifferentiated mesenchyme at the tip of the limb. This region of enhanced proliferation was first identified by morphological criteria (Saunders, 1948) to be approximately 250 μm thick and composed of undifferentiated cells, although its exact size remains controversial. During limb extension, however, the more proximal of these apical cells show diminished mitotic activity, and begin to condense in the core of the limb to form the cartilaginous elements that give rise to the skeleton (Fig. 15.2). The skeletal elements differentiate in a proximal-to-distal direction; the first cartilage to appear in the wing is the humerus, followed by the radius and ulna, and lastly the wrist elements and the digits (see Fig. 15.3; Saunders, 1948). Concomitantly, presumptive muscle cell precursors condense peripheral to the cartilage to form the muscle masses.

Limb Outgrowth Involves Reciprocal Interactions Between the Ectoderm and Mesoderm

The outgrowth of the limb involves continuous interactions between the AER and the subjacent mesoderm. The presence of the AER is essential for the continued proliferation of the underlying apical mesoderm. If the AER is removed, it does not regenerate, and the mesoderm beneath ceases to divide. As a result, limb outgrowth halts. Conversely, if a supernumerary AER is grafted to the side of a young limb, additional limb elements emerge under the influence of the second AER (Saunders and Gasseling, 1968; Saunders et al., 1976). When an AER is combined with limb mesoderm in tissue culture, the mesoderm is induced to proliferate (Fig. 15.4; Reiter and Solursh, 1982), whereas any limb ectoderm other than AER has no such effect on the mesoderm.

There exists some controversy about the role of the AER in limb outgrowth. It seems likely, however, that the AER emits a signal that

Figure 15.2

Stage 23 (~4 day) chick embryo limb bud, sectioned along the proximodistal axis, shows the zone of proliferating undifferentiated mesenchyme (U) at the distal tip and the more proximal regions where cartilage (C) and muscle cells (M) are beginning to differentiate. (From Summerbell and Lewis, 1975. Reprinted by permission of Company of Biologists, Ltd.)

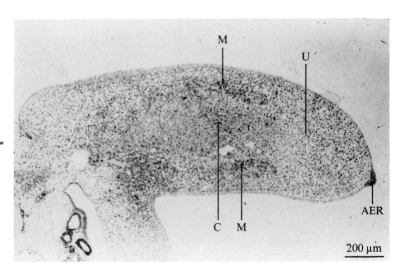

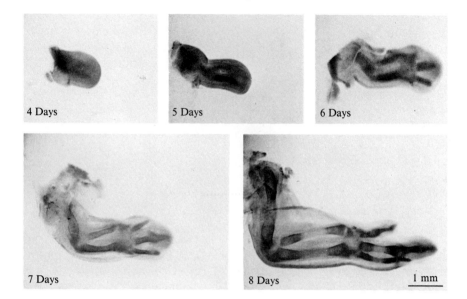

Figure 15.3

The progressive development of cartilage elements is revealed in these stained chick wing buds. In the earliest limb, only a humerus has been formed. As development proceeds, more distal elements progressively differentiate. (From Wolpert, 1978.)

stimulates cell division in the subjacent mesenchyme. The nature of the proliferative signal that is passed to the mesoderm is not known.

The mesoderm also plays a crucial role in limb outgrowth. The initial formation and continued existence of the AER are dependent upon the underlying mesoderm. If prospective limb mesoderm is grafted beneath flank ectoderm, an AER forms and a supernumerary limb develops (Saunders and Reuss, 1974). Similarly, if limb bud mesoderm is removed from an early

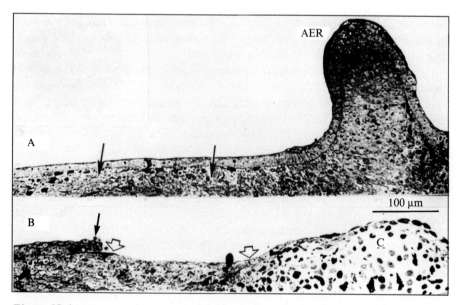

Figure 15.4

A, When chick limb ectoderm is combined with chick limb mesoderm in tissue culture, cells beneath the ectoderm (arrows) resemble limb mesenchyme. Furthermore, an outgrowth of mesenchyme forms beneath the AER. No cartilage differentiates from the mesoderm associated with the ectoderm. B, Cartilage (C) will differentiate from this same mesoderm, as long as there is no ectoderm within 200 μm (the distance between the two open arrowheads). The edge of the sheet of ectoderm is marked by an arrow. (From Solursh et al., 1981a.)

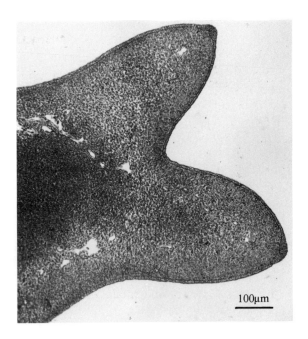

100μm

Figure 15.5

In this *eudiplodia* mutant chick embryo, an extra AER develops on the dorsal surface of the limb, resulting in a second limb axis that protrudes to the upper right. (From Goetinck, 1964.)

chick limb, leaving an ectodermal jacket that is then packed with nonlimb mesoderm, the AER regresses and the mesoderm ceases proliferating (Zwilling, 1961). The wing mesoderm acquires this inductive capacity by stage 11 (13-somite embryo, 40 to 45 hours) and loses it by stage 17 (31 somites; day 3); the leg mesoderm follows a similar pattern slightly later.

Thus, although the mesoderm is dependent upon the AER for continued proliferation, the AER is likewise dependent upon the limb mesoderm—not only for its initial formation but also for its maintenance. These findings suggest that the mesoderm produces a factor that is responsible for the presence and position of the AER. This putative substance has been named **apical ectoderm maintenance factor** (Zwilling and Hansborough, 1956), although no such molecule has been isolated.

A series of interesting chick mutations have demonstrated the importance of the mesodermal-ectodermal interaction in limb development. In one mutant, *wingless*, a normal wing fails to develop. In these embryos, a normal AER is seen on day 3 but soon regresses. If mutant limb mesoderm is combined with wild-type limb bud ectoderm, the AER similarly regresses, indicating that the mesoderm is unable to maintain the AER.

Conversely, another chick mutant, called *eudiplodia*, develops additional limb elements at its dorsal side. Histological examination reveals a dorsal extension of the AER (Fig. 15.5). If mutant mesoderm is combined with normal ectoderm, the AER will extend, resulting in extra digits. Thus, the mesoderm can determine the presence and shape of the AER.

The ability of the limb to develop is apparently dependent upon contact with the adjacent segmental plate, for, as Kieny (1971) demonstrated, prospective limb mesoderm taken from very young embryos (11-somite embryos; 33 to 38 hours) will only induce an AER and form a limb when grafted to the flank region if the associated segmental plate is also grafted.

15–3 Establishment of Axes in the Chick Limb

The fully formed limb involves the development of differences in three axes: the anteroposterior (A-P) axis (from thumb to little finger in humans; digit

II to IV in chick wing), the dorsoventral (D-V) axis (from back of the hand to palm in humans; top of the wing to bottom in chick) and the proximodistal (P-D) axis (from shoulder to hand in humans; girdle to wing tip in chick). These axes are diagrammed in Figure 15.6. It is not sufficient to understand what may control differentiation of different cell types within the limb (i.e., cartilage *versus* muscle); we must also determine what is responsible for the positioning of the various cell types along the three limb axes (i.e., bone in center and dermis at periphery).

The time at which the three limb axes are fixed has been determined by a variety of experimental perturbations. Harrison (1921) analyzed the timing of the establishment of the limb axes in a classical study using the salamander *Ambystoma* (Fig. 15.7). By removing the region of mesoderm that will give rise to the limb (the so-called **limb disk**) and transplanting it to the flank of a host salamander in a variety of orientations, he established that the A-P axis is determined first, followed by the D-V axis. The P-D axis is not specified until the limb is morphologically distinguishable.

For example, if the right forelimb disk from an early tailbud stage embryo is removed and rotated 180 degrees so that the A-P and the D-V axes are misaligned in relationship to the rest of the embryonic axis, the limb that grows out looks like a left arm; i.e., its D-V axis is normal, but its A-P axis is reversed compared with its position in the embryo. If a left forelimb disk is removed and grafted in place of a right forelimb, it can be oriented so that either the D-V axis or A-P axis is misaligned. When the D-V axis is reversed, a normal right limb develops. If the A-P axis is reversed, a left limb grows. These experiments show that, at these early stages of development, the A-P axis is already determined within the limb tissue, but the D-V axis can still be specified by the polarity of the surrounding tissue. At late tailbud stages, the same operations show that both the A-P

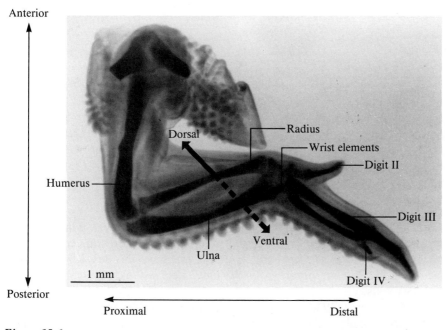

Figure 15.6

This stained and cleared 10-day chick wing reveals all the limb elements and their orientation along the anteroposterior and proximodistal axes. The dorsoventral axis is perpendicular to the plane of the page. (Adapted from Wolpert, 1978.)

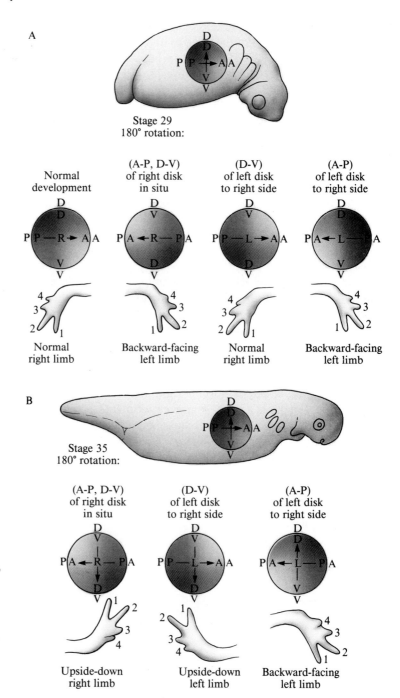

Figure 15.7

The timing of axial specification in *Ambystoma* was determined by grafting or rotating the limb disks as outlined in this diagram. When the experiments were performed on young embryos (A), only the A–P axis was specified. When they were done later (B), neither the A–P nor the D–V axis could be perturbed. (After Carlson, 1988.)

and D-V axes are now irreversibly determined and will develop independent of the surrounding embryonic tissue polarity. It is interesting to note that axial specification for some other organs (such as the developing ear) is acquired in the same order as for the limb; i.e., A-P axis first, followed by the D-V axis.

Hamburger (1938) extended these studies to the chick limb. He took prospective limb mesoderm from 16-somite (45 to 49 hours old) chick embryos and grafted it to the flank of a host embryo, where a supernumerary limb developed. If Hamburger rotated the graft so that its A-P axis was reversed compared with the host axis, the A-P axis of the resulting limb was also reversed. Likewise, reversing the D-V axis at the time of the graft produced a D-V reversed limb. Thus, even at a time when the limb is not yet visible, both of these two axes are fixed irreversibly in the presumptive limb mesoderm. The exception is the P-D axis, which is established as the limb begins to grow out. Axial specification is apparently the property of the mesoderm, not the ectoderm, because mesoderm denuded of its investing ectoderm and grafted to a host flank retains its original axis specification (Saunders and Reuss, 1974). We now examine what controls the regulation of axial organization in the chick limb.

Establishment of the Proximodistal Axis

Early experiments by John Saunders (1948), which were later confirmed by others (Summerbell, 1974; Rowe and Fallon, 1982), showed that the cartilaginous elements in the limb are laid down in the proximal-to-distal direction, because if the AER is removed at successively later stages (thereby stopping further limb outgrowth), progressively more distal limb elements will appear in the truncated limb (Fig. 15.8). For example, if the AER is removed from a wing bud at day 3, only a humerus will form in the stump. If the AER is amputated a day later, a radius and ulna (and perhaps some wrist elements) will also differentiate in the resulting wing.

The P-D polarity resides in the mesoderm and is not determined by the AER. If it were the AER itself that conferred pattern, one would expect that an older AER grafted onto a young limb bud would induce more distal elements. In a series of such experiments, Rubin and Saunders (1972) discovered that a young limb bud gives rise to the full complement of limb elements regardless of the age of the AER grafted onto it. Thus, the AER is important in maintaining mitotic activity in the apical tip mesoderm and in sustaining continued outgrowth but does not determine the positional properties of the mesoderm.

On the basis of Saunders' experimental results, Summerbell et al. (1973) formulated the following model for the establishment of the P-D axis in the chick limb. The fates of mesoderm cells within the limb are determined by how long they remain at the tip of the limb in a region called the **progress zone**. This is the area of proliferative activity at the limb tip first recognized by Saunders to be under the control of the AER. The thickness of the progress zone, which remains constant during limb outgrowth, is the distance

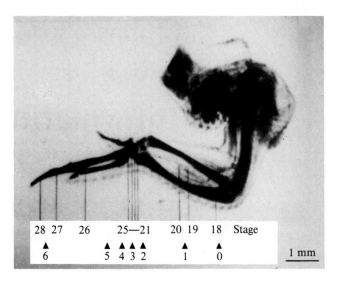

Figure 15.8

When the tip of a limb is severed, outgrowth ceases. The older the limb at amputation, the progressively more distal are the elements that develop from the stump. This figure summarizes what cartilage elements or portions of cartilage elements form after truncations at various stages. Numbers at the bottom indicate the number of cell cycles while in the progress zone. (From Summerbell and Lewis, 1975. Reprinted by permission of Company of Biologists, Ltd.)

over which the AER influences mitotic activity. They speculated that those cells that "escape" the progress zone first (i.e., those cells that are pushed out of the influence of the AER owing to the constant proliferation of cells in the progress zone) will become the most proximal limb elements. Cells that remain longer within the progress zone will become distal elements. Presumably, the developmental fates of mesoderm cells are acquired while they reside in the progress zone under the influence of the AER. Once a cell leaves the progress zone, its positional values (see Chap. 12 for a discussion of positional values) are fixed. Thus, a series of positional values is continuously generated in a proximal-to-distal direction as cells overflow the progress zone at its proximal boundary.

To test this hypothesis, Summerbell and Lewis (1975) removed the tip of a very young limb, leaving behind a stump containing material previously specified as humerus, and replaced it with a tip from an older four-day wing, which should have had only the more distal positional values (wrist and digits) in its progress zone (Fig. 15.9A). The result was a wing with the intermediate values (i.e., radius and ulna) missing. Alternatively, when the tip from a young limb (which should contain all or nearly all of the positional values in its progress zone) was grafted onto an old stump (which should have had most of the limb values except the most distal ones fixed), a

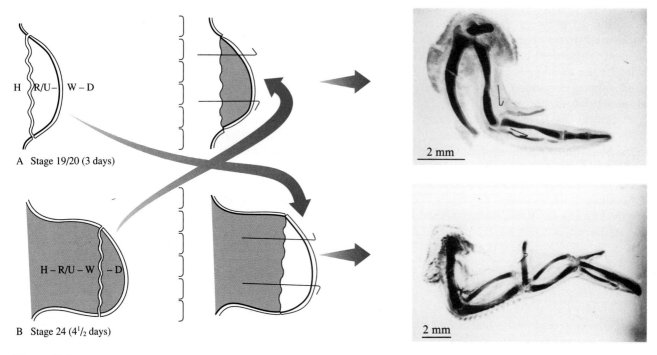

Figure 15.9

When the distal tips of "young" and "old" limb buds are exchanged, as diagrammed here, the resulting wings (seen at the right) show little or no regulation. When an old tip is grafted to a young stump (A), the radius and ulna are missing, whereas a young tip grafted to an old stump has duplicated humerus, radius, and ulna (B). These experiments support the hypothesis that positional values become fixed once cells are displaced from the growing tip. H: humerus; R/U: radius/ulna; W: wrist; D: digits. (After Summerbell et al., 1973. Reprinted by permission from *Nature* Vol. 244 pp. 492–496. Copyright © 1973 Macmillan Magazines Ltd.)

composite limb developed with all the intermediate values duplicated (see Fig. 15.9B). These experimental results have two important implications: (1) Once mesoderm cells have escaped the influence of the AER and left the progress zone, their fate is irreversibly fixed; and (2) the positional value of a cell is determined by the length of time it is under the control of the AER.

It is a matter of speculation how the progress zone might confer positional information on the mesoderm cells, but it has been suggested that the number of rounds of cell division that a cell undergoes in the progress zone determines whether it will be a proximal or distal element. The data of Summerbell et al. (1973) indicate that a cell that only divides once in the progress zone and then escapes the influence of the AER will have the most proximal value (i.e., humerus), whereas a cell that stays longer in the progress zone and undergoes more rounds of cell division will become a more distal element. Wolpert et al. (1979) have provided additional evidence to support this notion. Three-day chick embryos were irradiated with X-rays so that most of the dividing cells died. The progress zone from the irradiated embryo was then removed and grafted onto an unirradiated three-day chick limb stump. Those cells in the graft that lived resumed proliferation, but because so many cells had died during irradiation, it took several rounds of division before the progress zone had attained its normal width. At this point, the cells that began to escape the progress zone would have undergone several rounds of division more than normal and thus might be programmed to form more distal elements. In fact, the intermediate skeletal elements were missing in the resulting limb. Thus, the cells in the progress zone did, indeed, form more distal positional values, as predicted.

To date, the progress zone model remains controversial because at least some data suggest that positional values can be altered after cells have left the progress zone. Kieny and her co-workers (Kieny, 1964; Kieny and Pautou, 1976; Kieny, 1977) have performed experiments similar to those of Summerbell and Lewis, in which mesoderm has either been removed from a young wing bud or has had tissue added to it (see Fig. 15.10 for experimental design). The host and graft tissue were distinguished in some cases by using the chick-quail chimera technique developed by Le Douarin (see p. 265). When tissue that should form the radius and ulna was removed, at least portions of a new radius and ulna developed from the stump (i.e., from cells that should have formed a humerus). Conversely, when additional tissue was added to form a composite limb that should have contained a duplicated humerus and radius/ulna, a relatively normal wing developed in which only one large humerus formed, composed of host humerus and radius/ulna tissue and graft humerus cells (Fig. 15.10). These experiments show that positional values are not irreversibly fixed, because some regulation has occurred. In the former case, stump tissue was respecified to form more distal values, and in the latter case, stump tissue was respecified to form more proximal values (i.e., radius/ulna to humerus).

The disparity between the results from these two laboratories has not been completely explained, although the difference may be due, in part, to the ages at which the grafts were done. Summerbell (1977, 1981) has shown that regulation can occur if the deletion experiments are done on young embryos (less than three to four days), whereas positional values become fixed when the operations are performed on older limbs (older than three to four days) (Fig. 15.11).

Although the progress zone model remains controversial, it has served the worthy purpose of stimulating interesting research that may ultimately elucidate pattern specification in the limbs of higher vertebrates.

Stage 18–22 ($3–3^{1/2}$ days)

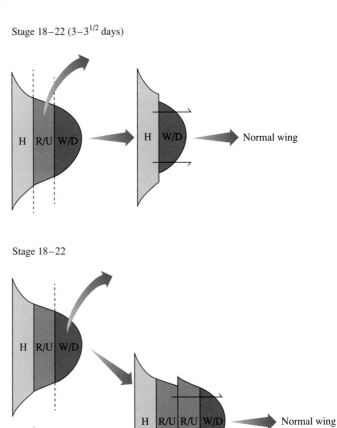

Figure 15.10

When Kieny and her colleagues constructed limbs in which the intermediate values were either deleted or duplicated, as outlined in this figure, limbs that displayed some or complete regulation were produced. These results are not consistent with the progress zone model, in which values are fixed once cells leave the progress zone. (Figure based on data in Kieny, 1977.)

The Anteroposterior Axis of the Limb: A Simple Gradient of Retinoic Acid

As mentioned in Chapter 12, a number of patterns are believed to be generated during development by gradients of morphogens. Evidence has accumulated that the A-P axis of the chick limb is determined by such a gradient. The first observations relating to the control of the A-P axial pattern of the limb were those of Saunders and Gasseling (1968). They noted that if they excised a small piece of tissue from the most posterior aspect of the chick limb that is at the junction with the body wall and grafted it to the anterior margin of another limb, a duplication of digits occurred in the limb receiving the graft. Furthermore, these duplicated digits were arranged in mirror symmetry with the host digits (Fig. 15.12). This region of the chick limb, whose position determines the A-P axis of the supernumerary digits, was named the **zone of polarizing activity (ZPA)**.

Wolpert and members of his research group (Tickle et al., 1975) speculated that a morphogen might diffuse from the ZPA across the limb to form a simple gradient that would have a sink (i.e., a low point in the gradient) at the anterior end of the limb. The particular digit that would

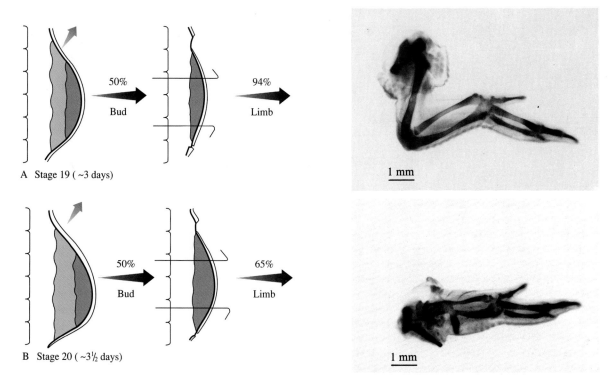

A   Stage 19 (~3 days)

B   Stage 20 (~3½ days)

Figure 15.11

The disparity in results between Kieny's and Wolpert's groups may be explained by the ages at which the experiments are done. If the middle portions of the limb are deleted from stage 19 (early day 3; A) limbs, a normal limb results, showing considerable regulation. If similar deletions are produced on a stage 20 (later day 3; B) limb, less regulation is observed, and the limb is missing its intermediate values (radius and ulna). (From Summerbell, 1981. Reprinted by permission of Company of Biologists, Ltd.)

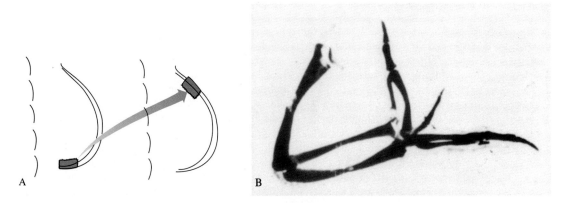

Figure 15.12

Grafting of a second ZPA produces a duplication of digits. A, A ZPA is removed from one chick embryo and grafted into the anterior margin of the right wing bud of a host embryo, as shown. B, The resulting limb shows the duplication of digits in a symmetrical array. The supernumerary parts are of left-handed symmetry. Digit 2, which lies in the mirror plane, is shared by the two sets of digits. (From Saunders and Gasseling, 1968. Copyright © 1968, the Williams & Wilkins Co., Baltimore.)

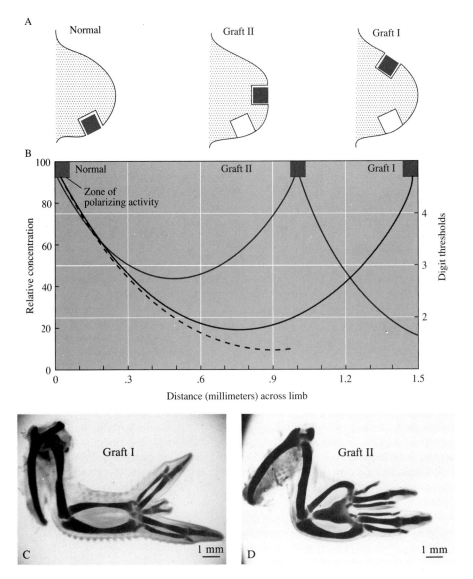

Figure 15.13

In the normal limb (A), the zone of polarizing activity (shaded box) is presumed to be the source of morphogen, which would theoretically diffuse across the limb to form a concentration curve, as seen in B, (dotted line). As the concentration of morphogen falls, the threshold for the next most anterior digit is reached. If a second ZPA is grafted into a limb near its anterior margin (A, graft I), the morphogen concentrations would be as predicted in B, and mirror-image duplications should result (specifically digits 432234 in anteroposterior order). Alternatively, if a second ZPA is grafted near the tip of the limb (A, graft II), the concentration of morphogen would be quite different (B), and one would predict a different distribution of digits (e.g., 234434). The experimental results were close to the predicted outcome (C, graft I; D, graft II). (A, and B, After Wolpert, 1978. C, and D, From Tickle et al., 1975. Reprinted by permission from *Nature* Vol. 254 pp. 199–202. Copyright © 1975 Macmillan Magazines Ltd.)

differentiate would depend upon the concentration of morphogen found at a particular position, with digit 4 forming at the highest level of the gradient, digit 2 at the lowest end, and digit 3 in the middle (Fig. 15.13B).

Tickle et al. (1975) grafted an additional ZPA into chick limbs in various positions and found that the patterns of duplication were compatible with a diffusing morphogen model. For example, if a ZPA is grafted at the anterior limb margin adjacent to the body wall, one might expect a morphogen gradient as seen for graft I in Figure 15.13A and B. Such a morphogen profile would result in the mirror-image duplication of all three digits. These were theoretical calculations on the part of Tickle and her colleagues. Although they had no idea what the absolute levels of the morphogen would be to form a specific digit, the pattern of digits they observed conforms to this model (see Fig. 15.13C).

However, if the additional ZPA is grafted near the apex of the limb, bringing it closer to the host ZPA, the concentration profile of the morphogen would be similar to that shown for graft II in Figure 15.13A and B. Now we would expect that the concentration of morphogen between the two ZPAs would never fall to the threshold needed to form digit 2, such that digit 3 would form between the two ZPAs. In addition, there is also limb mesoderm anterior to the grafted ZPA, such that a gradient can diffuse away in the anterior direction as well, thus producing digits 4, 3, and 2, in that order. This is exactly what was found experimentally (see Fig. 15.13D).

Further evidence for a diffusible morphogen came from studies by Summerbell (1979), who showed that if a barrier were inserted into the limb so as to divide it into anterior and posterior halves, digits formed at the posterior side in their normal order, but no digits developed on the anterior side. This study suggested that the putative morphogen originates on the posterior side of the limb.

The first clues as to the nature of the morphogen came from a study by Maden (1982), who found that if amputated salamander limbs were bathed in retinoic acid (a derivative of vitamin A), the regenerated limbs had duplicated limb parts (e.g., two sets of radius and ulna growing end-to-end). With this in mind, Tickle et al. (1982) and Summerbell (1983) showed that retinoic acid could bring about the same duplications of digits as could be produced by an additional ZPA. Insertion of a piece of adsorbent paper that had been impregnated with retinoic acid into the anterior ectoderm of the limb produced duplicated digits in mirror symmetry across the A-P axis (Fig. 15.14). At the very least, such experiments suggested that retinoic acid could mimic a natural morphogen.

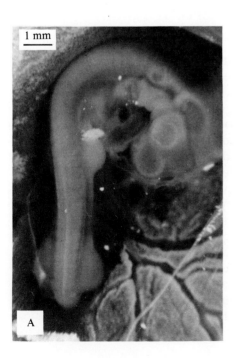

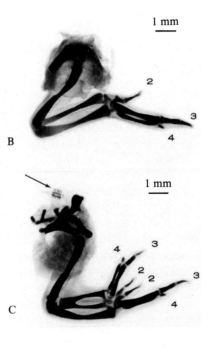

Figure 15.14

Retinoic acid can mimic the effects of an additional ZPA. A, A piece of newsprint is soaked in retinoic acid and inserted into the anterior margin of a chick limb, as shown here. B, A chick wing showing the normal distribution of digits. C, A chick limb with duplicated digits in mirror symmetry after receiving a graft of retinoic acid-soaked paper. The paper can still be seen in this preparation (arrow). (From Summerbell, 1983. Reprinted by permission of Company of Biologists, Ltd.)

Subsequently, Thaller and Eichele (1987) have demonstrated that retinoic acid is indeed present in the chick limb, and, significantly, it is also present in a gradient across the limb, with its high end at the posterior margin. This observation required a heroic effort on their part because they dissected, pooled, and analyzed over 5000 chick limb buds for this study. The identification of a gradient of retinoic acid in the chick limb lends credence to the possibility that it is responsible for specification of the A-P axis. Retinoic acid is a relatively large molecule (which would theoretically limit its diffusion), but it is lipophilic and thus probably diffuses in the extracellular spaces and is taken up through cell plasma membranes, rather than being passed from cell to cell through junctional complexes.

If retinoic acid is involved in patterning in the A-P axis of the limb, it is likely that it exerts its effects by regulating gene expression in cells of the limb. The mechanism of regulation of gene expression by retinoic acid is unclear (for review, see Summerbell and Maden, 1990). However, a receptor that binds to retinoic acid, forming a complex that binds to DNA, has been isolated from human cells (Petkovich et al., 1987). It is reasonable to assume that the retinoic acid-receptor complex interacts with specific promoters or enhancers and thus regulates gene expression (see Oliver et al., 1988).

There is some evidence that confounds the preceding results, however. For example, some mesoderm from embryonic regions other than the ZPA, such as lateral plate mesoderm, can also cause digit duplication. In addition, grafting a second ZPA into or adjacent to the host ZPA (thus effectively doubling the number of ZPA cells and, presumably, the morphogen concentration) does not have an effect on digit pattern (Iten et al., 1981; Javois and Iten, 1981). Finally, if anterior limb tissue is grafted to the posterior margin of the limb, supernumerary digits also develop (Iten and Murphy, 1980). An alternative explanation, which will be considered in more detail in the section on regeneration (Sect. 15–5), is that grafting of the ZPA to other regions of the limb places nonadjacent positional values next to each other, resulting in the intercalation of new values.

Thus, although current evidence strongly suggests that a gradient of retinoic acid specifies the A-P axis, sufficient conflicting data exist that must be explained before this hypothesis can be accepted at face value.

Determination of the Dorsoventral Axis

The D-V axis has received relatively little experimental attention compared with the other two axes. From the little work that has been done, it appears that this axis is specified by cells derived from both the mesoderm and the ectoderm, although at different stages in development.

When fragments of limb bud mesoderm or dissociated and centrifugally compacted limb bud cells are repacked into the ectodermal hull of a three- to four-day wing bud and then grafted to the flank of a host embryo, the skeleton and musculature of the distal elements have a D-V axis conforming to that of the ectoderm (MacCabe et al., 1973). Similarly, if intact mesodermal cores of three- to four-day chick embryo leg buds are recombined with the ectodermal hulls so that the D-V axis of the ectoderm is reversed, the musculature and skeleton will also be reversed along the D-V axis (MacCabe et al., 1974). The ectoderm can specify D-V axis as early as the 50- to 53-hour embryo, before the AER has even appeared. Before this stage, however, combinations of ectoderm and mesoderm produce a D-V axis according to the polarity of the mesoderm (Geduspan and MacCabe, 1987). As yet, there are no useful models for interpreting the specification of the D-V axis.

15–4 Establishment of Tissue Types Within the Limb

We have discussed models to account for the establishment of the major axes of the limb. Do these same models also account for the differentiation of specific cell types within the limb? For example, when a mesoderm cell leaves the progress zone with a particular positional value, does it also know whether it is to be a cartilage cell, a muscle cell, or a connective tissue cell as well? Current data suggest that cells differentiate according to cues in the limb bud that are different from those that may determine positional values.

Initially, as the limb bud grows out, it is composed of a morphologically homogeneous mesenchyme. By stage 26 (five days), distinct morphological regions develop within the limb (see Figs. 15.2 and 15.15). The central core becomes condensed and is bordered by a rich vascular bed. External to the vascular bed is a blood vessel-free zone that is approximately 150 μm wide. The core cells ultimately differentiate into cartilage, the nonvascularized zone beneath the ectoderm becomes connective tissue and the dermis of the skin, and the muscle masses develop around the vascular zone.

Determination of Cell Types Within the Limb

Several *in vitro* experiments confirm that the limb bud cells are pluripotent after they are laid down in the limb bud. Solursh et al. (1981b) removed cells from the limb core and placed these in culture where they all differentiated into cartilage cells. When they took cells from the vascular-free region directly beneath the ectoderm and cultured these cells, they also developed primarily into chondrocytes, even though they would have never done so in the embryo. How do we account for these results?

It seems probable that the limb bud ectoderm exercises considerable *negative* control over the cytodifferentiation of limb bud elements. If prospective cartilage cells are co-cultured with limb ectoderm, cartilage seldom (if ever) differentiates (see Fig. 15.4; Solursh et al., 1981a). The mesoderm behaves as if the ectoderm were secreting a diffusible factor that inhibits differentiation into chondrocytes because its ability to repress cartilage differentiation can be transmitted across a porous filter. Interestingly, this factor cannot diffuse across a filter 150 μm or greater in thickness, which is the precise dimension of the avascular sleeve in which chondrogenesis is absent in the embryonic limb bud.

The diffusible chondrostatic factor is thought to affect cell differentiation by virtue of its initial effect on cell shape (Zanetti and Solursh, 1984). In culture, cartilage differentiation is accelerated by making the cells more rounded. If the cells are prevented from rounding up (e.g., when they are plated on an adhesive substratum such as fibronectin), cartilage cells will not differentiate. A current hypothesis is that some extracellular matrix component produced by the ectoderm may cause cell flattening in the 150 μm subectodermal avascular zone and so repress cartilage development (Solursh, 1984).

It is thought that the vascular bed that forms near the core of the limb bud develops when it is not under the influence of the ectoderm because implanted ectoderm produces an avascular zone around itself. This may be due to the accumulation of hyaluronic acid, which is in high concentration in the avascular zone (Feinberg and Beebe, 1983) and, when implanted, stimulates an avascular region to form.

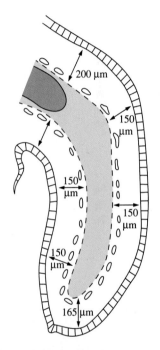

Figure 15.15

Diagram of a representative longitudinal section of a 4.5-day chick wing bud showing the distribution of cell types. The chondrogenic core is delineated by the dashed lines and is surrounded by a thick avascular sleeve that extends to the ectoderm. The vascular net and muscle masses form at the border of the chondrogenic core. The proximal cells that will form the humerus are shown in dark color, whereas the lighter color represents chondrocyte precursors that have not yet condensed. (From Solursh, 1984.)

The myogenic regions that develop in the vascular bed are derived from another tissue source altogether: the dermamyotomes of the somites. The myogenic precursors migrate into the limb after stage 14 (22 somites; 50 to 53 hours). This migration can be followed if a quail somite is grafted into a chick host and the chimeric embryo is allowed to develop (Fig. 15.16; Christ et al., 1977; Chevallier et al., 1977). The cells from the two sources can be distinguished by their different nuclear staining patterns using the Feulgen procedure. As soon as the myogenic precursors enter the limb, they are colocalized with the vascular bed. It has been speculated that the vascular bed guides the migration of the myogenic precursors, either by acting as a substratum for their migration or by producing a chemotactic factor (Solursh et al., 1987). The molecular biology of myogenesis is a fascinating story that will be discussed in more detail in Chapter 18.

Later, the muscle cells segregate into two major masses in the dorsal and ventral limb. The exact patterning of the limb musculature is dependent upon the cartilage model, which apparently establishes a pattern to be followed by the muscle cells. This conclusion is based on the observation that if somites that would normally give rise to wing muscles are grafted into the region opposite the future leg bud, these grafted cells will populate the leg and develop into normal leg musculature. Thus, muscle cells do not "know" what pattern they will adopt until after they arrive in the limb and receive local information (Chevallier et al., 1977).

15–5 Regeneration of the Urodele Limb

Some amphibians, in contrast to reptiles, birds, and mammals, have the ability to replace a severed limb. Although anurans can replace a limb only during the tadpole stage, urodeles can regenerate an amputated limb during either the larval or adult phase. The remarkable regenerative capacity of

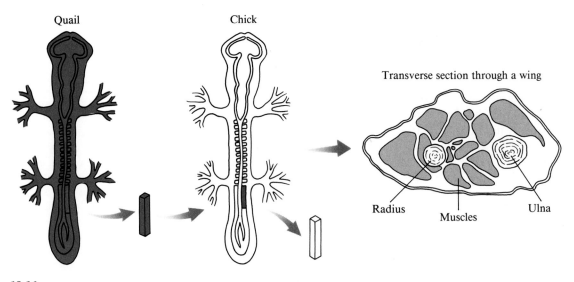

Figure 15.16

When the paraxial mesoderm in a chick embryo adjacent to the forelimb bud is replaced with quail mesoderm, the quail cells are only found in the muscle tissue of the wing, and none are found in the adjacent connective tissue. (After Walbot and Holder, 1987.)

the urodele limb makes it an excellent system for the study of regeneration. Furthermore, the principles that regulate patterning of the regenerating limb may be similar to those involved in patterning during initial limb development.

Regeneration has fascinated scientists for several generations. This fascination has produced a voluminous literature on the subject, dating back to the eighteenth century. Much of the philosophical framework for recent studies of regeneration was formulated by T. H. Morgan (1901). The major problems of regeneration identified by Morgan remain under active investigation today. These include the origins and developmental potentials of the cells that produce the regenerate, the roles of the adjacent tissues in determining the structure of the regenerated limb, and the reasons for the great variation in regenerative capabilities among animals.

Morgan recognized two primary morphogenetic processes for reconstitution of missing parts. One is **morphallaxis,** which involves reorganization of the remaining portions of the body to produce the missing structures. The other is **epimorphosis,** which occurs by proliferation of new tissue from the wound surface and progressive replacement of the missing parts. The well-known ability of the coelenterate *Hydra* to reconstitute itself out of a piece cut from a whole animal (see Chap. 12) is an example of morphallaxis. Limb regeneration is the most exhaustively studied example of epimorphosis, but this process actually may involve both mechanisms.

The favorite subject of investigators studying limb regeneration is the adult newt. Regeneration of newt limbs is illustrated in Figures 15.17 and 15.18. After the limb is amputated, repair of the wound begins with the spreading of epidermis from the edges of the wound to cover the open surface (see Fig. 15.18A and B). Once wound closure is accomplished, the epidermal cells proliferate to produce a multilayered mass of cells that forms

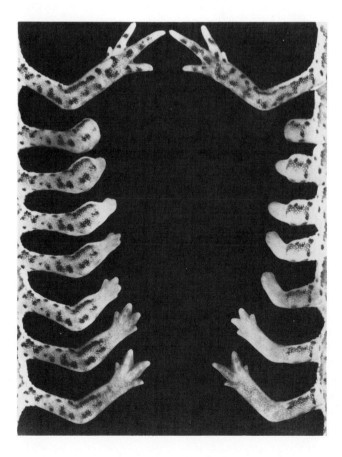

Figure 15.17

Adult salamander forelimbs (top) were amputated either at the level of the forelimb (left) or the upper arm (right). In both cases, a normal limb regenerated with the appropriate distal limb elements. (From Goss, 1969.)

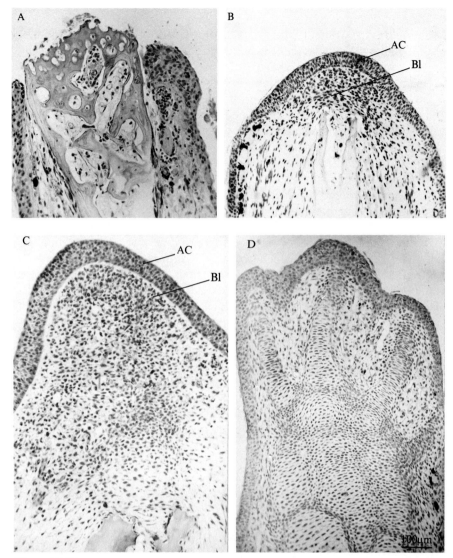

Figure 15.18

Stages in the regeneration of an adult salamander limb are demonstrated in these micrographs. A, Immediately after amputation, showing exposed bone. B, Sixteen days after amputation. The ectoderm has thickened into an apical epidermal cap (AC), and the blastema cells (B1) are accumulating beneath it. C, After 21 days, the blastema has grown quite large, and there is evidence of differentiation of cartilage within it. D, Skeletal rudiments are well developed in this 28-day regenerate. (From Saunders, 1982. Reprinted with permission of Macmillan Publishing Company from *Developmental Biology* by J. W. Saunders. Copyright © 1982 by Macmillan Publishing Co.)

a conical bulge at the tip of the limb. This structure is the **apical epidermal cap** (see Fig. 15.18B and C).

The cells beneath the wound epidermis subsequently lose their differentiated characteristics and detach from each other and the extracellular matrix to form a loose mesenchyme. Thus, former cartilage, bone, and connective tissue cells all appear identical to one another even when examined with the electron microscope (Hay, 1966). This process is called **dedifferentiation.**

The dedifferentiated cells form a mound beneath the apical cap called the **regeneration blastema.** The blastema proliferates rapidly and generates the new limb structures (see Fig. 15.18C). The first tissue to differentiate in the blastema is cartilage (see Fig. 15.18D). It appears initially at the ends of the persisting bone, which is completed by progressive addition to its distal end. Next, the more distal skeletal elements are added, just as they are in normal embryonic limb outgrowth. When the cartilaginous reconstruction is completed, the regenerated skeleton is transformed into bone.

Muscle is formed by both *de novo* appearance of muscles around the cartilage and terminal addition to persisting muscles. Blood vessels are not obvious in very early stages of reconstruction, but they later extend into the blastema from the stump and in the final regenerate reproduce the original pattern of vascularization.

Many nerve fibers are cut during amputation. Very soon after amputation, their axons grow into the wound and reconstruct the original nerve pattern. As we shall see later, the nerves play a significant role in controlling regeneration of the limb.

The Source and Role of Blastemal Cells

We have stated that blastemal cells arise from local dedifferentiation of stump tissue. An alternate source of blastemal cells could be reserve cells that are mobilized from elsewhere as a consequence of amputation. The origin of blastemal cells has been determined by use of X-rays. Irradiation with X-rays prevents regeneration, possibly because mitosis is impaired in irradiated tissue (Wertz and Donaldson, 1980). If a narrow portion of a limb is irradiated and amputated through that irradiated region, the limb is incapable of regeneration (Fig. 15.19B). However, if the same limb is severed through unirradiated portions, normal limb regeneration occurs (see Fig. 15.19A and C). Clearly, the cells necessary for regeneration must be supplied by, and proliferate from, a narrow region proximal to the level of amputation.

Given that the local cells support regeneration, which of these cells differentiates into the various limb tissues? The alternatives are listed in

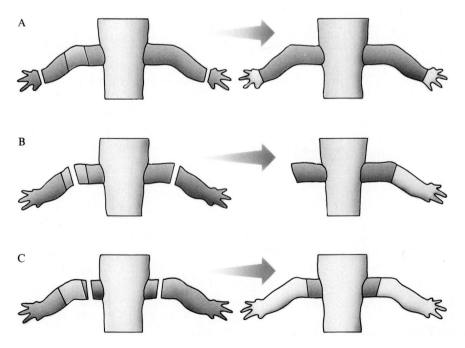

Figure 15.19

The cells that form the blastema are derived from a narrow region at the amputation site. This was demonstrated by irradiating a newt in a narrow region of the elbow (colorized region) so that cells in this region can no longer proliferate. If the limb is then amputated either distal (A) or proximal (C) to the irradiation site, a normal limb regenerates. If the limb is cut through the site of irradiation (B), no limb regenerates, showing that the mitotic cells of the blastema are derived from a local area at the amputation site. (After Saunders, 1982, based on Butler and O'Brien, 1942.)

Table 15–1. Do differentiated cells (e.g., muscle) lose their specialized properties, proliferate, and subsequently redifferentiate solely according to their previous differentiated state? Or do differentiated cells truly dedifferentiate to pluripotent stem cells that are capable of forming a variety of different cells? Finally, are there local populations of reserve cells that retain the embryonic property of pluripotency and can form the various differentiated cells of the limb? These alternatives are very difficult to test experimentally because the tissues of the limb are not pure populations of a single cell type. In addition to specialized cells, most tissues contain connective tissue cells, some of which may be pluripotent.

The standard experimental approach for examining the differentiation potential of tissues is to irradiate a host diploid animal, which limits the ability of the host cells to participate in regeneration, and then implant tissue from a triploid donor. The limb is then amputated so that it will regenerate, and a histological analysis is conducted on the regenerate. The presence of triploid nuclei in a tissue indicates that it is derived from the donor tissue and has participated in the regeneration of new tissues. In some experiments, the donor triploid cells are also labeled with ^{3}H-thymidine. Donor cells are detected in the regenerate by the presence of their triploid, radioactive nuclei.

When donor tissue is cartilage that has been cleaned of muscle and connective tissue, cells of the donor type are found in regenerated cartilage, perichondrium (the connective tissue layer that surrounds cartilage), connective tissue of joints, and fibroblasts. However, no donor cells are found in muscle or epidermis. On the other hand, when donor tissue is muscle, donor cells are found in all regenerated mesodermal derivatives (Steen, 1968; Namenwirth, 1974). Hence, the differentiation potential of cells in muscle tissue appears to be greater than that of cells in cartilage tissue. However, the presence of some perimuscular fibroblasts cannot be rigorously excluded as the source for some tissues.

Epidermis is not capable of producing mesodermal tissues, although skin dermis has the ability to produce several mesodermal tissues (Hay and Fischman, 1961; Namenwirth, 1974; Dunis and Namenwirth, 1977). Transplant experiments using triploid/diploid cell marking in the axolotl *Ambystoma mexicanum* show that 43% of the blastemal cells, in fact, originate

Table 15–1 Theoretical Possibilities for the Origin of Limb Regenerate Cells

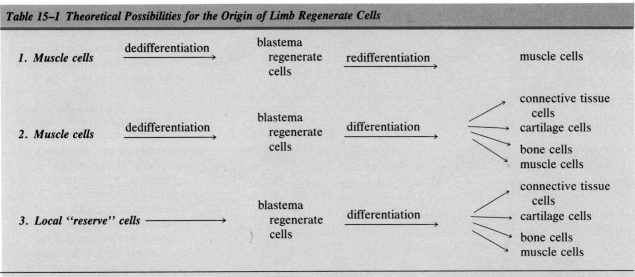

From Singer, M. 1973. Limb regeneration in the vertebrates. *Addison-Wesley Module in Biology*, no. 6. Addison-Wesley, Reading, MA, pp. 6-11.

from dermal fibroblasts, so that the dermis produces a disproportionate number of blastemal cells compared to the availability of different cell types at the plane of amputation (Muneoka et al., 1986). The dermis and muscle tissues both contain considerable numbers of fibroblasts. We must, therefore, consider the likelihood that many of the mesodermal derivatives of the regenerate are produced by fibroblastic cells.

Regulation of Regeneration

We have examined the cellular contributions that adult tissues make to the regenerating limb and now consider some of the factors that may regulate formation of the regenerate. These include (1) nerves of the stump and (2) the epidermis.

Role of Neurons

Neurons invade the regeneration blastema very soon after amputation and reconstitute the normal neural pattern. If the stump is denervated, however, regeneration is interrupted. Likewise, if nerve fibers do not regrow to the amputation site, the tissues degenerate. On the other hand, if axons are permitted to regrow into the stump, regeneration may be reinitiated, particularly if the amputation wound is once again disturbed. These results suggest that nerves promote regeneration. This interpretation is supported by observations that experimental deviation of a limb nerve to a skin wound at the base of the limb will promote formation of a supernumerary limb. The neural influence is thought to be due to a **trophic factor** released by neural tissue (for review, see Singer, 1978).

The neurotrophic effect is produced by all nerves (motor, sensory, or central), regardless of whether they have functional association with the central nervous system. There is a threshold number of nerve fibers that is necessary for regeneration. Below this threshold, regeneration does not occur (Singer, 1954). Above the threshold, regeneration occurs but is not affected by further increases in the amount of neural tissue. Nerves are thought to exert their effects on limb regeneration by promoting DNA synthesis and mitosis of mesodermal cells, which form the blastema (Mescher and Tassava, 1975; Loyd and Tassava, 1980). One possible nerve trophic factor is **glial growth factor,** which is produced by neural cells, is present in the blastema, and can stimulate mitotic activity in a denervated blastema (Brockes and Kinter, 1986). Once again, growth factors are implicated in a key role in a developmental process!

The neurotrophic effect is transitory, as demonstrated by sustained differentiation and morphogenesis of urodele limbs after denervation at stages following the onset of redifferentiation of the blastema. Growth of the denervated regenerates is retarded, however (Singer and Craven, 1948; Grim and Carlson, 1979). Apparently, nerves are necessary for the initiation of regeneration, but the redifferentiation and morphogenesis of the limb are independent of the nerves.

Vertebrate species that are incapable of limb regeneration have fewer nerve fibers per unit area of amputation wound than does *Triturus*. It is possible that the failure of these species to regenerate a limb is at least partially due to an inadequate nerve supply because some growth has been induced in the limb of the lizard, the frog, and the opossum by supplementing the nervous supply at the amputation site (Singer, 1974).

Role of Epidermis

Soon after amputation, skin epidermis rapidly covers the wound and proliferates to form the multilayered apical epidermal cap. Apical caps transplanted to the base of the blastema induce supernumerary limb regeneration. Thus, the apical epidermal cap may be necessary to permit regeneration. The specificity of the apical cap in this process is indicated by the fact that other epidermal grafts are without effect (Thornton, 1968).

One means for demonstrating the necessity for epidermis in promotion of regeneration is insertion of severed, amputated limbs into the coelomic cavity. In the coelom, a wound epidermis is not formed and regeneration fails to occur (Goss, 1956a). If, however, epidermal wound healing is allowed to occur before the stump is severed and inserted into the coelom, limb regeneration occurs (Goss, 1956b). Loyd and Tassava (1980) took advantage of this experimental procedure to compare the incorporation of ³H-thymidine and the level of mitosis in regenerating (epidermis intact) and nonregenerating (epidermis absent) stumps. They demonstrated that DNA synthesis and the level of mitosis are substantially higher when epidermis is present. These results suggest that, like the nerves, the wound epidermis stimulates cell division in the underlying tissue, thus promoting formation of the blastema and the subsequent regeneration of the limb.

The analogy between the apical epidermal cap of the regenerate and the apical ectodermal ridge of the chick embryonic limb bud is obvious. They both may play similar roles in permitting outgrowth of the limb by stimulating cellular proliferation, which provides the cells that differentiate proximally to produce definitive limb tissues.

Restoration of the Pattern of Cell Differentiation

One of the most intriguing aspects of regeneration is the fact that it reproduces the pattern of cell differentiation that is established during embryonic limb development. For example, if a limb is amputated, leaving behind a stump containing only a humerus, the regenerated structures will include a normal radius, ulna, wrist elements, and digits. The regenerated tissue contains neither duplicated nor missing values, and the limb elements are laid down in correct order. Furthermore, the limb always regenerates in the distal direction, regardless of the polarity of the stump (Fig. 15.20). This is known as the **rule of distal transformation.**

There is a second intriguing phenomenon that characterizes limb regeneration. A regenerating limb can also replace missing limb elements within the pattern, in addition to just replacing parts at a free cut surface. For example, if a blastema that has formed from the amputation of a hand is then grafted onto a stump that contains only a humerus, the resulting limb contains all the missing intermediate values (Fig. 15.21). This is a remarkable example of regulation and is termed **intercalary regeneration.** The regenerated intercalary elements are found to be of stump origin when labeled cells are used (Pescitelli and Stocum, 1980).

French et al. (1976) have developed a model, the **polar coordinate model**, to explain how the pattern is re-established during amphibian limb regeneration as well as in other regenerating systems. The model states that cells of the limb are thought to possess information about their positions. The positional information is presumed to be specified along polar coordinates. In such a system, a cell's position would be determined by two values: a circumferential and a radial component (Fig. 15.22A). The circumferential values (1 to 12) are arranged around a circle, similar to a clock face. These values will specify the position in the A-P and D-V axes. The radial values (A to E) represent the position along the P-D axis, with the outermost circle (A) representing the most proximal regions and the innermost circle (E) the distal parts. Note that this two-dimensional configuration of concentric circles can be expanded out to form a cone, which is

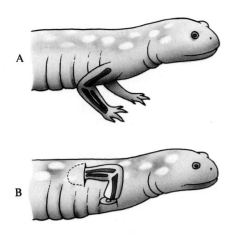

Figure 15.20

Urodele limbs can regenerate only in a distal direction even if the proximodistal polarity of the stump is reversed. In this case, an *Ambystoma* limb tip was amputated, and the limb was rotated and inserted into the flank skin. After the limb healed in place, it was severed at the shoulder, leaving two wounds (B). A normal leg regenerated from the right shoulder stump. The leg with reversed polarity regenerated a portion of the humerus, the forearm, and hand, thus only adding distal elements (C). (After Saunders, 1982.)

a more visual representation of a limb (see Fig. 15.22B). The tip of such a cone represents the digits, and its base represents the shoulder. It is currently believed that the dermal cells possess the positional information and are responsible for the regeneration of correct pattern.

There are several rules that have been proposed to govern the process of regeneration. The first is that if cells of disparate positional values, either circumferential or radial, are brought together either by grafting or during wound healing, then cell division is initiated, and this cell growth is necessary to generate the missing positional values. This is the process of **intercalation**.

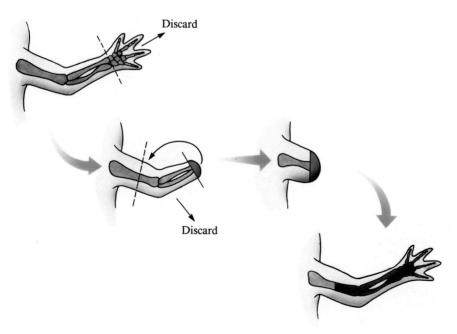

Figure 15.21

If elements in the middle of a limb are removed, the missing portions will be restored by intercalary regeneration. The tip of a urodele limb is first removed, producing a regeneration blastema. This blastema is next grafted to a limb freshly cut through the upper arm, thus resulting in a limb with proximal values fused with distal ones. The missing elements are replaced by intercalation. (After Walbot and Holder, 1987.)

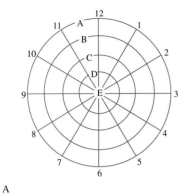

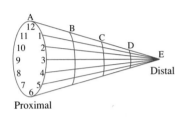

A

B

Figure 15.22

The polar coordinate model predicts that cells will have both circumferential values (1–12), which specify the anteroposterior and dorsoventral axes, and a radial value (A to E), which represents the proximodistal axis. These can be displayed on a clock face (A), and the radial values can be extended into a cone to more clearly depict the proximodistal axis (B). (After S. V. Bryant et al., 1981, Distal regeneration and symmetry. Science *212*: 993–1002. Copyright 1981 by the American Association for the Advancement of Science.)

As we have seen earlier, if the midvalues of a limb (e.g., a radius and ulna) are removed, they are regenerated to complete the missing limb structures.

Furthermore, if circumferential positional values are missing, intercalation occurs by the shorter, rather than the longer, of the two routes; i.e., if values 12 and 3 confront one another, 1 and 2 would be formed rather than 11, 10, 9, 8, 7, 6, 5, and 4 (Fig. 15.23). This is the **shortest intercalation rule.**

Regeneration of an amputated limb always produces parts that are distal to the cut, regardless of the level of the cut. This result implies that cell growth only generates more distal positional values. The polar coordinate model also predicts how this could occur (Fig. 15.24). After amputation, the wound heals by epidermis from the margins sealing the wound surface, followed by dermal fibroblasts migrating from the periphery to the wound center (Gardiner et al., 1986). This should cause dermal fibroblast cells that have different circumferential positional values to confront one another. Intercalation occurs, producing new cells whose circumferential positional values duplicate pre-existing cells. The model states that the same circumferential values cannot be duplicated at that P-D level. Therefore, new cells are forced to adopt a more distal positional value. This is called the **distalization rule.** Thus, as shown in Figure 15.24, amputation at A level generates new cells with a B-level value. The process of distalization continues until the most distal positional values are generated.

Evidence supporting the rule of distalization can be found in a variety of experimentally induced regenerated limbs. For example, Iten and Bryant (1975) were the first to demonstrate that if a limb blastema of the newt *Notophthalmus viridescens* were cut off, rotated, and reattached either to its own stump or to the limb stump on the opposite (contralateral) side, supernumerary limbs regenerated (Fig. 15.25). These remarkable supernumerary limbs appear to form at the point of maximal discontinuity between the axes. Figure 15.26 diagrams the relationship between positional values of a stump and a transplanted blastema that has been rotated so that the D-V axis is misaligned and the A-P axes are aligned. By the rule of intercalation, a complete circle of values should be reproduced at the point where D-V positional values are opposed. Thus, one would expect maximal distal growth at each of these points because intercalation creates a full circle of mismatched values. Figure 15.25C illustrates a regenerate that fulfills this prediction.

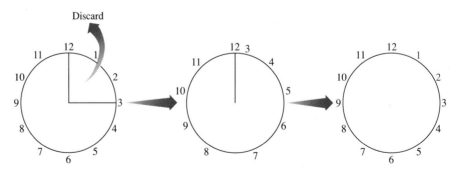

Figure 15.23

If tissue is removed so that nonadjacent positional values are brought together, the cells will be stimulated to divide and intercalate the missing values. The model further says that the values will be regenerated according to the shortest route. In this case, 1 and 2 will be regenerated to form a normal limb, rather than duplicating values 4 to 11.

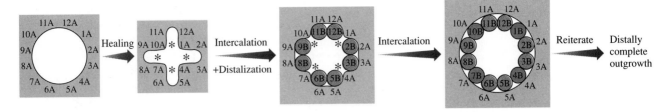

Figure 15.24

Model of distal outgrowth from a wound surface. Amputation removes B, C, D, and E levels of the pattern, leaving A (shaded). The wound edge is represented by the circle. During healing, different circumferential values are drawn into contact, resulting in circumferential intercalation (indicated by asterisks). The new cells have positional values identical to those of pre-existing adjacent cells, however, forcing the new cells to the next more distal level (B). The B level is completed by subsequent intercalation, and reiteration of the process continues until the most distal values (E) are generated. (After S. V. Bryant et al., 1981. Distal regeneration and symmetry. Science *212*: 993–1002. Copyright 1981 by the American Association for the Advancement of Science.)

The amount of distal outgrowth should be proportional to the total number of positional values at the base of an outgrowth. In the case of a full circle of values that is achieved at the amputation plane, many nonadjacent positional values are brought into contact during wound healing and thus maximal distal outgrowth can be achieved. However, one can construct limbs where there are fewer positional disparities. For example, double anterior or double posterior limbs can be constructed and then amputated (Fig. 15.27A). The positional values at the plane of amputation might be as depicted in Figure 15.28. In this case, there are fewer mismatches possible than with a full circle of values, and so there are fewer rounds of intercalation needed to correct the positional value disparity. In this case, one would predict the limb to taper off quickly and have fewer, but yet symmetrical, distal elements. This is what is often observed experimentally (see Fig. 15.27B).

The argument has been made that because urodeles have such remarkable regeneration capabilities, they represent an exceptional case on which to test models of patterning. Recently, Sessions et al. (1989) made chimeric grafts between urodeles and anurans, and the resulting limbs show that similar patterning mechanisms are at work in these two divergent animals. If a *Xenopus* (anuran) limb bud is grafted to an axolotl (urodele) stump so that their circumferential values are aligned, supernumerary digits are not formed. If, however, the grafts are rotated so that the A-P axes are misaligned, supernumerary digits do form. These results suggest that there is a pattern mechanism that is fundamentally the same among tetrapod limbs.

Finally, these mechanisms that regulate regeneration in tetrapod limbs may also be the same mechanisms that regulate pattern during development. Regeneration not only mimics embryonic limb development, but also regenerating and developing limb tissues can interact with each other to form a normal limb. Muneoka and Bryant (1982) grafted limb buds from a salamander tadpole onto regenerating blastema stumps so that their axes were properly aligned and a normal limb was produced (Fig. 15.29A). However, if their axes were misaligned, supernumerary digits would form, as has been observed with regenerating blastemas and stumps (see Fig. 15.29B). It is hoped that studies of regenerating and developing systems

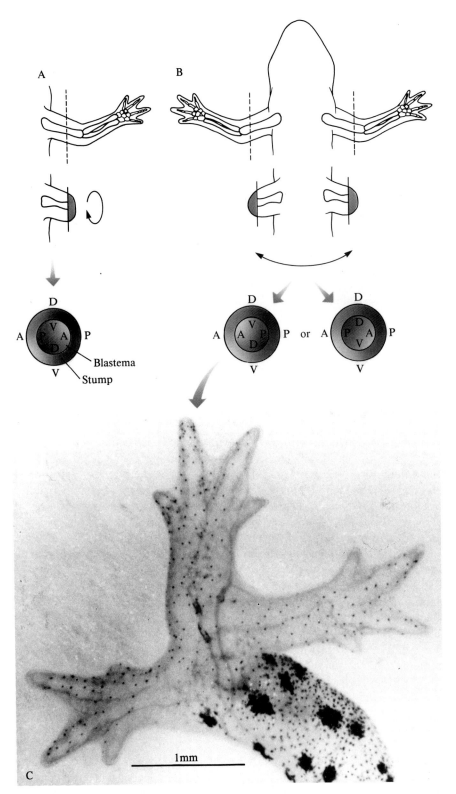

Figure 15.25

Blastemas can be removed and replaced on a limb stump so that the axes are misaligned. In A, both axes are misaligned, whereas in B, situations are shown in which either the anteroposterior or dorsoventral axes are misaligned. Such misalignment of circumferential positional values often results in supernumerary limbs (C). (A, and B, After Walbot and Holder, 1987. C, From Bryant, 1977.)

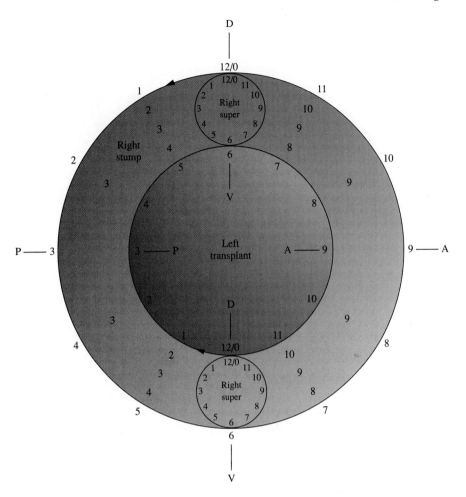

Figure 15.26

The polar coordinate model predicts both the number and position of supernumerary limbs produced by the misalignment of the anteroposterior or dorsoventral axes. Diagrammed here is a blastema from a left leg transplanted to the stump of a right leg and rotated so that the dorsoventral axis is at the point of maximum incongruity. The circumferential values are arrayed around each circle. The numbers between the circles are the values generated by the route of shortest intercalation between the confronted positional values of host and graft. Note that at the point of axis misalignment, the route of shortest intercalation can go in either direction, thus creating a complete circle of values. These circles are equivalent to the wound surface of an amputated limb; thus, distal outgrowth results at these two points in addition to outgrowth from the blastema. (From V. French et al., 1976. Pattern regulation in epimorphic fields. Science *193*: 969–971. Copyright 1976 by the American Association for the Advancement of Science.)

may elucidate the control of pattern in both situations and, moreover, outline principles common to pattern formation in other developing systems.

Earlier in this chapter, we learned that retinoic acid appears to play a role in specifying the A-P axis during development of the limb bud. Retinoic acid has equally dramatic effects on limb pattern during regeneration of the urodele (axolotl) limb. Maden (1982) found that if he amputated an adult axolotl limb and then bathed that animal in a solution of retinoic acid, the regenerated limb had a duplication of elements in the P-D axis. No matter

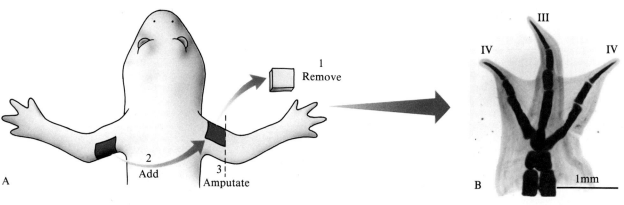

Figure 15.27

Symmetrical limbs can be created by grafting posterior half limbs in place of the
anterior half limb on the contralateral side (A). Such a graft, when amputated
through the forelimb, results in a symmetrical, tapered, sometimes truncated limb
(B). III: digit 3; IV: digit 4. (B, From Holder et al., 1980.)

at what level the limb was amputated, a complete limb could grow out from
the stump. For example, if a limb is amputated through the wrist, a complete
limb, including a humerus, radius/ulna, wrist, and hand, will grow out distal
to the amputation plane. Retinoic acid apparently resets the positional
information of the cells in the blastema to a more proximal value so that
the cells that would have formed distal structures now are "reprogrammed"
to form a proximal structure, the humerus (Maden, 1985).

Maden also found that the extent of proximal resetting is dependent
upon the concentration of retinoic acid or the length of exposure (Fig.
15.30). At lower concentrations, the degree of proximalization is reduced
so that if a limb is cut off at the wrist and exposed to low concentrations
of retinoic acid, a radius and ulna would develop at the stump, but not a
humerus.

Although the P-D axis is primarily affected by retinoic acid in adult
axolotls, both the A-P and P-D axes can be altered during regeneration in

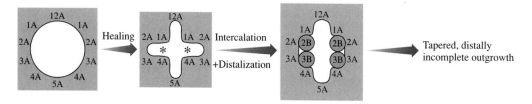

Figure 15.28

The outcome of double anterior or posterior limbs can be accounted for by the
polar coordinate model. In this case, there are fewer positional values that can
come together during wound healing, and therefore there is less intercalation. As a
result, distal outgrowth is limited to fewer radial values, and the limb is truncated.
Note that the limb will also be symmetrical. (After S. V. Bryant et al., 1981.
Distal regeneration and symmetry. *Science* 212: 993–1002. Copyright 1981 by the
American Association for the Advancement of Science.)

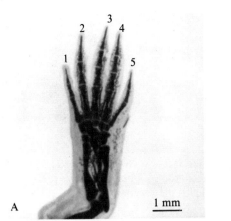

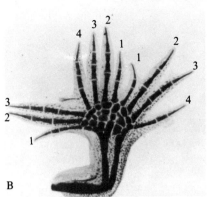

Figure 15.29

Regenerating and developing limbs may follow the same positional cues. Grafting a limb bud onto a regenerating limb stump results in a normal limb (A). If the limb bud is rotated and reattached to a regenerating limb stump so that the anteroposterior axes are reversed, supernumerary digits form (B), as they do when the same experiment is performed with regenerating blastemas and stumps only. (From Muneoka and Bryant, 1982. Reprinted by permission from *Nature* Vol. 298 pp. 369–371. Copyright © 1982 Macmillan Magazines Ltd.)

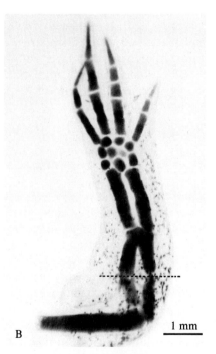

Figure 15.30

Regeneration of axolotl limbs amputated through the radius and ulna and treated with increasing concentrations of retinol palmitate, a derivative of retinoic acid. The dotted lines mark the amputation plane. A, Control showing the replacement of the radius (r), ulna (u), 8 carpals (c), and 4 digits. h: humerus. B, A medium dose that results in the radius and ulna being replaced, followed by a duplicated radius and ulna, then wrist and digits. C, At high doses, a complete limb beginning at the humerus emerges from the amputation plane. (From Maden, 1983a. Reprinted by permission of Company of Biologists, Ltd.)

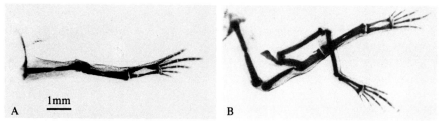

Figure 15.31

Effects of retinoic acid on larval frog leg regeneration. A, Normal frog's leg. Note that frogs have an extra long-bone segment compared to other vertebrates. B, After amputation and treatment with retinoic acid, the regenerated leg is duplicated along both the A–P and P–D axes. (From Maden, 1983b.)

developing frogs and toads. If a larval frog leg regenerates after bathing in high concentrations of retinoic acid, two complete limbs (the P-D duplication), which are mirror images (the A-P duplication), grow out from the amputation site (Maden, 1983b; Fig. 15.31).

It is not understood what the effects of retinoic acid are at the cellular level. It seems unlikely that retinoic acid is the normal agent that determines pattern because, at least in amphibians, gradients of this molecule have not been detected. Its importance in understanding pattern formation is clear, however. Because it produces similar responses in chick, salamander, and frogs, retinoic acid studies have shown that the patterning mechanisms may be similar across species boundaries. Moreover, retinoic acid is the only known external agent that can reset pattern. If we can determine what the effects of retinoic acid are at the cellular level, we will have in hand important clues as to what normally regulates limb pattern in both development and regeneration.

15–6 Chapter Synopsis

The limb forms from proliferating mesenchyme cells to give rise to a structure that is organized around three axes: the proximodistal (P-D), anteroposterior (A-P), and dorsoventral (D-V) axes. Thus, as the limb develops, not only must all the appropriate cell types differentiate within the limb, but they must be specified according to one of these axes. Because it has such distinct and readily distinguishable axes, the limb has served as a model system for studying pattern formation.

The positional information or values that specify placement along the P-D axis are thought to be acquired while the cells are rapidly dividing at the tip of the limb in a region that is called the progress zone. One model suggests that the greater the number of rounds of division mesoderm cells undergo in the progress zone, the more distal the positional values acquired by the cells. Mesenchyme cells at the tip are kept in a state of rapid division owing to the influence of a thickening of the ectoderm adjacent to this proliferative zone known as the apical ectodermal ridge. Although positional values are believed to be set at the limb tip, the cellular phenotype (e.g., bone versus connective tissue) is believed to be specified later, after the cells have exited the progress zone.

Experimental evidence suggests that the cells of the A-P axis receive their positional cues from a gradient of a morphogen that apparently has a source near the posterior wing margin in a region known as the zone of polarizing activity. Cells near the posterior margin are exposed to the highest concentration of the morphogen and are specified as posterior limb elements, whereas those at the opposite end see the lowest concentration of morphogen and become anterior limb elements. The morphogen is believed to be retinoic acid.

The limb of the urodele embryo has a remarkable ability to regenerate. Pattern specification during limb regeneration is explained in terms of the polar coordinate model, which suggests that cells within the limb obtain their positional cues from two sources: a circular array of values that specify position around the circumference of the limb and a radial array that specifies positions along the P-D axis. We do not know whether the same rules that specify positional values in regenerating systems can also apply to developing systems, and specifically whether the polar coordinate model is a better predictor of pattern during normal development than others currently supported.

References

Brockes, J. P., and C. R. Kinter. 1986. Glial growth factor and nerve-dependent proliferation in the regeneration blastema of urodele amphibians. Cell, *45*: 301–306.

Bryant, S. V. 1977. Pattern regulation in amphibian limbs. *In* D. A. Ede, J. R. Hinchliffe, and M. Balls (eds.), *Vertebrate Limb and Somite Morphogenesis*. Cambridge University Press, Cambridge, pp. 312–327.

Bryant, S. V., V. French, and P. J. Bryant. 1981. Distal regeneration and symmetry. Science, *212*: 993–1002.

Bryant, S. V., and L. E. Iten. 1976. Supernumerary limbs in amphibians: Experimental production in *Notophthalmus viridescens* and a new interpretation of their formation. Dev. Biol., *50*: 212–234.

Butler, E. G., and J. P. O'Brien. 1942. Effects of localized x-radiation on regeneration of the urodele limb. Anat. Rec., *84*: 407–413.

Carlson, B. M. 1988. *Patten's Foundations of Embryology*, 5th ed. McGraw-Hill Book Co., New York.

Chevallier, A., M. Kieny, and A. Mauger. 1977. Limb-somite relationship: origin of the limb musculature. J. Embryol. Exp. Morphol., *41*: 245–258.

Christ, B., H. J. Jacob, and M. Jacob. 1977. Experimental analysis of the origin of the wing musculature in avian embryos. Anat. Embryol., *150*: 171–186.

Dunis, D. A., and M. Namenwirth. 1977. The role of grafted skin in the regeneration of x-irradiated axolotl limbs. Dev. Biol., *56*: 97–109.

Feinberg, K. W., and D. C. Beebe. 1983. Hyaluronate in vasculogenesis. Science, *220*: 1177–1179.

French, V., P. J. Bryant, and S. V. Bryant. 1976. Pattern regulation in epimorphic fields. Science, *193*: 969–981.

Gardiner, D. M., K. Muneoka, and S. V. Bryant. 1986. The migration of dermal cells during blastema formation in axolotls. Dev. Biol., *118*: 488–493.

Geduspan, J. S., and J. A. MacCabe. 1987. The ectodermal control of mesodermal patterns of differentiation in the developing chick wing. Dev. Biol., *124*: 398–408.

Goetinck, P. F. 1964. Studies on limb morphogenesis. II. Experiments with the polydactylous mutant *eudiplodia*. Dev. Biol., *10*: 71–91.

Goss, R. J. 1956a. Regenerative inhibition following limb amputation and immediate insertion into the body cavity. Anat. Rec., *126*: 15–28.

Goss, R. J. 1956b. The regenerative responses of amputated limbs to delayed insertion into the body cavity. Anat. Rec., *126*: 283–298.

Goss, R. J. 1969. *Principles of Regeneration*. Academic Press, New York.

Grim, M., and B. M. Carlson. 1979. The formation of muscles in regenerating limbs of the newt after denervation of the blastema. J. Embryol. Exp. Morphol., *54*: 99–111.

Hamburger, V. 1938. Morphogenetic and axial self-differentiation of transplanted limb primordia of 2-day chick embryos. J. Exp. Zool., *77*: 379–400.

Harrison, R. G. 1921. On relations of symmetry in transplanted limbs. J. Exp. Zool., *32*: 1–136.

Hay, E. D. 1966. *Regeneration*. Holt, Rinehart and Winston, New York.

Hay, E. D., and D. A. Fischman. 1961. Origin of the blastema in regenerating limbs of the newt, *Triturus viridescens*. Dev. Biol., *3*: 26–59.

Holder, N., P. W. Tank, and S. V. Bryant. 1980. Regeneration of symmetrical forelimbs in the axolotl. Dev. Biol., *74*: 302–314.

Iten, L. E., and S. V. Bryant. 1975. The interaction between blastema and stump in the establishment of the anterior-posterior and proximal-distal organization of the limb regenerate. Dev. Biol., *44*: 119–147.

Iten, L. E., and D. J. Murphy. 1980. Pattern regulation in the embryonic chick limbs. Supernumerary limb formation with the anterior (non-ZPA) limb bud tissue. Dev. Biol., *75*: 373–385.

Iten, L. E., D. J. Murphy, and L. C. Javois. 1981. Wing buds with three ZPAs. J. Exp. Zool., *215*: 103–106.

Javois, L. C., and L. E. Iten. 1981. Position of origin of donor posterior chick wing bud tissue transplanted to an anterior host site determines the extra structures formed. Dev. Biol., *82*: 329–342.

Kieny, M. 1964. Etude du mécanisme de la régulation dans le développement du bourgeon de membre de l'embryon de poulet. II. Régulation des déficiences dans les chimères 'aile-patte' et 'patte-aile'. J. Embryol. Exp. Morphol., *12*: 357–371.

Kieny, M. 1971. Les phases d'activité morphogènes du mésoderme somatopleural pendant le développement précoce du membre chez l'embryon de poulet. Annales d'Embryologie et de Morphogenèse, *4*: 281–298.

Kieny, M. 1977. Proximo-distal pattern formation in avian limb development. *In* D. A. Ede, J. R. Hinchliffe, and M. Balls (eds.), *Vertebrate Limb and Somite Morphogenesis*. Cambridge University Press, Cambridge, pp. 87–103.

Kieny, M., and M.-P. Pautou. 1976. Régulation des excédents dans le développement du bourgeon de membre de l'embryon d'oiseau. Analyse expérimentale de combinaisons xénoplastiques caille/poulet. Wilhelm Roux's Archiv für Entwicklungsmechanik, *179*: 327–338.

Loyd, R. M., and R. A. Tassava. 1980. DNA synthesis and mitosis in adult newt limbs following amputation and insertion into the body cavity. J. Exp. Zool., *214*: 61–69.

MacCabe, J. A., J. Errick, and J. W. Saunders, Jr. 1974. Ectodermal control of the dorsoventral axis in the leg bud of the chick embryo. Dev. Biol., *39*: 69–82.

MacCabe, J. A., J. W. Saunders, Jr., and M. Pickett. 1973. The control of the anteroposterior and dorsoventral axes in embryonic chick limbs constructed of dissociated and reaggregated limb-bud mesoderm. Dev. Biol., *31*: 323–335.

Maden, M. 1982. Vitamin A and pattern formation in the regenerating limb. Nature (Lond.), *295*: 672–675.

Maden, M. 1983a. The effect of vitamin A on the regenerating axolotl limb. J. Embryol. Exp. Morphol., *77*: 273–295.

Maden, M. 1983b. The effect of vitamin A on limb regeneration in *Rana temporaria*. Dev. Biol., *98*: 409–416.

Maden, M. 1985. Retinoids and the control of pattern in limb development and regeneration. Trends Genet., *1*: 103–107.

Mescher, A. L., and R. A. Tassava. 1975. Denervation effects on DNA replication and mitosis during the initiation of limb regeneration in adult newts. Dev. Biol., *44*: 187–197.

Morgan, T. H. 1901. *Regeneration*. Macmillan, New York.

Muneoka, K., and S. V. Bryant. 1982. Evidence that patterning mechanisms in developing and regenerating limbs are the same. Nature (Lond.), *298*: 369–371.

Muneoka, K., W. F. Fox, and S. V. Bryant. 1986. Cellular contribution from dermis and cartilage to the regenerating limb blastema in axolotls. Dev. Biol., *116*: 256–260.

Namenwirth, M. 1974. The inheritance of cell differentiation during limb regeneration in the axolotl. Dev. Biol., *41*: 42–56.

Oliver, G. et al. 1988. A gradient of homeodomain protein in developing forelimbs of Xenopus and mouse embryos. Cell, *55*: 1017–1024.

Pescitelli, M. J., Jr., and D. L. Stocum. 1980. The origin of skeletal structures during intercalary regeneration of larval *Ambystoma* limbs. Dev. Biol., *79*: 255–275.

Petkovich, M. et al. 1987. A human retinoic acid receptor which belongs to the family of nuclear receptors. Nature (Lond.), *330*: 444–450.

Reiter, R. S., and M. Solursh. 1982. Mitogenic property of the apical ectodermal ridge. Dev. Biol., *93*: 28–35.

Rowe, D. A., and J. F. Fallon. 1982. The proximodistal determination of skeletal parts in the developing chick leg. J. Embryol. Exp. Morphol., *68*: 1–7.

Rubin, L., and J. W. Saunders, Jr. 1972. Ectodermal-mesodermal interactions in the growth of limb buds in the chick embryo: Constancy and temporal limits of the ectodermal induction. Dev. Biol., *28*: 94–112.

Saunders, J. W., Jr. 1948. The proximo-distal sequence of origin of the parts of the chick wing and the role of the ectoderm. J. Exp. Zool., *108*: 363–404.

Saunders, J. W., Jr. 1982. *Developmental Biology. Patterns/ Problems/ Principles.* Macmillan, New York.

Saunders, J. W., Jr., and M. T. Gasseling. 1968. Ectodermal-mesenchymal interactions in the origin of limb symmetry. *In* R. Fleischmajer and R. E. Billingham (eds.), *Epithelial-Mesenchymal Interactions.* 18th Hahnemann Symposium. Williams & Wilkins, Baltimore, pp. 78–97.

Saunders, J. W., Jr., M. T. Gasseling, and J. E. Errick. 1976. Inductive activity and enduring cellular constitution of a supernumerary apical ectodermal ridge grafted to the limb bud of the chick embryo. Dev. Biol., *50*: 16–25.

Saunders, J. W., Jr., and C. Reuss. 1974. Inductive and axial properties of prospective wing-bud mesoderm in the chick embryo. Dev. Biol., *38*: 41–50.

Searls, R. L., and M. Y. Janners. 1971. The initiation of limb bud outgrowth in the embryonic chick. Dev. Biol., *24*: 198–213.

Sessions, S. T., D. M. Gardiner, and S. V. Bryant. 1989. Compatible limb patterning mechanisms in urodeles and anurans. Dev. Biol., *131*: 294–301.

Singer, M. 1954. Induction of regeneration of the forelimb of the post metamorphic frog by augmentation of the nerve supply. J. Exp. Zool., *126*: 419–472.

Singer, M. 1973. Limb regeneration in the vertebrates. *Addison-Wesley Module in Biology*, no. 6. Addison-Wesley, Reading, MA.

Singer, M. 1974. Neurotropic control of limb regeneration in the newt. Ann. N. Y. Acad. Sci., *228*: 308–321.

Singer, M. 1978. On the nature of the neurotrophic phenomenon in urodele limb regeneration. Am. Zool., *18*: 829–841.

Singer, M., and L. Craven. 1948. The growth and morphogenesis of the regenerating forelimb of adult *Triturus* following denervation at various stages of development. J. Exp. Zool., *108*: 279–308.

Solursh, M. 1984. Ectoderm as a determinant of early tissue pattern in the limb bud. Cell Diff., *15*: 17–24.

Solursh, M., C. Drake, and S. Meier. 1987. The migration of myogenic cells from the somites at the wing level in avian embryos. Dev. Biol., *121*: 389–396.

Solursh, M., C. T. Singley, and R. S. Reiter. 1981a. The influence of epithelia on cartilage and loose connective tissue formation by limb mesenchyme cultures. Dev. Biol., *86*: 471–482.

Solursh, M. et al. 1981b. Stage- and position-related changes in chondrogenic response of chick embryonic wing mesenchyme to treatment with dibutyryl cyclic AMP. Dev. Biol., *83*: 9–19.

Steen, T. P. 1968. Stability of chondrocyte differentiation and contribution of muscle to cartilage during limb regeneration in the axolotl (*Siredon mexicanum*). J. Exp. Zool., *167*: 49–71.

Summerbell, D. 1974. A quantitative analysis of the effect of excision of the AER from the chick limb-bud. J. Embryol. Exp. Morphol., *32*: 227–237.

Summerbell, D. 1977. Regulation of deficiencies along the proximal distal axis of the chick wing-bud: a quantitative analysis. J. Embryol. Exp. Morphol., *41*: 137–159.

Summerbell, D. 1979. The zone of polarizing activity: Evidence for a role in normal chick limb morphogenesis. J. Embyrol. Exp. Morphol., *50*: 217–233.

Summerbell, D. 1981. Evidence for regulation of growth, size and pattern in the developing chick limb bud. J. Embryol. Exp. Morphol., *65* (Suppl.): 129–150.

Summerbell, D. 1983. The effect of local application of retinoic acid to the anterior margin of the developing chick limb. J. Embryol. Exp. Morphol., *78*: 269–289.

Summerbell, D., and J. H. Lewis. 1975. Time, place, and positional value in the chick limb-bud. J. Embryol. Exp. Morphol., *33*: 621–643.

Summerbell, D., J. H. Lewis, and L. Wolpert. 1973. Positional information in chick limb morphogenesis. Nature (Lond.), *244*: 492–496.

Summerbell, D., and M. Maden. 1990. Retinoic acid, a developmental signalling molecule. TINS, *13*: 142–147.

Thaller, C., and G. Eichele. 1987. Identification and spatial distribution of retinoids in the developing chick limb. Nature (Lond.), *327*: 625–628.

Thornton, C. S. 1968. Amphibian limb regeneration. Adv. Morphol., *7*: 205–249.

Tickle, C., D. Summerbell, and L. Wolpert. 1975. Positional signalling and specification of digits in chick limb morphogenesis. Nature (Lond.), *254*: 199–202.

Tickle, C. et al. 1982. Local application of retinoic acid to the limb bud mimics the action of the polarizing region. Nature (Lond.), *296*: 564–566.

Walbot, V., and N. Holder. 1987. *Developmental Biology*. Random House, New York.

Wertz, R. L., and D. J. Donaldson. 1980. Early events following amputation of adult newt limbs given regeneration inhibitory doses of X irradiation. Dev. Biol., *74*: 434–445.

Wolpert, L. 1978. Pattern formation in biological development. Sci. Am., *239*: 154–164.

Wolpert, L., C. Tickle, and M. Sampford (with an appendix by J. H. Lewis). 1979. The effect of cell killing by X-irradiation on pattern formation in the chick limb. J. Embryol. Exp. Morphol., *50*: 175–198.

Zanetti, N., and M. Solursh. 1984. Control of chondrogenic differentiation by the cytoskeleton. J. Cell Biol., *99*: 115–123.

Zwilling, E. 1961. Limb morphogenesis. Adv. Morphogen., *1*: 301–330.

Zwilling, E., and L. Hansborough. 1956. Interaction between limb bud ectoderm and mesoderm in the chick embryo. III. Experiments with polydactylous limbs. J. Exp. Zool., *132*: 219–239.

CHAPTER 16

Organogenesis: Gonad Development and Sex Differentiation

F. R. Lillie (1870–1947), a Professor at the University of Chicago, was one of the first to recognize that secondary sexual characteristics of the male are dependent upon a circulating factor, based upon his studies of the maculinized freemartin. He spent much of his later career directing the world-famous Marine Biological Laboratory in Woods Hole, Massachusetts. (Photograph courtesy of Marine Biological Laboratory Library, Woods Hole, MA.)

The development of the gonads (testis and ovary), their associated ducts, and the external genitalia is unique among organs because the same tissue will develop in dramatically different ways depending upon the sex of the individual. In this section, we first consider the development of the gonads and the origin of the tissues from which they are constructed. Then we consider what determines whether gonads and other sexual characteristics are male or female.

16–1 Development of the Vertebrate Gonads

The development of the gonads has been studied primarily through the use of preserved and sectioned material. Much of the early work, particularly on human specimens, relied on paraffin sections that provided low-resolution images. More recent studies of other animal species now employ plastic sections or electron microscopy in combination with experimental analysis, thereby providing more accurate details of gonadogenesis. Even so, many aspects of gonadogenesis are still subject to debate.

The gonad of the amniote initially appears as a thickening of the coelomic epithelium (splanchnic mesoderm) ventromedial to the developing mesonephric kidney, with which it is closely associated (Figs. 16.1A, 16.2A and B). This thickening is known as the **genital ridge** (also known as the germinal epithelium because it was once believed to be the source of the gametes). Subsequently, cells in the genital ridge continue to proliferate, causing the incipient gonad to bulge into the coelomic cavity (see Figs. 16.1B and 16.2B and C). The origin of the cells that make up the growing gonad is still controversial. Various sources considered are the coelomic epithelium, the undifferentiated intermediate mesoderm beneath the coelomic epithelium, and even the mesonephric kidney tubules, which are generally not functional in adult higher vertebrates. Recent experiments using chick/quail chimeras suggest that the initial thickening of the gonad is due to the proliferation of the mesoderm-derived coelomic epithelium (see Fig. 16.4).

As the coelomic epithelium proliferates, it sends disorganized, densely packed cords of cells into the mesenchyme (Merchant-Larios, 1979; Fargeix et al., 1981; see Figs. 16.1B and 16.2B and C). Immunocytochemistry and transmission electron microscopy show that the basal lamina of the coelomic epithelium is contiguous with the cords, further supporting the notion that the latter grow from the surface epithelium (Merchant, 1975; Paranko et al., 1983; Fig. 16.3). The epithelial cords are known as the **primitive sex cords** and are invaded at this time by the primordial germ cells (PGCs), the

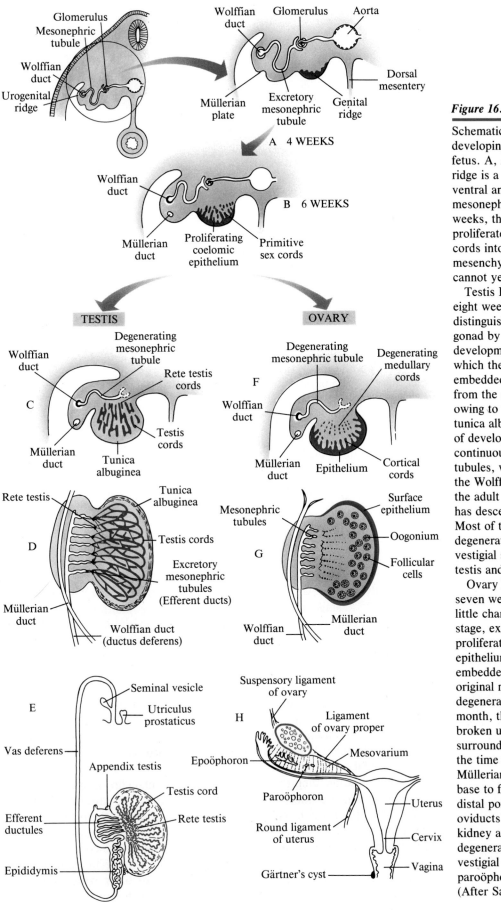

Figure 16.1

Schematic sections through the developing gonad of a human fetus. A, At four weeks, the genital ridge is a thickened epithelium ventral and medial to the mesonephric kidney. B, By six weeks, the coelomic epithelium has proliferated, sending epithelial cords into the adjacent mesenchyme. Ovaries and testes cannot yet be distinguished.

Testis Development (C-E). C, At eight weeks, the testis can be distinguished from the indifferent gonad by the continued development of the sex cords, in which the germ cells are now embedded. The cords separate from the coelomic epithelium owing to the formation of the tunica albuginea. D, By 16 weeks of development, the cords are continuous with the mesonephric tubules, which in turn connect with the Wolffian duct. E, Diagram of the adult features after the testis has descended into the scrotum. Most of the Müllerian duct has degenerated except for two vestigial structures, the appendix testis and the utriculus prostaticus.

Ovary Development (F-H). F, At seven weeks, the future ovary is little changed from the indifferent stage, except for the continued proliferation of the coelomic epithelium. The germ cells are embedded in the sex cords. The original medullary cords are degenerating. G, By the fifth month, the cortical epithelium has broken up to form clusters of cells surrounding each oogonium. H, By the time the ovary descends, the Müllerian ducts have fused at their base to form the uterus, while the distal portion becomes the paired oviducts. Most of the mesonephric kidney and Wolffian duct have now degenerated, except for the vestigial epoöphoron, paroöphoron, and Gartner's cyst. (After Sadler, 1985.)

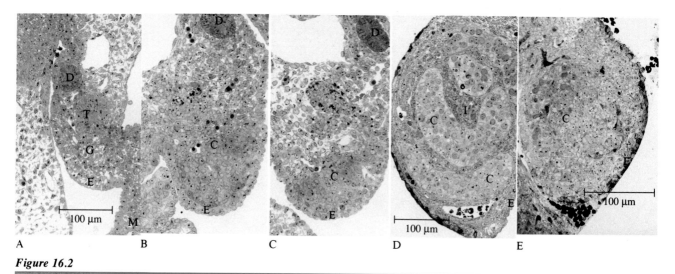

Figure 16.2

Sections of rat gonads during embryonic development. A, Indifferent stage (12 days, 10 hours). At this stage, the genital ridge (G) consists of a thickened coelomic epithelium (E) and underlying mesenchyme. The gonad develops in close association with the Wolffian duct (D) and mesonephric tubules (T). M: mesentery. B, and C, By 13 days, 9 hours, the indifferent gonad has expanded into the coelomic cavity owing to the proliferation of the coelomic epithelium (E), which generates unorganized sex cords (C). There is still little apparent difference between a future testis (B) and ovary (C). D, and E, By 15 days, 11 hours, the differences between the testis (D) and ovary (E) are now apparent. In the testis, the sex cords have organized into definitive testicular cords (C), which are invested in interstitial mesenchyme (I). Connective tissue forms a tunica albuginea beneath the surface epithelium (E), which separates the coelomic epithelium from the cords. In the ovary, the sex cords (C) are still relatively disorganized and maintain their connection with the coelomic epithelium (E). (From Paranko et al., 1983.)

precursors of the gametes, which migrate from a distant source into the genital ridge.

Initially, the presumptive gonads of both male and female embryos are identical, a stage consequently known as the **indifferent stage**. The "indifferent" gonads consist of the thickened coelomic epithelium, known as the **cortex**, and the core of the gonad composed of the sex cords and some mesenchyme, known as the **medulla**. Remarkable changes soon occur that are the first signs of overt sexual differentiation. If an embryo is destined to become a male, the primitive sex cords continue to grow, forming a mass of sex cords that thin at their distal tips to form the **rete testis** (see Figs. 16.1C and 16.2D). Mesenchymal cells invade between the sex cords, eventually also forming a layer of connective tissue beneath the coelomic epithelium, known as the **tunica albuginea**, which severs the cords from the surface epithelium of the testis. These cords are the forerunners of the seminiferous tubules, and their component cells differentiate into Sertoli cells, which comprise the cellular superstructure around which the spermatozoa differentiate (see Chap. 2). In addition to their roles in the maturation and nutrition of the spermatocytes, the Sertoli cells produce **Müllerian inhibiting substance (MIS)**, a growth factor that controls certain aspects of sex duct development (see p. 673). The germ cells remain embedded in the solid medullary cords until puberty, at which time the cords hollow out and the germ cells proliferate and differentiate into sperm.

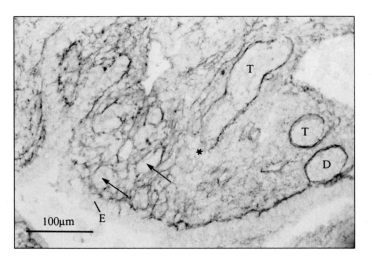

Figure 16.3

Section of a 13-day rat testis (similar stage as for Fig. 16.2B) stained with fibronectin antibody, which highlights the basement membranes of the epithelia. Note the well-developed basement membranes around the mesonephric tubules (T), Wolffian duct (D), and the primitive sex cords (arrows). This stain demonstrates that the cords are continuous with the coelomic epithelium (E), with which they share a common basement membrane. An * marks the mesenchyme, which is continuous between the mesonephric tubules and the sex cords. This mesenchyme may be derived from the degenerating mesonephric tubules. (From Paranko et al., 1983.)

The interstitial mesenchyme between the cords gives rise to Leydig cells, which produce the hormone testosterone. Recent studies, including experiments using chick/quail chimeras, suggest that the mesenchyme of the gonad is derived from the intermediate mesoderm—probably from Bowman's capsules of the degenerating mesonephric tubules (Zamboni and Upadhyay, 1982; Rodemer et al., 1986; Yoshinaga et al., 1988; Rodemer-Lenz, 1989; Fig. 16.4C).

If, on the other hand, the indifferent gonad is destined to become an ovary, the primitive sex cords and the mesenchyme of the medulla degenerate and will ultimately form connective tissue remnants in the hilum of the ovary. Instead, the coelomic epithelium continues proliferating to produce **secondary sex cords**, which maintain a continuous association with the ovarian cortex (see Figs. 16.1F and 16.2E). The cords will ultimately break up to form clusters of cells called primordial follicles, which consist of a primordial germ cell (now known as an oogonium) surrounded by a layer of follicle, or granulosa, cells derived from the sex cords (see Fig. 16.1G). These follicle cells are necessary for the arrest of the oocytes in meiosis and their survival. At puberty, the follicle cells are, in turn, enveloped by a single layer of cells derived from the mesenchyme, known as **thecal cells**, which produce steroid hormones.

16–2 Sex Ducts and External Genitalia

As the gonads develop, a set of ducts to transport the sperm or eggs also forms. In males, the **efferent ductules** connect the rete testis to the epididymis and the **vas deferens** (see Fig. 16.1E). In females, the ducts include the

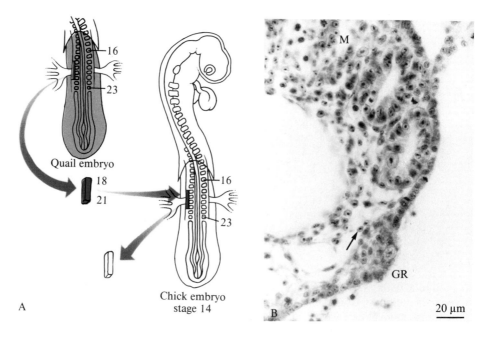

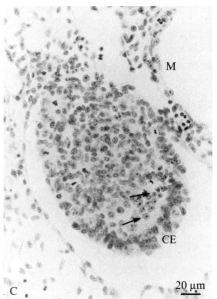

Figure 16.4

Contributions of the intermediate mesoderm to the gonad, as studied by construction of chick/quail chimeras. A, Schematic representation of the replacement of chick intermediate mesoderm (between somites 18 and 21), including the Wolffian duct, at the axial level of the genital ridge with quail intermediate mesoderm. B, Cross-section through the genital ridge (GR) of a host (chick) embryo two days after grafting quail intermediate mesoderm, which develops into the mesonephric kidney (M). The thickening genital ridge epithelium is composed entirely of host (chick) cells, although there are a few quail cells in the underlying loose mesenchyme (arrow). C, An indifferent gonad of a six-day chick embryo. The cords that develop from the coelomic epithelium (CE) are still of chick origin, but quail mesenchyme cells (arrows) from the mesonephric tubules are now beginning to invade the gonad. (A, After Rodemer et al., 1986. B, and C, From Rodemer-Lenz, 1989.)

oviduct, where fertilization occurs and through which the zygote is transported to the uterus, where it will implant (see Fig. 16.1H).

Two sources of tissue will contribute to these ducts: (1) the remnants of the mesonephric tubules and Wolffian duct (or mesonephric duct); and (2) a second duct, the **Müllerian duct** (also known as the paramesonephric duct), which develops *de novo* as an invagination of the coelomic epithelium. The remnants of the mesonephric kidney and ducts and the Müllerian duct are present in both sexes during the indifferent stage.

In embryos that develop a testis, the Müllerian duct begins to degenerate and is lost. The rete testis hooks up to the efferent ductules, which are remnants of the mesonephric tubules. These, in turn, retain their connection with the Wolffian duct, which becomes the vas deferens, and will carry mature sperm to the urethra and out of the body (see Figs. 16.1D and E and 16.5).

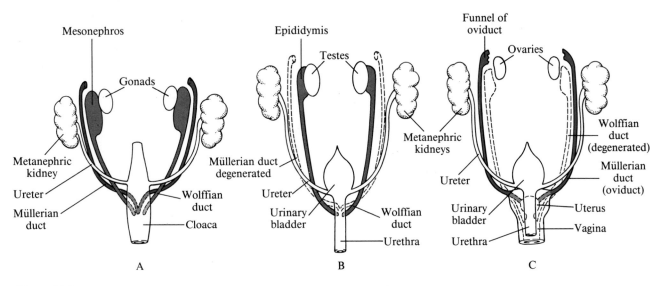

Figure 16.5

Frontal views summarizing the development of the sex ducts in association with the gonads in the indifferent stage (A) and in later-stage male (B) and female (C) embryos. (After Balinsky, 1981.)

The opposite scenario occurs in embryos that have formed an ovary: The mesonephric tubules and Wolffian duct degenerate to become connective tissue remnants, whereas the Müllerian duct is retained and undergoes further differentiation to form the oviducts and uterus (see Figs. 16.1G and H and 16.5).

Like the gonads and accompanying sex ducts, the external genitalia of both sexes have their origin in common primordia, the **genital swellings,** the **genital folds,** and the **genital tubercle** (Fig. 16.6). In human male embryos, the genital tubercle forms the glans penis and the shaft of the penis, the genital folds fuse to form the ventral shaft of the penis, and the genital swellings expand to form the scrotum. Conversely, in human female embryos, the genital swellings and folds remain paired to form the labia majora and labia minora, respectively.

16–3 Sex Determination

A question that has long fascinated biologists is what determines the sex of an individual. It appears that many different strategies have been adopted by different animals. For some, environmental conditions can determine the individual's sex. The sex of alligators, for example, is dictated by temperature. If alligator eggs are raised in the laboratory at temperatures of 30°C or lower, females will develop, whereas males are produced at 34°C or above (Ferguson and Joanen, 1982). The sex of the coral reef fish depends upon the number of males and females in a group. In these fish, a small group of females breeds with a single male. If for some reason the male is removed, a large female can turn into a male (Shapiro, 1984).

Sex is also determined by chromosome composition, although this is highly variable among different species. For example, *Drosophila* contain two sex chromosomes, X and Y, but sex is determined by the ratio of X chromosomes to the number of non-sex chromosomes (autosomes), which contain male determinants. For example, if a fly genome contains two X

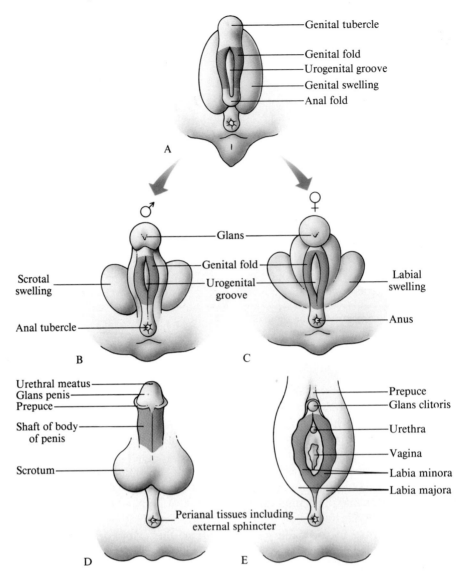

Figure 16.6

Summary of the development of the external genitalia in human male and female embryos from the indifferent stage. A, Indifferent stage (week 4). B, and D, Male at 9 and 12 weeks, respectively. C, and E, Female at 9 and 12 weeks, respectively. (After Hopper and Hart, 1985.)

chromosomes and a normal, diploid complement of autosomes (2X:2A), it will be female, but if it only contains one X chromosome, it will be male.

We now consider in detail sex determination in the mammals, emphasizing the human and the mouse, for which there is considerable evidence.

Control of Mammalian Sex Determination

Since the time of the Greeks, it was believed that sex in humans was determined by environmental factors such as nutrition and temperature at the time of conception. It was not until 1905 that sexual dimorphism in insects was observed to be based on specific sex chromosomes (Stevens, 1905; Wilson, 1905; Morgan, 1910). The existence of X and Y chromosomes in humans was demonstrated using cytological techniques a short time later (Painter, 1923).

In mammals, the normal number of sex chromosomes is two: XX individuals are female and are known as the homogametic sex, whereas XY individuals are male and are known as the heterogametic sex. During meiosis, mistakes can be made involving the sex chromosomes, resulting in **aneuploids** (not having the appropriate number of chromosomes). Genetic studies of such rare mutations involving the sex chromosomes in humans have revealed that the presence or absence of the Y chromosome determines the sex.

An example of an aneuploid condition involving the sex chromosomes is those men who carry an extra X chromosome (XXY), a condition known as **Klinefelter's syndrome**. These individuals are phenotypically male and are often mentally retarded. These men have normal testes, although they contain no spermatozoa. (The presence of two X chromosomes is incompatible with normal sperm production.) Extreme conditions can arise in which multiple (i.e., more than two) X chromosomes are found in an individual, but as long as there is still one Y chromosome, the individual will be male.

Conversely, individuals who are XO (i.e., one X chromosome and no Y chromosome) are phenotypically female, a condition known as **Turner's syndrome**. These individuals are generally underdeveloped sexually, are short in stature, and have only vestigial gonadal tissue as adults, although early in fetal life normal ovaries, sex ducts, and external genitalia are present.

Thus, presence of a Y chromosome confers maleness, regardless of the number of X chromosomes, whereas absence of a Y chromosome produces a female.

Such studies suggest that there is a gene or genes on the Y chromosome that direct the differentiation of a testis from the indifferent gonad, whereas the absence of the gene results in the default condition of differentiation of an ovary. However, the gene for such a "testis-determining factor" (termed *TDF* in humans, *Tdy* in the mouse) has remained elusive. To isolate this gene, the powerful tools of molecular biology are being combined with clinical studies that have identified "sex-reversed" individuals. An example of this approach is described here.

Very rare human defects arise in which the sex-chromosome complement does not match the sexual phenotype of the individual. For example, there are some males who are XX, whereas there are females who are XY. These rare individuals are usually discovered when they try to have children and are found to be sterile. In a series of studies, it was discovered that XX males contain a small bit of the Y chromosome translocated to one of the X chromosomes (Guellaen et al., 1984). Conversely, the XY females have lost varying pieces of the Y chromosome (Disteche et al., 1986). Molecular techniques have allowed investigators to determine exactly which portion of the Y chromosome must be present to give rise to a male (or, conversely, the smallest possible piece of the Y chromosome that must be missing to give rise to a female).

To accomplish this work, single-strand DNA probes were made that recognize various overlapping segments of the Y chromosome (Vergnaud et al., 1986). If one of the Y chromosome sequences hybridizes with the DNA from an XX male, then that portion of the Y chromosomal DNA is being carried in that man's genome. A precise map of the Y chromosome, called a deletion map, was constructed using these probes as landmarks (Fig. 16.7A).

Page and his colleagues (1987) analyzed the DNA from many XX males to find the smallest common piece of DNA from the Y chromosome

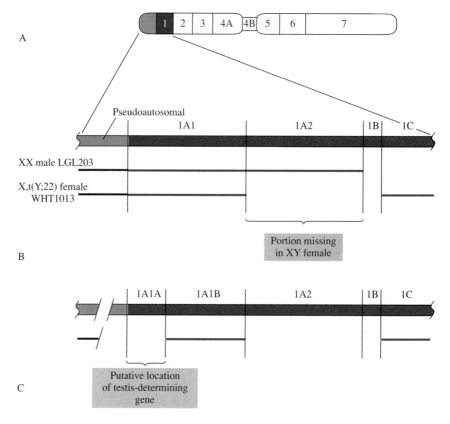

Figure 16.7

A, The human Y chromosome has been arbitrarily divided into sections based on deletion mapping. The *TDF* gene is found within region 1. B, Enlarged view of region 1, with bars indicating the portions of Y chromosome present in an XX male and in an XY female. These studies initially suggested that the TDF gene is located in region 1A2. C, Re-examination of XY female indicates that she lacks region 1A1A in addition to 1A2. The 1A1A region, which is also present in XX males, is the putative location of the testis-determining gene. (A, Adapted from Vergnaud et al., 1986. B, After Page et al., 1987. C, After Page et al., 1990.)

present in all these individuals. Conversely, they examined XY females to determine the smallest possible deletion from the Y chromosome that would produce sex reversal. They discovered that the piece of the Y chromosome that was translocated onto X chromosomes in XX males was the same piece that is lost from the Y chromosome in XY females.

Two patients were especially crucial in this analysis (see Fig. 16.7B). One male (patient LGL203) had lost all but 0.5% of the Y chromosome. This piece of DNA was located in the region of the Y chromosome termed 1A. Conversely, a female (patient WHT1013) appeared to retain 99.8% of the Y chromosome. She only lacked a small piece of DNA that spanned interval 1A2 (a portion of 1A) and 1B. Based on these patients, plus many others whose deletions were less extreme, it was determined that the testis-determining gene is located in the interval 1A2.

This type of analysis can be fraught with problems, however, and it is instructive to examine this particular line of research to see how these problems can be circumvented. Shortly after the study from Page's laboratory was published, another study (Palmer et al., 1989) examined a different group of human XX males and hermaphrodites; these individuals *lacked* the DNA from region 1A2, which—as we have discussed—was presumed to contain the gene for testis determination. This observation threw into doubt the identification of region 1A2 as the testis-determination region.

Page and colleagues examined in more detail female WHT 1013 and found that she had an additional deletion in region 1A1, suggesting that this could be the region containing the testis-determining gene (see Fig. 16.7C). Significantly, the small region missing from her Y chromosome (region 1A1A) is the very piece of the Y chromosome present in the XX males in the Palmer et al. (1989) study. Thus, attention has now turned to region 1A1A in the search for the testis-determining gene.

Two recent studies reinforce the notion that the testis-determining gene is in the 1A1A region. Sinclair et al. (1990) have searched this region and found only one sequence that could code for protein. They have sequenced the gene and found that it is conserved among a wide range of mammals. Most interestingly, it shares sequence homologies with a gene from yeast that codes for a DNA-binding protein. The authors have coined this candidate for the testis-determining gene *SRY*, for sex-determining region Y. Gubbay et al. (1990) have cloned the homologous gene from a mouse (the *Sry* gene) and shown that this gene is expressed in the gonadal tissue before testis differentiation, suggesting that it is present at the appropriate time to control differentiation. Furthermore, when mutations are introduced into this gene, the resulting XY mice are female.

Thus, the evidence is strong that the *SRY* gene (*Sry* in mice) is the elusive *TDF* gene. Given the history of this problem and the frequent intervals at which hypotheses for sex determination have been abandoned (see McLaren, 1990, for a short recapitulation), this current favorite child needs further documentation before it can be fully accepted. It would be reassuring, for example, if an XX mouse embryo were transfected with the *Sry* gene and it developed a testis. However, to date the *SRY* gene is the most promising candidate for the testis-determining gene.

It is unlikely that a single gene acts alone to control the phenotype of the gonads. In fact, a number of autosomal genes have been discovered that appear to influence gonad differentiation (Eicher and Washburn, 1983; Washburn and Eicher, 1983) or spermatogenesis (Burgoyne et al., 1986). These genes very likely act later in development, whereas the putative *TDF* gene (*SRY*?) is the first gene to act in a developmental cascade.

Secondary Sex Characteristics

If an embryo is to be male, the indifferent gonad develops into a testis; if it is to be female, the gonad stays in the indifferent stage a while longer and then begins to develop into an ovary (this has been termed **primary sex determination**). The differentiation of the sex ducts and external genitalia (formation of secondary sex characteristics) occurs subsequent to development of the gonads during embryogenesis, augmented by changes that occur during adolescence.

It was discovered primarily through the experimentation of Jost, a French biologist, that the development of the sex ducts and the external genitalia is dependent directly upon what type of gonad is present. He found that if testes are removed from 19-day-old rabbit fetuses, oviducts and a uterus will develop from the Müllerian duct, and the external genitalia are female (Jost, 1947, 1953). Similarly, removing the ovaries from a rabbit fetus still results in female duct and genital development. These data suggested that the testis produces some factor or factors that direct the development of the ducts and external genitalia in the male. In the absence of gonads, the female direction is always taken.

It had been suspected from the experiments of Lillie (1917) that circulating factors could control secondary sexual characteristics. His studies of freemartins (genetically female calves who are masculinized if they share the same circulatory system *in utero* with a male co-twin) suggested that some factor can pass through the circulatory system, causing the sex ducts to regress and the ovary to become an ovotestis.

Additional experiments by Jost (1947, 1953) provided the first evidence that the testis produces at least two factors that direct duct differentiation. If he grafted a testis near the ovary of a rabbit embryo in the indifferent stage, the vas deferens would develop on the operated side only and the Müllerian duct would regress (Fig. 16.8A). If instead he implanted a small crystal of testosterone in a female embryo, the vas deferens would develop, but the Müllerian duct would not regress (see Fig. 16.8B), indicating that regression of the Müllerian ducts is mediated by a factor from the testis that is distinct from testosterone. We now know that the vas deferens and external genitalia become male under the influence of testosterone produced by the Leydig cells and that the Müllerian ducts regress owing to MIS, which is produced by the Sertoli cells (Josso et al., 1976; Donahoe et al., 1987).

The role of testosterone in differentiation of secondary sex characteristics has been further demonstrated in individuals who are androgen insensitive, a syndrome known as **testicular feminization** (Fig. 16.9). These individuals are genetically XY, have a normal testis that produces normal levels of testosterone, but do not have functional cytoplasmic receptors for testosterone. Therefore, they are insensitive to the circulating hormone. They do not develop a vas deferens, and their external genitalia are female. This syndrome is also evidence of the fact that gonad differentiation is under genetic, rather than hormonal, control.

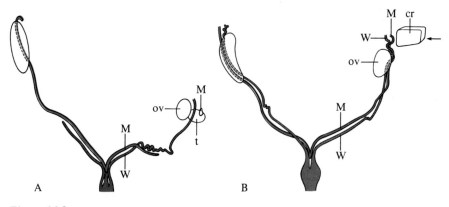

Figure 16.8

Schematic representations of the genital tracts of two 28-day-old rabbit fetuses. A, In this fetus, a testis (t) was grafted near the ovary (ov) on day 20. The Müllerian duct (M) has regressed, and the Wolffian duct (W) has been retained on the operated side. B, A crystal of testosterone (cr) was grafted next to the ovary of a 20-day-old rabbit fetus. The Wolffian ducts as well as the Müllerian ducts were retained. (After Jost, 1965.)

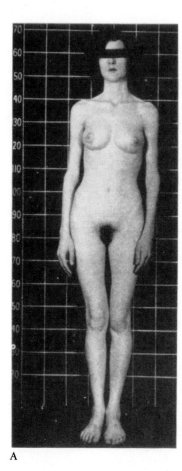

A

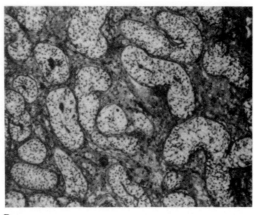

B

Figure 16.9

This 17-year-old girl (A) is an example of testicular feminization. Her chromosome constitution is XY, and her testes produce testosterone, but she lacks functional testosterone receptors. As a result, her external genitalia are female. B, A micrograph through the testis removed from this girl shows that it is composed of seminiferous tubules and differentiated Sertoli cells, but there are no sperm. (From Jones and Scott, 1958. Copyright © 1958, the Williams & Wilkins Co., Baltimore.)

   The regression of the Müllerian duct during differentiation of male sex characteristics apparently involves both cell death and a degeneration of the Müllerian epithelium into a mesenchyme (Fig. 16.10). As we have discussed, the duct regresses under the influence of MIS, a glycoprotein that is structurally very similar to another growth factor, TGF-β, which we have already considered (see Chap. 12; Donahoe et al., 1987). Müllerian inhibiting substance has been isolated and purified from a number of sources and, when applied directly to embryonic Müllerian ducts in *in vitro* assays, has been shown to cause the dissociation of the Müllerian duct cells into a mesenchyme, as occurs during normal male sex differentiation.

   Müllerian inhibiting substance has been implicated in a number of developmentally abnormal circumstances. In individuals who have testicular feminization, male sex ducts do not develop because their cells are insensitive to testosterone. However, a uterus and oviducts do not develop either because the normal testis does produce MIS and causes the Müllerian duct to regress. Thus, these individuals have no internal sex ducts.

   Müllerian inhibiting substance may also be involved in the masculinization of the freemartin. Freemartins usually have normal female external genitalia, but they have no oviduct and uterus and their ovaries are often smaller and sometimes show evidence of seminiferous tubules. Recently, Vigier et al. (1987) have treated cultured embryonic rat ovaries with purified bovine MIS. They find that MIS not only causes regression of the Müllerian duct but also influences the development of the ovary, in essence producing the freemartin condition in tissue culture. The effects of MIS on the ovary raise the interesting possibility that MIS may also play a role in at least the terminal stages of testis differentiation.

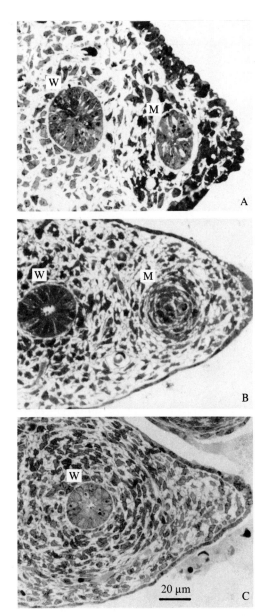

Figure 16.10

Sections through the urogenital ridge of male rat embryos, showing the regression of the Müllerian duct. A, At day 14, both the Wolffian (W) and Müllerian (M) ducts are intact epithelia. B, By early day 16, the dissolution of the Müllerian epithelium and an increase in mesenchyme around this duct are apparent. C, By day 17, there is no trace of the Müllerian duct. (From Trelstad et al., 1982.)

16–4 Origin of the Primordial Germ Cells

For some time it was believed that the cells that give rise to the gametes, the so-called primordial germ cells (pgcs), are derived from the coelomic epithelium that forms the genital ridge. However, we now know that germ cells develop elsewhere and then migrate and reach the genital ridge using a variety of mechanisms. Their origins and pathways of migration are considered in this section (see also Chap. 11).

Chick Germ Cell Migration

Primordial germ cells in chickens were first recognized by Swift (1914) using morphological criteria. These cells are larger than other embryonic cells and are recognizable in the hypoblast of the primitive streak-stage embryo at the border of the area pellucida and area opaca. This region was termed the **germinal crescent** (Figs. 16.11 and 16.12). If this region is excised or X-irradiated early in development, the resulting chicken is sterile, suggesting that these are indeed the precursors of the gametes. During development, the germ cells enter the space between the hypoblast and epiblast, where they invade the developing blood vessels and are carried through the blood vascular system. Swift found that the germ cells exit the blood vessels at the axial level of the developing genital ridges and migrate a short distance through the mesoderm until they reach the gonads, where they lodge in the coelomic epithelium (Fig. 16.13).

Recently, however, the origin of the germ cells from the hypoblast has been questioned, and it now seems clear that the germ cells first arise from the epiblast before appearing in the germinal crescent. This idea is based on several studies. First, Eyal-Giladi and her co-workers have constructed chick/quail chimeras in which the epiblast is from one species and the hypoblast from the other (Eyal-Giladi et al., 1981). Regardless of whether the epiblast is chick or quail, the germ cells are always shown to be of epiblastic origin, more specifically from the very central region of the area pellucida (Ginsburg and Eyal-Giladi, 1987). Antibodies have now been developed that recognize the germ cells at a stage earlier than has been possible by other techniques. These studies also show that germ cells form in the epiblast and then separate and move into the underlying hypoblast and mesoderm (Fig. 16.14; Pardanaud et al., 1987; Urven et al., 1988). The germ cells are then carried anteriorly by the movements of the hypoblast and mesoderm (DuBois, 1969; Ginsburg and Eyal-Giladi, 1986) until they are concentrated in the germinal crescent.

Early work of DuBois (1968) has suggested that germ cells move passively in the blood vascular system from the germinal crescent until they reach the region of the genital ridge. Here, their active locomotion through the mesentery to the gonads could be guided by chemotaxis. He showed, for example, that when chick germinal crescents or gonads from 3.5-day-

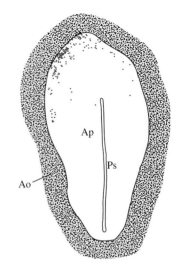

Figure 16.11

The distribution of chick primordial germ cells as determined in a whole blastodisc at the primitive streak (Ps) stage stained with periodic acid Schiff (PAS) reagent, which indicates the presence of polysaccharides. The germ cells are found in a crescent at the border between the area pellucida (Ap) and the area opaca (Ao). (After Fujimoto et al., 1976. Reprinted from Anatomical Record by permission of Wiley-Liss, a division of John Wiley and Sons, Inc. Copyright © 1976, Wiley-Liss.)

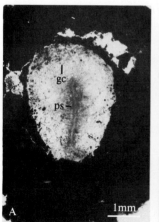
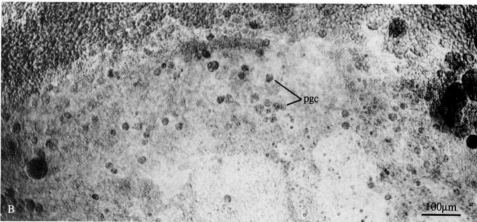

Figure 16.12

The chick germinal crescent. A, An intact blastodisc at the primitive streak (ps) stage removed from the yolk and stained to reveal the germ cells. Avian germ cells stain differentially with PAS. The area pellucida is the clear disc in the center. The germinal crescent (gc) is indicated at the margin of the area pellucida. B, A high magnification of the germinal crescent in A, showing the distribution of the primordial germ cells (pgc). (From Ginsburg and Eyal-Giladi, 1986. Reprinted by permission of Company of Biologists, Ltd.)

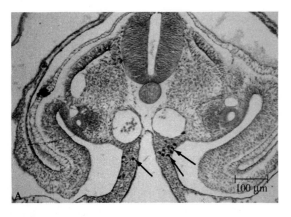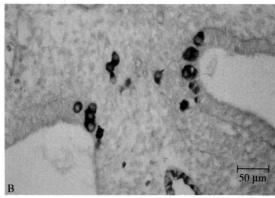

Figure 16.13

Migration of primordial germ cells from the dorsal aorta to the genital ridge. A, Section through a 3-day chick embryo stained with PAS, showing germ cells (arrows) in the mesentery, one cell leaving the dorsal aorta, and others embedded in the genital ridge. B, High magnification of germ cells identified by PAS staining in the mesentery and genital ridge epithelium. See Color plate 24 for color version. (Courtesy of L. Urven.)

old chick embryos are cultured in association with early gonads from embryos that have previously been sterilized, the germ cells migrate into the sterile gonadal tissue. This apparent attraction to the gonadal tissue could be due to adhesive molecules in the gonad trapping randomly migrating germ cells. However, he showed that when a sterile genital ridge and a germ cell-containing gonad are separated by a piece of vitelline envelope so that a diffusible chemotactic molecule can pass, the germ cells collect at the vitelline envelope immediately adjacent to the genital ridge (Fig. 16.15), providing evidence that germ cells are responding to an attractant.

Germ cell chemotaxis recently has been observed directly in tissue culture. Genital ridges from early chick embryos were cultured with germ cells isolated from the blood during their migratory stage. Under these circumstances, of those germ cells that actually adhered to the substratum, 154 of them migrated directly to the gonads, whereas only 11 moved to any of the other somatic tissues that were cultured as controls (Kuwana et al., 1986). A chemotactic molecule has not yet been isolated from the gonads, however.

Figure 16.14

A section through the epiblast (E) and hypoblast (H) of a 12-hour-old embryo stained with EMA-1 antibody, which recognizes germ cells at a stage earlier than is possible with PAS. Stained cells are caught in the process of separating from the epiblast (arrows). See Color plate 25 for color version. (From Urven et al., 1988. Reprinted by permission of Company of Biologists, Ltd.)

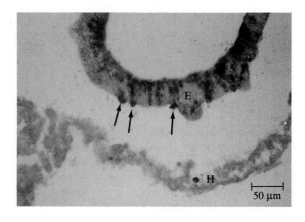

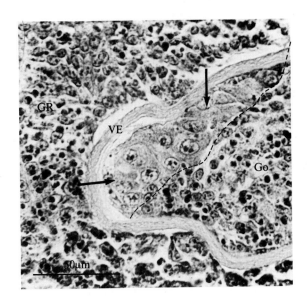

Figure 16.15

A sterile genital ridge (GR) was cultured in association with a germ cell-containing chick gonad (Go) with a vitelline envelope separating them. The germ cells (arrows) have accumulated at the vitelline envelope (VE) adjacent to the sterile genital ridge, suggesting that a diffusible factor can attract the germ cells. A dotted line demarks the border between the accumulated germ cells and the somatic tissue of the gonad. (From Dubois and Croisille, 1970.)

Mammalian Germ Cell Migration

As in the chick, mouse germ cells can be recognized by morphological criteria and their pathways subsequently traced, but more commonly mouse germ cells have been marked using a histological stain for alkaline phosphatase activity (Chiquoine, 1954). Recently, antibodies have been developed that recognize murine germ cells (Fig. 16.16); the migratory pattern identified using this marker is similar to that traced by the alkaline phosphate procedure (Hahnel and Eddy, 1986; Donovan et al., 1987).

Primordial germ cells can first be recognized in the extraembryonic mesoderm and endoderm of the yolk sac and allantois near the posterior end of the primitive streak (Fig. 16.17A). They then move into the developing gut, although it is not certain whether this is an active cell migration or passive movement due to the growth of the hindgut and yolk sac. However, it is clear that once they reach the level of the gonads, they migrate actively out of the gut and into the mesentery, where they then make their way into

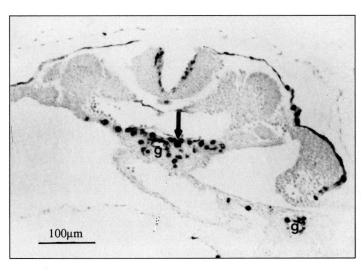

Figure 16.16

Germ cells in a 9-day mouse embryo stained with the antibody EMA-1. Here, the germ cells are found in the gut (g) and in transit (arrow) to the genital ridge. (From Hahnel and Eddy, 1986. Reprinted from Gamete Research by permission of Wiley-Liss, a division of John Wiley and Sons, Inc. Copyright © 1986, Wiley-Liss.)

Figure 16.17

Primordial germ cells in mammals are first recognized (A) in the yolk sac near the base of the allantois. From here, they migrate anteriorly through the endoderm of the gut and then into the mesentery at the level of the genital ridges, which they ultimately invade (B). (After Sadler, 1985. Copyright © 1985, the Williams & Wilkins Co., Baltimore.)

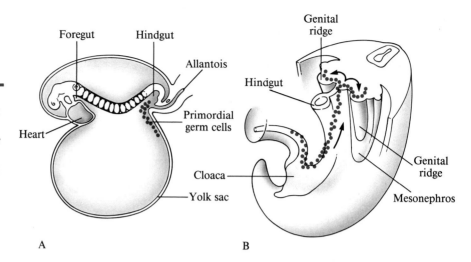

the genital ridges (see Fig. 16.17B; Chiquoine, 1954; Mintz and Russell, 1957), because not only do the germ cells have processes that look typically locomotory, but they also display active motility when they are explanted and maintained *in vitro* (Blandau et al., 1963; Donovan et al., 1986).

The primary origin of murine germ cells is apparently the epiblast, as we have already seen in birds. Copp et al. (1986) grafted small bits of ^{3}H-thymidine–labeled mouse posterior primitive streak into an unlabeled host—an extraordinary feat—and found that the future germ cells are derived from the epiblast that ingresses through the primitive streak. Conversely, endoderm grafted into this same region did not give rise to germ cells, negating the formerly widely-held view that germ cells are endoderm-derived by analogy to the anurans (see Chap. 11).

There is some indirect evidence that mouse germ cells are also attracted to the gonads by a chemotactic mechanism. If the posterior hind gut of a 9.5-day-old mouse embryo, which contains germ cells, is grafted into the coelomic cavity of a 2.5-day-old chick embryo, the mouse germ cells accumulate in that portion of the gut where it is opposed to the chick germinal ridges (Rogulska et al., 1971), and in some instances, the mouse germ cells even invade the chick gonad and mesonephros. This experiment also suggests that the chemotactic signal is conserved evolutionarily among disparate species.

It is unlikely, however, that chemotaxis exclusively guides murine germ cells, because a chemotactic gradient could probably not extend from the developing gonads all the way to the hindgut. Therefore, some substratum-mediated guidance mechanism must exist as well. Mouse germ cells have been shown to stick to fibronectin and laminin in culture (Donovan et al., 1987), and these matrix molecules are found in the migratory routes. It is not known, however, if they play a role in either guiding the cells or sustaining their migration.

Amphibian Germ Cell Migration

As we have seen in Chapter 11, primordial germ cells in anurans can be identified in cleaving embryos as those blastomeres that acquire germinal cytoplasm located at the vegetal pole of the egg (Fig. 16.18). As the embryo gastrulates, these cells can be traced to the floor of the archenteron (gut endoderm). The germ cells then relocate to the dorsal portion of the gut endoderm and move into the mesentery that suspends the gut and finally into the gonads (see Figs. 16.18 and 16.19). Primordial germ cell movement in *Xenopus* embryos is aligned with the underlying mesentery cells, sug-

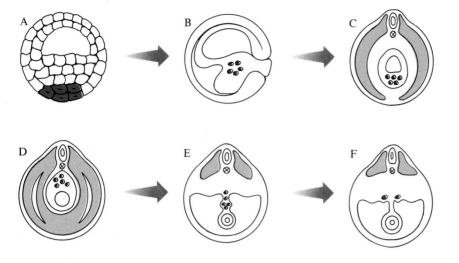

Figure 16.18

Diagrammatic representation of the
distribution of primordial germ
cells in anurans from their origin as
vegetal blastomeres until they
migrate through the dorsal
mesentery to the genital ridges. B,
Sagittal section; remainder are
cross-sections. (After Wylie et al.,
1982.)

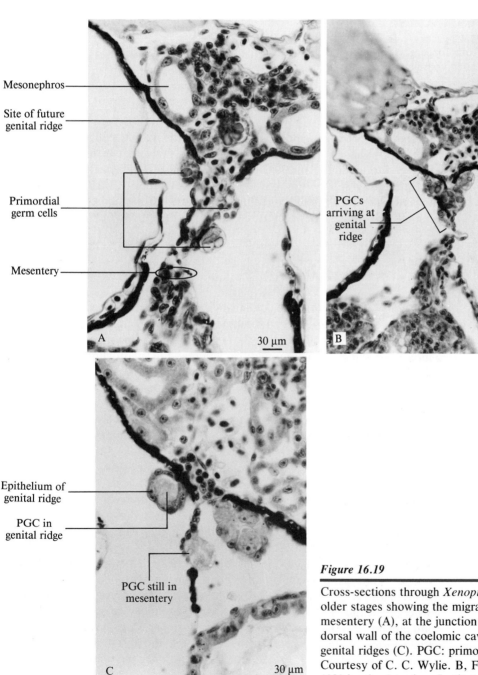

Mesonephros

Site of future
genital ridge

Primordial
germ cells

Mesentery

PGCs
arriving at
genital
ridge

30 µm

30 µm

Epithelium of
genital ridge

PGC in
genital ridge

PGC still in
mesentery

30 µm

Figure 16.19

Cross-sections through *Xenopus* embryos at successively
older stages showing the migration of the germ cells up the
mesentery (A), at the junction of the mesentery and the
dorsal wall of the coelomic cavity (B), and invading the
genital ridges (C). PGC: primordial germ cell. (A, and C,
Courtesy of C. C. Wylie. B, From Wylie, 1980. Copyrigh
1980 by the American Institute of Biological Sciences.)

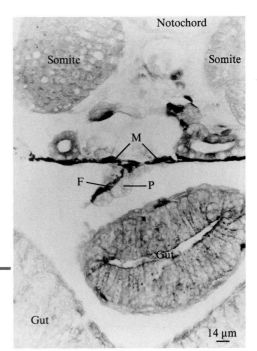

Figure 16.20

Cross-section through the abdominal region of a *Xenopus* embryo that has been stained with an antibody to fibronectin, showing the large, yolky primordial germ cells (P) associated with fibronectin-containing cells (F) in the mesentery. A few of the dark cells are darkly pigmented melanocytes (M). (From Heasman et al., 1981. Copyright © Cell Press.)

gesting that contact guidance may direct the germ cells dorsally to the gonads (Wylie et al., 1979). Fibronectin is at least one molecule that is believed to mediate frog PGC adhesion because (1) fibronectin is produced by mesentery cells when they are grown in culture, (2) frog PGCs adhere to and align on these fibronectin fibrils, and (3) antibodies to fibronectin inhibit PGC migration on cultured mesentery (Fig. 16.20).

16–5 Chapter Synopsis

The gonads arise from a thickening of the splanchnic mesoderm. During early developmental stages, a future testis and ovary cannot be distinguished from each other. Only later does the testis acquire the distinctive seminiferous tubules in which the sperm mature, whereas the ovary acquires follicles in which the oocytes differentiate. The primordial germ cells, from which the gametes form, arise from a distant source and must migrate into the genital ridge.

Whether an immature gonad will develop into an ovary or a testis depends upon the sex chromosome constitution. In humans, for example, the presence or absence of a Y chromosome determines the phenotypes of the gonad. If a Y is present, the gonad becomes a testis, whereas if the Y is absent, an ovary forms. There is believed to be a testis-determining gene on the Y chromosome, coined the *TDF* gene in humans and *Tdy* in mice; recent evidence suggests that this gene is a gene called *SRY* in humans and *Sry* in mice, which shares sequence homologies with a gene encoding a DNA-binding protein in yeast.

The other secondary sexual characteristics, including the sex ducts and external genitalia, are dependent upon the presence or absence of a testis. In the presence of a testis, male ducts and external genitalia form, whereas in the absence of the testis, the secondary sexual characteristics are female. Two factors produced by the testis are responsible for directing male differentiation. Testosterone promotes the differentiation of male-specific structures. In the absence of testosterone, the female phenotype is expressed. A second testis factor, Müllerian inhibiting substance, causes the Müllerian duct to regress in males.

References

Balinsky, B. I. 1981. *An Introduction to Embryology*, 5th ed. Saunders College Publishing, Philadelphia.

Blandau, R. J., B. J. White, and R. E. Rumery. 1963. Observations on the movements of the living primordial germ cells in the mouse. Fertil. and Steril., *14*: 482–489.

Burgoyne, P. S., E. R. Levy, and A. McLaren. 1986. Spermatogenic failure in male mice lacking H-Y antigen. Nature (Lond.), *320*: 170–172.

Chiquoine, A. D. 1954. The identification, origin, and migration of the primordial germ cells in the mouse embryo. Anat. Rec., *118*: 135–146.

Copp, A. J., H. M. Roberts, and P. E. Polani. 1986. Chimaerism of primordial germ cells in the early postimplantation mouse embryo following microsurgical grafting of posterior primitive streak cells *in vitro*. J. Embryol. Exp. Morphol., *95*: 95–115.

Disteche, C. M. et al. 1986. Small deletions of the short arm of the Y chromosome in 46,XY females. Proc. Natl. Acad. Sci. U.S.A., *83*: 7841–7844.

Donahoe, P. K. et al. 1987. Müllerian inhibiting substance: Gene structure and mechanism of action of a fetal regressor. Rec. Prog. Horm. Res., *43*: 431–467.

Donovan, P. J. et al. 1986. Migratory and postmigratory mouse primordial germ cells behave differently in culture. Cell, *44*: 831–838.

Donovan, P. J. et al. 1987. Studies on the migration of mouse germ cells. J. Cell Sci. Suppl., *8*: 359–367.

Dubois, R. 1968. La colonisation des ébauches gonadiques par les cellules germinales de l'embryon de Poulet, en culture *in vitro*. J. Embryol. Exp. Morphol., *20*: 189–213.

Dubois, R. 1969. Donneés nouvelles sur les localisation des cellules germinales primordiales dans le germe non incubé de Poule. C. R. Acad. Sci. (Paris), *269*: 205–208.

Dubois, R., and Y. Croisille. 1970. Germ-cell line and sexual differentiation in birds. Philos. Trans. R. Soc. Lond. (Biol.), *259*: 73–89.

Eicher, E. M., and L. L. Washburn. 1983. Inherited sex reversal in mice: identification of a new primary sex-determining gene. J. Exp. Zool., *228*: 297–304.

Eyal-Giladi, H., M. Ginsburg, and A. Farbarov. 1981. Avian primordial germ cells are of epiblastic origin. J. Embryol. Exp. Morphol., *65*: 139–147.

Fargeix, N., E. Didier, and P. Didier. 1981. Early sequential development in avian gonads. An ultrastructural study using selective glycogen labeling in the germ cells. Reprod. Nutr. Dévelop., *21*: 479–496.

Ferguson, M. W. J., and T. Joanen. 1982. Temperature of egg incubation determines sex in *Alligator mississippiensis*. Nature (Lond.), *296*: 850–853.

Fujimoto, T., A. Ukeshima, and R. Kiyofuji. 1976. The origin, migration and morphology of the primordial germ cells in the chick embryo. Anat. Rec., *185*: 139–154.

Ginsburg, M., and H. Eyal-Giladi. 1986. Temporal and spatial aspects of the gradual migration of primordial germ cells from the epiblast into the germinal crescent in the avian embryo. J. Embryol. Exp. Morphol., *95*: 53–71.

Ginsburg, M., and H. Eyal-Giladi. 1987. Primordial germ cells of the young chick blastoderm originate from the central zone of the area pellucida irrespective of the embryo-forming process. Development, *101*: 209–219.

Gubbay, J. et al. 1990. A gene mapping to the sex-determining region of the mouse Y chromosome is a member of a novel family of embryonically expressed genes. Nature (Lond.), *346*: 245–250.

Guellaen, G. et al. 1984. Human XX males with Y single-copy DNA fragments. Nature (Lond.), *307*: 172–173.

Hahnel, A. C., and E. M. Eddy. 1986. Cell surface markers of mouse primordial germ cells defined by two monoclonal antibodies. Gamete Res., *15*: 25–34.

Heasman, J. et al. 1981. Primordial germ cells of Xenopus embryos: The role of fibronectin in their adhesion during migration. Cell, *27*: 437–447.

Hopper, A. F., and N. H. Hart. 1985. *Foundations of Animal Development,* 2nd ed. Oxford University Press, New York.

Jones, H., and W. Scott. 1958. *Hermaphroditism, Genital Abnormalities and Related Endocrine Disorders*. Williams & Wilkins, Baltimore.

Josso, N., J.-Y. Picard, and D. Tran. 1976. The antimüllerian hormone. Rec. Prog. Horm. Res., *33*: 117–167.

Jost, A. 1947. Recherches sur la différenciation sexuelle de l'embryon de lapin. III. Rôle des gonades foetales dans la différenciation sexuelle somatique. Arch. Anat. Microscop. Morphol. Exptl., *36*: 271–315.

Jost, A. 1953. Problems of fetal endocrinology: The gonadal and hypophyseal hormones. Rec. Prog. Horm. Res., *8*: 379–413.

Jost, A. 1965. Gonadal hormones in the sex differentiation of the mammalian foetus. *In* R. L. DeHaan and H. Ursprung (eds.), *Organogenesis*. Holt, Rinehart and Winston, New York, pp. 611–628.

Koopman, P. et al. 1989. *Zfy* gene expression patterns are not compatible with a primary role in mouse sex determination. Nature (Lond.), *342*: 940–942.

Kuwana, T., H. Maeda-Suga, and T. Fujimoto. 1986. Attraction of chick primordial germ cells by gonadal anlage *in vitro*. Anat. Rec., *215*: 403–406.

Lillie, F. R. 1917. The freemartin, a study of the action of sex hormones in foetal life of cattle. J. Exp. Zool., *23*: 371–452.

McLaren, A. 1990. What makes a man a man? Nature (Lond.), *346*: 216–217.

Merchant, H. 1975. Rat gonadal and ovarian organogenesis with and without germ cells. An ultrastructural study. Dev. Biol., *44*: 1–21.

Merchant-Larios, H. 1979. Origin of the somatic cells in the rat gonad: An autoradiographic approach. Ann. Biol. Anim. Bioch. Biophys., *19*: 1219–1229.

Mintz, B., and E. S. Russell. 1957. Gene-induced embryological modifications of primordial germ cells in the mouse. J. Exp. Zool., *134*: 207–237.

Morgan, T. H. 1910. Sex-limited inheritance in *Drosophila*. Science, *32*: 120–122.

Page, D. C. et al. 1987. The sex-determining region of the human Y chromosome encodes a finger protein. Cell, *51*: 1091–1104.

Page, D. C. et al. 1990. Additional deletion in sex-determining region of human Y chromosome resolves paradox of X, t (Y;22) female. Nature (Lond.), *346*: 279–281.

Painter, T. 1923. Studies in mammalian spermatogenesis. II. The spermatogenesis of man. J. Exp. Zool., *37*: 291–335.

Palmer, M. S. et al. 1989. Genetic evidence that *ZFY* is not the testis-determining factor. Nature (Lond.), *342*: 937–939.

Paranko, J. et al. 1983. Morphogenesis and fibronectin in sexual differentiation of rat embryonic gonads. Differentiation, *23* (Suppl.): S72-S81.

Pardanaud, L., C. Buck, and F. Dieterlen-Lièvre. 1987. Early germ cell segregation and distribution in the quail blastodisc. Cell Differ., *22*: 47–60.

Rodemer, E. S., A. Ihmer, and H. Wartenberg. 1986. Gonadal development of the chick embryo following microsurgically caused agenesis of the mesonephros and using interspecific quail-chick chimaeras. J. Embryol. Exp. Morphol., *98*: 269–285.

Rodemer-Lenz, E. 1989. On cell contribution to gonadal soma formation in quail-chick chimeras during the indifferent stage of gonadal development. Anat. Embryol., *179*: 237–242.

Rogulska, T., W. Odeski, and A. Komar. 1971. Behaviour of mouse primordial germ cells in the chick embryo. J. Embryol. Exp. Morphol., *25*: 155–164.

Sadler, T. W. 1985. *Langman's Medical Embryology*, 5th ed. Williams & Wilkins, Baltimore.

Shapiro, D. Y. 1984. Sex reversal and sociodemographic processes in coral reef fishes. *In* G. W. Potts and R. J. Wootton (eds.), *Fish Reproduction: Strategies and Tactics*. Academic Press, London, pp. 103–118.

Sinclair, A. H. et al. 1990. A gene from the human sex-determining region encodes a protein with homology to a conserved DNA-binding motif. Nature (Lond.), *346*: 240–244.

Stevens, N. M. 1905. Studies in spermatogenesis with especial reference to the "accessory chromosome." Carnegie Institute Report 36 (Washington, D. C.).

Swift, C. H. 1914. Origin and early history of the primordial germ cells in the chick. Am. J. Anat., *15*: 483–516.

Trelstad, R. L. et al. 1982. The epithelial-mesenchymal interface of the male rat Müllerian duct: Loss of basement membrane integrity and ductal regression. Dev. Biol., *92*: 27–40.

Urven, L. E. et al. 1988. Analysis of germ line development in the chick embryo using an anti-mouse EC cell antibody. Development, *103*: 299–304.

Vergnaud, G. et al. 1986. A deletion map of the human Y chromosome based on DNA hybridization. Am. J. Hum. Genet., *38*: 109–124.

Vigier, B. et al. 1987. Purified bovine AMH induces a characteristic freemartin effect in fetal rat prospective ovaries exposed to it *in vitro*. Development, *100*: 43–55.

Washburn, L. L., and E. M. Eicher. 1983. Sex reversal in XY mice caused by a dominant mutation on chromosome 17. Nature (Lond.), *303*: 338–340.

Wilson, E. B. 1905. The chromosomes in relation to the determination of sex in insects. Science, *22*: 500–502.

Wylie, C. C. 1980. Primordial germ cells in anuran embryos: Their movement and its guidance. Bioscience, *30*: 27–31.

Wylie, C. C., A. P. Swan, and J. Heasman. 1982. The role of the extracellular matrix in cell movement and guidance. *In* J. D. Pitts and M. E. Finbow (eds.), *Functional Integration of Cells in Animal Tissues: British Society for Cell Biology Symposium 5*. Cambridge University Press, Cambridge, pp. 229–249.

Wylie, C. C. et al. 1979. Evidence for substrate guidance of primordial germ cells. Exp. Cell Res., *121*: 315–324.

Yoshinaga, K. et al. 1988. The development of the sexually indifferent gonad in the prosimian, *Galago crassicaudatus crassicaudatus*. Am. J. Anat., *181*: 89–105.

Zamboni, L., and S. Upadhyay. 1982. The contribution of the mesonephros to the development of the sheep fetal testis. Am. J. Anat., *105*: 339–356.

Organogenesis: Development of the Eye

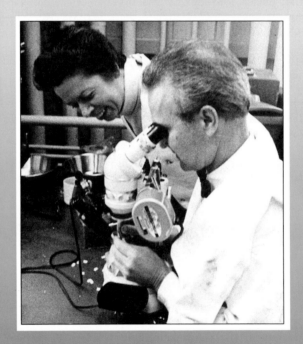

Jane (1922–1975) and Chris (b. 1922) Coulombre, shown here in their laboratory, worked together for over 20 years, first at Yale University and later at the National Institutes of Health, studying the mechanisms that control the development of the avian eye. Their seminal work employing the techniques of experimental embryology investigated many aspects of morphogenesis of the eye, including the role of intraocular pressure in controlling eye shape and cornea development, epitheliomesenchymal tissue interactions using the corneal epithelium as a model, and the role of extracellular matrix molecules in corneal morphogenesis. (Courtesy of Chris Coulombre.)

The eye is an extraordinary organ in terms of its function. A visual image of the world is focused on the retina, which in turn encodes this information and transmits it to the brain, where it is processed. The eye is equally amazing from a developmental perspective. The many parts that constitute the eye are derived from a variety of sources. For example, the retina is derived from an outpocket of the brain (the optic vesicle), the lens and portions of the cornea are derived from ectoderm, but the muscles that control the ability to focus are derived from the paraxial mesoderm, the optic vesicle, and neural crest. Thus, these many parts must come together in a process that is coordinated in time and in space to produce a functional organ.

In this chapter, we shall review what is known about eye development, focusing on the origins of various tissues in the eye and on the mechanisms that coordinate the assembly of these structures. The eye is a particularly useful system to show how the adult structures, which are extremely complex, are products of their developmental histories. In fact, many of the structures of the eye are so incongruous with their function that it is only through their embryology that the structures make sense. Therefore, we compare the adult histology of each component of the eye with its embryological origins and the developmental processes that produced it.

The eye also provides a classic example of sequential induction. Not only does one structure induce another (e.g., the lens induces the cornea), but sequential induction also occurs within the cornea.

Finally, the projection of nerve fibers from the optic retina to the tectum is a well-studied patterning event whose control mechanisms may well apply to other developmental systems.

17–1 Early Morphogenetic Events

The first visible evidence of the emergence of the eyes are two lateral expansions of the prosencephalon, the optic vesicles (Figs. 17.1A and B and 17.2A and B). As the optic vesicles grow distally toward the overlying ectoderm, they remain attached to the brain by way of a gradually thinning **optic stalk**.

When the optic vesicles contact the ectoderm, the latter thickens into the lens placode, which subsequently inpockets and separates from the ectoderm to form the **lens** (see Figs. 17.1C to E and 17.2C and D). The remaining ectoderm that covers the lens will become the outer layer of the **cornea**. The deeper layers of the cornea are derived from the head neural crest. Concomitant with lens morphogenesis, the optic vesicles also invag-

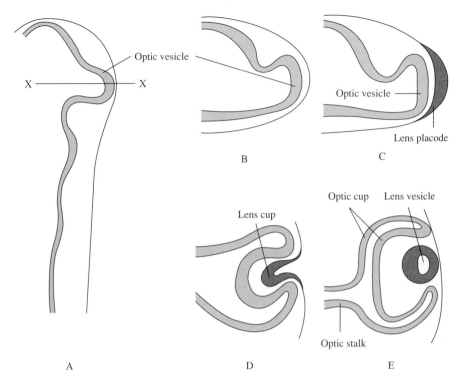

Figure 17.1

Development of the vertebrate eye. A, Dorsal view of the embryo as the optic vesicle first contacts the epidermis. X–X depicts the plane of section at the same stage in B and in progressively later stages in C to E. Sequential sections show the thickening of the lens placode (C), simultaneous invagination of the optic vesicle and lens placode (D), and separation of the lens vesicle from the surface (E). (After Coulombre, 1965.)

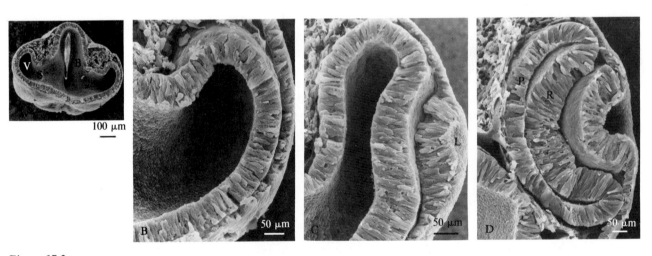

Figure 17.2

Cross-sections through the chick brain at the level of the developing eye, viewed by scanning electron microscopy. A, Low magnification revealing the evagination of the brain (B) to form the optic vesicles (V). S: optic stalk. B, High magnification of A showing the point of contact between the optic vesicle and overlying ectoderm. C, The lens placode (L) has formed and is beginning to invaginate, as is the optic vesicle. D, The optic vesicle is now a two-layered cup: The outer layer is the retinal pigmented epithelium (P), and the inner lining of the cup will become the neural retina (R). (From Hilfer and Yang, 1980. Reprinted from the Anatomical Record by permission of Wiley-Liss, a division of John Wiley and Sons, Inc. Copyright © 1980, Wiley-Liss.)

inate and form a double-layered cup. The inner layer of the cup will differentiate into the **neural retina**, the sensory layer of the eye. The neural retina will also produce the extracellular matrix component called the **vitreous body**, which fills the posterior cavity of the eye. The outer layer of the cup is the **retinal pigmented epithelium**. Cells of the retinal pigmented epithelium at the edge of the optic cup extend toward the lens and become the epithelium of the **iris**.

The neural crest-derived mesenchyme that surrounds the eye (Fig. 17.3) condenses around the pigmented retina and differentiates into two outer coats, the vascularized **choroid** and the connective-tissue **sclera**, which together serve to nourish and protect the eyeball (Fig. 17.4).

The muscles of the eye are derived from several sources. The **intrinsic muscles** in the iris and the ciliary body, which control the size of the pupil and the shape of the lens, respectively, are derived from the retinal pigmented epithelium and the neural crest, respectively. The **extrinsic ocular muscles**, which move the eye in its orbit, are derived from the head paraxial mesoderm (Noden, 1983).

The optic stalk is retained throughout development and is the migratory substratum for the neurons that extend from the neural retina back to the

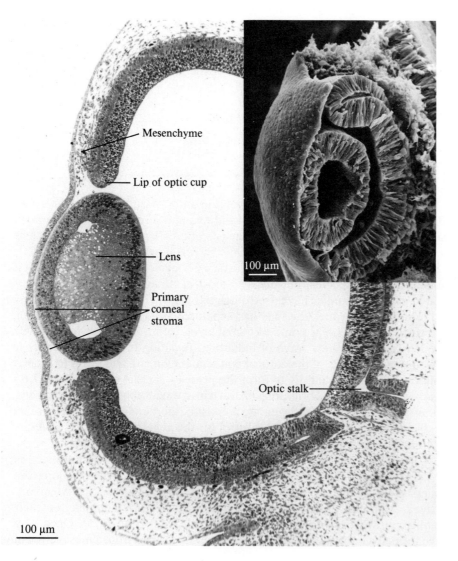

Mesenchyme

Lip of optic cup

Lens

Primary corneal stroma

100 μm

Optic stalk

100 μm

Figure 17.3

A low-power section through a stage 22 chick eye shortly after the early morphogenetic movements are complete. Note the extensive mesenchyme that surrounds the eye. Most of the mesenchyme is of neural crest origin and will later contribute to the corneal stroma and endothelium, the stroma of the iris, the connective tissue of the choroid and sclera, and to the ciliary ganglion and the ciliary muscles. (From Hay and Revel, 1969. Reprinted by permission of S. Karger AG, Basel.) The inset shows a similar stage and view by scanning electron microscopy. (Inset courtesy of Kathryn Tosney.)

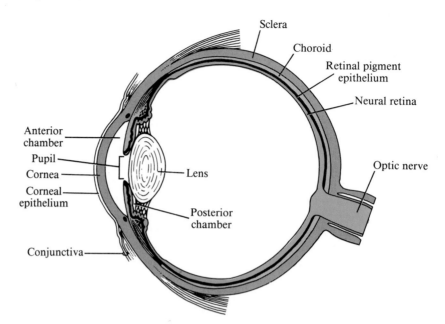

Figure 17.4

Schematic sagittal section through a vertebrate eye showing the relationships among the major structures. The large chamber between the lens and the neural retina contains the vitreous body. (Adapted from Ross et al., 1989.)

brain where they connect with the visual centers. These neurons together constitute the **optic nerve**. The final shape and components of the eye are diagrammed in Figure 17.4.

In the succeeding sections, we shall examine in detail what controls the morphogenesis and differentiation of the major elements of the eye.

17–2 The Cornea: Progressive Development Correlated with Changes in Extracellular Matrix

The cornea is the clear covering over the anterior surface of the eye (see Fig. 17.4). It serves both to protect the eye and also to focus light on the lens. In most species, the cornea consists of a collagenous stroma in which a few fibroblasts are embedded (Fig. 17.5). It is bounded by epithelial layers at both its anterior and posterior surfaces. Development of the cornea is similar in many species, but because most of the important morphological and biochemical studies have used chick tissues, we consider only the chick.

Immediately after the lens has separated from the surface ectoderm, the cornea consists of a single epithelial layer. Soon, however, contact with the lens initiates a new wave of induction, resulting in the first changes in the cornea. The cells of the corneal epithelium elongate and begin to secrete collagen types I, II, and IX to form the **primary corneal stroma** (Fig. 17.6A inset). These collagen fibrils are laid down in a remarkably regular orthogonal pattern (i.e., layers of collagen are deposited perpendicular to each other in the plane of the cornea). The control for the generation of such a precise and unusual arrangement is still unknown (Bard et al., 1988).

Soon after the primary stroma is established, cells from the margins of the optic cup (Noden, 1978; Johnston et al., 1979) migrate along the undersurface of the stroma and on the lens capsule. These cells, which are derived from the neural crest, use lamellipodia and filopodia for their motility, just as cells do in tissue culture (Bard and Hay, 1975). Once they have formed a monolayer of cells, they are transformed from a mesenchymal

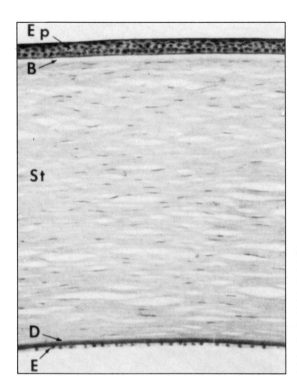

Figure 17.5

Micrograph of a human cornea revealing the anterior epithelium (Ep) and its associated basal lamina (B: Bowman's membrane), the posterior endothelium (E) and its basal lamina (D: Descemet's membrane), and the intervening stroma (St), which consists of a collagenous matrix in which fibroblasts are embedded. ×160. (Reproduced from D. G. Cogan and T. Kuwabara, 1973, The eye. *In* L. Weiss (ed.), *Histology*, 3rd ed. Reprinted by permission of McGraw-Hill Book Co.)

population to an epithelial layer called an endothelium, which will separate the corneal stroma from the anterior chamber of the eye (Hay and Revel, 1969; Bard et al., 1975; see Fig. 17.6A). Two days later, the endothelial cells, in turn, secrete hyaluronic acid into the primary stroma (Trelstad et al., 1974). Hyaluronic acid is a large space-forming molecule (see Chap. 9) that causes the primary stroma to swell (see Fig. 17.6B). The resultant expansion of the stroma may create spaces that make it easier for cells to invade. Consequently, mesenchymal cells (also derived from the neural crest; Fig. 17.7) that have paused at the margins of the optic cup invade the multilayered stroma (see Fig. 17.6B) and, in turn, secrete more collagen onto the primary template to create the secondary stroma (Bard and Higginson, 1977). The neural crest cells in this second wave of migration are referred to as stromal fibroblasts.

Once the secondary layer of collagen has been deposited, the cornea shrinks and subsequently becomes transparent—an attribute that is critical to normal eye function. There are apparently at least two controls for this process: The stromal fibroblasts produce hyaluronidase, which digests the hyaluronic acid in the cornea, causing it to shrink. Toole and Trelstad (1971) measured hyaluronidase activity in corneal extracts and found that activity was highest at the time that shrinkage began. A second factor involved in stromal condensation is the hormone thyroxine. If a thyroid gland from a 21-day-old chick is grafted to the chorioallantoic membrane of a 7-day-old embryo, the corneas of the embryo will condense prematurely and be transparent (Coulombre and Coulombre, 1964). Furthermore, if young (12- to 16-day old) chick embryos are injected with thyroxine, the corneas are clearer and more dehydrated than those in control chicks receiving saline injections. It has been proposed that thyroxine controls dehydration by making the endothelium more efficient in pumping solutes out of the cornea and into the anterior chamber (Masterson et al., 1977).

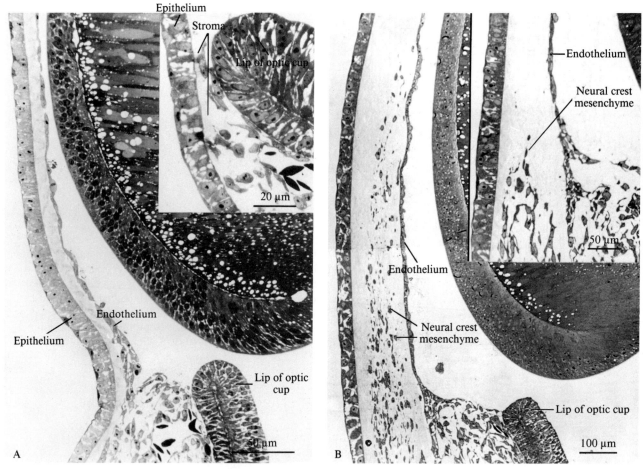

Figure 17.6

Development of the chick corneal stroma. A, Light micrograph of a section through the cornea of a chick embryo after the migration of the neural crest cells along the corneal stroma to form the endothelium. The inset shows the presumptive endothelial cells at the rim of the optic cup just as they initiate their migration. B, Light micrograph of a section through the cornea. The corneal stroma has thickened owing to the production of hyaluronic acid by the endothelial cells. This results in a second wave of mesenchyme migration, this time into the stroma. The inset shows some of the neural crest-derived fibroblasts poised at the threshold of the corneal stroma as they prepare to invade. (From Hay and Revel, 1969. Reprinted by permission of S. Karger AG, Basel.)

The changes in the corneal stroma and correlated morphogenetic events are summarized in Figure 17.8. This figure emphasizes the role of the extracellular matrix in the control of cell movement. All of these events and changes are required to generate the adult cornea. By the same token, physical constraints are as important as molecular changes in shaping the various components of the eye. The convex shape of the cornea, for example, is as critical as transparency in focusing light, and it acquires its specific curvature from intraocular pressure. By day 4 of development in the chick, the accumulation of the vitreous humor in the eye cavity begins to exert pressure against the wall of the eye. If, at day 4, a small capillary tube is inserted into the eye to drain away the vitreous (Coulombre, 1956), thus reducing intraocular pressure, the eye fails to enlarge. In addition to affecting the size of the eyeball, intraocular pressure also shapes the cornea.

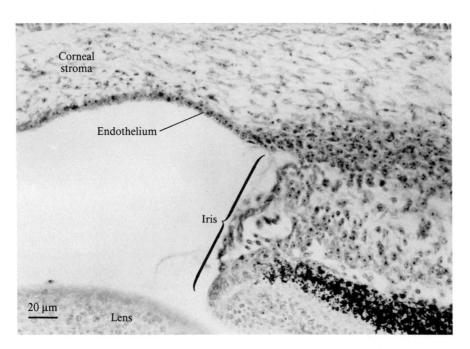

Figure 17.7

To identify the contribution of neural crest cells to various structures of the eye, chick head neural crest cells were replaced with quail neural crest before their migration from the neural tube. The quail cells (and therefore the neural crest cells) can be identified by their large deep-staining nucleoli, whereas the chick nucleoli stain uniformly. The stromal fibroblasts and endothelium of the cornea are derived from the neural crest, as are the mesenchymal cells of the iris. Note that the lens and the epithelium of the iris are chick cells and therefore not of neural crest origin. (From Noden, 1978.)

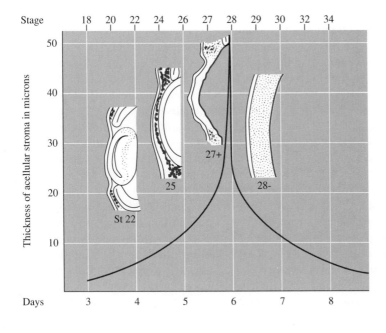

Figure 17.8

Drawings of successive stages of development of the chick cornea, demonstrating the changes in the thickness of the cornea. These changes are also plotted in the graph. St. 22: the cornea consists of the anterior epithelium and underlying primary stroma; St. 25: neural crest cells have migrated along the stroma to form the endothelium of the cornea; St. 27+: the endothelium secretes hyaluronic acid into the primary stroma, which causes it to thicken and stimulates a second invasion of neural crest cells into the stroma; St. 28–: stromal fibroblasts lay down a second layer of collagen and then secrete hyaluronidase into the matrix, causing it to shrink. (After Meier and Hay, 1973.)

Figure 17.9

The skull of an 18-day old embryonic chick stained with alizarin red S to reveal the ring of scleral ossicles encircling the cornea. (From Hall, 1981. Reprinted by permission of Company of Biologists, Ltd.)

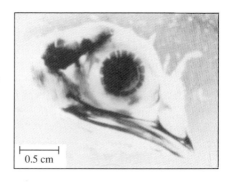

0.5 cm

As we shall see, a series of bony plates, called the scleral ossicles (Fig. 17.9), encircle the chick cornea. As the vitreous enlarges, the eye pushes against the ring of ossicles, where it is restrained from further expansion. But the cornea bulges out through the center of the ossicle ring, and in so doing the radius of curvature of the cornea is reduced (Coulombre and Coulombre, 1958). This bulging does not develop if the ocular pressure is reduced. Thus, this purely mechanical process can have a crucial role in morphogenesis.

17–3 The Sclera: Connective Tissue Covering of the Eye

The outer connective tissue layer of the mammalian eye is the sclera. The sclera is continuous with the more anterior cornea (see Fig. 17.4) and, like the cornea, is composed of a stroma of thick collagen fibers, although these are not arranged in orthogonal layers. This tough coat is responsible for maintaining the shape of the eyeball, a function critical for accurate vision as well as for protecting the eye. Because this layer is not dehydrated, it is white and not transparent like the cornea.

In nonmammalian vertebrates, the sclera is composed of two separate elements: (1) a layer of cartilage, the **scleral cartilage**, which surrounds the entire posterior surface of the eyeball; and (2) a series of bony plates that form more anteriorly around the cornea known as the **scleral ossicles**. Because much of the experimental work determining the origin of the scleral tissues, as well as defining the individual interactions, has been done in the chick, we consider the chick sclera in some detail.

From chick/quail grafting experiments, the connective tissues of the sclera (as well as the underlying choroid, which we shall not consider) are known to be derived from the cells of the neural crest (Noden, 1978; Johnston et al., 1979). At first, the neural crest cells form a loose mesenchyme around the pigmented retina (see Fig. 17.3). Soon, the cells condense and begin to develop into cartilaginous elements.

The differentiation of the scleral cartilage from the neural crest mesenchyme is directed, at least in part, by contact with the retinal pigmented epithelium. If neural crest cells are cultured with pieces of retinal pigmented epithelium, they differentiate into cartilage. Smith and Thorogood (1983) showed in transfilter experiments that direct contact is needed to produce cartilage. Subsequent experiments showed that collagen type II is present in the basement membranes of the pigmented epithelium just at the time when the inductive events occur, suggesting that collagen is important in this induction, although the actual role played by collagen remains to be determined (Thorogood et al., 1986).

Once the cartilage has been induced, its final shape and thickness are dependent upon mechanical factors, especially the shape of the eyeball, which develops under the influence of the expanding vitreous body (Coulombre, 1956).

The scleral ossicles form a series of overlapping bony plates (typically 14 in the chick) around the perimeter of the cornea (see Fig. 17.9). These underlie the **conjunctiva,** the epithelium that is continuous with the anterior corneal epithelium. Coulombre et al. (1962) showed that a single ossicle arises wherever the conjunctiva forms an invagination into the underlying mesenchyme. These ingrowths are known as the **scleral papillae**. The papillae appear beginning at the eighth day of development. Surgical removal of a papilla at an early stage results in the absence of the underlying ossicle. Such studies suggest that the conjunctiva exerts an inductive effect on the underlying neural crest cells. Hall (1981) explanted the conjunctiva into culture with neural crest cell mesenchyme isolated from embryos between 6 and 9 days of age, which resulted in production of scleral ossicles. The conjunctiva apparently has had its inductive effect by days 9 to 10 because at that time the mesenchyme can form bone when grown on the chick chorioallantoic membrane in the absence of the conjunctival membrane (Coulombre et al., 1962; Hall, 1981).

Curiously, the flattened shape of the bony ossicles may be prepatterned into the neural crest cell population. The information specifying shape of the ossicle apparently is not derived from the scleral papillae. Hall (1981) has combined neural crest cells that will form the sclera with mandibular (jaw) epithelium and still found that the unique flattened scleral ossicles form. Conversely, when conjunctival epithelium with its papillae was combined with mandibular mesenchyme (also of neural crest origin), the bones that differentiate are more typical of rounded jaw bones. Therefore, the shape of the various bones in the head seems to be patterned within the particular neural crest mesenchyme and not into the overlying epithelium. This experiment also suggests that this epitheliomesenchymal interaction is a permissive one (see p. 497) because several epithelia can substitute for the scleral papillae.

Clearly, much has yet to be learned about the information carried in the mesenchyme and in the overlying ectoderm and precisely when this information is acquired during development. Neural crest cells come in contact with many environments during their migration, and any or all of these could impart developmental information and pattern.

17–4 The Iris: Composite of Retina and Neural Crest

The iris, which separates the anterior and posterior chambers of the eye, is a pigmented disc with a hole in the center (the pupil). Because of the antagonistic actions of its dilator and sphincter muscles, the iris controls the amount of the light that enters the eye and thus functions as a diaphragm.

The bulk of the iris is a **spongy layer**, or **stroma**, which faces the anterior chamber of the eye (Fig. 17.10A). The stroma is filled with collagen, scattered fibroblasts, and varying numbers of melanocytes. The cells at the anterior border of the iris are more densely associated than in the rest of the stroma, but they are not assembled into an epithelium and have no junctional complexes between them. Within the stroma is the **sphincter muscle**, which is circumferentially oriented near the pupillary margin and is responsible for reducing the diameter of the pupil.

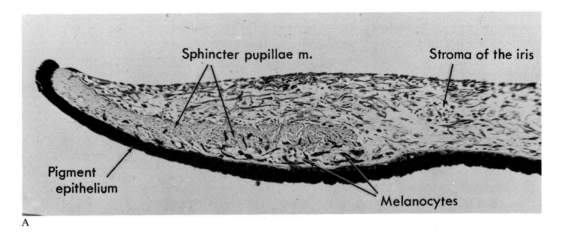

A

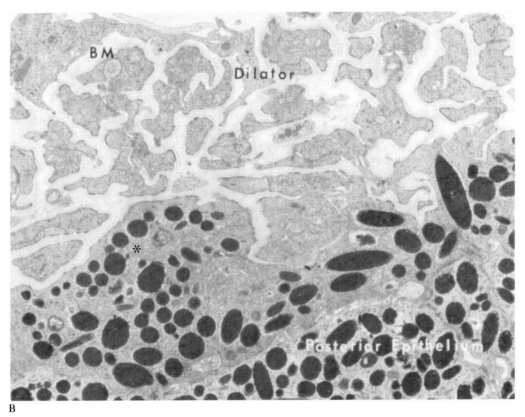

B

Figure 17.10

Structure of the iris. A, Transverse section through a human iris. The anterior
surface is composed of a stroma of loose connective tissue, fibroblasts, and
pigment cells. The posterior surface is bordered by a pigmented epithelium. B,
Electron micrograph through the pigmented epithelium of the iris. It is, in reality,
a two-layered epithelium. The posterior epithelium forms the posterior border of
the iris. The apical ends of the anterior epithelial cells (asterisk) are pigmented,
but long smooth muscle processes project from their basal surface and remain
trapped behind the basal lamina (BM; basement membrane). These form the
dilator muscle. ×18,000. (A, From Bloom and Fawcett, 1986. B, From Weiss,
1988. Reproduced with permission from Cell & Tissue Biology, 6th ed. by Leon
Weiss, 1988, Urban & Schwarzenberg, Baltimore-Munich.)

The posterior border of the iris is a double-layered pigmented epithelium, called the **epithelial layer** (see Fig. 17.10A and B). The layer of this epithelium closest to the anterior chamber generates a second intrinsic muscle, the **dilator muscle**, which opens the pupil and acts antagonistically with the sphincter muscle.

The color of the iris depends upon the number and arrangement of melanocytes in the spongy layer. If there are only a few melanocytes, the darkly pigmented posterior epithelium is seen through the colorless anterior portion, and the eye appears blue. With increasing numbers of melanocytes, the varying shades of gray or green develop. Large amounts of pigment result in brown color. Albino eyes appear pink because melanin pigment is absent, and one sees the red hemoglobin of the vascular system through an otherwise transparent iris.

The iris develops as an extension of the lips of the optic cup. Between days 4 and 6 in the chick embryo, the iris is little more than a thinning of the optic cup epithelium at its rim (Fig. 17.11A). By day 7, the iris is obvious as an extension of the double-walled optic cup. The epithelial layer closest to the lens is the posterior epithelium, and the layer facing the cornea is the anterior epithelium. This double-layered epithelium is overlaid with mesenchyme (see Fig. 17.11A and B), which is derived from the neural crest (Noden, 1978; see Fig. 17.7).

By day 7½, distinct epithelial buds have grown from the anterior epithelium at the margin of the iris and extend into the stroma (see Fig. 17.11B). These buds will continue to grow until day 10, when clusters of cells separate from the buds; muscle-specific antibodies show that they are muscle cells (see Fig. 17.11C). These are presumptive sphincter muscle cells. They will continue to arise from the margin until at least day 17 and are arranged circumferentially around the margin.

The dilator muscle is also derived from the anterior epithelium. By day 12 or 13, muscle processes begin to extend from the cells of the anterior epithelium into the stroma along the entire length of the anterior epithelium (Fig. 17.12). In the adult, the apical portion of the cells is typically epithelial in character and heavily pigmented, whereas the basal end is highly folded, contains smooth muscle elements, and is covered with basal lamina (see Fig. 17.10). The dilator muscle cell fibers are aligned radially away from the pupil; therefore, when they contract, the pupil dilates.

Most of the studies on the iris are strictly morphological; thus, the exact origin of the cells within it is subject to question. Studies of chick/ quail chimeras have identified unequivocally the iris stroma as being derived from neural crest mesenchyme, whereas the posterior and anterior epithelia are derived from neural epithelium. However, these same studies could not determine unequivocally whether the muscles are derived from the anterior epithelium or the neural crest because the quail marker becomes obscured in smooth muscle cells.

We are ignorant about most aspects of iris development, including what controls pigmentation, what induces the differentiation of the two intrinsic muscle groups from the anterior epithelium, and what initially stimulates the outgrowth of the iris from the lips of the optic cup.

17–5 *Origin and Development of the Lens*

As we discussed in Chapter 12, the lens forms from the surface ectoderm as the culmination of a series of inductive interactions. At first, the lens primordium is produced by an elongation of the surface ectoderm cells. The

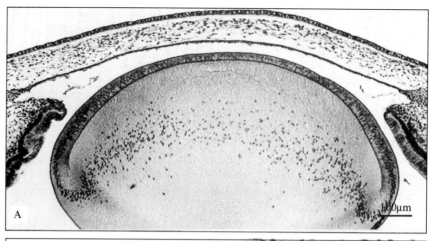

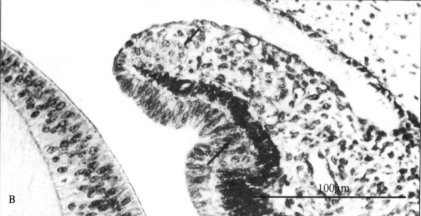

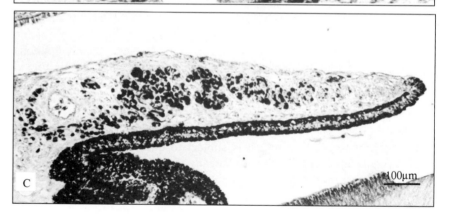

Figure 17.11

Iris development in the chick embryo. A, Section through the anterior eye region of a 6½-day chick embryo. At this time, the iris is only a small projection of the optic cup epithelium with associated neural crest-derived mesenchyme. B, As the iris extends, buds develop (arrow) from the anterior epithelium; the buds will eventually elaborate the sphincter muscle. C, An iris from a 13-day embryo stained with muscle-specific antibody to reveal the muscle cells in the stroma that were derived from the epithelial buds. (From Ferrari and Koch, 1984. Reprinted by permission of Company of Biologists, Ltd.)

elongation of lens placode cells is apparently achieved by the longitudinal alignment of microtubules (Byers and Porter, 1964), a mechanism suspected to occur in many other systems (see Chap. 9). In birds and mammals, the primordium then becomes a cup-shaped pocket that eventually constricts from the surface ectoderm and is transformed into a spherical **lens vesicle** (see Fig. 17.1E), which differentiates into the definitive lens. In amphibians and bony fishes, lens formation differs somewhat. In these groups, the inner layer of the epidermis thickens into a solid mass of cells; the cells rearrange themselves into a vesicle, which then undergoes differentiation.

Differentiation of the vertebrate lens begins with an elongation of the cells at the posterior side of the lens (Fig. 17.13). The cells first become columnar and are later transformed into long "fiber cells," which occupy

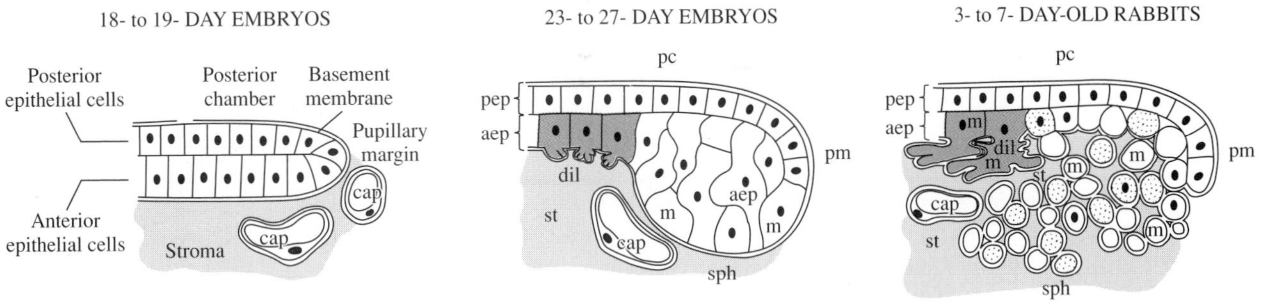

18- to 19- DAY EMBRYOS   23- to 27- DAY EMBRYOS   3- to 7- DAY-OLD RABBITS

Figure 17.12

Diagrammatic representation of the origin and development of the sphincter (sph) and dilator (dil; color) muscles in the iris of a rabbit. Note that the sphincter muscle cells (m) detach entirely from the anterior epithelial cells (aep), whereas the apical ends of the dilator muscle cells (m) remain within the epithelium and extend their muscle processes into the stroma (st). cap: capillary; pc: posterior chamber; pm: pupillary margin; pep: posterior epithelial cells. (After Tamura and Smelser, 1973. Reprinted from Archives of Ophthalmology, 1973, *89*: 332–339. Copyright 1973, American Medical Association.)

the center of the lens. This elongation step apparently is not due to alignment of microtubules because treatment with colchicine or nacodazole has no effect on the elongation process. Rather, Beebe and co-workers (1982) have evidence that elongation is due to an increase in cell volume. Because the lens is constrained around its circumference by an extracellular capsule, the lens fiber cells can get no wider as they swell; therefore they become longer (Fig. 17.14).

The cells of the outer, or anterior, face of the lens vesicle remain epithelial and form a sheet that covers the outer surfaces of the fiber cells (Fig. 17.15A). In the embryonic lens, this entire epithelial layer remains mitotically active, producing new cells that differentiate into fiber cells. As the animal matures, however, the central region of the anterior epithelium becomes mitotically quiescent, and mitosis is restricted to a zone that encircles the central region. The zone of cell division is called the **germinative region** (see Fig. 17.15B). As the cells divide, they are pushed to the equator

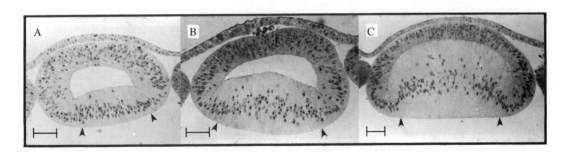

Figure 17.13

Sections through chick lenses at progressively more advanced stages of primary lens fiber cell elongation. Those fibers that elongate (between the arrowheads) increase substantially in volume, suggesting that this is the mechanism of elongation. Scale bars equal 100 μm. (From Beebe et al., 1982.)

Figure 17.14

Diagram of the forces (arrows) exerted on lens primary fiber cells during elongation. As the cells increase in volume, they press against each other and the lens capsule (c). Expansion is opposed in these two directions, resulting in cell elongation into the anterior lumen of the lens vesicle, where there is no competing force. Similar sorts of forces are probably involved in secondary fiber cell elongation as well. oc: optic cup. (After Beebe et al., 1982.)

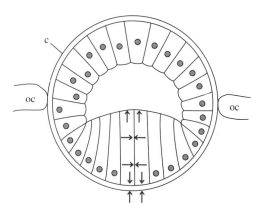

of the lens, where they begin to elongate. This elongation event—like the earlier elongation of the primary fiber cells—is probably due to pressure caused by increasing cell volume. The elongating cells are then pushed toward the center of the lens, where they undergo further elongation and terminal differentiation, losing their nuclei and acquiring a hard, transparent cytoplasm. The process of fiber cell formation continues throughout the life of the organism, as new fiber cells are added around the central core (called the **nucleus**), layer upon layer. Thus, the first fiber cells formed during embryogenesis constitute the central core, whereas newly formed fiber cells are found at progressively more peripheral locations.

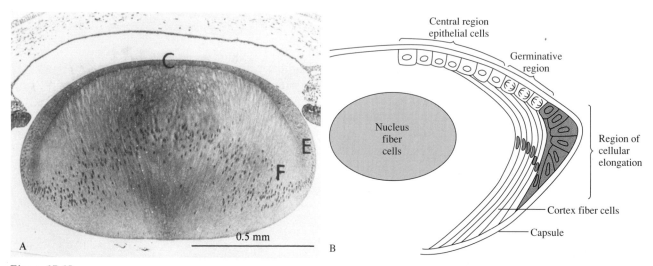

Figure 17.15

Structure of the vertebrate lens. A, Phase contrast micrograph of an axial section from a 6-day chick embryo. C: central epithelium; E: region of cellular elongation; F: lens fibers. B, Diagram of the vertebrate lens. The lens is composed of epithelial cells surrounded by an extracellular matrix capsule. A region of the anterior epithelium is mitotically active and is known as the germinative region. As these cells divide, they are pushed toward a zone at the equator of the lens, where they begin to elongate. As the cells elongate, they are displaced to the interior by the constantly proliferating and elongating cells at the periphery. The first lens fiber cells that formed occupy the center, or nucleus, of the lens. (A, From Piatigorsky et al., 1972. B, After Papaconstantinou, 1967.)

We have considered the various connective tissue elements of the eye, all of which are critical for its normal function. We now turn our attention to the development of the neural components of the eye: the neural retina and the optic nerve.

17–6 *The Neural Retina*

The neural retina is an extraordinarily complex arrangement of neurons and glial cells that receives sensory input, processes it, and transmits that information to the brain, where it is integrated. Because the eye forms from an outpocket of the brain, its development is reminiscent of our earlier discussion of brain development (see Chap. 7).

The retina in cross section (Fig. 17.16) shows a series of striations that represent (1) the cell bodies of various neurons (the nuclear layers) and (2) layers where cell processes that extend from cell bodies in adjacent nuclear layers synapse with each other (plexiform layers). The three nuclear layers, moving from the outer layer of the retina (closest to the retinal pigmented epithelium) to the inner layer of the retina (closest to the vitreal surface), are the **outer nuclear layer**, the **inner nuclear layer**, and the **ganglion cell layer**. These layers are separated by the **outer** and **inner plexiform layers**, respectively.

Within these layers, several well-defined neuronal types are found: (1) the rods and cones with their sensory ends extending into the space adjacent to the pigment layer and with their nuclei in the outer nuclear layer; (2) the bipolar cells, horizontal cells, and amacrine cells, with their nuclei in the inner nuclear layer; and (3) the ganglion cells, with their nuclei in the ganglion cell layer. These neurons synapse with each other in the plexiform layers. Thus, the rods and cones synapse with the bipolar cells and horizontal cells in the outer plexiform layer, and the bipolar cells and amacrine cells synapse with the ganglion cells in the inner plexiform layer. Finally, the ganglion cell axons are collected in bundles that comprise the **nerve fiber layer,** the innermost layer of the retina. These axons then enter the optic stalk and finally make synapses with cells in the brain.

The Müller glial cells, the major non-neuronal cell type, span the width of the retina, with their nuclei in the inner nuclear layer. These are the supportive cells of the retina and also modulate neuronal cell activity.

The retina is bounded at its inner and outer surfaces by **limiting membranes**, which are not plasma membranes at all. The outer limiting membrane represents the junctional complexes between the rod and cone cells and the Müller glial cells. The inner limiting membrane is a basal lamina formed by the Müller glial cells.

The neural retina layer is closely apposed to the outer pigmented retinal epithelium. During development, cell processes from the pigmented epithelial layer interdigitate with the outer segments of the rods and cones, and the ventricular space between them becomes obliterated. There are no differentiated junctions between these two layers, however; thus, it represents a weak point where the two layers of the retina can detach, causing blindness.

The presumptive neural retina of the optic vesicle is a simple columnar epithelium, as is the neural tube from which it has developed. The cells in the presumptive retina then undergo many rounds of cell division. The cell bodies migrate to the outer layer of the retina (the layer that is adjacent to the pigment layer), where they divide and then remigrate to the inner layer. This pattern of mitosis is reminiscent of mitosis in the neural tube, which was described in Chapter 7. This division eventually results in a thickening

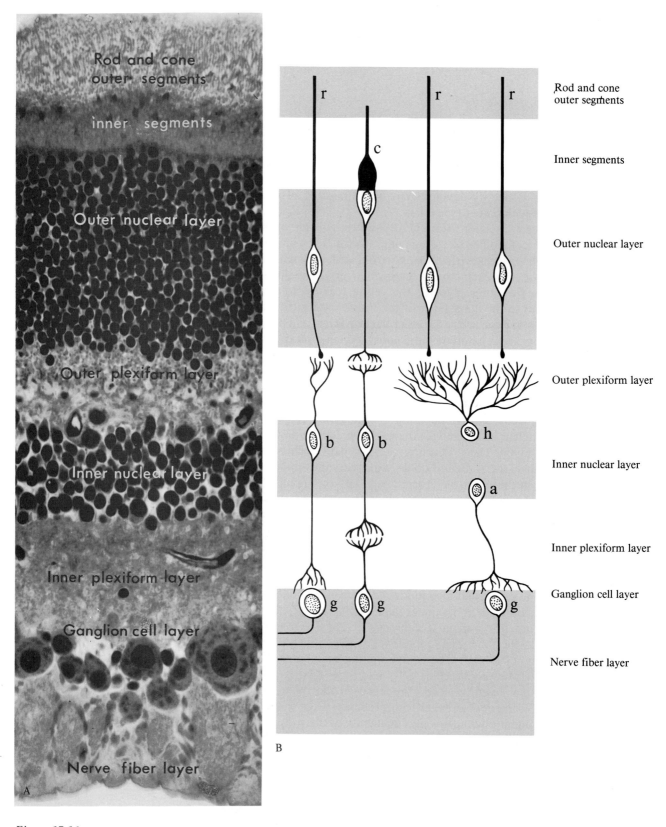

Figure 17.16

Structure of the retina. A, Section through the retina of a cat, showing the nuclear layers and plexiform (synaptic) layers. B, A scheme of the cellular organization and the synaptic connections between the various cell types in the vertebrate retina. r: rod; c: cone; b: bipolar cell; h: horizontal cell; a: amacrine cell; g: ganglion cell. (A, From Bloom and Fawcett, 1986. B, After Ramon y Cajal, 1972 [translation].)

of the retinal epithelium. As a result of cell division and cell movement, as well as cell differentiation, the retina becomes layered. Differentiation of the various cell types appears to progress in a wave from the inner layer to the outer layer (Coulombre, 1955; Mann, 1964; Blanks, 1982; Grun, 1982). In the mouse or chick, for example, the ganglion cell layer is the first to segregate at the inner surface of the retina. Next, the inner and outer nuclear layers segregate, with the amacrine and horizontal cells differentiating within them. Finally, the bipolar, photoreceptor (rod or cone), and Müller glial cells acquire their distinctive shapes and establish the appropriate synapses (Fig. 17.17). The molecular basis for the sorting out of cells into the various strata of the retina has not been established.

A major focus of recent research on the development of the eye concerns what controls the differentiation of the various cell types. Two alternative hypotheses can be posed. On the one hand, development could be mosaic, so that each cell type arises from a specific precursor cell. Alternatively, all cell types might be equivalent but differentiate according to where they find themselves within the retina (i.e., regulative development). Recent experiments, in which the precursor cells in the immature retina were labeled, have traced the lineages of the retinal cells, and all of these studies, which span several species, consistently suggest that development is regulative.

In the first of these studies (Turner and Cepko, 1987), a retrovirus vector containing the bacterial β-galactosidase gene was injected into neonatal rat eyes. The virus infected individual eye cells, and the vector was incorporated into the genomes of the host cells and inherited by each

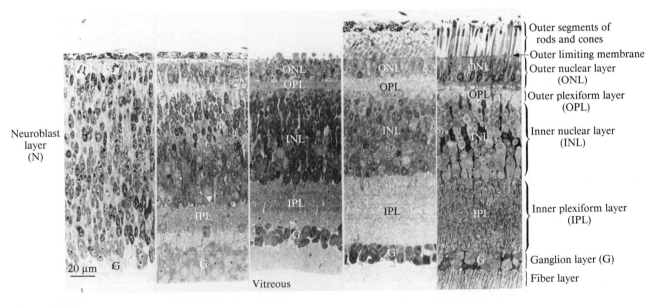

Figure 17.17

The retina is initially a pseudostratified layer of apparently undifferentiated cells. With time, they begin to differentiate and sort out to form many cell layers, as seen in this sequence of sections through the retinas of progressively older chicken embryos. The first cells to become overtly differentiated are the cells in the ganglion layers. Next, the neurons in the inner nuclear layer form processes that synapse in the plexiform layers. Finally, the photoreceptors at the outer surface of the retina differentiate. (From Sheffield and Fischman, 1970.)

daughter cell at mitosis, giving rise to clones of cells derived from the initial infected cells. Each clone could be identified by the labeling method described in the essay on page 703. In the neonatal rat retina, four cell types are known to retain mitotic ability: the Müller glial cells, horizontal cells, amacrine cells, and bipolar cells. Most clones, identified by staining for β-galactosidase, contained only a single cell type, but some clones contained up to three cell types and these could be in any combination, including a neuronal cell and a glial cell (Fig. 17.18). These results suggest that there are no specific precursors for each cell type and that precursor cells are at least pluripotent. Furthermore, stem cells do not seem to be committed early.

Similar studies in *Xenopus*, in which either rhodamine dextran (Wetts and Fraser, 1988) or horseradish peroxidase (Holt et al., 1988) was injected into individual neuroepithelial cells in the optic vesicle, showed that labeled clones arising from a single precursor could contain almost any combination of cell types. The clones were often arranged in columns across the width of the retina, spanning all three of the neuronal cell layers (Fig. 17.19).

These recent experiments suggest that cells differentiate according to the layer of the retina in which they find themselves and are not specified early by virtue of the precursor cell from which they arose.

Some recent experimental studies suggest that nearest neighbors may strongly influence differentiation. To test this hypothesis, kainate, which specifically kills some cells in the inner nuclear layer that possess glutamate receptors, was injected into *Rana* eyes (Reh, 1987). The eyes were examined for the location and type of mitotically active cells. There was a specific increase in the birth of the cell type killed and no change in the birth rate of others. Thus, the only cells that divided were those needed locally to replace the killed cells. A signal from adjacent mature cells appears to regulate the differentiation of the precursor cells.

Regular text continues on page 704.

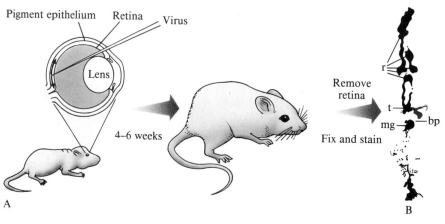

Figure 17.18

Retinal cell origins in the rat. A, A retroviral vector that expresses the bacterial β-galactosidase gene was injected into the space between the neural and pigmented retina in neonatal rats. The animals were killed at 4–6 weeks of age and the injected retinas fixed and processed for β-galactosidase activity. B, A section through an infected retina that has been processed for enzyme activity. A single infected cell has given rise to this radially-arrayed clone, which includes five rods (r), a bipolar cell (bp), and a Müller glial cell (mg). The labeled rod terminals are indicated (t) just above the bipolar cell. For color version, see Color plate 27. (After Turner and Cepko, 1987. Reprinted by permission from *Nature* Vol. 328 pp. 131–136. Copyright © 1987 Macmillan Magazines Ltd.)

Essay

Techniques for Tracing Cell Lineage

The analysis of cell lineage, which involves tracing the descendants or progeny from a precursor cell, has become the subject of increasing attention in developmental biology. The embryos of some animals, such as *Caenorhabditis elegans*, are sufficiently clear or have few enough cells that cell ancestry can be traced by observing the product of each cell division from the time of first cleavage (see Chap. 11). Even for the simplest organisms this has been an extremely difficult and painstaking task. For more complex and opaque embryos, such visual tracing is impossible.

Recently, several techniques have been developed that allow us to trace the progeny of single cells in more complex organisms.

The first of these techniques involves the injection of specially developed dyes that, once injected or taken up into a cell, cannot get out again and are not toxic to the cells. These include the fluorescein- and rhodamine-labeled dextrans (Gimlich and Braun, 1985). These have been used to advantage in tracing lineages in a variety of systems. They are visible without any biochemical reaction and therefore can be visualized in a living embryo as it develops. Furthermore, these dyes are very bright and will persist for some period of time in the cells. These dyes have also proved to be excellent markers for transplanted cells and can be used to mark individual cells and watch their migration in the embryo (see Fig. 6.9, p. 204).

The major disadvantage of fluorescent dyes, as well as other dyes such as horseradish peroxidase (HRP) (see essay on p. 280), is that they are eventually diluted beyond the level of detection owing to cytokinesis, especially in mammals, in which growth is so fast that such injection techniques must be limited to analysis of early developmental events.

Recently, however, retroviruses have been adapted to lineage studies to provide a stable, permanent marker. When a retrovirus infects a cell, its genome is integrated into the cell's DNA and is subsequently inherited by all that cell's progeny. Several research groups (e.g., Sanes et al., 1986; Price et al., 1987) have constructed viruses in which the viral structural genes have been replaced with an *Escherichia coli* β-galactosidase (*lac-Z*) gene, which becomes integrated into the host genome and is then expressed in the infected cell. The β-galactosidase can be detected by staining with the chromogenic substrate 5-bromo-4-chloro-3-indolyl-β-D-galactoside (X-gal), which forms a blue precipitate. Unlike injected dyes, which can be diluted out with cell division, the virus-infected cell and its progeny will permanently carry the *lac-Z* gene as a marker. Furthermore, the viral genomes have been engineered so that they cannot replicate and reinfect additional cells.

Sanes and colleagues (1986) injected mouse fetuses with their viral construct through the uterine wall and later sacrificed the embryos and examined them for the presence of β-galactosidase activity. Clusters of cells were found throughout the embryo, ranging in size from only a few cells to many hundreds. Each cluster is assumed to represent a clone of cells derived from a single virus-infected cell. In such clones, a variety of differentiated cell types arose, suggesting that precursor cells in a variety of tissues are pluripotent. Furthermore, the later the embryos were injected, the more limited the cell types in each clone, allowing the investigators to pinpoint when specific phenotypes were determined and segregated from the precursor stem cell.

Clonal analysis using viral constructs has now been used to study the development of such complex tissues as the central nervous system, including the neural retina and the optic tectum (see Fig. 1). Such markers will be invaluable in doing long-term clonal analysis in tissues that divide extensively. Because cells can be infected without directly injecting them, this may also remain the method of choice when injection of a single cell is not possible, either because it is too small or because it is inaccessible, as in the deep layers of the developing brain.

The developmental biologist now has available an impressive arsenal of techniques to study the control of cell differentiation.

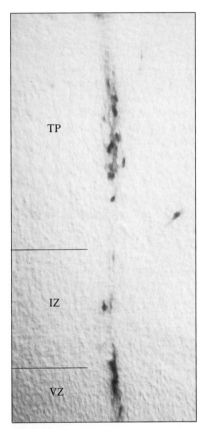

Figure 1

A chicken embryo was injected in the optic tectum with virus carrying the gene for β-galactosidase at stage 16 (51–56 hours of development) and fixed, sectioned, and stained at stage 34. A clone of cells derived from a single infected cell is revealed after staining with X-gal. The clone is arrayed radially across the tectum, suggesting that progeny are displaced vertically through the various layers, with minimal mixing laterally. Scale bar equals 50 μm. See Color plate 26 for color version. (From Gray et al., 1988.)

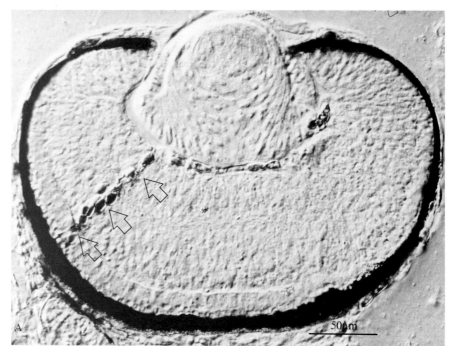

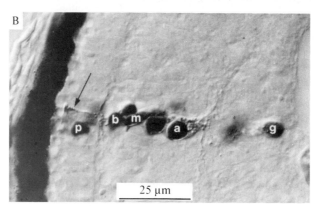

Figure 17.19

Retinal cell origins in *Xenopus*. A, A single cell in this *Xenopus* retina was injected with horseradish peroxidase and allowed to develop. All the labeled cells (arrowheads) are derived from a single precursor and are radially distributed through all the layers of the retina. B, Higher magnification of A. Differentiated cell types within this clone include photoreceptor cells (p), bipolar cells (b), Müller glial cells (m), amacrine cells (a), and ganglion cells (g). The arrow points to a rod cell in partial focus. (From Holt et al., 1988. Copyright © Cell Press.)

With the recent perfection of tissue culture systems in which different cell types of the retina can differentiate (Adler and Hatlee, 1989), we are now in a position to look at the environmental factors that control the precise distribution of cell types in the neural retina.

17–7 Retinotectal Projection

Without the appropriate connections established between the neurons of the retina and the brain, visual stimuli cannot be processed. The neurons of the retina arise from the neural epithelium and send out their axons along the optic stalk to the diencephalon. Considerable data and much controversy exist concerning (1) what cues guide the axons along specific routes in the brain, and (2) what directs synaptogenesis in the correct pattern to generate vision. In this section, we explore some of the elements that control this patterning. As we shall see, these are some of the same elements that control morphogenesis in other systems.

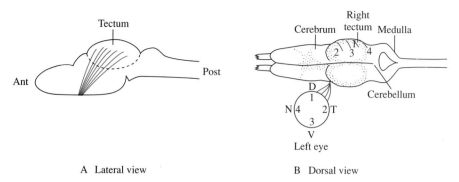

A Lateral view   B Dorsal view

Figure 17.20

Schematic drawings of the frog brain seen from lateral (A) and dorsal (B) perspectives. Retinal axons project from the eye to the opposite tectum in the pattern shown. D: dorsal; N: nasal; T: temporal; V: ventral.

Most of the experimental studies examining the development of the visual system have used lower vertebrates, often the amphibians *Xenopus laevis* and *Rana pipiens* or goldfish. In these animals, the retinal fibers extend to a region of the brain known as the **optic tectum**. This pattern of neuron distribution is thus known as the **retinotectal projection**. The tectum is a paired structure: All the fibers from the left retina project to the right tectum, and all the fibers from the right retina project to the left tectum (Fig. 17.20). Thus, the retinal ganglion axons must grow along the optic stalk and cross over to the opposite side of the brain before they climb up the side of the diencephalon to the tectum (Fig. 17.21).

Pathfinding along the Optic Tract

Before establishing the appropriate connections with the tectum, retinal axons must migrate from the eye through the optic tract. Recent experiments suggest that there are local positional cues in the embryonic brain that control navigation to the target. Harris (1989) removed part of the optic tract in the diencephalon through which the optic axons pass to reach the

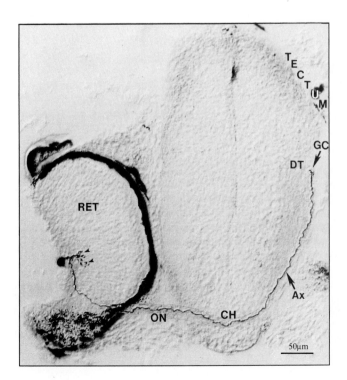

Figure 17.21

Cross-section through the diencephalon and retina of a stage 37/38 *Xenopus* embryo. The tectum is the region marked at the dorsal side of the brain. An individual retinal ganglion cell axon has been filled with Lucifer Yellow and immunostained with HRP so that its path from the retina (RET), through the optic nerve (ON), and up the diencephalon to the tectum can be visualized. Ax: axon; GC: growth cone; DT: dorsal optic tract; CH: chiasm. Arrowheads indicate dendrites. (Courtesy of C. Holt.)

tectum in *Xenopus* embryos and rotated these pieces by 90 degrees (Fig. 17.22A). When growing axons reached the rotated pieces, they deflected to either the right or left, depending upon the direction of rotation (see Fig. 17.22B to D). Eventually, the fibers still reached the tectum—but *only if they happened to pass near enough to it* (see Fig. 17.22D). Thus, there are guidance cues in the pathway that retinal axons take to get to the tectum. These are not strictly necessary because the axons can often reach their target by a circuitous route (Harris, 1986), but they clearly make axon migration more efficient and reinforce a sense of direction on a group of cells that must migrate and synapse in nearly perfect order with the tectum in order to assure normal vision. The nature of the cues that direct navigation through the brain is little understood and is being investigated actively.

Synapse with the Tectum

Once the retinal axons reach the tectum, they establish synapses in an extraordinarily regular pattern. Fibers from the dorsal retina synapse with the ventral tectum, those from the ventral retina synapse with the dorsal tectum, the anterior (nasal) retina to the posterior (caudal) tectum, and the

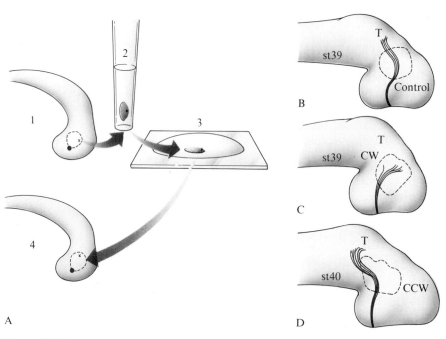

Figure 17.22

Cues in the optic tract guide retinal fibers to the tectum (T). A, A small piece of the diencephalon of a *Xenopus* embryo is cut out, incubated in a dye so that the grafted piece can be recognized at the end of the experiment, and then reinserted into the embryo—either in its normal position or rotated 90 degrees clockwise or 90 degrees counterclockwise. B, Fibers from the retina are traced as they pass through the graft region. This control shows the normal pathway of fibers. C, When the graft is rotated clockwise (CW), the fibers are deflected to the right. D, When the graft is rotated counterclockwise (CCW), the fibers are deflected to the left. If the fibers come close to the tectum as they do in D, they still innervate the tectum to form a normal projection. The factors in the pathway that guide the axons are unknown. (From Harris, 1989. Reprinted by permission from *Nature* Vol. 339 p. 218. Copyright © 1989 Macmillan Magazines Ltd.)

posterior (temporal) retina to the anterior (rostral) tectum (see Fig. 17.20). This produces a continuous point-to-point representation of the retina on the surface of the tectum that, in turn, produces an accurate map of the visual world.

This precise ordering of axon terminals on the tectum has been measured in several ways. In one, groups of fibers can be filled with dyes (see essay on p. 703) by injecting the ganglion cell bodies in the retina, and the dye-filled fibers can then be traced forward to their point of contact with the tectum.

A second method uses electrophysiological means to map the retinal axon terminals on the tectum. To do this, spots of light are focused on small regions of the retina of an anesthetized animal at the same time that an external electrode records activity from the exposed tectum. In this way, one can determine which optic nerve terminals on the tectum are electrically excited when a discrete spot on the retina is stimulated. By positioning the electrode in about 25 different positions and determining which part of the retina elicits a response, a **retinotectal map** is generated (Fig. 17.23; Jacobson, 1967). Such maps show that the arrangement of nearest neighbor neurons in the retina is preserved in their synapses to the tectum; i.e., neurons that are adjacent to each other in the retina also functionally innervate adjacent spots in the tectum. Thus, there is point-to-point specificity between retinal cells and their projection on the tectum.

Regeneration Experiments

Early experiments in which the optic nerve of a newt was crushed or severed demonstrated that the neurons could regrow and innervate the tectum in a near-perfect manner, re-establishing the correct retinotectal map and restoring normal vision (Matthey, 1925). Such regeneration experiments

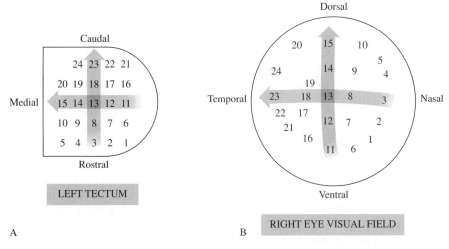

A

B

Figure 17.23

A retinotectal map of a *Xenopus* embryo assayed by electrophysiological methods. A, To construct a retinotectal map, an electrode is inserted into the tectum at the numbered sites. B, A visual field of the right eye. A spot of light is focused on the eye, which then evokes a response in the tectum. Each number on the visual field map marks the spot that evoked a response at that numbered site on the tectum. Note the one-to-one correspondence between the visual field and electrical activity of the axon terminals on the tectum. (After Fraser, 1985.)

suggested that individual neurons could recognize the correct cells on the tectum with which they synapse. Additional experiments reinforced the idea that certain "addresses" exist on the tectum that are recognized by particular retinal ganglion axons:

1. In a classic experiment, Sperry (1943) removed a newt eye and rotated it 180 degrees. With a rotated eye, the optic nerve will regenerate and reinnervate the tectum, but now the animal's visual field is upside down. If the newt is presented with a lure at the bottom of his visual field, it will leap up to catch the bait. No amount of training can reverse this visual impairment. Thus, the retinal cells have projected their axons back to cells on the tectum based on their original position in the retina.

2. If half of a goldfish eye is ablated and the severed optic nerve is allowed to regenerate, the retinal axons project initially only to the correct half of the tectum (Attardi and Sperry, 1963).

3. Finally, barricades can be placed in the path of the migrating fibers, or an eye can be grafted to an ectopic site. The appropriate retinal fibers will still find their way to the correct position on the tectum.

Such experimental evidence led to the development and gradual refinement of the **chemoaffinity hypothesis**, whose most vocal and influential proponent was Sperry (1963, 1965). Sperry argued that some sort of recognition mechanism must exist between retinal axons and the tectum that is molecular in nature. At its most extreme, his hypothesis stated that a different recognition signal must exist to account for each and every retinotectal connection. Specificity might possibly be based on complementary molecules present on each retinal axon and tectal cell that function as a lock and key.

Recent experimental work refutes such a rigid recognition system, although there is considerable evidence to support the notion that there is a gradient of one, or perhaps a few, molecules that may be responsible for the relative ordering of neurons across the tectum.

Some of the first work that argued against such rigid specificity in cell-surface labels was based on a continuation of the early work of Attardi and Sperry (1963). They showed that if half of a goldfish retina was ablated and the optic nerve was severed, the fibers from the remaining half-retina projected only to the correct portion of the tectum, leaving the rest of the tectum bare. Other workers subsequently found that, after many months, fibers from the half-retina will eventually spread across the entire tectum, forming a perfect, expanded "half-map" (Fig. 17.24A; Yoon, 1972; Schmidt et al., 1978). Similarly, if half a tectum is removed, a whole retina will project its axons to form a complete compressed map onto the half-tectum, given sufficient time (see Fig. 17.24B; Yoon, 1971; Schmidt et al., 1978). Such work refutes the idea that specific cues alone guide each neuron but supports the notion that there is relative affinity between the neurons and their target cells that allows them to order themselves across the space available.

Most of the studies on the control of retinal projection pattern that we have just discussed are regeneration experiments in which the optic nerve was severed or crushed and allowed to re-establish connections with the tectum. These can be done on mature and, therefore, larger animals that are easier to study and manipulate than embryos. However, a number of these experiments have revealed that regenerating axons may not follow the same cues as the original fibers during development. In fact, some evidence suggests that when retinal fibers establish connections with the tectum, they irreversibly alter it, allowing the regenerating fibers to find

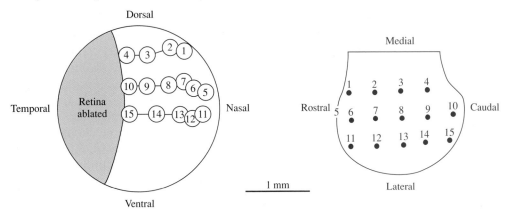

A Expansion of map

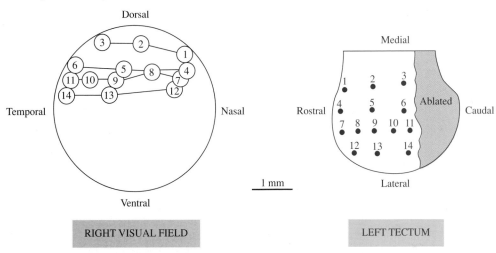

B Compression of map

RIGHT VISUAL FIELD

LEFT TECTUM

Figure 17.24

Plasticity in the retinotectal projection. A, When part of a goldfish retina was removed, the remaining ganglion cells projected fibers that map in the usual order across the entire tectum. B, Conversely, when part of the tectum was ablated, the nerve crushed, and allowed to regenerate, the whole visual field was compressed onto the remaining tectum, even though the usual sites of synapse on the tectum for some of the fibers have been removed. (A, After Schmidt et al., 1978. B, After Yoon, 1971. Reprinted from Experimental Neurology by permission of Wiley-Liss, a division of John Wiley and Sons, Inc. Copyright © 1971, Wiley-Liss.)

their original innervation site using a different mechanism. For example, as we have just seen, if part of a tectum is ablated, the retinal fibers will eventually form a normal, but compressed, map on the tectum, but only after some time. Similarly, if half a retina is removed, the remaining fibers will form an expanded map on the tectum, but again after a time lag. If, in either of these circumstances, the optic nerve is severed for a second time, the fibers regrow to the same spots immediately, without a time lag. These experiments suggest that the retinal fibers have made some indelible mark on the tectum, which then directs further regeneration (Schmidt, 1978).

Such experiments have made it clear that if we are to understand the cues that direct retinal fibers during development, we will have to study developing systems—not regenerating ones.

Developing Systems: Molecular Basis of Chemoaffinity

Recent work on developing systems also suggests that some sort of chemoaffinity has a role, perhaps even a dominant role, in establishing the retinotectal pattern. Willshaw and colleagues (1983) replaced one quarter to one third of the left eye rudiment of a young *Xenopus* larva with a similarly sized pie-shaped piece from a different position in the right eye rudiment. Various combinations included replacing a nasal (medial) piece with a temporal (lateral) piece and *vice versa*, or replacing a dorsal piece with a ventral piece (Fig. 17.25). The retinotectal projections were assessed using electrophysiological methods at various stages of development. The projections of the grafted piece became oriented according to their *origin*, rather than their position after grafting. These studies have been repeated recently by grafting small groups of *Xenopus* retinal cells that have been labeled with fluorescent dextrans into an unlabeled host retina, either in the correct position or in an abnormal position (Fraser et al., unpublished). In this way, the projections can be visualized. Once again, the retinal ganglion cells send their axons back to a position on the tectum that is consistent with the origin of the cells from the donor. These studies suggest that the retinal cells recognize some cue on the tectum to guide their innervation.

Retinal ganglion growth cones may recognize the appropriate target site because of an adhesive affinity. For example, dorsal retinal cells might adhere better to the ventral tectum, whereas ventral retinal cells might adhere better to the dorsal tectum. A variety of studies suggest that this is the case.

Figure 17.26 outlines the various assays that have been used to test adhesive preferences between retinal neurons and tectal cells. Retinal cells that have been labeled with radioactive isotopes can be dissociated and the single cells then added to tectal cells, which are either intact pieces of tectum or a monolayer of tectal cells in culture. The number of retinal cells that adhere to the tectal cells can be measured because of their radioactivity. Most such studies show that adhesive preferences reflect the normal retinotectal projections; i.e., dorsal retinal cells adhere preferentially to ventral tectum, whereas ventral retinal cells have a greater affinity with dorsal tectum (Barbera et al., 1973; Barbera, 1975; Gottlieb et al., 1976).

In addition to the demonstrated adhesive preferences, the axons of retinal cells from particular regions of the retina will also show a preference

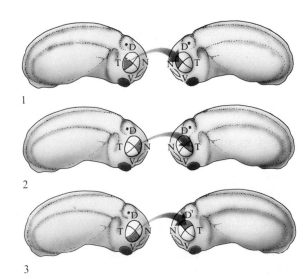

Figure 17.25

Schematic diagram of the operations replacing (1) nasal (N) with temporal (T), (2) temporal with nasal, and (3) dorsal (D) with ventral (V) eye segments from *Xenopus* embryos. (After Willshaw et al., 1983.)

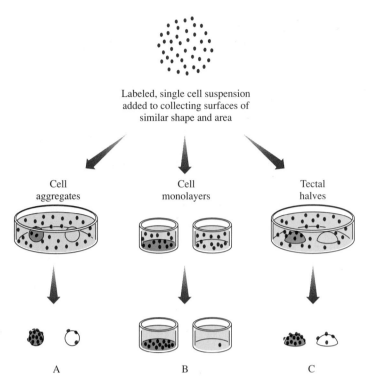

Labeled, single cell suspension
added to collecting surfaces of
similar shape and area

Cell
aggregates

Cell
monolayers

Tectal
halves

A

B

C

Figure 17.26

Schematic drawing of the variety of assays used to measure retinotectal affinity. Retinal cells are labeled with radioactive isotopes, dissociated into a single cell suspension, and then added to cultures that contain (A) aggregates of cells isolated from the tectum, (B) monolayers of cells isolated from the tectum, or (C) intact tecta cut in half. The number of retina cells sticking to the tectal tissue that comes from two different regions (i.e., dorsal vs. ventral) in any of these assays gives an estimate of the relative affinities between the two collecting surfaces. (After Roth and Marchase, 1976.)

to migrate over particular tectal cells in culture. For example, chick ganglion cells from the posterior (temporal) retina prefer to send their axons over tectal cells derived from the anterior tectum (Bonhoeffer and Huf, 1982), reflecting their natural topography in the animal. Furthermore, this preference is in a graded distribution along the anteroposterior tectal axis. Interestingly, nasal retinal axons will not distinguish anterior from posterior tectum. This migratory preference of axons derived from posterior retina probably depends upon some cell surface molecule because the same migratory preferences of retinal axons can be generated on isolated tectal plasma membranes (Fig. 17.27).

Such studies suggest that there are adhesive preferences that parallel the topography of at least the anteroposterior retinotectal map, but they have so far failed to yield much information about the *specific molecules* that may be involved. Another approach, using monoclonal antibody technology, has revealed at least one molecule that is present as a gradient across both the retina and the tectum; the molecule also seems to play a role in synapse formation.

Trisler et al. (1981) raised monoclonal antibodies against chick retinas and then screened the antibody-producing cells against intact retinas. By this means, they discovered an antigen, distributed in a gradient across the 14-day-old chick retina, that is 35 times more concentrated in the dorsal retina than in the ventral retina (Fig. 17.28). They named this antigen TOP, for toponymic, reflecting its unusual topographical distribution.

Trisler and Collins (1987) later showed that TOP is also present in a gradient across the developing chick tectum, but in the reverse polarity seen in the retina. Thus, the gradient of TOP is high in the dorsal retina and low in the ventral side, whereas TOP is 10 times higher in the ventral tectum than the dorsal tectum. This reversal of the gradient exactly correlates with the pattern of connectivity of the retinal axons with the tectum.

It is not yet clear if TOP has a role in the patterning of the retinotectal projection. However, Trisler et al. (1986) have injected antibody to TOP

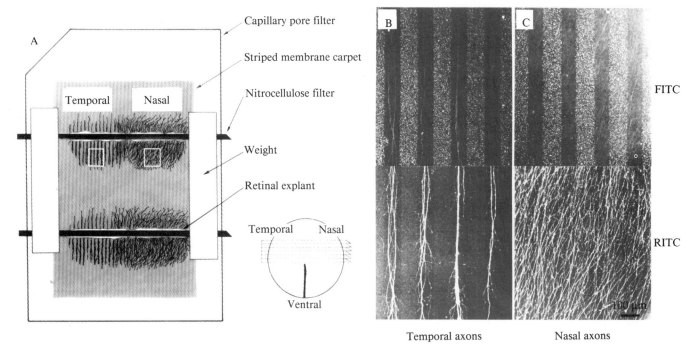

Figure 17.27

The protocol diagrammed here was developed to test the migratory ability of retinal ganglion axons on various portions of the tectum. A, Plasma membranes isolated from anterior and posterior tectum were deposited in alternating stripes to make a "carpet." Chick retina was placed on a nitrocellulose filter and strips were cut out as shown (bottom right), labeled with a red fluorescent dye (RITC: rhodamine), and laid across the tectal stripes so that nasal and temporal retinal axons will confront both anterior and posterior membranes. B, and C, Posterior tectal membranes are labeled with green fluorescence (FITC: fluorescein) and appear as speckled stripes in the top micrographs. The anterior tectal membranes are in the unlabeled intervening stripes. Temporal axons (B) grow only on the anterior tectal stripes, whereas nasal axons (C) do not distinguish between the anterior and posterior. (From Walter et al., 1987. Reprinted by permission of Company of Biologists, Ltd.)

into the developing chick retina in order to block the function of the TOP molecules. They found that the antibody prevented formation of synapses within the retina itself. Thus, the TOP molecule may be fundamentally important in connectivity between neurons and may ultimately be found to play a role in the development of synapses between retinal ganglion cell axons and the tectum as well.

Retinotectal Patterning Mechanisms in Addition to Chemoaffinity

As we discussed in Chapter 9, morphogenetic movements and pattern are often governed by multiple, and possibly redundant, cues. Multiple cues are likely to establish the precise pattern of retinotectal topography as well.

Several mechanisms other than chemoaffinity have been proposed to control the establishment of connections between ganglion cell axons and the tectum. The role of these other factors in the development and the refinement of the retinotectal projection has not been well studied until now for several reasons:

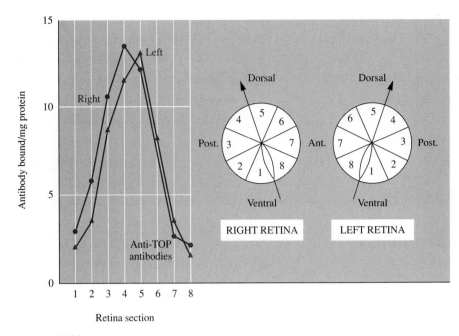

Figure 17.28

The distribution of the TOP molecule on retinal cells was determined by dividing the retina into 8 equal segments, dissociating the cells and adding to the cell suspension antibody to TOP that had been radioactively labeled. The numbers on the abscissa represent the similarly numbered segments of the retina. There is approximately 35 times more TOP in the dorsal segments of the retina than in the ventral segments. (Post.: posterior; Ant.: anterior) (After Trisler et al., 1981.)

1. Most of the studies that have identified other controlling factors have been done in regenerating systems. As we have already seen, the factors that control regeneration may be quite different from those governing the development of normal axonal patterning.

2. If there are redundant cues for controlling specificity, experiments that eliminate one factor may not alter the normal projection pattern because other remaining controls can compensate. Experiments that make an attempt to remove several possible cues at once will allow us to delineate all possible controls of retinotectal projection (e.g., Harris, 1984).

3. Until recently, it has not been possible to observe the actual behavior of an axon as it projects to the tectum and synapses with it. Now, individual (or groups of) ganglion cells can be labeled with fluorescent dyes (O'Rourke and Fraser, 1986), and with the use of sensitive video cameras, we can actually watch the behavior of these axons as they project and establish connections with the tectum (Fig. 17.29). There is no substitute in biology for actually observing a process as it occurs, rather than inferring behavior from the end result.

With these points in mind, we shall explore a few additionally proposed mechanisms for determining retinotectal connectivity, while acknowledging that much additional work will be required before they are either substantiated or refuted.

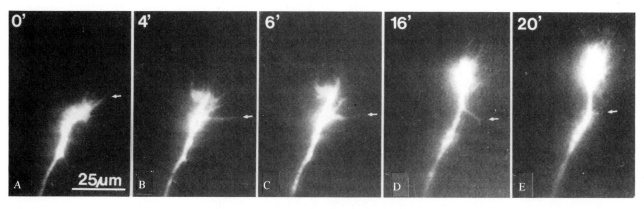

Figure 17.29

This retinal axon is labeled with the fluorescent dye DiI and observed migrating in the living intact brain of a *Xenopus* embryo. These are individual frames taken from a video recording showing the forward advance of the tip of the axon, the growth cone. Time intervals are indicated in the top left corner. (From Harris et al., 1987. Reprinted by permission of Company of Biologists, Ltd.)

Timing

One possible explanation for the correct ordering of retinal fibers on the tectum is that retinal ganglion cell neurons grow out in a developmental gradient and those fibers that arrive first at the tectum synapse with the first tectal cells they encounter. Those fibers that arrive later migrate to more distal sites that are not yet occupied. Thus, the maps may be constructed on a "first come, first served" basis (Bunt et al., 1978). Evidence from insects suggests that timing is indeed a plausible mechanism (Anderson, 1978). Is it also plausible for higher organisms?

Holt (1984) examined the timing mechanisms in the development of *Xenopus laevis*. To label the retinal ganglion cell neurons, she removed either the dorsal or ventral half of the eye of a tailbud (stage 26) embryo, incubated it in a solution of ³H-proline for 15 minutes, and then washed and regrafted the labeled eye back into the embryo (Fig. 17.30A). The fibers that grew from the labeled half of the eye could be traced using autoradiography. She found that retinal fibers leave the dorsal half of the eye first, enter the optic tract, and arrive at the tectum 6 hours ahead of the ventral retinal axons. Coincidentally, this pattern of outgrowth follows the establishment of functional connections between the retina and tectum. First, the dorsal retinal ganglion cells establish synapses with the ventromedial tectum (stage 37/38), followed by the ventral retinal cells synapsing with the dorsomedial tectum (stage 40) (see Fig. 17.30B). This developmental study thus suggests that timing could be one factor that organizes the retinotectal projection: The dorsal retinal axons occupy the first part of the tectum they encounter (the ventral tectum), and the ventral retinal axons must migrate past these fibers to the unoccupied, and still developing, dorsal tectum.

Holt then tested directly in developing animals whether the timing of outgrowth controls the spatial ordering of these connections. She labeled the dorsal half of an eye from a stage 21 embryo and grafted it in place of the dorsal half of a stage 27 eye (see Fig. 17.30C). In doing so, she replaced an older half eye with a younger half, so that the outgrowth of axons would be delayed with respect to the ventral half. Despite the fact that the dorsal fibers were the last to arrive in the tectum—at least 9 to 20 hours later than usual—a normal retinotectal map still developed (see Fig. 17.30D). These

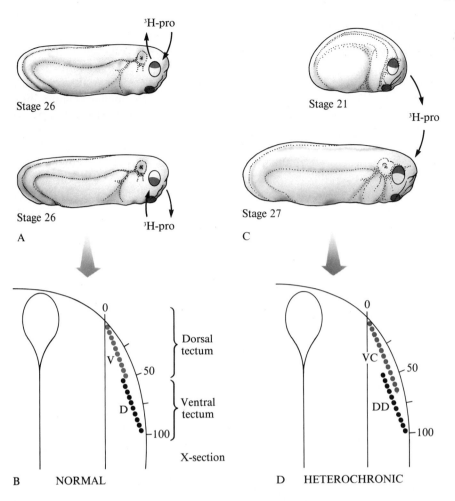

Figure 17.30

Role of timing in the correct projection of retinal cell axons to the tectum. A, To determine the order in which fibers arrive at the tectum, either a dorsal or ventral half eye primordium was removed from a tailbud-stage *Xenopus* embryo and labeled with ^{3}H-proline (^{3}H-pro). The eye was then reinserted into the embryo, and the retinal axons were allowed to grow to the tectum. The labeled fibers were identified using autoradiography. B, The distribution of fibers from the retina is mapped on this section through the tectum. The ventral retinal fibers (colored circles) synapse in the dorsal tectum, whereas the dorsal fibers (black circles) terminate in the ventral tectum. C, If a dorsal eye primordium from a young embryo (neurula stage) is grafted in place of a dorsal eye primordium from an older embryo, the outgrowth of its axons will be delayed in comparison to the ventral fibers. D, Despite the delay in their arrival at the tectum, the dorsal retinal axons (DD) still project to the ventral tectum in the same pattern observed during normal development. VC: ventral control retinal axons. (After Holt, 1984.)

results lend support to the hypothesis that retinal ganglion cells recognize at least the gross region of the tectum with which they will synapse and that timing probably does not play a major role in establishing a normal projection. Of course, as we have emphasized previously, other redundant and parallel controls may allow the establishment of appropriate connections in the absence of correct timing, so this experiment doesn't rule out any involvement of timing.

Refinement of the Initial Projection

Another proposed mechanism for establishing the retinotectal projection that clearly seems important in *regenerating* systems is that axons arborize (send out multiple branches) in an exuberant fashion over the surface of the tectum. The branches later are refined to a much smaller synaptic field. One means by which the excessive arbors could be recognized and "pruned" might be through neuronal activity. It has been hypothesized that two retinal ganglion cells will retain synapses on adjacent or the same tectal cells only if they are electrically stimulated at the same time. This coordinated "firing" would somehow reinforce and strengthen their synapses with the tectum. The mechanism of reinforcement is not known, although it clearly does occur. For example, it has been observed in the goldfish that if electrical activity is prevented by poisons during regeneration, the axons will terminate in large arborizations that never become refined with time (Meyer, 1983; Schmidt and Edwards, 1983).

During *normal development*, the role of refinement of terminal arborizations through neuronal activity is not so clear cut. In the first place, the initial map that the retina makes onto the tectum is well ordered (Holt and Harris, 1983), unlike the situation during regeneration in which many mistakes are made. In the second place, if axolotl embryos are poisoned with tetrodotoxin (to prevent the possibility of electrical cues) at the time when they are establishing their retinotectal projections, the retinotectal projections appear to be the same as in control animals (Harris, 1980).

A slightly different situation is seen in the development of the anteroposterior projection in *Xenopus* embryos (O'Rourke and Fraser, 1986). When a small group of cells in the temporal region of the retina is labeled with a fluorescein (green)-tagged dextran and a second group of cells in the nasal retina is labeled with a rhodamine (red) dextran, one can observe the overlap of these two projections in the same animal. This overlap represents the degree to which the map must ultimately be refined. At first, the axons from these two sources overlap entirely in their projection to the tectum (see Color plate 28A). This is an indication that individual arbors occupy a significant fraction of the anteroposterior dimension of the young tectum. Within a few days, the axons are seen to separate in their projections so that they synapse with different regions of the tectum (see Color plate 28B). If these embryos are poisoned so that there is no neuronal activity, or if they are raised in strobe light (which results in simultaneous firing of all retinal ganglion cells so that there is no coordinated activity), the gradual refinement of the anteroposterior projection is delayed by three days. These results suggest that although a normal map is achieved without electrical activity, such activity may help to accelerate the refinement of the projection, at least in *Xenopus*.

Summing Up

We are now poised to conduct many exciting experiments to probe this extraordinary patterning event, which has been the subject of intense investigation and considerable controversy in the past. At last, we are able to observe directly the behavior of individual axons during the developmental process, so that we will not have to infer behavior from fixed time points. In addition, we are now prepared to ask intelligent questions of the developing retinotectal projection based on the results of many experiments in regenerating systems. These experiments must be designed to perturb several possible mechanisms at the same time in order to identify all the redundant or overlapping controls exerted during the same developmental time frame.

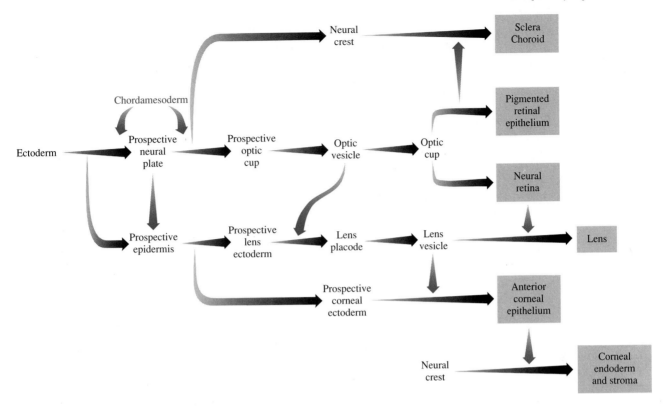

Figure 17.31

A schematic diagram of the many inductive events that occur during the development of the eye. The black arrows signify the pathway of development of a particular structure in the eye. The colored arrows signify inductive events.

17–8 Chapter Synopsis

The development of the eye is remarkable, given the many sources of tissue that must associate to form a structure that can only function within a very fine limit of tolerance. The exact placement of tissues is primarily a result of the many inductive events that occur throughout its development. The lens, for example, is positioned in the center of the eye cup because of the interaction of the retina with the overlying ectoderm. The scleral ossicles, which will ultimately produce the curvature of the cornea, are properly placed because of their induction by the conjunctiva. The numerous inductive events are summarized in the flow chart in Figure 17.31.

In most cases, we are not aware of the molecules mediating these events. In a few instances, however, such as in the cornea, we know that extracellular matrix molecules are crucial. The eye should be an important model system with which to study induction and the subsequent events that must occur at the level of the gene to control overt differentiation.

The development of the eye also encompasses a variety of morphogenetic movements. Epithelial sheets fold, such as in the formation of the lens or in the evagination of the eye cup, cells elongate, as in the growth of the lens fiber cells, and cells move actively from one region to another, as in the migration of corneal mesenchyme or the migration of neurons to the tectum.

Growth mediated by mitosis and cell death also figures prominently in the morphogenesis of the eye. In addition, the role of the expanding vitreous body, which is an extracellular matrix compartment, in the growth and shaping of the eyeball and the establishment of the proper curvature of the cornea is an aspect of morphogenesis that is unique to the eye.

Our challenge now is to understand how all of these mechanisms—morphogenetic movements, cell differentiation, and pattern—are so precisely orchestrated. Regulated gene expression as well as purely physical parameters will be at the center of this understanding.

References

Adler, R., and M. Hatlee. 1989. Plasticity and differentiation of embryonic retinal cells after terminal mitosis. Science, *243*: 391–393.

Anderson, H. 1978. Postembryonic development of the visual system of the locust, *Schistocerca gregaria*. II. An experimental investigation of the formation of the retina-lamina projection. J. Embryol. Exp. Morphol., *46*: 147–170.

Attardi, D. G., and R. W. Sperry. 1963. Preferential selection of central pathways by regenerating optic fibers. Exp. Neurol., *7*: 46–64.

Barbera, A. J. 1975. Adhesive recognition between developing retinal cells and the optic tecta of the chick embryo. Dev. Biol., *46*: 167–191.

Barbera, A. J., R. B. Marchase, and S. Roth. 1973. Adhesive recognition and retinotectal specificity. Proc. Natl. Acad. Sci. U.S.A., *70*: 2482–2486.

Bard, J. B. L., M. K. Bansal, and A. S. A. Ross. 1988. The extracellular matrix of the developing cornea: diversity, deposition and function. Development, *103* (Suppl.): 195–205.

Bard, J. B. L., and E. D. Hay. 1975. The behavior of fibroblasts from the developing avian cornea. Morphology and movement *in situ* and *in vitro*. J. Cell Biol., *67*: 400–418.

Bard, J. B. L., E. D. Hay, and S. M. Meller. 1975. Formation of the endothelium of the avian cornea: A study of cell movement *in vivo*. Dev. Biol., *42*: 334–361.

Bard, J. B. L., and K. Higginson. 1977. Fibroblast-collagen interactions in the formation of the secondary stroma of the chick cornea. J. Cell Biol., *74*: 816–829.

Beebe, D. C. et al. 1982. The mechanism of cell elongation during lens fiber cell differentiation. Dev. Biol., *92*: 54–59.

Blanks, J. C. 1982. Cellular differentiation in the mammalian retina. *In* J. G. Hollyfield and E. A. Vidrio (eds.), *The Structure of the Eye*. Elsevier Biomedical, New York, pp. 237–246.

Bloom, W., and D. W. Fawcett. 1986. *A Textbook of Histology*, 11th ed. W. B. Saunders, Philadelphia.

Bonhoeffer, F., and J. Huf. 1982. *In vitro* experiments on axon guidance demonstrating an anterior-posterior gradient on the tectum. EMBO J., *1*: 427–431.

Bunt, S. M., T. J. Horder, and K. A. C. Martin. 1978. Evidence that optic fibres regenerating across the goldfish tectum may be assigned termination sites on a "first come, first served" basis. J. Physiol. (Lond.), *241*: 89–90P.

Byers, B., and K. R. Porter. 1964. Oriented microtubules in elongating cells of the developing lens rudiment after induction. Proc. Natl. Acad. Sci. U.S.A., *52*: 1091–1099.

Cogan, D. G., and T. Kuwabara. 1973. The eye. *In* L. Weiss (ed), *Histology*, 3rd ed. McGraw-Hill Book Co., New York, pp. 963–1003.

Coulombre, A. J. 1955. Correlation of structural and biochemical changes in the developing retina of the chick. Am. J. Anat., *96*: 153–189.

Coulombre, A. J. 1956. The role of intraocular pressure in the development of the chick eye. I. Control of eye size. J. Exp. Zool., *133*: 211–225.

Coulombre, A. J. 1965. The eye. *In* R. L. DeHaan and H. Ursprung (eds.), *Organogenesis*. Holt, Rinehart and Winston, New York, pp. 219–251.

Coulombre, A. J., and J. L. Coulombre. 1958. The role of intraocular pressure in the development of the chick eye. IV. Corneal curvature. A. M. A. Arch. Ophthalmol., *59*: 502–506.

Coulombre, A. J., and J. L. Coulombre. 1964. Corneal development. III. The role of the thyroid in dehydration and the development of transparency. Exp. Eye Res., *3*: 105–114.

Coulombre, A. J., J. L. Coulombre, and H. Mehta. 1962. The skeleton of the eye. I. Conjunctival papillae and scleral ossicles. Dev. Biol., *5*: 382–401.

Ferrari, P. A., and W. E. Koch. 1984. Development of the iris in the chicken embryo. I. A study of growth and histodifferentiation utilizing immunocytochemistry for muscle differentiation. J. Embryol. Exp. Morphol., *81*: 153–167.

Fraser, S. E. 1985. Cell interactions involved in neuronal patterning: An experimental and theoretical approach. *In* G. M. Edelman, W. E. Gall, and W. M. Cowan (eds.), *Molecular Bases of Neural Development*. John Wiley & Sons, New York, pp. 481–507.

Gimlich, R. L., and J. Braun. 1985. Improved fluorescent compounds for tracing cell lineage. Dev. Biol., *109*: 509–514.

Gottlieb, D. I., K. Rock, and L. Glaser. 1976. A gradient of adhesive specificity in developing avian retina. Proc. Natl. Acad. Sci. U.S.A., *73*: 410–414.

Gray, G. G. et al. 1988. Radial arrangement of clonally related cells in the chicken optic tectum: Lineage analysis with a recombinant virus. Proc. Natl. Acad. Sci. U.S.A., *85*: 7356–7360.

Grun, G. 1982. *The Development of the Vertebrate Retina: A Comparative Survey*. Springer-Verlag, New York.

Hall, B. K. 1981. Specificity in the differentiation and morphogenesis of neural crest–derived scleral ossicles and of epithelial scleral papillae in the eye of the embryonic chick. J. Embryol. Exp. Morphol., *66*: 175–190.

Harris, W. A. 1980. The effects of eliminating impulse activity on the development of the retinotectal projection in salamanders. J. Comp. Neurol., *194*: 303–317.

Harris, W. A. 1984. Axonal pathfinding in the absence of normal pathways and impulse activity. J. Neurosci., *4*: 1153–1162.

Harris, W. A. 1986. Homing behaviour of axons in the embryonic vertebrate brain. Nature (Lond.), *320*: 266–269.

Harris, W. A. 1989. Local positional cues in the neuroepithelium guide retinal axons in embryonic *Xenopus* brain. Nature (Lond.), *339*: 218–221.

Harris, W. A., C. E. Holt, and F. Bonhoeffer. 1987. Retinal axons with and without their somata, growing to and arborizing in the tectum of *Xenopus* embryos: a time-lapse video study of single fibres *in vivo*. Development, *101*: 123–133.

Hay, E. D., and J.-P. Revel. 1969. *Fine Structure of the Developing Avian Cornea*. S. Karger, Basel.

Hilfer, S. R., and J.-J. W. Yang. 1980. Accumulation of CPC-precipitable material at apical cell surfaces during formation of the optic cup. Anat. Rec., *197*: 423–433.

Holt, C. E. 1984. Does timing of axon outgrowth influence initial retinotectal topography in *Xenopus*? J. Neurosci., *4*: 1130–1152.

Holt, C. E. 1989. A single-cell analysis of early retinal ganglion cell differentiation in *Xenopus*: From soma to axon tip. J. Neurosci., *9*: 3123–3145.

Holt, C. E. et al. 1988. Cellular determination in the *Xenopus* retina is independent of lineage and birth date. Neuron, *1*: 15–26.

Holt, C. E., and W. A. Harris. 1983. Order in the initial retinotectal map in *Xenopus*: A new technique for labelling growing nerve fibers. Nature (Lond.), *301*: 150–152.

Jacobson, M. 1967. Retinal ganglion cells: Specification of central connections in larval *Xenopus laevis*. Science, *155*: 1106–1108.

Johnston, M. C. et al. 1979. Origins of avian ocular and periocular tissues. Exp. Eye Res., *29*: 27–43.

Mann, I. 1964. *The Development of the Human Eye*. Grune & Stratton, New York.

Masterson, E., H. Edelhauser, and D. van Horn. 1977. The role of thyroid hormone in the development of the chick corneal endothelium and epithelium. Invest. Ophthalmol. Vis. Sci., *16*: 105–115.

Matthey, R. 1925. Recuperation de la vue aprés resection des nerfs optiques chez le triton. C. R. Soc. Biol., *93*: 904–906.

Meier, S., and E. D. Hay. 1973. Synthesis of sulfated glycosaminoglycans by embryonic corneal epithelium. Dev. Biol., *35*: 318–331.

Meyer, R. L. 1983. Tetrodotoxin inhibits the formation of refined retinotopography in goldfish. Dev. Brain Res., *6*: 293–298.

Noden, D. M. 1978. The control of avian cephalic neural crest cytodifferentiation. Dev. Biol., *67*: 296–312.

Noden, D. M. 1983. The embryonic origin of avian cephalic and cervical muscles and associated connective tissues. Am. J. Anat., *168*: 257–276.

O'Rourke, N. A., and S. E. Fraser. 1986. Dynamic aspects of retinotectal map formation revealed by a vital-dye fiber-tracing technique. Dev. Biol., *114*: 265–276.

Papaconstantinou, J. 1967. Molecular aspects of lens differentiation. Science, *156*: 338–346.

Piatigorsky, J., H. de F. Webster, and S. P. Craig. 1972. Protein synthesis and ultrastructure during the formation of embryonic chick lens fibers *in vivo* and *in vitro*. Dev. Biol., *27*: 176–189.

Price, J., D. Turner, and C. Cepko. 1987. Lineage analysis in the vertebrate nervous system by retrovirus-mediated gene transfer. Proc. Natl. Acad. Sci. U.S.A., *84*: 156–160.

Ramon y Cajal, S. 1972. *The Structure of the Retina*. Translated by S. A. Thorpe and M. Glickstein. Charles C Thomas, Springfield, IL.

Reh, T. A. 1987. Cell-specific regulation of neuronal production in the larval frog retina. J. Neurosci., *7*: 3317–3324.

Ross, M. H., E. J. Reith, and L. S. Romrell. 1989. *Histology: A Text and Atlas*. Williams & Wilkins, Baltimore.

Roth, S., and R. B. Marchase. 1976. An *in vitro* assay for retino-tectal specificity. *In* S. H. Barondes (ed.), *Neuronal Recognition*. Plenum Press, New York, pp. 227–248.

Sanes, J. R., J. L. R. Rubenstein, and J. F. Nicolas. 1986. Use of a recombinant retrovirus to study postimplantation lineages in mouse embryos. EMBO J., *5*: 3313–3342.

Schmidt, J. T. 1978. Retinal fibers alter tectal positional markers during expansion of the half-retinal projection in goldfish. J. Comp. Neurol., *177*: 279–300.

Schmidt, J. T., C. M. Cicerone, and S. S. Easter. 1978. Expansion of the half retinal projection to the tectum in goldfish: An electrophysiological and anatomical study. J. Comp. Neurol., *177*: 257–278.

Schmidt, J. T., and D. L. Edwards. 1983. Activity sharpens the map during the regeneration of the retinotectal projection in goldfish. Brain Res., *269*: 29–40.

Sheffield, J. B., and D. A. Fischman. 1970. Intercellular junctions in the developing neural retina of the chick embryo. Z. Zellforsch Mikrosk Anat., *194*: 405–418.

Smith, L., and P. Thorogood. 1983. Transfilter studies on the mechanism of epithelio-mesenchymal interaction leading to chondrogenic differentiation of neural crest cells. J. Embryol. Exp. Morphol., *75*: 165–188.

Sperry, R. W. 1943. Effect of 180° rotation of the retinal field on visuomotor coordination. J. Exp. Zool., *92*: 263–279.

Sperry, R. W. 1963. Chemoaffinity in the orderly growth of nerve fiber patterns and connections. Proc. Natl. Acad. Sci. U.S.A., *50*: 703–710.

Sperry, R. W. 1965. Embryogenesis of behavioral nerve nets. *In* R. L. DeHaan and H. Ursprung (eds.), *Organogenesis*. Holt, Rinehart and Winston, New York, pp. 161–171.

Tamura, T., and G. K. Smelser. 1973. Development of the sphincter and dilator muscles of the iris. Arch. Ophthalmol., *89*: 332–339.

Thorogood, P., J. Bee, and K. von Der Mark. 1986. Transient expression of collagen type II at epitheliomesenchymal interfaces during morphogenesis of the cartilaginous neurocranium. Dev. Biol., *116*: 497–506.

Toole, B. P., and R. L. Trelstad. 1971. Hyaluronate production and removal during corneal development in the chick. Dev. Biol., *26:* 28–35.

Trelstad, R., K. Hayashi, and B. Toole. 1974. Epithelial collagens and glycosaminoglycans in the embryonic cornea. J. Cell Biol., *62:* 815–830.

Trisler, C. D., M. D. Schneider, and M. Nirenberg. 1981. A topographic gradient of molecules in retina can be used to identify neuron position. Proc. Natl. Acad. Sci. U.S.A., *78:* 2145–2149.

Trisler, D., and F. Collins. 1987. Corresponding spatial gradients of TOP molecules in the developing retina and optic tectum. Science, *237:* 1208–1209.

Trisler, D., J. Bekenstein, and M. P. Daniels. 1986. Antibody to a molecular marker of cell position inhibits synapse formation in retina. Proc. Natl. Acad. Sci. U.S.A., *83:* 4194–4198.

Turner, D. L., and C. L. Cepko. 1987. A common progenitor for neurons and glia persists in rat retina late in development. Nature (Lond.), *328:* 131–136.

Walter, J. et al. 1987. Recognition of position-specific properties of tectal membranes by retinal axons *in vitro*. Development, *101:* 685–696.

Weiss, L. 1988. *Cell and Tissue Biology*. Urban & Schwarzenberg, Baltimore.

Wetts, R., and S. E. Fraser. 1988. Multipotent precursors can give rise to all major cell types of the frog retina. Science, *239:* 1142–1145.

Willshaw, D. J., J. W. Fawcett, and R. M. Gaze. 1983. The visuotectal projections made by *Xenopus* "pie slice" compound eyes. J. Embryol. Exp. Morphol., *74:* 29–45.

Yoon, M. G. 1971. Reorganization of retinotectal projection following surgical operations on the optic tectum in goldfish. Exp. Neurol., *33:* 395–411.

Yoon, M. G. 1972. Transposition of the visual projection from the nasal hemiretina onto the foreign rostral zone of the optic tectum in goldfish. Exp. Neurol., *37:* 451–462.

CHAPTER *18*

The Molecular Biology of Cell Differentiation

Paul Berg (b. 1926), a Professor at Stanford
University, shared the 1980 Nobel Prize in
Chemistry for development of DNA cloning
techniques with Herbert W. Boyer and Stanley
N. Cohen. These techniques have profoundly
changed the study and understanding of
developmental biology. (Courtesy of Stanford
University News and Publications Service.)

Every organ in a multicellular organism is composed of specialized cell types that perform unique functions. The acquisition of the abilities of cells to perform their ultimate specialized functions is cell differentiation. As we discussed in Chapter 10, every cell in the body retains the genome that is established at fertilization, and cell specialization is made possible by differential utilization of portions of the genome, resulting in the ability to synthesize a unique set of proteins. How does this selection process occur? As we learned in Chapter 14, differential gene expression can be regulated by the interactions of transcription factors (often called *trans*-**acting factors**; see the essay on p. 724 for a discussion of the classes of transcription factors) with regulatory elements (often called *cis*-**acting elements**), which are usually located in the flanking regions of genes. The transcription factors that interact with the common transcriptional elements in the promoters of Pol II genes (i.e., genes transcribed by DNA polymerase II) are relatively abundant in all cells that are transcriptionally active. However, the expression of tissue-specific genes requires not only these interactions but interactions between specialized sequence elements and tissue-specific transcription factors, which are typically present in limited amounts. In this chapter, we explore examples of such interactions in the regulation of specific genes in differentiating cells, and we examine the sequence of regulatory events that leads ultimately to differential gene expression.

The powerful experimental tools that are available to contemporary molecular biologists have spurred a vast increase in our knowledge in the past few years. The molecular regulation of cell differentiation has always been an exciting area of investigation, but it is now much more rewarding owing to the incredible experimental advantages that we enjoy. In the discussion that follows, some of these experimental advantages will become apparent.

The differentiation of individual cell types is likely to involve certain common principles. On the other hand, each cell type is the product of a unique set of circumstances. At present, we understand details about the differentiation of a relatively small number of cells in the most popular experimental organisms. By exploiting developmental systems that have advantages for experimental study, investigators have amassed an impressive amount of information about these systems. This, in turn, enables successive inquiries to probe at progressively greater depth into the molecular basis of cell differentiation. This **reductionist approach** to science has both advantages and pitfalls. The advantages are the incredible depth of knowledge that one can gain from concentrating on individual systems. Other systems can then be examined to determine if they employ the same principles. One danger

Regular text continues on page 725.

Essay

Eukaryotic Transcription Regulatory Proteins

Differential gene transcription is mediated by specific protein-DNA interactions that occur in the enhancer and promoter regions of genes. A variety of different transcription regulatory proteins has been discovered. Individual transcription regulatory proteins contain unique DNA binding domains that allow them to bind to nucleotide sequences that are specific to individual genes or groups of genes. However, they also contain elements that are shared by other transcription regulatory proteins, including those that enable protein-protein interactions. Comparisons of the structures of such proteins have revealed a small number of recurring structural features. This has allowed investigators to categorize transcription regulators according to these shared domains.

Zinc Finger Proteins

We have encountered these proteins several times in this book. The number of zinc fingers in such proteins can range from two to more than ten. The fingers are essential for DNA binding, although distinct elements may be responsible for specific DNA sequence interactions. Examples of zinc finger proteins are TFIIIA, the Krüppel protein of *Drosophila,* and steroid hormone receptors.

Helix-Loop-Helix (HLH) Proteins

These proteins share a region called the helix-loop-helix (HLH) domain, which contains a dimerization motif that allows them to bind with another HLH protein, and an adjacent basic region that mediates binding to specific DNA sequences (Murre et al., 1989). The HLH domain is proposed to consist of two helical regions that are separated by one or more β-loops. Examples of HLH proteins are *MyoD1* and related myogenesis regulatory factors (see pp. 731–740). This sequence is also found in the protein encoded by the *c-myc* proto-oncogene, certain immunoglobulin light-chain enhancer-binding proteins (not including Oct-1 and Oct-2; see p. 728), and a number of *Drosophila* proteins that are known to be involved in regulation of development.

Leucine Zipper Proteins

These proteins (see Fig. 1) contain four or five leucine residues that are spaced exactly seven amino acid residues apart, with one leucine for every two turns of the α-helix (Landschulz et al., 1988). Thus, the hydrophobic leucine residues will all protrude from the same side of the helix. Thus, two leucine zipper proteins can bind (i.e., "zip together") by lateral interactions between the leucine residues on the adjacent domains, thereby forming a dimer. Some leucine zipper proteins typically form homodimers, whereas others form heterodimers. For example, heterodimers are formed between Fos and Jun, which are the proteins encoded by the *fos* and *jun* proto-oncogenes (see essay on p. 380). The leucine-rich regions are adjacent to highly basic DNA-binding domains. The dimerization allows the DNA-binding domains of the two proteins to interact with adjacent stretches of nucleotides in the DNA. Thus, a dimer could span a longer stretch of nucleotides than a monomer. (A hypothetical model for the interaction between a leucine zipper dimer and DNA is shown in Color plate 29.)

Homeobox Proteins

A large number of proteins containing the homeobox have been discovered, and many of them are discussed in this book. An interesting group of proteins combines the homeobox with another conserved region called the POU box. These POU proteins are discussed on page 729.

The analysis of eukaryotic transcription regulators is in its infancy. Much remains to be learned about how they interact with one another and DNA to regulate the specific transcription of genes during development. This will remain an exciting area of investigation for the next several years.

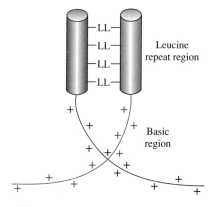

Figure 1

A model of a leucine zipper dimer.

is in extrapolation of information gained from one system to cell differentiation in general. Another is that focusing on only a few systems will preclude the investigation of other systems that would be illuminating; one of the beauties of the natural world is its amazing diversity. In the discussion that follows, we limit our consideration to a small number of particularly illustrative examples of gene regulation. We have chosen these examples either because they emphasize specific principles or because they have been particularly well suited for experimental exploitation. We should keep in mind that the principles discussed here do not exhaust nature's molecular repertoire; there are many surprises yet undiscovered concerning the regulation of gene expression in development.

18–1 Immunoglobulin Gene Expression

The interactions between *cis-* and *trans-*acting factors that result in tissue-specific gene expression are perhaps best understood for the immunoglobulin (Ig) genes that are expressed in B lymphocytes. These genes encode the antibodies (or **immunoglobulins**) that mediate the vertebrate immune response. Immunoglobulins are localized on the surfaces of B cells and serve as receptors for antigen. When antigen combines with the surface antibody, B cells are stimulated to proliferate and differentiate into clones of antibody-producing cells. The antibody they produce is released into the circulation for neutralization of the foreign antigen.

Each immunoglobulin molecule (Fig. 18.1) is a tetramer consisting of two identical light (L) chains and two identical heavy (H) chains. The great diversity among Ig molecules depends upon the production of diverse light and heavy chains, which combine to form tetramers. The L and H chains are composed of two segments each: The amino-terminal ends are quite variable among Ig molecules and are called the **variable (V) regions**, whereas the carboxy-terminal regions are the **constant (C) regions**. The light and heavy chain V regions form the combining site of an antibody molecule. Together they constitute the V domain, which recognizes the antigen and determines the specificity of an Ig molecule. The constant regions mediate the effector functions that produce the immune response.

Because an individual may encounter a wide variety of antigens in a lifetime, the potential to produce antibodies to counter these antigens must be immense. Recent estimates are that from 10^6 to 10^8 *different* antibodies can be produced by an individual (Tonegawa, 1983). How can the genome, which possesses the information to produce antibodies, cope with this

Regular text continues on page 728.

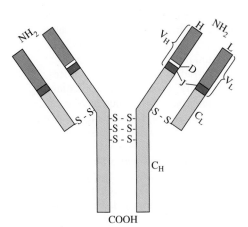

Figure 18.1

Model of an immunoglobulin molecule. Each immunoglobulin molecule is composed of two identical light chains and two identical heavy chains. Each chain consists of a V region (V_L or V_H) and a C region (C_L or C_H). The polypeptides are linked to one another by disulfide bonds.

Essay

Vertebrate Immunoglobulin Gene Rearrangement During Lymphocyte Differentiation

The variability among Ig molecules depends upon the association of diverse light and heavy chains. These diverse chains are generated by the combination of a limited number of constant (C) regions with many potential variable (V) regions. Dryer and Bennett (1965) proposed that the V and C regions of an Ig polypeptide are encoded by separate genes. However, each Ig chain is a covalently continuous polypeptide that is translated on a single mRNA molecule. This suggests that the DNA is rearranged by **somatic recombination** during B-cell development to bring a V gene into proximity with a C gene. Confirmation of Dreyer and Bennett's hypothesis has been provided primarily by experiments utilizing mice and, in particular, mouse **myeloma cells**. These cells are tumors of lymphoid cells. Because a lymphocyte produces a single species of immunoglobulin, a lymphocyte that becomes malignant can divide to produce a large number of progeny, each of which produces the same immunoglobulin. The myelomas can be maintained *in vitro*, providing an indefinite number of cells that are dedicated to the synthesis of a single molecular species and that can be subjected to molecular analysis.

There are two kinds of murine (i.e., mouse) light chains: κ and λ. All kappa (κ) chains have the same C region and one of several potential V regions. Lambda (λ) chains have one of two possible C regions but very little diversity among V regions. Distinct genomic regions encode the κ and λ chains; they are on separate chromosomes (i.e., unlinked). The heavy chains are polypeptides that may be composed of one of a large set of V regions and one of several C regions. The latter are classified into five types: C_μ, C_δ, C_γ, C_ϵ, and C_α. An Ig molecule is named for the type of heavy chain C region it contains; i.e., IgM, IgD, IgG, IgE, or IgA, respectively. The heavy chain genes are unlinked to either of the regions encoding the two kinds of light chains.

Individual B cells make antibodies with only a single type of antigen-combining site. Thus, during development, the cell becomes committed to producing either κ or λ light chains and expressing either the maternal or paternal allele at the light and heavy chain loci. The expression of only one of the alleles at Ig gene loci is called **allelic exclusion**. Because an antibody is composed of two light chains and two heavy chains, the production of only one light chain type and allelic exclusion ensure that homologous chains in a single Ig molecule have the same variable regions, and consequently the same antigenic specificity.

Evidence for somatic rearrangement of the V and C genes was first provided by a comparison of fragments of embryo and lymphocyte (myeloma cell) DNA produced by treatment with a restriction endonuclease. Hozumi and Tonegawa (1976) isolated full-length κ mRNA, which they used as a probe to identify both the V and C genes. They then prepared a fragment of the 3′ end of the mRNA, which is a specific probe for fragments of DNA containing the C gene. When embryo DNA was treated with the restriction enzyme *Bam* H1, two large fragments (9 and 6 kilobases [kb] long, respectively) were generated that hybridized with the full-length mRNA. However, only the 6 kb fragment hybridized with the C-gene probe. This must mean that the V gene is on the 9 kb fragment, whereas the C gene is only on the 6 kb fragment. Therefore, these genes are located at a distance from one another in embryo cell nuclei. By way of contrast, when myeloma DNA was restricted with *Bam* H1, a single 3.5 kb fragment was produced that hybridized with both probes. Thus, during lymphocyte differentiation, the V and C genes must be joined by recombination.

The V genes code for only a portion of the complete V region of an Ig chain. The remainder of the V region of λ chains is encoded by the **joining**, or **J**, **gene segment**, which is adjacent to the C gene, from which it is separated by an intron. During B-cell development, the V gene is translocated so that it abuts the J segment, resulting in a V-J-intron-C sequence. The situation is even more complicated for the κ- and heavy-chain genomic regions. The κ region contains five J segments (one of the five is nonfunctional). Any one of the V genes may be joined to any of the four functional J segments, thus increasing the diversity within κ-chain V regions. The diversity is further increased by variation in the exact site of V-J joining (Max et al., 1979; Sakano et al., 1979). The steps in the generation of κ light chain from the germ-line DNA, through somatic recombination, transcription, messenger RNA processing, and, finally, translation, are illustrated in Figure 1.

What is distinct about Ig genes and lymphocytes that allows these genes to undergo recombination in these cells? The gene segments that are capable of recombination are flanked by nucleotide sequences called **recombination signal sequences (RSSs)** that mediate recombination. The cellular machinery that recognizes the RSSs is unique to lymphocytes and is referred to as a **recombinase**. Recently, genes encoding recombinase activity in mice have been discovered. These genes are normally expressed exclusively in lymphocytes. However, when they are transfected into fibroblasts, the fibroblasts acquire the ability to recombine Ig genes (Oettinger et al., 1990). As more is learned about

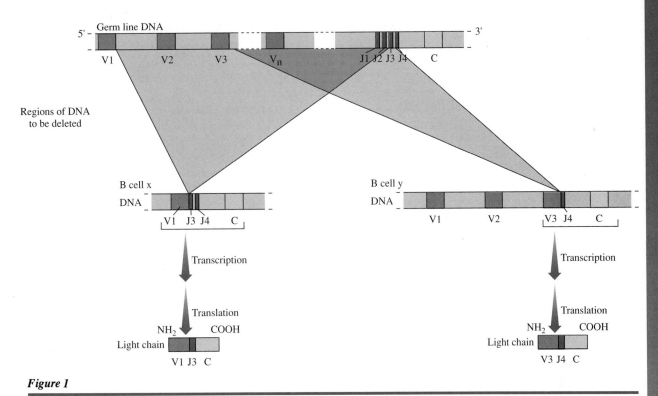

Figure 1

Summary of steps leading to synthesis of two different kappa light chains.

the recombinase mechanism, we should eventually be in the position to regulate gene recombination in cells.

Heavy-chain V regions consist of portions derived from the V gene, a J segment, and an additional segment called a **D** (for "diversified") **segment**. The heavy-chain genomic region consists of 100 to 200 different V segments, approximately 12 D segments, 4 functional J segments, and 8 C segments—one for each Ig class or subclass (Tonegawa, 1983).

During B-cell development, the heavy-chain genomic region is rearranged such that one each of the V segments, D segments, and J segments are conjoined to form a complete V region. As with the κ gene recombination, joining sites may be variable, thus increasing the diversity of potential V regions. The V region that is formed by recombination is initially expressed in all B cells along with C_μ to produce IgM. Later in B-cell development, the cells switch to production of one of the other heavy chain classes or subclasses. This is

called **heavy chain class switching**. Interestingly, the same V region is expressed in both the original and subsequent heavy chains. Thus, the antibody-binding site is not modified, even though the effector region changes.

Why is C_μ expressed first, and how does the cell effect a switch from expression of C_μ to one of the other C genes? The initial expression of C_μ is apparently due to its position in the heavy chain genomic region. The order of heavy chain C genes is $5'\text{-}C_\mu\text{-}C_\delta\text{-}C_{\gamma 3}\text{-}C_{\gamma 1}\text{-}C_{\gamma 2b}\text{-}C_{\gamma 2a}\text{-}C_\epsilon\text{-}C_\alpha\text{-}3'$ (Alt et al., 1982). The V region that is formed by recombination would, therefore, initially be adjacent to C_μ. Transcription of V-D-J-C_μ and translation of the messenger would result in IgM.

Two mechanisms have been reported for the subsequent class switch. One involves posttranscriptional processing of multiclass transcripts to remove the RNA encoding the inappropriate heavy chain, whereas the other involves rearrangement of the genome to move the V-D-J region into prox-

imity with the gene for a different heavy chain class and simultaneously to delete intervening genes (Tonegawa, 1983).

We have outlined several mechanisms that contribute to antibody diversity. Multiple V-region combinations and several combinations of variable and constant regions of either a light or heavy chain can be generated by somatic recombination. The production of various light and heavy chains results in additional diversity when they are assembled into a functional antibody molecule; i.e., a given light chain will form a different antibody when combined with heavy chain X than with heavy chain Y, and *vice versa*. There is one additional mechanism that we have not mentioned previously: **somatic mutation**, through which individual bases in the V region genes are modified (Weigert and Riblet, 1976). Clearly, the genome encoding the immunoglobulins is malleable. It will be exciting to learn how this malleability is regulated and harnessed during B-cell development.

incredible requirement? It is now evident that mechanisms exist to diversify the genetic information for antibody production and that this diversification is generated during B-cell development. This occurs by a rearrangement of Ig genes during early lymphocyte development, which culminates in formation of the definitive Ig genes (see the essay on p. 726). As a consequence of this rearrangement process, the genes encoding the antibody molecules differ among B lymphocytes and from those in the genome of the zygote. This is an obvious exception to the rule that the genome remains unaltered during development.

Before rearrangement, the enhancers and promoters of Ig genes are distant from one another. The rearrangement of the genes brings enhancers and promoters into proximity of one another, which facilitates regulation of Ig gene expression. The enhancers and promoters contain multiple nucleotide sequence elements (called **motifs**) that play discrete, but complementary, roles in regulating transcription when they are engaged by the appropriate *trans*-acting factors (Lenardo et al., 1987). Some of these factors are lymphoid-cell specific, whereas others are present in many cell types. When these various factors are bound to their respective motifs, they regulate transcription of the Ig genes. This is called a **combinatorial effect**. This is reminiscent of the regulation of the CyIIIa gene in sea urchin embryo aboral ectoderm, which involves the interaction of a number of regulatory factors with 5′ regulatory elements (see p. 584). One of the Ig gene sequence elements is an octanucleotide sequence called the **octamer motif** (ATTTGCAT), which is important for the function of the heavy chain enhancer and the heavy and light chain promoters. Variations of this sequence have also been found in the light chain enhancer (Currie and Roeder, 1989; Nelms and Van Ness, 1990).

What role do the octamers play in regulating Ig gene transcription? Transfection experiments have shown that the octamers are responsible for lymphoid-specific enhancement of transcription (Wirth et al., 1987). These experiments have involved transfecting lymphoid and nonlymphoid cells with hybrid genes in which the octamer sequence is situated upstream of a TATA box and a β-globin coding sequence. Expression of this hybrid gene was monitored by a nuclease protection assay (see Appendix), in which the fusion gene transcript was hybridized to a β-globin gene probe. Digestion of the hybrids with S1 nuclease (which specifically digests single-stranded nucleic acids) produces an 87-nucleotide fragment (Fig. 18.2A). Plasmids with the octamer in either orientation exhibited approximately 11-fold more transcription in lymphoid cells than plasmids lacking the octamer. However, this effect is *limited to lymphoid cells*; no such effect is seen in fibroblast cells (see Fig. 18.2B). Clearly, the octamer confers lymphoid-specific transcriptional enhancement on the fusion gene, implying a role for this sequence in lymphoid-specific expression of Ig genes. Mutations at any one of the first seven nucleotide positions within the octamer motif significantly curtailed transcription of the fusion gene. Hence, this portion of the sequence appears to be essential for conferring lymphoid-specific expression on the genes.

How does the octamer motif confer lymphoid-specific expression on Ig genes? Two octamer-binding proteins have been discovered; one (**Oct-1**) is ubiquitous (i.e., it is found in nuclei of a variety of cell types), whereas the other (**Oct-2**) is lymphoid-specific (Landolfi et al., 1986; Staudt et al., 1986). The function of Oct-2 as a transcriptional factor is demonstrated by its ability to activate transcription from a κ-light chain promoter *in vitro* (Fig. 18.3). In this experiment, genes with and without the κ-light chain promoter (which contains the octamer element) are transcribed *in vitro* in lymphoid nuclear extract. Nuclease protection assays reveal that a protected

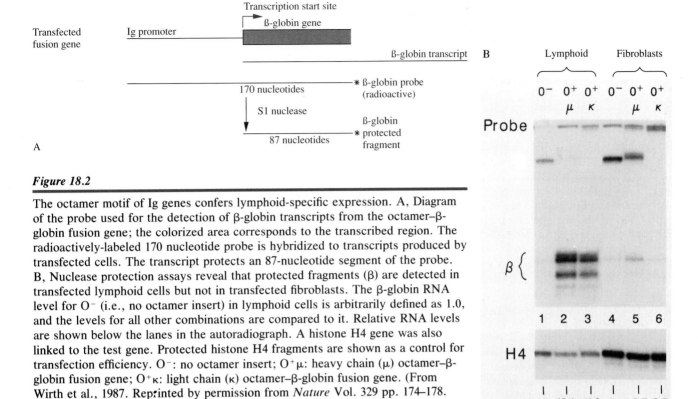

Figure 18.2

The octamer motif of Ig genes confers lymphoid-specific expression. A, Diagram of the probe used for the detection of β-globin transcripts from the octamer–β-globin fusion gene; the colorized area corresponds to the transcribed region. The radioactively-labeled 170 nucleotide probe is hybridized to transcripts produced by transfected cells. The transcript protects an 87-nucleotide segment of the probe. B, Nuclease protection assays reveal that protected fragments (β) are detected in transfected lymphoid cells but not in transfected fibroblasts. The β-globin RNA level for O⁻ (i.e., no octamer insert) in lymphoid cells is arbitrarily defined as 1.0, and the levels for all other combinations are compared to it. Relative RNA levels are shown below the lanes in the autoradiograph. A histone H4 gene was also linked to the test gene. Protected histone H4 fragments are shown as a control for transfection efficiency. O⁻: no octamer insert; O⁺μ: heavy chain (μ) octamer–β-globin fusion gene; O⁺κ: light chain (κ) octamer–β-globin fusion gene. (From Wirth et al., 1987. Reprinted by permission from *Nature* Vol. 329 pp. 174–178. Copyright © 1987 Macmillan Magazines Ltd.)

fragment resulting from *in vitro* transcription is produced in much greater abundance when the promoter is present than when it is absent. Depletion of Oct-2 from the nuclear extract reduces the production of transcripts. Furthermore, adding Oct-2 back to depleted extracts restored transcription to levels seen with the intact nuclear extract. Thus, Oct-2 is necessary for the octamer-dependent tissue-specific expression of the κ-light chain promoter (Scheidereit et al., 1987).

Both Oct-1 and Oct-2 have similarities to other DNA-binding transcriptional regulation proteins. The regions of these proteins that interact with the octamer element are similar to the *Antennapedia* homeodomain (Clerc et al., 1988; Ko et al., 1988; Müller et al., 1988; Scheidereit et al., 1988), which, as we learned in Chapter 14, interacts with DNA to promote specific gene expression. Another region is similar to a mammalian pituitary-specific transcription factor and the protein encoded by *unc-86*, a gene in *Caenorhabditis elegans* that has been implicated in cell lineage determination. This region has been called the **POU box** (**p**ituitary-**o**ctamer-**u**nc). The POU box in all these proteins is adjacent to the homeobox (which is called the **POU homeodomain**); the POU box and the POU homeodomain together are called the **POU domain** (Herr et al., 1988). The presence of the POU domain in these diverse transcriptional regulators implies that it is an important configuration for their common function in transcriptional regulation. In addition to their shared elements, each regulator must also have unique elements that confer specificity on them.

We have learned that development occurs progressively, with the establishment of germ layers, which then become regionalized and, ultimately, specialized into discrete tissues. These steps must necessarily

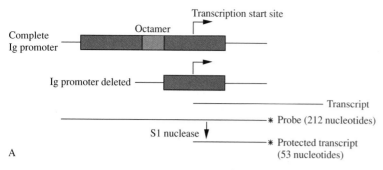

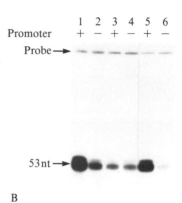

A

Figure 18.3

In vitro transcriptional activity of Oct-2. A, Schematic representation of the templates and the probe used for *in vitro* transcription and nuclease protection assay. The upper template contains the complete κ-light chain promoter, whereas it has been deleted in the lower template. The arrows show the transcription start sites. When the radioactively-labeled probe is hybridized to the transcripts and subjected to S1 nuclease digestion, the resultant protected transcript is 53 nucleotides long. Cloned inserts are shown in color and are flanked by portions of the plasmids in which they were inserted. B, Results of the nuclease protection assay. +: transcripts derived from templates with the promoter; −: transcripts derived from templates without the promoter. Lanes 1 and 2: Lymphoid nuclear extract; Lanes 3 and 4: Factor-depleted lymphoid nuclear extract; Lanes 5 and 6: Addition of purified factor to depleted extract restores enhanced κ transcription in the presence of the octamer sequence (compare lanes 5 and 3); this effect is totally dependent on the presence of the promoter (compare lanes 5 and 6). Correctly initiated 53 nucleotide transcripts are indicated (53 nt). (From Scheidereit et al., 1987. Copyright © Cell Press.)

involve shifts in the pattern of gene expression. Because the *trans*-acting factors that regulate gene expression are themselves the products of gene expression, the activation of their expression must precede the expression of the gene(s) that they regulate. Ultimately, we would hope to reconstruct the entire hierarchy of molecular events that begin during oogenesis and result in the spatial and chronological pattern of gene expression in the differentiated organism. Tissue-specific expression of the gene encoding Oct-2 must be established during lymphocyte differentiation. A number of cDNAs encoding Oct-2 have been isolated and used to demonstrate that Oct-2 mRNAs are, indeed, in highest abundance in lymphoid cells (Clerc et al., 1988; Müller et al., 1988; Scheidereit et al., 1988; Staudt et al., 1988). These cDNAs will be essential tools in tracing the sequence of events that culminates in differentiation of lymphocytes. For example, they will facilitate experiments that would explore the cellular events that lead to expression of the *oct-2* gene. Once the factors that regulate *oct-2* gene expression are known, the genes encoding them could be isolated and studied, and one more step in lymphoid differentiation will have been characterized.

18–2 *Muscle-Specific Gene Transcription*

Muscle cell development provides an excellent experimental system in which to study the molecular events involved in cell differentiation. In myogenesis, muscle cells develop from the mesoderm through a process beginning with determination of multipotent cells to develop into muscle precursor cells, called **myoblasts**. Myoblasts then fuse to form multinucleated myofibers that express muscle-specific genes, including those encoding cardiac and skeletal muscle actins, myosin heavy and light chains, tropomyosin, troponins I and T, acetylcholine receptor, and muscle-specific creatine kinase (Buckingham, 1977). Although we are on our way to understanding how each of these steps is controlled at the molecular level, a number of important pieces to the puzzle are still missing. Many of the missing pieces will fall into place quickly as research continues.

Our understanding of the initial steps of myogenesis has originated not from studies with embryos, but from work on tissue culture cells. Treatment of undifferentiated mouse embryonic fibroblast cells in culture with the chemical 5-azacytidine causes some of the fibroblasts to differentiate into myoblasts, which can fuse to form myotubes (Fig. 18.4). 5-Azacytidine is known to alter the pattern of methylation on cytidine residues in DNA, which may, in turn, alter the pattern of gene expression (see the essay on p. 732 for a discussion of the possible role of methylation in regulation of gene expression). Examination of the myogenic lineages led to the discovery that expression of the gene *MyoD1* (for *Myo*blast *D*etermination) is sufficient to cause conversion of the fibroblasts into muscle cells. Subsequently, it was learned that transfection of *MyoD1* DNA into a variety of fibroblast cell lines is sufficient to convert them into a myogenic phenotype (Davis et al., 1987; Weintraub et al., 1989). The conversion of fibroblasts into myogenic cells can be monitored by using immunofluorescence to stain for the presence of muscle-specific proteins in transfected cells. Figure 18.5 shows an elongated former fibroblast that had been transfected with *MyoD1* and stained for the muscle-specific protein myosin heavy chain (MHC). An alternative to transfection with DNA is to infect cells with an RNA virus

Regular text continues on page 733.

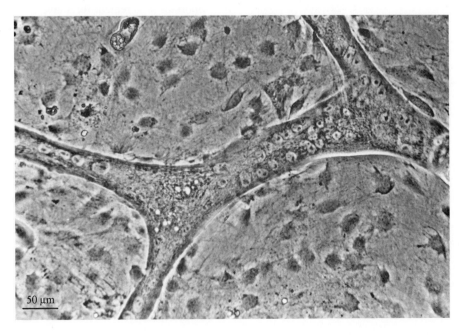

50 μm

Figure 18.4

Formation of myotubes from mouse fibroblasts after treatment with 5-azacytidine. (From Taylor and Jones, 1979. Copyright © Cell Press.)

Essay

DNA Methylation

As we discussed in Chapter 10, the genomes of eukaryotes are largely unchanged during development. However, cytidine residues in the dinucleotide sequence CpG are subject to postreplication methylation of cytosine to produce 5-methylcytosine (m^5C). Clusters of CpGs, which are known as **CpG islands**, are found in association with many genes (Bird, 1986; Gardiner-Garden and Frommer, 1987). *De novo* methylation of these sequences during development can interfere with transcription of the associated genes. For example, evidence suggests that cytosine methylation is linked to X-chromosome heterochromatization (see Chap. 10 for a discussion of X-chromosome heterochromatization): CpG islands on mammalian X chromosomes become methylated after X chromosome heterochromatization (Monk, 1986). Conversely, genes on inactive X chromosomes can be reactivated experimentally by treatment with the nucleoside analogue 5-azacytidine, which causes demethylation of cytosine residues in DNA (Lock et al., 1986).

5-Azacytidine is a powerful experimental tool for assessing the effects of methylation on gene expression. In addition to reactivation of quiescent genes on X chromosomes, it has been shown that treatment of cultured fibroblasts with this drug can induce their *in vitro* differentiation (Taylor and Jones, 1979; Jones and Taylor, 1980; Taylor and Jones, 1982). One of the most striking *de novo* differentiation events that is induced by 5-azacytidine is myogenesis, which is discussed in this chapter.

Methylation is detected experimentally by a comparison of the cleavage patterns of the restriction enzymes *Hpa* II and *Msp* I. Both of these enzymes recognize sites containing CCGG sequences. However, *Hpa* II cannot cleave Cm⁵CGG, whereas *Msp* I cleaves both CCGG and Cm⁵CGG. Thus, *Msp* I is used to identify CCGG sites in genes, and methylation of these sites is detected by comparison with the digestion pattern obtained with *Hpa* II.

In a typical experiment, total DNA is restricted with these enzymes, after which the DNA fragments are resolved by agarose gel electrophoresis and transferred by Southern blotting to nitrocellulose. The DNA fragments containing genes of interest are then visualized by hybridization to a labeled probe. The results of such an experiment are shown in Figure 1.

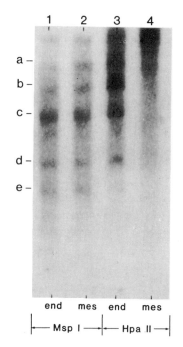

Figure 1

Methylation of the α-fetoprotein (α-FP), gene in the mouse yolk sac. DNA was restricted with either *Hpa* II or *Msp* I and analyzed by Southern blotting using α-FP cDNA as a probe. Tracks 1 and 2 show *Msp* I-restricted DNA from yolk sac endoderm and mesoderm respectively, while tracks 3 and 4 show *Hpa* II-restricted DNA from these same tissues. (From Andrews et al., 1982.)

The figure compares the restriction patterns obtained for the α-fetoprotein (α-FP) gene from yolk sac endoderm and mesoderm (see Chap. 8 for a discussion of the mammalian yolk sac). The α-FP gene is expressed in endoderm but not in the mesoderm. *Msp* I cleaves the α-FP genes from both tissues into five fragments (labeled a–e). The same five fragments are generated with *Hpa* II treatment of endoderm DNA. However, treatment of mesoderm DNA with *Hpa* II releases only one fragment. These contrasting results demonstrate that the CCGG sites are methylated in mesoderm but hypomethylated (i.e., the level of methylation is reduced) in endoderm—the tissue in which the gene is expressed.

Results with a number of genes indicate that in cells expressing a tissue-specific gene, CG sites are hypomethylated, whereas the highest levels of methylation are found in tissues in which the gene is not expressed (Razin and Riggs, 1980). These observations support the hypothesis that a hypomethylated state may be necessary for tissue-specific gene expression during development. One of the most intriguing consequences of this hypothesis for regulation of gene expression is that it satisfies the important requirement for stable transmission of the pattern of gene expression of a cell to its progeny (Wigler et al., 1981; Stein et al., 1982).

Methylation of the appropriate cytosine residues occurs enzymatically after replication of DNA, conserving the methylation pattern of the parental strands in each daughter strand (Riggs, 1975; Holliday and Pugh, 1975; Bird, 1978). This is represented in Figure 2. Figure 2 illustrates another important property of methylated CG sequences; i.e., the methylation pattern is symmetrical on the two strands of the DNA double helix. The symmetrical methylation makes possible the clonal inheritance of the exact methylation pattern at cell division, presumably by

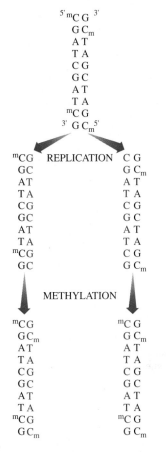

```
        5' mC G 3'
           G Cm
           A T
           T A
           C G
           G C
           A T
           T A
          mC G
        3' G Cm 5'
```

```
mCG   REPLICATION   C G
G C                 G Cm
A T                 A T
T A                 T A
C G                 C G
G C                 G C
A T                 A T
T A                 T A
mCG                 C G
G C                 G Cm
```

METHYLATION

```
mCG                 mC G
G Cm                G Cm
A T                 A T
T A                 T A
C G                 C G
G C                 G C
A T                 A T
T A                 T A
mCG                 mC G
G Cm                G Cm
```

Figure 2

Restoration of the parental methylation pattern after DNA replication. Note that of the three CpG dinucleotides, only those methylated before replication show methylation in their complementary strands after replication.

means of a methylase enzyme that recognizes the half-methylated sites that exist after replication and restores the fully methylated symmetrical pattern.

The exact mechanism whereby methylated CpG islands exert their inhibitory effect on transcription is not known; one possibility is that they bind with nuclear transcriptional inhibitors that specifically recognize methylated CpG residues (Antequerra et al., 1989; Meehan et al., 1989).

How is methylation regulated during development: Are genes that are expressed only during early development methylated to prevent their further expression, or does activation of previously inert genes involve a reduction in gene methylation level to facilitate transcription? In fact, both may occur. **Developmental methylation** is proposed to be a secondary event that helps to maintain transcriptional inertia of genes after they are first inactivated by other mechanisms (Gautsch and Wilson, 1983; Lock et al., 1986). **Developmental hypomethylation** would involve reduction in methylation levels of tissue-specific genes in tissues in which they are to be expressed. The latter phenomenon has indeed been

detected for δ-crystallin genes during lens development in chick embryo lens cells (Sullivan et al., 1989). (δ-Crystallin is the predominant protein synthesized in the developing chick lens.) The δ-crystallin gene undergoes a specific hypomethylation event in lens cells (as detected by *Msp* I-*Hpa* II DNA digestion analysis) concurrently with the onset of high-level accumulation of δ-crystallin transcripts (Grainger et al., 1983; Sullivan et al., 1989). However, it is uncertain whether the hypomethylation event itself activates transcription of the gene or is involved in maintaining a high level of transcription after it is first activated by other mechanisms (i.e., by enhancer-binding *trans*-acting factors), as is apparently the case for the κ light chain immunoglobulin gene (Kelley et al., 1988). Such a shift away from dependence on enhancer-driven transcription after activation of specific gene expression would eliminate the need for cells to continue to produce enhancer-binding *trans*-acting factors to maintain high-level transcription of tissue-specific genes; their commitment to transcribe these genes would be maintained by hypomethylation.

containing a *MyoD1* gene that will be strongly expressed in the infected cells, which causes dramatic changes in cell morphology (compare Figs. 18.6A and B). As shown in Figure 18.6C, infected human fibroblasts produce MHC. Infected rat fibroblasts become elongate and fuse to become large multinucleated myotubes (see Fig. 18.6D, upper frame). As a control, rat fibroblasts infected with an RNA virus containing a nonrelevant gene under the control of a viral promoter do not differentiate (see Fig. 18.6D, lower frame). In addition to fibroblasts, a number of diverse differentiated cell lines can be caused to express muscle-specific genes. Figure 18.7 illustrates immunofluorescent staining for MHC in transfected melanoma (A), neuroblastoma (C), and fat (D) cells. All three cell types express this muscle-specific marker. Figure 18.7E shows the conversion of normal fat cells with fat globules (upper frame) to cells that resemble muscle (lower frame).

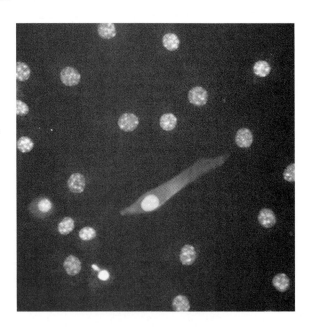

Figure 18.5

Conversion of a fibroblast to muscle by transfection with the *MyoD1* gene. A single transfected cell is shown against a background of untransfected cells. The nuclei of the untransfected cells are stained with the DNA-specific stain DAPI. Expression of MyoD1 and myosin heavy chain protein in the transfected cell are demonstrated by immunofluorescent staining. The MyoD1 is localized to the nucleus of the transfected cell, whereas myosin heavy chain protein is present in the cytoplasm. Note the elongate appearance of the cell. See Color plate 30 for color version. (Courtesy of S. J. Tapscott from the front cover of Science, 4 August, 1989. Copyright 1989 by the American Association for the Advancement of Science.)

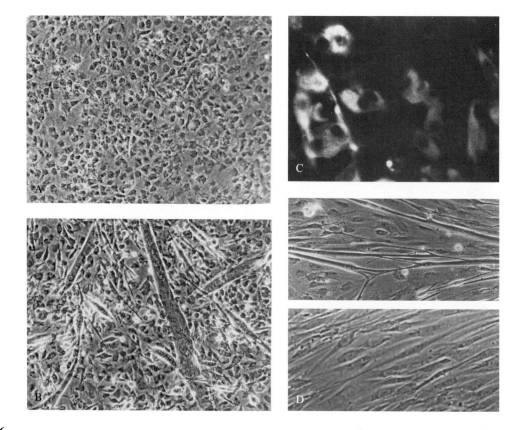

Figure 18.6

Conversion of fibroblasts to muscle by infection with RNA viruses containing the *MyoD1* gene. Cultured fibroblasts were infected with RNA viruses containing either a *MyoD1* gene (B) or a non-relevant gene (A). In C, infected human fibroblasts were stained for myosin heavy chain protein. In D (upper frame), *MyoD1*-infected rat fibroblasts have formed large multinucleated myotubes, whereas the control-infected rat fibroblasts (D, lower frame) retain a typical fibroblast morphology. × 140. (A, and B, Courtesy of H. Weintraub. C, and D, From Weintraub et al., 1989.)

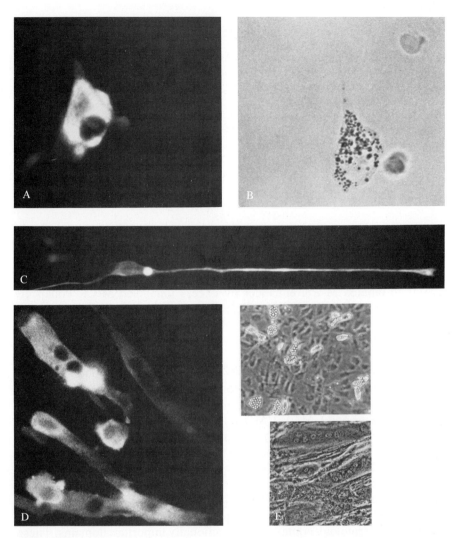

Figure 18.7

Muscle-specific gene expression in transfected differentiated cells. Melanoma (A and B), neuroblastoma (C), and fat (D and E) cells were transfected with *MyoD1* DNA. Cells in A, C, and D were stained for myosin heavy chain protein. B, The same melanoma cell as shown in A displays pigment granules in addition to myosin heavy chain protein. E, Normal fat cells with fat globules (upper frame) are converted to muscle (lower frame). ×160. (A, B, and C, From Weintraub et al., 1989. D, and E, Courtesy of H. Weintraub.)

The protein (MyoD1) encoded by *MyoD1*, is a DNA-binding nuclear protein that apparently regulates expression of certain muscle-specific genes by binding to their enhancer elements. Specific binding of MyoD1 to the enhancer region of the muscle-specific creatine kinase gene (M-CK) is shown by the mobility shift assay in Figure 18.8. MyoD1 protein incubated with a labeled M-CK gene enhancer fragment results in formation of a DNA-protein complex that decreases the electrophoretic mobility of the fragment. The specificity of the interaction is shown by competition for binding of

Figure 18.8

Mobility shift assay, showing the specific interaction between the muscle-specific creatine kinase enhancer and MyoD1 protein. MyoD1 protein was incubated with radioactively labeled M-CK enhancer and subjected to electrophoresis and autoradiography (lane 1). A significant portion of the radioactivity migrates more slowly than the free enhancer fragments, indicating that it is bound to MyoD1 protein. The specificity of the binding is indicated by adding unlabeled M-CK enhancer to the labeled fragments before the addition of MyoD1 protein (lane 2). The unlabeled fragments competed with the labeled fragments for binding to MyoD1, reducing the number of radioactive fragments associated with the MyoD1. Unlabeled chicken skeletal muscle α-actin (sk. actin) promoter fragments, however, do not compete for binding (lane 3). (From Lassar et al., 1989. Copyright © Cell Press.)

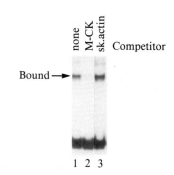

MyoD1 to labeled M-CK enhancer by excess unlabeled M-CK enhancer, but not by an unrelated enhancer (skeletal muscle α-actin promoter). Similar experiments have shown that MyoD1 also binds specifically to a rat myosin light chain gene enhancer as well as to the promoters of a number of some other muscle-specific genes.

Recently, additional genes involved in myogenic differentiation have been identified. Each of them has the capacity to convert fibroblasts into myoblasts in transfection experiments. They are *myd*, *myf-5*, *mrf-4*, myogenin, herculin, and CMD1. CMD1 is apparently the avian homologue of *MyoD1* (Lin et al., 1989). The myogenin, myf-5, mrf-4, and herculin proteins are intriguing because each contains a region that is strikingly similar to a 68 amino acid long region of MyoD1 that is required to initiate myogenesis in transfection experiments (Tapscott et al., 1988; Murre et al., 1989; Rhodes and Konieczny, 1989; Miner and Wold, 1990). Thus, these proteins apparently belong to a family of skeletal muscle-specific transcription factors. The region of homology contains two functionally important domains. These are the **helix-loop-helix (HLH)** domain, which is essential for dimerization of these so-called HLH proteins, and an adjacent **basic region**, which binds DNA specifically. As discussed in the essay on page 724, a number of other proteins that are not specific to muscle are also HLH proteins.

Dimerization (either through the formation of homodimers or heterodimers) is essential for the function of these proteins in high-affinity binding to DNA, but it lacks specificity. In fact, the dimerization region of MyoD1 can be replaced with the comparable region of another HLH protein without affecting its ability to bind DNA or regulate transcription of muscle-specific genes. That specificity resides in the basic region because modifications to that region can either abolish the ability of MyoD1 to bind DNA or eliminate the ability to activate transcription of muscle-specific genes (Davis et al., 1990). Presumably, functional activity of these proteins requires the juxtaposition of the basic regions of the monomers which is caused by dimerization.

The binding of MyoD1 to the enhancer of the muscle-specific creatine kinase gene (M-CK) is stronger when MyoD1 dimerizes with the Ig enhancer-binding HLH protein E12 than when it dimerizes with itself (Davis et al., 1990). Because E12 is present in muscle cells, E12/MyoD1 heterodimers may be responsible for regulating high-level expression of M-CK and other muscle-specific genes. On the other hand, another gene has been discovered that encodes an HLH protein that lacks the basic region (Benezra et al., 1990). This protein, which has been named **Id** (for "inhibitor of DNA binding'') may bind to MyoD1 and other HLH proteins and, because it lacks a basic region, *prevent* their interaction with gene enhancers before muscle differentiation, thus inhibiting transcription of muscle-specific proteins. A reduction in Id levels would allow differentiation to occur. Clearly, the interactions among the HLH proteins are complex and may hold the key to regulating expression of the muscle-specific genes.

Because *MyoD1* is sufficient to promote development along the myogenic pathway, one would predict that expression of this gene would be restricted to skeletal muscle tissue during development. This is the case for mouse embryos (Davis et al., 1987). Presumably, production of MyoD1 is at least one of the triggers for mesoderm cells to differentiate into muscle during early embryonic development. Recently, the *Xenopus* homologue of *MyoD1* has been detected, cloned, and used to examine a possible role for *MyoD1* in early development of this organism. Northern blot analysis of RNA from eggs and early embryos (Fig. 18.9) shows that eggs have a small amount of *MyoD1* transcript and the amount remains low until gastrulation, at which time it rapidly increases in abundance. Transcript abundance peaks in the late neurula stage before declining. The same blot was also probed for α-cardiac actin transcripts. As we learned in Chapter 12, α-cardiac actin

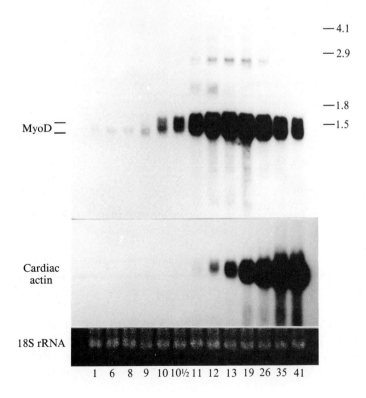

Figure 18.9

Accumulation of *MyoD1* transcripts during early *Xenopus* development is shown in an RNA blot probed with *Xenopus MyoD1* cDNA. Two sizes of *MyoD1* transcript are seen. Only the smaller size transcript is seen before the gastrula stage. The blot was then re-probed with a *Xenopus* α-cardiac actin probe. Equal loading of RNA on the gel was monitored by staining for 18S rRNA with ethidium bromide. Developmental stages are shown as both stage descriptors (blastula, gastrula, etc.; sw. tadpole: swimming tadpole) and arbitrary stage numbers as defined by Nieuwkoop and Faber (N&F), 1967. (From Hopwood et al., 1989.)

is a valuable marker of muscle cell differentiation in early embryonic mesoderm (Sturgess et al., 1980; Mohun et al., 1984). Clearly, *MyoD1* transcript accumulation precedes that of α-cardiac actin. Two techniques have been utilized to determine the sites of *MyoD1* transcript accumulation in the embryo. Northern blot analyses of RNA from dissected embryos demonstrate quite clearly that the transcripts accumulate preferentially in the somites of neurula-stage embryos, along with α-cardiac actin transcripts (Fig. 18.10). *In situ* hybridizations give more precise spatial information; they show that both transcripts accumulate exclusively in the somites of neurulae (Fig. 18.11A and B). Because the Northern blot analyses show that *MyoD1* transcripts accumulate as early as the gastrula stage, *in situ* hybridizations were also done to show the sites of localization of these transcripts in gastrulae. As shown in Figure 18.11C and D, *MyoD1* transcripts accumulate in the involuting mesoderm, which presumably corresponds to the prospective somite mesoderm.

As we discussed in Chapter 12, mesoderm is induced to develop by an interaction between animal and vegetal hemisphere cells in the equatorial region of the blastula. One would predict that the expression of *MyoD1* is

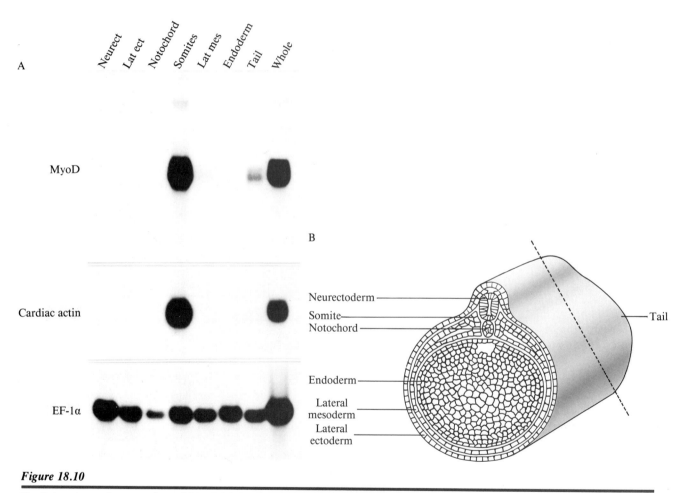

Figure 18.10

Spatial pattern of expression of *MyoD1* in *Xenopus* neurulae. A, Northern blot of RNA extracted from dissected tissues probed initially for *MyoD1* and then re-probed for cardiac actin and for EF-1α. The latter shows the relative amounts of total RNA in each lane (see page 611 for a discussion of the use of this control). Neurect: neural ectoderm; Lat ect: lateral ectoderm; Lat mes: lateral mesoderm. B, Diagram showing the locations of tissues dissected for analysis in A. (From Hopwood et al., 1989.)

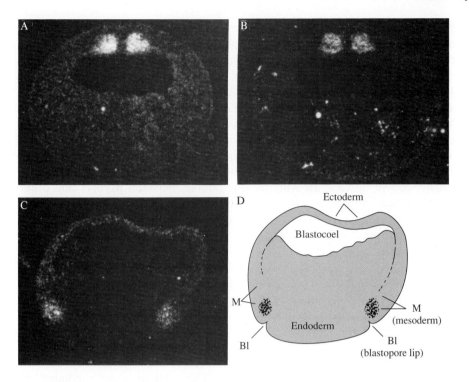

Figure 18.11

In situ hybridization analyses showing the locations of *MyoD1* and cardiac actin transcripts in early *Xenopus* embryos. A, Transverse section of neurula hybridized with cardiac actin probe. Transcripts are localized exclusively in the somites. B, Equivalent transverse section hybridized with the *Xenopus MyoD1* probe. Transcript localization is identical to that of cardiac actin. C, Gastrula hybridized with *MyoD1*. Labeling is restricted to the involuting mesoderm. The apparent labeling of the animal ectoderm is considered to be non-specific. D, Diagram showing the locations of the labeled cells in C. (From Hopwood et al., 1989.)

a consequence of mesoderm induction and that it might serve as a mediator of induction of muscle differentiation by activating other muscle-specific genes. To examine the possibility that *MyoD1* expression is a consequence of mesoderm induction, animal, vegetal, and equatorial portions of blastula-stage embryos were either cultured in isolation or animal and vegetal portions were combined and cultured. When control embryos had reached the blastula stage, RNA was extracted from the explants and analyzed for *MyoD1* expression by Northern blot analysis. As shown in Figure 18.12, neither animal nor vegetal portions alone accumulated *MyoD1* transcripts, but equatorial fragments and animal-vegetal conjugates both expressed *MyoD1*. Clearly, *MyoD1* expression is activated by induction. Is *MyoD1* involved in mediating, either directly or indirectly, the effects of induction by activating the transcription of other muscle-specific genes? This possibility has been examined by injecting *MyoD1* RNA into early embryos, separating animal caps at the late blastula stage, and culturing them until control embryos reached the late neurula stage (Hopwood and Gurdon, 1990). Because isolated animal caps will not normally form muscle or express muscle-specific genes, this procedure assesses the ability of *MyoD1* to promote muscle differentiation. In fact, the α-cardiac actin and skeletal

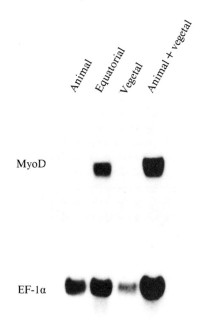

Figure 18.12

Accumulation of *MyoD1* transcripts in response to mesoderm induction. Blastulae were dissected into animal, equatorial zone, and vegetal pieces and either cultured in isolation or as animal-vegetal conjugates until control embryos became late gastrulae. The Northern blot was probed with *Xenopus MyoD1* cDNA and then re-probed with EF-1α. *MyoD1* transcripts are seen only in equatorial zone explants and in animal-vegetal conjugates. (From Hopwood et al., 1989.)

muscle actin genes were expressed as shown by Northern blot analysis, but muscle did not differentiate. These experiments show that additional factors are necessary for complete and stable myogenesis.

The advantages of myogenesis for studying gene control of development have stemmed primarily from the ability to monitor the sequence of events of myogenesis *in vitro* and the abundance of muscle-specific proteins, which has facilitated the cloning of genes for muscle-specific proteins. We have now entered a new era, in which the discoveries made with the *in vitro* systems are being exploited to learn how muscle cell differentiation is regulated at the molecular level during embryonic development. One challenge will be to learn whether interactions between MyoD1 and other HLH proteins are involved in regulation of myogenesis and—if so—how the synthesis and activity of these proteins are regulated by developmental events, such as embryonic induction.

18–3 Regulation of Transcription by Vertebrate Steroid Hormones

Vertebrate steroid hormones (which include the sex hormones and adrenal steroids) can evoke changes in patterns of protein synthesis in cells of certain steroid-responsive **target organs**. These changes are mediated, at least in part, by modifications in transcriptional patterns. This conclusion was originally based upon the fact that steroid responses are inhibited by α-amanitin and actinomycin D (O'Malley and Means, 1974; Edelman, 1975) and is confirmed by modern molecular biology techniques. Steroids can not only regulate gene expression in differentiated target cells but they can also *cause* differentiation of certain cells through the activation of new patterns of gene expression during development.

The mode of transcriptional regulation by steroid hormones has been investigated in considerable detail. Steroid target cells have specific receptor molecules that combine with the hormones to form **hormone-receptor complexes**, which mediate the effects of the hormone. The hormone-receptor complexes interact with *cis*-acting regulatory elements called **hormone response elements** (or **HREs**) in the 5′ flanking regions of hormone-inducible genes. In the absence of the hormone, the receptors are unable to interact with the HREs. Thus, even if a cell has receptors, the receptors will remain inactive until the appropriate hormone enters the cell. When receptors bind to hormone, they undergo structural modifications that facilitate interaction with DNA and mediation of the effects of the hormone.

The function of the hormone response elements in regulation of transcription has been demonstrated by ligating them to otherwise hormone-non-responsive genes, which become hormone-responsive as a consequence (Lee et al., 1981; Robins et al., 1982; Slater et al., 1985). The HREs are position- and orientation-independent and therefore behave like enhancers (Green and Chambon, 1988).

The binding of hormone-receptor complexes to HREs has been demonstrated by gel retardation assays (Kumar and Chambon, 1988). In the experiment shown in Figure 18.13, HeLa cells (human tissue culture cell line) were transiently transfected with vectors containing a human estrogen receptor gene. The transfected cells will produce estrogen receptors encoded by these exogenous genes. Extracts of transfected cells were incubated with estrogen response elements (EREs) of the *Xenopus* vitellogenin gene, which had been labeled with ^{32}P. (Recall that in Chap. 3 we discussed the control of vitellogenin synthesis in *Xenopus* liver cells by estrogen.) If the receptors

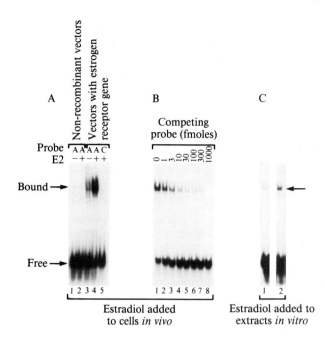

Figure 18.13

Binding of estrogen-receptor complexes to estrogen response elements. A, HeLa cells were transfected with either non-recombinant vector or vector containing the human estrogen receptor gene. The cells were treated *in vivo* either with (+) or without (−) 10^{-8} M estradiol (an estrogen; E2). Extracts were made of these cells and incubated with radiolabeled DNA probes representing the wild-type (Probe A, lanes 1–4) and mutant (Probe C, lane 5) EREs. The free and bound probes were resolved by gel electrophoresis and visualized by autoradiography. B, Extracts from HeLa cells transfected with the recombinant vector and treated with estradiol were incubated with a subsaturating level of radiolabeled probe and increasing amounts (as indicated) of unlabeled probe A (representing the wild-type ERE). C, Extracts from HeLa cells transfected with the recombinant vector were incubated with the ERE probe. Retarded complex is not formed if estrogen is not added to the extract (lane 1) but is formed if estrogen is added (arrow, lane 2). (From Kumar and Chambon, 1988. Copyright © Cell Press.)

bind to the EREs, they will retard their migration during gel electrophoresis. As shown in Figure 18.13A, no retarded complex is seen when cells were transfected with the nonrecombinant vector (lanes 1 and 2), and very little retarded complex is seen after transfection with recombinant vector unless estrogen had been added to the cell culture medium. Thus, estrogen-receptor complexes had bound to the EREs. No complex had formed when a mutated ERE was added (lane 5), indicating that the nucleotide sequence of the ERE specifies binding. The specificity of retardation is demonstrated by adding cold competitor ERE probe, which decreases the binding of labeled probe (see Fig. 18.13B). Mobility shift of labeled ERE was also demonstrated when estrogen was added to *extracts* of transfected HeLa cells; as with the *in vivo* experiments, formation of the retarded complex is dependent upon addition of estrogen to the extracts (see Fig. 18.13C). These experiments indicate that the estrogen receptor binds specifically to the response element and that the binding is hormone-dependent.

The steroid receptor proteins are a very large family of proteins that differ in their affinities to particular hormones (Evans, 1988). Each of them contains regions that mediate (1) binding to the appropriate steroid hormone, (2) DNA binding, and (3) transcriptional activation. The DNA binding region possesses two zinc fingers. The zinc finger region is responsible for determining specificity of the target gene. This specificity has been demonstrated experimentally by exchanging the zinc finger regions of the estrogen and glucocorticoid receptors, which switches receptor specificity; i.e., an estrogen receptor with a glucocorticoid receptor zinc finger region activates transcription of glucocorticoid-responsive genes rather than of estrogen-responsive genes (Green and Chambon, 1987). Recent experiments have shown that change of specificity can even be achieved by making subtle changes in the amino acid sequence within the zinc finger region (Danielson et al., 1989; Mader et al., 1989; Umesono and Evans, 1989).

The steroid receptors provide a highly specific mechanism to select genes for transcriptional activation in response to steroid. The exact mechanism of transcriptional activation by hormone-receptor complexes is not well understood. One possibility is that the binding of hormone-receptor complexes to HREs may facilitate the interactions of other regulatory *trans-acting* factors (both generalized and tissue-specific) with their target sequences, thus facilitating formation of active transcription complexes (Ham and Parker, 1989). We shall discuss an example of such factors later in this chapter.

As mentioned previously, steroids can activate cell differentiation during development. An example is the tubular gland cells of the chick oviduct. Estrogen is required for the differentiation of the tubular gland cells, the cells that produce the egg-white proteins (ovalbumin, conalbumin, lysozyme, and ovomucoid). The major egg-white protein is ovalbumin, which makes up as much as 50% of the protein that is synthesized by the gland cells. Because estrogen is not present in measurable amounts in immature chicks, the effects of the hormone on gland cell differentiation and ovalbumin synthesis can be studied by daily administration of estrogen, which causes extensive proliferation of oviduct epithelial cells and their differentiation into tubular gland cells (Fig. 18.14).

Ovalbumin synthesis is first detectable in the oviduct about 18 hours after hormone administration and reaches a plateau after about ten days (Palmiter, 1975). The increased ovalbumin synthesis promoted by steroids is the direct result of an accumulation of ovalbumin messenger RNA in the tubular gland cells of hormone-treated tissues. This is illustrated by Northern blot analysis in Figure 18.15 and by *in situ* hybridization, which shows that the ovalbumin transcripts in the oviduct are localized to the differentiating tubular gland cells (Rempel, 1990).

Does the accumulation of ovalbumin messenger reflect an increased rate of ovalbumin mRNA synthesis, or could it be due to a stabilization of transcripts? Ovalbumin mRNA synthetic rates have been determined by hybridization of radioactively labeled messenger RNA to cloned ovalbumin cDNA. Chronic injection of estrogen leads to high levels of ovalbumin messenger synthesis that remain constant as long as hormone administration is continued. The synthetic rate drops rapidly during estrogen withdrawal to become undetectable by 60 hours. Re-administration of hormone causes a rapid return to the maximal level of synthesis (Swaneck et al., 1979a,b).

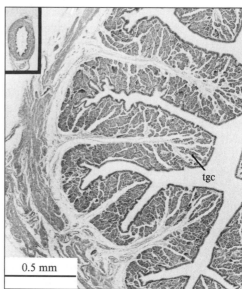

Figure 18.14

Light micrographs of cross-sections of an unstimulated chick oviduct (inset) and a portion of an oviduct from a chick stimulated for 14 days with estradiol. Both oviduct sections are shown at the same magnification. The growth of the oviduct in response to estradiol is readily apparent in the size differences of the oviducts before and after stimulation. Tubular gland cell (tgc) differentiation is apparent in the highly convoluted lining of the lumen of the oviduct. (From Rempel and Johnston, 1988. Reprinted by permission of Company of Biologists, Ltd.)

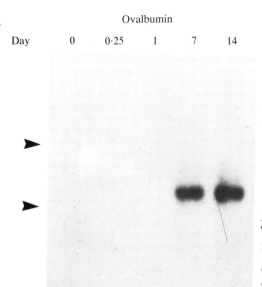

Ovalbumin

Day   0   0·25   1   7   14

Figure 18.15

Accumulation of ovalbumin transcripts as illustrated by Northern blotting. Immature chicks were injected daily with 1 mg estradiol for 14 days and sacrificed at day 0 (unstimulated), 6 hours, day 1, day 7, and day 14. RNA was isolated from the oviducts and resolved by electrophoresis. Ovalbumin transcripts were visualized by hybridization to a radiolabeled ovalbumin gene probe and autoradiography. Arrowheads indicate the positions of 28S and 18S ribosomal RNA. (From Rempel and Johnston, 1988. Reprinted by permission of Company of Biologists, Ltd.)

Thus, the steroids exert their effects on the level of ovalbumin gene transcription, and their continued action is necessary to sustain transcription of this gene.

As discussed previously, transcriptional activation by hormones may require the activities of a variety of regulatory factors in addition to the steroid receptors. In fact, the participation of a multitude of transcriptional factors in regulation and facilitation of transcription appears to be the norm in eukaryotic gene expression. Recently, one of the additional factors involved in ovalbumin gene expression has been described. It is **chicken ovalbumin upstream promoter transcription factor (COUP-TF)**. COUP-TF has been shown to bind to the ovalbumin promoter by the technique of **DNase footprinting analysis** (Fig. 18.16A). This technique involves first labeling DNA at only one end and then digesting it with the enzyme deoxyribonuclease I under conditions in which the enzyme will make a single cut in the DNA molecule, producing fragments of varying length. Hence, when the DNA fragments are separated by electrophoresis and the gel is subjected to autoradiography, one will see bands corresponding to cleavage at each nucleotide position. The presence of protein on the DNA will, however, prevent the enzyme from cutting the DNA at the protected site. Hence, no fragments corresponding to this site will be produced, resulting in a gap (which is called a "footprint") in the sequence on the gel.

COUP-TF purified from chicken oviduct tissue binds to the ovalbumin promoter, producing a footprint, as shown in Figure 18.16B. The location of the footprint corresponds exactly to the so-called COUP sequence, which has been shown to be essential for transcription of the ovalbumin gene (Knoll et al., 1983; Elbrecht et al., 1985; Pastorcic et al., 1986). Gel retardation assays demonstrate that the binding between COUP-TF and the promoter element is specific. When unlabeled promoter DNA fragments with or without the COUP sequence were added as competitor to labeled COUP sequence, only the fragments containing the COUP sequence were effective competitors (Fig. 18.17).

Steroid responsiveness thus depends upon the presence of a multitude of factors that can mediate the effects of the steroid on gene expression.

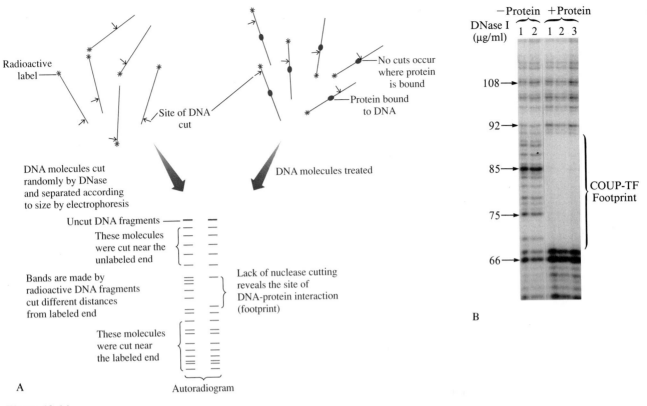

Figure 18.16

DNase I footprinting analysis of sites of protein binding to DNA. A, Diagram of technique. B, DNase I footprinting analysis of COUP-TF. Purified COUP-TF was added to the lanes indicated. The numbers on the left indicate the base locations in the ovalbumin gene upstream sequence. The blank region in lanes with added COUP-TF correspond to the COUP sequence. (B, From Bagchi et al., 1987. Copyright © 1987 American Society for Microbiology.)

Differentiation of tubular gland cells must therefore be preceded by production of these factors. Various genes encoding steroid receptors and other *trans*-acting regulatory proteins that operate in hormone-responsive tissue (such as COUP; Wang et al., 1989) have now been cloned, making it possible to begin to reconstruct the sequence of events leading to the development of steroid-responsive cells, including the tubular gland cells of the oviduct. As with other cell types that we have discussed in this chapter, the final step in differentiation involves the expression of the genes encoding cell-specific proteins (e.g., ovalbumin in the case of tubular gland cells); this must be preceded by production of the *trans*-acting factors that regulate activity of those genes, which is preceded by production of the regulators of the regulators, and so forth. As we work back in time in this sequence of events, we will ultimately reconstruct the entire sequence of molecular events that begins with oogenesis and that leads to terminal differentiation through a succession of regulatory events that control the utilization of the genetic repertoire.

18–4 Globin Gene Switching

Our final example of regulation of gene expression during development is β-globin gene expression in human erythroid cells. We have chosen this

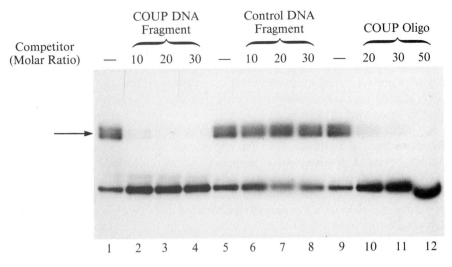

Figure 18.17

Effects of competitor DNA fragments on the binding of COUP-TF to the chicken ovalbumin promoter as demonstrated by gel retardation assay. Different amounts of unlabeled competitor fragments were added to the labeled DNA before the protein was added. Lanes 1, 5, and 9: no competitor DNA; lanes 2–4: COUP DNA fragment added as competitor; lanes 6–8: control DNA fragment lacking COUP sequence added as competitor; lanes 10–12: synthetic double-stranded COUP oligonucleotide added as competitor. The arrow indicates the position of the COUP-TF–DNA complex. (From Bagchi et al., 1987. Copyright © 1987 American Society for Microbiology.)

example because of the intriguing changes in β-globin gene expression that occur during development. The human genome contains a cluster of five functional β-globin genes that are arranged in the order $5'–\epsilon–{}^{G}\gamma–{}^{A}\gamma–\delta–\beta–3'$. The expression of each of these genes is restricted to a particular developmental stage and reflects the changing sites of production of erythrocytes as development proceeds. In the embryo, the ϵ gene is expressed in primitive erythrocytes, which are derived from extraembryonic **blood islands** located in the yolk sac. In the early fetus, ${}^{G}\gamma$ and ${}^{A}\gamma$ are expressed in erythrocytes produced by the fetal liver, whereas the δ and β genes are expressed in adult erythrocytes, which are produced in the bone marrow. Thus, not only must erythroid cells become specialized for globin synthesis, but determination must be made as to *which* of these β-globin genes is to be expressed. These changes in globin gene expression during development are called **globin gene switching**. An understanding of the mechanisms that discriminate which β-globin gene is to be expressed will be instructive in clarifying how differential gene expression during development is regulated.

Because mice also undergo changes in sites of erythropoiesis during development, the introduction of human β-globin genes with variable portions of the 5′ flanking regions into mouse embryos has enabled investigators to identify the sequences that are involved in developmental regulation of human β-globin gene expression. The use of such so-called **transgenic mice** to study developmental regulation of gene expression is a powerful means to examine gene regulatory interactions (see Appendix). These gene transfer experiments have demonstrated that the adult human β-globin gene with its immediate 5′ and 3′ flanking regions is expressed in definitive adult erythrocytes and not in embryonic erythrocytes (Magram et al., 1985; Townes et al., 1985; Behringer et al., 1987). The level of expression of such exogenous β-globin genes is, however, quite low in relation to expression of endogenous β-globin genes. Thus, although sequences adjacent to the gene can control developmental stage specificity, they are insufficient to produce the high levels of transcription that are typical of the endogenous β-globin genes; other, more distant sequences are necessary to enhance the level of gene expression. Recent investigations have characterized sequences that fit these two categories.

The identification of putative β-globin regulatory sequences has stemmed from two lines of evidence. One line of evidence is derived from a technique called **DNase I hypersensitive mapping**. Certain chromatin regions are especially sensitive to digestion by the enzyme DNase I. These include enhancer regions and sequences around the promoters of developmentally regulated genes in developmentally committed cells (Weisbrod, 1982). Such DNase I hypersensitive sites in the vicinity of the β-like globin gene cluster in erythroid cells help to identify *cis*-acting sequences that are likely involved in regulation of transcription of these genes. When such analyses are done with the β-like globin gene cluster in human erythroid cells, two classes of hypersensitive sites are seen: major and minor. The so-called minor hypersensitive sites are near the 5′ ends of transcribed β-like globin genes and are digested by relatively high concentrations of DNase I. For example, a minor hypersensitive site is close to the promoter region of the transcriptionally active β-globin gene in erythrocyte precursors of adult human bone marrow. The inactive ε, $^G\gamma$, $^A\gamma$, and δ globin genes do not possess such sites in these cells. Conversely, the β-globin gene in primitive erythrocytes lacks a minor hypersensitive site (Tuan et al., 1985). These minor hypersensitive sites thus represent regions involved in regulation of developmental stage specificity.

The major hypersensitive sites are sensitive to low levels of enzyme and are found far upstream and far downstream of the β-like globin gene cluster in erythroid cells, *regardless of which β-like globin gene is being expressed* (Tuan et al., 1985). The major hypersensitive sites are, therefore, erythroid-cell specific but independent of the developmental stage, whereas the minor hypersensitive sites are not only erythroid-cell specific but appear only at specific developmental stages.

Another line of evidence stems from the study of a human genetic disorder known as **β-thalassemia**, which causes a deficiency in β-globin gene expression. Some individuals with β-thalassemia have been shown to have deletions far upstream of normal β-globin genes; these deletions encompass the major hypersensitive sites (Curtin et al., 1989). This suggests that one or more of these hypersensitive sites may be a regulator of β-globin gene expression. Recently, Curtin et al. (1989) have examined the putative role of one of these regions in directing high-level expression of the β-globin gene. They linked a small fragment (882 base pairs) of DNA that encompasses one of the major hypersensitive sites to a human β-globin gene in either orientation (i.e., 5′-to-3′ or 3′-to-5′) and transfected mice with these fusion genes. As a control, they also utilized human β-globin genes lacking this far-upstream element. They then examined expression of these various genes in erythropoietic tissue (fetal liver) and nonerythropoietic tissue (brain) from day-16 transgenic fetuses. At this stage of development, the fetal liver is the primary site of definitive erythropoiesis, producing only adult globin. (The mouse differs from the human in that the adult pattern of gene expression is established in the fetal liver.)

The expression of human β-globin genes in transgenic mice is shown in Figure 18.18. This figure shows the results of ribonuclease protection assays of fetal brain and fetal liver RNA from mice containing three gene constructs: (A) the normal human β-globin gene without upstream regulatory

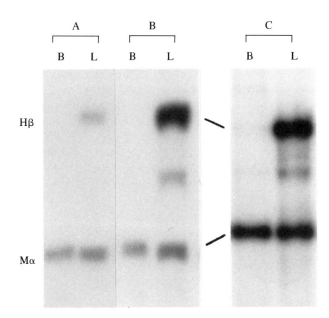

Figure 18.18

Tissue specificity of β-globin gene expression from transgenic mice. Ribonuclease protection assays were conducted on RNA from fetal brain (B lanes) and liver (L lanes) using human β-globin (Hβ) and mouse α-globin (Mα) probes. A, Human β-globin transgenic mouse. B, Mouse possessing fusion gene with far-upstream element in genomic orientation fused to β-globin gene. C, Transgenic mouse having fusion gene with far-upstream element in reverse orientation fused to β-globin gene. (From Curtin et al., 1989.)

elements; (B) the β-globin gene linked to the upstream element in the normal genomic orientation; and (C) the β-globin gene linked to the upstream element in the opposite orientation. For comparison, an assay was also conducted using a mouse α-globin gene probe. Normally, α- and β-globin genes are expressed to the same extent. However, only low-level expression of the human β-globin gene without the upstream element can be seen in the fetal liver. Among seven transgenic mice, the average level of expression per gene was 9.9% of the mouse α-globin gene expression. On the other hand, the upstream element dramatically enhanced expression of the β-globin gene in fetal liver, regardless of orientation (averaging 139.5% and 202% of the mouse α-globin gene expression for the genomic orientation and the opposite orientation, respectively). Human β-globin gene expression was not seen in the brain from any of the three constructs, illustrating the tissue specificity of β-globin gene expression in transgenic mice.

The characteristics of the far-upstream element of the β-globin gene are consistent with those of an enhancer of promoter function (i.e., tissue specificity and long-distance stimulation of transcription). The enhancer and globin promoter sequences apparently work synergistically in erythroid tissue. The enhancer is thought to "open" the chromosomal domain containing the β-like gene cluster in erythroid tissue, allowing for high-level transcription of the genes within the cluster. Their expression at any particular developmental stage is, however, dependent upon the function of the promoters that flank the individual genes (Ryan et al., 1989).

The observation that an enhancer-like sequence can facilitate high-level β-globin gene transcription opens up the prospect of **somatic gene therapy** to correct β-thalassemia. *In vitro* infection of mouse bone marrow cells with retroviruses carrying the human β-globin gene without the far-upstream elements results in low-level erythroid-cell–specific expression of the introduced β-globin gene (Dzierzak et al., 1988). It has recently been proposed that virally mediated introduction of a gene construct in which the enhancer-like sequence is linked to the β-globin gene would make correction of β-thalassemia in humans a possibility (Friedmann, 1989; Talbot et al., 1989). The scenario that is envisioned is to remove bone marrow cells from affected individuals, infect them with recombinant retroviruses, and return them to the patient. Permanent stem cells transformed with the β-globin gene construct should allow for the production of circulating red blood cells with normal levels of β-globin.

Figure 18.19

Erythroid cell-specific expression of GF-1 mRNA. RNAs from several mouse cell lines and tissues were probed by Northern analysis with GF-1 cDNA. The GF-1 mRNA is detected only in erythroid cells; the RNA from all non-erythroid cells is negative. The erythroid cells used in this experiment are mouse erythroleukemia cells (MEL). MEL cells are a committed erythroid precursor cell line, which can be induced to undergo erythropoietic differentiation by adding dimethylsulfoxide (DMSO). GF-1 RNA is detected in both the induced and uninduced MEL cells. The non-erythroid cells and tissue are: fibroblasts, B cells, T cells, macrophages, myotubes, kidney, stomach, uterus, and heart/lung. The positions of 18S and 28S rRNA are shown. (From Tsai et al., 1989. Reprinted by permission from *Nature* Vol. 339 pp. 446–451. Copyright © 1989 Macmillan Magazines Ltd.)

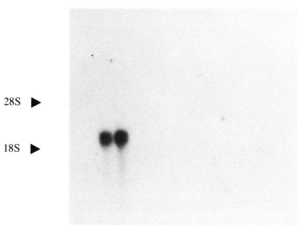

The operation of both the enhancer of the β-like gene cluster and the promoters flanking the individual genes are dependent upon the presence of the appropriate *trans*-acting factors to enable them to function. The identification of these factors, the cloning of their genes, and discovery of the basis for their developmental regulation are important goals for improving our understanding of development of erythropoietic tissue. In fact, investigators have begun that process. A DNA-binding protein that is restricted to erythroid cells and that recognizes sites in promoters and enhancers of genes with erythroid-specific expression has been discovered. This protein, which is called **GF-1**, is not restricted to any particular stage of erythropoiesis. GF-1 binding sites are characteristic of the promoters and enhancers of nearly all genes that are expressed specifically in erythroid cells (including the enhancer of the β-globin gene cluster). Mutations of the GF-1 binding sites of erythroid-specific genes affect the expression of these genes, strongly implicating GF-1 binding in regulation of erythroid-specific gene expression.

A mouse GF-1 cDNA has now been cloned by Tsai et al. (1989), thus providing a potent molecular tool for the study of regulation of gene expression during erythropoiesis. Nucleotide sequencing of the cDNA has revealed that the protein has two zinc fingers. Northern blot analyses using the GF-1 cDNA as a probe have revealed that expression of the GF-1 gene is strictly limited to erythroid tissue (Fig. 18.19). The mouse GF-1 cDNA also cross-hybridizes to an mRNA in human erythroid cells, demonstrating that this mode of regulation of erythroid gene expression has been conserved during mammalian evolution (Tsai et al., 1989). The GF-1 cDNA will be an invaluable reagent for uncovering the sequence of events that is involved in determination of the erythroid lineage during development.

18–5 Chapter Synopsis

Tissue- and stage-specific gene expression are regulated by the interactions of transcription regulators with regulatory elements of genes. This chapter includes examples of both spatial and chronological regulation of gene transcription during cell differentiation.

Immunoglobulin genes, which undergo reorganization during B-cell differentiation, are regulated by a variety of factors that bind to Ig gene regulatory motifs. The octamer motif is one of the sequences that confers lymphoid-specific expression on Ig genes. The factor Oct-2, which is a member of the POU family of transcription factors, interacts with the octamer to promote lymphoid-specific Ig gene expression.

Our current understanding of muscle-specific gene transcription during development is based upon experiments initiated with cultured fibroblasts, which can be converted to myoblasts by transfection with the gene *MyoD1*. This gene encodes the transcription regulator MyoD1, which is a member of the helix-loop-helix family of proteins containing a dimerization motif and a basic DNA-binding motif. Dimerization of MyoD1 with other HLH proteins can modulate its ability to activate muscle-specific gene expression. Hence, regulation over the composition of HLH protein dimers may be an important factor in regulation of muscle-specific gene expression. *MyoD1* is a candidate for involvement in regulation of myogenesis during development. *MyoD1* transcripts accumulate in mesoderm of *Xenopus* gastrulae before overt differentiation of muscle as a consequence of mesoderm induction. The accumulation of *MyoD1* transcripts before the onset of muscle differentiation is consistent with a possible role for *MyoD1* in regulating the onset of muscle-specific gene expression during development.

Steroid hormones regulate transcription by the formation of complexes with receptor molecules. These complexes interact with enhancer-like elements of hormone-responsive genes to activate specific gene transcription. Estrogen activation of tubular gland cell differentiation in chick oviducts is an example of hormone-induced cell differentiation. Estrogen promotes transcription of ovalbumin genes in tubular gland cells. One of the factors involved in mediating steroid regulation of ovalbumin gene transcription is the chicken ovalbumin upstream promoter transcription factor (COUP-TF), which binds to the ovalbumin promoter.

β-Globin gene expression undergoes changes during both human and mouse development that are called globin gene switching. Various experimental approaches have revealed regulatory elements that restrict β-globin gene expression to erythroid tissue and determine *which* β-globin gene is to be expressed at any particular stage.

18–6 Conclusions

The study of developmental biology is now in a very exhilarating period of discovery. Almost every new issue of journals covering developmental biology carries exciting new information about the regulation of embryonic development. The pace of discovery exceeds our wildest expectations of only a few years ago. We hope that this textbook captures the spirit and flavor of this dynamic period in developmental biology. We will have succeeded if we have engendered in our readers a fascination and enthusiasm for studying the processes of change that typify development.

The utilization of genetic information for the synthesis of proteins and the subsequent utilization of those proteins to form exquisite functional entities in a stereotypic three-dimensional relationship is the essence of the developmental process. Although investigators have identified many of the genomic sequences involved in regulating cell differentiation and are beginning to reconstruct the regulatory hierarchies that control their utilization, we still only have a rudimentary understanding of how that information is used to form an embryo. The construction of even the simplest embryonic structures is beyond our current level of understanding. However, with the rapid recent progress in developmental biology, this challenge will soon be met.

In addition to the initial processes of development, reprises of developmental events can occur later in life that can have both beneficial and deleterious results to the organism. Regeneration, which recapitulates many of the processes in the initial development of a structure, can replace an amputated structure. On the other hand, inappropriate expression of proto-oncogenes during later life can lead to cancer. As we learn more about the regulation of initial development, we will simultaneously acquire the knowledge that will lead to controlling both regeneration and cancer. Furthermore, as we learn how to target genes to cells that are deficient in particular gene products and to manipulate their expression, we will be well on the way toward repairing congenital defects and metabolic diseases. The future is bright and very exciting.

References

Alt, F. W. et al. 1982. Immunoglobulin heavy-chain expression and class switching in a murine leukaemia cell line. Nature (Lond.), *296*: 325–331.

Andrews, G. K., M. Dziadek, and T. Tamaoki. 1982. Expression and methylation of the mouse α-fetoprotein gene in embryonic, adult and neoplastic tissues. J. Biol. Chem., *257*: 5148–5153.

Antequera, F., D. Macleod, and A. P. Bird. 1989. Specific protection of methylated CpGs in mammalian nuclei. Cell, *58*: 509–517.

Bagchi, M. K. et al. 1987. Purification and characterization of chicken ovalbumin gene upstream promoter transcription factor from homologous oviduct cells. Mol. Cell. Biol., *7*: 4151–4158.

Behringer, R. R. et al. 1987. Two 3′ sequences direct adult erythroid-specific expression of human β-globin genes in transgenic mice. Proc. Natl. Acad. Sci. U.S.A., *84*: 7056–7060.

Benezra, R. et al. 1990. The protein Id: A negative regulator of helix-loop-helix DNA binding proteins. Cell, *61*: 49–59.

Bird, A. P. 1978. Use of restriction enzymes to study eukaryotic DNA methylation: II. The symmetry of methylated sites supports semi-conservative copying of the methylation pattern. J. Mol. Biol., *118*: 49–60.

Bird, A. P. 1986. CpG-rich islands and the function of DNA methylation. Nature (Lond.), *321*: 209–213.

Buckingham, M. E. 1977. Muscle protein synthesis and its control during the differentiation of skeletal muscle cells *in vitro*. Int. Rev. Cytol., *15*: 269–332.

Clerc, R. G. et al. 1988. The B-cell specific Oct-2 protein contains POU box- and homeo box-type domains. Genes Dev., *2*: 1570–1581.

Currie, R. A., and R. G. Roeder. 1989. Identification of an octamer-binding site in the mouse kappa light-chain immunoglobulin enhancer. Mol. Cell. Biol., *9*: 4239–4247.

Curtin, P. T. et al. 1989. Human β-globin gene expression in transgenic mice is enhanced by a distant DNase I hypersensitive site. Proc. Natl. Acad. Sci. U.S.A., *86*: 7082–7086.

Danielson, M., L. Hinck, and G. M. Ringold. 1989. Two amino acids within the knuckle of the first zinc finger specify DNA response element activation by the glucocorticoid receptor. Cell, *57*: 1131–1138.

Davis, R. L., H. Weintraub, and A. B. Lassar. 1987. Expression of a single transfected cDNA converts fibroblasts to myoblasts. Cell, *51*: 987–1000.

Davis, R. L. et al. 1990. The MyoD DNA binding domain contains a recognition code for muscle-specific gene activation. Cell, *60*: 733–746.

Dreyer, W. J., and J. C. Bennett. 1965. The molecular basis of antibody formation. A paradox. Proc. Natl. Acad. Sci. U.S.A., *54*: 864–869.

Dzierzak, E. A., T. Papayannopoulou, and R. C. Mulligan. 1988. Lineage-specific expression of a human β-globin gene in murine bone marrow transplant recipients reconstituted with retrovirus-transduced stem cells. Nature (Lond.), *331*: 35–41.

Edelman, I. S. 1975. Mechanism of action of steroid hormones. J. Steroid Biochem., *6*: 147–159.

Elbrecht, A. et al. 1985. Identification by exonuclease footprinting of a distal promoter-binding protein from HeLa cell extracts. DNA, *4*: 233–240.

Evans, R. M. 1988. The steroid and thyroid hormone receptor superfamily. Science, *240*: 889–895.

Friedmann, T. 1989. Progress toward human gene therapy. Science, *244*: 1275–1281.

Gardiner-Garden, M., and M. Frommer. 1987. CpG islands in vertebrate genomes. J. Mol. Biol., *196*: 261–282.

Gautsch, J. W., and M. C. Wilson. 1983. Delayed *de novo* methylation in teratocarcinoma suggests additional tissue-specific mechanisms for controlling gene expression. Nature (Lond.), *301*: 32–37.

Grainger, R. M. et al. 1983. Is hypomethylation linked to activation of δ-crystallin genes during lens development? Nature (Lond.), *306*: 88–91.

Green, S., and P. Chambon. 1987. Oestradiol induction of a glucocorticoid-responsive gene by a chimaeric receptor. Nature (Lond.), *325*: 75–78.

Green, S., and P. Chambon. 1988. Nuclear receptors enhance our understanding of transcription regulation. Trends Genet., *4*: 309–314.

Ham, J., and M. G. Parker. 1989. Regulation of gene expression by nuclear hormone receptors. Curr. Top. Cell Biol., *1*: 503–511.

Herr, W. et al. 1988. The POU domain: A large conserved region in the mammalian *pit-1*, *oct-2* and *C. elegans unc-86* gene products. Genes Dev., *2*: 1513–1516.

Holliday, R., and J. E. Pugh. 1975. DNA modification mechanisms and gene activity during development. Science, *187*: 226–232.

Hopwood, N. D., and J. B. Gurdon. 1990. Activation of muscle genes without myogenesis by ectopic expression of MyoD in frog embryo cells. Nature (Lond.), *347*: 197–200.

Hopwood, N. D., A. Pluck, and J. B. Gurdon. 1989. MyoD expression in the forming somites is an early response to mesoderm induction in *Xenopus* embryos. EMBO J., *8*: 3409–3417.

Hozumi, N., and S. Tonegawa. 1976. Evidence for somatic rearrangement of immunoglobulin genes coding for variable and constant regions. Proc. Natl. Acad. Sci. U.S.A., *73*: 3628–3632.

Jones, P. A., and S. M. Taylor. 1980. Cellular differentiation, cytidine analogs and DNA methylation. Cell, *20*: 85–93.

Kelley, D. E. et al. 1988. The coupling between enhancer activity and hypomethylation of κ immunoglobulin genes is developmentally regulated. Mol. Cell. Biol., *8*: 930–937.

Knoll, B. J. et al. 1983. Definition of the ovalbumin gene promoter by transfer of an ovalglobin fusion gene into cultured cells. Nucleic Acids Res., *11*: 6733–6754.

Ko, H.-S. et al. 1988. A human protein specific for the immunoglobulin octamer DNA motif contains a functional homeobox domain. Cell, *55*: 135–144.

Kumar, V., and P. Chambon. 1988. The estrogen receptor binds tightly to its responsive element as a ligand-induced homodimer. Cell, *55*: 145–156.

Landolfi, N. F. et al. 1986. Interaction of cell-type-specific nuclear proteins with immunoglobulin V_H promoter region sequences. Nature (Lond.), *323*: 548–551.

Landschulz, W. H., P. F. Johnson, and S. L. McKnight. 1988. The leucine zipper: A hypothetical structure common to a new class of DNA binding proteins. Science, *240*: 1759–1764.

Lassar, A. B. et al. 1989. MyoD is a sequence-specific DNA binding protein requiring a region of *myc* homology to bind to the muscle creatine kinase enhancer. Cell, *58*: 823–831.

Lee, F. et al. 1981. Glucocorticoids regulate expression of dihydrofolate reductase cDNA in mouse mammary tumour virus chimaeric plasmids. Nature (Lond.), *294*: 229–232.

Lenardo, M., J. W. Pierce, and D. Baltimore. 1987. Protein-binding sites in Ig gene enhancers determine transcriptional activity and inducibility. Science, *236*: 1573–1577.

Lin, Z.-y. et al. 1989. An avian muscle factor related to MyoD1 activates muscle-specific promoters in nonmuscle cells of different germ-layer origin and in BrdU-treated myoblasts. Genes Dev., *3*: 986–996.

Lock, L. F. et al. 1986. Methylation of the mouse *hprt* gene differs on the active and inactive X chromosomes. Mol. Cell. Biol., *6*: 914–924.

Mader, S. et al. 1989. Three amino acids of the oestrogen receptor are essential to its ability to distinguish an oestrogen from a glucocorticoid-responsive element. Nature (Lond.), *338*: 271–274.

Magram, J., K. Chada, and F. Costantini. 1985. Developmental regulation of a cloned adult β-globin gene in transgenic mice. Nature (Lond.), *315*: 338–340.

Max, E. E., J. G. Seidman, and P. Leder. 1979. Sequences of five potential recombination sites encoded close to an immunoglobin κ constant region gene. Proc. Natl. Acad. Sci. U.S.A., *76*: 3450–3454.

Meehan, R. R. et al. 1989. Identification of a mammalian protein that binds specifically to DNA containing methylated CpGs. Cell, *58*: 499–507.

Miner, J. H., and B. Wold. 1990. Herculin, a fourth member of the *MyoD* family of myogenic regulatory genes. Proc. Natl. Acad. Sci. U.S.A., *87*: 1089–1093.

Mohun, T. J. et al. 1984. Cell type-specific activation of actin genes in the early amphibian embryo. Nature (Lond.), *311*: 716–721.

Monk, M. 1986. Methylation and the X chromosome. Bioessays, *4*: 204–208.

Müller, M. M. et al. 1988. A cloned octamer transcription factor stimulates transcription from lymphoid-specific promoters in non-B cells. Nature (Lond.), *336*: 544–551.

Murre, C., P. Schonleber McCaw, and D. Baltimore. 1989. A new DNA binding and dimerization motif in immunoglobulin enhancer binding, *daughterless*, *MyoD*, and *myc* proteins. Cell, *56*: 777–783.

Nelms, K., and B. Van Ness. 1990. Identification of an octamer-binding site in the human kappa light-chain enhancer. Mol. Cell. Biol., *10*: 3843–3846.

Nieuwkoop, P. D., and J. Faber. 1967. *Normal Table of Xenopus laevis (Daudin)*. North Holland Publishing Company, Amsterdam.

Oettinger, M. A. et al. 1990. RAG–1 and RAG–2, adjacent genes that synergistically activate V(D)J recombination. Science, *248*: 1517–1523.

O'Malley, B. W., and A. R. Means. 1974. Female steroid hormones and target cell nuclei. Science, *183*: 610–620.

Palmiter, R. D. 1975. Quantitation of parameters that determine the rate of ovalbumin synthesis. Cell, *4*: 189–197.

Pastorcic, M. et al. 1986. Control of transcription initiation *in vitro* requires binding of a transcription factor to the distal promoter of the ovalbumin gene. Mol. Cell. Biol., *6*: 2784–2791.

Razin, A., and A. D. Riggs. 1980. DNA methylation and gene function. Science, *210*: 604–610.

Rempel, S. A. 1990. Steroid hormone regulation of c-*myc* gene expression in proliferating chick oviduct. Ph.D. thesis, University of Calgary, Calgary, Alberta, Canada.

Rempel, S. A., and R. N. Johnston. 1988. Steroid-induced cell proliferation *in vivo* is associated with increased c-*myc* proto-oncogene transcript abundance. Development, *104*: 87–95.

Rhodes, S. J., and S. F. Konieczny. 1989. Identification of MRF4: a new member of the muscle regulatory factor gene family. Genes Dev., *3*: 2050–2061.

Riggs, A. D. 1975. X-inactivation, differentiation, and DNA methylation. Cytogenet. Cell Genet., *14*: 9–25.

Robins, D. M. et al. 1982. Regulated expression of human growth hormone genes in mouse cells. Cell, *29*: 623–631.

Ryan, T. M. et al. 1989. A single erythroid-specific DNase I super-hypersensitive site activates high levels of human β-globin gene expression in transgenic mice. Genes Dev., *3*: 314–323.

Sakano, H. et al. 1979. Sequences at the somatic recombination sites of immunoglobulin light chain genes. Nature (Lond.), *280*: 288–293.

Scheidereit, C., A. Heguy, and R. G. Roeder. 1987. Identification and purification of a human lymphoid-specific octamer-binding protein (OTF-2) that activates transcription of an immunoglobulin promoter in vitro. Cell, *51*: 783–793.

Scheidereit, C. et al. 1988. A human lymphoid-specific transcription factor that activates immunoglobulin genes is a homeobox protein. Nature (Lond.), *336*: 551–557.

Slater, E. P. et al. 1985. Glucocorticoid receptor binding and activation of a heterologous promoter by dexamethasone by the first intron of the human growth hormone gene. Mol. Cell. Biol., *5*: 2984–2992.

Staudt, L. M. et al. 1986. A lymphoid-specific protein binding to the octamer motif of immunoglobulin genes. Nature (Lond.), *323*: 640–643.

Staudt, L. M. et al. 1988. Cloning of a lymphoid-specific cDNA encoding a protein binding the regulatory octamer DNA motif. Science, *241*: 577–580.

Stein, R. et al. 1982. Clonal inheritance of the pattern of DNA methylation in mouse cells. Proc. Natl. Acad. Sci. U.S.A., *79*: 61–65.

Sturgess, E. A. et al. 1980. Actin synthesis during the early development of *Xenopus laevis*. J. Embryol. Exp. Morphol., *58*: 303–320.

Sullivan, C. H. et al. 1989. Developmental regulation of hypomethylation of δ-crystallin genes in chicken embryo lens cells. Mol. Cell. Biol., *9*: 3132–3135.

Swaneck, G. E. et al. 1979a. Absence of an obligatory lag in the induction of ovalbumin mRNA by estrogen. Biochem. Biophys. Res. Comm., *88*: 1412–1418.

Swaneck, G. E. et al. 1979b. Effect of estrogen on gene expression in chicken oviduct: Evidence for transcriptional control of ovalbumin gene. Proc. Natl. Acad. Sci. U.S.A., *76*: 1049–1053.

Talbot, D. et al. 1989. A dominant control region from the human β-globin locus conferring integration site-independent gene expression. Nature (Lond.), *338*: 352–355.

Tapscott, S. J. et al. 1988. MyoD1: A nuclear phosphoprotein requiring a *myc* homology region to convert fibroblasts to myoblasts. Science, *242*: 405–411.

Taylor, S. M., and P. A. Jones. 1979. Multiple new phenotypes induced in 10T1/2 and 3T3 cells treated with 5-azacytidine. Cell, *17*: 771–779.

Taylor, S. M., and P. A. Jones. 1982. Changes in phenotypic expression in embryonic and adult cells treated with 5-azacytidine. J. Cell. Physiol., *111*: 187–194.

Tonegawa, S. 1983. Somatic generation of antibody diversity. Nature (Lond.), *302*: 575–581.

Townes, T. M. et al. 1985. Erythroid-specific expression of human β-globin genes in transgenic mice. EMBO J., *4*: 1715–1723.

Tsai, S.-F. et al. 1989. Cloning of cDNA for the major DNA-binding protein of the erythroid lineage through expression in mammalian cells. Nature (Lond.), *339*: 446–451.

Tuan, D. et al. 1985. The ''β-like-globin'' gene domain in human erythroid cells. Proc. Natl. Acad. Sci. U.S.A., *82*: 6384–6388.

Umesono, K., and R. M. Evans. 1989. Determinants of target gene specificity for steroid/thyroid hormone receptors. Cell, *57*: 1139–1146.

Vinson, C. R., P. B. Sigler, and S. L. McKnight. 1989. Scissors-grip model for DNA recognition by a family of leucine zipper proteins. Science, *246*: 911–916.

Wang, L.-H. et al. 1989. COUP transcription factor is a member of the steroid receptor superfamily. Nature (Lond.), *340*: 163–166.

Weigert, M., and R. Riblet. 1976. Genetic control of antibody variable regions. Cold Spring Harbor Symp. Quant. Biol., *41*: 837–846.

Weintraub, H. et al. 1989. Activation of muscle-specific genes in pigment, nerve, fat, liver, and fibroblast cell lines by forced expression of MyoD1. Proc. Natl. Acad. Sci. U.S.A., *86*: 5434–5438.

Weisbrod, S. 1982. Active chromatin. Nature (Lond.), *297*: 289–295.

Wigler, M., D. Levy, and M. Perucho. 1981. The somatic replication of DNA methylation. Cell, *24*: 33–40.

Wirth, T., L. Staudt, and D. Baltimore. 1987. An octamer oligonucleotide upstream of a TATA motif is sufficient for lymphoid-specific promoter activity. Nature (Lond.), *329*: 174–178.

Wright, W. E., D. A. Sassoon, and V. K. Lin. 1989. Myogenin, a factor regulating myogenesis, has a domain homologous to MyoD1. Cell, *56*: 607–617.

Appendix: Techniques of Molecular Biology

Application of the techniques of molecular biology has revolutionized the study of developmental biology. Such experiments utilize elegant, but simple, technology. We briefly review some of the most important techniques used in molecular biology research as applied to developmental biology.

I. Electrophoresis

Electrophoresis is an extremely valuable technique used to separate complex mixtures of macromolecules. Most macromolecules are electrically charged and will, therefore, migrate in an electrical field; this migration is electrophoresis. The molecules are placed in an electrical field where they migrate to either the positive electrode (**anode**) or the negative electrode (**cathode**) (i.e., negatively charged molecules to the anode and positively charged molecules to the cathode). The greater the charge, the more rapid the migration. Mobility is also affected by other physical properties of the molecules—large molecules move more slowly than small ones, and certain shaped molecules migrate more readily than others. In a complex mixture of macromolecules, the combination of these different variables causes molecules to migrate at different rates and results in a sorting out of different components. Electrophoresis is quite versatile and can be adapted for a number of different separation requirements, including resolution of complex mixtures of proteins or nucleic acids.

A sample containing a solution of macromolecules is normally placed on a semisolid support medium, an electrical field is applied, and the molecules migrate through the medium. This is called **zone electrophoresis**, since the molecules separate and migrate as discrete **zones** or **bands**. The support medium may act as a molecular sieve that assists in separating molecules by differentially impeding their movement. Proteins and nucleic acids are usually separated on a gel composed of either **polyacrylamide** or **agarose**. The gels are made with buffers that are selected to produce a pH that facilitates molecular separation. After separation, the molecules are localized by staining the gel with a specific reagent, by scanning the gel with ultraviolet light (to localize nucleic acids), or, if the sample is radioactive, by determining the radioactivity distribution in the gel. In actual practice, many variations of the electrophoretic technique are used for separation of different types of macromolecules.

Gel Electrophoresis of Nucleic Acids

Separation of nucleic acids is achieved mainly by molecular sieving, because the charge-to-mass ratio is nearly the same for all nucleic acids. This means that small molecules will migrate faster than large ones. Either RNA molecules or DNA fragments can be separated in this way. Nucleic acid electrophoresis uses either polyacrylamide or agarose gels. Separation of DNA fragments by polyacrylamide gel electrophoresis is so accurate that a mixture of fragments that differ from one another by only one nucleotide can be resolved into unique bands on a gel.

SDS-Gel Electrophoresis

This procedure is used to separate proteins according to their molecular weights. Electrophoresis is done in polyacrylamide gels containing the detergent sodium dodecyl sulfate (SDS). Before electrophoresis, the protein mixture is treated with SDS and a reducing agent, such as mercaptoethanol. The latter reagent reduces disulfide bonds, causing polypeptide subunits to dissociate and erasing secondary structure. The SDS binds to the polypeptides and forms complexes that migrate as though they have uniform shapes. The charge on the complexes is determined solely by the SDS. Since all polypeptides treated in this way have similar charge-to-mass ratios, and shape differences are eliminated, the only factor that influences migration is molecular weight—due to molecular sieving. A plot of the distance migrated versus log molecular weight gives a straight line (Fig. A.1). The molecular weight of any protein can be calculated by subjecting it and proteins of known molecular weight to electrophoresis, plotting molecular weight versus distance migrated for the knowns, and determining the molecular weight for the unknown by interpolation.

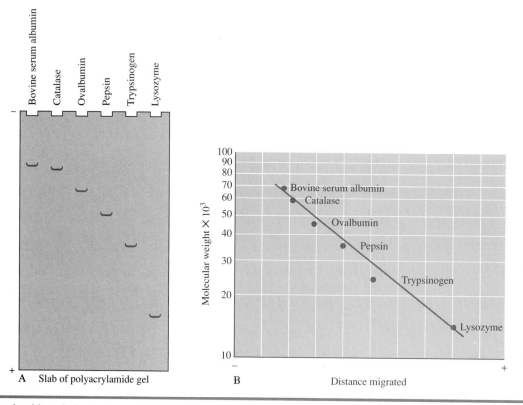

Figure A.1

SDS-polyacrylamide gel electrophoresis. A, Drawing of an SDS-polyacrylamide slab gel stained to show the relative migrations of a number of polypeptides. Each polypeptide is initially loaded onto the gel in the wells shown at the top. B, Semilogarithmic plot of molecular weight vs. distance migrated.

Two-Dimensional Gel Electrophoresis of Proteins

When a mixture of proteins is separated in one-dimensional gel electrophoresis, each band on the gel may correspond to a mixture of proteins that migrate the same distance under a single set of conditions. The resolution of electrophoresis can be increased tremendously if the separated proteins are electrophoresed subsequently under a different set of conditions and at a direction that is perpendicular to the first dimension. The most commonly used procedure for two-dimensional electrophoresis of proteins is one developed by O'Farrell (1975).

In the O'Farrell procedure, proteins are separated initially by isoelectric focusing. This technique is based upon the fact that a protein will not migrate in an electrical field at the pH at which it is electrically neutral. This pH is called the **isoelectric point (pI)**. Since the isoelectric points of proteins differ from one another, proteins can be separated by electrophoresing a mixture of proteins in a narrow tube of polyacrylamide containing a stable pH gradient, with the lowest pH at the anode and the highest pH at the cathode. The proteins migrate in the gel until they reach a pH in the gradient that corresponds to their pI. They then lose electrophoretic mobility and become focused in a narrow zone on the gel.

After isoelectric focusing, the tube gel is then placed across the top of a polyacrylamide slab gel where it is subjected to SDS electrophoresis. In the second dimension, the components in each zone on the isoelectric focusing gel migrate into the slab gel and separate according to molecular weight. Two-dimensional electrophoresis is a powerful technique for resolving individual polypeptides in a large, heterogeneous protein mixture. An example of the type of results obtained with this technique is shown in Figure 13.10.

II. DNA Renaturation

The basic organization of eukaryotic genomes has been determined largely by the technique of **DNA reassociation, reannealing,** or **renaturation**. This technique takes advantage of the complementary relationships between nucleotide pairs on the two strands of the DNA double helix. When the strands are separated (**denatured**) by experimental means, they will recognize one another and reassociate under appropriate experimental conditions. Denaturation of DNA is accomplished either by raising the pH above 11 with alkali or by heating a solution of short segments of purified DNA in a dilute salt solution. The short segments are produced by shearing the DNA with ultrasound or high pressure. Denaturation generally occurs at 80 to 90° C as a result of the disruption of the hydrogen bonds that join the two strands. The midpoint of this transition is T_m, or the **melting temperature** of DNA. When a solution of denatured DNA is subsequently cooled, renaturation or reannealing occurs at about 20 to 25° C below the T_m. Likewise, if denaturation is caused by high pH, reannealing can be produced by lowering the pH.

Annealing between two strands of DNA depends upon the complementarity between bases of the two strands. Although denatured DNA of a single species will renature readily, denatured DNA of two species will renature only if they have nucleotide sequences in common, and the extent of renaturation depends upon the degree of similarity between nucleotide sequences. Renaturation results from a random collision between two complementary strands. Thus, the **concentration of the DNA** and the **duration of the reaction** must be sufficient to allow the matching to occur. The renaturation reaction is controlled by the product of these two factors, and this is expressed as $C_o t$.

$$C_o t = \text{moles of nucleotides} \times \text{seconds/liter}$$

$C_o t$ values vary over several orders of magnitude, and therefore $C_o t$ is expressed by a logarithmic scale. An ideal $C_o t$ curve is shown in Figure A.2, in which the percent of reassociated DNA is plotted against $C_o t$. The reassociation properties of a given type of DNA are described by the $C_o t$ at 50% reassociation. This is called $C_o t\frac{1}{2}$. Since reassociation is the result of random collisions between single strands of DNA, the reaction rate is inversely proportional to the *number* of different sequences present. Thus, the larger the size of the genome, the more difficult it is for complementary strands to find one another, because each sequence is diluted by all the others. Renaturation will therefore occur more readily with a small genome than with a large genome. For example, the *E. coli* genome contains 4.2×10^6 nucleotide pairs, and the MS-2 virus genome contains 4×10^3 nucleotide pairs. Therefore, the concentration of any particular sequence is reduced a thousandfold in *E. coli* as compared to MS-2, requiring a $C_o t\frac{1}{2}$ approximately 10^3 greater (Fig. A.3).

Figure A.2

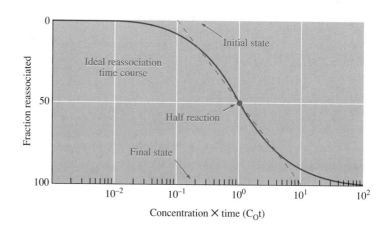

Ideal $C_o t$ curve. The percent reassociation is plotted against the product of total concentration and time on a logarithmic scale. (After R. J. Britten and D. E. Kohne. 1968. Repeated sequences in DNA. Science, *161*: 529–540. Copyright 1968 by the American Association for the Advancement of Science.)

Figure A.3

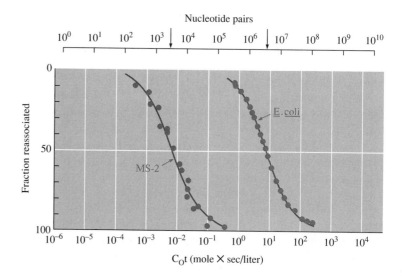

Reassociation of double-stranded nucleic acids from MS-2 virus and *E. coli*. (After R. J. Britten and D. E. Kohne. 1968. Repeated sequences in DNA. Science, *161*: 529–540. Copyright 1968 by the American Association for the Advancement of Science.)

$C_o t\frac{1}{2}$ also indicates the combined length of *different* sequences present in a genome. This measure is called the **complexity** of the genome, which is usually expressed in nucleotide pairs. If a genome consists of sequences A, B, and C, for example, the complexity is the sum of the different sequences present; i.e., $A + B + C$. If each of the preceding sequences were 10^2 base pairs in length, the complexity would be 3×10^2. For *E. coli*, every sequence is represented once in the genome. Therefore, its complexity is identical with its genome size; i.e., 4.2×10^6 nucleotide pairs.

When reassociation of two genomes is compared, each should have a $C_o t\frac{1}{2}$ proportional to its complexity. *E. coli* is usually the standard used for such comparisons. Thus, the complexity of MS-2 (4×10^3 nucleotide pairs) is considerably less than that of *E. coli*. However, such straightforward comparisons are not possible with advanced eukaryotic organisms. We shall use the mammalian genome as an example. Because the mammalian genome is considerably larger than that of *E. coli*, it should be more complex and have a $C_o t\frac{1}{2}$ commensurate with its complexity. For example, the calf haploid genome contains 3.2×10^9 nucleotide pairs. Since the $C_o t\frac{1}{2}$ of *E. coli* DNA is approximately 10 (see Fig. A.3), the $C_o t\frac{1}{2}$ of calf DNA should be approximately 10,000. However, eukaryotic DNA reanneals in a more complicated way than does prokaryotic DNA.

As shown in Figure A.4, a portion of calf thymus DNA is very rapidly reassociating, with a $C_o t\frac{1}{2}$ of 0.03 and low complexity. This rapid renaturation indicates that a portion of the DNA is present in numerous copies. Repetition of nucleotide sequences increases their concentration and consequently causes more rapid reannealing. There is a clear separation between this DNA and a slow annealing, high complexity fraction. The latter fraction has a $C_o t\frac{1}{2}$ of 3×10^3, which is close to the predicted value for calf DNA present in single copies. Because the reassociation of the rapidly renaturing DNA is 100,000 times as rapid as the single copy DNA, the nucleotide sequences in the repetitive DNA are repeated on the average of 100,000 times. There is an extremely broad spectrum of sequence repetition in eukaryotic DNA. This is illustrated very clearly by the $C_o t$ curve for mouse

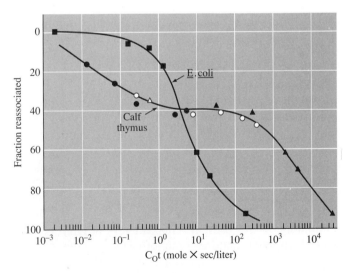

Figure A.4

Reassociation of calf thymus DNA compared to *E. coli* DNA. (After R. J. Britten and D. E. Kohne. 1968. Repeated sequences in DNA. Science, *161*: 529–540. Copyright 1968 by the American Association for the Advancement of Science.)

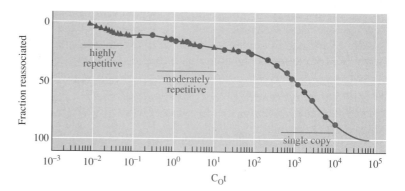

Figure A.5

Reassociation of mouse DNA. (After McConaughy and McCarthy, 1970).

DNA (Fig. A.5). Approximately 70% of the DNA has a $C_0t\frac{1}{2}$ value corresponding to high complexity, unique sequence DNA. An intermediate component, which corresponds to about 15% of the DNA, apparently contains sequences repeated 1,000 to 100,000 times, whereas a rapidly annealing component, which constitutes 10% of the genome, seems to consist of sequences repeated 1,000,000 times.

The presence of both single copy DNA and sequences with variable levels of repetition is a universal property of eukaryotic genomes—including both plants and animals. There is considerable variation among organisms as to the relative amounts of the different categories of DNA (i.e., single copy, moderately repetitive, and highly repetitive). However, there are some generalizations that can be made about the functions that these portions of the genome perform in cell metabolism.

The high complexity (or single copy) DNA contains most of the structural genes (i.e., the genes that encode messenger RNA). Although it is unclear whether all unique sequences are structural genes, it can be said with a great deal of certainty that most structural genes are present as single copies. This conclusion has important repercussions for regulation of gene expression because it means that if a cell becomes specialized for the production of a protein, it can rely upon only a single template (two in a diploid cell) for production of the messenger RNA for that protein.

Moderately repetitive DNA is composed of two categories of nucleotide sequences: (1) identifiable genes that occupy a limited number of chromosomal sites (histone genes, ribosomal RNA genes, and transfer RNA genes) and (2) sequences that do not fit the classic concept of genes and are scattered throughout the genome. These sequences are interspersed between the unique sequences. It is now thought that most, if not all, unique sequences are flanked on either side by moderately repetitive DNA. The function of interspersed repetitive DNA, if any, is unknown. As we discuss in Chapters 3 and 13, transcripts with interspersed moderately repetitive and single copy sequences are synthesized during oogenesis in the sea urchin and *Xenopus*. These transcripts may result from transcription that is initiated on single copy sequences and proceeds into the flanking moderately repetitive regions.

Highly repetitive DNA consists of short, repetitive nucleotide sequences that do not code for protein. This category of DNA is usually called **satellite DNA** (which has no known function).

III. RNA-DNA Hybridization

Because RNA is produced as a complementary copy of DNA, it can be annealed with single-stranded DNA to form an RNA-DNA hybrid. The formation of hybrids is dependent upon a random association of RNA-DNA fragments. Thus, like DNA-DNA reassociation, hybridization can be characterized by the time required to form an RNA-DNA duplex and by the concentration of the reactants to determine whether the mRNA hybridizes with unique or repetitive DNA. A substantial fraction of total mRNA has been found to hybridize almost exclusively with nonrepetitive DNA. Hence, most of the mRNA molecules consist of transcripts from this type of DNA (see, e.g., Goldberg et al., 1973).

The number of different gene transcripts is calculated from the **RNA sequence complexity**. As with DNA complexity, RNA complexity is the total length of different sequences, as measured in nucleotides (Davidson, 1976). One method of measuring the complexity of structural gene transcripts is determination of the extent of hybridization of RNA to total single copy DNA (Galau et al., 1974), the class of DNA that contains most structural genes (see previous discussion).

A variation of the technique of RNA-DNA hybridization is *in situ* hybridization. In this procedure, RNA in tissue sections or in isolated cellular components is hybridized to radioactively-labeled probes representative of particular gene sequences. The labeled DNA that has hybridized to the RNA is detected by autoradiography. Autoradiography is one of the most commonly used techniques in cellular and molecular biology. It involves using photographic emulsion to detect radioactivity. In the case of *in situ* hybridization, the emulsion is placed over a cytological preparation. Particles emitted from the radioisotope activate the emulsion so that silver grains are produced when the emulsion is developed by normal darkroom procedures. The silver grains are seen as tiny black dots when observed with the microscope and are located immediately over the source of radiation. An example of *in situ* hybridization is seen in Figure 3.31, page 84. Autoradiography is used to detect radioactive macromolecules in a variety of other applications. A commonly used procedure is to employ X-ray film to detect labeled nucleic acids or proteins after analytical procedures such as electrophoresis. The film is placed over the source of the radioactivity, and when the film is developed, a black region on the clear film indicates the presence of radioactive macromolecules.

IV. cDNA

One of the most valuable techniques in molecular biology is the preparation of **cDNA** (**complementary DNA**; Fig. A.6), which is a single strand of DNA made by copying mRNA with the enzyme **reverse transcriptase**. This enzyme, which is isolated from RNA tumor viruses, uses RNA to make a complementary DNA copy. cDNA can be used as a molecular probe to detect specific mRNA sequences, or it can be made double-stranded and cloned by techniques that we describe in the next section.

V. Recombinant DNA

No single achievement in molecular biology has had a greater impact on developmental biology than the discovery of procedures to clone eukaryotic DNA by inserting it into bacterial plasmids. This technique has made it possible to produce virtually unlimited amounts of specific DNA sequences. **Plasmids** are small, circular DNA molecules that replicate independently in bacteria, such as the common laboratory organism *E. coli*. The plasmid is a vehicle (or **vector**) used to carry the eukaryotic DNA, which is inserted

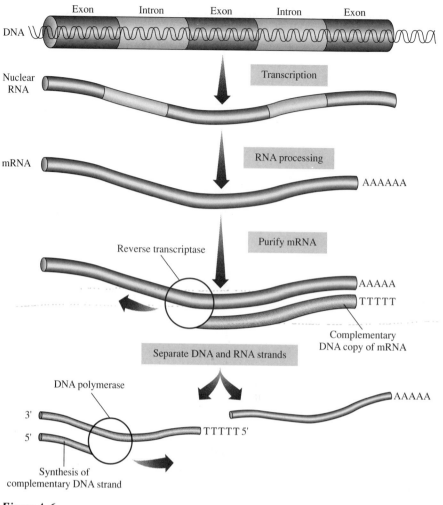

Figure A.6

Construction of cDNA from a messenger RNA molecule. Reverse transcriptase is used to make a complementary copy of the RNA, and DNA polymerase is used to make a complementary copy of the single-stranded cDNA molecule. (After Villee et al., 1989.)

into the plasmid by **enzymatic recombination** (another commonly used *E. coli* vector is the bacteriophage λ). Essentially, this technique involves breaking the circular plasmid DNA and inserting the desired DNA. The latter becomes a "passenger" carried by the plasmid vehicle and is replicated together with the plasmid. Insertion of the passenger DNA utilizes a class of enzymes called **restriction endonucleases**. These are enzymes isolated from microorganisms that cut DNA internally at sequence-specific sites. These "restriction sites" usually consist of four or six base pairs and have **palindromic symmetry**; i.e., the base sequence on one strand is the inverse of the other (see Table A–1).

The names of restriction enzymes are based upon a shorthand notation that identifies the microorganism from which they are derived. For example, *Eco* R1 is derived from *E. coli*. Some commonly used restriction enzymes and their restriction sites are shown in Table A–1 (Roberts, 1982). The enzymes listed here produce staggered cuts in the double-stranded DNA, which leave single-stranded ends on the fragment (Fig. A.7). These ends are "sticky" in the sense that they could anneal with one another. Thus, a

Table A–1 Three commonly used restriction enzymes and their restriction sites

Enzyme	Sequence[1]
Eco R1	5' G A A T T C 3' 3' C T T A A G 5'
*Hin*d III	5' A A G C T T 3' 3' T T C G A A 5'
Hpa II	5' C C G G 3' 3' G G C C 5'

[1] Arrows indicate cleavage sites of the enzymes.

single fragment produced in this way could become circular by the annealing of its two complementary sticky ends. However, if two different molecules are treated with the same restriction enzyme, the sticky ends of each can anneal, producing a single, circular recombinant DNA molecule.

The utilization of this technique to produce recombinant plasmids is illustrated in Figure A.8. Both the plasmid and the eukaryotic DNA are treated with the same restriction enzyme, producing single-stranded ends in each. The plasmid is now a linear molecule. The two kinds of DNA are combined, and the sticky ends anneal, causing formation of a circular, recombinant molecule. Another enzyme, **DNA ligase**, is used to seal gaps remaining in the two strands of the hybrid molecule. After formation of the hybrid plasmid, it is introduced into a host *E. coli* cell, where it can replicate. This process is called **transformation**: The DNA is added to a culture of bacteria, and it is taken up by a small proportion of the cells. The plasmids are constructed in a way that gives the transformed cells a selective advantage. Thus, the small minority of transformed cells can be selected from the nontransformed cells. The most common way to do this is to place

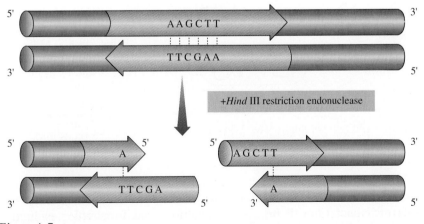

Figure A.7

Cleavage of DNA with *Hin*d III. Note that the restriction site has the same base sequence in reverse order on the two DNA strands. Cleavage of DNA with this enzyme leaves complementary "sticky ends." (After Villee et al., 1989.)

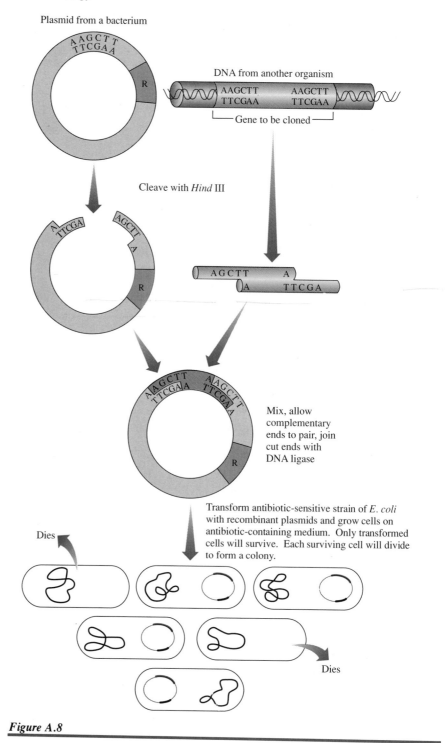

Plasmid from a bacterium

DNA from another organism

AAGCTT
TTCGAA

AAGCTT
TTCGAA

Gene to be cloned

Cleave with *Hind* III

AGCTT

A

A

TTCGA

Mix, allow
complementary
ends to pair, join
cut ends with
DNA ligase

Transform antibiotic-sensitive strain of *E. coli*
with recombinant plasmids and grow cells on
antibiotic-containing medium. Only transformed
cells will survive. Each surviving cell will divide
to form a colony.

Dies

Dies

Figure A.8

Scheme of events in cloning. (After Villee et al., 1989.)

a gene for resistance to an antibiotic in the plasmid. Thus, only the
transformed bacteria will be able to multiply and form colonies in the
presence of the antibiotic. Every cell in a colony contains the same passenger
DNA. Replication of this DNA by plasmids results in "cloning" of the
DNA. The amplified foreign DNA can be recovered from the plasmids by
treating them with a restriction enzyme.

Construction of Gene Libraries

How does an investigator clone a specific gene of interest? Genes to be cloned are usually isolated from a **gene library**, which is a full set of cloned DNA from either the total DNA of an organism (**genomic library**) or the set of cDNA molecules copied from all the mRNAs in a cell (**cDNA library**) (Fig. A.9). Each kind of cloning procedure has its advantages: Genomic libraries represent a random sample of all the genes in a particular organism; consequently, genomic libraries are usually the same regardless of the cell type from which the DNA was obtained. These clones contain not only the coding sequence of genes, but also the flanking regulatory sequences and the introns, which are missing from mRNA. A cDNA library, alternatively, represents the unique set of transcripts found in a particular cell type. Hence, cDNA libraries provide the means to clone genes that are expressed

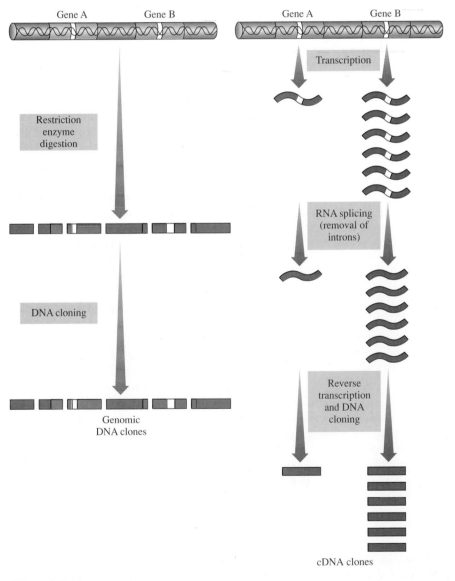

Figure A.9

Comparison of genomic cloning and cDNA cloning. Transcripts of gene B are enriched in the cDNA cloning procedure because gene B is transcribed frequently and gene A is transcribed infrequently.

at restricted times during development or in specialized cell types. These cloned sequences lack both the nontranscribed sequences of genes and introns (the sequences that are transcribed, but removed by processing). Thus, they contain the uninterrupted protein coding sequence as well as nontranslated regions that might be important in posttranscriptional regulation. Thus, if one wishes to deduce the amino acid sequence of a protein or produce the protein by expressing it in bacteria or yeast, a cDNA clone is preferable. cDNA clones are often used to obtain the corresponding genomic clone by probing a genomic library (see following discussion). With the genomic clone in hand, the investigator can analyze the putative regulatory sequences and their role in controlling expression of the gene during development.

One of the most important factors that has contributed to the rapid progress made in contemporary molecular biology has been sharing of libraries and specific cloned genes among investigators. This avoids the necessity to repeat time-consuming library construction or cloning procedures that have been done by other investigators.

Probing a Library

Selecting a single desired sequence from a mixed population (a genomic library might have a million different sequences) can be a daunting task. One way to do this is to use a radioactive DNA probe for *in situ* hybridization to the desired gene. The probe can be a synthetic DNA oligonucleotide designed to match the predicted nucleotide sequence of a portion of the gene or a sequence with presumed similarity to a portion of the desired gene. For example, a gene cloned from *Drosophila* might be used to find a related gene in a human genomic library.

A common strategy for preparing and probing a genomic library is outlined in Figure A.10. Genomic libraries are usually prepared in λ bacteriophage. Genomic fragments and λ DNA are prepared by treating the DNA with a restriction endonuclease and mixing the two populations of DNA. After treatment with DNA ligase, the recombinant DNA is coated with bacteriophage proteins prepared from infected *E. coli* cells. The resultant infectious λ bacteriophage are grown on a lawn of *E. coli*, where they produce plaques. Individual plaques contain genetically identical phage particles.

In order to probe the library, nitrocellulose filters are placed on the bacterial lawn to make a replica of the Petri plate. After this, the filter is treated with alkali to lyse the phage and denature the DNA. The filter is then incubated with the radioactive DNA probe, which will hybridize to complementary sequences, and the positions of plaques with the desired gene are determined by autoradiography. Once identified, the plaque on the plate is used as a source of the desired gene, which is removed from the bacteriophage genome by treatment with restriction endonuclease and inserted into a plasmid vector for cloning.

In some cases, it is necessary to identify a gene in a library without having a DNA probe. In that case, one can use an expression vector, in which the λ genome contains a strong promoter that will result in transcription of the library genes. A replica filter can then be probed with an antibody to the protein encoded by the desired gene. The presence of the antibody-antigen complexes (and, hence, the plaques containing the gene) can be detected by a color reaction on the filter.

Once suspected positive clones have been selected from a library, additional procedures may be necessary to confirm that the genes encode the desired protein. One way to do this is through a technique called **hybrid**

A

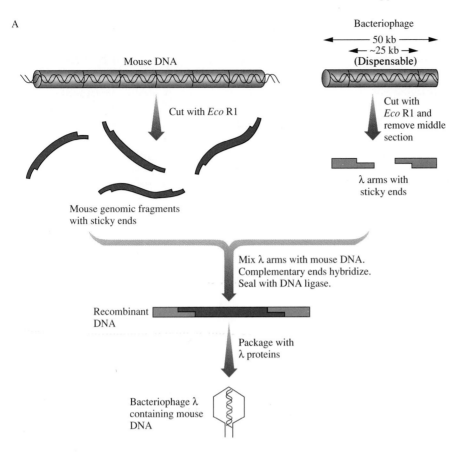

B

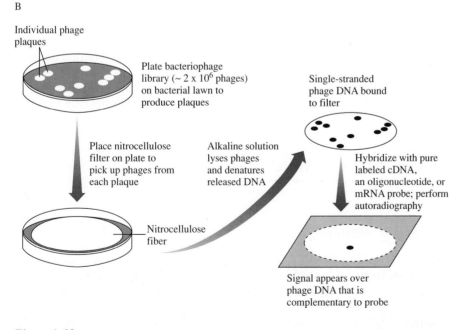

Figure A.10

Preparation (A) and probing (B) of a mouse genomic library. (After Villee et al., 1989, and Darnell et al., 1990.)

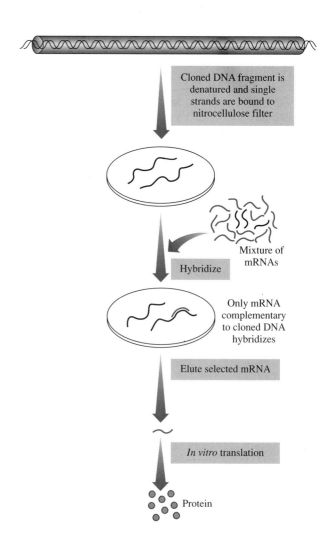

Figure A.11

The technique of hybrid selection. (After Alberts et al., 1989.)

selection (Fig. A.11). The cloned fragment is denatured and immobilized on a filter that is used to select mRNA by RNA-DNA hybridization. The filter is incubated with an RNA population that would contain the messenger for the protein. RNA that hybridizes with the DNA is translated *in vitro*, and the protein is analyzed to confirm that it is the correct protein (e.g., by use of a specific antibody). This confirms the identity of the gene.

VI. Analyses of Cloned Genes

Restriction Mapping

One of the first orders of business after cloning a gene is to construct a **restriction map** of the gene. The restriction sites in a gene are reflective of the nucleotide sequence of the gene and provide a signature that is unique to that gene. That "signature" can be revealed by cutting cloned DNA with a small number of restriction enzymes, followed by electrophoresis to separate the fragments according to their lengths. The sizes of these fragments allow the investigator to infer the positions of the restriction sites in the gene and the distances between them (Fig. A.12). The fragments can be recovered and subcloned for sequencing or other analytical purposes.

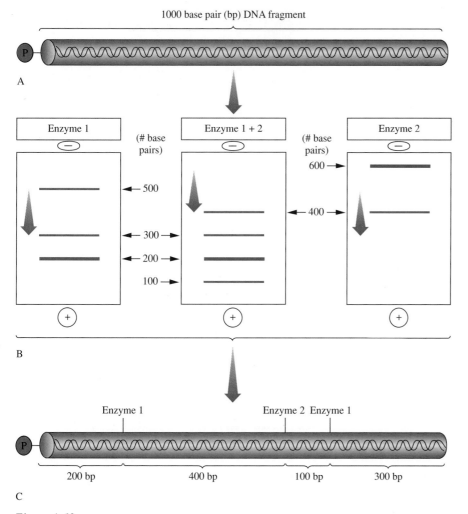

Figure A.12

Restriction mapping of a cloned DNA fragment. Two restriction enzymes are used in this example, both separately and together. A, The 1000 base-pair DNA fragment that is to be subjected to restriction nuclease digestion. The fragment is labeled with ^{32}P on one end to provide a reference point. B, After digestion, the sizes of the fragments are determined by electrophoresis. The location of the radioactive fragment is determined by autoradiography. (The radioactive fragments are shown in color.) C, The locations of the restriction sites. (After Villee et al., 1989.)

DNA Sequencing

The precision of polyacrylamide gel electrophoresis of DNA fragments, which allows for the resolution of fragments that differ in length by a single nucleotide, is exploited for DNA sequencing. The most widely-used sequencing procedure is the Sanger dideoxy method, which uses DNA polymerase to extend a single-stranded primer of end-labeled DNA (Fig. A.13). The single strand of DNA to be sequenced is hybridized to the primer, and four separate reaction mixtures for elongation of the primer are prepared. In addition to the four deoxynucleoside triphosphates (dATP, dCTP, dGTP, and dTTP), each reaction mixture contains a low concentration of a modified form of one of the nucleotides (a dideoxynucleoside triphosphate), which stops the reaction when it is incorporated into the elongating

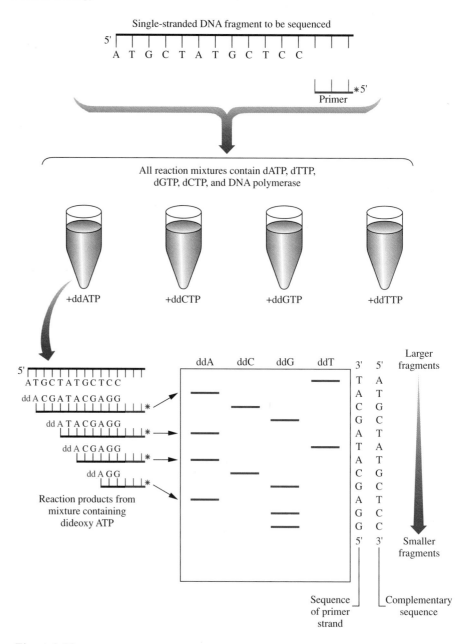

Figure A.13

The Sanger dideoxy procedure for sequencing DNA. (After Villee et al., 1989.)

chain. Thus, one reaction contains dideoxy ATP, one contains dideoxy CTP, and so on. Each reaction will produce chains that have terminated prematurely wherever a dideoxynucleotide has inserted. Reaction conditions are controlled so that a dideoxynucleotide will have been substituted at every occurrence of that nucleotide in the sequence. This produces a nested set of terminated chains, each with a common origin but ending at a different nucleotide. When the products of all four reactions are separated by electrophoresis in adjacent lanes, followed by autoradiography, four parallel ladders of bands that represent each occurrence of the individual nucleotides will be seen on the film. The nucleotide sequence of the primer (which is complementary to that of the strand being sequenced) can be read directly from the film.

VII. Southern Blotting

This technique, which is named for the individual who developed it (Southern, 1975), is one of the most powerful tools used in molecular biology research. It is a method for eluting electrophoretically separated fragments of DNA from polyacrylamide or agarose gels and identifying specific fragments by hybridization to a radioactive probe. The eluted DNA fragments are bound to nitrocellulose paper, which retains the exact pattern of separated fragments and is a suitable medium for the hybridization reactions to take place.

Before transferring the DNA to nitrocellulose, the DNA is denatured (i.e., treated with sodium hydroxide to make it single-stranded) while in the gel. The gel is then placed on filter paper that is soaking in buffer. A sheet of nitrocellulose is placed over the gel, and dry blotting paper is laid over the nitrocellulose. Capillary action draws the buffer through the gel, transferring an exact replica of the DNA pattern in the gel into the nitrocellulose.

A radioactive single-stranded probe (RNA or DNA) for a specific gene is then hybridized to complementary single-stranded DNA fragments on the nitrocellulose paper. Unhybridized probe is washed off, and autoradiography is conducted. Thus, a replica of the band or bands containing the specific DNA is formed on the autoradiograph. Southern blotting is used when one needs to detect specific DNA fragments from a large, heterogeneous mixture of fragments, as, for example, after restriction enzyme treatment of DNA. The mixture of fragments is resolved into a smear of overlapping bands of fragments of different size by electrophoresis. The band corresponding to a specific gene can then be detected from the smear of overlapping bands by transfer to nitrocellulose, followed by hybridization with a radioactive probe for that gene.

VIII. Northern Blotting (RNA Blot Analysis)

This technique is a modification of the Southern blotting technique, which monitors the level of RNA abundance by hybridization analysis between RNA that has been electrophoresed and blotted onto a filter and a labeled single-stranded probe. This is a particularly useful procedure for following the accumulation of individual messengers as development occurs. An example of Northern analysis is shown in Figure 3.42B on page 100. Another means for monitoring messenger accumulation during development is the nuclease protection assay. Figure 3.42C utilizes this technique, which is discussed in detail on page 101.

IX. Polymerase Chain Reaction (PCR)

This is a new procedure that is rapidly assuming a key role in the arsenal of molecular biologists. It allows for the selective amplification of DNA sequences from a complex mixture of DNA sequences. Thus, if a miniscule amount of a particular sequence is present, very large amounts of it can be obtained for analysis. Furthermore, if at least part of a sequence is known, this technique can be used to amplify the entire sequence by more than a millionfold. Finally, a modification of this technique can be used to detect messenger RNA that is present in minute amounts—an advancement that has obvious advantages for investigators studying the expression of low abundance mRNA during development.

As shown in Figure A.14, this technique involves denaturing a DNA fragment by heating it and then adding excess amounts of synthetic DNA oligonucleotides that are complementary to the 3′ ends of the sequence to be amplified (i.e., the oligonucleotides hybridize to the opposite strands of denatured DNA at the 3′ ends of the sequence to be amplified). The oligonucleotide fragments serve as primers for *in vitro* DNA synthesis, which is catalyzed by a DNA polymerase. When the reaction is complete, the reaction mixture is heated to denature the newly-formed DNA duplexes, and another round of synthesis is initiated by cooling the mixture to allow the primers (remember that they are present in excess) to bind to the strands and the polymerase to catalyze chain extension. Every time the process is repeated (each cycle requires only about 5 minutes), the number of copies of the sequence is doubled. Thus, by repeating the process 20 to 30 times, a large amount of DNA can be produced in a very short period of time.

An adaptation of this technique that is of particular interest to developmental biologists is **RT-PCR**, which involves isolating RNA from cells of interest, making cDNA from the RNA, and using sequence-specific

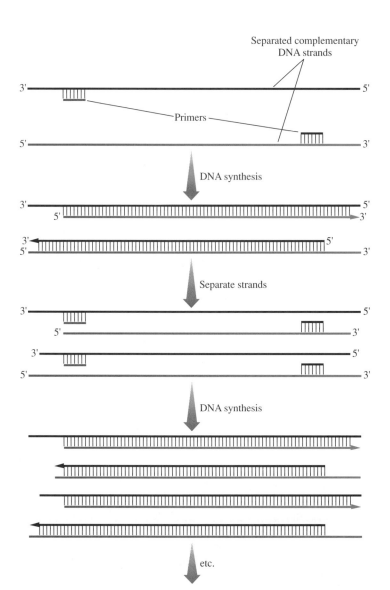

Figure A.14

The polymerase chain reaction. Two rounds of synthesis are shown here.

primers to amplify particular cDNA (Rappolee et al., 1990). The amplified fragments can then be identified by techniques such as sequencing, restriction analysis, and Southern blotting. The technique is so sensitive that mRNA from a single cell can be detected.

X. Transgenic Animals

Insertion of cloned genes into the genome of a host organism would allow that gene to be replicated at cell division and regulated by host cell regulatory mechanisms. Furthermore, if the foreign gene could be integrated into the genome of the host's germ line, the gene would be transferred to successive generations and a new strain of the organism would be produced. Organisms that have been altered in this way are called **transgenic**. Transgenic organisms are valuable for examining the expression of cloned genes during development, but the technique also has potential value for correcting genetic defects.

Germ-line transformation of *Drosophila* is discussed on page 606. Mice have also been used extensively for transgenesis (Fig. A.15). Insertion of

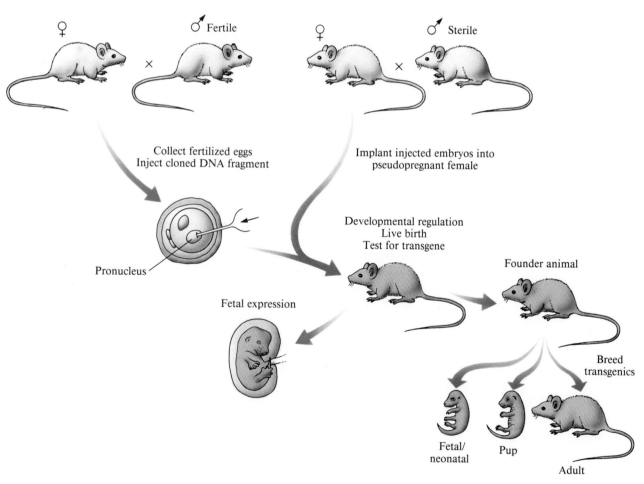

Figure A.15

Production of transgenic mice. Cloned genes are injected into one of the pronuclei of a fertilized egg, and the egg is transferred to a foster mother. If the gene is incorporated into the germ line, a strain of mice containing the transgene can be established. (After Villee et al., 1989.)

cloned genes is accomplished by injecting the DNA into one of the two pronuclei of a fertilized egg with a micropipette. The injected embryos are transplanted into pseudopregnant mice. The mice born to the foster mother are tested for expression of the transgene, and selected mice are bred to test for the presence of the transgene in their germ cells. As discussed on page 745, this can be a powerful technique for studying the functions of gene regulatory sequences during development.

References

Alberts, B. et al. 1989. *Molecular Biology of the Cell*, 2nd ed. Garland, New York.

Britten, R. J., and D. E. Kohne. 1968. Repeated sequences in DNA. Science, *161*: 529–540.

Darnell, J., H. Lodish, and D. Baltimore. 1990. *Molecular Cell Biology*, 2nd ed. Scientific American Books, New York.

Davidson, E. H. 1976. *Gene Activity in Early Development*, 2nd ed. Academic Press, New York.

Galau, G. A., R. J. Britten, and E. H. Davidson. 1974. A measurement of the sequence complexity of polysomal messenger RNA in sea urchin embryos. Cell, *2*: 9–20.

Goldberg, R. D. et al. 1973. Nonrepetitive DNA sequence representation in sea urchin embryo messenger RNA. Proc. Natl. Acad. Sci. U.S.A., *70*: 3516–3520.

McConaughy, B. L., and B. J. McCarthy. 1970. Related base sequences in the DNA of simple and complex organisms. VI. The extent of base sequence divergence among the DNAs of various rodents. Biochem. Genet., *4*: 425–446.

O'Farrell, P. H. 1975. High resolution two-dimensional electrophoresis of proteins. J. Biol. Chem., *250*: 4007–4021.

Rappolee, D. A. et al. 1990. The expression of growth factor ligands and receptors in preimplantation mouse embryos. *In* S. Heyner and L. M. Wylie (eds.), *Early Embryo Development and Paracrine Relationships*. Alan R. Liss, New York, pp. 11–25.

Roberts, R. J. 1982. Restriction and modification enzymes and their recognition sequences. Nucleic Acids Res., *10*: r117–r144.

Southern, E. M. 1975. Detection of specific sequences among DNA fragments separated by gel electrophoresis. J. Mol. Biol., *98*: 503–517.

Villee, C. A. et al. 1989. *Biology*, 2nd ed. Saunders College Publishing, Philadelphia.

Glossary

A23187 *See* ionophore.

abdominal A **and** *abdominal B* Homeotic selector genes of *Drosophila*.

abdominal histoblast nests Imaginal cells in the insect larva that will form adult abdominal structures.

aboral ectoderm Single layer of cells in the sea urchin larva, forming the body wall on the side opposite the mouth.

accessory cells Nongerm cells (follicle cells and nurse cells) in the ovary that are intimately associated with the oocytes.

acetylcholinesterase Enzyme that appears primarily in muscle; used as a histological marker for muscle cell differentiation in tunicate embryos.

acini *See* pancreatic acini.

acron Anterior unsegmented region of the *Drosophila* head.

acrosin Protease exposed on the sperm head after the acrosome reaction; may assist sperm digestion through the zona pellucida.

acrosomal granule Golgi-derived material that forms the matrix of the acrosome.

acrosomal process Filamentous thread of cytoplasm containing microfilaments; extended from the head of marine invertebrate sperm to the egg surface during fertilization.

acrosomal vesicle Precursor of the mammalian acrosome.

acrosome Membrane-enclosed organelle containing hydrolytic enzymes and other proteins; undergoes exocytosis during fertilization and assists the sperm in penetrating the extracellular coats of the egg.

acrosome reaction Response of sperm upon contact with eggs; causes the plasma membrane and the outer acrosomal membrane to vesiculate and be shed, thus releasing the enzymes of the acrosome.

α-actinin Actin-binding protein that helps to form flexible cross-links between actin microfilaments.

actinomycin D Inhibitor of transcription.

activation program Sequence of metabolic and morphological changes in the egg after fertilization, leading to cleavage and embryonic development.

activins Growth factors in the TGF-β family that are involved in mesoderm induction in *Xenopus* and in controlling the proliferation of erythroid cells in adult mammals.

adherens junctions Junctions between epithelial cells at their apical margins; characterized by insertion of actin filaments into the junction. Adhesion is mediated by the protein A-CAM.

adhesive hierarchy Scale of relative adhesiveness of cell types.

aequorin Luminescent protein isolated from jellyfish; when it binds Ca²⁺, light is produced.

afferent neurons Sensory fibers that carry sensations such as pain, touch, and proprioception from the periphery toward the central nervous system.

alecithal eggs Eggs that have little or no yolk.

allantois 1. Bulge from the mammalian yolk sac endoderm that pushes into the connecting stalk mesoderm; will give rise to the vasculature of the umbilical cord and connect the fetal blood vessels with those of the placenta. 2. Extension of the hindgut in birds and reptiles that serves as a repository for excretory wastes; fuses with the chorion to form the chorioallantoic membrane.

allelic exclusion Expression of only one of the alleles at a gene locus.

α-amanitin Inhibitor of RNA polymerases II and III.

Ambystoma mexicanum Species of urodele amphibian; also known as the axolotl.

amiloride Inhibitor of Na⁺/H⁺ flux.

amnion Fluid-filled sac surrounding the embryos of higher vertebrates that is responsible for keeping the embryo moist and protecting it from shock.

amnioserosa Extraembryonic tissue that protects the developing *Drosophila* embryo.

amniotes Higher vertebrates (birds, reptiles, and mammals) that possess an amnion in the embryonic stage.

Amphioxus Cephalochordate.

amplification Repeated replication of genes, such as the ribosomal RNA genes in some amphibian oocytes.

androgen-binding protein (ABP) Protein that binds androgens in the fluid of seminiferous tubules and presumably transports them to the germ cells to promote cell differentiation.

androgenesis Development promoted exclusively by paternal nuclei.

androgens Steroid hormones produced by the interstitial cells of the testis; principal promoters of germ cell differentiation in male vertebrates; responsible for male secondary sex characteristics.

aneuploidy Some number of chromosomes other than an exact multiple of the haploid number.

animal cap Pigmented animal hemisphere of the amphibian blastula.

animal pole Pole of the egg at which the maturation divisions take place and where the polar bodies are formed.

animalized embryo Larva of a sea urchin egg showing exaggerated development of animal hemisphere derivatives.

animalizing factors Hypothetical factors that are most concentrated at the animal pole of the sea urchin embryo, gradually decreasing toward the vegetal pole, which promote ectodermal development.

annulate lamellae Groups of parallel stacks of membranes in oocytes; contain pore complexes and bear a striking similarity to the nuclear envelope.

annulus Structure at the end of the sperm middle piece.

anode Positive electrode in an electrical field.

Antennapedia (Antp) Dominant homeotic mutation in *Drosophila*, causing legs to arise from the region of the head that would normally produce antennae.

antennapedia complex Cluster of homeotic selector genes, containing *Antennapedia*, *deformed*, and *sex combs reduced*, as well as the *bicoid* and *fushi tarazu* genes; controls head and anterior thorax development.

anteriobithorax Homeotic selector gene of *Drosophila* that causes the anterior compartment of the metathorax to develop like the anterior compartment of the mesothorax.

anterior nuclear fossa Specialized invagination of the anterior nuclear membrane of sea urchin sperm; contains globular actin, which polymerizes into filamentous actin during the formation of the acrosomal process.

anteroposterior axis Body axis extending from the anterior to the posterior pole of a bilaterally symmetric embryo (or animal).

antibody Specialized proteins (immunoglobulins) that bind selectively to foreign molecules or organisms.

antrum Large fluid-filled cavity in the mammalian Graafian follicle.

anucleolate mutant (0-nu) Mutant of *Xenopus laevis* that lacks nucleolar organizers and nucleoli, cannot manufacture ribosomes, and is lethal in the homozygous condition.

apical ectoderm maintenance factor Putative factor produced by mesoderm; thought to be responsible for the presence and position of the apical ectodermal ridge.

apical ectodermal ridge (AER) Ridge of thickened ectoderm that runs along the length of the anteroposterior margin of the limb buds of most higher vertebrates.

apical epidermal cap Multilayered mass of cells that forms a conical bulge at the tip of an amputated urodele limb.

apical tuft Tuft of cilia formed by animal pole blastomeres in *Cerebratulus* embryos.

approximation Process during fertilization in which the pronuclei migrate toward one another, become closely apposed, but do not fuse.

archencephalon Forebrain.

archenteron Cavity formed by the endoderm during gastrulation; will later become the gut lumen.

area opaca Opaque area of the avian embryo comprised of cells that contact the underlying yolk mass directly; forms a ring that surrounds the area pellucida.

area pellucida Translucent area of the avian embryo comprised of the cells that lie directly above the subgerminal cavity.

artificial parthenogenesis Activation of the egg by chemical or physical means carried out in the laboratory.

arylsulfatase Enzyme encoded by a gene in sea urchins that is expressed exclusively in the aboral ectoderm.

Ascaris Genus of nematode that undergoes chromatin diminution during cleavage.

asters Microtubules that radiate from the centrioles (or centrosomes) toward the periphery of the cell.

autonomic nervous system Motor fibers innervating the smooth muscles of the visceral organs and various glands; also known as the visceral nervous system.

autoradiography Experimental method for detecting radioactivity by which energy of emissions from a radioactive source are used to convert silver halide to silver grains in a photographic emulsion.

axolotl *See Ambystoma mexicanum.*

axoneme Motor apparatus in the sperm tail; consists of two central microtubules surrounded by an array of nine doublet microtubules.

axons Long processes of neurons that transmit nervous impulses either toward other neurons or to effector organs such as muscles.

5-azacytidine Nucleoside analogue that causes demethylation of cytosine residues in DNA.

Balbiani rings Large puffs of polytene chromosomes of *Chironomus tentans.*

Barr body Darkly staining heterochromatic X chromosome lying near the nuclear envelope in interphase nuclei of female mammals.

basal decidua Portion of the uterine decidua lying beneath the embryo; forms the maternal component of the placenta.

basal disc Foot of *Hydra.*

basal lamina Dense sheet of extracellular matrix underlying the base of epithelia.

basal layer of epidermis Deepest layer of the epidermis, which proliferates to produce keratinocytes.

bicoid (*bcd*) Maternal-effect gene that is involved in specification of the anterior region of the *Drosophila* embryo.

bilateral cleavage Cleavage type in which only one plane divides the embryo into symmetrical halves.

bindin Protein associated with the acrosomal process of sea urchin sperm after the acrosome reaction; promotes the species-specific attachment of the acrosomal process to the vitelline envelope.

bithorax Homeotic selector gene of *Drosophila* that causes the anterior compartment of the metathorax to develop like the anterior compartment of the mesothorax.

bithorax gene complex Cluster of homeotic selector genes, including *Ultrabithorax, abdominal A,* and *abdominal B*; specifies segments posterior to the prothorax.

Bithynia Genus of snail whose embryo has a small polar lobe.

blastocoel Fluid-filled cavity that forms in the embryo after the morula stage.

blastocyst Stage of early mammalian development in which a layer of cells surrounds an internal cavity.

blastoderm Epithelial layer that surrounds the blastocoel.

blastodisc Small yolk-free zone at the animal pole of telolecithal eggs. Cleavage is restricted to this region.

blastomeres Cells of a cleavage-stage embryo.

blastula Stage of early development that is characterized by the presence of an epithelial layer of cells surrounding a blastocoel.

blood islands Aggregate of cells in the early vertebrate embryo that produces primitive erythrocytes.

blood-testis barrier Barrier formed by tight junctions between adjacent Sertoli cells; seals the seminiferous tubule and prevent many substances of the blood plasma from entering the tubule.

body folds Folds that undercut a flat embryo to produce a cylindrical body.

bone morphogenetic protein (BMP) Protein related to TGF-β that causes mesenchyme to differentiate into bone cells.

bottle cells Epithelial cells that temporarily become bottle-shaped, owing to the contraction of their apical margins and the expansion of their basal margins; found at the site of initiation of gastrulation in amphibian embryos.

bride of sevenless (*boss*) *Drosophila* mutation that interrupts eye development.

bristle coat Coat of clathrin on the cytoplasmic side of an endocytotic vesicle.

cadherins Calcium-dependent intercellular adhesion molecules.

Caenorhabditis elegans Species of nematode that is valuable for the study of the roles of genes in development.

calcium wave Chain reaction of intracellular Ca^{2+} release and uptake that accompanies the cortical reaction.

CAMs Cell adhesion molecules.

cap Methylated inverted nucleotide at the 5' end of mRNA.

capacitation Changes to mammalian sperm that occur in the female reproductive tract, enabling sperm to fertilize the egg.

cappuccino *Drosophila* egg-shape gene.

capsular decidua Portion of the uterine decidua that covers the embryo and lines the uterine cavity.

α-cardiac actin Actin that is restricted to muscle in the early *Xenopus* embryo (and adult heart muscle) and serves as a marker of muscle-specific gene expression.

cardiac jelly Gelatinous hyaluronic acid-rich matrix in the embryonic heart between the endocardium and myocardium.

CAT *See* chloramphenicol acetyl transferase.

cathode Negative electrode in an electrical field.

cdc genes Genes that control the cell division cycle of fission yeast.

cDNA DNA made by copying mRNA with the enzyme reverse transcriptase.

cDNA library Set of cDNA molecules copied from all the mRNAs in a cell.

ced Genes in *Caenorhabditis elegans* involved in regulating programmed cell death.

cell cycle Period between the formation of a cell by the division of its parent cell and the formation of two new cells by cell division.

cell determination Specification of the developmental pathway of a cell.

cell differentiation Process by which cells acquire their final structural and functional characteristics.

cell-free translation system All the components required for protein synthesis except mRNA. When mRNA is added, protein synthesis is initiated.

cellular blastoderm stage Stage after simultaneous cytokinesis of the syncytial blastoderm of the insect embryo.

cellular oncogenes *See* proto-oncogenes.

central nervous system (CNS) Portion of the nervous system consisting of the brain and spinal cord; integrates nervous function of the whole organism.

centrolecithal eggs Eggs of arthropods in which the yolk is centrally located and is surrounded by a thin coat of cytoplasm. The nucleus is in a central island of cytoplasm.

Cerebratulus Genus of nemertine worm used to study the role of ooplasmic determinants in development.

chemoaffinity hypothesis Proposal that the precision in patterning of retinotectal connections depends upon molecular recognition between retinal axons and specific regions of the tectum.

chemotaxis Directed movement of cells toward the source of a diffusible chemical.

chiasmata Cytological indication of exchanges between chromatids during crossing-over.

chimera Embryo containing cells from two or more animals.

Chironomus tentans Midge (a dipteran insect) that has polytene chromosomes that form particularly large puffs, called Balbiani rings.

chloramphenicol acetyl transferase (CAT) Bacterial enzyme not found in eukaryotes. The CAT gene is used as a reporter gene when its transcription is driven by a heterologous promoter.

chondrocytes Cellular component of cartilage.

chordamesoderm Presumptive notochord and somite mesoderm; functions as the ''organizer'' of the vertebrate embryonic axis.

chordoplasm Region of cytoplasm in tunicate eggs that becomes the notochord.

chorioallantoic membrane Fusion product of the chorion and allantois in bird and reptile embryos; has a respiratory role.

chorion 1. Outermost extraembryonic membrane of avian embryos, which abuts the egg shell; functions as the major vehicle of gas exchange between the embryo and the environment. 2. In mammals, the chorion becomes a portion of the placenta, in which it acquires the functions of supplying nutrition and waste removal in addition to respiration. 3. Tough capsule surrounding eggs of insects and fish.

chorion frondosum Disk-shaped portion of the mammalian chorion surrounding the umbilical cord; forms the fetal component of the placenta.

chorion laeve Portion of the mammalian chorion in which chorionic villi are lost.

choroid Vascularized layer of neural crest origin that nourishes and protects the eyeball.

chromatin DNA-protein complex of which chromosomes are comprised.

chromatin diminution Loss of a portion of the genome during development.

chromocenter Aggregation of the centromeres of polytene chromosomes.

chromomeres Local compacted domains along chromatids.

chromosomal vesicles Vesicles composed of nuclear envelope material; form around each chromosome during the reformation of the nuclear envelope at the end of karyokinesis.

ciliary body Muscle that controls the shape of the lens; derived from the neural crest.

ciliary ganglion Parasympathetic ganglion found in the posterior portion of the eye near the iris and ciliary muscle.

cis-**acting elements** Regulatory elements usually located in the flanking regions of the genes they control.

clathrin Protein that coats the cytoplasmic surface of an endocytotic vesicle.

cleavage Mitotic division of the zygote; occurs without intervening cell growth.

cleavage-arrested embryos Embryos treated with cytochalasin B to arrest cytokinesis. Nuclei continue to divide within the cytoplasm of the cleavage-arrested cells.

cleavage furrow Constriction of the egg surface that splits the egg or blastomere into two parts during telophase.

cleavage-stage (CS) histones Histones assembled into chromatin during the very early cleavage stages in sea urchins; synthesized by oogenic mRNAs and, for a short time after fertilization, by mRNAs transcribed from the zygotic genome.

Clepsine Genus of leech that was used for the first cell lineage study.

cloaca Posterior end of the endodermal tube into which the ducts of the excretory and reproductive systems empty.

cloacal plate Point of contact between the proctodeum and the posterior end of the endodermal tube.

clomiphene Drug used to regulate follicular growth and ovulation.

clone 1. Population of cells all derived from the same precursor cell. 2. DNA molecules produced by replication of a single DNA molecule. 3. Genetically identical individuals derived from a single individual.

cloning 1. Production of genetically identical copies of individuals, as through serial nuclear transplantation. 2. Production of identical copies of DNA sequences by recombinant DNA technology.

close contact Class of adhesion between cell and substratum in which the cell's plasma membrane comes within 30 to 50 nm of the substratum and appears gray in an interference reflection image.

c-MOS Protein product of the c-*mos* proto-oncogene; kinase enzyme responsible for both activation of maturation and metaphase arrest in oocytes.

coated vesicles Cytoplasmic vesicles that are coated with the protein clathrin.

codon Triplet of bases in mRNA that either specifies the placement of an amino acid in protein or that punctuates protein synthesis.

coelom Main body cavity lined by mesoderm. All the viscera will be suspended in this cavity.

coelomic sac Mesodermal precursor to the coelom that buds off from the archenteron of sea urchin embryos after gastrulation.

colcemid Drug that depolymerizes microtubules.

colchicine Drug that depolymerizes microtubules.

collagen Glycoprotein that is the most abundant extracellular matrix molecule. Collagen molecules, which are rich in glycine and resistant to proteases, are composed of three polypeptide chains that wrap around each other to form a triple helix that has a striated appearance in the microscope.

collagenase Enzyme that digests collagen.

combinatorial effect Aggregate effect of multiple regulatory factors with gene regulatory elements in regulating transcription.

comb plates Locomotory organs derived from the E cell lineage of ctenophore cydippid larvae.

commissures Transverse tracts of neurons in the insect.

compaction Process in which mammalian blastomeres become tightly packed into a compact morula.

complexity *See* sequence complexity.

conjunctiva Epithelium that is continuous with the anterior corneal epithelium.

connecting piece Major structural element in the sperm neck that articulates with a concave depression in the base of the sperm head.

connecting stalk Attachment between the posterior end of the mammalian embryo and the cytotrophoblast; will contribute ultimately to connective tissue of the umbilical cord.

constant regions Regions at the carboxy-terminus of immunoglobulin molecules that do not vary among antibody molecules within a class of Ig molecules; responsible for the specialized functions of each Ig class.

contact guidance Alignment and migration of cells along discontinuities in the substratum.

contact inhibition When one cell contacts another in culture, the ruffling lamellipodium of the contacting cell ceases protrusive activity, forms a new process elsewhere, and migrates away from the cell it contacted.

contractile ring Thickened region of the egg cortex within the cleavage furrow; pinches the egg or blastomere into two parts.

convergent extension Convergence of an epithelial sheet toward a central site, followed by its extension along a single axis through forceful intercalation of the cells of the epithelium.

coordinate genes Maternal-effect genes of *Drosophila* that establish the anterior and posterior regions.

cornea Clear covering over the anterior surface of the eye; protects the eye and focuses light on the lens.

corona-penetrating enzyme Mammalian acrosomal enzyme that helps sperm penetrate between corona radiata cells.

corona radiata Single layer of elongated cells with fine processes that radiate toward the ovulated mammalian egg.

corpus allatum Gland in insects that produces juvenile hormone.

cortex 1. Gel-like cytoplasmic layer just below the egg plasma membrane. 2. Thickened coelomic epithelium of developing gonads.

cortical granules Membrane-enclosed vesicles in the egg cortex that contain acid mucopolysaccharides and protein. Contents of cortical granules are extruded into the perivitelline space after fertilization.

cortical reaction Wave of exocytosis that occurs as the cortical granules fuse with the egg plasma membrane and release their contents after sperm–egg fusion.

cortical rotation Rotation of the cortex with respect to the internal cytoplasm in amphibian eggs.

C_0t Value that describes a DNA renaturation reaction; product of the concentration of the DNA $\times$ the duration of the reaction.

$C_0t^{1/2}$ The C_{ot} value needed to allow 50% DNA renaturation.

cranial nerves Nerves emerging from the brain stem and consisting of preganglionic fibers of the parasympathetic system, together with sensory and motor fibers.

crossing-over Physical exchange of material between chromatids during meiotic prophase; results in recombination of genes in a linkage group.

crural plexus Aggregation of motor neurons from thoracic level 7 and lumbosacral levels 1 through 3.

δ-crystallin Predominant protein synthesized in the developing chick lens.

cumulus oophorus Follicle cells that do not detach from the zona pellucida during ovulation of the mammalian egg but remain associated with the egg.

cupule Specialized extracellular structure surrounding the animal pole of the jellyfish egg. A cupule protein acts as a sperm attractant.

CyI Cytoskeletal actin gene of sea urchins that is initially expressed throughout the embryo but continues to be expressed in oral ectoderm and gut while becoming silenced in aboral ectoderm.

CyIIa Cytoskeletal actin gene of sea urchins that is expressed relatively late in development, exclusively in the vegetal plate derivatives and in skeletogenic mesenchyme.

CyIIb Cytoskeletal actin gene of sea urchins that is initially expressed throughout the embryo but continues to be expressed in oral ectoderm and gut while becoming silenced in aboral ectoderm.

CyIIIa Cytoskeletal actin gene of sea urchins that is expressed exclusively in the aboral ectoderm.

cyclic AMP Intracellular second messenger.

cyclin Protein that accumulates during interphase and is destroyed by proteolysis after mitosis during each cell cycle; one of the components of maturation promotion factor.

cycloheximide Inhibitor of protein synthesis.

cydippid larva Larva of ctenophores.

Cynops pyrrhogaster Species of amphibian lacking the fast block to polyspermy.

cystoblast Cell that establishes the oocyte-nurse cell clone of *Drosophila*.

cytochalasin B Drug that blocks actin polymerization and thus microfilament formation.

cytogenetics   Discipline that examines the effects of chromosomal structure on expression of the genome.

cytokinesis   Division of the cytoplasm during mitosis.

cytoplasmic localization   Localized differences in cytoplasmic regions of the egg.

cytoskeleton   Array of cytoplasmic fibers composed of microtubules, microfilaments, and intermediate filaments; organizes the cytoplasm and generates motility.

cytostatic factor (CSF)   Factor responsible for metaphase arrest in the unfertilized egg of *Xenopus*; has been shown to be identical to the c-MOS protein.

cytotrophoblast   Inner cellular layer of the trophectoderm.

D600   Drug that inhibits Ca²⁺ transport.

decidua   Uterine endometrium during pregnancy.

dedifferentiation   Loss of differentiated characteristics.

deep cells   Mass of individual cells in the fish blastula, sandwiched between an overlying enveloping layer and the underlying yolk syncytial layer. During gastrulation, the deep cells migrate on the yolk syncytial layer and converge to form the embryonic shield, which will become the embryo proper.

deep zone   Ring of presumptive mesodermal cells in the *Xenopus* embryo originating below the surface on the vegetal edge of the involuting marginal zone.

deformed   Homeotic selector gene of *Drosophila*.

denaturation   Separation of the two strands of the DNA double helix.

dendrites   Processes of neurons that receive nervous impulses.

denticles   Projections on the ventral cuticle of *Drosophila* larvae that are organized in segmentally repeated rows.

dermatome   Epithelial portion of the somite that will contribute to the dermis of the skin.

dermis   Layer of skin beneath the epidermis; derived from the mesoderm.

desmin filaments   Class of intermediate filaments that appear in skeletal, cardiac, and smooth muscle.

desmosomes   Punctate "weld" spots along the lateral sides of epithelial cells; hold cells together.

determinative development   *See* mosaic development.

deuterocephalon   Hindbrain.

developmental field   Group of cells that is destined to become a specific structure.

dextral cleavage   Right-handed (clockwise) displacement of micromeres during spiral cleavage.

diacylglycerol (DAG)   One of the two breakdown products of phosphatidyl inositol-4, 5-bisphosphate; activates protein kinase C.

diakinesis   Final phase of prophase I of meiosis. During this stage, the chromosomes shorten, the nuclear envelope breaks down, and the chromosomes begin moving toward the metaphase plate.

Dictyostelium   Slime mold; often used in the study of cell motility and chemotaxis.

dideoxynucleoside triphosphate   Modified nucleoside used in DNA sequencing reaction.

diencephalon   Posterior subdivision of the prosencephalon.

differential cellular adhesiveness   Hypothesis that the relative positions of cells in mixed aggregates result from random motility of the cells and quantitative differences in adhesiveness between them.

differential gene expression   Expression of different genes in different cells and/or at different times during development.

differentiation induction factor (DIF)   Phenyl hexone that induces isolated *Dictyostelium* amoebae in culture to differentiate into stalk cells.

dilator muscle   Iris muscle that opens the pupil.

discoidal cleavage   Cleavage of eggs in which the embryo proper is formed from the blastodisc, with the remainder of the zygote forming the yolk sac.

disphermy   Fertilization of eggs by two sperm.

distalization rule   New cells in a regenerate are forced to adopt a more distal positional value.

DNA ligase   Enzyme used to seal gaps in DNA.

DNA methylation   Postreplication methylation of cytosine to produce 5-methylcytosine (m⁵C); interferes with transcription.

DNase footprinting analysis   Technique for locating sites on genes where gene regulatory proteins are bound.

DNase I hypersensitive mapping   Means of identifying *cis*-acting sequences that are likely to be involved in regulation of transcription. Such regions are especially sensitive to digestion by the enzyme DNase I.

dorsal   Maternal-effect gene involved in dorsoventral axis formation in the *Drosophila* embryo.

dorsal closure   Expansion of the mesoderm and ectoderm of the *Drosophila* embryo around the midgut and fusion at the dorsal midline.

dorsal hairs   Fine extensions of the cuticle that characterize the dorsal side of the *Drosophila* larva.

dorsal lip of the blastopore   Site of initiation of gastrulation in the amphibian embryo. The dorsal lip, which forms at the site of the gray crescent, forms the dorsal margin of the blastopore.

dorsalized embryos   Embryos with duplicated or exaggerated dorsal structures.

dorsalized larvae   Larvae with exaggerated dorsal characteristics.

dorsoventral axis   Body axis extending from the dorsal to the ventral pole of a bilaterally symmetric embryo (or animal).

dosage compensation   Random heterochromatization of either the paternal or maternal X chromosome during development of female mammals.

double abdomens   Larvae of the insect *Smittia* that were irradiated with ultraviolet light in the early embryonic stage to produce a mirror-image duplication of the posterior segments.

double gradient model   Model used to explain cell determination in sea urchin embryos, which proposes opposite gradients of animalizing and vegetalizing factors initiated at the animal pole and vegetal pole, respectively, of the cleaving embryo.

double method   Procedure for inducing artificial parthenogenesis in sea urchins; involves an initial treatment to activate the Ca²⁺ flux followed by a second treatment to activate cleavage.

Drosophila melanogaster   Fruit fly that was used to unravel the basic concepts of transmission genetics and has more recently been used to identify genes that regulate development.

dynamic instability Rapid shortening and lengthening of microtubules.

dynein 1 Protein comprising the arms associated with microtubule protofilaments that possesses ATPase activity; responsible for converting chemical energy into mechanical movement.

early histones Histones that are encoded by both oogenic and zygotic mRNAs in sea urchins. Early histone mRNA is stored in the female pronucleus.

early responses Events of the activation program occurring within the first few minutes after fertilization.

easter Maternal-effect gene involved in dorsoventral axis formation in the *Drosophila* embryo.

ecdysone Steroid hormone produced by the prothoracic gland of insects that (1) stimulates the fat body to produce vitellogenin and (2) promotes growth and molting in insects.

ectoderm Germ layer that gives rise to the epidermis and nervous tissues.

ectoplasm Region of cytoplasm in tunicate eggs that becomes epidermis and neural tissues.

efferent ductules Ducts in the male that connect the rete testis with the vas deferens; remants of the mesonephric tubules.

efferent neurons Motor fibers that carry messages from the central nervous system toward the periphery to communicate with a muscle or gland.

egg chamber Cluster of oocyte and associated nurse cells in polytrophic ovaries of insects.

egg-shape genes Genes in *Drosophila* that initiate dorsoventral polarization exclusively during oogenesis. Mutant alleles at these loci cause changes in the shape of the egg and its shell.

EGTA Calcium chelator.

ejaculation Explosive release of semen from the penis.

electron microscope Instrument with high resolving power that enables biological material to be observed at very high magnifications.

electrophoresis Technique used to separate complex mixtures of macromolecules by differential migration through an electrical field.

embryology Study of changes in the form or shape of animals during their embryonic phase.

embryonic induction Determination of cell fate as a result of interactions with neighboring cells.

embryonic polarization Process in which bilateral symmetry is superimposed on the egg's radial symmetry.

embryonic shield Blastodisc of the mammalian embryo at the primitive streak stage.

endocardial cushion cells Cells that detach from the endocardium and become the valves and septa of the heart.

endocardium Inner lining of the heart.

endocytosis Uptake of an exogenous substance via a vesicle that pinches off from the plasma membrane.

endoderm Germ layer that gives rise to the respiratory organs, gut, and the gut accessory glands.

endodermal plate Flattened plate of presumptive endoderm cells in the vegetal hemisphere of the *Amphioxus* embryo that invaginates to form the archenteron.

endoplasm 1. Region of cytoplasm in tunicate eggs that becomes gut. 2. Yolky internal cytoplasm of eggs.

endoreplication Replication of DNA without further cell division, which leads to polyploidy.

endosomes Uncoated cytoplasmic vesicles.

energid Small cytoplasmic islands containing nuclei within the central yolk mass of insect eggs and embryos.

engrailed Segment polarity mutant of *Drosophila* in which the posterior region of each segment is replaced by a mirror image of the anterior region of the adjacent segment.

ependymal layer Layer of the central nervous system closest to the lumen; derived from the germinal epithelium.

epiblast Layer of cells above the blastocoel in the discoidal embryo.

epiboly Process in which epithelial cells flatten perpendicular to their apicobasal axes, accompanied by the lateral expansion of the sheet; causes the superficial layers of the blastula to spread and completely surround the inner portions of the embryo.

epibranchial placodes Ectodermal thickenings that will give rise to some of the neurons in the cranial sensory ganglia.

epidermal growth factor (EGF) Growth factor whose binding to its receptor activates the receptor's tyrosine kinase activity, which is a signal for mitosis.

epididymis Duct in which sperm are stored before their forcible release at ejaculation.

epigenesis Hypothesis that the adult organism gradually develops from a rather formless egg.

epimere Dorsal portion of the myotome, which will form the epaxial muscles.

epimorphosis Mode of regeneration involving proliferation of new tissue from the wound surface and progressive replacement of the missing parts.

epitheliomesenchymal interactions Inductive interactions that occur between the epithelium and mesenchyme in organ rudiments.

equatorial cleavage Cleavage that is perpendicular to the animal–vegetal axis in the plane of the equator of the egg.

equatorial segment Portion of the mammalian acrosome that remains intact during the acrosome reaction and is the site of initial contact between the sperm and egg at fertilization.

erythroblasts Mitotically active precursors of erythrocytes.

erythrocytes Mature, non-proliferative red blood cells.

estrogen Steroid hormone produced by follicle cells in the ovary; regulates reproduction and secondary sexual characteristics in the female.

estrogen response elements (EREs) *cis*-acting regulatory elements in the 5′ flanking regions of estrogen-inducible genes.

Eucidaris tribuloides Species of primitive sea urchin having a slower rate of cleavage and different origin for mesenchyme cells than typical sea urchins.

Eudiplodia Chick mutant in which supernumerary dorsal limb elements develop.

Euromatus Genus of polychaete annelid with moderately telolecithal eggs.

evagination Bending of an epithelium away from a lumen or the surface of the embryo.

even-skipped (*eve*) Pair-rule mutant of *Drosophila* that lacks the even-numbered parasegments.

exocrine cells Cells derived from the pancreatic acini that produce the various pancreatic enzymes.

exocytosis Extrusion of the contents of vesicles such as the acrosome or cortical granules.

exogastrulation Abnormal condition in which the archenteron forms inside-out.

exon Gene sequence that is represented in mRNA after processing of a nuclear transcript.

extracellular matrix (ECM) Meshwork secreted into the spaces between cells and composed of collagen, proteoglycans, and other glycoproteins.

extraembryonic coelom Cavity between layers of extraembryonic mesoderm.

extraembryonic membranes *See* allantois, amnion, chorion, and yolk sac.

extrinsic ocular muscles Muscles that move the eye in its orbit; derived from the head paraxial mesoderm.

exuperantia (*exu*) Maternal-effect gene that affects head and thorax development in *Drosophila*.

F-actin Filamentous (polymerized) actin.

Fab′ fragments Univalent fragments of antibodies.

facultative heterochromatin Condensation of one chromosome of a chromosome pair during development (such as for female mammalian X chromosomes).

fascicles Bundles of longitudinal nerve axons in the insect.

fasciclins Cell surface glycoproteins identified by antibodies that disrupt the patterning of the nervous system.

fasciculation Process by which nerve axons come together to form bundles.

fast block to polyspermy Block that is mediated by the fertilization potential.

fate map Map that shows what each part of an egg or early embryo will become at a later stage in development.

feedback inhibition Situation in which the product of an activity inhibits the level of that activity; e.g., a hormone produced by the action of another hormone inhibits the release of the latter hormone.

female pronucleus Nucleus of egg after completion of maturation and fertilization but before syngamy.

fertilization Union of sperm and egg to form a zygote.

fertilization acid Efflux of H^+ from the fertilized sea urchin egg.

fertilization cone Extension of sea urchin egg cytoplasm that forms around the entering sperm head.

fertilization envelope Elevated vitelline envelope that has undergone physical changes as a result of fertilization.

fertilization potential Electrical depolarization of the egg plasma membrane; mediates the fast block to polyspermy.

fertilization wave Change in optical properties at the surface of the sea urchin egg that accompanies the cortical reaction.

fibroblasts Cells derived from connective tissue.

fibroblast growth factor (FGF) Family of growth factors that are involved in mesoderm induction in *Xenopus*.

fibronectin Adhesive glycoprotein that mediates extracellular matrix assembly and cell attachment to the extracellular matrix.

fibrous sheath Two longitudinal columns that are connected by a series of hemispherical ribs in the sperm principal piece.

filamin Actin-binding protein that helps to form flexible cross-links between actin microfilaments.

filopodia Long, fine locomotory extensions of cells.

fimbrin Actin-binding protein that bundles microfilaments into parallel arrays.

focal contact Class of adhesion between cell and substratum in which the cell's plasma membrane comes within 10 to 20 nm from the substratum and appears black in an interference reflection image.

folded gastrulation Gene of *Drosophila* that is involved in regulating gastrulation movements.

follicle Functional unit in the ovary that consists of the oocyte and its surrounding follicle cells.

follicle cells Somatic cells surrounding the germ cells in the female gonad.

follicle-stimulating hormone (FSH) Pituitary gonadotropic hormone in vertebrates that (1) in the male regulates spermiogenesis by acting directly on the Sertoli cells and (2) in the female promotes growth and development of the oocyte.

fos Oncogene that encodes a nuclear protein that is a transcriptional regulator.

freemartin Genetically female calf who is masculinized if it shares the same circulatory system *in utero* with a male co-twin.

fucoserraten Hydrocarbon molecule secreted by *Fucus* eggs that attracts sperm.

Fucus serratus Species of brown alga that is often used in the study of fertilization and polarity.

Fundulus Genus of teleost fish with transparent embryos that are frequently used to study morphogenesis.

fushi tarazu (*ftz*) Pair-rule mutant of *Drosophila* that lacks the odd-numbered parasegments.

fusion gene Composite gene constructed of elements of two separate genes, usually combining regulatory sequences of one gene with coding sequences of another.

G-actin Globular (unpolymerized) actin.

G_1 **phase** Phase in the cell cycle between the completion of cell division and the initiation of DNA synthesis.

G_2 **phase** Phase in the cell cycle between the completion of DNA synthesis and the next cell division.

G-protein *See* GTP-binding protein.

β-galactosidase Bacterial enzyme encoded by the *lac-Z* gene, which is a widely used reporter gene.

ganglia Aggregates of the cell bodies of neurons.

ganglion cell layer Layer of the retina containing nuclei of the ganglion cells.

ganglion mother cells Undifferentiated neural cells that give rise repeatedly to neurons during nervous system development.

gap genes Class of segmentation genes of *Drosophila* that are expressed beginning at the syncytial blastoderm stage. Mutations in gap genes eliminate a large block of adjacent segments in the anterior, middle, or posterior regions of the embryo.

gap junctions Junctions that mediate communication between two adjacent cells by allowing inorganic ions and small molecules to pass between the cells.

gastrulation Process by which cells of the blastoderm are translocated to new positions in the embryo, producing the three primary germ layers.

gel retardation analysis *See* mobility-shift analysis.

gelsolin Ca^{2+}-dependent actin-fragmenting protein.

gene batteries Diverse, unlinked genes that are co-expressed.

genetic dissection Process by which the gene-controlled steps in a developmental process can be reconstructed by careful analysis of a series of mutations.

genital folds Form the ventral shaft of the penis in the male and the labia minora in the female.

genital swellings Form the scrotum in the male and the labia majora in the female.

genital tubercle Forms the glans penis and shaft of the penis in the male and the glans clitoris in the female.

genome DNA complement in the nucleus, mitochondrion, or chloroplast of a cell.

genomic library Total cloned nuclear DNA of an organism.

germ band Concentration of cells on the ventral side of the yolk mass that will form most of the *Drosophila* embryo.

germ band extension Expansion of the germ band of the *Drosophila* embryo, first posteriorly, then around the posterior pole and spreading in an anterior direction along the dorsal surface of the embryo.

germ band shortening Contraction of the extended germ band of the *Drosophila* embryo around the posterior pole to return to the ventral surface of the embryo.

germ cells Cells that will form gametes.

germ layers The three primitive cell layers (ectoderm, endoderm, and mesoderm); precursors to all the organ rudiments.

germ line proliferation defective (*glp-1*) *Caenorhabditis* mutation that causes a deficiency in germ cell production.

germ-line transformation Method for inserting exogenous genes into the *Drosophila* germ line genome utilizing the P-element.

germ plasm Region of the egg containing the determinants of the germ cell line.

germinal crescent Region in the chick embryo hypoblast at the border of the area pellucida and area opaca that contains primordial germ cells.

germinal epithelium Mitotically active epithelial layer of a proliferating tissue, such as the ovary or the neural tube.

germinal granules Specially staining granules localized in the vegetal cytoplasm of frog eggs; show similarities to the polar granules of insect eggs and may be involved in germ cell determination.

germinal vesicle Name given to the enlarged nucleus of the oocyte.

germinative region Zone of cell division in the developing lens epithelium.

GF-1 DNA-binding protein that is restricted to erythroid cells and that recognizes sites in promoters and enhancers of genes with erythroid-specific expression.

glial cells *See* neuroglia.

glial filaments Class of intermediate filaments that are found in glial cells.

glial growth factor Growth factor produced by neural cells; can stimulate mitotic activity in a denervated limb blastema.

globin gene switching Changes in globin gene expression during development.

glomerulus Tuft of capillaries that filters the blood in the mesonephric and metanephric kidneys.

glomus Vascular ridge near the pronephric tubule that filters the blood.

glue proteins Insect salivary gland secretory products that attach the pupal case to the substrate.

glycocalyx Extracellular layer composed of the carbohydrate side chains of glycolipids and integral membrane glycoproteins as well as extracellular proteoglycans and glycoproteins.

glycosaminoglycans (GAGs) Class of negatively charged sugar chains that are constituents of the extracellular matrix; made up of dimeric subunits, consisting of two alternating classes of monosaccharides: the amino sugars and uronic acids.

gonadal ridges Thickenings of the dorsal lining of the coelomic cavity; precursors of the gonads.

gonadotropic hormones Peptide hormones released from the pituitary gland that regulate gonadal function in vertebrates.

gonadotropin releasing hormone (GnRH) Brain neurohormone that regulates the gonadotropin secretory functions of the pituitary.

gooseberry Segment polarity mutant of *Drosophila* in which the posterior region of each segment is replaced by a mirror image of the anterior region of the adjacent segment.

Graafian follicle Large follicles in mammalian ovaries; characterized by the presence of a large fluid-filled antrum.

grandchildless Maternal-effect mutations in *Drosophila* in which homozygous females produce embryos with deficiencies in pole plasm, pole cell formation, and germ cell development. Thus, a first generation of progeny is produced, but it is sterile, and there is no second generation.

granulofibrillar material (GFM) String-like electron-dense material found associated with mitochondria in the amphibian oocyte.

granulosa cells Follicle cells in the mammalian ovary.

gray crescent Region of intermediate pigmentation in the marginal zone of the amphibian egg caused by a shift in the pigmented egg cortex toward the site of sperm entry; marks the future site of the dorsal lip of the blastopore.

gray matter   *See* mantle layer.

growth cone   Migratory tip of an axon.

growth control cascade   Intracellular transmission of cell division signals caused by binding of growth factors to their plasma membrane receptors.

growth factors   Proteins secreted into the extracellular spaces of an embryo that cause changes in cell growth and differentiation; thought to be inducers in amphibian embryos.

GTP-binding protein (G protein)   Protein located on the cytoplasmic side of the egg plasma membrane; stimulates the activity of phospholipase C.

GTP cap   Unhydrolyzed stable cap of GTP at one end of a microtubule that will serve as a primer for the addition of tubulin dimers.

gurken   *Drosophila* egg-shape gene.

gynogenesis   Development promoted exclusively by maternal nuclei.

hairy   Pair-rule mutant of *Drosophila* that is defective in parasegment-sized regions that are out of register with the parasegments.

halteres   Balancing organs located on the metathorax of *Drosophila*; maintain the fly in an upright position while flying.

haptotaxis   Directed movement of cells along an adhesive gradient.

Hbox1   Gene of sea urchins containing a homeobox sequence related to the *Antennapedia* homeobox of *Drosophila*; initially expressed throughout the aboral ectoderm, but later restricted to a small area of ectoderm that lies at the extreme aboral end of the pluteus.

heat shock proteins   Specialized set of proteins synthesized during a sudden exposure to high temperature.

heat shock response   Induction of heat shock protein synthesis by a sudden increase in temperature.

heavy chain class switching   Change in production of immunoglobulin heavy chain class during development.

heavy meromyosin (HMM)   Proteolytic fragment of myosin that contains the actin-binding portion of the molecule; used to determine the polarity of actin microfilaments.

HeLa cells   Human tissue culture cell line derived from a cervical carcinoma.

helix-loop-helix (HLH) proteins   Transcription regulatory proteins that contain (1) a dimerization motif, which allows them to bind with another HLH protein, and (2) an adjacent basic region, which mediates binding to specific DNA sequences.

hemolymph   Fluid ("blood") that circulates through the body cavity of insects.

Hensen's node   Thickening at the anterior end of the primitive streak where presumptive notochord cells accumulate.

hepatic diverticulum   Endodermal primordium of the liver.

heterochromatin   Highly condensed chromatin that is inactive in transcription.

heterokaryon   Hybrid cell formed by fusion of cells from different species.

heterosynthesis   External synthesis of a cellular component.

histones   Basic proteins complexed with DNA in chromatin.

Hoechst 33258   Fluorescent dye that stains DNA.

holoblastic cleavage   Complete cleavage.

Holtfreter sandwich   *In vitro* culture system developed by Holtfreter to study induction; composed of the presumptive inducer sandwiched between two sheets of ectoderm.

homeobox   DNA sequence that encodes a homeodomain; discovered during the cloning and sequencing of the *Drosophila* homeotic selector genes.

homeodomain   Conserved DNA-binding sequence found in certain transcription factors; originally discovered in proteins encoded by *Drosophila* homeotic selector genes.

homeotic mutations   Mutations in *Drosophila* that result in substitutions of one body part for another.

homeotic selector genes   Genes of *Drosophila* whose expression distinguishes between alternative differentiated states; term applied to the genes of the bithorax and antennapedia complexes, which specify segment identity.

homolecithal eggs   Eggs with a small amount of evenly distributed yolk.

homophilic binding   Binding of two like molecules to each other.

homunculus   Name given to a tiny creature that was once thought to be curled up in the sperm head; was proposed to be a preformed embryo.

hormone response elements (HREs)   *cis*-acting regulatory elements in the 5′ flanking regions of hormone-inducible genes.

horseradish peroxidase (HRP)   Heme protein isolated from the roots of the horseradish that, when reacted with a suitable substrate, produces a readily visible reaction product; commonly used as an intracellular marker of axons.

Hox genes   Homeobox genes in mice related to the *Antennapedia* gene of *Drosophila*.

Hpa II   Restriction enzyme that cleaves sites in DNA containing CCGG sequences, but not Cm⁵CGG sites; used in conjunction with *Msp* I to detect methylation sites in DNA.

humerus   Bone of the upper arm.

hunchback   Gap mutant of *Drosophila* that lacks head segments.

hyaline layer   Layer composed of material released from the cortical granules that surrounds the sea urchin embryo; maintains blastomere adherence and participates in morphogenetic changes in the developing embryo.

hyaluronic acid (HA)   Glycosaminoglycan composed of alternating N-acetylglucosamine and glucuronic acid; characterized by its extremely large size and hydrophilic nature.

hyaluronidase   Enzyme that hydrolyzes hyaluronic acid, a cell surface polysaccharide.

hybrid gene   *See* fusion gene.

hybrid selection Technique used to confirm that a cloned gene encodes a desired protein.

hybridization 1. Annealing of RNA and DNA via complementary base pairing. 2. Fertilization of eggs by sperm of another species.

Hynobius nebulosus Species of amphibian with voltage-dependent fast block to polyspermy.

hyperplasia Cell proliferation.

hypoblast Layer of cells below the blastocoel in the discoidal embryo.

hypomere Ventral portion of the myotome; will form the hypaxial muscles.

hypostome Mouth and ring of tentacles of *Hydra*.

hypothalamus Region of the brain that produces factors that regulate production of pituitary hormones.

Id "Inhibitor of DNA binding protein"; may bind to MyoD1 and related proteins and prevent their interaction with gene enhancers, thus inhibiting transcription of muscle-specific genes.

Ilyanassa Genus of marine snail with embryos having a large polar lobe.

imaginal cells Cells that remain undifferentiated in the insect larva but differentiate to form specific adult structures after metamorphosis.

imaginal discs Discrete packages of undifferentiated cells in the insect larva. Disc cells remain undifferentiated until metamorphosis, when they form specific elements of the adult exoskeleton.

immotile cilia syndrome Syndrome that is apparently inherited as an autosomal recessive mutation and results in immotile sperm. The normal allele of the gene that causes immotile cilia syndrome is apparently responsible for either the synthesis of the dynein protein or the attachment of the dynein arms to the doublets.

immunocontraception Use of immunization against components of either eggs or sperm to prevent conception.

immunofluorescence microscopy Technique for detecting a cell protein by visualizing with the fluorescence microscope antibodies to that protein that have been coupled to a fluorescent dye.

immunoglobulins Proteins that function as antibodies in vertebrates.

imprinting Differential programming of the maternal and paternal genomes in mammals that restricts their potential after fertilization.

in situ **hybridization** Technique for identifying specific RNA or DNA molecules by hybridization to a labeled probe, followed by autoradiography.

in vitro **fertilization** Technique for fertilization of eggs outside the reproductive tract. When done on mammals, the zygote is maintained *in vitro* before the embryo is implanted into the uterus.

indeterminate development *See* regulative development.

indifferent stage Stage during development when the presumptive gonads of both male and female embryos are identical.

indirect immunofluorescence Use of a fluorescent antibody to detect another antibody bound to an antigen.

induction Alteration of cell fate as a result of interactions with neighboring cells.

ingression Process whereby individual epithelial cells change shape, lose contact with other epithelial cells, and migrate into the blastocoel.

inhibin Growth factor secreted by the Sertoli cells of the mammalian testis and the corpus luteum of the mammalian ovary; suppresses the secretion of FSH from the pituitary.

initiation factors Proteins required during the initiation phase of protein synthesis.

inner cell mass Inner cluster of cells of the mammalian blastocyst; will form the embryo proper.

inner limiting membrane Inner boundary of the retina; basal lamina formed by the Müller glial cells.

inner nuclear layer Layer of the retina containing nuclei of the bipolar cells, horizontal cells, and amacrine cells.

inositol trisphosphate (IP₃) One of the two products of breakdown of phosphatidyl inositol-4, 5-bisphosphate; can cause release of Ca^{2+} from the endoplasmic reticulum.

inside–outside hypothesis Proposal that the relative positions of cells in the early mouse embryo determine whether they will form trophectoderm or inner cell mass.

instars Feeding stages between molts during the larval growth period of insects.

instructive induction Interaction in which the inducing tissue gives instructions that commit cells to a new pathway of development.

integral membrane protein Membrane protein that extends across the lipid bilayer.

integrins Family of transmembrane proteins that mediate cell attachment to the substratum; function as receptors for extracellular matrix proteins and interact with elements of the cytoskeleton inside the cell. *See also* RGD sequence.

intercalary regeneration Replacement of missing elements within a pattern.

intercalation 1. Expansion process whereby cells from different layers lose contact with their neighbors and rearrange into a single layer, which consequently spreads laterally, owing to an increase in surface area. 2. Generation of missing positional values during regeneration when cells of disparate positional values are brought together after amputation.

interchromomeric chromatin Thin strands of chromatin between the chromomeres on the axes of lampbrush chromosomes.

interdigital necrotic zones Regions of cell death between developing digits.

interference reflection microscopy Specialized microscopic technique used to visualize contacts between the cell and the substratum. The closer the cell comes to the substratum, the darker the image of that contact.

intermediate filaments Components of the cytoskeleton with a mean diameter of 10 nm and composed of one of five classes of subunit protein.

intermediate mesoderm Region of mesoderm between the

somites and the lateral plate mesoderm that gives rise to components of the urogenital system.

intermediate placode   Ectodermal thickening that will give rise to some of the neurons in the cranial sensory ganglia.

interspecific hybrids   Embryos (or organisms) produced by fertilization between species.

interstitial cells   *See* Leydig cells.

interstitial cell-stimulating hormone (ICSH)   Another name for luteinizing hormone.

intrinsic ocular muscles   Muscles in the iris that control the size of the pupil; derived from the retinal pigmented epithelium.

intron   Gene sequence that is deleted from RNA during processing of a nuclear transcript.

invagination   Buckling and folding of an epithelium into an interior space.

involuting marginal zone   Vegetal portion of the marginal zone of the *Xenopus* embryo that turns inside the embryo during involution.

involution   Process by which an expanding epithelium turns over on itself and continues to spread in the opposite direction along its basal margin.

ionophore A23187   Molecule that inserts into the egg plasma membrane or internal membrane system of a cell and transports Ca^{2+} into the cytoplasm.

iris   Colored area of the eye surrounding the pupil.

islets of Langerhans   Cells derived from the pancreatic mesoderm that are endocrine in function and produce the hormones insulin, glucagon, and somatostatin.

isoelectric point (pI)   pH at which a protein is electrically neutral and will not migrate in an electrical field.

isolecithal eggs   Eggs that have modest quantities of evenly distributed yolk.

jelly coat   Outermost extracellular layer of eggs of many species. In sea urchins, the jelly coat contains several small peptides and large acidic polysaccharides that are involved in fertilization.

jun   Oncogene that encodes a nuclear protein that is a transcriptional regulator.

juvenile hormone   Gonadotropic hormone that is secreted by the corpus allatum of insects such as *Drosophila*; promotes oocyte differentiation and stimulates the ovary to produce ecdysone, which stimulates the fat body to produce vitellogenin.

karyokinesis   Division of the nucleus.

karyoplast   Nucleus surrounded by a cell membrane.

karyotype   Chromosomes spread for microscopic examination of chromosome number and morphology.

keratin   Protein subunit of keratin filaments.

keratin filaments   Class of intermediate filaments (also called tonofilaments) that are restricted to epithelial cells arising from both ectoderm and endoderm; often stretch across an epithelial cell from one desmosome to another and form a superstructure within the cell.

keratinocytes   Skin cells of the adult epidermis.

keratohyalin granules   Granules in epidermal cells that contain the highly phosphorylated protein profilaggrin, which is used to aggregate keratin filaments.

kinase   Phosphorylating enzyme.

Klinefelter's syndrome   Condition in humans caused by an extra X chromosome. These individuals are phenotypically male, have testes lacking spermatozoa, and are often mentally retarded. These men have normal testes, although they contain no spermatozoa.

knirps   Gap mutant of *Drosophila* that lacks posterior abdominal segments.

Krüppel   Gap mutant of *Drosophila*; lacks thorax and anterior abdominal segments.

labeled pathway hypothesis   Proposal that different axons possess different chemical cues on their surfaces, and these direct the fasciculating behavior of each and every axon.

lac-Z   Gene encoding β-galactosidase in *E. coli*; widely used reporter gene.

lacunae of placenta   Cavities formed as the syntrophoblast erodes its way through the uterine wall; they fill up with maternal blood and lymph. Contents of the lacunae provide the initial nourishment and oxygen to the embryo.

lamella   Thin, fan-shaped region at the anterior end of a migrating fibroblast that propels the cell forward.

lamellipodium   Thin, flat extension of the leading lamella of a migrating fibroblast; as the lamellipodium protrudes forward, parts of it may lift up off the substratum and fold back on itself, forming a ruffling lamellipodium.

laminin   Adhesive glycoprotein component of the basal lamina that mediates epithelial cell adhesion to the extracellular matrix of the basal lamina and is also thought to stimulate motility of a number of embryonic cells.

lamins   Proteins that form a fibrillar network (the nuclear lamina) on the inner surface of the nuclear envelope.

lampbrush chromosomes   Large chromosomes with prominent lateral loops found in oocytes of some species, notably in amphibians.

laryngotracheal groove   Midventral furrow in the floor of the posterior pharyngeal endoderm from which the lungs and trachea are derived.

late histones   Histones that are synthesized from the 16-cell stage through the pluteus and adult stages in sea urchins.

late responses   Events of the egg activation program occurring after the first few minutes following fertilization. These events succeed the early responses.

lateral plate mesoderm   Lateral-most portion of the mesoderm layer of the vertebrate embryo; splits to form splanchnic mesoderm and somatic mesoderm, separated by the coelom.

lectins   Proteins that have multiple carbohydrate binding sites and can therefore bind to the carbohydrate moieties of cell surface glycoproteins.

left–right axis   Body axis extending from the left to the right side of a bilaterally symmetric embryo (or animal).

lens   Spherical body in the eye that focuses light on the retina.

lens placode   Ectodermal thickening that will give rise to the lens.

lens vesicle Spherical vesicle derived from the lens placode; differentiates into the definitive lens.

leptonema First stage of prophase I.

leucine zipper proteins Proteins containing leucine residues that are spaced exactly seven amino acid residues apart, with one leucine for every two turns of the α-helix. The hydrophobic leucine residues will all protrude from the same side of the helix so that two leucine zipper proteins can bind by lateral interactions between the leucine residues on the adjacent domains, thereby forming a dimer. Leucine zippers are found in certain transcription regulatory proteins.

Leydig cells (interstitial cells) Androgen-producing cells derived from the interstitial mesenchyme between the sex cords in the vertebrate testis; localized in the connective tissue between the seminiferous tubules.

limb disc Mesoderm that will give rise to the *Ambystoma* limb.

lineage Developmental history of an individual embryonic cell, including the region of cytoplasm that a blastomere inherits from the egg.

lipovitellin Lipophosphoprotein component of amphibian yolk platelets.

lithium chloride Salt that (1) causes vegetalization of the sea urchin embryo and (2) anteriorizes *Xenopus* embryos, possibly by reducing Xhox3 transcript levels.

lobopodia Finger-like protrusions of the anterior end of migrating cells; can shorten, thereby pulling the cell body forward.

Lucifer yellow Dye that can diffuse from cell to cell only through gap junctions.

luteinizing hormone (LH) Pituitary gonadotropic hormone that (1) in the vertebrate male regulates production of androgens by the interstitial cells and (2) in the female causes the growth of the follicle, prepares the egg for ovulation, and stimulates steroid production by the follicle cells. A surge of LH triggers ovulation in mammals.

Lymnaea Genus of freshwater snail used to study the genetic control of spiral cleavage.

Lyon hypothesis Random heterochromatization of either the paternal or maternal X chromosome during development of female mammals, which results in dosage compensation.

Lytechinus Genus of sea urchin commonly used in developmental biology research.

M phase Period of cell division.

macromeres Large blastomeres.

maculae adherens *See* desmosomes.

male pronucleus Nucleus of sperm after entry into the egg but before syngamy.

manchette Parallel array of microtubules that surrounds the nucleus in spermatids.

mantle layer Intermediate cell layer of the central nervous system. Axons extend from cell bodies in the mantle layer into the marginal layer, which is peripheral. Mantle layer axons are devoid of myelin.

marginal layer Axon-rich outer layer of the central nervous system. Glial cells may cover axons in this layer and produce myelin membranes, which give the axons a white appearance. Thus, the marginal layer is known as the white matter.

marginal zone Region of intermediate pigmentation between the pigmented animal hemisphere and the unpigmented vegetal hemisphere of the amphibian egg.

masked mRNA hypothesis Proposal that oogenic mRNA is complexed with inhibitory proteins that prevent its interaction with the translational machinery.

maternal mRNA *See* oogenic mRNA.

maternal-effect genes Genes that are transcribed during oogenesis and exert their phenotypic effects after fertilization.

matrix Protein and associated RNA that coat lampbrush chromosome loops.

maturation 1. Resumption of meiosis of the oocyte at the completion of oogenesis. 2. Changes to sperm that occur within the epididymis and that cause sperm to become potentially capable of fertilizing eggs.

maturation promotion factor (MPF) Cytoplasmic factor composed of p34^{cdc2} and cyclin that promotes oocyte maturation and may be a universal regulator of the cell cycle; functions by promoting the $G_2 \rightarrow$ M-phase transition.

medaka Fish with large, clear eggs that are useful for studying the cortical reaction.

medulla Core of the gonad composed of the sex cords and some mesenchyme.

melanocytes Pigment cells derived from the neural crest that contain the brown pigment melanin.

melanosomes Membrane-enclosed vesicles containing the brown pigment melanin.

menstrual cycle Ovarian cycle of primates.

mercaptoethanol Reagent that reduces disulfide bonds, causing polypeptide subunits to dissociate and erasing secondary structure

meridional cleavage Cleavage that passes through the animal–vegetal axis.

meroblastic cleavage Incomplete cleavage.

merogone Fragment of sea urchin egg formed by splitting an egg, such as by centrifugation.

meroistic oogenesis Oogenesis in insects with nurse cells.

mesencephalon Midbrain.

mesenchyme blastula Stage of sea urchin development during which primary mesenchyme ingression occurs.

mesoderm Germ layer that gives rise to the muscles, bones, and connective tissue of most organs.

mesodermalizing factors Factors such as guinea pig bone marrow that induce mesoderm and occasionally posterior neural tube structures.

mesolecithal eggs Eggs with a moderate amount of yolk.

mesomeres Intermediate-sized blastomeres.

mesonephric duct *See* Wolffian duct.

mesonephric kidney Functional kidney of adult amphibians and fishes and in the embryos of higher vertebrates. In birds, reptiles, and mammals, it regresses and is replaced by the metanephric kidney.

mesonephric tubules Tubules in the mesonephric kidney through which waste products are channeled.

mesothorax Middle thoracic segment of *Drosophila*; contains the second pair of legs on its ventrolateral margin and the wings on its dorsolateral margin.

messenger ribonucleoprotein (mRNP) particles Messenger RNA complexed with protein.

metallothioneins Small cysteine-rich proteins that bind heavy metals.

metamorphosis Process in which larval tissues are destroyed and replaced by cells destined to form the adult tissues and organs.

metanephric blastema Condensation of mesenchyme from which the tubular excretory units of the kidney arise.

metanephric diverticulum Outgrowth of the mesonephric kidney; primordium of the ureter and renal pelvis.

metanephric kidney Permanent kidney of higher vertebrates.

metathorax Posterior-most thoracic segment of *Drosophila*; contains the third pair of legs on its ventrolateral margin and halteres on its dorsolateral margin.

metencephalon Anterior subdivision of the rhombencephalon.

methylation *See* DNA methylation.

microfilaments Contractile cytoskeletal actin filaments of 6-nm diameter.

micromeres Small blastomeres.

micropyle Narrow channel in the chorion or eggshell of some species through which sperm must swim to fertilize the egg.

microtubule associated proteins (MAPs) Proteins that associate with the surface of the microtubule and stabilize it against sudden disassembly.

microtubule organizing center (MTOC) Point of initiation of growth for microtubules; may also specify the number, distribution, and length of the microtubules that emerge.

microtubules Components of the cytoskeleton composed of hollow cylindrical rods, 25 nm in diameter, formed of 13 rows of solid tubulin protofilaments that run parallel to the microtubule long axis.

microvilli Long finger-like projections of the plasma membrane; contain cytoplasm with F-actin.

mid-blastula transition Abrupt change in the duration and synchrony of cell cycles, cell motility, and transcriptional activity that occurs at the end of the cleavage period in amphibian embryos.

Miller technique Technique for preparing chromatin for visualization of nascent transcripts.

Millipore filters Filters with very small, anastomosing pores; used to separate tissues in *in vitro* induction experiments.

mitochondrial cloud Cluster of mitochondria and granulofibrillar material in the amphibian oocyte.

mitotic apparatus Cage-like structure consisting of spindle fibers with attached chromosomes, paired centrioles at either pole, and (in animals) the asters, which emanate from the centrioles toward the periphery.

MM3 Homeobox-containing gene of *Xenopus*.

mobility-shift analysis Experiments in which radio-labeled subfragments of a gene are incubated with putative DNA-binding proteins, followed by electrophoresis and autoradiography. If the subfragment contains sequences that interact stably with proteins in the nuclear extract, the mobility of the DNA in the gel will be retarded by the protein.

molting Process in which the old larval insect exoskeleton is shed, the body grows, and a new exoskeleton is elaborated to cover the larger larva.

morphallaxis Mode of regeneration involving reorganization of the remaining portions of the body to produce the missing structures.

morphogenesis Process by which embryonic form and structure are achieved.

morphogens Molecules that direct morphogenesis.

morula Cleavage-stage embryo formed of a solid cluster of blastomeres resembling a mulberry.

MOS *See* c-MOS.

mosaic analysis Technique for inducing a patch of mutant tissue in an otherwise normal wild-type environment.

mosaic development Pattern of development in which the fates of blastomeres become restricted during the first few cleavages. Mosaic embryos cannot compensate for blastomeres that are removed or destroyed, so the embryo lacks the structures derived from the missing blastomeres.

motifs Nucleotide sequence elements that play discrete roles in regulating transcription when they are engaged by the appropriate *trans*-acting factors.

mRNA transit time Interval required for an mRNA to be translated into a polypeptide.

Msp I Restriction enzyme that cleaves sites in DNA containing CCGG sequences, whether methylated or not; used in conjunction with *Hpa* II to detect methylation sites in DNA.

msp130 Gene of sea urchins that encodes a cell surface glycoprotein that is restricted to skeletogenic mesenchyme cells only after they have ingressed into the blastocoel.

mucosa Lining of the digestive tube in vertebrates; derived from the endoderm plus underlying connective tissue and muscles.

Müller glial cells Major non-neuronal cell type in the retina; supportive cells of the retina and modulators of neuronal cell activity.

Müllerian ducts Form the vertebrate oviducts and uterus; develop as an invagination of coelomic epithelium.

Müllerian inhibiting substance (MIS) Growth factor related to TGF-β; causes regression of the Müllerian duct and influences gonadal development.

multivesicular bodies (MVBs) Structures containing small membranous vesicles within a membrane-enclosed lumen. MVBs in amphibian oocytes contain yolk precursors incorporated by endocytosis.

myc Oncogene that encodes a nuclear protein that is involved in the processing of nuclear RNA molecules.

myelencephalon Posterior subdivision of the rhombencephalon.

myelin Specialized plasma membrane produced by glial cells for sheathing neurons.

myeloma Tumor of lymphoid cells.

myoblasts Muscle precursor cells that fuse to form multinucleated myofibers.

myocardium Muscle layer of the heart.

MyoD1 Myoblast determination gene that encodes a DNA-binding nuclear protein that regulates expression of certain muscle-specific genes by binding to their enhancer elements.

myoplasm Region of cytoplasm in tunicate eggs that is yellow (*see also* yellow crescent) in *Styela* eggs and orange in *Boltenia* eggs; localized in a crescent-shaped region in the posterior part of the egg and segregated to muscle cells during cleavage.

myosin Cytoplasmic protein that interacts with actin to cause contraction.

myotome Dorsolateral portion of the somite; will develop into the musculature of the trunk and the limbs.

nanos (nos) Maternal-effect gene that affects posterior development in *Drosophila*.

nasal epithelium Sensory lining of the nose, which also serves as the origin of the neurons of cranial nerve I.

nasal placode Ectodermal thickening that will give rise to the nasal epithelium.

N-CAM Calcium-independent neural-cell adhesion molecule.

Neanthes Genus of polychaete annelid with small isolecithal eggs.

necrotic zones Regions of cell death in the mesoderm that are involved in limb morphogenesis.

nerve fiber layer Layer of the retina containing bundles of ganglion cell axons; innermost layer of the retina. These axons enter the optic stalk and make synapses with cells in the brain.

nerve growth factor (NGF) Diffusible substance that promotes the outgrowth of nerve fibers from spinal and sympathetic ganglia of chick embryos.

neural crest cells Individually migrating cells that separate from the dorsal neural tube and come to lie on top of it. They then migrate laterally and ventrally to give rise to a variety of cell types scattered throughout the body.

neural groove Groove flanked by the neural folds; extends along the entire mid-dorsal line of the vertebrate embryo.

neural plate Primordium of the nervous system formed by the flattening and thickening of the dorsal ectoderm.

neural retina Sensory layer of the eye; derived from the inner layer of the optic cup.

neural tube Rudiment of the central nervous system formed by fusion of the neural folds.

neuralizing factors Factors such as guinea pig liver that induce anterior brain structures.

neurofilaments Class of intermediate filaments found in nerve cells.

neuroglia Cells derived from the neural epithelium that can act as supportive cells of the nervous system, phagocytes of the central nervous system, or that wrap themselves around axons, completely ensheathing them in multiple layers of myelin.

neuromeres Segments of the anterior neural tube of vertebrates.

neurons Elongate cells containing a nucleus and processes emanating from it that transmit nervous impulses.

neurotransmitters Chemicals that transmit electrical impulses across synapses.

neurotrophic factor Growth factor released by nerves; promotes limb regeneration.

neurula Embryo during the phase of development when the neural folds are being formed.

neurulation Formation and inward displacement of the neural tube.

Nile blue sulfate Vital dye used to mark areas of the egg or embryo surface.

nocodazole Drug that binds to tubulin subunits and prevents polymerization; inhibitor of microtubule function.

noninvoluting marginal zone Animal portion of the marginal zone of the *Xenopus* embryo that spreads in front of the animal cap but does not involute during gastrulation.

Northern blotting Method for eluting electrophoretically separated RNA molecules from polyacrylamide or agarose gels and identifying specific fragments by hybridization to a radioactive probe.

notochord Stiff rod of mesodermal origin that occupies a mid-dorsal position in the chordate embryo flanked by somite mesoderm.

Notophthalmus viridescens Species of newt often used in limb regeneration experiments.

notoplate Ventral portion of the neural plate in contact with the notochord.

nuage Fibrous substance associated with the nuclear envelope of primordial germ cells and gametes.

nuclear pores Sites containing a complex of proteins located where the inner and outer membranes of the nuclear envelope join; thought to be involved in nucleocytoplasmic communication.

nuclear transplantation Transfer of the nucleus from one cell to another.

nuclease protection assay Technique used to detect specific transcripts; derives its name from the fact that double-stranded nucleic acids are protected from digestion by S_1 nuclease.

nucleolar organizer Gene locus that contains the 18S and 28S ribosomal RNA genes.

nucleolus Site of ribosomal RNA synthesis in the nucleus.

Nucleopore filters Filters with very small, straight pores; used to separate tissues in *in vitro* induction experiments.

nucleus of the lens Central core of the lens.

nurse cells Cells of germ line origin that are associated with oocytes via cytoplasmic bridges; synthesize RNA and protein for transport into the oocyte.

Oct-1 and Oct-2 Octamer-binding proteins.

octamer motif Octanucleotide sequence (ATTTGCAT) that is an important regulatory sequence in the immunoglobulin heavy chain enhancer and the heavy and light chain promoters.

olfactory pits Derivatives of the nasal placodes; form the olfactory epithelium of the nose and neurons of cranial nerve I.

oligodendrocytes Glial cells that produce myelin membranes in the central nervous system.

oligolecithal eggs Eggs with a small amount of evenly distributed yolk.

ommatidia Repeating units in insect eyes containing a stereotypic arrangement of photoreceptor cells.

oncogenes *See* proto-oncogenes.

one gene–one enzyme hypothesis Proposal that each individual gene encodes a specific enzyme.

oocyte maturation Phase between the resumption of meiosis and metaphase II in the oocyte.

oogenesis Formation of ova (or eggs) from oogonia.

oogenic mRNA Messenger RNA transcribed by the maternal genome during oogenesis.

oogonia Female germ cells in the proliferative stage.

ooplasm Oocyte or egg cytoplasm.

ooplasmic determinants Intrinsic factors in eggs that control embryonic determination and are partitioned unequally to different blastomeres during cleavage.

ooplasmic segregation Cytoplasmic movements that cause distinct cytoplasmic regions to become redistributed after fertilization.

optic nerve Neurons that carry visual stimuli from the neural retina to the brain.

optic stalk Attenuated portion of optic vesicle; migratory substratum for the neurons that extend from the neural retina back to the brain where they connect with the visual centers.

optic tectum Region of the brain in lower vertebrates to which retinal fibers extend.

optic vesicles Lateral outgrowths of the prosencephalon that are the precursors of the eyes and optic nerves.

oral ectoderm Single layer of cells in the sea urchin larva, forming the body wall on the side where mouth is located; also gives rise to the larval nervous system.

oral plate Point of contact between the stomodeum and the anterior end of the endodermal tube.

organizer *See* primary organizer.

oskar (osk) Maternal-effect gene that affects posterior development in *Drosophila.*

otic placode Ectodermal thickening from which the inner ear will develop.

otic vesicle Derivative of the otic placode; will form the inner ear.

outer dense fibers A set of nine fibers parallel to the axoneme doublets in the sperm tail; proposed to play a role in flagellar flexibility.

outer limiting membrane Outer boundary of the retina made up of the junctional complexes between the Müller glial cells and the rod and cone cells.

outer nuclear layer Layer of the retina containing nuclei of rods and cones.

ova deficient (o) Maternal-effect mutation in the axolotl that causes progeny to arrest during gastrulation.

ovalbumin Major egg white protein in chickens.

ovariole Lobe of telotrophic insect ovary.

oviduct Duct in the female that transports the egg from the ovary.

oviparous species Animals that produce eggs that develop outside the organism.

ovulation Release of the oocyte from the mature follicle.

ovum (egg) Fully differentiated female germ cell.

PDGF *See* platelet-derived growth factor.

P-element Small DNA transposon that contains sequences allowing it (and any DNA linked to it) to be inserted into the *Drosophila* genome.

P granules Granules in the *Caenorhabditis elegans* egg that are segregated into the posterior cytoplasm of the egg, localize to the P cell lineage during cleavage, and finally enter the primordial germ cells.

pachynema Stage of prophase I that is characterized by a shortening of the chromosomes.

pair-rule genes Zygotic segmentation genes in *Drosophila*; mutations in these genes show defects in all or part of alternate parasegments, leaving the larva with half the number of normal segments.

palindromic symmetry Base sequence on one strand of DNA that is the inverse of the sequence on the other.

palisading Thickening of an epithelial sheet by elongation of cells from a cuboidal to columnar shape.

Panagrellus redivivus Nematode related to *Caenorhabditis elegans.*

pancreatic acini Blind pockets of endoderm in the pancreas rudiment that differentiate into exocrine cells and produce the various pancreatic enzymes.

Paracentrotus Genus of sea urchin used in developmental biology research.

paramesonephric ducts *See* Müllerian ducts.

parasegments Series of bulges along the *Drosophila* embryo that divide the embryo into 14 distinct regions along its anteroposterior axis and reflect different domains of gene expression. Parasegments are a half-segment out-of-frame with larval segments.

parasympathetic ganglia Aggregates of cell bodies of nerves of the parasympathetic autonomic nervous system.

parasympathetic nervous system Portion of the autonomic nervous system that contains chiefly cholinergic fibers; innervates the smooth muscles of organs, glands, and blood vessels.

paraxial mesoderm Thick band of mesoderm immediately adjacent to the neural tube; segments into somites.

parietal decidua Portion of the uterine decidua that is unperturbed by the invading embryo.

parthenogenesis Process by which eggs are activated and develop in the absence of a paternal genetic contribution.

Patella Genus of limpet used for studies on cell determination.

pattern formation Establishment of ordered spatial arrangement of differentiated cells.

pattern formation genes Genes involved in establishing the basic body plan of the organism; first discovered in *Drosophila.*

pericardial cavity Cavity enveloping the heart.

pericardium Somatic layer of lateral plate mesoderm that lines the pericardial cavity.

periderm Temporary protective covering of the skin of primate embryos.

peripheral nervous system (PNS) Nervous tissue peripheral to the central nervous system that mediates communication of all tissues of the body with the central nervous system; composed of nerves and glial cells that are derived from the central nervous system, the neural crest, and epidermal placodes.

periplasm Thin peripheral layer of the cytoplasm in the insect egg.

peritoneal cavity Main body cavity of the adult vertebrate; derived from the coelom.

perivitelline space Area between the egg plasma membrane and the vitelline envelope.

permissive induction Interaction in which responding cells are already determined and poised to differentiate and simply require a signal from the inducing tissue to allow them to express their potential.

pharyngeal arches Columns of mesenchyme between the pharyngeal pouches/grooves.

pharyngeal clefts Regions of contact between pharyngeal pouches and grooves that perforate to form the gill slits.

pharyngeal grooves Inward projections of the ectodermal pockets in the pharyngeal region that approach the pharyngeal pouches.

pharyngeal pouches Evaginations of the pharynx that approach the overlying ectoderm. In advanced vertebrates, individual pouches form specialized derivatives.

pharynx Expanded anterior end of the endodermal tube.

phorbol ester Substance that mimics diacylglycerol.

phosphatidyl inositol Plasma membrane lipid whose derivatives are involved in signal transduction.

phosphatidylinositol-4, 5-bisphosphate (PIP₂) Phosphorylated derivative of phosphatidyl inositol; cleavage of PIP_2 produces inositol trisphosphate (IP₃) and diacylglycerol (DAG), which can stimulate cell mitosis.

phospholipase C Enzyme that cleaves the membrane lipid phosphatidylinositol-4, 5-bisphosphate (PIP₂) into inositol trisphosphate (IP₃) and diacylglycerol (DAG).

phosvitin Phosphoprotein component of amphibian yolk platelets.

photocytes Light-producing cells derived from the M cell lineage of ctenophore cydippid larvae.

photoreactivation Reversal of the effects of ultraviolet light on embryos by treating them with visible light, which reverses damage to nucleic acids.

physiological polyspermy Natural fertilization by several sperm, only one of which combines with the female pronucleus to form a zygote nucleus.

pigmented epithelium Pigmented layer of the retina; derived from the outer layer of the optic cup.

pipe Maternal-effect gene involved in dorsoventral axis formation in the *Drosophila* embryo.

pituitary gland Endocrine gland that lies below the diencephalon of the brain; most important endocrine organ of vertebrates, producing a variety of hormones, including the gonadotropic hormones, which regulate the reproductive system.

placenta Highly developed organ in mammals that serves the nutritional and respiratory functions of the fetus.

placodes Plate-like thickenings of the head ectoderm that are rudiments of specialized structures, including the lens, inner ear, and neurons.

plasmids Small, circular DNA molecules that replicate independently in bacteria, such as *E. coli*; vectors used for cloning DNA.

plasmin Product of plasminogen activator activity.

platelet-derived growth factor (PDGF) Growth factor that activates the growth control cascade by stimulating tyrosine kinase activity.

Pleurodeles Genus of urodele amphibians with very large lampbrush chromosomes in oocyte nuclei.

pluripotent Potential of nuclei to promote development of a variety of diverse organs and tissues.

pluteus Swimming larva of sea urchins.

Poisson buckling Stretching of an elastic sheet such that it forms two parallel ridges that roll up and meet to form a tube.

pol I genes Genes transcribed by RNA polymerase I.

pol II genes Genes transcribed by RNA polymerase II.

pol III genes Genes transcribed by RNA polymerase III.

polar bodies Cells with small amounts of cytoplasm produced at the end of meiosis I and II of oogenesis.

polar coordinate model Explanation to account for reestablishment of pattern during regeneration; cells are thought to possess positional information that is specified along polar coordinates.

polar granules Specially staining granules in the pole plasm of *Drosophila* eggs that may contain germ cell determinants.

polar lobe Spherical cytoplasmic protrusion that forms at the vegetal pole of some annelid and mollusk eggs during cleavage; has been shown to contain factors that affect axis determination.

polarity Organization along an axis.

polarization *See* embryonic polarization.

pole plasm Specialized cytoplasm at the posterior pole of insect eggs; contains the germ cell determinants.

poly (A) Tracts of up to 250 adenylate residues at the 3′ ends of messenger RNA molecules.

polymerase chain reaction (PCR) Technique for selective amplification of DNA sequences from a complex mixture of DNA sequences.

polyploidy Presence of multiple haploid sets of chromosomes in the genome.

polyspermy Fertilization by more than one sperm.

polytene chromosomes Large chromosomes of the salivary glands and certain other tissues of dipteran insects; result from repeated DNA replication of aligned homologues.

polytrophic ovaries Insect ovaries in which the nurse cells are intimately connected to the oocyte.

positional information Ability of cells to interpret their location in a developing structure.

posterior midgut invagination Cavity that forms at the posterior end of the ventral furrow of the *Drosophila* embryo into which the pole cells enter.

postganglionic fibers Motor fibers of the autonomic nervous system that have their cell bodies in ganglia and transmit impulses to smooth muscles or glands.

postvitellogenesis Phase of oogenesis after yolk deposition.

potency Developmental potential of a nucleus or cell.

preformation Concept that embryos are preformed in the gametes of one of the sexes.

preganglionic fibers Motor fibers of the autonomic nervous system that originate in the central nervous system and synapse on cell bodies in ganglia.

prevalent mRNA class Messengers that are present in

abundance (1000 or more copies per cell) and account for most of the total mass of mRNA.

previtellogenesis Phase of oogenesis before yolk deposition.

primary corneal stroma Collagenous matrix produced by the corneal epithelium.

primary egg envelopes Noncellular egg envelopes that are produced by the oocyte during oogenesis.

primary induction Neural induction by the chordamesoderm.

primary mesenchyme Cells of micromere origin that ingress from the vegetal plate of the sea urchin blastula and enter the blastocoel to form the larval skeleton.

primary oocytes Female germ cells in the first meiotic division.

primary organizer Dorsal lip of the amphibian blastopore; responsible for coordinating the development of the body plan during gastrulation.

primary sex determination Development of the indifferent gonad into testis or ovary.

primary spermatocytes Male germ cells in the first meiotic division.

primitive groove Depression in the midline of the primitive streak.

primitive sex cords Epithelial cords emanating from the coelomic epithelium into mesenchyme of the vertebrate gonad.

primitive streak Long cleft in the blastodisc through which cells ingress during gastrulation in birds and mammals.

primordial germ cells (PGCs) Precursors of the vertebrate gametes; migrate from a distant source into the genital ridge.

primordial yolk platelets (PYPs) Precursors of the definitive yolk platelets in amphibian oocytes.

proacrosomal granules Membrane-enclosed granules in the Golgi complex of mammalian spermatids; contain the precursors of the acrosomal material.

proctodeum Inpocketing of the ectoderm at the posterior end of the vertebrate embryo that contacts the posterior end of the endodermal tube to form the cloacal plate.

progesterone Steroid hormone produced by follicle cells in the ovary; regulates the reproductive cycle.

programmed cell death Cell death that functions in morphogenesis.

progress zone Area of proliferative activity at the tip of the avian embryo limb.

promoter Nucleotide sequence of a gene that contains the transcription start site and directs where transcription should begin.

pronephric duct Duct that carries waste products from the pronephric kidney to the cloaca; becomes the functional excretory duct of the mesonephric kidney.

pronephric kidney First kidney to develop in vertebrates. In most vertebrates, it has a brief and transitory appearance and then begins to regress as the mesonephric kidney forms posterior to it.

pronephric tubules Tubules in the pronephric kidney through which waste products are channeled.

pronucleus Sperm or egg nucleus after fertilization. The pronuclei undergo a number of changes that facilitate formation of the zygote nucleus.

prosencephalon Forebrain, which is later subdivided into the anterior telencephalon and the posterior diencephalon.

protamines Chromatin-associated proteins that replace histones during spermiogenesis of some species.

protease Enzyme that digests protein.

protein kinase C (PKC) Enzyme that can phosphorylate proteins and activate the Na^+/H^+ pump, which increases cytoplasmic pH—both of which can stimulate cell division.

protein kinases Enzymes that phosphorylate proteins.

proteoglycans Large, space-filling extracellular matrix molecules that are composed of glycosaminoglycans bound to a core protein.

prothorax Anterior-most thoracic segment of *Drosophila*; contains the first pair of legs on its ventrolateral margin.

proto-oncogenes Genes that regulate cell growth and certain developmental processes. Overexpression or mutation of proto-oncogenes is responsible for the development of cancer cells from normal cells.

puffing Unfolding of bands of polytene chromosomes in a region of intense transcriptional activity.

pupal stage Phase during which the larval insect is converted into an adult by metamorphosis.

radial cleavage Cleavage in which the mitotic apparatus of each blastomere is oriented in a plane parallel or perpendicular to the plane of the egg's animal–vegetal axis and in which any plane passing through the animal–vegetal axis divides the embryo into symmetrical halves.

radius One of the bones of the forearm.

Rana pipiens. Leopard frog, which has been valuable in experimental embryology, particularly in nuclear transplantation experiments.

rare mRNA class Messengers that are present in only a few copies each per cell.

reaction-diffusion model Class of model of patterning involving two morphogens (an activator and an inhibitor) that diffuse from a single source to form separate gradients.

read-through transcription Transcription that continues past normal termination signals.

reannealing *See* renaturation.

reassociation *See* renaturation.

recombination signal sequences (RSSs) Nucleotide sequences that mediate somatic gene recombination.

reconstitution Reformation of new sponges from aggregates of cells.

reductionist approach Philosophy that the whole is understood most completely by studying its parts and the nature of the sum of the parts.

regeneration Replacement of a severed structure.

regeneration blastema Mound of cells at the tip of an amputated limb; rapid proliferation of these cells generates a new limb.

regulative development Pattern of development in which cell fates are restricted late in development. Regulative embryos can compensate for blastomeres that are removed or destroyed early in development.

renaturation Restoration of the DNA double helix by reassociation between the two strands by complementary base pairing.

reporter gene Gene encoding a protein that is readily detected; fusion of the reporter sequence to a promoter enables the investigator to detect readily the products of promoter activity.

resact (sperm respiratory activating peptide) Small peptide in the sea urchin egg jelly coat that acts as a sperm attractant.

residual bodies Lobules of cytoplasm that accumulate in the neck regions of spermatids at the completion of spermiogenesis.

restriction endonucleases Enzymes isolated from microorganisms that cut DNA internally at sequence-specific sites.

restriction map Locations of restriction endonuclease cleavage sites of a gene.

retinoic acid Morphogen controlling pattern formation in chick and amphibian limbs.

retinotectal projection Pattern of neuron distribution from the retina to the tectum of the brain of lower vertebrates.

reverse transcriptase Enzyme isolated from RNA tumor viruses that uses RNA to make a complementary DNA copy.

RGD sequence Domain of integrins consisting of the three amino acids Arg-Gly-Asp; functions as the binding site for extracellular matrix proteins.

rhodamine dextran Fluorescent vital dye.

rhombencephalon Hindbrain, which gives rise to the anterior metencephalon and the posterior myelencephalon.

rhombomeres Segments of the vertebrate hindbrain.

Rhynchosciara Genus of midge (diperan insect) that undergo selective amplification of genes during development.

ribonucleotide reductase Enzyme that catalyzes the reduction of uridine to thymidine and thus is involved in DNA synthesis.

ribosomal RNA Major nucleic acid component of ribosomes; comprised of 5S, 18S, and 28S RNA.

ribosomes Ribonucleoprotein complexes on which protein synthesis occurs in the cytoplasm. The major RNA components are: 18S, 28S, and 5S RNA.

ring canals Intercellular bridges that are formed by incomplete cytokinesis in the insect oocyte-nurse cell complex.

RNA polymerase Class of enzymes that transcribe DNA.

rotational cleavage Cleavage in which one mitotic apparatus (or cell) rotates to a position perpendicular to the other during the first cleavage division.

rough (ro) *Drosophila* mutation that interrupts eye development.

rudiments Precursors of organs or tissues.

rudimentary (r) Maternal-effect mutation in *Drosophila* that affects wing size and is involved in pyridine metabolism.

ruffling lamellipodia *See* lamellipodium.

rule of distal transformation Refers to the pattern of urodele limb regeneration; the limb always regenerates in the distal direction, regardless of the polarity of the stump.

runt Pair-rule mutant of *Drosophila* that is defective in parasegment-sized regions that are out of register with the parasegments.

S unit Unit dependent upon the molecular weight and shape of a particle; used to describe RNA molecules, such as 18S RNA, 28S RNA, and 5S RNA.

S_1 nuclease Enzyme that specifically digests single-stranded nucleic acids; used in nuclease protection assays.

satellite DNA Highly repetitive DNA consisting of short, repetitive nucleotide sequences that do not code for protein.

Schizosaccharomyces pombe Fission yeast, which have been used extensively in cell cycle research.

Schwann cells Glial cells that produce myelin membranes in the peripheral nervous system.

sciatic plexus Aggregation of motor axons that emerge from lumbosacral levels 4 through 8.

sclera Outer connective-tissue layer that nourishes and protects the eyeball and is responsible for maintaining the shape of the eyeball.

scleral cartilage Layer of cartilage that surrounds the posterior surface of the eyeball in nonmammalian vertebrates; derived from the neural crest.

scleral ossicles Series of bony plates that form around the cornea in nonmammalian vertebrates; derived from the neural crest.

scleral papillae Invaginations of the conjunctiva into the underlying mesenchyme in the developing chick eye; thought to induce formation of scleral ossicles from the mesenchyme.

sclerotome Mesenchyme derived from the ventromedial wall of the somite; gives rise to vertebrae and ribs.

Scolopos Genus of polychaete annelid with large isolecithal eggs.

sea urchins Group of echinoderms used extensively in research on fertilization, egg activation, and early development.

secondary egg envelopes Noncellular egg envelopes that are produced by the follicle cells that surround the oocyte.

secondary mesenchyme Cells that constitute the second wave of mesenchyme in the sea urchin embryo; they enter the blastocoel by ingression from the blunt end of the archenteron and differentiate into larval muscle, pigment cells, and the coelomic sac.

secondary oocytes Female germ cells in the second meiotic division.

secondary sex cords Cords derived from proliferation of the coelomic epithelium in the vertebrate ovary; ultimately break up to form primordial follicles.

secondary spermatocytes Male germ cells in the second meiotic division.

segmental plate Paraxial mesoderm in amniote embryos before overt segmentation. All future somites will develop from the segmental plate in an anteroposterior wave.

segmentation genes Zygotically expressed genes of *Drosophila* that mediate development along the anteroposterior axis after the anterior and posterior regions of the embryo are established.

segment-polarity genes Zygotic segmentation genes of *Drosophila* that begin to be transcribed between the cellular blastoderm stage and gastrulation; mutations in these genes remove a part of each segment and replace it with a duplication of the remaining part of the segment.

segments Repetitive body units of *Drosophila* and other arthropods.

seminiferous growth factor (SGF) Growth factor produced by mammalian Sertoli cells; stimulates somatic cell proliferation and blood vessel production in the testes during fetal and postnatal development and regulates protein secretion by the Sertoli cells themselves in the adult.

seminiferous tubules Tubules in the testis that are circumscribed by radially distributed Sertoli cells, with which the germ cells remain associated during spermatogenesis.

Sendai virus When killed with ultraviolet light, these viruses bind to cell membranes and facilitate cell fusion.

sequence complexity 1. Measure of the number of different mRNA species in a cell. 2. Measure of the number of different sequences present in a genome.

serial nuclear transplantation Sequential transfer of nuclei derived from a single transplanted nucleus to enucleated eggs, resulting in the perpetuation of a clone of genetically identical individuals.

Sertoli cells Somatic cells derived from the sex cords; form a cellular superstructure that supports sperm differentiation.

sevenless (sev) *Drosophila* mutation that interrupts eye development.

sex combs reduced Homeotic selector gene of *Drosophila*.

shortest intercalation rule Intercalation of missing values during regeneration occurs by the shortest route.

sialic acid Highly negatively charged sugar molecule.

sinistral cleavage Left-handed (counterclockwise) displacement of micromeres during spiral cleavage.

skeletogenic mesenchyme Micromere derivatives in the sea urchin embryo that enter the blastocoel to form the skeleton of the pluteus larva.

slow block to polyspermy Block that is mediated by the cortical reaction.

slug Migratory aggregate of *Dictyostelium* cells.

SM50 Gene of sea urchins encoding a 50 kDa spicule matrix protein that is expressed early in the skeletogenic mesenchyme.

small micromeres Subset of micromeres in the sea urchin embryo that may contribute to formation of the coelomic sacs.

Smittia Genus of chironomid insect that has been valuable for the study of anterior determinants.

snake Maternal-effect gene involved in dorsoventral axis formation in the *Drosophila* embryo.

sodium acetate Reagent that lowers intracellular pH in eggs.

somatic cell hybridization Fusion of somatic cells.

somatic cells Nongerm cells.

somatic gene therapy Introduction of genes into somatic cells to correct genetic defects.

somatic nervous system Motor neurons (under voluntary control) that innervate and control the skeletal muscles.

somatic recombination DNA rearrangement during development of vertebrate B cells.

somatopleure Combination of ectoderm and somatic mesoderm.

somites Blocks of tissue in the trunk derived from the originally unsegmented paraxial mesoderm.

somitomeres Circular swirls of the mesodermal cells in the segmental plate; reflect the pattern of prospective somites in the trunk.

sorting out Segregation of a mixture of dissociated cells according to cell type and re-establishment of former associations.

Southern blotting Method for eluting electrophoretically separated fragments of DNA from polyacrylamide or agarose gels to a filter and identifying specific fragments by hybridization to a radioactive probe.

Spec 1 Spec genes that are expressed early and exclusively in the aboral ectoderm in the sea urchin *Strongylocentrotus purpuratus*.

Spec 2 Spec genes that are expressed in the aboral ectoderm of *Strongylocentrotus purpuratus* several hours after Spec 1 expression.

Spec proteins Family of ectoderm-specific Ca^{2+}-binding proteins related to calmodulin in *Strongylocentrotus purpuratus* embryos.

specific cellular adhesiveness Hypothesis stating that cell adhesion is a property of specific cell surface macromolecules that allow cells to recognize like cells and adhere.

sperm-activating peptide (speract) Small peptide that stimulates sperm motility.

sperm aster Complex of long microtubules that radiate from the paired sperm centrioles in the egg cytoplasm after fertilization.

sperm maturation Physiological, morphological, and biochemical changes that occur in the epididymis; enables spermatozoa to acquire the potential ability to fertilize eggs.

spermatids Haploid cells resulting from meiosis in the male; differentiate into spermatozoa.

spermatogenesis Formation of sperm from spermatogonia.

spermatogonia Male germ cells in the proliferative stage.

spermatozoa (sperm) Fully differentiated male germ cells.

spermiation Release of individual sperm from their syncytial network at the completion of spermiogenesis.

spermiogenesis Process of differentiation of the male germ cells (spermatozoa).

Sphaerechinus Genus of sea urchins that is used in interspecies hybridization experiments.

sphincter muscle Circumferentially oriented muscle of the iris; responsible for reducing the diameter of the pupil.

spicules Mineralized components of the sea urchin larval skeleton.

spinal nerves Nerves emerging from the spinal cord that contain preganglionic axons of the sympathetic system and sensory and motor fibers of the somatic nervous system.

spindle Structure composed of microtubules that organizes chromosomes on the metaphase plate and moves the chromosomes to the opposite poles during anaphase.

spiral cleavage Cleavage in which the mitotic apparatus is oriented obliquely, rather than parallel, to the long axis of the egg or blastomere.

spiralians Animal species that exhibit spiral cleavage.

spire *Drosophila* egg-shape gene.

Spisula Genus of clam that has been valuable for the study of differential translation of oogenic mRNA during oocyte maturation.

splanchnopleure Combination of endoderm and splanchnic mesoderm.

src Gene initially detected in the Rous sarcoma virus that can transform normal cells into cancer cells. A cellular homologue of *src* exists in eukaryotic genomes.

SRY Candidate testis-determining gene in humans.

Sry Candidate testis-determining gene in mice.

staggerer Mutation in mice that causes defects in the connection of some of the neurons in the cerebellum, resulting in abnormalities in their motion and balance; correlated with a defect in the processing of N-CAM.

staufen (*stau*) Maternal-effect gene that affects anterior development in *Drosophila*.

stem cells Cells that divide to produce either (1) more stem cells or (2) cells that differentiate.

stomodeum Inpocketing of the ectoderm at the anterior end of the embryo that contacts the anterior end of the endodermal tube to form the oral plate.

stratum corneum Outer layers of epidermis consisting of flattened squamous cells that have lost their nuclei and are tightly packed with bundles of keratin intermediate filaments.

stratum granulosum Intermediate layer of the epidermis whose cells contain keratohyalin granules.

stratum spinosum Intermediate layer of the epidermis that consists of polyhedral cells that are held together by developing desmosomes.

stress bundles Bundles of microfilaments that extend along the length of the cell.

stroma Extracellular matrix between cells or between cell layers.

Strongylocentrotus Genus of sea urchin that is commonly used in developmental biology research.

structural genes Genes that are transcribed into messenger RNA, which is transported into the cytoplasm, where it is translated into protein.

Styela Genus of tunicate having eggs with pigmented cytoplasmic regions. These regions have been useful for the study of cytoplasmic localization.

subgerminal cavity Cavity in the early avian embryo that lies between the blastoderm and the yolk mass.

superficial cleavage Cleavage that occurs only in a thin layer of superficial cytoplasm; characteristic of some arthropods.

swallow (*sww*) Maternal-effect gene that affects anterior development in *Drosophila*.

sympathetic ganglia Aggregates of cell bodies of nerves of the sympathetic autonomic system; form a chain along the spinal cord.

sympathetic nervous system Portion of the autonomic nervous system that contains chiefly adrenergic fibers; innervates the smooth muscles of organs, glands, and blood vessels.

synapse Specialized junction by which a neuron communicates with other neurons or with effector organs.

synapsis Pairing of homologous chromosomes during prophase I of meiosis.

Synapta Genus of sea cucumber that shows radial holoblastic cleavage.

syncytial blastoderm stage Stage of insect development in which nuclear division occurs without cytokinesis in the periplasm.

syncytium Multinucleate cell.

synkaryon Zygote nucleus.

syntrophoblast Outer syncytial layer of the trophectoderm.

talpid Mutation in chickens that prevents the appearance of necrotic zones between digits in the embryo with the consequence that a web of mesenchyme continues to connect the digits.

TATA box Nucleotide sequence in the core promoter that directs RNA polymerase to initiate transcription about 30 nucleotides downstream from it.

TDF Hypothetical gene for testis-determining factor in humans.

Tdy Hypothetical gene for testis-determining factor in the mouse.

telencephalon Anterior subdivision of the prosencephalon that gives rise to the cerebral hemispheres.

teloblasts Stem cells that compose the growth zone from which the segmented region of the annelid body develops.

telolecithal eggs Eggs containing a large amount of yolk. The cytoplasm in such eggs is restricted to a thin layer covering the yolk at one end of the egg.

telotrophic ovaries Insect ovaries in which the nurse cells are in a cluster at one end of the ovary and are connected via long trophic cords to the oocytes.

telson Terminal unsegmented region of the *Drosophila* abdomen.

territory Contiguous region of the sea urchin blastoderm that is characterized by a distinct pattern of gene expression and unique cell fates.

tertiary egg envelopes Noncellular egg envelopes that are added as the egg passes through the reproductive tract after ovulation.

testicular feminization Condition found in individuals who are androgen insensitive. These individuals are genetically XY, have a normal testis that produces normal levels of testosterone, but lack functional cytoplasmic receptors for testosterone.

TFIIIA   One of the transcription factors responsible for synthesis of 5S RNA and binds to 5S RNA after its synthesis; has a prominent zinc finger element.

TGF-β   *See* transforming growth factor-β.

β-thalassemia   Human genetic disorder that causes a deficiency in β-globin gene expression.

thecal cells   Mesenchyme-derived cells enveloping follicle cells in the ovary; produce steroid hormones.

three-signal model   Proposal that three distinct inductive signals are involved in mesoderm induction in *Xenopus*.

thymidine   One of the four nucleosides found in DNA. Because thymidine is not found in RNA, radioactively-labeled thymidine is frequently used specifically to monitor DNA synthesis.

thyroid diverticulum   Primordium of the thyroid that forms in the floor of the pharynx between the first and second pharyngeal pouches.

tight junctions   Junctions that prevent molecules from diffusing into the paracellular spaces.

tissue culture   Technique for growing living tissues outside of the organism in culture.

tissue-type plasminogen activator (tPA)   Serine protease that cleaves plasminogen to plasmin and degrades directly some components of the extracellular matrix; tPA activity increases in ovulatory follicles of mammals.

T$_m$ Temperature at which the double-stranded DNA is denatured into two individual strands; often called the melting temperature.

Toll   Maternal-effect gene involved in dorsoventral axis formation in the *Drosophila* embryo.

tonofilaments   *See* keratin filaments.

TOP   Antigen present in a gradient across the chick retina.

torpedo   *Drosophila* egg-shape gene.

totipotent   Complete potential of a nucleus to support the development of all cell types that make up the organism.

trans-acting factors   Transcription factors.

transcription factors   Proteins that interact with gene regulatory elements to influence transcription; often called *trans*-acting factors.

transdetermination   Alteration in imaginal disc cell fate.

transfection   Introduction of a gene via a viral infection.

transfer RNA (tRNA)   Class of small RNA molecules that read the genetic code in the ribosome and bear amino acids for insertion into the growing peptide.

transferrin   Iron-binding protein produced by the Sertoli cells in vertebrate testes.

transformation   Introduction of DNA into a cell.

transforming growth factor-β (TGF-β)   Family of growth factors that is involved in mesoderm induction in *Xenopus*.

transgene   Exogenous gene in a transgenic organism.

transgenic organism   Organism having an exogenous gene inserted into its genome.

transition proteins   Proteins that replace somatic histones during mammalian spermiogenesis before being replaced themselves by protamines.

translational control   Process by which gene expression is modulated at the level of mRNA translation.

translational potentiator   Developmentally regulated factor that is necessary to unwind double-stranded regions of mRNA so that translation can proceed.

transposable elements   Mobile elements in the genome.

trefoil stage   Spiralian embryo during telophase of first cleavage, when the polar lobe is connected to the future CD blastomere by a narrow neck of cytoplasm, giving the embryo a three-lobed appearance.

triploid   Genome containing three haploid chromosome sets.

Triton X-100   Nonionic detergent.

Triturus   Genus of urodele amphibians (called newts) used in experimental embryology and for the study of lampbrush chromosomes and nucleoli during oogenesis.

trochoblast cells   Ciliated cells in the embryo of *Patella*, a limpet.

trochophore   Larva of spiralians.

tropharium   Syncytial complex of nurse cells in the telotrophic insect ovary.

trophectoderm   Outer epithelium of the mammalian blastocyst; will give rise to portions of the placenta.

trophic cords   Long cords that connect nurse cells to the oocytes in telotrophic ovaries.

trophic factor   Growth factor.

tubulin   Proteins of which microtubule protofilaments are comprised.

tunica albuginea   Layer of connective tissue beneath the coelomic epithelium derived from mesenchymal cells that invade between the sex cords; severs the cords from the surface epithelium of the testis.

Turing model   Class of model in which spatial patterns arise after minor perturbations in morphogen gradients that had been essentially homogeneous.

Turner's syndrome   Condition in humans caused by having one X and no Y chromosome. These individuals are phenotypically female, generally underdeveloped sexually, short in stature, and have only vestigial gonadal tissue as adults.

twisted gastrulation   Gene of *Drosophila* that is involved in regulating gastrulation movements.

ulna   One of the bones of the forearm.

Ultrabithorax (Ubx)   Dominant homeotic selector gene mutation in *Drosophila;* causes halteres to be replaced by a second pair of wings.

unc-86   Gene in *Caenorhabditis elegans* that has been implicated in cell lineage determination.

ureter   Duct of the metanephric kidney.

uterus   Organ in female mammals in which the embryo implants.

uvomorulin   Adhesive glycoprotein localized on the surfaces of mammalian blastomeres that stabilizes cell–cell associations.

v-erb-B   Oncogene found in the avian erythroblastosis virus that encodes a protein that is identical to the cytoplasmic and membrane domains of the epidermal growth factor (EGF) receptor.

v-*ras* Viral oncogene that encodes a G-protein that regulates cell division.

v-*sis* Oncogene isolated from the Simian sarcoma virus that encodes a protein that is similar to PDGF and binds to PDGF receptors.

variable regions Amino-terminal ends of immunoglobulin molecules, which vary considerably from molecule to molecule; form the combining site of an antibody molecule.

vas deferens Duct in the male that carries mature sperm to the urethra and out of the body.

vector Vehicle (such as bacterial plasmid) into which DNA is inserted for cloning.

vegetal body Granular RNA-containing organelle in the small polar lobe of the *Bythinia* embryo.

vegetal plate Cells of the sea urchin swimming blastula that invaginate during gastrulation and form the gut and secondary mesenchyme.

vegetal pole Pole of the egg that lies 180 degrees opposite the site where maturation divisions occur and polar bodies are formed.

vegetalized pluteus Larva of a sea urchin egg showing exaggerated development of vegetal hemisphere derivatives.

vegetalizing factors Hypothetical factors that are most concentrated at the vegetal pole of the sea urchin embryo, gradually decreasing toward the animal pole, which promote development of mesoderm.

veliger Larval stage of some mollusks.

velum Ciliated structure required for locomotion and feeding of the veliger larva of mollusks.

ventral furrow Structure formed during gastrulation of the *Drosophila* embryo by invagination of blastoderm cells. After invagination, the folded epithelium separates from the blastoderm remaining on the surface of the embryo and forms a tube of mesodermal cells beneath the ventral midline.

ventralized embryos Embryos lacking dorsal structures.

ventralized larvae Larvae with exaggerated ventral characteristics.

verapamil Drug that inhibits Ca^{2+} transport.

vernix Material that is smeared on fetal primate skin and consists of periderm cells mixed with sebaceous secretions from the skin; protects the developing epidermis from abrasion as well as damage from immersion in amniotic fluid.

Vg1 mRNA that is localized to the vegetal pole in *Xenopus* oocytes. Sequence analysis has shown that the protein encoded by Vg1 is related to transforming growth factor-β (TGF-β).

villi of placenta Branched processes derived from the cytotrophoblast; contain blood vessels and function in exchange between the mother and fetus.

villin Ca^{2+}-dependent actin-fragmenting protein.

vimentin filaments Class of intermediate filaments found primarily in mesodermally-derived connective tissues such as bone and cartilage.

viral oncogenes Viral counterparts of proto-oncogenes.

Viral oncogenes have been identified that substitute for, and mimic, various elements in the growth response pathway.

visceral arches *See* pharyngeal arches.

visceral nervous system *See* autonomic nervous system.

vitelline duct Synonym for yolk stalk.

vitelline envelope Thin glycoprotein layer that surrounds the egg plasma membrane.

vitellogenesis Phase of oogenesis during which yolk is deposited.

vitellogenin Yolk precursor.

vitreous body Extracellular matrix component that fills the posterior cavity of the eye.

white matter *See* marginal layer.

wingless 1. Chick mutant in which a normal wing fails to develop; defect is due to failure of mutant mesoderm to maintain the AER. 2. *Drosophila* segmentation gene.

Wolffian (mesonephric) duct Duct derived from the pronephric duct; serves as the duct for waste products from the mesonephric kidney. In higher vertebrates, the Wolffian duct will form the vas deferens.

Wolffian ridge Longitudinal ridge caused by proliferation of somatic mesoderm cells under the ectoderm along the length of the amniote embryo.

X-gal Substrate of β-galactosidase that forms a blue precipitate.

Xenopus laevis South African clawed frog; valuable experimental organism for studies of oogenesis and embryonic development.

Xhox3 Homeobox-containing gene of *Xenopus*; appears to be involved in mediating anteroposterior polarity during *Xenopus* development.

XTC-MIF *Xenopus* tissue culture cell mesoderm-inducing factor (recently identified as activin A), which is secreted into the culture medium by XTC cells.

yellow crescent Subequatorial region in the posterior aspect of the *Styela* egg that contains the myoplasm; segregated to the muscle cell lineage during cleavage.

yolk Lipo- and glyco-proteins stored in the egg cytoplasm; metabolized during embryogenesis as a source of energy.

yolk platelets Yolk-containing organelles of amphibia.

yolk plug Large endodermal cells of the amphibian gastrula that are encircled by the blastopore.

yolk sac 1. Extraembryonic membrane composed of endoderm and splanchnic mesoderm that intimately surrounds and digests the yolk of the chick embryo. 2. Extraembryonic membrane below the mammalian embryo derived from extraembryonic mesoderm and endoderm; vestigial during later development.

yolk stalk Connection between the embryo and the yolk sac.

zinc finger DNA-binding element found in many proteins that regulate transcription; formed by the binding of cysteine and histidine residues to zinc atoms.

zona pellucida   Noncellular layer containing three glycoproteins (ZP-1, ZP-2, and ZP-3); surrounds the mammalian oocyte and functions in species-specific sperm binding.

zona reaction   Alterations in the zona pellucida of the mammalian egg at fertilization.

zone of polarizing activity (ZPA)   Junction of the posterior margin of the developing chick limb with the body wall; determines the anteroposterior axis of the limb.

zonula adherens   *See* adherens junctions.

zonula occludens   *See* tight junctions.

zygonema   Stage of prophase I when synapsis occurs.

zygote   Fertilized egg.

zygotic mRNA   Messenger RNA transcribed by the zygotic genome after fertilization.